Green Energy and Technology

For further volumes:
http://www.springer.com/series/8059

Acknowledgement

The Editor wishes to acknowledge with gratitude all the contributing authors for their enthusiasm, cooperation, and expertise as exemplified in their fields of interest. It is their great effort and contribution that makes the publication of this book possible.

The Editor is most grateful to Mr. Benjamin Bunes for his critical reading and accuracy reviewing of all the chapters, and Dr. Xiaomei Yang for her expert assistance in the formatting and editing of the book. Their enormous efforts are wholeheartedly appreciated.

The Editor is also indebted to Claire Protherough, senior editorial assistant at Springer, for her constant support and guidance in preparation of the book. Special thanks also go to the Editor's students and colleagues for their generous help in many ways so as to allow the Editor to allocate time working on the book.

Lastly, but not least, the Editor's deep appreciation goes to his family, particularly his daughter, Susu, for their understanding of his numerous absences during evenings, weekends, and even holidays when working on the book. The Editor would also like to dedicate this book to his father on the occasion of his seventy-fifth birthday.

Contents

Ling Zang
Editor

Energy Efficiency and Renewable Energy Through Nanotechnology

Springer

Dr. Ling Zang
Department of Materials Science and Engineering
University of Utah
122 S. Central Campus Drive
Salt Lake City, UT 84112
USA
e-mail: lzang@eng.utah.edu

ISSN 1865-3529
ISBN 978-0-85729-637-5
DOI 10.1007/978-0-85729-638-2
e-ISSN 1865-3537
e-ISBN 978-0-85729-638-2
Springer London Dordrecht Heidelberg New York

British Library Cataloguing in Publication Data
A catalogue record for this book is available from the British Library

Cover design: eStudio Calamar, Berlin/Figueres

Printed on acid-free paper

Springer is part of Springer Science+Business Media (www.springer.com)

Preface

Emerging nanotechnology has transformed how people from different fields think, work and interact to fabricate and optimize new materials and processes at the nanometer scale, which, in turn, improves performance when employed in real devices. These nanomaterials and nanostructures often provide unprecedented efficiency or function compared to conventional bulk phase materials manufactured by traditional techniques. Reflecting the rapid growth of nanotechnology research and its potential impact to the increasing energy demand, this book "Energy Efficiency and Renewable Energy Through Nanotechnology" summarizes leading-edge research in energy-related nanoscience and nanotechnology.

Although there are a number of research review books already available covering various specific topics of nanotechnology, none tightly correlates nanotechnology with energy issues in such a general, thorough manner that makes this book not only suitable as a desk reference for researchers, but also a knowledge resource for the nonexpert public. This book includes 26 chapters covering broad topics of nanotechnology and nanoscience research that are relevant to energy efficiency and renewable energy. The chapters are categorized into five major parts: Electricity Generation via Inorganic and Organic Solar Cells, Electricity Generation via Fuel Cells and Piezoelectric Materials, Clean Fuels Generation and Environmental Remediation by Sun Light, Energy Storage, and Energy Efficiency and Saving. All chapters were written by world-leading experts who research energy-related nanotechnology.

The aim of this book is to provide readers with basic knowledge and information in energy efficiency and renewable energy, and the approaches using nanotechnology, thus inspiring their interest in related research and development. This book is suitable for a broad range of readership: (1) academia and higher education for general information and public outreach of leading edge techniques and improvements in energy efficiency and renewable energy; (2) government and federal agencies for making energy policies and administration; (3) industrial R&D labs as reference book for developing energy research, strategizing and planning; and (4) the general public, including scientists and engineers, for personal interests and self-education in the emergence of nanotechnology and its long-term impact on meeting the rising energy demand. Moreover, this book can also be used as a primary or supplementary textbook for third/fourth-year undergraduate and graduate students, for a special topic course related to energy and nanotechnology, which has become popular in both undergraduate and graduate curricula.

While the Editor is responsible for the selection of the topics and contributing authors, it is the primary responsibility of the authors to ensure the accuracy and appropriateness of the contents covered in their respective chapters. Despite the great effort that the Editor and authors have put forth, this book may not be error-free. Any comments or suggestions to improve the book for the next edition are welcome.

Acknowledgement

The Editor wishes to acknowledge with gratitude all the contributing authors for their enthusiasm, cooperation, and expertise as exemplified in their fields of interest. It is their great effort and contribution that makes the publication of this book possible.

The Editor is most grateful to Mr. Benjamin Bunes for his critical reading and accuracy reviewing of all the chapters, and Dr. Xiaomei Yang for her expert assistance in the formatting and editing of the book. Their enormous efforts are wholeheartedly appreciated.

The Editor is also indebted to Claire Protherough, senior editorial assistant at Springer, for her constant support and guidance in preparation of the book. Special thanks also go to the Editor's students and colleagues for their generous help in many ways so as to allow the Editor to allocate time working on the book.

Lastly, but not least, the Editor's deep appreciation goes to his family, particularly his daughter, Susu, for their understanding of his numerous absences during evenings, weekends, and even holidays when working on the book. The Editor would also like to dedicate this book to his father on the occasion of his seventy-fifth birthday.

Contents

Part I
Electricity Generation via Inorganic and Organic Solar Cells

Enhancement of Si-Based Solar Cell Efficiency via Nanostructure Integration

Junshuai Li and Hong-Yu Yu

Abstract Solar cells are considered one of the most promising clean and renewable energy sources. Si wafer-based solar cells currently dominate the photovoltaic (PV) market with over 80% of the market share, largely owing to the available and rich manufacturing processes developed for the integrated circuit industry. However, the relatively high cost of the PV modules using Si wafer solar cells compared to conventional fossil fuels-based energy restricts its wide adoption for the civil electricity supply. How to effectively lower the costs of PV modules becomes one of the most important scientific and technical topics, especially considering the current world-wide efforts to combat climate change due to the "greenhouse" gas emissions when consuming carbon-based fossil energy. Two methodologies are generally pursued to realize this goal: one is to utilize low-grade raw materials and the other is by increasing the power conversion efficiency (PCE). In this chapter, the approaches to lower the costs and enhance the PCE of the Si-based solar cells by incorporating various Si nanostructures (e.g., nanodots, nanowires, nanocones and nanoholes) are presented, with details on the preparation techniques and their optical and electrical characteristics. The possible mechanisms of PCE improvement using these Si nanostructures are discussed in terms of enhanced light absorption and photogenerated carrier collection.

J. Li · H.-Y. Yu (✉)
School of EEE, Nanyang Technological University,
50 Nanyang Avenue, Singapore 639798, Singapore
e-mail: hyyu@ntu.edu.sg

L. Zang (ed.), *Energy Efficiency and Renewable Energy Through Nanotechnology*,
Green Energy and Technology, DOI: 10.1007/978-0-85729-638-2_1,

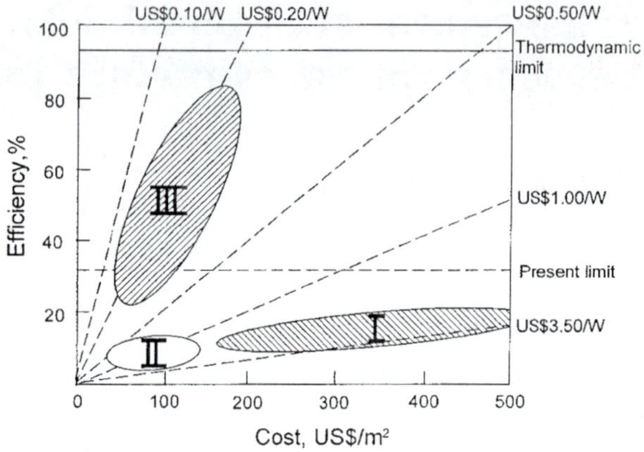

Fig. 1 Efficiency-cost trade-off for the three generations of solar cell technologies: wafers (region I), thin films (region II) and advanced thin films (region III). (Reprinted from Green [3], with permission from John Wiley and Sons; http://onlinelibrary.wiley.com/journal/10.1002/(ISSN)1099-159X)

1 Introduction

Nowadays, the negative impact of human activities to the environment receives tremendous attention, especially on the increased global temperature resulted from the emission of "greenhouse" gases such as CO_2 during production of carbon-based fossil energy. To combat climate change, clean and renewable energy sources need to be rapidly developed. Solar cell is considered as one of the ideal candidates, which directly converts solar energy into electricity without any "greenhouse" gas emissions [1].

At present, the first-generation solar cells based on Si wafers dominate, with more than 80% of the photovoltaic (PV) market [2]. However, the relatively high cost of Si wafer-based PV modules compared to other electricity sources severely restricts wide adoption of solar cells for civil utilities. To address this concern, both second and third generation solar cells are being actively pursued, as indicated in Fig. 1 [3]. Second generation solar cells are the thin film-based solar cells, which utilize inexpensive raw materials and cost-effective manufacturing techniques. Third generation solar cells are also under active research and introduce advanced physical concepts such as band gap engineering, down/up conversions for efficiently utilizing the ultraviolet/infrared photons, which are expected to significantly boost the power conversion efficiency (PCE) and reduce PV prices [4–6].

Among various materials and structures, Si-based nanostructures, such as nanodots (NDs) [7], nanowires (NWs) [8], nanocones (NCs) [9] and nanoholes (NHs) [10] are particularly promising for enhancing the PCE (and the cost per kilowatt hour) due to their unique optical and electrical properties. For example, the spectral response of the corresponding solar cells can be adjusted by incorporating SiNDs with different sizes [11], owing to quantum confinement effect on the energy gaps [12]. Accordingly, solar energy can be more efficiently utilized as compared to Si wafer solar cells with a fixed band gap ($\sim$1.12 eV). For Si-NWs, the decoupling between the light trapping and photogenerated carrier collection enables the use of low-grade raw materials, which is beneficial to reducing the manufacturing cost without sacrificing the PCE [8, 13]. To date, the development of nanofabrication techniques has enabled researchers to prepare various nanostructures with controllable size, shape and spatial distribution in a low-cost manner, potentially leading to further cost reduction of the resultant PV modules. For example, Si-NDs have already been successfully prepared in SiO_2, Si_3N_4 and SiC matrix materials-based thin films by sputtering or plasma enhanced chemical vapor deposition (PECVD) followed by a post-annealing treatment [14–16]. Si-NWs, NCs, and NHs can also be produced in large areas, compatible with solar cell production via direct synthesis or dry/wet etching approaches [17–22].

In this chapter, the current research and development efforts on cell PCE enhancement via integration of these Si nanostructures are presented. The corresponding mechanisms to PCE improvement are also discussed.

2 Si Nanostructure-Based Solar Cells

Before turning to Si nanostructure-based solar cells, the energy losses in a standard single junction solar cell are briefly summarized. Figure 2 schematically illustrates the various energy loss processes [23]. When the incident light strikes solar cells, a portion of the photons are reflected back. For the photons entering the solar cells, only those with the energy above E_g (the energy band gap of the semiconductor materials) can be absorbed and generate electron–hole pairs. Those photons with the energy below E_g pass through the device without any contribution to the resulting PCE, as marked by process ①. For the photons absorbed by the solar cells, prior to extraction of the photogenerated electron–hole pairs to the load, the main energy loss occurs in process ②, i.e., the thermalization loss, which is due to the rapid thermal relaxation for the electron–hole pairs activated by high-energy (E) photons, in the form of releasing the energy of (E–E_g) to generate phonons. The other energy loss mechanisms include junction and contact voltage losses (③ and ④), and recombination loss ⑤. Both processes ① and ② account for $\sim$50% of the solar energy loss and are related to the fixed band gap of the semi-conductor (accordingly photons with too little or too much energy cannot be effectively utilized). To address these concerns, the multijunction (tandem) configuration is actively studied. PCE as high as $\sim$86.8% has been theoretically predicted for the multijunction configuration with an infinite stack of independently operated cells under direct sunlight exposure [5]. The Si-NDs system is an ideal candidate for the multijunction configuration due to its varied band gap value caused by the inclusion of dots with different sizes. The other design to efficiently utilize the high-energy photons is referred to as the "hot carrier" cell, where the idea is to quickly extract the "hot" carriers, i.e., the high-energy photogenerated carriers, before relaxation occurs [24]. To collect the "hot" carriers, the cell thickness must be thin due to the short carrier lifetime (less than $\sim$1 ps) [5].

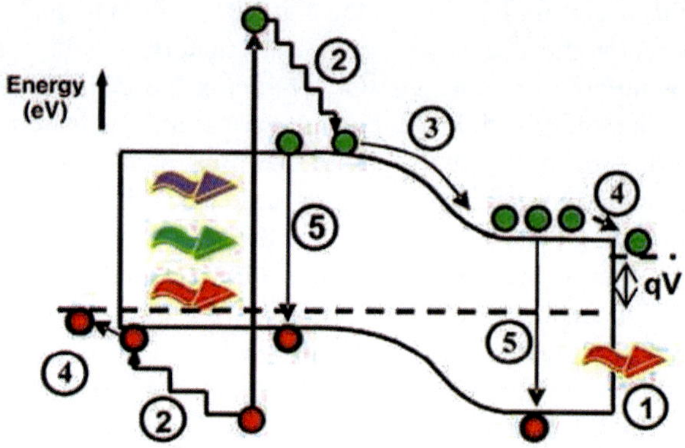

Fig. 2 Energy loss processes in a standard single junction solar cell: ① non-absorption of sub-band gap photons; ② thermalization loss; ③ and ④ junction and contact voltage losses; ⑤ recombination loss. (Reprinted from Conibeer [23], copyright 2007, with permission from Elsevier)

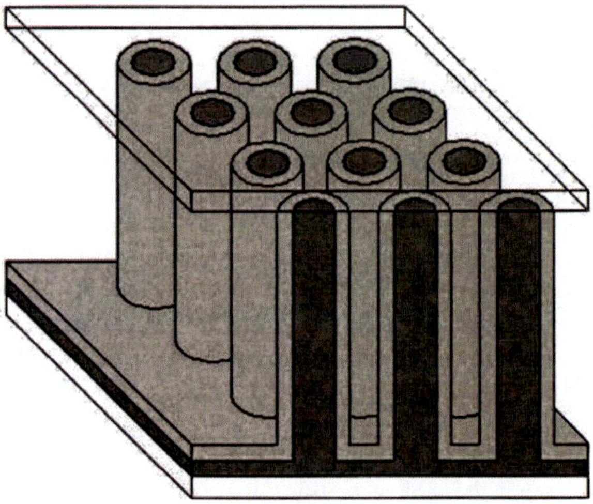

Fig. 3 Schematic of the radial p–n junction Si-NWs solar cell. Light is absorbed along the Si-NWs axis, the photogenerated carrier collection is along the radial direction of the Si-NWs. (Reprinted with permission from [8], copyright 2005, American Institute of Physics)

On the other hand, thin cell thicknesses are not able to absorb long-wavelength light. It is believed that this concern can be addressed through introducing the Si-NW, Si-NC and Si-NH structures, in which photogenerated carrier collection and light trapping can be decoupled, as shown in Fig. 3. Furthermore this type of solar cell design is beneficial to reduce the junction contact ③ and recombination losses ⑤ by controlling the Si nanostructure's dimension.

It is worth mentioning that the energy losses discussed above are only applicable to the absorbed and transmitted solar energy. The reflected portion is still considerable, which comes from the large difference of the refractive indices between air and solid Si [25]. An antireflection coating (ARC) can suppress light reflection to a certain extent; however, the effective light antireflection for the main energy range of the solar spectrum from $\sim$1 to 4 eV is challenging when using one layer of ARC with a low cost and without complicating the solar cell

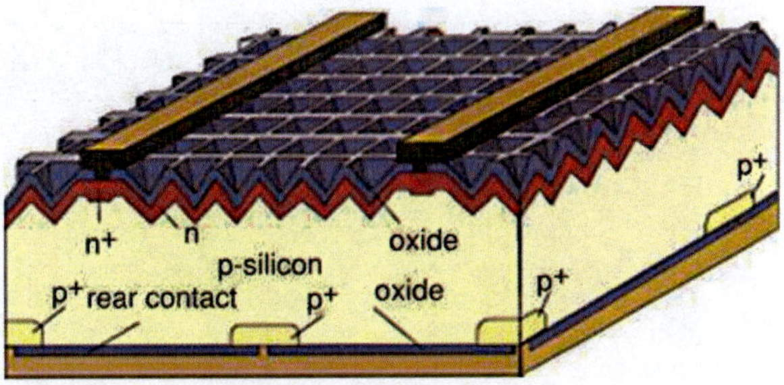

Fig. 4 Schematic of the state of the art Si wafer-based solar cell with "inverted pyramid" surface texture. (Reprinted from Green [29], with permission from John Wiley and Sons; http://onlinelibrary.wiley.com/journal/10.1002/(ISSN)1099-159X)

manufacturing [26]. For Si wafer-based solar cells, the well-known method to effectively reduce the light reflection is by surface texturing in micrometer scales, which elongates the optical path via multiple optical reflection [27, 28]. The state of the art Si wafer solar cell using the "inverted pyramid" surface texture (shown in Fig. 4) has demonstrated a PCE as high as 25.0%, and a short circuit current density of ~42.7 mA/cm^2, which is ~95% of the theoretical value of ~45 mA/cm^2 (under the radiation condition of the revised AM 1.5G) [29]. However, the technique developed for Si wafer-based solar cells cannot be directly transferred to thin film solar cells due to thickness limitations. Fortunately, the nanoscale surface texturing by Si-NWs, Si-NCs and Si-NHs arrays can address this important issue. Different from microscale surface textured Si wafer solar cells, the mechanism for enhancing the light absorption in the Si nanostructure-textured thin film solar cells can be understood based on wave optics [30, 31]. More detailed discussion on this will be provided later in this chapter.

2.1 Si-Nanodot Solar Cells

2.1.1 Preparation of Si-NDs with Controllable Size

It is reported that Si-NDs can be prepared either by direct deposition using a PECVD system or by annealing sputtered or PECVD Si-rich oxides (SiO_{2-x}), nitrides (Si_3N_{4-x}) or carbides (SiC_{1-x}) [14–16]. The sizes of the Si-NDs can be controlled by adjusting the deposition parameters. In this section, both approaches are briefly discussed.

Figure 5 shows TEM images of samples of Si-NDs in silicon nitride matrices, which are deposited on p-type Si (100) wafers at a substrate temperature of 300°C using a conventional PECVD system [11]. Nitrogen-diluted 5% SiH_4 and NH_3 are used as the precursor gases, and the flow rate of SiH_4 is fixed at 190 sccm for all samples. For the samples shown in Fig. 5a–c, the working pressure during deposition is 1.0 Torr, and the flow rate of NH_3 increases from 10 to 90 sccm. The flow rate of NH_3 is 90 sccm for the sample shown in Fig. 5d and the working pressure is 0.5 Torr. It is clear that the Si-NDs' sizes are reduced by decreasing the flow ratio

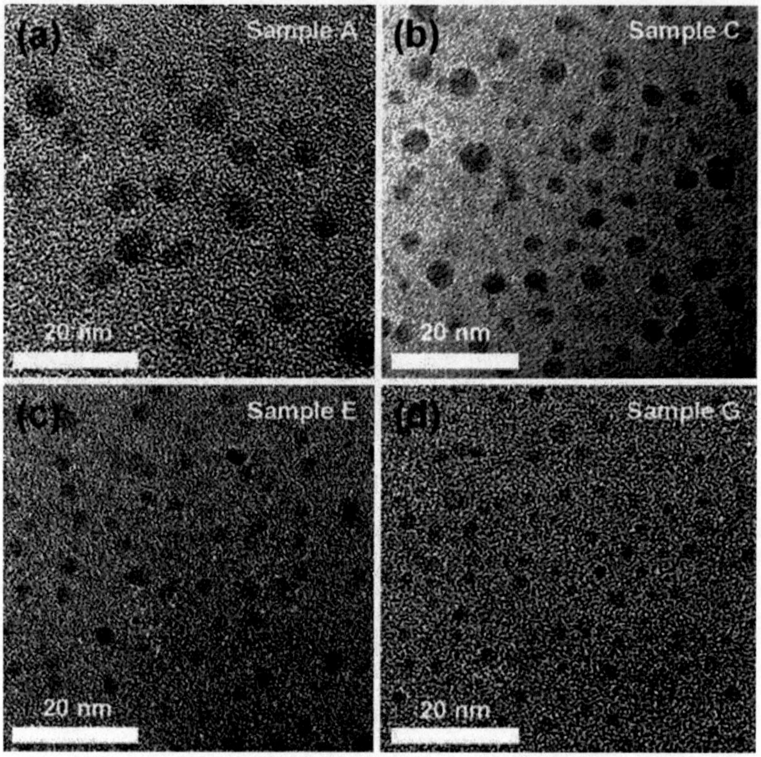

Fig. 5 TEM images of the in situ grown Si-NDs in the PECVD silicon nitride matrices. **a–c** The working pressure is 1 Torr, and the flow rate of NH_3 increases from 10 to 90 sccm. **d** The flow rate of NH_3 is 90 sccm, and the working pressure is 0.5 Torr. (Reprinted with permission from [11], copyright 2006, American Institute of Physics)

of SiH_4 to NH_3 and by decreasing the working pressure, demonstrating size control. The average diameters of the four samples are 4.9, 3.7, 3.2 and 2.9 nm, respectively. In this approach, the Si-NDs are formed during the gas phase transport process from the Si-contained radicals activated by plasma.

Different from the growth mechanism in the aforementioned direct deposition, the formation of Si-NDs through annealing Si-rich oxides follows the chemical reaction, as depicted in Eq. 1 [14]. Figure 6 shows the crystal size distribution and the corresponding plane-view TEM images after annealing the Si-rich oxide films, which were deposited using PECVD on Si (100) wafers at a substrate temperature of 300°C [32]. The precursor gases were SiH_4 (purity >99.99%) and N_2O. The crystal radius increases with an increasing annealing temperature from 1100 to 1250°C or increasing Si composition in the SiO_{2-x} films.

$$SiO_{2-x} \rightarrow \frac{x}{2}Si + \left(1 - \frac{x}{2}\right)SiO_2 \tag{1}$$

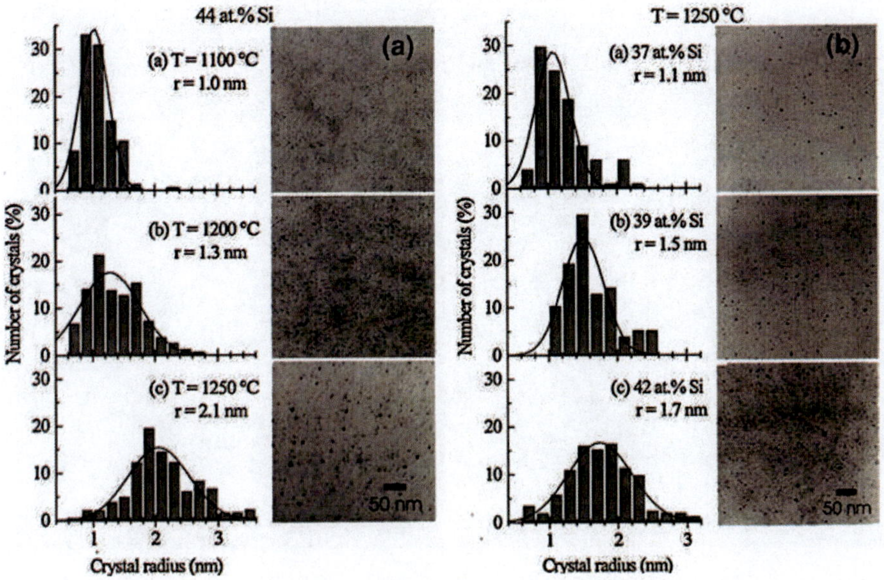

Fig. 6 Crystal size distribution and the corresponding plane-view TEM images for the Si-NDs prepared by furnace annealing the SiO_{2-x} films at different temperatures for 1 h at N_2 (**a**), and for the as-deposited films with various Si compositions at 1250°C for 1 h at N_2 (**b**). (Reprinted with permission from [32], copyright 2000, American Institute of Physics)

It was also reported that Si-NDs size can be controlled by forming a sandwich structure, such as $SiO_2/SiO_{2-x}/SiO_2$ [14, 33, 34]. Figure 7 shows cross-sectional TEM images of the SiO_{2-x}/SiO_2 superlattice before (a) and after (b) furnace annealing at 1100°C for 1 h in a N_2 atmosphere [14]. Amorphous SiO_{2-x}/SiO_2 is first prepared using reactive evaporation of SiO powders in an oxygen atmosphere with a Si-wafer substrate temperature of 100°C. The thickness of SiO_{2-x} or SiO_2 single layer is ~2.8–3.2 nm for the as-prepared sample, as indicated in Fig. 7a. After furnace annealing, the Si-NDs are segregated in the Si-rich SiO_{2-x} layers, and confined by the adjacent SiO_2 layers, as shown in Fig. 7b. The mean size estimated from TEM measurements is 3.3 ± 0.5 nm, consistent to the original thickness of the SiO_{2-x} layers. For this approach, it should be noted that the thickness of the Si-rich layers should not exceed 4 nm to achieve uniform size distribution of the Si-NDs. Otherwise, the precipitation of Si atoms will transit from the two-dimensional (2D) diffusion to 3D diffusion, resulting in a wider size distribution [7].

2.1.2 Optical and Electrical Characteristics of Si-NDs in Dielectric Matrices

As mentioned earlier, the energy band gap of Si-NDs is tunable, owing to the quantum confinement effect, and this implication is important for solar cell

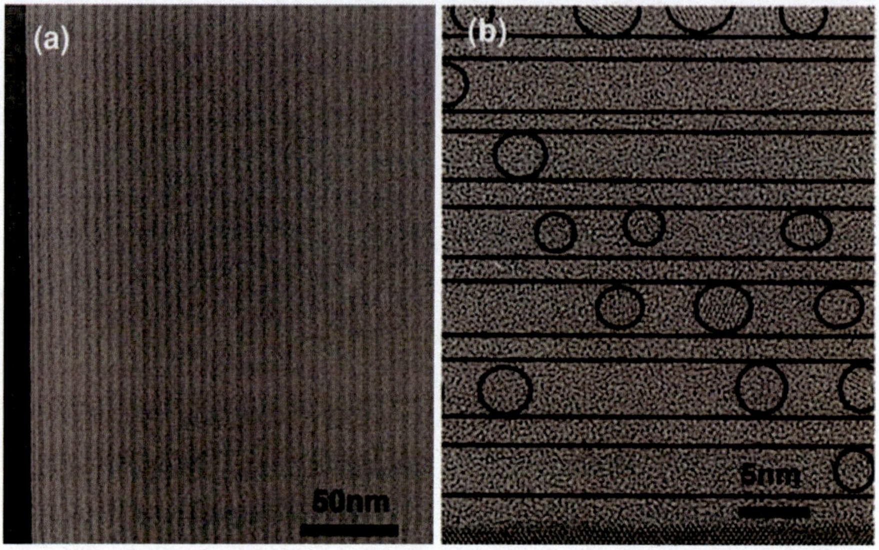

Fig. 7 TEM images of the SiO_{2-x} ($x \sim 1$)/SiO_2 superlattices: **a** as-prepared superlattice (the darker regions indicate the SiO_{2-x} sublayers); **b** the same sample after annealing at 1100°C for 1 h under N_2 atmosphere. (Reprinted with permission from [14], copyright 2002, American Institute of Physics)

applications. It favors the selective absorption of photons with different energies to minimize the energy relaxation of "hot" carriers, hence boosting the resultant PCE. Accordingly, understanding how the optical and electrical characteristics depend on Si-NDs size and distribution is necessary to guide material preparation and structure design for solar cells based on this technology. In this section, the optical and electrical characteristics of Si-NDs in various dielectric matrices are introduced, with emphasis on solar cell applications.

The optical characteristics of Si-NDs directly depend on the effective band gap, which is further determined by the Si-NDs size and the surrounding matrix [7]. Note that the band structure of the Si-NDs becomes almost direct when the Si-NDs' sizes are small enough (in general, this value should be below 7 nm for quantum confinement effect to dominate) [7]. Figure 8 shows the room temperature photoluminescence (PL) spectra of the Si-NDs, which are grown in situ in the silicon nitride films using PECVD with 10% SiH_4, diluted in argon and pure (>99.9999%) NH_3, on Si (100) wafers at 250°C [35]. It is interesting that the peak position in the PL spectra can be modulated from infrared to ultraviolet with decreasing size of the Si-NDs. By correlating the size distribution of the Si-NDs with the corresponding PL peak, the relationship between the Si-ND size and band gap (Fig. 9) can be empirically established, which is described by Eq. 2:

$$E = 1.16 + \frac{11.8}{d^2}, \tag{2}$$

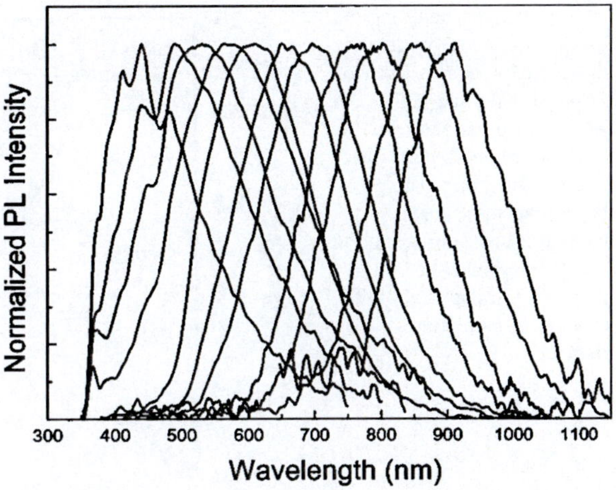

Fig. 8 Room temperature PL spectra of the Si-NDs grown in situ in silicon nitride films. (Reprinted with permission from [35], copyright 2004, American Institute of Physics)

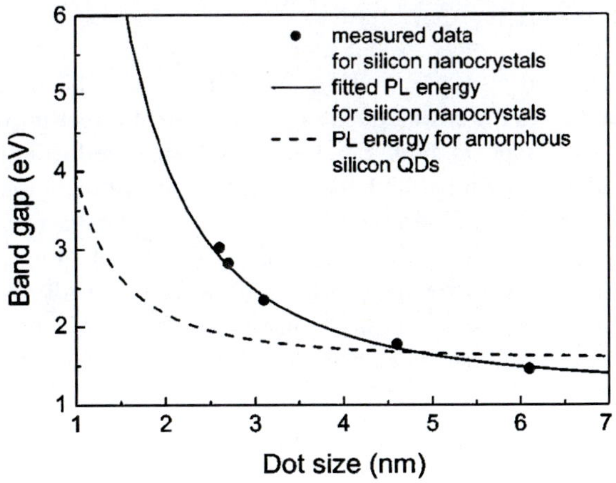

Fig. 9 PL peak energy of Si-NDs grown in situ in silicon nitride films as a function of dot size. The *solid line* indicates the energy band gap calculated using Eq. 2. (Reprinted with permission from [35], copyright 2004, American Institute of Physics)

where E is the band gap in eV and d is the Si-ND's diameter in nm.

Here it should be mentioned that Eq. 2 is based on effective mass theory with the assumption of an infinite potential barrier [35]. A more commonly used expression is shown in Eq. 3 [36]:

$$E = E_{\text{bulk}} + \frac{C}{d^2}, \tag{3}$$

where E and E_{bulk} are the energy band gaps of the NDs and the bulk material (with the same composition as the NDs) in eV, d is the ND's diameter in nm, and C is the confinement parameter. For different materials and surrounding environments, the effect on the energy band gap can be reflected by the variation in E_{bulk} and C. Figure 10 summarizes the values of the band gap of Si-NDs in silicon oxide and nitride matrices from PL measurements [7]. The data follows the trend

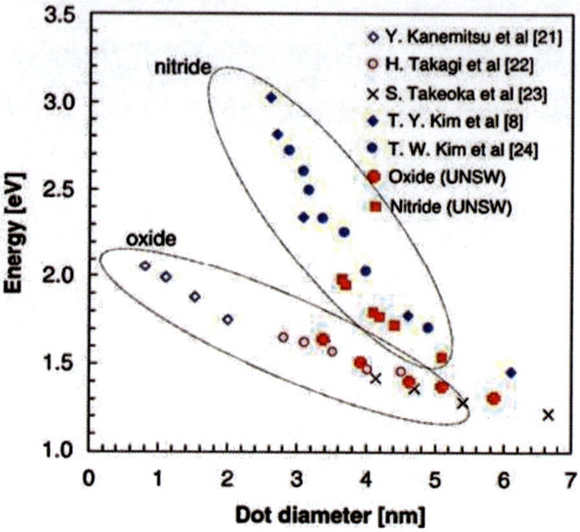

Fig. 10 Energy band gap of Si-NDs in silicon oxide and nitride matrices as a function of the Si-ND's diameter. The data are summarized from the PL measurements performed by different groups (note that the reference numbers marked in this figure given in [7]). (Reprinted from Conibeer [7], copyright 2008, with permission from Elsevier)

depicted by Eq. 3, and the sharp increase in band gap with decreasing diameters indicates the strong quantum confinement effect in Si-NDs in silicon nitride films.

To achieve high efficiencies in solar cells, both effective utilization of the incident photons and the efficient collection of the photogenerated carriers are important. Different from a standard Si wafer-based solar cell, the carrier transport in Si-ND solar cells strongly depends on tunneling between the neighboring dots through a barrier formed by the matrix materials.

The transmission probability (*T*) of a carrier between the two states with the same level can be approximated by a simplified formula [37]:

$$T \approx 16\exp\left\{-d\sqrt{\frac{8m^*}{\hbar^2}\Delta E}\right\}, \tag{4}$$

where d is the barrier width, i.e., the spacing between the neighboring dots, m^* is the effective mass of the carrier, ΔE is the energy difference between the conduction band edges (for electrons) of the matrix material and the Si-NDs, or between the energy level in the Si-NDs and the valence band edge (for holes) of the matrix material, and $\hbar$ is the reduced Plank constant. It is obvious that the transmission probability exponentially decreases with increasing $\left[d(m^*\Delta E)^{1/2}\right]$. Figure 11 schematically illustrates the energy band diagram for bulk Si in contact with SiC, Si_3N_4 and SiO_2 [6]. Among SiO_2, Si_3N_4, and SiC, the energy band offset between bulk Si to SiC is smallest, indicating an enhanced T for Si-NDs embedded in SiC. On the other hand, it is also noted that the formation of Si-NDs in SiC is more challenging than in SiO_2 due to the weak polarity or Si–Si length in Si-rich carbides. As indicated in Eq. 4, another way to increase T is by reducing the Si-NDs' size, which increases the energy level of carriers in Si-NDs and therefore

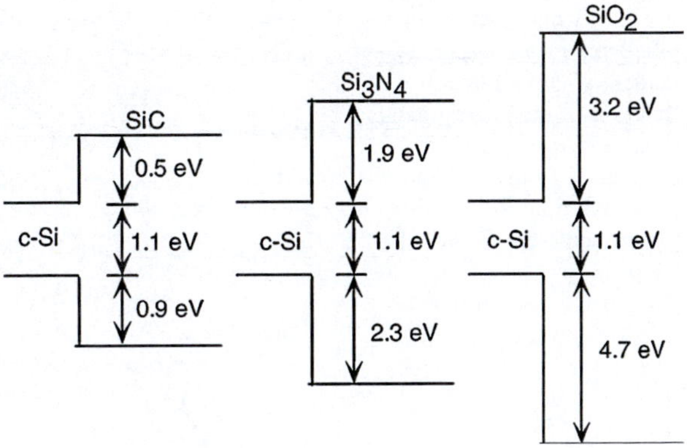

Fig. 11 Schematic of the energy band diagrams for bulk Si contacted to SiC, Si_3N_4 and SiO_2. (Reprinted from Conibeer [6], copyright 2006, with permission from Elsevier)

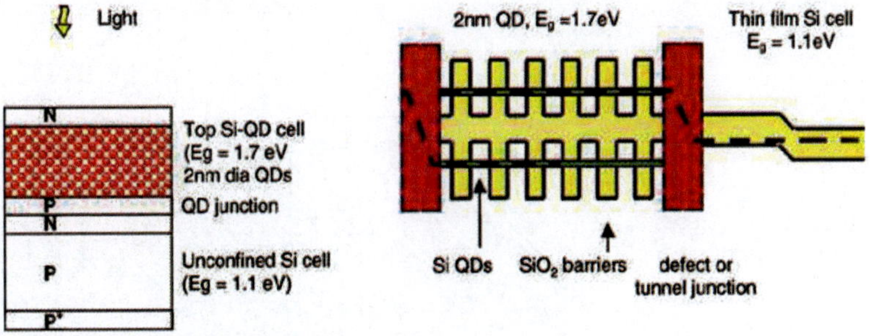

Fig. 12 Schematic of a Si-based tandem solar cell with a Si-ND *top* cell and a Si *thin* film *bottom* cell (*left*), and the energy band diagram of the tandem solar cell (*right*). (Reprinted from Conibeer [7], copyright 2008, with permission from Elsevier)

decreases ΔE. Last but not least, to enhance the resonance tunneling current, uniform size among all Si-NDs is a key factor, supported by theoretical calculations [38]. For the tunneling distance d between neighboring Si-NDs, a slight variation does not have significant impact on T.

2.1.3 Research Status of Si-ND-Based Solar Cells

Figure 12 schematically shows the structure of a typical Si-ND-based tandem solar cell, consisting of a Si-ND-based top cell and a Si thin film-based bottom cell [7]. Due to quantum confinement, the energy band gap of ~1.7 eV is achieved for Si-NDs with the diameter of ~2 nm in the SiO_2 matrix, as indicated in Fig. 10. This cell can capture the photons with the energies greater than 1.7 eV and convert them into photogenerated carriers. Photons with the energies below 1.7 eV pass through and are absorbed by the bottom cell. The expected overall efficiency is enhanced. As shown in Fig. 13, the efficiency of this tandem cell can reach ~35%,

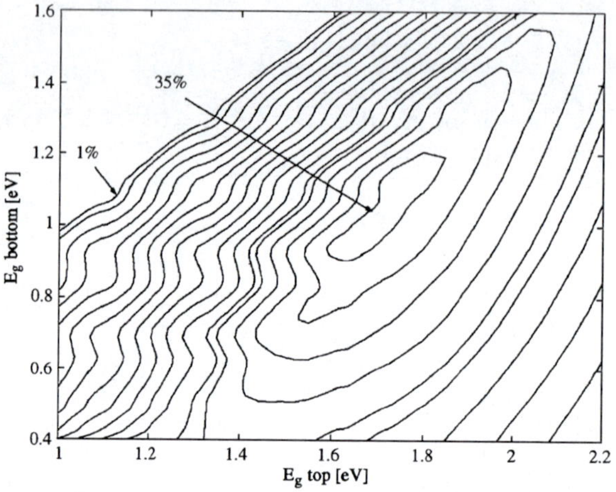

Fig. 13 Upper limit of the efficiency of p–i–n/p–i–n tandem solar cell as a function of the energy gaps $E_{g,top}$ and $E_{g,bottom}$. (Reprinted from Meillaud [39], copyright 2006, with permission from Elsevier)

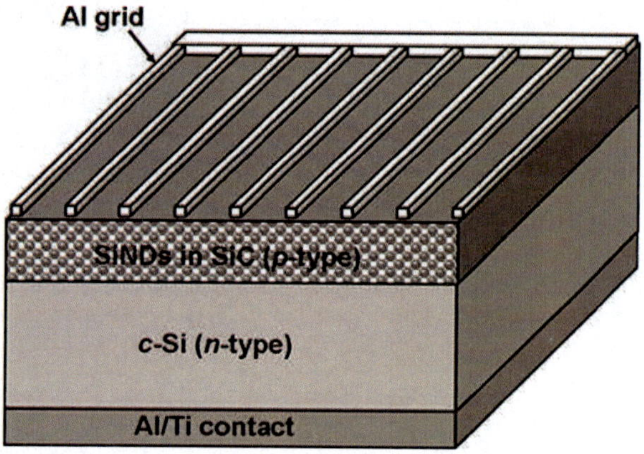

Fig. 14 Schematic of a p-type Si-NDs layer/n-type Si wafer heterojunction solar cell. (Reprinted from Song et al. [40], copyright 2008, with permission from Elsevier)

which is an improvement over the 29% calculated for a single junction Si wafer solar cell [39].

Currently, Si-ND-based solar cells are still in the early phase of research [40–44], with the majority of the work concentrating on the preparation and optical/electrical characterization of high quality materials with controllable size and/or size distribution. Thus, only preliminary experimental results related to the Si-ND layer/Si wafer heterojunction devices are summarized in this section, which is followed by guiding proposals to manufacture high efficiency Si-NDs-based solar cells.

Figure 14 shows the schematic of a typical Si-ND layer/Si wafer heterojunction device [40]. The top layer ($\sim$0.8 μm thick) is an Al metal grid, which serves as the top electrode, permitting incident light to pass through and forming an Ohmic contact with the underlying p-type Si-ND layer. The p-type Si-ND layer with Si-ND diameters of $\sim$3–5 nm is prepared by annealing the $Si_{1-x}C_x/SiC$ ($x \sim 0.1 - 0.15$) multilayers at 1100°C for 9 min in a N_2 environment and are

Fig. 15 Illuminated *I–V* curve of a p-type Si-ND layer/n-type Si wafer heterojunction solar cell under AM 1.5G. The *solid line* indicates a fitting curve using a two-diode model. The *inset* shows the pseudo *I–V* without considering the series resistance. (Reprinted from Song et al. [40], copyright 2008, with permission from Elsevier)

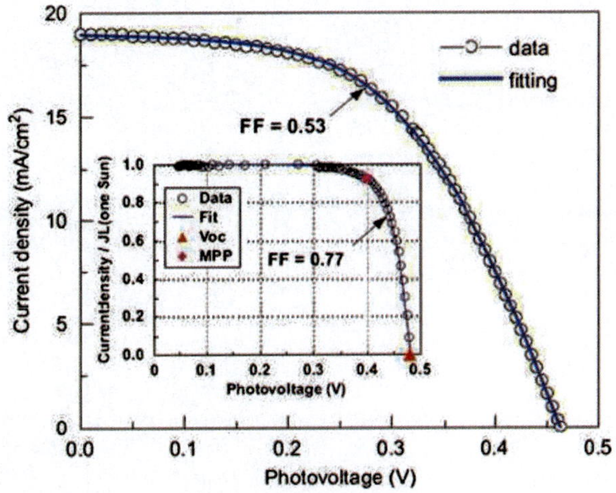

deposited using magnetron co-sputtering from Si and SiC targets (the SiC target is boron-doped). The respective thicknesses of the as-deposited $Si_{1-x}C_x$ and SiC layers are ~6 and ~2.5 nm, and the total thickness is ~160 nm. After annealing and dopant activation, the resulting Si-ND layer has a resistivity of ~10^{-1}–10^{-2} Ω cm. The resistivity of the n-type Si (100) wafer is ~2–9 Ω cm (a doping concentration of ~0.4–2.5 × 10^{15} cm^{-3}). The Ohmic back contact is built by evaporating a layer of 30 nm thick Ti film, followed by a 1.0 μm-thick Al layer. It is worth noting that there is no surface texturing or antireflection coating in the resulting solar cells.

Figure 15 shows the illuminated *I–V* characteristics of the above heterojunction solar cell at the standard AM 1.5G, i.e., the illumination condition of 100 mW/cm^2 at 25°C. The open circuit voltage, V_{oc}, short circuit current density, J_{sc} and fill factor are ~463 mV, 19 mA/cm^2 and 0.53, respectively, resulting in a PCE of ~4.66%. From the comparison with the pseudo *I–V* curve obtained by the *Suns-*V_{oc} method without considering the series resistance (shown in the inset of Fig. 15), the degradation of the fill factor is mainly attributed to the high series resistance, ~4.72 Ω cm^2, which probably comes from the imperfect Ohmic contact between the top electrode and the Si-ND layer. The low V_{oc} is due to carrier recombination, especially in the junction region, which is reflected by the relatively high ideality factor of ~1.24 extracted from the dark *I–V* measurement in the intermediate bias voltage of ~0.1–0.4 V. Meanwhile, the high light reflection (as shown in Fig. 16) of the solar cell leads to the lower external quantum efficiency (EQE) and J_{sc} compared to the surface-textured devices. One notes that the internal quantum efficiency (IQE) in the high-energy region of the solar spectrum, such as ~400 nm, is higher than that of the conventional Si wafer solar cell, which is attributed to the enlarged energy gap of the Si-ND layer. The following results validate this statement [41].

Devices consisting of an n-type Si-NDs layer/p-type Si (100) wafer (resistivity ~5–20 Ω cm) heterojunction were also prepared [41]. Si and SiO_2 targets were used

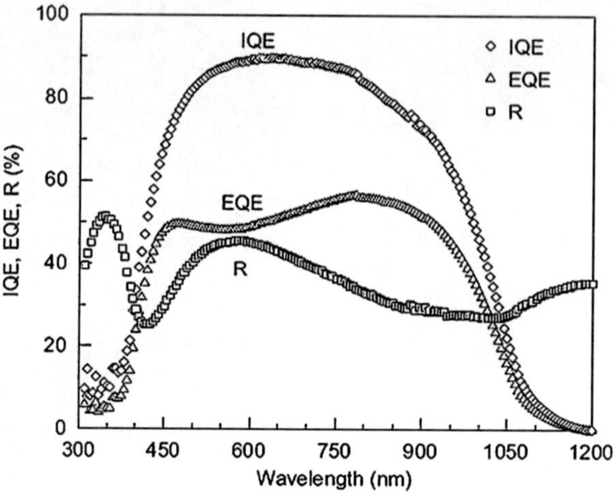

Fig. 16 Reflectance (*R*), external quantum efficiency (EQE) and internal quantum efficiency (IQE) for the same solar cell shown in Fig. 15. (Reprinted from Song et al. [40], copyright 2008, with permission from Elsevier)

Table 1 Parameters of the illuminated n-type SiNDs layer/p-type Si wafer heterojunction solar cells at AM 1.5G, i.e., ~100 mW/cm^2 at 25°C

No.	Thickness and number of $SiO_{0.89}/SiO_2$ multilayers	V_{oc} (mV)	J_{sc} (mA/cm^2)	FF (%)	PCE (%)
1	3 nm/2 nm, 15	555.6	29.8	63.8	10.6
2	4 nm/2 nm, 25	540.3	25.0	76.8	10.4
3	5 nm/2 nm, 25	517.9	27.9	72.3	10.5
4	8 nm/1 nm, 25	470.8	18.6	65.1	5.7

Reprinted with permission from [41], copyright 2008, Institute of Physics

to deposit $SiO_{0.89}$ and SiO_2 multilayers, and a P_2O_5 target was utilized for n-type doping in the Si-rich oxide layers. Si-ND formation and dopant activation were accomplished by the furnace annealing at 1100°C for 1.5 h in a N_2 environment. Al electrodes were evaporated on the top (via a shadow mask) and bottom of the solar cells, and then annealed at 400°C for 30 min in N_2 to improve the contact.

The electrical parameters of the solar cells with $SiO_{0.89}/SiO_2$ multilayers with varying thicknesses are collected under the illumination condition of AM 1.5G and summarized in Table 1. It is worth noting that V_{oc} increases with decreasing thickness of the $SiO_{0.89}$ sublayer, in line with the resulting size of the Si-NDs. It is believed that the enlarged energy band gap of the Si-ND layer with the decreased $SiO_{0.89}$ thickness can be correlated to the increased V_{oc}. It is also seen that the J_{sc} for device-1, with the thinnest Si-ND layer, is highest, indicating that the majority of photocurrent is not from this layer. As the thickness of the Si-ND layer increases, the larger tunneling distance coupled with the accumulated defects in this layer leads to the reduction of J_{sc}.

Figure 17 shows the IQE curves for the four samples. It is clearly shown that the IQE is reduced in the high-energy region with the increase of the Si-NDs' size, and the peak shifts toward low energy (or long wavelength). This figure provides

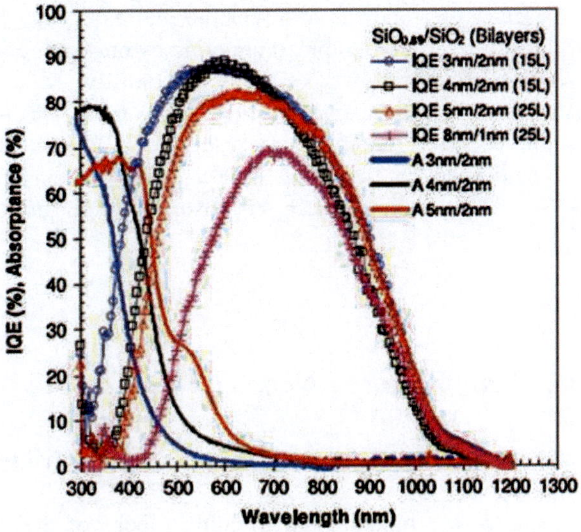

Fig. 17 IQE of the n-type SiNDs layer/p-type Si wafer heterojunction solar cells with different thicknesses and numbers of the $SiO_{0.89}/SiO_2$ multilayers. The light absorption of the SiNDs layer for the first three devices is also included. (Reprinted with permission from [41], copyright 2008, Institute of Physics)

evidence that the enlarged energy band gap for the SiNDs layer leads to the higher IQE for the Si-ND layer/Si wafer heterojunction solar cell in the higher energy region of the solar spectrum compared to the Si wafer homojunction devices [40]. As seen from the same figure, the light absorption in the Si-NDs layers mostly occurs in the high-energy region of the solar spectrum. With the increase of Si-ND size, the absorption edge shifts toward the low-energy regime due to the decrease in the effective band gap.

Following the brief summary of the current research efforts on SiNDs layer/Si wafer heterojunction solar cells, some guidelines to achieve the high-efficiency and cost-effective Si-ND-based solar cells are proposed below:

(i) *Close-packed Si-NDs with uniform size distribution.* To efficiently collect photogenerated carriers, the Si-NDs must be closely packed for the high resonant tunneling efficiency between the neighboring Si-NDs. On the other hand, how to prepare close-packed Si-NDs by controlling the nucleation sites is not only critical, but challenging.

(ii) *Effective high doping of the Si-NDs layer.* The doping of Si-NDs is difficult from the aspects of energy and kinetics due to small-size effects [45, 46]. To realize a high-quality p–n junction for effectively extracting the photogenerated carriers, high-dose doping in the Si-NDs emitters is necessary.

(iii) *High-quality Ohmic contact with the Si-ND layer.* As discussed previously, the series resistance severely affects the illuminated *I–V* curve by decreasing the fill factor. Accordingly, exploring suitable metal electrodes and metallization process is highly demanded.

(iv) *Optimized conjunction between neighboring cells.* The low V_{oc} for the above devices results from carrier recombination, especially in the conjunction regions, can be attributed to the defects created during high-temperature annealing and the different thermal expansion factors of the various materials. Optimizing the conjunction by employing low-temperature processes is desirable for lowering the defect density.

(v) *Antireflection design for suppressing light reflection.* It is clear that the light reflection is another major source of loss in the aforementioned devices, evidenced by their low EQEs. Therefore, an appropriate surface texturing or antireflection coating is also essential for high-efficiency Si-ND-based solar cells.

2.2 Stand-Alone Si Nanowire-Based Solar Cells

2.2.1 Preparation of Si-NWs Array with Controllable Dimension

As one of the fundamental building blocks of the "nanoworld", Si-NWs have been extensively studied in recent years [47–53]. The methods to prepare Si-NWs can generally be categorized into two classes: one is the "top-down" method [54, 55] and another is the "bottom-up" approach [56, 57]. The former commonly involves preparing Si-NWs using various etching methods or combining different patterning techniques. The latter case is generally based on the vapor–liquid–solid (VLS) mechanism [58] to grow Si-NWs with the assistance of various catalysts such as Au, Cu, Fe, Al, etc. [59–62]. On the other hand, it needs to be pointed out that neither approach is perfect. For instance, by using the reactive ion etch (RIE) with electron beam or photolithography, one can fabricate highly uniform Si-NWs, which are commonly used in integrated circuit devices such as transistors [63, 64]. However, the high cost/low throughput of such a process strictly restricts its application in solar cell industry. The VSL growth of Si-NWs with an area of over 1 cm^2 can be realized even on glass substrates [17]. However, the unavoidable doping from the catalysts during growth can introduce deep energy level defects, which severely affect the electrical characteristics of the resulting Si-NWs [65]. On the other hand, for solar cell applications, the critical factor to be considered for the preparation of Si-NWs is the cost. With low costs, efficient light trapping and photogenerated carrier collection also needs to be studied. In this section, we highlight the approaches to prepare the Si-NWs with low costs and high controllability in terms of dimension and spatial distribution.

Electron beam or photolithography is acknowledged as a key factor leading to the high cost of preparing Si-NWs for solar cell applications. The alternative patterning methods to replace them have been actively pursued. Recently, the Cui group from Stanford University reported using a Langmuir–Blodgett (LB) [66] assembled monolayer of the SiO_2 particles to serve as the "nanosphere" lithographic mask. Using this approach, uniform and dense patterning can be achieved over a large area as shown in Fig. 18 [19].

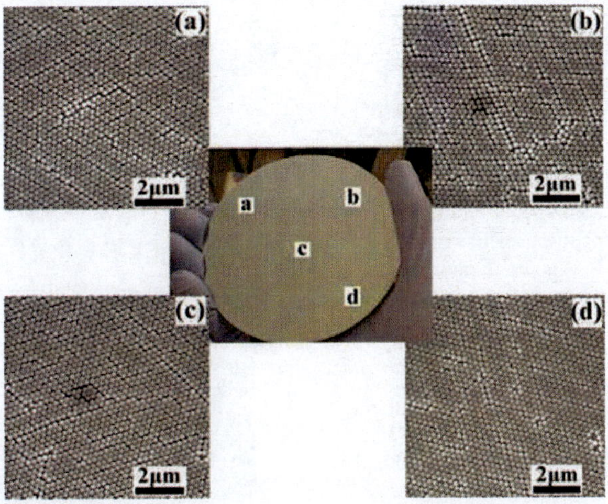

Fig. 18 SEM images of the close-packed SiO_2 nanosphere (diameter of ~200 nm) monolayer prepared by the Langmuir–Blodgett method on a 4 in. Si (100) wafer. Figures **a–d** demonstrate the four different regions corresponding the symbols marked in the wafer, indicating a high uniformity over the large area. (Reprinted with permission from [19], copyright 2008, American Institute of Physics)

After forming the close-packed SiO_2 nanosphere monolayer, reactive ion etching (RIE) can be performed to selectively reduce the dimension of the SiO_2 nanospheres using O_2 and CHF_3, as demonstrated in Fig. 19a, b [19]. Then Cl_2-based RIE is used to anisotropically etch away the underlying Si material with SiO_2 nanospheres acting as a hard mask, creating a highly-ordered Si-NWs array (see Fig. 19c). The residual SiO_2 at the tip of the Si-NWs can then be removed by HF acid. Figure 19d shows the tilted SEM image of the resulting Si-NWs array. It is clear that the array's periodicity is transferred from the SiO_2 nanospheres in the as-prepared monolayer. The Si-NWs' diameters can be modulated by adjusting the RIE parameters. Moreover, the dimension of the Si-NWs can be further scaled to below 100 nm with additional thermal oxidation and subsequent HF etching. It is worth mentioning that this approach can be utilized to manufacture Si-NWs arrays on polycrystalline Si or amorphous Si thin films [9].

Except the RIE method, which requires relatively expensive equipment, another more cost-effective wet etching approach has also been used to fabricate high-quality Si-NWs arrays [67]. Figure 20 shows the schematic process flow and the corresponding SEM images in preparing a highly-ordered Si-NWs array using electroless wet etching to replace the anisotropic RIE. Si wafers are first cleaned in a boiled solution of NH_4OH: H_2O_2: H_2O = 1:1:5 for 1 h to achieve a hydrophilic surface. The close-packed SiO_2 nanosphere monolayer is then prepared via the method described by Lu et al. [68] from a SiO_2 colloidal solution, as shown in Fig. 20a. The SiO_2 nanospheres' dimensions can be reduced by chemical etching in HF solution (see Fig. 20b). This is followed by catalyst metal deposition, such as Ag or Au, using vacuum evaporation. The area on the Si wafer not covered by SiO_2 nanospheres is exposed to the catalyst metal layer as shown in Fig. 20c. After removing the SiO_2 nanospheres by ultrasonication in water for 2–3 min, the metal thin film with the periodic nanopore array is achieved (see Fig. 20d). After immersing the sample into the HF/H_2O_2 solution, the highly-ordered Si-NWs array was created by the selective electroless etching of the Si underneath the catalyst metal, as observed in Fig. 20e. The Si-NWs' length can be easily adjusted by

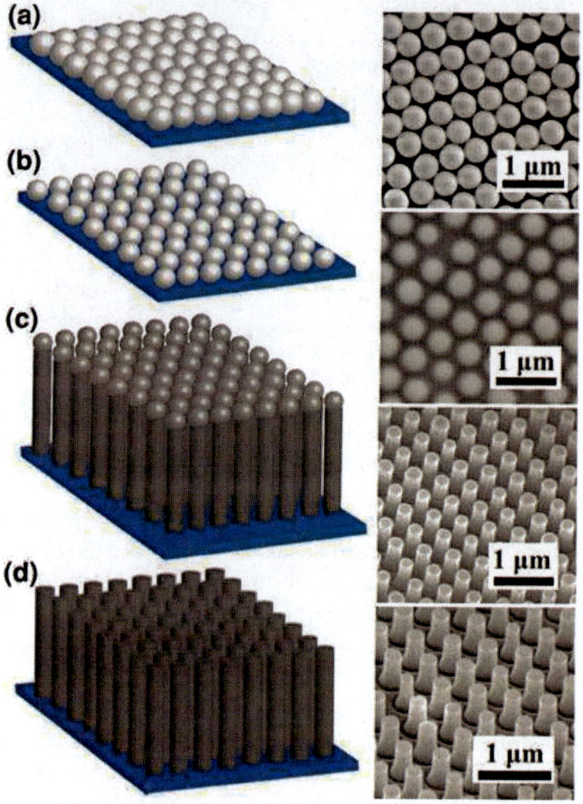

Fig. 19 Schematic (*left*) and the corresponding SEM image (*right*) of each step used to prepare a highly-ordered Si-NW array with the controllable periodicity and Si-NW diameter using RIE combined with the SiO_2 nanosphere Langmuir–Blodgett patterning method. **a** Deposition of the silica nanoparticles by Langmuir–Blodgett. **b** Shrinking of the mask by isotropic RIE of SiO_2. **c** Anisotropic etching of Si into pillars by RIE. **d** Removal of the residual mask by HF etching. (Reprinted with permission from [19], copyright 2008, American Institute of Physics)

changing the etching time, and the array's periodicity and Si-NWs diameter are controlled by the diameters of the starting and reduced SiO_2 nanospheres. In general, the etching mechanism is based on the following chemical processes [69]:

With Ag nanoparticles:

$$H_2O_2 + 2H^+ \rightarrow 2H_2O + 2h^+,$$

$$2H^+ + 2e \rightarrow H_2 \uparrow,$$

With Si:

$$Si + 4h^+ + 4HF \rightarrow SiF_4 + 4H^+,$$

$$SiF_4 + 2HF \rightarrow H_2SiF_6,$$

Overall reaction:

$$Si + H_2O_2 + 6HF \rightarrow 2H_2O + H_2SiF_6 + H_2 \uparrow .$$

There are also other approaches based on the "top-down" paradigm to create Si-NWs arrays. Generally speaking, to integrate Si-NWs arrays in solar cells, the

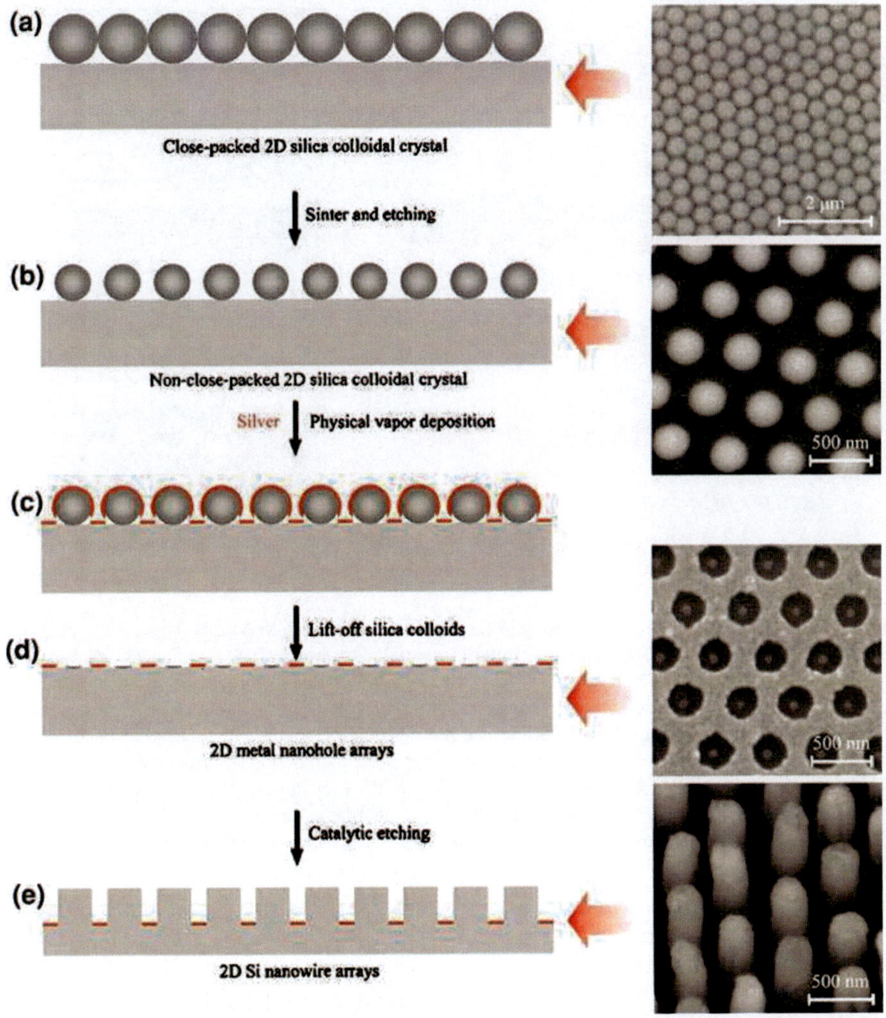

Fig. 20 Schematic of the process flow (*left*) and the corresponding SEM image (*right*) of each step used in preparing a highly-ordered Si-NW array with controllable periodicity and diameter using electroless etching combined with a self-assembled SiO_2 nanosphere monolayer severing as the mask. **a** Deposition of monolayer silica colloidal crystal template on silicon surface; **b** fabrication of 2D non-close-packed silica colloidal crystals on silicon surface; **c** deposition of silver layer on silicon surface through the non-close-packed colloidal crystal template; **d** formation of regular silver nanohole arrays by removing silica colloids by brief ultrasonication in water; and **e** formation of Si-NWs by catalytic etching. The corresponding SEM micrographs on the right show the monolayer silica colloidal crystal template (**a**), the 2D nonclose-packed silica colloidal crystal template (**b**), the silver film with periodic nanohole arrays (**d**), and ordered Si-NW arrays produced using catalytic silver film with periodic nanoholes (**e**). (Reprinted with permission from [67], copyright 2007, American Institute of Physics)

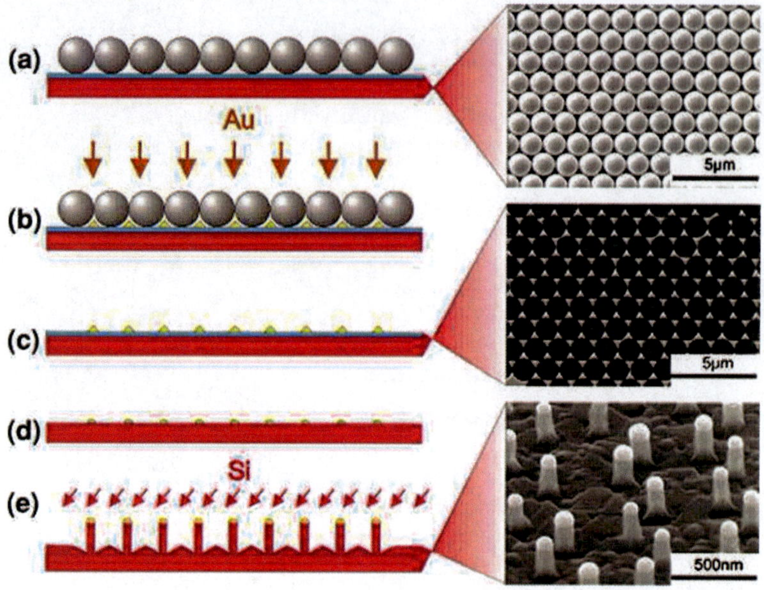

Fig. 21 Schematic of the process flow (*left*) and the corresponding SEM image (*right*) to prepare highly-ordered Si-NW arrays with controllable spatial distribution and size using VLS-based "bottom-up" growth combined with a pre-patterned Au catalyst. **a** Deposition of a mask of polystyrene particles on a Si-(111) substrate covered by a 2-nm-thick oxide layer (*blue*), **b** deposition of gold by thermal evaporation, **c** removal of the spheres, **d** thermal annealing and cleaning step to remove the oxide layer, and **e** Si deposition and growth of nanowires by MBE. The first two SEM images correspond to the use of PS nanospheres having the diameter of ~1320 nm, and the third corresponds to the PS nanospheres with the diameter of 600 nm. (Reprinted with permission from [70], copyright 2005, American Chemical Society)

trade-off between the manufacturing cost and the enhanced PCE should be seriously taken into account.

Next let us turn to the "bottom-up" approaches to grow Si-NWs arrays. Through manipulating the spatial and size distribution of the catalysts, the controllable periodicity and diameter can be achieved using the VLS paradigm. Because VLS growth itself is cost-effective, the final manufacturing cost strongly depends on the lithographic process to pattern the catalysts. Here several low-cost lithographic processes are discussed.

Figure 21 illustrates an example process to prepare a Si-NWs array by the VSL-based "bottom-up" technique [70]. Similar to the patterning process used in the aforementioned "top-down" etching technique, a close-packed monolayer of polystyrene (PS) nanospheres with desired diameters is deposited by spin coating, as shown in Fig. 21a. Then, an Au layer with a thickness between 10 and 20 nm is deposited via thermal evaporation. Au nanoislands on the substrate are formed due to the triangular interspacing between neighboring PS nanospheres, as demonstrated in Fig. 21b. After clearing the PS nanospheres by ultrasonication in CH_2Cl_2 for 2 min, the well-arranged Au nanoislands with a hexagonal pattern are formed

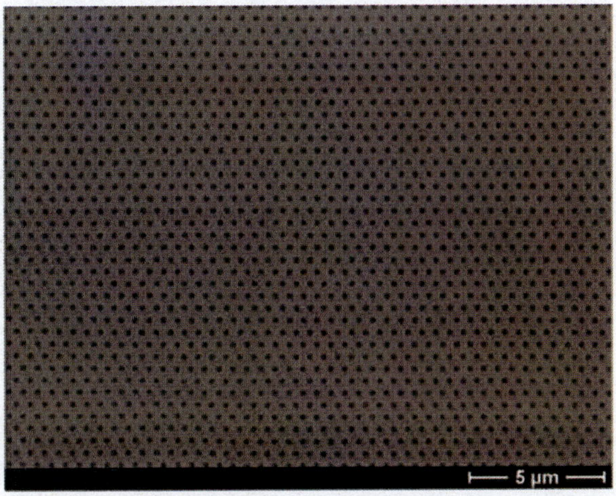

Fig. 22 Plane-view SEM image of monodomain AAO with a pore diameter of 180 nm and interpore distance of 500 nm. (Reprinted with permission from [73], copyright 2003, American Chemical Society)

(see Fig. 21c). The sample can be transferred to the Si-NWs growth system, an ultrahigh vacuum MBE chamber in this example. The sample is heated to 810°C for 10 min to transform the triangular Au nanoislands into hemispheres as schematized in Fig. 21d. Then the Si-NWs growth is performed at a sample temperature of 525–570°C. The Si-NWs' lengths are controlled by the deposition time. As indicated in Fig. 21e, the resulting Si-NWs array transfers the pattern from the hexagonal Au nanoparticles. The spatial distribution and diameter of the Si-NWs can be modulated by varying the diameter of the PS nanospheres and the thickness of the Au film.

Anodic aluminum oxide (AAO) with hexagonally arranged nanopores (see Fig. 22) has been widely utilized in preparing aligned arrays of various nanostructures of different materials [71–74]. AAO can be manufactured using the electrochemical anodization of pure aluminum over a large area based on the dissipation structure mechanism [75]. The nanopore size and spacing can be tuned by adjusting the electrochemical anodization conditions. For instance, the nanopore size and density can be controlled by varying the applied voltage [71, 76]. After the formation of the AAO layer, it can be transferred and bonded with other substrates (e.g., Si wafers) [74, 77]. Then the catalyst (Au, Ag, Cu, etc.) nanoparticle array can be created on the substrate through the AAO layer. Figure 23 shows a typical evaporated Au nanoparticle array formed with the assistance of an AAO mask [77].

Upon obtaining the Au nanoparticle decorated substrate, the Si-NWs array can be grown based on the VLS mechanism. Figure 24 shows the Si-NWs array grown on the Si (111) substrate (shown in Fig. 23), with $SiCl_4$ as the precursor gas at 900°C. The resulting Si-NWs array follows the Au nanoparticle pattern, with good size control.

In addition, it is also noted that nano-imprint lithography is actively studied to pattern the catalyst metal layer and holds great potential to prepare large area Si-NWs arrays with low cost and high controllability [78]. Furthermore, the preparation of large-area and highly-ordered Si-NWs arrays embedded in polymer substrates has been successfully realized by transferring the VLS grown Si-NWs array, which paves the way for flexible Si-NWs-based solar cells [79].

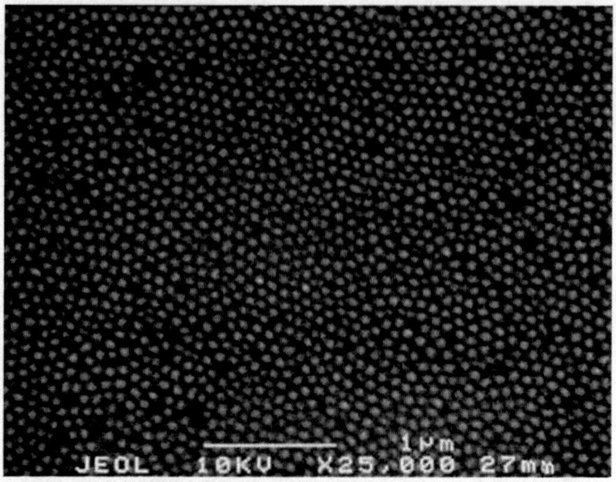

Fig. 23 Au nanoparticle array deposited through an AAO mask manufactured from an aluminum plate having multidomains. The average diameter, spacing and height of the nanoparticles are ~53, 100 and 5 nm, respectively. (Reprinted with permission from [77], copyright 2006, American Chemical Society)

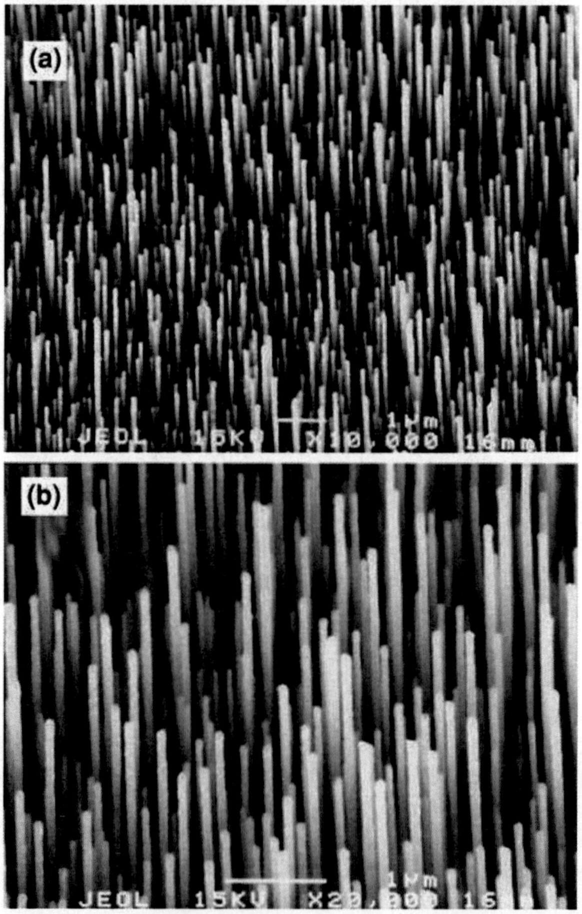

Fig. 24 Si-NW array with different magnifications grown on the substrate shown in Fig. 23. **a** Low magnification and **b** high magnification SEM images of vertically aligned, diameter-controlled Si-NWs grown from ordered Au dots on Si(111) substrates. The average diameter of the Si-NWs is ~72 nm. (Reprinted with permission from [77], copyright 2006, American Chemical Society)

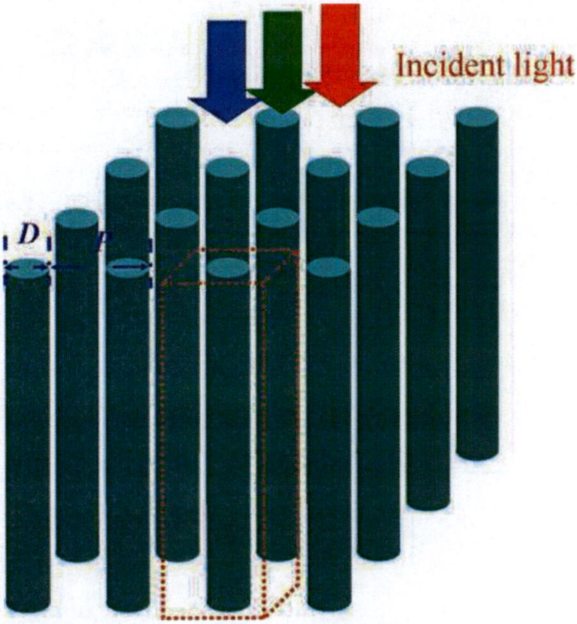

Fig. 25 Schematic of the stand-alone Si-NWs array studied by simulation. (Reprinted with permission from [87], copyright 2009, American Institute of Physics)

2.2.2 Optical and Electrical Characteristics of Stand-Alone Si-NWs Arrays

One of the most attractive advantages of using Si-NWs arrays in solar cells is the decoupling between light trapping and photogenerated carrier extraction [8]. In other words, Si-NWs can be grown long enough to effectively trap the incident photons and thin enough to efficiently collect the photogenerated carriers, enabling the utilization of low-grade raw materials, reducing the manufacturing cost. For either effective light absorption or efficient photogenerated carrier collection, the trade-off between the manufacturability (e.g., cost, controllability, reproducibility) and the resulting PCE should be considered. For instance, Si-NWs tend to break down if they are too thin or too long, which eventually leads to degraded photogenerated carrier extraction. Electrode contact and the light absorption capability would also be negatively impacted. These factors can explain why the present Si-NW-based solar cells have low PCEs, below 1% [80–82]. Therefore, providing the optimal microstructural parameters, such as array periodicity, Si-NWs diameter/length, is critical to the fabrication of Si-NWs-based solar cells.

Enhanced light absorption by Si-NWs arrays has been demonstrated empirically, especially in the high-energy regime of the solar spectrum [18, 50]. It was also observed that periodic and highly-ordered Si-NWs arrays are desirable for such a purpose [83]. Recently, the Chen group from Massachusetts Institute of Technology studied the optical characteristics, including light reflection, transmission and absorption, for a periodic stand-alone Si-NWs array (schematically shown in Fig. 25) with variable Si-NWs length/diameter at a fixed array periodicity of 100 nm using the transfer matrix method (TMM) [84]. Figure 26 shows

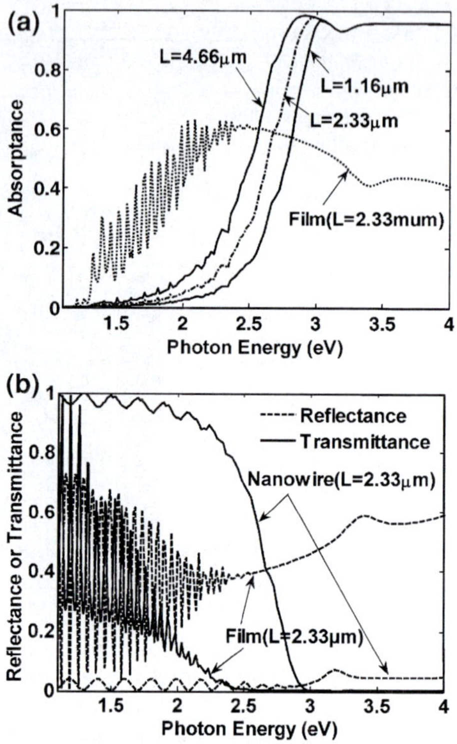

Fig. 26 **a** Absorption spectra of the Si-NW arrays as a function of the Si-NW length, and **b** reflection and transmission spectra of the 2330 nm long Si-NW array. The array periodicity and Si-NW diameter are fixed at 100 and 50 nm, respectively. The spectra of the 2330 nm thick Si film serve as the reference. (Reprinted with permission from [84], copyright 2007, American Chemical Society)

the optical characteristics of the Si-NWs array as a function of Si-NWs length. The diameter of the Si-NWs are fixed at 50 nm, and the incident light, parallel to the Si-NWs axes, varies from 1 to 4 eV, covering the majority of the solar spectrum [85]. It is clearly seen that the light absorption of a stand-alone Si-NWs array is above 95% in the high-energy regime (>2.8 eV) of the solar spectrum, much higher than that of the Si thin film counterpart. However, the light absorption of the Si-NWs array sharply decreases to nearly zero when the energy of the incident light is below ~2.8 eV for an array of Si-NWs with lengths of 1160 nm. Even for the 4660 nm length Si-NW array, the light trapping capability is inferior to that of a 2330 nm thick Si film below the energy of ~2.6 eV. As indicated in Fig. 26b, the poor light absorption for Si-NWs arrays in the low-energy region mainly stems from high light transmission in this energy regime. Low light reflection in the entire energy range, i.e., 1–4 eV, demonstrates the excellent antireflection function of Si-NWs arrays.

To evaluate the overall light trapping capability for solar cell applications, the ultimate efficiency (η) is calculated according to the following formula [30, 84, 86]:

$$\eta = \frac{\int_{E_g}^{\infty} \frac{E_g \times I(E) \times \alpha(E)}{E} \mathrm{d}E}{\int_0^{\infty} I(E) \mathrm{d}E}, \tag{5}$$

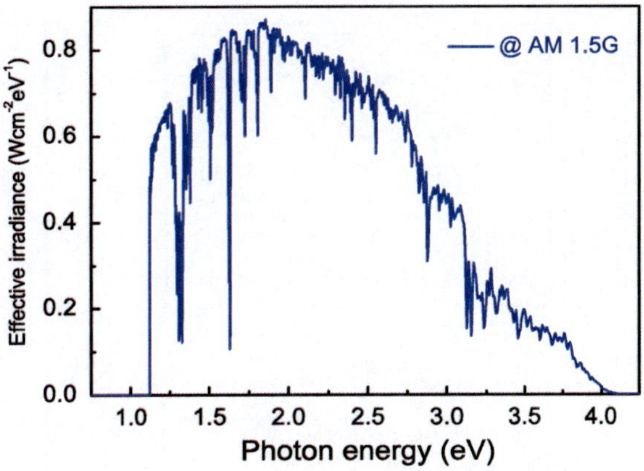

Fig. 27 Effective irradiance of the solar spectrum under AM 1.5G for single band gap Si solar cells. The effective irradiance is defined as the product of the irradiated power density and $(E_g/h\nu)$

where E_g is the band gap for the material (~1.12 eV for Si), E is the photon energy, $I(E)$ is the solar energy density spectrum, and $\alpha(E)$ is the absorption spectrum. From Eq. 5, one notes that η is defined as the optical–electrical conversion capability, provided that a photon with the energy above E_g, which can be trapped by the solar devices, is converted into one electron–hole pair with energy equal to E_g, and the electron–hole pair can be completely extracted for external output. In other words, when calculating the ultimate efficiency, the internal quantum efficiency is assumed to be 100%. The calculated η for the Si-NWs arrays with the lengths of 1160, 2330 and 4660 nm are ~4.4, ~5.8 and ~7.8%, respectively. However, for the 2330 nm thick Si film, the ultimate efficiency reaches ~15.5%, almost twice that of the best Si-NWs array. Thus, it is concluded that strong light absorption only in the high-energy regime of the solar spectrum does not make Si-NWs arrays competitive with Si thin films.

This can be understood from the potential contribution of the solar spectrum to a single junction solar cell. Figure 27 depicts the effective irradiance, or the product of the irradiated power (under AM 1.5G) per cm^2 and $(E_g/h\nu)$ (note that for the photons with energy below E_g, this value is 0, i.e., no generation of electron–hole pairs), which illustrates the potential maximum power conversion capability. It is clearly shown that the effective irradiance for Si solar cells in the low-energy regime of the solar spectrum (~2 eV) plays a more profound role than in the high-energy regime to enhance the PCE. Due to the poor light absorption in the low-energy regime, it is reasonable that the ultimate efficiency for the 4660 nm Si-NW array is lower than that of the 2330 nm thick Si film. Similar to the case of increasing the Si-NWs' lengths (as demonstrated in Fig. 26a), increasing the Si-NWs diameters from 50 to 80 nm also leads to a redshift of the absorption edge [84]. However, the limited redshift still does not improve the overall light absorption of the Si-NWs array to be comparable to that of Si thin films with the same thickness. As discussed below, the poor light absorption for the Si-NWs

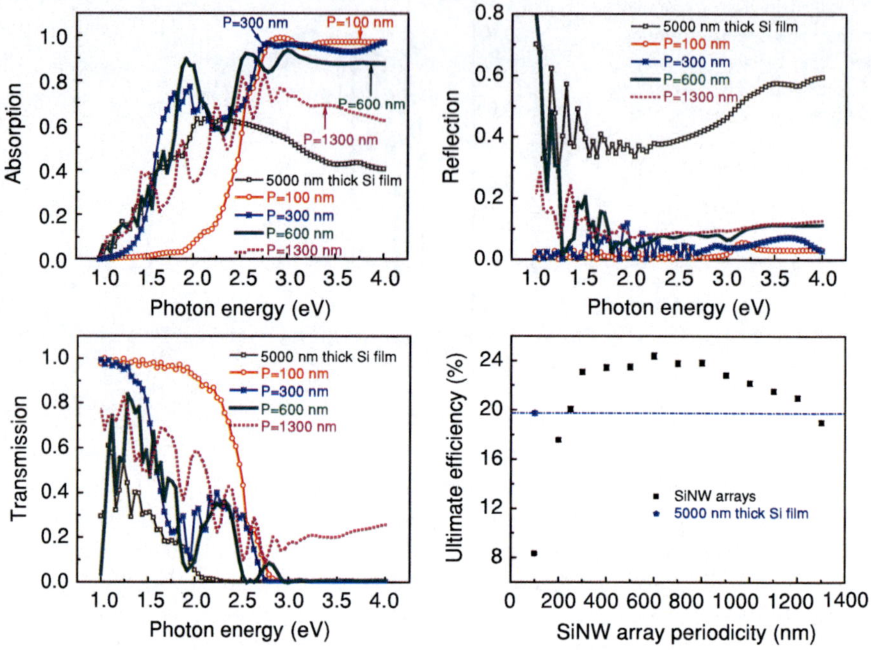

Fig. 28 **a** Absorption, **b** reflection, and **c** transmission spectra of the Si-NW arrays with different periodicities, *P*, of 100, 300, 600 and 1300 nm. The ultimate efficiency, η, of the Si-NW arrays as a function of *P* is summarized in **d**. The length of the Si-NWs in the array is 5000 nm, and the ratio of the Si-NW diameter to the array periodicity is fixed at 0.5. The 5000 nm thick Si film serves as a reference. (Reprinted with permission from [87], copyright 2009, American Institute of Physics)

arrays discussed by Chen et al. is due to their relatively small periodicity of 100 nm.

More recently, our group conducted a study on the impact of the Si-NWs array's periodicity to the optical characteristics using the finite element method (FEM) and presented the underlying physics responsible for the observations [87]. Figure 28a–c shows the light absorption, reflection and transmission spectra of the Si-NWs arrays as a function of periodicity. The Si-NWs length is set to 5000 nm, and the ratio of the Si-NWs diameter (*D*) to the array periodicity (*P*) is fixed at 0.5. One notes that the shift of the light absorption edge is much more sensitive to the array periodicity.

When the array periodicity is increased to 600 nm, the light absorption for the Si-NWs array is higher than that of the reference sample, i.e., the Si film with the same thickness of 5000 nm, almost in the entire energy range of 1–4 eV. As indicated by the transmission spectrum in Fig. 28c, the evident redshift of the light absorption edge can be attributed to the significantly suppressed light transmission in the lower energy region of the solar spectrum with an increased array periodicity. However, as the array periodicity is further increased, the

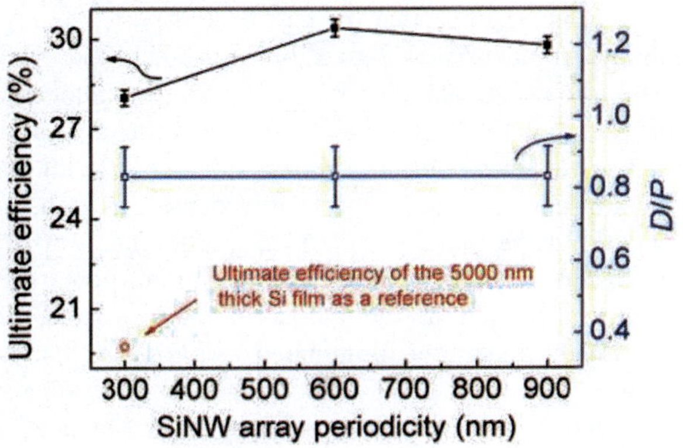

Fig. 29 Maximum ultimate efficiency and the corresponding *D*/*P* for the Si-NW arrays as a function of the array periodicity ranging from 300 to 900 nm. The value for the 5000 nm thick Si film is shown as a reference. (Reprinted with permission from [87], copyright 2009, American Institute of Physics)

spacing between neighboring Si-NWs also increases, resulting in increased light transmission, as indicated by the transmission spectrum of the Si-NWs array with a periodicity of 1300 nm. Moreover, light reflection (see Fig. 28b) is gradually enhanced due to the increased cross-sectional area of the Si-NWs. Accordingly, the light absorption is degraded when the array periodicity becomes too large. Figure 28d summarizes the calculated ultimate efficiencies, η. In agreement with the light absorption change with the array periodicity, η first increases significantly when increasing the array periodicity from 100 to 300 nm. Beginning with the Si-NWs array with the periodicity of 250 nm, η already becomes larger than that of the reference Si film with the same thickness of 5000 nm ($\sim$19.7%). As the array periodicity is further increased to 600 nm, η slightly increases to a maximum value of $\sim$24% due to the combined effects of the suppressed light transmission in the low-energy regime and the increased light reflection. The slight decrease in η can be ascribed to light transmission in the high-energy regime in addition to the increased light reflection when the array periodicity is above 600 nm.

Our further study reveals that the maximum ultimate efficiency for each periodicity ranging from 300 to 900 nm can be achieved when the ratio of the Si-NWs diameter to array periodicity (*D*/*P*) is around 0.8 [87]. At this *D*/*P* value, η of $\sim$30.5% is achievable for the Si-NWs array with the array periodicity of 600 nm, as demonstrated in Fig. 29. More meaningfully, the window of the array periodicity and *D*/*P* is wide enough to realize a higher light trapping capability compared to the Si film counterpart, which gives the manufacturing side more process to choose from. It is interesting to note that the Lewis and Atwater group from California Institute of Technology has recently reported the synthesis of Si-NWs with 10 μm minority carrier diffusion lengths by Cu-catalyzed VLS growth [88]. Accordingly, it is believed that the resulting high PCE of the Si-NWs array-based on an optimized optical design is achievable.

Kayes et al. have explored the carrier transport properties of stand-alone Si-NWs via simulation and discussed the relationship between the illuminated *I*–*V* characteristics and geometrical parameters, such as the length and diameter, and the defect state densities in the quasi-neutral and depletion regions [8]. In their study, it is assumed that the carrier transport is along the radial direction for the radial p–n junction configuration. It is found that for a given Si-NWs length, the short circuit current density, J_{sc}, is almost independent of the minority carrier diffusion length if the minority carrier diffusion length is larger than the Si-NWs radius. However, the J_{sc} in the solar cells with the conventional planar p–n junction configuration significantly decreases with the minority carrier diffusion length. For the radial junction case, different from J_{sc}, V_{oc} shows dependence on the Si-NWs' length, and decreases with an increase in the Si-NWs' length. This can be attributed to the increased junction area, and thus a decreased shunt resistance. Under the illuminated condition of the AM 1.5 spectrum, it is also reported that the Si-NWs-based solar cell can reach a PCE of ~11%. The Si-NWs have diameters of 100 nm and lengths between 20 and 500 μm, the minority-electron diffusion length is 100 nm (comparable to the Si-NWs diameters), and the recombination center densities of 7×10^{18} cm^{-3} in the quasi-neutral region and 10^{14} cm^{-3} in the depletion region. However, for the Si wafer counterpart with the same material parameters, the PCE is only ~1.5% and saturates when the thickness is over 450 nm. This study points out the great potential of stand-alone Si-NWs-based solar cells. Here it is noted that the above study does not account for any interaction between the incident light and the studied Si-NWs array (and hence the enhanced light absorption). After considering the enhanced light absorption for the optically optimized Si-NWs array, the PCEs of Si-NWs-based solar cells should become much more competitive with that of planar Si solar cells.

2.2.3 Research Status of Stand-Alone Si-NWs-Based Solar Cells

In 2007, the Lewis and Atwater group demonstrated a Si-NWs array-based photoelectrochemical cell [89]. The highly-ordered and vertically-aligned n-type (resistivity of ~0.32 Ω cm) Si-NWs array (see Fig. 30) with wires of diameter of ~2 μm, length of ~20 μm, and an array periodicity of ~7 μm was grown on a degenerately doped n-type Si (111) wafer via VLS growth. The Si-NWs array is immersed into a 1,1′ dimethylferrocene $(Me_2Fc)^{+/0}$ redox system in CH_3OH to form the p–n junction. This method can easily realize uniform coverage of the transparent electrode over the Si-NWs surface. The measured V_{oc} and J_{sc} of the Si-NWs array-based solar cell under AM 1.5G are 389 ± 18 mV and 1.43 ± 0.14 mA/cm^2, which is much better than that of the control sample without the Si-NWs array (V_{oc} : 232 ± 8 mV; J_{sc} : 0.28 ± 0.01 mA/cm^2).

Also in 2007, Tsakalakos et al. of General Electric's Global Research Center, USA fabricated the first Si-NWs-based all-inorganic solar cell [80]. Figure 31a, b shows the schematics of the Si-NWs-based solar cell configuration and the SEM images of the Si-NWs under different views (the cross-sectional view in the top-left). The p-type (estimated doping concentration of ~10^{18} cm^{-3}) Si-NWs with diameters of 109 ± 30 nm and length of ~16 μm were grown without orientation on a stainless steel foil coated by 100 nm thick Ti_2N using the Au catalyzed VLS growth (the precursor gas is the mixture of SiH_4, H_2, HCl and $B(CH_3)_3$). Here the

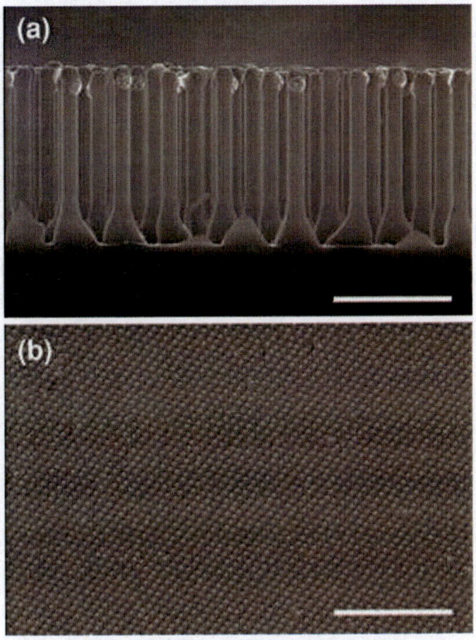

Fig. 30 **a** Cross-sectional (scale bar: 15 μm), and **b** 45° view (scale bar: 85.7 μm) SEM images of an Si wire array prepared by the VLS growth combined with the prepatterned Au catalyst. (Reprinted with permission from [89], copyright 2007, American Chemical Society)

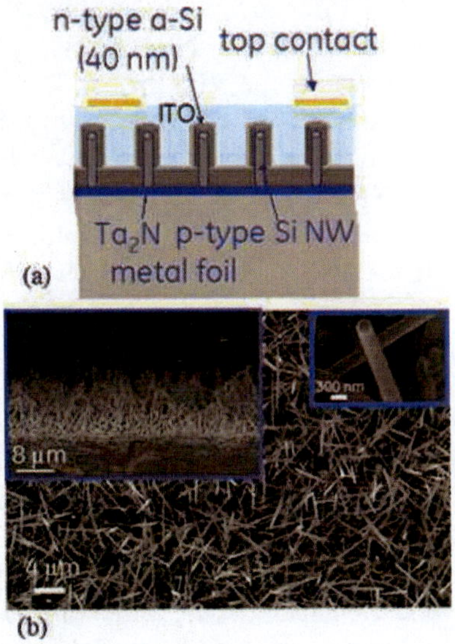

Fig. 31 **a** Schematic of the Si-NW-based all-inorganic solar cell using a stainless steel foil as the substrate, and **b** the SEM images of the Si-NWs under different views. (Reprinted with permission from [80], copyright 2007, American Institute of Physics)

Ti_2N film serves as the back electrode and also prevents interdiffusion between the Si and the steel substrate. As indicated by the SEM images in Fig. 31b, the synthesized Si-NWs randomly distribute on the substrate owing to the unpatterned Au catalyst. The p–n junction is introduced through depositing the 40 nm thick n-type a-Si:H layer by PECVD. This is followed by the ITO layer deposition with

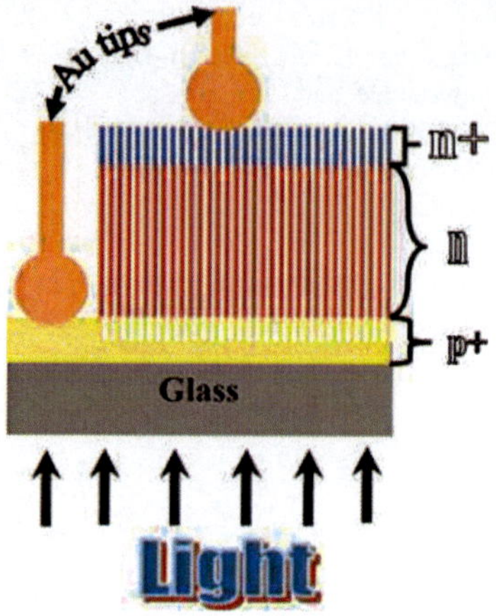

Fig. 32 Schematic of the Si-NW-based solar cell prepared by the electroless etching of the corresponding μc-Si (n^+)/μc-Si (n)/μc-Si (p^+) structure on glass. (Reprinted with permission from [90], copyright 2009, American Chemical Society)

the thickness of ~200 nm to electrically connect the Si-NWs. Then the top finger electrode consisting of 50 nm thick Ti and 2000 nm thick Al is prepared by shadow evaporation. The PCE of the fabricated solar cells with the size of 1×1 cm^2 was measured at only ~0.1% under AM 1.5G although the enhanced light absorption is obvious. The low PCE can be ascribed to the poor photogenerated carrier separation capability and poor electrical contact, both related to the randomly grown Si-NWs array.

Except for the VLS grown Si-NWs array, the low-cost "top-down" approaches also attract a lot of attention to fabricate stand-alone Si-NWs-based solar cells. Sivakov et al. fabricated Si-NWs-based solar cells by etching the corresponding $\mu c - \mathrm{Si}\ (n^+)/\mu c - \mathrm{Si}\ (n)/\mu c - \mathrm{Si}\ (p^+)$ structure on glass substrates using the electroless chemical etching solution prepared by mixing 0.02 M $AgNO_3$ and 5 M HF with a volume ratio of 1:1 [90]. Figure 32 shows the schematic of the cell structure with a superstrate configuration [91]. The $\mu c - \mathrm{Si}\ (n^+)$, $\mu c - \mathrm{Si}\ (n)$ and $\mu c - \mathrm{Si}\ (p^+)$ layers with the respective thicknesses of 300, 2000, and 200–400 nm were prepared by electron beam evaporation and the laser annealing, with respective doping concentrations of 5×10^{19}, 6×10^{16}, and 5×10^{19} cm^{-3}. As indicated by the cross-sectional SEM image shown in Fig. 33a, the resulting Si-NWs are vertically aligned on the glass substrate. The length and diameter of the Si-NWs vary from 2300 to 2500 nm, and from 20 to 100 nm respectively, as estimated from the SEM and TEM measurements. Figure 33b shows the measured optical characteristics, indicating excellent light trapping capability, especially in the high-energy regime of the solar spectrum compared to the Si thin film counterpart. The *I–V* characteristics of these solar cells are recorded by contacting the Au tips with the radius of 450 μm onto the Si-NWs top surface and the p^+ Si-film layer (see Fig. 32). The V_{oc} is recorded in the range of 410–450 mV and the J_{sc} varies from 13.4 to 40.3 mA/cm^2 when measuring different points on the same sample. Despite the error in calculating the Au tip area (hence the J_{sc}), the PCE of the cell is in the range of 1.7–4.4%. Further enhancement of the PCE can be

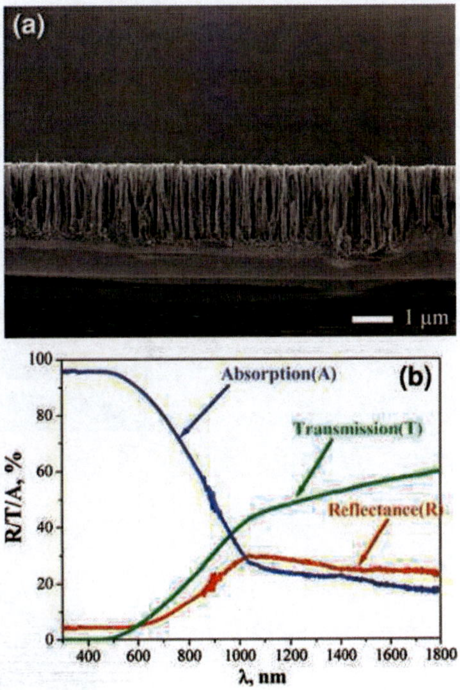

Fig. 33 **a** Cross-sectional SEM image of vertically aligned Si-NWs on glass after electroless etching, and **b** the optical characteristics of the above Si-NW array. (Reprinted with permission from [90], copyright 2009, American Chemical Society)

expected by optimizing the Si-NWs' structural parameters, such as diameter, array periodicity, and the electrode configuration.

More recently, the Yang group at the University of California at Berkeley reported Si-NWs-based solar cells with the controllable Si-NWs diameters and array periodicity using the RIE-based "top-down" technique combined with a self-assembled SiO_2 monolayer mask [92]. In their study, to mimic the PV response of the stand-alone Si-NWs layer, a highly doped Si wafer was used as the substrate to minimize the PV contribution from the substrate and an epitaxial Si thin layer on the wafer is employed to manufacture the Si-NWs array. After the formation of the Si-NWs array, boron diffusion was performed to form radial p–n junctions, followed by finger metal electrode deposition. Figures 34a, b respectively show the SEM image of the resulting Si-NWs array (Si-NWs with diameters of $\sim$390 nm, lengths of $\sim$5 μm, and an array periodicity of $\sim$530 nm), and the optical image of the large-scale Si-NWs-based solar cells on the same substrate after isolation. The illuminated *I–V* characteristics recorded under AM 1.5G demonstrate a PCE of $\sim$4.83% (V_{oc}: 525 ± 2 mV; J_{sc}: 16.45 ± 0.19 mA/cm^2; FF: 0.559 ± 0.002) for the solar cell with the absorber consisting of the above Si-NWs array and a 3 μm thick underlying Si layer, which is $\sim$20% higher than that of the Si ribbon solar cell with the same total thickness, i.e., 8 μm [93]. Here it should be noted that the electrode configuration is yet to be optimized for Si-NWs-based solar cells. They further declare that the

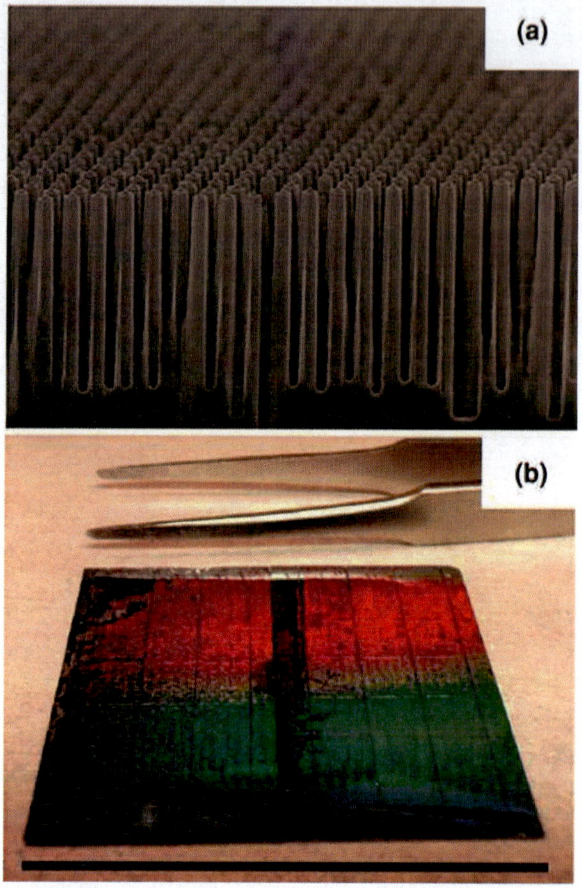

Fig. 34 **a** Titled SEM image of the Si-NWs array after forming the radial p–n junction (the scale bar: 1 μm), and **b** the tilted optical image of the Si-NW array-based solar cells on the same substrate after isolation (the scale bar: 4 cm). (Reprinted with permission from [92], copyright 2010, American Chemical Society)

light absorption mainly occurs in the Si-NWs array, and an optical path length enhancement factor of ~73 can be achieved, which is much larger than the randomized scattering limit (~25 without a back reflector) [94].

Despite the great potential held by Si-NWs-based solar cells, the currently reported PCE record from the literature is still relatively low compared to mainstream Si wafer cells. To further improve the PCE of stand-alone Si-NWs-based solar cells, the following points need to be addressed:

(i) *Structural optimization of the Si-NWs*. Si-NWs-based solar cells make it possible to design the light trapping and photogenerated carrier collection processes due to the decoupling between them. However, for practical operation, the trade-off between light trapping and carrier collection should be considered when designing high-efficiency devices. For example, the surface defect states are critical, especially for Si-NWs prepared by the etching approaches [50] because they lead to a decreased V_{oc} [8]. How to achieve compromise between the enhanced light absorption and the degraded V_{oc} while increasing the Si-NWs' lengths needs to be investigated via simulations coupling both optical and electrical aspects.

(ii) *Optimized top transparent electrodes*. Optimal electrode configuration is another key factor to realizing high-efficiency solar cells. Especially for Si-NWs-based solar cells, high contact resistance needs to be addressed through optimization of nanoscale design and manufacturing processes. The conformal deposition of the top transparent electrode to uniformly cover the Si-NW surface is necessary to effectively extract the photogenerated carriers.

(iii) *Reducing the surface defect density*. Although the radial p–n junction based solar cells allow for the usage of low-grade materials with high bulk defect densities, the defect state density on Si-NWs surfaces needs to be minimized. Surface defects can trap photogenerated carriers, act as recombination zones and degrade the solar performance [95, 96]. Therefore effectively reducing the number of Si-NWs surface defect density remains important to obtaining high efficiency.

2.3 *Si Thin Film Solar Cells Textured with Si Nanostructures*

In this section, surface texturing of Si thin film solar cells using three types of Si nanostructures, including Si-NWs (or Si nanopillars (Si-NPs)), Si nanocones (Si-NCs), and Si nanoholes (Si-NHs), are discussed for efficiency boosting. The preparation of Si-NWs arrays was introduced in Sect. 2.2.1. Therefore, the fabrication techniques of the Si-NCs and Si-NHs-textured surfaces are to be discussed in detail in this section.

2.3.1 Preparation of Si Thin Films Textured with Nanaostructures (Si-NCs and Si-NHs)

Different from the "bottom-up" paradigm in preparing stand-alone Si-NWs on foreign substrates, the fabrication of the Si nanostructure-textured Si thin films is mainly based on "top-down" processes, which are able to meet the solar cells manufacturing requirements, i.e., low cost, large scale, and high throughput. Several representative methods used for the preparation of Si-NC- and Si-NH-textured surfaces will be discussed. Hsu et al. prepared highly-ordered Si-NC-textured surfaces using C_2ClF_5/SF_6-based isotropic RIE on the corresponding Si-NWs arrays, which were formed by the Cl_2-based anisotropic RIE of Si wafers combined with a monolayer of SiO_2 nanospheres as a Langmuir–Blodgett mask (for more details, refer to Sect. 2.2.1) [19]. Figure 35 illustrates the fabrication process and the corresponding SEM images. In this method, the Si-NCs' structural parameters, such as height, base diameter, and array periodicity can be controlled

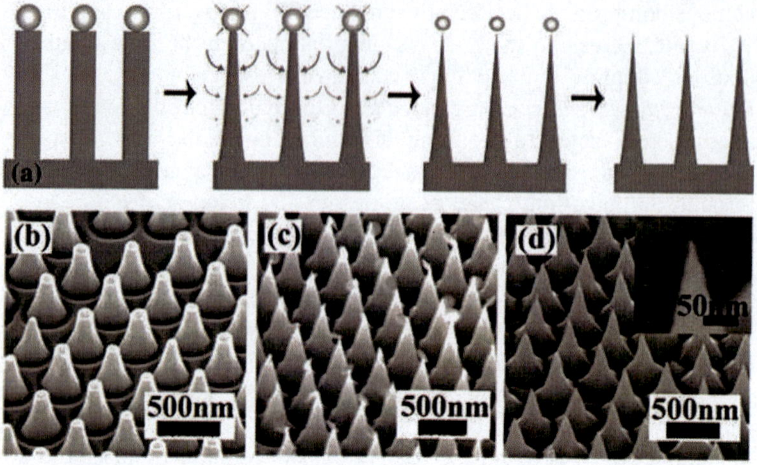

Fig. 35 **a** Schematic of the fabrication process of a highly-ordered Si-NC-textured surface, **b–d** show the SEM images corresponding to the intermediate and final steps. (Reprinted with permission from [19], copyright 2008, American Institute of Physics)

during the preparation of the Si-NWs array. It is interesting to note that this method has been successfully applied to fabricate Si-NC-textured a-Si:H films, paving the way for its application in Si thin film-based solar cells [9].

Another approach to make Si-NC-textured surfaces is reported based on one-step self-assembly processes without involving the mask preparation [20, 97–99]. Figure 36 schematically shows the process flow using the "self-masking method" [20]. During the process, a plasma with precursor gases of SiH_4, CH_4, Ar, and H_2 can generate and deposit SiC nanoparticles on the substrate surface as demonstrated in Fig. 36a, b. This is followed by the introduction of Ar and H_2 plasma to etch the Si substrate and obtain the Si-NC array using the SiC nanoparticles as the hard mask. It is expected that Si-NC arrays can be made in large scales using this method. Figure 37a, b shows the as-prepared Si-NC texturing on single-crystal and polycrystalline Si substrates, respectively. It is noted that during the processes, the substrate temperature was maintained below 250°C, which facilitates the usage of low-temperature and low-cost substrates. In Fig. 37a, a Si-NC density of $\sim 1.5 \times 10^{11}/cm^2$ and aspect ratio of 50 are shown. For this approach, the structural parameters of the Si-NC array, such as the Si-NCs' base diameter, spacing, and height can be adjusted through varying the size and density of the SiC nanoparticles via modifying the plasma conditions, and the substrate temperature.

Analogous to macrohole-textured surfaces in Si wafer-based solar cells, the excellent antireflection property of the Si nanoholes array-decorated surfaces makes Si-NHs promising in Si thin film-based solar cells. For the sake of cost reduction, only the fabrication approaches based on cost-effective mask or maskless processes are discussed here. Li et al. developed the laser nanoimprinting technique to create large area Si-NH-textured surfaces [100]. After coating the

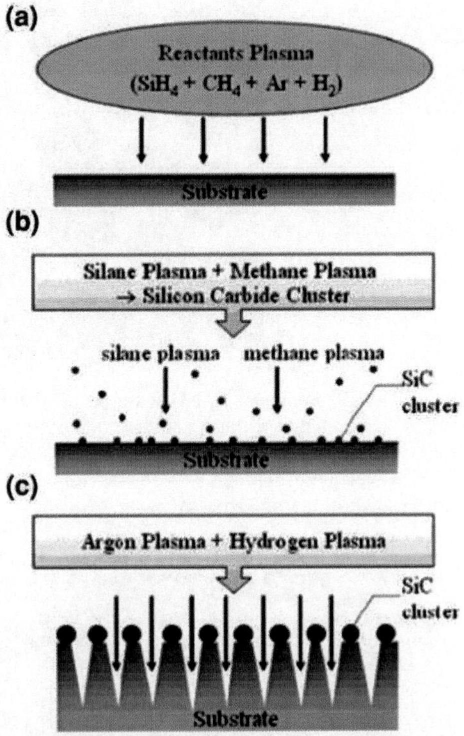

Fig. 36 Schematic of the procedure for preparing the Si-NC-textured surface using the one-step self-assembly method. **a** The reactive gases are composed of silane, methane, argon, and hydrogen; **b** the SiC nanosized clusters are formed from the reaction of SiH_4 and CH_4 plasma and uniformly distributed on the substrate surface; and **c** the unmasked region is etched by Ar and H_2 plasma, whereas the region masked by SiC caps protects the substrate from etching, and hence creates the conical tips. (Reprinted with permission from [20], copyright 2004, American Chemical Society)

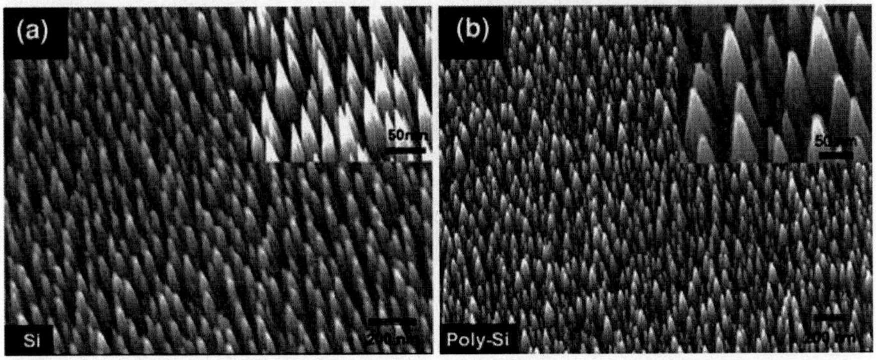

Fig. 37 Tilted SEM images of the Si-NC array on **a** single-crystal and **b** polycrystalline Si substrates using the one-step self-assembling method depicted in Fig. 36. (Reprinted with permission from [20], copyright 2004, American Chemical Society)

sample with a monolayer of self-assembled SiO_2 nanospheres, a focused laser beam is directed onto the SiO_2 nanospheres. Due to the extremely high temperature at the contact point between the SiO_2 and substrate, which results from the optical resonance and near-field effects [101], the substrate in the vicinity of the

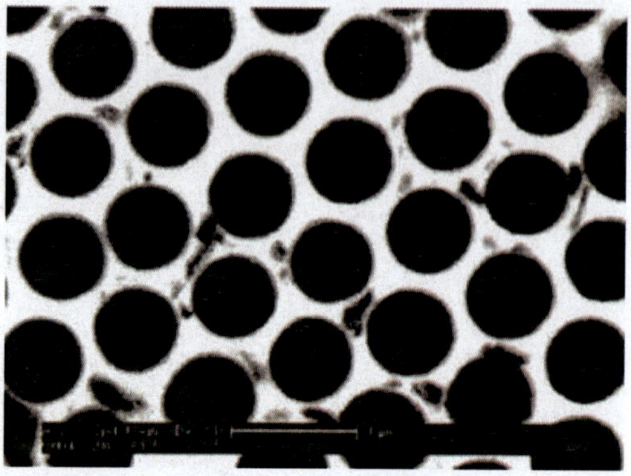

Fig. 38 SEM image of Si-NH-textured Si (100) wafer prepared by the laser nanoimprinting technique combined with the self-assembled SiO_2 nanosphere monolayer (the scale bar is 1 μm). The average diameter of the SiO_2 nanospheres is 970 nm, and the fluence of the single laser pulse (KrF, wavelength: 248 nm) is 1 J/cm^2. (Reprinted with permission from [100], copyright 2004, American Institute of Physics)

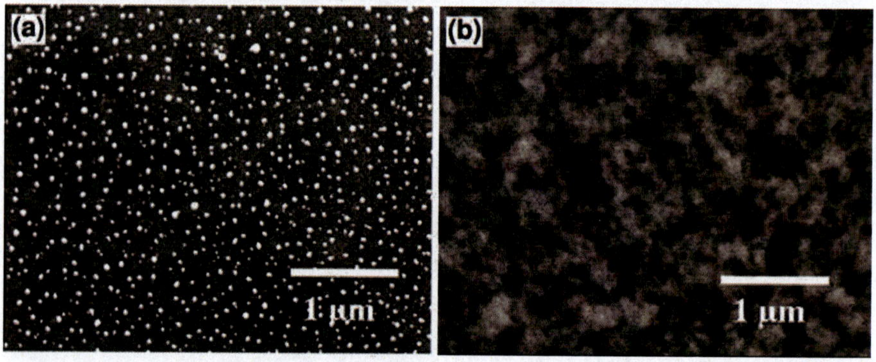

Fig. 39 SEM images of **a** the Ag nanoparticles on the Si wafer surface, and **b** the SiNHs array decorated surface after wet etching in the HF/H_2O_2 solution. (Reprinted from Tsujino et al. [102], copyright 2006, with permission from Elsevier)

nanospheres melts, forming hemisphere-shaped holes into the substrate. Figure 38 shows the SEM image of the Si-NH-textured Si (100) wafer prepared using SiO_2 nanospheres with diameters of ~970 nm with the fluence of a single laser pulse (KrF, wavelength: 248 nm) set at 1 J/cm^2. The Si-NH size and depth can be modulated by modifying the SiO_2 nanosphere size and the laser fluence. This method is applicable to Si thin films on glass and plastic substrates due to the localization of the melting.

Next, the randomly distributed Si-NH-textured surface is discussed, which can be prepared in a low cost and high throughput manner. The formation of the Si-NHs can be realized in the HF/H_2O_2 solution with the catalysis of noble metals such as Ag, Au, etc (see Sect. 2.2.1). Figure 39 shows the SEM images of (a) the

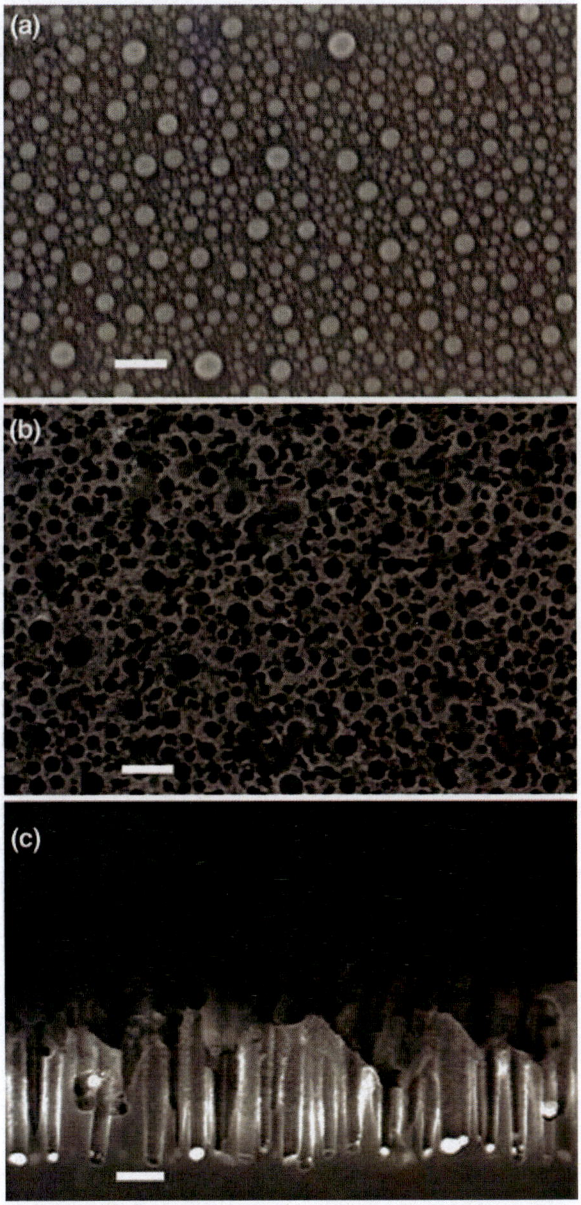

Fig. 40 SEM images of **a** the Ag nanoparticles on the Si wafer surface prepared by laser annealing of the Ag film, **b** the top and **c** cross-sectional views of the resulting Si-NH array after wet etching in the HF/H_2O_2 solution. In **c**, the white particles at the bottom of the Si-NHs are the residual Ag nanoparticles. The scale bar in the pictures is 200 nm

Ag nanoparticles on the multicrystalline Si wafer surface, and (b) the resulting randomly distributed Si-NH-textured surface after the wet etching in a mixture of 10% HF and 30% H_2O_2 (10:1 v/v) for ~5 min [102]. The Ag nanoparticles are prepared using the electroless plating [103]. The dimensions of the Si-NHs are consistent with the Ag nanoparticles sizes and can be modulated by varying the Ag

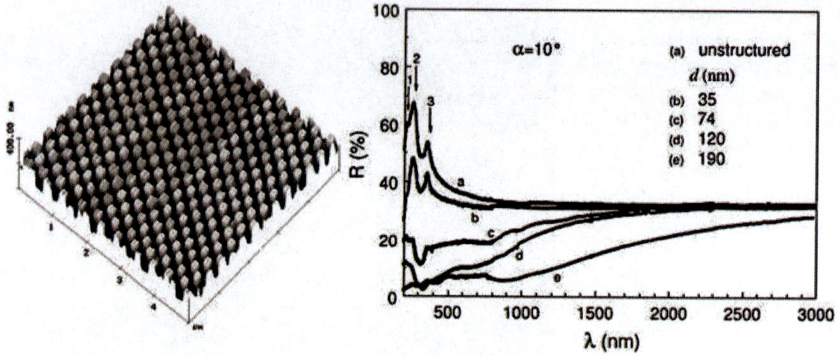

Fig. 41 AFM image of the Si-NW-textured surface (array periodicity: 300 nm) (*left*), and the reflectivity, *R* (*right*) of the Si-NW-textured surface with the same array periodicity of 300 nm and varying Si-NW array depths, *d*. The measurements were performed using unpolarized light at the incident angle of 10°. (Reprinted with permission from [64], copyright 2000, Institute of Physics)

nanoparticles size and distribution. The Si-NHs' depth can also be easily controlled by the etching time.

Our group has also developed a method to form the Ag nanoparticles by employing rapid laser annealing on an as-deposited Ag thin film [104]. The SEM image, as shown in Fig. 40a demonstrates that the Ag nanoparticles can be synthesized with a high density. Figure 40b, c shows the top and cross-sectional views of the etched Si surface using the HF/H_2O_2 mixture. It is clear that the resulting Si-NH array strictly follows the Ag nanoparticles pattern, which can be modified by the thickness of the Ag film, laser pulse energy, etc. Similar to the laser nanoimprinting technique, this approach is also applicable to Si thin films on glass or plastic due to the localized thermal effect of the laser annealing process.

2.3.2 Optical and Electrical Characteristics of Si Nanostructure-Textured Si Thin Films

It is well documented that the key advantage of the Si nanostructure texturing in thin film-based solar cells is its excellent antireflection property. It is thus expected that thicker texturing layers with large aspect ratios will lead to more efficient antireflection, as indicated in Fig. 41 [64]. However for solar cell applications, as discussed previously, other aspects must also be considered. The effective extraction of photogenerated carriers relies on optimized electrical contacts, which further depend on the conformal deposition of the electrodes on the Si nanostructure. Therefore, the thickness of the Si nanostructure layers is a compromise, also partially owing to the manufacturing concerns.

Different from the light multireflection mechanism based on geometrical optics in microscale surface textures for light absorption enhancement, the interaction

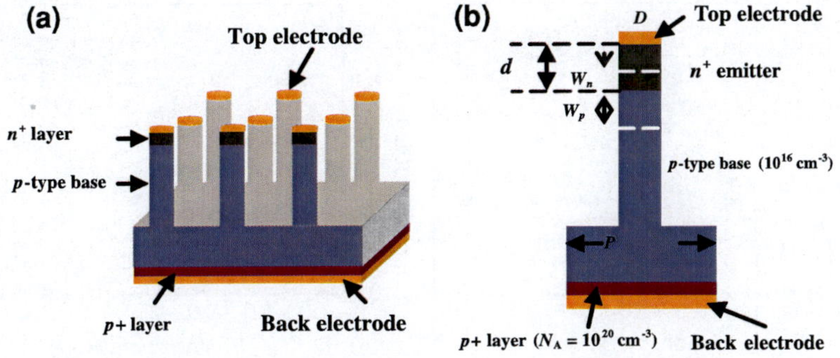

Fig. 42 **a** Schematic of the Si-NW-textured thin film-based solar cell, and **b** the unit used for the calculation of the optical and electrical characteristics

between the incident light and nanoscale surface textures follows elemental optical processes based on wave optics, such as scattering [30]. Enhanced scattering will significantly elongate the optical path length, hence increasing the light trapping capability, i.e., increasing the light absorption. Our group has systematically studied the optical characteristics of Si nanostructure-textured Si thin films, including Si-NWs, Si-NCs and Si-NHs [30, 31, 105, 106]. The results indicate that there is a critical value for the thickness of all the Si nanostructure textures. When the thickness is beyond this value, the light trapping capability becomes nearly saturated. From calculations, the optimized critical thicknesses for Si-NWs, Si-NCs and Si-NHs are ~1000, 400, and 2000 nm, respectively. Furthermore, light absorption is strongly affected by array periodicity. Figure 42 schematically shows the Si thin film-based solar cell configuration with a Si-NWs-textured surface. During the calculation, the thickness of the Si thin film was fixed at 800 nm. The optically optimized Si-NWs length was set as 1000 nm, which is also acceptable for the high-quality electrode preparation. The ratio of the Si-NWs diameter (*D*) to *P* of 0.5 was taken from our previous studies [30, 31].

Figure 43a, c shows the light absorption, reflection, and transmission spectra of the studied structure (see Fig. 42a) with different Si-NWs array periodicities. As expected, the light absorption is significantly enhanced when incorporating the Si-NWs array into the device. More interestingly, the sample with a *P* of 100 nm shows a consistent absorption with the 800 nm thick Si film, and the absorption becomes stronger when the energy is above ~2.2 eV. In the low-energy region of the solar spectrum, the wavelength of the incident light is much longer than the *P* of the Si-NWs array. Accordingly, incident light can easily penetrate through the Si-NWs array and reach the underlying Si layer. This statement is further evidenced by the reflection and transmission spectra for both samples in the corresponding energy region (see Fig. 43b, c). It was also observed that the "deviation" point of the absorption spectra for the Si film with and without Si-NWs array shifts towards the low-energy regime with increasing *P*. The deviation point for the sample with *P* of 200 nm is around 1.5 eV.

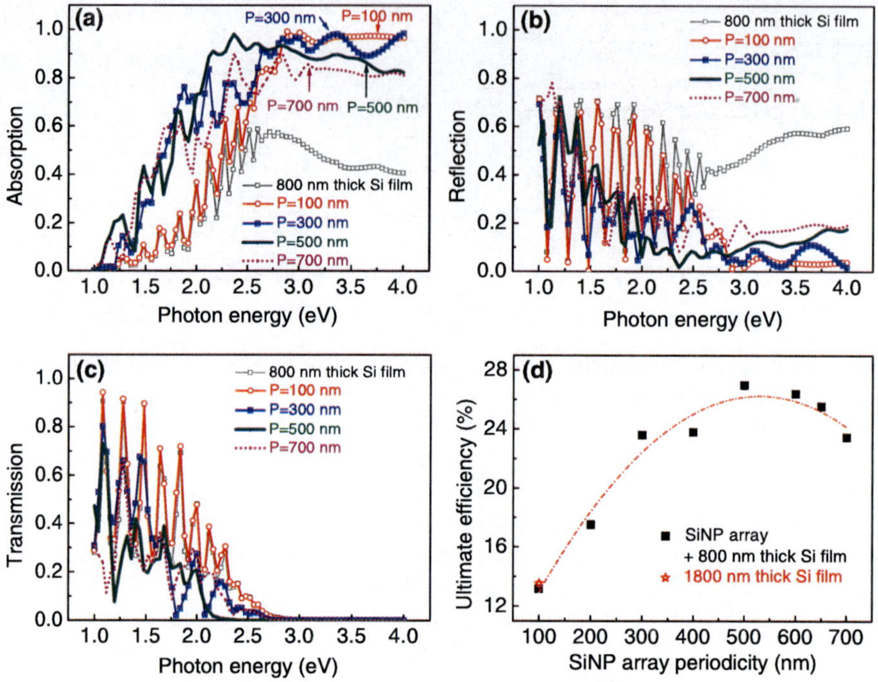

Fig. 43 **a** Absorption, **b** reflection, and **c** transmission spectra of the Si-NW (*length*: 1000 nm) textured Si *thin* films (thickness: 800 nm) as a function of the array periodicity, *P*. The ultimate efficiency is summarized in **d**. *D*/*P* is held constant at 0.5. The 800 and 1800 nm *thick* Si films serve as the reference. (Reprinted with permission from [30], copyright 2009, American Institute of Physics)

In the high-energy regime, the wavelength of the incident light is comparable with *P*. Thus, the scattering of the incident light is significantly enhanced, resulting in an elongated optical path and, therefore, enhanced light absorption, which is verified by the excellent light absorption of ~95% in the energy region above ~2.9 eV for the sample with *P* of 100 nm. As *P* increases, the incident light scattered by the structure shifts toward longer wavelengths and the reflection in the short wavelength range increases accordingly. As shown in Fig. 43a, the absorption peak shifts to ~2.5 eV with increasing *P* from 100 to 500 nm. On the other hand, decreased light absorption in the high-energy region was observed. However, the energy density is relatively weak in this regime of the solar spectrum and the decreased light absorption in this energy region is effectively compensated by the shift of the absorption edge. Further increasing *P* to 700 nm, the light reflection becomes so strong in the broad range that it cannot be compensated by the absorption edge shift, resulting in lower light absorption. Figure 43d shows the ultimate efficiency as a function of *P*, and the 1800 nm film serves as the reference. For the 800 nm thick Si film with the Si-NWs-decorated surface, the ultimate

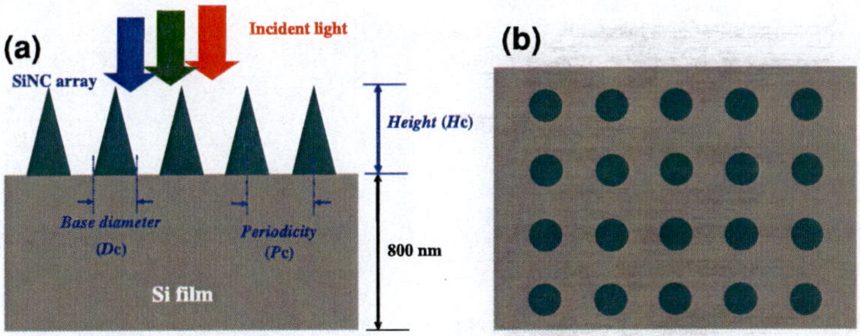

Fig. 44 Schematic of the **a** cross-sectional and **b** plane views of the Si-NC-textured thin film

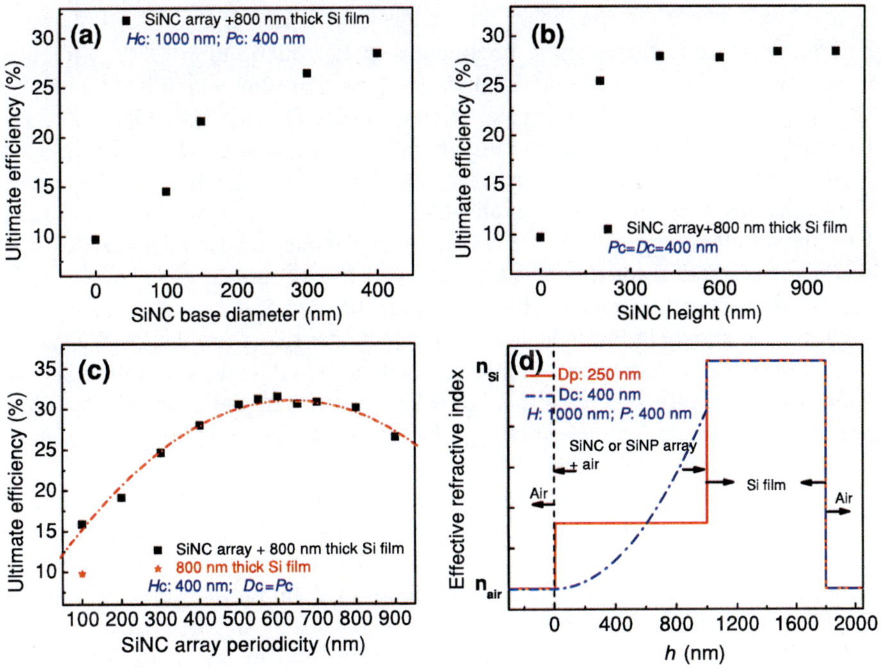

Fig. 45 Ultimate efficiency of the Si-NC-textured thin film as a function of **a** base diameter, D_c, **b** Si-NC height, H_c, and **c** array periodicity, P_c. **d** shows a comparison of the spatial distribution of the effective refractive indices for the Si-NC- and Si-NW-textured thin films

efficiency first increases with P, reaching a maximum of $\sim$27% when P is $\sim$500 nm, more than 200% of that of the 1800 nm thick Si film.

Following a similar methodology, the optical characteristics of the Si-NC- and Si-NH-textured thin films are also investigated using the FEM method [31]. Figure 44 shows the schematic of cross-sectional and top views of a Si thin film (800 nm) textured by a Si-NC array. This study indicates that for effective anti-reflection, the base diameter, D_c, of the Si-NCs should be equal to the array

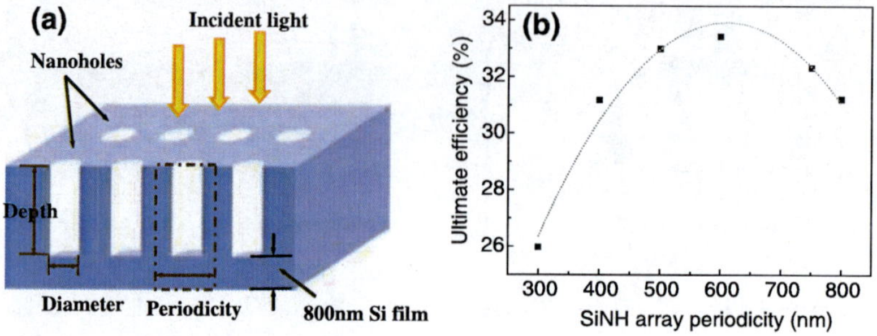

Fig. 46 **a** Schematic of the Si-NH structure for computational simulations, and **b** ultimate efficiency of the structure as a function of the array periodicity

periodicity, P_c, due to the more continuous spatial distribution of the effective refractive index between air and the underlying Si thin film, which is reflected by the continuously increasing ultimate efficiency with D_c (see Fig. 45a). As demonstrated in Fig. 45b, the ultimate efficiency becomes nearly saturated when the Si-NC height exceeds 400 nm. Further, it is believed that the conformal deposition of the top transparent electrode is realizable for this Si-NC height. The variation of the ultimate efficiency with P_c demonstrates a similar trend compared with the Si-NWs case, and a maximum value of $\sim$31.5% occurs at a P_c of 600 nm, although the total thickness is only 1200 nm (400 nm Si-NC + 800 nm Si thin film layer), as shown in Fig. 45c. This is much greater than the value of $\sim$27% for the optimized Si-NWs-textured Si thin film with a total thickness of 1800 nm, due to the continuous effective refractive index between air and the underlying Si layer (see Fig. 45d), enabling more efficient reduction of the incident light reflection, as suggested by the Fresnel theory [25].

For the Si-NH-textured surface (schematically shown in Fig. 46a), based on the similar mechanism of wave optics, it was found that the solar energy absorption could be optimized when the dimensions of the Si-NH array are set as follows: array periodicity of $\sim$600 nm (Fig. 46b), depth of 2000 nm, and the ratio of the Si-NH diameter to array periodicity of $\sim$87.5%.

Following the optical study, the electrical behaviors in the Si nanostructure-textured thin film-based solar cells are discussed in terms of the minority carrier diffusion length, doping concentration, and junction depth. The discussion is based on the optically optimized Si-NWs-textured (length: 1000 nm; *P*: 500 nm; *D*: 250 nm) Si thin film (thickness: 800 nm) [107]. In Sect. 2.2.2, the carrier transport in Si-NWs-based solar cells was briefly discussed for a radial p–n junction. Here, for the convenience of comparison with the Si thin film solar cells, the traditional planar p–n junction configuration is under consideration, as shown in Fig. 42. The p-type base has a light doping of 10^{16} cm^{-3}. A thin p^+ layer of 50 nm with the doping concentration of 10^{20} cm^{-3} is used to form a high-quality Ohmic contact with the back electrode. During the calculation, Shockly-Reed-Hall and Auger recombinations are considered in the lightly- and heavily-doped

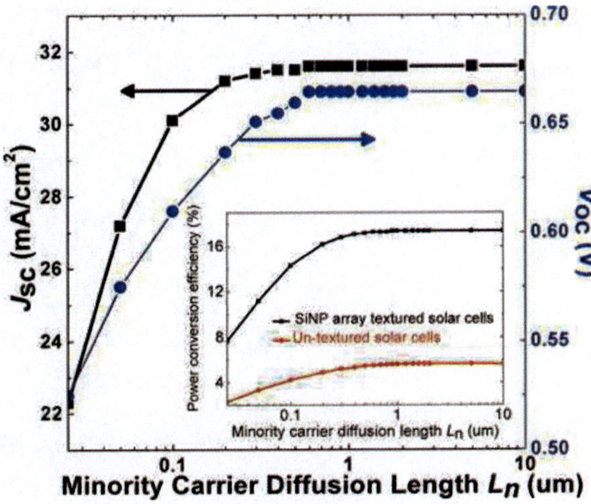

Fig. 47 Short circuit current density, J_{sc}, and open circuit voltage, V_{oc}, of the solar cell with the 1000 nm long Si-NW array (*P*: 500 nm; *D*: 250 nm) and an 800 nm thick Si film as a function of the minority carrier (electron) diffusion length, L_n. The *inset* shows the predicted PCE. The data of the thin film solar cells having the thickness of 1800 nm without texturing serve as the reference

regions, respectively [108, 109]. The illumination condition is AM 1.5G, i.e., ~100 mW/cm^2. The carrier generation rate (*G*) under illumination is expressed by the following formula (6) [110]:

$$G = \frac{1}{2\hbar\omega}\text{re}\{\nabla \cdot \mathbf{P}\} = \frac{\varepsilon_i|\mathbf{E}|^2}{2\hbar}, \tag{6}$$

where **P** is the Poynting vector, **E** is the electric field, $\hbar$ is the reduced Plank constant, ω is the angular frequency, ε_i is the imaginary part of the material's permittivity. It is noted that Eq. 6 applies to photons with the energies greater than E_g. For the photons with energies below E_g, there is no carrier generation.

Figure 47 depicts the J_{sc} and V_{oc} of the Si-NWs-textured thin film-based solar cells as a function of the minority carrier (electron) diffusion length, L_n. The doping concentration in the n$^+$ thin layer (50 nm) was set to 10^{20} cm^{-3}. It is clear that both J_{sc} and V_{oc} increase with increasing L_n, and become saturated when L_n is above ~0.5 μm. The predicted PCE is summarized in the inset of Fig. 47, and also compared with that of the Si thin film solar cells having the same thickness of 1800 nm without surface texturing. Corresponding to the dramatically enhanced light absorption of the Si-NWs-textured thin film solar cells, the predicted PCE of ~17.3% is achievable when $L_n = 0.6\,\mu$m, much larger than the value of ~5.62% for the Si thin film solar cell with L_n of 1.4 μm. Another interesting point is that the PCE of the thin film solar cells without texturing saturates when L_n exceeds ~1.2 μm, larger than the value of ~0.5 μm. This indicates that the intense light absorption in the Si-NWs-textured solar cell is much closer to the top surface as compared to the case of the devices without texturing, which in the meanwhile facilitates the extraction of the minority carriers by the top electrode.

The open circuit voltage and PCE at different emitter doping concentrations are summarized in Fig. 48. The width of the emitter is held constant at 50 nm and the

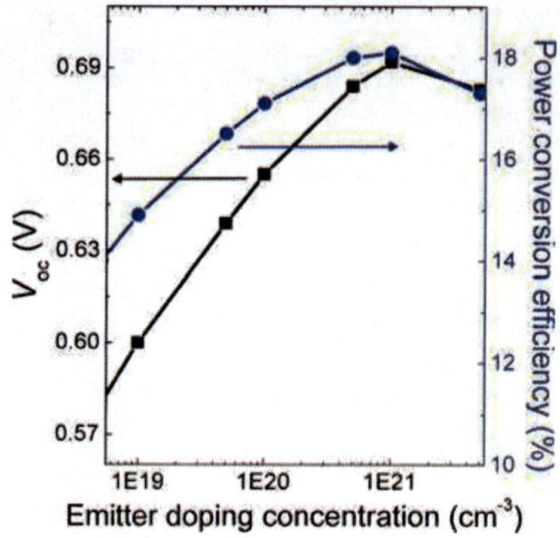

Fig. 48 V_{oc} and PCE of solar cells with a 1000 nm long Si-NW array (*P*: 500 nm; *D*: 250 nm) and an 800 nm thick underlying Si film as a function of the emitter dopant concentration, N_D

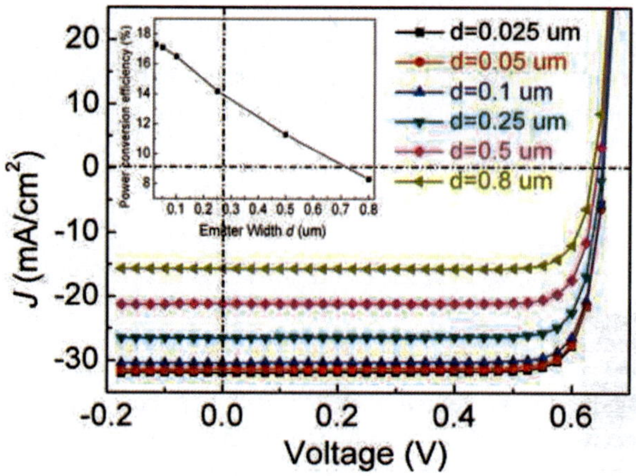

Fig. 49 *J–V* characteristics of a solar cell with the 1000 nm long Si-NWs in an array (*P*: 500 nm; *D*: 250 nm) and an 800 nm thick underlying Si film as a function of emitter width, *d*

minority-electron diffusion length of 0.6 μm is used. As depicted in Fig. 48, the V_{oc} first increases due to the widened Fermi level difference in the p–n junction with the increase of N_D. When N_D is above 10^{21} cm^{-3}, the V_{oc} then decreases due to increased carrier recombination stemming from enhanced Auger recombination. The predicted PCE shows the similar change trend with N_D, and a maximum of 18.1% is achieved for N_D of 10^{21} cm^{-3}. Here it is worth noting that for practical operation, it is difficult to achieve such a high doping concentration, and hence the emitter doping concentration of 10^{20} cm^{-3} is recommended, although there is a slight decrease of ~0.8% in PCE.

Figure 49 shows the *J–V* characteristics of the Si-NWs-textured Si thin film-based solar cells with different emitter widths, *d*. The doping concentration in the emitter and the minority carrier diffusion length are fixed at 10^{20} cm^{-3} and 0.6 μm, respectively. The J_{sc} decreases with increasing *d*, especially for *d* values above

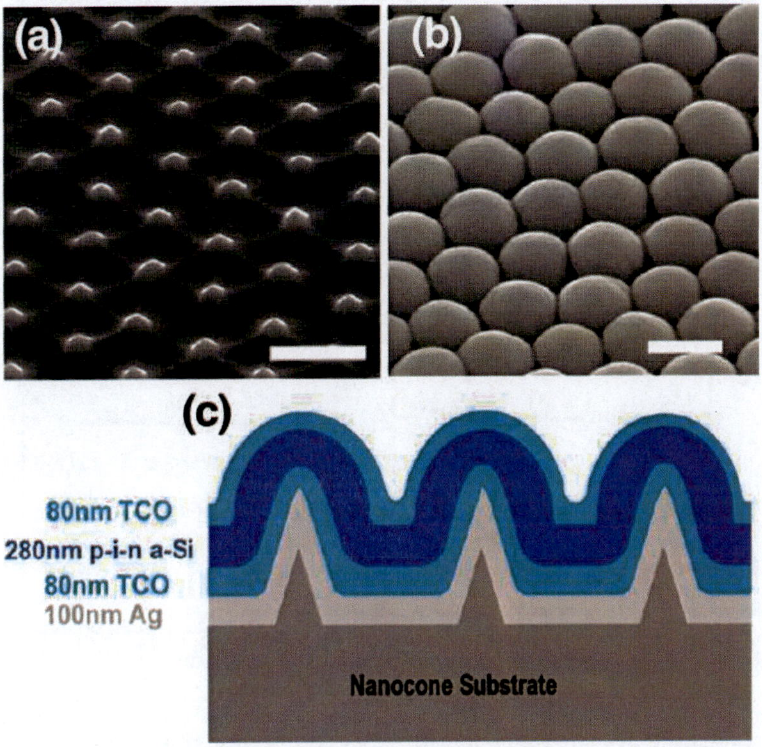

Fig. 50 Structure of the a-Si:H film-based solar cell coated onto the nanocone-textured quartz substrate. SEM images taken at 45° on **a** the nanocones array textured substrate, and **b** nanodome-shaped solar cell (the scale bar represents 500 nm). The schematic of the cross section of the solar cell is shown in (**c**). (Reprinted with permission from [112], copyright 2009, American Chemical Society)

0.1 μm, which indicates that the high-energy photons absorbed inside the emitter may not contribute to the photocurrent. This is because the photogenerated carriers fail to cross the p–n junction, and recombine in the emitter layer. The PCE and V_{oc} thus decrease with increasing d, as it is directly related to J_{sc}. Considering the practical manufacturing issues, emitter width between 0.05 and 0.1 μm is thus recommended.

From the above discussion, the conclusion can be drawn that the Si nano-structure-textured thin film solar cells are superior in both light absorption and carrier collection, compared to their thin film counterparts. Here it is also worth mentioning that for high PCE in this type of solar cell, minimizing the surface defect density is critical. As calculated, the predicted PCE for the optimized structure decreases from 17.3 to 15.2%, and further to 6.1% when increasing the surface recombination velocity from 1 to 10^3 and further to 10^5 cm/s [111].

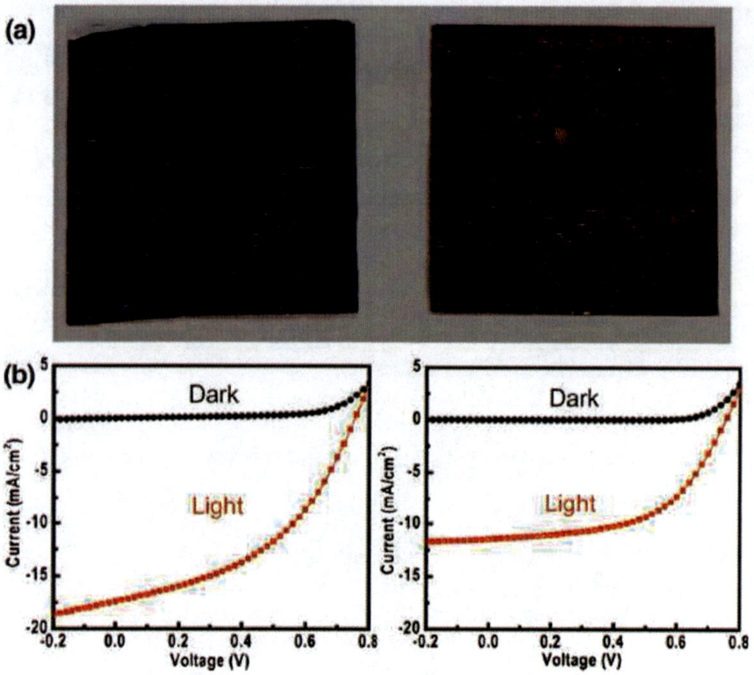

Fig. 51 **a** Photographs, and **b** dark and illuminated (AM 1.5G) *J–V* characteristics of the nanocones array incorporated solar cell (*left*) and the reference flat solar cell (*right*). (Reprinted with permission from [112], copyright 2009, American Chemical Society)

2.3.3 Research Status of Si Thin Film Solar Cells Textured with Si Nanostructures

Motivated by the excellent light trapping capability and efficient photogenerated carrier collection, Si nanostructure-textured Si thin film-based solar cells have attracted much attention. Figure 50 shows the structure of an a-Si:H film-based solar cell coated onto a NC-textured quartz substrate, fabricated by RIE using the SiO_2 nanosphere Langmuir–Blodgett monolayer mask [112]. The NCs (Fig. 50a) have base diameters of 100 nm, heights of 150 nm, and form an array with a periodicity of 450 nm. Figure 50b shows the solar cell morphology after depositing a 100 nm thick Ag back reflector, 80 nm thick TCO, 280 nm thick n–i–p a-Si:H cell (the thicknesses of the n, i and p layer are 20, 250 and 10 nm, respectively), and another 80 nm thick TCO layer onto the NCs in sequence. Figure 50c shows the cross-section of the solar cell.

Owing to the NC-textured substrate, the solar cell demonstrates superb ght absorption compared to the planar devices, which is clear from the darker appearance shown in Fig. 51a. From the *J–V* measurements (see Fig. 51b), the NC array-based solar cells can achieve a record high J_{sc} of $\sim$17.5 mA/cm^2 [113], although the contact and junction of the device were not optimized as observed

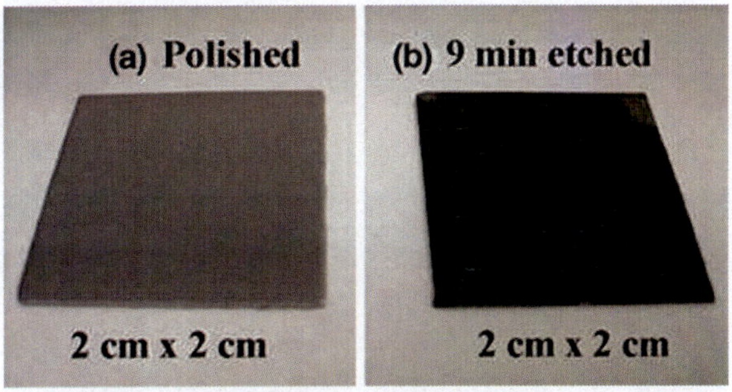

Fig. 52 Photographs of the samples **a** without and **b** with the Si-NH texturing. (Reprinted from Nishioka et al. [114], copyright 2009, with permission from Elsevier)

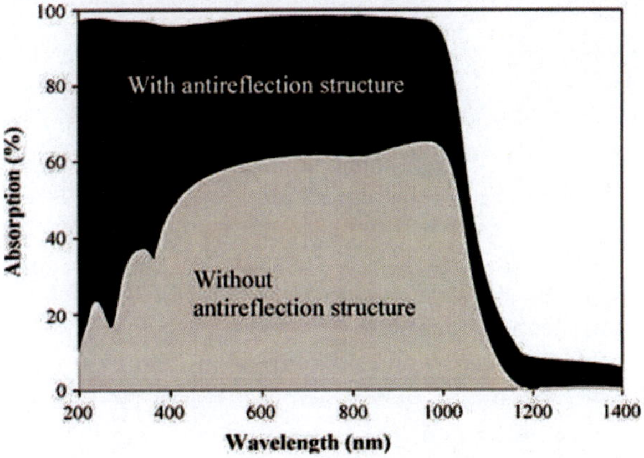

Fig. 53 Absorption spectra of the samples corresponding to those shown in Fig. 52. (Reprinted from Nishioka et al. [114], copyright 2009, with permission from Elsevier)

from the *J–V* curve shape. On the other hand, the J_{sc} of the reference solar cell is only ~11.4 mA/cm^2, owing to its relatively poor light trapping capability. The reported PCE for the NC array-based solar cell is 5.9% (V_{oc}: 750 mV; FF: 0.45), which is ~25% higher than that of the reference flat solar cell (PCE ~4.7% with V_{oc} of 760 mV and FF of 0.54).

Nishioka et al. reported solar cells with the surface textured by Si-NH arrays, where the arrays were fabricated by immersing Ag nanoparticle (3–5 nm in diameter) coated Si in a HF/H_2O_2 solution [114]. Figure 52 shows photographs of the samples with and without the Si-NH array. The dark appearance for the etched sample indicates significantly suppressed light reflection. Corresponding to the efficient antireflection, the absorption in the Si-NH-textured sample is greatly enhanced, as shown in Fig. 53. The *J–V* measurements at AM 1.5G show that the sample textured by the Si-NH array has a J_{sc} of ~31.25 mA/cm^2, much larger compared to ~24.94 mA/cm^2 for the device without surface texturing. Here it is noted that the FF decreased to 0.557 from 0.596 after texturing, which can be attributed to poor electrical contact.

Recently, a research group from National Renewable Energy Laboratory developed Si-NH surface texturing by directly immersing the sample into a Au-containing solution (0.4 mM $HAuCl_4$ plus a mixture of HF, H_2O_2 and

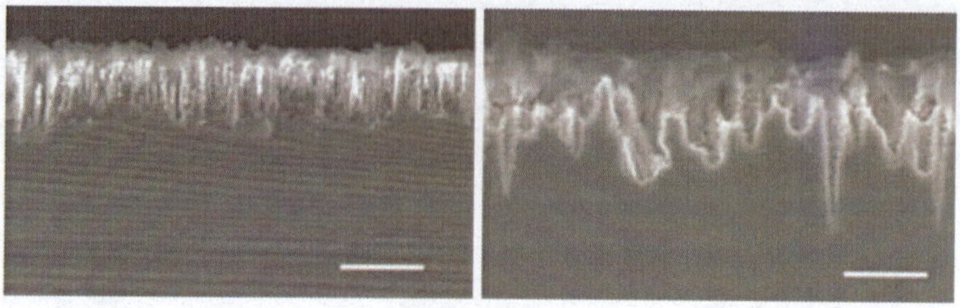

Fig. 54 Cross-sectional SEM images of the Si-NH-textured samples with etching durations of 3 min (*left*), and 6 min (*right*). The scale bar denotes 500 nm. (Reprinted with permission from [10], copyright 2009, American Institute of Physics)

H_2O (1:5:2)) [10]. Figure 54 shows the cross-sectional SEM images of the samples with etching durations of 3 min (left) and 6 min (right). The Si nanoholes were randomly distributed on the substrate surface. The optical measurement indicates that the light reflection is close to zero in the range from 350 to 1000 nm. Excellent light absorption is suggested by the high resulting PCE of ~16.8% (V_{oc}: 612 mV; J_{sc}: 34.1 mA/cm^2; FF: 0.806) for the sample after introducing the Si-NH texturing, which is higher than the PCE of 13.9% for the solar cell without an antireflection coating.

Finally, we would like to point out a pressing concern that needs to be addressed for Si nanostructure-textured Si thin film-based solar cells. A high surface defect density can be created during the etching process when fabricating the nanostructures, which can severely affect the performance of the resulting solar cells. Furthermore, a rough surface on the nanostructure makes it difficult to deposit the top transparent electrode in the conformal manner necessary for efficiently collecting the photogenerated carriers.

3 Summary

In this chapter, Si nanostructure-based solar cells including Si-ND, Si-NWs and Si nanostructure-textured (Si-NWs, Si-NCs, and Si-NHs) Si thin film-based solar cells have been discussed in terms of the structure/device preparation, optical and electrical characteristics, and latest device research status. Excellent photon management is the key advantage for these solar cells compared to their planar counterparts, such as Si wafer and thin film solar cells. Preparation of high-quality nanomaterials and structures suitable for solar cells has been greatly assisted by the recent developments in nanofabrication technologies. To obtain high-efficiency Si nanostructure-based solar cells, the bottleneck is on collecting the photogenerated carriers, which relies on high-quality electrical contacts and junctions.

References

1. Würfel P (2005) Physics of solar cells: from principles to new concepts. Wiley-VCH, Weinheim
2. Pagliaro M, Palmisano G, Ciriminna R (2008) Flexible solar cells. Wiley-VCH, Weinheim
3. Green MA (2001) Third generation photovoltaics: ultra-high conversion efficiency at low cost. Prog Photovolt Res Appl 9:123–135
4. Green MA (2003) Third generation photovoltaics: advanced solar energy conversion. Springer, Berlin
5. Green MA (2002) Third generation photovoltaics: solar cells for 2020 and beyond. Physica E 14:65–70
6. Conibeer G, Green MA, Corkish R et al (2006) Silicon nanostructures for third generation photovoltaic solar cells. Thin Solid Films 511:654–662
7. Conibeer G, Green MA, Cho EC et al (2008) Silicon quantum dot nanostructures for tandem photovoltaic cells. Thin Solid Films 516:6748–6756
8. Kayes BM, Atwater HA, Lewis NS (2005) Comparison of the device physics principles of planar and radial p–n junction nanorod solar cells. J Appl Phys 97:114302-11
9. Zhu J, Yu Z, Burkhard GF et al (2009) Optical absorption enhancement in amorphous silicon nanowire and nanocone array. Nano Lett 9:279–282
10. Yuan HC, Yost VE, Page MR et al (2009) Efficient black silicon solar cell with a density-graded nanoporous surface: optical properties, performance limitations, and design rules. Appl Phys Lett 95:123501-3
11. Kim TW, Cho CH, Kim BH et al (2006) Quantum confinement effect in crystalline silicon quantum dots in silicon nitride grown using SiH_4 and NH_3. Appl Phys Lett 88:123102-3
12. van Buuren T, Dinh LN, Chase LL et al (1998) Changes in the electronic properties of Si nanocrystals as a function of particle size. Phys Rev Lett 80:3803–3806
13. Tian B, Zheng X, Kempa TJ et al (2007) Coaxial silicon nanowires as solar cells and nanoelectronic power sources. Nature 449:885–890
14. Zacharias M, Heitmann J, Scholz R et al (2002) Size-controlled highly luminescent silicon nanocrystals: A SiO/SiO_2 superlattice approach. Appl Phys Lett 80:661–663
15. Scardera G, Puzzer T, Conibeer G et al (2008) Fourier transform infrared spectroscopy of annealed silicon-rich silicon nitride films. J Appl Phys 104:104310-7
16. Song D, Cho EC, Conibeer G et al (2007) Fabrication and electrical characteristics of Si nanocrystal/*c*-Si heterojunctions. Appl Phys Lett 91:123510-3
17. Kayes BM, Filler MA, Putnam MC et al (2007) Growth of vertically aligned Si wire array over large areas ($>1\ cm^2$) with Au and Cu catalysts. Appl Phys Lett 91:103110-3
18. Stelzner T, Pietsch M, Andrä G et al (2008) Silicon nanowire-based solar cells. Nanotechnology 19:295203-4
19. Hsu CM, Connor ST, Tang MX et al (2008) Wafer-scale silicon nanopillars and nanocones by Langmuir–Blodgett assembly and etching. Appl Phys Lett 93:133109-3
20. Hsu CH, Lo HC, Chen CF et al (2004) Generally applicable self-masked dry etching technique for nanotip array fabrication. Nano Lett 4:471–475
21. Huang MJ, Yang CR, Chiou YC et al (2008) Fabrication of nanoporous antireflection surfaces on silicon. Sol Energy Mater Sol Cells 92:1352–1357
22. Branz HM, Yost VE, Ward S et al (2009) Nanostructured black silicon and the optical reflectance of graded-density surfaces. Appl Phys Lett 94:231121-3
23. Conibeer G (2007) Third-generation photovoltaics. Mater Today 10:42–50
24. Conibeer G, Ekins-Daukes N, Guillemoles JF et al (2009) Progress on hot carrier cells. Sol Energy Mater Sol Cells 93:713–719
25. Boutry GA (1948) Augustin Fresnel: his time, life and work 1788–1827. Sci Prog 36:587–604
26. Xi JQ, Schubert MF, Kim JK et al (2007) Optical thin-film materials with low refractive index for broadband elimination of Fresnel reflection. Nat Photon 1:176–179

27. Arndt RA, Allison JF, Haynos JG et al (1975) Optical properties of the COMSAT non-reflective cell. In: Proceedings of 11th IEEE photovoltaic specialists conference, Scottsdale, May, pp 40–43
28. Zhao J, Wang A, Green MA et al (1998) 19.8% efficient "honeycomb" textured multicrystalline and 24.4% monocrystalline silicon solar cells. Appl Phys Lett 73:1991–1993
29. Green MA (2009) The path to 25% silicon solar cell efficiency: history of silicon cell evolution. Prog Photovolt Res Appl 17:183–189
30. Li JS, Yu HY, Wong SM et al (2009) Si nanopillar array optimization on Si thin films for solar energy harvesting. Appl Phys Lett 95:033102-3
31. Li JS, Yu HY, Wong SM et al (2009) Surface nanostructure optimization for solar energy harvesting in Si thin film based solar cells. In: IEEE technical digest, international electron devices meeting, pp 547–550
32. Iacona F, Franzò G, Spinella C (2000) Correlation between luminescence and structural properties of Si nanocrystals. J Appl Phys 87:1295–1303
33. Tsybeskov L, Hirschman KD, Duttagupta SP et al (1998) Nanocrystalline-silicon superlattice produced by controlled recrystallization. Appl Phys Lett 72:43–45
34. Zhang RJ, Chen YM, Lu WJ (2009) Influence of nanocrystal size on dielectric functions of Si nanocrystals embedded in SiO_2 matrix. Appl Phys Lett 95:161109-3
35. Kim TY, Park NM, Kim KH et al (2004) Quantum confinement effect of silicon nanocrystals *in situ* grown in silicon nitride films. Appl Phys Lett 85:5355–5357
36. Park NM, Choi CJ, Seong TY et al (2001) Quantum confinement in amorphous silicon quantum dots embedded in silicon nitride. Phys Rev Lett 86:1355–1357
37. Boer KW (1990) Survey of semiconductor physics. van Nostrand Reinhold, New York
38. Jiang CW, Green MA (2006) Silicon quantum dot superlattices: modeling of energy bands, densities of states, and mobilities for silicon tandem solar cell applications. J Appl Phys 99:114902-7
39. Meillaud F, Shah A, Droz C et al (2006) Efficiency limits for single-junction and tandem solar cells. Sol Energy Mater Sol Cells 90:2952–2959
40. Song D, Cho EC, Conibeer G et al (2008) Structural, electrical and photovoltaic characterization of Si nanocrystals embedded SiC matrix and Si nanocrystals/*c*-Si heterojunction devices. Sol Energy Mater Sol Cells 92:474–481
41. Cho EC, Park S, Hao X et al (2008) Silicon quantum dot/crystalline silicon solar cells. Nanotechnology 19:245201-5
42. Stupca M, Alsalhi M, Saud TA et al (2007) Enhancement of polycrystalline silicon solar cells using ultrathin films of silicon nanoparticle. Appl Phys Lett 91:063107-3
43. Kim SK, Cho CH, Kim BH et al (2009) Electrical and optical characteristics of silicon nanocrystal solar cells. Appl Phys Lett 95:143120-3
44. Perez-Wurfl I, Hao X, Gentle A et al (2009) Si nanocrystal p–i–n diodes fabricated on quartz substrates for third generation solar cell applications. Appl Phys Lett 95:153506-3
45. Cantele G, Degoli E, Luppi E et al (2005) First-principles study of n- and p-doped silicon nanoclusters. Phys Rev B 72:113303-4
46. Erwin SC, Zu L, Haftel MI et al (2005) Doping semiconductor nanocrystals. Nature 436:91–94
47. Westwater J, Gosain DP, Tomiya S et al (1997) Growth of silicon nanowires via gold/silane vapor–liquid–solid reaction. J Vac Sci Technol B 15:554–557
48. Hochbaum AI, Fan R, He R et al (2005) Controlled growth of Si nanowire array for device integration. Nano Lett 5:457–460
49. Schmidt V, Senz S, Gösele U (2005) Diameter-dependent growth direction of epitaxial silicon nanowires. Nano Lett 5:931–935
50. Tsakalakos L, Balch J, Fronheiser J et al (2007) Strong broadband optical absorption in silicon nanowire films. J. Nanophoton 1:013552-10
51. Hochbaum AI, Chen R, Delgado RD et al (2008) Enhanced thermoelectric performance of rough silicon nanowires. Nature 451:163–168

52. Chan CK, Peng H, Liu G et al (2008) High-performance lithium battery anodes using silicon nanowires. Nat Nanotechnol 3:31–35
53. Pan C, Wu H, Wang C et al (2008) Nanowire-based high performance "micro fuel cells": one nanowire, one fuel cell. Adv Mater 20:1644–1648
54. Peng KQ, Yan YJ, Gao SP et al (2002) Synthesis of large-area silicon nanowire array via self-assembling nanoelectrochemistry. Adv Mater 14:1164–1167
55. Huang Z, Fang H, Zhu J (2007) Fabrication of silicon nanowire array with controlled diameter, length, and density. Adv Mater 19:744–748
56. Wang Y, Schmidt V, Senz S et al (2006) Epitaxial growth of silicon nanowires using an aluminium catalyst. Nat Nanotechnol 1:186–189
57. Shimizu T, Xie T, Nishikawa J et al (2007) Synthesis of vertical high-density epitaxial Si (100) nanowire array on a Si (100) substrate using an anodic aluminum oxide template. Adv Mater 19:917–920
58. Wagner RS, Ellis WC (1964) Vapor–liquid–solid mechanism of single crystal growth. Appl Phys Lett 4:89–90
59. Westwater J, Gosain DP, Usui S (1998) Si nanowires grown via the vapor–liquid–solid reaction. Phys Stat Sol (a) 165:37–42
60. Yao Y, Fan S (2007) Si nanowires synthesized with Cu catalyst. Mater Lett 61:177–181
61. Morales AM, Lieber CM (1998) A laser ablation method for the synthesis of crystalline semiconductor nanowires. Science 279:208–211
62. Ke Y, Weng X, Redwing JM et al (2009) Fabrication and electrical properties of Si nanowires synthesized by Al catalyzed vapor–liquid–solid growth. Nano Lett 9:4494–4499
63. Chen W, Ahmed H (1993) Fabrication of high aspect ratio silicon pillars of <10 nm diameter. Appl Phys Lett 63:1116–1118
64. Hadobás K, Kirsch S, Carl A et al (2000) Reflection properties of nanostructure-array silicon surfaces. Nanotechnology 11:161–164
65. Bullis WM (1966) Properties of gold in silicon. Solid-State Electron 9:143–168
66. Dabbousi BO, Murray CB, Rubner MF et al (1994) Langmuir–Blodgett manipulation of size-selected CdSe nanocrystallites. Chem Mater 6:216–219
67. Peng K, Zhang M, Lu A et al (2007) Ordered silicon nanowire array via nanosphere lithography and metal induced etching. Appl Phys Lett 90:163123-3
68. Lu Y, Xiong H, Jiang X et al (2003) Asymmetric dimers can be formed by dewetting half-shells of gold deposited on the surfaces of spherical oxide colloids. J Am Chem Soc 125:12724–12725
69. Li X, Bohn PW (2000) Metal-assisted chemical etching in HF/H_2O_2 produces porous silicon. Appl Phys Lett 77:2572–2574
70. Fuhrmann B, Leipner HS, Höche HR et al (2005) Ordered array of silicon nanowires produced by nanosphere lithography and molecular beam epitaxy. Nano Lett 5:2524–2527
71. Masuda H, Fukuda K (1995) Ordered metal nanohole array made by a two-step replication of honeycomb structures of anodic alumina. Science 268:1466–1468
72. Che G, Lakshmi BB, Fisher ER et al (1998) Carbon nanotubule membranes for electrochemical energy storage and production. Nature 393:346–349
73. Choi J, Sauer G, Nielsch K et al (2003) Hexagonally arranged monodisperse silver nanowires with adjustable diameter and high aspect ratio. Chem Mater 15:776–779
74. Mei X, Kim D, Ruda HE et al (2002) Molecular-beam epitaxial growth of GaAs and InGaAs/GaAs nanodot array using anodic Al_2O_3 nanohole array template masks. Appl Phys Lett 81:361–363
75. Nasir ME, Allsopp DWE, Bowen CR et al (2010) The fabrication of mono-domain highly ordered nanoporous alumina on a wafer scale by a guided electric field. Nanotechnology 21:105303-6
76. Jessensky O, Müller F, Gösele U (1998) Self-organized formation of hexagonal pore array in anodic alumina. Appl Phys Lett 72:1173–1175
77. Lombardi I, Hochbaum AI, Yang P et al (2006) Synthesis of high density, size-controlled Si nanowire array by porous anodic alumina mask. Chem Mater 18:988–991

78. Guo LJ (2007) Nanoimprint lithography: methods and material requirements. Adv Mater 19:495–513
79. Plass KE, Filler MA, Spurgeon JM et al (2008) Flexible polymer-embedded Si wire array. Adv Mater 21:325–328
80. Tsakalakos L, Balch J, Fronheiser J et al (2007) Silicon nanowire solar cells. Appl Phys Lett 91:233117-3
81. Garnett EC, Yang P (2008) Silicon nanowire radial p–n junction solar cells. J Am Chem Soc 130:9224–9225
82. Kalita G, Adhikari S, Aryal HR et al (2009) Silicon nanowire array/polymer hybrid solar cell incorporating carbon nanotubes. J Phys D Appl Phys 42:115104-5
83. Street RA, Qi P, Lujan R et al (2008) Reflectivity of disordered silicon nanowires. Appl Phys Lett 93:163109-3
84. Hu L, Chen G (2007) Analysis of optical absorption in silicon nanowire array for photovoltaic applications. Nano Lett 7:3249–3252
85. Air Mass 1.5 Spectra (2010) American society for testing and materials. http://rredc.nrel.gov/solar/spectra/am1.5/#1962. Accessed 9 March 2010
86. Shockley W, Queisser HJ (1961) Detailed balance limit of efficiency of p–n junction solar cells. J Appl Phys 32:510–519
87. Li JS, Yu HY, Wong SM et al (2009) Design guidelines of periodic Si nanowire array for solar cell application. Appl Phys Lett 95:243113-3
88. Putnam MC, Turner-Evans DB, Kelzenberg MD et al (2009) 10 μm minority-carrier diffusion lengths in Si wires synthesized by Cu-catalyzed vapour–liquid–solid growth. Appl Phys Lett 95:163116-3
89. Maiolo JR III, Kayes BM, Filler MA et al (2007) High aspect ratio silicon wire array photoelectrochemical cells. J Am Chem Soc 129:12346–12347
90. Sivakov V, Andrä G, Gawlik A et al (2009) Silicon nanowire-based solar cells on glass: synthesis, optical properties, and cell parameters. Nano Lett 9:1549–1554
91. van den Donker MN, Gordijn A, Stiebig H et al (2007) Flexible amorphous and microcrystalline silicon tandem solar modules in the temporary superstrate concept. Sol Energy Mater Sol Cells 91:572–580
92. Garnett E, Yang P (2010) Light trapping in silicon nanowire solar cells. Nano Lett 10:1082–1087
93. Yoon J, Baca AJ, Park SI et al (2008) Ultrathin silicon solar microcells for semitransparent, mechanically flexible and microconcentrator module designs. Nat Mater 7:907–915
94. Campbell P, Green MA (1987) Light trapping properties of pyramidally textured surfaces. J Appl Phys 62:243–249
95. Gray JL (2003) The physics of the solar cell. In: Luque A, Hegedus S (eds) Handbook of photovoltaic science and engineering. Wiley, Chichester
96. Kelzenberg MD, Turner-Evans DB, Kayes BM et al (2008) Single-nanowire Si solar cells. In: Proceedings of the 33rd IEEE photovoltaic specialists conference, pp 144–149
97. Bai XD, Zhi CY, Liu S et al (2003) High-density uniformly aligned silicon nanotip array and their enhanced field emission characteristics. Solid State Commun 125:185–188
98. Wang Q, Li JJ, Bai XD et al (2005) Field emission properties of carbon coated Si nanocone array on porous silicon. Nanotechnology 15:2919–2922
99. Hsu CH, Huang YF, Chen LC et al (2006) Morphology control of silicon nanotips fabricated by electron cyclotron resonance plasma etching. J Vac Sci Technol B 24:308–311
100. Li LP, Lu YF, Doerr DW et al (2004) Parametric investigation of laser nanoimprinting of hemispherical cavity array. J Appl Phys 96:5144–5151
101. Huang SM, Hong MH, Luk'yanchuk BS et al (2002) Pulsed laser-assisted surface structuring with optical and near-field enhanced effects. J Appl Phys 92:2495–2500
102. Tsujino K, Matsumura M, Nishimoto Y (2006) Texturization of multicrystalline silicon wafers for solar cells by chemical treatment using metallic catalyst. Sol Energy Mater Sol Cells 90:100–110

103. Gorostiza P, Díaz R, Servat J et al (1997) Atomic force microscopy study of the silicon doping influence on the first stages of platinum electroless deposition. J Electrochem Soc 144:909–914
104. Wang F, Yu HY, Wang XC et al (2010) Maskless fabrication of large scale Si nanohole array via laser annealed metal nanoparticles catalytic etching for photovoltaic application. J Appl Phys 108:024301-3
105. Li JS, Yu HY, Wong SM et al (2010) Si nanocone array optimization on crystalline Si thin films for solar energy harvesting. J Phys D Appl Phys 43:255101-7
106. Wang F, Yu HY, Li JS et al (2010) Optical absorption enhancement in nanopore textured-silicon thin film for photovoltaic application. Opt Lett 35:40–42
107. Wong SM, Yu HY, Li JS et al (2010) Design high-efficiency Si nanopillar-array-textured thin film solar cell. IEEE Electron Device Lett 31:335–337
108. Shockley W, Read WT (1952) Statistics of the recombinations of holes and electrons. Phys Rev 87:835–842
109. Shibib MA, Lindholm FA, Fossum JG (1979) Auger recombination in heavily doped shallow-emitter silicon p–n-junction solar cells, diodes and transistors. IEEE Trans Electron Devices ED 26:1104–1106
110. Kelzenberg MD, Putnam MC, Turner-Evans DB et al (2009) Predicted efficiency of Si wire array solar cells. In: Proceedings of the 34th IEEE photovoltaic specialists conference, pp 391–396
111. Li JS, Wong SM, Li YL et al (2010) High-efficiency crystalline Si thin film solar cells with Si nanopillar array textured surfaces. In: Proceedings of the 35th IEEE photovoltaic specialists conference
112. Zhu J, Hsu CM, Yu Z et al (2010) Nanodome solar cells with efficient light management and self-cleaning. Nano Lett 10:1979–1984
113. Green MA, Emery K, Hishikawa Y et al (2009) Solar cell efficiency tables (version 34). Prog Photovolt Res Appl 17:320–326
114. Nishioka K, Sueto T, Saito N (2009) Formation of antireflection nanostructure for silicon solar cells using catalysis of single nano-sized silver particle. Appl Surf Sci 255:9504–9507

Organic and Hybrid Solar Cells Based on Small Molecules

Luiz C. P. Almeida, Jilian N. de Freitas, Flavio S. Freitas and Ana F. Nogueira

Abstract In this chapter, the recent literature involving small molecule-based organic solar cells (OSCs) will be reviewed. The number of papers published in the fields of organic semiconductor and OSCs has grown exponentially in the past decade. Such growth is stimulated by the exciting properties of these materials, combined with the possibility to produce colored, flexible, transparent and cheap solar cells. The main focus of this review is to give an overview and a perspective of the recent advances in this area, highlighting the most interesting results, novel materials as well as their limitations and challenges. This chapter will explore the properties and applications of several classes of small organic molecules, as electron donors and acceptors, dyes, and hole transport materials. Different architectures and techniques will be also discussed in the assembly of double, heterojunction, and multilayer films.

1 Introduction

"Size is not important". This well-known adage is heard everywhere when the matter is size (and in most cases, in a positive perspective). In the field of organic semiconductors (OSs) and in particular those involving organic field-effect transistors, organic light-emitting devices and organic solar cells (OSCs), size has been demonstrated to be an irrelevant factor. Indeed, the best performing devices are those based on a limited number of low-molecular weight materials.

L. C. P. Almeida · J. N. de Freitas · F. S. Freitas · A. F. Nogueira (✉)
Chemistry Institute, University of Campinas-UNICAMP,
P.O. Box 6154, Campinas, SP 13083-970, Brazil
e-mail: anaflavia@iqm.unicamp.br

L. Zang (ed.), *Energy Efficiency and Renewable Energy Through Nanotechnology*, Green Energy and Technology, DOI: 10.1007/978-0-85729-638-2_2,

There are currently two major classes of OSs: the low-molecular weight materials, or small molecules, and π-conjugated polymers such as poly(thiophenes), poly(fluorine), and poly(phenylene vinylenes) derivatives. They have many features in common, such as their excitonic nature (excitons are electrically neutral quasi-particles consisting of bound electron–hole pairs formed after photoexcitation), low dielectric constants, localized charge carriers (electrons and holes), optically and electrically anisotropic properties, high extinction coefficients, narrow absorption bands, and more disordered structures compared to inorganic semiconductors.

Organic semiconductors, both small molecules and polymers, show great promise for photoconversion through their synthetic variability, low-temperature processing (similar to that applied to plastics), and the possibility of producing lightweight, flexible, easily manufactured, and inexpensive solar cells.

Although OSCs have been known for more than 50 years, they have become the subject of active research only in the past 20 years. This new generation of solar cells is also referred to as nanostructured solar cells since at least one component and/or the morphology is in the nanoscale range. Low-temperature processing of organic small molecules and polymers from the vapor-phase or from solution have a crucial advantage over silicon technology since the high-temperature processing requirements of the latter increases the production cost and limits the range of substrates on which they can be deposited. Additionally, OSs can be easily applied using low-cost methods, such as the high-speed, roll-to-roll technique. Unfortunately, despite significant advances, the power-conversion efficiency of OSCs remains low, with maximum values in the range of 6–7%. When at least one component is replaced by an inorganic counterpart, these solar cells are referred to as hybrid devices. Dye-sensitized solar cells (DSSC) fit well in this class. For this kind of solar cell, the efficiency is $\sim$11%, but has remained at this plateau in the last years.

In order to enhance their competitiveness with other technologies, efficiency and long-term stability are crucial in the field of OSC. The photocurrent in these solar cells is limited by the light-harvesting capability of the individual molecules or polymers in the device. Small band gap molecules have been intensively studied to overcome these drawbacks, but it is a complicated matter. Morphology is also important in this context because it impacts charge transport and an intimate contact between donor and acceptor materials in a nanoscale range is difficult to achieve due to phase separation. A better understanding of these processes at the material level, particularly those in layer-to-layer interfaces, which determine the open-circuit voltage (V_{OC}), is critical and remains the subject of active research. For DSSCs, photoelectrochemical devices, an additional factor is the presence of a liquid electrolyte, which can hamper large-area production.

In this chapter, the application of small organic molecules applied to organic or hybrid solar cells (including DSSCs) will be reviewed. A detailed description of the chemical, physical, and electrical properties of OSs, or the state of art of OSCs based on π-conjugated polymers is not the focus of this chapter. For this purpose, the reader is encouraged to see references [1–7].

Although many small molecules have been known and studied for decades, only a small fraction have been used successfully in OSCs. This reflects the diversity in charge-carrier mobilities, exciton diffusion lengths, thin film morphology, energy levels, band gap, absorption coefficient and ambient and thermal stability. In this chapter, we present the most commonly used donor and acceptor materials used as active layers in OSCs and the most important contributions in this field will be highlighted.

This chapter will be divided into five sections. The first two deal with small molecules applied to OSCs, separated by fabrication techniques: physical and solution methods. The third and fourth sections deal with liquid crystals (LCs) and three-component solar cells, respectively, using diverse techniques. The last section involves a description of small molecules in dye-sensitized solar cells as sensitizer and hole transport materials.

2 Solution-Processable OSC

Solution-processable OSCs (SPOSC), prepared using solution processing techniques including spin-coating, casting, roll-to-roll, etc., have attracted increasing attention in academia and industry because of their potential advantages. Such advantages include easy fabrication, low-cost, low-weight, large-area production, and mechanical flexibility. On the other hand, films formed from blend solutions tend to phase separate. The scale of the phase separation depends on the solvent, solubility of the materials and other parameters associated with deposition, such as the speed and temperature of the spin-coating process. The nanomorphology of the active layer film is very important to device efficiency. Also, the materials used in solution-processed solar cells must fulfill the physical conditions of high charge-carrier mobilities, suitable values of highest occupied molecular orbital (HOMO) and lowest unoccupied molecular orbital (LUMO) energy levels, strong absorption in the visible region and several processing conditions, such as: (i) the materials must be soluble in common organic solvents, (ii) the solubility has to be high enough to enable the deposition of smooth thin films, or even thick films, (iii) the morphology of the spin-casted films should have the desired structure concerning phase separation on the exciton diffusion length-scale and the percolation path to the electrodes.

In this section, we present a review of recent reports on the use of small molecule-based OSC assembled using solution-processing methods. This kind of solar cell is analogous to the bulk-heterojunction devices based on polymer and fullerenes, which are not in the scope of this review. Concerning the efficiencies, OSCs based on small processable molecules are considered to be less efficient, but, recent reports have demonstrated the potential of the small molecules and efficiencies are catching up to those of the "standard" bulk-heterojunction devices.

The first achievement in the area of SPOSC was to make the small molecules soluble in organic solvents. Before their use in organic or hybrid solar cells, most small molecules must undergo several chemical modifications in order to be made appropriate for a solution-deposition method. Even the buckminsterfullerene, C_{60}, had to be converted into its well-known soluble derivative, [6, 6]-phenyl-C_{61}-butyric acid methyl ester (PCBM). Most of the significant achievements have been realized with chemically modified molecules.

For example, perylene and its derivatives have been combined with different molecules and polymers in SPOSCs. Li et al. [8] investigated a blend of perylene tetracarboxydiimide as the acceptor and the polymer poly[*N*-(20-decyltetradecyl)carbazole]-2,7-diyl as the donor. The bulk-heterojunction device based on this combination afforded 0.63% of overall power-conversion efficiency (η) under low light intensity illumination (10 mW cm^{-2}). Sharma et al. [9] reported the fabrication of photovoltaic devices using a bulk-heterojunction layer of a small molecule named compound T (Fig. 1a), containing a central *p*-phenylenevinylene unit, an intermediate thiophene moiety, and terminal 4-nitrophenyl-cyano-vinylene as the donor, and a perylene–pyrene bisimide (Fig. 1b) as the acceptor. At the optimum blend ratio (donor:acceptor 1:3.5 wt%), η was $\sim$1.9%. The efficiency was further increased to $\sim$3.2% when the device was annealed (100°C for 5 min) and a thin ZnO layer was incorporated between the active layer and the Al electrode.

Squaraine dyes are another class of molecules which have been the subject of many recent investigations, owing to their unique photochemical/photophysical properties. The attraction to these dyes includes their broad absorption spectrum (550–900 nm), facile synthetic access, a wide variety of possible structures and oxygen/moisture stability. These properties enable an active layer deposition under ambient conditions, in contrast to the inert atmosphere required for most conducting polymers. New soluble squaraine derivatives (Fig. 1c) were used as long-wavelength absorbers and donor components in SPOSC, and the effects of core modification on active layer film morphology and photovoltaic response were investigated [10]. Both linear and branched alkyl chain substituents provide solubility, but each gave rise to different effects on the solid state organization [11]. SPOSCs with different squaraine:PCBM ratios were fabricated by spin-coating the blends from chloroform or *o*-dichlorobenzene. The optimum annealing conditions were found to be 70°C for 1 h, as evidenced by the increased phase separation observed in atomic force microscopy (AFM) images. Devices fabricated with a 1:3 squaraine:PCBM ratio spin-cast from chloroform exhibited higher carrier mobility and improved performance, with a short-circuit current (J_{SC}) of 5.70 mA cm^{-2} and an efficiency of 1.24%. This value was 1.5 times higher than the result obtained using *o*-dichlorobenzene and was attributed to microstructure evolution effects that occur during a more rapid film growth and drying rate when using chloroform [10].

Recently, Winzenberg et al. [12] reported the polycyclic aromatic template dibenzo[*b,def*]chrysene as a promising candidate for use in organic electronic devices. These compounds have an advantage over other well-studied small molecules, such as pentacenes, because they do not undergo cycloaddition reactions with fullerenes. The bulk-heterojunction solar cells made from a

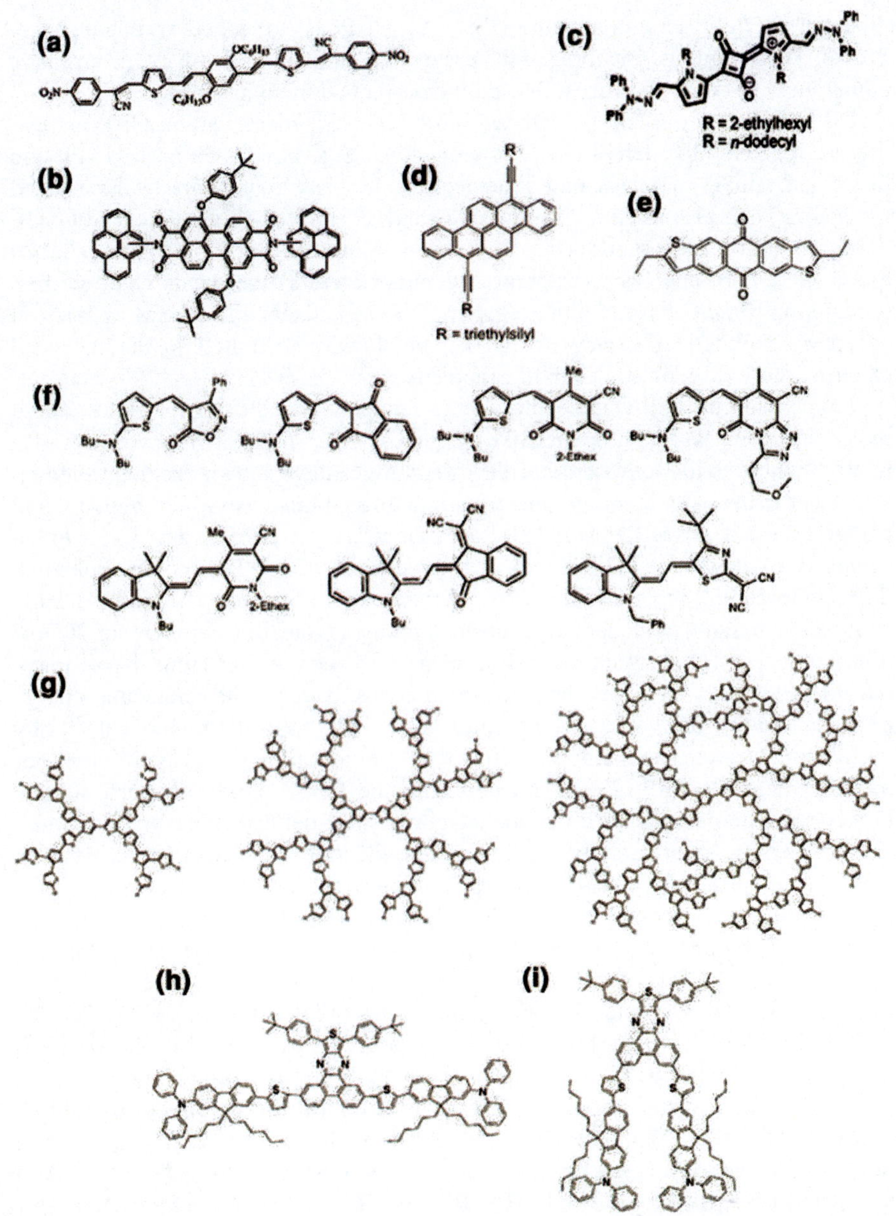

Fig. 1 Examples of chemical structures of selected processable small molecules **a** compound T, **b** perylene-pyrene (PPI) bisimide, **c** squaraine, **d** dibenzo[*b,def*]chrysene, **e** anthradithiophene, **f** merocyanine, **g** dendritic thiophene, **h**, **i** dibenzo[*f,h*]thieno[3,4-*b*]quinoxaline [9, 10, 13–16]

dibenzo[*b,def*]chrysene compound (Fig. 1d) and PCBM reached an efficiency of 2.25%. These findings encouraged the use of other unexplored polycyclic aromatic compounds in solution-processable bulk-heterojunction solar cells.

Other molecules, such as, anthradithiophene [13], merocyanine [14], or dendritic thiophene [15] derivatives, shown in Fig. 1e–g, have been used in solution processed, small molecule-based devices in the last few years. Amine-based molecules were also explored as possible candidates for highly efficient SPOSCs. Two isomeric compounds (Fig. 1h–i) containing a dibenzo[*f,h*]thieno[3,4-*b*] quinoxaline core and two peripheral arylamines were synthesized. The bulk-heterojunction SPOSC based on these materials as sensitizers and PCBM as electron acceptor exhibited efficiency of 1.70%, which was attributed to the balanced electron and hole mobility found in the active layer film [16].

Triphenylamine (TPA) molecules are also potential candidates for OSCs. These molecules possess 3D, propeller-like geometry, glass-forming properties, and a relatively high oxidation potential. TPA derivatives have shown excellent thermal and electrochemical stability, electron-donating ability, isotropic optical, and charge-transport properties [17, 18]. Research efforts have led to a progress in the synthesis of new molecules which mainly consist of a TPA moiety linked to different acceptor moieties, including dicyanovinyl, perylene, benzothiadiazole (BT), or 2-pyran-4-ylidenemalononitrile, aiming at the development of donor-π-acceptor (D-π-A) structures. Molecules with D-π-A structure have many advantages, such as a lower bandgap which arises from the intramolecular charge transfer between the donor and acceptor, and easily controlled energy levels by introducing acceptor and donor moieties with different pull–push electron abilities into the molecules [19, 20]. Compared with the D-π-A type polymers, soluble D-π-A small molecules have the advantages of well-defined molecular structure, monodispersity, and relatively simple and reproducible synthesis and purification, making them promising materials for application in solution-processing solar cells.

New solution-processable molecules, shown in Fig. 2, based on the combination of TPA donor units and BT acceptor units have been synthesized recently [21–24]. He et al. [21] compared the properties of devices based on the star-shaped molecule S(TPA-BT) and the linear molecule L(TPA-BT) (see Fig. 2). The S(TPA-BT) film showed a broader and stronger absorption band in the range of 440–670 nm, lower band gap, higher hole mobility and better film-forming properties (high quality uniform spin-cast film) than those of the corresponding linear L(TPA-BT). The devices assembled with PCBM as acceptor and S(TPA-BT) reached a J_{SC} of 4.2 mA cm^{-2}, V_{OC} of 0.81 V, FF of 0.39 and 1.3% of efficiency, while those based on L(TPA-BT) showed J_{SC} of 1.25 mA cm^{-2}, V_{OC} of 0.84 V, FF of 0.34 and 0.35% of efficiency. The same group later investigated a molecule with TPA as core and benzothiadiazole-(4-hexyl) thiophene as arms, S(TPA-BT-4HT). The end group of 4-hexylthiophene in S(TPA-BT-4HT) instead of the TPA end group in S(TPA-BT) was intended to further improve the film morphology and photovoltaic properties. The SPOSC devices were fabricated by spin-coating the blended solution of S(TPA-BT-4HT) and [6, 6]-phenyl-C_{71}-butyric acid methyl ester ($PC_{71}BM$) (1:3, w/w) and by using a Mg/Al electrode.

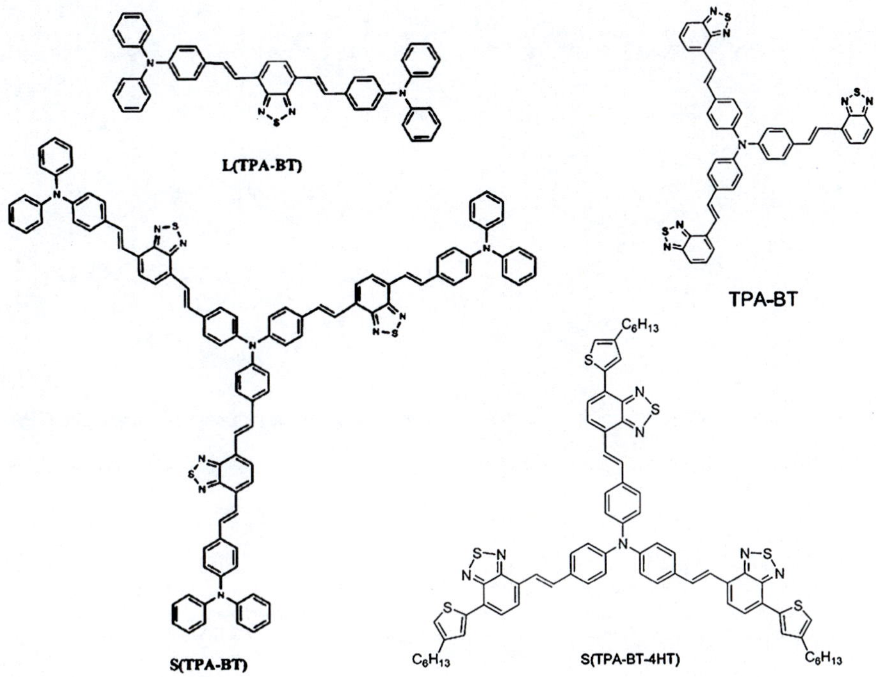

Fig. 2 Examples of linear and star-shaped molecules consisting of a TPA moiety linked to different acceptor moieties [21, 22]

The obtained J_{SC} was 8.58 mA cm^{-2}, V_{OC} of 0.85 V and FF of 32.7%, resulting in maximum efficiency of 2.4% [22].

D-π-A structures containing electron-accepting sulfonyldibenzene cores and electron-donating TPA dendrons have also been synthesized. Since the dendrimers were highly soluble in common organic solvents, they were used to assemble bulk-heterojunction SPOSCs in combination with PCBM, by spin-coating [25]. The cell based on dendrimer G0 (Fig. 3) showed an efficiency of only 0.34%. This might be related to the absorption characteristics of the molecule, which strongly absorbs light only below 450 nm. Devices fabricated with other molecules with similar structures (G1 and G2, Fig. 3) had even lower performances. Although the larger molecules G1 and G2 absorb photons at longer-wavelengths when compared to G0, their extinction coefficients are lower than that of G0.

2-{2,6-*bis*-[2-(4-styryl)-vinyl]-pyran-4-ylidene}malononitrile (DCM)-type organic dyes have also been combined with TPA. DCM dyes are known as low-molecular weight red-emitting materials and most of them have a D-π-A molecular structure that contains (dicyanomethylene)pyran (PM) as the electron acceptor. The strong electron-withdrawing PM group can lower the LUMO energy level and extend the absorption band such that it nearly overlaps with the entire visible spectrum when combined with strong electron-donating units. Moreover, the strong intermolecular dipole–dipole interaction or intermolecular π-stacking of

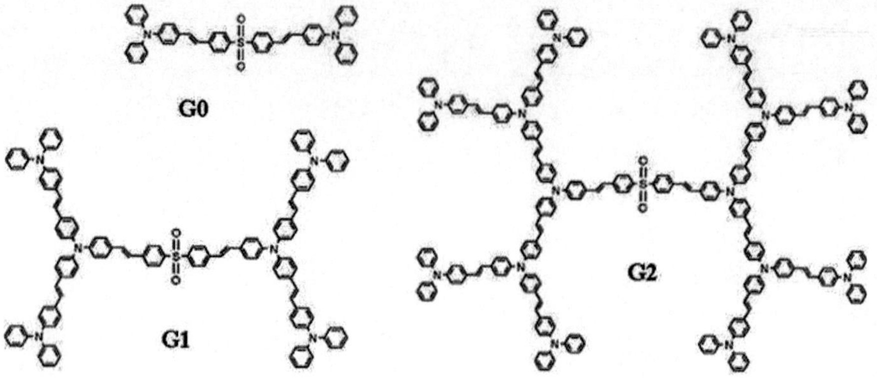

Fig. 3 Chemical structures of D-π-A molecules containing sulfonyldibenzene cores and TPA dendrons [25]

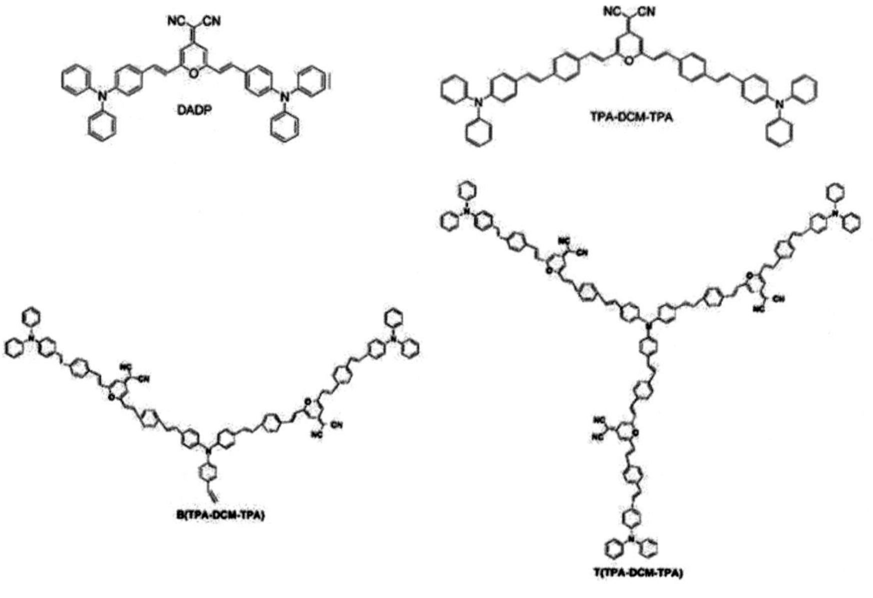

Fig. 4 Examples of D-π-A small molecules used in SPOSC, containing TPA groups linked to PM groups [26–28]

DCM-type organic dyes may be beneficial to the charge-carrier transportation. The combination of electron-rich TPA and electron deficient PM groups via π-conjugated spacers is also an interesting combination, since they can effectively reduce the band gap and produce special physical and photoelectric properties.

He et al. [26] reported the synthesis of a symmetric D-π-A small molecule TPA-DCM-TPA where two TPA groups are linked by divinylbenzene bridges at both ends of the PM group (Fig. 4). This material was applied in SPOSC and gave an efficiency of 0.79%.

Another symmetrical dye molecule, DADP (Fig. 4), was also used in a SPOSC [27]. The optimized bulk-heterojunction devices based on the combination of this dye and PCBM as the acceptor exhibited a J_{SC} of 4.16 mA cm^{-2}, V_{OC} of 0.98 V, FF of 0.37, and efficiency of 1.50%. The efficiency of the device based on DADP was almost twice that of the device based on TPA-DCM-TPA, although there is only a small difference between their molecular structures: DADP has a shorter distance between the TPA group and the PM group than TPADCM-TPA. This small structural difference results in a lower-lying HOMO energy level and a higher hole mobility in DADP, leading to increased efficiency.

Zhao et al. [28] synthesized bi-armed B(TPA-DCM-TPA) and tri-armed T(TPA-DCM-TPA) molecules, shown in Fig. 4. Although the B(TPA-DCM-TPA) and T(TPA-DCM-TPA) films show broad and strong absorption band in a wavelength range of 300–750 nm, when B(TPA-DCM-TPA) was used as the electron donor in an OSC with PCBM as the electron acceptor, the devices delivered a power-conversion efficiency of only 0.73%.

Phthalocyanines (PC) and porphyrins (PP) have also been successfully used in SPOSCs. In fact, the first efficient OSC based on small molecules reported by Tang [29] had PC as the donor material. Several examples of binary blends involving PP have also been reported, designed for narrow-band absorption at the end of the visible spectrum [30, 31]. However, higher efficiencies were found for copper phthalocyanine/C_{60} p–n junction devices [32], due to the longer exciton diffusion length in PC films as compared to PP films, which is reported to be as high as 68 nm [33].

The solution processing of the small molecule subnaphthalocyanine (SubNc) was developed by Ma et al. [34]. Due to the high solubility, low tendency to aggregate and strong light absorption in the visible region, amorphous SubNc films with high charge transporting and light-harvesting properties were prepared via solution casting. By using SubNc as the donor and C_{60} as the acceptor in a planar heterojunction SPOSC, the authors reported a power-conversion efficiency of 1.5%, with a V_{OC} of 0.55 V, J_{SC} of 5.6 mA cm^{-2}, and FF of 0.49. This device performance was considered high for a planar heterojunction device based on solution processable small molecules, and was assigned to the contribution of triplet excitons from SubNc.

Soluble 1,4:8,11:15,18:22,25-tetraethano-29*H*,31*H*-tetrabenzo[*b*,*g*,*l*,*q*]porphyrin (CP), which can be thermally converted to a highly insoluble and crystalline tetrabenzoporphyrin (BP) donor (Fig. 5), was used with a new fullerene acceptor *bis*(dimethylphenylsilylmethyl)C_{60} (SIMEF, Fig. 5) in a three-layered p–*i*–n structure, in which the *i*-layer possessed a defined interdigitated structure, formed by spontaneous crystalline phase separation during thermal processing [35]. This device showed a power-conversion efficiency of 5.2%, a value that is among the best reported for solution-processed small-molecule OSCs.

Huang et al. [36] synthesized two soluble alternating porphyrin-dithienothiophene conjugated copolymers where the units are linked by a single bond (I) or triple bond (II), as shown in Fig. 6. The absorption spectrum of a thin film of I exhibited a sharp Soret band at 450 nm and two weak Q-bands at 563–619 nm,

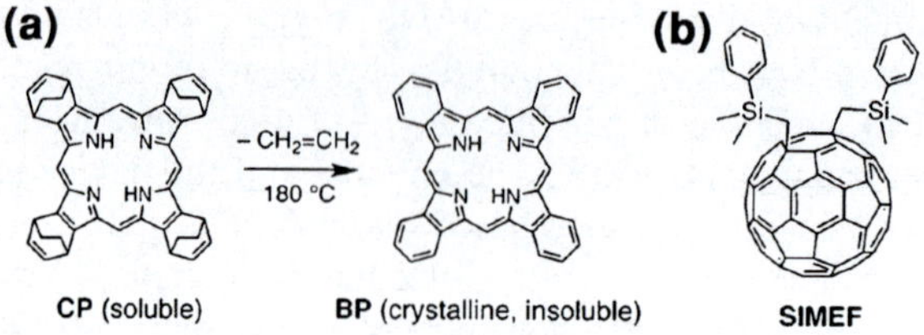

Fig. 5 Donor and acceptor materials used in SPOSC: **a** thermal retro-Diels–Alder conversion of CP (donor precursor) to BP (donor) at 180°C; **b** SIMEF (acceptor). "Reprinted with permission from Matsuo et al. [35]. Copyright 2009 American Chemical Society"

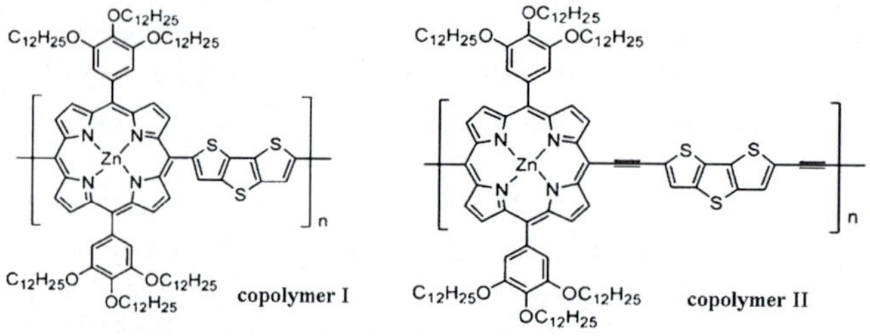

Fig. 6 Chemical structures of porphyrin-dithienothiophene copolymers containing single (I) or triple bonds (II) [36]

while II exhibited a sharp Soret band at 491 nm and a strong Q-band at 760 nm. The field-effect hole mobilities were measured to be 2.1×10^{-4} cm^2 V^{-1} s^{-1} for the two copolymers. Solar cells based on blends of these copolymers with PCBM were prepared by spin-coating an *o*-dichlorobenzene solution, followed by drying at 80°C for 30 min. Despite the high charge-carrier mobilities, the efficiencies obtained were only 0.30 and 0.15% for devices using II:PCBM or I:PCBM as active layer. Although the authors do not comment on the low efficiency values obtained, they attribute the difference observed to the stronger Q-band absorption shown by the copolymer II.

Recently, efficiencies of 2.3 and 3.0% were reported for devices combining the electron-poor diketopyrrolopyrrol-containing low-dimensional oligothiophene 2,5-di-(2-ethylhexyl)-3,6-*bis*-(500-*n*-hexyl-[2, 20, 50, 200]terthiophen-5-yl)-pyrrolo [3,4-c]pyrrole-1, 4-dione (SMDPPEH, Fig. 7) as donor, with PCBM or $PC_{71}BM$ as acceptors [37–39]. Aiming at further exploring film morphology formation and its effects on device performance as a function of donor–acceptor (DA) interactions, Tamayo et al. [40] fabricated solar cells using SMDPPEH and methanofullerene derivatives having alkyl substituents with different lengths (as shown in Fig. 7). The authors observed that the absorption, crystallinity, film morphology, and device performance characteristics of both as-cast and annealed blended films of diketopyrrolopyrrole-based donors and fullerene acceptors are significantly affected by the length of the alkyl substituent attached to the methanofullerene. The domain size of these structures increased when the alkyl chain attached to the

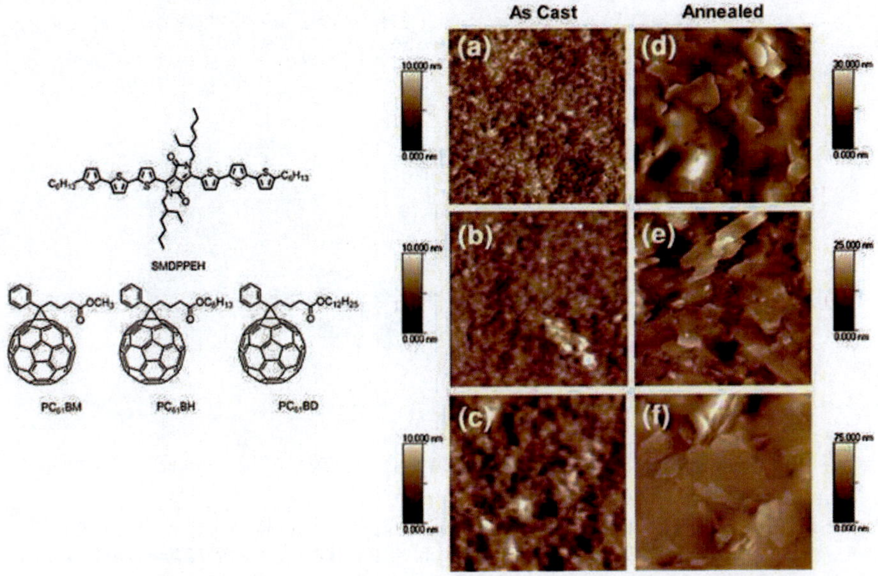

Fig. 7 Chemical structures of the donor SMDPPEH and the methanofullerene derivatives acceptors $PC_{61}BM$, $PC_{61}BH$ and $PC_{61}BD$, which contain alkyl substituents with different lengths. AFM topographic images of as-cast (**a**, **b**, **c**) and annealed (**d**, **e**, **f**) SMDPPEH:PC61BX (50:50) from (2% w/v) solution where X = M = methyl (**a** and **d**), X = H = *n*-hexyl (**b** and **e**), X = D = *n*-dodecyl (**c** and **f**). Images for the as-cast and annealed films are 1 μm × 1 μm and 2 μm × 2 μm in size, respectively. Ref. [40]—reproduced by permission of The Royal Society of Chemistry (http://dx.doi.org/10.1039/B912824G)

fullerene acceptor was increased. Furthermore, annealing the blended films led to varying degrees of phase separation and the most pronounced phase separation was observed for the fullerene containing the largest alkyl chain. The different morphologies observed are shown in Fig. 7, and are believed to have resulted from the differences in hydrophobicity of the methanofullerenes, which affects their interaction with the relatively more polar donor material. Thus, the simple variation of the alkyl chain length results in devices with power-conversion efficiencies between 1.5 and 3%.

To date, OSCs exhibiting fill factor (FF) values exceeding 50% have only been realized with fullerenes [41–43], TiO_2 [44], or CdSe [45] as electron acceptors. However, the V_{OC} of these devices seldom exceeds 0.7 V, which is in part due to the high electron affinity of the electron-accepting phase, while an V_{OC} exceeding 1 V would be desirable for high power generation. On the other hand, V_{OC} values reaching 1.5 V have been obtained when blending suitable electron-donating and accepting polymers [46–48], but the FF of these devices was below 40% in most cases.

In this context, 2-vinyl-4,5-dicyanoimidazole (Vinazene) has been used as a precursor to design a novel family of electron-accepting materials in which, by changing the central aromatic unit, the energy of the LUMO can be tuned to sufficiently low values. Ooi et al. [49] employed the small-molecule

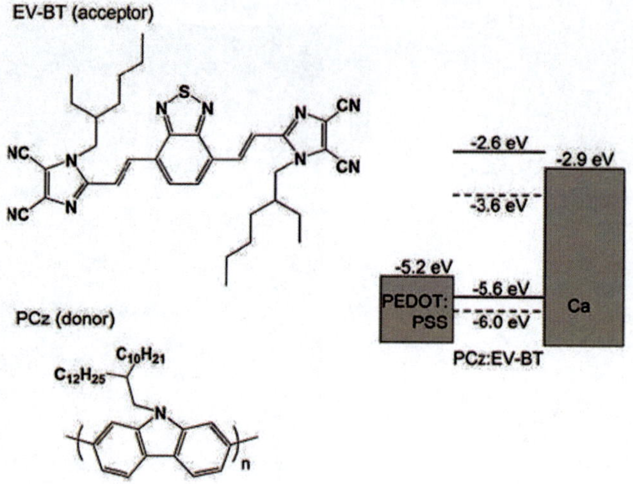

Fig. 8 Chemical structure of the donor polymer PCz and the acceptor molecule EV-BT. The *dashed lines* indicate the HOMO and LUMO levels of EV-BT and the *solid lines* those of PCz. Ooi et al. [49]—reproduced by permission of The Royal Society of Chemistry (http://dx.doi.org/10.1039/B813786M)

electron-acceptor EV-BT (Fig. 8) based on Vinazene in a SPOSC. This material has a LUMO level of 3.6 eV and strongly absorbs light in the visible region up to 520 nm. These properties make this material attractive when compared to the most widely used acceptor, PCBM. This Vinazene derivative was incorporated into bulk-heterojunction devices using a poly(2,7-carbazole) (PCz) as electron donor. This material provided absorption in a complementary range region and its HOMO level was found to be 5.6 eV. The devices were fabricated by spin-coating PCz:EV-BT films from chloroform solutions and the influence of blend composition and annealing temperature were investigated using different weight ratios of PCz:EV-BT. The most efficient device was obtained using 70% EV-BT followed by annealing at 80°C, which presented a high V_{OC} of 1.36 V and FF of 49%, but a low J_{SC} of 1.14 mA cm^{-2}, yielding an efficiency of only 0.75%.

Schubert et al. [50] investigated devices comprised of another Vinazene acceptor, 4,7-*bis*[2-(1-hexyl-4,5-dicyano-imidazol-2-yl)vinyl] benzo[c][1, 2, 5]-thiadiazole, (HV-BT, in Fig. 9), and poly(2,5-dimethoxy-1,4-phenylenevinylene-2-methoxy-5-(2-ethylhexyloxy)1,4-phenylenevinylene) (MEH-PPV) as the electron donor. Since HV-BT is soluble in common organic solvents, it can also be deposited by thermal evaporation; the authors were able to vary the device preparation scheme, systematically varying the heterojunction topology. Figure 9 shows the scheme of these devices, AFM images of the topology, current–voltage curve (J–V) and incident photon to current efficiency (IPCE) curves. The as-prepared bulk-heterojunciton blend provided relatively low FF and IPCE values of 0.26 and 4.5%, respectively, attributed to significant recombination of geminate pairs and free carriers in a highly intermixed blend morphology. In the all-solution processed bilayer device, the FF

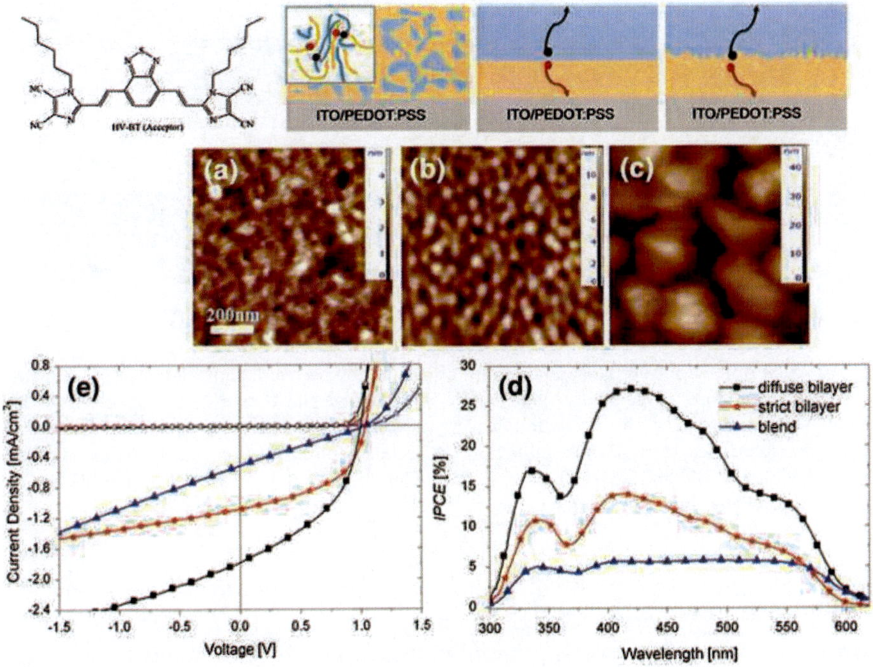

Fig. 9 Chemical structure of HV-BT and schematic depicting the interface topology at the donor/acceptor heterojunction for a blend device (*left*), the discrete bilayer with a vacuum processed acceptor layer deposited on top of the solution-cast polymer film (*center*), and the all-solution-processed bilayer device (*right*). A closed view of geminate pair recombination is shown in the *inset* of the left drawing. AFM topography images of **a** as-prepared 1:1 MEH-PPV:HV-BT blend spin-coated from chlorobenzene, **b** high-vacuum deposited HV-BT film on top of a MEH-PPV interlayer, and **c** pure HV-BT film spin-coated on top of the MEH-PPV interlayer. The **d** ICPE and **e** J–V curves measured under 100 mW cm^{-2}, AM 1.5 illumination, of an as-prepared 1:1 MEHPPV: HV-BT blend (*blue triangles*), a strict bilayer (*red stars*) with an evaporated HV-BT layer, and a diffuse bilayer (*black squares*) solar cell with solution-processed HV-BT on top of a MEH-PPV interlayer. "Reprinted with permission from Schubert et al. [50]. Copyright (2009), American Institute of Physics"

and IPCE dramatically increased to 0.43 and 27%, respectively. The FF increases further to 0.57 in devices comprised of thermally deposited Vinazene layers when there is virtually no interpenetration at the donor/acceptor interface. The AFM images in Fig. 9 suggests that both the solution-cast and evaporated HV-BT layers consist of crystallites. Such nanocrystalline morphology allows for the photogenerated electrons to quickly migrate into the acceptor phase away from the hetero-interface, thereby reducing the mutual Coulomb binding energy to the hole remaining in the donor phase. A high V_{OC} of about 1.0 V was nearly the same for all devices. This suggests that the heterojunction topology does not affect the energetics of the active layer or at the electrodes.

In a more recent work, Inal et al. [51] reported the photovoltaic properties of HV-BT combined with poly(3-hexylthiophene) (P3HT) (Fig. 10). The P3HT/

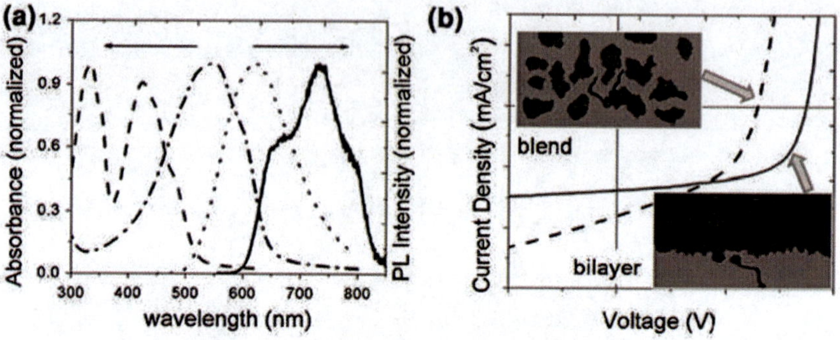

Fig. 10 **a** Normalized absorbance and photoluminescence spectra of thin films of P3HT (*dash-dotted* and *solid line*, respectively) and of HV-BT (*dashed* and *dotted line*, respectively). PL spectra were recorded at the excitation wavelengths of absorption maxima of each material. **b** J–V characteristics for bulk-heterojunction (*solid line*) or bilayer-type (*dashed line*) devices prepared with HV-BT and P3HT. Inal et al. [51]. Copyright Wiley–VCH Verlag GmbH & Co. KGaA. Reproduced with permission

HV-BT blend covers a broader spectral range (from 290 to 820 nm) than the HV-BT/MEH-PPV system. The photoluminescence (PL) intensity from the individual components was largely reduced in the blend and calculations showed that at least 95% of the excited states on both the donor and the acceptor are quenched in the non-annealed blend. The solar cells were fabricated with as-prepared 1:1 donor/acceptor mixtures and exhibited higher FF and IPCE when compared to earlier studies on blends of HV-BT with polymers [50, 52, 53]. This was attributed to the high molecular weight of the polymer sample used, which might assist the rapid motion of holes moving away from the heterojunction, rendering the photovoltaic properties less sensitive to the nanomorphology of the blend. No improvements were observed upon annealing the samples. On the other hand, a significant improvement of FF and V_{OC} was observed when inducing phase separation at a longer length-scale, i.e., in solution processed bilayer devices, comprising a layer of P3HT and a top-layer of HV-BT. The optimized device exhibited a power-conversion efficiency of close to 1%, and the efficiency of such cells was mainly determined by the device architecture [51].

Another interesting solution-based deposition method used to obtain thin films is the electrostatic layer-by-layer (LBL) adsorption technique, in which molecular-level control can be achieved. This technique was used to assemble thin films of PP and PC combined with several different organic molecules [54–58] and inorganic nanoparticles [59–63]. Although the properties of these films are usually well characterized, only a few reports can be found showing the use of these materials in photovoltaic devices.

Ultrathin films serving as a light-harvesting and hole transporting materials were fabricated via LBL deposition of the water-soluble copper(II) phthalocyanine-3,4′,4″,4‴-tetrasulfonic acid tetrasodium salt (CuPCTS) and poly(diallyldimethylammonium chloride) (PDDA) [64]. The absorption characteristics of these films

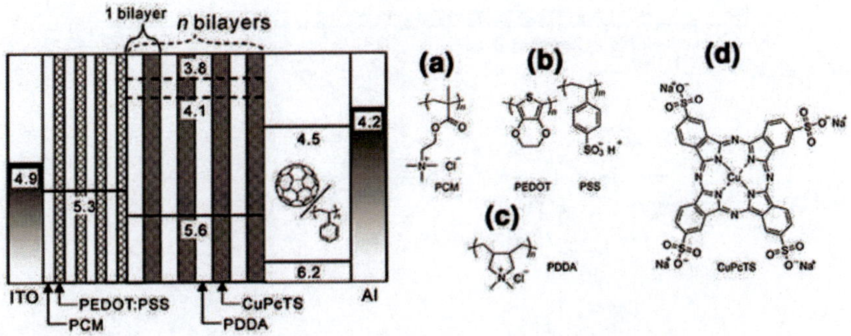

Fig. 11 Illustration of the photovoltaic cell with a triple-layered structure ITO|PCM/PEDOT:PSS|(PDDA/CuPcTS)n|C_{60}|Al. The thickness of PCM/PEDOT:PSS layer was 100 nm (20 bilayers) and C_{60} layer was 50 nm, which were prepared by LBL deposition of PCM and PEDOT:PSS, and spin-coating from the blend solution of C_{60}: PS (4: 1 wt%) in *o*-dichlorobenzene, respectively. The thickness of PDDA/CuPCTS layer was varied from 0 ($n = 0$) to 14.4 nm ($n = 6$). Numerical values represent the Fermi-level energies for ITO (4.9 eV), PEDOT:PSS (5.3 eV), Al (4.2 eV), and HOMO and LUMO levels for CuPCTS (5.6 and 3.8–4.1 eV), and C_{60} (6.2 and 4.5 eV). The chemical structures of polyelectrolytes used in the LBL deposition: **a** PCM (polycation), **b** PEDOT:PSS (polyanion), **c** PDDA (polycation), and **d** CuPCTS. "Reprinted from Benten et al. [64], copyright (2009), with permission from Elsevier." (http://www.sciencedirect.com/science/journal/00406090)

indicated that the CuPCTS forms dimers or oligomers, and their molecular planes were oriented parallel to the substrate. Triple-layered OSCs were developed by combining CuPCTS with a poly(3,4-ethylenedioxythiophene):polystyrenesulfonate (PEDOT:PSS) hole-transport layer and a C_{60} electron-transport layer, displayed in Fig. 11. The J_{SC} increased with increasing CuPCTS film thickness up to ca. 10 nm, as a result of exciton generation in each CuPCTS layer and transport to the interface with the C_{60} layer. The best J_{SC} value of 0.114 mA cm^{-2} reported is low compared to devices assembled using other solution processing methods and is probably caused by a poor light absorption inherent from the small layer thicknesses.

Generally, the photocurrent of devices assembled using the LBL technique are still on the order of microamperes, but the self-assembly properties obtained using this technique might be very useful for the future development of well-organized and highly efficient devices.

3 Evaporated Small Molecules OSC

3.1 Growth Techniques

Among the various growth techniques capable of producing thin films of small-molecular weight materials for OSCs, two techniques deserve special attention: vacuum thermal evaporation (VTE) [65] and organic vapor-phase deposition

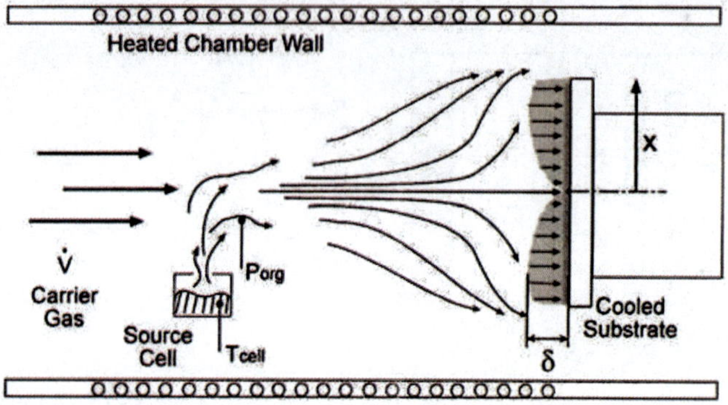

Fig. 12 Schematic diagram of the OVPD method. Organic materials are first evaporated into a carrier gas stream, which transports the molecules toward a cold substrate, where the vapor condenses into a solid film. When the substrate is rotated, the thickness of the boundary layer becomes uniform. "Reprinted with permission from Shtein et al. [66]. Copyright [2001], American Institute of Physics"

(OVPD) (Fig. 12) [66]. Thermal sublimation in vacuum is the most frequently used means for depositing small molecules that are insoluble in the majority of solvents [65]. The VTE procedure involves placing a purified organic material in a baffled tantalum or tungsten boat which is located between the electrodes in a vacuum chamber with a pressure of 10^{-6}–10^{-7} Torr. When current is passed through the boat or crucible, the temperature is increased beyond the sublimation point of the organic material, and it is evaporated, depositing everywhere on the chamber walls, as well as on the target substrate. A quartz crystal microbalance is used to monitor the growth rate (typically 0.5–3 Å s^{-1}) and the thickness of the film. VTE has the advantage of being able to form films with high uniformity, high degree of purity, good structural control, and "run-to-run" reproducibility. Additionally, one particular advantage of VTE is its ability to grow several OS layers without delamination or dissolution of the previous layers during subsequent deposition steps [67]. However, this technique presents several drawbacks such as lack of control over film thickness, uniformity, and dopant concentration over large areas needed for many applications. Moreover, this technique requires a relatively high material consumption and the initial setup costs for equipment are also very high.

To overcome the drawbacks associated with the VTE technique, OVPD was been introduced as an alternative method for depositing thin films of small molecular weight materials and allowed greater control over doping. Besides, this technique is suitable for fast, particle-free, uniform deposition in large-area substrates [66, 68].

The process of OVPD differs from VTE; here the organic material is thermally evaporated into an inert gas stream such as nitrogen or argon, which then transports the vapor in the hot-walled reactor vessel and toward a cooled substrate where deposition occurs [69]. In this respect, OVPD allows one to adjust multiple parameters such as reactant concentration, carrier gas flow rate, deposition rate, substrate temperature, chamber temperature, and chamber pressure. Thus, the deposition efficiency and film morphology are controlled [70, 71].

In this technique, the heated chamber walls avoid material deposition, allowing for a more efficient use of source materials compared with VTE [72]. Even with such differences, both VTE and OVPD techniques are used to manufacture OSCs based on small molecules [73, 74]. In the next section, different types of solar cell architectures using evaporated small molecules and recent results from many research groups are presented.

3.2 Organic Solar Cells Architectures

Besides manipulating the chemical and physical properties of small molecules, the choice of the growth technique and the selection of the best donor and acceptor combination are also important factors in controlling the device's architecture. These additional factors should be taking into account when attempting to improve the performance of solar cells [72]. In the next section, several device architectures using small molecules are described, highlighting the most important contributions in this field.

3.2.1 Double-Layer Solar Cells

The first successful OSC to reach a power-conversion efficiency of 1% was introduced by Tang at Kodak in 1986 [29]. Tang's solar cell consisted of a double layer of ~300 Å copper phthalocyanine (CuPC) as the electron-donating material and perylene tetracarboxylic derivative (~500 Å), as the electron-accepting material. Both materials were deposited by sequential thermal evaporation onto an indium tin oxide (ITO) substrate. On top, an opaque Ag layer was evaporated (Fig. 13). One of the most important conclusions derived from Tang's experiment is that the increase in thickness of the organic material layers led to a reduction in efficiency. Hence, the authors inferred that only excitons generated in the proximity of the interface between the donor and the acceptor materials were able to generate free charge carriers. Such a phenomenon reflected on the values of J_{SC}.

In the work reported by Tang, different metal contacts such as In, Al, Cu, and Ag were tested in the role of a top electrode. These metals, with different work functions produced solar cells with V_{OC} values differing by only 50 mV. It was assumed that V_{OC} is strongly dependent on the choice of the particular pair of organic layers.

Aiming to improve the efficiency of OSC, the group of Prof. Forrest [75] introduced an additional layer of bathocuproine (BCP) in double heterojunction solar cells. In this pioneering work, evaporated films of CuPC and 3,4,9,10-perylenetetracarboxylic *bis*-benzimidazole (PTCBI) were employed as the electron

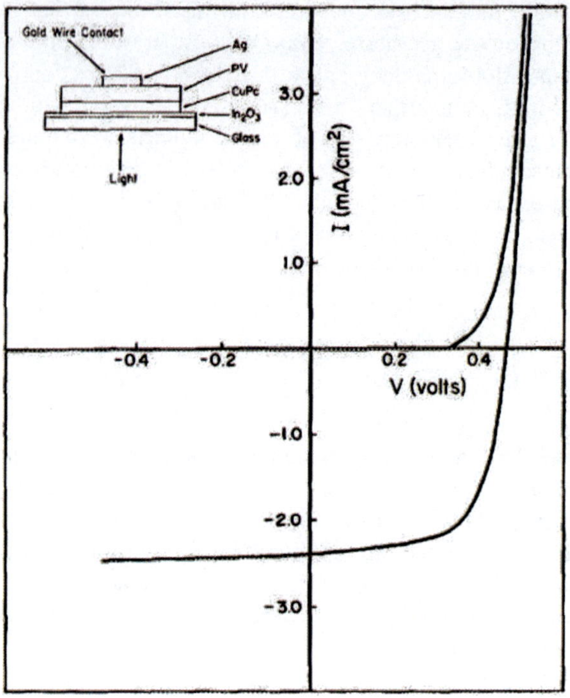

Fig. 13 Device configuration and current–voltage characteristics of an ITO/CuPC (250 Å)/perylene tetracarboxylic derivative (450 Å)/Ag solar cell. "Reprinted with permission from Tang [29]. Copyright [1986], American Institute of Physics"

donor and acceptor, respectively. The BCP and Ag cathode were also evaporated and the final cell architecture was ITO/CuPC/PTCBI/BCP/Ag. 2.4% efficient solar cells were produced using this double heterojunction approach. This device architecture demonstrates that the control over the exciton diffusion can lead to a significant increase in the number of charge carriers. One year after this work, the same group reported an external power-conversion efficiency of 3.6% employing the same architecture [32]; here, the acceptor molecule was replaced by C_{60}. Both reports highlighted the strong influence of BCP in these devices.

The introduction of BCP molecule represents one of the greatest advances in small molecule-based solar cells assembly by evaporation techniques. This wide band gap molecule acts as an exciton blocking layer (EBL) between the electron acceptor material and the cathode, as shown in Fig. 13.

The HOMO–LUMO offsets between the acceptor and the BCP provide reflection of excitons at the interface of these materials, preventing recombination [75]. In this context, Vogel et al. [76] presented a detailed PL study on the function of the BCP layer insertion in small-molecule solar cell (Fig. 14a). In the case of sample 1, the weak PL was attributed to the fast nonradiative recombination (i.e., exciton quenching) at the Al-C_{60} interface. In contrast to sample 1, significant PL was observed in sample 2, clearly showing the PL spectrum of the C_{60}. Sample 3, containing a BCP layer between C_{60} and Al, showed the strongest PL and consequently the lowest exciton quenching yield. These results corroborate

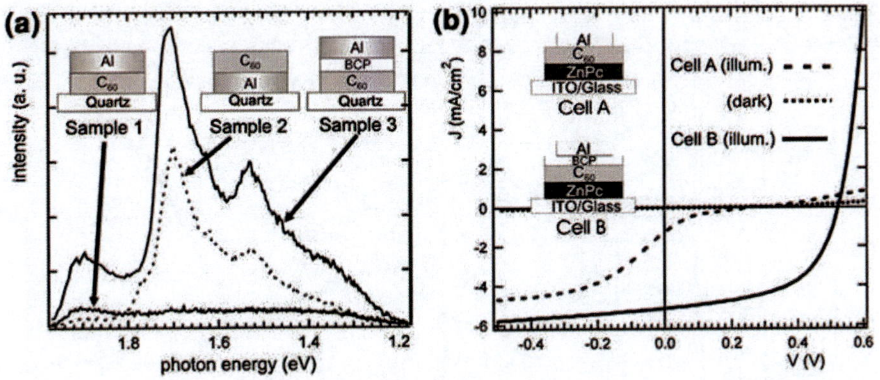

Fig. 14 **a** Photoluminescence spectra of C_{60} films at 10 K with different interfaces with Al. The direction of excitation and PL is indicated in the *inset* sample schemes. **b** Current density–voltage characteristics of photovoltaic cells with sequentially deposited ZnPC and C_{60} films with and without BCP buffer layer. Photovoltaic cell parameters are the following: cell A: $J_{SC} = 1.2$ mA cm^{-2}, $V_{OC} = 0.12$ V, FF = 0.12, and $\eta = 0.0\%$; cell B: $J_{SC} = 5.2$ mA cm^{-2}, $V_{OC} = 0.52$ V, and FF = 0.5, and $\eta = 1.5\%$. "Reprinted with permission from Vogel et al. [76]. Copyright [2006], American Institute of Physics"

previous reports that BCP may in fact be preventing exciton recombination at the Al-C_{60} interface [76].

Vogel et al. [76] further analyzed the function of BCP layer by interpretation of the photocurrent–voltage (*J*–*V*) in double-layer photovoltaic cells. Two types of devices were prepared using zinc phthalocyanine (ZnPC) as the donor and C_{60} as the acceptor, with and without BCP layers. These devices are represented by cell A and cell B in Fig. 14b, respectively. The short-circuit current density (J_{SC}) was reduced from 5.2 mA cm^{-2} in cell B to 1.2 mA cm^{-2} in cell A. In contrast, at a bias of −0.5 V, the photocurrent in cell A is only about 20% lower. Such strong dependence of the photocurrent on the voltage shows that exciton quenching explains only a small part of the photocurrent loss in cell A, since excitons are neutral and therefore independent of applied bias. Important theoretical calculations of electric field distribution and exciton diffusion were carried out by Breyer et al. [77]. The authors show that even complete quenching of excitons at the Al-C_{60} interface does not reduce the photocurrent by more than about 13% compared to an exciton-reflecting interface. Therefore, the larger part of the photocurrent increases in cell B is mainly attributed to the improved efficiency of electron transport from C_{60} to the Al electrode via the BCP layer. In other words, the most important function of BCP is to establish an Ohmic contact between Al and C_{60}.

BCP has been typically used as an EBL material in small molecule-based solar cells. However, its large energy gap and low conductivity make it unsuitable for use in thick layer devices because the increase in the series resistance degrades device's performance.

Interestingly, Rand et al. [41] introduced an EBL composed of *tris*(acetyl-acetonato) ruthenium(III) (Ru(acac)$_3$). Its functionality results from an energy-level

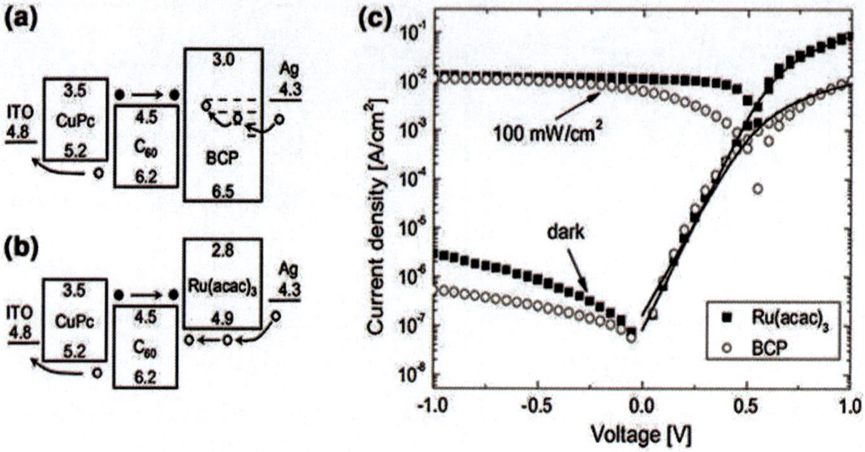

Fig. 15 Schematic energy-level diagram and proposed photovoltaic process for double-heterostructure devices using either **a** BCP or **b** $Ru(acac)_3$ EBL. Holes are shown as *open circles* and electrons as *filled circles*. Energy levels are given in units of electron-volts (eV). **c** Current-density–voltage (J–V) characteristics, in the dark and under 1 sun (100 mW cm^{-2}) intensity of simulated AM1.5G, for the organic photovoltaic cells with the following structure: ITO/CuPC(200 Å)/C_{60}(400 Å)/EBL(200 Å)/Ag(1000 Å). The EBL consists of either BCP (*open circles*) or $Ru(acac)_3$ (*filled squares*). The efficiencies for BCP and $Ru(acac)_3$ based devices are 1.1 and 2.7%, respectively. The *solid lines* are fits to the J–V characteristics based on the modified ideal diode equation (Rand et al. [41]). Copyright Wiley–VCH Verlag GmbH & Co. KGaA. Reproduced with permission

alignment more favorable than BCP, as seen in Fig. 15a, b. This is reflected in the remarkable differences observed in photocurrent–voltage characteristics as shown in Fig. 15c.

3.2.2 Bulk-Heterojunction Solar Cells

The DA interface can be considered to be the heart of the small-molecule solar cells. It is at this interface that strongly bound photogenerated excitons are dissociated to generate photocurrent. The external quantum efficiency (η_{EQE}) of a photovoltaic cell based on exciton dissociation at a DA interface is $\eta_{EQE} = -\eta_A \times \eta_{ED} \times \eta_{CC}$ [78], where η_A is the absorption efficiency, η_{ED} is the exciton diffusion efficiency that corresponds to the fraction of photogenerated excitons that reach the DA interface before recombining, and η_{CC} is the carrier collection efficiency and corresponding to the probability that a free carrier, generated at a DA interface by dissociation of an exciton, reaches its corresponding electrode. When the total thickness, L, in bilayer DA solar cells is of the order of the optical absorption length, L_A, we can assume η_A, described in Eq. 1 [78], to be higher than 50% if optical interference effects are ignored, and $\eta_{CC} \approx 100\%$. Nevertheless, since the exciton diffusion length (L_D) is typically

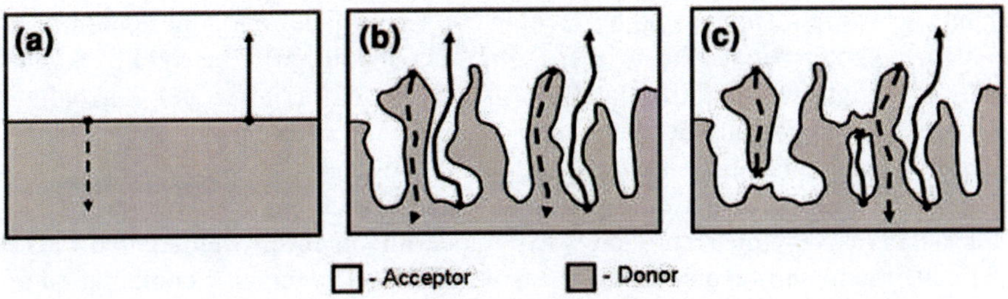

Fig. 16 Representation of donor/acceptor interface architecture possibilities: **a** a double layer, formed between thin films of donor and acceptor materials; **b** an optimal bulk heterojunction, where there is complete phase separation of donor to one side and acceptor to the other side of the device structure; and **c** a non-ideal bulk heterojunction, where isolated regions of donor and/or acceptor phases prevent the collection of photogenerated charges. *Dashed* and *solid lines* correspond to hole and electron transport, respectively

an order of magnitude smaller than L_A, a large fraction of the photogenerated excitons are not used for photocurrent generation as shown in Fig. 16. Therefore, limited η_{EQE} and hence limited η are observed for double-layer solar cells.

$$\eta_A = 1 - \exp(-L/L_A) \tag{1}$$

The η_{EQE} of an organic double-layer solar cell is often limited by a short exciton diffusion length. To overcome this drawback, the bulk-heterojunction concept has been introduced [78–82], resulting in an improvement in both η_{EQE} and in the η.

Bulk or mixed heterojunction is a mixture of the donor and acceptor materials in a bulk volume, exhibiting a DA phase separation of about 10–20 nm which provides an interpenetrating DA network with large interface area [83]. Compared with the planar double layer cells introduced by Tang, in which the donor and acceptor phases are completely separated from each other (Fig. 16a), the bulk heterojunction has both materials intimately intermixed. It expands the photocurrent generation capability of the device by increasing the excitons' probability of reaching a nearby DA interface where they can dissociate [72].

In polymer photovoltaic cells, the exciton diffusion bottleneck has been overcome by the introduction of bulk heterojunctions [84, 85] (Fig. 16c). In a bulk heterojunction, the DA interface is highly folded so that photogenerated excitons can find an interface within a distance L_D of their generation site. In an optimal architecture [86], the width of the phases in the interdigitated structure should be on the order of 2 L_D to guarantee high probability of dissociation for the excitons generated in the bulk of the material. At the same time, this optimal architecture provides low resistance pathways for charge transport. In practice, however, achieving such a structure is difficult. Recently, optimized polymer bulk-heterojunction cells have reached power-conversion efficiency of 7% [87]. In this work, a blend of a novel semiconducting polymer based on alternating ester substituted thieno[3,4-b]thiophene and benzodithiophene units and $PC_{71}BM$. The polymer exhibits a low bandgap, providing an efficient absorption around the region with the highest photon flux of the solar spectrum (about 700 nm). The rigid backbone results in a high hole mobility, and the side chains on the ester and benzodithiophene enable good solubility in organic solution and suitable miscibility with the fulleride acceptor. The polymer chain is found to be stacked on the substrate in the face-down conformation from grazing-incidence wide-angle X-ray scattering studies. This is very different from the polymer alignment in well-studied P3HT solar cell system and favors charge transport [87]. This synergistic combination of properties lead to an excellent photovoltaic effect (and efficiency record!).

Remarkable efforts have been conducted to perform bulk-heterojunction devices prepared by co-deposition of the donor and acceptor materials yielding η values falling short of those attainable in optimized bilayer cells using the same materials [32, 79, 82, 88–91]. Co-deposited donor and acceptor small molecules have been used to prepare bulk-heterojunction photovoltaic devices [78, 80, 92–96] reaching power-conversion efficiency of approximately 3.5%. Using this approach, the deposition of the mixed donor and acceptor phase is usually followed by a decrease in charge-carrier mobility, i.e., by an increase in the series resistance (R_S) of devices [92, 97]. This behavior is different from that observed in polymer bulk-heterojunction structures [98]. The charge-carrier mobility is critically dependent on the composition and morphology in both polymer and small molecule bulk-heterojunction structures. From space charge limited current mobility measurements, Rand et al. [92] demonstrated that mixed layers presented lower charge-carrier mobilities than neat films.

Aiming to accomplish a nearly ideal structure in the thermally co-evaporated thin film, Peumans et al. [78] carried out an annealing treatment of the CuPC:PTCBI mixture (300–500 K). Such post-treatment was responsible for inducing phase separation of the two materials and an increase in η_{CC}.

Donor and acceptor mixed layers can also be deposited via the OVPD growth technique to directly form a bulk-heterojunction architecture in which the two layers are phase-separated [99] (Fig. 16b). By this method, it is possible to control the growth mode of the CuPC film on ITO such that, crystalline needles of CuPC extend out of the CuPC film. After growth of the CuPC film, the OVPD process was successfully applied to fill the spaces within the rough CuPC film. This contrasts with growth induced by ultra high vacuum, in which the VTE process results in voids in the film [99] (Fig. 16a). This architecture resulted in an increase in device efficiency of the CuPC/PTCBI solar cell from 1.1 to 2.7%. The photovoltaic parameters for all architectures based on ITO/CuPC/PTCBI/BCP/Ag solar cells are shown in Table 1.

Achieving a balance between the absorption needed for photocurrent generation and a good charge transport in bulk-heterojunction devices is challenging. Therefore, maintaining a good charge transport and a low R_S are important factors in creating efficient OSCs [100] with high FFs, and reduced recombination probability of the photogenerated charges within the mixture. This can be achieved using a hybrid architecture: combined double layer and bulk heterojunction approaches. This configuration consists of a bulk-heterojunction DA layer

Table 1 Comparison of performance of several ITO/CuPC/PTCBI/BCP/Ag photovoltaic cell structures characterized under 1 sun simulated AM1.5G illumination

Device	J_{SC} (mA cm^2)	V_{OC} (V)	FF	η (%)	R_S (Ω cm^2)
Double layer using VTE[a]	6	0.49	0.49	1.1 ± 0.1	30 ± 10
Annexed bulk hetero junction[b]	9	0.50	0.40	1.4 ± 0.1	60 ± 10
Double layer using OVPD	5	0.4S	0.47	1.1 ± 0.1	18.2 ± 0.5
Controlled bulk OVPD heterojunction	11	0.49	0.58	2.7 ± 0.1	2.2 ± 0.1

R_S is the specific series resistance [99]
[a] Ref. [75]
[b] Ref. [78]

Table 2 Comparison of several OSC using different architectures, under approximately 1 sun simulated AM 1.5G illumination, where P_0 is the incident optical power density [101]

Device	P_0 (suns)	J_{SC}/P_0 (mA W^{-1})	FF	V_{oc} (V)	η (%)
Double layer[a]	1.3	11.8 ± 0.5	0.61	0.51	3.7 ± 0.2
Bulk heterojunction[b]	0.9	15.4 ± 0.7	0.46	0.50	3.5 ± 0.2
Double layer/bulk heterojunction	1.2	15 ± 0.6	0.61	0.54	5.0 ± 0.3

[a] Ref. [100]
[b] Ref. [93]

sandwiched between homogeneous donor and acceptor layers [101]. Because each homogeneous layer has a thickness of approximately L_D, excitons are generated in the entire heterojunction structure with a high probability of diffusing to a nearby DA interface. This architecture provides both high η_{ED} of a bulk heterojunction and efficient charge collection characteristic of a double-layer device [101].

Using this approach, Xue et al. [101] fabricated solar cells consisting of the donor CuPC, and acceptor C_{60}, obtaining a power-conversion efficiency of 5% under 1–4 sun of simulated AM1.5G illumination. The authors also compared these devices with the bulk heterojunction and double-layer solar cells. The photovoltaic parameters of these solar cells are summarized in Table 2. Remarkably, this innovative concept afforded an increase in FF values and further resulted in efficient charge transport in the active layer.

3.2.3 Tandem Solar Cells

In solar cells with only one band gap, the Shockley–Queisser limit [102] can be reduced due to mainly two loss mechanisms [103]: thermalization losses and losses via sub-band gap transmission of photons. Aiming to overcome these limitations, tandem or multi-junction architectures have been investigated and appear to be promising solutions. When two or more donor materials with non-overlapping absorption spectra are used in a tandem solar cell, a broader range of the solar spectrum can be covered. In the last years, several approaches for organic tandem cells have been employed, depending on the materials used for the active layer and the separation or recombination layer(s). In general, these approaches can be divided in three main categories: (i) tandem OSCs where both the bottom (in front of the light illumination) and the top (back) cells are based on small molecules growth by vacuum-deposition techniques; (ii) hybrid tandem OSCs in which the bottom cell is processed from solution while the top cell is made of vacuum-deposited small molecules; and (iii) fully solution-processed tandem OSCs where both the bottom and top cells are deposited from solution. In this section, only the first approach will be described, together with some promising results described recently in the literature.

The use of small molecules presents a great advantage for tandem architectures because different layers of donor and acceptor (or mixed layer) materials can be evaporated or co-evaporated with sharp interfaces on top of each other, without affecting the previously evaporated layer. The disadvantage, however, comes from the relative low evaporation rate of active materials, which limits the processing speed in large-area applications [104].

Tandem solar cells can theoretically be comprised of an infinite number of sub-cells connected in series. According to Kirchhoff's law, this type of connection implies that the voltage across the entire cell is equal to the sum of the voltage across each sub-cell. The V_{OC} of these devices containing series sub-cells, in the case of loss-free connections, is given by [105, 106]:

$$V_{OC,tot} = V_{OC1} + V_{OC2} + V_{OC2} + K \tag{2}$$

On the other hand, on contrary to what is often mentioned, the short-circuit current of the organic tandem solar cells is not equal to the smallest short-circuit current of the sub-cells, but depends strongly on the FF values of the respective devices [106]. This behavior, illustrated in Fig. 17a, results in the combination of a cell with lower J_{SC} and a significantly higher FF and a cell with higher J_{SC} and extremely low FF. According to Kirchhoff's law, a J–V characteristic of the tandem cell with a $J_{SC} = \text{Min}\ (J_{SC1}, J_{SC2})$ is attained. Oppositely, Fig. 17b shows the combination of a cell with extremely low FF and lower J_{SC} and another with very good FF and higher J_{SC}, leading to a resultant tandem device with a $J_{SC} = \text{Max}(J_{SC1}, J_{SC2})$. A description of the various cases was presented by Hadipour et al. [105].

The tandem cell architecture was originally introduced by Hiramoto et al. [107]. This architecture was constructed from two identical bilayers based on evaporated small molecules, where each bilayer was an organic DA junction consisting of 50 nm of metal-free phthalocyanine (H_2PC) and 70 nm of perylene tetracarboxylic derivative. These two bilayers were separated by a thin interstitial layer (2 nm) of Au, in order to establish an Ohmic contact. The organic tandem cell resulted in almost double the V_{OC} (0.78 V) compared to single cell, in which the V_{OC} was 0.44 V. This result showed that a thin interstitial Au layer was able to accomplish an effective charge recombination site for electrons arriving from the perylene tetracarboxylic derivative of the back cell with the holes coming from the H_2PC of

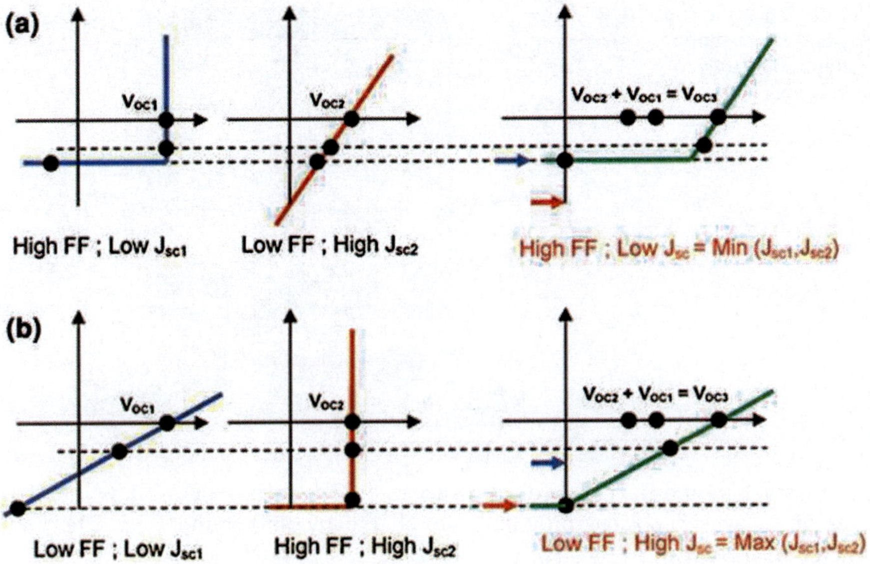

Fig. 17 **a** The combination of a cell with lower J_{SC} and significantly higher FF and a cell with higher J_{SC} and extremely low FF results in a tandem cell with a $J_{SC} = \mathrm{Min}(J_{SC1}, J_{SC2})$. **b** The combination of a cell with extremely low FF and lower J_{SC} and another with very good FF and higher J_{SC} leads to a tandem device with a JSC = Max (JSC1, JSC2). Ameri et al. [106]—reproduced by permission of The Royal Society of Chemistry

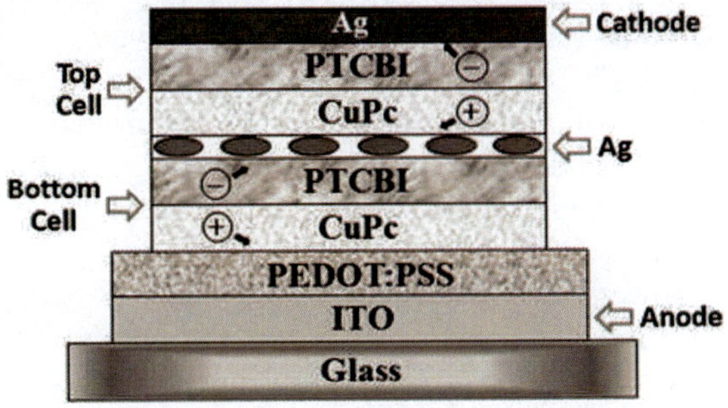

Fig. 18 Schematic structure of an organic tandem solar cell based on the small molecules CuPC as donor and PTCBI as acceptor performed by Yamikov and Forrest [109]. The 0.5 nm Ag separation layer provides recombination sites for the electrons and holes (http://dx.doi.org/10.1039/B817952B)

the front cell. In general, this interstitial layer is sufficiently thin to efficiently supply charge recombination sites, and also is not thick enough to absorb light on its way to the back cell, nearest the reflecting cathode [108].

Yamikov and Forrest [109] reported in 2002 the first tandem solar cells by stacking two, three, or five vacuum-deposited thin heterojunction consisting of

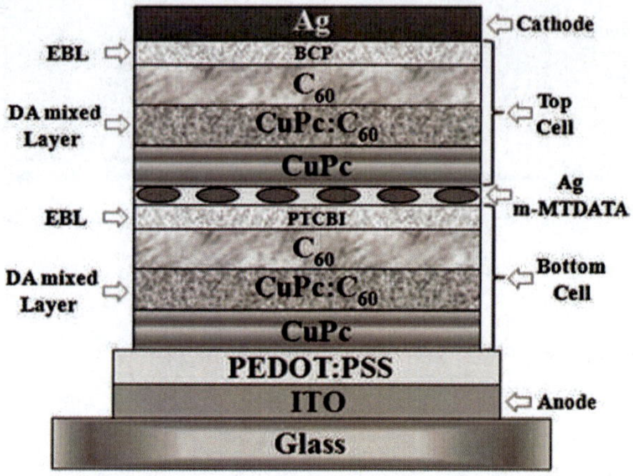

Fig. 19 Schematic structure of hybrid double-bulk-heterojunction organic tandem solar cell

CuPC, as donor, and PTCBI, as electron acceptor. Between the DA bilayers, a thin layer of Ag was deposited providing the charge recombination sites. The final structure can be seen in Fig. 18.

The power-conversion efficiencies reached by the two and three-stacked solar cells were $\eta = 2.5\%$ and $\eta = 2.3\%$, with $V_{OC} = 0.93$ and 1.2 V, respectively. These values are twofold higher compared to a single junction cell based on identical materials ($\eta = 1.1\%$). Rand et al. [110] found that the optical field enhancement, due to surface plasmon generation on the metal clusters located at the interstitial layers, was responsible for the higher efficiency. In contrast, the five stacked cell showed a considerable decrease in η (1%) compared to devices containing two or three-stacked bilayers. This observation comes from the reduction in light absorption by the first bilayers, that being the main limitation of multiple-heterojunction solar cells. Triyana et al. performed similar approach using CuPC and PTCBI as donor and acceptor materials, respectively, in combination with ultrathin Ag and Au interlayers to produce multiple-junction solar cells with two and three stacks [111, 112].

To improve the power-conversion efficiency in tandem solar cells, Xue et al. [113] applied several modifications to the device's structure. The use of C_{60} as acceptor materials, with a longer exciton diffusion length ($L_D \sim 40$ nm) compared to the acceptor PTCBI [114], is the first modification. The second is the use of an evaporated mixed DA layer or bulk-heterojunction structure sandwiched between neat donor and acceptor layers. Thin layers of PTCBI and BCP were employed as EBLs in the bottom and top sub-cells, respectively. These modifications gave rise to a highly efficient double bulk-heterojunction structure. It was suggested that one EBL may be acting as a protection for the hot metallic particles during the thermal evaporation process. In such device, ultrathin Ag interstitial layer are recombination site which have a thickness of 5 Å buried in a 50 Å thick 4, 4′, 4″-*tris*(3-methyl-phenyl–phenyl-amino)triphenylamine (*m*-MTDATA) p-doped with 5 mol% tetrafluoro-tetracyano-quinodimethane. This structure with a mixed DA layer sandwiched between homogenous donor and acceptor layers is named hybrid

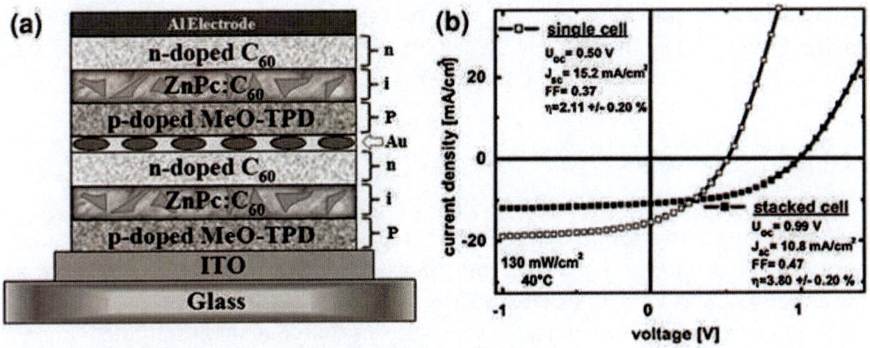

Fig. 20 **a** Concept of a stacked p–i–n OSC with active layers sandwiched between p- and n-type wide-gap transport layers. **b** J–V characteristics of single and tandem p–i–n solar cells under 130 mW cm^{-2} simulated AM 1.5 solar illumination. The single cell is identical to the bottom cell in the tandem configuration (Cell A) and prepared simultaneously. The performance parameters are given. "Reprinted with permission from Drechsel et al. [115]. Copyright [2005], American Institute of Physics"

double-bulk heterojunction [101, 113], and is schematically represented in Fig. 19. Another important finding concerns the effect of different layers thickness that resulted in 5% efficiency. It demonstrates the undercover potential of such approach and that the device's optimization may even augment the efficiency value.

Dreshsel et al. [115] introduced an effective means to space the sub-cells apart by employing a p–i–n heterojunction architecture, as depicted in Fig. 20. In this case, the *N,N,N′,N′*-tetrakis(4-methoxyphenyl)-benzidine (p-doped MeO–TPD) and the n-type C_{60} wide-gap transport layers (which optimally do not absorb the incident light) are used to spatially separat the mixed ZnPC:C_{60} (i-type materials) layers that generate photocurrent.

The photocurrent–voltage characteristics for the single and tandem p–i–n solar cells are shown in Fig. 20b. The power-conversion efficiency is 2.1% for the single device and 3.8% for the tandem cell. It is very common to observe that optimized organic tandem cells have lower J_{SC} values than optimized single cells.

More recently, Yu et al. [116] assembled organic tandem solar cells without metallic nanoclusters between the sub-cells. In contrast, they fabricated organic tandem cells employing all-organic units by continuous deposition. These all-organic connecting units were heterojunction films, which have a better transparency and a lower sublimation temperature than Au or Ag. A tunneling mechanism was suggested as an explanation to why the organic heterojunction became an effective charge recombination center. In an optimized tandem solar cell comprising a tin phthalocyanine dichloride ($SnCl_2PC$)/copper hexadecafluorophthalocyanine ($F_{16}CuPC$) heterojunction as the connecting unit, the V_{OC} is almost twice that as a sub-cell, reaching up to 1.04 V, and 60% enhanced IPCE ($\eta = 1.8\%$). Furthermore, the all-organic connecting units can be continuously deposited, which provides an easy way to fabricate these devices and avoids damage to the organic films.

4 Liquid Crystals

If the nanomorphology of the active layer could be controlled on a molecular scale, the efficiency of charge separation and transport would be expected to increase substantially, improving the performance of these devices. Since most organic photovoltaic materials are amorphous solids or polymers with limited charge mobilities and exciton diffusion lengths, an approach to achieve the necessary enhancement in efficiency is the use of self-assembled materials into large domains of crystalline or liquid crystalline order. In this context, the use of discotic liquid crystalline materials and composites might reach this goal because of their capacity to self-organize into columnar stacks, maintaining high charge-carrier mobility while providing a well-distributed interface between the donor and acceptor semiconductors.

The supramolecular assemblies of aromatic disc-shaped molecules, which lead to the formation of discotic LCs, were discovered in 1977 by Chandrasekhar and colleagues [117]. These disc-shaped molecules exhibit liquid crystalline properties at room temperature (RT), with LC to liquid transition (clarification) temperatures (CT) usually above 150°C. The discotic molecules generally consist of an aromatic core surrounded by aliphatic chains. Due to strong π–π interaction between the cores and weak interaction between the flexible aliphatic chains, the molecules can stack one over the other, forming columns, when slowly cooled from above CT to RT. Thus, well-oriented large domains can be obtained. These columns can be arranged in several ways, forming columnar hexagonal, columnar rectangular, columnar oblique, columnar helical, or columnar plastic phases [118]. Typical column–column distance in the columnar hexagonal phase is 2–4 nm depending on the aliphatic chain length and the core–core distance within the column is around 0.35 nm with a length of a few tens of nanometers.

The strong intracolumnar interaction and weak intercolumnar interaction contribute to the quasi-one-dimensional electrical conductivity along the columns. Long exciton diffusion lengths (a few 100 nm) and high charge-carrier mobilities (up to 0.1 $cm^2\ V^{-1}\ s^{-1}$) have been found in the highly ordered liquid crystalline discotic molecules [119]. The mobility along the columns is superior to typical mobility of amorphous organic films and can reach values found for amorphous silicon [120–122]. Provided the distance between the single discs is short, electrons and excitons readily move along the columnar axes. Both p-type discotic LC, such as triphenylene, dibenzopyrene, and hexabenzocoronene (HBC) (Fig. 21a–c), and n-type discotic LC, such as tricycloquinazoline, anthraquinone, and perylene (Fig. 21d–f), are known [123]. Moreover, liquid crystalline phases are self-repairing and thereby minimize defects that can act as recombination sites. While the aliphatic side groups affect the melting point, solubility, and film-forming

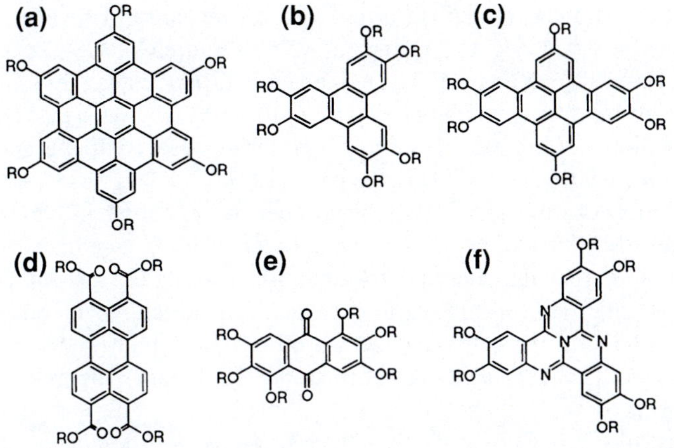

Fig. 21 Chemical structure of p-type (**a–c**) and n-type (**d–f**) discotics liquid crystals [123]

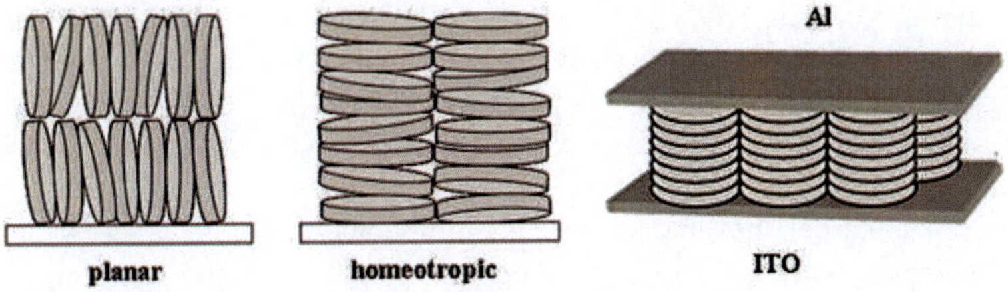

Fig. 22 Schematic representation of the planar (edge-on) and homeotropic (face-on) orientations of a columnar mesophase and ideal arrangements of the discotic stacks in a device configuration. The LC stacks should be perpendicular to the surface substrate and be surrounded by the electron-accepting material to obtain an ideal structure

properties, they have a negligible effect on the electrochemical oxidation/reduction potentials and optical properties, which depend on the core structure and the π-conjugation.

After the deposition of a LC molecule onto a substrate, the alignment can be planar (columns parallel to the substrate) or homeotropic (columns perpendicular to the substrate), as depicted in Fig. 22, depending on the type of the material, the substrate, and the experimental conditions (film thickness, cooling rate, etc.) [124]. Sometimes, after spin-coating or vacuum sublimation of a film with thickness below a critical value (typically a few hundred nanometers), the material clears upon heating to form an isotropic liquid that rapidly destabilizes into isolated droplets. This droplet formation can be avoided in many cases by an appropriate ITO surface treatment (e.g., annealing at high temperature) combined with a rapid cooling of the organic film through the liquid-to-LC phase transition [124, 125].

Grelet et al. [124] showed that uniform vertical orientation of the columnar axis on the substrate can be achieved in submicron open films by controlling the growth kinetics of the columnar domains during thermal annealing. However, when the film thickness was reduced to the typical value required in OSC (50–100 nm), dewetting occurred, leading to inhomogeneous films. Later, the same group showed that by treating the substrate surface either by UV ozone or by nitrogen plasma, thin films of 50 nm of a columnar mesophase could be stabilized in homeotropic orientation on an ITO electrode [126].

HBC liquid crystalline derivatives have been widely used in efficient photovoltaic diodes [127]. The disk-like aromatic cores of HBC assemble face-on into columns, allowing for high charge-carrier mobilities along the discotic cores [128, 129]. The potential to form controlled, organized structures in optoelectronic devices is evident by the self-organization of HBCs in both "edge-on" heterotropic alignment [130] as well as homeotropic alignment perpendicular to the substrate [131].

The photovoltaic behavior of three hexa-peri-hexabenzocoronene derivatives with different-sized alkyl side chains was investigated [132]. Increasing the side chain length dilutes the HBC chromophore core, decreasing the amount of light absorbed by the film. Also, differential scanning calorimetry and X-ray analysis showed that, at RT, the HBC derivative with the 2-ethyl-hexyl side chain is in a crystalline state, while the HBC containing 2-hexyl-decyl or 2-decyl-tetradecyl substituents are in the so-called "plastic crystalline state". The HBC with the shortest side chain was proven to be the best donor for perylenediimide, showing IPCE of 12% (at 470 nm).

Fluorenyl hexa-peri-hexabenzocoronene functionalized with a series of thiophene dendrons, were synthesized using the Suzuki–Miyaura coupling [133]. Ordered structures were observed in blends of these materials and fullerene acceptor materials. The larger thiophene dendritic substituent attached to the HBC derivative broadened its absorption profile, and also altered the morphology. A power-conversion efficiency of 2.5% was achieved for a device containing the compound depicted in Fig. 23a with $PC_{71}BM$ as the acceptor material [133].

Schmidt-Mende et al. [134] built a photovoltaic solar cell using hexadodecylphenylhexabenzocoronene as the hole-transporting layer (Fig. 23b). A chloroform solution of this LC and the crystalline dye *N,N*(*bis*(1-ethylpropyl)-3,4,9,10-perylene-tetracarboxdiimide (PTCDI, Fig. 23c) was spin-coated onto ITO. PTCDI is a LC dye with a long chain hydrocarbon moiety bound to the nitrogen in the imide group, which self assembles in a similar manner to other perylene diimides [135]. The device exhibited IPCE up to 34% and power efficiencies of up to ~2%. The efficient photoinduced charge transfer and facile charge transport through vertically segregated perylene and HBC were considered responsible for the high efficiencies. The same group investigated later the performance of solar cells assembled with blends of perylene diimide and other HBC derivatives, spin-coated directly from solution [136]. The use of different HBC derivatives as hole conductors showed lower efficiencies, attributed to the different film morphology originated when spin-casting these materials.

In a different approach, Schmidtke et al. [137] employed the elastomer polydimethylsiloxane (PDMS) to control the film morphology and phase separation of

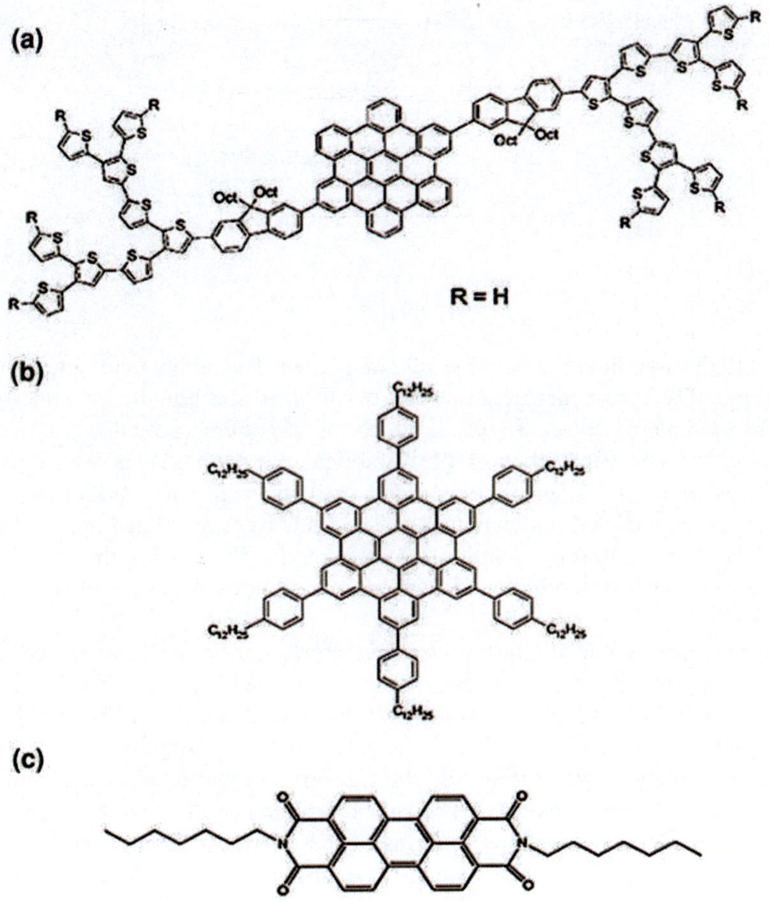

Fig. 23 Examples of HBC derivatives used as donors (**a**, **b**) and PTCDI dye used as an acceptor (**c**) in LC-based SPOSCs [133, 134]

blends containing PTCDI and hexadodecylsubstituted-HBC. The PDMS stamp formed a flat, top-surface, which was utilized during the annealing step and then removed. The authors observed that the annealed devices showed a modest increase in IPCE compared to that of the as-spun devices, whereas the PDMS-annealed devices presented a twofold increase in IPCE over the as-spun films, reaching 29.5% at 460 nm. The AFM images showed different textures and patterns in each case. The annealing processes resulted in increased roughness and polycrystalline features, which suggest higher charge-carrier mobilities associated with the crystalline phases of the HBC derivative. Furthermore, the increased vertical stratification increased the V_{OC} in these devices [137].

Porphyrine-based molecules may also present liquid crystalline properties. Gregg et al. [138] studied in 1990 the photovoltaic effects of symmetrical cells

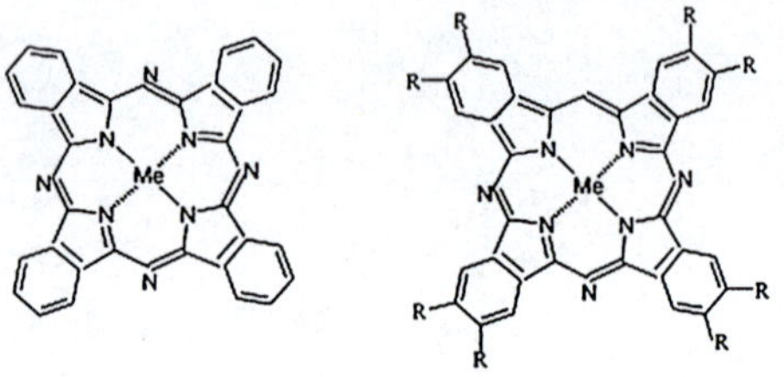

Fig. 24 Chemical structures of CuPC1, CuPC2, and liquid crystal CuPC3 [140]

filled with discotic liquid crystalline PP complexes. The authors did not study the charge mobility in the mesophase itself, but utilized the liquid crystalline properties to promote macroscopic order, which, upon cooling, provided polycrystalline films. In 1999, Petritsch et al. [139] fabricated a double-layer device using a liquid crystalline PC. They heated the PC until it reached its clarification point (292°C) and then cooled it slowly to RT. The films were fabricated in air by spin-coating from a chloroform solution. On top of the PC layer a thin layer of a perylene derivative was sublimated. The assembled devices presented efficiencies up to 0.5%.

Recently, Levitsky et al. [140] presented the preparation of a solar cell based on n-type nanoporous Si filled with copper phthalocyanine (CuPC1, in Fig. 24) and its derivatives (CuPC2, in Fig. 24), including a discotic LC form (CuPC3, in Fig. 24). The conversion efficiencies were between 0.01 and 0.02% (at 30 mW cm^{-2}) when the CuPC-derivatives were used. For CuPC1, on the other hand, conversion efficiencies up to 2% were observed. The striking difference in conversion efficiency of hybrid devices was explained in terms of the interfacial area between organic and inorganic components. The critical factor in this case was believed to be the average distance between the CuPC core and the Si surface. The relatively long alkyl chains in CuPC2 and CuPC3 hindered the charge transfer in such a way that the LC organization in CuPC3 could not compensate for the low charge transfer caused by the long alkyl chains.

The properties of discotic LC can be modulated by doping with either electron-rich or electron-deficient molecules into the supramolecular, ordered phase. The doping should be kept at an optimum concentration while retaining the liquid crystalline phase and introducing sufficient electron or hole concentrations into the liquid crystalline medium to increase the conductivity. Trinitrofluorenone and several inorganic dopants, such as iodine, aluminum chloride, nitrosonium tetrafluoroborate and gold nanoparticles have been introduced in the columnar liquid crystalline matrix [141–145].

The versatility of LCs allows the application of these molecules in other types of solar cells as well, including silicon-based devices. For example, LC-based thermography has been successfully used for the investigation of various thermal phenomena in a wide range of applications, including the detection of shunts in polycrystalline silicon solar cells [146].

5 Ternary Component-Based OSC

For OSCs, effective absorption over a wide wavelength range is an important goal, which usually is not sufficient when using a simple DA pair. As an alternative, the inter or intralayer cascaded energy transfer concept has been proposed by Koeppe et al. [147] and Liu et al. [148], utilizing multiple photoactive materials in a single cell. The application of this method, however, is limited by the subtle balance of hole and electron mobilities in the bulk heterojunctions. Also, the interface formation and morphology are difficult to control in the presence of a third component in the active layer. In this section, recent reported results for three-component-based solar cells with small organic molecules are reviewed.

Figure 25a illustrates the basic ideas underlying the design of the multiple-heterojunction system with antenna effects: (i) a p–n heterojunction, containing an electron donor and an acceptor is responsible for exciton separation; (ii) antenna layers, i.e., energy donors with wide bandgap and large exciton diffusion length are introduced as sensitizers; (iii) excitons in the wide-gap antenna layers enter p- or n-type layers via efficient energy transfer, and then diffuse to and dissociate at the p–n junction; and (iv) efficient charge transport and collection of photogenerated charge carriers can be realized if no significant barriers for holes and/or electrons from the p–n junctions to the energy donors/antenna layers exist.

Using this concept for the purpose of improving photon harvesting in OSC, Hong et al. [149] combined two hole-conducting materials, pentacene and ZnPC, and electron conducting C_{60} to construct three-component heterojunctions. Figure 25c shows the absorption spectra of the three active materials and their molecular structures. Pentacene and ZnPC have strong absorption mainly in the longer wavelength range (500–800 nm). In pentacene/ZnPC/C_{60} multi-heterojunction cells (see scheme in Fig. 25b), some of the excitons in pentacene might reach the ZnPC/C_{60} interface, where efficient exciton separation occurs and contributes to the photocurrent. The V_{OC} of this device was slightly higher than that of the ZnPC/C_{60} cell, which suggests a higher carrier concentration under the same illumination density, due to sensitization effects of pentacene for the ZnPc interlayer.

PC and PP [150–153] have also been added to polymer/fullerene systems to improve light-harvesting. For example, solar cells with multilayer structure containing P3HT, PCBM, and CuPC have been reported (Fig. 26) [151]. In these devices, the CuPC layer was thermally evaporated onto the substrate and the P3TH:PCBM layer was deposited by spin-coating. For solar cells with optimized layer thicknesses, enhanced light absorption was responsible for the high J_{SC} (12.5 mA cm^{-2}) and high efficiency (4.1%), a consequence of a second optical interference peak in the multilayer structure.

Cyanoporphyrins have also been incorporated into the active layer of P3HT:PCBM solar cells, which were obtained by spin-casting the blends from chloroform solutions [154]. Before annealing, the greatest PP contribution to the spectral response occurred when both P3HT and PCBM were present. Upon

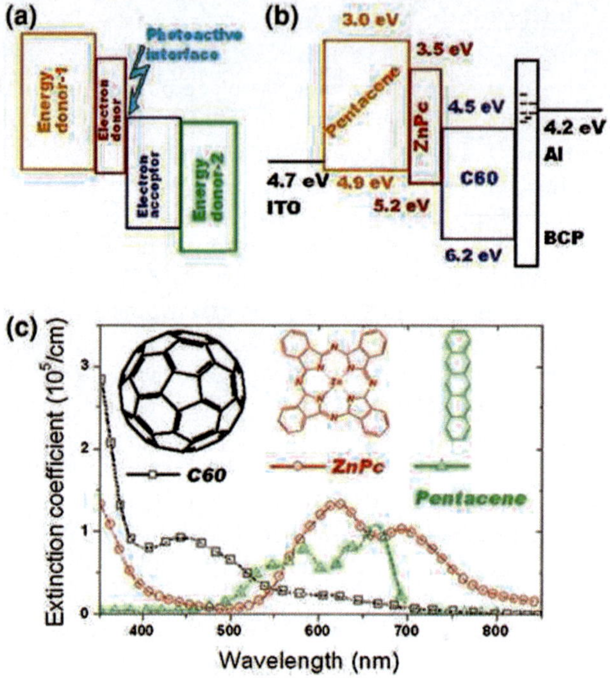

Fig. 25 **a** Concept of multiple-heterojunction design. **b** Energy diagram of OSC based on pentacene/ZnPC/C_{60} multiple heterojunctions. **c** Extinction coefficient curves of the three materials and their molecular structures. "Reprinted with permission from Hong et al. [149]. Copyright [2009], American Institute of Physics"

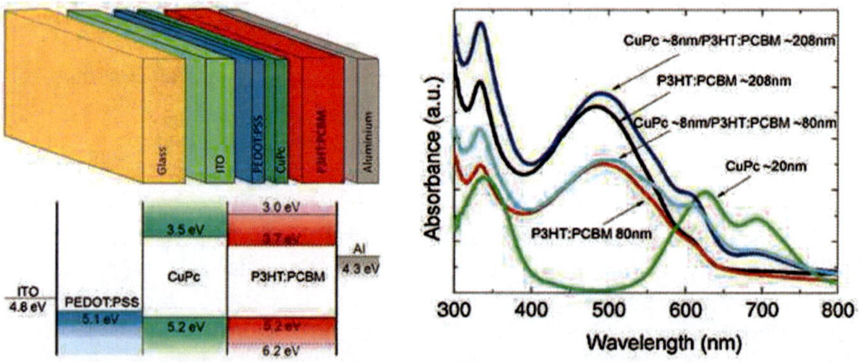

Fig. 26 Schematic illustration of the device structure and energy diagram for a OSC of ITO/PEDOT:PSS/CuPC/P3HT:PCBM/Al. The absorption spectra of P3HT:PCBM (80 nm), P3HT:PCBM (208 nm), CuPC (20 nm), CuPC (8 nm)/P3HT:PCBM (80 nm), and CuPC (8 nm)/P3HT:PCBM (208 nm) structures. "Reprinted with permission from Zhang et al. [151]. Copyright [2008], American Institute of Physics"

annealing (140°C for 4 min), the photocurrent generated by the PP was lost, due to aggregation, except in devices where only a small amount of polymer was present. For devices with low PCBM content, the PP did not contribute to the photocurrent and hindered the photocurrent generation by P3HT at the Soret band. Overall, this data suggests that the PP tends to interact with the polymer to quench photocurrent generation, but interacts preferentially with PCBM when it is present in the film.

In order to improve the exciton diffusion length, both high-mobility compounds and triplet materials with long exciton lifetimes can be introduced in the device. An efficient bulk-heterojunction OSC based on the triplet material 2,3,7,8,12, 13,17,18-octaethyl-21*H*,23*H*porphineplatinum (II) was demonstrated by Shao and Yang [155]. Schulz and Holdcroft [156] and Yang et al. [157] also showed enhanced photovoltaic responses due to singlet-to-triplet exciton conversion in conjugated polymer/iridium complex-based cells. Li et al. [158] fabricated ITO/PEDOT:PSS/P3HT:Pt dendrimer/C_{60}/Al cells, in which the charge generating structure consisted of a film of P3HT blended with triplet platinum dendrimer complex deposited by spin-coating and a thermally evaporated fullerene layer. In this platinum dendrimer, the platinum porphyrin core acts as heavy metal center, which induces intersystem crossing in the host polymer [157, 159], and the external carbazole groups enhance the conductive properties. In particular, the alkyl chains provide good solubility for the dendrimers in organic solvents and excellent film-forming properties by spin-coating. The cells showed a poor efficiency of 0.70%, which was attributed to poor mobility in the Pt dendrimer film.

The construction of supramolecular assemblies is also interesting and promising for the future development of photovoltaics. In this perspective, recent developments of supramolecular systems for light energy conversion, which are mainly composed of PC dyes and nanocarbon materials, such as fullerenes and carbon nanotubes were reported. The water-soluble CuPC derivative 3,4,4,4 tetrasulphonic acid tetra sodium salt copper phthalocyanine (TS-CuPC) was blended with concentrated dispersions of acid-treated carbon nanotubes to form stable solutions with excellent film-forming properties [160]. The application of this nanocomposite material as hole-extracting electrode and donor layer in bilayer OSC using C_{60} as electron acceptor was demonstrated. The interaction between surface-oxidized multi-walled carbon nanotubes (o-MWCNTs) and TS-CuPC was also investigated [161]. The compatibility between the two components was shown to result from hydrogen-bonding interactions and ground-state charge-transfer interactions. The self-organization of the o-MWCNT decorated PC molecules into extended aggregates of 1D linearly stacked PC polymers was observed to occur after the spin-coating deposition. The hybrid material was incorporated into an organic photovoltaic cell at the interface between the P3HT:PCBM bulk-heterojunction layer and the ITO electrode. This extra-layer increased the light-harvesting and facilitated the hole extraction, enhancing the η.

Despite the interesting concepts underlying the introduction of a third component to the OSC, more efforts are necessary in the search for better material combinations, since higher efficiencies are still found for two-component-based devices.

6 Small Molecules in Dye-sensitized Solar Cells

The possibility of exploring the solar energy with reduced costs became a reality after the report of an efficient DSSC by O'Regan and Grätzel [162]. Since then these devices have attracted significant attention from the scientific and industrial community because DSSCs use cheaper materials and production processes than the analogous Si-based photovoltaics. These solar cells can achieve up to 10.4% of certified solar power efficiency [163] and their stability data indicates at least 10 years of use in outdoor applications [164]. Generally, DSSCs are assembled using nanocrystalline TiO_2, organic or inorganic dyes (normally Ruthenium (II) coordination compounds), and a liquid electrolyte. The liquid component is considered a drawback for large-area production as a consequence of potential liquid leakage and contamination concerns; a sealed and secured cell is essential. Also, the iodine usually employed is capable of attacking the transparent conducting oxide substrate typically used in fabrication. Solid or quasi-solid electrolytes, such as polymers and gels, have been applied as alternatives to the liquid component [165–167]. However, all-solid-state solar cells are only achievable when p-type semiconductors are employed, such as inorganic hole conductors, organic low-molecular weight molecules or conducting polymers [168–173, 218–220].

In this scenario, small organic molecules are also promising candidates to replace the liquid electrolyte for several reasons: easy fabrication, low cost, and versatile deposition. When applied as hole conductors (or hole transporting materials, HTM) in DSSCs, better penetration is achieved due to their ability to fill the TiO_2 pores more effectively.

Small organic molecules can also be used to replace the sensitizer dye in DSSCs. In fact, in recent years, several groups have concentrated their efforts in the search of novel organic dyes with more extended light absorption toward the infrared. Besides, the small organic dye can replace the expensive metal complexes based on the rare metal Ruthenium [174–176]. It is also possible that the sensitizer dye acts simultaneously as both absorber and HTM in this kind of solar cell, although the efficiency is low.

In this section we will focus mainly on the attempts to replace the liquid electrolyte in DSSC by small organic molecules as HTMs. Small molecules are powerful sensitizers, particularly perylene, indoline and arylamine derivates. Due to the great versatility and performance demonstrated in DSSC, a brief review is provided for this application.

6.1 Hole-Conductor Materials

Energy conversion in a DSSC is based on the injection of an electron from a photoexcited state of the sensitizer dye (typically a bipyridine Ruthenium (II) complex) into the conduction band of the nanocrystalline semiconductor (TiO_2 is

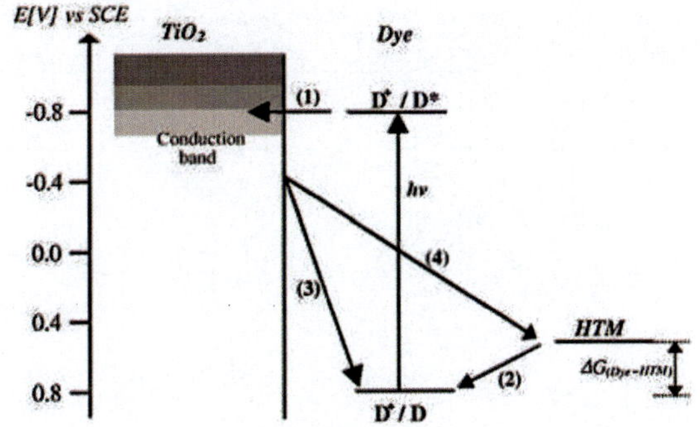

Fig. 27 Schematic diagram of the processes occurring at the dye-sensitized TiO_2/ organic HTM heterojunction (Haque et al. [217]). Copyright Wiley–VCH Verlag GmbH & Co. KGaA. Reproduced with permission

by far the most employed oxide semiconductor), as depicted in Fig. 27. These cells also employ a liquid electrolyte (usually an iodide/triiodide redox-active couple dissolved in an organic solvent) to reduce the dye cation (viz., regenerate the ground state of the dye). Regeneration of iodide ions, which are oxidized in this reaction to triiodide, is achieved at a platinum counter electrode.

A p-type semiconductor can replace the liquid electrolyte if it is able to accept holes efficiently from the excited state of the dye cation. The main processes that occur at the TiO_2/dye/HTM interfaces are represented in Fig. 27. Cell operation using HTM is analogous to liquid or polymer electrolyte-based DSSCs. However, after dye excitation and electron transfer (reaction 1 in Fig. 27), the ground state of the dye is regenerated by the HTM (reaction 2): electrons from the HOMO of the HTM regenerate the ground state of the dye molecules instead of the redox couple of the electrolyte. The oxidized HTM material is then reduced at the counter electrode (in most cases a nanometric layer of gold). The main difference relies on the kind of transport between the electrodes. In comparison to the ionic transport in the DSSC which uses liquid or gel polymer electrolyte, the HTM cell transport is typically electronic. The losses are represented by the electron recombination with the dye cations (reaction 3) and with HTM (reaction 4). The recombination reactions are important because they limit the efficiency of these solar cells: at open-circuit and short-circuit conditions, the recombination is 10 and 100 times faster than in the cells with liquid electrolyte, respectively [177].

The hole-transfer reaction is limited by the thermodynamic driving force, defined as $\Delta G_{\text{dye-HTM}}$ [169]. In order to achieve an 85% charge-transfer yield, the energy difference between the HOMO of the HTM and the HOMO of the dye ($\Delta G_{\text{dye-HTM}}$) must be at least 0.2 eV. According to the energy diagram, the HOMO position of the HTM must lie above the ground state of the sensitizer dye.

Triarylamine derivates are so far the most important class of HTM applied in DSSC. The first successful triarylamine derivate applied in DSSC was the amorphous compound 2,2′,7,7′-tetrakis-(*N*,*N*-di-*p*-methoxyphenylamine)9,9′-spirobifluorene (referred as spiro-OMeTAD, structure 1, Table 3), introduced by the group of Prof. M. Grätzel in 1998 [168]. In 2007, the efficiency of the solid-state DSSC based on spiro-OMeTAD reached 5.1% [178]. Until this present day, this molecule has been unsurpassed in terms of hole conduction in DSSC. However,

Table 3 Characteristics of some organic HTM based on triarylamine derivatives applied in DSSCs

	Structure	Mw	Hole mobility/ $cm^2\ V^{-1}\ s^{-1}$	HOMO level/ eV	References
1	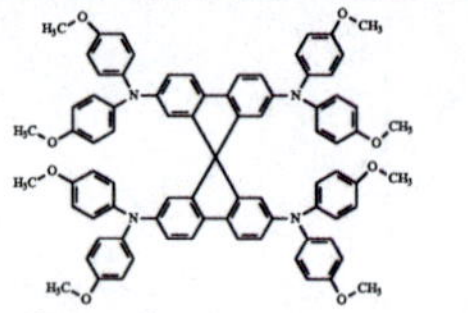	1230	2×10^{-4} ([a])	−4.77	[170]
2	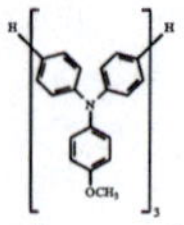	821	4.86×10^{-4} ([b])	−4.97	[169]
3	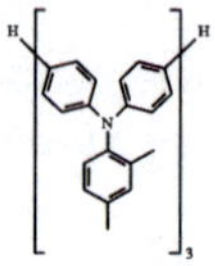	815	8.07×10^{-4} ([b])	−5.0	[169]
4	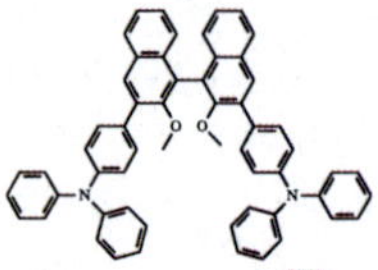	800	([c])	−5.26	[171]
5	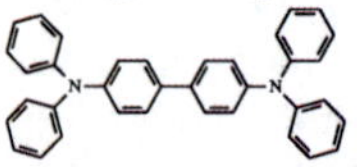	488	10^{-3} ([a])	([c])	[172, 173]
6	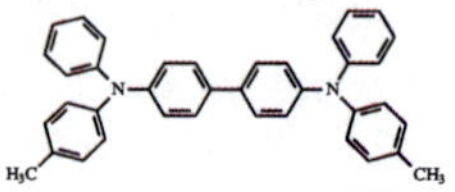	516	$\sim 10^{-3}$ ([a])	−5.13	[172, 173]
7	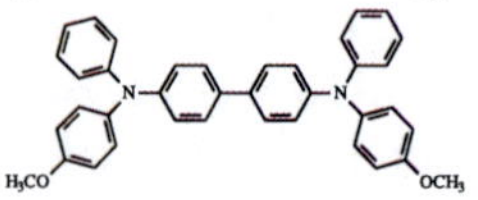	548	$\sim 10^{-3}$ ([a])	−5.06	[172, 173]
8	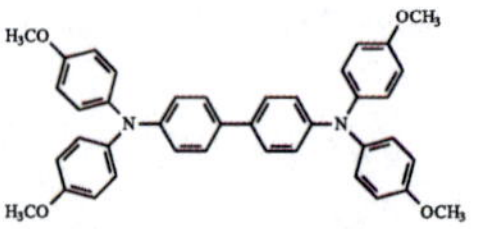	608	$\sim 10^{-3}$ ([a])	−4.97	[173]
9	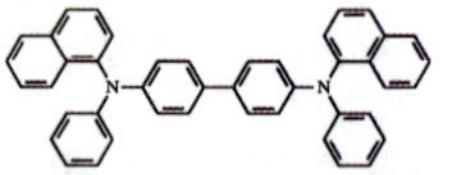	588	$\sim 10^{-3}$ ([a])	−5.20	[173]
10	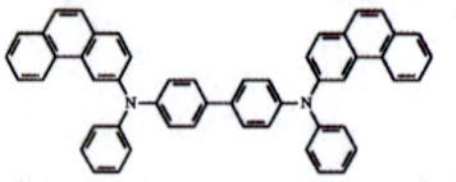	688	$\sim 10^{-3}$ ([a])	−5.18	[173]
11	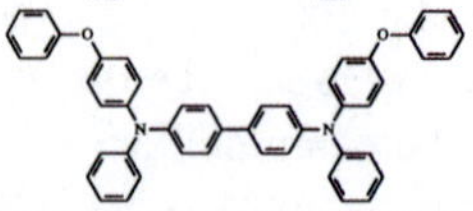	640	$\sim 10^{-3}$ ([a])	−5.12	[173]

(continued)

Table 3 (Continued)

	Structure	Mw	Hole mobility/ $cm^2\ V^{-1}\ s^{-1}$	HOMO level/ eV	References
12	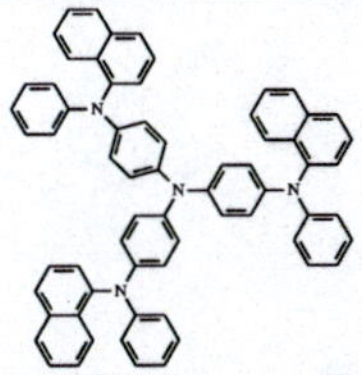	896	([c])	([c])	[173]
13	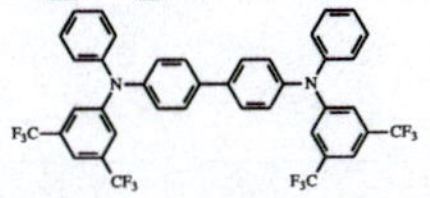	760	([c])	−5.45	[173]

[a] Determined using the time-of-flight technique
[b] Field-effect transistor (FET) mobilities
[c] Not available

other types of organic HTMs have been synthesized and have demonstrated the potential to replace liquid electrolytes, as well. Table 3 summarizes the structures and some characteristics of small organic HTM based on tryarylamine derivatives applied in DSSC.

Interestingly, the charge-transport properties of the spiro-OMeTAD showed that its hole mobility is independent of film thickness from 0.135 to 4 μm [170]. Thus, hole mobility is a non-dispersive characteristic and this indicates that the photocurrent may be dependent on the rate of interfacial hole transfer, not as function of the thickness. In fact, Durrant et al. [169] found that the hole transfer yield (related to the efficiency of reaction 2 presented in Fig. 27) is directly proportional to the photocurrent (Fig. 28).

Although a high hole mobility is a strong consideration when choosing between potential HTMs, other important requirements must be taken into. It is well-established that good film-formation ability, low tendency toward crystallization, and excellent pore filling and HOMO energy value strongly influence the overall conversion efficiency of DSSCs [173]. As observed in the performance of solar cells using the hole conductors 1–3 as displayed in Table 3, the best efficiency was achieved with spiro-OMeTAD (2.8%), although this molecule presented the lowest hole mobility of all. Despite the higher hole mobility of the other triarylamine oligomers, they are poor pore fillers [179] and have significant variations in the hole transfer yields.

Similar conclusions involving the HTM properties and their relation to device performance were obtained out by Karthikeyan and Thelakkat using several synthesized HTMs based on triphenyldiamines with mobility of the order of $10^{-3}\ cm^2\ V^{-1}\ s^{-1}$ [173]. They calculated the charge-transfer rate for these molecules and observed their dependence on the HOMO energy offset. The maximum charge-transfer rate was realized for the energy gap of 0.79 eV. The results give significant information for the design of novel dyes and HTM.

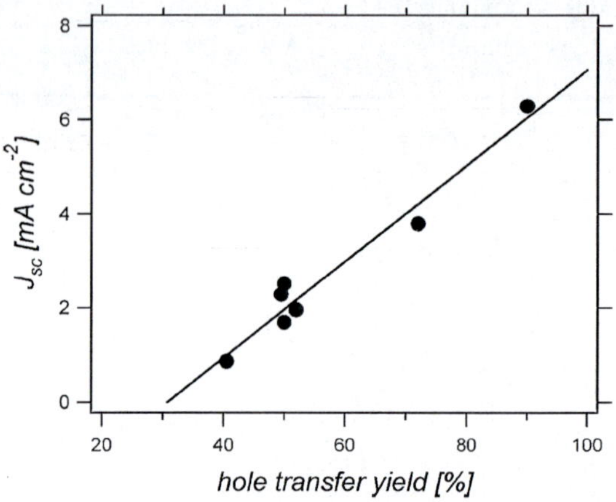

Fig. 28 Short-circuit photocurrent density (J_{SC}) as a function of the hole transfer yield, determined using Transient Absorption Spectroscopy—TAS (Kroeze et al. [169]). Copyright Wiley–VCH Verlag GmbH & Co. KGaA. Reproduced with permission

Another class of HTM is macrocyclic aromatic compounds such as PP and PC. They have high stability and high optical absorption, covering a great extension of the solar spectrum. Some techniques have showed that PP and PC films are frequently p-type semiconductors [180–184]. Few reports demonstrate the usage of this class of small molecules in transporting holes after dye excitation. The combination of TiO_2 and PC or PP is limited [185–190], and the efficiency is low. For the TiO_2/ZnPC/Au cells, $V_{OC} = 0.376$ V, $J_{SC} = 0.142$ mA cm^{-2}, and FF = 0.34 (under simulated AM 2 conditions) were achieved [185]. Other reports were found for zinc (ZnPC) [185, 189, 190], copper (CuPC) [190], palladium (PdPC) [185], lead (PbPC) [188], and iron (II) phthalocyanines (FePC) [187]. The best result was obtained for an ITO/TiO_2/ZnTCPP/Hg solar cell (where ZnTCPP is zinc-tetra(4-carboxyphenyl) porphyrin), which showed a V_{OC} of 0.7 V, J_{SC} of 0.22 mA cm^{-2}, and FF of 0.25 [182].

For the most part, these molecules, PP and PC are more intensively explored as dyes [191–198] . The dual function of this class of small molecules, as sensitizer and as HTM, is rarely investigated [182]. A possible explanation for the limited use of PP and PC as HTMs (and dye-HTM combination) in DSSCs comes from low efficiency and this may be related to the difficulty of filling the pores of the TiO_2 electrodes, annihilation of the molecule excited state by energy transfer due to aggregation, and poor hole mobility after deposition. In fact, for this class of molecules, the values of hole mobility is closely related to device performance.

Intensity-modulated photocurrent spectroscopy indicates that the transport of holes in these small molecules is not governed by the electric field, but driven by diffusion [182]. Mobility values of 5 and 2.5×10^{-4} cm^2 V^{-1} s^{-1} were found for columnar stacks of PP [199] and a palladium phthalocyanine (PdPC) [186]. In this latter case the solar cell presented a $V_{OC} = 0.46$, $J_{SC} = 30$ μA cm^{-2}, FF = 0.35 and $\eta = 0.025\%$ under 20 mW cm^{-2}. However, in the majority reports, the mobility values fall between 10^{-10} and 10^{-7} cm^2 V^{-1} s^{-1} [181, 182].

It is also important, however, to call attention to the difficulty in preparing such films. This may explain the large discrepancy between the reported mobilities

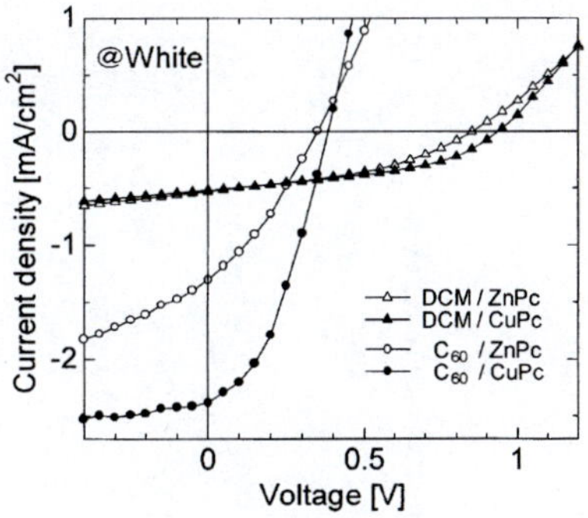

Fig. 29 I–V curves of DCM and C_{60} based devices as an intermediate layer, which is sandwiched between TiO_2 and ZnPC or CuPC layer under white light illumination. "Reprinted from Ohmori et al. [190], copyright (2006), with permission from Elsevier"

values derived from current–voltage [200, 201], time-of-flight [202] and microwave-conductivity measurements [199], since only vacuum sublimation, electropolymerization and dispersion in other HTMs seems to be a successful method to achieve homogeneous and stable films [183].

Nevertheless, the conversion efficiency can be substantially increased when dyes or electron acceptors such as C_{60} and 4-(dicyano-methylene)-2-methyl-6-(4-dimethylaminostyryl)-4H-pyran (DCM) are added as n-type material in ITO/TiO_2/n-type layer/PC/Au cells, where PC refers to ZnPC or CuPC [190]. The hole mobililties were estimated as 3.8×10^{-3} and 2.6×10^{-5} $cm^2\ V^{-1}\ s^{-1}$ for CuPC and ZnPC, respectively. The largest energy difference between the HOMO level of PC and LUMO level of DCM leads to an increase in the V_{OC} of the device. The best solar cell performance was achieved with C_{60} combined with CuPC (Fig. 29), which causes an increase of J_{SC}, likely due to a higher carrier mobility, i.e., C_{60} acts as an electron acceptor, increasing the electron transport.

Pentacene molecules have also been applied as HTM in DSSC, however, the efficiency of the solar cells is lower compared with triarylamine derivates, possibly due to a fast recombination of the charge carriers at the interfaces [203].

6.2 Small Molecules as Sensitizers

In DSSCs, the photocurrent values are limited by the absorption spectrum of the sensitizer dye. Tuning the optical and electronic properties of small organic molecules by modifying their chemical structures has become an important step in overcoming this issue and developing solar cells with improved light-harvesting capabilities. In fact, the number of publications involving the synthesis of novel organic dyes and their application as sensitizers in DSSCs has grown rapidly, motivated by the high efficiency values reported recently (Fig. 30). The best result achieved to date reached 9.8% of efficiency using a TPA derivative and a liquid electrolyte [204].

Among the organic molecules sensitizers, cyanine [174], coumarin [205], porphyrins [191–194], phthalocyanine [192, 195–198], perylene derivatives [206–209],

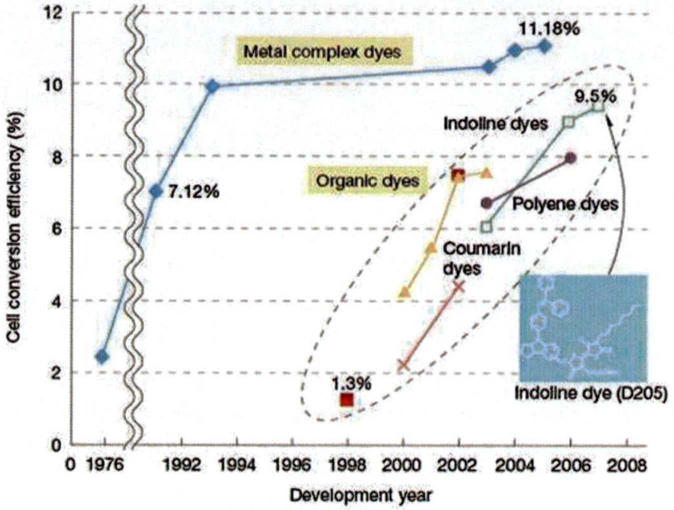

Fig. 30 Ruthenium complexes developed by the Grätzel Group in 1991 achieved an efficiency of 10% quite rapidly and they are currently at 11%. The use of organic dyes began with extremely low efficiencies but has shown a fast growth in the last years. The D205 developed by Mitsubishi Paper Mills in 2007 achieves an efficiency of 9.5% (from Tetsuo Nozawa Nikkei Electronics Asia, July 2008) (http://www.sciencedirect.com/science/journal/00406090)

indoline [175, 176] and arylamine derivates dyes [204, 210, 211] have received particular attention. In this section we will focus on perylene, indoline, and arylamine dyes, since recent reviews on the use of cyanine [174], coumarin [174], porphyrin [192, 193], and phthalocyanine [192] dyes in solar cells can be found elsewhere. The perylene derivatives present high stability and high versatility attained by use of different functional groups. Indoline/arylamine dyes have shown the highest efficiency for organic dyes in DSSCs [176, 204].

Photophysical studies are a powerful tool to monitor the excited state of these molecules and to correlate the effect of the substituent groups in molecular dynamics. For example, the degree of dye aggregation can be modulated by substituents like alkyl side-chains with free rotation capacities. Long alkyl chains [206] and 2,6-diisopropylphenyl [207] have demonstrated promising results and are shown in Fig. 31.

The presence of two alkyl chains in perylene (Fig. 31) showed that a free rotation capacity prevents aggregation, enhancing the photoelectron transfer injection from the dye to the titania film. Icli et al. [206] showed that the presence of two chains instead of one long or small chains with many substituents prevents charge recombination, resulting in photocurrent values up to 9 mA cm^{-2} and an efficiency of 1.61%. When the substituent group is larger and possesses strong electron-donating properties (i.e., two pyrrolidines at the perylene core) as in perylene b (Fig. 31), the first oxidation potential is shifted in the negative direction, improving the power-conversion efficiency up to 2% [207].

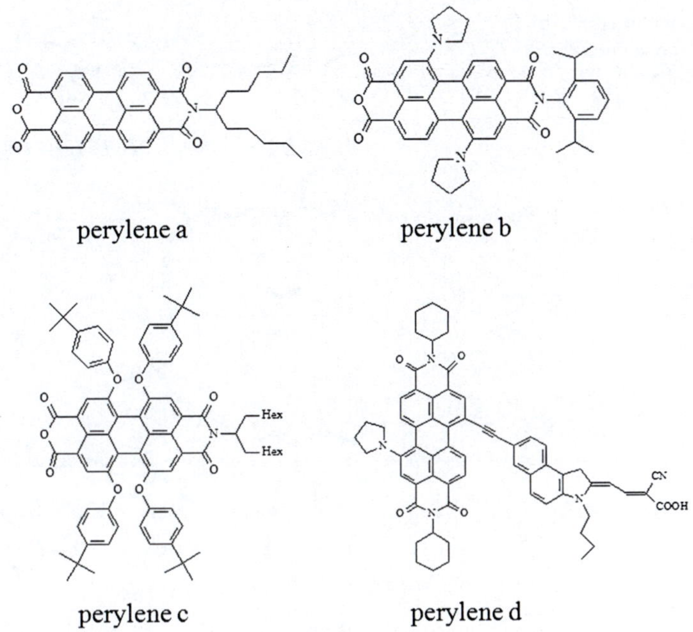

Fig. 31 Structures of perylene derivates [206–209]

In addition, the presence of four O-aryl groups makes the perylene sufficiently electron donating when applied as TiO_2 sensitizer. The conjugation system between the perylene core and the benzimidazole was tested and leads to a significant bathochromic shift of the absorption maximum and to an enhanced molar absorption coefficient. However, the electron injection was low, with efficiencies below 1%. The absence of benzimidazole and the presence of alkyl chain (perylene c, Fig. 31) led to a decrease in the dye recombination, reaching an efficiency of 2.29% [208]. An interesting report by Jin et al., demonstrated that the presence of two imide groups (perylene d, Fig. 31) in a perylene dye deteriorated device performance. Such molecules contain a strong electron-withdrawing group, resulting in ineffective electron transfer to the carboxylic groups anchored on the TiO_2 [209].

Other organic dyes have also shown promise in the DSSC field. Indoline dyes are a class of molecules with high sensitization. The presence of a rhodanine ring in the indoline dyes (Fig. 32) contributed to a red shift in the absorption spectrum, resulting in devices with a $V_{OC} = 693$ mV, $J_{SC} = 18.50$ mA cm^{-2}, FF = 0.624, and $\eta = 8.00\%$ [175]. Solid-state DSSCs using spiro-OMeTAD and D102 dye have been reported. The devices showed efficiency higher than 4% [212]. The introduction of a *n*-octyl substituent onto the rhodanine ring (D205 in Fig. 32) has pushed the conversion efficiency up to 9.52% under 100 mW cm^{-2}. This is the highest value reported so far for an indoline dye-based DSSC, compared to the same cells using Ruthenium (II) complexes [175].

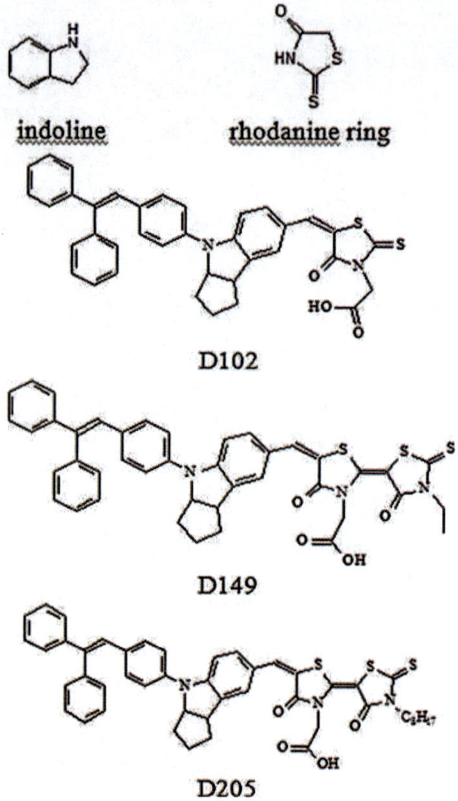

Fig. 32 Molecular structures of indoline, rhodanine ring, and indoline dyes D102, D149, D205 [175, 176, 213]

The high efficiency values can be explained by a red shift caused by attaching a second rhodanine unit to the indoline structure, extending its π-conjugation [213]. Although the chromophoric units of these two dyes are identical and the IPCEs are very close [175], the best behavior of D205 (Fig. 33) is a result of the extension of the alkyl chain on the terminal rhodanine moiety from ethyl to octyl. The successful combination of the *n*-octyl chain in the dye with the adsorption chenodeoxycholic acid (CDCA) onto the oxide resulted in blocking the charge recombination between I_3^- and electrons injected in the nanocrystalline-TiO_2 electrodes [176].

However, the D205 indoline does not seem to be stable for use in outdoor photovoltaic devices [214]. Thus, arylamine derivates dyes have been synthesized to improve the resistance to degradation over light soaking at full solar intensity and elevated temperatures [204, 215, 216]. These dyes contain an arylamine (electron donor) and a cyanoacrylate group (electron acceptor) connected by one or several thiophene moieties acting as a π-conducting bridge. During light excitation, electrons are transferred from the arylamine through the thiophene bridge to the surface-bound cyanoacrylate, producing an efficient and rapid electron injection from the excited state of the sensitizer into the conduction band of the TiO_2 [214].

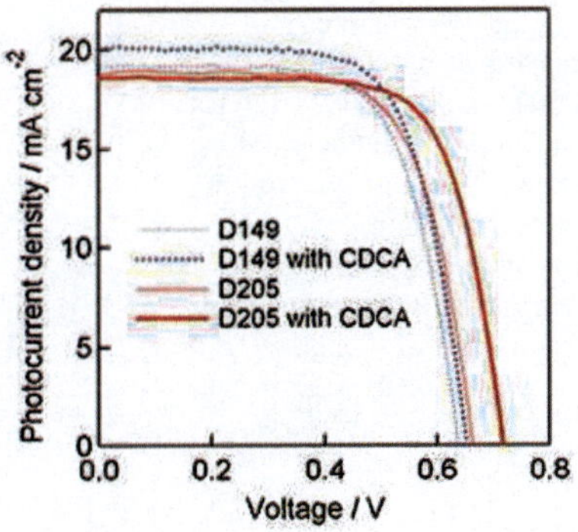

Fig. 33 Current density vs. voltage characteristics for DSSC with indoline dyes (D149 and 205) as sensitizers with/without chenodeoxycholic acid (CDCA) under AM1.5 simulated sunlight (100 mW cm^{-2}) illumination. Ito et al. [176]—reproduced by permission of The Royal Society of Chemistry

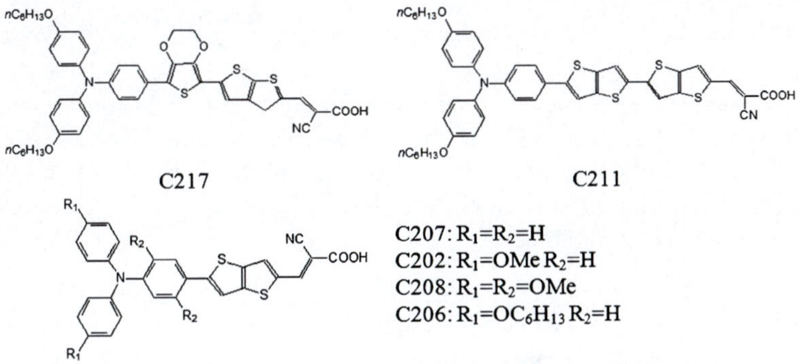

Fig. 34 Structures of the arylamine derivates dyes [204, 210, 211] (http://dx.doi.org/10.1039/B809093A)

Due to the strong coupling of the excited state wave function with the Ti (3d,t_{2g}) orbitals [214], the record efficiency of 9.8% is reached using organic dyes that belongs to an arylamine derivate dye named C217, shown in Fig. 34. Such a remarkable dye combines an electron donor and an electron acceptor with an electron-rich 3,4-ethylenedioxythiophene unit that has a small torsion angle with respect to the adjoining phenyl fragment, ensuring efficient electronic communication between donor and acceptor units [204].

An analogous thiophene π-conducting bridge (thienothiophene and bisthienothiophene) has also been demonstrated (C207, C202, C208, C206 e C211 in Fig. 34) [210, 211], showing that the addition of one more thienothiophene units both increases HOMO level slightly and drives down the LUMO, narrowing the gap [210]. Comparing the dyes in Fig. 35, the band gap can be further shifted into the infrared regime when two electron-donating methoxy groups (C208) are replaced at R_1 positions [211].

Although DSSCs employing indolines and arylamines have demonstrated improved and impressive results in recent years, the overall power-conversion efficiencies still remain slightly below the values achieved with expensive

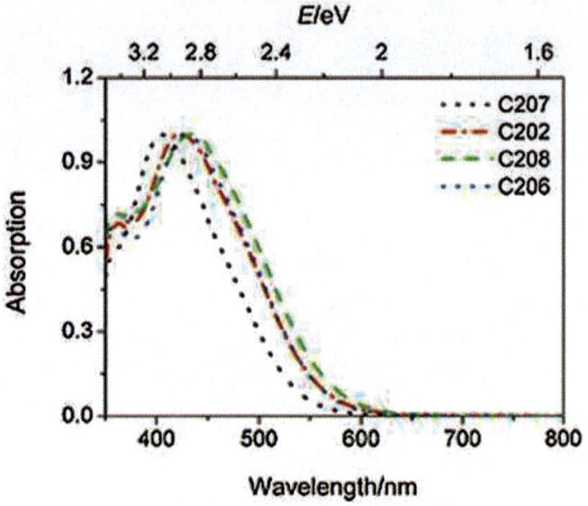

Fig. 35 Normalized absorption spectra of C207, C202, C208 and C206 dyes anchored the nanocrystalline titania film. The absorption from the titania and glass substrate has been subtracted. "Reprinted with permission from Xu et al. [211]. Copyright 2008 American Chemical Society"

ruthenium complexes. However, scientists working in this field foresee that the most powerful solar cells will be those based on organic dyes. Theoretical chemical calculation, for example, is a powerful tool that can help us design and select new dyes with more suitable absorption spectra, and as consequence, increased light-harvesting abilities.

7 Conclusions and Perspectives

Any emerging technology represents a challenging task and this is no different in the development of efficient, stable OSCs based on small molecules. This field requires a multidisciplinary contribution that involves strong cooperation between creative synthetic chemists, solid-state physicists, theoreticians, and device engineers.

The efficiency of OSCs based on small molecules is currently inferior to those employing conjugated polymers. This is partially attributed to the contact interfaces. A polymer adsorbed on a surface can only desorb if a substantial number of its segments desorb simultaneously, which is improbable and slow. However, interface quality and stability in small molecule devices have been lacking. Thus, more stable interfaces are required.

In addition to searching for novel donor and acceptor molecules with improved solubility, light-harvesting capability and charge mobility, the most challenging and critical issue is to achieve a suitable morphology without phase separation in solution-processed OSCs. It is very common to find FF values below 50% in solar cells using the solution approach. Tryarilamines, vinazene, and diketopyrrolo-pyrrol-based molecules are among the most promising candidates for solution processable solar cells. In the case of vinazene molecules, values of open-circuit voltage exceeding 1 V were obtained. Efficiency is expected to increase with better control over phase separation and the use of self-assembly techniques may be interesting in this context.

Solar cells based on evaporated small molecules offer an alternative to SPOSCs as OS films and/or multilayers can be easily deposited with high purity, high crystallinity, and without pinholes. VTE and OVPD are the most-used techniques, providing morphological control in the double layer, bulk heterojunction and tandem approaches. The introduction of an EBL has allowed efficiencies up to 4%. Besides those advantages, the major drawback is the cost associated with the production method, especially for large-area solar cells.

Organic small molecules have also found their place as components in hybrid solar cells, which include dye-sensitized solar cells. In this type of device, organic molecules can act as both the dye and hole transporting materials. In fact, the field of DSSCs has grown intensively in the area of novel small organic molecules with ability to absorb long wavelengths. The indoline dyes represent the DSSC-ruthenium free class that has delivered the highest efficiencies to date.

As HTMs, triarylamine derivatives, especially Spiro-OMeTAD, are the most studied class of p-type semiconductors. The replacement of the liquid electrolyte by an HTM is not trivial. Problems associated with poor electrode filling, HOMO position (which controls the charge transfer yield) and mobility still need to be solved.

For OSCs in general, the challenges are many. As solar cell efficiencies improve, however, long-term device stability becomes a more relevant and important issue to consider. Novel donor and acceptor molecules with improved optical, electronic, and transport properties are desired. Control over morphology is experimentally difficult and efforts in this direction continue to receive increasing attention. Although most OSCs employ amorphous films, partly because it is relatively easy to deposit amorphous pinhole-free films that adhere well to substrates and to other films, a tendency to produce more crystalline and/or aligned OSs has been observed. New cell architectures, the design of an appropriate DA combination based on theoretical investigations, more ordered structures and stable interfaces have been pointed out as strategies to push the efficiency of these solar cells toward a competitive level. The future is bright for OSCs and based on recent achievements there is no doubt that their place is certainly guaranteed in the photovoltaic market.

Acknowledgments The authors thank CNPq, Fapesp (2009/15428-0, 2008/53059-4) and INEO (National Institute for Organic Electronics) for financial support and scholarships. The authors also acknowledge Jason Guy Taylor and Prof. Roy Bruns for English revision.

References

1. Nelson J (2004) The physics of solar cells. Imperial College Press, London
2. Brabec CJ, Dyakonov V, Parisi J, Sariciftci NS (2003) Organic photovoltaics: concepts and realization. Springer, Berlin

3. Sun S-S, Sariciftci NS (2005) Organic photovoltaics. Taylor & Francis Group, Boca Raton
4. Würfel P (2009) Physics of solar cells. Wiley–VCH, Weinheim
5. Archer MD, Hill R (2005) Clean electricity from photovoltaics. Imperial College Press, London
6. Archer MD, Nozik AJ (2008) Nanostructured and photoelectrochemical systems for solar photon conversion. Imperial College Press, London
7. Brabec C, Dyakonov V, Scherf U (2009) Organic photovoltaics. Wiley–VCH, Weinheim
8. Li JL, Dierschke F, Wu JS, Grimsdale AC, Mullen K (2006) Poly(2, 7-carbazole) and perylene tetracarboxydiimide: a promising donor/acceptor pair for polymer solar cells. J Mater Chem 16:96–100
9. Sharma GD, Suresh P, Mikroyannidis JA, Stylianakis MM (2010) Efficient bulk heterojunction devices based on phenylenevinylene small molecule and perylene–pyrene bisimide. J Mater Chem 20:561–567
10. Silvestri F, Irwin MD, Beverina L, Facchetti A, Pagani GA, Marks TJ (2008) Efficient squaraine-based solution processable bulk-heterojunction solar cells. J Am Chem Soc 130:17640–17641
11. Balakrishnan K, Datar A, Baddo T, Huang J, Otiker R, Yen M, Zhao J, Zang L (2006) Effect of side-chain substituents on self-assembly of perylene diimide molecules: morphology control. J Am Chem Soc 128:7390–7398
12. Winzenberg KN, Kemppinem P, Fanchini G, Bown M, Collins GE, Forsyth GM, Hegedus K, Singh TB, Wattkins SE (2009) Dibenzo[b, def]chrysene derivatives: solution-processable small molecules that deliver high power-conversion efficiencies in bulk heterojunction solar cells. Chem Mater 21:5701–5703
13. Lloyd MT, Mayer AC, Subramanian S, Mourney DA, Herman DJ, Bapat AV, Anthony JE, Malliaras GGJ (2007) Efficient solution-processed photovoltaic cells based on an anthradithiophene/fullerene blend. J Am Chem Soc 129:9144–9149
14. Kronenberg NM, Deppisch M, Würthner F, Lademann HWA, Deing K, Meerholz K (2008) Bulk heterojunction organic solar cells based on merocyanine colorants. Chem Commun 6489–6491
15. Ma C-Q, Fonrodona M, Schikora MC, Wienk MM, Janssen RAJ, Bauerle P (2008) Solution-processed bulk-heterojunction solar cells based on monodisperse dendritic oligothiophenes. Adv Funct Mater 18:3323–3331
16. Velusamy M, Huang J-H, Hsu Y-C, Chou H-H, Ho K-C, Wu P-L, Chang W-H, Lin JT, Chu C-W (2009) Dibenzo[f, h]thieno[3, 4-b] quinoxaline-based small molecules for efficient bulk-heterojunction solar cells. Org Lett 11:4898–4901
17. Shirota Y (2000) Organic materials for electronic and optoelectronic devices. J Mater Chem 10:1–25
18. Shirota Y, Kageyama H (2007) Charge carrier transporting molecular materials and their applications in devices. Chem Rev 107:953–1010
19. Roncali J (1997) Synthetic principles for bandgap control in linear pi-conjugated systems. Chem Rev 97:173–205
20. Roncali J (2005) Linear-conjugated systems derivatized with C60-fullerene as molecular heterojunctions for organic photovoltaics. Chem Soc Rev 34:483–495
21. He C, He Q, Yi Y, Wu G, Bai F, Shuai Z, Li Y (2008) Improving the efficiency of processable organic photovoltaic devices by a star-shaped molecular geometry. J Mater Chem 18:4085–4090
22. Zhang J, Yang Y, He C, He Y, Zhao G, Li Y (2009) Solution-processable star-shaped photovoltaic organic molecule with triphenylamine core and benzothiadiazole-thiophene arms. Macromolecules 42:7619–7622
23. Wu G, Zhao G, He C, Zhang J, He Q, Chen X, Li Y (2009) Synthesis and photovoltaic properties of a star-shaped molecule with triphenylamine as core and benzo[1, 2, 5]thiadiazol vinylene as arms. Sol Energy Mater Sol Cells 93:108–113

24. Li W, Du C, Li F, Zhou Y, Fahlman M, Bo Z, Zhang F (2009) Benzothiadiazole-based linear and star molecules: design, synthesis, and their application in bulk heterojunction organic solar cells. Chem Mater 21:5327–5334
25. Li K, Qu J, Xu B, Zhou Y, Liu L, Peng P, Tian W (2009) Synthesis and photovoltaic properties of a novel solution-processable triphenylamine-based dendrimers with sulfonyldibenzene cores. New J Chem 33:2120–2127
26. He C, He Q, Yang X, Wu G, Yang C, Bai F, Shuai Z, Wang L, Li Y (2007) Synthesis and photovoltaic properties of a solution-processable organic molecule containing triphenylamine and DCM moieties. J Phys Chem C 111:8661–8666
27. Xue L, He J, Gu X, Yang Z, Xu B, Tian W (2009) Efficient bulk-heterojunction solar cells based on a symmetrical D–A–D organic dye molecule. J Phys Chem C 113:12911–12917
28. Zhao G, Wu G, He C, Bai F-Q, Xi H, Zhang H-X, Li Y (2009) Solution-processable multiarmed organic molecules containing triphenylamine and DCM moieties: synthesis and photovoltaic properties. J Phys Chem C 113:2636–2642
29. Tang CW (1986) Two layer organic photovoltaic cell. Appl Phys Lett 48:183–185
30. Takahashi K, Asano M, Imoto K, Yamaguchi T, Komura T, Nakamura J, Murata K (2003) Sensitization effect of porphyrin dye on the photocurrent of Al/polythiophene schottky-barrier cells. J Phys Chem B 107:1646–1652
31. Takahashi K, Nakajima I, Imoto K, Yamaguchi T, Komura T, Murata K (2003) Sensitization effect by porphyrin in polythiophene/perylene dye two-layer solar cells. Sol Energy Mater Sol Cells 76:115–124
32. Peumans P, Forrest SR (2001) Very-high-efficiency double-heterostructure copper phthalocyanine/C-60 photovoltaic cells. Appl Phys Lett 79:126–128
33. Stübinger T, Brütting W (2001) Exciton diffusion and optical interference in organic donor–acceptor photovoltaic cells. J Appl Phys 90:3632–3641
34. Ma B, Woo CH, Miyamoto Y, Frechet JMJ (2009) Solution processing of a small molecule, subnaphthalocyanine, for efficient organic photovoltaic cells. Chem Mater 21:1413–1417
35. Matsuo Y, Sato Y, Niinomi T, Soga I, Tanaka H, Nakamura E (2009) Columnar structure in bulk heterojunction in solution-processable three-layered p–i–n organic photovoltaic devices using tetrabenzoporphyrin precursor and silylmethyl[60]fullerene. J Am Chem Soc 131:16048–16050
36. Huang X, Zhu C, Zhang S, Li W, Guo Y, Zhan X, Liu Y, Bo Z (2008) Porphyrin-dithienothiophene-conjugate copolymers: synthesis and their applications in field-effect transistors and solar cells. Macromolecules 41:6895–6902
37. Tamayo AB, Walker B, Nguyen TQ (2008) A low band gap, solution processable oligothiophene with a diketopyrrolopyrrole core for use in organic solar cells. J Phys Chem C 112:11545–11551
38. Tamayo AB, Dang XD, Walker B, Seo J, Kent T, Nguyen TQ (2009) A low band gap, solution processable oligothiophene with a dialkylated diketopyrrolopyrrole chrmophore for use in bulk heterojunction solar cells. Appl Phys Lett 94:103301-1–103301-3
39. Walker B, Tamayo AB, Dang X-D et al (2009) Nanoscale phase separation and high photovoltaic efficiency in solution-processed, small-molecule bulk heterojunction solar cells. Adv Funct Mater 19:3063–3069
40. Tamayo A, Kent T, Tantitiwat M, Dante MA, Rogers J, Nguyen T-Q (2009) Influence of alkyl substituents and thermal annealing on the film morphology and performance of solution processed, diketopyrrolopyrrole-based bulk heterojunction solar cells. Energy Environ Sci 2:1180–1186
41. Rand BP, Li J, Xue J, Holmes RJ, Thompson ME, Forrest SR (2005) Organic double-heterostructure photovoltaic cells employing thick *tris*(acetylacetonato)ruthenium(III) exciton-blocking layers. Adv Mater 17:2714–2718
42. Li G, Shrotriya V, Huang J, Yao Y, Moriarty T, Emery K, Yang Y (2005) High-efficiency solution processable polymer photovoltaic cells by self-organization of polymer blends. Nat Mater 4:864–868

43. Peet J, Kim JY, Coates NE, Moses D, Heeger AJ, Bazan GC (2007) Efficiency enhancement in low-bandgap polymer solar cells by processing with alkane dithiols. Nat Mater 6:497–500
44. Goh C, Scully SR, McGehee MD (2007) Effects of molecular interface modification in hybrid organic–inorganic photovoltaic cells. J Appl Phys 101:114503-1–114503-12
45. Huynh WU, Dittmer JJ, Alivisatos AP (2002) Hybrid nanorod-polymer solar cells. Science 295:2425–2427
46. Kietzke T, Egbe DAM, Horhold HH, Neher D (2006) Comparative study of M3EH-PPV-based bilayer photovoltaic devices. Macromolecules 39:4018–4022
47. Koetse M, Sweelssen J, Hoekerd K, Schoo H (2006) Efficient polymer: polymer bulk heterojunction solar cells. Appl Phys Lett 88:083504-1–083504-3
48. McNeill CR, Abrusci A, Zaumseil J, Wilson R, McKiernan MJ, Burroughes JH, Halls JJM, Greenham NC, Friend RH (2007) Dual electron donor/electron acceptor character of a conjugated polymer in efficient photovoltaic diodes. Appl Phys Lett 90:193506-1–193506-3
49. Ooi ZE, Tam TL, Shin RYC, Chen ZK, Kietzke T, Sellinger A, Baumgarten M, Mullen K, de Mello JC (2008) Solution processable bulk-heterojunction solar cells using a small molecule acceptor. J Mater Chem 18:4619–4622
50. Schubert M, Yin C, Castellani M, Bange S, Tam TL, Sellinger A, Hörhold H-H, Kietzke T, Neher D (2009) Heteojunction topology versus fill factor correlations in novel hybrid small-molecular/polymeric solar cells. J Chem Phys 130:094703-1–094703-9
51. Inal S, Castellani M, Sellinger A, Neher D (2009) Relationship of photophysical properties and the device performance of novel hybrid small-molecular/polymeric dolar cells. Macromol Rapid Commun 30:1263–1268
52. Kietzke T, Shin RYC, Egbe DAM, Chen ZK, Sellinger A (2007) Effect of annealing on the characteristics of organic solar cells: polymer blends with a 2-vinyl-4, 5-dicyanoimidazole derivative. Macromolecules 40:4424–4428
53. Schin RYC, Kietzke T, Sudhakar S, Dodabalapur A, Chen Z-K, Sellinger A (2007) N-type conjugated materials based on 2-vinyl-4, 5-dicyanoimidazoles and their use in solar cells. Chem Mater 19:1892–1894
54. Martinson ABF, Massari AM, Lee SJ, Gurney RW, Splan KE, Hupp JT, Nguyen ST (2006) Organic photovoltaics interdigitated on the molecular scale. J Electrochem Soc 153:A527–A535
55. Smith ARG, Ruggles JL, Yu A, Gentle IR (2009) Multilayer nanostructured porphyrin arrays constructed by layer-by-layer self-assembly. Langmuir 25:9873–9878
56. Yui T, Kameyama T, Sasaki T, Torimoto T, Takagi K (2007) Pyrene-to-porphyrin excited singlet energy transfer in LBL-deposited LDH nanosheets. J Porphyr Phthalocyanines 11:428–433
57. Li LS, Jia QX, Li ADQ (2002) Effects of organic self-assembled polymer and metal phthalocyanine multilayers on the surface photovoltaic properties of indium tin oxide and titanium oxide. Chem Mater 14:1159–1165
58. Alencar WS, Crespilho FN, Santos MRMC, Zucolotto V, Oliveira ON Jr, Silva WC (2007) Influence of film architecture on the charge-transfer reactions of metallophthalocyanine layer-by-layer films. J Phys Chem C 111:12817–12821
59. Arakawa T, Munaoka T, Akiyama T, Yamada SW (2009) Effects of silver nanoparticles on photoelectrochemical responses of organic dyes. J Phys Chem C 113:11830–11835
60. Nishiyama F, Yokoyama T, Kamikado T, Yokoyama S, Mashiko S (2006) Layer-by-layer growth of porphyrin supramolecular thin films. Appl Phys Lett 88:253113-1–253113-3
61. Zhang B, Mu J, Li XQ (2006) Linear assemblies of aged CdS particles and cationic porphyrin in multilayer films. Appl Surf Sci 252:4990–4994
62. Alencar WS, Crespilho FN, Martins MVA, Zucolotto V, Oliveira ON, Silva WC (2009) Synergistic interaction between gold nanoparticles and nickel phthalocyanine in layer-by-layer (LbL) films: evidence of constitutional dynamic chemistry (CDC). Phys Chem Chem Phys 11:5086–5091

63. Pradhan B, Bandyopadhyay A, Pal AJ (2004) Molecular level control of donor/acceptor heterostructures in organic photovoltaic devices. Appl Phys Lett 85:663–665
64. Benten H, Kudo N, Ohkita H, Ito S (2009) Layer-by-layer deposition films of copper phthalocyanine derivative; their photoelectrochemical properties and application to solution-processed thin-film organic solar cells. Thin Solid Films 517:2016–2022
65. Forrest SR (1997) Ultrathin organic films grown by organic molecular beam deposition and related techniques. Chem Rev 97:1793–1896
66. Shtein M, Gossenberger HF, Benziger JB, Forrest SR (2001) Material transport regimes and mechanisms for growth of molecular organic thin films using low-pressure organic vapor phase deposition. J Appl Phys 89:1470–1476
67. Ling MM, Bao ZN (2004) Thin film deposition, patterning, and printing in organic thin film transistors. Chem Mater 16:4824–4840
68. Burrows PE, Forrest SR, Buma T, Fenter P, Sapochak LS, Schwartz J, Ban VS, Forrest JL (1995) Organic vapor phase deposition: a new method for the growth of organic thin films with large optical non-linearities. J Cryst Growth 156:91–98
69. Baldo M, Deutsch M, Burrows P, Gossenberger H, Gerstenberg M, Ban V, Forrest SR (1998) Organic vapor phase deposition. Adv Mater 10:1505–1514
70. Shtein M, Peumans P, Benziger JB, Forrest SR (2003) Micropatterning of small molecular weight organic semiconductor thin films using organic vapor phase deposition. J Appl Phys 93:4005–4016
71. Shtein M, Mapel J, Benziger JB, Forrest SR (2002) Effects of film morphology and gate dielectric surface preparation on the electrical characteristics of organic-vapor-phase-deposited pentacene thin-film transistors. Appl Phys Lett 81:268–270
72. Rand BP, Genoe J, Heremans P, Poortmans J (2007) Solar cells utilizing small molecular weight organic semiconductors. Prog Photovoltaics 15:659–676
73. Meiss J, Leo K, Riede MK, Uhrich C, Gnehr W-M, Sonntag S, Pfeiffer M (2009) Efficient semitransparent small-molecule organic solar cells. Appl Phys Lett 95:213306
74. Rusu M, Wiesner S, Mete T, Blei H, Meyer N, Heuken M, Lux-Steiner MC, Fostiropoulos K (2008) Organic donor, acceptor and buffer layers of small molecules prepared by OVPD technique for photovoltaics. Renew Energy 33:254–258
75. Peumans P, Bulovic V, Forrest SR (2000) Efficient photon harvesting at high optical intensities in ultrathin organic double-heterostructure photovoltaic diodes. Appl Phys Lett 76:2650–2652
76. Vogel M, Doka S, Breyer C, Lux-Steiner MC, Fostiropoulos K (2006) On the function of a bathocuproine buffer layer in organic photovoltaic cells. Appl Phys Lett 89:163501
77. Breyer C, Vogel M, Mohr M, Johnev B, Fostiropoulos K (2006) Influence of exciton distribution on external quantum efficiency in bilayer organic solar cells. Physica Status Solidi (b) 243:3176–3180
78. Peumans P, Uchida S, Forrest SR (2003) Efficient bulk heterojunction photovoltaic cells using small-molecular weight organic thin films. Nature 425:158–162
79. Dittmer JJ, Lazzaroni R, Leclère P, Moretti P, Granström M, Petritsch K, Marseglia EA, Friend RH, Brédas JL, Rost H, Holmes AB (2000) Crystal network formation in organic solar cells. Sol Energy Mater Sol Cells 61:53–61
80. Gebeyehu D, Maennig B, Drechsel J, Leo K, Pfeiffer M (2003) Bulk-heterojunction photovoltaic devices based on donor–acceptor organic small molecule blends. Sol Energy Mater Sol Cells 79:81–92
81. Pannemann C, Dyakonov V, Parisi J, Hild O, Wöhrle D (2001) Electrical characterization of zinc-phthalocyanine-fullerene photovoltaic cells. Synth Metals 121:1585
82. Geens W, Aernouts T, Poortmans J, Hadziioannou G (2002) Organic co-evaporated films of a PPV-pentamer and C-60: model systems for donor/acceptor polymer blends. Thin Solid Films 403:438–443
83. Günes S, Neugebauer H, Sariciftci NS (2007) Conjugated polymer-based organic solar cells. Chem Rev 107:1324–1338

84. Granström M, Pertritsch K, Arias AC, Lux A, Andersson MR, Friend RH (1998) Laminated fabrication of polymeric photovoltaic diodes. Nature 395:257–260
85. Yu G, Gao J, Hummelen JC, Wudl F, Heeger AJ (1995) Polymer photovoltaic cells: enhanced efficiencies via a network of internal donor–acceptor heterojunctions. Science 270:1789–1791
86. Hiramoto M, Yamaga T, Danno M, Suemori K, Matsumura Y, Yokoyama M (2006) Design of nanostructures for photoelectric conversion using an organic vertical superlattice. Appl Phys Lett 88:213105
87. Liang Y, Xu Z, Xia J, Tsai ST, Wu Y, Li G, Ray C, Yu L (2010) For the bright future—bulk heterojunction polymer solar cells with power conversion efficiency of 7.4%. Adv Mater 22:1–4
88. Feng W, Fujii A, Lee S, Wu H, Yoshino K (2000) Broad spectral sensitization of organic photovoltaic heterojunction device by perylene and C60. J Appl Phys 88:7120–7123
89. Tsuzuki T, Shirota Y, Rostalski J, Meissner D (2000) The effect of fullerene doping on photoelectric conversion using titanyl phthalocyanine and a perylene pigment. Sol Energy Mater Sol Cells 61:1–8
90. Hiramoto M, Suemori K, Yokoyama M (2002) Photovoltaic properties of ultramicrostructure controlled organic co-deposited films. Jpn J Appl Phys 41:2763–2766
91. Halls JJM, Friend RH (1997) The photovoltaic effect in a poly(p-phenylenevinylene)/perylene heterojunction. Synth Metals 85:1307–1308
92. Rand BP, Xue J, Uchida S, Forrest SR (2005) Mixed donor acceptor molecular heterojunctions for photovoltaic applications. I. Material properties. J Appl Phys 98:124902
93. Uchida S, Xue J, Rand BP, Forrest SR (2004) Organic small molecule solar cells with a homogeneously mixed copper phthalocyanine: C60 active layer. Appl Phys Lett 84:4218–4220
94. Heutz S, Sullivan P, Sanderson BM, Schultes SM, Jones TS (2004) Influence of molecular architecture and intermixing on the photovoltaic, morphological and spectroscopic properties of CuPc-C60 heterojunctions. Sol Energy Mater Sol Cells 83:229–245
95. Sullivan P, Heutz S, Schultes SM, Jones TS (2004) Influence of codeposition on the performance of CuPc-C-60 heterojunction photovoltaic devices. Appl Phys Lett 84:1210–1212
96. Xue J, Rand BP, Uchida S, Forrest SR (2005) Mixed donor acceptor molecular heterojunctions for photovoltaic applications. II. Device performance. J Appl Phys 98:124903
97. Opitz A, Bronner M, Bruetting W (2007) Ambipolar charge carrier transport in mixed organic layers of phthalocyanine and fullerene. J Appl Phys 101:063709
98. Mihailetchi VD, Koster LJA, Blom PWM, Melzer C, de Boer B (2005) Compositional dependence of the performance of poly(p-phenylene vinylene): methanofullerene bulk-heterojunction solar cells. Adv Funct Mater 15:795–801
99. Yang F, Shtein M, Forrest SR (2005) Controlled growth of a molecular bulk heterojunction photovoltaic cell. Nat Mater 4:37–41
100. Xue J, Uchida S, Rand BP, Forrest SR (2004) 4.2% efficient organic photovoltaic cells with low series resistances. Appl Phys Lett 84:3013–3015
101. Xue J, Rand BP, Uchida S, Forrest SR (2005) A hybrid planar-mixed molecular heterojunction photovoltaic cell. Adv Mater 17:66–70
102. Shockley W, Queisser HJ (1961) Detailed balance limit of efficiency of p–n junction solar cells. J Appl Phys 32:510
103. Schueppel R, Timmreck R, Allinger N, Mueller T, Furno M, Uhrich C, Leo K, Riede M (2010) Controlled current matching in small molecule organic tandem solar cells using doped spacer layers. J Appl Phys 107:044503
104. Hadipour A, De Boer B, Blom PWM (2008) Organic tandem and multi-junction solar cells. Adv Funct Mater 18:169–181
105. Hadipour A, De Boer B, Blom PMW (2008) Device operation of organic tandem solar cells. Org Electron 9:617–624

106. Ameri T, Dennler G, Lungenschmied C, Brabec CJ (2009) Organic tandem solar cells: a review. Energy Environ Sci 2:347–363
107. Hiramoto M, Suezaki M, Yokoyama M (1990) Effect of thin gold interstitial-layer on the photovoltaic properties of tandem organic solar cell. Chem Lett 19:327–330
108. Forrest SR (2005) The limits to organic photovoltaic cell efficiency. MRS Bull 30:28–32
109. Yakimov A, Forrest SR (2002) High photovoltage multiple-heterojunction organic solar cells incorporating interfacial metallic nanoclusters. Appl Phys Lett 80:1667–1669
110. Rand BP, Peumans P, Forrest SR (2004) Long-range absorption enhancement in organic tandem thin-film solar cells containing silver nanoclusters. J Appl Phys 12:7519–7526
111. Triyana K, Yasuda T, Fujita K, Tsutsui T (2004) Effects of different materials used for internal floating electrode on the photovoltaic properties of tandem type organic solar cell. Jpn J Appl Phys Part I 43:2352–2356
112. Triyana K, Yasuda T, Fujita K, Tsutsui T (2004) Organic tandem and multi junction solar cells. Thin Solid Films 447:198–202
113. Xue J, Uchida S, Rand BP, Forrest SR (2004) Asymmetric tandem organic photovoltaic cells with hybrid planar-mixed molecular heterojunctions. Appl Phys Lett 85:5757–5759
114. Peumans P, Yakimov A, Forrest SR (2003) Small molecular weight organic thin-film photodetectors and solar cells. J Appl Phys 93:3693–3723
115. Drechsel J, Männig B, Kozlowski F, Pfeiffer M, Leo K (2005) Efficient organic solar cells based on a double p–i–n architecture using doped wide-gap transport layers. Appl Phys Lett 86:244102
116. Yu B, Zhu F, Wang H, Li G, Yan D (2008) All-organic tunnel junctions as connecting units in tandem organic solar cell. J Appl Phys 104:114503
117. Chandrasekhar S, Sadashiva BK, Suresh KA (1977) Liquid-crystals of disc-like molecules. Pramana 9:471–480
118. Kumar S (2006) Self-organization of disc-like molecules: chemical aspects. Chem Soc Rev 35:83–109
119. Adam D, Schuhmacher P, Simmerer J, Haussling L, Siemensmeyer K, Etzbacj KH, Ringsdorf H, Haarer D (1994) Fast photoconduction in the highly ordered columnar phase of a discotic liquid-crystal. Nature 371:141–143
120. van de Craats AM, Warman JM, Fechtenkotter A, Brand JD, Harbison MA, Müllen K (1999) Record charge carrier mobility in a room-temperature discotic liquid-crystalline derivative of hexabenzocoronene. Adv Mater 11:1469–1472
121. Sergeyev S, Pisula W, Geerts Y (2007) Discotic liquid crystals: a new generation of organic semiconductors. Chem Soc Rev 36:1902–1929
122. Woon KL, Aldred MP, Richards GJ, Vlachos P, Mehl GH, Kelly SM, O'Neill M (2006) Electronic charge transport in extended nematic liquid crystals. Chem Mater 18:2311–2317
123. Kumar S (2002) Discotic liquid crystals for solar cells. Curr Sci India 82:256–257
124. Grelet E, Bock H (2006) Control of the orientation of thin open supported columnar liquid crystal films by the kinetics of growth. Europhys Lett 73:712–718
125. Archambeau S, Seguy I, Jolinat P, Farenc J, Destruel P, Nguyen TP, Bock H, Grelet E (2006) Stabilization of discotic liquid organic thin films by ITO surface treatment. Appl Surf Sci 253:2078–2086
126. Charlet E, Grelet E, Brettes P, Bock H, Saadaoui H, Cisse l, Destruel P, Gherardi N, Seguy I (2008) Ultrathin films of homeotropically aligned columnar liquid crystals on indium tin oxide electrodes. Appl Phys Lett 92:024107-1–024107-3
127. Schmidt-Mende L, Fechtenkotter A, Mullen K, Moons E, Friend RH, MacKenzie JD (2001) Self-organized discotic liquid crystals for high-efficiency organic photovoltaics. Science 293:1119–1122
128. Lemaur V, Da Silva Filho DA, Coropceanu V, Lehmann M, Greerts Y, Piris J, Debije MG, van de Craats AM, Senthilkumar K, Siebbeles LDA, Warman JM, Bredas JL, Cornil J (2004) Charge transport properties in discotic liquid crystals: a quantum-chemical insight into structure–property relationships. J Am Chem Soc 126:3271–3279

129. van de Craats AM, Warman JM (2001) The core-size effect on the mobility of charge in discotic liquid crystalline materials. Adv Mater 13:130–133
130. Tracz A, Jeszka JK, Watson MD, Pisula W, Müllen K, Pakula T (2003) Uniaxial alignment of the columnar super-structure of a hexa (alkyl) hexa-peri-hexabenzocoronene on untreated glass by simple solution processing. J Am Chem Soc 125:1682–1683
131. Pisula W, Tomovi Z, El Hamaoui B, Watson MD, Pakula T, Müllen K (2005) Control of the homeotropic order of discotic hexa-peri-hexabenzocoronenes. Adv Funct Mater 15:893–904
132. Li J, Kastler M, Pisula W, Willem J, Robertson F, Wasserfallen D, Grimsdale AC, Wu J, Müllen K (2007) Organic bulk-heterojunction photovoltaics based on alkyl substituted discotics. Adv Funct Mater 17:2528–2533
133. Wong WWH, Ma C-Q, Pisula W, Yan C, Feng X, Jones DJ, Müllen K, Janssen RA, Bäuerle P, Holmes AB (2010) Self-assembling thiophene dendrimers with a hexa-peri-hexabenzocoronene core-synthesis, characterization and performance in bulk heterojunction solar cells. Chem Mater 22:457–466
134. Schmidt-Mende L, Fechtenkotter A, Mullen K, Friend RH, MacKenzie JD (2002) Efficient organic photovoltaics from soluble discotic liquid crystalline materials. Physica E 14:263–267
135. Liu S-G, Sui G, Cormier RA et al (2002) Self-organizing liquid crystal perylene diimide thin films: spectroscopy, crystallinity, and molecular orientation. J Phys Chem B 106:1307–1315
136. Schmidt-Mende L, Watson M, Müllen K, Friend RH (2003) Organic thin film photovoltaic devices from discotic materials. Mol Cryst Liq Cryst 396:73–90
137. Schmidtke JP, Friend RH, Kastler M, Müllen K (2006) Cobtrol of morphology in efficient photovoltaic diodes from discotic liquid crystals. J Chem Phys 124:175704-1–175704-6
138. Gregg BA, Fox MA, Bard AJ (1990) Photovoltaic effect in symmetrical cells of a liquid-crystal porphyrin. J Phys Chem 94:1586–1598
139. Petristch K, Friend RH, Lux A, Rozenberg G, Moratti SC, Holmes AB (1999) Liquid crystalline phthalocyanines in organic solar cells. Synth Metals 102:1776–1777
140. Levitsky IA, Euler WB, Tokranova N, Xu B, Castracane J (2004) Hybrid solar cells based on porous Si and copper phthalocyanine derivatives. Appl Phys Lett 85:6245–6247
141. Boden N, Bushby RJ, Clements J (1993) Mechanism of quasi-one-dimensional electronic conductivity in disotic liquid-crystals. J Chem Phys 98:5920–5931
142. Boden N, Bushby RJ, Cammidge AN, Clements J, Luo R, Donovan KJ (1995) Transient photoconductivity and dark conductivity in discotic liquid crystals. Mol Cryst Liq Cryst 261:251–257
143. Kumar S, Pal SK, Kumar PS, Lakshminarayanan V (2007) Novel conducting nanocomposites: synthesis of triphenylene-covered gold nanoparticles and their insertion into a columnar matrix. Soft Matter 3:896–900
144. Kumar PS, Kumar S, Lakshminarayanan V (2008) Electrical conductivity studies on discotic liquid crystal-ferrocenium donor–acceptor systems. J Phys Chem B 112:4865–4869
145. Kumar PS, Kumar S, Lakshminarayanan V (2009) Hybrid organic/inorganic nanocomposite as a quasi-one-dimentsional semiconductor under ambient conditions. J Appl Phys 106:093701-1–093701-6
146. Schmidt J, Dierking I (2001) Localization and imaging of local shunts in solar cells using polymer-dispersed liquid crystals. Prog Photovoltaics 9:263–271
147. Koeppe R, Bossart O, Calzaferre G, Sariciftci NS (2007) Advanced photon-harvesting concepts for low-energy gap organic solar cells. Sol Energy Mater Sol Cells 91:986–995
148. Liu YX, Summers MA, Edder C, Frechet JMJ, McGehee MD (2005) Using resonance energy transfer to improve exciton harvesting in organic–inorganic hybrid photovoltaic cells. Adv Mater 17:2960–2964
149. Hong ZR, Lessmann R, Maennig B, Huang Q, Harada K, Riede M, Leo K (2009) Antenna effects and improved efficiency in multiple heterojunction photovoltaic cells based on pentacene, zinc phthalocyanine, and C60. J Appl Phys 106:064511-1–064511-6
150. Imahori H, Fukuzumi S (2004) Porphyrin- and fullerene-based molecular photovoltaic devices. Adv Funct Mater 14:525–536

151. Zhang C, Tong SW, Jiang C, Kang ET, Chan DSH, Zhu CX (2008) Efficient multilayer organic solar cells using the optical interference peak. Appl Phys Lett 93:043307-1–043307-3
152. Dastoor PC, McNeill CR, Frohne H, Foster CJ, Dean B, Fell CJ, Belcher WJ, Campbell WM, Officer DL, Blake IM, Thordarson P, Crossleu MJ, Hush NS, Reimers JR (2007) Understanding and improving solid-state polymer/C60-fullerene bulk-heterojunction solar cells using ternary porphyrin blends. J Phys Chem C 111:15415–15426
153. Burke KB, Belcher WJ, Thomsen L, Watts B, McNeill CR, Ade H, Dastoor PC (2009) Role of solvent trapping effects in determining the structure and morphology of ternary blend organic devices. Macromolecules 42:3098–3103
154. Belcher WJ, Wagner KI, Dastoor PC (2007) The effect of porphyrin inclusion on the spectral response of ternary P3HT:porphyrin:PCBM bulk heterojunction solar cells. Sol Energy Mater Sol Cells 91:447–452
155. Shao Y, Yang Y (2005) Efficient organic heterojunction photovoltaic cells based on triplet materials. Adv Mater 17:2841–2844
156. Schulz GL, Holdcroft S (2008) Conjugated polymers bearing iridium complexes for triplet photovoltaic devices. Chem Mater 20:5351–5355
157. Yang C-M, Wu C-H, Liao H-H, Lai KY, Cheng HP, Horng SF, Meng HF, Shy JT (2007) Enhanced photovoltaic response of solar cell by singlet-tot-triplet exciton conversion. Appl Phys Lett 90:133509-1–133509-3
158. Li Y, Mastria R, Li K, Fiore A, Wang Y, Cingolani R, Manna L, Gigli G (2009) Improved photovoltaic performance of bilayer heterojunction photovoltaic cells by triplet materials and tetrapod-shaped colloidal nanocrystals doping. Appl Phys Lett 95:043101-1–043101-3
159. Guo FQ, Kim YG, Reynolds JR, Schanze KS (2006) Platinum-acetylide polymer based solar cells: involvement of the triplet state for energy conversion. Chem Commun 1887–1889
160. Hatton RA, Blanchard NP, Miller AJ, Silva SRP (2007) A multi-wall carbon nanotube-molecular semiconductor composite for bi-layer organic solar cells. Physica E 37:124–127
161. Hatton RA, Blanchard NP, Stolojan V, Miller AJ, Silva SRP (2007) Nanostructured copper phthalocyanine-sensitized multiwall carbon nanotube films. Langmuir 23:6424–6430
162. O'Regan B, Grätzel M (1991) A low-cost, high-efficiency solar cell based on dye-sensitized colloidal TiO2 films. Nature 353:737–740
163. Grätzel M (2001) Measured under standard air mass 1.5 reporting conditions, PV calibration. Laboratory of the National Energy Research Laboratory (NREL), Golden
164. Hinsch A, Kroon JM, Späth M, van Roosmale JAM, Bakker NJ, Sommeling P, van der Burg N, Kinderman R, Kern R, Ferber J, Schill C, Schubert M, Meyer A, Meyer T, Uhlendorf I, Holzbock J, Niepmann R (2000) In: Proceedings of the 16th European photovoltaic solar energy conference and exhibition, Glasgow
165. Nogueira AF, Durrant JR, De Paoli M-A (2001) Dye-sensitized nanocrystalline solar cells employing a polymer electrolyte. Adv Mater 13:826–830
166. De Freitas JN, Nogueira AF, De Paoli M-A (2009) New insights into dye-sensitized solar cells with polymer electrolytes. J Mater Chem 19:5279–5294
167. Freitas FS, De Freitas JN, Ito BI, De Paoli M-A, Nogueira AF (2009) Electrochemical and structural characterization of polymer gel electrolytes based on a PEO copolymer and an imidazolium-based ionic liquid for dye-sensitized solar cells. Appl Mater Interfaces 1:2870–2877
168. Bach U, Lupo D, Comte P, Moser JE, Weissortel F, Salbeck J, Spreitzer H, Grätzel M (1998) Solid-state dye-sensitized mesoporous TiO2 solar cells with high photon-to-electron conversion efficiencies. Nature 395:583–585
169. Kroeze JE, Hirata N, Schmidt-Mende L, Orizu C, Ogier SD, Carr K, Grätzel M, Durrant JR (2006) Parameters influencing charge separation in solid-state dye-sensitized solar cells using novel hole conductors. Adv Func Mater 16:1832–1838
170. Poplavskyy D, Nelson J (2003) Nondispersive hole transport in amorphous films of methoxy-spirofluorene-arylamine organic compound. J Appl Phys 93:341–346

171. Zhao Y, Chen W, Zhai J, Sheng X, He Q, Wei T, Bai F, Jiang L, Zhu D (2007) Solid-state dye-sensitized photovoltaic device with newly designed small organic molecule as hole-conductor. Chem Phys Lett 445:259–264
172. Hagena J, Schaffrath W, Otschik P, Fink R, Bacher A, Schmidt H-W, Haarer D (1997) Novel hybrid solar cells consisting of inorganic nanoparticles and an organic hole transport material. Synth Metals 89:215–220
173. Karthikeyan CS, Thelakkat M (2008) Key aspects of individual layers in solid-state dye-sensitized solar cells and novel concepts to improve their performance. Inorg Chim Acta 361:635–655
174. Tian H, Meng F (2005) Solar cells based on cyanine and polymethine dyes. In: Sun S-S, Sariciftci NS (eds) Organic photovoltaics, 1st edn. Taylor & Francis, Boca Raton
175. Horiuchi T, Miura H, Sumioka K, Uchida S (2004) High efficiency of dye-sensitized solar cells based on metal-free indoline dyes. J Am Chem Soc 126:12218–12219
176. Ito S, Miura H, Uchida S, Takata M, Sumioka K, Liska P, Comte P, Péchy P, Grätzel M (2008) High-conversion-efficiency organic dye-sensitized solar cells with a novel indoline dye. Chem Commun 5194–5196
177. O'Regan B, Lenzmann F (2004) Charge transport and recombination in a nanoscale interpenetrating network of n-type and p-type semiconductors: transient photocurrent and photovoltage studies of TiO2/Dye/CuSCN photovoltaic cells. J Phys Chem B 108:4342–4350
178. Snaith HJ, Moule AJ, Klein C, Meerholz K, Friend RH, Grätzel M (2007) Efficiency enhancements in solid-state hybrid solar cells via reduced charge recombination and increased light capture. Nano Lett 7:3372–3376
179. Schmidt-Mende L, Grätzel M (2006) TiO2 pore-filling and its effect on the efficiency of solid-state dye-sensitized solar cells. Thin Solid Films 500:296–301
180. Boschloo GK, Goossens A (1996) Electron trapping in porphyrin-sensitized porous nanocrystalline TiO2 electrodes. J Phys Chem 100:19489–19494
181. Wienke J, Schaafsma TJ, Goossens A (1999) Visible light sensitization of titanium dioxide with self-organized porphyrins: organic P–I–N solar cells. J Phys Chem B 103:2702–2708
182. Savenije TJ, Goossens A (2001) Hole transport in porphyrin thin films. Phys Rev B 64:115323
183. Miyairi K, Itoh E, Hashimoto Y (2003) Photovoltaic properties of double layer devices consisting of titanium dioxide and porphyrin dispersed hole transporting material layer. Thin Solid Films 438–439:147–152
184. Simon J, Andre JJ (1998) Molecular semiconductors, photoelectrical properties and solar cells. Springer, Berlin
185. Gregg BA (1996) Bilayer molecular solar cells on spin-coated TiO2 substrates. Chem Phys Lett 258:376–380
186. Signerski R, Jarosz G, Koscielska B (2009) On photovoltaic effect in hybrid heterojunction formed from palladium phthalocyanine and titanium dioxide layers. J Non-Crystalline Solids 355:1405–1407
187. Sharma GD, Kumar R, Roy MS (2006) Investigation of charge transport, photo generated electron transfer and photovoltaic response of iron phthalocyanine (FePc):TiO2 thin films. Sol Energy Mater Sol Cells 90:32–45
188. Tracey SM, Hodgson SNB, Ray AK (1998) Sol–gel derived TiO2/lead phthalocyanine photovoltaic cells. J. Sol–Gel Sci. Techn 13:219–222
189. Kajihara K, Tanaka K, Hirao K, Soga N (1996) Photovoltaic effect in titanium dioxide/zinc phthalocyanine cell. Jpn J Appl Phys 35:6110–6116
190. Ohmori Y, Itoh E, Miyairi K (2006) Photovoltaic properties of phthalocyanine based p–n diode evaporated onto titanium dioxide. Thin Solid Films 499:369–373
191. Tachibana Y, Haque SA, Mercer IP, Durrant JR, Klug DR (2000) Electron injection and recombination in dye sensitized nanocrystalline titanium dioxide films: a comparison of ruthenium bipyridyl and porphyrin sensitizer dyes. J Phys Chem B 104:1198–1205

192. Imahori H, Umeyama T, Ito S (2009) Large-aromatic molecules as potential sensitizers for highly efficient dye-sensitized solar cells. Acc Chem Res 42:1809–1818
193. Campbell WM, Burrell AK, Officer DL, Jolley KW (2004) Porphyrins as light harvesters in the dye-sensitised TiO2 solar cell. Coord Chem Rev 248:1363–1379
194. Lee CY, Hupp JT (2010) Dye sensitized solar cells: TiO2 sensitization with a bodipy-porphyrin antenna system. Langmuir 26:3760–3765
195. Nazeeruddin MK, Humphry-Baker R, Grätzel M, Murrer BA (1998) Efficient near IR sensitization of nanocrystalline TiO2 films by ruthenium phthalocyanine. Chem Commun 719–720
196. Yum J-H, Jang S-R, Humphry-Baker R, Grätzel M, Cid J-J, Torres T, Nazeeruddin MK (2008) Effect of coadsorbent on the photovoltaic performance of zinc pthalocyanine-sensitized solar cells. Langmuir 24:5636–5640
197. Shen L, Zhu G, Guo W, Tao C, Zhang X, Liu C, Chen W, Ruan S, Zhong Z (2008) Performance improvement of TiO2/P3HT solar cells using CuPc as a sensitizer. Appl Phys Lett 92:073307
198. Morandeira A, López-Duarte I, O'Regan B, Martínez-Díaz MV, Forneli A, Palomares E, Torres T, Durrant JR (2009) Ru(II)-phthalocyanine sensitized solar cells: the influence of co-adsorbents upon interfacial electron transfer kinetics. J Mater Chem 19:5016–5026
199. Schouten PG, Warman JM, De haas MP, Fox MA, Pan HL (1991) Charge migration in supramolecular stacks of peripherally substituted porphyrins. Nature 353:736–737
200. Lawrence MF, Huang Z, Langford CH, Ordonez I (1993) Photocurrent generation and charge transport in tin dioxide/ion-exchange polymer-zinc meso-tetraphenylporphyrin/gold cells. J Phys Chem 97:944–951
201. Taleb T, Nasr C, Hotchandani S, Leblanc RM (1996) Effect of temperature on capacitance of Al/microcrystalline chlorophylla/Ag sandwich cells. J Appl Phys 79:1701
202. Ioannidis A, Lawrence MF, Kassi H, Cote R, Dodelet JP, Leblanc RM (1993) Field dependence of hole mobilities in chloro-aluminum phthalocyanine. Chem Phys Lett 205:46–50
203. Senadeera GKR, Jayaweera PVV, Perera VPS, Tennakone K (2002) Solid-state dye-sensitized photocell based on pentacene as a hole collector. Sol Energy Mater Sol Cells 73:103–108
204. Zhang G, Bala H, Cheng Y, Shi D, Lv X, Yu Q, Wang P (2009) High efficiency and stable dye-sensitized solar cells with an organic chromophore featuring a binary π-conjugated spacer. Chem Commun 2198–2200
205. Hara K, Tachibana Y, Ohga Y, Shinpo A, Suga S, Sayama K, Sugihara H, Arakawa H (2003) Dye-sensitized nanocrystalline TiO2 solar cells based on novel coumarin dyes. Sol Energy Mater Sol Cells 77:89–103
206. Zafer C, Kus M, Turkmen G, Dincalp H, Demic S, Kuban B, Teoman Y, Icli S (2007) New perylene derivative dyes for dye-sensitized solar cells. Sol Energy Mater Sol Cells 91:427–431
207. Shibano Y, Umeyama T, Matano Y, Imahori H (2007) Electron-donating perylene tetracarboxylic acids for dye-sensitized solar cells. Org Lett 9:1971–1974
208. Fortage J, Séverac M, Houarner-Rassin C, Pellegrin Y, Blart E, Odobel F (2008) Synthesis of new perylene imide dyes and their photovoltaic performances in nanocrystalline TiO2 dye-sensitized solar cells. J Photochem Photobiol A Chem 197:156–169
209. Jin Y, Hua J, Wu W, Ma X, Meng F (2008) Synthesis, characterization and photovoltaic properties of two novel near-infrared absorbing perylene dyes containing benzo[e]indole for dye-sensitized solar cells. Synth Metals 158:64–71
210. Zhang G, Bai Y, Li R, Shi D, Wenger S, Zakeeruddin SM, Grätzel M, Wang P (2009) Employ a bisthienothiophene linker to construct an organic chromophore for efficient and stable dye-sensitized solar cells. Energy Environ Sci 2:92–95
211. Xu M, Li R, Pootrakulchote N, Shi S, Guo J, Yi Z, Zakeeruddin SM, Grätzel M, Wang P (2008) Energy-level and molecular engineering of organic D-π-A sensitizers in dye-sensitized solar cells. J Phys Chem C 112:19770–19776

212. Schmidt-Mende L, Bach U, Humphry-Baker R, Horiuchi T, Miura H, Ito S, Uchida S, Grätzel M (2005) Organic dye for highly efficient solid-state dye-sensitized solar cells. Adv Mater 17:813–815
213. Kuang D, Uchida S, Humphry-Baker R, Zakeeruddin SM, Grätzel M (2008) Organic dye-sensitized ionic liquid based solar cells: remarkable enhancement in performance through molecular design of indoline sensitizers. Angew Chem Int Ed 47:1923–1927
214. Grätzel M (2009) Recent advances in sensitized mesoscopic solar cells. Acc Chem Res 42(11):1788–1798
215. Yum J-H, Hagberg DP, Moon S-J, Karlsson KM, Marinado T, Sun L, Hagfeldt A, Nazeeruddin MK, Grätzel M (2009) A light-resistant organic sensitizer for solar-cell applications. Angew Chem Int Ed 48:1576–1580
216. Choi H, Baik C, Kang SO, Ko J-J, Kang M-S, Nazeeruddin MK, Grätzel M (2008) Highly efficient and thermally stable organic sensitizers for solvent-free dye-sensitized solar cells. Angew Chem Int Ed 47:327–330
217. Haque SA, Park T, Holmes AB, Durrant JR (2003) Transient optical studies of interfacial energetic disorder at nanostructured dye-sensitised inorganic/organic semiconductor heterojunctions. ChemPhysChem 4:89–93
218. O'Regan B, Schwartz DT (1998) Large enhancement in photocurrent efficiency caused by UV illumination of the dye-sensitized heterojunction TiO2/RuLL'NCS/CuSCN: initiation and potential mechanisms. Chem Mater 10:1501–1509
219. Tennakone K, Kumara G, Kumarasinghe AR, Wijayantha KGU, Sirimanne PM (1995) A dye-sensitized nano-porous solid-state photovoltaic cell. Semicond Sci Technol 10:1689–1693
220. Tennakone K, Senadeera GKR, De Silva D, Kottegoda IRM (2000) Highly stable dye-sensitized solid-state solar cell with the semiconductor 4CuBr 3S(C4H9)2 as the hole collector. Appl Phys Lett 77:2367

Organic Solar Cells with Inverted and Tandem Structures

De Wei Zhao, Aung Ko Ko Kyaw and Xiao Wei Sun

Abstract During the past decade, organic solar cells have attracted great attention due to their wide applicability and potentially low-cost fabrication from printing at low temperature on flexible substrates. Although the technologies of small molecule and polymer solar cells have advanced significantly, the efficiency and stability still need to be improved to fulfill the commercial requirements. In principle, the primary way to improve device performance is to introduce new materials with the properties of broad absorption range, high charge-carrier mobility, and long-term stability. On the other hand, the device performance can be also enhanced by optimizing device structures. In this chapter, we will discuss the recent progress in organic solar cells with inverted and tandem structures, two effective approaches to improve device performance. We will review various interfacial and intermediate layers employed in solar cells based on these concepts. The stability of the devices with these structures in ambient environment will also be discussed.

D. W. Zhao and A. K. K. Kyaw contribute equally to this chapter.

X. W. Sun (✉)
Department of Applied Physics, College of Science, Tianjin University,
300072 Tianjin, China
e-mail: exwsun@ntu.edu.sg

X. W. Sun
Tianjin Key Laboratory of Low-Dimensional Functional Material Physics and
Fabrication Technology, Tianjin University, 300072 Tianjin, China

D. W. Zhao · A. K. K. Kyaw
School of Electrical and Electronic Engineering, Nanyang Technological University,
Singapore 639798, Singapore

L. Zang (ed.), *Energy Efficiency and Renewable Energy Through Nanotechnology*,
Green Energy and Technology, DOI: 10.1007/978-0-85729-638-2_3,

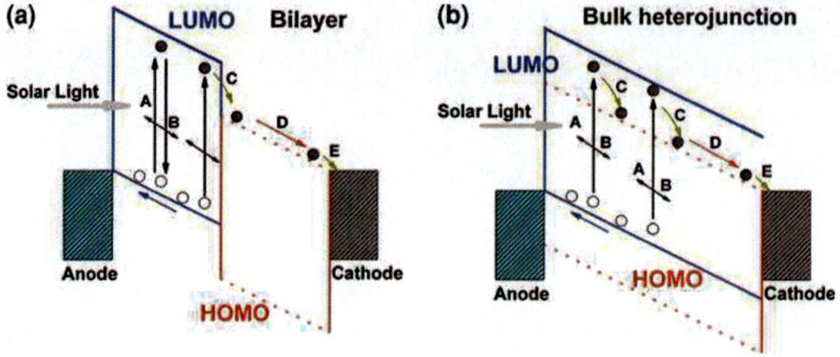

Fig. 1 The basic working principle of OSCs. The energy level diagrams of **a** bilayer; **b** bulk heterojunction. The *solid blue line* is presenting the energy levels of donor material, and the *dash red line* is presenting the energy levels of acceptor material. In the diagrams, *A*, *B*, *C*, *D*, and *E* corresponds to light absorption, exciton formation and diffusion, exciton dissociation into free charge carriers, charge transport in their corresponding layers or pathways, and charge collection by the electrodes, respectively

1 Introduction to Organic Solar Cells

1.1 Introduction

Organic solar cells (OSCs) are attracting much attention due to their potential as a low-cost and flexible energy conversion device. Light conversion to electricity is carried out through the following processes: (A) light absorption, (B) exciton formation and diffusion, (C) exciton dissociation into free charge carriers, (D) charge transport, and (E) charge collection by the electrodes [1, 2] (Fig. 1). Therefore, efficiency improvement can be achieved by enhancing the efficiencies of these processes in both bilayer and bulk heterojunction (BHJ) based OSCs.

In order to increase light absorption, a few classes of organic materials with broad absorption ranges have been developed [3–5], increasing the number of excitons formed in the photoactive layer. Improvement can also be achieved simply by increasing the thickness of photoactive layer; however, the thickness is ultimately limited by the short exciton diffusion length and low charge-carrier mobility. For the improvement of the diffusion/dissociation efficiency, BHJ devices, which blend donors and acceptors, are generally used. In such devices an ultra-fast charge transfer (~45 fs) [6] between the LUMOs of donors and acceptors occurs at the donor/acceptor interface. Moreover, the formation of the nanoscale morphology facilitates the charge transport in interpenetrating networks [7–9], correlated with the phase separation between the two components. The phase separation mainly depends on the crystallization of polymer or fullerene [9], determined to a great extent by the solvents used [10–13], the weight ratios of polymer to fullerene [7, 14], the concentration of the blend solution [8], and postannealing treatment [1, 15, 16]. To improve charge collection, buffer layers

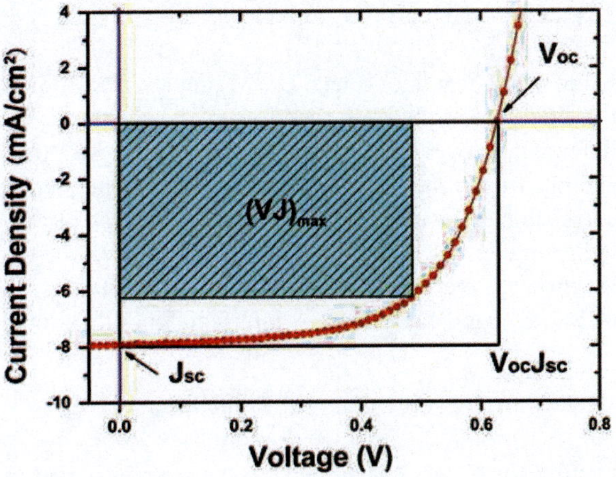

Fig. 2 The typical *I–V* characteristic of an OSC under 100 mW/cm^2 illumination

inserted between active layer and electrodes help align the energy levels. On the anode side, PEDOT:PSS is the most commonly used hole-transporting layer. P-type-like transition metal oxides have also been introduced to improve the ITO anode/active layer contact, such as NiO [17], MoO_3 [18–20], V_2O_5 [18], and WO_3 [21] due to their desirable work functions and good hole-transporting behaviors. On the cathode side, modification by Ca [22], LiF [23], TiO_x [4, 24, 25], ZnO [26, 27], and Cs_2CO_3 [28, 29] have been reported to improve the electron transport and extraction. Therefore, not only the light absorption but also the charge transport plays a crucial role in the efficiency improvement.

Compared to the conventional device structure, inverted and tandem structures have been shown to improve device performance for the following reasons: improved charge transport, better interface stability, and enhanced light harvesting by means of stacking multiple cells with complementary absorption spectra [19, 20, 30]. In this chapter, we will review OSCs based on these two structures.

1.2 Main Parameters

The *I–V* characteristic of a typical OSC is shown in Fig. 2. Generally, the power conversion efficiency (PCE) of a solar cell is calculated by:

$$\mathrm{PCE} = \frac{V_{\mathrm{oc}} J_{\mathrm{sc}} \mathrm{FF}}{P_{\mathrm{in}}}, \quad \mathrm{FF} = \frac{(VJ)_{\mathrm{max}}}{V_{\mathrm{oc}} J_{\mathrm{sc}}}$$

where V_{oc} is the open-circuit voltage, J_{sc} is the short-circuit current density, FF is the fill factor, $(VJ)_{max}$ is the maximum product of V and J, and P_{in} is the incident light power. Generally, the standard incident light is defined by measurement temperature (25°C), spectral irradiance (AM 1.5G), and total irradiance (100 mW/cm^2) with a spectral intensity distribution matching that of the Sun on the Earth's surface at an angle of 48.2°.

1.2.1 Open-Circuit Voltage (V_{oc})

In the metal–insulator–metal (MIM) device, the V_{oc} is considered to be determined by the difference of the work functions of the electrodes used [31]. On the other hand, the difference between HOMO of donor and LUMO (i.e., the bandgap) of acceptor is another important factor on which V_{oc} is dependent [32], showing a nearly linear correlation. However, charge-carrier losses at electrodes can lower the V_{oc} [33], indicating V_{oc} is sensitive to the energy levels of organic materials, and the wok functions of electrodes, as well as their interfaces. Hence, a good contact between active layer and electrodes, by interfacial modification, is critical for a reasonable V_{oc}.

1.2.2 Short-Circuit Current Density (J_{sc})

Ideally, the J_{sc} is determined by the photoinduced charge-carrier density and the charge-carrier mobility within organic materials:

$$J_{sc} = ne\mu E$$

where n is the density of charge carriers, e is the elementary charge, μ is the charge-carrier mobility, and E is the electric field. Assuming the 100% efficiency for the photoinduced charge generation in a BHJ, n is the number of absorbed photons per unit volume.

Understanding the optical absorption profile in the active layer is necessary to obtain a high J_{sc}. Currently, optical modeling of light electric field distribution has been widely used in OSCs [34–36], considering the interference effect in thin film devices. This model provides a convenient computation which facilitates the estimation of exciton diffusion ranges and optimization of film thickness. A matched spectrum of organic materials with solar spectrum is needed to absorb a large fraction of solar light. Therefore, low band gap organic materials covering a broad spectral range are required [3, 5, 37–39]. Moreover, J_{sc} is also dependent on the temperature, demonstrating thermally activated hopping transport in the polymer/fullerene BHJ devices. Generally, organic semiconductors exhibit low charge-carrier mobility, within the range from 10^{-5} to 1 cm^2/V s, limiting the thickness and the charge transport in the active layer. The BHJ structure can raise the interfacial areas between donors and acceptors, as well as enhance exciton dissociation. Only when the BHJ forms fine nanoscale morphology, will this structure be beneficial to the charge transport and contribute to photocurrent generation. Thus, the achievements in low band gap materials, high charge-carrier mobility, and controllable nanoscale morphology, will enhance the J_{sc}.

1.2.3 Fill Factor (FF)

The FF reflects the practical fraction of charge carriers reaching the electrodes when the built-in potential is lower than the V_{oc}. There is a drastic competition between charge recombination and charge transport [36]. As a result, the product of the lifetime τ and mobility μ determines the distance that charge carriers can drift under an electric field E,

$$d = \mu \times \tau \times E$$

Hence, $\tau \times \mu$ should be maximized to reduce charge recombination [36]. Furthermore, the series and parallel resistances in a device largely influence the FF. Generally speaking, the series resistance should be minimized, which can be overcome by using highly conductive semitransparent substrates and introducing buffer layers for interfacial modification; on the other hand, it is necessary to free the devices from "short-circuits" in order to maximize the parallel resistance, which can be implemented by electrode modification.

1.2.4 Incident Photon-to-Current Conversion Efficiency

Another parameter related to the photocurrent is IPCE, a spectral response representing the ratio of the number of electrons generated under the short-circuit condition to the number of incident photons. It characterizes the capability of the device converting the photons to electrons, where J_{sc} can be yielded from the integral of the product of IPCE and AM 1.5G solar spectrum. It is defined by,

$$\text{IPCE}(\lambda) = \frac{1,240 J_{sc}(\lambda)}{\lambda P_{in}(\lambda)}$$

where $P_{in}(\lambda)$ is the incident light power at a particular wavelength λ, $J_{sc}(\lambda)$ is the short-circuit current density at that wavelength, and λ is the wavelength of light in nm.

2 Inverted Organic Solar Cells

2.1 Importance of Inverted Structure

The BHJ concept is typically implemented in the common device, where the BHJ active layer is sandwiched between a PEDOT:PSS-coated ITO anode and a low work function (LWF) metal cathode. However, the conventional structure has drawbacks in the stability and lifetime of device. Neugebauer et al. [40] demonstrated the rapid degradation of the device in air just after 8–12 h using MDMO-PPV/PCBM and C_{60} systems. Schuller et al. [41] developed devices with increased air-stability by encapsulation with a second glass. They showed that the performance of the OSC with a structure of ITO/PEDOT:PSS/MDMO-PPV:PCBM/Ca/Ag degrades less than 20% after more than 1,000 h of operation under 1/3 sun at 85°C; however, at least 25,000 h of operation is needed for a viable product [42]. One limiting parameter associated with device stability is the LWF metal cathode. LWF metals, such as Li, Ca, and Al, are air-sensitive and can be easily oxidized, increasing the series resistance at the organic layer/electrode interface and degrading the device performance. Moreover, diffusion of oxygen into the active layer through pinholes and grain boundaries in the cathode causes the degradation of the active layer, leading to device instability in air [43]. Also, the interface between ITO and PEDOT:PSS is not stable, owing to indium contamination into the polymer layer and the acidic nature of PEDOT:PSS which etches ITO [44, 45]. The degradation of the ITO/PEDOT:PSS interface is faster upon exposure to air because water is absorbed by PEDOT:PSS film and an aqueous acid environment is formed due to the reaction between water and PSS [44].

To circumvent the problem associated with LWF metal electrode, a TiO_x layer was inserted between organic active layer and Al electrode, serving as a shielding and scavenging layer, preventing the penetration of oxygen and moisture into the active layer [46]. The lifetime of non-encapsulated devices exposed to air was improved by nearly two orders of magnitude. Such a buffer layer also acts as an optical spacer and helps increase the photo-generated current [25, 26]. Despite a better structure for device stability, it still cannot avoid the ITO/PEDOT:PSS interfacial degradation. On the other hand, to overcome the ITO/PEDOT:PSS interface problem, Shrotriya et al. [18] replaced PEDOT:PSS with transition metal oxides (MoO_3 and V_2O_5) as a hole-selective layer, demonstrating the PCEs of 3.1 and 3.33% for the devices with 3 nm V_2O_5 and 5 nm MoO_3 film thicknesses, respectively, comparable to that of the device (3.18%) with 25 nm PEDOT:PSS. Unfortunately, this approach could not eliminate the usage of LWF metal (Ca). An intelligent way to address both top electrode and ITO/PEDOT:PSS interface issues is to construct an inverted structure, where the charge collection of electrode is opposite to that of conventional device structure. The reversed polarity of charge collection allows the use of a high work function (HWF) metal as top electrode as well as the removal of ITO/PEDOT:PSS interface. HWF metals such as Au and Ag are less sensitive to air, yielding prolonged interface stability between the active layer and metal electrode. Moreover, HWF metals offer the possibility of using non-vacuum-processed techniques such as lamination [47, 48], printing [49, 50], and coating [51, 52], which decreases production costs and brings closer the ultimate aim of roll-to-roll manufacturing. In addition, it has been reported that charge collection is also enhanced in the inverted geometry owing to inherent vertical phase separation with a donor-enriched top surface and an acceptor-enriched bottom surface in the blend polymer. Such a vertical concentration distribution of donor and acceptor disfavors charge transport in conventional geometry [49]. Another advantage of inverted structure is that if an n-type metal oxide with large-surface area is employed with a BHJ, the interface between the metal oxide and light absorbing polymer can provide additional exciton dissociation sites and generate additional photocurrent [50]. Hence, inverted geometry contains the advantages of device stability, design flexibility, increased robustness, and higher photocurrents. So far, the inverted structure has been employed in both polymer and small molecule OSCs [53, 54]. At the time the inverted structure was introduced, the PCE of devices with P3HT:PCBM blend were as low as 1.4% [51]. Through optimization of the electron selective buffer layer via increasing its conductivity and aligning the energy levels at the polymer/electrode interfaces, the

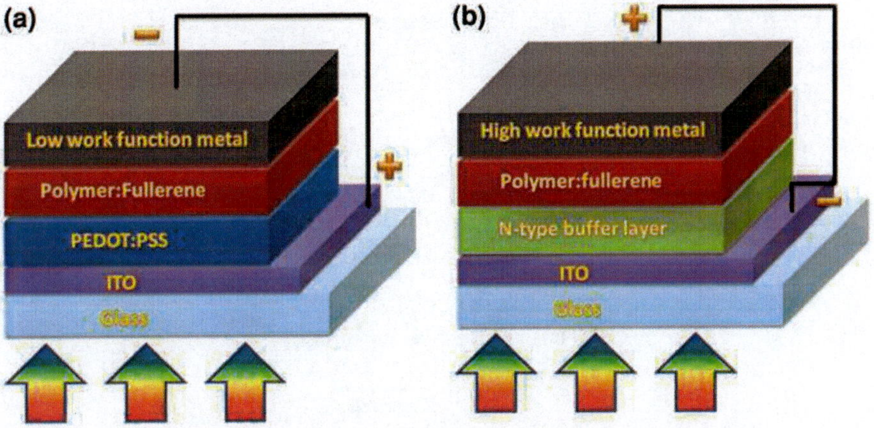

Fig. 3 Typical device architecture for **a** conventional and **b** inverted cell

PCE of inverted device with P3HT:PCBM active layer recently achieved 4.2%, comparable to that of regular device [52].

2.2 Charge Collection in Conventional and Inverted Devices

The energy conversion process of incident photons into electricity in an inverted structure device is the same as that in a conventional OSC, as discussed in Sect. 1. The only difference between two structures is the process of charge collection. Typically, a conventional structure device consists of ITO/PEDOT:PSS/polymer:fullerene blend/LWF metal from bottom to top (Fig. 3a), whereas an inverted structure device is composed of ITO/n-type material/polymer:fullerene blend/HWF metal (Fig. 3b). In principle, ITO is able to collect either holes or electrons from the polymer:fullerene network because its work function (from −4.5 to −4.9 eV) is between the HOMO and LUMO values of these organic materials (Fig. 4a). In the conventional device, ITO is modified by p-type PEDOT:PSS, which has a HWF so that electrons are blocked and only holes are collected by ITO anode (Fig. 4b). Since the electrons must be extracted from the top electrode, air-sensitive LWF metal is applicable as the top cathode. On the other hand, if ITO is modified by n-type material with a LWF, the holes will be blocked and electrons will be exclusively collected by ITO electrode (Fig. 4c). Collecting the holes at the other side allows the usage of HWF metal as top electrode without losing the asymmetric electrode structure and concurrently eliminating the PEDOT:PSS/ITO interface issue. Applying this concept in an inverted device, the ITO is coated with n-type material which can be transition metal oxides [55, 56], an alkali-metal compound [52], an ultra-thin LWF metal [53], or an organic material [42], instead of hole-transporting material. In some modified inverted devices, a hole-transporting layer is inserted between the active layer and the HWF electrode to effectively block the excitons and adjust the optical field distribution in the active layer [53, 54].

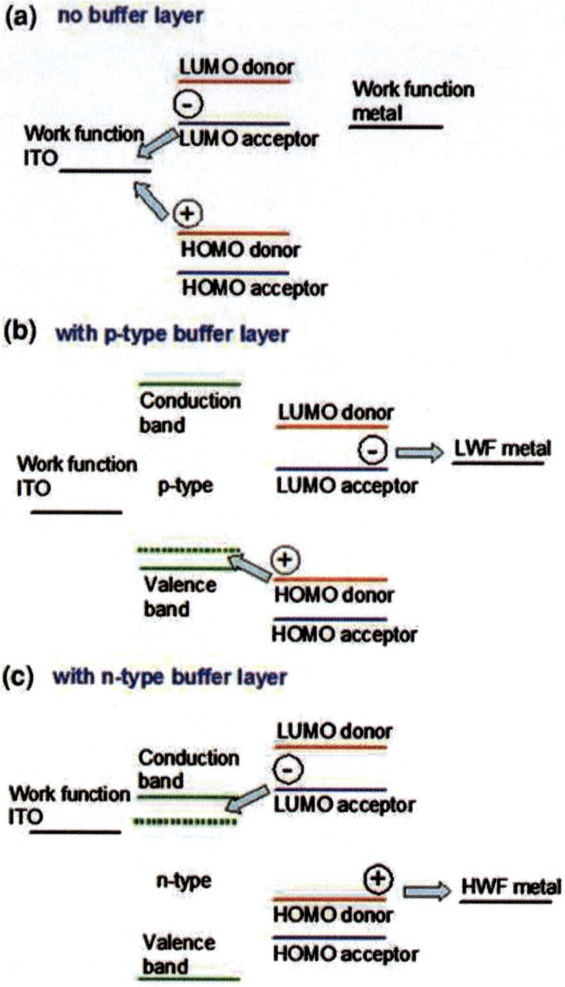

Fig. 4 Schematic of charge carrier extraction in OSCs **a** without buffer layer, **b** with p-type buffer layer, and **c** with n-type buffer layer at the bottom electrode

2.3 Bottom Illuminated Inverted Structure

Owing to the advantages in device stability and design flexibility compared to conventional structure, the inverted geometry has attracted significant attention in OSC research. To date, many efforts have been made to improve the efficiency, stability, and ease of processing via altering the materials and sequence of functional interfacial layers and studying the influence of interfacial layers on the performance of the device. Based on the previous research, inverted structures OSCs can be categorized into two groups: bottom illuminated and top illuminated, determined by the direction of light illumination. The implementation of bottom illuminated structure was achieved by Sahin et al. [42] in 2005. Typically, the bottom illuminated inverted structure employs a HWF metal as the hole collecting

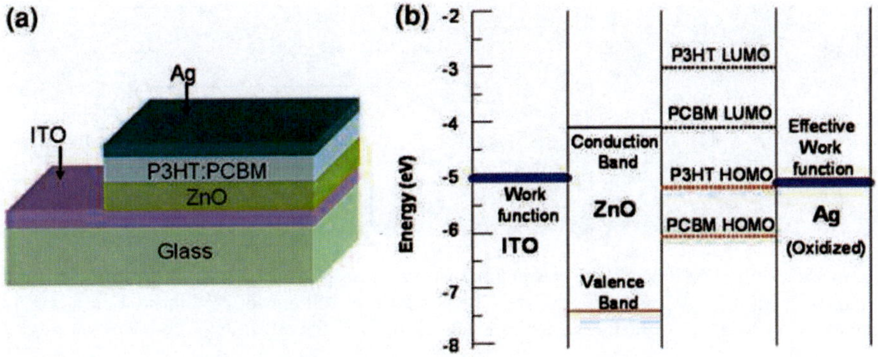

Fig. 5 **a** Device structure of White's inverted device and **b** estimated band diagram for the device prior to any Fermi- or vacuum-level alignment of the layers [56]

top electrode while ITO modified with a LWF material is used as the bottom electrode to effectively collect the electrons from the active layer and allow the light to enter the active layer, similar to the structure shown in Fig. 3b.

2.3.1 Transition Metal Oxide as Interfacial Layer

Transition metal oxides offer a unique opportunity to fabricate the devices with large area and at a low cost via solution-processing, which is similar to the solution-processing of organic material. The organic materials' electrical and optical properties, such as the control of work function, charge injection, and high transparency in the visible range, play important roles in their applications in optoelectronic devices. The modification of organic layer/electrode interface by inserting a transition metal oxide has been very beneficial to organic light-emitting diodes (OLEDs) [54] and organic thin film transistors (OTFTs) [57].

ZnO is one of the transition metal oxides whose work function is suitable to collect electrons as well as block holes from photoactive layer. White et al. demonstrated solution-processed ZnO as a functional interfacial layer in the inverted device [56]. The PCE of 2.58% (Certified NREL measurements 4 days after fabrication) was achieved with a structure of ITO/ZnO/P3HT:PCBM/Ag (Fig. 5). The solution-processed ZnO film after annealing at 300°C for 5 min in air results in crystalline ZnO with a work function of −4.3 eV which is close to its conduction band of −4.1 eV. The V_{oc} of 0.556 V for these devices is similar to that commonly reported in a conventional device with ITO/PEDOT:PSS/P3HT:PCBM/Al structure, implying that there is no significant energy loss during the electron transferring from the PCBM to the ZnO layer. Based on the finding that V_{oc} value of MDMO-PPV:ZnO nanoparticle blend device [58] and that of MDMO-PPV:PCBM [12] are similar, the authors also suggested that LUMO of PCBM is close to the conduction band of ZnO at −4.1 eV, rather than commonly reported value of −3.7 eV [18, 59]. Similarly, the work function of Ag is shifted

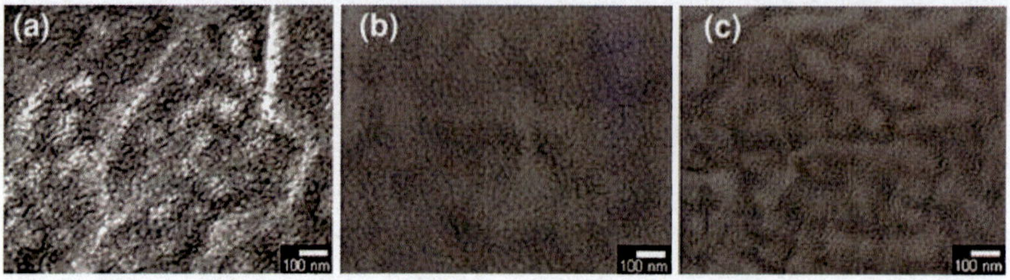

Fig. 6 Field emission scanning electron microscopy (FESEM) images of ZnO films derived from **a** 0.75 M, **b** 0.5 M, and **c** 0.3 M sol. The bar scale is 100 nm [60] (With kind permission from Springer Science + Business Media: Kyaw et al. [60], figure 3)

away from the vacuum level instead of documented value at −4.5 eV due to the exposure to oxygen. The matched energy level between LUMO of PCBM and conduction band of ZnO as well as that between HOMO of P3HT and shifted work function of Ag provides efficient electron and hole extractions at the electrodes, resulting in a high quantum efficiency of 85% from 500 to 550 nm. The high rectification of the diode (on the order of 10^5) is attributed to the efficient hole blocking by ZnO at P3HT/ZnO interface.

Despite an initial realization of ZnO as interfacial layer, White et al. did not reveal how the ZnO layer influences the performance of inverted cells. We conducted a systematic study on the effect of solution-processed ZnO film and its properties on the performance of the device, employing a device structure of FTO/ZnO/P3HT:PCBM/Au where ZnO film was fabricated by the sol–gel technique [60]. One important criterion to be utilized as an interfacial layer in organic devices is to ensure a pinhole-free film to prevent short-circuits in the device. We demonstrated that conformal and pinhole-free film can be achieved by manipulating the concentration of sol. As shown in Fig. 6, ZnO colloids are not closely packed in the film derived from 0.75 M sol, creating nano-sized gaps on the film. These gaps are favorable for polymer to infiltrate the ZnO film and cause direct contact between polymer layer and FTO, leading to short-circuits. On the other hand, the grains produced from both 0.5 and 0.3 M are smaller than that from 0.75 M, resulting in conformal and closely packed films. Therefore, the *I–V* characteristics of the devices with ZnO layer derived from 0.5 and 0.3 M sols do not show any short-circuit behavior, but do demonstrate the photovoltaic effect.

We also showed that room-temperature surface activation effectively removes unwanted chemical groups capped in the ZnO colloids without further increasing the annealing temperature. Generally, zinc-oxo-acetate oligomers are formed during the hydrolysis and condensation of the sol–gel process due to the presence of OH^- groups and acetate (CH_3COO^-) groups from a zinc acetate dihydrate precursor [61]. Ammonia treatment is able to remove the $ZnAc_2$ coated on the surface of the ZnO film by forming dissolvable $Zn(NH_3)_4^{2-}$ ions [62]. The removal of Ac ions by ammonia treatment was verified by XPS. In high resolution C(1s) spectra (Fig. 7), two distinct peaks are observed at binding energies around 288.6 and 284.6 eV for as-deposited film. One peak at higher binding energy

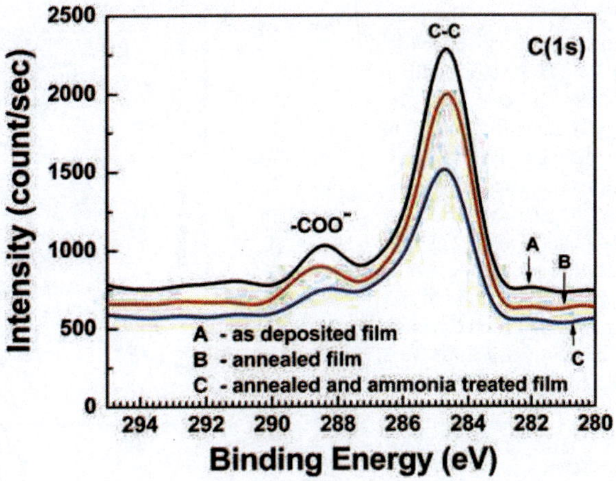

Fig. 7 XPS high resolution C (1s) spectra of as-deposited film, annealed film, and both annealed and ammonia treated sol–gel derived ZnO film [60] (With kind permission from Springer Science + Business Media: Kyaw et al. [60], figure 5)

corresponds to the carbon atom bonded to carbonyl oxygen atom in the acetate group and the other peak associates with adventitious carbon atoms. The intensities of both peaks are attenuated after low-temperature annealing while the acetate group diminishes strongly and the intensity of carbon reduces to half after additional ammonia treatment. As a result, the PCE of the device with ZnO interfacial layer derived from 0.3 M sol increases significantly from 0.82 to 1.55% (Fig. 8). The *I–V* curves (in semilogarithmic scale) of the cells in the dark (the inset of Fig. 8) show that the device with ammonia treated ZnO film has a higher current in the forward direction and a lower leakage current in the reverse direction. Finally, our optimized device adopted with a sol–gel derived ZnO electron selective layer and MoO_3 hole selective layer achieves a maximum PCE of 3.09% [63].

Hau et al. also reported that inverted cells with spin-coated ZnO nanoparticle as an electron selective layer and PEDOT:PSS as a hole selective layer have PCEs of 3.78 and 3.58% on glass and plastic substrates, respectively [64]. Despite a similar FF, the J_{sc} of inverted device is higher than those of regular devices due to the additional acceptor–donor interface between ZnO and P3HT, which generates additional photocurrent. In addition to solution-processed ZnO film, flexible inverted cell was recently fabricated with an atomic layer deposited (ALD) ZnO interfacial layer [65]. Being deposited layer by layer at atomic scale by introducing reactant gas into the reaction zone separately and sequentially, a conformal and pinhole-free film can be achieved at low temperature process (<100°C). The PCE of the device with ZnO grown at 80°C is higher than those with ZnO grown at 45°C, which is attributed to the lower resistivity of ZnO layer and a more hydrophobic surface, beneficial to the adhesion of polymer films. The PCE of the device using ALD ZnO (grown at 80°C) as the electron selective layer has achieved 4.18%, higher than previously reported values; however this technique has the drawbacks of high equipment cost and slow deposition rate. In addition to polymer:fullerene-based OSCs, an inverted architecture was also implemented in a

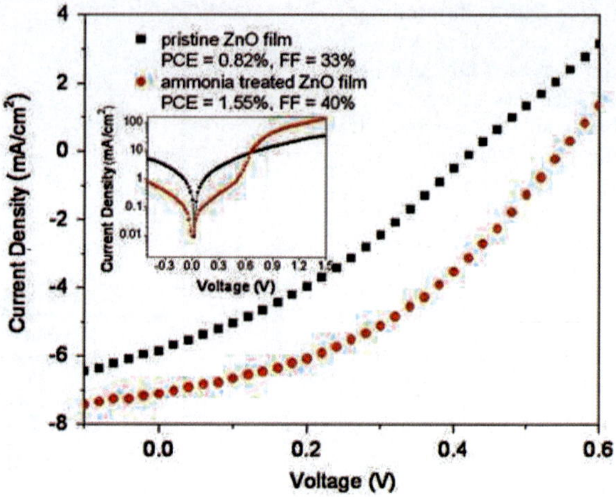

Fig. 8 The *I–V* characteristics of inverted devices (FTO/ZnO/P3HT:PCBM/Au) with ammonia treated and pristine (only annealed) ZnO film under 100 mW/cm^2. ZnO films were derived from 0.3 M sol. *Inset*: dark *I–V* curves of the respective devices in semilogarithmic scale [60] (With kind permission from Springer Science + Business Media: Kyaw et al. [60], figure 6)

small molecule-based OSC, employing solution-processed ZnO as the interfacial layer. Liu et al. fabricated an inverted small molecule OSC with an ITO/ZnO/CuPc:C_{60}/CuPc/PEDOT:PSS/Ag structure, showing a PCE of 0.31% using a ratio of 4:1 for CuPc:C_{60} [66]. It is worthwhile to note that, the devices without ZnO interfacial layer do not show good rectification behavior in dark in both BHJ and small molecule heterojunction devices (Figs. 9a and 10). For the inverted device without ZnO interfacial layer, since the energy barrier height for electron injection into the active layer is quite large, a rise in dark current can be observed only at a sufficiently high positive bias and the current is quite small, compared to that of the device with ZnO layer. Moreover, in the *J–V* curve under illumination (Figs. 9b and 10), the device without ZnO shows a rise of current at low positive (even at negative voltage in the small molecule device), a small V_{oc} because ITO has a larger work function than Ag, the flat-band condition is met at a negative bias voltage, and electron transport to ITO is not favored at positive bias voltage. The LWF interfacial layer, therefore, is a key factor to obtain a rectification behavior of diode in the dark and photovoltaic effect under illumination.

TiO_x is another transition metal oxide applicable as an interfacial layer in an inverted structure since its conduction band (−4.4 eV) [25] is well matched with LUMO level of PCBM (−4.3 eV) [67] and its valence band (−7.5 eV) [25] also effectively blocks the hole collection to ITO. Waldauf et al. [55] investigated an inverted OSC using solution-processed TiO_x as the electron selective layer, comparing the performance of regular device (ITO/PEDOT:PSS/P3HT:PCBM/LiF/Al) and inverted device (ITO/TiO_x/P3HT:PCBM/PEDOT:PSS/Ag). In contrast to the literature [64], the J_{sc} of inverted device is lower than that of regular one. However, this group claims that the FF of the inverted device is higher than that of the regular device because TiO_x serves as an efficient hole-blocking layer, leading to a lower leakage current. They prepared the solution of P3HT:PCBM blend in low-cost *o*-xylene solvent rather than chlorinated-based solvents, considering that *o*-xylene is more beneficial to inverted devices, based on the electron

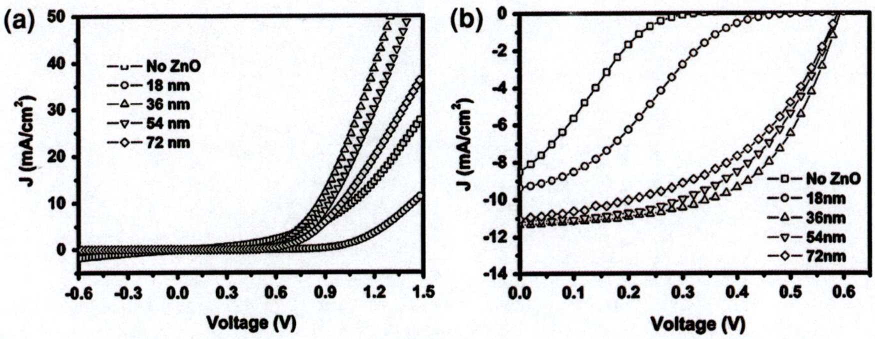

Fig. 9 The *J–V* characteristics of inverted OSCs (ITO/ALD-grown ZnO/P3HT:PCBM/MoO_3/Ag) with different ZnO thicknesses deposited at 45°C, **a** in the dark, and **b** under AM 1.5G illumination [65] (Reproduced by permission of The Royal Society of Chemistry (RSC) for the European Society for Photobiology, the European Photochemistry Association, and the RSC) (http://dx.doi.org/10.1039/B921396A)

and hole mobility extracted from OTFT characterization. In the blend layer prepared from *o*-xylene, the hole mobility is 25 times higher than the electron mobility. Although balanced transport is required to reduce space charge building and charge recombination, the group speculates that for the inverted device, where light is illuminated from bottom and hole is collected by top, higher mobility of hole is necessary due to a longer distance travel. No other groups, however, continue the investigation on the effect of *o*-xylene as the solvent on inverted device performance. Nanostructure n-type oxides, such as ZnO nanorods [68, 69] and TiO_2 nanotubes [50, 70] are also employed as an electron selective layer in inverted devices, which will be discussed in detail in double-heterojunction device, in (Sect. 2.5).

While n-type oxides serve as an interfacial layer at the bottom electrode for electron collection, p-type-like oxides, which have been utilized for hole injection in OLEDs [71, 72], anode buffer layers in conventional OSCs, and intermediate layer in tandem OSCs [17, 18], are introduced as a hole selective layer and exciton blocking layer in inverted device. Although PEDOT:PSS layer can be employed as a hole-transporting layer in an inverted cell [73, 74], its hygroscopic nature is likely to form insulating patches due to the water adsorption, degrading device performance [73, 75]. We demonstrated that V_{oc} and FF of the inverted device (FTO/ZnO/P3HT:PCBM/Ag) increases from 0.567 to 0.616 V and from 47 to 57%, respectively, by inserting a thermal-evaporated p-type-like MoO_3 layer between the active layer and the top electrode, as shown in Fig. 11 [63]. The p-type-like MoO_3 serves as an exciton blocking layer because the energy band gap (~3.0 eV) of MoO_3 is higher than the exciton energy (~1.9 eV) of P3HT, thereby suppressing exciton quenching at the organic/metal interface and leading to the increase in FF. Tao et al. also adopted thermal-evaporated MoO_3 as a hole selective interfacial layer in an inverted device together with sol–gel processed TiO_2 electron selective layer [76]. They studied three different metals (Au, Ag,

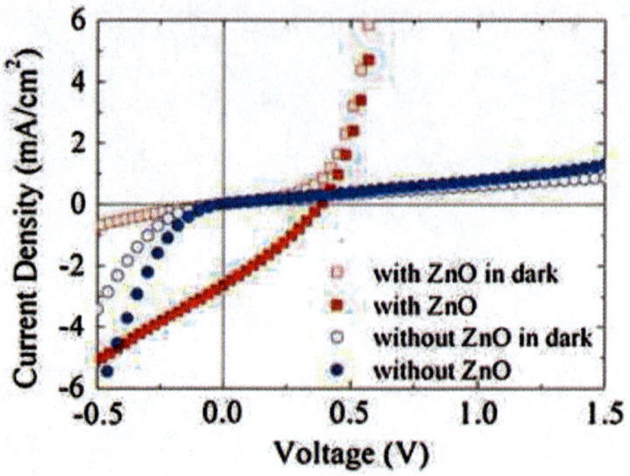

Fig. 10 The *J–V* characteristics of inverted small molecule devices with ZnO (ITO/ZnO/CuPc:C_{60}/PEDOT:PSS/Ag) and without ZnO (ITO/CuPc:C_{60}/PEDOT:PSS/Ag) in the dark and under illumination using an AM 1.5G solar simulator [66] (Reprinted from [66], with permission from Elsevier) (http://www.sciencedirect.com/science/journal/00092614)

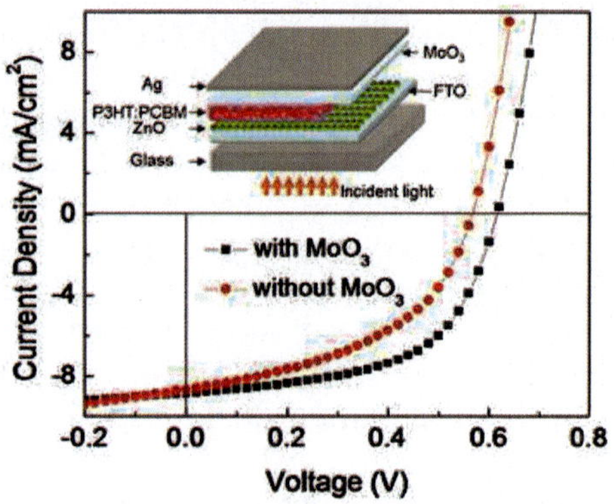

Fig. 11 The *J–V* characteristics of devices with and without MoO_3 p-type-like buffer layer. Device structure is shown in the *inset*. ZnO film was derived from solution-processing with sol concentration 0.1 M. (Reprinted with permission from Ref. [63]. Copyright 2008 American Institute of Physics)

and Al) as top electrodes and found that there is a considerable difference in the V_{oc} of the devices without MoO_3 present. However, the variation in the V_{oc} for the device with MoO_3 is within 20 mV, suggesting that the V_{oc} of device is independent of the work function of top electrode when an anode buffer layer is included. This group also revealed that the device with MoO_3 has a higher V_{oc} than corresponding devices without MoO_3, which is consistent with our finding. Schmidt et al. [77] demonstrated thermally evaporated thick MoO_3 as a protective layer in a semitransparent inverted OSC. The control device (ITO/TiO_2/P3HT:PCBM/MoO_3/Al) and semitransparent device (ITO/TiO_2/P3HT:PCBM/MoO_3/sputtered-ITO) were fabricated and the effect of the thickness of MoO_3 layer was investigated. Surprisingly, the characteristics of the control devices vary slightly when MoO_3 thickness changes from 5 to 40 nm, yet a significant difference is observed for semitransparent devices by varying MoO_3 thickness. The *J–V* characteristics of the semitransparent device with 5 nm-thick MoO_3 shows an S-shape with a low FF, whereas the device with 40 nm-thick MoO_3 exhibits an increased FF of 60% with no increased series resistance. The authors claimed that 5 nm MoO_3 is not thick enough to protect the active layer from the impact of sputtering ITO top electrode and therefore the charge extraction at the organic/MoO_3 and MoO_3/ITO interfaces is deteriorated and FF is declined. In contrast, the damage from sputtering ITO is well protected by 40 nm-thick MoO_3, resulting in a high FF.

Alternatives to MoO_3 include other widely utilized p-type-like oxides, such as V_2O_5 and WO_3. Thermally fevaporated V_2O_5 was applied as a hole injection layer in the inverted OSCs, complemented with Cs_2CO_3 [59] and TiO_2 nanotubes [70] as the electron selective layer. Due to the HWF of V_2O_5, a reasonable value of V_{oc} (~0.59 V) is achieved in the inverted cell even though a LWF Al is used as the top electrode [70]. Like MoO_3, V_2O_5 was employed as a protective layer in a transparent OSC, which is important for realizing tandem OSCs for further efficiency improvement [59]. Besides thermally evaporated V_2O_5, solution-processed V_2O_5, which is more compatible with solution-processed polymer OSCs, was also demonstrated by Huang et al. [68]. In addition to a higher V_{oc} and FF due to suppression of leakage current at the organic layer/top electrode interface and HWF, V_2O_5 acts as an optical spacer which adjusts high optical field distribution across the active layer, leading to 6% increase in IPCE at 550 nm and a 5% increase in J_{sc}. It also serves as a barrier for preventing oxygen and water from entering the active layer owing to V_2O_5's relative insensitivity to water and stability in air. A thermally evaporated WO_3 as a hole selective interfacial layer in inverted OSC was realized by Tao et al. [78]. Similarly, WO_3 has multiple function: efficient hole collection, suppression of leakage current, and enhancement in built-in potential, all of which result in higher J_{sc}, V_{oc}, and FF, compared to devices without WO_3. Similar to MoO_3, inserting WO_3 buffer layer makes the influence of the work function of the electrode on V_{oc} diminutive. J_{sc}, however, depends on the top electrode due to different properties of reflection of light for different metals. The group also demonstrated the transparent OSC by applying WO_3 (10 nm)/Ag (13 nm)/WO_3 (40 nm) where the bottom 10 nm WO_3 serves as the buffer layer and the top Ag (13 nm)/WO_3 (40 nm) serves as the transparent electrode.

2.3.2 Alkali-Metal Compound as Interfacial Layer

Alkali and alkali-earth metal compounds are considered as candidates for electron injection materials due to their LWF and insensitivity to moisture and oxygen. Alkali metal (Li^+, Na^+, K^+, Rb^+, and Cs^+) acetates and fluorides have revealed the enhanced electron injection in OLEDs since 2001 [79]. Cs_2CO_3, easier to handle in comparison with other commonly used alkali-metals, was introduced by Toshinori et al. [75] from Canon as an electron injection layer for top emission OLEDs. It was also implemented as an interfacial layer in an n-channel OTFT for the reduction in the energy barrier of electron injection, the prevention of unfavorable chemical interaction between organic layer and metal electrode, and the reduction in contact resistance [80].

It is believed that Cs_2CO_3 was first demonstrated in OSCs by Li et al. [59]. By interchanging the positions of Cs_2CO_3 (1 nm) and V_2O_5 (10 nm) below and

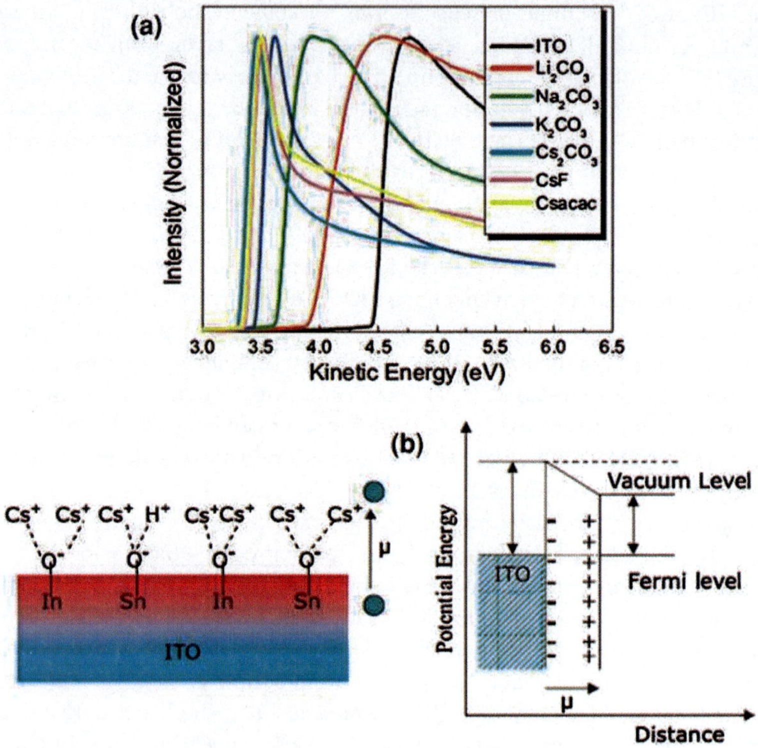

Fig. 12 **a** Evolution of secondary electron edge with different alkali metal compounds on ITO, **b** illustration of the formation of dipole layer on ITO due to alkali metal compound and its effect on reducing the work function of ITO [81] (Copyright Wiley–VCH Verlag GmbH & Co. KGaA. Reproduced with permission)

above the active layer, a conventional cell was transformed into inverted cell with a PCE of 2.25%. In the conventional cell, V_{oc} and FF are improved significantly, however J_{sc} is reduced by inserting 1 nm Cs_2CO_3, suggesting that physical damage may reduce the photocurrent. However, in the inverted cell, Cs_2CO_3 is deposited onto ITO and all parameters are improved. Cs_2CO_3 can be deposited by both solution-processing and thermal evaporation, without significantly influencing the PCE. Although LiF thin film enhances V_{oc} and FF in conventional cells [23], it results in low V_{oc} and FF in inverted cell due to the presence of an anti-diode. An interesting observation related to alkali compounds in inverted OSCs is that the same metal salt (Cs) with different anions (carbonate, fluoride, and acetate) results in almost the same device characteristic because buffer layers containing the same metal salt have similar work functions. However, different metal salts with the same CO_3 anion yield different device characteristics owing to different work functions (Fig. 12a) [81]. A strong dipole layer is formed at the interface of ITO (or FTO) and alkali metal due to the electron donating ability of metal species, which alters the work function of ITO (or FTO) (Fig. 12b). Among the

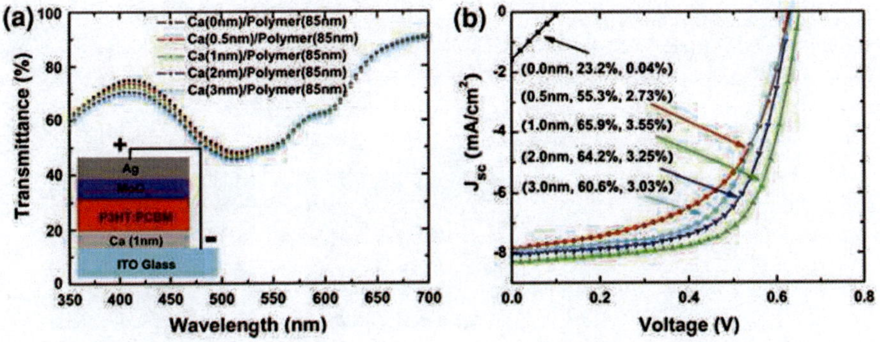

Fig. 13 **a** The transmittance spectra of the films Ca (x nm)/P3HT:PCBM (85 nm) with $x = 0$, 0.5, 1, 2 and 3 nm. The *inset* shows the device structure of inverted organic solar cell, **b** The *J–V* characteristics of inverted cells with 3 nm MoO_3 and different thickness of Ca under 100 mW/cm^2 illumination. The corresponding Ca thickness, FF, and PCE are shown in the format of (Ca thickness, FF, PCE) (Reprinted with permission from Ref. [53]. Copyright 2009 American Institute of Physics)

alkali-metal compounds (Cs, K, Na, Li metal with CO_3 anions), Cs_2CO_3 reduces the work function the most due to formation of the strongest dipole moment, leading to higher J_{sc} and V_{oc} and a lower series resistance, compared to other compounds. Higher J_{sc} and lower series resistance are related to the enhanced interfacial charge transfer rate at the cathode and higher V_{oc} corresponds to the lower work function of cathode before Fermi level pinning.

2.3.3 Ultra-Thin LWF Metal as Interfacial Layer

LWF metals such as Ca and Mg are inserted between active layer and top electrode in a conventional OSC to enhance the charge collection and built-in potential of the device [82]. Similarly, being sandwiched between an ITO electrode and the active layer in an inverted cell, LWF metals can serve as an interfacial layer that efficiently collects electron. Recently, we have investigated the impact of thermally evaporated, ultra-thin Ca in inverted cell using a structure of ITO/Ca/P3HT:PCBM/MoO_3/Ag (the inset of Fig. 13a) [53]. Without Ca, the cell is a hole-only device, exhibiting nearly no photovoltaic effect. When Ca is inserted, the PCE significantly increases and achieves 3.55% with 1 nm-thick Ca (Fig. 13b). The work function of Ca (−2.9 eV) lowers the work function of ITO, thereby increasing the V_{oc} of device considerably due to Fermi level pinning between Ca and PCBM via surface states. As a result, an Ohmic contact is favored between the Ca and PCBM [83]. In contrast, a rectifying contact between Ca and P3HT is formed, which blocks the hole collection on the ITO side [83]. It is worth mentioning that Ca layer is not oxidized because Ca is deposited in a vacuum of 9.0×10^{-5} Pa and subsequent layers protect it from moisture and oxygen. This is indirectly verified by a poorly performing device with Ca being replaced by CaO

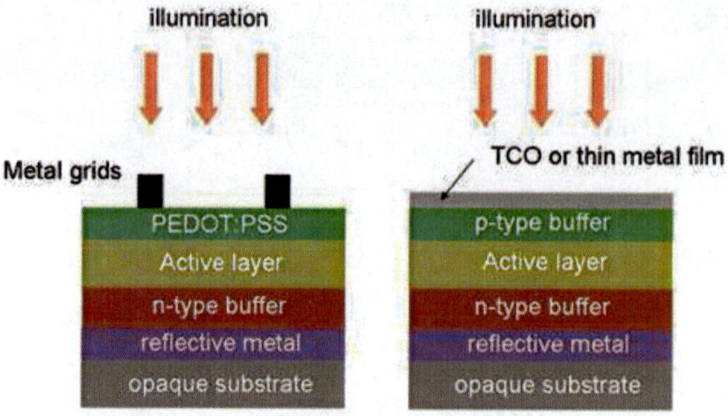

Fig. 14 Structures of typical top illuminated inverted OSCs

(oxidizing Ca by dry air). Similarly, other LWF metals like Mg result in photovoltaic effect comparable to Ca. On the other hand, when HWF metal Ag (−4.5 eV) is applied to the ITO surface, an V_{oc} of only 0.29 V is yielded. This clearly reveals that LWF metals are also suitable to be adopted as electron selective interfacial layer for inverted cell.

2.4 Top Illuminated Inverted Structure

In contrast to bottom illuminated devices, the reflective metal electrode is embedded in the device and either a transparent conductive oxide or a thin metal or metal grid is employed as the top electrode in a top illuminated device (Fig. 14). The top illuminated structure is appealing in solar cell research because it allows for the usage of low-cost metal foil substrates and plastic foils with opaque metal coatings. Moreover, it can eliminate the need for expensive ITO if a metal grid or thin metal film is used as top electrode.

It is believed that the first top illuminated inverted structure was introduced by Glatthaar et al. in 2005 [51]. The top illuminated inverted cell was formed by Al (80 nm), Ti (20 nm), P3HT:PCBM (250 nm), PEDOT:PSS (250 nm), and an Au-grid (50 nm) sequentially placed on a glass substrate. Electrons are extracted from the bottom Al electrode and holes from the top Au-grid electrode. Herein, a thin Ti layer, which has high electron mobility even if oxidized, is used to prevent the formation of highly insulating oxide at the surface. The combination of highly conductive PEDOT:PSS and Ag-grid allows the light to enter the active layer without sacrificing sheet resistance. Despite a low PCE of 1.4% with $J_{sc} = 4.6$ mA/cm^2, $V_{oc} = 0.58$ V, and FF = 53%, the result was promising and the device structure was significant at the point of time. Unfortunately, such a device structure was not well-known in inverted solar cell research and no similar work was reported until recently. Chen et al. [84] rejuvenated this structure on a stainless-steel (SS) substrate. Its device structure is photoresist (PR) and SiO_x coated SS foil/Ag(100 nm)/ITO(100 nm)/Cs_2CO_3/P3HT:PCBM/MoO_3(15 nm)/Al-grid/ITO. PR is used to reduce the roughness of SS foil and SiO_x is to prevent

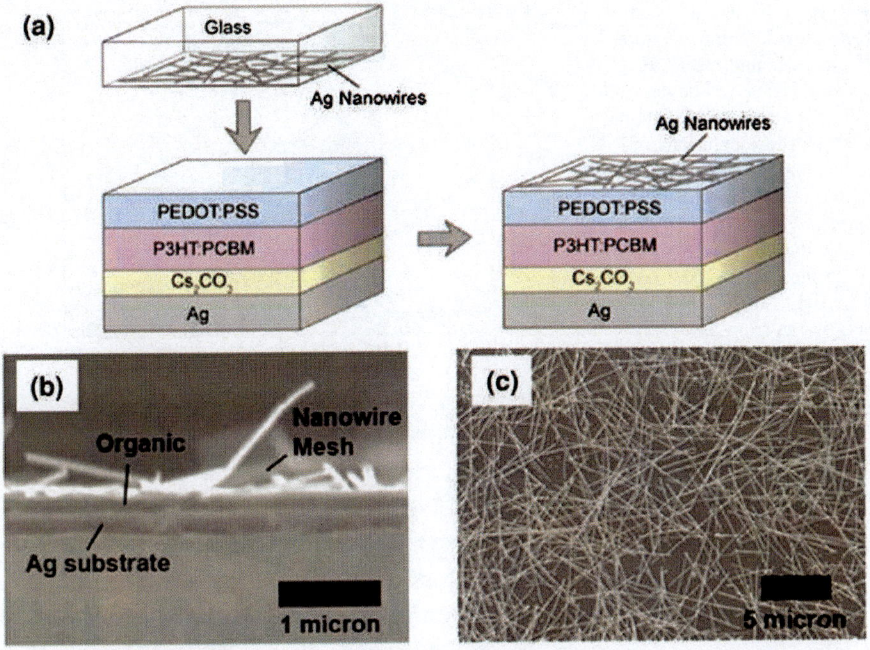

Fig. 15 **a** Device structure of all solution-processed top illuminated inverted cell with laminated Ag electrode and illustration of lamination process. Scanning electron micrograph (SEM) image of **b** cross-section and **c** top view of the device. (Reprinted ("Adapted" or "in part") with permission from Ref. [47]. Copyright 2010 American Chemical Society)

PR from damage associated with solution process. While PR and SiO_x serve as non-functional buffer layers, Ag plays an important role in reducing sheet resistance and serves as a light reflector. The reason for inserting ITO between Ag and Cs_2CO_3 is not clear. Al-grid with 10% shadow fraction embedded between MoO_3 and ITO further reduces the sheet resistance of the device without sacrificing much in optical loss. Another example of using Cs_2CO_3 as electron selective layer in top illuminated inverted structure is a device composed of an opaque Ag substrate/ Cs_2CO_3/P3HT:PCBM/PEDOT:PSS/Ag nanowires (Fig. 15) [47]. The device is fabricated from the Ag substrate up to the PEDOT:PSS using solution-processing. To deposit the Ag nanowires as the top electrode, Ag nanowire meshes were pressed with a glass substrate to flatten the nanowire mesh first and then laminated onto the previously deposited PEDOT:PSS surface. The advantage of this device is that all solution-processed inverted cells can be realized without an ITO electrode, potentially reducing cost. The top Ag nanowire electrode has a transparency, similar to ITO with a sheet resistance of 10 Ω/square. A PCE of 2.5% with $J_{sc} = 10.59$ mA/cm^2, $V_{oc} = 0.51$ V, and FF = 46% was achieved. However, as the case in bottom illuminated device, the device without Cs_2CO_3 layer results in low J_{sc}, V_{oc}, and FF. Another all solution-processed top illuminated inverted cell was fabricated on a granular surface-nickelized polyimide (NiPI), which serves as

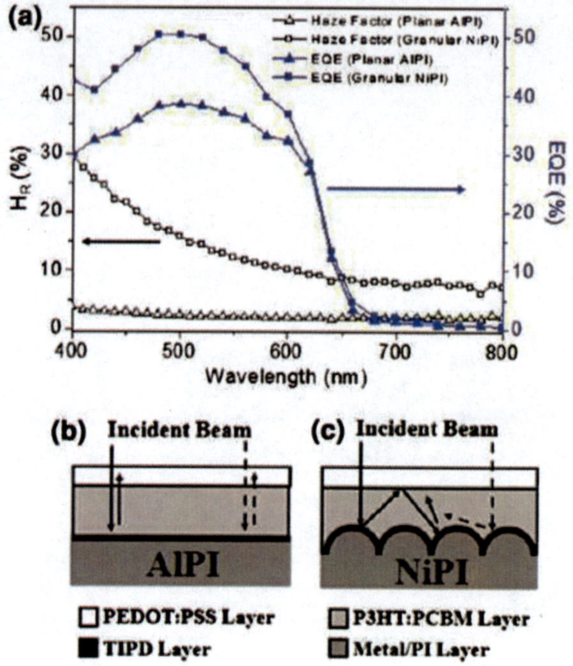

Fig. 16 **a** Reflection haze factor for different structure of cathodes and the EQE spectra of inverted devices on different structure of cathode. Cartoon representations of light illuminated on devices with different structures of cathodes, **b** planar AlPI and **c** granular NiPI [85] (Reprinted from [85], Copyright 2009 with permission from Elsevier) (http://www.sciencedirect.com/science/journal/15661199)

both substrate and cathode [85]. Different from other inverted cells, titanium (diisopropoxide)bis(2,4-pentanedionate) (TIPD) is used as the electron selective interfacial layer based on the previous finding that TIPD interface significantly reduces the interface resistance between the cathode and the active layer [86]. A high-thickness layer (minimum 200 nm) of P3HT:PCBM, about twice as thick as those used in conventional devices, is used as the active layer to prevent short-circuit due to the imposed morphology (granular structure) of the anode. High conductive PEDOT:PSS with an Ag grid is used as the top electrode. Despite the fact that the work function of Ni is −5.3 eV [87], and that of NiPI is reduced to −3.9 eV, compatible with the conduction band of TIPD (−3.9 eV) and lower than the work function of Al. One advantage of employing a granular anode over a planar anode is improved light-trapping in the active layer. Since the refractive index of P3HT:PCBM (1.6–2.2) is higher than that of PEDOT:PSS (1.1–1.6) [88, 89], the tilted light reflected from the granular NiPI is likely to be trapped by P3HT:PCBM layer rather than refracted into the PEDOT:PSS layer due to total internal reflection as illustrated in Fig. 16. The reflective haze factor measurement and external quantum efficiency (EQE) test, in which the reflective haze factor and EQE obtained from granular NiPI film are higher than that obtained from planar AlPI film, thus verify that a granular anode improves the light scattering and trapping in such device structure. This ITO-free inverted cell on flexible granular NiPI film has a PCE of 2.4%.

The top illuminated inverted structure was also implemented in small molecule OSCs. Tong et al. [90] demonstrated an inverted small molecule OSC on a reflective substrate, employing a structure of quartz substrate/Ni/CuPc/PTCBI/BCP/ITO. The top ITO electrode was deposited by sputtering. This device with optimized thickness of acceptor (10 nm) and donor (35 nm) gives a PCE of 0.74% ± 0.03%. In fact, this device is more similar to a conventional device because holes are extracted by the bottom electrode and electrons are collected by the top electrode. Only the direction of illumination is reversed. The actual inverted structure, in which electrons are collected by the bottom electrode and holes are collected by the top electrode, was also fabricated with quartz/Ag/BCP/PTCBI/CuPc/ITO, yet this device structure did not work, probably due to the damage caused by sputtering ITO directly on CuPc surface.

An ITO-free top illuminated small molecule BHJ OSC was realized using 100 nm Al as the bottom electrode, 30 nm of 10 wt% ZnPC (10 nm)/ZnPC:C_{60} (25 nm, weight ratio of 1:1)/C_{60} (40 nm) as the active layer, 7 nm of BPhen as the exciton-blocking and electron-transporting layers, and finally ultra-thin Al and Ag with different combinations of thickness as the top electrode [91]. The novelty of this device is the fabrication of cell without ITO or a metal grid. Likewise previous small molecule OSCs, although it was called an inverted OSC, its structure resembles a conventional structure. By varying the thicknesses of Al and Ag top electrodes, optimized device performance was achieved. Al acts as a surface-mediating layer to ensure good adhesion of Ag and formation of a closed, flat Ag layer with a smooth morphology, but causes optical loss due to high reflectivity. The optical transmittance does not suffer much from the Ag layer and it serves as the main conductive component of the top electrode. However, at a thickness below a certain coalescence threshold, depending on evaporation rate, substrate, and pressure [92], some isolated Ag clusters, islands or hillocks that act as charge traps center form, leading to unbalanced charge-carrier extraction. The optimized thickness of the top electrode, therefore, is a combination of 1 nm Al and 14 nm Ag, which results in an improved morphology of metal contact without causing optical loss, leading to a PCE of 2.21% with $J_{sc} = 7.9$ mA/cm^2, $V_{oc} = 0.519$ V, and FF = 53.86%.

2.5 Inverted Organic Solar Cells with Nanostructure Metal Oxides or Double-Heterojunction Structure

As discussed in Sect. 2.3.1, n-type transition metal oxides serve as an interfacial layer between a polymer:fullerene blend and an ITO electrode to effectively collect electrons and block holes. In addition to planar thin film, nanostructure metal oxide can also be employed as an interfacial layer in inverted cells. The inverted OSC with a nanostructure metal oxide, also referred to as a double-heterojunction device, owing to the formation of two acceptor–donor heterojunctions: one junction is between polymer and fullerene interface and the other one is between polymer and n-type metal oxide. Despite an efficient dissociation of photogenerated excitons due to the electron transfer from semiconducting polymer to fullerene, the BHJ still suffers from poor charge collection due to the inherently random penetrating network morphology and the problem of segregation of the donor and acceptor phases. On the other hand, an ordered heterojunction (e.g., a well-ordered, mesoporous TiO_2 infiltrated with polymer) provides a straightforward pathway for electron transport/collection and enhances the hole transport due to the vertical confinement of the polymer chains in the pores if the pore size is appropriate (ca. 20–30 nm) [93]. A double-heterojunction, as a result of combining features of BHJ and ordered heterojunction, not only has advantages of ordered network morphology but also enhances the exciton dissociation sites due to the large-surface area of the nanostructures, providing additional acceptor–donor junctions between polymer and n-type metal oxide. Light harvesting can be improved due to an increased amount of active layer; however, there is a negligible drawback in exciton recombination due to the infiltration of polymers inside nanostructure with pore size comparable to the exciton diffusion length.

However, the challenge in realizing such a device is to fabricate the nanostructures with a highly ordered and controlled pore size since the structure of polymer chain infiltrated inside the nanostructure significantly depends on both alignment and pore size. Coakley et al. revealed that the charge-carrier mobility of the polymer is increased as much as 20-fold due to the vertical channel confinement in the pores along the direction perpendicular to the substrate [94]. However, it should be noted that the polymer chains cannot twist and a high-degree of π-stacking between polymer chains cannot exists (which would allow for the delocalization of excitons over multiple chain) due to the constrictions imposed by the strongly confining pores (ca. 8 nm) [93]. Moreover, very small pores allow only isolated chains of polymer rather than coiled interacting chains, resulting in low polaron yield or polaron lifetime [95]. Larger pore sizes allow more film-like environments with aggregated and coiled polymer chains. According to preliminary study, it is believed that straight pores and larger pore size will be required for double-heterojunction devices to achieve good hole transport in the films.

Mor et al. [50] demonstrated an inverted OSC with nanostructures by infiltrating P3HT:PCBM blend into TiO_2 nanotubes with a length of 270 nm and a pore size of 50 nm, which is large enough to allow the formation of coiled interacting chains as in bulk polymer film. TiO_2 nanotubes were fabricated on FTO glass while Au was employed as the top contact. PEDOT:PSS was used as the hole collecting and electron blocking layer as shown in Fig. 17. With this structure, a PCE of 4.07% with $V_{oc} = 0.641$ V and FF = 51.1% was achieved. The high J_{sc} (12.4 mA/cm^2) and high EQE (maximum 80% at 538 nm) are most likely due to the additional exciton dissociation sites at the P3HT:TiO_2 interfaces and large P3HT:PCBM interface area, since the blend layer is infiltrated into large-surface area nanostructures instead of planar layers. However, the fraction of the contribution from the P3HT:TiO_2 interface cannot be determined since the experimental result of photocurrent generated from the P3HT:TiO_2 only device (without PCBM) was not

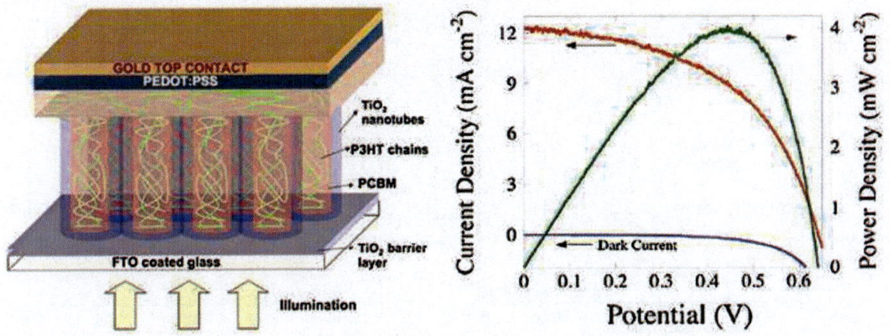

Fig. 17 Inverted device structure using TiO_2 nanotubes as electron selective layer (*left*) and *J–V* characteristic of the device (*right*) (Reprinted with permission from Ref. [50]. Copyright 2007 American Institute of Physics)

disclosed. Another inverted cell with TiO_2 nanotubes was reported by Yu et al., employing an ITO/TiO_2 nanotubes/P3HT:PCBM/V_2O_5/Al structure [70]. With TiO_2 nanotubes having 15 nm pores prepared by anodizing 100 nm Ti film on ITO for 27 min at 20 V and then annealing at 500°C for 1 h, the resulting inverted cell obtained J_{sc} = 10.96 mA/cm^2, V_{oc} = 0.59 V, and FF = 42%, leading to a PCE of 2.71%. A very short anodization time results in a low J_{sc} because residual Ti film exists due to incomplete formation of nanotubes, causing a high series resistance in the cell. It is also possible that the low carrier mobility of the polymer resulting from the lack of high-degree π-stacking of polymer chain inside small pores is responsible for inferior device performance. Over anodization also deteriorates the device performance because the weakly acidic electrolyte over-etches the ITO electrode. In fact, this inverted cell should be fabricated on FTO rather than ITO because the annealing process of TiO_2 nanotubes at high temperature has a negative impact on the resistivity of ITO but FTO is not susceptible. This may be one of the reasons for that the device performance is lower than Mor's reported values despite a similar fabrication process and device structure.

In addition to TiO_2, ZnO nanostructures are also applied in inverted devices. In fact, the realization of inverted device with ZnO nanostructure was earlier than TiO_2. Olson et al. reported an inverted device with ITO/ZnO nanofiber/P3HT:PCBM/Ag architecture, where ZnO fibers were fabricated from a hydrothermal method, with a PCE of 2% with a J_{sc} of 10 mA/cm^2, V_{oc} of 0.475 V, and FF of 43% [69]. However, the device without PCBM (i.e., ITO/ZnO nanofiber/P3HT/Ag hybrid) shows a J_{sc} of 2.2 mA/cm^2, V_{oc} of 0.44 V, and FF of 56%, leading to a PCE of 0.53%, implying that about 20% of photogenerated current in the inverted device is contributed from the P3HT (donor):ZnO (acceptor) heterojunction. Takanezawa et al. fabricated an inverted cell with the same architecture, which achieved a PCE of 2.7%. The J_{sc} of the device with ~160 nm ZnO nanorods is ~7 mA/cm^2 and that of the device under the same condition without PCBM is ~1.2 mA/cm^2, suggesting that about 17% of photocurrent is contributed from P3HT:ZnO heterojunction [96]. Based on these observations, it can be

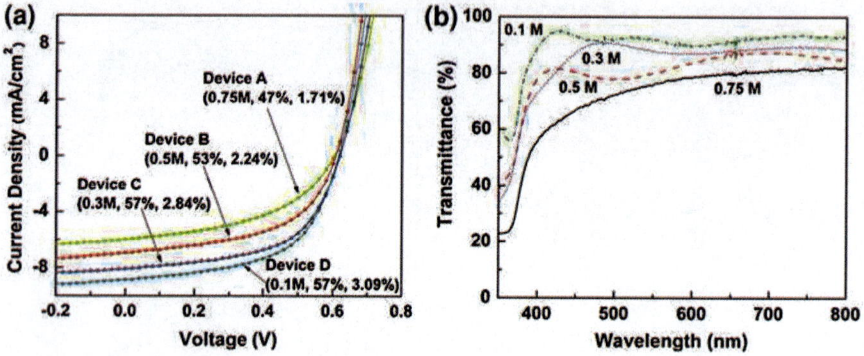

Fig. 18 **a** The *J–V* characteristics of inverted devices with ZnO films derived from different sols under simulated solar irradiation of AM 1.5G. The corresponding sol concentration, FF and PCE are shown in the format of (sol concentration, FF, PCE). **b** The transmission spectra of ZnO films derived from different sols (Reprinted with permission from Ref. [63]. Copyright 2008 American Institute of Physics)

concluded that nanostructure oxides serve as an electron selective layer and an acceptor, which allows the formation of additional heterojunction although the photocurrent primarily comes from polymer:fullerene heterojunction.

2.6 Power Efficiency Improving Techniques

The performance of inverted OSCs can be improved in two main aspects. One is manipulating the active layer so that the orientation of vertical phase separation is beneficial to the charge transport to respective electrodes. The other is engineering the interfaces between the organic layer and electrode to promote efficient charge collection, to reduce the contact resistance, and to minimize the charge trapping at the interfaces. Many efforts have been made on latter part, yet work on manipulating the active layer has been lacking. In this section, a few techniques that improve the efficiency by interface engineering will be discussed.

The interfacial buffer layers, both electron and hole selective layers, must be as transparent as possible to reduce the light losses (both incident light and reflected light from the reflective metal electrode). We demonstrated an efficiency improvement of the inverted device by manipulating the optical transmittance of the electron selective ZnO buffer layer [63]. The transmittance of a crystalline thin film depends on the crystal size, which is controlled by the concentration of precursors in the sol [97]. Therefore, we prepared sols with different concentrations of precursors to control the transmittance of the film. Devices A, B, C, and D refer to the devices with ZnO film derived from 0.75, 0.5, 0.3, and 0.1 M sol, respectively. It can be clearly seen that the trends in the current density (Fig. 18a) and IPCE (not shown here) are consistent with the variation in the transmittance of

ZnO films used in the devices (Fig. 18b). Since the ZnO films were controlled to obtain the same thickness except for device A, whose thickness is higher than the others due to its large grain size, the variation in the transmittance is not due to different thicknesses, but is caused by the grain size of the film.

The surfaces of commonly used buffer layers such as transition metal oxides and alkali-metal compounds are not as smooth as that of PEDOT:PSS, but, generally, a smooth surface is necessary for acceptable performance of organic devices [54]. Interface modification to reduce the roughness of buffer layer is a useful technique to improve the efficiency. Steim et al. [98] reported that the series resistance is reduced from 3 to 1 Ω and the shunt resistance increases from 14 to 31 kΩ, leading to an increase in FF from 55 to 64% when inserting an ultrathin organic interfacial layer, polyoxyethylene tridecyl ether (PTE) between ITO and TiO_x buffer layer. The PTE layer reduces the roughness (distance between peaks and valleys) of the TiO_x buffer layer, providing more intimate contacts with the active layer. The fact that built-in voltage (V_{bi}) of the devices with and without PTE is the same suggests that the ultrathin, insulating PTE layer does not alter the work function of ITO. Hence, it is a solely passive interfacial layer and its ability to reduce the roughness of buffer layer increases the FF of the device. Liao et al. [52] also found that the PCE of the inverted device is improved while increasing the contact angle of the electron selective buffer layer. Since the smooth surface generally has higher contact angle for hydrophilic surfaces [99], the improvement in Liao's inverted device is attributed to the smooth surface of buffer layer. Moreover, an improvement in device performance was observed by modifying TiO_2 buffer layer with a functionalized self-assembled monolayer (SAM), which makes the morphology of the buffer layer smoother and increases the contact angle of the surface [100].

Manipulating the work function of the electrodes is another technique to promote efficient charge collection and to increase the PCE. The work function manipulation can be achieved by intrinsic or extrinsic doping of the buffer layer, or using SAM that creates dipoles at the interfaces. Liao et al. [52]. demonstrated that the PCE of inverted cell was improved by decreasing the work function of the Cs_2CO_3 buffer layer from 3.45 to 3.06 eV via low-temperature annealing, which decomposes Cs_2CO_3 into Cs_2O intrinsically doped with Cs_2O_2. We also reported that charge collection at the cathode is improved and J_{sc} increases by extrinsically doping ZnO with indium, which lowers the work function and the resistivity of the buffer layer [101]. Lowering the work function decreases the barrier height for the electrons to be extracted by the electrode and reduces the electron injection barrier for diodes as illustrated in Fig. 19, and therefore improves the charge collection efficiency.

A dipolar SAM, which can effectively increase or decrease the work function of the electrode corresponding to the direction of the dipole moment, is widely used in other organic electronic devices [102] and conventional OSCs [27, 103, 104] to tune the work function of ITO or metal electrode, yet it is rarely adopted in inverted OSCs to tune the work function of the electrodes. Hau et al. utilized a SAM and inserted it between a TiO_2 buffer layer and an organic active layer, but

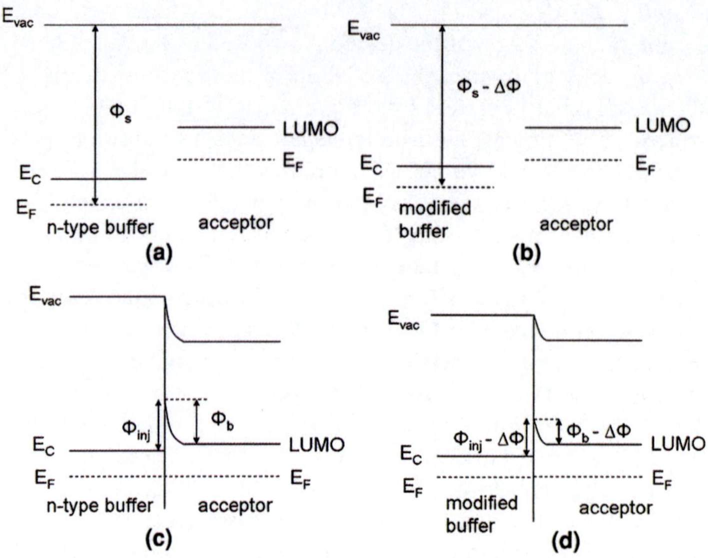

Fig. 19 Schematic energy band diagram of **a** typical n-type buffer layer and acceptor and **b** work-function-tuned (lower) buffer layer and acceptor before Fermi level alignment of the layers. **c** and **d** Energy band diagram after Fermi level alignment. Φ_s, Φ_b and Φ_{inj} represent the work function of buffer layer, the barrier for electron to be transported to electrode and the electron injection barrier under forward bias respectively. E_{vac}, E_c, and E_f refer to the energy of vacuum level, conduction band and Fermi level, respectively

the main purpose was to reduce the charge recombination rather than to tune the work function of the electrode [100]. The PCE of inverted cell increased from 2.8 to 3.78% by adding the C_{60}-based SAM because the C_{60}'s electroactive functional group reduced the recombination losses at the interface by passivation of inorganic trap states and promoted photoinduced charge transfer at the interface.

Chen et al. reported that the PCE of their inverted cell was slightly lower than that of their best conventional cell due to a lower V_{oc} and FF. However, their device's J_{sc} was higher than conventional cell (11.1 vs. 10.6 mA/cm^2) and the EQE was higher in the whole spectrum [49]. In addition, the maximum EQE of more than 80% was reported [50, 56]. Based on these facts, it was speculated that the spontaneous vertical phase separation of the BHJ blend resulted in a donor-enriched top surface and an acceptor-enriched bottom contact, which accounts for the enhancement of the charge collection efficiency [49]. Therefore, manipulating the orientation of vertical segregation in an acceptor–donor blend is an alternative and promising technique to further improve the efficiency of inverted cells. The segregation in the blend can be manipulated by the careful control of driving forces such as entropic and enthalpic forces [105], crystallization [106], and surface energy of the materials [107]. Utilizing surface energy as a driving force, Wei et al. reported self-organized buffer layers for conventional OSCs, in which a small amount of PCBM with a fluorocarbon chain (F-PCBM) is mixed in P3HT:PCBM solution. Owing to the low surface energy of fluorocarbon, F-PCBM spontaneously migrates to the surface of organic layer during spin-coating, forming a monolayer of F-PCBM on the surface, which serves as a buffer layer for the electron extraction. This self-organized F-PCBM layer also creates a dipole moment pointing away from top Al electrode, which further reduces the work function of Al top electrode, resulting in an increase in FF from 64.4 to 70% and J_{sc} from 8.72 to 9.51 mA/cm^2. Mimicking this approach and adopting the technologies for the segregation of the blend materials, the manipulation of segregation of BHJ blend to obtain a donor-enriched top surface and an acceptor-enriched bottom contact is achievable, leading to further efficiency improvement of inverted cells.

2.7 *Air-Stability Study of Inverted Organic Solar Cell*

Principally, the inverted structure provides better air-stability than a conventional structure due to a less air-sensitive top metal electrode. However, it is necessary to investigate the extent of air-stability that an inverted cell provides compared to a conventional cell without aid of encapsulation. Sahin et al. [42] first demonstrated that the inverted cell has a better air-stability than the conventional cell. The inverted cell (ITO/Perylene/MEH-PPV:PCBM/CuPc/Au) shows a PCE of 0.14% under illumination with 74.5 mW/cm^2 in the first day measurement and decreases by only 27% of the first day value after 2 weeks despite the presence of oxygen and humidity in the environment. In contrast, such cells using an Al top cathode with a LiF interlayer (conventional architecture) yields an initial PCE of 1.3% under 100 mW/cm^2 illumination in an N_2 filled glove box, but the PCE drops by 91% after 24 h of being kept in oxygen environment and diminishes completely after only a few days in air. Kuwabara et al. also studied the stability of P3HT:PCBM-based inverted cells by comparing an inverted cell (ITO/TiO_x/P3HT:PCBM/PEDOT:PSS/Au) with a conventional cell (ITO/PEDOT:PSS/P3HT:PCBM/Al), carrying out the test in an ambient atmosphere under continuous 100 mW/cm^2 illumination [74]. The PCE of the conventional cell drops to 50% of its maximum value after light illumination for 10 h, whereas the inverted cell without encapsulation maintains its efficiency under continuous light illumination for 20 h, demonstrating its stability not only in ambient environment but also under continuous illumination (Fig. 20). Hau et al. also conducted the air-stability study of inverted cells employing ZnO nanoparticles (NP) as an electron selective layer and PEDOT:PSS with Ag as a top contact [64].

The shelf lifetime, which is defined as the degradation time of PCE from the original value to half of it, of conventional device (ITO/PEDOT:PSS/P3HT:PCBM/LiF/Al) is only one day and the device is completely degraded after 4 days, showing the rapid degradation of conventional device (Fig. 21a). After 4 days, a large decrease in dark current density at 2 V was observed compared to its initial value (the inset of Fig. 21a), suggesting an increase in series resistance is part of the degradation mechanism. In contrast, the inverted device (ITO/ZnO NP/

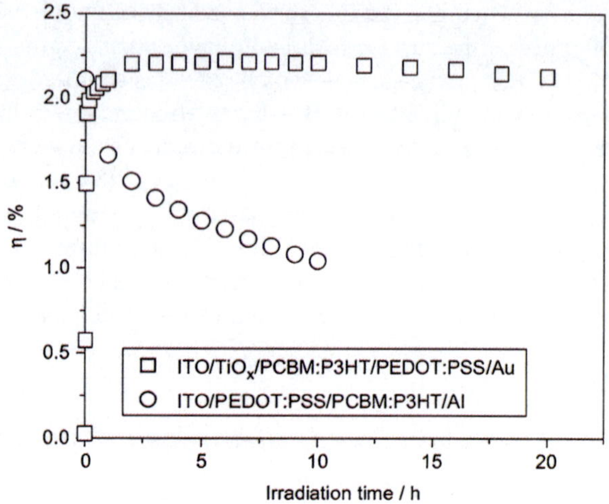

Fig. 20 The variation in η (PCE) of inverted device (ITO/TiO$_x$/P3HT:PCBM/PEDOT:PSS/Au) and conventional device (ITO/PEDOT:PSS/P3HT:PCBM/LiF/Al) with irradiation time. The cells were exposed in air during the irradiation [74] (Reprinted from [74], Copyright 2008 with permission from Elsevier) (http://www.sciencedirect.com/science/journal/09270248)

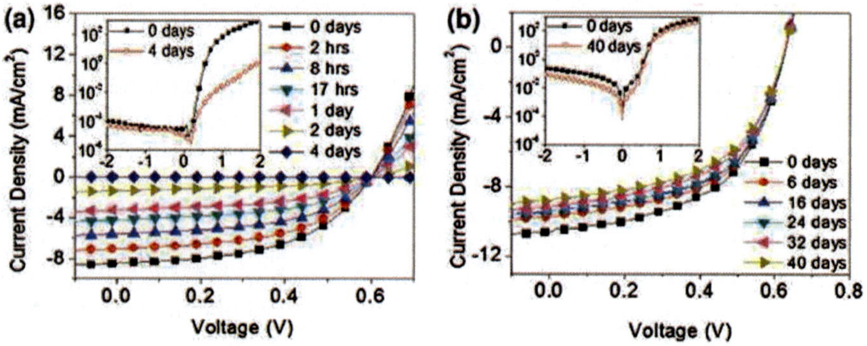

Fig. 21 **a** *J–V* characteristics of unencapsulated (a) conventional cells over a period of our days and **b** inverted cell over a period of 40 days in air under ambient conditions. *Inset*: dark *J–V* characteristics of the respective cells. (Reprinted with permission from Ref. [64]. Copyright 2008 American Institute of Physics)

P3HT:PCBM/PEDOT:PSS/Ag) without encapsulation is relatively stable in air over a long period of time, leading to only a 20% drop in PCE after 40 days (Fig. 21b). A decrease in dark current density due to storage in air was not observed (the inset of Fig. 21b). The improvement in the stability of small molecule OSCs by adopting an inverted structure was reported by Liu et al. [66]. Two air-stability tests were performed; devices were kept in the dark for one test and kept under 100 mW/cm^2 illumination for the other. The former test revealed that the shelf lifetime of an inverted cell (ITO/ZnO/CuPc:C$_{60}$/CuPc/PEDOT:PSS/Ag) kept in air is 912 h whereas that of the conventional cell (ITO/CuPc/C60/BCP/Al) kept in air is only 256 h. At 912 h, the PCE of conventional cell drops by 99.5% and the cell is almost completely degraded. In the latter test, continuous illumination also degrades the conventional device faster while the degradation of

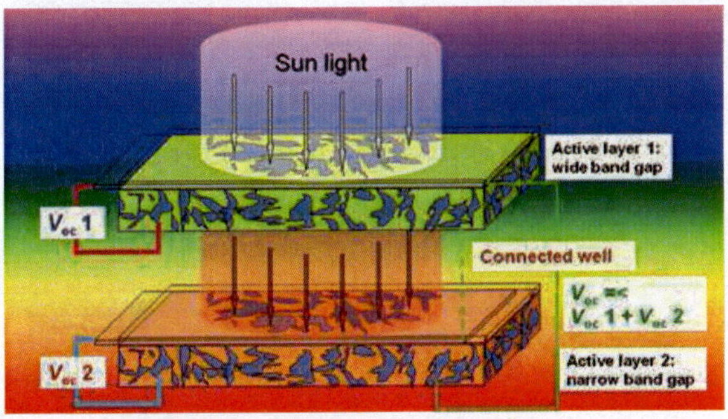

Fig. 22 General device structure of a tandem OSC

inverted cell is negligible. An initial increase in the PCE of the inverted cell was observed, which can be attributed to the improvement in mobility of ZnO due to the filling of electron traps under light illumination. All of these stability tests show that the degradation rate for inverted cells is much slower (more than one order of magnitude) than conventional cells, although the degradation is not totally evitable due to the air-vulnerable nature of organic materials.

3 Tandem Organic Solar Cells

3.1 Introduction

Although the PCEs of conventional and inverted OSCs have been enhanced remarkably [1, 3, 5, 16], improvement is still needed. The limitations of current OSCs lie mainly in the narrow absorption range, short exciton diffusion length, and relatively low charge mobility of organic semiconductors, limiting the photocurrent generation and charge transport in the devices. Due to wide band gaps, a large fraction of solar light is lost in most organic semiconductors. Therefore, a thick layer is required to absorb sunlight as much as possible, but the low charge mobility and short exciton diffusion length limit the thickness of active layer. In order to improve the absorption of the solar irradiation by relatively thin active layers, materials with low band gaps and broad absorption ranges have been designed and applied [3, 5, 108, 109]. When two active layers with complementary absorption ranges are combined, a full coverage of solar spectrum can be realized. Hence, the performance of tandem OSCs is expected to improve by stacking different active layers with complementary absorption spectra and a proper thickness.

A tandem structure [19, 20, 30, 110] consisting of two complementary absorbers in series, as shown in Fig. 22, offers several merits: (1) summation of the V_{oc}'s of individual cells; (2) a broad absorption spectrum obtained by using materials with complementary absorption spectra, ensuring a high J_{sc}; (3) higher optical density over a wider fraction of the solar spectrum than that of a single cell without increasing the internal resistance [30], guaranteeing a reasonable FF.

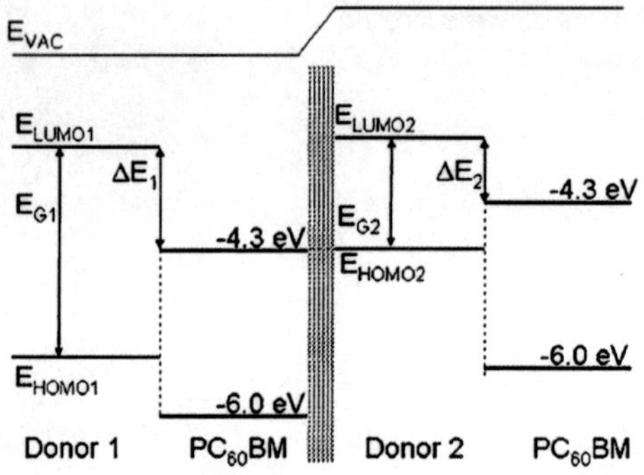

Fig. 23 Scheme of the energy levels of a tandem cell considered in the prediction [111] (Copyright Wiley–VCH Verlag GmbH & Co. KGaA. Reproduced with permission)

In this section, we will discuss device design and operation of tandem OSCs, and their theory and limits. Then we will review the devices developed in detail, concentrating on the intermediate layers employed and device performance. Finally, the lifetime or stability of tandem OSCs will be taken into account.

3.2 Device Design and Operation: Theory and Limits

3.2.1 Design Rules

Just like efficiency prediction in single OSCs, a simple model has been developed to predict tandem cell efficiency with estimates approaching a promising 15% [111], assuming a few ideal properties of materials such as the HOMO, LUMO, and band gap, which will assist the realization of highly efficient tandem cells. In addition, a very important assumption for all calculations is a constant EQE for the absorbed wavelengths. Figure 23 shows the scheme of the energy levels of tandem cells considered in this prediction. Dennler et al. considered three cases: the first is a tandem cell that has a P3HT-based bottom cell and variable donor-based top cells; and the second is a tandem cell that has a PCPDTBT:PC_{60}BM-based top cell and variable donor-based bottom cells. From these two cases, conclusions have been drawn that the realization of tandem cell by stacking two sub-cells in series can only enhance the efficiency of the corresponding single cells only in the situation that the performance of the donor materials based devices is achieved under their maximum potential capability. This has been proven by Kim's work [30]. Moreover, the third calculation is carried out by fixing $E_{\mathrm{LUMO1}} = E_{\mathrm{LUMO2}} = -4$ eV, the case of PCBM as the acceptor and varying the donor materials. The maximum efficiency of 15% was proposed to be a bottom cell with a donor E_g of 1.6 eV and a top cell with a donor E_g of 1.3 eV.

For more specific calculations, Dennler et al. introduced rigorous coupled wave analysis (RCWA) and transfer matrix formalism (TMF) to perform the optical calculations [112]. It has been presented that the selection of donor materials in the corresponding cells plays an important role. And the best device performance can be obtained by stacking PCPDTBT:PC_{60}BM as the bottom cell and P3HT:PC_{70}BM as the top cell in order to achieve a well-matched photocurrent. This conclusion is in accordance with the work performed by Kim et al. [30]. Two points that are worth mentioning here are: (1) the interference effect is obvious in the top cell, but, on the contrary, no strong interference effect appears in the bottom cell; (2) the proper acceptors must be carefully chosen in order to optimize the tandem device performance, since the $PC_{60}BM$ and $PC_{70}BM$ acceptors have significantly different absorption ranges that could lead to the balanced photocurrent in the tandem cell.

Moreover, Persson et al. [113] unified both optical and electrical modelings under monochromatic or polychromatic light in tandem cells, consisting of APFO 3:PCBM (1:4) as the bottom cell and APFO Green 1:BTPFC70 (1:4) as the top cell, connected by an intermediate layer PDMS (a "separator"). The spatial absorption profile in the cell was calculated by unifying coherent and incoherent light addition in the incident and reflected lights. The thickness of each active layer has been optimized by means of optical electric field distribution, energy dissipation at a given wavelength, and accumulated dissipation, taking light absorption and charge transport into account. Such a combined model provides a simple and promising way to optimize the tandem OSCs.

3.2.2 Device Operation

If an intermediate layer is ideal, i.e., the holes from one cell should efficiently recombine with the electrons from the other cell, there is no voltage loss across this intermediate layer. Therefore, the V_{oc} of the tandem cell in a series connection is the summation of the V_{oc}'s of the individual cells ($V_{oc,\ tandem} = V_{oc,\ 1} + V_{oc,\ 2} + V_{oc,\ 3,\ldots}$).

As reviewed by Ameri et al. [114, 115], the FF of the individual cells determines the overall efficiency of the tandem cell. Figure 24 shows the comparison of two cases using different combinations of the individual cells. In one case, shown in Fig. 24a, one cell has a lower J_{sc} and much higher FF whereas the other cell has a high J_{sc} and much low FF, resulting in the J_{sc} of the tandem cell equal to the smaller of the two ($J_{sc,\ tandem} = \text{Min}\ [J_{sc,\ 1}, J_{sc,\ 2}]$) [111]. In the other case, shown in Fig. 24b, one cell has a much lower J_{sc} and FF while the other cell has a much higher J_{sc} and FF, leading to an extremely low FF and the J_{sc} of the tandem cell ($J_{sc,\ tandem} = \text{Max}\ [J_{sc,\ 1}, J_{sc,\ 2}]$) [114]. These calculations can predict the performance of tandem cell by knowing performance of both sub-cells [115], providing a direction for the optimization of tandem cell in order to approach the optimal device performance.

3.3 Realization of Tandem Organic Solar Cells: Different Types

In tandem OSCs, the intermediate layer is crucial and should possess the following properties: (1) low electrical resistance; (2) high transparency in the visible and infrared ranges; (3) low barriers for both electron and hole extractions; (4) easy-fabrication process; and (5) protection for the previously deposited active layer in

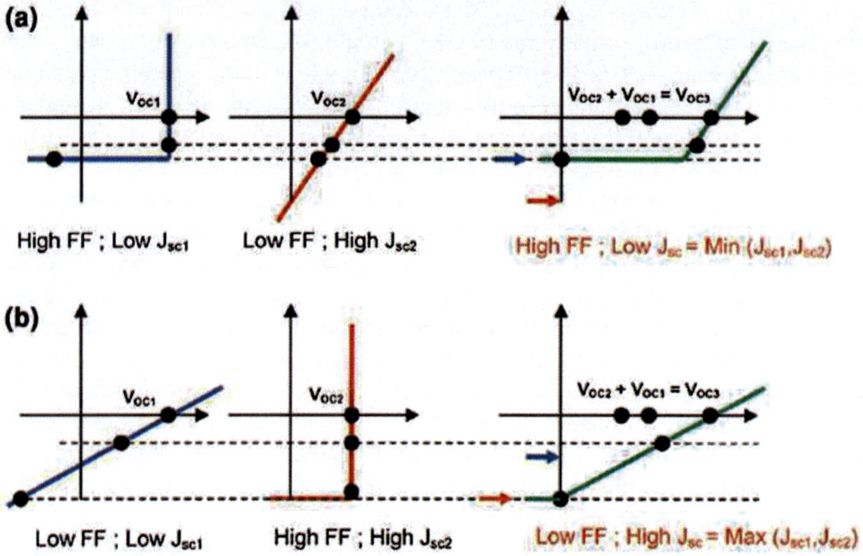

Fig. 24 **a** A case in which a tandem cell, consisting of a cell with lower J_{sc} and a much higher FF and the other cell with a high J_{sc} and much lower FF, has a J_{sc} of the tandem cell equal to the minimum of the two cells ($J_{sc,\,tandem} = \mathrm{Min}\,[J_{sc,\,1}, J_{sc,\,2}]$. **b** A case in which a tandem cell, consisting of one cell with a much lower J_{sc} and FF as well as the other cell with much higher J_{sc} and FF, has a J_{sc} of the tandem cell equal to the maximum of the two cells ($J_{sc,\,tandem} = \mathrm{Max}\,[J_{sc,\,1}, J_{sc,\,2}]$ [114] (Reproduced by permission of The Royal Society of Chemistry) (http://dx.doi.org/10.1039/B817952B)

a solution-processed tandem cell. Herein, we focus on the functional intermediate layer and discuss the realization of tandem cells by using various combinations of multiple layers to connect sub-cells. Based on their different compositions (active layers) and intermediate layers with their corresponding fabrication techniques, tandem OSCs can be divided into three types:

(a) Small molecule/small molecule tandem solar cells, in which thermally evaporated, low-weight small molecules are used as the active materials in both bottom and top cells.
(b) Polymer/small molecule tandem solar cells, in which a solution-processed polymer BHJ is used as the active medium in the bottom cell and thermally evaporated small molecules are active in the top cell.
(c) Polymer/polymer solution-processed tandem solar cells, in which solution-processed polymer BHJs are used as the active materials in both bottom and top cells by spin-coating.

3.3.1 Small Molecule/Small Molecule Tandem Solar Cells

For small molecule/small molecule tandem cells, the whole process is completed in a high-vacuum evaporation system, ensuring the sequential multiple layers are deposited with high-quality interfaces. The devices consist of multiple layers of

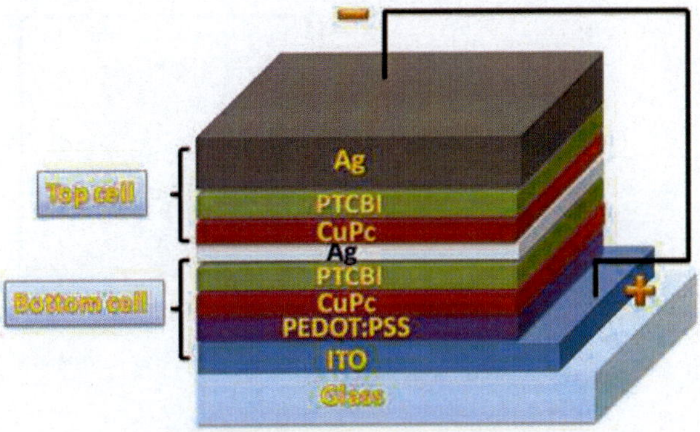

Fig. 25 The structure of the tandem OSC with 0.5 nm Ag nanocluster as recombination center, made of CuPc and PTCBI as donor and acceptor, respectively [110]

donors and acceptors or their mixtures, without damaging any previously deposited layers.

In 1990, Hiramoto et al. [116] reported the first small molecule/small molecule tandem OSCs with phthalocyanine (H_2Pc) (50 nm) and a perylene tetracarboxylic derivative (Me–PTC) (70 nm) as the active layers. An ultrathin Au (2 nm) layer was inserted to provide Ohmic contact between the sub-cells. This tandem cell exhibits a doubled V_{oc} of 0.8 V, compared to the V_{oc} of the single cell (0.4 V), demonstrating that the Au layer acts as an effective recombination center for the holes from the top cell and the electrons from the bottom cell.

Yakimov and Forrest [117] presented the first multiple heterojunction-based tandem cells with Ag interfacial metallic nanoclusters as the recombination centers by stacking two, three, or five cells. The heterojunction cells are made of CuPc and PTCBI as the donor and acceptor materials, respectively. The device structure is shown in Fig. 25. It is mentioned that the built-in voltage of such cell is determined by the difference in the Fermi levels of the constituent organic materials. The cell without the Ag cluster exhibits no tandem effect. When the nanoclustered Ag is incorporated between two sub-cells, the V_{oc} is summated to be 0.9 V, where the V_{oc}'s of individual cells is 0.45 V. With the Ag thickness increasing, the V_{oc} does not change but the J_{sc} decreases dramatically. Thick Ag lessens the photocurrent generation because the amount of light reaching the top cell is reduced, resulting in a decrease of the photocurrent in top cell. This demonstrates that 0.5 nm discontinuous nanocluster Ag is optimal for the performance of tandem cell, serving as the recombination center. It is noted that the final current generated in the tandem cell depends on how efficient charges can recombine at the nanoclustered Ag intermediate layer [110]. Any more current generated in one cell than the other one will induce unbalanced charges at the intermediate layer, leading to the degradation of performance. Finally, the maximum PCE of double-tandem cell achieves 2.5% with $V_{oc} = 0.9$ V under 100 mW/cm^2.

Triyana [118] reported triple-tandem cells comprised of materials with different absorption ranges as the active layers. Similarly, the V_{oc} is summated.

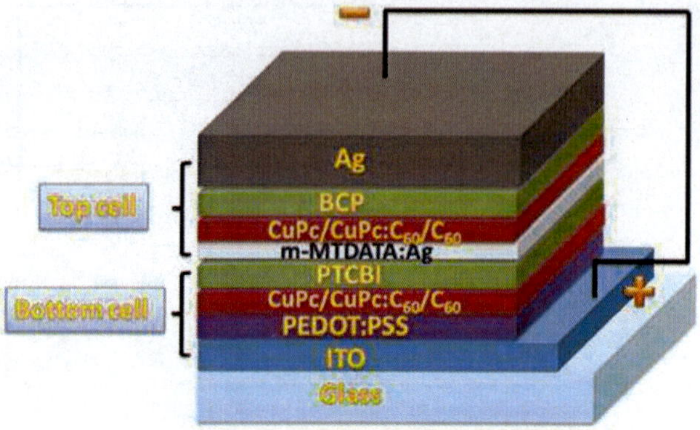

Fig. 26 The structure of hybrid PM-heterojunction tandem OSC [120]

The photocurrent is still low since the materials (phthalocyanine/perylene, and CuPc/PTCBI) have short exciton diffusion lengths. However, the PCE is improved due to more balanced photogenerated carriers by using appropriate material for the second heterojunction component. Cheyns [119] presented double-tandem cells, consisting of pentacene and C_{60} as the donor and acceptor, respectively, in order to improve photon harvesting. The authors compared different metal-based intermediate layers. The results indicate that thin Al causes S-shaped curves in *I–V* characteristics due to a barrier for charge extraction inside the device, however, by incorporating a thin metal layer of Au or Ag, the S-shaped curves disappear and the V_{oc} increases.

In order to further improve the efficiency of tandem cells, Xue and Forrest developed a few fabrication technique to enhance the performance of single [121] and tandem cells [120]. The structure of their tandem cell is shown in Fig. 26. This tandem cell has a symmetric spectral response from the individual cells, since the front cell has an absorption peak in the long-wavelength range and the back cell efficiently absorbs in the shorter-wavelength range. As a result, the PCE of such an "asymmetric" tandem cell reaches nearly 6% under 100 mW/cm^2 illumination. This approach makes efficiencies in excess of 7% a possibility. It is worth highlighting some points here: (1) C_{60} is applied as the acceptor instead of PTCBI due to its longer exciton diffusion length ($\sim$40 nm) and high charge-carrier mobility; (2) a mixed donor–acceptor layer is used, sandwiched between pure donor and acceptor layers, a hybrid planar-mixed heterojunction (PM-HJ) [121], in order to increase the donor/acceptor interfaces for the enhancement of exciton dissociation; and (3) BCP and PTCBI are used in both the electron-transporting layer and the blocking layer in the top cell and the bottom cell, respectively, for the interfacial modifications, as shown in Fig. 26. The blocking layer functions to: (1) block excitons from diffusing to the interfaces of organic layer/metal electrode and consequently quenching; (2) prevent damage of Ag evaporation onto the top cell; (3) reduce the diffusion of Ag particles into active layer, which act as quenching centers; and (4) provide an optical spacer for the entire device to tune the optical field distribution. Ag nanoclusters are also inserted in m-MTDATA layer, a p-doped organic semiconductor. Doping this intermediate layer increases its conductivity and reduces Ohmic losses due to non-Ohmic contacts. The highly conductive transporting layer can adjust the optical field distribution in the device by controlling its thickness, and furthermore the increase of the thickness can make the devices stable and the probability of short-circuit low [110].

Yu et al. [122] presented an all-organic intermediate layer in small molecule/small molecule tandem cells based on ZnPc:C_{60} heterojunction. The $SnCl_2Pc$/$F_{16}CuPc$ intermediate layer has advantages in higher transparency and lower sublimation temperature, serving as an effective recombination center. First of all, only $F_{16}CuPc$ is inserted between two sub-cells. Although the V_{oc} of the tandem cell approaches 0.78 V, it is still smaller than the combined V_{oc} of the single cells (0.54 V) due to energy loss, primarily caused by the formation of an interface barrier at the $F_{16}CuPc/C_{60}$ interface. In order to improve this interface, an n-type organic material $SnCl_2Pc$ is introduced between C_{60} and $F_{16}CuPc$, leading to the direct transport of electrons from $SnCl_2Pc$ layer to $F_{16}CuPc$ layer and reducing the energy loss. As a result, the V_{oc} of 1.04 V, nearly the doubled V_{oc} of the single cell, was obtained with an optimal $SnCl_2Pc$ thickness of 3 nm. Finally, the tandem cell has a PCE of 1.81% with a $J_{sc} = 3.64$ mA/cm^2, FF = 48%, and $V_{oc} = 1.04$ V. The explanation for the improved intermediate layer is that the electrons are accumulated in $F_{16}CuPc$ layer in both heterojunctions of $SnCl_2Pc/F_{16}CuPc$ and ZnPc/$F_{16}CuPc$, resulting in high electron density and improving the barrier height as well as the tunneling probability. Therefore, the charge carriers recombine via tunneling at the ZnPc/$F_{16}CuPc$ interface.

3.3.2 Polymer/Small Molecule Tandem Solar Cells

This type of tandem cell is constructed by solution and high-vacuum processes. In general, such a tandem cell can be easily realized by stacking two sub-cells with complementary absorption ranges without damaging the previously deposited active layer.

Dennler [123] presented tandem OSCs with a bi-layer of P3HT and PCBM as the bottom cell and small molecules ZnPc/C_{60} in multiple layers as the top cell, as shown in Fig. 27. In this tandem cell, the bottom cell mainly absorbs light in the range from 375 to 630 nm and the top cell has an absorption band ranging from 600 to 800 nm. As a result, the absorption of this tandem cell covers a broad range compared to those covered by individual cells. A thin layer of Au (1 nm) is used as the intermediate layer. It is worth mentioning that the bottom cell is made of a stratified bilayer or diffused bilayer due to different solubility of P3HT and PCBM in chlorobenzene and dichloromethane, respectively. Such sequential double layers can assist exciton dissociation at the P3HT/PCBM interfaces and charge transport. The usage of ZnPc, C_{60}, and their mixture ZnPc:C_{60} is similar to that of CuPc:C_{60} system; however, ZnPc absorbs light in a longer wavelength range than CuPc. This makes the absorption spectrum much more complementary in such a

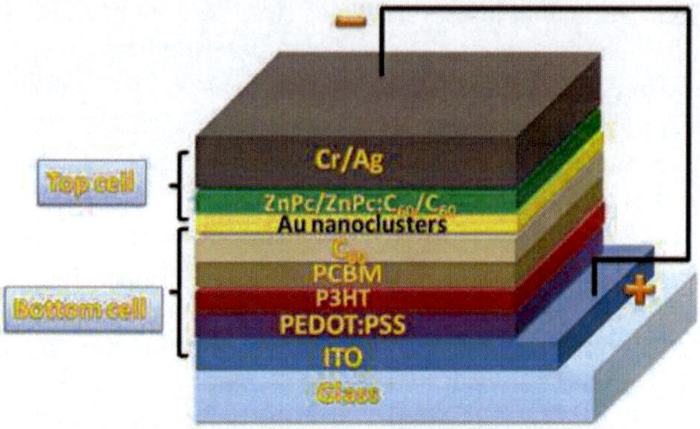

Fig. 27 The structure of polymer/small molecule tandem OSC, where the bottom cell consists of a stratified bilayer of P3HT and PCBM, and the top cell is fabricated by the evaporation of ZnPc, C_{60} and their mixture [123]

tandem cell. One advantage of this method is that the fabrication of top cell has no effect on the previously deposited bottom cell. Therefore, the V_{oc} of tandem cell (1.02 V) reaches the sum of that of bottom cell (0.55 V) and that of top cell (0.47 V). In addition, the J_{sc} decreases to 4.8 mA/cm^2 compared to that of bottom cell (8.5 mA/cm^2) and that of top cell (9.3 mA/cm^2); correspondingly, the FF reduces to 45% from 55 and 50%, respectively. Finally, the PCE of 2.3% was attained, compared to single reference cells (2.6 and 2.2%). Although the enhanced V_{oc} is realized, the performance of the tandem cell is still limited by the low photocurrent and reduced FF.

Colsmann [124] reported tandem OSCs comprised of a P3HT/PCBM heterojunction as the bottom cell and a small molecule CuPc/C_{60} layer as the top cell with a new intermediate layer of BPhen:Li/Au/MTDATA:F4-TCNQ, which was fabricated by using two doped organic semiconductor layers and a thin noble metal interlayer. This intermediate layer functions well, making the summation of V_{oc} achievable; however, the performance of the tandem cell is low due to the poor individual single cells. Through optimization, 60 nm-thick active layer of bottom cell is the best available, and either doped layer is mandatory for the operation of the tandem cell. In such a tandem cell, the total current is lower than those of individual cells, explained by the large spectral overlap of both sub-cells and thus the competition for photon harvesting. This approach is similar to the technique employed in tandem OLEDs, where doped transporting layers are used as the intermediate layer. In addition, the high transparency of intermediate layer is rather crucial for tandem OSCs.

Janssen [125] used a thin metal oxide WO_3 (3 nm) layer instead of Au (10 nm) layer as part of the intermediate layer of Al (1 nm)/WO_3 (3 nm) to connect the bottom (P3HT:PCBM) and top cells (CuPc:C_{60}). The structure is shown in Fig. 28. The results show that the intermediate layer with 3 nm WO_3 has much higher transparency than that with 10 nm Au (only 50% in the spectral region for the

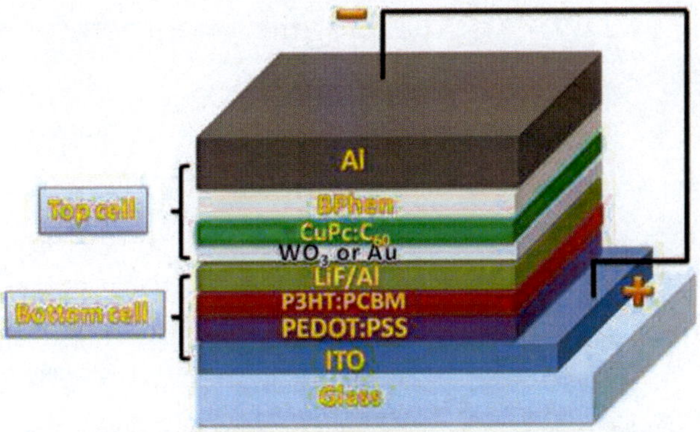

Fig. 28 The structure of the tandem cell with different intermediate layers: Al/Au and Al/WO_3 [125]

optimum absorption of top cell), although both sub-cells exhibit the summation of V_{oc}. The 200 nm bottom cell of P3HT:PCBM, optimized for the best performance, is thick enough to absorb its corresponding light. Therefore, the maximum PCEs of 3.0% at 160 mW/cm^2 and up to 4.6% at 16 mW/cm^2 were obtained in the tandem cell with the Al/WO_3 intermediate layer, higher than those of tandem cells with Al/Au intermediate layers. Additionally, a reduced series resistance of 6 Ω cm^2 for WO_3 cell is found compared to that of 8 Ω cm^2 for Au cell.

We developed a new and effective intermediate layer, combining the metal nanocluster and metal oxide layer together [19]. This intermediate layer consists of 1 nm Al and 15 nm MoO_3, connecting the P3HT:PCBM active layered bottom cell and CuPc:C_{60} in multiple layers as the top cell. Figure 29a shows the structure of the tandem cell. Figure 29b shows the absorption spectra of P3HT:PCBM, CuPc/CuPc:C_{60}/C_{60} and P3HT:PCBM/LiF/Al/MoO_3/CuPc/CuPc:C_{60}/C_{60} films. Obviously, the absorption spectrum of tandem film is broadened since the bottom film mainly covers the visible range from 400 to 650 nm and the top film absorbs complementarily from 650 to 750 nm. More importantly, the transmittance spectrum of Al (1 nm)/MoO_3 (15 nm) intermediate layer is almost 98% in the range from 350 to 900 nm, as shown in Fig. 29b. Hence, this intermediate layer with high transparency satisfies the optical requirement as outlined in the previous section.

The *I–V* characteristics of bottom single, top single, and tandem cells in the polymer/small molecule tandem structure are shown in Fig. 30a under 100 mW/cm^2. The PCEs of the bottom and top single cells are 2.11 and 1.68%, respectively. With the Al (1 nm)/MoO_3 (15 nm) intermediate layer, the tandem cell has a PCE of 2.82% with $V_{oc} = 1.01$ V, $J_{sc} = 6.05$ mA/cm^2, and FF = 46.2%. The V_{oc} (1.01 V) of the tandem cell is the sum of the V_{oc}'s of the bottom single cell (0.63 V) and top single cell (0.45 V), which demonstrates that the tandem cell is connected electrically by this Al/MoO_3 intermediate layer. The matched J_{sc} of bottom and top single cells is important for tandem cells in series. From Fig. 30a,

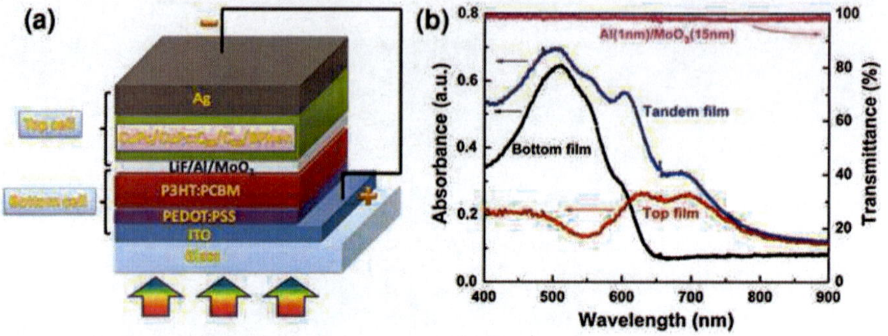

Fig. 29 **a** The resulting structure of the polymer/small molecule tandem cell. **b** The absorption spectra of the bottom film P3HT:PCBM (120 nm), top film CuPc(7.5 nm)/CuPc:C_{60} (12.5:12.5 nm)/C_{60}(27.5 nm) and tandem film P3HT:PCBM(120 nm)/LiF(0.5 nm)/Al(1 nm)/MoO_3(15 nm)/CuPc(7.5 nm)/CuPc:C_{60}(12.5:12.5 nm)/C_{60}(27.5 nm) films. **b** The transmittance spectrum of the Al(1 nm)/MoO_3(15 nm) intermediate layer (Reprinted with permission from Ref. [19]. Copyright 2008 American Institute of Physics.)

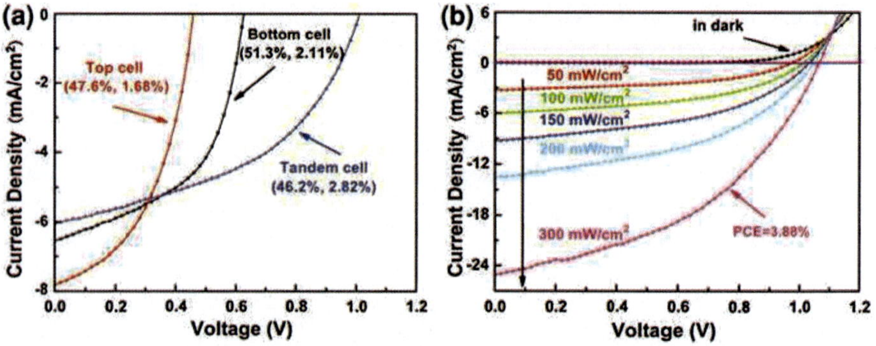

Fig. 30 **a** The *I–V* characteristics of bottom single, top single cells and polymer/small molecule tandem cell under 100 mW/cm^2. **b** The *I–V* characteristics of polymer/small molecule tandem cell under different illuminations. The corresponding FF and PCE are shown in the format of (FF, PCE) (Reprinted with permission from Ref. [19]. Copyright 2008 American Institute of Physics)

the J_{sc} (7.83 mA/cm^2) of top single cell is slightly higher than that (6.54 mA/cm^2) of bottom one, demonstrating that the nearly matched photocurrent density between the bottom and top cells can be achieved in this tandem cell. Overall, the PCE of the tandem cell is higher than either of the individual single cells due to the increase of the V_{oc}. Figure 30b shows the *I–V* characteristics of the polymer/small molecule tandem cell under different illuminations. Clearly, its V_{oc} increases with the intensity of illumination because of the increased V_{oc}'s of both bottom and top single cells. Despite FF reduction from 47.2% (at 50 mW/cm^2) to 43.3% (at 300 mA/cm^2) monotonically, the maximum PCE reaches 3.88% at 300 mW/cm^2 primarily due to the enhanced V_{oc}.

In this tandem cell, MoO_3 has two functions: (1) hole transport and (2) exciton blocking. The holes collected/transported by MoO_3 and electrons generated from the bottom cell are recombined in the Al/MoO_3 intermediate layer (1 nm Al, most likely in nanocluster form). MoO_3 has been used as a hole-injection layer for OLEDs [71, 126] and a replacement of PEDOT:PSS in polymer OSCs [18] for its good hole-transporting property. Meanwhile, MoO_3 has the other function of exciton blocking. Due to the higher energy band gap of MoO_3 (3.0 eV) than the exciton energy of CuPc (1.9 eV), excitons formed in the top cell will be blocked by MoO_3. Only after exciton dissociation (at the CuPc/C_{60} interfaces) can holes be collected by MoO_3 layer, contributing to photocurrent. More importantly, this intermediate layer can also act as the protection layer for the previously deposited polymer active layer. This function will be discussed in the following section.

3.3.3 Polymer/Polymer Solution-Processed Tandem Solar Cells

Polymer OSCs are believed to have the most potential in the commercial application due to their large area, easy fabrication, and low cost. Generally speaking, polymer OSCs are prepared by using simple techniques such as spin-coating, printing, and doctor-blading. For their applications in tandem structure, the main problem is that the previously deposited active layer can be easily dissolved or damaged by the process to deposit the top cell since the polymers can be dissolved in most organic solvents. Therefore, this requires an intermediate layer with a special property of protecting the previously deposited layers while providing optical and electrical connections. This issue has been overcome to a certain degree, which will be reviewed in the following section.

ITO/PEDOT as Intermediate Layer

Kawano [127] reported a tandem OSC made of two identical BHJ MDMO-PPV/PCBM as the active layers, in which ITO/PEDOT acts as the intermediate layer, as shown in Fig. 31.

The intermediate layer of ITO is deposited by dc magnetron sputtering in 1 Pa of argon without substrate heating, which prevents the bottom active layer from damage by sputtering. Sputtering is followed by spin-coating layers of PEDOT and polymer for the top cell. However, it was not revealed how the PEDOT was treated in the intermediate layer since PEDOT must be dried under high temperature. The authors compared two thicknesses of sputtered ITO (20 and 100 nm). The results demonstrate that the thinner layer has lower series resistance (20 Ω) than the thicker one (27 Ω).

The tandem cell with a 20 nm ITO layer has $V_{oc} = 1.34$ V, $J_{sc} = 4.1$ mA/cm^2, FF = 56%, and PCE = 3.1%. The V_{oc} is the sum of those of bottom (0.48 V) and top cells (0.73 V), although the V_{oc} of bottom cell is lower than expected due to the HWF of its ITO cathode. Therefore, it is necessary to optimize the intermediate

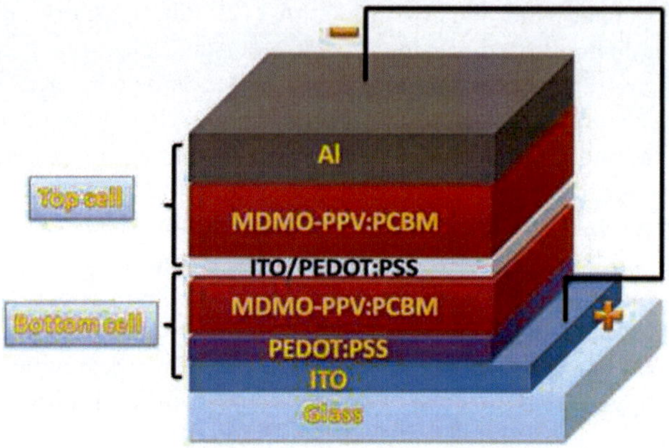

Fig. 31 The structure of tandem cell with ITO/PEDOT as the intermediate layer [127]

layer to obtain proper energy level alignment between the two sub-cells, which may be implemented using LWF materials and thin metal film or clusters, taking the structural continuity and chemical and mechanical stability against the solution process into account [127].

Metals/PEDOT as Intermediate Layer

Hadipour et al. [128] first reported a solution-processed tandem OSCs made of two BHJ cells with complementary absorption spectra, in which PFDTBT is a wide band gap donor polymer used for the bottom cell and PTBEHT is a low band gap donor polymer used for the top cell, while PCBM is used as the acceptor in both. As a result, the light in the short wavelength is absorbed by the bottom cell and that in the long wavelength is absorbed mainly by the top cell. The absorption spectra of the used materials (PFDTBT [128] and PTBEHT [108, 128]) are shown in Fig. 32.

These two sub-cells are connected by an intermediate layer of Al/Au/PEDOT:PSS, and the corresponding structure of the tandem cell is shown in Fig. 33. The LiF (0.5 nm)/Al (0.5 nm) layer acts as the cathode for the bottom cell, supplying an Ohmic contact between PCBM and LiF/Al to extract electrons, and the Au (15 nm)/PEDOT:PSS (60 nm) layer acts as the anode for the top cell, providing a stable and Ohmic contact for hole extraction. Then, the extracted electrons and holes recombine in this intermediate layer. More importantly, the Au layer is employed to protect the bottom cell from being dissolved when the subsequent PEDOT:PSS and top cell layers are deposited by spin-coating, a key point in solution-processed tandem cells. In addition, the intermediate layer is fabricated as thin as possible to allow maximum light transmission. Through careful optical and current optimization, the most desirable thicknesses are 110 nm (PFDTBT:PCBM, 1:4) for the bottom cell and 90 nm (PTBEHT:PCBM, 1:4) for

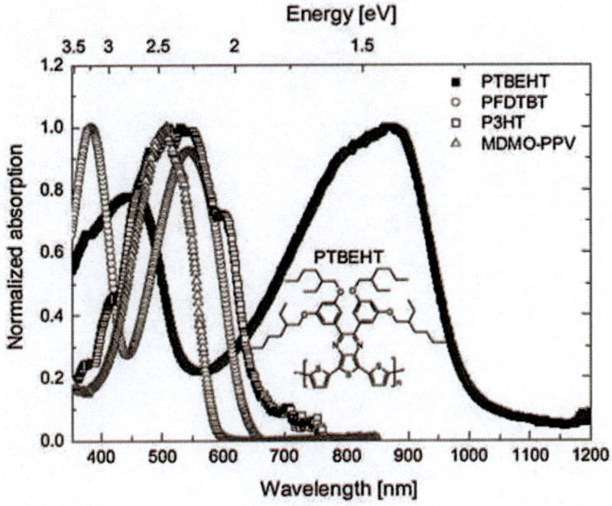

Fig. 32 The absorption spectra of the materials used in the tandem cell [128] (Copyright Wiley–VCH Verlag GmbH & Co. KGaA. Reproduced with permission)

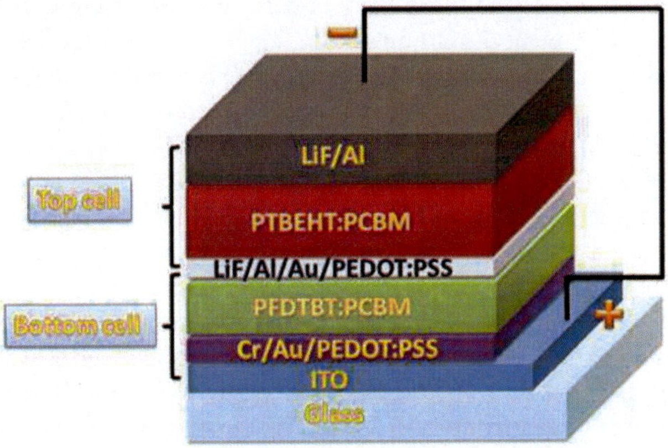

Fig. 33 The structure of the tandem cell consisting of PFDTBT:PCBM as the bottom cell and PTBEHT:PCBM as the top cell, connected by Al/Au/PEDOT:PSS intermediate layer [128]

the top cell, yielding a PCE of 0.57% with $V_{oc} = 1.5$ V, $J_{sc} = 0.9$ mA/cm^2, and FF = 55%.

Metal Oxides/PEDOT as Intermediate Layer

Gilot et al. [129] presented fully solution-processed tandem OSCs with intermediate layers of ZnO (nanoparticles)/PEDOT:PSS. These tandem cells have the

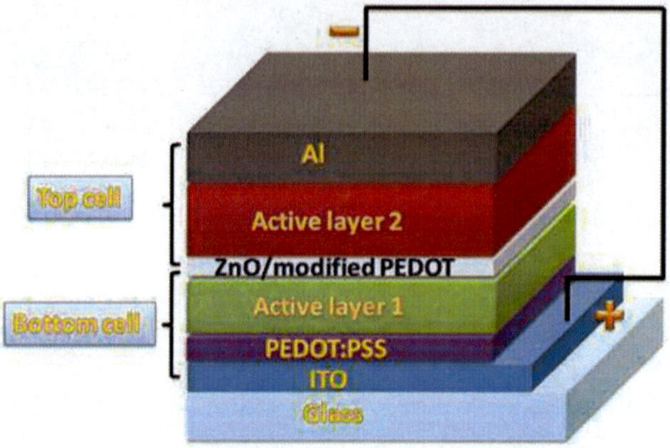

Fig. 34 The structure of double-tandem cell with ZnO/PEDOT:PSS intermediate layer. The active layers are made of MDMO-PPV/PCBM and P3HT/PCBM for the bottom and top cells, respectively [129]

combination of MDMO-PPV/PCBM and P3HT/PCBM as the active layers. The structure of a tandem cell is shown in Fig. 34.

It is worth mentioning that these tandem cells are fabricated without the use of vacuum deposition. For the ZnO fabrication, ZnO nanoparticles were prepared in acetone and spin-coated onto the active layer. ZnO is a wide band gap metal oxide and usually a high-conductivity, n-type semiconductor. PEDOT:PSS is a water-based solution with a slight acidity. The PEDOT:PSS must be modified to be neutral because it can easily dissolve the ZnO layer during spin-coating. The ZnO layer serves as the electron-transporting layer and PEDOT:PSS as the hole-transporting layer and the two form the interface at which electrons and holes recombine. Generally, such recombination is poor due to the high offset of work functions; instead an ohmic contact between these two layers is desired. In this work, the authors expose ZnO layer to UV lights for a few seconds to provide an n-type doping. Then, the V_{oc} of triple-tandem cell with MDMO-PPV/PCBM as active layers is increased from 1.40 to 1.92 V. It was found that the V_{oc} of the tandem cell was lower than the sum of those of individual cells (0.84 V), which might be attributed to large voltage drop across the interface of ZnO and PEDOT:PSS. Although the V_{oc} is much higher than those of the individual cells, the overall efficiency is still poor. This approach provides a new method to fabricate fully solution-processed tandem cells with broad complementary absorption spectra.

Moreover, another n-type metal oxide TiO_x was also introduced to fabricate tandem cells with a PCE of 6.7% by Kim et al. [30]. In such a tandem cell, a highly transparent TiO_x layer is used as the cathode for the bottom cell and is combined with PEDOT PH 500 to form the intermediate layer. This tandem cell is also fully solution-processed and the TiO_x/PEDOT PH 500 intermediate layer serves as the recombination center and protects the previously deposited active layer.

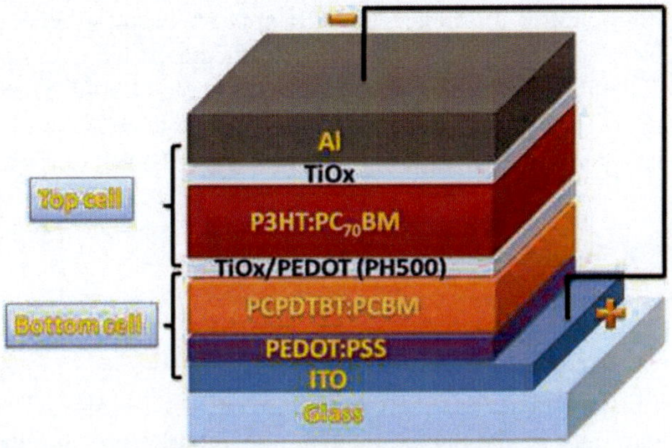

Fig. 35 The structure of the tandem cell with TiO_x/PEDOT as the intermediate layer [110]

The tandem cell, shown in Fig. 35, consists of a 130 nm thick layer of PCPDTBT/PCBM (1:3.6) as the bottom cell and a 170 nm thick layer of P3HT/$PC_{70}BM$ (1:0.7) as the top cell, with complementary absorption spectra. The absorption of this tandem cell covers the entire visible range and a small part of infrared range of solar spectrum.

For the fabrication of intermediate layer, TiO_x was prepared in a methanol solution by means of sol–gel chemistry and spin-coated onto the active layer of the bottom cell to form its cathode. The highly conductive PEDOT PH 500 was applied as the anode of the top cell. This tandem cell achieves the highest performance, with PCE = 6.7% under 20 mW/cm^2 and 6.3% under 100 mW/cm^2. The fine performance is attributed to high photocurrent generation, facilitated by these two complementary donors as photoactive absorbing layers in both sub-cells.

Sista and Yang et al. [130] recently reported a highly efficient tandem cell with P3HT:$PC_{70}BM$ as the bottom cell and PSBTBT:$PC_{70}BM$ as the top cell, connected by an improved intermediate layer of Al/TiO_2:Cs/PEDOT:PSS 4083. In this tandem cell, the bottom cell (P3HT:$PC_{70}BM$) absorbs the light ranging from 400 to 600 nm, and the top cell (PSBTBT:$PC_{70}BM$) absorbs light in the range from 600 to 800 nm, a complementary absorption spectrum. It is worth mentioning that the donor polymer PSBTBT is newly synthesized, showing a PCE of 5.5% for a single cell. For the intermediate layer, ultra-thin Al is thermally evaporated for improving both wettability and electrical contact between P3HT:$PC_{70}BM$ and TiO_2. TiO_2, which forms a densely packed network of nanocrystals, serves as the protective layer for the previously deposited active, hole-blocking, and electron collection layers. Moreover, with this intermediate layer exposed to UV light, the TiO_2/PEDOT:PSS 4083 interface transforms from a Schottky to an Ohmic contact, leaving no energy barrier at the TiO_2/PEDOT:PSS 4083 interface and facilitating efficient electron tunneling. Furthermore, the authors compared the tandem cells with PEDOT:PSS 4083 and PEDOT:PSS PH500 as part of an intermediate layer, indicating that the tandem device with PH500 gives a higher J_{sc} than that with

PEDOT:PSS 4083 and no difference in EQE appears. This is because the active area for PH500-based tandem cell is larger than the electrode overlap due to high conductivity of PH500 films. Therefore, the active area should be carefully concerned when PH500 is used as part of the intermediate layer. Overall, the PCE of tandem cell is up to 5.84% with a $J_{sc} = 7.44$ mA/cm^2, FF = 63.2%, and $V_{oc} = 1.25$ V.

Metal/Metal Oxide as Intermediate Layer

As mentioned above, metallic thin films can be used as semitransparent intermediate layers in tandem cells, such as Ag [120], Au [123], and Al/Au [128]. However, such an intermediate layer alone is not suitable for solution-processings because either the thickness is insufficient to protect the previously deposited layer or the light loss is high. The combination of n-type metal oxides (e.g., TiO_x [30] and ZnO [129]) and PEDOT:PSS has been used as intermediate layers. Disadvantages of this kind of intermediate layer are that the fabrication of the PEDOT layer is always carried out in an oxygen and moisture environment (outside the glove box) with baking and being modified to be neutral is required due to the acidic nature of PEDOT [129], which is harmful to the previously deposited layers [36]. Hence, it is desirable to replace the PEDOT layer with a functional layer serving as the anode for the top cell and protecting the previously deposited layers. Some p-type-like metal oxides, such as NiO, MoO_3, V_2O_5, and WO_3, with HWFs and good hole transport behavior, are proper candidates. These oxide films can generally be deposited by thermal evaporation, which does not damage the previously deposited organic layers.

Accordingly, we applied the Al (1 nm)/MoO_3 (15 nm) intermediate layer in polymer/polymer tandem cells (both double-tandem and triple-tandem). This intermediate layer overcomes the challenges of stacking solution-processed polymer layers repeatedly by preventing the existing polymer layers from dissolving and reducing the light loss induced by the added intermediate layer. We used an identical polymer blend (P3HT:PCBM) as the active layers of the double- and triple-tandem cells, aiming to increase the light absorption depth and shorten the charge transport length. The thicknesses of the active layer and MoO_3 layers need to be carefully adjusted to optimize the optical field distributed in each cell, resulting in maximum light absorption and matched J_{sc} in each sub-cell. The resulting structure of the solution-processed triple-tandem OSC is shown in Fig. 36.

Figure 37a compares the I–V characteristics of all cells covered in the solution-processed triple-tandem cell under 100 mW/cm^2 irradiation. The PCEs of the first (70 nm), second (85 nm), and third (50 nm) single cells are 1.94%, 2.64%, and 1.76%, respectively. The double-tandem cell has a PCE of 2.19% with V_{oc} of 1.19 V and J_{sc} of 3.71 mA/cm^2. The triple-tandem cell achieves a V_{oc} of 1.73 V with $J_{sc} = 2.41$ mA/cm^2, FF = 48.4%, and PCE = 2.03%. The V_{oc} almost triples that of a single cell (0.62 V), suggesting a relatively good Ohmic contact between

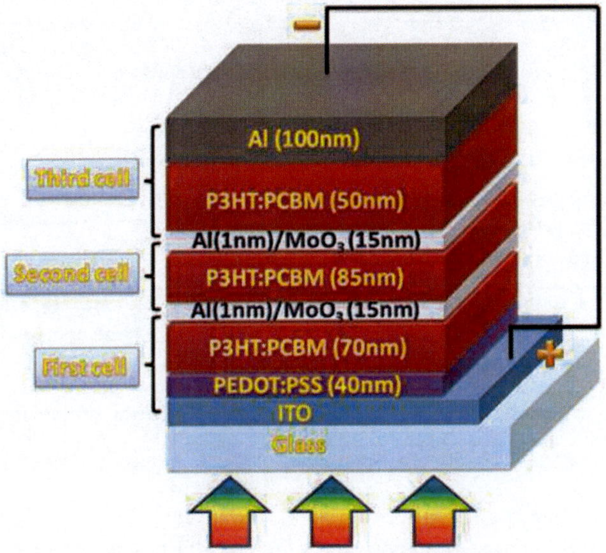

Fig. 36 The structure of the polymer–polymer triple-tandem OSC [20]

the Al and MoO_3 layers [129]. The transmittance spectra of the first (70 nm), double-tandem (70 nm/85 nm), and triple-tandem (70 nm/85 nm/50 nm) cells are shown in Fig. 37b. Obviously, the triple-tandem cell fully absorbs the solar light in the main range of P3HT/PCBM blend, indicating there is appropriate light incident on each cell. As a result, the J_{sc} (2.41 mA/cm^2) of the triple-tandem cell is reasonable. It is worth mentioning that the FF shrinks as the number of the cells rises due to an increase in the series resistance in the multi-tandem cell; however, the FF is still high here compared to that of another reported triple-tandem cell [129]. The resulting PCE of triple-tandem cell (2.03%) is comparable to those of the single cells and the double-tandem cell. Additionally, this high V_{oc} can drive a red light-emitting diode (LED) easily with only 0.1 cm^2 of active area. Since the V_{oc} of the triple-tandem cell can increase from 1.73 V (under 100 mW/cm^2) to 1.87 V (under 300 mW/cm^2), the LED grows brighter.

Additionally, Sakai et al. [131] reported a similar intermediate layer consisting of a LiF (0.5 nm), ITO (20 nm), and MoO_3 (10 nm), which connects the P3HT:PC_{60}BM layer of the bottom cell and P3HT:PC_{70}BM layer of the top cell, enhancing the light absorption due to the different absorption ranges of the two types of PCBMs. This intermediate layer also prevents the solutions used to fabricate the top cell from leaking into the previously deposited bottom active layer. In addition, the theoretical and experimental results have been compared and are in good agreement. The tandem cell obtains a PCE of 5.16% with a $J_{sc} = 6.14$ mA/cm^2, FF = 73.7%, and $V_{oc} = 1.14$ V. It is worth noting two points: one is the extremely high FF, the highest value achieved by a tandem cell,

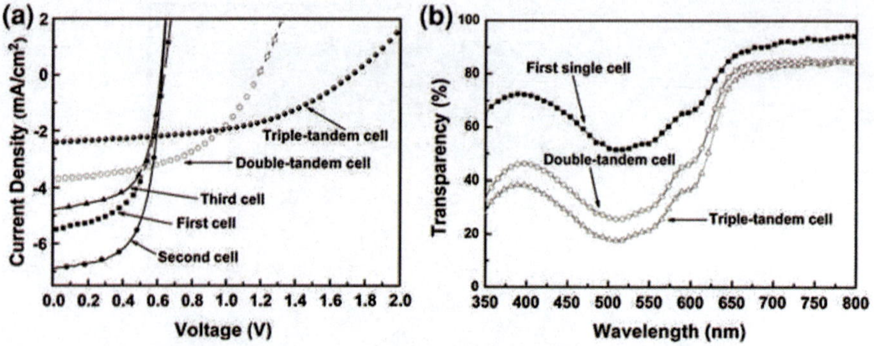

Fig. 37 **a** The *I–V* characteristics of first, second, third single, double- and triple-tandem cells under 100 mW/cm^2 irradiation. **b** The transmittance spectra of the first single film, double-tandem film, and triple-tandem film [20]

likely due to good charge transport and efficient charge-carrier recombination at the intermediate layer, as explained by the authors; the other is the somewhat low V_{oc}, less than the sum of the individual cells' V_{oc}'s, which might be caused by V_{oc} loss in the bottom cell.

As reviewed above, it was found out that most of intermediate layers involve PEDOT. Since PEDOT needs to be dried and treated outside a glove box and at high temperature, this process damages previously deposited layers, causing the degradation of the tandem cell. Therefore, a superior intermediate layer, which can be processed in a nitrogen environment without high temperature treatment, is necessary for the efficiency improvement of tandem OSCs. Moreover, good lifetime and stability are also required for commercial applications.

3.3.4 Lifetime and Stability of Tandem Organic Solar Cells

Quite few researches have focused on lifetime or stability of tandem cells. As mentioned above, increasing the thickness of each layer can stabilize the devices and reduce the probability of a short-circuit [110]. Moreover, the lifetime and stability of OSCs are influenced by chemical, physical, and mechanical degradations [22].

Franke [132] discussed the long-term stability measurements of tandem OSCs consisting of mixed ZnPc and C_{60} as the active layer with a PCE of 4%. The results indicate that these cells are stable upon exposure to a halogen light since their PCE only decreases by 3% or less under continuous 185 mW/cm^2 illumination at 50°C for more than 1,400 h, as shown in Fig. 38. Subsequently, the cell degrades fast, which can be attributed to the effect of halogen light exposure on the encapsulation adhesive. The authors also compared the degradation of those

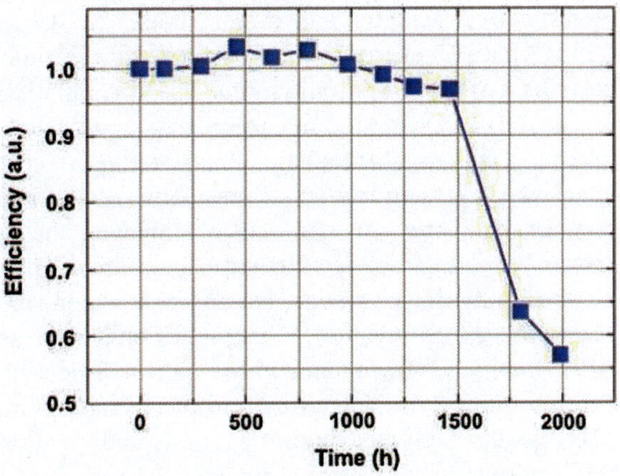

Fig. 38 Change of the normalized PCE of ZnPc:C_{60} tandem cell over time [132] (Reprinted from [132], Copyright 2008 with permission from Elsevier) (http://www.sciencedirect.com/science/journal/09270248)

tandem cells stored in the dark environment at 85°C, showing a much faster degradation likely due to the stability variation of hole-transporting layer with a low glass transition temperature.

Moreover, Kim [30] presented tandem OSCs with promising stability as well. The fact that the PCE decreases from 6.5 to 5.5% after 3,500 h storage in N_2 confirms that such a tandem structure exhibits robustness and reasonable stability. Even though exposed continuously to 100 mW/cm^2 illumination, the tandem cell keeps 70% of its original efficiency after 40 h and over 60% after 100 h.

4 Conclusion

In order to further improve the efficiency of OSCs, materials with low band gaps, which can absorb a larger percentage of the solar spectrum, are required. Additionally, adopting an effective structure of a solar cell is crucial to maximize the device performance based on specially designed donor and acceptor materials. A tandem structure consisting of two or more cells with complementary absorption spectra is able to enhance the light harvesting by overcoming the intrinsic limits of organic semiconductors, such as low charge-carrier mobility and short exciton diffusion length. On the other hand, the inverted structure, in which the charge collection is opposite to that in conventional structure, provides stable and robust device.

Many efforts have been made to understand the charge-carrier extraction behavior in inverted structures and to improve efficiency by employing various interfacial layers between the organic active layer and electrodes. The polarity of the charge collection is controlled by the interfacial layers, which are also responsible for efficient charge collection at electrode. For further efficiency improvement, matching of work functions of both electrodes with the energy levels of respective acceptor and donor should be stressed to minimize Ohmic losses during charge transfer. Work function can be tuned by doping a buffer layer or utilizing a SAM with dipole moments. Manipulating the orientation of segregation in the BHJ blend is another promising approach to improve the performance of inverted devices. Moreover, incorporating nanostructure metal oxides as the interfacial layer can further increase efficiency owing to additional heterojunctions between nanostructure and donor. However, the device must overcome charge-carrier recombination at the surface of metal oxides via charge trapping sites created by hydroxyl groups, leading to low contribution to the final photocurrent. Therefore, the passivation of trap states on the surface of nanostructure by SAM or other methods will enhance the performance of double-heterojunction cells. Through tuning the work function of electrodes, manipulating the segregation in BHJ blends, and incorporating passivated nanostructure metal oxide, inverted OSCs with higher efficiency can be achieved.

The intermediate layer, which connects sub-cells optically and electrically, plays a significant role in the realization and efficiency improvement of tandem cells. It should provide high transparency, high conductivity, Ohmic contact, and protection of the lower layers. Many types of tandem OSCs have been applied and developed using a variety of fabrication techniques. The choice of intermediate layer is not merely material exploration; it accounts for efficiency improvement of tandem cell, where the intermediate layer serves as an efficient recombination center for charge carriers extracted from the sub-cells. It is the reason that the photocurrent in a tandem cell is somewhat low, primarily resulting from the large overlapping of absorption spectra of the materials used for both sub-cells. This issue must be solved by chemical structure design and material synthesis, aiming to synthesize excellent organic materials with broad absorption range and high carrier mobility. Therefore, through carefully tuning the band gaps of donors used in sub-cells and the thickness of each cell, a balanced photocurrent can be implemented. Moreover, an optical spacer with a high transparency and high conductivity inserted between two sub-cells can also adjust the optical field distribution across the entire tandem cell, for efficient light absorption in each sub-cell. Hence, combining these considerations will improve the performance of tandem OSCs.

In summary, inverted and tandem structures implemented in OSCs are aiming to fulfill the commercial requirements. They have advanced much in the past few years and have achieved an exciting and potential level. Such devices will drive the realization of flexible and large-area OSCs with low costs.

Appendix

Most common abbreviations used throughout the chapter.

Abbreviations	Full expressions
OSCs	Organic solar cells
OLEDs	Organic light-emitting diodes
OTFTs	Organic thin film transistors
OLEDs	Organic light-emitting diodes
BHJ	Bulk heterojunction
HOMO	Highest occupied molecular orbital
LUMO	Lowest unoccupied molecular orbital
V_{bi}	Built-in voltage
E_g	Energy band gap
PEDOT:PSS	Poly(3,4-ethylenedioxythiophene): poly(styrenesulfonate)
ITO	Indium tin oxide
NiO	Nickel(II) oxide
MoO_3	Molybdenum oxide
V_2O_5	Vanadium oxide
WO_3	Tungsten trioxide
LiF	Lithium fluoride
TiO_x	Titanium oxide
ZnO	Zinc oxide
Cs_2CO_3	Cesium carbonate
J_{sc}	Short-circuit current density
V_{oc}	Open-circuit voltage
FF	Fill factor
PCE	Power conversion efficiency
AM 1.5G	Air Mass 1.5 Global
I-V	Current-Voltage
IPCE	Incident photon-to-electron conversion efficiency
LWF	Low work function
HWF	High work function
MDMO-PPV	Poly[2-methoxy-5-(3,7-dimethyloctyloxy)]-1,4-phenylenevinylene)
MEH-PPV	Poly(2-methoxy-5-(2'-ethylhexyloxy)-1, 4-phenylenevinylene)
P3HT	Poly(3-hexylthiophene)
PPV	Polyphenylenevinylene
PCPDTBT	Poly[2,6-(4,4-bis-(2-ethylhexyl)-4*H*-cyclopenta[2,1-*b*;3,4-*b'*]dithiophene)-*alt*-4,7-(2,1,3-benzothiadiazole)]
PFDTBT	Poly((2,7-(9,9-dioctyl)-fluorene)-*alt*-5,5-(4,7-di-2-thienyl-2,1,3-benzothiadiazole)
PFDTBT	Poly((2,7-(9,9-dioctyl)-fluorene)-*alt*-5,5-(4,7-di-2-thienyl-2,1,3-benzothiadiazole)
PTBEHT	Poly{5,7-di-2-thienyl-2,3-bis(3,5-di(2-ethylhexyloxy)phenyl)-thieno[3,4-*b*]pyrazine}

(continued)

(continued)

Abbreviations	Full expressions
PSBTBT	Poly[(4,40-bis(2-ethylhexyl) dithieno[3,2-b:20,30-d]silole)-2,6-diyl-alt-(2,1,3 benzothiadiazole)-4,7-diyl]
PDMS	Polydimethylsiloxane
PTE	Polyoxyethylene tridecyl ether
$PC_{60}BM$	1-(3-methoxycarbonyl)-propyl-1-phenyl-(6,6)C_{61}
$PC_{70}BM$	1-(3-methoxycarbonyl)-propyl-1-phenyl-(6,6)C_{71}
CuPc	Copper phthalocyanine
ZnPc	Zinc phthalocyanine
$SnCl_2Pc$	Tin phthalocyanine dichloride
$F_{16}CuPc$	Copper hexadecafluorophthalocyanine
PTCBI	3,4,9,10 perylenetetracarboxylic bisbenzimidazole
MTDATA	4,4′,4″-*tris*(*N*-3-methylphenyl-*N*-phenyl-amino)triphenylamine
F_4-TCNQ	(2,5-cyclohexadine-1,4-diylidene)-dimalononitrile
TIPD	Titanium (diisopropoxide)bis(2,4-pentanedionate)
C_{60}	Fullerene
BPhen	Bathophenanthroline
SEM	Scanning electron microscope
FE-SEM	Field emission scanning electron microscope
XPS	X-ray photoelectron spectroscopy
SAM	Self-assembled monolayer

References

1. Ma WL, Yang CY, Gong X et al (2005) Thermally stable, efficient polymer solar cells with nanoscale control of the interpenetrating network morphology. Adv Funct Mater 15:1617–1622
2. Reyes-Reyes M, Kim K, Carroll DL (2005) High-efficiency photovoltaic devices based on annealed poly(3-hexylthiophene) and 1-(3-methoxycarbonyl)-propyl-1-phenyl-(6, 6)C-61 blends. Appl Phys Lett 87:083506
3. Bundgaard E, Krebs FC (2007) Low band gap polymers for organic photovoltaics. Sol Energy Mater Sol Cells 91:954–985
4. Park SH, Roy A, Beaupre S et al (2009) Bulk heterojunction solar cells with internal quantum efficiency approaching 100%. Nat Photonics 3:297–303
5. Hou JH, Chen HY, Zhang SQ et al (2008) Synthesis, characterization, and photovoltaic properties of a low band gap polymer based on silole-containing polythiophenes and 2, 1, 3-benzothiadiazole. J Am Chem Soc 130:16144–16145
6. Brabec CJ, Zerza G, Cerullo G et al (2001) Tracing photoinduced electron transfer process in conjugated polymer/fullerene bulk heterojunctions in real time. Chem Phys Lett 340:232–236
7. Chirvase D, Parisi J, Hummelen JC et al (2004) Influence of nanomorphology on the photovoltaic action of polymer–fullerene composites. Nanotechnology 15:1317–1323
8. Hoppe H, Niggemann M, Winder C et al (2004) Nanoscale morphology of conjugated polymer/fullerene-based bulk-heterojunction solar cells. Adv Funct Mater 14:1005–1011
9. Hoppe H, Sariciftci NS (2006) Morphology of polymer/fullerene bulk heterojunction solar cells. J Mater Chem 16:45–61

10. Zhang FL, Jespersen KG, Bjorstrom C et al (2006) Influence of solvent mixing on the morphology and performance of solar cells based on polyfluorene copolymer/fullerene blends. Adv Funct Mater 16:667–674
11. Chen FC, Tseng HC, Ko CJ (2008) Solvent mixtures for improving device efficiency of polymer photovoltaic devices. Appl Phys Lett 92:103316
12. Rispens MT, Meetsma A, Rittberger R, et al. (2003) Influence of the solvent on the crystal structure of PCBM and the efficiency of MDMO-PPV: PCBM 'plastic' solar cells. Chem Commun 2116–2118
13. Moule AJ, Meerholz K (2008) Controlling morphology in polymer–fullerene mixtures. Adv Mater 20:240–245
14. Jin SH, Naidu BVK, Jeon HS et al (2007) Optimization of process parameters for high-efficiency polymer photovoltaic devices based on P3HT: PCBM system. Sol Energy Mater Sol Cells 91:1187–1193
15. Yang XN, Loos J, Veenstra SC et al (2005) Nanoscale morphology of high-performance polymer solar cells. Nano Lett 5:579–583
16. Kim K, Liu J, Namboothiry MAG et al (2007) Roles of donor and acceptor nanodomains in 6% efficient thermally annealed polymer photovoltaics. Appl Phys Lett 90:163511
17. Irwin MD, Buchholz B, Hains AW et al (2008) p-Type semiconducting nickel oxide as an efficiency-enhancing anode interfacial layer in polymer bulk-heterojunction solar cells. Proc Natl Acad Sci USA 105:2783–2787
18. Shrotriya V, Li G, Yao Y et al (2006) Transition metal oxides as the buffer layer for polymer photovoltaic cells. Appl Phys Lett 88:073508
19. Zhao DW, Sun XW, Jiang CY et al (2008) Efficient tandem organic solar cells with an Al/MoO3 intermediate layer. Appl Phys Lett 93:083305
20. Zhao DW, Sun XW, Jiang CY et al (2009) An efficient triple-tandem polymer solar cell. IEEE Electron Device Lett 30:490–492
21. Chan MY, Lee CS, Lai SL et al (2006) Efficient organic photovoltaic devices using a combination of exciton blocking layer and anodic buffer layer. J Appl Phys 100:094506
22. Jorgensen M, Norrman K, Krebs FC (2008) Stability/degradation of polymer solar cells. Sol Energy Mater Sol Cells 92:686–714
23. Brabec CJ, Shaheen SE, Winder C et al (2002) Effect of LiF/metal electrodes on the performance of plastic solar cells. Appl Phys Lett 80:1288–1290
24. Roy A, Park SH, Cowan S et al (2009) Titanium suboxide as an optical spacer in polymer solar cells. Appl Phys Lett 95:013302
25. Kim JY, Kim SH, Lee HH et al (2006) New architecture for high-efficiency polymer photovoltaic cells using solution-based titanium oxide as an optical spacer. Adv Mater 18:572–576
26. Gilot J, Barbu I, Wienk MM et al (2007) The use of ZnO as optical spacer in polymer solar cells: theoretical and experimental study. Appl Phys Lett 91:113520
27. Yip HL, Hau SK, Baek NS et al (2008) Self-assembled monolayer modified ZnO/metal bilayer cathodes for polymer/fullerene bulk-heterojunction solar cells. Appl Phys Lett 92:193313
28. Chen FC, Wu JL, Hsieh KH et al (2008) Polymer photovoltaic devices with highly transparent cathodes. Org Electron 9:1132–1135
29. Chen FC, Wu JL, Yang SS et al (2008) Cesium carbonate as a functional interlayer for polymer photovoltaic devices. J Appl Phys 103:103721
30. Kim JY, Lee K, Coates NE et al (2007) Efficient tandem polymer solar cells fabricated by all-solution processing. Science 317:222–225
31. Parker ID (1994) Carrier tunneling and device characteristics in polymer light-emitting-diodes. J Appl Phys 75:1656–1666
32. Brabec CJ, Cravino A, Meissner D et al (2001) Origin of the open circuit voltage of plastic solar cells. Adv Funct Mater 11:374–380
33. Malliaras GG, Salem JR, Brock PJ et al (1998) Photovoltaic measurement of the built-in potential in organic light emitting diodes and photodiodes. J Appl Phys 84:1583–1587

34. Pettersson LAA, Roman LS, Inganas O (1999) Modeling photocurrent action spectra of photovoltaic devices based on organic thin films. J Appl Phys 86:487–496
35. Peumans P, Yakimov A, Forrest SR (2003) Small molecular weight organic thin-film photodetectors and solar cells. J Appl Phys 93:3693–3723
36. Moliton A, Nunzi JM (2006) How to model the behaviour of organic photovoltaic cells. Polym Int 55:583–600
37. Hagemann O, Bjerring M, Nielsen NC et al (2008) All solution processed tandem polymer solar cells based on thermocleavable materials. Sol Energy Mater Sol Cells 92:1327–1335
38. Kroon R, Lenes M, Hummelen JC et al (2008) Small bandgap polymers for organic solar cells (polymer material development in the last 5 years). Polym Rev 48:531–582
39. Zhang FL, Bijleveld J, Perzon E et al (2008) High photovoltage achieved in low band gap polymer solar cells by adjusting energy levels of a polymer with the LUMOs of fullerene derivatives. J Mater Chem 18:5468–5474
40. Neugebauer H, Brabec C, Hummelen JC et al (2000) Stability and photodegradation mechanisms of conjugated polymer/fullerene plastic solar cells. Sol Energy Mater Sol Cells 61:35–42
41. Schuller S, Schilinsky P, Hauch J et al (2004) Determination of the degradation constant of bulk heterojunction solar cells by accelerated lifetime measurements. Appl Phys A Mater Sci Process 79:37–40
42. Sahin Y, Alem S, de Bettignies R et al (2005) Development of air stable polymer solar cells using an inverted gold on top anode structure. Thin Solid Films 476:340–343
43. Krebs FC, Norrman K (2007) Analysis of the failure mechanism for a stable organic photovoltaic during 10000 h of testing. Prog Photovolt Res Appl 15:697–712
44. de Jong MP, van Ijzendoorn LJ, de Voigt MJA (2000) Stability of the interface between indium-tin-oxide and poly(3, 4-ethylenedioxythiophene)/poly(styrenesulfonate) in polymer light-emitting diodes. Appl Phys Lett 77:2255–2257
45. Wong KW, Yip HL, Luo Y et al (2002) Blocking reactions between indium-tin oxide and poly (3, 4-ethylene dioxythiophene): poly(styrene sulphonate) with a self-assembly monolayer. Appl Phys Lett 80:2788–2790
46. Lee K, Kim JY, Park SH et al (2007) Air-stable polymer electronic devices. Adv Mater 19:2445–2449
47. Gaynor W, Lee JY, Peumans P (2010) Fully solution-processed inverted polymer solar cells with laminated nanowire electrodes. ACS Nano 4:30–34
48. Bernards DA, Biegala T, Samuels ZA et al (2004) Organic light-emitting devices with laminated top contacts. Appl Phys Lett 84:3675–3677
49. Chen LM, Hong ZR, Li G et al (2009) Recent progress in polymer solar cells: manipulation of polymer: fullerene morphology and the formation of efficient inverted polymer solar cells. Adv Mater 21:1434–1449
50. Mor GK, Shankar K, Paulose M et al (2007) High efficiency double heterojunction polymer photovoltaic cells using highly ordered TiO2 nanotube arrays. Appl Phys Lett 91:152111
51. Glatthaar M, Niggemann M, Zimmermann B et al (2005) Organic solar cells using inverted layer sequence. Thin Solid Films 491:298–300
52. Liao HH, Chen LM, Xu Z et al (2008) Highly efficient inverted polymer solar cell by low temperature annealing of Cs_2CO_3 interlayer. Appl Phys Lett 92:173303
53. Zhao DW, Liu P, Sun XW et al (2009) An inverted organic solar cell with an ultrathin Ca electron-transporting layer and MoO_3 hole-transporting layer. Appl Phys Lett 95:153304
54. Kim SY, Baik JM, Yu HK et al (2005) Highly efficient organic light-emitting diodes with hole injection layer of transition metal oxides. J Appl Phys 98:093707
55. Waldauf C, Morana M, Denk P et al (2006) Highly efficient inverted organic photovoltaics using solution based titanium oxide as electron selective contact. Appl Phys Lett 89:233517
56. White MS, Olson DC, Shaheen SE et al (2006) Inverted bulk-heterojunction organic photovoltaic device using a solution-derived ZnO underlayer. Appl Phys Lett 89:143517
57. Chu CW, Li SH, Chen CW et al (2005) High-performance organic thin-film transistors with metal oxide/metal bilayer electrode. Appl Phys Lett 87:193508

58. Beek WJE, Wienk MM, Janssen RAJ (2005) Hybrid polymer solar cells based on zinc oxide. J Mater Chem 15:2985–2988
59. Li G, Chu CW, Shrotriya V et al (2006) Efficient inverted polymer solar cells. Appl Phys Lett 88:253503
60. Kyaw AKK, Sun XW, Jiang CY (2009) Efficient charge collection with sol–gel derived colloidal ZnO thin film in photovoltaic devices. J Sol–Gel Sci Technol 52:348–355
61. Znaidi L, Illia G, Benyahia S et al (2003) Oriented ZnO thin films synthesis by sol–gel process for laser application. Thin Solid Films 428:257–262
62. Liu XZ, Luo YH, Li H, et al. (2007) Room temperature fabrication of porous ZnO photoelectrodes for flexible dye-sensitized solar cells. Chem Commun 2847–2849
63. Kyaw AKK, Sun XW, Jiang CY et al (2008) An inverted organic solar cell employing a sol–gel derived ZnO electron selective layer and thermal evaporated MoO_3 hole selective layer. Appl Phys Lett 93:221107
64. Hau SK, Yip HL, Baek NS et al (2008) Air-stable inverted flexible polymer solar cells using zinc oxide nanoparticles as an electron selective layer. Appl Phys Lett 92:253301
65. Wang JC, Weng WT, Tsai MY et al (2010) Highly efficient flexible inverted organic solar cells using atomic layer deposited ZnO as electron selective layer. J Mater Chem 20:862–866
66. Liu JP, Wang SS, Bian ZQ et al (2009) Inverted photovoltaic device based on ZnO and organic small molecule heterojunction. Chem Phys Lett 470:103–106
67. Scharber MC, Wuhlbacher D, Koppe M et al (2006) Design rules for donors in bulk-heterojunction solar cells—towards 10% energy-conversion efficiency. Adv Mater 18:789–794
68. Huang JS, Chou CY, Liu MY et al (2009) Solution-processed vanadium oxide as an anode interlayer for inverted polymer solar cells hybridized with ZnO nanorods. Org Electron 10:1060–1065
69. Olson DC, Piris J, Collins RT et al (2006) Hybrid photovoltaic devices of polymer and ZnO nanofiber composites. Thin Solid Films 496:26–29
70. Yu BY, Tsai A, Tsai SP et al (2008) Efficient inverted solar cells using TiO_2 nanotube arrays. Nanotechnology 19:255202
71. Tokito S, Noda K, Taga Y (1996) Metal oxides as a hole-injecting layer for an organic electroluminescent device. J Phys D Appl Phys 29:2750–2753
72. You H, Dai YF, Zhang ZQ et al (2007) Improved performances of organic light-emitting diodes with metal oxide as anode buffer. J Appl Phys 101:026105
73. Hau SK, Yip HL, Leong K et al (2009) Spraycoating of silver nanoparticle electrodes for inverted polymer solar cells. Org Electron 10:719–723
74. Kuwabara T, Nakayama T, Uozumi K et al (2008) Highly durable inverted-type organic solar cell using amorphous titanium oxide as electron collection electrode inserted between ITO and organic layer. Sol Energy Mater Sol Cells 92:1476–1482
75. Toshinori H, Seishi M, Takashi M et al (2004) Novel Electron-Injection Layers for Top-Emission OLEDs. SID Symposium Digest of Technical Papers 35:154–157
76. Tao C, Ruan SP, Zhang XD et al (2008) Performance improvement of inverted polymer solar cells with different top electrodes by introducing a MoO_3 buffer layer. Appl Phys Lett 93:193307
77. Schmidt H, Flugge H, Winkler T et al (2009) Efficient semitransparent inverted organic solar cells with indium tin oxide top electrode. Appl Phys Lett 94:243302
78. Tao C, Ruan SP, Xie GH et al (2009) Role of tungsten oxide in inverted polymer solar cells. Appl Phys Lett 94:043311
79. Ganzorig C, Suga K, Fujihira M (2001) Alkali metal acetates as effective electron injection layers for organic electroluminescent devices. Mater Sci Eng B Solid State Mater Adv Technol 85:140–143
80. Chu CW, Sung CF, Lee YZ et al (2008) Improved performance in n-channel organic thin film transistors by nanoscale interface modification. Org Electron 9:262–266

81. Huang JS, Li G, Yang Y (2008) A semi-transparent plastic solar cell fabricated by a lamination process. Adv Mater 20:415–419
82. Na SI, Oh SH, Kim SS et al (2009) Efficient organic solar cells with polyfluorene derivatives as a cathode interfacial layer. Org Electron 10:496–500
83. Blom PWM, De Jong MJM, Vleggaar JJM (1996) Electron and hole transport in poly(p-phenylene vinylene) devices. Appl Phys Lett 68:3308–3310
84. Chen FC, Wu JL, Lee CL et al (2009) Flexible polymer photovoltaic devices prepared with inverted structures on metal foils. IEEE Electron Device Lett 30:727–729
85. Hsiao YS, Chen CP, Chao CH et al (2009) All-solution-processed inverted polymer solar cells on granular surface-nickelized polyimide. Org Electron 10:551–561
86. Tan ZA, Yang CH, Zhou EJ et al (2007) Performance improvement of polymer solar cells by using a solution processible titanium chelate as cathode buffer layer. Appl Phys Lett 91:023509
87. Skriver HL, Rosengaard NM (1992) Surface-energy and work function of elemental metals. Phys Rev B 46:7157–7168
88. Monestier F, Simon JJ, Torchio P et al (2008) Optical modeling of organic solar cells based on CuPc and C-60. Appl Opt 47:C251–C256
89. Pettersson LAA, Ghosh S, Inganas O (2002) Optical anisotropy in thin films of poly(3, 4-ethylenedioxythiophene)-poly(4-styrenesulfonate). Org Electron 3:143–148
90. Tong XR, Bailey-Salzman RF, Wei GD et al (2008) Inverted small molecule organic photovoltaic cells on reflective substrates. Appl Phys Lett 93:173304
91. Meiss J, Riede MK, Leo K (2009) Towards efficient tin-doped indium oxide (ITO)-free inverted organic solar cells using metal cathodes. Appl Phys Lett 94:013303
92. Sennett RS, Scott GD (1950) The structure of evaporated metal films and their optical properties. J Opt Soc Am 40:203–211
93. Coakley KM, McGehee MD (2004) Conjugated polymer photovoltaic cells. Chem. Mater. 16:4533–4542
94. Coakley KM, Srinivasan BS, Ziebarth JM et al (2005) Enhanced hole mobility in regioregular polythiophene infiltrated in straight nanopores. Adv Funct Mater 15:1927–1932
95. Cadby AJ, Tolbert SH (2005) Controlling optical properties and interchain interactions in semiconducting polymers by encapsulation in periodic nanoporous silicas with different pore sizes. J Phys Chem B 109:17879–17886
96. Takanezawa K, Hirota K, Wei QS et al (2007) Efficient charge collection with ZnO nanorod array in hybrid photovoltaic devices. J Phys Chem C 111:7218–7223
97. O'Brien S, Koh LHK, Crean GM (2008) ZnO thin films prepared by a single step sol–gel process. Thin Solid Films 516:1391–1395
98. Steim R, Choulis SA, Schilinsky P et al (2008) Interface modification for highly efficient organic photovoltaics. Appl Phys Lett 92:093303
99. Muller B, Riedel M, Michel R et al (2001) Impact of nanometer-scale roughness on contact-angle hysteresis and globulin adsorption. J Vac Sci Technol B 19:1715–1720
100. Hau SK, Yip HL, Acton O et al (2008) Interfacial modification to improve inverted polymer solar cells. J Mater Chem 18:5113–5119
101. Kyaw AKK, Sun XW, Zhao DW, et al. Improved inverted organic solar cells with a sol–gel derived indium-doped zinc oxide buffer layer. IEEE J Sel Top Quantum Electron. doi: 10.1109/JSTQE.2009.2039200
102. Appleyard SFJ, Day SR, Pickford RD et al (2000) Organic electroluminescent devices: enhanced carrier injection using SAM derivatized ITO electrodes. J Mater Chem 10:169–173
103. Khodabakhsh S, Sanderson BM, Nelson J et al (2006) Using self-assembling dipole molecules to improve charge collection in molecular solar cells. Adv Funct Mater 16:95–100

104. Kim JS, Park JH, Lee JH et al (2007) Control of the electrode work function and active layer morphology via surface modification of indium tin oxide for high efficiency organic photovoltaics. Appl Phys Lett 91:112111
105. Krishnan RS, Mackay ME, Duxbury PM et al (2007) Self-assembled multilayers of nanocomponents. Nano Lett 7:484–489
106. Goffri S, Muller C, Stingelin-Stutzmann N et al (2006) Multicomponent semiconducting polymer systems with low crystallization-induced percolation threshold. Nat Mater 5:950–956
107. Wei QS, Nishizawa T, Tajima K et al (2008) Self-organized buffer layers in organic solar cells. Adv Mater 20:2211–2216
108. Wienk MM, Turbiez MGR, Struijk MP et al (2006) Low-band gap poly(di-2-thienylthienopyrazine): fullerene solar cells. Appl Phys Lett 88:153511
109. Zhang FL, Mammo W, Andersson LM et al (2006) Low-bandgap alternating fluorene copolymer/methanofullerene heterojunctions in efficient near-infrared polymer solar cells. Adv Mater 18:2169–2173
110. Hadipour A, de Boer B, Blom PWM (2008) Organic tandem and multi-junction solar cells. Adv Funct Mater 18:169–181
111. Dennler G, Scharber MC, Ameri T et al (2008) Design rules for donors in bulk-heterojunction tandem solar cells-towards 15% energy-conversion efficiency. Adv Mater 20:579–583
112. Dennler G, Forberich K, Ameri T et al (2007) Design of efficient organic tandem cells: on the interplay between molecular absorption and layer sequence. J Appl Phys 102:123109
113. Persson NK, Inganas O (2006) Organic tandem solar cells—modelling and predictions. Sol Energy Mater Sol Cells 90:3491–3507
114. Ameri T, Dennler G, Lungenschmied C et al (2009) Organic tandem solar cells: a review. Energy Environ Sci 2:347–363
115. Hadipour A, de Boer B, Blom PWM (2008) Device operation of organic tandem solar cells. Org Electron 9:617–624
116. Hiramoto M, Suezaki M, Yokoyama M (1990) Effect of thin gold interstitial-layer on the photovoltaic properties of tandem organic solar-cell. Chem Lett 327–330
117. Yakimov A, Forrest SR (2002) High photovoltage multiple-heterojunction organic solar cells incorporating interfacial metallic nanoclusters. Appl Phys Lett 80:1667–1669
118. Triyana K, Yasuda T, Fujita K et al (2005) Tandem-type organic solar cells by stacking different heterojunction materials. Thin Solid Films 477:198–202
119. Cheyns D, Gommans H, Odijk M et al (2007) Stacked organic solar cells based on pentacene and C-60. Sol Energy Mater Sol Cells 91:399–404
120. Xue JG, Uchida S, Rand BP et al (2004) Asymmetric tandem organic photovoltaic cells with hybrid planar-mixed molecular heterojunctions. Appl Phys Lett 85:5757–5759
121. Xue JG, Rand BP, Uchida S et al (2005) A hybrid planar-mixed molecular heterojunction photovoltaic cell. Adv Mater 17:66–71
122. Yu B, Zhu F, Wang HB et al (2008) All-organic tunnel junctions as connecting units in tandem organic solar cell. J Appl Phys 104:114503
123. Dennler G, Prall HJ, Koeppe R et al (2006) Enhanced spectral coverage in tandem organic solar cells. Appl Phys Lett 89:073502
124. Colsmann A, Junge J, Kayser C et al (2006) Organic tandem solar cells comprising polymer and small-molecule subcells. Appl Phys Lett 89:203506
125. Janssen AGF, Riedl T, Hamwi S et al (2007) Highly efficient organic tandem solar cells using an improved connecting architecture. Appl Phys Lett 91:073519
126. You H, Dai YF, Zhang ZQ et al (2007) Improved performances of organic light-emitting diodes with metal oxide as anode buffer. J Appl Phys 101:3
127. Kawano K, Ito N, Nishimori T et al (2006) Open circuit voltage of stacked bulk heterojunction organic solar cells. Appl Phys Lett 88:073514
128. Hadipour A, de Boer B, Wildeman J et al (2006) Solution-processed organic tandem solar cells. Adv Funct Mater 16:1897–1903

129. Gilot J, Wienk MM, Janssen RAJ (2007) Double and triple junction polymer solar cells processed from solution. Appl Phys Lett 90:143512
130. Sista S, Park MH, Hong ZR et al (2010) Highly efficient tandem polymer photovoltaic cells. Adv Mater 22:380–383
131. Sakai J, Kawano K, Yamanari T et al (2010) Efficient organic photovoltaic tandem cells with novel transparent conductive oxide interlayer and poly (3-hexylthiophene): fullerene active layers. Sol Energy Mater Sol Cells 94:376–380
132. Franke R, Maennig B, Petrich A et al (2008) Long-term stability of tandem solar cells containing small organic molecules. Sol Energy Mater Sol Cells 92:732–735

Organic Solar Cells and Their Nanostructural Improvement

Serap Günes

Abstract Organic solar cells comprised of organic semiconductors have attracted considerable attention in the areas of photonics and electronics during the last decade. Organic semiconductors are a less expensive alternative to inorganic semiconductors. Organic molecules and conjugated polymers as organic semiconductors can be processed by simple techniques that are not available to crystalline inorganic semiconductors. The flexibility in the synthesis of organic molecules allows for the alteration of molecular weight, band gap, energy levels, and structural order, which makes organic semiconductors unique. The conversion of sunlight to electricity by organic solar cells is very interesting and promising since organic solar cells offer the possibility of fabricating large area, light-weight, cost-effective, flexible devices using simple and environmental friendly techniques. Also, organic solar cells can be integrated into wide variety of structures and products in ways not possible for conventional solar cells. As a clean renewable energy source organic solar cells are rapidly developing. A power conversion efficiency over 7–8% was reported recently. However, this value must be improved to compete with conventional solar cells. On the other hand, there is a considerable progress in the evolution of organic solar cells from pure scientific research to a possible industrial application. Recent efforts are devoted to the investigation of operating mechanisms, new synthesis routes, new device architectures, stability of the organic materials, lifetime, encapsulation, etc. If comparable or even slightly lower efficiencies than those of conventional technologies can be achieved, the cost-effectiveness and versatility of organic compounds will make organic solar cells more favorable. In this chapter, an overview on principles of operation, critical parameters, nanomorphology, charge transport and mobility,

S. Günes (✉)
Faculty of Arts and Science, Department of Physics, Davutpasa Campus,
Yildiz Technical University, 34210, Esenler, Istanbul, Turkey
e-mail: sgunes@yildiz.edu.tr

L. Zang (ed.), *Energy Efficiency and Renewable Energy Through Nanotechnology*,
Green Energy and Technology, DOI: 10.1007/978-0-85729-638-2_4,

stability, possible routes for improvement, and the recent status and future aspects of organic solar cells will be discussed.

1 Introduction

A solar cell converts sunlight into electricity. Harvesting energy directly from sunlight using photovoltaic (PV) technologies is a viable way to eliminate the disadvantages of decreased availability of fossil fuel sources and the long-term effects of CO_2 emission. The recognition of the potential of PV technology led to the rapid growth of production of terrestrial solar cell modules [1]. Current production is highly dominated by crystalline silicon (Si) modules, which represent 94% of the market. A maximum theoretical power conversion efficiency (PCE) of almost 31% under direct air mass (AM) 1.5 sunlight is expected for silicon-based devices [1, 2]. Although the industry is currently based on Si, significant material challenges and high manufacturing costs might not allow Si-based devices to meet the long-term goals of PV technology. Thus, new high efficiency or low-cost technologies such as multi-junction and organic-based devices are rapidly growing [1]. Organic photovoltaics (OPV) have the potential for low-cost device production.

An organic solar cell consists of an organic layer which realizes the basic steps in PV conversion such as light absorption, charge carrier generation, charge carrier transport and extraction or injection of charge carriers through the contacts [3]. PV cells based on organic compounds have potential cost-effectiveness, flexibility, and easy processing [4]. Two main approaches have been considered in the research of OPVs: bulk heterojunction (BHJ), which is represented in an ideal case as a bicontinuous composite of donor and acceptor phases [5, 6], and donor–acceptor bilayers achieved by vacuum deposition of organic molecules [7, 8].

BHJ type devices employ conjugated polymers which can be solution processed. This is seen as an advantage over the vacuum deposition since processing in solution gives the ability to process the active layer from solution in a single step using various techniques such as ink jet printing, spin coating, and roller casting [9]. In polymer-based BHJ solar cells, which are achieved by blending a polymer donor and acceptor, the most common donor polymers are poly[2-methoxy-5-(3,7-dimethyloctyloxy)-1,4-phenylene vinylene] (MDMO-PPV] [10, 11], regioregular poly(3-hexylthiophene) (RR-P3HT) [12–16], and poly [2-methoxy-5-(2'-ethylhexyoxy)-1,4-phenylene vinylene] (MEH-PPV) [17–19]. The most widely used acceptor material of choice has been [6,6]-phenyl C_{61}-butyric acid methyl ester (PCBM) [20]. Several small organic materials, such as zinc pthalocyanine (ZnPc) [21, 22] and copper pthalocyanine, [23] have also been used as donors in bilayer heterojunction solar cells.

Optimized organic devices have PCE of 5–8% [24–26]. Although there has been significant improvement in organic solar cell PCEs, they still have not reached that of conventional solar cells [27]. In order to compete with the conventional inorganic cells, a PCE of more than 10% is desirable [4, 28–31]. Various approaches to organic solar cell designs have been employed such as using novel organic materials [16, 32] or a combination of organic and inorganic (hybrids) [33, 34] as active layer materials, tandem cells [35] and also, various techniques have been employed to improve the performance of organic solar cells such as low band gap polymers [36, 37] polymer–polymer blends [38, 39], new device structures [40], etc. Improving the PCE is the primary interest of the current research on organic solar cells. The efficiencies can be further improved by tailoring the materials and also engineering of the device structures. However, one should note that stability and life times are also important issues [41, 42].

In this chapter, an overview of basic operation principles, organic solar cell materials, and possible routes for performance improvement of organic solar cells will be overviewed.

2 Basics of Organic Photovoltaics

2.1 Organic Photovoltaic Materials

Research on organic solar cells focuses on two types of materials: one is solution processed, such as semiconducting polymers/molecules, and the other is vacuum processed, such as small organic molecules. Polymers decompose under excessive heat and have too large molar mass for evaporation. Therefore, most polymer-based PV elements are solution processed at low temperatures [43].

The discovery of conducting polymers opened up a way of achieving a new generation of polymers: materials that exhibit the electrical and optical properties of metals or semiconductors and retain the attractive mechanical properties and processing advantages of polymers [44]. Saturated polymers in which all of the four valence electrons are used up in covalent bonds are not very interesting as electronic materials since they are insulators. However, in conjugated polymers the electronic configuration is fundamentally different. The chemical bonding in conjugated polymers leads to one unpaired electron (the π electron) per carbon atom and π bonding leads to electron delocalization along the back bone of the polymer which provides the "high way" for charge mobility along the backbone of the polymer chain. The chain symmetry such as the number and kind of atoms within the repeat unit determines the electronic structure in conducting polymers. As a result, such polymers can exhibit semiconducting or even metallic properties [44]. Materials that have an extended delocalized π electron system can absorb sunlight and create photogenerated charge carriers and also transport these charge carriers, which makes them interesting as optoelectronic materials [43]. Conjugated polymers have become the materials of choice not only for solar cells but also for various optoelectronic applications, such as light emitting diodes (OLEDs) [45–47] and field-effect transistors (OFETs) [48, 49].

Organic materials for use in PV devices require a good chemical stability and high optical absorption in the visible range with respect to the AM1.5G spectrum [50]. Efficiencies of the first hole conducting polymers based on conjugated polymers were less than satisfactory [51]. Previously, higher efficiencies were achieved by switching to different classes of donor type conjugated polymers and by mixing them with suitable electron acceptors like fullerenes [52–56]. High performance organic solar cells employ polymer-fullerene blends in which the polymer acts as a donor whereas the fullerene acts as an acceptor. The Buckminsterfullerene, C_{60}, is a strong acceptor. A soluble derivative of C_{60}, namely PCBM (1-(3-methoxycarbonyl) propyl-1-phenyl [6,6] C_{61}), which has been widely used in polymer/fullerene solar cells due to its solution processability owing to their side-chain substitution was synthesized by Wudl et al. [57]. Currently, a PVC of 7.4% is reported. The main reason for the increase in the performance is the much greater absorption of C_{70} in the visible region relative to that of C_{60}. The high symmetry of C_{60} renders low-energy transitions formally dipole forbidden which results in a weak absorption of light in the visible region [9]. C_{70} has an asymmetric structure with significantly stronger absorption across the visible range [11].

Most semiconducting polymers are hole conductors (electron donors). Important reprensentatives of hole conducting donor-type semiconducting polymers are derivatives of phenylene vinylene backbones such as poly[2-methoxy-5-(3,7-dimethyloctyloxy)]-1,4-phenylenevinylene) (MDMO-PPV), thiophene chains such as poly(3-hexylthiophene) (P3HT) and fluorene backbones such as (poly(9,9′-dioctylfluorene-co-bis-N,N′-(4-butylphenyl-1,4 phenylenediamine) (PFB) and PCDTBT. Light harvesting by PCDTBT is better than that of P3HT due to its smaller band gap.

Phthalocyanine and perylene have commonly found applications in thin-film organic solar cells [58]. Phthalocyanine is a reprensentative of the *p*-type, hole conducting materials that work as electron donors, whereas perylene and its derivatives show *n*-type, electron conducting behavior, and serve as the electron acceptor materials. Small organic molecule semiconductors are generally suitable for evaporation/sublimation techniques since their solubility in common organic solvents is limited. Fig. 1.

2.2 Operation Principles

There are four important steps for the conversion of solar illumination into photocurrent in organic solar cells: (i) absorption of a photon to create an exciton, (ii) diffusion of the exciton to a donor–acceptor interface, (iii) charge transfer of an exciton into an electron in the acceptor and a hole in the donor, and (iv) collection of the charges at the electrodes [59].

The absorption spectrum of the organic material defines the spectral range over which the cell will respond to light. This depends on the chemical structure of the polymer or molecule. Most organic dyes or molecules are strongly absorbing and a

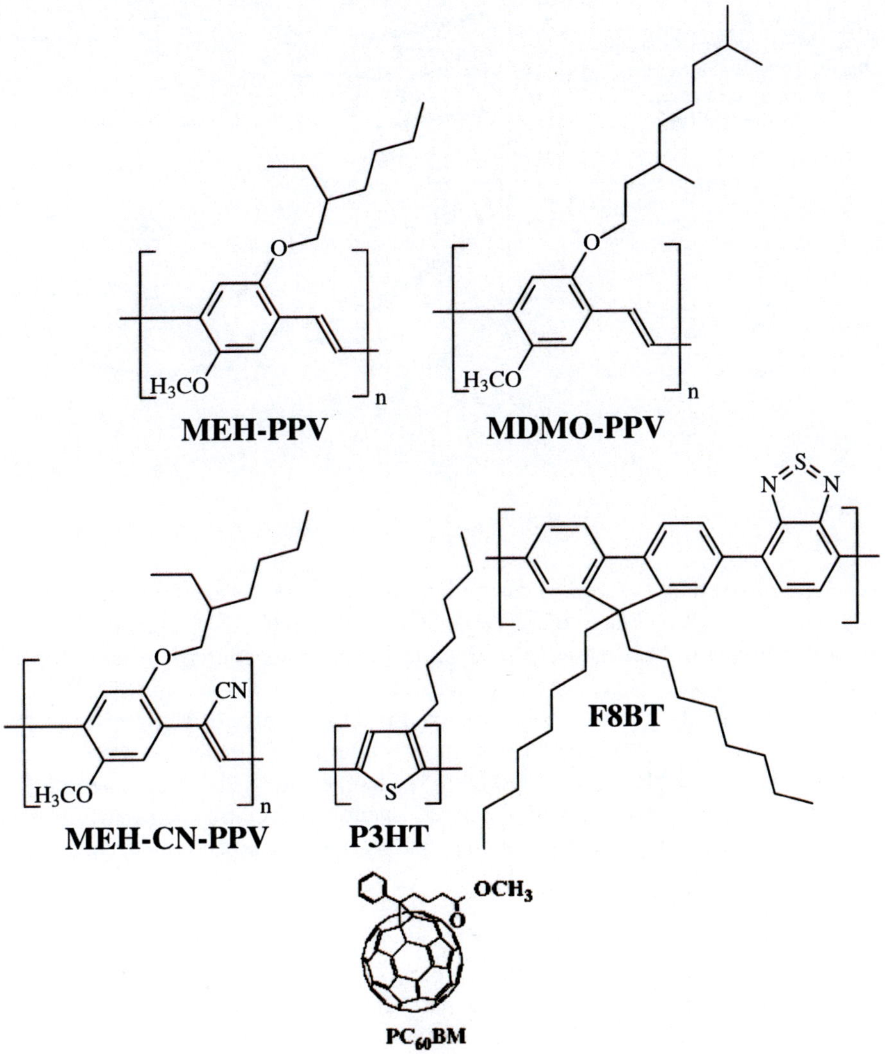

Fig. 1 Some examples of the representative donors and acceptors used in organic solar cells

film only a few 100 nm thick is sufficient to absorb significant proportion of light if it falls within the absorption band of the material [60]. The first process in photoconversion is related to the absorption. The absorption efficiency is governed by the absorption spectra of the organic semiconductors, their thickness and also, the device architecture.

In organic semiconductors, optical excitations lead to electron–hole pairs (excitons) that are bound at the room temperature [61]. The efficiency of exciton diffusion in organic solar cells is determined by the exciton diffusion length and

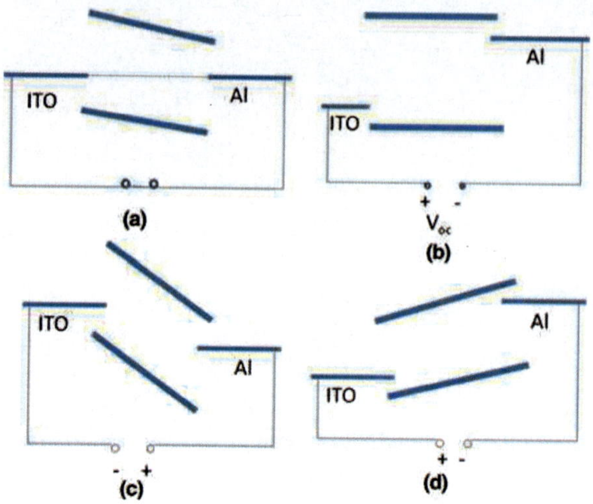

Fig. 2 Metal-insulator–metal (MIM) models of an organic diode: **a** short circuit, **b** open circuit, **c** reversed bias, and **d** forward bias

the morphology of the donor–acceptor interface [59]. Once excitons are created they can diffuse over a length of approximately 5–15 nm [62–64]. Then they decay either radiatively or non-radiatively. Excitons must be separated into free charge carriers within their lifetime for PV purposes. For efficient dissociation of excitons, strong electric fields are utilized, via externally applied fields or via interfaces. At an interface where abrupt changes of the potential occur, there are strong local electrical fields possible ($E = -\text{grad U}$) [65]. Here, excitons can dissociate into an electron in one phase and a hole in the other. Exciton dissociation can also occur at bulk trap sites, leading to one trapped carrier and one potentially free carrier [66, 67]. Conversion of the excess photon energy above singlet exciton into the vibrational heat bath of a polymer segment is also considered a main source of the energy required for charges to escape from a potential well formed by a superposition of the Coulomb and external electric fields [68].

Once, the electrons and holes are separated, and the free charge carriers are created, these free charge carriers are transported to the electrodes [69]. To understand the rectifying behavior of a semiconductor device, the metal–insulator–metal (MIM) model is used. In Fig. 2, a single layer cell with ITO and Al electrodes is shown. Figure 2a shows the short-circuit conditions. The current delivered by a solar cell under zero bias is called the short-circuit current (I_{sc}). In the MIM picture, the built-in potential is equal to the difference in the work functions of the metal electrodes. Exciton dissociation and charge transport are driven by the built-in potential. Figure 2b shows the open-circuit case. The voltage where the current equals to zero is called open-circuit voltage (V_{oc}). V_{oc} is equal to the difference between the metals' work functions and balances the built-in potential. Thus, the current is zero since there is no net driving force acting on the charge carriers. Figure 2c shows the negative bias case. The diode works as a photodetector. Under illumination, the charge carriers drift to the appropriate electrodes. Figure 2d shows the forward bias

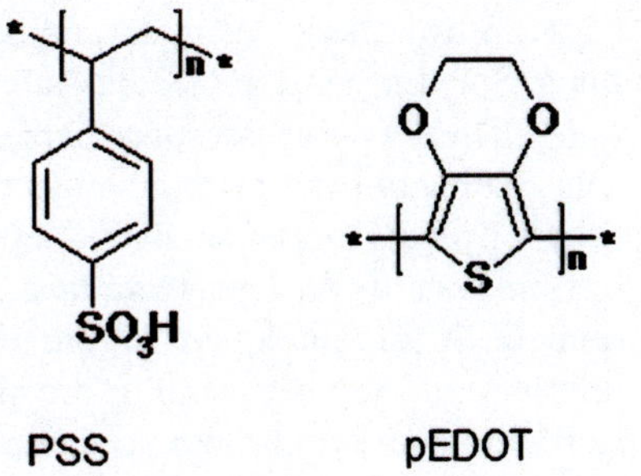

Fig. 3 Chemical structure of PEDOT and PSS

case. In the case of an applied forward bias larger than the V_{oc} the contacts start injecting charges into the semiconductor. If these charges recombine radiatively, the diode works as a LED.

2.3 Device Architecture

As substrates, transparent and conducting electrodes (for example, glass or plastic covered with indium-tin-oxide (ITO)) are used. As a transparent conductive electrode ITO allows light to pass through the cell. However, ITO is not the ideal conductive material due to the following problems: release of oxygen and tin into the organic layer, poor transparency in the blue region and complete crystallization of ITO films, which requires high-temperature processing [70, 71]. Also, the increasing cost of indium prevents large-scale use of ITO in low-cost PV devices [70]. Therefore, alternatives for ITO electrode such as carbon nanotube network electrodes are being investigated [72].

The substrate electrode can be structured by etching. On the transparent conducting substrate, PEDOT:PSS, poly(ethylene-dioxythiophene) doped with polystyrene-sulphonic acid is commonly coated from an aqueous solution. This PEDOT:PSS layer improves the surface quality of the ITO electrode (reducing the probability of shorts) and facilitates hole injection/extraction. Furthermore, the work function of this electrode can be changed by chemical/electrochemical manipulation of the PEDOT layer [73]. PEDOT:PSS is the most promising organic-based electrode material (see Fig. 3) and currently, various modifications of PEDOT:PSS with even greater conductivities are being investigated as electrodes for organic devices [70, 74–76].

As already mentioned, there are two major classes of organic semiconductors: low molecular weight materials and polymers. They have in common a conjugated π electron system formed by the p_z orbitals of sp^2 hybridized C atoms in the molecules. Compared to the σ bonds that form the backbone of the molecules, the π bonding is significantly weaker. Therefore, the lowest electronic excitations of conjugated molecules are the π-π^* transitions with an energy gap leading to light absorption or emission in the visible spectral range [77]. An important difference between two classes of materials lies in the way they are processed to form thin films. Small organic molecules are usually deposited by sublimation or evaporation whereas conjugated polymers can only be processed from solution by using spin coating or printing techniques [77]. Side chain functionalization is principally used for processing of semiconducting polymers from solution in organic solvents or from water [44]. The active layers are then coated depending on the class of the semiconductor employed in the device. Spin coating, doctor blading, screen printing, and ink jet printing are the most common wet processing techniques. Evaporation of two or more organic molecules at once is called coevaporation and it can be applied to small molecules to create interpenetrating donor–acceptor networks or to achieve molecular doping [78, 79].

The top electrode is, in general, a metal with an underlayer of ultrathin lithiumfluoride. The exact nature of this LiF underlayer is unknown, but certainly such thicknesses (ca 0.6 nm) cannot form a closed layer. The exact role of the LiF underlayer is under controversial debate in the literature [80–82]. Photoelectron spectroscopy studies showed that the metal workfunction can be considerably reduced by evaporation of such LiF layers [83].

The most widely used device configurations in organic solar cells are summarized below.

2.3.1 Single Layer Organic Solar Cells

The first organic solar cells were based on single layers sandwiched between two metal electrodes of different work functions. One of the electrodes must be (semi-) transparent, generally ITO, but a thin metal layer can also be used. The other electrode is generally aluminium (calcium, magnesium, gold and others are also used).

Photoexcitations in an organic semiconductor lead to formation of bound electron–hole pairs (excitons). Excitons created in an organic semiconductor must dissociate into free charge carriers according to the PV requirements. The only way to break apart the excitons in the devices consisting of pristine polymers is to use an electric field [84]. If the organic semiconductor is doped, band bending can occur, leading to a Shottky contact at the metal-electrode interface. As a result of the limited exciton diffusion length, only photoexcitations generated close to the depletion region of the Schottky contact can lead to separated charge carriers. Therefore, only a small region contributes to photocurrent generation in pristine polymer-based devices.

The early efforts to realize organic solar cells using conjugated polymers focused on polyacetylene [85] and polythiophene [86]. Previously, visible light emission was observed at Shottky diodes made from semiconducting polymers [45, 87]. The observation of a dual-function device, both as a light emitting diode and a photodiode under reverse bias, using poly[2-methoxy-5-(2′-ethyl-hexyloxy)-1,4-phenylene vinylene] (MEH-PPV) made PPV a material of interest in single-layer organic optoelectronics [88]. However, the PCE of ITO/PPV/Al based

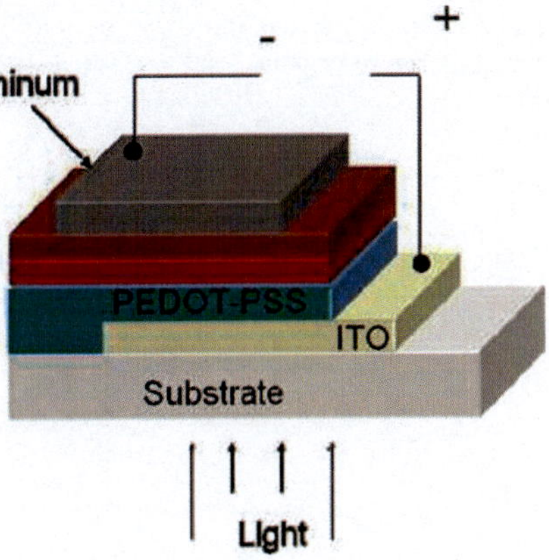

Fig. 4 Bilayer heterojunction configuration of organic solar cells

devices were reported as 0.07 for light intensities of 1 mW/cm^2 [89]. The intrinsic field created by the asymmetry of the work function of electrodes was not large enough to generate significant power. Single-layer conjugated polymer-based devices lacked a significant PV effect.

The device architecture plays an important role in the overall efficiency of organic solar cells. In a single-layer device, since only a small region contributes to the photocurrent generation, the efficiency of such devices is rather low [90].

2.3.2 Bilayer Heterojunction Organic Solar Cells

The major problem in single-layer organic solar cells is inefficient charge generation in conjugated polymers. To overcome this limitation, the donor/acceptor approach has been suggested [91]. A bilayer device is prepared by stacking a donor and an acceptor material (see Fig. 4). The effective interaction between the donor and the acceptor takes place at the geometric interface in the bilayer heterojunction organic solar cells. Many conjugated polymers in their undoped, semiconducting states are electron donors upon photoexcitation [91]. The photophysics studies on conjugated polymers and fullerenes reveal that there is ultrafast, reversible and metastable electron transfer from conjugated polymers to Buckminster fullerenes in solid films [18, 52–54].

Several devices have been constructed using this effect [18, 58]. Tang et al. demonstrated PV activity in small molecular bilayers that were vacuum deposited [58]. They fabricated bilayer organic solar cells by vacuum depositing copper phthalocyanine and a perylene tetracarboxylic derivative. A PVC of about 1% was achieved under simulated AM2 illumination. A novel feature of this device was that the charge-generation efficiency was relatively independent of the bias voltage, resulting in cells with fill factor values as high as 0.65. The interface between the two organic materials, rather than the electrode/organic contacts, was found to be crucial in determining the PV properties of the cell [58]. In a study by

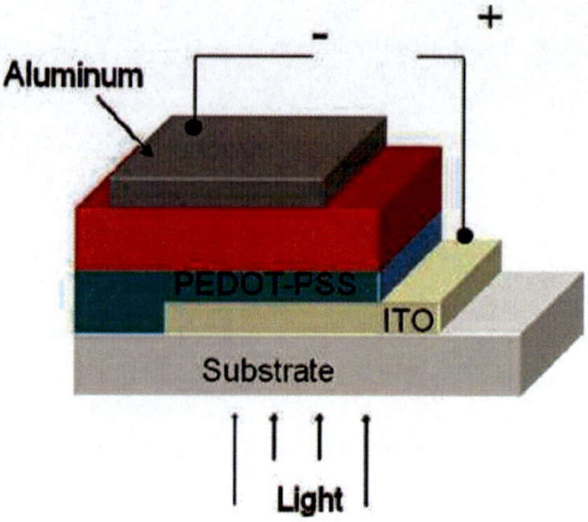

Fig. 5 Bulk heterojunction configuration of organic solar cells

Yamashita et al. the characteristics of a bilayer heterojunction diode consisting of C_{60}/tetratiafulvalene (TTF) were reported. The device under dark conditions was almost insulative, however, upon light illumination, remarkable rectification was observed [92]. Halls et al. measured the PV properties of heterojunctions consisting of a layer of the molecular pigment bis(phenethylimido)perylene, sublimed onto a film of the conjugated polymer poly(*p*-phenylenevinylene) [PPV], and sandwiched between electrodes of ITO and aluminum. Quantum yields (electrons collected per incident photon) of up to 6%, fill factors of 0.6 and open-circuit voltages approaching 1 V were obtained, representing significant improvements over single-layer PPV devices [93]. Efficiencies of 3.6% were reported for vacuum-deposited copper phthalocyanine/C_{60} thin-film double-heterostructure PV cells incorporating an exciton-blocking layer (EBL) [94].

Although significant improvement compared to single-layer devices was achieved in bilayer heterojunction solar cells, the PVC of bilayer heterojunction solar cells is limited due to the following reasons: (i) Efficient charge separation occurs only close to the donor/acceptor interface whereas the photoexcitations far from the donor/acceptor heterojunction will recombine and (ii) PV conversion efficiency is limited by the amounts of photons absorbed in the region where charge separation takes place [91].

2.3.3 Bulk Heterojunction Organic Solar Cells

In a bulk heterojunction organic solar cell, donor and acceptor components are mixed to form an interpenetrating network at the nanometer scale (see Fig. 5). The bulk heterojunction concept maximizes the donor–acceptor contact area. If the phase separation between the donor and the acceptor can be controlled to form an interpenetrating network through the control of morphology a high interfacial area within a bulk material can be achieved [91].

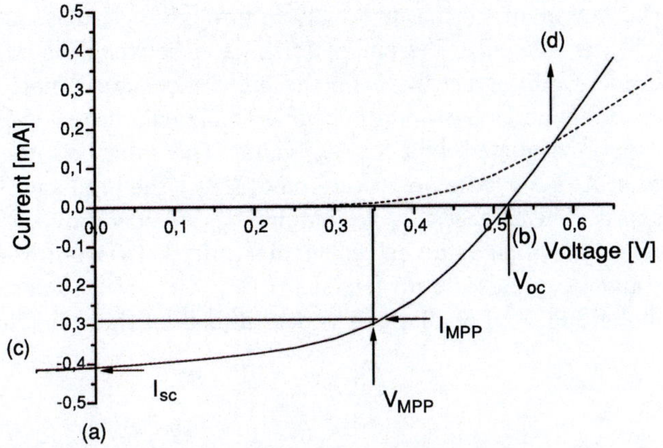

Fig. 6 Current–voltage (*I–V*) curves of an organic solar cell (*dark, dashed*; illuminated, full line)

Early studies focus on polymer-fullerene bulk heterojunction solar cells. The bulk heterojunction is presently the most widely used photoactive layer. One of the most promising combinations of materials is a blend of a semiconducting polymer and a fullerene derivative as the acceptor.

The study by Shaheen et al. was a breakthrough on bulk heterojunction solar cells with efficiencies reaching 2.5% under simulated AM1.5G illumination using MDMO-PPV as a donor and PCBM as an acceptor. In these cells, the photoactive layer was sandwiched between two electrodes with different work functions. Since then several studies on the morphology, electronic structure, and charge transport of MDMO-PPV/PCBM-based devices were performed [95–100].

P3HT has been another choice of material as a donor in bulk heterojunction cells. The high charge carrier mobility and lower band gap compared to MDMO-PPV make this material advantageous over MDMO-PPV. P3HT/PCBM blends provided an increased PV performance as compared to MDMO-PPV/PCBM solar cells [14, 101]. Recently, 5% efficiency has been achieved using P3HT as the donor and PCBM as the acceptor. By introducing a TiO_x layer as the optical spacer they were able to achieve polymer solar cells with PCE approximately 50% higher than similar devices fabricated without an optical spacer [102].

2.4 Current–Voltage Characteristics of an Organic Solar Cell

The characteristic parameters of an organic solar cell can be deduced from current–voltage (*I–V*) curves. The *I–V* characteristics of a solar cell in the dark and under illumination are shown in Fig. 6.

In the dark there is almost no current flow until the start to inject at forward bias. When a cell is illuminated, the *I–V* curve is shifted down by the short-circuit current, I_{sc}. The maximum current that can run through the cell is determined by the I_{sc} [102]. V_{oc} is related to the energetic relation between the donor and the acceptor. The energy difference between the highest occupied molecular orbital (HOMO) of the donor and the lowest unoccupied molecular orbital (LUMO) of the acceptor is closely correlated with the V_{oc} value. The solar cell operates in the fourth quandrant. At the maximum power point (MPP), the product of current and voltage is the largest. The largest power output (P_{max}) is determined by the point where the product of voltage and current is maximized. Division of P_{max} by the product of I_{sc} and V_{oc} yields the fill factor, FF.

The photovoltaic PCE of a solar cell is determined by the following formula:

$$\eta_e = \frac{V_{oc} * I_{sc} * \mathrm{FF}}{P_{in}} \tag{1}$$

$$\mathrm{FF} = \frac{I_{mpp} * V_{mpp}}{I_{sc} * V_{oc}} \tag{2}$$

where V_{oc} is the open-circuit voltage, I_{sc} is the short-circuit current, FF is the fill factor and P_{in} is the incident light intensity.

The light intensity at 1,000 W/m^2 with a spectral intensity distribution matching that of the sun on the earth's surface at an incident angle of 48.2°, which is called the AM1.5G spectrum is accepted as standard for solar cell testing [104].

An experimentally accessible value is the external quantum efficiency or incident photon to current efficiency (IPCE). IPCE is defined as the number of photogenerated charge carriers contributing to the photocurrent per incident photon [102]. IPCE is calculated using the following formula:

$$\mathrm{IPCE} = \frac{1240 * I_{sc}}{\lambda * P_{in}} \tag{3}$$

where λ (nm) is the incident photon wavelength, I_{sc} (μA/cm^2) is the photocurrent of the device and P_{in} (W/m^2) is the incident power.

2.4.1 Power Conversion Efficiency Reports on Organic Solar Cells

Organic solar cell research has grown rapidly during the last decades and now it is close to the level of the commercial applications. It became crucial [103] to accurately determine the efficiency values to enable a fair comparison of the results from several groups for the healthy development of this technology. Since the PCE is the representative parameter to evaluate the performance of a PV cell, the measurement of this value should be accurately determined and reported and also should be reproducible to be able to compare different devices [105].

The performance of PV cells is commonly described in terms of their efficiencies with respect to the standart reporting conditions (SRC) defined by the temperature, spectral irradiance, and total irradiance [105, 106]. The SRC for the performance of the PV cells are as follows: 1,000 W/m^2 irradiance, AM 1.5 global reference spectrum, and 25°C cell temperature.

Shrotriya et al. have already described what must be done for accurate measurement and characterization of organic solar cells [105]. They performed experiments on polymer/fullerene bulk heterojunction solar cells and small organic molecule-based bilayer cells. They considered the effects of the spectral-responsivity, light source calibration and spectral mismatch factor and also the device area. They first investigated the effect of the light bias intensity on the spectral responsivity. They observed that the responsivities of all the cells show a slight dependence on the light bias intensity, although the behavior was different for different material systems. However, they showed that the light bias dependence of the responsivity for all test cells was constant with respect to the wavelength, which suggested that the mismatch factor calculation would be independent of light bias intensity. They also observed relatively weak dependence of the external quantum efficiency on light bias intensity. They counted the response of the cell to the chopped light as an important factor that has to be considered when measuring the spectral response of the PV device.

A solar simulator replicates the solar spectrum and is used for testing PV devices. Reference cells are used to set the intensity of a light source to a particular test condition (e.g., one sun) for *I–V* measurements. The relative spectral responsivities of the test and reference cells are an important factor in the solar simulator calibration procedure. For crystalline solar cells, the reference cell is made of the same materials and technology as the test device. The primary interest in a reference cell is the stability in the reference cell's calibration value. For this reason most thin-film organic and inorganic devices use a Si reference cell that may have a filter to improve the spectral match. However, for polymer and small organic molecule organic solar cells, it is extremely difficult to fabricate reference cell from the same materials. The reasons for this are the lack of the consistent reproducibility and the poor life times of these devices. Therefore, for the purpose of light source calibration for organic solar cell testing, it is important to select a reference cell whose spectral response matches that of the actual test cells as closely as possible to minimize the spectral error [105].

Another parameter to consider is the lamp's age, which affects the actual irradiance. As a result, the spectral mismatch changes with the age of the solar simulator's lamp. The spectral mismatch factor accounts for deviations in the spectral output of the solar simulator with respect to the standard AM1.5G spectrum and deviations in the spectral response with respect to that of the reference cell [10, 107].

It is also essential to correctly measure the device area in order to accurately determine the current density through the device. Usually, device area is chosen as the area defined by the shadow mask used for evaporating the top contact. An important factor that can result in significant errors in the estimation of the area is the shadow effect arising from evaporating successive layers from multiple sources [105].

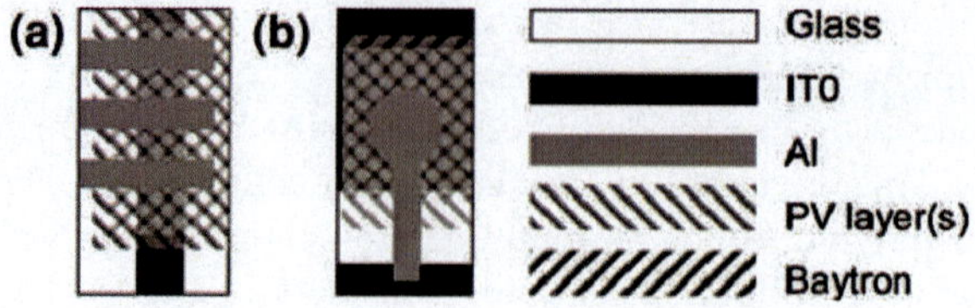

Fig. 7 Two common layouts for OPV devices with unpatterned PEDOT:PSS (Baytron) and photoactive layers: **a** In devices with the crosses layout, the Al electrode overlaps with the photoactive and PEDOT:PSS layers outside the patterned ITO region, **b** in this layout, outside the nominal area, the photoactive layer is in contact with unpatterned PEDOT:PSS/ITO only. Reproduced with permission from Cravino et al. [108], Copyright Wiley-VCH Verlag GmbH&Co. KGaA

Besides device area, device design can also affect the determination of the efficiency of organic solar cells. Cravino et al. demonstrated that the shape and size of a cell, which is referred to as the cell layout, might be very important in organic solar cells (see Fig. 7) [108]. They investigated the two common layouts for organic solar cells with patterned PEDOT:PSS layers and photoactive layers, as shown in the Fig. 7. They considered the results of the geometrically resolved photocurrent measurements and concluded that the most critical region in the crossed layout is the region where the Al electrode overlaps with the both photoactive and PEDOT:PSS layers.

The relationship between the performance and the electrode geometry of organic solar cells was also investigated by Kim et al. [109]. They used the two most commonly used electrode geometries: (i) island-type and (ii) crossbar-type, as shown in the figure. They found that the commonly used PV cells can produce a significant error in PCE measurement if the beam diameter of the illuminated light is larger than the area of electrodes. They concluded that the larger PCE observed using such a configuration is due to the excess photocurrent generated from a parasitic organic solar cell structure, where the conductive PEDOT:PSS layer acts as an effective anode. This can also explain why the excess photocurrent is proportional to the conductivity of the PEDOT:PSS layer and increases with the illuminated area. Kim et al. suggested two solutions to prevent the error in characterization: (i) making the area of light illumination equal to that of the active electrode of the PV cells and (ii) island-type cathode design. Considering the technical difficulty of the former method they suggested that the latter approach would be more convenient and practical (see Figs. 8 and 9) [109].

There are also other challenges in PCE measurements since there are many other parameters such as processing conditions [14, 16, 110, 111], solvent selection [10], and the presence of oxygen and moisture [50, 112], which will be discussed later.

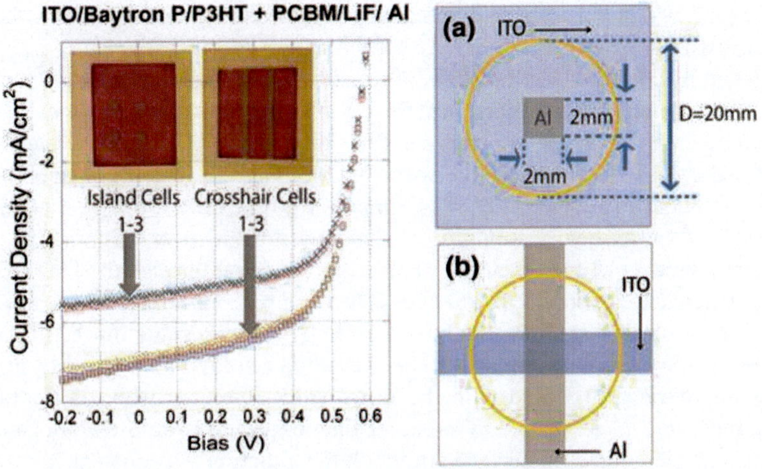

Fig. 8 *I–V* curves of OPV cells having (**a**) island-type electrode geometry and (**b**) crossbar-type electrode geometry _illumination diameter: 20 mm, intensity: AM1.5G 100 mW/cm^2. For the device characterization having the island-type electrode, the anode contact was made using a probe and the cathode contact was made using a gold wire. Reprinted with permission from Kim et al. [109] Copyright (2008), American Institute of Physics

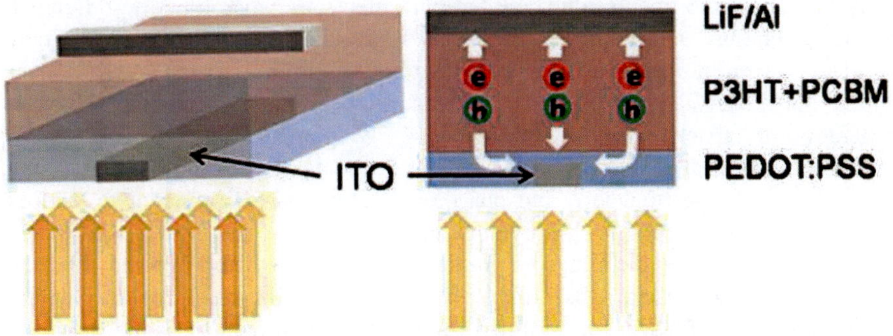

Fig. 9 *Left* Characterization scheme of OPV cells having crossbar-type electrode geometry under illumination larger than the overlapped area of the crossbar-type electrodes. *Right* Excess current generation in PEDOT:PSS/Al device where there is no ITO. Reprinted with permission from Kim et al. [109]. Copyright (2008), American Institute of Physics

2.5 Characteristic Parameters of an Organic Solar Cell

2.5.1 Open-Circuit Voltage

One of the key parameters of PV devices is the open-circuit voltage (V_{oc}), which is the voltage for which the current in the external circuit equals to zero [113]. Since the efficiency of an organic solar cell is directly proportional to the V_{oc}, improvement of the V_{oc} is crucial to achieve higher efficiencies. However, there is a controversial debate in the literature on the nature of the V_{oc}.

A generally accepted estimate for the built-in potential is given by the V_{oc}, which underestimates the built-in potential at room temperature and converges to the correct value at low temperatures [98, 106]. In a MIM device the V_{oc} is determined by the difference in the work functions of two metal contacts [84].

The PV effect in diode structures formed with thin films of PPV sandwiched between ITO and either aluminium (Al), magnesium (Mg) or calcium (Ca) was studied by Marks et al. [107]. Under illumination incident through the ITO contact, they measured large open-circuit voltages. They compared the difference in the work function between ITO and the metal in question by taking a value for the ITO work function of 4.8 eV. They found that the measured photovoltages for Mg and Ca devices approximately scaled with the metal work function, whereas the photovoltage measured in Al devices was larger than expected from the work function difference of the contact materials. Although the situation was unclear, they commented that the surface layer of the Al reacts with PPV to form covalent bonds across the vinylene linkages and introduces a non-conjugated barrier layer [107].

The magnitude of V_{oc} can be attributed to the electrode work function difference [108]. However, other research studies showed that there are other parameters that contribute to the V_{oc}, such as dark currents [114], Fermi level pinning [91] and chemical potential gradients [115, 116].

A scaling of the open-circuit voltage with electrode work function difference has also been observed in bilayer devices by Ramsdale et al. [116]. However, they observed an additional intensity-dependent contribution from the active layer within the device. This additional contribution was attributed to photoinduced generation of carriers.

Built-in potential is an essential parameter of PV devices and influences charge dissociation, charge transport, and charge collection [98]. Compared to devices made from pristine conjugated polymers, the nature of the thin-film devices made by mixing fullerenes and conjugated polymers is completely modified and, thus, the V_{oc} of the corresponding cells differ. Therefore, the MIM model or the Shottky junctions [117, 118] that successfully explain the situation in pristine polymer-based devices cannot satisfactorily explain the nature of V_{oc} in BHJ solar cells [98].

Brabec et al. investigated the critical parameters influencing the built-in potential in conjugated polymer/fullerene based devices. They analyzed the open-circuit voltage of the corresponding devices as a function of the acceptor strength [98]. They observed that the V_{oc} of the corresponding devices correlated directly with the acceptor strength of the fullerenes, whereas it was rather insensitive to variations of the work function of the negative electrode. They discussed their observations within the concept of Fermi level pinning between fullerenes and metals via surface charges.

Gadisa et al. measured and compared V_{oc} of solar cells based on series of conjugated polythiophene polymers. In every cell, they blended donor polymer with an electron acceptor fullerene molecule. They constructed devices in a sandwich structure with ITO/metallic polymer (PEDOT:PSS) acting as the anode and Al or LiF/Al acting as a cathode. Comparing the V_{oc} of all the cells they showed that this important PV parameter systematically varied with the polymer. The variation of photovoltage was attributed to the variation of the oxidation potential of the donor conjugated polymers after due consideration of the different injection conditions in the varying polymers [119].

The influence of an altered doping level of the hole collecting electrode was investigated by Frohne et al. [73] and they observed that the V_{oc} can be influenced by the electrochemical potential of the PEDOT:PSS.

The morphology of the active layer in BHJ solar cells also affect the V_{oc}. Liu et al. introduced a new term, which is the product of cross-sectional area physically occupied by C_{60} (c) that is related to the concentration of C_{60} in the composite and the product of f, which is the morphology induced interfacial factor and they concluded that the observed V_{oc}'s can be used to estimate cf for PV devices fabricated with different solvent [120].

The V_{OC}'s of BHJ solar cells based on PCBM as the electron acceptor and poly[2-methoxy-5(3′,7′-dimethyloctyloxy)-*p*-phenylene vinylene] (OC_1C_{10}-PPV) as the electron donor were investigated by Mihailetchi et al. [97]. They demonstrated that for non-ohmic contacts, the experimental V_{OC} was determined by the work function difference of the electrodes. A total variation of more than 0.5 V of the V_{OC} was observed by variation of the negative electrode (cathode) work function. They added that for ohmic contacts the V_{OC} was governed by the LUMO and HOMO levels of the acceptor and donor, respectively, which pin the Fermi levels of the cathode and anode. According to this work, the band bending created by accumulated charges at the ohmic contact produced considerable loss in V_{OC} of 0.2 V at room temperature. They concluded that the experimentally observed voltage loss in V_{OC} of 0.38 V due to the presence of ohmic contacts at both interfaces strongly limited the maximum open-circuit voltage of OC_1C_{10}-PPV/PCBM solar cells [97].

In another study, by Mihailetchi et al., they observed that an increase in the workfunction of the metal top electrode led to a reduction of the open-circuit voltage, short-circuit current, and PVC of organic bulk-heterojunction solar cells. They demonstrated that the photocurrent obtained from an active layer comprised of a blend of OC_1C_{10}-PPV and PCBM, with lithium fluoride-topped aluminum, silver, gold, or palladium electrodes, showed a universal behavior when scaled against the effective voltage across the device. Model calculations confirmed that the dependence of the photocurrent on the effective voltage was responsible for the observed variation in performance of each different electrode. Consequently, for any given metal, only the device's open-circuit voltage was required in order to be able to predict the remaining solar cell parameters [99].

Koster et al. investigated the V_{oc} of polymer:fullerene BHJ solar cells as a function of light intensity at different temperatures [113]. The devices consisted of PPV derivative as the hole conductor and PCBM as the electron conductor. The observed photogenerated current and V_{oc} varied with classical *p–n* junction based models. They showed that the V_{oc}, when plotted as a function of light intensity, had a slope equal to kT/q. They explained the influence of light intensity and recombination strength on V_{oc} by a model based on the notion that the quasi-Fermi levels were constant throughout the device, including both drift and diffusion of charge carriers.

Scharber et al. investigated the relation between the energy levels of the donor–acceptor blend and the V_{oc} of 26 different BHJ devices [30]. They derived a simple relation between the energy level of the HOMO of the polymer and the V_{oc}, which was used to estimate the maximum efficiency of BHJ solar cells. Based on the model, they determined the ideal material parameters for a conjugated polymer-PCBM device. They demonstrated that for a total of 26 different BHJ solar cells, there was a linear relation between V_{oc} and the conjugated polymer oxidation potential. They estimated the V_{oc} of a conjugated polymer-PCBM solar cell using the following formula [30]:

$$V_{oc} = (1/e)\left(\left|E^{\mathrm{Donor}}\mathrm{HOMO}\right| - \left|E^{\mathrm{PCBM}}\mathrm{LUMO}\right|\right) - 0.3\,\mathrm{V} \tag{4}$$

where e is the elementary charge and using −4.3 eV for the PCBM LUMO energy. The value of 0.3 V in Eq. 4 is an empirical factor. They found that the MIM model is not applicable to BHJ devices. The observed deviation of V_{oc} from V_{BI} (HOMO donor minus LUMO of the acceptor), which they related to the working principle of the BHJ and they suggested that it could be minimized by optimizing materials, the active layer thickness, and the charge carrier mobility [30].

2.5.2 Short Circuit Current

Since I_{sc} also directly affects the efficiency, it is an essential parameter of the solar cells. Theoretical calculations predict that the development of novel donor materials is required to reach efficiencies of 10% or more [30]. New materials should also be considered to overcome the limitations of MDMO-PPV/PCBM or P3HT/PCBM-based devices. One important problem is limited absorption. It was calculated that 240 nm-thick P3HT is only capable of absorbing 21% of the sun's photons [121] and only in the wavelength range between 350 and 650 nm. Conjugated polymers have absorption coefficients in the order of 10^5 cm^{-1} [122]. The limitation in the absorption is due to the mismatch of the absorption of the organic semiconductor and the solar spectrum (see Fig. 10).

A polymer having a 1.1 eV band gap can absorb 77% of all the solar irradiation. However, majority of semiconducting polymers have band gaps higher than 2 eV, which limits the possible absorption [122]. This limitation led researchers to investigate low band gap polymers, which are achieved by shifting the polymer absorption spectrum into the near-infrared region.

The synthesis and application of low band gap polymers that absorb light above 600 nm in organic solar cells have been reported by several groups [124–130]. The band gap is defined as the difference between the HOMO and LUMO energy levels in polymers, neglecting Coulombic interactions. Low band gap polymers are defined as those having band gaps below approximately 2 eV [131]. One of the

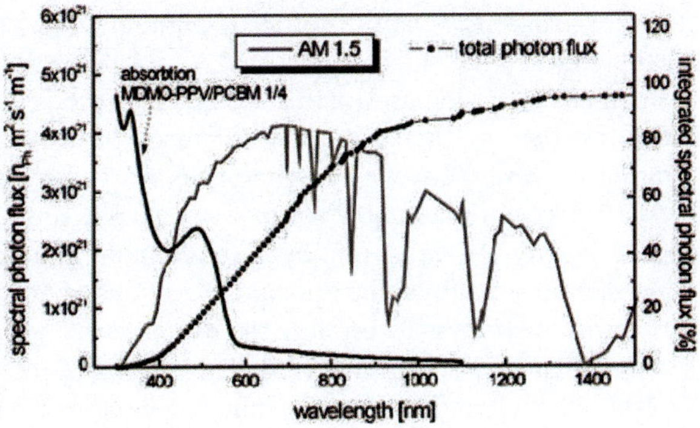

Fig. 10 AM1.5G spectrum compared to the absorption profile of a MDMO-PPV:PCBM (1:4) film. Circles show the total photon flux, i.e., the percentage of photons available for a material with a certain bandgap. Winder and Sariciftci [123] Reproduced by permission of the Royal Society of Chemistry (http://dx.doi.org/10.1039/B306630D)

limiting parameters in plastic solar cells is the mismatch of the absorption of organic materials and the terrestrial solar spectrum [130]. The optical band gap of the generally used CPs in organic solar cells have values around 2.0–2.2 eV [130]. The use of low band gap polymers expands the spectral region of bulk heterojunction solar cells and is a viable route to enhance the number of photons absorbed [130]. An ideal band gap of 1.3–1.6 eV for a bulk heterojunction device is described in a study by Scharber et al. [30].

Another way to overcome the barrier to absorbing more photons is increasing the layer thicknesses. However, one should note that an increase in the layer thickness may be limited by the charge carrier mobility and lifetime. At present, bulk heterojunction polymer solar cells are typically fabricated with an active layer thickness of between 80 and 100 nm. This active layer thickness has traditionally been chosen based on convenience and empirical results. However, it was studied by Moule et al. that active layer thickness has an effect on the short-circuit current and efficiency of BHJ polymer solar cells [132]. They demonstrated that the performance of these devices was highly dependent on the active layer thickness and, using a model for optical interference, they showed that such effects were responsible for the variations in performance as a function of active layer thickness. They also showed that the ideal composition ratio of the donor and acceptor materials was not constant, but depended on the active layer thickness in a predictable manner. Also, they confirmed that their results are not material specific and that high efficiency solar cells can be fabricated with active layer thickness greater than 100 nm [132].

The optical properties of BHJ solar cells were modelled by Hoppe et al. [133]. They showed that upon illumination of BHJ solar cells using MDMO:PPV/PCBM with the standard AM1.5G solar spectrum, the short-circuit current can be determined for any given internal quantum efficiency as a function of the active layer thickness. Also, the depth profile of photoinduced charge-generation rates were calculated.

Peumans et al. demonstrated a method for efficient photon harvesting in organic thin films, thereby increasing the efficiency of organic PV cells [134]. By incorporating an EBL between the photoactive organic layers and the metal cathode, they achieved an external PVC of 2.4%±0.3% from vacuum-deposited ultrathin organic bilayer PV cells employed in a simple light trapping geometry.

As mentioned above, the optical absorptions of conjugated polymers are strong, but even at maximum absorption, the penetration depth of light into these materials is in the range of 10–100 nm [135]. Therefore, the generation of excited states over 10–100 nm deep and harvesting, which occurs over a much shorter distance, must be combined. Making thicker films to collect more light by absorption decreases the field and reduces the collection efficiency. One way of bringing these requirements together is to trap light in the polymer layers by diffraction into guided modes in the thin polymer films [135]. This approach has been used to enhance light trapping and absorption in silicon solar cells in the energy range where optical absorption range in silicon is low [136]. Since optical absorption in conjugated polymers is high, Roman et al. used a similar approach to trap light in thinner polymer films [135]. They used an elastomeric mold to transfer a submicron grating pattern from a commercially available grating template to the active polymer layer in a PV device. They concluded that the grating function improved the optoelectronic properties of photodiodes [135].

Niggemann et al. also investigated two novel cell concepts: light trapping with diffraction gratings and buried nano electrodes [137]. In the buried electrode approach, they used one planar electrode and the other electrode was oriented perpendicular to the substrate surface, forming a lamellar structure. In this case low mobility charge carriers could be collected at the lamellar electrodes and high mobility charge carriers could be collected at the planar electrode. In their study, they discussed the potential of this approach and the initial experimental results [137].

2.5.3 Fill Factor

The fill factor of a device depends on charge dissociation, charge carrier transport, and recombination [138]. Effective cell design variables have large impact on the FF of an OPV device. The characteristic properties of an organic layer such as morphology and thickness, the regioregularity of the conjugated polymer, and the two interfaces between the electrodes and the blend layer have a large impact on the FF by affecting the series resistance (R_s), and the shunt resistance (R_{sh}). The R_s can be calculated from the inverse slope of the *I–V* curve in the first quadrant and is closely correlated with the intrinsic resistance, morphology, and thickness of the semiconductor layer whereas the R_{sh} is correlated with the amount and character of the impurities and defects in the active organic semiconductor layer since impurities and defects cause charge recombination and leakage current, as described by

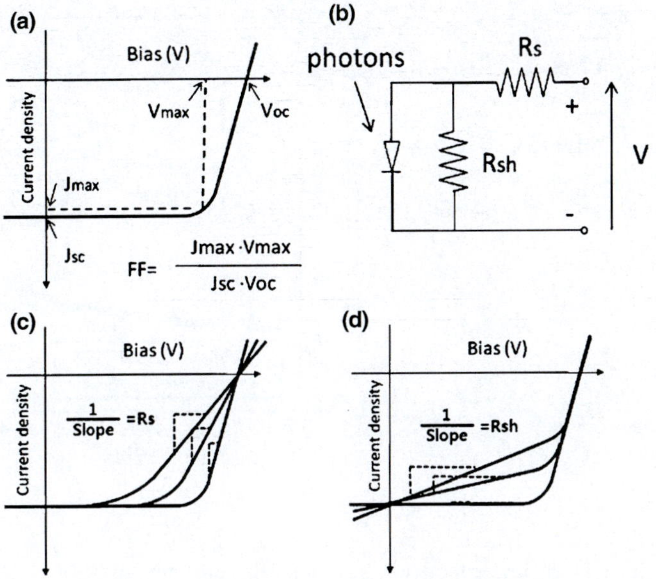

Fig. 11 **a** Definition of FF. J_{max} is the current density at the maximum of *JV* in the fourth quadrant and V_{max} is the bias at the maximum of *JV* in the fourth quadrant. **b** Circuit model of a photovoltaic device. R_s is the series resistance and R_{sh} is the shunt resistance. **c** Impact of the variation of the series resistance (*R*s) on the FF. The indicated inverse slope represents *R*s. **d** Impact of the variation of the shunt resistance (R_{sh}) on the FF. The indicated inverse slope represents R_{sh}. Reprinted with permission from Kim et al. [139]. Copyright (2009), American Chemical Society

Kim et al. [139] (see Fig. 11). R_{sh} determines the inverse slope of the *I*–*V* curve in the fourth quadrant.

The FF is determined using Eq. 2, which is basically the ratio of the product of maximum voltage and maximum current defined by the MPP and the product of the V_{oc} and I_{sc}.

Gupta et al. studied the fill factor in organic solar cells and observed that the shape of the *I*–*V* characteristics in the power generating fourth quadrant strongly depends on the quality of polymer-cathode interface and does not always reflect the bulk characteristics [138]. They pointed out that a partial metal coverage or a chemically modified layer can reduce the FF drastically. They concluded that defects at the polymer-metal interface give rise to charge carrier accumulation due to inefficient collection. They suggested that a conformal coating of metal that follows the polymer surface undulations is a preferable way to improve the FF [138].

Effective variables to control the FF of organic solar cells were studied by Kim et al. [139]. They reported relationships between the FF and various design parameters. Their device structure was ITO/PEDOT:PSS/P3HT+PCBM/LiF/Al.

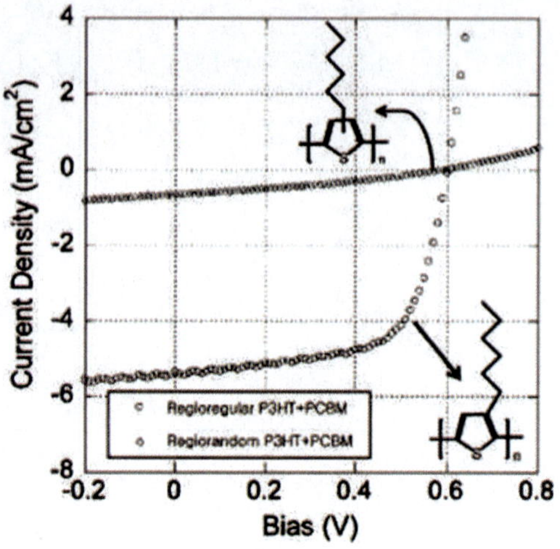

Fig. 12 Effect of regioregularity of conjugated polymers on the fill factor. Reprinted with permission from Kim et al. [139], Copyright (2009), American Chemical Society

They used the island-type electrode geometry to prevent additional charge collection observed in the crossbar-type device configuration, as described in Ref. [109]. They investigated the effects of characteristic properties of the organic layer to the FF, including morphology, thickness, the regioregularity of the conjugated polymer, and the two interfaces between the electrodes and the blend layer [139]. Their results showed that when the crystallinity of the blend layer was increased by thermal annealing, R_s decreased. When the regioregular P3HT was investigated, they observed that R_s was also reduced by one order of magnitude. They pointed out that the higher crystallinity induced by thermal annealing and regioregularity enhance the efficiency of the inter-and intramolecular charge transport [139] (see Fig. 12).

They observed that as the thickness of the blend layer was increased, R_s increased and R_{sh} decreased, which they attributed to the increased distance the charges must travel until they reach the electrodes, possibly increasing the resistivity and charge recombination (see Fig. 13).

They also studied the quality of two interfaces between the blend layer and the electrodes. They found that when a less conducting cathode was used, the FF decreased because of more charge recombination and leakage current, which was confirmed by the decrease in R_{sh} [139] (see Fig. 14).

They also proved that R_{sh} and the resulting FF were largely affected by the efficiency of charge extraction through the cathode, which they investigated by controlling the amount of the photoinduced charges in the blend layer through various illumination conditions. They observed that as the illumination intensity increased, R_{sh} and FF gradually decreased, indicating that there is more charge recombination in the blend layer. Finally, they showed that the nature of the interface between the Al anode and the blend layer also impacts the FF [139].

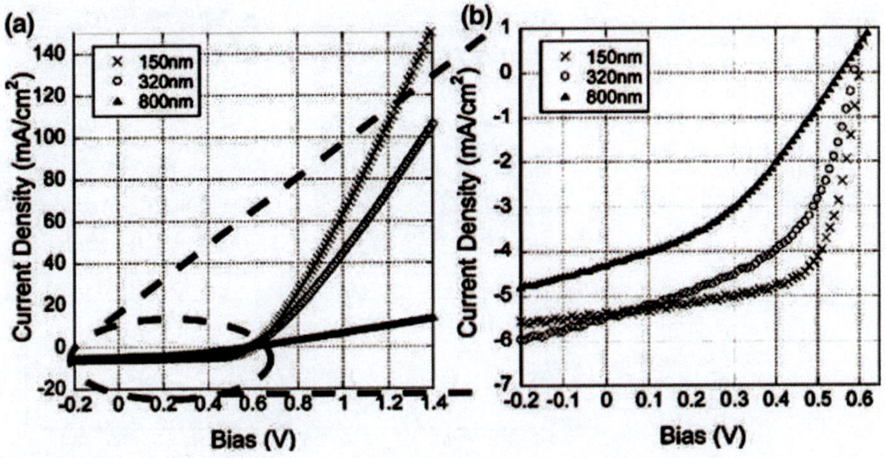

Fig. 13 Thickness effect on the fill factor. Reprinted with permission from Kim et al. [139]. Copyright (2009), American Chemical Society

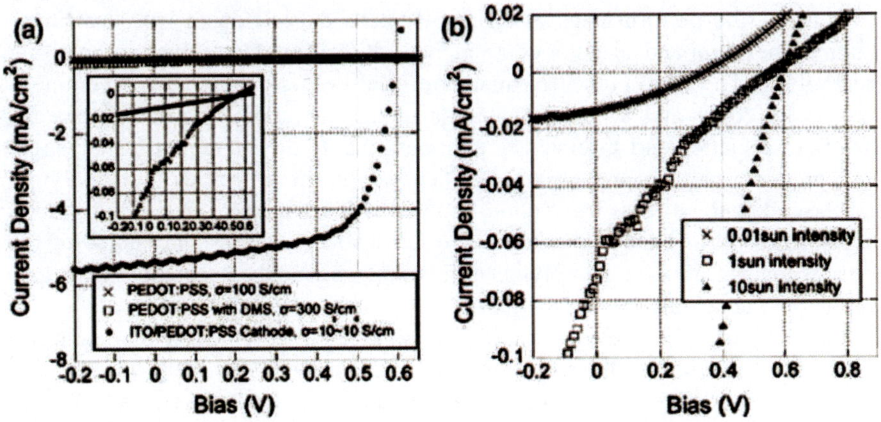

Fig. 14 **a** Effect of anode conductivity, **b** effect of illumination intensity on the fill factor. Reprinted with permission from Kim et al. [139]. Copyright (2009), American Chemical Society

3 Nanomorphology and Charge Transport in Organic Solar Cells

3.1 Nanomorphology

The BHJ concept requires blending two organic materials, one of which is a donor (generally a conjugated polymer) and the other is an acceptor (generally a fullerene derivative). Even if these donors and acceptors have an ideal electronic

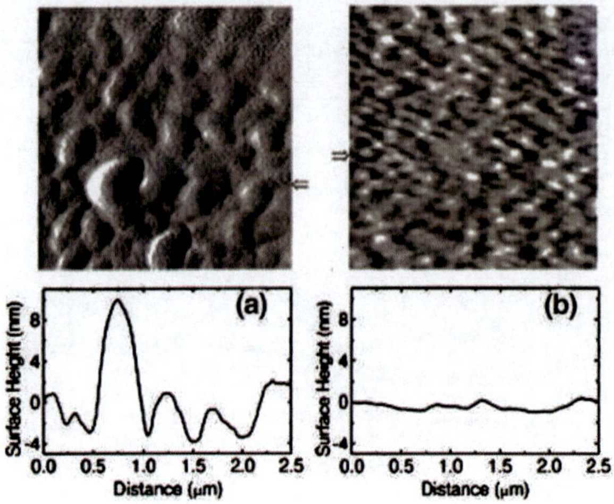

Fig. 15 AFM images showing the surface morphology of MDMO-PPV:PCBM (1:4 by wt) blend films with a thickness of approximately 100 nm and the corresponding cross-sections. **a** Film spin coated from a toluene solution. **b** Film spin coated from a chlorobenzene solution. Reprinted with permission from Shaheen et al. [10]. Copyright (2001), American Institute of Physics

relationship, the performances of BHJ solar cells depend on the physical interaction of the donor and acceptor components, which is evidenced by the composite morphology. The ideal bulk heterojunction solar cell is defined as a bicontinous network of donor and acceptor materials with a maximum interfacial area for exciton dissociation and a domain size proportional to the exciton diffusion length [9]. The morphology of the active layer depends on the properties of the polymer and the fullerene, such as interaction between two components (miscibility), and on influences associated with the device fabrication, such as the choice of the solvent, concentration of the polymer-fullerene components, deposition technique, and thermal annealing [140].

The PCE of OPV devices based on conjugated polymer/methanofullerene blends are dramatically affected by the molecular morphology [10] (see Fig. 15). Shaheen et al. reported one of the earliest studies [10]. They fabricated a device with a PVC of 2.5% under AM1.5G illumination by structuring the blend to be a more intimate mixture that contains less phase segragation of methanofullerenes, simultaneously increasing the degree of interactions between conjugated polymer chains. They showed that the choice of the proper casting solvent may lead to high PCEs. Shaheen et al. fabricated devices using MDMO:PPV and PCBM. Figure 15 shows the atomic force microscopy (AFM) images of the surfaces of MDMO-PPV:PCBM blend films spin coated using either toluene or chlorobenzene.

Measurement of the mechanical stiffness and adhesion properties of the surface indicated that the vertical features on the AFM images have a chemical composition different than the surrounding valleys. Since these features were not observed in pristine films of MDMO-PPV, they assigned them to be phase-segregated regions that contain a different fullerene concentration. In the case of chlorobenzene as the solvent they observed much more uniform mixing of the constituents. Their results indicated that spin coating the active layer blend from chlorobenzene has the effect of simultaneously enhancing the morphological microstructures of both components that form the interpenetrating networks [10].

The conformation of the polymer chains can be controlled by the selection of the organic solvents, by the concentration of the polymer solution, and by the rotational speed of the spin-casting process. The resulting morphology in the thin film plays an important role in controlling both the material and the device characteristics [120]. Liu et al. fabricated OPV devices by blending the conjugated polymer, poly (2-methoxy-5-(2′-ethylhexyloxy)-1,4-phenylenevinylene) (M3EH-PPV) with the buckminsterfullerene, C_{60} [120]. They showed that the photocurrent and V_{oc} show a strong dependence on the polymer processing conditions. They found that the PV devices fabricated with tetrahydrafuran (THF) or chloroform (non-aromatic solvents) have smaller photocurrents under the same reverse bias as well as higher open-circuit voltages than the devices fabricated with xylene, dichlorobenzene, or chlorobenzene (aromatic solvents). The device performance dependence on the processing solvent was attributed to the different solvation induced polymer morphology [120].

The combination of AFM, scanning electron microscopy (SEM), and transmission electron microscopy (TEM) were applied to MDMO-PPV/PCBM blends to investigate the nanomorphology of chlorobenzene and toluene cast blends [95, 96, 100].

Van Duren et al. used a technique to resolve the morphology of spin cast films of MDMO-PPV:PCBM blends in three dimensions on a nanometer scale and related the results to the performance of the corresponding solar cells [95]. They employed AFM, TEM, and depth profiling using dynamic time of flight secondary ion mass spectrometry (TOF–SIMS) techniques. They showed that phase separation is not observed up to 50 wt% PCBM and nanoscale phase separation sets in for concentrations of more than 67 wt% PCBM. They observed that the performance of the corresponding PV devices features a strong increase in PCE when the phase separated network develops, with sharp increases in photocurrent and fill factor between 50 and 67 wt% PCBM. They concluded that as the phase separation sets in, enhanced electron transport and a reduction of bimolecular charge recombination provide the conditions for improved performance [95].

Martens et al. also observed phase separation in MDMO-PPV:PCBM based organic solar cells using both TEM and AFM. They observed a two-phase system that consists of PCBM-rich domains embedded in a matrix consisting of MDMO-PPV and PCBM. They showed that changing the solvent influences the size of the phase-separated PCBM-rich domains. They also showed that the composition of the matrix is determined by the choice of the solvent. They studied this effect by changing the ratio of PCBM relative to MDMO-PPV [96].

The relation between nanoscale morphology and the associated device properties in conjugated polymer/fullerene bulk heterojunction solar cells was also investigated by Hoppe et al. [100]. They performed measurements on devices based on blends of MDMO-PPV and PCBM, spin cast either from toluene or chlorobenzene solutions. They applied SEM and AFM techniques to characterize the nanoscale morphology. In addition, they manipulated the morphology via

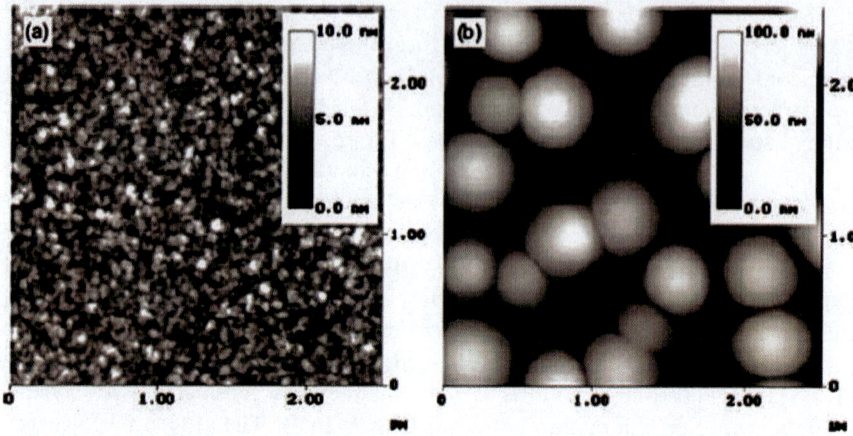

Fig. 16 AFM topography scans of MDMO-PPV/PCBM 1:4 (by weight) blended films spin cast from (**a**) chlorobenzene, and (**b**) toluene solution. The toluene cast films shows a ten-fold greater height variation compared to the chlorobenzene cast one. Features of a few hundred nanometers in width are visible in (**b**), while features in (**a**) are around 50 nm. Reproduced with permission from Hoppe et al. [100]. Copyright Wiley-VCH Verlag GmbH&Co. KGaA.

annealing, to increase the extent of phase separation in the thin film blends and to identify the distribution of materials. They confirmed the demixing of the materials under thermal treatment using photoluminescence (PL) measurements. They observed PL luminescence of PCBM clusters with sizes up to a few 100 nm which indicated a photocurrent loss in films of the coarser phase-separated blends cast from toluene. They concluded that the scale of phase separation in the toluene cast films depends strongly on the ratio of MDMO-PPV to PCBM, as well as on the total concentration of the casting solution. They also observed small beads of 20–30 nm diameter, attributed to MDMO-PPV, in blend films cast from both toluene and chlorobenzene (see Figs. 16, 17 and 18) [96].

The degree and length scale of phase separation between the donor and the acceptor phases has also an influence on the efficiency of BHJ solar cells [141]. As previously described, the common operation principle of a BHJ solar cell requires a photogenerated exciton in the donor or acceptor phase to diffuse to the phase boundary, where charge transfer takes place. These freely or bound carriers are subsequently separated and the free charges are then transported to the contacts through the percolating paths in different phases [141]. The critical length scale for the phase separation is the exciton diffusion length, which in disordered organic materials is on the order of 5 nm [141, 142].

Recently, Maturova et al. presented a model for the device performance of BHJ solar cells that takes into account the phase separated morphology [141]. It was applied to devices with MDMO-PPV:PCBM as the active material. They found that the characteristic features of the *I*–*V* characteristics and their evolution with morphology can be reproduced without taking into account exciton diffusion or field dependent dissociation of electrostatically bound charge transfer states.

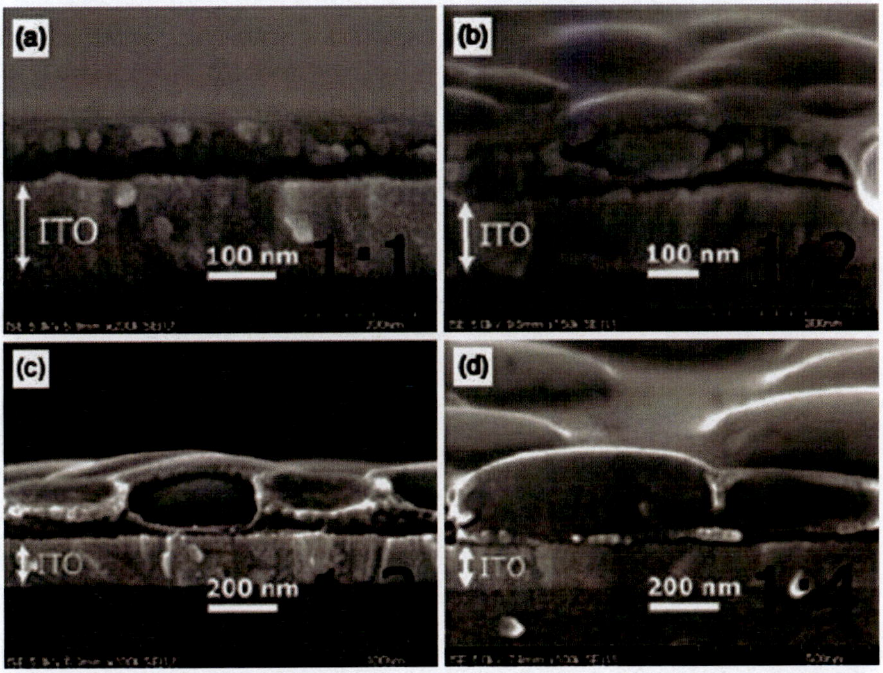

Fig. 17 SEM side views (cross-sections) of MDMO-PPV/PCBM blend films cast from toluene with various weight ratios of MDMO-PPV and PCBM. For the ratios 1:4, 1:3 and 1:2 (**b–d**), the nanoclusters, in the form of discs, are surrounded by another phase, called the skin, that contains smaller spheres of about 20–30 nm diameter. For the 1:1 film, only these smaller spheres are found. Reproduced with permission from Hoppe et al. [100]. Copyright Wiley-VCH Verlag GmbH&Co. KGaA.

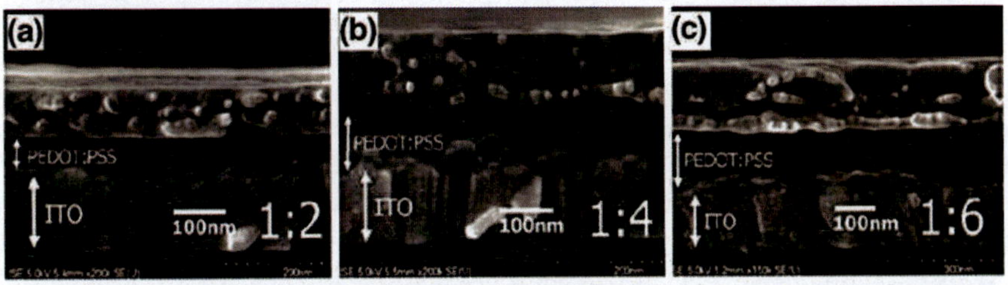

Fig. 18 SEM side views of MDMO-PPV:PCBM blend films spin-cast from chlorobenzene solutions with various ratios (by weight) of MDMO-PPV to PCBM on top of PEDOT:PSS coated ITO glass. In all samples nanospheres of about the same size (20 nm) are found. Reproduced with permission from Hoppe et al. [100]. Copyright Wiley-VCH Verlag GmbH&Co. KGaA.

They showed that the lateral transport of free photogenerated electrons from the donor-rich to the acceptor-rich phase is the critical process. They fabricated devices by blending MDMO-PPV with PCBM and tuned the length scale of phase separation from ca 40 nm to over 500 nm by changing the solvent and varying the

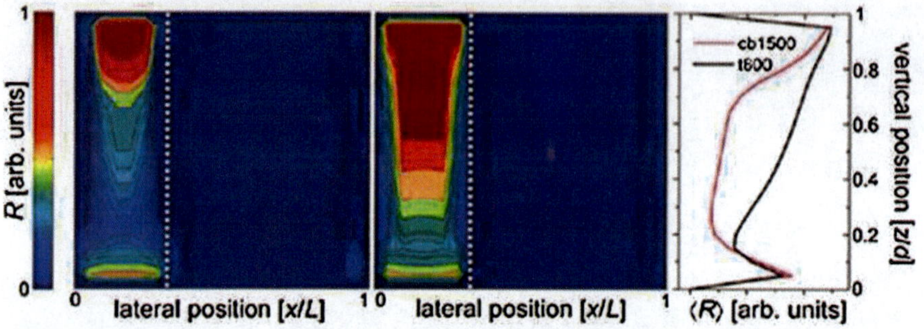

Fig. 19 Calculated recombination rates R under short circuit conditions inside devices spin-cast from (*left*) chlorobenzene at 1,500 rpm and (*middle*) toluene at 800 rpm versus normalized vertical and lateral position. The *dotted white line* indicates the phase boundary. The color scale is the same for both devices. The *right panel* shows the averaged recombination rate for each device. Reprinted with permission from Maturova et al. [141]. Copyright (2009), American Chemical Society

spin speed. They showed that the device characteristics of polymer:PCBM BHJ solar cells are largely determined by the lateral transport of the free photogenerated electrons from the polymer-rich donor phase to the PCBM-rich acceptor phase. They pointed out that under low fields, the nanometer length scale of phase separation strongly influences the probability of escaping bimolecular recombination, leading to an increased short circuit current for finer phase separation whereas under high fields, the lateral transport is suppressed, and a transition to a vertical electron extraction pathway occurs [141] (see Fig. 19).

Regioregular P3HT (RR-P3HT) is a promising candidate for OPV research due to its stability and absorption in the red region. Currently, one of the most promising organic solar cells, both in terms of efficiency and long-term stability is the system based on regioregular P3HT as the electron donor and PCBM as the acceptor, with efficiencies of 4–5% [16, 110]. An essential step to achieve high efficiency is to subject P3HT/PCBM solar cells to an annealing treatment, either at elevated temperature or during slow solvent evaporation [143]. As a result of heating, the morphological structure of the organic active layer can be improved by reducing the free volume and the density of defects at the interface during evaporation of the solvent [15, 144] and by enhancing interchain interactions [15, 145]. It is commonly expected that morphology development in a P3HT/PCBM system depending on the film preparation method, annealing treatment should create and stabilize a nanoscale interpenetrating network with high crystalline order and favorable concentration gradients of both components through the thickness of a photoactive layer [143].

The method of thermal annealing varies in the literature. Structural, optical, and electrical properties of poly(3-dodecylthiophene) (P3DT) have been studied after heat treatment at various temperatures by Nakazono et al. [146]. They showed the crystallinity and crystallite size of P3DT increased after heat treatment. They also found that the electrical conductivity decreased with increasing the heating

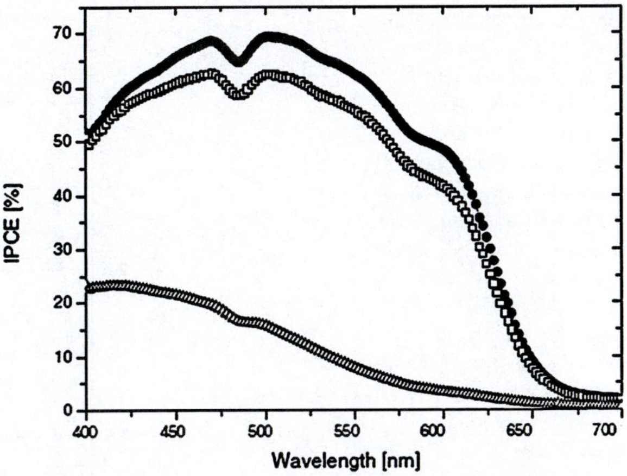

Fig. 20 IPCE of P3HT-PCBM solar cells: as-produced solar cell (*open triangles*), annealed solar cell (*open squares*), and cell simultaneously treated by annealing and applying an external voltage (*filled circles*). Reproduced with permission from Padinger et al. [14]. Copyright Wiley-VCH Verlag GmbH&Co. KGaA.

temperature. According to the results obtained they suggested that the interchain hopping of charge carrier between neighboring π conjugation planes is the dominant process in conductivity and carrier mobility in P3DT [146].

Studies on thermal annealing under chloroform vapor on polythiophene copolymers indicate a strong increase in the external quantum efficiency of photogenerated charge carriers [147].

Using a tempering cycle at elevated temperatures in which an external voltage is simultaneously applied, an increase in short-circuit current density and external quantum efficiency was demonstrated [14] (see Fig. 20). Padinger et al. presumed that the enhancement in the V_{oc} and the FF of post-production treated devices compared to untreated devices results partly from the burning of the shunts, while they presumed that the increase in I_{sc} results from an increase of the charge carrier mobility [14] (see Fig. 21).

The results of the studies on the electrical and physical modifications to P3HT upon thermal annealing showed that at temperatures higher than 100°C, the content of O_2 could no longer be detected due to a dedoping process, and the organic absorber became protected against deterioration [148].

The influence of nanomorphology on the PV action of polymer-fullerene composites was studied by Chirvaze et al. [15]. They analyzed the consequences of thermal annealing on ITO/PEDOT:PSS/P3HT:PCBM/Al PV devices. They realized that inspite of a considerable increase in the short-circuit current density due to heat treatment, a large amount of PCBM in the P3HT:PCBM-based devices do not automatically imply a high-energy conversion efficiency. In their experimental study, the PCBM concentration in the BHJ P3HT:PCBM composites was optimized to approximately 50%, the critical value which stems from the mutual interplay between an efficient generation of mobile charge carriers via photoinduced charge transfer and an efficient carrier transport, less affected by the recombination processes [15] (see Fig. 22).

Fig. 21 Influence of the duration of the post-production treatment on the white light efficiency (800 W/m^2) for different post-production treatment methods: annealing (*dotted line*) and annealing plus external voltage (*solid line*). Reproduced with permission from Padinger et al. [14] Copyright Wiley-VCH Verlag GmbH&Co. KGaA.

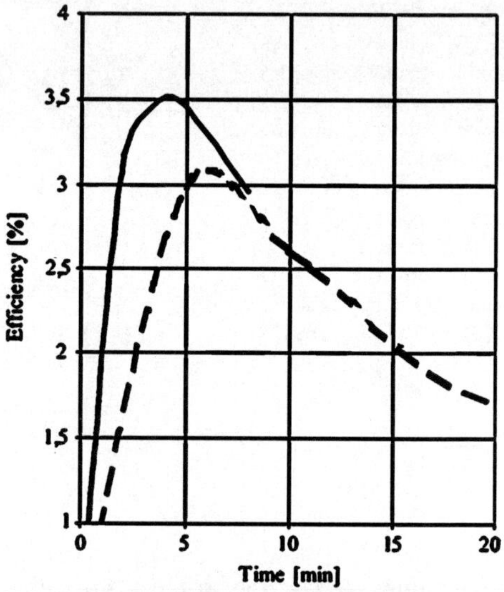

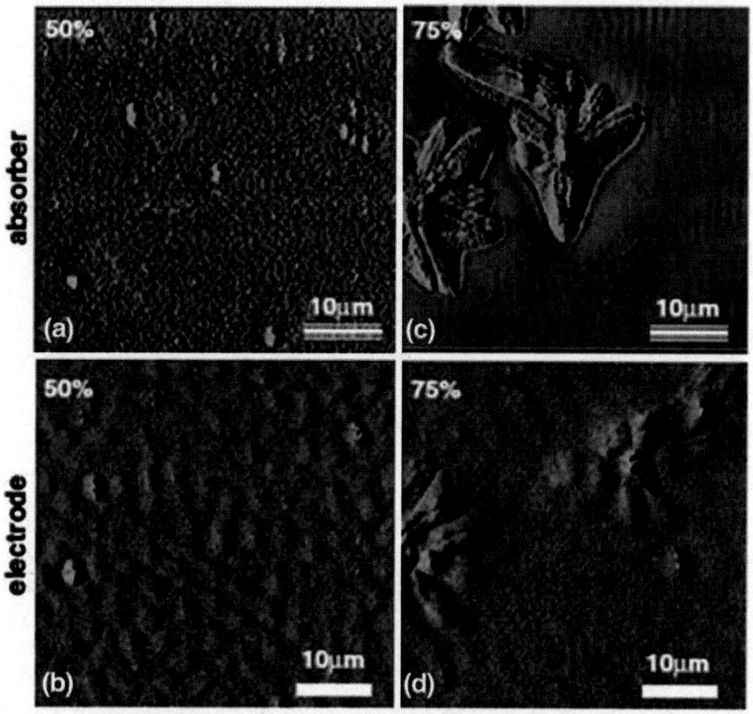

Fig. 22 AFM images in tapping mode (phase) of the P3HT:PCBM absorber surface (**a**, **c**) as well as of the aluminium electrode (**b**, **d**) for a PCBM concentration of 5 and 75%, respectively. Scan area 50 × 50 μm^2. Reprinted with permission from Chirvaze et al. [15]. Copyright (2004), American Institute of Physics

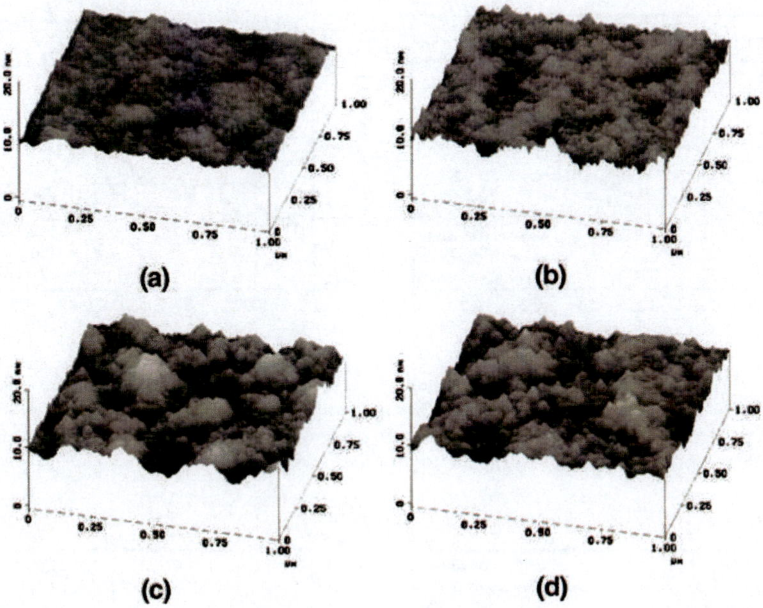

Fig. 23 AFM height images of the surface of the active layer consisting of P3HT:PCBM with a 1:1 weight ratio **a** before annealing, and after annealing at **b** 70°C, **c** 110°C, and **d** 150°C. The annealing time for all the films was 10 min. The P3HT:PCBM films were spin-coated on top of PEDOT:PSS-coated ITO glass. Reprinted with permission from Li et al. [149]. Copyright (2005), American Institute of Physics

The effects of annealing temperature and time on the device performance for devices annealed before and after cathode deposition were studied by Li et al. [149] (see Fig. 23). They showed that thermal annealing shows significant improvement in the performance of both types of annealing conditions, with post production annealing being slightly better. They found that the best annealing condition for the device is post-production treatment at 110°C for 10 min. By keeping the optimized thermal annealing condition and varying the active layer thickness, they fabricated devices with PCEs up to 4% [149] (see Figs. 24 and 25).

The effects of solvent and annealing on the performance of P3HT:PCBM-based organic solar cells were studied by Al Ibrahim et al. [150]. They fabricated P3HT:PCBM devices using two different solvents. They showed that the P3HT:PCBM films cast from chlorobenzene solution absorbed more red light than the films cast from chloroform solution. After thermal annealing, the films cast from chloroform showed higher absorption than the films cast from chlorobenzene. They observed that solar cells made from P3HT:PCBM chlorobenzene solution showed no change in the white light PVC after annealing. They concluded that the solar cells processed from P3HT:PCBM chloroform solution showed higher PCEs after annealing [150].

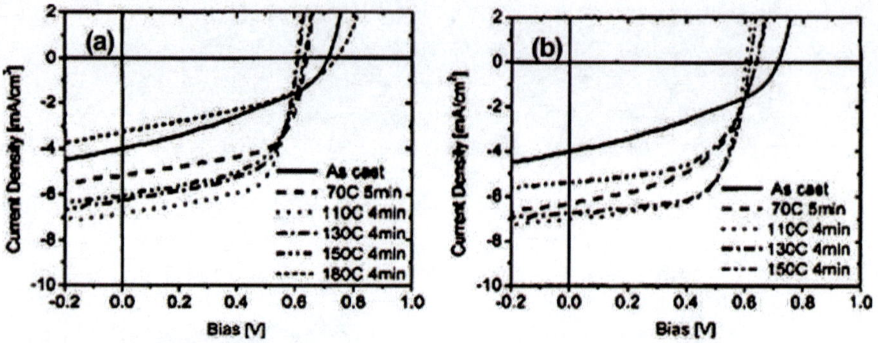

Fig. 24 *I-V* characteristics under an illumination of 100 mW/cm^2 AM1.5G for devices that have undergone annealing **a** pretreatment and **b** post-production treatment. The different curves represent different annealing temperatures ranging from room temperture to 180°C. Reprinted with permission from Li et al. [149]. Copyright (2005), American Institute of Physics

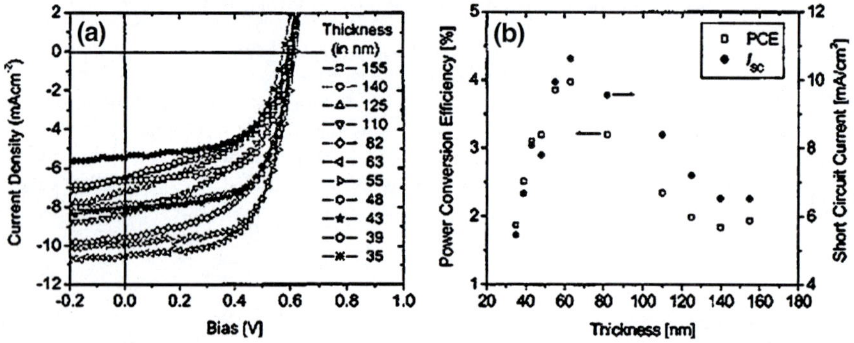

Fig. 25 **a** *I-V* characteristics under illumination for photovoltaic device based on P3HT:PCBM with a 1:1 weight ratio and varying active layer thickness, and **b** short-circuit current density and PVC as a function of active layer thickness. All the devices here were annealed at 110°C for 10 min post-production. Reprinted with permission from Li et al. [149]. Copyright (2005), American Institute of Physics

Erb et al. studied the correlation between structural and optical properties of composite P3HT/PCBM films [151]. They investigated the structural properties of P3HT:PCBM films using grazing-incidence X-ray diffraction (XRD) and determined the size and orientation of crystalline P3HT nanodomains. They did not detect PCBM crystallites in thin films by XRD. Upon annealing, they observed that the P3HT crystallinity increases, leading to an increase in the optical absorption and spectral photocurrent in the low photon energy region. They demonstrated a direct relation between efficiency and P3HT crystallinity [151].

Savenije et al. investigated the influence of various thermal treatment steps in the morphology and photoconductive properties of a non-contacted, 50 nm thick blend of P3HT and PCBM [152]. They characterized the films using TEM and electrodeless time resolved microwave conductivity (TRMC) techniques. They observed that after annealing the film for 5 min at 80°C, the TEM images showed the formation of crystalline fibrils of P3HT due to a more ordered packing of the polymer chains. They found that the thermal treatment results in a large increase of photoconductivity due to an enhancement of hole mobility in these crystalline P3HT domains from 0.0056 cm^2/Vs for the non-annealed sample to 0.044 cm^2/Vs for the annealed sample. They demonstrated that further annealing of the sample at 130°C results in the formation of three different substructures within the heterojunction: a PCBM:P3HT blend with PCBM-rich clusters, a region depleted of PCBM and large PCBM single crystals. A tenfold rise in decay time of the photoconductivity was explained by the authors by the formation of PCBM-rich clusters and large PCBM single crystals, resulting in an increased diffusional escape probability for mobile charge carriers and hence reduced recombination [152].

Yang et al. used TEM and electron diffraction to study the changes in morphology of P3HT:PCBM-based devices (see Fig. 26) [153]. They showed that thermal annealing produces and stabilizes a nanoscale interpenetrating network with crystalline order for both components. They demonstrated that P3HT forms long, thin conducting nanowires in a rather homogeneous, nanocrystalline PCBM film. They explained the increase in the PCE observed in these devices as a result of both the improved crystalline nature of films and increased but controlled demixing between the two constituents after annealing [153].

Van Bavel et al. performed an analysis of the three dimensional (3D) nanoscale organization of P3HT/PCBM photoactive layers before and after annealing treatments with nanometer resolution in all three dimensions by applying the technique of electron tomography (ET), with the aim to identify the critical morphology parameters contributing to the improved performance of annealed devices;they identified the critical morphology parameters contributing to the improved performance of P3HT/PCBM solar cells after thermal or solvent assisted annealing [143]. They explained that after spin-coating from solution, few aggregates of P3HT are present in the film and the overall crystallinity of P3HT is quite low. However, they showed that after annealing, reorganization of the P3HT/PCBM morphology takes place: many highly crystalline and long (up to several microns) P3HT nanorods form over the volume of the films. These nanorods form a genuine 3D network, which was proved by ET, and serve as physical barriers to PCBM diffusion by suppressing a large-scale phase separation at any point in the film preparation process. The authors concluded that all of the morphological changes in the P3HT/PCBM system after thermal or solvent assisted annealing result in improved photon absorption, better exciton dissociation, and more efficient and balanced transport of free charges to the electrodes, and are reflected in higher values of FF and J_{sc}, and hence better overall efficiency of corresponding solar cell devices (see Figs. 27 and 28) [143].

Another ET study on imaging the 3D nanostructure of a polymer cell was performed by Andersson et al. [154]. In ET, three-dimensional reconstructions are

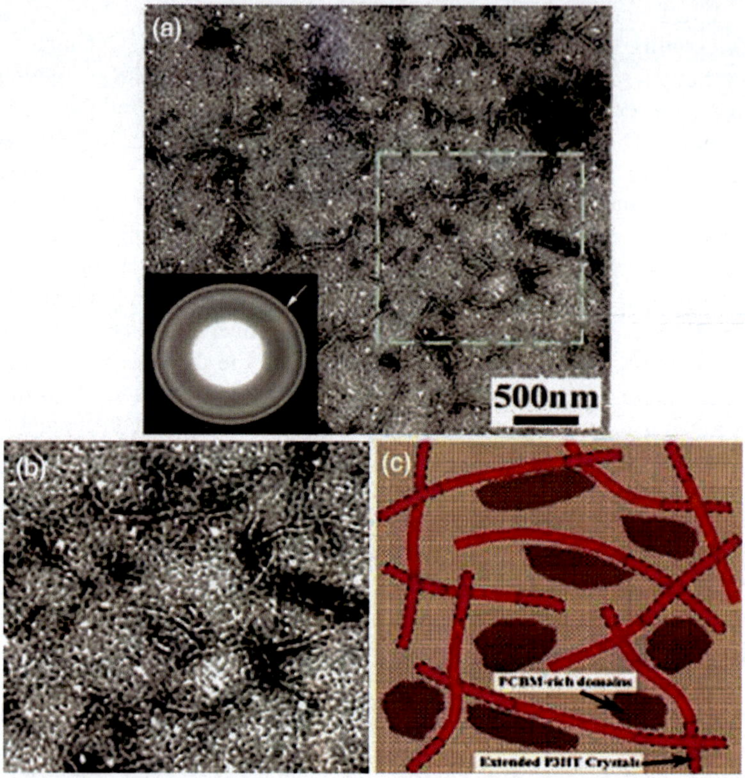

Fig. 26 BF TEM images show the overview (**a**), the zoom-in (**b**) and the corresponding schematic representation (**c**) of the thermal annealed photoactive layer. The inset in (**a**) is the corresponding SAED pattern. The *arrow* is to indicate the increased intensity of (020) Debye-Scherer ring from P3HT crystals compared to the SAED pattern. For Fig. 3c the *dash line* bordered regions represent the extension of existing P3HT crystals in the pristine film or newly developed PCBM-rich domain during the annealing step. Reprinted with permission from Yang et al. [153]. Copyright (2005), American Chemical Society

made from micrographs obtained from transmission electron microscopy. The authors investigated the blends of poly [2,7-(0,9-dioctyl-fluorene)-alt-5,5-(4′,7′-di-2-thienyl-2′,1′3′-benzothia-diazole)] (APFO-3) and PCBM. They used this method to supply a 3D representation of the morphology of the film, where domains with different scattering properties can be distinguished [154].

Marsh et al. employed sub-picosecond transient absorption (TA) spectroscopy on operating P3HT:PCBM devices to probe the effect of annealing on charge transfer dynamics and nanoscale morphology [155]. They removed the effect of high excitation densities which could otherwise dominate using their measurement configuration. The authors demonstrated that the charge transfer in pristine P3HT:PCBM devices proceeds on a sub-picosecond time scale. According to the authors, in annealed devices, annealing results in diffusion limited charge

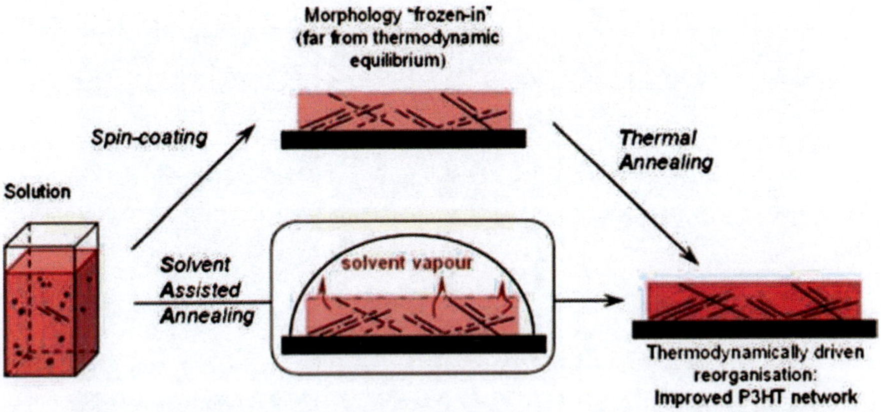

Fig. 27 Morphology development in P3HT/PCBM photoactive layers depending on film preparation method. Reprinted with permission from Van Bavel et al. [143]. Copyright (2009), American Chemical Society

generation with a half time of ca. 3 ps, complete only after 30 ps. The authors address this as a result of exclusion of PCBM molecules and ordering of P3HT domains and correlated this with the improved PV efficiency [155].

Keawprajak et al. investigated the effect of a crystallizable solvent on the morphology and performance of BHJ solar cells based on a 1:1 weight-ratio RR-P3HT/PCBM blend, using chlorobenzene with different concentrations of 1,3,5-trichlorobenzene (TCB) as crystallizable aromatic solvents [156]. The authors demonstrated that the decelerated release of the residual solvent during the solidification of the blend film improves the self organization, and the surface of the film becomes smoother with the addition of TCB, which in turn improves organic/cathode contact. The authors concluded that the polymer blend film formed with a TCB crystallizable solvent exhibited better crystallinity, optical absorption, and polymer morphology than the polymer blend film formed without the TCB crystallizable solvent [156].

Polyfluorene copolymers are well known for their high charge carrier mobility, good processability, and high absorption coefficients. Among this class of polymers, poly[9,9′-dioctyl-fluorene-co-bithiophene] (F8T2) has excellent properties in both hole transport and thermotropic liquid crystallinity to allow better packing of the polymer via self assembly. The nanomorphological changes on the PV performance based on a F8T2:PCBM BHJ system were demonstrated by Huang et al. [157].

In summary, obtaining a desired morphology in polymer-fullerene composite solar cells is critical for device performance. Optimizing the electronic structure is not alone sufficient and there is not a direct recipe to apply to new material systems; therefore, morphology optimization should be performed for any combination of materials, where solvent selection is just one parameter among many [11].

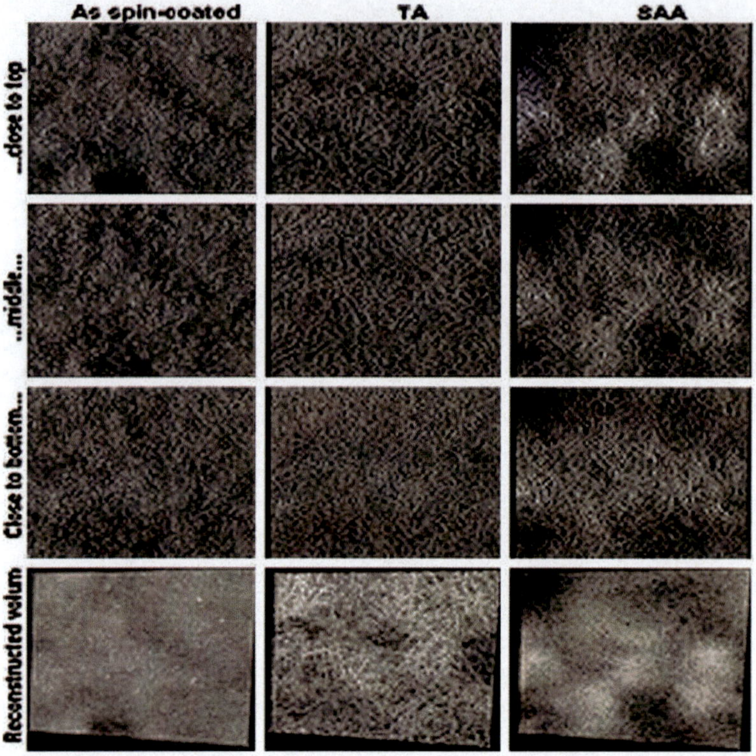

Fig. 28 Results of EM applied to P3HT/PCBM photoactive layers: as spin-coated, thermally annealed at 130°C for 20 min (TA) and solvent assisted annealed for 3 h (SAA). The first three rows contain slices taken out of a reconstructed volume of the corresponding film. All slices are lying in the horizontal (X,Y) plane of the film at different depths (Z location): one slice close to the top of the film (i.e., close to the electron collecting LiF/Al electrode), another one in the middle of a film, and the third one close to the bottom of the film (near the hole collecting PEDOT:PSS/ITO electrode). The dimensions of the slices are around 1,700 nm × 1,700 nm. The images in the fourth row are snapshots of the corresponding film's reconstructed volume, i.e., a stack of all of the slices through the whole thickness of a film, with dimensions of around 1,700 nm × 1,700 nm × 100 nm. Reprinted with permission from Van Bavel et al. [143].Copyright (2009), American Chemical Society

3.2 Charge Transport and Mobility

Charge carrier transport and mobility in organic semiconductors are of critical importance for the device performance of organic solar cells. There have been several studies on modeling the charge transport through organic solar cells. Depending on the degree of order, the charge carrier transport mechanism in organic solar cells falls between two extreme cases: band or hopping transport [77, 158]. For many systems, such as organic semiconductors it is not easy to determine which type of transport exists. Band transport is typically observed in highly

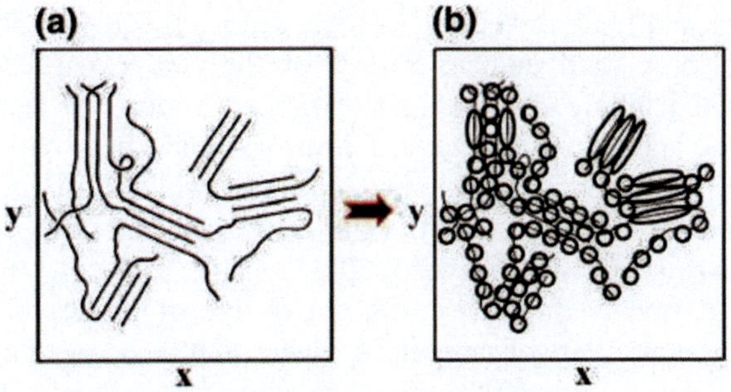

Fig. 29 **a** Schematic description of polymer chains and **b** an illustration of how a long chain breaks into small conjugation units (or sites). Reproduced with permission from Tessler et al. [159]. Copyright Wiley-VCH Verlag GmbH&Co. KGaA.

purified molecular crystals at temperatures that are not too high [77]. For organic crystals, for example, it has been demonstrated that band transport dominates in the low *T* regime, while at room temperature hopping is the prevailing transport mechanism [158]. A characteristic feature of band transport, the temperature dependence follows power law behavior at lower temperatures. However, in the presence of traps, significant deviations from such behavior are observed [77]. In the other extreme case of an amorphous organic solid, hopping transport prevails, which leads to much lower mobility values. Instead of a power law temperature dependence, an activated behavior appears and the mobility also depends on the applied electric field. Furthermore, space charge and trapping effects as well as details of charge carrier injection mechanism have to be considered for describing electrical transport in organic solids [77]. Many studies based on the charge transport and mobility of organic semiconducting materials agree that most OPV materials investigated so far are disordered in nature. Introducing disorder into the lattice and breaking the crystal symmetry results in the wave functions becoming localized and the formation of energy states in the forbidden band gap. Therefore, former models describing the conductivity phenomenon in metals or other types of materials can no longer explain the transport under such circumstances. Since OPV materials are disordered in nature, the physical theories relating to the charge carriers through disordered media should be taken into account [159].

Tessler et al. suggested that a conjugated polymer cannot be considered a single electronic wire but rather as a wire broken into subunits by chemical or physical defects that are called conjugation units and the number of monomers making this unit is called the conjugation length (see Fig. 29) [159]. In this case, there are now many small electronic units that can also be thought of as small molecules distributed across the film. The only role that was played by the polymer chain was to determine the distribution of the units ("molecules") in the film. With this simple picture, one can consider that there is a distribution of sites that can host charge and enable it to move across a film. Thus, one needs to analyze the charge transport across sites that are distributed in real space as well as in energy space according to their studies [159].

The key quantity that characterizes charge transport is the carrier mobility [160]. In the absence of any external potential, transport is purely diffusive and is generally described by a simple diffusion equation. The application of an external electric field induces a drift of the charge carriers; the mobility can then be alternatively defined as the ratio between the velocity and the amplitude of the applied electric field. Diffusion should be seen as a local displacement of the charge around an average position, while drift induces a displacement of the average position. Drift is the effect that dominates the migration of the charges across an organic layer [160].

Even in the best cases, the charge mobilities of organic semiconductors are several orders of magnitude lower than those of inorganic PV materials. The low mobilities arise from the localization of electronic states on individual molecules or segments of molecules. Relatively weak intermolecular interactions mean that charge transport is best described by hopping transport rather than band transport [161]. There is extensive evidence from temperature and field dependent conductivity measurements that charge transport in both conjugated polymers and small molecules occurs by a hopping mechanism [77].

The requirement of balanced hole and electron mobilities is of critical importance for the device performance of organic solar cells. Low mobilities limit the performance of organic solar cells by enhancing the probability of charge recombination, limiting the charge separation yield, and increasing resistive losses [161]. Nelson et al. demonstrated that packing disorder can reduce mobility by several orders of magnitude for a disordered molecular semiconductor relative to the crystal. They mentioned that the effect of disorder on mobility is amplified by anisotropy in the orbitals involved in the charge transfer; side chains strongly influence mobility and transport in conjugated polymers is a function of both the extension of polymer chains and order in chain packing [161].

Charge carrier mobilities can be determined experimentally by using various techniques such as time of flight (TOF), field effect transistor (FET) configuration, charge extraction under linearly increasing voltage (CELIV), steady state current density–voltage response, admittance spectroscopy, and TRMC [160, 162]. Results from methods that measure mobilities under macroscopic distances ($\approx$1 mm) are often dependent on the purity and order in the material. Methods that measure mobilities over microscopic distances are less dependent on these characteristics [160].

As previously described, after photoinduced electron transfer at the donor/acceptor interface and dissociation, the electrons are localized in the PCBM phase whereas the holes remain in the PPV polymer chains. Then, the free electrons and holes must be transported via percolated PCBM and PPV pathways toward electrodes [163]. For pure PCBM, the electron mobility was reported as 2×10^{-7} m^2/Vs [164], which is 4,000 times higher than the hole mobility in pristine MDMO-PPV, reported as 5×10^{-11} m^2/Vs [165]. The hole mobility of pristine MDMO-PPV was rather low. Also, the results showed that usually the electron mobility and the hole mobility are unbalanced.

One of the ways to improve the charge carrier mobility in conjugated polymers is increasing the regioregularity of the conjugated backbone, as reported by Lutsen et al. [166]. However, fully regioregular MDMO-PPV lacked solubility, which is quite important for OPV cells. Later, a soluble regiospecific (RS) MDMO-PPV was synthesized and the room temperature TOF mobility of this 70:30 RS-MDMO-PPV was found to be higher than the commercially available regiorandom RRa-MDMO-PPV [167].

Mozer et al. investigated the hole mobility of P3HT by the TOF technique and observed a negative electric field dependence of mobility for the first time in conjugated semiconducting polymers [168].

The question of interest was whether or not the mobility values of pristine materials changed upon blending. Melzer et al. demonstrated that the hole mobility in a 20:80 weight ratio MDMO-PPV:PCBM blend was enhanced by more than two orders of magnitude compared to the pure polymer value using space charge limited current (SCLC) measurements [169]. The difference between the electron and hole mobility was reduced by only a factor of 10, resulting in a more balanced transport [163].

The 2.5% efficient MDMO-PPV:PCBM-based solar cells make use of 80 wt% PCBM in the blend, a material which barely contributes to the absorption. Mihailetchi et al. demonstrated that one of the main reasons for the relatively large amount of PCBM being required is the enhancement of the hole transport in MDMO-PPV, which is the slower carrier, when blended with PCBM [170]. An insufficient hole mobility will also lead to a build up of space-charge in the solar cell, which affects the fill factor and efficiency [171]. Therefore, an intrinsically higher hole mobility in the blend permits one to reduce the amount of PCBM and inhibits the occurrence of space charge, which will further increase the magnitude of the photogenerated current and enhance the PCE [170].

Among the factors that may influence charge transport in polymer-molecule or polymer–polymer blends are the different morphologies adopted by the two components, especially in the case where one is a high molecular weight polymer and the other is a small conjugated molecule [172]. Frost et al. studied the influence of polymer blend morphology on charge transport and photocurrent generation in polymer blends using Monte Carlo simulations. They reported that different morphologies, such as well dispersed, loosely coiled polymer chains (heterophilic) or aggregated, tightly coiled chains (homophilic) can be produced by varying the interaction energies between the chains with different types of process treatments, such as solvent or annealing. Frost et al. pointed out that the morphology strongly influences charge transport characteristics; that is, homophilic blends show a lower threshold for charge percolation and more dispersive transport than heterophilic blends, consistent with a higher probability of charge trapping in configurational traps. They explained the experimental observations, such as the effect of annealing on charge transients in some conjugated polymers, by their model in terms of process-induced changes in polymer morphology, whereas monomer-based transport models cannot explain the data [172]. They concluded that the morphology influences both transport and charge recombination such that the optimum composition and morphology is a sensitive function of charge recombination and separation rates and faster transport, achieved through optimizing the blend morphology does not necessarily lead to higher photocurrent generation [172].

Pivrikas et al. reviewed the charge transport and recombination in polymer/fullerene organic solar cells [173]. They studied the transport and recombination in two polymer/fullerene mixtures: RR-P3HT/PCBM and MDMO-PPV/PCBM. They showed that the carrier mobility is strongly dependent on the film morphology and disorder. Therefore, charge carrier mobility can change orders of magnitude depending on film preperation techniques, solvents used for film deposition, time and temperature of the thermal treatment, etc. [173].

Shah et al. used a model to address the interplay between anisotropic charge transport and the orientational and/or crystalline ordering of donor molecules [174]. They presented the influence of the degree of ordering, the impact of ordering parallel or perpendicular to the electrodes, and the role of device thickness. They used a drift–diffusion model, which accounted for the effects of the anisotropic nature of charge transport in conjugated polymers and the donor–acceptor density inhomogeneties, to determine the PV performance. They tried to find answers for the following questions: "Is there an optimal degree of alignment influencing the efficiency of the solar cells?", "Which among the parallel or perpendicular orientations (relative to the electrodes) of the molecules are more favorable for optimizing the efficiencies?", and "What is the influence of anisotropic transport characteristics upon the optimal domain widths of the donor and acceptors?". They considered "simulated morphologies" where alternating lamellae of donor and acceptor are aligned perpendicular to the two electrodes, forming an ordered heterojunction structure. They assumed that the donor molecules are aligned only either paralel or perpendicular to the interface to specifically elucidate the role of anisotropic charge transport. They characterized the degree of alignment by a scalar order parameter. According to their model, for perpendicularly aligned donors there exists an optimal degree of alignment to extract the best device efficiencies. In contrast, for the case of parallely oriented donor molecules, they observed an opposite effect where stronger alignment is always beter due to accompanying enhancement in transport to the electrodes. They observed that the magnitudes of short-circuit current density (J_{sc}) are much smaller in the parallel orientation, reflecting the fact that the transport of excitons to the interface and subsequent dissociation is greatly diminished in this arrangement. For the interplay of the *D–A* size with the anisotropic charge transport characteristics, Shah et al. commented that for the perpendicularly oriented donors, increasing the anisotropy enhances the exciton transport. In contrast, parallel-oriented donors are less affected by such anisotropies [174].

Ballantyne et al. used the TOF method to study the effect of P3HT molecular weight (MW) (M_n = 13–121 kDa) on the charge mobility in pristine and PCBM blend films using highly RR P3HT [175]. They observed a constant 10^{-4} cm^2/Vs mobility value as the molecular weight increased from 13–18 kDa, but the mobility value decreased by one order of magnitude as molecular weight further increased from 34–121 kDa. They showed that the decrease in charge mobility observed in the blend films is accompanied by a change in surface morphology, and leads to a decrease in the performance of PV devices made from these blends. They attributed the decrease in mobility to the change in packing of the polymer chains. They mentioned that it is possible that high molecular weight chains become tangled, causing twisting of the polymer backbone, decreased intra-chain transport by creating more traps, and/or reduction of inter-chain charge hopping by allowing less overlap of conjugated segments. Combining their results, they proposed that there is an optimal P3HT molecular weight for charge transport and device performance between 13 and 34 kDa. Their modeling of the device behavior indicated that the effect of MW on device performance could not be explained solely in terms of differences in charge mobility. They mentioned that other factors, such as the effect of blend morphology on interface resistance, the effect of film thickness on transport, or the non-linear effect of transport on the net charge-generation rate, could be partly responsible for the variations in the efficiency [175].

In organic solar cells, the mobility simultaneously controls both the carrier extraction and the losses via carrier recombination [176]. Mandoc et al. demonstrated that the balance between carrier losses by extraction and by recombination leads to a distinct optimal carrier mobility with regard to the efficiency of organic solar cells [176]. They stated that for low mobilities, recombination losses limit the performance, whereas efficient extraction at high mobilities leads to a reduction in open-circuit voltage. The authors concluded that the maximal performance of the BHJ solar cells is governed by balancing transport and recombination and that both extremes of too low or too high mobility contribute to the loss of efficiency through different mechanisms. Therefore, an optimized carrier mobility is an important condition that must be fulfilled to obtain highly efficient organic solar cells [176].

In summary, low mobilities limit the performance of organic solar cells since the probability of charge recombination increases [161]. It should also be noted that using different methods to characterize the mobility may lead to very different mobility results. For example, when the hole mobility results of pristine P3HT are examined, the measured hole mobility ranges from 10^{-8} m^2/Vs in TOF and SCLC up to 10^{-5} m^2/Vs in FETs [177]. RR-P3HT self-organizes into a crystalline structure and, owing to the π–π stacking direction, the charge (hole) transport is extremely efficient. Since in FET measurements the current travels in the plane of the film (parallel to the substrate), the anisotropy in the polymer-chain orientation contributes strongly to the difference in the measured mobility [177]. A different molecular weight of P3HT, the presence of PCBM, and/or the application of a thermal treatment will also affect the measured electron and hole mobilities [1, 178, 179]. Therefore, relevant values for charge carrier mobilities can only be obtained when measured in the same configuration and experimental conditions as used in an operational solar cell device [177].

4 Current Status of Solution Processed Conjugated Polymer-Based Organic Solar Cells

The first attempts to create organic solar cells were made by sandwiching a single layer of organic material between two electrodes of different work functions [180]. In these cells, the performance strongly depends on the nature of the electrodes.

In 1986, Tang realized a breakthrough by fabricating a bilayer organic PV cell using copper phthalocyanine and a perylene tetracarboxylic derivative [58]. A PVC of about 1% has been achieved under simulated AM2G illumination. A novel feature of the device was that the charge-generation efficiency was relatively independent of the bias voltage, resulting in cells with fill factor values as high as 0.65. The interface between the two organic materials, rather than the electrode/organic contacts, was found to be crucial in determining the PV properties of the cell [58]. Although this efficiency was a breakthrough at that time, the performance of this PV device was still limited since the only light absorption within a very thin layer close to the interface contributes to the photocurrent generation.

The carrier collection efficiency and PVC of polymer PV cells were improved by blending of the semiconducting polymer with C_{60} or its functionalized derivatives by Yu et al. [17]. Composite films of MEH-PPV and fullerenes exhibited efficiencies better than those that have been achieved with devices made with pure MEH-PPV by over two orders of magnitude. The efficient charge separation results from photoinduced electron transfer from the MEH-PPV (donor) to the C_{60} (acceptor); the high collection efficiency results from a bicontinuous network of internal donor–acceptor heterojunctions [17].

Halls et al. also showed that the interpenetrating network formed from a phase-segregated mixture of two semiconducting polymers provides both the spatially distributed interfaces necessary for efficient charge photo-generation, and the means for separately collecting the electrons and holes [5].

Various combinations of donor and acceptor materials have been used to fabricate BHJ solar cells. The most promising and successful combination of materials is a blend of a semiconducting polymer and a fullerene derivative as the acceptor. It was demonstrated that sub-picosecond charge transfer takes place at the interface of these materials, which ensures efficient charge generation [52].

A breakthrough on solution processed organic solar cells employing MDMO-PPV as the donor and PCBM as the acceptor was realized by Shaheen et al. [10]. They showed that the PCE of OPV devices based on a conjugated polymer/methanofullerene blend is dramatically affected by molecular morphology. By structuring the blend to be a more intimate mixture that contains less phase segregation of the methanofullerenes and simultaneously increasing the degree of interactions between conjugated polymer chains, they fabricated a device with a PVC of 2.5% under AM1.5G illumination. This was a nearly three-fold enhancement over previously reported values. The crucial step that improved the device performance was the use of a special solvent that improves the nanoscale morphology for charge generation and transport [10].

Although the efficiency of single-layer devices was improved by simple mixing the *p*- and *n*-type materials, this solution bears a new challenge. In the BHJ approach, the donor and acceptor components are mixed in a bulk volume so that each donor–acceptor interface is within a distance smaller than the exciton diffusion length of each absorbing site [43]. The donor and acceptor phases must form a bicontinuous and interpenetrating network. Therefore, bulk heterojunction devices are much more sensitive to the nanoscale morphology in the blend [43]. As previously discussed, to organize the donor and acceptor in a nanometer scale is a big challenge.

Further improvement in OPV efficiency was achived when the PPV was replaced with P3HT. P3HT gave a further increase in the hole mobility and the PVC was improved to 3.5% [14].

Recently, polymer PV cells with PCEs approaching 5% were fabricated using titanium oxide (TiO_x) as an optical spacer by Kim et al. [102]. In this study, solar cells with a TiO_x layer (deposited via a sol–gel process) between the active layer and the electron-collecting aluminum electrode exhibited approximately a 50% enhancement in short-circuit current compared to similar devices without the optical spacer, as a result of modification of the spatial distribution of the light intensity inside the device [102].

Konarka Technologies demonstrated a 5.21% power conversion efficient plastic solar cell with an active area of 1.024 cm^2 (NREL certificate for Konarka solar cell from 11-20-2006) [181].

Park et al. demonstrated 6.1% efficient bulk heterojunction solar cells with internal quantum efficiency reaching 100% [25]. They fabricated devices using alternating copolymer, poly[*N*-9″-hepta-decanyl-2,7-carbazole-alt-5,5-(4′,7′-di-2-thienyl-2′,1′,3′-benzothiadiazole) (PCDTBT), in bulk heterojunction composites with the fullerene derivative [6,6]-phenyl C_{70}-butyric acid methyl ester ($PC_{70}BM$). The PCDTBT/$PC_{70}BM$ solar cells exhibited a J_{SC} of 10.6 mA cm^{-2}, an V_{OC} of 0.88 V, a *FF* of 0.66 and a PCE of 6.1% under AM1.5G. The nearly 100% internal quantum efficiency implies that essentially every absorbed photon results in a separated pair of charge carriers and all photogenerated carriers are collected by the electrodes [25]. PCBM, which amounts to 80% of the photoactive layer, has a very low absorption coefficient in the visible region of the spectrum due to its symmetrically forbidden HOMO–LUMO transition. When the C_{60} moiety of [60]PCBM is replaced by a less symmetric fullerene, these transitions become slightly allowed and an increase in light absorption/harvesting is expected [11, 182].

Chen et al. showed that the open-circuit voltage of polymer solar cells constructed based on a low-bandgap polymer, PBDTTT, can be tuned, step by step using different functional groups to achieve values as high as 0.76 V [183]. This increased open-circuit voltage combined with a high short-circuit current density results in a polymer solar cell with a PVC as high as 6.77%, as certified by the National Renewable Energy Laboratory [183].

The demonstration of PCEs close to 8% under calibrated AM1.5G conditions demonstrates the potential of this technology. The fabrication using suitable printing methods for the production of such organic solar cells is regarded as the next important milestone [181].

In terms of research trends, the main direction has been toward the achievement of high PCEs under simulated sunlight [184]. However, other important aspects include stability, cost and processing [184]. Long operational lifetimes of solar cells are required in real-life application. Polymer solar cells degrade under illumination and in the dark. Therefore, methods for enhancing the stability through the choice of better active materials, encapsulation, etc., are important [184].

5 Stability

Organic solar cells degrade under illumination and in the dark. Long operation lifetimes of solar cell devices are required in real-life applications and understanding the degradation mechanisms is of great importance for successful implementation of this promising technology [184]. For the technological development of organic solar cells, the important aspects aside from PCE are the stability, cost and processing [185, 186].

Organic materials are, by nature, more sensitive to chemical degradation, e.g., from oxygen and water, than inorganic materials. Diffusion of oxygen and water into the OPV device, photochemistry and photooxidation of polymers, degradation of a function of polymer preparation, and chemical degradation of the ITO electrode are seen as chemical degradation mechanisms whereas morphological stability can be seen as a physical degradation mechanism for OPV cells [184]. In a recent review by Jorgensen et al. these mechanisms are discussed [184]. The following points cause problems for the stability of devices [184]:

- Oxygen from the atmosphere will oxidize the organic layer, especially when the device is illuminated.
- Materials such as P3HT are less sensitive to this degradation. Annealing can lead to further structural changes when the devices are heated.
- Aluminium, which is the most commonly used metal electrode, reacts with oxygen and water and perhaps with the organic material due to its low work function.
- The transparent electrode, which is commonly ITO, is also not stable. Indium atoms pass to the organic layer and distribute themselves into the polymer layer.
- PEDOT:PSS layer is hygroscopic in nature and may bind to water.

These possible factors for degradation have various degrees of influence on the device lifetime.

Due to the degradation mechanisms mentioned above, polymer solar cells cannot operate in air for more than few hours. Therefore, appropriate encapsulation may help increase the lifetimes of these devices [112]. Dennler et al. reported a shelf lifetime study of MDMO-PPV:PCBM solar cells encapsulated in a new flexible and transparent poly(ethylene naphthalate) (PEN)-based ultra-high barrier

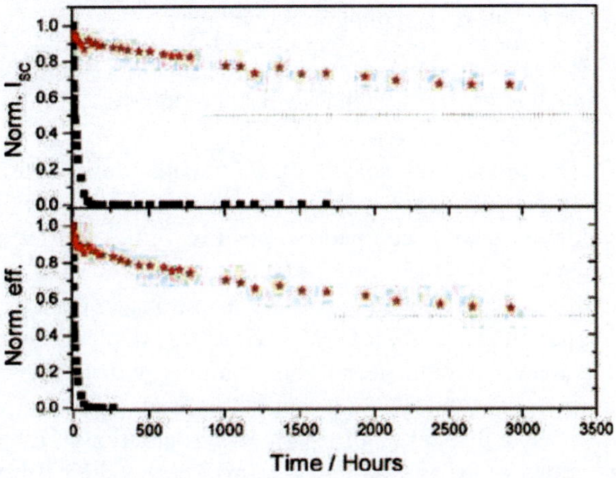

Fig. 30 Normalized I_{SC} and efficiency of solar cells encapsulated with uncoated PET (*black*) and with the flexible gas barrier material (*red*) versus storage time in the dark under ambient air. Reprinted from [112] Copyright (2006) with permission from Elsevier

material fabricated entirely by plasma enhanced chemical vapor deposition (PECVD) [112]. They demonstrated that encapsulation raises the shelf lifetime (defined as the time performace degrades to 50% of the initial efficiency) of 30–57 mm^2 solar cells from a few hours to over 3,000 h, or more than 4 months [112].

Figure 30 shows the evolution of the normalized V_{OC} and FF of a typical solar cell versus storage time. In the case of PET encapsulation, both V_{OC} and FF decreased by 50% within less than 200 h. They observed that aging of the cells results in the development of a back diode that tends to drastically reduce FF. However, in the case of the flexible barrier encapsulation, V_{OC} remains at about 90% of its initial value after 3,000 h, as does FF (see Fig. 30) [112].

However, 3,000 h of device stability is not sufficient for this type of device in the market. MDMO-PPV is known to be particularly unstable and sensitive to aging. However, new materials, such as polythiophenes, that are much less sensitive to moisture and oxygen and allow even better PV conversion are becoming available [187]. The combination of such materials with this flexible gas barrier encapsulation might constitute a very promising advance in this field [112].

6 Technological Aspects of Organic Solar Cells

The technologically important features of organic solar cells can be summarized as following [185]:

- flexibility and transparency.
- potential to be manufactured in a continuous printing process.
- large area coating.
- easy integration in a wide variety of devices.
- cost-effectiveness.

- ecological and economic advantages.

These features are very important for commercialization. However, as in inorganic PV cells, OPV cells should also fullfill the basic requirements for renewable energy production [185].

The production aspects of OPV cells were reviewed by Brabec et al. [185]. They pointed out that there are three essential parameters for every solar technology to take its competitive position in the market: efficiency, lifetime, and cost per watt peak [185]. From efficiency point of view, a device efficiency of ~10% and a module efficiency of ~5% are regarded as critical market entry values. Higher efficiencies for OPV devices will demand novel materials, either low band gap materials or material combinations with better matched electronic levels, to improve I_{sc} and V_{oc}, respectively. From lifetime point of view, for low-cost devices, a life time of 3–5 years (operational life time of 3,000–5,000 h) is regarded as the market entry point. These values were derived from typical usage lifetimes of electronic goods that could potentially be powered by solar cells. According to Brabec et al., printing techniques are the necessary tool for significant cost reduction for PVs. Under the assumption that the material costs (including packaging and substrates) will further decrease with scale up, organic solar cells with a cost structure significantly below 1 USD/Wp can be realized [185]. Brabec et al. foresee that the first applications will most likely be in consumer electronics: calculators, watches, sensors and innovative applications such as mobile phone chargers.

Compared to the efficiencies over 20% achieved by Si-PVs, the efficiency of OPV cells seem poor. However, other advantages, particularly the low cost, flexibility and easy integration into a wide variety of systems make them competitive for a range of applications such as packaging, smart fabrics, and signalization. Organic PVs utilize cheap and easy processing techniques, such as inkjet printing or spin coating, on large, flexible plastic substrates. The flexibility of the final product offers low weight and easy installation.

One company has already announced that plastic solar panels will be available for portable electronics, smart fabrics and building integrated PVs in the first quarter of 2011; they also added that customers interested in protypes of plastic solar panels for portable electronics testing can obtain samples by mid-2010.

7 Conclusions

Recent milestones and improved efficiencies achieved by organic solar cells opened up a possibility for OPV devices to take their place in the market. The biggest advantages of organic-plastic solar cells are their low cost, flexibility and ability to be wrapped around structures or even applied like paint. As predicted by Scharber et al., PCEs of 10% are within the achievable range [30] and 15% can be achieved using the tandem solar cell concept [29]. Both of these efficiencies designate the development of novel donor materials.

Three key parameters will determine the success of organic solar cells: price, efficiency and lifetime. Since they are manufactured using rather simple techniques, their cost can be less than Si-based PV devices. The efficiencies achieved recently are considerable and encouraging, and the research continues for further improvement. Lifetime is a critical issue since organic materials are rather sensitive to oxygen and humidity. Since, full commertialization will take time, it is not easy to determine the product's lifetime of plastic solar cells.

The first commercial applications seem to be portable electronics and smart fabrics. It seems that the initial goal is not to replace the existing technology, such as solar panels used on rooftops.

From the scientific point of view, new donor and acceptor materials and possibly device concepts are needed for further efficiency improvement of organic solar cells. Several strategies can be summarized as the synthesis and development of low band gap materials for improved matching with the solar spectrum, synthesis and development of new electron accepting materials, tandem solar cells, improvement of hybrid solar cell and dye-sensitized solar cell concepts.

The ideal structure of BHJ solar cells requires that the two phases of donors and acceptors within the bulk heterojunction be spaced with an average separation of around 10–20 nm, equal or less than the exciton diffusion length. The two phases must be separated in percolated (ideally straight) pathways to ensure high mobility charge carrier transport with reduced recombination. Such a well organized nanostructure is not easy to obtain in classical polymer mixtures due to disorder. However, self organization of the organic semiconducting polymers (molecules) can be introduced using an amphiphilic structure di-block copolymer, which results in a self-organized phase [188], using supramolecular fullerene architectures by quadruple hydrogen bonding [189] and using uniaxially aligned, thin films of discotic columnar liquid crystalline materials [190].

Research on organic solar cells is highly interdisciplinary, promising, and challenging. The interdisciplinarity has already led to scientific and technological advancement since the organic solar cells are close to commercialization and is expected to lead to further achievements in this field.

References

1. Ginley D, Green MA, Collins R (2008) Solar energy conversion toward 1 Terawatt. MRS Bull 33:355–363
2. Shockley W, Queisser HJ (1961) Detailed balance limit of efficiency of *p–n* junction solar cells. J App Phys 32:510–519
3. Poortmans J (2007) Organic solar cells: linking nanoscale to gigawatts? Prog Photovolt Res App 15:657–658
4. Minnaert B, Burgelman M (2007) Efficiency potential of organic bulk heterojunction solar cells. Prog Photovolt Res App 15:741–748
5. Halls JM, Walsh CA, Greenham NC et al (1995) Efficient photodiodes from interpenetrating polymer networks. Nature 376:498–500

6. Yu G, Heeger AJ (1995) Charge separation and photovoltaic conversion in polymer composites. J App Phys 78:4510–4515
7. Xue J, Uchida S, Rand BP et al (2004) 4.2% Efficient organic photovoltaic cells with low series resistances. Appl Phys Lett 84:3013–3015
8. Peumans P, Yakimov A, Forrest SR (2003) Small molecular weight organic thin-film photodetectors and solar cells. J App Phys 93:3693–3723
9. Thompson BC, Frechet JMJ (2008) Polymer-fullerene composite solar cells. Angew Chem Int Ed 47:58–77
10. Shaheen SE, Brabec C, Padinger F et al (2001) 2.5% efficient organic plastic solar cells. Appl Phys Lett 78:841–843
11. Wienk M, Kroon JM, Verhees WJH et al (2003) Efficient methano[70]fullerene/MDMO-PPV bulk heterojunction photovoltaic cells. Angew Chemie Int Ed 42:3371–3375
12. Chen T, Rieke RD (1992) The first regioregular head-to-tail poly(3-hexylthiophene-2, 5-diyl) and a regiorandom isopolymer: nickel versus palladium catalysis of 2(5)-bromo-5(2)-(bromozincio)-3-hexylthiophene polymerization. J Am Chem Soc 114:10087–10088
13. McCullough RD, Lowe RD, Jayaraman M et al (1993) Design, synthesis, and control of conducting polymer architectures: structurally homogeneous poly(3-alkylthiophenes). J Org Chem 58:904–912
14. Padinger F, Rittberger RS, Sariciftci NS (2003) Effects of post production treatment on plastic solar cells. Adv Func Mat 13:85–88
15. Chirvaze D, Parisi J, Hummelen JC et al (2004) Influence of nanomorphology on the photovoltaic action of polymer–fullerene composites. Nanotechnology 15:1317–1323
16. Ma WL, Yang CY, Gong X et al (2005) Thermally stable, efficient polymer solar cells with nanoscale control of the interpenetrating network morphology. Adv Func Mat 15:1617–1622
17. Yu G, Gao J, Hummelen JC et al (1995) Enhanced efficiencies via a network of internal donor-acceptor heterojunctions. Science 270:1789–1791
18. Sariciftci NS, Braun D, Zhang C et al (1993) Semiconducting polymer-buckminster fullerene heterojunctions: diodes, photodiodes and photovoltaic cells. Appl Phys Lett 62:585–587
19. Alem S, Bettignies R, Nunzi JM et al (2004) Efficient polymer based interpenetrated network photovoltaic cells. Appl Phys Lett 84:2178
20. Hummelen JC, Knight BW, LePeq F et al (1995) Preparation and characterization of fulleroid and methanofullerene derivatives. J Org Chem 60:532–538
21. Drechsel J, Maennig B, Kozlowski F et al (2004) High efficiency organic solar cells based on single or multi PIN structures. Thin Solid Films 451–452:515–517
22. Drechsel J, Maennig B, Gebeyehu D et al (2004) MIP-type organic solar cells incorporating phthalocyanine/fullerene mixed layers and doped wide-gap transport layers. Org Electron 5:175–186
23. Peumans P, Uchida S, Forrest SR (2003) Efficient bulk heterojunction photovoltaic cells using small-molecular-weight organic thin films. Nature 425:16–158
24. Peet J, Kim JY, Coates NE et al (2007) Efficiency enhancement in low-bandgap polymer solar cells by processing with alkane dithiols. Nat Mater 6(7):497–500
25. Park SH, Roy A, Beaupre S et al (2009) Bulk heterojunction solar cells with internal quantum efficiency approaching 100%. Nat Photonics 3(5):297–302
26. Kim Y, Cook S, Tuladhar S et al (2006) A strong regioregularity effect in self-organizing conjugated polymer films and high-efficiency polythiophene: fullerene solar cells. Nat Mater 5:197–203
27. Green M, Emery E, Hisikawa Y et al (2007) Solar cell efficiency tables. Prog Photovolt Res Appl 15:425–430
28. Dennler G, Scharber MC, Brabec CJ (2009) Polymer-fullerene bulk heterojunction solar cells. Adv Mater 21:1323–1338

29. Dennler G, Scharber MC, Ameri T et al (2008) Design rules for donors in bulk-heterojunction tandem solar cells?Towards 15% energy-conversion efficiency. Adv Mater 20:579–583
30. Scharber MC, Mühlbacher D, Koppe M et al (2006) Design rules for donors in bulk heterojunction solar cells-towards 10% energy conversion efficiency. Adv Func Mat 18:789–794
31. Servaites JD, Yeganeh S, Marks TJ et al (2010) Efficiency enhancement in organic photovoltaic cells: consequences of optimizing series resistance. Adv Func Mat 20:97–104
32. Wang X, Liu D, Li J (2010) Organic photovoltaic materials and thin film solar cells. Front Chem China 5:45–60
33. Huynh WU, Dittmer JJ, Alivisatos AP (2002) Hybrid nanorod polymer solar cells. Science 295:2425–2427
34. Coakley KM, McGehee MD (2003) Photovoltaic cells made from conjugated polymers infiltrated into mesaporous Titania. Appl Phys Lett 83:3380–3383
35. Kim JY, Lee K, Coates NE et al (2007) Efficient tandem polymer solar cells fabricated by all-solution processing. Science 317:222–225
36. Wienk MM, Turbiez MGR, Struijk MK et al (2006) Low-band gap poly(di-2 thienylthienopyrazine):fullerene solar cells. Appl Phys Lett 88:153511–153514
37. Hou J, Chen HY, Zhang S et al (2008) Synthesis, characterization, and photovoltaic properties of a low band gap polymer based on silole-containing polythiophenes and 2, 1, 3-benzothiadiazole. J Am Chem Soc 130:16144–16145
38. Koetse MM, Sweelsson J, Hoekerd KT et al (2006) Efficient polymer:polymer bulk heterojunction solar cells. Appl Phys Lett 88:083504–083507
39. Tada K, Hirohata, Hosada M et al (1997) Donor polymer (PAT6)—acceptor polymer (CNPPV) fractal network photocells. Synth Met 85:1305–1306
40. Liao H, Chen LM, Xu Z et al (2008) Highly efficient inverted polymer solar cell by low temperature annealing of CS_2CO_3 interlayer. Appl Phys Lett 92:173303–173306
41. Katz EA, Gevorgyan S, Orynbayev MS et al (2007) Outdoor testing and long-term stability of plastic solar cells. Eur Phys J Appl Phys 36:307–311
42. Hauch JA, Schilinsky P, Choulis SA et al (2008) Flexible organic P3HT:PCBM bulk heterojunction modules with more than 1 year outdoor lifetime. Solar Energy Mater Solar Cells 92:727–731
43. Hoppe H, Sariciftci NS (2004) Organic solar cells: an overview. J Mater Chem 19:1924–1945
44. Heeger AJ (2001) Nobel lecture: semiconducting and metallic polymers: the fourth generation of polymeric materials. Rev Mod Phys 73:681–700
45. Borroughes JH, Bradley DDC, Brown AR et al (1990) Light emitting diodes based on conjugated polymers. Nature 347:539–541
46. Berggren M, Gustafsson G, Ingaenes O et al (1994) Green electroluminescence in Poly-(3-cyclohexylthiophene) light-emitting diodes. Adv Mater 6:488–490
47. Gadisa A, Person E, Andersson MR et al (2007) Red and near infrared polarized light emissions from polyfluorene copolymer based light emitting diodes. Appl Phys Lett 90:113510–113513
48. Bao Z, Dodabalapur A, Lovinger A et al (1996) Soluble and processable regioregular poly(3-hexylthiophene) for thin film field-effect transistor applications with high mobility. Appl Phys Lett 69:4108–4111
49. Sirringhaus H, Tessler N, Friend RH (1998) Integrated optoelectronic devices based on conjugated polymers. Science 280:1741–1744
50. Neugebauer H, Brabec C, Hummelen JC, Sariciftci NS (2000) Stability and photodegradation mechanisms of conjugated polymer/fullerene plastic solar cells 61:35–42
51. Kanicki J (1986) Handbook of conducting polymers Marcel Dekker, New York, p 544
52. Sariciftci NS, Smilowitz L, Heeger AJ et al (1992) Photoinduced electron transfer from a conducting polymer to buckminsterfullerene. Science 258:1474–1476

53. Smilowitz L, Sariciftctci NS, Wu R et al (1993) Photoexcitation spectroscopy of conducting-polymer–C_{60} composites: photoinduced electron transfer. Phys Rev B 47:13835–13842
54. Kraabel B, Hummelen JC, Vacar D et al (1996) Subpicosecond photoinduced electron transfer from conjugated polymers to functionalized fullerenes. J Chem Phys 104:4267–4274
55. Morita S, Zakhidov AA, Yoshino K (1992) Doping effect of Buckminsterfullerene in conducting polymer: change of absorption spectrum and quenching of luminescene. Solid State Commun 82:249–252
56. Yoshino K, Yin XH, Morita S et al (1993) Enhanced photoconductivity of C_{60} doped Poly(3-alkylthiophene). Solid State Commun 85:85–88
57. Wudl F (1992) The chemical properties of Buckminsterfullerene (C60) and the birth and infancy of fulleroids. Acc Chem Res 25:157–161
58. Tang CW (1986) Two layer organic photovoltaic cell. App Phys Lett 43:183–186
59. Heremans P, Cheyns D, Rand BP (2009) Strategies for increasing the efficiency of heterojunction organic solar cells: material selection and device architecture. Acc Chem Res 42:1740–1747
60. Archer M, Hill R (2001) Clean electricity from photovoltaics. Imprerial College Press, London
61. Würfel P (2007) Photovoltaic principles and organic solar cells. Chimia 61:770–774
62. Haugeneder A, Neges M, Kallinger C et al (1999) Exciton diffusion and dissociation in conjugated polymer/fullerene blends and heterostructures. Phys Rev B 59:15346–15351
63. Theander M, Yartsev A, Zigmantas D et al (2000) Photoluminescence quenching at a polythiophene/C_{60} heterojunction. Phys Rev B 61:12957–12963
64. Stübinger T, Brütting W (2001) Exciton diffusion and optical interference in organic donor/acceptor photovoltaic cells. J Appl Phys 90:3632–3642
65. Mozer A, Sariciftci NS (2006) Conjugated polymer photovoltaic devices and materials. C R Chimie 9:568–577
66. Gregg BA, Hanna MC (2003) Comparing organic to inorganic photovoltaic cells: theory, experiment, and simulation. J Appl Phys 93:3605–3614
67. Popovic ZD, Hor A, Loutfy R (1988) A study of carrier generation mechanism in benzimidazole perylene/tetraphenyldiamine thin film structures. Chem Phys 127:451–457
68. Arkhipov V, Emelianova EV, Baessler (1999) Hot eciton dissociation in a conjugated polymer. Phys Rev Lett 82:1321–1324
69. Koster LJA, Smits ECP, Mihailetchi VD et al (2005) Device model for the operation of polymer/fullerene bulk heterojunction solar cells. Phys Rev B 72:085205–085214
70. Na S, Kim S, Jo J et al (2008) Efficient and flexible ITO free organic solar cells using highly conductive polymer anodes. Adv Mater 20:4061–4067
71. Cui J, Wang A, Edleman NL et al (2001) Indium tin oxide alternatives—high work function transparent conducting oxides as anodes for organic light-emitting diodes. Adv Mater 13:1476–1480
72. Rowell MW, Topinka MA, McGehee MA et al (2006) Organic solar cells with carbon nanotube network electrodes. Appl Phys Lett 88:233506–233509
73. Frohne H, Shaheen S, Brabec C et al (2002) Influence of anodic work function on the performance of organic solar cells. Chem Phys Chem 3:795–799
74. Zhang F, Johansson M, Andersson MR et al (2002) Polymer photovoltaic cells with conducting polymer anodes. Adv Mater 14:662–665
75. Huang J, Miller PF, Wilson JS et al (2005) Investigation of the effects of doping and post-deposition treatments on the conductivity, morphology, and work function of poly(3, 4-ethylenedioxythiophene)/poly(styrene sulfonate) films. Adv Func Mater 15:290–296
76. Hiramoto M, Fujiwara H, Yokoyama M et al (1991) Three-layered organic solar cell with a photoactive interlayer of codeposited pigments. Appl Phys Lett 58:1062–1065
77. Brütting W (2006) Introduction to physics of organic semiconductors in physics of organic semiconductor. pp 1–12, Wiley-VCH, Weinheim. 10.1002/3527606637

78. Geens W, Aernouts T, Poortmans J et al (2002) Organic co-evaporated films of a PPV-pentamer and C_{60}: model systems for donor/acceptor polymer blends. Thin Solid Films 403–404:438–443
79. Tsuzuki TT, Shirota J, Rostalski J et al (2000) The effect of fullerene doping on photoelectric conversion using titanyl phthalocyanine and a perylene pigment. Sol Energy Mater Sol Cells 61:1–8
80. Brabec CJ, Shaheen SE, Winder C et al (2002) Effect of LiF/metal electrodes on the performance of plastic solar cells. App Phys Lett 80:1288–1291
81. Hung LS, Tang CW, Mason NG (1997) Enhanced electron injection in organic electroluminescence devices using an Al/LiF electrode. App Phys Lett 70:152–155
82. Jabbour GE, Kawahe Y, Shaheen S et al (1997) Highly efficient and bright organic electroluminescent devices with an aluminum cathode. Appl Phys Lett 71:1762–1765
83. Jong D, Friedlein MP, Osikowicz RW et al (2006) Ultraviolet photoelectron spectroscopy of polymers. Mol Cryst Liq Cryst 455:193–203
84. Parker I (1994) Carrier tunneling and device characteristics in light emitting diodes. J Appl Phys 75:1656–1667
85. Horowitz G (1990) Organic semiconductors for new electronic devices. Adv Mater 2:287–292
86. Glenis S, Tourillon G, Garnier F (1984) Electrochemically grown polythiophene and Poly (3-methylthiophene) organic photovoltaic cells. Thin Solid Films 111:93–103
87. Braun D, Heeger AJ (1991) Visible light emission from semiconducting polymer diodes. Appl Phys Lett 58:1982–1985
88. Yu G, Zhang C, Heeger AJ (1994) Dual function semiconducting polymer devices light emitting and photodetecting diodes. Appl Phys Lett 64:154–1540
89. Antoniadis H, Hsieh BR, Abkowitz MA et al (1994) Photovoltaic and photoconductive properties of aluminum/poly(p-phenylene vinylene) interfaces. Synt Metals 62:265–271
90. Hoppe H, Sariciftci NS (2008) Polymer solar cells. Adv Poly Sci Springer Berlin Heidelberg 1–86
91. Brabec CJ, Sariciftci NS, Hummelen JC (2001) Plastic solar cells. Adv Func Mater 11:15–26
92. Yamashita Y, Takashima W, Kaneto K (1993) Characteristics of heterojunction diode of C_{60}/tetratiafulvalene (TTF). Jpn J Appl Phys 32:L1017–L1020
93. Halls JJM, Friend R (1997) The photovoltaic effect in a poly(p-phenylenevinylene)/perylene heterojunction. Synth Met 85:1307–1308
94. Peumans P, Forrest SR (2001) Very-high-efficiency double-heterostructure copper phthalocyanine/C_{60} photovoltaic cells. Appl Phys Lett 79:126–129
95. Van Duren JKJ, Yang XN, Loos J et al (2004) Relating the morphology of Poly(p-phenylene vinylene)/methanofullerene blends to solar-cell performance. Adv Func Mat 14:425–434
96. Martens T, Dhaen J, Munters T et al (2003) Disclosure of the nanostructure of MDMO-PPV:PCBM bulk hetero-junction organic solar cells by a combination of SPM and TEM). Synth Met 138:243–247
97. Mihailetchi VD, Blom PWM, Hummelen JC et al (2003) Cathode dependence of the open-circuit voltage of polymer:fullerene bulk heterojunction solar cells. J Appl Phys 94:6849–6855
98. Brabec CJ, Cravino A, Meissner D et al (2001) Origin of the open circuit voltage of plastic solar cells. Adv Func Mat 11:374–380
99. Mihailetchi VD, Koster LJA, Blom PWM (2004) Effect of metal electrodes on the performance of polymer:fullerene bulk heterojunction solar cells. Appl Phys Lett 85:970–973
100. Hoppe H, Niggemann M, Winder C et al (2004) Nanoscale morphology of conjugated polymer/fullerene based bulk heterojunction solar cells. Adv Func Mat 14:1005
101. Schilinsky P, Waldauf C, Brabec CJ (2002) Recombination and loss analysis in polythiophene based bulk heterojunction photodetectors. Appl Phys Lett 81:3885–3888

102. Kim JY, Kim SH, Lee HH et al (2006) New architecture for high efficiency polymer photovoltaic cells using solution based titanium oxide as an optical spacer. Adv Mat 18:572–576
103. Spanggaard H, Krebs F (2004) A brief history of the development of organic and polymeric photovoltaics. Sol Energy Mater Sol Cells 83:125–146
104. Rostalski J, Meissner D (2000) Monochromatic versus solar efficiencies of organic solar cells. Sol Energy Mater Sol Cells 61:87–95
105. Shrotriya V, Li G, Yao Y et al (2006) Accurate measurement and characterization of organic solar cells. Adv Func Mat 16:2016–2023
106. Dutta U, Chatterjee P (2004) The open circuit voltage in amorphous silicon p-i-n solar cells and its relationship to material, device and dark diode parameters. J Appl Phys 96:2261–2272
107. Marks RN, Halls JJM, Bradley DDC et al (1994) The photovoltaic response in poly(p-phenylene vinylene) thin film devices. J Phys Condens Matter 6:1379–1394
108. Cravino A, Schilinsky P, Brabec C (2007) Characterization of organic solar cells: the importance of device layout. Adv Func Mat 17:3906–3910
109. Kim M-S, Kang M-G, Guo L-J et al (2008) Choice of electrode geometry for accurate measurement of organic photovoltaic cell performance. Appl Phys Lett 92:133301–133304
110. Li G, Shrotriya V, Huang J et al (2005) High-efficiency solution processable polymer photovoltaic cells by self-organization of polymer blends. Nat Mater 4:864–868
111. Miller S, Fanchini G, Lin Y-Y et al (2008) Investigation of nanoscale morphological changes in organic photovoltaics during solvent vapor annealing. J Mater Chem 18:306–312
112. Dennler G, Lungenschmied C, Neugebauer H et al (2006) A new encapsulation solution for flexible organic solar cells. Thin Solid Films 511:349–353
113. Koster LJA, Mihailetchi VD, Ramaker R et al (2005) Light intensity dependence of open circuit voltage of polymer:fullerene solar cells. Appl Phys Lett 86:123509–123512
114. Malliaras GG, Salem JR, Brock PJ et al (1998) Photovoltaic measurement of the built-in potential in organic light emitting diodes and photodiodes. J Appl Phys 84:1583–1588
115. Gregg B (2003) Exitonic solar cells. J Phys Chem B 107:4688–4698
116. Ramsdale CM, Barker JA, Arias AC et al (2002) The origin of the open circuit voltage in polyfluorene-based photovoltaic devices. J Appl Phys 92:4266–4271
117. Antoniadis H, Hsieh BR, Abkowitz MA et al (1993) Photovoltaic properties of poly (p-phenylenevinylene/aliminum interfaces. Polymer Preprints. Polymer preprints 34:490-491
118. Karg S, Riess W, Dyakonov V et al (1993) Electrical and optical characterization of poly(phenylene-vinylene) light emitting diodes. Synth Met 54:427–433
119. Gadisa A, Svensson M, Andersson MR et al (2004) Correlation between oxidation potential and open circuit voltage of composite solar cells based on blends of polythiophenes/ fullerene derivative. Appl Phys Lett 84:1609–1612
120. Liu J, Shi Y, Yang Y (2001) Solvation induced morphology effects on the performance of polymer-based photovoltaic devices. Adv Func Mat 11:420–424
121. Soci C, Hwang IW, Moses D et al (2007) Photoconductivity of a low-bandgap conjugated polymer. Adv Func Mat 17:632–636
122. Nunzi JM (2002) Organic photovoltaic materials and devices. C R Physique 3:523–542
123. Winder C, Sariciftci NS (2004) Low band gap polymers for photon harvesting in bulk heterojunction solar cells. J Mat Chem 14:1077–1086
124. Campos L, Tontcheva A, Günes S et al (2005) Extended photocurrent spectrum of a low band gap polymer in a bulk heterojunction solar cell. Chem Mater 17:4031–4033
125. Perzon E, Wang X, Admassic S et al (2006) An alternating low band-gap polyfluorene for optoelectronic devices. Polymer 47:4261–4268
126. Cravino A, Loi M, Scharber M et al (2003) Spectroscopic properties of PEDOTEHIITNa novel soluble low band-gap conjugated polymer. Synth Met 137:1435–1436
127. Wienk M, Struijk M, Janssen R (2006) Low band gap polymer bulk heterojunction solar cells. Chem Phys Lett 422:488–491

128. Bundgaard E, Krebs F (2006) Low-band-gap conjugated polymers based on thiophene, benzothiadiazole, and benzobis(thiadiazole). Macromolecules 39:2823–2831
129. Bundgaard E, Krebs F (2007) Large-area photovoltaics based on low band gap copolymers of thiophene and benzothiadiazole or benzo-bis(thiadiazole). Sol Energy Mater Sol Cells 91:1019–1025
130. Winder C, Matt G, Hummelen JC et al (2002) Sensitization of low bandgap polymer bulk heterojunction solar cells. Thin Solid Films 403–404:373–379
131. Bundgaard E, Krebs F (2007) Low band gap polymers for organic photovoltaics. Sol Energy Mater Sol Cells 91:954–985
132. Moule AJ, Bonekamp BJ, Meerholtz K (2006) The effect of active layer thickness and composition on the performance of bulk heterojunction solar cells. J Appl Phys 100:094503–094510
133. Hoppe H, Sariciftci NS, Meissner D (2003) Modeling the optical absorption within conjugated polymer/fullerene-based bulk-heterojunction organic solar cells. Sol Energy Mater Sol Cells 80:105–113
134. Peumans P, Bulovic V, Forrest SR (2000) Efficient photon harvesting at high optical intensities in ultrathin organic double-heterostructure photovoltaic diodes. Appl Phys Lett 76:2650–2653
135. Roman LS, Inganaes O, Nyberg T et al (2000) Trapping light in polymer photodiodes with soft embodded gratings. Adv Mat 12:189–195
136. Delley B, Kiess H (1994) 2-dimensional simulation of high efficiency silicon solar cells. Sol Energy Mater Sol Cells 33:1–10
137. Niggemann M, Glatthaar M, Gombert A et al (2004) Diffraction gratings and buried nano-electrodes-architectures for organic solar cells. Thin Solid Films 451–452:619–623
138. Gupta D, Mukhopadhyay S, Narayan K S (2010) Fill factors in organic solar cells. solar energy materials and solar cells. 94:1309–1313
139. M-Su Kim, B-Gi Kim, Kim J (2009) Effective variables to control the fill factor of organic photovoltaic cells. Appl Mater Interf 1:1264–1269
140. Barrau S, Andersson V, Zhang F et al (2009) Nanomorphology of bulk heterojunction organic solar cells in 2D and 3D correlated to photovoltaic performance. Macromolecules 42:4646–4650
141. Maturova K, Van Bavel SS, Wienk MM et al (2009) Morphological device model for organic bulk heterojunction solar cells. Nano Lett 9:3032–3037
142. Markov DE, Hummelen JC, Blom PWM et al (2005) Dynamics of exciton diffusion in poly(*p*-phenylene vinylene)/fullerene heterostructures. Phys Rev B 72:045216–045221
143. Van Bavel SS, Sourty E, De With G et al (2009) Three dimensional nanoscale organization of bulk heterojunction polymer solar cells. Nano Lett 9:507–513
144. Ahn T, Sein HHHL (2002) Effect of annealing of polythiophene derivative for polymer light-emitting diodes. Appl Phys Lett 80:392–395
145. Brown PJ, Thomas DS, Köhler A et al (2003) Effect of interchain interactions on the absorption and emission of poly(3-hexylthiophene). Phys Rev B 67:064203–064219
146. Nakazono M, Kawai T, Yoshino K et al (1994) Effects of heat treatment on properties of poly(3-alkyl)thiophene. Chem Mater 6:864–870
147. Zhang F, Svensson M, Andersson M et al (2001) Soluble polythiophenes with pendant fullerene groups as double cable materials for photodiodes. Adv Mat 13:1871–1874
148. Mattis B A, Chang P C, Subramarian V (2003) Effect of thermal cycling on the performance of poly(3-hexyl)thiophene transistors. Mater Res Soc Symp Proc 771 L.10.35.1
149. Li G, Shrotriya V, Yang Y (2005) Investigation of annealing effects and film thickness dependence of polymer solar cells based on poly (3-hexylthiophene). J Appl Phys 98:043704–043709
150. Al-Ibrahim M, Ambacher O, Sensfuss S (2005) Effects of solvent and annealing on the improved performance of solar cells based on poly(3-hexyl)thiophene. Fullerene 86:201120–201123

151. Erb T, Zhokhavets U, Gobsh G et al (2005) Correlation between structural and optical properties of composite polymer/fullerene films for organic solar cells. Adv Func Mat 15:1193–1196
152. Savenije TJ, Kroeze JE, Yang X et al (2005) The effect of thermal treatment on the morphology and charge carrier dynamics in a polythiophene-fullerene bulk heterojunction. Adv Func Mat 15:1260–1266
153. Yang X, Loos J, Veenstra SC et al (2005) Nanoscale morphology of high-performance polymer solar cells. Nanoletters 5:579–583
154. Andersson V, Herland A, Masich S et al (2009) Imaging of the 3D nanostructure of a polymer solar cell by electron tomography. Nano Lett 9:853–855
155. Marsh RA, Hodgkiss JM, Seifried SA et al (2010) Effect of annealing on P3HT:PCBM charge transfer and nanoscale morphology probed by ultrafast spectroscopy. Nano Lett 10:923–930
156. Keawprajak A, Piyakulawat P, Klamchuen A et al (2010) Influence of crystallizable solvent on the morphology and performance of P3HT:PCBM bulk heterojunction solar cells. Sol Energy Mater Sol Cells 94:531–536
157. Huang JH, Ho ZY, Kekuda D et al (2009) Effects of nanomorphological changes on the performance of solar cells with blends of poly[9, 9'-dioctyl-fluorene-co-bithiophene] and a soluble fullerene. Nanotechnology 20:025202–025211
158. Ortmann F, Bechstedt F, Karsten H (2010) Charge transport in organic crystals: interplay of band transport, hopping and electron-phonon scattering. New J Phys 12:023011. doi: 10.1088/1367-2630/12/2/023011
159. Tessler N, Preezant Y, Rappaport N et al (2009) Charge tranpsort in disordered organic materials and its relevance to thin film devices. Adv Mater 21:2741–2761
160. Coropceanu V, Cornil J, Filho DAS et al (2007) Charge tranport in organic semiconductors. Chem Rev 107:926–952
161. Nelson J, Kwiatkowski JJ, Kirkpatrick J et al (2009) Modeling charge transport in organic photovoltaic materials. Acc Chem Res 42:1768–1778
162. Karl N (2003) Charge carrier transport in organic semiconductors. Synth Met 133–134:649–657
163. Blom PWM, Mihailetchi VD, Koster LJA et al (2007) Device physics of polymer:fullerene bulk heterojunction solar cells. Adv Mater 19:1551–1566
164. Mihailetchi VD, van Duren JKJ, Blom PWM et al (2003) Electron transport in a methanofullerene. Adv Func Mat 13:43–46
165. Blom PWM, de Jong MJM, van Munster MG (1997) Electric-field and temperature dependence of the hole mobility in poly(p-phenylene vinylene). Phys Rev B 55:R656–R659
166. Lutsen L, Adriaensens, Becker H et al (1999) New synthesis of a soluble high molecular weight poly(arylene vinylene): poly[2-methoxy-5-(3, 7-dimethyloctyloxy)-p-phenylene vinylene] polymerization and device properties. Macromolecules 32:6517–6525
167. Mozer AJ, Denk P, Scharber MC et al (2004) Novel regiospecific MDMO–PPV copolymer with improved charge transport for bulk heterojunction solar cells. J Phys Chem B 108:5235–5242
168. Mozer A, Sariciftci NS (2004) Negative electric field dependence of charge carrier drift mobility in conjugated, semiconducting polymers. Chem Phys Lett 389:438–442
169. Melzer C, Koop E, Mihailetchi VD et al (2004) Hole transport in poly(phenylene vinylene)/methanofullerene bulk-heterojunction solar cells. Adv Func Mater 14:865–870
170. Mihailetchi VD, Koster LJA, Blom PWM et al (2005) Compositional dependence of the performance of poly(p-phenylene vinylene): methanofullerene bulk-heterojunction solar cells. Adv Func Mater 15:795–801
171. Mihailetchi VD, Wildeman J, Blom PWM (2005) Space-charge limited photocurrent. Phys Rev Lett 94:126602–126606
172. Frost JM, Cheynis F, Tuladhar SM et al (2006) Influence of polymer-blend morphology on charge transport and photocurrent generation in donor-acceptor polymer blends. Nano Lett 6:1674–1681

173. Pivrikas A, Sariciftci NS, Juska G et al (2007) A review of charge transport and recombination in polymer/fullerene organic solar cells. Prog Photovolt Res Appl 15:677–696
174. Shah M, Pryamitsyn V, Ganesan V (2009) Effect of anisotropic charge transport on device characteristics of polymer solar cells. Appl Phys Lett 95:194101–194104
175. Ballantyne AM, Chen L, Dane J et al (2008) The effect of poly(3-hexyl)thiophene (P3HT) molecular weight on charge transport and the performance of polymer:fullerene solar cells. Adv Func Mater 18:2373–2380
176. Mandoc MM, Koster LJA, Blom PWM (2007) Optimum charge carrier mobility in organic solar cells. Appl Phys Lett 90:133504–133507
177. Mihailetchi VD, Xie H, Boer de B et al (2006) Charge transport and photocurrent generation in poly(3-hexylthiophene):methanofullerene bulk-heterojunction solar cells. Adv Func Mater 16:699–708
178. Kline RJ, McGehee MD, Kadnikova EN et al (2003) Controlling the field effect mobility of regioregular polythiophene by changing the molecular weight. Adv Mater 1519–1522
179. Goh C, Kline RJ, McGehee MD et al (2005) Molecular-Weight-Dependent Mobilities in Regioregular Poly (3-hexyl-thiophene) Diodes,. Appl Phys Lett 86:122110–122112
180. Chamberlain GA (1983) Organic solar cells: a review. Solar Cells 8:47–83
181. Hoth CN, Choulis SA, Schilinsky P et al (2007) High photovoltaic performance of inkjet printed polymer:fullerene blends. Adv Mater 19:3973–3978
182. Arbogast JW, Foote CS (1991) Photophysical properties of C70. J Am Chem Soc 113:8886–8889
183. Chen H-Y, Hou J, Zhang S et al (2009) Polymer solar cells with enhanced open-circuit voltage and efficiency. Nat Photonics 3:649–653
184. Jorgensen M, Norrman K, Krebs FC (2008) Stability and degradation of polymer solar cells. Sol Energy Mater Sol Cells 92:686–714
185. Brabec CJ, Hauch JA, Schilinsky P et al (2005) Production aspects of organic photovoltaics and their impact on the commercialization of devices. MRS Bull 30:50–52
186. Krebs FC (2005) Alternative PV: large scale organic photovoltaics. Refocus 6:38–39
187. Schuller S, Schilinsky P, Hauch J et al (2004) Determination of the degradation constant of bulk heterojunction solar cells by accelerated lifetime measurements. Appl Phys A 79:37–40
188. Jenekhe S, Chen XL (1998) Self-assembled aggregates of rod-coil block copolymers and their solubilization and encapsulation of fullerenes. Science 279:1903
189. Rispens MT, Sanchez L, Beckers EHA et al (2003) Supramolecular fullerene architectures by quadruple hydrogen bonding. Synth Met 135:801–803
190. Tracz A, Jeszka JK, Watson MD et al (2003) Uniaxial alignment of the columnar superstructure of a hexa (Alkyl) hexa-*peri*-hexabenzocoronene on untreated glass by simple solution processing. J Am Chem Soc 125:1682–1683

On the Importance of Morphology Control for Printable Solar Cells

Svetlana S. van Bavel and Joachim Loos

Abstract Polymer and hybrid solar cells have the potential to become the leading technology of the twenty-first century to convert sunlight to electrical energy because they can be easily processed from solution printing devices in a roll-to-roll fashion with high speed and low-cost. The performance of such devices critically depends on the nanoscale organization of the photoactive layer, which is composed of at least two functional materials, the electron donor and the electron acceptor forming a bulk-heterojunction; however, control of its volume morphology still is a challenge. The main requirements for the morphology of efficient photoactive layers are nanoscale phase separation for a large donor/acceptor interface area and hence efficient exciton dissociation, short and continuous percolation pathways of both components leading through the layer thickness to the corresponding electrodes for efficient charge transport and collection, and high crystallinity of both donor and acceptor materials for high charge mobility. In this chapter we review recent progress of our understanding on how the efficiency of a bulk-heterojunction printable solar cell largely depends on the local nanoscale volume organization of the photoactive layer.

S. S. van Bavel · J. Loos
Department of Chemical Engineering and Chemistry, Eindhoven University of Technology, P.O. Box 513, 5600 MB, Eindhoven, The Netherlands

S. S. van Bavel · J. Loos
Dutch Polymer Institute, P.O. Box 902, 5600 AX, Eindhoven, The Netherlands

J. Loos (✉)
School of Physics and Astronomy, University of Glasgow,
Glasgow, G12 8QQ, Scotland, UK
e-mail: joachim.loos@glasgow.ac.uk

L. Zang (ed.), *Energy Efficiency and Renewable Energy Through Nanotechnology*,
Green Energy and Technology, DOI: 10.1007/978-0-85729-638-2_5,

1 Introduction

Besides conventional silicon-based solar cells, in recent years, an alternative type of solar cells has been intensively studied, viz. thin-film devices printable from purely organic or hybrid solutions using semiconducting polymers, such as poly phenylene-vinylene, polythiopene, or polyfluorine, for light absorption and charge transport [1, 2]. Despite comparatively low efficiencies of about 7–8% [3] achieved so far with modelling studies that predict 10–11% efficiencies are attainable [4] and rather low stabilities, presently 1 or 2 years at most [5], such printable solar cells (PSCs) have a distinct advantage over inorganic counterparts, viz. their fast and low-cost manufacturing process: they can be fabricated by processing polymers, eventually together with other organic and/or inorganic materials, in solution and depositing them by printing or coating in a roll-to-roll fashion like newspapers. Thanks to the speed and ease of this manufacturing process, the energy payback time of PSCs may, according to some estimates, be reduced to months to about a year [6, 7]. Additional advantages include the low weight and flexibility of organic materials, enabling fast and easy applications on curved surfaces and thus freedom of design.

PSCs are still in the research and development phase; however, the first commercial products were recently introduced to the market. To bring them closer to the stage of practical and efficient devices, several issues still must be addressed, including further improvement to efficiency and stability. These, in turn, are determined to a large extent by the morphological organization of the photoactive layer, i.e., the layer where light is absorbed and converted into electrical charges. Thus, the general scope of this chapter is to examine how the morphology formation of ultra-thin (100–200 nm) photoactive layers prepared via solvent-based techniques can be controlled and manipulated and to establish relationships between the three-dimensional morphological organization of photoactive layers and the performance of corresponding PSCs.

2 Morphology Requirements of Photoactive Layers in PSCs

There is a principal difference in operation of solar cells based on inorganic semiconductors and organic (polymer) semiconductors, governed by a differing magnitude of exciton binding energy (an exciton is a bound electron–hole pair) in these materials. In many inorganic semiconductors, the exciton binding energy is small compared to the thermal energy (kT) at room temperature and therefore free charges are directly created under ambient conditions upon absorption of a photon of light [8]. An organic semiconductor, on the other hand, typically possesses an exciton binding energy that exceeds kT by roughly more than an order of magnitude [9]. As a consequence, excitons do not directly split into free charges in organic semiconductors and an additional mechanism is required to achieve this.

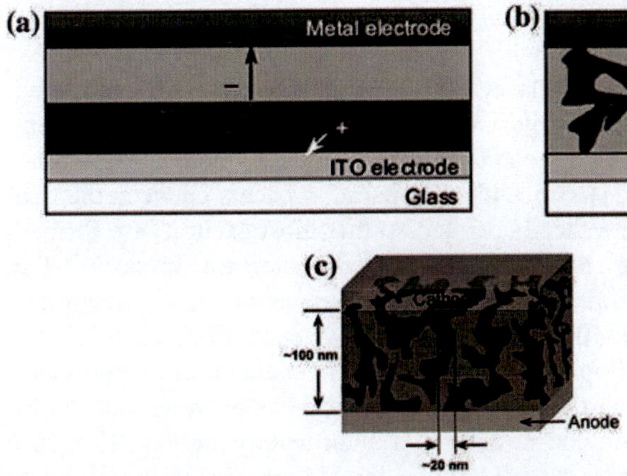

Fig. 1 **a** Schematic representations of a Tang double layer cell and **b** a bulk-heterojunction structure; **c** schematic three-dimensional representation of a bulk-heterojunction structure (electron donor and acceptor constituents in different colors). The common device structure is depicted here, with a photoactive layer sandwiched between an electron collecting electrode (typically metal, such as Al) and a hole collecting transparent electrode of indium-doped tin oxide (ITO) (**c** Reprinted with permission from [26]. Copyright 2007 American Chemical Society)

A successful way to dissociate excitons formed in organic semiconductors into free charges is to use a combination of two materials: an electron donor (the material with low ionization potential) and electron acceptor (with large electron affinity). At the donor/acceptor interface, an exciton can dissociate into free charges by rapid electron transfer from the donor to acceptor [10, 11]. Afterward, both charge carriers move to their respective electrodes when electrode materials are chosen with the proper work functions [1]. Significant photovoltaic effects of organic semiconductors applying the heterojunction approach were first demonstrated by Tang in the 1980s (Fig. 1a) [12]. A thin-film, two-layer organic photovoltaic cell was fabricated that showed a power conversion efficiency of about 1% and a large fill factor (FF) of 0.65 under simulated AM 2 illumination.

The external quantum efficiency, η_{EQE}, of a photovoltaic cell based on exciton dissociation at the donor/acceptor interface is $\eta_{EQE} = \eta_A \times \eta_{ED} \times \eta_{CC}$, where η_A is the light absorption efficiency, η_{ED} is the exciton dissociation efficiency, which is the fraction of photogenerated excitons that dissociate into free charge carriers at a donor/acceptor interface, and η_{CC} is the carrier collection efficiency, which is the probability that a free carrier generated at a donor/acceptor interface by dissociation of an exciton reaches its corresponding electrode [13]. Donor/acceptor interfaces can be very efficient in separating excitons; in some systems, it has been shown that the forward reaction, the charge generation process takes place on the femtosecond time scale, while the reverse reaction, the charge recombination step, occurs in the microsecond range [14]. The typical exciton diffusion length in most organic semiconductors is, however, limited to 5–20 nm [15–18]. Consequently, acceptor/donor interfaces have to be within this diffusion range for efficient exciton dissociation into free charges; however, efficient light absorption only can be guaranteed for photoactive layers with thickness larger than 200 nm.

Independently, Yu et al. and Halls et al. addressed the problem of limited exciton diffusion length by intermixing two conjugated polymers with different electron affinities [19, 20] or a conjugated polymer with C_{60} molecules or their methanofullerene derivatives [21]. Since phase separation occurs between the two constituents, a large internal interface is created so that most excitons are formed near the interface and are thus able to dissociate at the interface. In case of the polymer/polymer intermixed film, evidence for the success of the approach has been found in the observation that the photoluminescence from each of the polymers was quenched. This implies that the excitons generated in one polymer within the intermixed film reach the interface with the other polymer and dissociate before decaying. This device structure, a bulk-heterojunction (Fig. 1b), provides a route through which nearly all photogenerated excitons in the film can split into free charge carriers. At present, bulk-heterojunction structures are the main candidates for high-efficiency polymeric solar cells.

Nanoscale phase separation between the donor and acceptor components is not the only requirement for the morphology of photoactive layers of bulk-heterojunction PSCs. Once free charges are formed upon exciton dissociation, they should be transported through the donor and acceptor phases toward the corresponding electrodes: holes through the donor phase to the hole collecting (positive, usually indium-doped tin oxide, ITO) electrode and electrons through the acceptor phase to the electron collecting (negative, e.g., aluminium) electrode. Thus, the nanoscale phases of donor and acceptor must form continuous, and preferably short (to minimize charge recombination), percolation pathways leading to the electrodes.

Additionally, the transport of charge carriers can be enhanced if donor and/or acceptor (ideally both) are characterized by high mesoscopic order and crystallinity [22, 23]. Also, the transport and collection of charges should be facilitated in the case where there is enrichment of the acceptor material at the site of a photoactive layer close to a negative (metal) electrode and enrichment of the donor material close to a positive electrode [24, 25]. Such favorable concentration gradients of donor and acceptor materials through the thickness of the active layer should ensure that the percolation pathways leading to electrodes are short and limit possibilities of charge recombination.

The efficiency of a bulk-heterojunction PSC is thus largely dependent on the local nanoscale organization of the photoactive layer in all three dimensions [26–28]. The key requirements for efficient PSCs, including those dealing with photoactive layer morphology, are listed in Fig. 2, together with the parameters of device performance (J_{SC}, FF, V_{OC}, and η).

The short circuit current, J_{SC}, generated by a solar cell is found at the end of the whole chain: it is determined by the external quantum efficiency, i.e., efficiency of all basic processes of the PSC operation as discussed above, viz. light absorption, exciton dissociation at the donor/acceptor interface, and transport and collection of free charges at the electrodes. The meaning of the FF is more specific: a higher FF

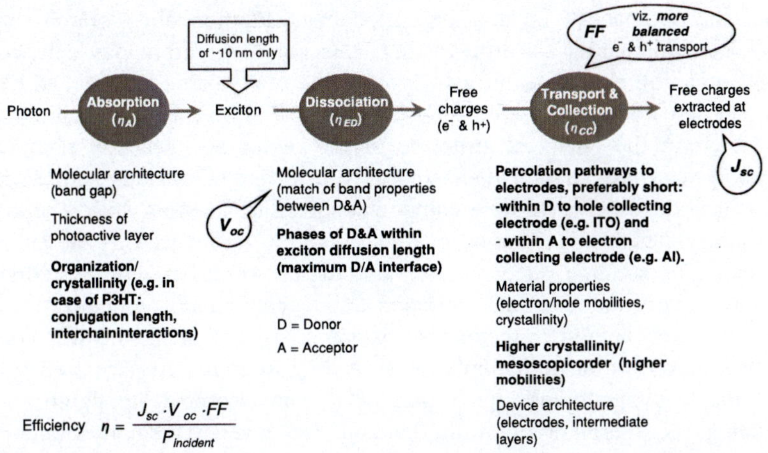

Fig. 2 The key factors determining the power conversion efficiency (η) of bulk-heterojunction PSCs, together with parameters of solar cell device performance are: short circuit current density J_{SC}, open circuit voltage V_{OC}, and FF. All three basic processes: light absorption (characterized by efficiency η_A), exciton dissociation (η_{ED}), and transport and collection of charges (η_{CC}) should be efficient in order to get efficient PSCs. The efficiency determining factors are listed under each step; those dealing with photoactive layer morphology are shown in *bold* (Reprinted with permission from [27]. Copyright 2010 Wiley–VCH)

implies an improved balance of electron and hole transport, low traps, and negligible space-charge effects [4, 29].

The open circuit voltage, V_{OC}, scales with the energy difference between the lowest unoccupied molecular orbital of the acceptor material and the highest occupied molecular orbital (HOMO) of the donor [30, 31]. Lower values of V_{OC} obtained experimentally are attributed to band bending created by accumulated charges at each electrode, with associated losses of $\sim$0.2 eV per electrode [30, 32]. Recent studies indicated, however, that V_{OC} may be determined by the single occupied molecular orbital of the donor (i.e., the low-energy polaronic level) rather than its HOMO level [33]. It has also been reported that V_{OC} is dependent on the probability of exciton dissociation into free charges [34], presence of electron traps [35], and mobility of free charges [36], i.e., parameters that may be influenced by morphology. However, the impact of these parameters on V_{OC} can usually be neglected [37, 38] and, for the present discussion, V_{OC} is considered as a purely material property, independent of the photoactive layer morphology [39].

Looking to the morphology requirements mentioned above, the following parameters have been identified as the most significant for their influence on the nanoscale volume organization in the photoactive layers of bulk-heterojunction PSCs: the chemical structure of donor and acceptor materials, the solvent(s) used for processing, concentration in solution, the ratio between donor and acceptor, and post-production treatments such as thermal annealing or exposure to solvent vapor [26, 40]. Some examples of how these parameters affect photoactive layer morphology formation and device performance are considered below.

Since the photoactive layer is deposited from solution, for research studies, mostly via spin-coating or drop-casting, the volume morphology-determining parameters can be classified into two groups: thermodynamic aspects and kinetic effects that play a role during thin-film formation process. Thermodynamic aspects are reflected in the chemical structure of the donor and acceptor compounds determining to a large extent their solubility in different solvents and the interaction (miscibility) between these compounds taken in a certain ratio. The kinetic aspects are related to the duration of film formation, influenced by the solvent's boiling point, by solution viscosity, etc., with the rate of ordering or crystallization in the case of crystallizing materials and thus accompanied reorganization, and with post film formation treatments such as annealing that enable the diffusion and crystallization of one or both compounds, leading to enhanced phase separation.

Both thermodynamic and kinetic parameters show comparable significance in determining the morphology of the photoactive layer. Thermodynamics will, however, drive (and kinetics may limit) eventual morphological reorganization after the films have formed, and thus determine the long-term stability of the photoactive layer morphology and corresponding solar cell devices.

3 Analyzing the Volume Organization of the Bulk-Heterojunction Photoactive Layer

The thin-film nature of the photoactive layers with typical thicknesses of 100–200 nm and the need for local morphology information makes high-resolution microscopy techniques essential tools for morphology characterization. Transmission electron microscopy (TEM), including conventional imaging and electron diffraction operation modes [26, 41–44], scanning electron microscopy (SEM) [40, 45] and scanning probe microscopy (SPM), atomic force microscopy (AFM) in particular [43, 45–50], have proven their utility for characterization of the morphology of photoactive layers.

The main difference between TEM, and SPM and SEM is that TEM provides mostly morphological information on the lateral organization of thin-film samples by acquisition of two-dimensional (2D) projections through the whole volume of the photoactive layer, whereas SPM and SEM are probing the topography or mainly phase demixing at the sample surface.

As discussed above, the performance of a PSC strongly depends on the three-dimensional organization of the compounds within the photoactive layer. Donor and acceptor materials should form co-continuous networks with nanoscale phase separation to be able to effectively dissociate excitons into free electrons and holes, and to guarantee fast charge carrier transport from any place within the active layer to the electrodes.

Attempts to gain more information on the three-dimensional organization of polymer solar cells have been described by applying techniques for cross-sectional preparation of TEM and SEM samples. Besides use of a focused ion beam (FIB) to

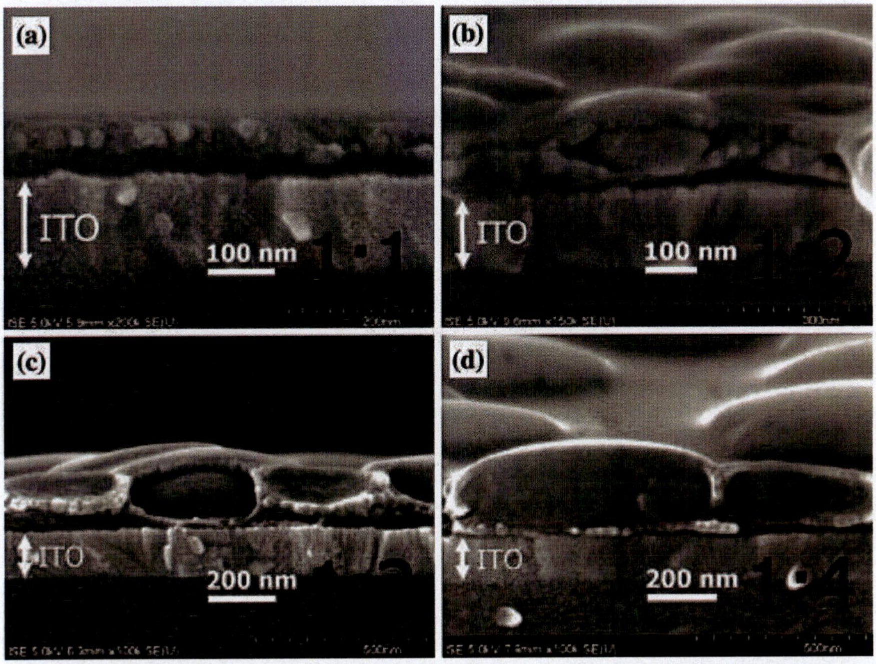

Fig. 3 SEM side views (cross-sections) of MDMO-PPV/PCBM blend films cast from toluene with various weight ratios of MDMO-PPV and PCBM. For ratios 1:4, 1:3, and 1:2 (**b–d**) the nanoclusters, in the form of discs are surrounded by another phase, called the skin, that contains smaller spheres of about 20–30 nm diameter. For the 1:1 film (**a**), only these smaller spheres are found (Reprinted with permission from [45]. Copyright 2004 Wiley–VCH)

section whole devices prepared on glass substrates [51, 52], conventional cross-sectional ultra-microtoming of the photoactive layer is used [43, 45]. Looking from the side perpendicular to the photoactive layer lateral plane, the organization of segregated phases of the bulk-heterojunction can be analyzed in the thickness direction of the film. Similar to the top view, the cross-sectional side view provides information on the sizes and shapes of phases. For instance, Sariciftci et al. demonstrated by cross-sectional SEM investigations that in case of the system with poly [2-methoxy-5-(3′, 7′-dimethyloctyloxy)-1, 4-phenylene-vinylene] (MDMO-PPV) as a donor and the fullerene derivative [6]-phenyl-C_{61}-butyric acid methyl ester (PCBM) as acceptor, the PCBM-rich phases always are imbedded in the MDMO-PPV matrix independent of their size (Fig. 3) [45]. Most likely, this feature reduces the electron injection capability from the PCBM to the electrode.

In principle, 3D morphology information can be obtained by cutting the sample into thin sections and subsequently imaging the remaining sample or the sections one by one with AFM or SEM. Using a FIB/SEM combination for FIB slicing and SEM viewing is referred to as "slice-and-view". For photoactive layers of PSCs these approaches are, however, of limited use due to the high risk of complete sample destruction as a result of local heating by FIB, and due to the relatively low lateral resolution of SEM, and depth resolution as determined by the thickness of removed slices in the range of 10–100 nm at most.

The technique that circumvents these limitations and does provide 3D morphology information with nanometer resolution in all three dimensions is electron tomography (ET), also referred to as transmission electron microtomography and 3D TEM [53–58]. Besides conventional morphology characterization, in this chapter, we focus on advanced ET analysis to reveal volume information on the local nanoscale organization of the photoactive layer of PSCs.

4 On the Volume Organization of Bulk-Heterojunction Photoactive Layers

So far, we have discussed the importance of morphology control in the bulk-heterojunction photoactive layer of high performance PSCs. In this part, we discuss the important morphology issues reported in the literature. Further, we introduce recent morphology studies we have performed with the help of ET on three bulk-heterojunction systems: MDMO-PPV as a donor and the fullerene derivative PCBM as acceptor for the first bulk-heterojunction system for which the influence of nanoscale volume organization on the performance of corresponding devices was investigated in great detail, poly(3-hexylthiophene) (P3HT)/PCBM as the currently most successful electron donor/acceptor combination in terms of efficiency and long-term stability, and finally P3HT as electron donor and ZnO as acceptor in a hybrid system, which we have performed in-depth quantification of the nanoscale volume organization and related it to the performance of corresponding devices.

4.1 MDMO-PPV/PCBM

Intensive morphology studies have been performed on polymer/fullerene systems, in which PCBM is applied [37, 59]. PCBM is by far the most widely used electron acceptor, and the most successful PSCs have been obtained by mixing it with the donor polymers like MDMO-PPV and other PPV derivatives [38, 46], with poly (3-alkylthiophene)s such as regioregular poly(3-hexylthiophene) (P3HT) [23, 60–63] or a less studied combination with polyfluorenes [64–67] and other amorphous semiconducting polymers such as poly[2, 6-(4, 4-bis-(2-ethylhexyl)-4H-cyclopenta [2, 1-b; 3, 4-b′]dithiophene)-alt-4, 7-(2,1,3-benzothiadiazole)] (PCPDTBT) [68–71].

The influence of the solvent used for processing was first observed in MDMO-PPV/PCBM system when a strong increase in power conversion efficiency was obtained by changing the solvent from toluene (0.9% efficiency) to chlorobenzene (2.5% efficiency) [46]. The improved performance of MDMO-PPV/PCBM cells in case of using chlorobenzene (or o-dichlorobenzene) as a solvent rather than

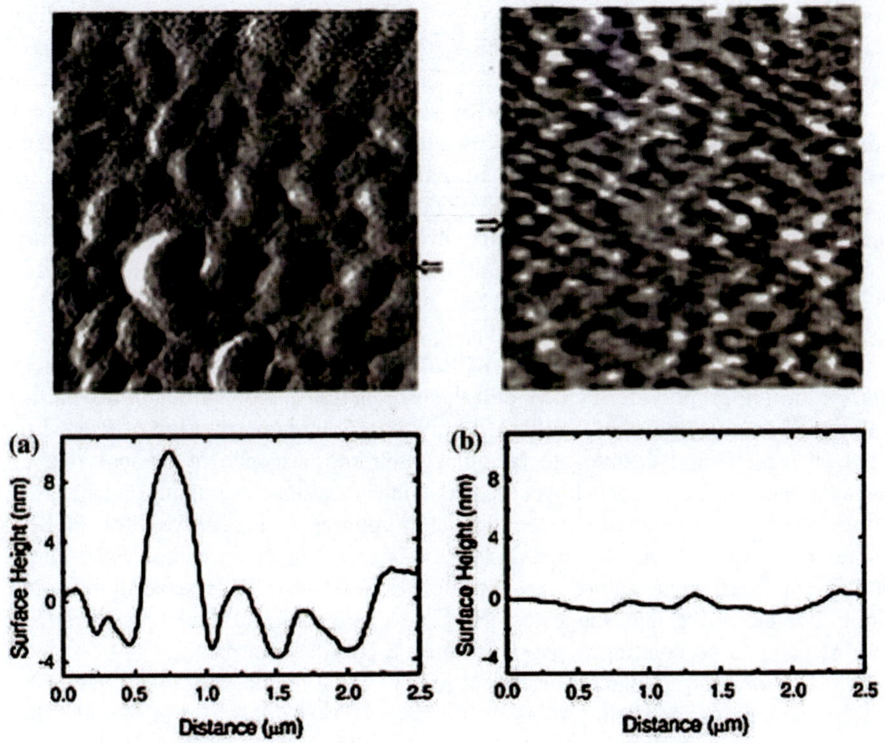

Fig. 4 AFM images showing the surface morphology of MDMO-PPV/PCBM (1:4 by wt) blend films with a thickness of approximately 100 nm and the corresponding cross-sections. **a** Film spin coated from a toluene solution. **b** Film spin coated from a chlorobenzene solution. The images show the first derivative of the actual surface heights. The cross-sections of the true surface heights for the films were taken horizontally from the points indicated by the *arrow* (Reprinted with permission from [46]. Copyright 2001 American Institute of Physics)

toluene was found to be due to a smaller (more favorable) scale of phase separation (Fig. 4), viz. smaller PCBM-rich domains in the MDMO-PPV-rich matrix, formed during spin-coating as a result of higher solubility of PCBM in chlorobenzene [42, 45, 46, 72].

The evaporation rate of a solvent during film formation is also of importance. Even when a good solvent (e.g., chlorobenzene) is used for MDMO-PPV/PCBM but its evaporation is slowed by using lower spin speed during spin-coating or by using drop-casting instead of spin-coating, coarse phase segregation is observed in the resulting films similar to faster spin-coating from a less favorable solvent like toluene [42]. Since a film has then a longer time to form and kinetic factors become less limiting, thermodynamically driven re-organization, viz. large-scale PCBM crystallization, takes place. Not surprisingly, thermal annealing of MDMO-PPV/PCBM boosts PCBM crystallization even further leading to formation of bulky PCBM crystals. Besides annealing conditions, the kinetics of their formation was also found to depend on a type of spatial confinement, i.e., in a free-standing film, PCBM clusters are formed much faster than in a film sandwiched between two substrates [73].

Besides the solvent used and the evaporation rate applied, the overall compound concentration and the ratio between two compounds in solution are important parameters controlling morphology formation. High compound concentrations induce large-scale phase segregation in MDMO-PPV/PCBM during formation of the film [74]. The maximum solubility of PCBM was determined to be roughly 1 wt% in toluene and 4.2 wt% in chlorobenzene (at room temperature), so for concentrations above these critical levels, aggregation of PCBM is anticipated in the solvent and is enhanced further during film formation [26].

For the systems of MDMO-PPV/PCBM and MEH-PPV/PCBM, as for most other amorphous polymer/PCBM bulk-heterojunctions, the optimal ratio of the compounds was found to be 1:4 [21], despite a very low contribution of PCBM to light absorption and despite the fact that photoluminescence of the polymer is already quenched for much lower PCBM concentrations (less than 5%) [11]. A rather abrupt improvement in the device properties was observed for PCBM contents of around 67%, accompanied by the onset of phase separation [48]. Thus, it was concluded that charge transportation rather than charge separation is the limiting factor here and suggested that, only above this critical concentration, PCBM forms a percolating network within the polymer matrix.

Most of the morphological studies carried out so far on the MDMO-PPV/PCBM system were performed using AFM [43, 45–50], TEM [42–44], and SEM and TEM imaging of cross-sections of the active layers [40, 43, 45]. They showed that at compositions of >60 wt% PCBM, the phase separation sets in with domains of rather pure PCBM distributed in a homogeneous MDMO-PPV/PCBM matrix. It was speculated that MDMO-PPV/PCBM films are built by interpenetrating networks at high PCBM loadings [48] but, up to now, there was no unambiguous experimental evidence for existence of any networks or percolations in MDMO-PPV/PCBM photoactive layers, owing largely to the lack of appropriate tools that allow a proper visualization of the morphological organization in all three dimensions.

To visualize the 3D morphology of MDMO-PPV/PCBM films (prepared from solutions in chlorobenzene), we applied ET. The tilt series were acquired via bright field conventional TEM (BF-CTEM) and annular dark-field scanning TEM (ADF-STEM) modes, so that (denser) PCBM-rich domains look dark (BF-CTEM) or bright (ADF-STEM), respectively. After 3D reconstruction, detailed information about form and size of PCBM-rich domains in three dimensions was obtained. In case of MDMO-PPV/PCBM films with 80 wt% PCBM, ET showed that the PCBM-rich domains are interlinked by thin PCBM-rich strands, all together forming a percolating nanoscale network (see the snapshots of the resulted volume reconstructions in Fig. 5).

The presence of a nanoscale percolating network formed by PCBM-rich strands as revealed by ET in BF-CTEM and ADF-STEM imaging modes presents two advantages for the performance of corresponding MDMO-PPV/PCBM solar cell devices. First, these interlinking PCBM domains form additional interfaces where

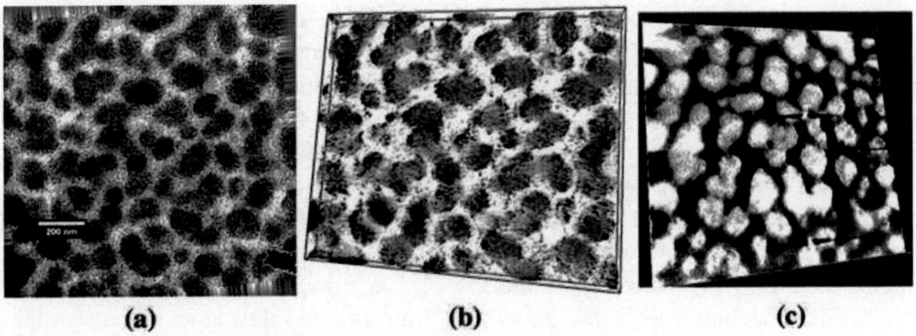

Fig. 5 **a** A BF-CTEM image of the MDMO-PPV/PCBM film (with 80 wt% PCBM), where dark PCBM-rich domains are visible, next to snapshots of volume reconstructions as obtained by **b** BF-CTEM, and **c** ADF-STEM. In the volume reconstruction thin PCBM-rich strands connecting PCBM-domains can be discerned (Reprinted with permission from [28]. Copyright 2010 Wiley–VCH)

more excitons can dissociate into free charges and, second, such a 3D network of PCBM, surrounded by a 3D network of polymer-rich material, creates efficient percolation pathways for transport of both electrons and holes from any place within the photoactive layer to the top and bottom electrodes. This optimum morphology results in higher values of short circuit current density (J_{sc}) and *FF* and thus higher efficiency of devices made of 80 wt% PCBM and 20 wt% MDMO-PPV [48]. No such beneficial morphology was observed (neither with BF-CTEM nor with ADF-STEM) in MDMO-PPV/PCBM layers with at most 60 wt% PCBM.

In general, thermal annealing is a useful way to probe the morphological stability of photoactive layers. Apart from accelerating thermodynamically favorable changes in the layer's morphology, mild annealing also mimics practical conditions since solar cells can easily heat up during operation to temperatures of around 60°C. Obviously, long-term stability of PSCs based on MDMO-PPV/PCBM is rather poor, due to the tendency of PCBM to crystallize by forming micron-sized clusters in amorphous MDMO-PPV. Such large-scale crystallization implies that exciton dissociation becomes rather inefficient, and the quality of a percolating network of PCBM deteriorates, too. The formation of large PCBM crystals can, however, be largely suppressed by choosing a polymer having a higher glass transition temperature, T_g (e.g., 138°C) [75] than that of MDMO-PPV, so that diffusion of PCBM molecules in the blends is hindered (another example of interplay between thermodynamics and kinetics in these systems).

4.2 P3HT/PCBM

The P3HT/PCBM system, where both components can crystallize, differs in its behavior and morphological organization from MDMO-PPV/PCBM blends. The type of crystalline morphology formed by regioregular P3HT ranges from well-dispersed nanorods to well-developed spherulites, depending on solution processing conditions [77]. Typically, P3HT crystallizes in thin-films by forming crystalline nanorods with widths of around 15–25 nm, thicknesses of just a few nanometers, and lengths of hundreds of nanometers or even a few microns [23, 26, 74, 76].

The best devices with power conversion efficiencies exceeding 5% and rather stable morphologies, are obtained for P3HT/PCBM ratios of around 1:1 after an annealing treatment, either at elevated temperature or during slow solvent evaporation (solvent assisted annealing) [30, 51, 77, 78]. Similar results are also attained by adding high boiling point additives like alkyl thiols into the solution of P3HT/PCBM because this slows down the film formation during spin-coating due to a longer solvent evaporation time, analogous to solvent annealing [48, 79].

Various reasons have been named to account for morphology changes causing efficiency improvement in P3HT/PCBM films upon annealing, such as the increased crystallinity of P3HT [80], favorable dimensions of (long and thin) P3HT crystals [76], suppressed formation of bulky PCBM clusters due to presence of P3HT crystals [23, 63], improved light absorption of the P3HT/PCBM films as a result of morphological changes in P3HT [47], improved hole mobility and hence more balanced hole and electron transport in P3HT/PCBM films [29, 81, 82]. This list, however, long it may seem, is not complete as it does not include details on morphological organization throughout the volume of the photoactive layer, such as the quality of percolating networks of nanocrystalline P3HT and PCBM and the exact scale of phase separation.

In our studies, we have applied ET (among other techniques) to analyze the 3D nanoscale organization of the P3HT/PCBM photoactive layers before and after annealing treatments and found correlations between the observed morphology and parameters of corresponding device performance [41, 83]. Electron tomography was applied to P3HT/PCBM in the BF-CTEM mode, with the resulting contrast due to the density difference between the polymer and the PCBM. Crystalline P3HT nanowires are actually very unstable and loose their crystallinity fast because the polymer crystals are under exposure to the electron beam (as shown by electron diffraction studies) but they remain morphologically stable for a much longer time, certainly long enough to acquire a tilt series (Fig. 6). The crystallinity of the nanowires and their morphological stability were confirmed by conical dark-field TEM imaging combined with the BF-TEM imaging. In the dark-field TEM images, the image formation is governed by the diffraction contrast as opposed to the mass-thickness and/or phase contrast in BF-CTEM. For this investigation we can conclude that even though the P3HT crystals are destroyed, their morphological integrity is sustained during tilt-series acquisition; thus, amorphous but contrast-rich structures remain in place of the original P3HT crystals are visualized in the P3HT/PCBM system by ET in BF-CTEM mode. Dark-field TEM was also used to identify the size (width) of the P3HT crystals and confirms similar sizes for ET data.

After reconstructing the morphology of as spin-coated and annealed samples, only for the latter a genuine network is visible within the photoactive layer connecting any part of the P3HT nanowires with the respective hole collecting

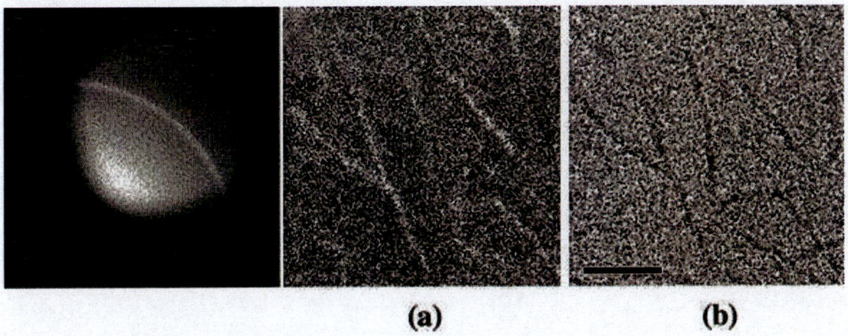

Fig. 6 **a** Dark-field TEM images of the P3HT nanowires: those structures get visualized that give rise to the shown part of the (020) P3HT-ring in the electron diffraction pattern (*left*). **b** BF-CTEM images of the same locations of the specimen after the crystals have been destroyed (i.e., no Debye-Scherer ring is any longer visible in the electron diffraction pattern). The scale bar is 200 nm (Reprinted with permission from [28]. Copyright 2010 Wiley–VCH)

electrode. By going slice by slice through the reconstructed volume of the annealed P3HT/PCBM films, the amount of P3HT nanowires can be quantified in relation to the actual *z*-position within the photoactive layer (Fig. 7, performed on a thermally annealed P3HT/PCBM layer). As evident from Fig. 7, there is enrichment of P3HT nanowires in the lower part of the photoactive layer close to the hole collecting electrode. The same trend was observed for P3HT/PCBM layers obtained by solvent assisted annealing for 3 h. In this gradient analysis, only initially crystalline P3HT nanowires are accounted for, as explained above (see Fig. 6). Amorphous P3HT should be located around the crystalline P3HT nanowires, which is generally the case in folded chain or chain stacked (for low molecular weights) polymer crystals. As this amorphous P3HT is in all probability mixed with PCBM, it has insufficient contrast with the rest of PCBM to be visualized by BF-CTEM. Vertical segregation of initially crystalline P3HT in the annealed films, as showed by ET, should thus apply to all P3HT, including amorphous parts. Correspondingly, there should be enrichment of PCBM close to the top (electron collecting) electrode. The situation when more crystalline P3HT is located in the lower part of the film is expected to be beneficial for collection of holes as it suggests better percolation networks made of crystalline P3HT and PCBM.

Further, when comparing the volume organization of photoactive layers with different thicknesses, the basic conclusion is that an optimal morphological organization of a photoactive layer in all three dimensions is more crucial for high efficiency of solar cell devices than absorption alone. We have obtained the best device performance using moderately thick (100 nm) P3HT/PCBM photoactive layers characterized by high overall crystallinity of P3HT, namely more numerous and more perfect crystalline P3HT nanowires forming a genuine 3D network, and by enrichment of P3HT close to the hole collecting electrode. Thicker films (200 nm) absorb more light but show less favorable morphological organization in photoactive layers, i.e., lower crystallinity of P3HT, especially next to the hole collecting electrode, and as a result produce poorly performing solar cell devices.

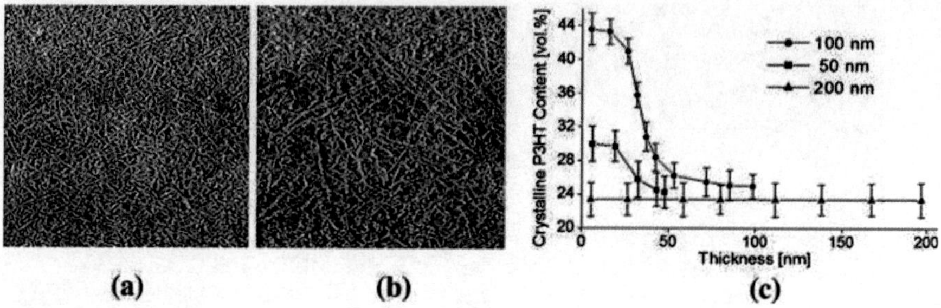

Fig. 7 Results of electron tomography: quantification of the crystalline P3HT nanowires distribution through the thickness of the thermally annealed P3HT/PCBM photoactive layer. Images **a** and **b** are original slices taken out of the reconstructed volume of the film, with slice in **a** located close to the bottom of the film and **b** close to the top of the film. Dimensions of the slices are 1700 nm × 1700 nm. **c** The relative area (volume) occupied by P3HT in each slice is determined for all slices through the whole volume of the P3HT/PCBM film and plotted depending on a slice position and thickness of the photoactive layer (Reprinted with permission from [41] and [83]. Copyright 2009 American Chemical Society)

What exactly causes vertical segregation of crystalline P3HT is not clear at the moment. Previous studies based on the modeling of data obtained by variable-angle spectroscopic ellipsometry (VASE) [84] and X-ray photoelectron spectroscopy (XPS) [85] suggested that PCBM, and not P3HT, was preferentially concentrated on the bottom of the P3HT/PCBM films due to the high surface energy of the PEDOT:PSS layer. The experimental approach and conclusions drawn in these studies do raise some questions though.

In general, however, we do not doubt that the unfavorable gradient (more of PCBM below and more of P3HT on top of the film) is possible. We have observed it with ET, too, e.g., in P3HT/PCBM samples prepared from P3HT grades with different molecular weight distributions, lower regioregularity, or different type and amount of impurities, and by using different film preparation conditions.

It should be noted that there are also SIMS (secondary ion mass spectrometry) studies of MDMO-PPV/PCBM [48] and polyfluorene/PCBM blends [86], which show a homogeneous distribution of both polymer and PCBM throughout the film. It may be interesting to apply SIMS to P3HT/PCBM blends to follow the vertical distribution of the (total) P3HT and PCBM.

Compared to XPS and modeling VASE data, the technique of ET has a number of advantages to study vertical organization in P3HT/PCBM films. First of all, this technique does not require any assumptions on morphological organization (like in modeling of VASE or ellipsometry data), nor is there a risk of introducing such artefacts as carbon and oxygen contaminations in case of surface-sensitive techniques like XPS. Next, ET does provides some numbers on how P3HT is distributed in the film and it also allows direct observation of the initially crystalline P3HT nanowires with high accuracy at any depth of the film. A minor disadvantage of this technique is that a few nanometers of the top and bottom of the films are usually difficult to interpret but, for this, AFM may provide relevant data.

The fact that ET revealed different sorts of vertical gradients in P3HT/PCBM films of different thickness deposited on the same PEDOT:PSS substrate dismisses the proposed role of the underlying substrate's surface energy for the formation of vertical gradients. On the basis of the findings of this study, we infer that the existence and type of composition gradients through the thickness of the active layer are largely determined by the kinetic aspects of film formation due to different solution viscosities, different times for the solvent to evaporate and eventual differences in local solvent concentration. These aspects have a direct impact on how long the (macro)molecules are mobile in the given solution/dispersion, on the eventual precipitation of components depending on local variations in solvent concentration, and thus on formation of nuclei and subsequent growth and distribution of (nano)crystals throughout the active layer.

Besides composition gradients, ET data can also be used to quantify the overall degree of crystallinity of P3HT in the P3HT/PCBM films. Based on density values of P3HT and PCBM and a 1:1 weight ratio of these components in the photoactive layer, P3HT should occupy ca. 58% of the total volume of the thermally annealed layer. From the plot presented in Fig. 7, it can be estimated that approximately 35% of the layer volume is actually made up of crystalline P3HT nanowires, which indicates a high crystallinity of P3HT of about 60%. A comparable degree of P3HT crystallinity was also obtained for P3HT/PCBM after solvent assisted annealing (data not shown). For spin-coated films, this information was not directly accessible due to poor contrast in the corresponding 3D datasets but, based on electron diffraction data (where the intensity of the peak attributed to (020) reflections of P3HT crystals increased threefold upon annealing) the degree of P3HT crystallinity after spin-coating is estimated at around 20%.

4.3 P3HT/ZnO

The examples considered above are polymer/fullerene PSCs but, in general, all the parameters influencing morphology formation are also valid for polymer/polymer systems [87–91] and hybrid systems, where semiconducting polymers such as P3HT are combined with inorganic materials such as ZnO, TiO_2, or CdSe [92–95]. A potential advantage of all-polymer systems is improved absorption compared to systems using poorly absorbing fullerenes. Hybrid solar cells form an attractive alternative because of their high dielectric constant (facilitating carrier generation), high carrier mobility in the inorganic semiconductors, and the thermal morphological stability of the photoactive layers.

Ultimately, hybrid cells offer the prospect of direct control over the morphology of the blend by first constructing an inorganic scaffold with the proper layout and dimensions [96–98] and then filling the scaffold with the organic material. But until now this approach failed and the best hybrid solar cells were made by simultaneous deposition of the two components. This often involves tricky processing, due to the differing natures of the materials involved. These drawbacks

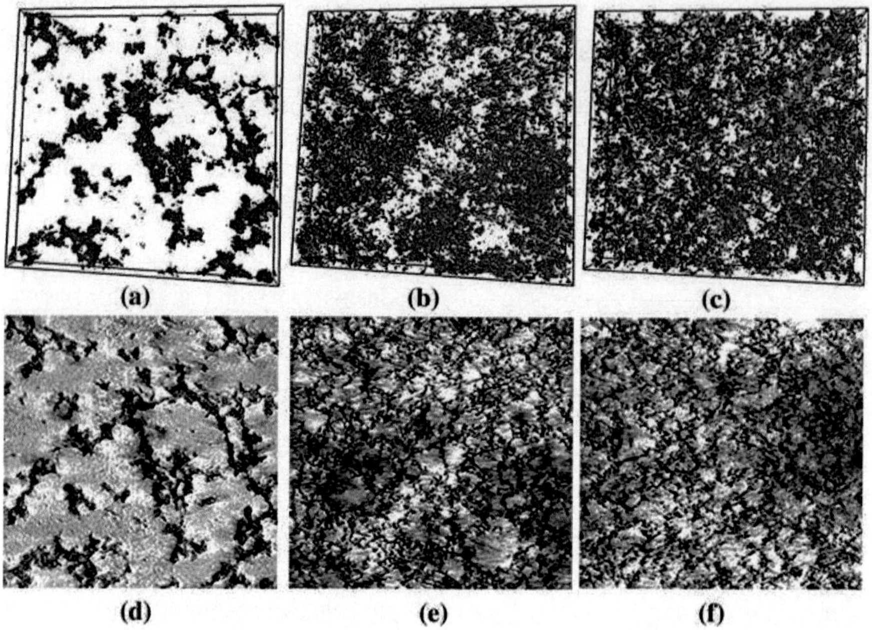

Fig. 8 **a–c** Reconstructed volumes of the P3HT/ZnO layers, obtained by electron tomography in the BF-CTEM mode, and corresponding 2D slices taken from these volume datasets: with thickness of 57 nm (**a**, **d**), 100 nm (**b**, **e**) and 167 nm (**c**, **f**). The lateral size of the datasets in **a–c** and correspondingly the size of the slices in **d–f** is ca. 700 nm × 700 nm. In the volumes in **a–c**, the threshold was applied to the raw data such that ZnO appears grey and P3HT transparent (Reprinted with permission from [101]. Copyright 2009 Nature Publishing Group)

can largely be circumvented by the in situ generation of the inorganic semiconductor inside the organic material [99, 100]. In this process a soluble organometallic precursor is deposited from solution together with the semiconducting polymer. During and after this deposition, the precursor is converted by reacting with moisture from the surrounding atmosphere to an inorganic network inside the polymer film.

Here we describe and analyze in situ prepared P3HT/ZnO solar cells [101]. Electron tomography was applied to characterize the three-dimensional morphology of the P3HT/ZnO photoactive layers and the resulting 3D datasets were segmented and quantified in great detail. The elaborate 3D morphology quantification combined with photophysical characterization and device performance data provide detailed insight to the role of the nanoscale 3D morphology in creating and transporting charges in the bulk-heterojunction photoactive layers.

To examine the morphology of the ZnO/P3HT films, ET was applied to freestanding films of different thicknesses (57, 100, and 167 nm) obtained by spincoating onto PEDOT:PSS. The snapshots of the reconstructed volume of these films are shown in Fig. 8a–c. A threshold was applied to the raw data as obtained by ET in such a way that ZnO appears grey in the volume and P3HT looks transparent. Obviously, there is a large difference between the three films obtained, with finer phase separated domains observed in thicker films. The thinnest film displays large domains both for the ZnO and the P3HT and these domains are substantially larger than the exciton diffusion length of 5–20 nm. This is consistent with the poor device performance of thinner films.

In order to quantify the relevant morphological parameters, an extensive statistical analysis of the 3D datasets provided by ET was performed. The original 3D data of the bulk of the film were binarized to decide which voxels ("volume pixels") are ZnO and which are P3HT. The threshold for this binarization has a major impact on the final outcome, and hence error margins were estimated by applying the two extremes for this threshold.

First of all, the volume fraction of ZnO was determined from the 3D datasets (see Column 1 in the Table 1). In the two thickest layers, the estimated ZnO volume fraction (ca. 21 vol.%) is close to the expected value of 19 vol.%, based on the ratio of diethylzinc and P3HT in the spin-coating solution. The ZnO content in the thinnest layer is significantly lower at 13 vol.%. This is rationalized by a comparatively large fraction of the diethylzinc evaporating during spin-coating with the higher spin speed applied for this thin layer. The low ZnO content of the thinnest film partially accounts for the relatively large observed P3HT domains.

Next, spherical contact distances, defined as the distance from a certain voxel of one material to the nearest voxel of the other material, were determined for these three films. Cyclic boundary conditions were applied for this quantification, i.e., the datasets were extended by mirroring the bulk part of the film (i.e., about 60% of the total thickness). Because excitons are mostly generated inside the P3HT, we focused on the distance distribution from P3HT to ZnO. Figure 9 shows the probability to find P3HT at a certain shortest distance to a ZnO domain. For the 100 and 167 nm thick films, most P3HT lies well within the shortest distance of 10 nm from ZnO. On the other hand, the 57 nm thick sample displays a large amount of polymer at shortest distances as high as 25 nm from an interface with ZnO. This analysis substantiates that coarser phase separation is present in thinner layers.

The efficiency of the charge carrier generation was then estimated by modeling the exciton diffusion through the P3HT phase. For this, the three-dimensional exciton diffusion equation was solved. As the result, an estimated fraction of excitons formed within P3HT that reach the interface with ZnO was obtained (Column 2 in Table 1). Assuming that excitons efficiently dissociate into free charges at the interface with ZnO [97], the numbers obtained coincide with the efficiency of charge generation.

Besides charge carrier generation, carrier collection is also essential for solar cell operation. Efficient collection relies on continuous pathways for both carriers (Fig. 9c, d). Viewing the large volume excess of polymer in the blend, we see that connectivity of this material will not be a limiting factor. The fraction of ZnO voxels that is interconnected via other ZnO voxels to the top of the investigated slab is quite high, at values well over 90% for all three layers despite the low ZnO content (see Column 4 in Table 1). The connectivity is smaller for thicker layers, likely because larger distances have to be crossed.

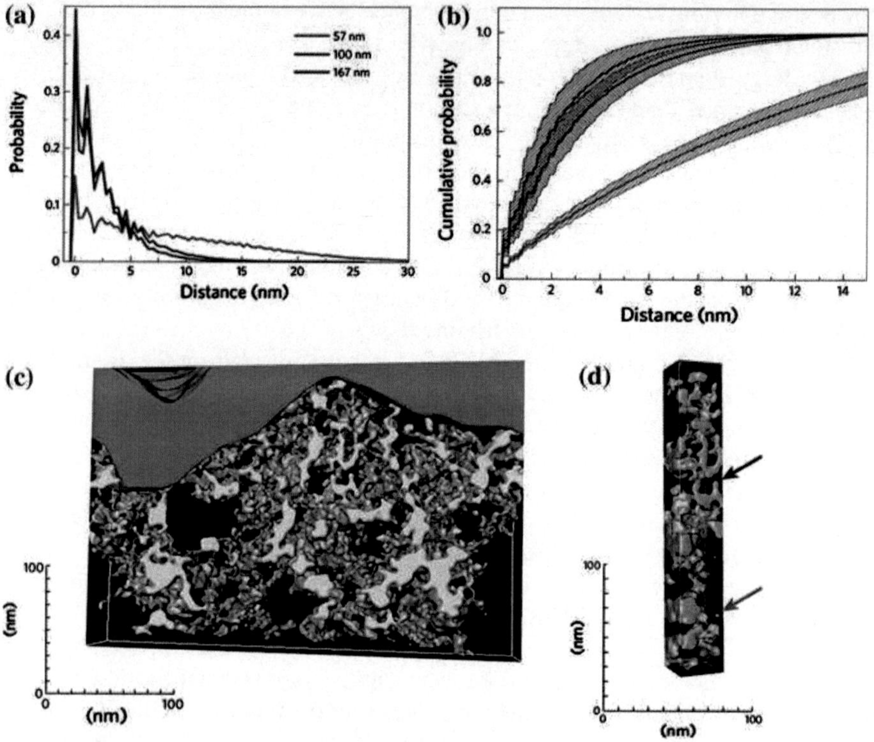

Fig. 9 Statistical analysis of the 3D morphology: **a** Distribution of the probability to find a P3HT voxel at a certain distance from a ZnO domain for mixed P3HT/ZnO films of different thickness, calculated from the 3D-datasets displayed in Fig. 8. **b** Cumulative probability to have P3HT within a shortest distance to ZnO. The error margins indicated are obtained from the two most extreme thresholds possible for the binarization of the 3D data; color code as in a. **c** Reconstructed volume of a cross-section of the P3HT/ZnO device with grey ZnO domains in transparent P3HT matrix. **d** The part of this volume with the *green arrow* indicating an isolated ZnO domain and the *red arrow* indicating a ZnO domain connected to the top, but not through a strictly rising path (Reprinted with permission from [101]. Copyright 2009 Nature Publishing Group)

Mere continuity of ZnO-phase may not be enough to effectively collect the charges. Within a continuous phase, pathways may exist that are discontinuous in the direction of the collecting electrode. Due to the macroscopic electric field over the active layer of the device, charges may be trapped inside those cul-de-sacs and thus not collected (Fig. 9d). Therefore, we also determined the fraction of ZnO connected to the top through a strictly rising path (Column 5 in Table 1). The calculated unidirectionally connected fraction of ZnO is still very high (93%) for the 57 nm thin layer but reduces for thicker layers, dropping to 80% for the 167 nm thick film.

The volume fraction of ZnO (with respect to the total volume of the photoactive layer) connected to the top electrode is, however, still higher in the thicker layers than in the thin 57 nm layer, where the total ZnO content is significantly lower

Table 1 Calculated volume fraction, fraction of excitons reaching the interface with ZnO before decaying, fraction of excitons quenched at the electrodes (and thus not contributing to the current generation) and percolation connectivity of P3HT/ZnO layers, all inferred from the electron tomography data

Film thickness (nm)	ZnO volume fraction (vol.%)	Without electrodes	With electrodes		ZnO connected to top (%)	ZnO monotonously connected to top (%)	Volume fraction of ZnO monotonously connected to top (vol.%)
		Excitons reaching the interface (%)	Excitons reaching the interface (%)	Excitons quenched by electrodes (%)			
57	13 ± 4	40 ± 3	32 ± 1	32 ± 1	96 ± 2	93 ± 2	12 ± 4
100	21 ± 8	78 ± 5	73 ± 5	7 ± 1	94 ± 5	85 ± 10	18 ± 8
167	21 ± 8	83 ± 5	79 ± 6	4 ± 1	92 ± 5	80 ± 12	17 ± 9

Quantification "without electrodes" was performed on the homogeneous bulk (ca. 60% of the total thickness) of the films extended by mirroring part of the morphology until the correct thickness was obtained. Quantification "with electrodes" was obtained by using two perfectly quenching planes as boundary conditions. The connectivity of ZnO was calculated for the bulk dataset and extrapolated to the correct thickness

(Column 6 in Table 1, which is the product of values in Column 1 and Column 5). This implies that electron transport through thin P3HT/ZnO layers may be less efficient. However, as described above, there are also fewer charges to be transported in thin layers due to the unfavorably large phase separation. Since the FF values (that reflects the balanced transport of free charges) are comparable in all films, charge transport on the whole is unlikely to be less efficient in thin-films.

Looking at the combined effects of charge carrier generation and collection, we can conclude that the relative poor performance of thin P3HT/ZnO solar cells is related to inefficient charge generation as a result of low ZnO content, to the coarse phase separation, and to the exciton losses impaired by the electrodes. For thicker photoactive layers charge generation is much more efficient, owing to a much more favorable phase separation.

Thicker devices show superior efficiencies, but still the internal quantum efficiency (IQE) only reaches 50%. Since in thicker layers most excitons (around 80%) reach the P3HT/ZnO interface where they can dissociate into free charges, the IQE is most probably limited by inefficient charge transport. Electron transport may be limited by a low volume fraction of ZnO, whereas hole transport may be inefficient due to low hole mobilities in P3HT.

5 Summary and Outlook

As evident from the above discussion, there is a complex interplay between different aspects that determine photoactive layer morphology during film formation, its eventual reorganization during post-production treatments and its long-term stability. Due to this complexity and the fact that the desired structure should form spontaneously by deposition from solution (to retain low-cost manufacturing), the optimal morphology is explored in practice by time-consuming optimization. However, increasing the understanding of the underlying structure–property relationships should make direct manipulations possible in the future.

What complicates the matter is that the ideal photoactive layer morphology is characterized by different length-scales in the volume of the film (see Fig. 1c): for efficient charge separation, it should have phases of donor and acceptor materials on the order of 10–20 nm (in two dimensions) and, for charge collection, it should have percolation pathways through the whole thickness of the film, i.e., a few hundred nanometres in thickness direction of the photoactive layer. The requirement of donor and acceptor phases of different length-scales in three dimensions makes the control of spontaneous morphology formation very challenging, especially in case of thicker photoactive layers. For optimal light absorption, the layer should ideally be at least 200 nm thick, whereas it is often observed that thinner films of 100–150 nm perform better in bulk-heterojunction PSCs even though they effectively absorb less light. Poor performance of thicker layers is typically attributed to enhanced recombination of free charges resulting from imperfect percolation pathways.

Several attempts have been made to promote formation of such a well-organized structure through the whole volume of the photoactive layer by using an amphiphilic primary structure like diblock copolymers [102] or dyad structures [103, 104], but performances of the resulting solar cells have been lower than with conventional approaches thus far. Moreover, creation of nanostructures already in the solvent prior to deposition might be a route to split structure formation from the film deposition process, which probably allows for better morphology control of the photoactive layer [105].

A key requisite to control and optimize the morphology is access to reliable quantitative data sets reflecting the local nanoscale and the overall organization of the photoactive layer in all three dimensions, allowing for straightforward correlation of the volume morphology with device performance. In this respect, we have introduced ET as a versatile tool to reveal volume data on the organization of bulk-heterojunction photoactive layers of PSCs with nanometer resolution. Basically, ET is able to provide information on size of separated phases, connectivity of the phases in three dimensions, crystallinity, compositional gradients, and the overall network organization in the whole volume of the photoactive layer. Ultimately, we have demonstrated that correlation between device processing, device properties, and volume morphology of the photoactive layer is possible.

The next step toward a better understanding of relations between the nanoscale volume organization and the ultimate performance of PSCs is further controlled over the complete device architecture. In this respect, sectioning of whole device by FIB is a potential route allowing for straightforward preparation of complex multilayered specimens for morphology investigations. Besides gaining additional information on interfacial organization, investigation of whole devices is the only accurate way to study lifetime dependent chemical and physical aging. Certainly volume morphology information can be very helpful to understanding the aging mechanisms involved, and chemical volume mapping can tell us about the local oxidation or diffusing electrode material, for example.

Moreover, it should be said that, besides better morphology control, device performance can also be optimized through smart device architecture, e.g., by applying hole blocking layers [106], optical spacers to enhance light absorption in a layer of the same thickness [107, 108], and by using the tandem cell architecture [109–111], where two photovoltaic cells are added in series. In a tandem cell, it is possible to combine two, or more, thinner (more efficient) active layers using semiconductors with different bandgaps for more efficient light harvesting. Besides, since individual cells are added in series, the open circuit voltage of a tandem cell is directly increased to the sum of the V_{OC} values of individual cells.

Acknowledgments The authors would like to use this opportunity to thank René Janssen, Martijn Wienk, Jan Kroon, Sjoerd Veenstra, Volker Schmidt, and Xiaoniu Yang for helpful discussions. Supported by the Dutch Polymer Institute (DPI), the Royal Dutch Academy of Sciences (KNAW), and the Chinese Academy of Sciences (Grant 2009J2-28: Visiting Professorship for Senior International Scientists).

References

1. Gunes S, Neugebauer H, Sariciftci NS (2007) Chem Rev 107:1324
2. Hoppe H, Sariciftci NS (2008) Adv Polym Sci 214:1
3. Liang Y, Xu Z, Xia J et al (2010) Adv Mater 22:1
4. Koster LJA, Mihailetchi VD, Blom PWM (2006) Appl Phys Lett 88:09351
5. Jorgensen M, Norrman K, Krebs FC (2008) Sol Energy Mater Sol Cells 92:686
6. Dennler G, Brabec CJ (2008) Socio-economic impact of low-cost PV technologies. In: Brabec CJ, Dyakonov V, Scherf U (eds) Organic photovoltaics. Wiley–VCH, Weinheim, p 531
7. Roes AL, Alsema EA, Blok K et al (2009) Prog Photovolt Res Appl 17:372
8. Bube RH (1992) Photoelectronic properties of semiconductors. Cambridge University Press, Cambridge
9. Pope M, Swenberg CE (1999) Electronic processes in organic crystals and polymers. Oxford University Press, Oxford
10. Sariciftci NS, Smilowitz L, Heeger AJ et al (1992) Science 258:1474
11. Haugeneder A, Neges M, Kallinger C et al (1999) Phys Rev B 59:15346
12. Tang CW (1986) Appl Phys Lett 48:183
13. Peumans P, Yakimov A, Forrest SR (2003) J Appl Phys 93:3693
14. Smilowitz L, Sariciftci NS, Wu R et al (1993) Phys Rev B 47:13835
15. Yoshino K, Hong YX, Muro K et al (1993) Jpn J Appl Phys Part 2 32:L357
16. Halls JJM, Pichler K, Friend RH et al (1996) Appl Phys Lett 68:3120
17. Savenije TJ, Warman JM, Goossens A (1998) Chem Phys Lett 287:148
18. Kroeze JE, Savenije TJ, Vermeulen MJW et al (2003) J Phys Chem B 107:7696
19. Yu G, Heeger AJ (1995) J Appl Phys 78:4510
20. Halls JJM, Walsh CA, Greenham NC et al (1995) Nature 376:498
21. Yu G, Gao J, Hummelen JC et al (1995) Science 270:1789
22. Sirringhaus H, Brown PJ, Friend RH et al (1999) Nature 401:685
23. Yang X, Loos J, Veenstra SC et al (2005) Nano Lett 5:579
24. Kim Y, Choulis SA, Nelson J et al (2005) Appl Phys Lett 86:063502
25. Heutz S, Sullivan P, Sanderson BM et al (2004) Sol Energy Mater Sol Cells 83:229

26. Yang X, Loos J (2007) Macromolecules 40:1353
27. van Bavel SS, Veenstra S, Loos J (2010) Macromol Rapid Commun 31:1835
28. van Bavel SS, Loos J (2010) Adv Mater 20:3217
29. Mihailetchi VD, Xie HX, de Boer B et al (2006) Adv Funct Mater 16:699
30. Mihailetchi VD, Blom PWM, Hummelen JC et al (2003) J Appl Phys 94:6849
31. Brabec CJ, Cravino A, Meissner D et al (2001) Adv Funct Mater 11:374
32. Scharber MC, Mühlbacher D, Koppe M et al (2006) Adv Mater 18:78
33. Cravino A (2007) Appl Phys Lett 91:243502
34. Koster LJA, Mihailetchi VD, Ramaker R et al (2005) Appl Phys Lett 86:123509 (the expression in question has the following form: $V_{oc} = E_{gap}/q - (kT/q) * \ln[C^*(1 - P)/P]$, where P is the dissociation probability of excitons into free charges; unless P is approaching zero, the dependency of V_{oc} on P can be neglected)
35. Mandoc MM, Kooistra FB, Hummelen JC et al (2007) Appl Phys Lett 91:263505
36. Mandoc MM, Koster LJA, Blom PWM (2007) Appl Phys Lett 90:133504 (very high carrier mobilities (above 10^{-3} m^2 V^{-1} s^{-1}) affect the difference between the quasi-Fermi levels and lead to the reduction of V_{oc}. For comparison, mobilities of holes and electrons in annealed P3HT/PCBM blends are around 10^{-8} m^2 V^{-1} s^{-1} and 10^{-7} m^2 V^{-1} s^{-1}, respectively, ref. 30)
37. Wudl F (1992) Acc Chem Res 25:157
38. Mozer A, Denk P, Scharber M et al (2004) J Phys Chem B 108:5235
39. Vandewal K, Tvingstedt K, Gadisa A et al (2009) Nat Mater 8:904
40. Hoppe H, Sariciftci NS (2006) J Mater Chem 16:45
41. van Bavel SS, Sourty E, de With G et al (2009) Nano Lett 9:507
42. Yang X, van Duren JKJ, Janssen RAJ et al (2004) Macromolecules 37:2152
43. Martens T, D'Haen J, Munters T et al (2003) Synth Met 138:243
44. Hoppe H, Drees M, Schwinger W et al (2005) Synth Met 152:117
45. Hoppe H, Niggemann M, Winder C et al (2004) Adv Funct Mater 14:1005
46. Shaheen SE, Brabec CJ, Sariciftci NS et al (2001) Appl Phys Lett 78:841
47. Chirvase D, Parisi J, Hummelen JC et al (2004) Nanotechnology 15:1317
48. van Duren JKJ, Yang X, Loos J et al (2004) Adv Funct Mater 14:425
49. Zhong H, Yang X, de With G et al (2006) Macromolecules 39:218
50. Alexeev A, Loos J (2008) Org Electron 9:149
51. Loos J, van Duren JKJ, Morrissey F et al (2002) Polymer 43:7493
52. van Duren JKJ, Loos J, Morrissey F et al (2002) Adv Funct Mater 12:665
53. Weyland M (2002) Top Catal 21:17
54. Weyland M, Midgley PA (2004) Mater Today 7:32
55. Jinnai H, Spontak RJ (2009) Polymer 50:1067
56. Möbus G, Inkson BJ (2007) Mater Today 10:18
57. Cormack AM (1963) J Appl Phys 34:2722
58. Radermacher M (1980) PhD thesis, Department of Physics, University of Munich, Munich, Germany
59. Hummelen JC, Knight BW, LePeq F et al (1995) J Org Chem 60:532
60. Padinger F, Rittberger RS, Sariciftci NS (2003) Adv Funct Mater 13:85
61. Waldauf C, Schilinsky P, Hauch J et al (2004) Thin Solid Films 451–452:503
62. Al-Ibrahim M, Ambacher O, Sensfuss S et al (2005) Appl Phys Lett 86:201120
63. Ma W, Yang C, Gong X et al (2005) Adv Funct Mater 15:1617
64. Svensson M, Zhang F, Veenstra SC et al (2003) Adv Mater 15:988
65. Yohannes T, Zhang F, Svensson M et al (2004) Thin Solid Films 449:152
66. Slooff LH, Veenstra SC, Kroon JM et al (2007) Appl Phys Lett 90:43506
67. Inganäs O, Zhang F, Andersson MR (2009) Acc Chem Res 42:1731
68. Kroon R, Lenes M, Hummelen JC et al (2008) Polym Rev 48:531
69. Peet J, Kim JY, Coates NE et al (2007) Nat Mater 6:497
70. Qin R, Li W, Li C et al (2009) J Am Chem Soc 131:14612
71. Chu T-Y, Alem S, Verly PG et al (2009) Appl Phys Lett 95:063304

72. Rispens MT, Meetsma A, Rittberger R et al (2003) Chem Commun 2116
73. Yang X, Alexeev A, Michels MAJ et al (2005) Macromolecules 38:4289
74. Merlo JA, Frisbie CD (2004) J Phys Chem B 108:19169
75. Bertho S, Janssen G, Cleij TJ et al (2008) Sol Energy Mater Sol Cells 92:753
76. Ihn KJ, Moulton J, Smith P (1993) J Polym Sci Polym Phys 31:735
77. Li G, Shrotriya V, Huang J et al (2005) Nat Mater 4:864
78. Reyes-Reyes M, Kim K, Carroll D (2005) Appl Phys Lett 87:083506
79. Peet J, Soci C, Coffin RC et al (2006) Appl Phys Lett 89:252105
80. Zhao Y, Yuan GX, Roche P et al (1995) Polymer 36:2211
81. Savenije TJ, Kroeze JE, Yang X et al (2005) Adv Funct Mater 15:1260
82. Koster LJA, Mihailetchi VD, Lenes M et al (2008) Performance improvement of polymer:fullerene solar cells due to balanced charge transport. In: Brabec CJ, Dyakonov V, Scherf U (eds) Organic photovoltaics. Wiley–VCH, Weinheim, p 283
83. van Bavel SS, Sourty E, de With G et al (2009) Macromolecules 42:7396
84. Campoy-Quiles M, Ferenczi T, Agostinelli T et al (2008) Nat Mater 7:158
85. Xu Z, Chen L-M, Yang G et al (2009) Adv Funct Mater 19:1227
86. Nilsson S, Bernasik A, Budkowski A et al (2007) Macromolecules 40:8291
87. van Hal PA, Christiaans MPT, Wienk MM et al (1999) J Phys Chem B 103:4352
88. Halls JJM, Arias AC, MacKenzie JD et al (2000) Adv Mater 12:498
89. Stalmach U, de Boer B, Videlot C et al (2000) J Am Chem Soc 122:5464
90. Zhang F, Jonforsen M, Johansson DM et al (2003) Synth Met 138:555
91. Veenstra SC, Verhees WJH, Kroon JM et al (2004) Chem Mater 16:2503
92. Saunders BR, Turner ML (2008) Adv Colloid Interface Sci 138:1
93. Beek WJE, Wienk MM, Janssen RAJ (2004) Adv Mater 16:1009
94. Huynh WU, Dittmer JJ, Alivisatos AP (2002) Science 295:2425
95. Kuo CY, Tang WC, Gau C et al (2008) Appl Phys Lett 93:033307
96. Wang H, Oey CC, Djurisic AB et al (2005) Appl Phys Lett 87:023507
97. Ravirajan P, Peiro AM, Nazeeruddin MK et al (2006) J Phys Chem B 110:7635
98. Zhu R, Jiang C-Y, Ramakrishna S (2009) Adv Mater 21:994
99. van Hal PA, Wienk MM, Kroon JM et al (2003) Adv Mater 15:118
100. Beek WJE, Slooff LH, Kroon JM et al (2005) Adv Funct Mater 15:1703
101. Oosterhout SD, Wienk MM, van Bavel SS et al (2009) Nat Mater 8:818
102. Jenekhe S, Chen XL (1998) Science 279:1903
103. Nishizawa T, Lim HK, Tajima K et al (2009) Chem Commun 2469
104. Yang C, Lee JK, Heeger AJ et al (2009) J Mater Chem 19:5416
105. Moule AJ, Meerholz K (2009) Adv Funct Mater 19:3028
106. Hayakawa A, Yoshikawa O, Fujieda T et al (2007) Appl Phys Lett 90:163517
107. Hansel H, Zettl H, Krausch G et al (2003) Adv Mater 15:2056
108. Gilot J, Barbu I, Wienk MM et al (2007) Appl Phys Lett 91:113520
109. Kim JY, Lee K, Coates NE et al (2007) Science 317:222
110. Gilot J, Wienk MM, Janssen RAJ (2007) Appl Phys Lett 90:143512
111. Hadipour A, de Boer B, Wildeman J et al (2006) Adv Funct Mater 16:1897

Development of Low Band Gap Polymers for Roll-to-Roll Coated Polymer Solar Cell Modules

Eva Bundgaard and Frederik Krebs

Abstract Organic photovoltaics (OPV) have the potential for low production cost. Additionally, there has been an increase in both efficiency and stability of small-area OPV devices prepared in research laboratories worldwide and, consequently, attention on OPVs has increased tremendously. In this chapter we describe the challenges of OPVs and give suggestions on how these can be overcome. Design and synthesis of a new group of materials and low band gap polymers are described. Problems and possible solutions of OPV stability are shortly discussed. Furthermore, the latest technology to manufacture large-area OPV modules is described along with production of large-area modules by roll-to-roll printing of a low band gap polymer.

1 Introduction

1.1 Organic Photovoltaics

In the past, there has been a tremendous focus on renewable energy sources, such as wind, hydro, biomass, geothermal and solar. This is mainly to find alternative energy sources that are CO_2 neutral since United Nations have decided that a 20% reduction in the CO_2 emissions by 2020 is necessary [1, 2]. Furthermore, due to the growth of human population on Earth and its increasing demand for energy, it is

E. Bundgaard (✉) · F. Krebs
Risø National Laboratory for Sustainable Energy, The Technical University of Denmark, Frederiksborgvej 399, DK 4000 Roskilde, Denmark
e-mail: evbu@risoe.dtu.dk

L. Zang (ed.), *Energy Efficiency and Renewable Energy Through Nanotechnology*, Green Energy and Technology, DOI: 10.1007/978-0-85729-638-2_6,

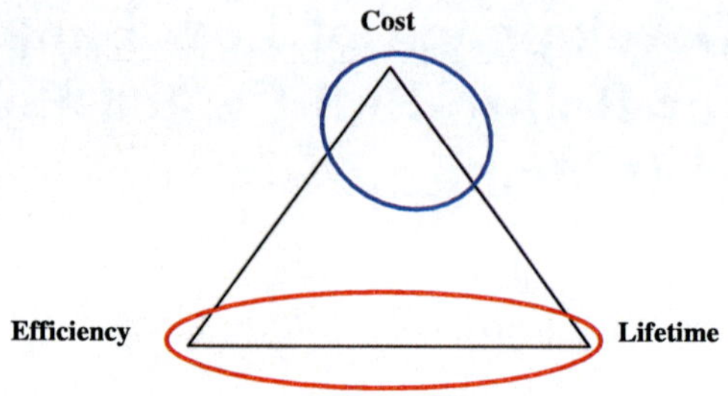

Fig. 1 Brabec triangle [4]: red represents inorganic PV and blue represents OPV

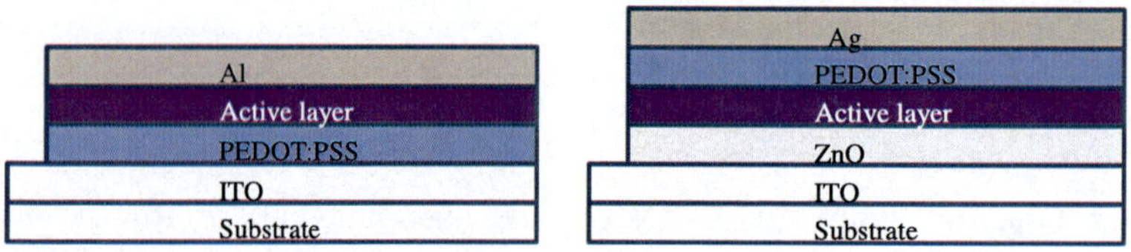

Fig. 2 Illustration of OPV devices, normal structure (*left*) and inverted (*right*)

likely that fossil fuels (which are the primary energy sources we depend on today) will be exhausted within the next 30–50 years.

The sun delivers more energy to the Earth's surface than we can consume and thus this renewable energy source is impossible to ignore. To harvest the energy from the sun and turn it into electricity, photovoltaics (PV) are used. There are three generations of photovoltaics. In the first generation, referred to as inorganic PV, the active material is crystalline silicon. In the second generation, referred to as thin film PV, the active layer is typically based on CdS or $CuIn(Ga)Se_2$ type materials. Finally the third generation, referred to as organic PV or OPV, the active layer is based on polymers or other organic molecules. There are advantages and disadvantages for all three generations. Inorganic PVs are relatively expensive to produce, mostly due to high temperatures and the need for very clean conditions; however, they have a very high efficiency and are very stable [3]. Thin film PVs frequently involve toxic materials and, eventhough they have demonstrated very high efficiencies and a potentially low cost of production, the controlled fabrication is still a challenge [3]. Polymer-based PVs currently have a low efficiency and poor stability. However, they have a low production cost and can be printed or coated using roll-to-roll (R2R) methods. Thus, a simple comparison shows that the inorganic photovoltaics dominate the areas of efficiency and stability, whereas organic photovoltaics can take the lead in production cost (Fig. 1) [4].

Within the last 5 years the stability and efficiency of small-area organic photovoltaic (OPV) prepared in research laboratories have increased and they now live fully up to the role as a competitor to inorganic PV and thin film PV at the laboratory level. In this chapter we will therefore focus on this type of PV. For further information on OPV (see [4–11]).

llustrations of two typical OPVs are shown in Fig. 2. The structure is a bulk heterojunction, where the polymer is mixed with a soluble fullerene,

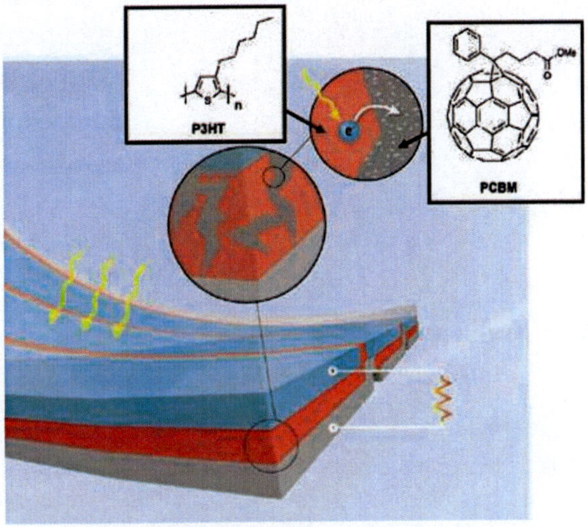

Fig. 3 Processes in OPV: from sunlight to electricity

e.g., [6,6]-phenyl-C_{61}-butyric acid methyl ester (PCBM), sandwichied between a transparent electrode, typically indium tin oxide (ITO) and poly(3,4-ethylenedioxythiophene):poly(styrene sulfonate) (PEDOT:PSS), and a back electrode, Al. Recently, an inverted structure for the OPV device was developed, where the transparent electrode is ITO and ZnO and the back electrode is Ag. The device structure is referred to as inverted due to the inversion of the way the current flows in the cell [12].

The typical processes in a bulk heterojuntion OPV is demonstrated in Fig. 3. The polymer absorbs a photon and an exciton (electron–hole pair) is created. This exciton diffuses to the donor/acceptor interface (i.e., the polymer/PCBM interface), where an electron is transferred from the donor to the acceptor (i.e., from polymer to PCBM). Dissociation of the exciton into free carriers, transport of the free carriers to the electrodes and charge collection at the electrodes complete the working cycle of the OPV. These processes determine how efficient the cell operates, i.e., losses in these processes by charge recombination or non-absorbed photons will result in a decrease in current and, hence, the efficiency will be lower.

1.2 Challenges of OPV

In the field of OPV, there are three focus areas within research groups worldwide. Those are: efficiency, stability and production.

So far, power conversion efficiency (PCE) has received the most attention. The literature shows a focus on device structure and device optimization to improve efficiencies mostly for OPV devices based on poly(3-hexyl-thiophene) (P3HT) and PCBM. Since the limit in efficiency for P3HT has been reached, the focus has shifted during the past few years toward designing polymers with a lower band gap

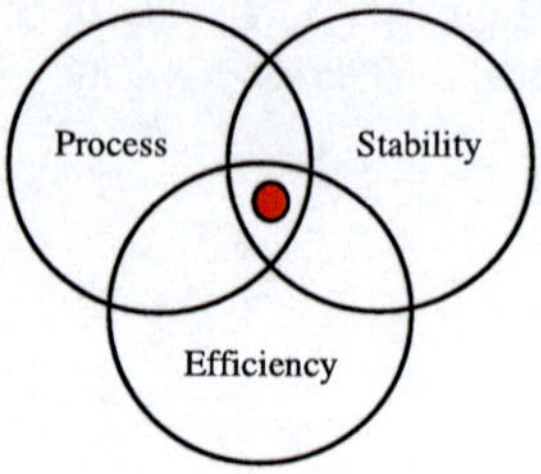

Fig. 4 Unification challenge of organic photovoltaic [17]

for potential improvement of the PCE. Low band gap polymers are expected to enable a significant increase in the efficiency of the solar cell due to increased absorption in the visible spectrum [13–16]. These types of polymers will be described in detail in the next section.

The focus on stability is increasing and more research groups are carrying out lifetime studies either by long time studies or accelerated studies. The stability can be increased by encapsulating the OPV with layers that protects the cell from water/moisture and oxygen or by design of the polymer [17]. In the third part of this chapter we briefly describe the problems and possible solutions for prolonging the stability of flexible R2R coated OPV modules.

Processing is a rather new topic, since most research has been carried out on small-area devices, e.g., <0.1 cm^2. However, the OPVs have the advantage that they can be printed on flexible substrates, making the processing of modules both faster and lower in cost compared to inorganic photovoltaics. There are several methods and techniques that need to be studied and in the last part of this chapter, we describe in detail the methods and the preparation of modules based on a low band gap polymer. The overall challenge for scientists working in the field of organic solar cells is to unite their knowledge into a single material such that the resulting device structure yields efficient and stable OPV modules that are easy to produce at a low cost (Fig. 4).

In the following three parts of this chapter, we look upon the challenges separately. We discuss in detail (1) the low band gap materials, the type of polymer that is believed to increase the efficiency of OPV, (2) stability issues, with a focus on thermocleavable side chains and (3) the processing of large area OPV modules and the different techniques that have been developed over the past years.

2 Low Band Gap Polymers

2.1 What and Why?

Within the past few years the focus within OPV research has been drifting more and more toward low band gap polymers. But what are these polymers and why are they so interesting?

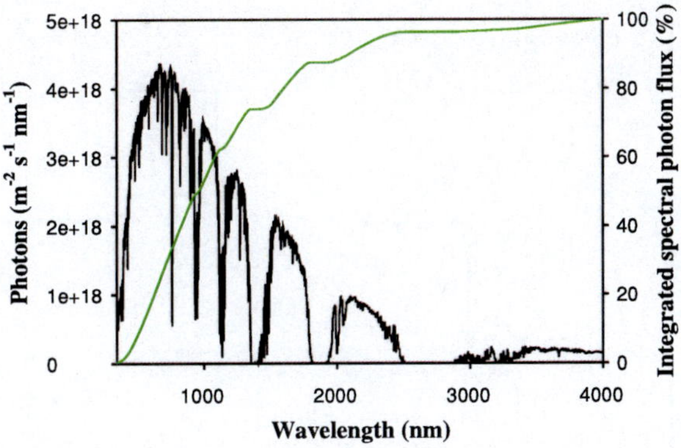

Fig. 5 Solar spectrum (AM1.5G) shown as number of photons (*black*) and integrated photon flux (*green*)

Low band gap polymers are loosely defined as polymers that absorb light with wavelengths longer than 600 nm, i.e. they have a band gap below 2 eV. Low band gap polymers are believed to have the ability to increase the efficiency of OPV devices since they have the ability to harvest more photons from the sun than the classes of polymers developed previously (e.g., MEHPPV, MDMOPPV and P3HT) [13, 15, 16].

In Fig. 5, the solar spectrum (AM1.5G) is shown as number of photons together with the integrated photon flux. This indicates that regular P3HT, which absorbs light up to ~700 nm can absorb at most 27.6% of incident photons. If all photons are harvested and converted into electrons this corresponds to a maximum current of 17.6 mA cm^{-2}. However, a low band gap polymer that absorbs light to 900 nm, for example, the portion of photons that can be absorbed increases to 46.7%, corresponding to a maximum current of 29.8 mA cm^{-2} [15]. These examples highlight the importance of low band gap polymers; however, in practice it is far more complicated than just lowering the band gap. When the band gap is lowered several parameters may change.

In Fig. 6, one can see that if the band gap is decreased by shifting the position of the lowest unoccupied molecular orbital (LUMO) of the polymer to lower energies, it approaches the LUMO of the acceptor (PCBM) and, as a result, it may not be energetically favored for the electron to transfer from the donor to the acceptor, meaning that recombination of electron and hole is favored instead (scenario B) [13, 15]. Furthermore, the open circuit voltage (V_{OC}) is reduced if the band gap is lowered by increasing the energy of the highest occupied molecular orbital (HOMO) of the polymer, since the V_{OC} is determined by the difference between the HOMO of the donor and LUMO of the acceptor in a simple pn-junction (scenario C) [13, 15]. This shows the importance of both designing the polymer and choosing an appropriate acceptor (e.g., PCBM or $PC_{70}PM$) so the energy levels of HOMO and LUMO are aligned optimally.

The different acceptors that are often used in OPV devices are shown in Fig. 7, along with an energy diagram showing their relative energy levels. The energy level alignment can, in principle, also be optimized by addition of another polymer instead of the fullerene based acceptor, i.e., combining a low band gap donor

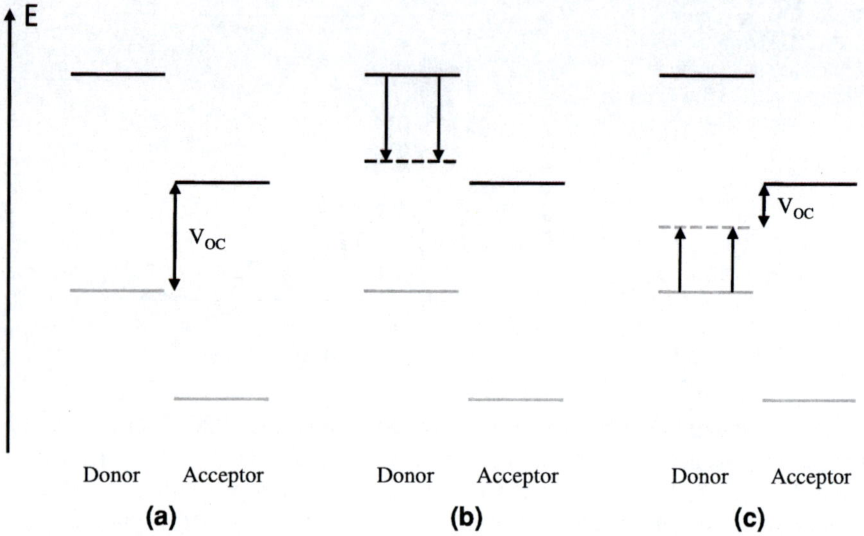

Fig. 6 Consequences of lowering the band gap of the donor. Gray lines denote the LUMO and black lines represent the HOMO levels

polymer with an acceptor polymer in a polymer–polymer OPV device [18]. This approach has not been pursued by many.

The band gap of a polymer is affected by several factors, such as donor–acceptor units, intramolecular interactions and molecular weight. The band gap is affected by the molecular weight (M_W) of the polymer, since an increase in M_W decreases the band gap due to a longer conjugation length [13].

Conjugated polymers with an alternating single double bond structure normally have two resonance structures: the aromatic and the quinoid forms. The aromatic form is normally lower in energy than the quinoid form and, hence, it is the dominating form in the polymer backbone. The band gap of the polymer is lowered if the energy difference between the two is decreased, i.e., the difference between double and single bond length is decreased [13]. An example of a polymer where the quinoid form is more stable than the aromatic is poly(isothianaphtalene) (PITN). Here the thiophene ring loses its aromaticity when going to the quinoid form, however, the benzene ring gains aromaticity and stabilizes the quinoid structure. Another way to lower the band gap of the polymer is to use donor–acceptor alternation, which increases the double bond character between the units, thereby stabilizing the quinoid form of the polymer back bone [13, 15]. The donor unit, often thiophene, is electron-rich and therefore increases the electron density between the units or donating electrons to the acceptor, which is electron poor. Copolymers with donor and acceptor alternating units are the most dominating low band gap polymers, as seen in Fig. 8. There are a few details one needs to keep in mind when designing these types of polymers. The energy levels of the acceptor should fit those of the donor. An example of optimizing the

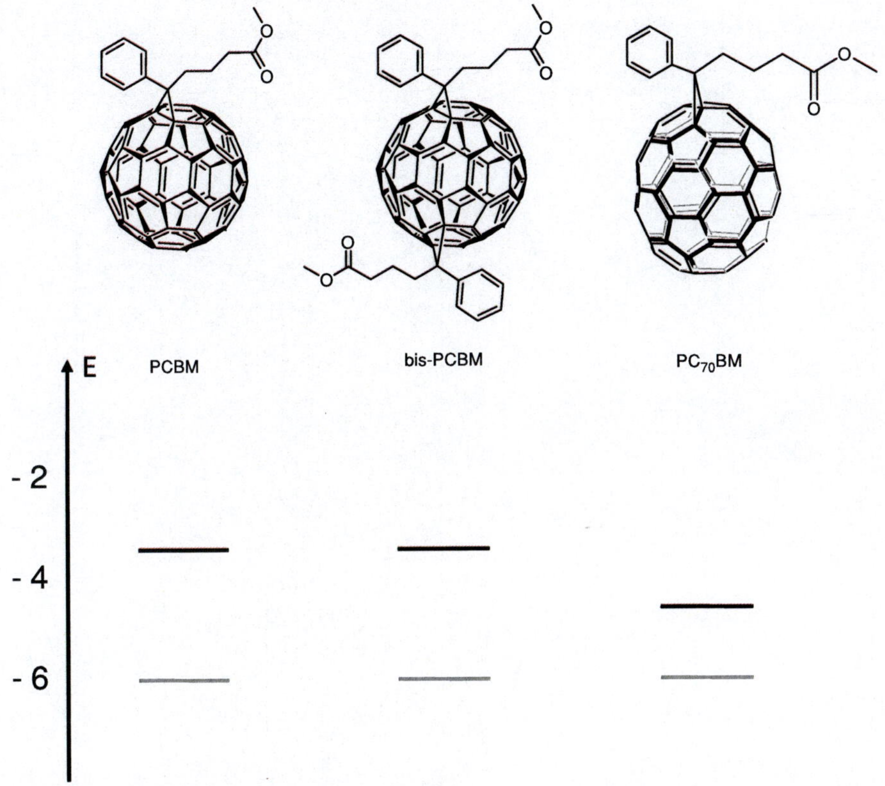

Fig. 7 Different acceptors used in OPV devices along with their corresponding energy diagram [19–22]. For other examples of fullerenes see [15]

polymer and lowering its band gap was shown for copolymers of benzothiadiazole and thiophenes, where the number of thiophenes was varied from 1 to 4. It was found that the band gap decreased with an increasing number of thiophenes [23]. And using benzo-bis-thiadiazole as the acceptor unit decreased the band gap to 0.6 eV [23]. Furthermore, when the band gap of the polymer is lowered, the energy level alignment with the acceptor in OPV device should be taken into account as described above.

Other design methods to decrease the band gap of a polymer are retention of backbone planarity, thereby keeping the conjugation length as long as possible. Thus, the torsion angle between units should be kept to a minimum by decreasing the steric hindrance in the molecule, e.g., using smaller side chains. [13]. Side groups can have another effect on the band gap of the polymer. When an electron donating group (e.g., alkoxy or alkyl) is attached to a ring, it pushes electrons into the aromatic system, making it easier to remove an electron from the HOMO due to a raise in the energy level. When an electron accepting group (e.g., fluorine or carbonyl) is attached, it pulls electrons away from the aromatic system, lowering the LUMO of the polymer and, therefore, the band gap [13].

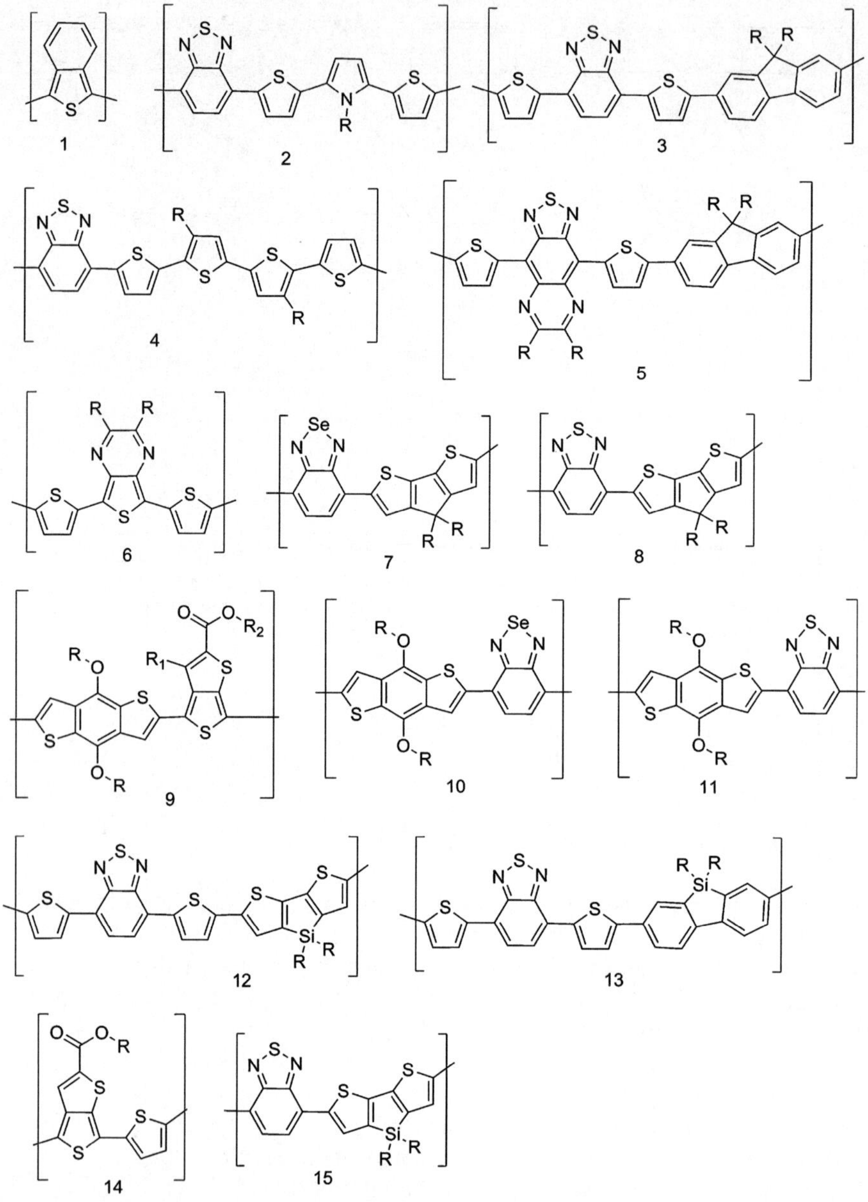

Fig. 8 Examples of low band gap polymers reported in the literature. Their band gap and photovoltaic data is reported in Table 1 along with references

Substituents can also influence the band gap by intermolecular effects. For P3HT the hexyl side chains ensure the polymer orders in a lamellar structure in film. This is clearly seen in UV-vis spectrum of a solution and film of P3HT; the band gap is decreased for the film [15].

Table 1 Band gap and photovoltaic data for the low band gap polymers shown in Fig. 8. The data for P3HT are given at the end of the table as a reference

Polymer	Band gap (eV)	I_{SC} (mA cm^{-2})	V_{OC} (V)	FF (%)	Efficiency (%)	Reference
1	1.00	0.05	0.55	30	0.008	[16,24–27]
2	1.60	3.10	0.72	37	1.0	[28–31]
3	1.78	4.66	1.04	46	2.2	[32, 33]
4	1.65	3.59	0.61	46	1.0	[23, 34]
5	1.27	3.40	0.58	35	0.7	[35–39]
6	1.20	3.50	0.56	58	1.1	[40]
7	1.30	5.00	0.52	34	0.9	[41]
8	1.73	16.2	0.62	55	5.5	[42–44]
9A	1.61	14.7	0.70	64	6.6	[45]
9B	1.63	9.20	0.76	45	3.1	[21]
10	1.52	1.05	0.55	32	0.2	[46]
11	1.70	2.97	0.68	44	0.9	[46]
12	1.51	10.7	0.62	52	3.4	[47]
13	1.82	9.50	0.90	51	3.4	[48]
14	1.30	–	–	–	1.0	[20]
15	1.45	12.7	0.68	55	5.1	[49]
P3HT	1.99	10.6	0.61	67	4.4	[50–52]

Design of the polymer is thus of great importance to control the band gap and one should bare the following in mind during the design process:

- Donor and acceptor units should be chosen to "fit" each other
- Choice of side chains (solubility, electron donating/accepting, steric hindrance and intermolecular interactions)
- Synthesis of polymers that ensures high molecular weight

Examples of low band gap polymers reported in the literature are given in Fig. 8. In Table 1, the corresponding band gap and photovoltaic data is summarized.

3 Stability and Degradation of OPV

3.1 Degradation of OPV

Stability of OPV devices is a relatively new research topic in the field of OPV. The lifetime of a device has gone from minutes to years with in the past decade. This is largely due to a better understanding of the degradations paths that an OPV device can follow. The types of degradation include, but are not limited, to the following [16]:

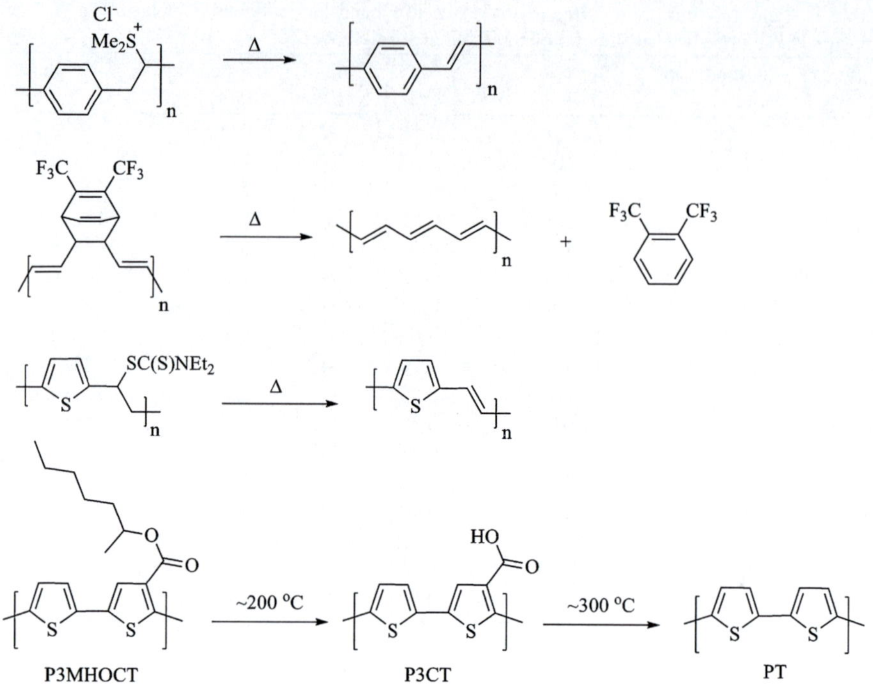

Fig. 9 Thermocleaving of polymers

- Reaction with either water or oxygen (all layers)
- Degradation of electrodes and PEDOT:PSS, e.g., diffusion of Al electrode into the active layer
- Photooxidation, this is especially for polymers with an alkoxy side chain
- Impurities in the polymer, e.g., Pd particles from the catalyst used in the synthesis of the polymer
- Morphology

This indicates that the stability of OPV devices can be improved with careful considerations to design of both the polymer and the device. However, other ways to increase the stability of the devices have also been reported, e.g., encapsulation layers such as PET or glass [53, 54].

3.2 Materials to Ensure Higher Stability

There have been reports on polymers with increased stability by way of thermocleavable side chains [55]. Historically conjugated polymers were prepared by thermal routes, as shown in Fig. 9. PPVs and polyacetylenes were prepared by the Wessling [56–59] and Durham [60–62] routes. Recently, the dithiocarbamate route was developed [63–67]. Common to these three approaches, the conjugated polymer backbone is formed in the last thermal step. A different route employs tertiary ester groups that are cleaved upon heating to form the acid and further heating results in decarboxylation and formation of the rigid polymer backbone as shown in Fig. 9. The advantage of the latter technique is that the conjugated backbone is present throughout the whole process and device films are active in photovoltaic devices before and after thermocleavage.

The thermocleavable side chains have also been studied on low band gap polymers where a dithienylthienopyrazine was coupled with different donor groups to produce polymers with band gaps between 1.17 and 1.37 eV. Photovoltaic performance was carried out and the highest efficiency was measured to be 1.21%. However, with a material loss upon heating to 200°C where the alkyl chain of the ester was removed, the performance decreased and no further thermocleaving was observed before decomposition at 400°C. The decrease in performance was ascribed to phase segregation in the active layer upon heating [68]. However, in a similar study where the donor was thiophene it was found that the decay in device stability was similar for both oxygen and inert atmospheres, indicating stability toward oxygen when aluminium electrodes are used [69].

4 Manufacture of Large Area OPV Modules

4.1 Production Methods

The typical laboratory polymer solar cell is prepared by spin-coating solutions of the materials onto a rigid glass substrate covered by a layer of ITO. The standard device employs a spincoated layer of the hole conducting PEDOT:PSS from an aqueous dispersion followed by spincoating of the active materials from an organic solvent such as chlorobenzene. The metallic electron collecting back electrode is applied by evaporation in a high vacuum. This approach has proven highly successful for materials screening, testing and development of the polymer solar cell technology in academia, requiring a relatively small investment in equipment (metal evaporator and a source meter). It should be noted that the device geometry with ITO/PEDOT:PSS as the transparent hole collecting front electrode requires that the evaporated back electrode is the electron collector, implying that it must be a low work function metal, such as aluminium or calcium. This is problematic in terms of device stability and, as a result, most academic device work is performed in an inert glovebox environment with very low humidity and oxygen content. In order to take those devices outside the laboratory there are strict requirements for the encapsulation that often requires active getter materials that remove oxygen and water from the enclosed compartment. An example of an encapsulated device is shown in Fig. 10, where a device prepared on glass is sealed with a glass fiber-filled, thermosetting epoxy against a milled aluminium back plate [53].

Fig. 10 A rigid encapsulation of a 10 cm^2 low band gap laboratory device prepared on a glass substrate with a milled aluminum back plate. The sealing of the device was achieved with a glass fiber-filled, thermosetting epoxy

An additional problem with the devices prepared on rigid glass substrates and evaporated metal electrodes is the relatively slow batch process with which they are prepared. In the typical laboratory experiment, a batch in the order of 10 devices is prepared over the course of several hours including cleaning the substrates, spin-coating, masking and removal of unwanted material, evaporation and encapsulation and an optimistic production time for such a device is around an hour. While this is clearly not prohibitive for laboratory work and development, it is impossible to implement this as a cost-effective method in a competitive solar cell market. While it is possible to envisage the scaling of device production on rigid glass substrates, the poor performance and stability makes it unlikely to be successful and a different approach must be sought to create a financially viable manufacturing process for polymer solar cells.

4.2 Other Film Forming Techniques

The most favored academic film forming technique, spin-coating, described above, is limited by the fact that it provides no control of the pattern of the formed film (zero-dimensional). Spin-coating is, however, very successful because it enables very good control over the film thickness, allows for preparation of very thin films and is highly reproducible. An additional reason for the success of spin-coating is that it is a non-equilibrium film forming technique, which allows the experimenter to prepare films in cases where wetting is not possible in an equilibrium situation. In many cases experiments have been successful without the experimenter ever knowing what peril (s)he faced! There are many other film forming techniques [70] available, some of which are suited for particular purposes while others are less suitable for polymer solar cells. The most well-known film forming techniques are:

- ink jet printing,
- electro/magnetographic printing,

- offset printing,
- screen printing,
- rotary screen printing,
- gravure printing,
- pad printing,
- flexographic printing,
- slot-die coating,
- curtain coating,
- slide coating,
- spray coating and
- knife coating.

Ink jet and electrographic coating are unique because they employ a digital master and provide full two-dimensional patterning of the printed area. Only ink jet printing has been used successfully in the context of polymer solar cells. Offset printing, screen printing, rotary screen printing, gravure printing, pad printing, flexographic printing all provide full two-dimensional patterning, but require a master plate with the desired layout. The complexity of the master is simplest and lowest in cost for screen printing, mid-range for rotary screen, flexographic and offset printing, and quite expensive for gravure printing, in which an engraved roller is required. They are all contact techniques, meaning that the application of the ink to the substrate is made through physical contact between the master and the substrate. It is currently debated whether this type of method can be used to apply the active layer. They have all been explored in the context of polymer solar cells, but none have been truly successful. Slot-die coating is a non-contact technique that allows for one-dimensional patterning. Curtain coating, slide coating, spray coating and knife coating are all non-contact zero-dimensional techniques, but curtain and slide coating allow for exceptionally high speeds and multilayer formation. In addition to the above techniques, several more exist that are derived from this overall set of printing and coating techniques. There is currently no clear view on which techniques are the most suitable and each is plausible, pending dedicated development of ink systems that suit the requirements of each individual technique. A few facts have, however, been influential on the development of the different techniques in the context of polymer solar cells. The most important factor is scale, as some of these techniques simply cannot be made to work on a small scale. This is possibly also the reason for the success of the few techniques that work on a small scale. The second factor is ink usage because some techniques require enormous amounts of ink before printing can even be started. Flexographic and gravure printing require that the rollers be continuously bathed and this can easily mean that liters of ink are required. Finally, research has focused on film formation of the active layers. It is anticipated that a mature polymer solar cell technology is a fully-printed, multilayer structure and it is likely that several different film forming techniques will enter the final process, with each being chosen because of its particular advantage for a specific layer.

4.3 Roll-to-Roll Printed Modules Based on a Low Band Gap Polymer

A frequently-highlighted attribute of polymer solar cells is their flexibility. While this is true, it is often misunderstood that flexibility is a prerequisite for success in application. While this may be the case in a few instances, it is unlikely that the success of the technology rests on a flexible product. The flexibility is, however, beneficial during manufacture to decrease costs and increase throughput. Many of the film forming techniques mentioned above are R2R compatible and this is viewed as a prerequisite for a printing or coating technique to be valuable. In order to prepare a low band gap polymer solar cell by R2R coating it is necessary to have a device geometry that enables the R2R processing of all layers.

The most successful process described so far is ProcessOne [71] that employs an inverted geometry (see Fig. 2), where ITO-covered PET is employed as the substrate. The ITO is patterned by screen printing and etch resist followed by etching, stripping and washing away the resist. The ITO is converted into the electron collecting electrode by slot-die coating a thin ZnO layer on top of the ITO to give a patterned PET/ITO/ZnO substrate that serves as a platform for testing virtually any active layer materials combination. The active layer material is slot-die coated onto the PET/ITO/ZnO composite electrode. The device is completed by slot-die coating of PEDOT:PSS on top as the hole collecting electrode. The PEDOT:PSS electrode is improved by screen printing a full silver electrode or a silver grid electrode on top depending on the sheet resistivity of the PEDOT:PSS and on whether semi-transparency is required. The complete five-layer device is thus prepared by a combination of screen printing and slot-die coating. Selected steps of the process are shown in Fig. 11. The low band gap polymer yielded semi-transparent modules with a deep blue, semitransparent color.

The IV curves of one module of indoor measurements under a solar simulator and outdoor measurements (Fig. 12) are shown in Fig. 13 and the photovoltaic data is summarized in Table 2. It shows a small decrease in current when measured outdoors and, even though the voltage is increased outdoors, the overall efficiency is higher under the simulated sun, which has a higher intensity (1000 versus 915 W m^{-2}).

All modules were characterized by a R2R procedure under a solar simulator and the average efficiency of the modules was 0.3%. However, it is clear that after annealing under the sun for about 30 min, the efficiency of the modules increased to 0.5–0.6%. Lifetime studies were carried out at 55 and 85°C and clearly showed the effect temperature has on the stability of the modules, i.e., the stability decreases at higher temperatures (85 versus 55°C).

5 Summary and Outlook

In this chapter, we have described the possible solutions to some challenges in the area of OPV. We have described in detail how the efficiency is believed to increase for low band gap polymers and provided examples from the literature.

Fig. 11 The slot-die coating of the low band gap polymer onto PET/ITO/ZnO. The wet film is shown immediately after the slot-die coating head (*top right*). The dry film is shown as it exits the oven (*top left*). The completed and laminated devices are shown below where the semitransparent blue color of the devices is visible

Fig. 12 Outside testing of one of the completed modules on a solar tracking platform for general testing of OPV modules and panels. The inset shows the low band gap device

Additionally, we have shown that device stability can be improved using thermocleavable side chains. And finally, production methods were described, along with the results of production of large-area modules based on low band gap polymers. The challenge for OPV devices is to combine the efficiency, stability and production into a single material.

Therefore, it is of great importance to design the optimal material. This can be accomplished by using donor and acceptor units that fit together. Another factor one must bear in mind is the choice of side chain, which can have an effect on:

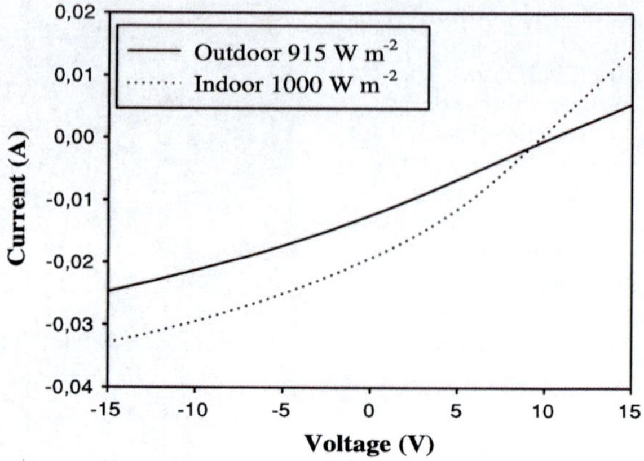

Fig. 13 IV curves for large area modules based on low band gap polymers measured outdoors and indoors

Table 2 Photovoltaic data of a module where measurements were performed outdoors and indoors under a sun simulator

	I_{SC} (mA cm^{-2})	V_{OC} (V)	FF (%)	η (%)
Indoor	0.20	9.5	29	0.55
Outdoor	0.13	10.3	26	0.38

(1) the band gap, by intermolecular ordering or withdraw/donating groups, (2) the production method, by ensuring solubility and (3) the stability, by thermocleavable ester groups.

It is clear that polymer solar cells hold great potential as a novel type of low-cost photovoltaic technology. There are, however, several developments needed before one can realize polymer solar cells as a competitive photovoltaic technology. The largest cost limitation currently is the transparent ITO electrode and there are currently few performing alternatives. The elimination of ITO is anticipated to enable a cost reduction in the range of 20–35%. The operational stability of the devices must be improved significantly in order for polymer solar cells to reach beyond the crowded thin film photovoltaic market. Finally, the power conversion efficiency is likely to require improvements to the 10–15% range before the technology can become pervasive.

Acknowledgments This work was supported by the Danish Agency for Science Technology and Innovation (FTP, ref. 274-08-0057). We would like thank Ole Hagemann and Jan Alstrup at Risø DTU for technical support.

References

1. Energy Information Administration (EIA), International Energy Outlook 2006, report # DOE/EIA-0484 (2006)
2. Politiken.dk, http://politiken.dk/eu/article261249.ece article in Danish published in 2007

3. Kippelen B, Brédas J-L (2009) Organic photovoltaics. Energy Environ Sci 2:251–261
4. Brabec CJ (2004) Organic photovoltaics: tehnology and market. Sol Mater Energy Sol Cells 83:273–292
5. Helgesen M, Søndergaard R, Krebs FC (2009) Advanced materials and processes for polymer solar cell devices. J Mater Chem 20:36–60
6. Spanggaard H, Krebs FC (2004) A brief history of the development of organic and polymeric photovoltaics. Sol Mater Energy Sol Cells 83:125–146
7. Coakly KM, McGehee MD (2004) Conjugated polymer photovoltaic cells. Chem Mater 16:4533–4542
8. Hoppe H, Sariciftci NS (2004) Organic solar cells: an overview. J Mater Res 19:1924–1945
9. Special issue: (2004) The development of organic and polymer photovoltaics. Sol Energy Mater Sol Cells (83):2–3
10. Special issue: (2005) Organic-based photovoltaics. MRS Bull. 30(1)
11. Brabec CJ, Sariciftci NS, Hummelen JC (2001) Plastic solar cells. Adv Funct Mater 11:15–26
12. Waldauf C, Morana M, Denk P, Schilinsky P, Coakley K, Choulis SA, Brabec CJ (2006) Highly efficient inverted organic photovoltaics using solution based titanium oxide as electron selective contact. Appl Phys Lett 89:233517
13. Kroon R, Lenes M, Hummelen JC, Blom PWM, De Boer B (2008) Small bandgap polymers for organic solar cells (polymer material development in the last 5 years. Pol Rev 48:531–582
14. Chen J, Cao Y (2009) Development of novel conjugated donor polymers for high-efficiencyefficiency bulk-heterojunction photovoltaic devices. Acc Chem Res 42:1709–1718
15. Bundgaard E, Krebs FC (2007) Low band gap polymers for organic photovoltaics. Sol Energy Mater Sol Cells 91:954–985
16. Winder C, Sariciftci NS (2004) Low band gap polymersLow band gap polymers for photon harvesting in bulk heterojunctions solar cells. J Mater Chem 14:1077–1086
17. Jørgensen M, Norrman K, Krebs FC (2008) Stability/degradation of polymer solar cells. Sol Energy Mater Sol Cells 92:686–714
18. McNeill CR, Abrusci A, Zaumseil J, Wilson R, McKiernan MJ, Burroughes JH, Halls JJM, Greenham NC, Friend RH (2007) Dual electron donor/electron acceptor character of a conjugated polymer in efficient photovoltaic diodes. Appl Phys Lett 90(1–3):193506
19. Lenes M, Wetzelaer G-JAH, Kooistra FB, Veenstra SC, Hummelen JC, Blom PWM (2008) Fullerene bisadducts for enhanced open-circuit voltages and efficiencies in polymer solar cells. Adv Mater 20:2116–2119
20. Yao Y, Liang Y, Shrotriya V, Xiao S, Yu L, Yang Y (2007) Plastic near-infrared photodetectors utilizing low band gap polymer. Adv Mater 19:3979–3983
21. Liang Y, Feng D, Wu Y, Tsai S-T, Li G, Ray C, Yu L (2009) Highly efficient solar cell polymers developed via fine-tuning of structural and electronic properties. J Am Chem Soc 131:7792–7799
22. Kim JY, Lee K, Coates NE, Moses D, Nguyen T-Q, Dante M, Heeger AJ (2007) Efficient tandem polymer solar cells frabricated by all-solution processing. Science 317:222–225
23. Bundgaard E, Krebs FC (2006) Low band gap conjugated polymers based on thiophene, benzothiadiazole and benzobis(thiadiazole). Macromolecules 39:2823–2831
24. Wudl F, Kobayshi M, Heeger AJ (1984) Poly(isothianaphthene). J Org Chem 49:3382–3384
25. Cava MP, Laksshmikatham MV (1975) Nonclassical condensed thiophenes. Acc Chem Res 8:139–144
26. Kobayshi M, Colerani N, Boysel M, Wudl F, Heeger AJ (1985) The electronic and electrochemical properties of poly(isothianaphthene). J Chem Phys 82:5717–5723
27. Henckens A, Knipper M, Polec I, Manca J, Lutsen L, Vanderzande D (2004) Poly(thienylene vinylene) derivates as low band gap polymerslow band gap polymers for photovoltaic applications. Thin Solid Films 451–452:572–579
28. Dhanabalan A, van Duren JKJ, van Hal PA, van Dogen JLJ, Janssen RAJ (2001) Synthesis and characterzation of a low bandgap conjugated polymer for bulk heterojunction photovoltaic cells. Adv Funct Mater 11:255–262

29. van Duren JKJ, Dhanabalan A, van Hal PA, Janssen RAJ (2001) Low-bandgap polymer photovoltaic cells. Synth Meter 121:1587–1588
30. Brabec CJ, Winder C, Sariciftci NS, Hummelen JC, Dhanabalan A, van Hal PA, Janssen RAJ (2002) A low-bandgap semiconducting polymer for photovoltaic devices and infrared emitting diodes. Adv Funct Mater 12:709–712
31. Winder C, Mühlbacher D, Neugebauer H, Sariciftci NS, Brabec C, Janssen RAJ, Hummelen JK (2002) Polymer solar cells and infrared light emitting diodes: dual function low band gap polymer. Mol Cryst Liq Cryst 385:[213]\93–[220]\100
32. Svensson M, Zhang F, Veenstra SC, Verhees WJH, Hummelen JC, Kroon JM, Inganäs O, Andersson MR (2003) High-performance polymer solar cells of an alternating polyfluorene copolymer and a fullerene derivative. Adv Mater 15:988–991
33. Shi C, Yao Y, Yang Y, Pei Q (2006) Regioregular copolymers of 3-alkoxythiophene and their photovoltaic application. J Am Chem Soc 128:8980–8986
34. Bundgaard E, Shaheen SE, Krebs FC, Ginley D (2007) Bulk heterojunctions based on a low band gap copolymer of thiophene and benzothiadiazole. Sol Energy Mater Sol Cells 91:1631–1637
35. Chen M, Perzon E, Andersson MR, Marcinkevicius S, Jönsson SKM, Fahlman M, Berggren M (2004) 1 micron wavelength photo- and electroluminescence from a conjugated polymer. Appl Phys Lett 84:3570–3572
36. Wang X, Perzon E, Oswald F, Langa F, Admassie S, Andersson MR, Inganäs O (2005) Enhanced photocurrent spectral responses in low-bandgap polyfluorene and C_{70}-derivative-based solar cells. Adv Funct Mater 15:1665–1670
37. Perzon E, Wang X, Zhang F, Mammo W, Delgado JL, de la Cruz P, Inganäs O, Langa F, Andersson MR (2005) Design, synthesis and properties of low band gap polyfluorene for photovoltaic devices. Synth Met 154:53–56
38. Chen M, Perzon E, Robisson N, Jönsson SKM, Andersson MR, Fahlman M, Berggren M (2204) Low band gap donor-acceptor-donor polymers for infra-red electroluminescence and transistors. Synth Met 146:233–236
39. Perzon E, Wang X, Admassie S, Inganäs O, Andersson MR (2006) An alternating low band-gap polyfluorene for optoelectronic devices. Polymer 47:4261–4268
40. Wienk MM, Turbiez MGR, Struijk MP, Fonrodona M, Janssen RAJ (2006) Low band gap poly(di-2-thienylthienopyrazine):fullerene solar cells. Appl Phys Lett 88(1–3):153511
41. Hou J, Chen TL, Zhang S, Chen H-Y, Yang Y (2009) Poly[4, 4-bis(2-ethylhexyl) cyclopenta[2, 1-b:3, 4-b']dithiophene-2, 6-diyl-alt-2, 1, 3-benzoselenadiazole-4, 7-diyl], a new low band gap polymer in polymer solar cells. J Phys Chem C 113:1601–1605
42. Peet J, Kim JY, Coates NE, Ma WL, Moses D, Heeger AJ, Bazan GC (2007) Efficiency enhancement in low-bandgap polymer solar cells by processing with alkane dithiols. Nat Mater 6:497–500
43. Mühlbacher D, Scharber M, Morana M, Zhu Z, Waller D, Gaudiana R, Brabec C (2006) High photovoltaic performance of a low bandgap polymer. Adv Mater 18:2884–2889
44. Mühlbacher D, Scharber M, Morana M, Zhu Z, Waller D, Gaudiana R, Brabec C (2006) High photovoltaic performance of a low bandgap polymer. Adv Mater 18:2884–2889 correction
45. Hou J, Chen H-Y, Zhang S, Chen RI, Yang Y, Wu Y, Li G (2009) Synthesis of a low band gap polymer and its application in highly efficient polymer solar cells. J Am Chem Soc 131:15586–15587
46. Hou J, Park M-H, Zhang S, Yao Y, Chen L-M, Li J-H, Yang Y (2008) Bandgap and molecular energy level control of conjugated polymer photovoltaic materials based on benzo[1, 2-b:4, 5-b']dithiophene. Macromolecules 41:6012–6018
47. Huo L, Chen H-Y, Hou J, Chen TL, Yang Y (2009) Low band gap dithieno[3,2-b:2',3'-d]silole-containing polymers, synthesis, characterization and photovoltaic application. Chem Comm (37):5570–5572
48. Wang E, Wang L, Lan L, Luo C, Zhuang W, Peng J, Cao Y (2008) High-performance polymer heterojunction solar cells of a polysilafluorene. Appl Phys Lett 92:033307-1–033307-3

49. Hou J, Chen H-Y, Zhang S, Li G, Yang Y (2008) Synthesis, characterization and photovoltaic properties of a low band gap polymer based on silole-containing polythiophenes and 2, 1, 3-benzothiadiazole. J Am Chem Soc 130:16144–16145
50. Ma W, Yang C, Gong X, Lee K, Heeger AJ (2005) Thermally stable, efficient polymer solar cells with nanoscale control of the interpenetrating network morphology. Adv Funct Mater 15:1617–1622
51. Li G, Shrotriya V, Huang J, Yao Y, Moriarty T, Emery K, Yang Y (2005) High-efficiencyefficiency solutionproccessable polymer photovoltaic cells by self-organization of polymer blends. Nat Mater 4:854–868
52. Kim Y, Cook S, Tuladhar SM, Choulis SA, Nelson J, Durrant JR, Bradley DDC, Giles M, Mcculloch I, Ha C-S, Ree M (2006) A strong regioregularity effect in self-organizing conjugated polymer films and high-efficiencyefficiency polythiophenes:fullerene solar cells. Nat Mater 5:197–203
53. Krebs FC, Alstrup J, Spanggaard H, Larsen K, Kold E (2004) Production of large-area polymer solar cells by industrial silk screen printing, lifetime considerations and lamination with polyethyleneterephthalate. Sol Energy Mater Sol Cells 83:293–300
54. Krebs FC (2006) Encapsulation of polymer photovoltaic prototypes. Sol Energy Mater Sol Cells 90:3633–3643
55. Krebs FC, Spanggaard H (2005) Significant improvement of polymer solar cell stabilitystability. Chem Mater 17:5235–5237
56. Gagnon DR, Capistran JD, Karasz FE, Lenz RW, Antoun S (1987) Synthesis, doping, and electrical conductivity of high molecular weight poly(p-phenylene vinylene). Polymer 28:567–573
57. Garay RO, Mayer B, Karasz FE, Lenz RW (1995) Synthesis and characterization of poly[2, 5-bis(triethoxy)-1, 4-phenylene vinylene]. J Polym Sci, Part A: Polym Chem 33:525–531
58. Lenz RW, Han CC, Stengersmith J, Karasz FE (1988) Preparation of poly(phenylene vinylene) from cycloalkylene sulfonium salt monomers and polymers. J Polym Sci, Part A: Polym Chem 26:3241–3249
59. Wessling RA (1985) The polymerization of xylene bisdialkyl sulfonium salts. J Polym Sci, Polym Symp 72:55–66
60. Bott DC, Brown CS, Chai CK, Walker NS, Feast WJ, Foot PJS, Calvert PD, Billingham NC, Friend RH (1986) Durham poly acetylene: preparation and properties of the unoriented material. Synth Met 14:245–269
61. Feast WJ, Winter JN (1985) An improved synthesis of polyacetylene. J Chem Soc, Chem Comm 4:202–203
62. Furlani A, Napoletano C, Russo MV, Feast WJ (1986) Stereoregular polyphenylacetylene. Polym Bull 16:311–317
63. Henckens A, Colladet K, Fourier S, Cleij TJ, Lutsen L, Gelan J, Vanderzande D (2005) Synthesis of 3, 4-diphenyl-substituted poly(thienylene vinylene), low band-gap polymers via the dithiocarbamate route. Macromolecules 38:19–26
64. Nguyen LH, Gunes S, Neugebauer H, Sariciftci NS, Banishoeib F, Henckens A, Cleij T, Lutsen L, Vanderzande D (2006) Precursor route poly(thienylene vinylene) for organic solar cells: photophysics and photovoltaic performance. Sol Energy Mater Sol Cells 90:2815–2828
65. Banishoeib F, Adriaensens P, Berson S, Guillerez S, Douheret O, Manca J, Fourier S, Cleij TJ, Lutsen L, Vanderzande D (2007) The synthesis of region-regular poly(3-alkyl-2, 5-thienylene vinylene) derivates using lithium bis(trimethylsilyl)amide (LHMDS) in the dithiocarbamate precursor route. Sol Energy Mater Sol Cells 91:1026–1034
66. Banishoeib F, Henckens A, Fourier S, Vanhooyland G, Breselge M, Manca J, Cleij TJ, Lutsen L, Vanderzande D, Nguyen LH, Neugebauer H, Sariciftci NS (2008) Synthesis of poly(2, 5-thienylene vinylene) and its derivates: low band gap materials for photovoltaics. Thin Solid Films 516:3978–3988
67. Girotto C, Cheyns D, Aernouts T, Banishoeib F, Lutsen L, Cleij TJ, Vanderzande D, Genoe J, Poortman J, Heremans P (2008) Bulk heterojunction organic solar cells based on soluble poly(thienylene vinylene) derivates. Org Electron 9:740–746

68. Helgesen M, Krebs FC (2010) Photovoltaic performance of polymers based on dithienylthienopyrazines bearing thermocleavable benzoate esters. Macromolecules 43: 1253–1260
69. Helgesen M, Krebs FC (2008) Thermocleavable low band gap polymerslow band gap polymers and solar cells therefrom with remarkable stabilitystability toward oxygen. Macromolecules 41:8986–8994
70. Krebs FC (2009) Fabrication and processing of polymer solar cells: A review of printing and coating techniques. Sol Energy Mater Sol Cells 93:394–412
71. Krebs FC, Gevorgyan SA, Alstrup J (2009) A roll-to-roll process to flexible polymer solar cells: model studies, manufacture and operational stabilitystability studies. J Mater Chem 19:5442–5451

Nanoarchitectured Electrodes for Enhanced Electron Transport in Dye-Sensitized Solar Cells

Tao Xu

Abstract The invention of dye-sensitized solar cell (DSSC) provided a promising alternative to Si-based photovoltaic devices. The first generation of DSSCs was constructed on nanoparticle wide bandgap semiconductor photoanodes. However, despite its unmatched success to date, the nanoparticle-based photoanode suffers from exceedingly slow electron transport due to the intrinsic defect states in the nanoparticle network, which eventually limits any further advancement in the device efficiency. Recent efforts have been directed toward developing ordered electron transport pathways using a variety of pseudo-1D photoanodes that exhibit enhanced charge transport and greater material versatility. Further exploration and optimization of these alternative nanoarchitectured photoanodes may eventually lead to device performance exceeding the current state-of-the-art.

1 Introduction

The global energy consumption is projected to double by 2050 from 13 terawatts (TW) at present to about 30 TW, largely due to population growth and enhanced industrial development world-wide [1]. The current reliance on fossil fuels (coal, petroleum, and natural gas) incurs serious problems as model for the sustainable energy future, such as scarcity and the disastrous impact on environment associated with their exploitation and utilization. Along with the increasing concern of

T. Xu (✉)
Department of Chemistry and Biochemistry, Northern Illinois University,
DeKalb, IL 60115, USA
e-mail: txu@niu.edu

L. Zang (ed.), *Energy Efficiency and Renewable Energy Through Nanotechnology*,
Green Energy and Technology, DOI: 10.1007/978-0-85729-638-2_7,

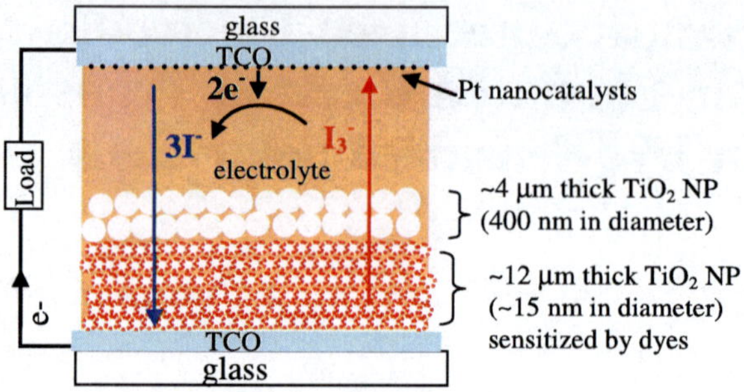

Fig. 1 A classic configuration of a TiO_2 nanoparticle-based DSSC

global climate change, cutting greenhouse gas emissions from energy production has become a major priority. Nuclear power can supply large amounts of energy in a carbon-neutral fashion. However, the scale is enormous: in order to cover the ever-growing energy demands nearly 1 gigawatt (GW) nuclear power plant must be commissioned world-widely every day for the next decade. Solar energy is currently thought to hold immense potential as a reliable clean energy source. The sun deposits 120,000 TW of power onto the Earth's surface, which can in 1 min provide the entire human energy need for a whole year [1]. Several approaches are being developed to tap into this vast, environmentally safe energy resource in an economically competitive fashion: photovoltaic cells, solar fuels generation (inspired by natural photosynthesis), and solar thermal conversion.

Photovoltaic devices convert solar photonic energy directly into electric energy in an environmental friendly way. Thus, lowering the cost while improving the efficiency of photovoltaic devices has become an attractive goal for both scientific and economic purposes. Dye-sensitized solar cells (DSSCs) have attracted much attention due to their potentially low cost and simple fabrication process in comparison to silicon-based photovoltaics. However, compared to the progressively increasing efficiency of silicon-based photovoltaic technology, the efficiency of DSSCs has been stagnate since the ground-breaking work by O'Regan and Grätzel [2], who reported the first TiO_2 nanoparticle-based DSSC with an efficiency of 7%. After that, the device efficiency was soon improved to 10% in 1993 by Grätzel et al. [3]. Ever since then, in nearly two decades, Grätzel's pioneering work has aroused tremendous research effort to understand the basic photophysics, electrochemistry, and materials chemistry relevant to this fascinating device, but only relatively incremental improvements in efficiency have been realized so far, reaching the current record of 11.2% [4, 5]. A brief schematic and the working principle of a TiO_2 nanoparticle-based DSSC are illustrated in Fig. 1.

Incident solar light first penetrates the device through a transparent conducting oxide (TCO), on which a thick layer (typically 10–15 μm) of interconnected semiconducting nanoparticles (NP), such as TiO_2, are coated to provide a large internal surface area for anchoring the light-harvesting dye molecules, whose lowest unoccupied molecular orbital (LUMO) matches the conduction band edge of the semiconducting NP. The dye molecules are excited by the incident photons, leading to electron-hole pairs (excitons), from which the electrons are

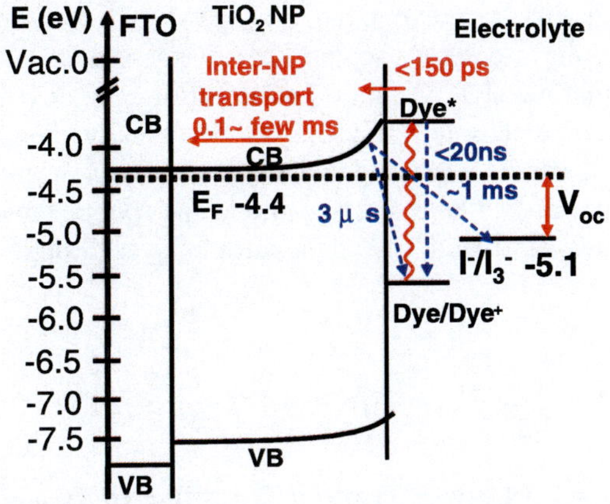

Fig. 2 Energetic and kinetic processes in DSSCs on an idealized energy level diagram

quickly injected into the conduction band of the semiconducting NP and subsequently collected by the TCO anode, while the holes in the highest unoccupied molecular orbital are refilled by electrons from the cathode via redox species, e.g., an I^-/I_3^- couple in an electrolyte that interpenetrates the nanoparticle network [6].

2 Device Energetics and Kinetics

The schematic representation of the energetic and kinetic diagram of a TiO_2 nanoparticle-based DSSC is illustrated in Fig. 2.

In the most studied and most efficient devices to date (Fig. 1), the photoanode consists of two layers of TiO_2 NP subsequently coated on a TCO glass. The first layer is directly attached to the TCO anode and is composed of a 12 μm thick film of transparent 10–20 nm diameter TiO_2 NP, which is sintered from the paste of TiO_2 at 450°C in air. A doctor blade is typically used to control the thickness of the paste. This first layer of TiO_2 nanoparticle film provides a large internal surface area (characterized by a roughness factor, defined as the total film area per unit projected substrate area) of ~1,000 for the anchoring of solar-absorbing dye molecules to yield high light absorption in the 400–800 nm region, covering ~60% of the solar spectrum. Furthermore, this first layer is treated with $TiCl_4$ or other organometallic TiO_2 precursors to form a blocking TiO_2 layer on the exposed TCO to enhance the shunt resistance of the device. A second layer of much larger (400 nm diameter) TiO_2 particles with thickness of ~4 μm is deposited atop the first layer. This second layer of TiO_2 particles is used to scatter the red and near-IR photons that are not absorbed by the dyes on the first layer of TiO_2 back into the first layer [7, 8]. This photoanode must be sintered prior to dye-loading in order to enhance the adhesion between TiO_2 particles. Typically, ruthenium-based dye molecules, e.g., [Ru(4,4′-dicarboxy-2,2′-bipyridine)2(NCS)$_2$] (N_3 dye), are used as solar absorbers and were anchored to the surface of TiO_2 NP via carboxylate groups, followed by desiccation in dark to prevent the TiO_2-catalyzed photo-bleaching of dye molecules. Then, using a polymer sealant, the dye-sensitized photoanode is attached to a piece of platinized TCO glass that acts as a cathode, (i.e., the so-called counter electrode, dark electrode, or back electrode). Finally, the electrolyte containing I^-/I_3^- as redox shuttles is infiltrated in the space between the photoanode and the cathode through a pre-drilled hole on the cathode and spontaneously interpenetrates the matrix of the dye-sensitized TiO_2 nanoparticle layer through capillary force.

2.1 Forward Electron Transport

2.1.1 Electron Transfer from Dye to TiO_2

Following the light absorption, an excited dye rapidly injects an electron from the LUMO of the dye molecule to the conduction band of the TiO_2, as illustrated by the reaction below.

$$D + h\upsilon \rightarrow D^* \quad (1)$$

$$D^* \rightarrow D^+ + e^-_{\mathrm{TiO_2}} \quad (2)$$

Here, the LUMO level of the dye must be higher than the conduction band edge of the semiconductor in order to provide a driving force for the photo-electron injection. With the help of this driving force, the charge injection efficiency (quantum yield) of Ru-based dyes can reach nearly unit. [4] The injected hot electrons undergo a rapid decay through electron–phonon interaction to the conduction band edge of the TiO_2 in a time scale of less than 1 ps [4] and this energy is not harvested, causing a loss of photovoltage of approximately $0.3 \sim 0.4$ V.

2.1.2 Electron Transport in TiO_2 Nanoparticle Network

Then, the injected photoelectrons diffuse through the sintered TiO_2 nanoparticle network and are collected by the TCO anode. The nature of electron transport in the TiO_2 nanoparticulate matrix is well-understood through time-resolved photocurrent and photovoltage measurements [9, 10] and simulation studies [11, 12]. In a typical DSSC, there is a strong electrostatic coupling between the electrons in the TiO_2 nanoparticulate matrix and the nearby counter ions in the electrolyte. Such an electron–ion coupling can screen off any macroscopic electric fields necessary for drift transport found in many solid-state solar cells. Hence, the electron transport in wet, illuminated nanoparticle network in DSSCS proceeds through a trap-limited ambipolar diffusion process [13]. In details, the electrons in the conduction band of TiO_2 undergo a random walk through the TiO_2 nanoparticulate film, during which the electrons frequently interact with a distribution of defective trap states (due to vacancies, crystal dislocations etc.) on the surface or in the bulk of the TiO_2 NP. [14, 15]. Under steady-state, one Sun radiation, an injected electron in the TiO_2 network can encounter, on average, one million trapping/detrapping events prior to either percolating to the electron-collecting TCO electrode or recombining with an oxidizing species, mainly the I_3^- and dye$^+$ cation in the electrolyte in the proximity of the moving electrons [14]. τ_{diff}, the time taken for a free electron to diffuse across the TiO_2 film to the collecting TCO anode, is given by $\tau_{diff} = d/D_0$, where d is the thickness TiO_2 nanoparticle film and D_0 is the diffusion coefficient of free electrons in the TiO_2 nanoparticle film. For $d = 10$ μm and $D_0 = 10^{-4}$ cm^2 s^{-1} at room temperature found in the nanoparticle film, electrons should reach the anode in 0.1 to a few microseconds depending on the voltage. In comparison, the diffusion coefficient of free electrons for single crystal anatase is ~0.5 cm^2 s^{-1} [16], 3–4 orders of magnitude faster than that through a TiO_2 nanoparticle film. In fact, injected electrons appear to become trapped on a time scale of picoseconds to nanoseconds [17].

2.1.3 Electron Transfer from Cathode to Dye Molecules

Upon electron injection from the LUMO of dye to the TiO_2 NP, the dye molecules loose an electron in their HOMO orbitals and are in an oxidative state. Thus, prompt refilling of an electron in the HOMO of the dye molecules is highly demanded. The function of the redox shuttle is to convey electrons from the counter electrode to the oxidized dyes to complete the electrochemical circuit. To date, the most efficient redox shuttle is the iodide/triiodide redox couple (I^-/I_3^-). The reaction between the oxidized dye and iodide leads to the formation of diiodide radicals ($I_2^{-\bullet}$). The pathway for the reduction of the oxidized dye (D^+) by iodide is given by the following reactions [18]:

$$D^+ + I^- \rightarrow (D\cdots I) \tag{3}$$

$$(D\cdots I) + I^- \rightarrow D + I_2^{-\bullet} \tag{4}$$

$$2I_2^{-\bullet} \rightarrow I_3^- + I^- \tag{5}$$

Although several alternatives to the I^-/I_3^- mediator have been developed for DSSCs, the performance of the iodide/triiodide couple remains unrivaled. Understanding the physical mechanisms that influence the device parameters is essential for further development. One of the key parameters that govern the attainable power from a DSSC is the open-circuit voltage (V_{oc}), which is the maximum voltage the device can generate. The maximum value of V_{oc} is determined by the energy gap between the quasi-Fermi level in the semiconductor (i.e., TiO_2 for a traditional TiO_2 nanoparticle-based DSSC) under illumination and the redox potential of the I^-/I_3^- in electrolyte. The V_{oc} of a typical TiO_2-based DSSC is 0.75–0.85 V. Considering the 1.8 eV optical gap of the N3 dye, for example, only half of the energy available in a DSSC can be utilized. A significant loss of V_{oc} is caused by the overpotential required for interfacial electron transfer from the excited dye to the TiO_2. Besides, another major loss in V_{oc} is the nearly 500 mV overpotential of I^-/I_3^- (with respect to the HOMO of the dye molecules), which seems to be a necessary driving force for I^- to transfer its electron to the dye$^+$ species at an acceptable rate. This overpotential may also be a consequence of both mechanistic complexity and the unusually large reorganization energy ($\sim$0.7 eV) for I^-/I_3^- [18].

The regeneration of I_3^- to I^- requires the mass flow of I_3^- from the voids of TiO_2 matrix to the counter electrode. The transport of triiodide is mostly driven by diffusion. Under a low concentration of triiodide or in a viscous electrolyte, the mass-transport of I_3^- can be a rate-limiting step in the DSSCs. The diffusion coefficient of I_3^- in the electrolyte is less than 10^{-4} cm^2 s^{-1} at room temperature [19]. Depletion of triiodide at the counter electrode causes an overpotential, lowering the voltage output of the cell. In order to have rapid regeneration of the oxidized dye, high iodide concentration is required. In nonviscous electrolytes such as acetonitrile, an iodide concentration of 0.3 M is sufficient to assure prompt regeneration. When a high concentration of triiodide is present, a Grotthus mechanism, illustrated below, may increase the observed diffusion coefficient [20].

$$I_3^- + I^- \rightarrow I^- \cdots I_2 \cdots I^- \rightarrow I^- + I_3^- \tag{6}$$

2.2 Back Electron Transfer

If the cell is illuminated with an open circuit under steady-state conditions, no current can be collected from the cell. The forward electron transfer, namely the electron injection from dye to semiconductor must, therefore, be exactly offset by the back electron transfer from semiconductor to the oxidized dye (D^+) or to I_3^-, as shown in the following equations [17]:

$$e^-_{TiO_2} + D^+ \rightarrow D \tag{7}$$

$$2e^-_{TiO_2} + I_3^- \rightarrow 3I^- \tag{8}$$

The refilling of D^+ by the electrons from TiO_2 (reaction 7) is a major competitive process against the forward electron in reactions 3–5, in which it is desired for D^+ to be refilled by electrons donated from I^-. Thus, in an efficient DSSC, the loss of electrons in TiO_2 must be prevented so that electrons can accumulate in the TiO_2 to build-up the output voltage. This can be achieved in two ways with regard to the two possible electron leaking pathways in TiO_2. First, the dye regeneration via forward electron transfer in reactions 3–5 needs to be much faster than the back electron transfer from TiO_2 to D^+ (reaction 7). Transient absorbance measurements in the absence of I_3^- indicate that the back electron transfer from TiO_2

to D^+ (reaction 7) exhibits a strong dependence on the electron concentration in the TiO_2 [21]. Second, the back electron transfer from TiO_2 to I_3^- (reaction 8) needs to be suppressed. However, under high light intensity, the regeneration of the dye via reactions 3–5 can no longer compete effectively with reaction 7 due to following side reactions [22].

$$I_3^- + h\upsilon \rightarrow I_2^{-\bullet} + I^\bullet \tag{9}$$

$$I^\bullet + I^- \rightarrow I_2^{-\bullet} \tag{10}$$

The absorption band of I_3^-/I^- redox electrolyte overlaps with solar spectrum in the red and yellow regions (450–600 nm). It was reported that over 13% depletion in the photocurrent can occur when the concentration of triiodide ionic liquid electrolyte reaches 1 M, mainly resulting from strong photoexcitation of the triiodide (reactions 9–10) [23]. In TiO_2-based DSSCs, the diiodide radicals $I_2^{-\bullet}$ produced in reactions 9–10 are expected to accept electrons from the TiO_2 conduction band or the electron-collecting TCO anode, leading to an extra recombination pathway.

The device performance is determined by several key parameters. The overall efficiency (η) of a solar cell is calculated from $\eta = (FF \times |J_{sc}| \times V_{oc})/I$, where J_{sc} is the short-circuit photocurrent density, FF is the fill factor of the cell and I is the light intensity ($I = 0.1$ W cm^{-2} for one sun at AM 1.5G). In addition to maximizing the overlap of the absorption window of dyes and the solar spectrum through the discovery of new dyes, J_{sc} can also be improved by thickening the TiO_2 nanoparticle film for increased optical density. However, a dilemma arises from the fact that the electron diffusion length, typically 10 μm in TiO_2 nanoparticle networks, limits the useful TiO_2 nanoparticle film thickness.

As shown in Fig. 3, Ito et al. systematically studied the thickness dependence of TiO_2 nanoparticulate films with respect to each device performance parameter, including η, FF, J_{sc}, and V_{oc} [24]. It was revealed that by increasing the thickness of the TiO_2 film from ~5 to ~25 μm, the J_{sc} nearly monotonically increases by 96% until 15 μm and saturates or even slightly declines thereafter, presumably due to the increase in surface area. The open-circuit voltage, however, decreases monotonically with increasing film thickness. The fill factor also decreases when thickness increases. The η of the cell reflects the overall device performance and increases linearly with thickness increases until 15 μm, and saturates thereafter. The decrease in V_{oc} is associated with the increase of back electron transfer from the conduction band of the TiO_2 electrode to the I_3^- ions because thicker TiO_2 films have increased difficulty in the mass flow of the redox species in the electrolyte, leading to a high concentration of I_3^- ions in the film. Thus, it is suggested that optimum η can be attained with around a 15 μm TiO_2 thickness for N719 dye.

As a matter of fact, the initial transparent metal oxide nanoparticle film was an extremely successful starting point for DSSCs in terms of the simple device fabrication and low cost. However, despite the remarkable performance of nanoparticle films in conventional DSSCs, this photoanode geometry has several apparent disadvantages. The primary weakness of the nanoparticle-based

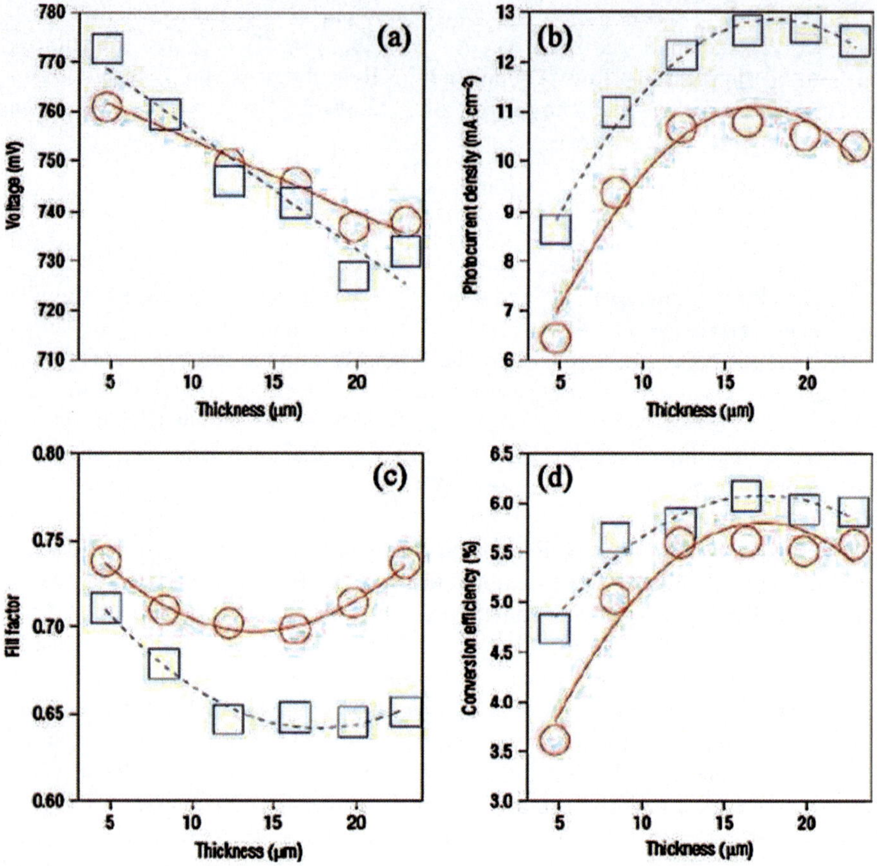

Fig. 3 Relationship between the photovoltaic characteristics of bifacial DSSCs with TiO_2/SiO_2 (3 mm thick) porous electrodes and the thickness of the porous TiO electrodes. Blue squares and red circles represent front- and rear-side irradiation, respectively. **a** Open-circuit photovoltage, V_{OC}. **b** Short-circuit photocurrent density, J_{SC}. **c** Fill factor, FF. **d** Photopower energy conversion efficiency, η. The lines represent best fit [24]. Copyright 2005, reprinted with permission from Nature Publishing Group

photoanode is the extremely slow electron diffusion coefficient, $D_n \leq 10^{-5}$ cm^2 s^{-1}, which is several orders of magnitude slower than that of TiO_2 and ZnO single crystals, as discussed previously. As a direct consequence of the slow electron transport in the TiO_2 nanoparticle layer, the kinetics redox mediator must be even slower to allow the successful traverse of the electrons across 10 ~ 14 um thick TiO_2 nanoparticle film in milliseconds However, the slow kinetics of the redox species also leads to the slow dye regeneration, which have to be compensated by the large driving force namely, the energy gap between dye HOMO and redox potential (refer to Fig. 2), leading to low V_{oc} (<0.8 V). In other words, the current state-of-the-art DSSCs achieve high photon-to-electron conversion efficiency, i.e., quantum yield at a significant loss of the photovoltage that is used as driving force for the regeneration of dyes. Thus, exploration of innovative photoanode architectures with prospect of enhanced electron transport kinetics dominates the current effort toward highly efficient DSSCs.

3 One-Dimensional Nanoarchitectured Photoanodes

Despite the remarkable performance and design breakthroughs in DSSCs using semiconducting nanoparticulate films as the photoelectrodes and iodide/triiodide as the redox shuttle, there is no doubt that additional efforts are desired to improve the parameters relevant to the efficiency. The incident photon-to-current efficiency (IPCE), a key metric of the overall device efficiency (η), is a combined measure of the efficiencies of three critical cell processes: light-harvesting efficiency (η_{LH}), charge separation efficiency (η_{CS}), and charges collection efficiency (η_{CC}) at the appropriate electrodes. IPCE is expressed by the following equation [3]:

$$\mathrm{IPCE} = \eta_{LH} \cdot \eta_{CS} \cdot \eta_{CC} \tag{11}$$

To achieve a high IPCE at a certain absorption wavelength, it first requires that nearly every photon of this wavelength is absorbed ($\eta_{LH} \sim 1$). Second, virtually each photoexcited dye molecule must inject an electron into the conduction band of the semiconductor ($\eta_{CS} \sim 1$). Finally, at least at short circuit, the injected electron that percolates the nanoparticle network must be havested at the transparent photoanode without being scavenged by the oxidized dye or nearby oxidizing redox species (I_3^-) in solution ($\eta_{CC} \sim 1$).

Enhancing η_{CC} is believed to be the most challenging part in this rather complicated photoelectrochemical system, since the journey of electrons traversing the photoanode can last milliseconds up to second, during which the electrons are exposed to various oxidizing species nearby. For comparison, the duration of electron-collecting time in silicon-based photovoltaics is typically in the range of tens of microseconds, and charge collection in efficient solid-state organic solar cells is on a similar time scale [25].

Despite the extremely slow nature of such trap-mediated diffusion-dominated transport, electron collection still remains favored over recombination because of the even slower multi-electron kinetics of I_3^- reduction on oxide surfaces [18]. Due to a simple kinetic competition between electron collection and its capture by I_3^-, IPCEs will be severely reduced when electron recombination is accelerated (lifetime shortened). It is not surprising that many efforts are directed toward developing novel photoanode architectures with the prospect of rapid charge collection. Those semiconducting nanostructures typically are one-dimensional semiconducting nanoarchitectures, including nanowires, nanotubes, or nanochannels that are aligned vertically with respect to the TCO glass [26]. Fabrication of these transformative electrode architectures has been supported by several attractive synthetic techniques; among them are (a) surfactant-controlled crystallization, (b) electrochemically controlled film anodization, and (c)

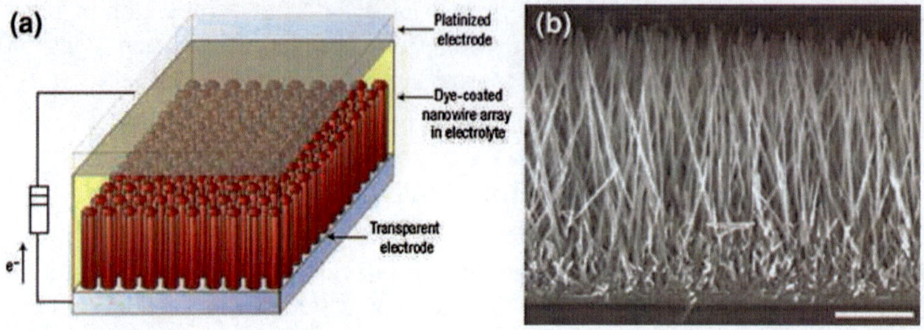

Fig. 4 The nanowire dye-sensitized cell, based on a ZnO wire array. **a** Schematic diagram of the cell. Light is incident through the bottom electrode. **b** Typicalscanning electron microscopy cross-section of a cleaved nanowire array on FTO. The wires are in direct contact with the substrate, with no intervening particle layer. Scale bar, 5 μm [27]. Copyright 2005, reprinted with permission from Nature Publishing Group

templated atomic-layer deposition (ALD) [26]. Pseudo 1-D nanostructures, particularly with single or polycrystalline domains have received the most intense investigations with the prospect of improved electron diffusion length because percolating electron transfer in 1-D nanostructures can follow a directed pathway with higher degree of order than the random fractal-like assembly of NP, which can lead to significant scattering of free electrons at the particle–particle interfaces.

3.1 Nanowires Photoanodes

One promising approach to boost the electron diffusion length in the anode is to substitute the nanoparticle film with an array of oriented, single-crystalline nanowires. Electron transport in single-crystalline wires is expected to be several orders of magnitude faster than percolation through a random polycrystalline network. In principle, when a sufficiently dense array of long, thin nanowires is used as a platform to load the dye molecules, it should increase the DSSC's dye-loading (and thus its absorption of red light) while simultaneously maintaining very efficient carrier collection. Furthermore, the rapid electron transport provided by a nanowire anode would be very favorable for cell designs that use non-standard electrolytes, such as polymer gels or solid inorganic phases, in which recombination rates are high compared with the liquid electrolyte cell.

A DSSC photoanode based on an array of aligned semiconducting nanowires was introduced in 2005 by Yang's group [27]. Figure 4 shows the ordered nanowire DSSC and illustrates how this topology can improve the understanding and performance of DSSCs and other types of excitonic solar cells.

The photoanode was prepared on a TCO glass substrate onto which a layer of ZnO NP were spin coated as seeds for the growth of ZnO nanowires. Preferential growth of the [0001] crystal face from solution affords fairly high aspect ratio single crystal nanowires perpendicular to the TCO. The diameters of the nanowires are 50–70 nm and the packing density of the nanowires is $\sim$35 nanowires/μm^2 [28]. In order to further enhance the anisotropic growth, poly(ethyleneimine) was incorporated in the growth solution, allowing nanorods with aspect ratios in exceeding 125. In addition, if the radius of the 1-D *n*-type nanowires is large enough, an upward band bending at the semiconductor surface can occur, which suppresses the adverse back electron transfers (charge recombination) [27, 29]. This is because the Fermi level of an *n*-type semiconductor is typically higher than the redox potential of the electrolyte. To equilibrate the two electron levels, electrons flow from the semiconductor into the electrolyte. As a result, there is a built-in circular electric field from the surface of the semiconductor nanowires toward their centers. This internal electric field pulls the injected electron toward the center of the wire and reduces the interception of the electrons by the electrolyte around the surface of the wire. The suppression of back electron transfer improves the current density of the cells.

However, compared to nanoparticle films (roughness factors >1000) the nanowires arrays are notably smooth (roughness factor <200), leading to a considerably decreasing J_{SC} and, therefore, reducing energy conversion efficiency to 1.5%, like the nanowire-based DSSC first reported [27]. In addition, the solution growth of nanowires has been limited primarily to ZnO. This is unfortunate because ZnO photoanodes show consistently lower performance than similar TiO_2 devices, owing primarily to the instability of ZnO in acidic dye solution. For this reason, one must reduce the dye-loading time on ZnO-based photoanode to lessen the dissolution of ZnO in the acidic Ruthenium dye solutions and the subsequent formation of the Zn^{2+}/dye complexes, which can diminish the light absorption [30–33]. Nonetheless, despite their modest success in DSSCs to date, nanowire photoanodes still possess several attractive features including low cost, scalability, and accelerated electron transport.

The instability of ZnO nanowires in acidic dye solution is perhaps the most inconvenient for the preparation of ZnO-based DSSC. Through a low temperature, nonpolar solvent/hydrophilic solid substrate interfacial reaction under hydrothermal conditions, Grimes' group reported the fabrication of densely packed single crystal TiO_2 nanowire arrays supported on FTO coated glass substrates (Fig. 5) [34]. Dye-sensitized solar cells fabricated using such nanowire arrays demonstrated very encouraging photoelectric conversion coefficiencies, 5.02% for a 2–3 μm long nanowire array when using N719 dye as solar absorber.

3.2 Nanotube Photoanodes

To conquer the apparent weakness of the reduced roughness factor found in nanowire arrays, a second route to low-dimension DSSC photoanodes has been established with nanotubes. Self-organized nanotube layers (NTs) provide

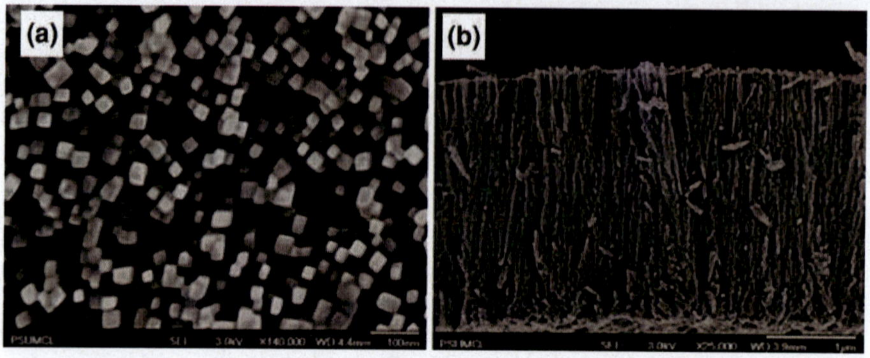

Fig. 5 FE-SEM images of vertically oriented self-organized TiO_2 nanowire array grown on FTO coated glass at 180°C for 24 h: **a** Top-view images; **b** Cross-sectional FE-SEM image of the same array, mechanically fractured [34]. Copyright 2008, American Chemical Society

considerably augmented surface roughness factors, mainly resulting from the inner surfaces of the nanotubes. The use of nanotubes for DSSCs provides a very high surface area with a continuous cylindrical morphology to minimize the number of detrimental grain boundaries that one electron must pass when traveling to the electron-collecting TCO anode.

3.2.1 TiO_2 Nanotubes

Grimes's group reported the use of highly-ordered TiO_2 nanotube-arrays (Fig. 6), made by potentiostatic anodization of a starting titanium film, in both front- and back-side illuminated DSSCs [35].

A key advantage of the self-aligned TiO_2 nanotube photoanode is its high degree of ordering. In addition, the structural parameters of the nanotubes architecture can be conveniently controlled by the selection of the electrochemical conditions, including the solution concentration, anodization voltage and duration, etc. In contrast to solution-phase nanorod growth, the electrochemical anodization of select metallic films produces an array of metal oxide nanotubes with a tunable roughness reaching over 1000.

Backside illuminated DSSCs using an array of nanotubes 6.2 μm long as the photoanode show an AM1.5 short-circuit current density of 10.6 mA cm^{-2}, 0.82 V open-circuit potential and a fill factor of 0.51, which yields a solar conversion efficiency of 4.4%. The front-side illuminated DSSCs, with a negative electrode consisting of a transparent nanotube-array of 3600 nm thick, showed an AM1.5 photocurrent of 10.3 mA cm^{-2}, 0.84 V open-circuit potential, 0.54 fill factor, and 4.7% efficiency. Open-circuit photovoltage decay measurements suggest superior electron transport in the titania nanotube-array-based DSSCs with an electron lifetime exceeding 1 s at the voltage of the maximum power point. Thus, it is possible to achieve greater efficiency by increasing the lengths of the

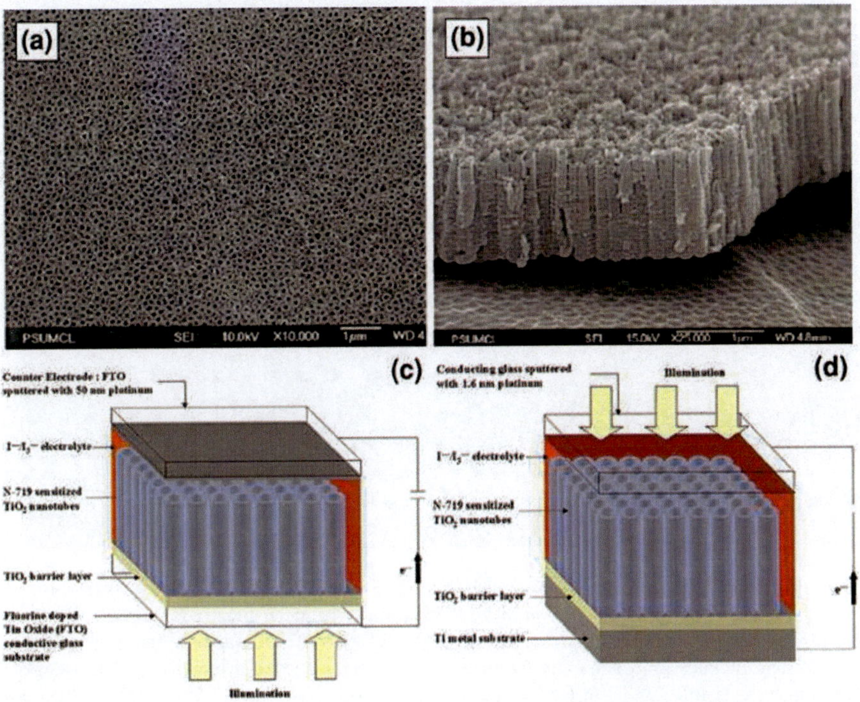

Fig. 6 FE-SEM images of illustrative TiO_2 nanotube-array: **a** Top view; and **b** Cross-sectional images from mechanically fractured samples, **c** and **d** Illustration of back-side illuminated nanotube-array dye solar cell architecture [35]. Copyright 2006, Institute of Physics

nanotubes in the array on the negative electrode since the amount of the absorbed dye appears to be the limiting factor.

In fact, Grimes et al. further demonstrated the synthesis of TiO_2 nanotube-arrays 1000 um in length with a free-standing nanotube-array membrane thickness of over 2 mm by anodic oxidation of a 1.0 mm thick Ti film as shown in Fig. 7 [36]. Such thick nanotube-array films can be readily transformed into mechanically robust membranes for DSSC applications. Because the nanotube has a morphology similar to that of a test tube with a closed bottom, there may be difficulty in effective dye-loading in the inner surface of the nanotubes due to the trapped air bubbles during the dye-loading. Plus, the long tubes correspondingly increase the transport length of redox shuttles. Therefore, further work is needed to solve these problems.

3.2.2 ZnO Nanotubes by Atomic-Layer Deposition

Using atomic-layer deposition, Hupp's group fabricated ZnO nanotubes by conformably coating anodic aluminum oxide (AAO) templates [26, 37–39]. In this technique, alternating exposures to reactive gas precursors deposit films of metal oxides, sulfides, or nitrides. Inert gas is used as purging gas between the precursor

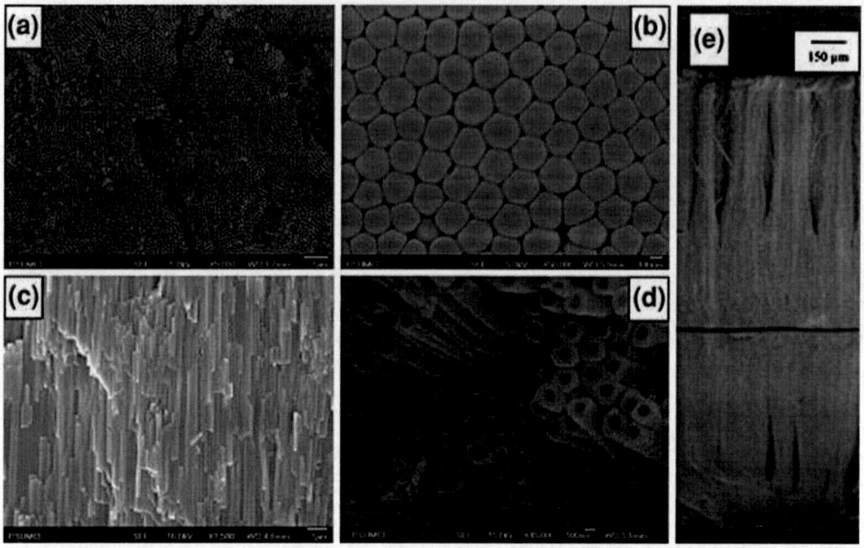

Fig. 7 FE-SEM images of (**a**) top side of TiO_2 nanotube-array film, (**b**) back-side or barrier layer side of nanotube-array film, and (**c, d**) cross-sectionalimage of mechanically fractured nanotube-array film showing its tubular nature. (**e**) Cross-sectional view of self-standing titania membrane over 2 mm in thickness, mechanically fractured for imaging, achieved by anodizing both sides of a 1.0 mm thick Ti foil sample at 60 V for 216 h in 0.6 wt% NH4F and 3.5% water in ethylene glycol. The membrane consists of two, back-to-back nanotube-arrays no less than 1000 μm in length [36]. Copyright 2007, American Chemical Society

gases. The self-limiting nature of the layer-by-layer growth technique makes ALD exceptionally suitable for high aspect ratio nanofabrication applications. Hupp's group reported a transparent array of polycrystalline ZnO nanotubes, the wall thickness of which can be controlled with angstrom resolution as shown in Fig. 8 [26]. An optimal ZnO tube thickness grown within and upon commercially available AAO yields DSSCs with 1.6% efficiency, limited primarily by the modest roughness factor of the commercial membranes (<450).

The fill factor (0.64) and V_{OC} (729 mV) of ZnO nanotube devices exceed those of any other ZnO photoanode reported in the literature [38, 39]. The effective diffusion coefficient for electrons in ZnO nanotube-array photoanodes was three orders of magnitude greater than in any other DSSC photoanode reported to date. Photovoltage decay measurements and analysis of IPCEs with increasing I_3^- concentration are consistent with the values of charge lifetime, τ_n, and charge collection time, τ_d, as derived from electrochemical impedance spectroscopy. Despite the relatively long lengths of the tubes constituting the photoanode (64 μm), electrons were efficiently and rapidly extracted throughout the anode. In addition, one can expect that efficient charge collection will be feasible in ZnO nanotube-based DSSCs that employ redox couples other than I_3^-/I^-.

The most remarkable advantage of the templated photoanodes grown by ALD is that a wide range of metal oxides can be synthesized. Thus, high surface area

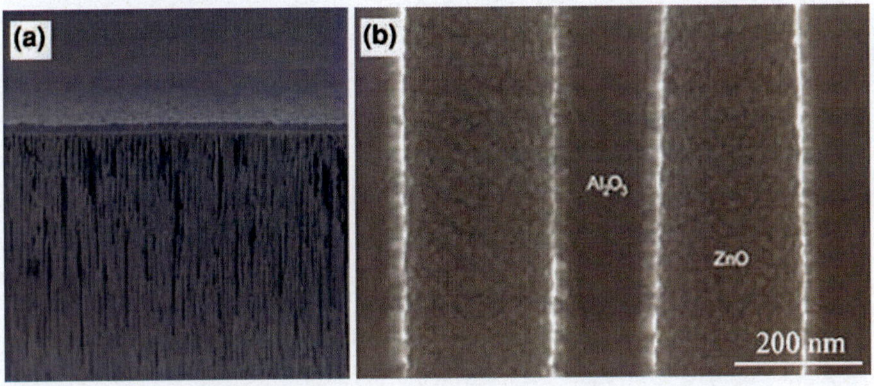

Fig. 8 **a** Cross-sectional SEM image of an anodic aluminum oxide template coated with ZnO by atomic-layer deposition. On average, the pores of the 64 μm thick membranes are 215 nm in diameter and spaced 330 nm on center. **b** Magnified Cross-sectional SEM image of commercial AAO membrane pores coated with 20 nm of ZnO by ALD [26]. Copyright 2008, with permission from Elsevier

photoanodes based on new semiconducting metal oxides with tunable band structures can be introduced to explore more basic science in DSSCs. In contrast, the hydrothermal growth of nanorod arrays has been reported for only a limited portion of metal oxides. Similarly, fabricating high surface area nanoparticle photoelectrodes of different metal oxides is a technical challenge that has not yet been overcome for several oxides of interest (e.g., NiO) [40]. In addition, ALD also opens a new window for multilayer metal oxide devices to be utilized in DSSCs. Due to the high reproducibility and well-controlled structural parameters of the nanoporous template, ALD is also particularly useful for precisely comparing semiconducting metal oxides independent of photoanode geometry and roughness, which are independently determined by the template.

A large effective surface area, which enables significant light absorption from moderate extinction dyes, is a prerequisite for efficient DSSCs. Nanowires, although in principle can have very high surface area mainly by increasing the length of nanorod, the collapse of the long nanowires becomes inevitable due to surface tension [41]. Similar to nanoparticle systems whose surface area can be readily tuned by the size of NP and the thickness of the film, the AAO templates allow for large and controllable surface areas. Unlike NP, however, AAO templates allow for independent control of the porosity, with 50% porous membranes feasible [42]. The large porosities and the straight pore direction may enable thick membranes to accommodate large current densities by overcoming the mass-transport of redox species that is apparently a limiting factor in thick nanoparticle films due to the highly disordered flow pathway and dead volumes. On the contrary, the mass-transport in the straight nanochannels in AAO will suffer less from the flow resistance. In addition, the alternative architectures are one dimensional, or pseudo-one dimensional, which should result in faster electron transport than in three-dimensional nanoparticle films. Finally, both hydrothermal growth and ALD are expected to yield more pristine semiconductors with larger polycrystalline domains compared to nanoparticle films, resulting in fewer surface trap states and grain boundaries or particle–particle junctions.

4 Advancing beyond Current 1-Dimensional Nanoarchitectured Photoanodes

Despite the tremendous work spent on developing new 1-D nanoarchitectured photoanodes to date, no DSSCs utilizing such an anode have achieved an efficiency exceeding those of conventional TiO_2 nanoparticle-based DSSCs. This is because many other device parameters are often interlinked, which can offset or reduce the improvements available through the new features. One particular problem is the diametric opposing effect resulting from increasing the length of the ZnO nanowires. On one hand, longer wires exhibit higher short-circuit current densities due to the increased surface area and thus higher dye-loading. On the other hand, longer wires lead to higher series resistance, lowering the fill factor. Narrower, thus denser, nanowires appear as a potential approach to overcome this problem. However, if the Debye–Hückel screening length exceeds the wire radius, reducing the diameter of the wires can eliminate the upward band bending at the wire's surface, an advantage of 1-D semiconductor is presented above. Typically, depending on the carrier density and the electrolyte, the width of the depletion layer can extend to tens of nanometers into the ZnO wires [27].

Another fundamental bottleneck that substantially impedes the advantages of 1-D semiconductor photoanodes is the slow hole transport by the redox species through mass-transport in the electrolyte. In all DSSCs using I^-/I_3^- as the redox shuttle, the cathode is essentially a planar-platinized TCO that is separated from the semiconductor layer by the electrolyte. Pt is an catalyst for efficient reduction of I_3^- to I^-. The Pt cathode is typically 20–40 μm from the top of the semiconductor layer, defined by a polymer spacer as sealer. The diffusion coefficient of I_3^- in the electrolyte is less than 10^{-4} cm^2 s^{-1} at room temperature [19], which is 2–3 orders of magnitude slower than the electron diffusion coefficient in 1-D semiconductor nanowires (>10^{-2} cm^2 s^{-1} for ZnO nanowires, for example). Thus, the synchronism of charge carrier (both electrons and holes) transport cannot be established. As a result, many adverse back electron transfers will take place, including the recombination of the electron and the semiconductors with D^+ and I_3^- as well as the formation of dye$^+$-I_3^- complex.

These dilemmas make it particularly necessary to explore further electrode architectures to drastically improve this fascinating photoelectrochemical device. The aims are (1) to increase the effective semiconductor thickness in the conventional nanoparticle-based DSSC without exceeding the electron diffusion length, (2) to increase the surface roughness of 1-D nanostructured semiconductors without significantly increasing the length, and thus the series resistance; (3) to alter the route of I^-/I_3^- diffusion pathways to coincide with the fast electron

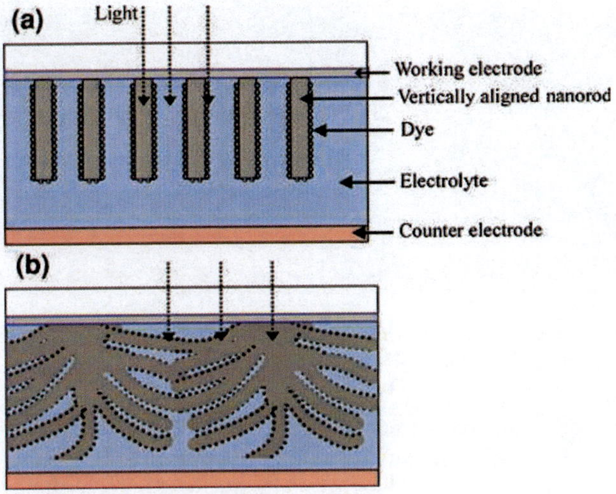

Fig. 9 Schematic of upright nanorod (**a**) and flower-like arrays under illumination (**b**) [43]. Copyright 2007, American Institute of Physics

transport in the 1-D ordered semiconductor-based photoanode; and (4) to enhance the interfacial rectifying effect to suppress the back electron transfer.

4.1 Branched One-Dimensional ZnO Nanowires

Besides maintaining the high electronic conductivity attributed to the high crystallinity in 1-D nanowire-based photoanode, renovation on the morphology of 1-D semiconductors also plays a very important role in enhancing dye-loading and light harvesting. In Fig. 9a and b schematically show two nanostructured 1-D arrays: upright nanorods and nanoflowers, respectively [43].

In principle, a portion of the incident photons could possibly fall on the voids between neighboring semiconducting nanorods, missing the chance to be absorbed by the dye molecules anchored to the semiconductor surface. Thus, the simple upright nanorod array may not be the optimal structure for light harvesting. Even with the light shining vertically on the nanorods, the absorption is not complete because the light might only pass through one thin layer of the dye-nanorod interface at the very apex of the rod. Therefore, the light loss may be significant for the upright ZnO nanorod arrays, i.e., low η_{LH}. In the nanoflower morphology as shown in Fig. 9b, the random branches of the nanoflowers provide a larger surface area resulting from the extra branches, and also an improved photon-dye interaction; meanwhile and more importantly, the nanoflowers maintain the good electron transportation featured by 1-D nanostructures.

Based on the above discussions, Jiang et al. reported a DSSC employing a ZnO-nanoflower film as the photoanode (Fig. 10), fabricated by hydrothermal

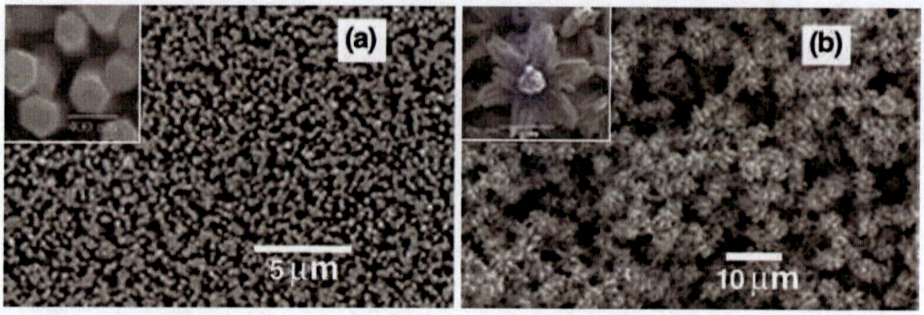

Fig. 10 SEM images of the (**a**) ZnO-nanorod array and (**b**) nanoflower film grown on FTO substrates by hydrothermal synthesis [43]. Copyright 2007, American Institute of Physics

decomposition and compared its performance to that of the DSSC with an upright ZnO-nanorod film as the photoanode [44]. The nanoflower photoanode exhibited enhanced dye-loading and light harvesting, while retaining good electron conductivity as in the upright nanorod photoanode. With the nanoflower ZnO photoanode, a DSSC with power conversion efficiency of 1.9% was attained, and a 90% enhancement over the control DSSC using the upright ZnO-nanorod array.

In addition, the nanoflower-based DSSC shows a greater fill factor (0.53) than that of the nanorod-based cell (0.36). The low fill factor of nanorod-based DSSC is attributed to the recombination of charges at the interface between ZnO nanorod and I^-/I_3^- electrolyte. As shown by the IV curves, the shunt resistance $dV/dIV = 0$ becomes larger for nanoflower-based cell than that for nanorod-based one, demonstrating the suppressed charge recombination in ZnO-nanoflower based DSSC.

Hsieh et al. reported the DSSC based on branched ZnO nanowires on conductive glass substrates via a solvothermal method (Fig. 11) [45].

The 1-D branched nanostructures provide a directed conduction pathway instead of a disordered interparticle hopping pathway as in nanoparticle-based DSSC. Furthermore, the short-circuit current density and the energy conversion efficiency of the branched ZnO nanowire DSSCs are 4.27 mA cm^{-2} and 1.51%, respectively, which are twice as high as those in bare ZnO nanowire devices. This improvement was a product of the enlargement of the internal surface area within the photoelectrode as suggested by the significantly enhanced IPCE spectra compared with the bare ZnO nanowires.

Overall, these 1-D branched nanostructures could simultaneously offer a direct conduction pathway and achieve higher dye adsorption to significantly enhance the overall energy conversion efficiency of the DSSCs.

4.2 Branched One-Dimensional TiO_2 Nanotubes

Self-organized TiO_2 NTs are used as photoanodes to enhance the light-to-electricity conversion efficiency predominantly due to a continuous cylindrical

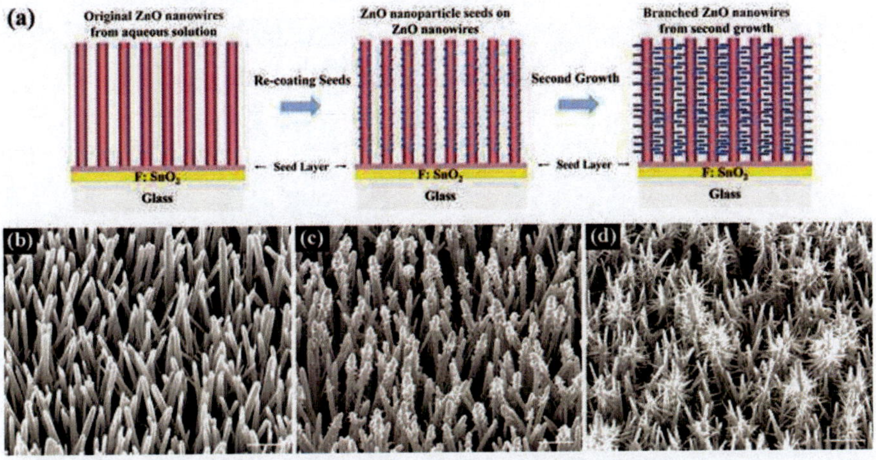

Fig. 11 **a** The schematic growth procedure from the original ZnO nanowires to the branched ZnO nanowires, **b** before and **c** after recoating a seed layer of the original ZnO nanowires obtained from a solvothermal method, and **d** the branched ZnO nanowires after second growth; scale bar, 1 μm [44]. Copyright 2008, American Chemical Society

morphology accompanied by reasonably high surface area. Frank's group reported that the light-harvesting efficiencies of NT-based DSSCs were higher than those of NP-based DSSCs owing to stronger internal light-scattering effects inside the tubes [45]. Such improvement agrees well with the three key factors for efficient light harvesting and conversion, as mentioned in the beginning of this section, including (i) high surface area of nanostructured oxide films for dye adsorption, (ii) highly efficient electron transfer from photoexcited dyes to the conduction band of TiO_2, and (iii) the fast forward transport of charge carriers in the conduction band of semiconductor and in the redox electrolyte without a minimized recombination loss of charge carriers.

Attempts to improve the TiO_2 nanotube geometry toward higher conversion efficiency have been reported. An effective improvement was achieved by reducing disorder at the junction of the semiconductor nanotube-TCO anode that is associated with the electrochemical growth of nanotubes in nonaqueous electrolytes. An increase in the efficiency from 1.6 to 1.9% under AM1.5G illumination was reported [46].

Schmuki et al. showed that a new version of stratified TiO_2 nanotubes with a bamboo-type morphology, in which the nanotubes are interlinked with nodes as found in bamboo [47]. This feature can considerably boost the conversion efficiency of nanotube-based DSSCs. Figure 12 schematically illustrates the different generations of solar cells from the conventional nanoparticle-based DSSC to the nanotube-based and to the branched-NT-based solar cells [47].

Like conventional smooth-walled tubes, the bamboo-type TiO_2 nanotubes were prepared electrochemically via controlled anodization of Ti in an electrolyte consisting of HF. The nodes that connect the nanotubes, however, were formed

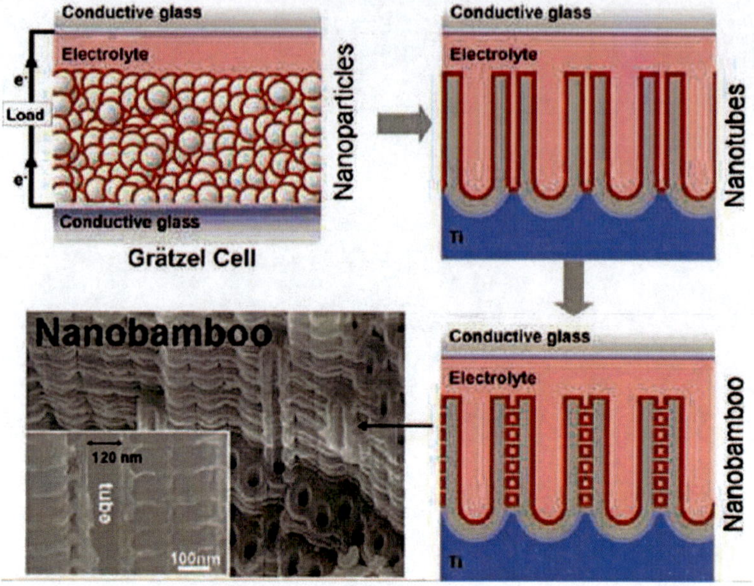

Fig. 12 Schematic representation of dye-sensitized TiO_2 solar cells based on nanoparticle (NP), nanotube (NT), and bamboo-type nanotube (B-NT) electrodes. The inset shows a tilted SEM image of the nanobamboo morphology and a high magnification inset shows the inner side of the B-NT structure open over the entire length [47]. Copyright 2008, American Chemical Society

under appropriate alternating voltage cycling, which features a switch between growth conditions for tube and for a compact layer. Essentially, at each voltage step a compact connecting layer is formed in between the tubes and the bamboo-type structure can be grown.

The amorphous structure of the as-anodized self-organized TiO_2 nanotubes can be converted into the anatase structure upon annealing in air at 450°C. DSSCs prepared by using bamboo-type TiO_2 nanotubes as photoanodes exhibit notable enhancement over regular TiO_2 nanotubes in terms of J_{sc} and, thus, efficiency η ($\sim$3%), while V_{oc} and FF are similar. Hence, the key improvement can be attributed to the extra surface area for dye-loading provided by the bamboo rings.

4.3 *One-Dimensional Semiconductor on a Rough Metallic Electron-Collecting Anode*

It would be a formidable challenge to retain the state-of-the-art strategy of enhancing the electron transfer afforded in ordered 1-D semiconductor structures, but simultaneously reducing the series resistance of the semiconductor without a significant loss of surface roughness compared to 1-D nanowires. Rather than

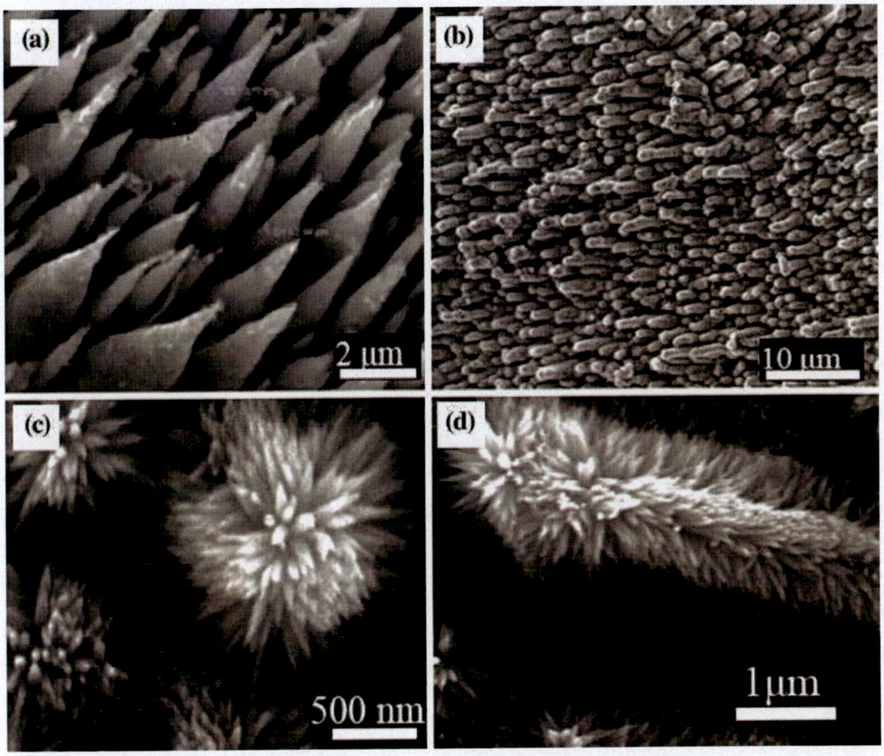

Fig. 13 **a** High magnification SEM image of Zn microtip array formed by anodization of a Zn foil; **b** A representative large-area SEM image of ZnO-nanotips grown on an array of Zn-microtips; **c** A magnified SEM top-view image of ZnO-nanotips grown on Zn-microtips; **d** A magnified SEM top-view image of the ZnO-nanotips grown on the Zn-microtips [48]. Copyright 2009, American Chemical Society

placing all the roughness onto the semiconductor layer, Xu's group reported a strategy to allocate part of the roughness onto the collecting anode [48]. A rough collecting anode provides a larger surface area to accommodate more individual 1-D nanoscale semiconductors.

Hence, for the same surface roughness factor (SRF, defined as the ratio of the total semiconductor surface area to the nominal planar substrate area), the length of the 1-D semiconductor on a coarsened anode can be shorter than that of conventional 1-D semiconductors supported on a planar anode. Xu et al. demonstrated this strategy by employing a Zn-microtip/ZnO-nanotip core–shell hierarchical nano-architecture as the anode/semiconductor component of the DSSC (Fig. 13) [48].

Figure 14 illustrates the schematic diagram of a DSSC based on the Zn-microtip/ZnO-nanotip core–shell hierarchy nano-architectured electrode, and its *J–V* characteristics under one Sun (AM1.5G) illumination.

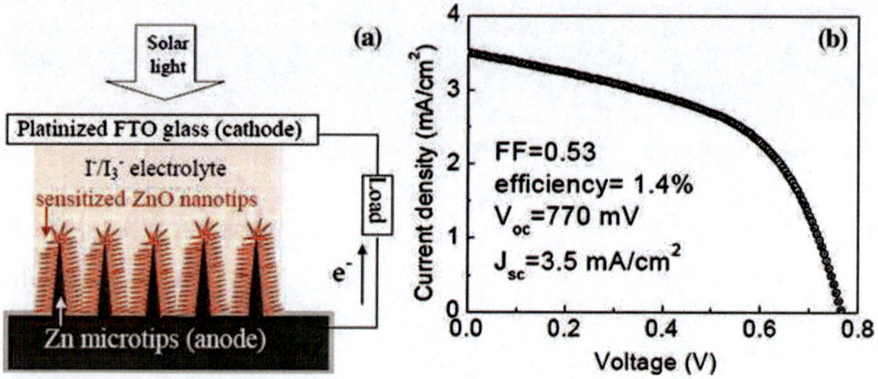

Fig. 14 **a** A schematic diagram of the DSSC based on Zn-microtip/ZnO-nanotip core–shell hierarchy nano-architectured electrode; **b** A representative current density–voltage (J–V) curve of the DSSC using Zn-microtip|ZnO-nanotip as the anode|semiconductor component [48]. Copyright 2009, American Chemical Society

The measured J_{sc} for the Zn-microtip/ZnO-nanotip nanoarchitecture-based DSSC was 3.5 mA μm^{-2}. This value is close to that of the planar anode-supported ZnO nanowires with the same SRF, which indicates that the dye-loading amounts (1.1×10^{-8} mol cm^{-2}) are comparable [27].

Remarkably, two key parameters of this device, the fill factor and open-circuit voltage show pronounced improvements over those of DSSCs based on planar anode-supported ZnO nanowires. The fill factors previously reported for DSSCs based on planar anode-supported ZnO nanowires are typically between 0.3 and 0.4 [27, 43, 44, 49, 50]. In contrast, the fill factor of this device is consistently greater than 0.5. This trend agrees with the reported enhancement of the fill factor through the reduction of the series resistance in ZnO nanowires using interconnected ZnO nanowires. For the same roughness factor (SRF = 60), the series resistance of shot ZnO-nanotips ($\sim$0.3 μm) is greatly reduced due to the much shorter tip length compared to the planar supported ZnO nanowires (6–7 μm). This 20-fold reduction in the length of ZnO-nanotips results in a series resistance $R_s = dV/dJ_{(J=0)}$ as low as 28 Ωcm^2. For the same SRF, this R_s value is significantly lower than that of DSSCs based on planar anode-supported ZnO nanowires, which fall in the 46–106 Ωcm^2 range [44, 49]. Therefore, the enhanced fill factor can be partially contributed to the reduced series resistance in the shorter ZnO-nanotips.

The V_{oc}'s of the Zn-microtip/ZnO-nanotip DSSCs were measured consistently around 770 mV, greater than any literature reported values for long ZnO nanowire-based DSSCs, which are typically below 700 mV. The maximum value of V_{oc} is determined by the energy gap between the quasi-Fermi level in ZnO under illumination and the redox potential of the I^-/I_3^- electrolyte [51]. The cause of the enhancement of V_{oc} in our Zn-microtip/ZnO-nanotip DSSC can be elucidated by comparing its energy diagram with that of a planar supported ZnO nanowire DSSC. Figure 15a is the highly idealized energy diagram of a planar FTO anode-supported ZnO nanowire-based DSSC.

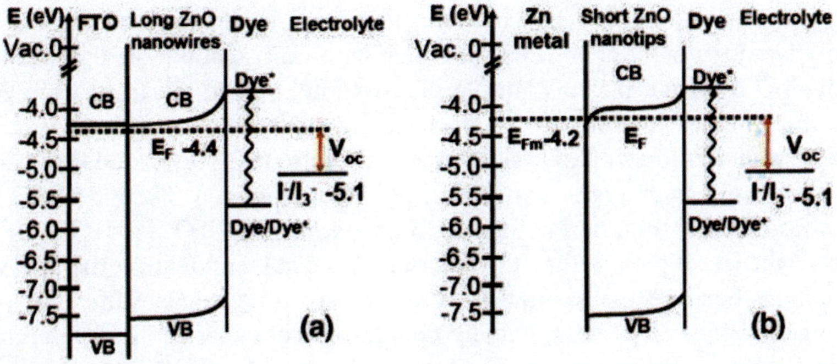

Fig. 15 Highly idealized energy diagrams of DSSCs using **a** a planar FTO-supported long ZnO nanowires as the anode/semiconductor component; **b** Zn-microtip/ZnO-nanotip as the anode/semiconductor component [48]. Copyright 2009, American Chemical Society

The upward band bending at the ZnO-electrolyte interface is due to the higher Fermi level of *n*-type ZnO compared to the redox potential of the electrolyte, and hence electrons will flow from the electrode into the electrolyte until equilibrium is established [29]. Consequently, there is a positive space charge region in the center of the wires, which are reflected by the upward bending at the wire surface. This electron depletion layer in ZnO nanowires can extend into the semiconductor a few nanometers up to tens of nanometers, depending on the carrier density and the dielectric constant of the semiconductor, as well as the electrolyte. For example, even at a fairly high-carrier density of 10^{18} cm^{-3}, the diameter of a ZnO nanowire should at least be 20 nm to attain a potential drop of 0.2 V cross the depletion layer in a carbonate propylene electrolyte (0.1 M $LiClO_4$) [29].

Figure 15b is the energy diagram of the DSSC using the Zn-microtip/ZnO-nanotip core–shell nano-architecture as the anode/semiconductor component. Considering that a large portion of a ZnO nanotip has a radius above 20 nm, an upward band banding should also exist at the ZnO-nanotip/electrolyte interface, similar to the case of the planar anode-supported ZnO nanowires. Besides, an added feature in this device is a metal/*n*-type semiconductor Ohmic junction established at the interface between Zn and ZnO-nanotips. This is because the Fermi level of Zn (−4.2 eV) [52, 53], is higher than that of ZnO (−4.4 eV, average value reported as the Fermi level for ZnO) [54–56]. To achieve thermal equilibrium in this junction, electrons must flow from the Zn metal into the lower energy states of the conduction band in the *n*-type semiconductor, i.e., ZnO, which makes the ZnO in the junction region more n-type.[17] As a consequence, the Fermi level of the entire ZnO-nanotip is raised since the length of our ZnO-nanotips is so short that a considerable section of the wire is affected by the electron accumulation layer. The width of the junction (w) can be estimated by $w = \sqrt{2\varepsilon_0\varepsilon_r V_b/(eN_d)}$ [57], where ε_0 is the vacuum permittivity, ε_r (=10) is the dielectric constant of ZnO nanowires [29], V_b is the amount of band bending, assumed to be 0.2 V, e is the elementary charge, and N_d is the carrier density of ZnO, and we assume $N_d = 10^{17}$ cm^{-3}. Thus, w is estimated to be approximately 50 nm. As a result, the V_{oc}, that is, the energy gap between the quasi-Fermi level of ZnO and the electrolyte redox potential is widened, which takes into account the upshifted band edge of ZnO by about 0.15 eV due to the accumulation of electrons from the Zn metal. In contrast, if there is an electron accumulation layer at the interface between the much longer ZnO nanowires and the planar FTO anode, it can only affect a relatively small portion (at the root of the nanowires). Therefore, this effect is not considered in Fig. 6a for long ZnO nanowires on FTO.

Kieven et al. reported that the shorter the ZnO nanowires, the better the photovoltage because the shorter electron collection distance suffers less from the competing kinetic process, namely charge recombination [58]. Therefore, the enhanced open circuit voltage in this device can also be contributed from the fact that the short ZnO nanotips provide a short electron collection distance for the metal, Zn. Such a short collection distance makes charge transport much more favorable than the charge recombination process.

The realigned band structure at the metal–semiconductor interface affects the Fermi level of the semiconductor and provides us with a potential opportunity to improve the open-circuit voltage of the cells, a key parameter governing the attainable power from a DSSC.

4.4 Semimetallic One-Dimensional Photoanodes

While employing of highly-ordered self-organized anodic TiO_2 anotube as photoanodes for DSSCs, the inadequate conductivity due to the semiconducting nature of TiO_2 impedes its broader and more efficient use as photoanode with prospect of a fast electron transport, a crucial factor to achieve high efficient energy conversion. Schmuki et al. demonstrated the fabrication of robust semimetallic TiO_2 nanotubes to overcome this problem [59]. They adopted carbon-thermal reduction treatment that converts TiO_2 into an oxy carbide compound with significantly enhanced conductivity.

First, the well-established self-organizing electrochemical anodization method is used to produce TiO_2 NTs as described previously. Then, the TiO_2 nanotubes were treated in acetylene at a high temperature to trigger the desired carbonization reaction, which converts the semiconducting anatase phase into carbon-rich Magnéli-type phases with pronouced semimetallic conductivity. This entire synthetic process can be processed without loss of the ordered nanotubular morphology, as illustrated in Fig. 16 [59].

Solid-state electrical measurements, as shown in Fig. 17, unambiguously show a notable enhancement in the conductivity of the TiO_xC_y nanotubes in comparison to C-doped TiO_2 nanotubes and undoped anatase nanotubes. The carbon-modified nanotubes exhibit an Ohmic I–V behavior, for which the resistance is in orders of magnitude lower than that for the carbon-doped or pure-anatase TiO_2 nanotubes. A rough estimate yields approximate conductivity values of 3×10^5 S m^{-1} for TiO_xC_y, 9×10^4 S m^{-1} for C-doped TiO_2, and 10^3 S m^{-1} for anatase TiO_2 nanotubes.

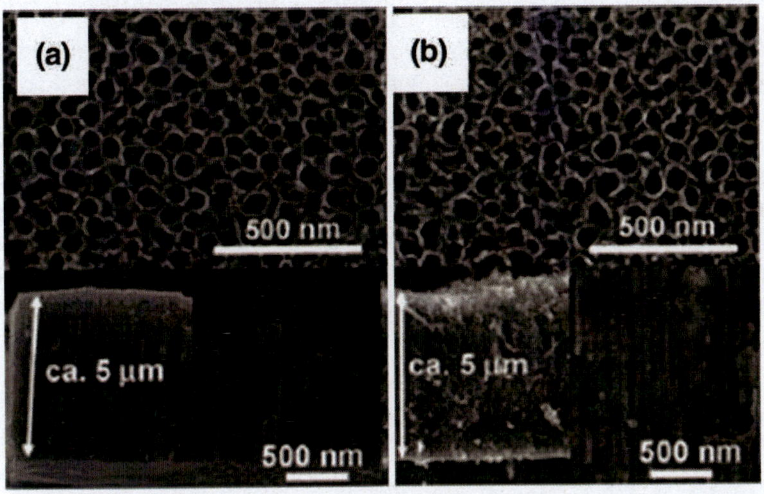

Fig. 16 SEM images of the nanotubes after **a** anodization formation and **b** after a thermal acetylene treatment at 8508C for 10 min. The lower insets are the corresponding cross-sectional views [59]. Copyright 2009, Wiley–VCH Verlag GmbH & Co. KGaA. Reproduced with permission

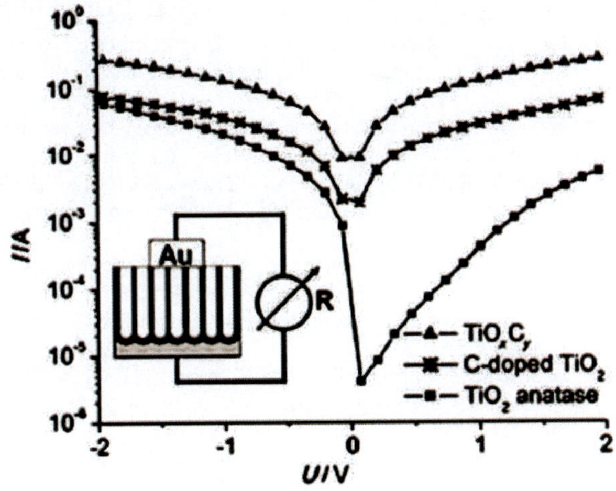

Fig. 17 I–V characteristics of 2-point solid-state measurements of carbonized tubes and of anatase TiO_2 nanotube layers and C-doped layers. Inset: a simplified sketch of the 2-point measurement arrangement [59]. Copyright 2009, Wiley–VCH Verlag GmbH & Co. KGaA. Reproduced with permission

5 Conclusion and Outlook

Since O'Regan and Grätzel reported highly efficient TiO_2 nanoparticle-based DSSCs, many efforts have been dedicated to explore transformative photoanodes with enhanced electron transport through innovations in morphology and interface engineering of the photoanodes. However, compared to the 11% overall energy conversion efficiency in the conventional configuration of TiO_2 nanoparticle-based DSSCs, no new photoanode has produced a device with an efficiency above this record value. Variations in the photoanode have, instead of improving the efficiency, lead to degraded overall photovoltaic performance due to many antithetic coupling among various energetic and kinetic parameters of the device. In order to push DSSCs toward practical photovoltaics, it is necessary to alter not only the photoanode, but also other major device components simultaneously, including the solar absorbing dyes and the redox shuttles, which may potentially lead to DSSCs with efficiencies exceeding 16% [60]. For example, for I^-/I_3^- based redox species, future effort should also be conducted toward exploring new cathode architectures that can provide alternative pathways to facilitate the mass flow of redox species so that the transportation of holes can be enhanced to match the fast electron transport found in the 1-D nanoarchitectured photoanodes.

References

1. Basic Research Needs for Solar Energy Utilization, DoE Report of Basic Energy Sciences Workshop on Solar Energy Utilization August 18–21, 2005
2. O' Regan B, Grätzel M (1991) A low-cost, high-efficiency solar cell based on dye-sensitized colloidal TiO_2 films. Nature 353:737–740
3. Nazeeruddin MK, Kay A, Rodicio I et al (1993) Conversion of light to electricity by cis-X2bis(2, 2'-bipyridyl-4, 4'-dicarboxylate)ruthenium(II) charge-transfer sensitizers (X = Cl-, Br-, I-, CN-, and SCN-) on nanocrystalline titanium dioxide electrodes. J Am Chem Soc 115:6382–6390
4. Nazeeruddin MK, De Angelis F, Fantacci S et al (2005) Combined experimental and DFT-TDDFT computational study of photoelectrochemical cell ruthenium sensitizers. J Am Chem Soc 127:16835–16847
5. Chiba Y, Islam A, Watanabe Y et al. (2006) Dye-sensitized solar cells with conversion efficiency of 11.1%. Jap J Appl Phys 45: L638–L640
6. Grätzel M (2003) Applied physics—solar cells to dye for. Nature 421:586–587
7. Gratzel M (2005) Solar energy conversion by dye-sensitized photovoltaic cells. Inorg Chem 44:6841–6851
8. Gratzel M (2005) Dye-sensitized solid–state heterojunction solar cells. MRS Bull 30:23–27
9. Fisher AC, Peter LM, Ponomarev EA et al (2000) Intensity dependence of the back reaction and transport of electrons in dye-sensitized nanocrystalline TiO_2 solar cells. J Phys Chem B 104:949–958
10. Oekermann T, Zhang D, Yoshida T, Minoura H (2004) Electron transport and back reaction in nanocrystalline TiO_2 fi lms prepared by hydrothermal crystallization. J Phys Chem B 108:2227–2235
11. Nelson J (1999) Continuous-time random-walk model of electron transport in nanocrystalline TiO_2 electrodes. Phys Rev B 59:15374–15380
12. van de Lagemaat J, Frank AJ (2001) Nonthermalized electron transport in dye-sensitized nanocrystalline TiO_2 films: transient photocurrent and random-walk modeling studies. J Phys Chem B 105:11194–11205
13. Kopidakis N, Schiff EA, Park NG et al (2000) Ambipolar diffusion of photocarriers in electrolyte-fi lled, nanoporous TiO_2. J Phys Chem B 104:3930–3936
14. Benkstein KD, Kopidakis N, van de Lagemaat J (2003) Influence of the percolation network geometry on electron transport in dye-sensitized titanium dioxide solar cells. J Phys Chem B 107:7759–7767
15. Kopidakis N, Benkstein KD, van de Lagemaat J et al (2003) Transport-limited recombination of photocarriers in dye-sensitized nanocrystalline TiO_2 solar cells. J Phys Chem B 107: 11307–11315

16. Kavan L, Grätzel M, Gilbert SE et al (1996) Electrochemical and photoelectrochemical investigation of single-crystal anatase. J Am Chem Soc 118:6716–6723
17. Peter L (2009) "Sticky electrons" transport and interfacial transfer of electrons in the dye-sensitized solar cell. Acc Chem Res 42:1839–1847
18. Boschloo G, Hagfeldt A (2009) Characteristics of the iodide/triiodide redox mediator in dye-sensitized solar cells. Acc Chem Res 42:1819–1826
19. Asano T, Kubo T, Nishikitani Y (2005) Short-circuit current density behavior of dye-sensitized solar cells. Jpn J Appl Phys 44:6776–6780
20. Zistler M, Wachter P, Wasserscheid P et al (2006) Comparison of electrochemical methods for triiodide diffusion coefficient measurements and observation of non-stokesian diffusion behaviour in binary mixtures of two ionic liquids. Electrochim Acta 52:161–169
21. Haque SA, Tachibana Y, Klug DR (1998) Charge recombination kinetics in dye-sensitized nanocrystalline titanium dioxide films under externally applied bias. J Phys Chem B 102: 1745–1749
22. Roy JC, Hamill WH, Williams RR (1955) Diffusion kinetics of the photochemical and thermal dissociation-recombination of trihalide ions. J Am Chem Soc 77:2953–2957
23. Kubo W, Kambe S, Nakade S et al (2003) Photocurrent-determining processes in quasi-solid-state dye-sensitized solar cells using ionic gel electrolytes. J Phys Chem B 107:4374–4381
24. ITO S, Zakeeruddin SM, Comte P et al (2008) Bifacial dye-sensitized solar cells based on an ionic liquid electrolyte. Nature Photonics 2:693–698
25. Li G, Shrotriya V, Huang JS et al (2005) High-efficiency solution processable polymer photovoltaic cells by self-organization of polymer blends. Nat Mater 4:864–868
26. Martinson ABF, Hamann TW, Pellin MJ et al (2008) New architectures for dye-sensitized solar cells. Chem Eur J 14:4458–4467
27. Law M, Greene LE, Johnson JC et al (2005) Nanowire dye-sensitized solar cells. Nat Mater 4:455–459
28. Greene LE, Law M, Goldberger J et al (2003) Low-temperature wafer-scale production of ZnO nanowire arrays. Angew Chem Int Ed 42:3031–3034
29. Mora-Seró I, Fabregat-Santiago F, Denier B et al (2006) Determination of carrier density of ZnO nanowires by electrochemical techniques. Appl Phys Lett 89:203117
30. Guillén E, Casanueva F, Anta JA et al (2008) Photovoltaic performance of nanostructured zinc oxide sensitised with xanthene dyes. J Photochem Photobio A Chem 200:364–370
31. Keis K, Lindgren J, Lindquist SE et al (2000) Studies of the adsorption process of Ru complexes in nanoporous ZnO electrodes. Langmuir 16:4688–4694
32. Horiuchi H, Katoh R, Hara K (2003) Electron injection efficiency from excited N3 into nanocrystalline ZnO films: effect of (N3 – Zn^{2+}) aggregate formation. J Phys Chem B 107:2570–2574
33. Sayama K, Tsukagoshi S, Hara K et al (2002) Photoelectrochemical properties of j aggregates of benzothiazole merocyanine dyes on a nanostructured TiO_2 film. J Phys Chem B 106: 1363–1371
34. Feng X, Shankar K, Varghese OK et al (2008) Vertically aligned single crystal TiO_2 nanowire arrays grown directly on transparent conducting oxide coated glass: synthesis details and applications. Nano Lett 8:3781–3786
35. Paulose M, Shankar K, Varghese OK et al (2006) Application of highly-ordered TiO_2 nanotube-arrays in heterojunction dye-sensitized solar cells. J Phys D Appl Phys 39:2498–2503
36. Paulose M, Prakasam HE, Varghese OK et al (2007) TiO_2 nanotube arrays of 1000 μm length by anodization of titanium foil: phenol red diffusion. J Phys Chem C 111:14992–14997
37. Martinson ABF, Elam JW, Hupp JT et al (2007) ZnO nanotube based dye-sensitized solar cells. Nano Lett 7:2183–2187
38. Martinson ABF, Elam JW, Liu J et al (2008) Radial electron collection in dye-sensitized solar cells. Nano Lett 8:2862–2866
39. Martinson ABF, Goes MS, Fabregat-Santiago F (2009) Electron transport in dye-sensitized solar cells based on ZnO nanotubes: evidence for highly efficient charge collection and exceptionally rapid dynamics. J Phys Chem A 113:4015–4021

40. Irwin MD, Buchholz DB, Hains AW (2008) *p*-Type semiconducting nickel oxide as an efficiency-enhancing anode interfacial layer in polymer bulk-heterojunction solar cells. Proc Natl Acad Sci 105:2783–2787
41. Liang Y, Zhen C, Zou D et al (2004) Preparation of free-standing nanowire arrays on conductive substrates. J Am Chem Soc 126:16338–16339
42. Ko S, Lee D, Jee S et al (2006) Mechanical properties and residual stress in porous anodic alumina structures. Thin Solid Films 515:1932–1937
43. Jiang CY, Sun XW, Lo GQ et al (2007) Improved dye-sensitized solar cells with a ZnO-nanoflower photoanode. Appl Phys Lett 90:263501
44. Cheng HM, Chiu WH, Lee CH et al (2008) Formation of branched ZnO nanowires from solvothermal method and dye-sensitized solar cells applications. J Phys Chem C 112: 16359–16364
45. Zhu K, Neale NR, Miedaner A et al (2007) Enhanced charge-collection efficiencies and light scattering in dye-sensitized solar cells using oriented TiO_0 nanotube arrays. Nano Lett 7:69–74
46. Zhu K, Vinzant TB, Neale NR et al (2007) Removing structural disorder from oriented TiO_2 nanotube arrays: reducing the dimensionality of transport and recombination in dye-sensitized solar cells. Nano Lett 7:3739–3754
47. Kim D, Ghicov A, Albu SP et al (2008) Bamboo-type TiO_2 nanotubes: improved conversion efficiency in dye-sensitized solar cells. J Am Chem Soc 130:16454–16455
48. Yang Z, Xu T, Ito Y et al (2009) Enhanced electron transport in dye-sensitized solar cells using short ZnO nanotips on a rough metal anode. J Phys Chem C 113:20521–20526
49. Du Pasquier A, Chen HH, Lu YC (2006) Dye sensitized solar cells using well-aligned zinc oxide nanotip arrays. Appl Phys Lett 89:253513
50. Chen HH, Du Pasquier A, Saraf G et al (2008) Dye-sensitized solar cells using ZnO nanotips and Ga-doped ZnO films. Semicond Sci Technol 23:045004
51. Peter LM (2007) Characterization and modeling of dye-sensitized solar cells. J Phys Chem C 111:6601–6612
52. Anderson PA (1940) The contact difference of potential between barium and zinc the external work function of zinc. Phys Rev 57:122–127
53. Reinaudi L, DelPopolo M, Leiva E (1997) Work function calculation for thick metal slabs with local pseudopotentials. Surf Sci 372:L309–L314
54. Sun Z, Wang C, Yang J, Zhao B, Lombardi JR (2008) Nanoparticle metal—semiconductor charge transfer in ZnO/PATP/Ag assemblies by surface-enhanced Raman spectroscopy. J Phys Chem C 112:6093–6098
55. Katoh R, Furube A, Barzykin AV, Arakawa H, Tachiya M (2004) Kinetics and mechanism of electron injection and charge recombination in dye-sensitized nanocrystalline semiconductors. Coord Chem Rev 248:1195–1213
56. Kamiya T, Tajima K, Nomura K, Yanagi H, Hosono H (2008) Interface electronic structures of zinc oxide and metals: first-principle study. Physica Status Solidi (a) 205:1929–1933
57. Benda V, Gowar J, Grant DA (1999) Power semiconductor devices: theory and applications. Wiley, New York, pp 62–65
58. Kieven D, Dittrich T, Belaidi A, Tornow J, Schwarzburg K, Allsop N, Lux-Steiner M (2008) Effect of internal surface area on the performance of ZnO/In2S3/CuSCN solar cells with extremely thin absorber. Appl Phys Lett 92:153107
59. Hahn R, Schmidt-Stein F, Salonen J, Thiemann S, Song Y, Kunze J, Lehto V, Schmuki P (2009) Semimetallic TiO_2 nanotubes. Angew Chem Int Ed 48:7236–7239
60. Hamann TW, Jensen RA, Martinson ABF, Ryswyk HV, Hupp JT (2008) Advancing beyond current generation dye-sensitized solar cells energy. Environ Sci 1:66–78

Dye-Sensitized Solar Cells Using Natural Dyes and Nanostructural Improvement of TiO_2 Film

Shoji Furukawa

Abstract The characteristics of the dye-sensitized solar cells using natural dyes, such as those of red-cabbage, curcumin, and red-perilla, and synthesized dyes, such as NKX-2553, NKX-2677, and D149, in which a precious metal is not contained, will be explained. The largest conversion efficiency obtained is over 1% for the dye-sensitized solar cell using the dye of red-cabbage. This value is relatively small. However, the cost performance (defined by [conversion efficiency]/[cost of dye]) is more than 50 times greater than that of the dye-sensitized solar cell using Ruthenium complex. Therefore, when the cost of FTO and ITO substrates, oxide semiconductor, electrolyte solution, and opposite electrode becomes very low, dye-sensitized solar cells fabricated using natural dyes may become more pervasive, although the physical device becomes larger than one using a Ruthenium complex. The effects of pH of the dye solution on the characteristic of the dye-sensitized solar cells will be also described. The conversion efficiencies of the dye-sensitized solar cells using the dye of red-cabbage, red-perilla, NKX-2553, and NKX-2677 become larger when the pH value is low. It is expected that this technique will be used in future dye-sensitized solar cell systems.

1 Introduction

Clean energy has become a focal point for many researchers. Among the clean energy, solar cells are quite promising device because their cost is relatively low. However, traditional crystalline silicon solar cells require much energy in order to

S. Furukawa (✉)
Graduate School of Computer Science and Systems Engineering,
Kyushu Institute of Technology, Kitakyushu-shi 804-8550, Japan
e-mail: furukawa@cse.kyutech.ac.jp

L. Zang (ed.), *Energy Efficiency and Renewable Energy Through Nanotechnology*,
Green Energy and Technology, DOI: 10.1007/978-0-85729-638-2_8,

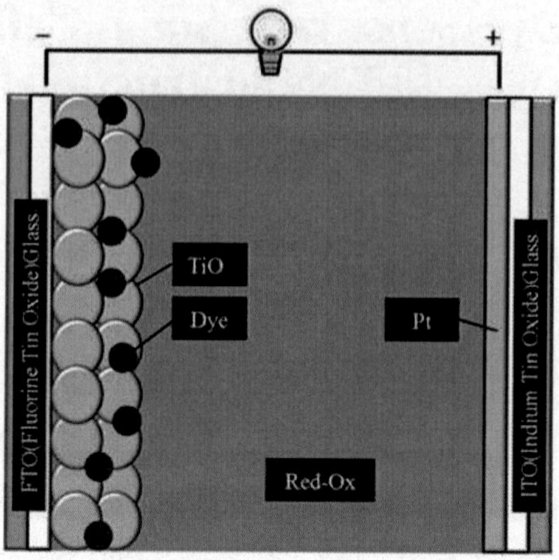

Fig. 1 Fundamental structure of dye-sensitized solar cell

fabricate the wafers. Therefore, recently, dye-sensitized solar cells are expected as cheap energy conversion devices [1, 2].

Most researchers working in the field of dye-sensitized solar cells are using a Ruthenium complex as the dye because such devices can have a high conversion efficiency. To date, the highest conversion efficiency obtained experimentally is over 12%. However, Ruthenium is a rare metal, so its cost is very high.

Our group has fabricated various kinds of dye-sensitized solar cells using natural dye, such as those of red-cabbage, curcumin, and red-perilla [3–5]. The cost of these natural dyes is very low. The largest conversion efficiency of dye-sensitized solar cells using these dyes is about 1%, which is smaller than solar cells fabricated using a Ruthenium complex. However, the cost performance (defined by [conversion efficiency]/[cost of dye]) of the dye-sensitized solar cell using the dye of red-cabbage is more than 50 times larger than that of the dye-sensitized solar cell fabricated using a Ruthenium complex. Therefore, when the cost of FTO substrate, oxide semiconductor, electrolyte solution, and opposite electrode becomes very low, dye-sensitized solar cells using natural dyes may be a more cost effective solution, although such devices will be physically larger than those using a Ruthenium complex.

In this chapter, the characteristics of the dye-sensitized solar cells using natural dyes and synthesized dyes in which precious metal is not contained are explained. The effects of pH of the dye solution on the characteristic of the dye-sensitized solar cell are also described.

2 Structure of Dye-Sensitized Solar Cell

Figure 1 shows the fundamental structure of a dye-sensitized solar cell. A glass sheet coated by the transparent and electrically conductive film (FTO) is used as the cathode. On the cathode, an oxide semiconductor film, such as TiO_2, is coated and the dye is adsorbed onto its surface. The anode is prepared by sputtering Pt film onto another ITO substrate. Then, the electrolyte solution is inserted between the cathode and anode.

When sunlight irradiates the cathode substrate, most of it propagates to the dye. This is because glass is transparent to visible light, and the band gap of the oxide semiconductor is large. Then, the electronic states of the electrons in the dye molecules jumps to higher levels due to energy absorbed from the sunlight. The excited electrons transfer to the conduction band of the oxide semiconductor, then move through the transparent oxide to the cathode. After arriving at the conducting film on the anode, the electrons move to the dye surface through the iodine ions in the electrolyte solution and back to the dye, creating electric current and completing the circuit.

3 Fabrication Process for Dye-Sensitized Solar Cells

In order to fabricate the dye-sensitized solar cells, an oxide semiconductor paste was prepared [6–11]. In our laboratory, two kinds of TiO_2 were used in fabrication of the paste: P-25, with an average particle size of 26 nm, and PC-101, with an average particle size of 20 nm. The two kinds of TiO_2, P-25 (0.14 g) and PC-101 (0.06 g), were first placed in a shaker with HNO_3 (0.6 ml). Then, polyethylene glycol (PEG) was also added to the shaker in order to obtain a porous oxide semiconductor film. The characteristics of the porous oxide semiconductor film depend on the molecular weight of PEG. In our work, two kinds of PEGs were used: PEG with a molecular weight of 500,000 and PEG with a molecular weight of 2,000,000. The TiO_2 paste was coated onto the FTO glass substrate, which had been cleaned by ethyl alcohol and acetone. Then, the substrate was placed in an electric heater, and annealed at 450°C for 1 h.

After depositing the porous oxide semiconductor film onto the FTO glass substrate, the dye was adsorbed onto the oxide semiconductor film surface by dipping the substrate into the dye solution. In our study, six types of dyes were used. They were dyes of red-cabbage, curcumin, red-perilla, NKX-2553, NKX-2677, and D149. The former three dyes are natural and the latter three are synthetic. All dyes contain no precious metal. The chemical structures of the natural dyes are shown in Fig. 2, and the molecular structures of NKX-2553, NKX-2677, and D149 are shown in Fig. 3. The molar density of the red-cabbage, curcumin, and red-perilla dye solutions was 0.5 mM. The solvent of the solutions of red-cabbage and red-perilla was water, and the solvent for the curcumin solution was ethyl alcohol.

The electrolyte solution was prepared using I_2, LiI, 1-propyl-2,3 dimethylimidazolium iodide (DMPImI), 4-*tert*-butyl pyridine (TBP), and propylene carbonate (PC). The weights of I_2, LiI, DMPImI, TBP, and PC were 0.0317, 0.330, 0.342, 0.189, and 3.00 g, respectively. The electrolyte solution was inserted between the two electrodes.

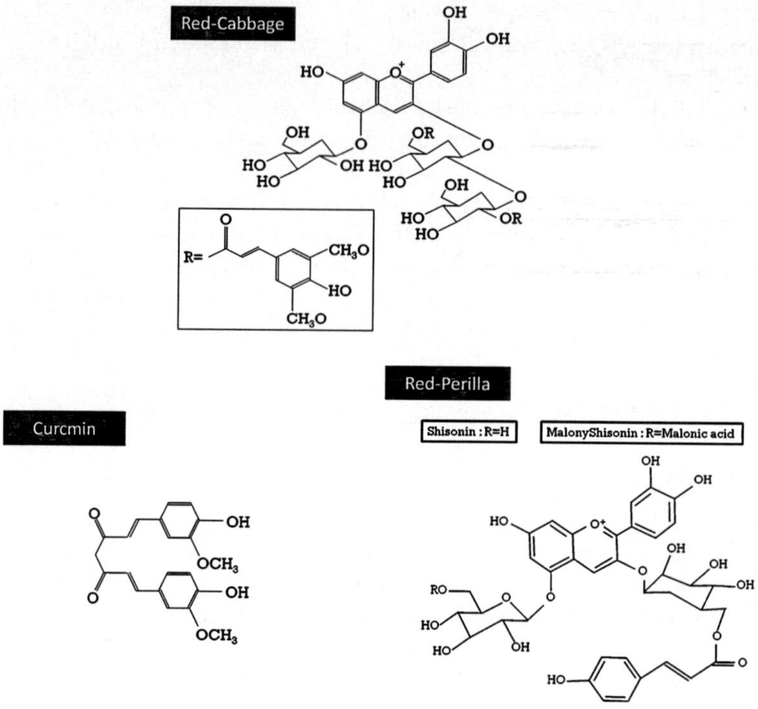

Fig. 2 Chemical structures of the dyes from red-cabbage, curcumin, and red-perilla

Photocurrent-voltage characteristics of the dye-sensitized solar cells were measured using a halogen lamp as a light source. The power of the light source was 50 mW/cm^2. The conversion efficiency obtained using the halogen lamp was about 5% larger than that obtained while using the standard light source that was constructed in our laboratory. In our standard light source system, visible and infrared components were partly eliminated by an optical filter.

4 Characteristics of Dye-Sensitized Solar Cells Using Natural Dyes

4.1 Characteristics of Dye-Sensitized Solar Cells using Dye of Red-Cabbage

Figure 4 shows the photocurrent-voltage characteristics of the dye-sensitized solar cells using the dye of red-cabbage. Five solar cells have been fabricated, and the numbers in Fig. 4 correspond to the sample number. In this case, the oxide semiconductor paste was prepared using the PEG with the molecular weight of 500,000.

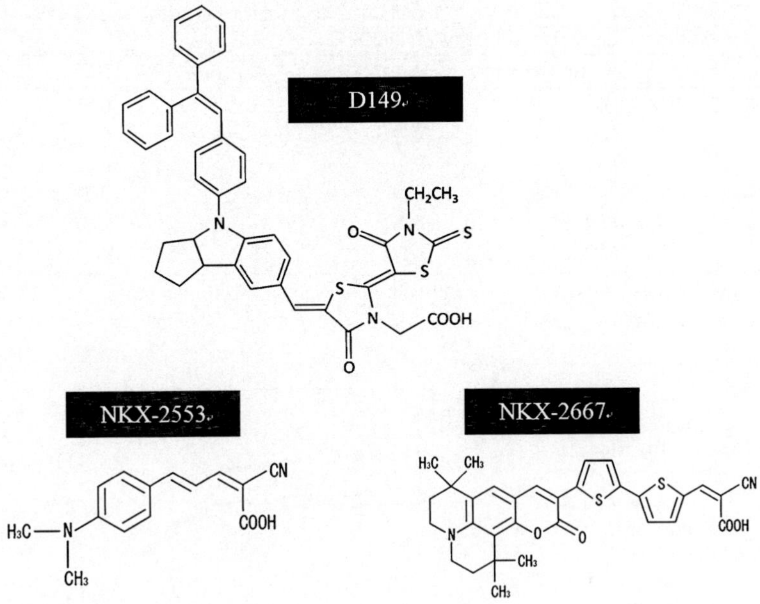

Fig. 3 Molecular structures of NKX-2553, NKX-2677, and D149

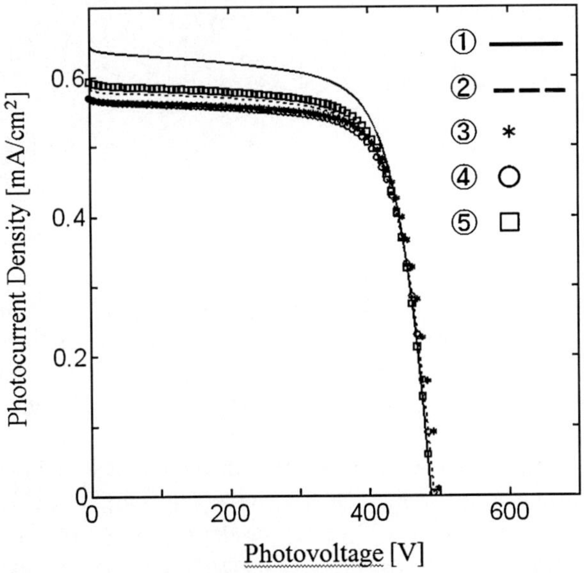

Fig. 4 Photocurrent-voltage characteristics of the dye-sensitized solar cells using dye of red-cabbage. The numbers correspond to the sample number. The oxide semiconductor paste was fabricated using the PEG whose molecular weight was 50,000

Table 1 shows the summary of the characteristics of the solar cells fabricated using the dye of red-cabbage (molecular weight of PEG is 500,000). The average fill factor was 0.72, and the average open circuit voltage was 0.49 V. In this case, the average conversion efficiency was 0.42%.

Table 1 Summary of the characteristics of the solar cells fabricated using dye of red-cabbage and the PEG with the molecular weight of 500,000

Sample number	J_{SC} (mA/cm^2)	V_{OC} (V)	P_{MAX} (mW)	F.F.	η (%)
Sample 1	0.64	0.48	0.22	0.70	0.44
Sample 2	0.59	0.49	0.20	0.71	0.41
Sample 3	0.57	0.50	0.20	0.72	0.41
Sample 4	0.57	0.49	0.20	0.72	0.40
Sample 5	0.59	0.48	0.21	0.72	0.42
Average	0.59	0.49	0.21	0.72	0.42

J_{SC} short circuit current density; V_{OC} open circuit voltage; P_{MAX} maximum power; *F.F.* fill factor; η conversion efficiency

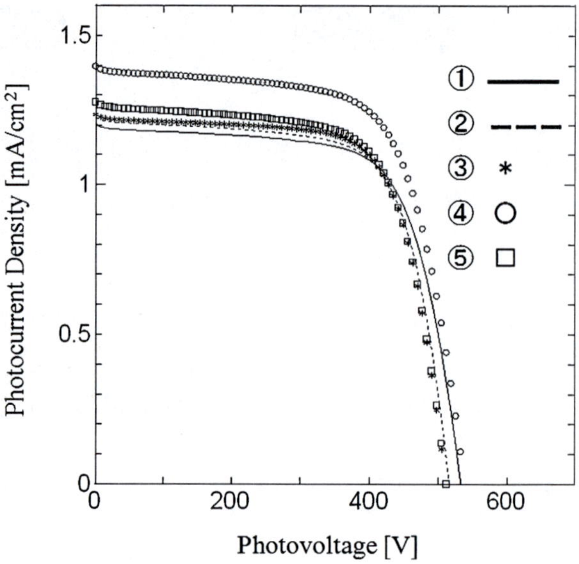

Fig. 5 Photocurrent-voltage characteristics of the dye-sensitized solar cells using dye of red-cabbage. The numbers written in the figure are the sample numbers. In this case, the oxide semiconductor paste was fabricated using the PEG whose molecular weight was 2,000,000

Similar to Figs. 4, 5 show the photocurrent-voltage characteristics of the dye-sensitized solar cells using the dye of red-cabbage. In this case, the oxide semiconductor paste was fabricated using the PEG whose molecular weight was 2,000,000. Five solar cells were fabricated and the numbers written in Fig. 5 correspond to the sample number.

Table 2 shows the summary of the characteristics of the solar cells fabricated using the dye of red-cabbage (molecular weight of PEG is 2,000,000). In this case, the average short circuit current density is approximately twice that of the device fabricated with the PEG with the smaller molecular weight. The average conversion efficiency also doubled to 0.90%.

In order to investigate the difference in conversion efficiencies between the two devices, the oxide semiconductor surfaces were characterized by scanning electron microscopy (SEM). This study showed that the pores were uniformly dispersed in the oxide semiconductor film prepared using the PEG with a molecular weight of

Table 2 Summary of the characteristics of the solar cells fabricated using dye of red-cabbage and the PEG whose molecular weight is 2,000,000

Sample number	J_{SC} (mA/cm^2)	V_{OC} (V)	P_{MAX} (mW)	F.F.	η (%)
Sample 1	1.2	0.53	0.44	0.68	0.87
Sample 2	1.2	0.51	0.44	0.69	0.87
Sample 3	1.2	0.50	0.44	0.70	0.88
Sample 4	1.4	0.53	0.50	0.67	1.00
Sample 5	1.3	0.51	0.44	0.68	0.89
Average	1.3	0.52	0.45	0.69	0.90

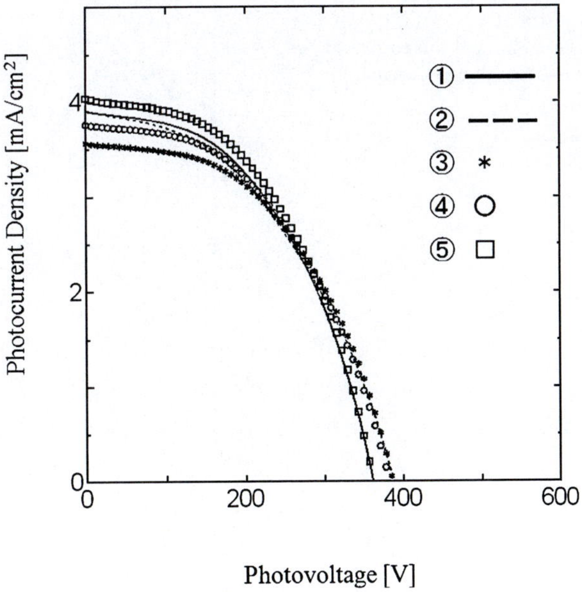

Fig. 6 Photocurrent-voltage characteristics of the dye-sensitized solar cells using dye of red-cabbage with an improved electrolyte solution

2,000,000. On the other hand, some large particles were observed in the oxide semiconductor film prepared using the PEG with a molecular weight of 500,000. This change in morphology is hypothesized to be the cause of the difference in efficiencies between the two devices.

In order to improve the characteristics of the dye-sensitized solar cells using the dye of red-cabbage, the electrolyte solution was changed [12]. The photocurrent-voltage characteristics of the solar cells fabricated using the dye of red-cabbage are shown in Fig. 6. Table 3 shows the summary of the characteristics of these devices (the pH is controlled by adding citric acid) with the improved electrolyte solution. The highest conversion efficiency was approximately 1.4%. The improved electrolyte solution is considered to be the cause of the improved performance.

Table 3 Summary of the characteristics of the dye-sensitized solar cells fabricated using red-cabbage dye solutions (whose pH is controlled by adding citric acid) with improved electrolyte solution

Sample number	J_{SC} (mA/cm^2)	V_{OC} (V)	P_{MAX} (mW)	F.F.	η (%)
Sample 1	3.9	0.39	0.66	0.44	1.3
Sample 2	4.1	0.36	0.71	0.49	1.4
Sample 3	3.6	0.39	0.67	0.49	1.3
Sample 4	3.9	0.36	0.68	0.48	1.4
Sample 5	3.8	0.38	0.68	0.47	1.4
Average	3.9	0.37	0.68	0.47	1.4

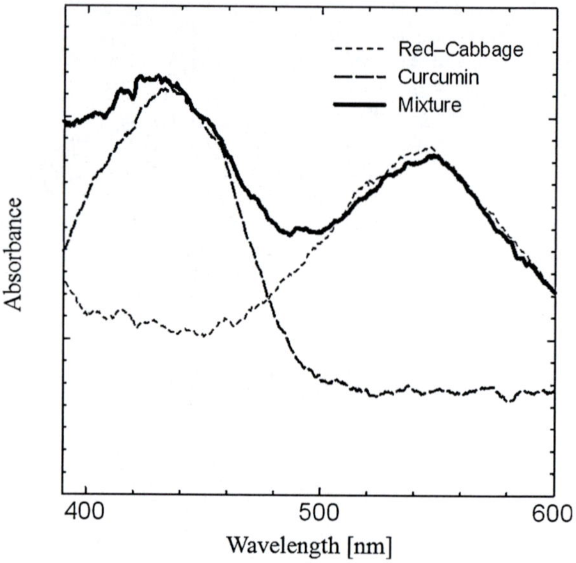

Fig. 7 Optical absorption spectra of the red-cabbage, curcumin, and mixed dye of red-cabbage and curcumin solutions

4.2 Characteristics of Dye-Sensitized Solar Cells Using Mixed Dye of Red-Cabbage and Curcumin

Figure 7 shows the optical absorption spectra of the red-cabbage, curcumin, and the mixture of the two dyes in solution. As shown in Fig. 7, the dyes of red-cabbage and curcumin have absorption peaks at about 550 and 440 nm, respectively. When the both dyes are mixed, optical absorption occurs in wide band, as shown in Fig. 7. Therefore, it is expected that the conversion efficiency of the solar cell increases when using the dye mixture.

Figure 8 shows the photocurrent-voltage characteristics of the dye-sensitized solar cells fabricated using the dye of red-cabbage, curcumin, and mixed dye of red-cabbage and curcumin. As shown in Fig. 8, the characteristics of the dye-sensitized solar cell using the mixed dye of red-cabbage and curcumin is better than those fabricated using the dye of red-cabbage or curcumin.

Fig. 8 Photocurrent-voltage characteristics of the dye-sensitized solar cells using the dye of red-cabbage, curcumin, and mixed dye of red-cabbage and curcumin

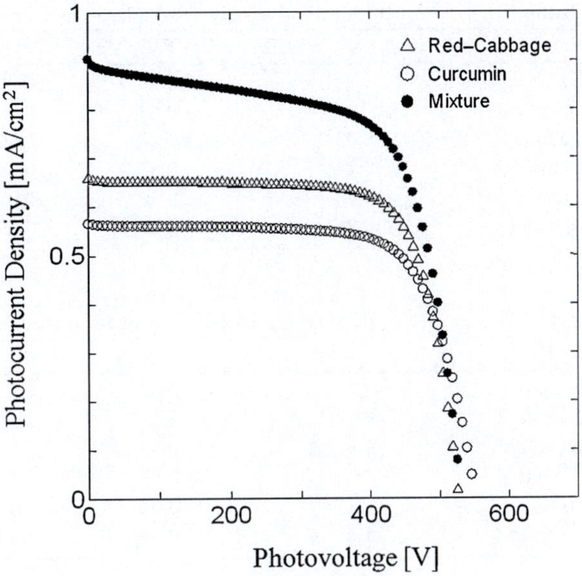

Table 4 Summary of the characteristics of the dye-sensitized solar cells fabricated using the dye of red-cabbage, curcumin, and mixed dye of red-cabbage and curcumin

Red-cabbage:curcumin	J_{SC} (mA/cm^2)	V_{OC} (V)	P_{MAX} (mW)	F.F.	η (%)
1:0	0.69	0.52	0.25	0.70	0.50
100:1	0.73	0.52	0.27	0.71	0.53
90:1	0.78	0.50	0.27	0.69	0.54
80:1	0.78	0.50	0.27	0.69	0.54
70:1	0.81	0.53	0.30	0.69	0.60
60:1	0.74	0.52	0.27	0.71	0.54
50:1	0.74	0.50	0.25	0.69	0.51
0:1	0.53	0.53	0.20	0.72	0.41

Table 4 shows the summary of the characteristics of the dye-sensitized solar cells using the dyes of red-cabbage, curcumin, and the mixture. The conversion efficiency of the dye-sensitized solar cell using the mixed dye of red-cabbage and curcumin becomes greater than those of the solar cells fabricated using the dye of red-cabbage or curcumin.

Our recent data showed a large conversion efficiency of about 1.3% for the dye-sensitized solar cell fabricated using the dye of curcumin with the improved electrolyte solution [12].

4.3 Characteristics of Dye-Sensitized Solar Cells Using Dye of Red-Perilla

Table 5 shows the characteristics of the dye-sensitized solar cells using the dye of the red-perilla. The average conversion efficiency was 0.17%, which is smaller

Table 5 Characteristics of the dye-sensitized solar cells fabricated using the dye of red-perilla

Sample number	J_{SC} (mA/cm^2)	V_{OC} (V)	P_{MAX} (mW)	F.F.	η (%)
Sample 1	0.20	0.48	0.06	0.63	0.12
Sample 2	0.27	0.48	0.09	0.68	0.17
Sample 3	0.33	0.48	0.10	0.65	0.21
Sample 4	0.24	0.53	0.09	0.70	0.18
Sample 5	0.26	0.50	0.09	0.68	0.18
Average	0.26	0.49	0.09	0.67	0.17

Table 6 Summary of the characteristics of the dye-sensitized solar cells using NKX-2553

Sample number	J_{SC} (mA/cm^2)	V_{OC} (V)	P_{MAX} (mW)	F.F.	η (%)
Sample 1	4.2	0.63	1.5	0.56	3.0
Sample 2	4.3	0.63	1.5	0.54	2.9
Sample 3	4.3	0.63	1.5	0.54	2.9
Sample 4	4.3	0.63	1.5	0.55	3.0
Sample 5	4.7	0.62	1.6	0.54	3.2
Average	4.4	0.63	1.5	0.55	3.0

than that of the solar cell fabricated using the dye of the red-cabbage. As shown in Fig. 2, the chemical structure of the main part of the red-perilla dye is the same as that of the red-cabbage dye. However, the chemical structure of other part is different. The OH groups in the dyes are believed to react with the OH groups on the oxide semiconductor film surface. This reaction causes H_2O and covalent bonding between the dye and the oxide semiconductor via oxygen atom. The number of OH groups of the dye of red-perilla is smaller than that of the dye of red-cabbage (see Fig. 2). This may cause a lower conversion efficiency of the solar cell using the dye of red-perilla.

5 Characteristics of Dye-Sensitized Solar Cells Using Synthetic Dyes

In this section, the characteristics of dye-sensitized solar cells using the synthetic dyes containing no precious metal, such as Ruthenium, will be presented.

5.1 Characteristics of Dye-Sensitized Solar Cells Using NKX-2553

Table 6 shows the summary of the characteristics of the dye-sensitized solar cells using NKX-2553. Five solar cells were fabricated. The average conversion efficiency obtained was 3.0%, which is larger than that of the solar cells fabricated

Table 7 Summary of the characteristics of the dye-sensitized solar cells using NKX-2677

Sample number	J_{SC} (mA/cm^2)	V_{OC} (V)	P_{MAX} (mW)	F.F.	η (%)
Sample 1	5.9	0.61	1.8	0.51	3.7
Sample 2	5.7	0.61	1.8	0.51	3.5
Sample 3	5.8	0.61	1.8	0.51	4.0
Sample 4	6.0	0.60	1.7	0.47	3.4
Sample 5	5.4	0.61	1.8	0.55	3.7
Average	5.8	0.61	1.8	0.51	3.6

using natural dyes. However, the cost of the synthesized dyes is high, so the cost performances of the dye-sensitized solar cells fabricated using synthetic dyes is smaller than those of the solar cells fabricated using natural dyes.

5.2 *Characteristics of Dye-Sensitized Solar Cells Using NKX-2677*

Table 7 shows the summary of the characteristics of the dye-sensitized solar cells using NKX-2677. Five solar cells were fabricated. The average conversion efficiency obtained was 3.6%, again larger than those found in solar cells fabricated using natural dyes or NKX-2553. Similar to the solar cell using NKX-2553, the cost performance is smaller than those of the solar cells fabricated using natural dyes.

5.3 *Characteristics of Dye-Sensitized Solar Cells using Mixture of NKX-2553 and D149*

Figure 9 shows the optical absorption spectra of NKX-2553, D149, and the mixed dye of NKX-2553 and D149. The ratio of the mixture is 1:1. The vertical axis indicates normalized absorbance.

As shown in Fig. 9, NKX-2553 and D149 have absorption peaks at about 450 nm and 540 nm, respectively. By mixing NKX-2553 and D149, the absorption in the range from about 400 to 500 nm increases, suggesting an increase of conversion efficiency of the solar cell.

In our study, the FTO substrate on which the oxide semiconductor film was fabricated was dipped into the dye solution, and various dye-sensitized solar cells were fabricated by changing dipping time. Figure 10 shows the relation between conversion efficiency and dipping time obtained at 12, 40, 60, and 80°C. The ratio of the mixture of NKX-2553 and D149 is 1:1. As shown in Fig. 10, the maximum conversion efficiency was obtained at a short dipping time when the dipping temperature was high. However, the largest conversion efficiency was obtained at a low temperature of 12°C. The largest conversion efficiency was 5.4%.

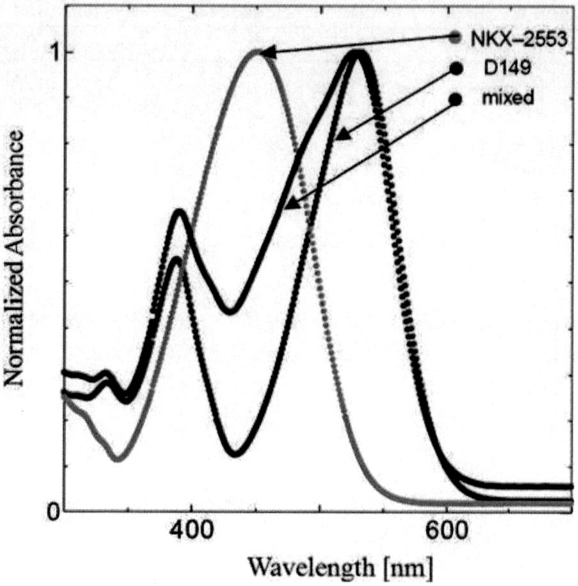

Fig. 9 Optical absorption spectra of NKX-2553, D149, and the mixed dye of NKX-2553 and D149. The ratio of the mixture is 1:1. The vertical axis indicates normalized absorbance

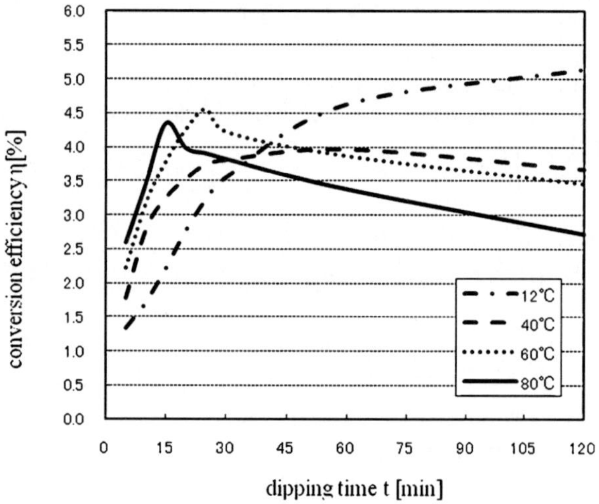

Fig. 10 Relation between conversion efficiency and dipping time obtained at 12, 40, 60, and 80°C. The ratio of the mixture of NKX-2553 and D149 is 1:1

6 Effects of pH of Dyes on Characteristics of Dye-Sensitized Solar Cells

6.1 Effects of pH of Dye on Characteristics of Dye-Sensitized Solar Cells Using Dye of Red-Cabbage

Figure 11 shows the optical absorption spectra of various red-cabbage dye solutions in which the pH is controlled by adding citric acid or sodium hydroxide.

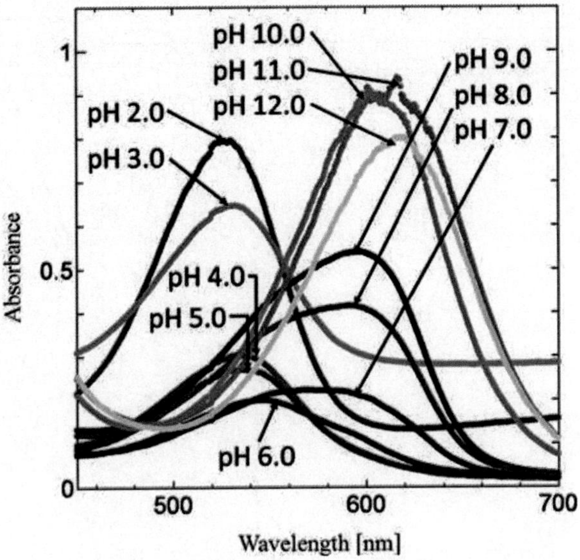

Fig. 11 Optical absorption spectra of various red-cabbage dye solutions whose pH is controlled by adding citric acid or sodium hydroxide

Table 8 Summary of the characteristics of the dye-sensitized solar cells fabricated using various red-cabbage dye solutions in which the pH is controlled by adding citric acid or sodium hydroxide

pH	J_{SC} (mA/cm^2)	V_{OC} (V)	P_{MAX} (mW)	F.F.	η (%)
2.0	2.7	0.38	0.56	0.56	1.1
3.0	2.4	0.39	0.52	0.57	1.0
4.0	2.1	0.39	0.49	0.59	0.97
5.0	2.0	0.40	0.44	0.55	0.89
6.0	1.4	0.42	0.34	0.59	0.68
7.0	0.92	0.42	0.22	0.57	0.45
8.0	0.91	0.42	0.22	0.57	0.43
9.0	0.96	0.42	0.25	0.61	0.50
10.0	0.93	0.44	0.26	0.63	0.51
11.0	0.38	0.47	0.09	0.50	0.19
12.0	0.15	0.44	0.03	0.45	0.06

In the case of pH = 7.0, no addition was made. As shown in Fig. 11, the strength of the absorption peak became increased when pH was 2.0, 3.0, 4.0, 5.0, 8.0, 9.0, 10.0, 11.0, and 12.0. When pH was low, the strength of the peak at about 530 nm increased, whereas when pH was high, the strength of the peak at about 610 nm increased. Therefore, it is expected that the conversion efficiency will increase by adding citric acid or sodium hydroxide into the dye solution.

Table 8 shows the summary of the characteristics of the dye-sensitized solar cells fabricated using various red-cabbage dye solutions whose pH is controlled by adding citric acid or sodium hydroxide. The largest conversion efficiency (1.1%) was obtained when pH was 2.0.

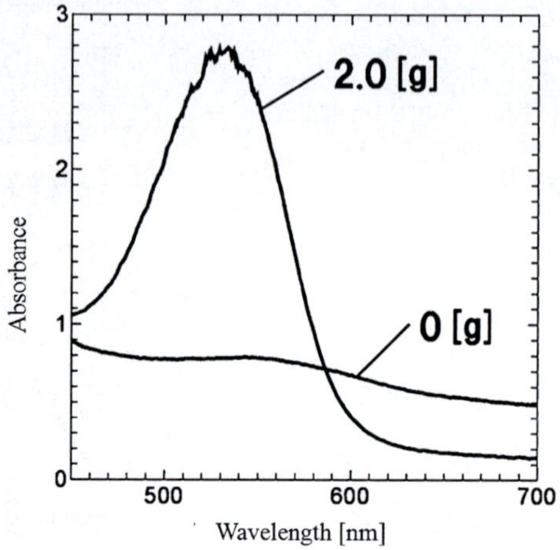

Fig. 12 Optical absorption spectra of the red-perilla dye solutions with unmodified pH and a pH controlled by adding citric acid (2 g)

According to the chemical structure shown in Fig. 2, the basic structures of the dyes of red-cabbage and red-perilla are both anthocyanin. As shown in Fig. 2, there are three OH groups in the fundamental anthocyanin structure. However, one of the OH groups changes to oxygen (double bond) under neutral conditions. As discussed in "Nanomorphology and charge transport in organic solar cells", it is considered that these OH groups in the dyes react with the OH groups which terminate the oxide semiconductor film surface. This reaction causes the strong covalent bond between the dye and the oxide semiconductor. In the acidic case, there are three OH groups in the fundamental structure. This may cause an increase in the number of the strong covalent bond between the dye and the oxide semiconductor, resulting in an increase of the conversion efficiency.

6.2 Effects of pH of Dye on Characteristics of Dye-Sensitized Solar Cells Using Dye of Red-Perilla

Figure 12 shows the optical absorption spectra of the red-perilla dye solutions with unmodified pH and a pH modified by adding citric acid (2 g). The strength of the absorption peak at around 530–540 nm is very much increased by adding citric acid.

Table 9 shows the summary of the characteristics of the dye-sensitized solar cells fabricated using red-perilla dye solutions with modified pH. The average conversion efficiency is 0.27% which is ~60% larger than that of solar cells fabricated using the red-perilla dye solution with an unmodified pH (0.17%, see Table 5). This increase of the conversion efficiency is proposed to be caused by an increase of the optical absorption strength shown in Fig. 12.

Table 9 Summary of the characteristics of the dye-sensitized solar cells fabricated using red-perilla dye solutions whose pH is controlled by adding citric acid (2 g)

Sample number	J_{SC} (mA/cm^2)	V_{OC} (V)	P_{MAX} (mW)	F.F.	η (%)
Sample 1	0.39	0.50	0.13	0.68	0.26
Sample 2	0.40	0.51	0.14	0.67	0.27
Sample 3	0.40	0.51	0.14	0.67	0.27
Sample 4	0.43	0.48	0.14	0.68	0.28
Sample 5	0.40	0.48	0.13	0.69	0.26
Average	0.40	0.50	0.13	0.68	0.27

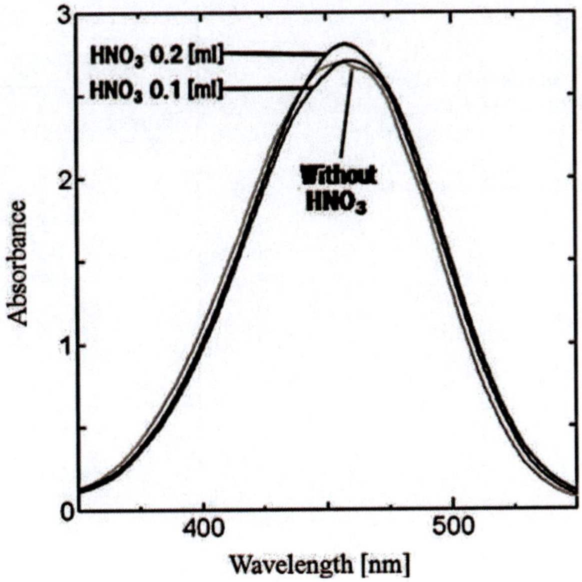

Fig. 13 Optical absorption spectra of NKX-2553 solutions with and without HNO_3. The former specimen was prepared by adding NHO_3 (0.1 or 0.2 ml) into NKX-2553 solution (25 ml)

6.3 Effects of pH of Dye on Characteristics of Dye-Sensitized Solar Cells Using NKX-2553

Figure 13 shows the optical absorption spectra of NKX-2553 solutions with and without HNO_3. The former solutions were prepared by adding NHO_3 (0.1 or 0.2 ml) into NKX-2553 solution (25 ml). As shown in Fig. 13, the strength of the absorption peak at about 460 nm increased slightly by adding 0.2 ml of HNO_3.

Table 10 shows the summary of the characteristics of the dye-sensitized solar cells fabricated using NKX-2553 with and without HNO_3 (0.1 ml). Five solar cells were fabricated. The average conversion efficiency is 3.3% which is larger by 0.3 point (about 9%) than that of the solar cells fabricated using NKX-2553 without HNO_3 (see Table 6). The reason for this increase of the conversion efficiency is not clear. However, it may be related to a small increase of the optical absorption strength at about 460 nm shown in Fig. 13.

Table 10 Summary of the characteristics of the dye-sensitized solar cells fabricated using NKX-2553 with and without HNO_3 (0.1 ml)

Sample number	J_{SC} (mA/cm^2)	V_{OC} (V)	P_{MAX} (mW)	F.F.	η (%)
Sample 1	4.7	0.62	1.6	0.55	3.2
Sample 2	4.9	0.64	1.7	0.54	3.3
Sample 3	4.9	0.64	1.7	0.54	3.4
Sample 4	4.3	0.64	1.5	0.55	3.0
Sample 5	5.0	0.62	1.7	0.55	3.4
Average	4.8	0.63	1.6	0.55	3.3

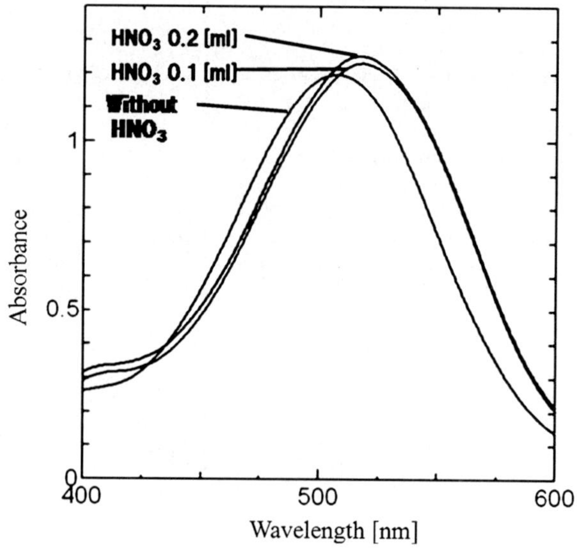

Fig. 14 Optical absorption spectra of NKX-2677 solutions with and without HNO_3. The former specimen was prepared by adding NHO_3 (0.1 or 0.2 ml) into NKX-2677 solution (25 ml)

6.4 Effects of pH of Dye on Characteristics of Dye-Sensitized Solar Cells Using NKX-2677

Figure 14 shows the optical absorption spectra of NKX-2677 solutions with and without HNO_3. The former specimen was prepared by adding NHO_3 (0.1 or 0.2 ml) into NKX-2677 solution (25 ml). As shown in Fig. 14, the strength of the absorption peak at about 520 nm was increased by adding HNO_3 (0.1 or 0.2 ml).

Table 11 shows the summary of the characteristics of the dye-sensitized solar cells fabricated using NKX-2677 with and without HNO_3 (0.1 ml). Five devices were fabricated. The average conversion efficiency is 4.1% which is larger by 0.5 point (about 15%) than that of the solar cells fabricated using NKX-2677 without HNO_3 (see Table 7). The reason of this increase of the conversion efficiency is considered to be an increase of the optical absorption strength at about 520 nm shown in Fig. 14.

Table 11 Summary of the characteristics of the dye-sensitized solar cells fabricated using NKX-2677 with and without HNO_3 (0.1 ml)

Sample number	J_{SC} (mA/cm^2)	V_{OC} (V)	P_{MAX} (mW)	F.F.	η (%)
Sample 1	6.5	0.60	2.0	0.50	4.0
Sample 2	6.7	0.60	2.0	0.49	4.0
Sample 3	7.0	0.61	2.0	0.48	4.1
Sample 4	7.2	0.61	2.1	0.47	4.1
Sample 5	7.4	0.61	2.1	0.47	4.2
Average	7.0	0.61	2.0	0.48	4.1

7 Summary

In this chapter, the characteristics of the dye-sensitized solar cells using natural dyes, such as those found in red-cabbage, curcumin, and red-perilla, and synthesized dyes, such as NKX-2553, NKX-2677, and D149, with no precious metal have been discussed.

The obtained largest conversion efficiency is greater than 1% for the dye-sensitized solar cell using the dye of red-cabbage. This value is small. However, the cost performance is more than 50 times larger than that of the dye-sensitized solar cell using Ruthenium complex. Therefore, when the cost of FTO and ITO substrates, oxide semiconductor, electrolyte solution, and opposite electrode becomes very low, dye-sensitized solar cells fabricated using natural dyes may become the most cost effective type of solar cell available.

The effects of the dye solution's pH on the characteristics of the dye-sensitized solar cells have been also described. The conversion efficiencies of the dye-sensitized solar cells using the dyes of red-cabbage, red-perilla, NKX-2553, and NKX-2677 become larger when the pH value is low. It is expected that this technique will be used for future dye-sensitized solar cell systems.

References

1. Gratzel M (2003) Dye-sensitized solar cells. J Photochem Photobiol C: Photochem Rev 4:145
2. Law M, Greene LE, Johnson JC, Saykally R, Yang P (2005) Nanowire dye-sensitized solar cells. Nat Mater 4:455–459
3. Furukawa S, Okada K, Ohno T, Sato H, Ohta H, Iino H, Arakawa H, Yasuda T (2007) Trans Mater Res Soc Jpn 32(2):321
4. Furukawa S, Yamauchi S, Iino H, Iwamoto T, Kuwada K (2008). In: The 5th international symposium on organic molecular electronics, 6C-4, 22–23 May, Himeji
5. Furukawa S, Iino H, Iwamoto T, Kukita K, Yamauchi S (2009) Characteristics of dye-sensitized solar cells using natural dye. Thin Solid Films 518:526–529
6. Shen Q, Arae D, Toyoda T (2004) J Photochem Photobol A: Chem 164:75
7. Sabataityte J, Oja I, Lenzmann F, Volobujeva O, Krunks M, Chim CR (2006) 9:708
8. Kang M, Hak Kim J, Won J, Soo Kang Y (2006) Dye-sensitized solar cells based on crosslinked poly(ethylene glycol) electrolytes. J Photochem Photobiol A: Chem 183:15

9. Shaheer Akhtar M, Chun J-M, Yang O-B (2007) Electrochem Commun 9:2833
10. Lee K, Suryanarayanan V, Ho K (2007) Sol Energy Mater Sol Cells 91:1416
11. Lan Z, Wu J, Lin J, Huang M, Yin S, Sato T (2007) Electrochim Acta 52:6673
12. Kukita K, Furukawa S (2010) Annual meeting of the Institute of Electrical Engineers of Japan, 17–19 March, Tokyo (In Japanese)

Nanotube- and Nanorod-Based Dye-Sensitized Solar Cells

Yung-Eun Sung, Soon Hyung Kang and Jae-Yup Kim

Abstract Considerable efforts have been devoted to the design and synthesis of low-dimensional, nanostructured materials due to their morphology-dependent performances. In particular, one-dimensional (1-D) TiO_2 nanostructures, including nanorods (NRs), nanowires (NWs), and nanotubes (NTs), have attracted considerable interest due to their unique characteristics. In dye-sensitized solar cell (DSSC) operation, 1-D nanostructure-based photoanodes can contribute to rapid electron transport, ensuring efficient charge collection by the conducting substrate in competition with recombination. Relying on the ordering of 1-D TiO_2 nanomaterial, the conversion efficiency of DSSCs was affected because electron collection is determined by trapping/detrapping events at the site of the electron traps, such as defects, surface states, grain boundaries, and self-trapping. This point has promoted research on self-ordered, 1-D photoanodes stretched on a substrate with enhanced electron transport properties due to their desirable features: highly decreased intercrystalline contacts and a structure with a specified directionality. In this literature review, the preparation of various 1-D nanomaterials from disordered to ordered states and their electron dynamics in the application of DSSCs are reviewed.

Y.-E. Sung (✉)
World Class University (WCU) Program of Chemical Convergence for Energy & Environment (C2E2), School of Chemical & Biological Engineering,
Seoul National University, Seoul 151-744, Korea
e-mail: ysung@snu.ac.kr

S. H. Kang
Department of Chemistry Education, Chonnam National University,
Gwangju 500-757, Korea
e-mail: skang@jnu.ac.kr

J.-Y. Kim
School of chemical and Biological Engineering and Interdisciplinary Program
in Nano Science and Technology, Seoul National University,
Seoul 151-742, Korea

L. Zang (ed.), *Energy Efficiency and Renewable Energy Through Nanotechnology*,
Green Energy and Technology, DOI: 10.1007/978-0-85729-638-2_9,

1 Introduction

Titanium dioxide (TiO_2) is a typical wide-band gap (3.2 eV) semiconductor that has been studied extensively in areas such as synthesis, deposition methods, crystal structure, optical properties, phase stability, photosensitivity,and morphological changes [1, 2]. In particular, among the various morphological phases, 1-dimensional (1-D) TiO_2 nanostructures such as nanorod (NR)/nanowire (NW) and nanotube (NT) have been widely investigated on account of their unique microstructure and promising features, such as a high-aspect-ratio, high surface area, higher surface area/volume ratio, increased number of delocalized carriers, and improved charge transport afforded by dimensional anisotropy with the conventional properties [3–5]. Their remarkable properties have led to their use in a variety of applications including dye-sensitized solar cells (DSSCs), photocatalysts,and photochromic devices [6, 7]. In particular, the DSSC is a photoelectrochemical cell that converts visible light energy into electricity and is considered an alternative to silicon solar cells due to its attainable high efficiency, environmental intimacy, and cost effectiveness.

Figure 1 shows the operating principle of a DSSC consisting of a semiconductor photoanode (electron transporting layer), dye (a photosensitizer), electrolyte containing I^-/I_3^- redox couples (an electron transfer mediator), and Pt-coated counter electrode (a catalyst for electron transfer). The DSSC features dye molecules chemisorbed onto the surface of a semiconductor nanocrystal filled with an electrolyte. Under light illumination, charge separation occurs at the dye/semiconductor interface, where photoexcited dyes inject electrons into the conduction band of the semiconductor, after which the original state of the dye is restored by electron donation from the electrolyte containing the I^-/I_3^- redox system. The iodide is regenerated, in turn, by reduction of triiodide at the counter electrode, with the circuit being completed through the external load. The theoretical maximum voltage under the illumination corresponds to the difference between the redox potential of the redox mediator and the Fermi level of the semiconductor photoanode. Along with these processes, as side reactions, electrons in the conduction band of the semiconductor may be recombined with the oxidized dye molecules or electron acceptor species from the electrolyte. Therefore, the following aspects of the photoanode deserve important consideration. Firstly, injected electrons diffuse through thousands of particles before they reach the transparent conductive oxide (TCO) substrate, emphasizing the importance of good electronic inter-particle connection. Furthermore, recombination plays a critical role because of the extremely large TiO_2/dye/electrolyte interface area and the proximity of electrons to the oxidized redox species (the hole carriers), which can limit the photovoltaic performance. In its original form, the DSSC used the mesoporous nanocrystalline TiO_2 film in replacement of the previous flat TiO_2 electrode to enhance the light-harvesting dye. Adsorption resulted from the highly expanded surface area, with a reported conversion efficiency of 7% [8]. Subsequently, to improve the conversion efficiency, both the size (<20 nm) and phase (anatase, rather than rutile) of the TiO_2 nanoparticles (NPs) were optimized. The TiO_2 film

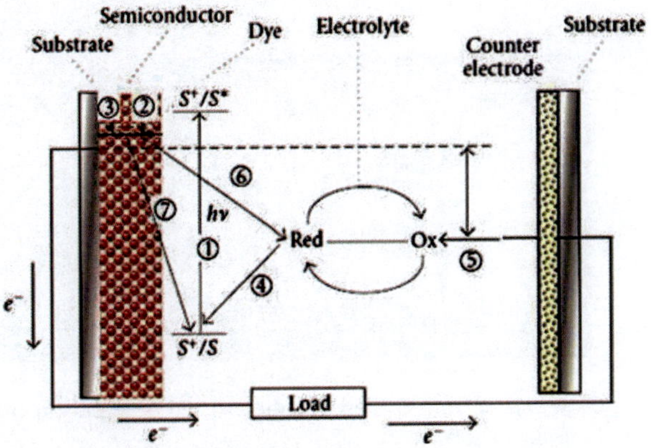

Fig. 1 Simple representation and operating principal of dye-sensitized solar cells (DSSCs)

thickness has a significant effect on the light-harvesting efficiency and charge recombination in that the conversion efficiency increases linearly with increasing TiO_2 thickness due to the increased dye uptake. However, above a limited thickness, the efficiency begins to degrade as a result of dominant charge recombination and mass transport limitations [9]. This is easily explained by the electron diffusion length in TiO_2 films, which is the distance that photoinjected electrons transport through a nanoporous network before reacting with the cations of the redox electrolyte in the DSSC system. In general, an electron diffusion length of approximately 15–20 μm in traditional electrodes composed of nano-sized TiO_2 particles (e.g., several orders of magnitude smaller than those in bulk single crystal TiO_2) is believed to limit the power conversion efficiency. This is because electron collection is determined by trapping/detrapping events along the site of the electron traps (defects, surface states, grain boundaries, self-trapping, etc.) [10, 11]. This point has focused research on photoanodes comprised of variously shaped nanoporous materials (rod, tube, wire, and tetra pad, etc.) with enhanced electron transport properties due to highly decreased intercrystalline contacts and structure with specified directionality. As a result, various methods for synthesizing 1-D TiO_2 particles have been reported [12–14].

However, the preparation of 1-D TiO_2 nanomaterials strongly influences their peculiar properties due to their enhancement or retardation in the fabrication process of the photoanode layer for DSSCs. Figure 2 shows the simple scheme to describe the disordered and ordered state of 1-D nanomaterials on the TCO substrate for the application of DSSCs. The disordered state of 1-D nanomaterial-based photoanode also provides additional grain boundaries or trapping sites, while the ordered state of 1-D nanomaterial-based photoanode shows straight electron percolation.

Herein, several approaches to prepare disordered (surfactant assisted, oriented attachment, sol–gel reaction, and hydrothermal process) and ordered (electrochemical anodization and hydrothermal process) 1-D TiO_2 nanomaterials and their electron dynamic properties for the application of DSSCs are summarized.

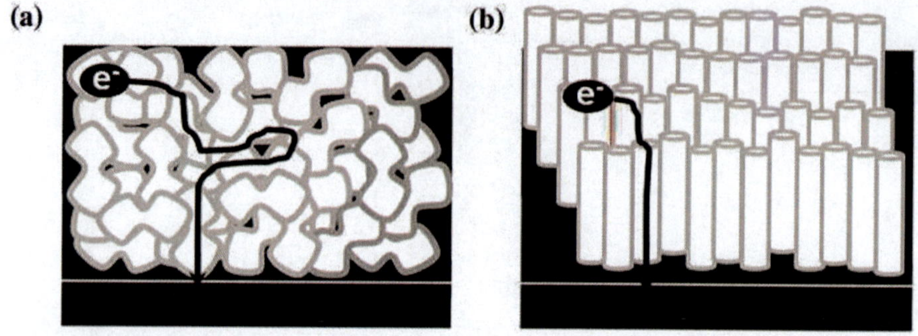

Fig. 2 Simple scheme to describe the ordered **a** and disordered **b** states of 1-D nanomaterials based on photoanode in the DSSCs

2 Preparation of Randomly Oriented TiO_2 Nanorod and Nanotube Films for Dye-Sensitized Solar Cells

NR-shaped TiO_2 nanocrystals are believed to have exceptional properties and have been considered an alternative to NPs. NRs have a higher surface area-to-volume ratio than NPs, thereby supplying a higher density of active sites for surface reactions as well as a high interfacial charge carrier transfer rate. Furthermore, increased delocalization of carriers in rods, where they can move freely throughout the length of the NRs, is expected to reduce the e^-/h^+ recombination probability. However, this is partially compensated by the traps in the surface sites to ensure more efficient charge separation [15]. Finally, NRs can potentially improve charge transport in the photoanodes of DSSCs when an ordered orientation of 1-D inorganic electron transporters is employed. Therefore, NRs offer direct electrical pathways for photogenerated electrons and can increase the electron transport rate, which in turn may improve the performance of DSSCs. Accordingly, a photoanode composed of 1-D TiO_2 NRs is a promising approach to which the following methodologies for synthesizing anisotropic TiO_2 nanomaterials have been applied: surfactant-assisted methods, oriented attachment, sol–gel reactions, and hydrothermal reactions.

2.1 Surfactant-Assisted Method

As a stabilizing agent, surfactants prevent aggregation of synthesized NPs and control their size and shape. With a large excess of surfactant, such as oleic acid, lateral expansion of the crystal lattice must be suppressed to achieve anisotropic crystal growth [16]. Oleic acid plays two roles: stabilizing solvent and chemical modifier to control the hydrolysis rate of the precursor. For better tuning of the growth rate of the different planes and control over the shape evolution of the anatase nanocrystals, the use of mixture surfactants that selectively bind to different crystal faces,and elimination of high-energy facets has been encouraged.

In this method, rod formation is usually realized when the surface free energies of the various crystallographic planes differ significantly. The use of different surface ligands that bind selectively to specific surface planes has been demonstrated as a suitable approach for rod formation in a controlled manner. As a representative example, Weller et al. reported the controlled growth of TiO_2 nanocrystals by modulation of the hydrolysis rate, using oleic acid as a stabilizing surfactant at 80°C [17]. In the present study, oleic acid served as an adsorbing-chelating ligand, restricting the growth rate along some crystallographic directions.

Moreover, the chemical modification of titanium alkoxide has been proven as a reasonable strategy to tune the reactivity of the precursor to water, manipulating the nanocrystal's growth kinetics and providing shape control over the resulting NPs. This synthetic route is a unique tool for easily functionalizing the TiO_2 nanocrystals surface with different capping ligands. Furthermore, Alivisatos's group reported a surfactant-mediated shape evolution of TiO_2 anatase nanocrystals in non-aqueous media. By using a surface selective surfactant, such as carboxylic acid, which binds strongly to the anatase (001) facets, modulation of the surface energies of the different crystallographic faces controlled the shape [18].

Recently, the Adachi group reported DSSCs using hydrothermally synthesized, single-crystalline TiO_2 NRs with diameters of 20–30 nm and lengths greater than 100 nm [19]. They synthesized TiO_2 NRs using selective surfactants, since the highest-energy facets were eliminated during crystal growth so that the progressive addition of a selective surfactant yielded a sequence of shapes. In the case of TiO_2 anatase, nucleation formed truncated octagonal bi-pyramidal seeds, exposing eight equivalent (101) faces and two equivalent (001) faces. The surface free energy of the (001) faces is nearly 1.4 times larger than that of the (101) faces. Therefore, the shape evolution in TiO_2 anatase was realized by modifying the surface free energy and growth rate of the nucleus with the surface adhesion of ligands. In addition, by controlling the cetyltrimethylammonium bromide and poly(ethylene oxide)$_{100}$–poly(propylene oxide)$_{65}$–poly(ethylene oxide)$_{100}$ as a surfactant and triblock co-polymer(F127), the NR shape and length were modulated. For the application of DSSCs, the TiO_2 NR film was sintered at a high temperature, while retaining the rod shape. This contributed to the achievement of a high conversion efficiency (7.29%) of light-to-electricity (J_{sc}: 13.1 mA/cm^2, V_{oc}: 0.767 V, fill factor: 0.728) with a thickness of 16 μm. This achievement was attributed to the increased rate of electron transport resulting from the high crystalline anatase NR and high dye loading.

2.2 *Oriented Attachment Method*

Oriented attachment involves the spontaneous self-organization of adjacent particles so that they share a common crystallographic orientation, followed by joining of these particles at a planar interface [20]. Bonding between the particles reduces the overall energy by removing the surface energy associated with unsatisfied bonds. Figure 3 shows how oriented attachment gives rise to

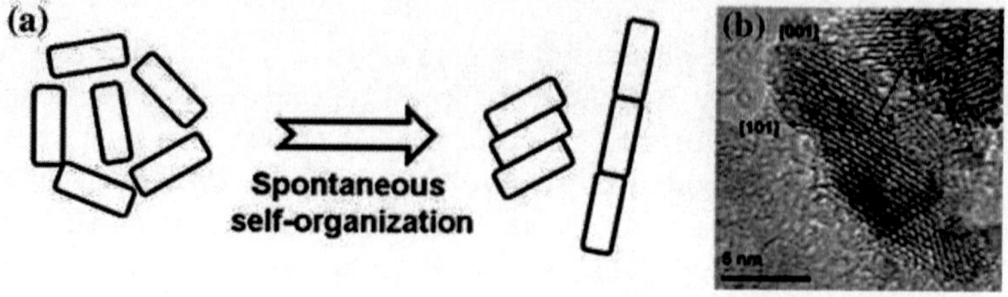

Fig. 3 **a** Simple schematic drawing of the oriented attachment mechanism, **b** TiO_2 nanorods synthesized by oriented attachment process

homogenous single crystals or to crystals separated by twin boundaries or other planar defects. This issue of morphological evolution is of direct importance to material science because growth and morphological evolution dramatically modify physical properties and surface reactivity.

Penn et al. reported the formation of anisotropic TiO_2 nanocrystals during hydrothermal treatment of TiO_2 nanocrystals under acidic conditions by oriented attachment [21]. Oriented attachment is important in nominally dry aggregates and in periodically wet environments. The particle movement needed to achieve orientation within random aggregates is provided by Brownian motion and other physical and energetic effects [22]. In addition, the researchers predicted that the particles are in contact with a solution under pH conditions near the isoelectric point. The organic molecules may hinder or modify oriented attachment by preventing contact between the faces on which adsorption has selectively occurred. Furthermore, they showed the evolution of the chains of particles, starting with the attachment of two primary crystallites. The attachment appears to occur most commonly on (112), less commonly on (001), and rarely on (101). This mechanism effectively serves to reduce the overall energy by eliminating the surfaces at which the crystallites join. In the most common case of (112) attachment, the highest surface energy face is eliminated.

Moreover, Hyeon et al. reported highly crystalline TiO_2 nanocrystals with various shapes and crystal structures in aqueous media at room temperature by oriented attachment [23]. In particular, the poly(ethylene glycol)-block-poly(propylene glycol)-block-poly(ethylene glycol) (PEO-PPO0-PEO) (P123) template controls the hydrolysis and condensation of the titanium tetraisopropoxide precursor. Under various reaction conditions, TiO_2 nanocrystals with different sizes and shapes were synthesized. With 1 M NaCl, spherical anatase nanocrystals with a particle size of 6 nm were produced, while in 1 M CH_3COOH solution, short anatase NRs with an average size of 4×20 nm were produced. Additionally, the addition of 1 M HCl induced the formation of long rutile NRs with an average size of 6×50 nm. These results reveal the essential presence of salt or acid to synthesize TiO_2 nanocrystals with unique shapes and crystal structures. In the DSSC, the photoanode comprised of oriented attachment. TiO_2 NRs showed the following two main advantages: (1) confirmation of high surface area directly proportional to the light-harvesting yield (dye uptake) resulted from the NRs synthesized from the necking of truncated NPs by recovering the low surface area of the general TiO_2 NRs, and (2) fast electron

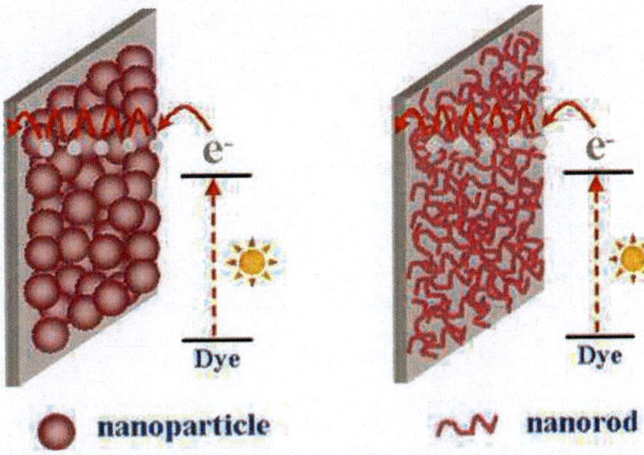

Scheme 1 TiO_2 photoanodes consisting of NPs and NRs in the configuration of DSSCs (Reprinted with permission from Ref. [24]. Copyright 2008 Wiley Interscience)

transport rate and degraded charge recombination from the decreased intercrystalline contacts between grain boundaries and specific directionality of NRs, bringing about the improved charge collection efficiency. To demonstrate the distinguished marks, the NP- and NR-based photoanodes for DSSCs were prepared in the same conditions (light-harvesting efficiency and charge injection yield). Only the charge transport ability of both samples was varied for comparison [24].

Scheme 1 describes the TiO_2 photoanodes consisting of the NPs and NRs in the configuration of DSSCs. At the same dye uptake ($n = 2.8 \times 10^{16}$), the deposited thickness variation between NP (5.47 μm) and NR (4.87 μm) films was small at approximately 10%. This indirectly indicates that NRs have a large surface area. From the photocurrent-voltage (*J–V*) measurements, a NP-based DSSC showed an V_{oc} of 0.68 V, a J_{sc} of 6.9 mA/cm^2, a FF of 0.71, and an efficiency (η) of 3.36%, while the NR-based DSSC exhibited results of 0.7 V, 11.7 mA/cm^2, 0.6, and 4.95%, respectively, under one sun condition.

To measure the electron transport and charge recombination characteristics between NP- and NR-based photoanodes in DSSCs, the stepped light-induced measurements of photocurrent and voltage (SLIM-PCV) was performed under front side illumination [25]. Then, truncated NRs with a particle size of about 6 nm were prepared as a control sample using the same synthesis tool.

Figure 4 shows the electron diffusion coefficients (*D*) and lifetimes (τ) of the NP-and NR-based DSSCs as a function of J_{sc}, as analyzed by SLIM-PCV. The *D* values of both samples showed a power-law dependence on the light intensity, indicating an aspect controlled by similar mechanistic factors [26]. The *D* value of the NR film was slightly higher than that of the NP film, showing that NR films have more favorable properties for electron transport than NP films. Considering the condition where small TiO_2 NPs (<10 nm) strongly bind to each other to decrease the unstable surface energy during thermal treatment, this minor improvement may be caused by the reduction of electron loss in the grain boundaries resulting from the necking of NPs, as well as by the increase of average crystallite size. From the parallel slope of both samples, similar traps were distributed over the whole range of the film because the slope indicates a steeper trap-state distribution. Figure 4b

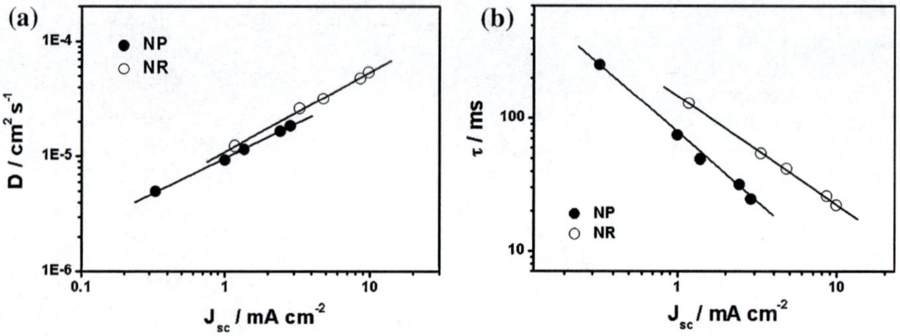

Fig. 4 Electron diffusion coefficients (*D*) and electron lifetimes (τ) for dye-sensitized NP and NR cells as a function J_{sc} controlled by ND filter from 635 nm laser illumination (Reprinted with permission from Ref. [24]. Copyright 2008 Wiley Interscience)

shows the values of τ as a function of J_{sc}. Over the light intensity range, the τ values of the NR films become an order of magnitude higher than those of the NP films. The lower lifetime of the NP films might be due to the combined effect of the downward band-edge shift, as confirmed by the relatively low V_{oc} value, and the increased charge recombination rate, as supported by dark current measurements (not shown here). This is mainly illustrated by the effects of the surface state, which leads to intraband gap states and enormous electron loss between the grain boundaries that experience tens of thousands of trapping/detrapping events during their transit through the film [27, 28]. The phenomenon indicates that more efficient DSSCs will be realized with a thicker TiO_2 NR layer than that in the NP film because both the enhanced electron transport due to the geometry effect and the increased electron lifetime due to the suppression of charge recombination contribute to increasing charge collection efficiency.

In a similar context, Adachi's group reported single-crystal-like TiO_2 NWs prepared by the oriented attachment mechanism using surfactant-assisted processes at a low temperature for highly efficient DSSCs [29]. The crystal growth direction of the oriented attachment was controlled by changing the molar ratio of acetylacetone to Ti, regulating the adsorption of surfactant molecules via control of the reaction rate and the surface energy. They showed that most of the aggregated particles form a wire shape with a single-crystalline (sc) structure. The oriented attachment occurred mainly in the (101) direction and formed a network structure. They suggested that the use of a (101) exposure plane with a network structure of single-crystal-like anatase NWs is one of the most promising methods because the intercrystalline titania contacts are greatly decreased by using a single-crystal-like network structure in comparison with a porous titania thin film composed of accumulated nanosized particle. This might be useful for easier electron transfer through the titania layer and suppression of back reaction of photoinjected electrons with I_3^-. This also influenced the 4-fold greater adsorption of ruthenium dye compared to P25. To summarize, a high light-to-electricity conversion rate of

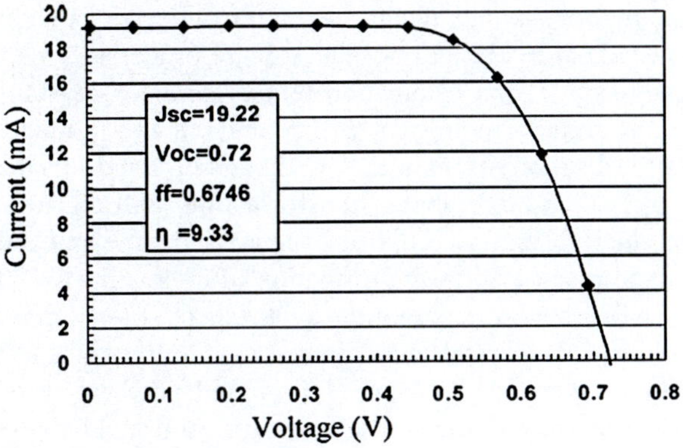

Fig. 5 Illustration of the photocurrent-voltage characteristics of the cell with titania thin film composed of a TiO_2 network structure of single-crystal-like anatase NWs (Reprinted with "Adapted" or "in part" permission from Ref. [29]. Copyright 2004 American Chemical Society)

9.33% (J_{sc}: 19.2 mA/cm^2, V_{oc}: 0.72 V, and FF: 0.675) was obtained for the cell with a TiO_2 network of single-crystalline anatase NWs (Fig. 5).

Furthermore, the hydrothermal process with the transparent TiO_2 sol prepared by peptization of anatase TiO_2 precipitates was suggested to dispersible and NR-like TiO_2 nanocrystals by an oriented attachment mechanism [30]. After further hydrothermal treatment of the mixture of NR-like TiO_2 nanocrystals and the original sol, branched and nanoring-like nanostructures with diameters of 6 nm in the anatase phase were obtained. When they were with the oriented crystallographic plane, more complex structures, such as nanoring-like and *T*-type TiO_2 nanostructures, were formed. By counting the number of the primary crystallites formed in the nanoring-like structures, about 20% of the nanocrystals were orientated and attached to the closed and nearly closed nanorings. They showed 2-fold higher ruthenium dye adsorption than commercial TiO_2 nanocrystals ST-01, giving a DSSC efficiency of 5.75%.

2.3 Sol–Gel Method

The sol–gel process is a wet-chemical technique regarded as a powerful approach for preparing dispersed NPs by tailoring the structure of a primary NP in which metal atoms are uniformly distributed. Such methods are initiated as soluble precursor molecules and are hydrolyzed to form a dispersion of colloidal particles (sol). As the reaction proceeds, the bonding of sol particles forms the integrated network of NPs (gel). The gel is typically heated to produce the desired materials. During this process, the metal alkoxides and metal chlorides as a precursor undergo various forms of hydrolysis and polycondensation reactions. To obtain homogenous macromolecular oxide networks for qualified nanomaterials in sol–gel processing, several factors in the reaction mixture (e.g., water, hydrolysis rate, precursor ratio etc.) must be controlled [31]. The key parameter to form 1-D TiO_2 NRs and NTs with the desired nanocrystalline structure and shape is the control of the hydrolysis rate, due to the high reactivity of titanium precursors such as $TiCl_4$ and alkoxide [32]. In normal sol–gel processing, gelization and calcinations are necessary. If the calcination temperature for crystallization is not high enough, some organic molecules will remain in the product, leading to the incomplete crystallization of titania. On the other hand, if the temperature is too high, aggregation and phase transformation will occur, resulting in phase-impure product. Therefore, control over critical experimental parameters in the sol–gel reaction can lead to the formation of crystalline 1-D TiO_2 NR and NT.

Niihara et al. reported the synthesis of sol–gel derived TiO_2 NTs [33]. Needle-shaped TiO_2 crystals with diameters of 8 nm and lengths of 100 nm were obtained when sol–gel derived fine TiO_2-based powders were treated chemically with a 5–10 M NaOH aqueous solution for 20 h at 110°C, followed by the addition of HCl and distilled water in the solution. The following formation mechanism of titania NTs has been proposed: the crystalline raw material is first converted into an amorphous product through alkali treatment, after which the treatment with a distilled water and HCl aqueous solution results in the formation of titania NTs. This report is the first evidence that oxide NTs can be obtained easily via the sol–gel process without requiring molds for replication or templates.

Yanagida et al. reported the sol–gel synthesis of TiO_2 NTs with large aspect ratio and large specific surface area from P25 NP [34]. They optimized the fabrication conditions of DSSCs, i.e., the pH of the starting paste, sintering temperature for TiO_2 electrodes, electrolyte compositions, and $TiCl_4$ treatment, to give a conversion efficiency (7.1%) that was higher than that of the DSSC made of P25 (6.2%) due to the higher open-circuit voltage. To investigate comprehensively the high efficiency obtained by using the TiO_2 NT-based photoanode, the electron dynamic properties (electron diffusion coefficient (D) and lifetime (τ)) were examined (Fig. 6).

The D value of TiO_2 NTs was almost comparable with that of P25. The trend of diffusion coefficients was explained by the charge trap site density and distribution in the mesoporous electrode. The trap sites may have been formed mainly by grain boundaries and crystal defects. As for the grain boundaries, the number of boundaries across the TiO_2 film may not have decreased greatly because the orientation of NTs perpendicular to the conductive substrate had not been achieved in the TiO_2 NTs electrodes. On the other hand, the τ value in TiO_2 NTs electrode was about 3 times greater than that in P25, indicating that the probability of recombination between the electron in TiO_2 and the I_3^- in the electrolyte is smaller than that of P25. Considering both factors affecting the electron behavior, TiO_2 NTs were found to have longer diffusion length and thus more favorable electron transport property than P25.

More recently, the non-hydrolytic sol–gel method was successfully used for the synthesis of 1-D TiO_2 NRs by simultaneous modulation of the phase and size [35]. The synthesis of TiO_2 NRs was achieved with this reaction via the continuous delivery of two titanium precursors using two separate syringe pumps. By varying the injection rate of the precursors, the TiO_2 NRs simultaneously underwent phase transformation and length elongation. Furthermore, the morphology of the TiO_2 NRs evolved into a branched shape and their length was increased. In addition,

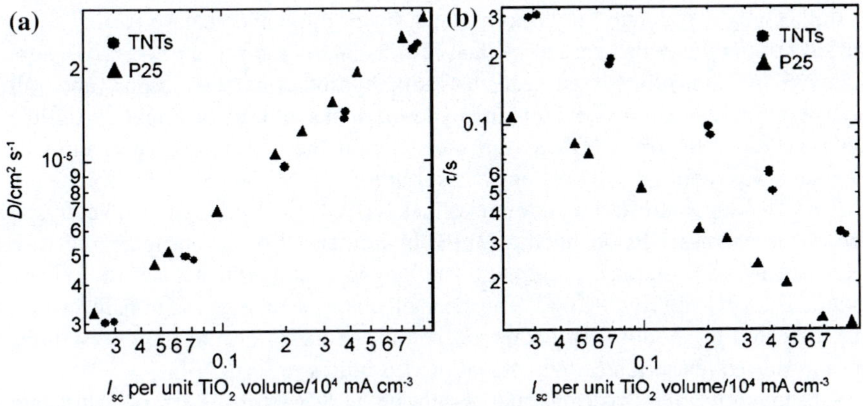

Fig. 6 Electron diffusion coefficients **a** and electron lifetimes **b** in the electrodes constructed with TiO_2 NTs (*circles*) and P25 (*triangles*) as a function of short-circuit photocurrent (I_{sc}) per unit TiO_2 volume (Ref. [34]-Reproduced by permission of the PCCP Owner Societies)

the crystalline phase of TiO_2 NRs was simultaneously affected by the injection rate. Pure anatase and pure rutile NRs were produced at fast and slow injection rates, respectively. Herein, the TiO_2 NRs were successfully employed as the photoanode for DSSCs. Particularly, the photoconversion efficiency (3.83%) of the mixture comprised of star-shaped rutile TiO_2 NRs and a small fraction of anatase NRs was comparable to that of P25 (4.1%).

2.4 Hydrothermal Method

Hydrothermal synthesis has been considered the most powerful technique to prepare single-crystalline 1-D TiO_2 nanomaterials that rely on the solubility of the precursor in hot water under high pressure. In particular, this method enables the growth of crystalline phases, which have unstable melting points, and materials showing high vapor pressure at their melting points can also be grown. Normally, the hydrothermal process is performed in an autoclave (steel pressure vessel) with Teflon® liners under controlled temperature and pressure. The hydrothermal process of TiO_2 nanomaterials under various experimental conditions (temperature, pH, and additives) yielded different morphologies and structural forms such as anatase, rutile, and even brookite. In general, the TiO_2 nanopowder is used as a precursor to form a suspension sol in the alkaline solution, and is subsequently hydrothermal treated to form TiO_2 NTs. On the other hand, alkaline titanate NTs were grown hydrothermally in alkaline solution, by exchanging the alkaline ions with protons to convert the alkaline titanate NTs to hydrogen titanate NTs. Afterward, TiO_2 NTs were produced by the thermal dehydration in air at high temperature or the hydrothermal reaction of hydrogen titanate NTs.

Initially, the production of uniformly nanosized rutile and anatase particles was carried out via the hydrothermal method [36]. Anatase and rutile phase TiO_2 were achieved by controlling the precursor ratio, hydrothermal conditions, and salt incorporation [37–39]. In addition, the hydrothermal synthesis of single-crystalline anatase TiO_2 NRs from NTs as the precursor in the absence of surfactants or templates was reported. In the strong basic solution (10 M NaOH), the pH of the NT suspension was controlled at different values by HNO_3 concentration. The crystal size of the anatase NPs obtained from the hydrothermal treatment increased with pH, and NRs with aspect ratios up to 6 and long axes along the anatase (001) were obtained at a pH slightly below 7. The mechanism of the tube-to-rod transformation was indicated by the local shrinkage of the tube walls to form anatase crystallites and the subsequent oriented attachment of crystallites to form NRs.

Subsequently, the hydrothermal synthesis (130°C for 72 h) of trititanate ($H_2Ti_3O_7$)-type NTs, with diameters of 9 nm and lengths of 100 to several hundreds of nanometers, using a single alkali treatment was suggested. In this method, TiO_2 reacts with a NaOH solution to form a highly disordered intermediate phase containing Ti, O, and Na [40, 41]. Single sheets of the trititanate $Ti_3O_7{}^{2-}$ started to grow inside the disordered phase. Afterward, when such trititanate sheets grew two-dimensionally, they simultaneously rolled up into NTs. When this reconstruction took place three-dimensionally, they suggested that $H_2Ti_3O_7$ plates with several trititanate layers were formed. The utilization of trititanate NTs with a multiwall structure as an electrode material in DSSCs was reported. These layered trititanate NTs showed highly efficient DSSC performance (7.5%: V_{oc} of 0.78 V, J_{sc} of 12.8 mA/cm^2, FF of 0.75) with a thickness of 14.4 μm. The incident photon to current efficiency (IPCE) at a wavelength of 520 nm was 61%. This indicated that the physicochemical characteristics of the trititanate NTs, including the high surface area, afforded a higher photocurrent density and a network structure useful for easier electron transfer through the titania layer, thereby suppressing the charge recombination between photoinjected electrons from the dye to the conduction band of the TiO_2 electrode and the $I_3{}^-$ ions in the electrolyte.

Aydil et al. hydrothermally synthesized randomly oriented, anatase TiO_2 NWs on a titanium substrate using 10 M NaOH as an alkaline source [42]. During the hydrothermal reaction, the top surface of the titanium foil transformed to $Na_2Ti_2O_4(OH)_2$ NTs. Subsequently, the $Na_2Ti_2O_4(OH)_2$ NTs were converted by an ion exchange reaction to $Na_2Ti_2O_4(OH)_2$ NTs, which were in turn converted to polycrystalline anatase NWs through a topotactic transformation. The prepared TiO_2 NW film showed a photoconversion efficiency of about 1.8% (J_{sc}: 5.6 mA/cm^2, V_{oc}: 0.62 V, FF: 0.52). Additionally, IPCE as a function of wavelength typically reached 50% at 530 nm (the maximum absorbance of the dye). Furthermore, the researchers investigated the electron transport time constant using intensity modulated photocurrent/voltage spectroscopy (IMPS/VS) and photocurrent decay methods. Figure 7 shows the charge recombination and transport time constants for TiO_2 NW DSSCs as a function of J_{sc} (light intensity) [43].

The magnitude of the electron transport time (10^{-2}–10^{-3} s) and its dependence on the illumination intensity were similar to those reported for DSSCs made from

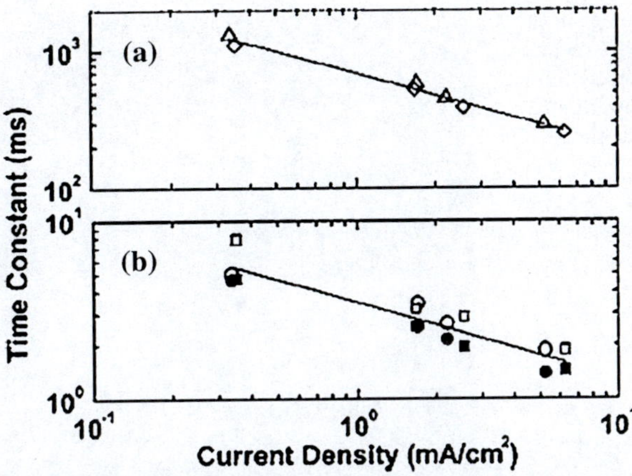

Fig. 7 **a** Charge recombination (Δ, ◇) and **b** transport (○, ●, □, ■) time constants for TiO_2 NW DSSCs as a function of J_{sc} (light intensity). In **b**, the open (○, □) and filled (●, ■) symbols represent transport times extracted from IMPS and photocurrent decay, respectively. Data for two cells (○, ●), and (□, ■) are shown to illustrate reproducibility. Similarly, recombination time constant data are shown for two different cells (◇, Δ) in **a** (Reprinted with permission from Ref. [43]. Copyright 2007, American Institute of Physics)

TiO_2 NPs, which exhibited a power-law dependence on J_{sc}. This indicated that the electron capture and release by surface traps determine transport times even in NWs, although the ratio of the electron recombination time to the electron collection time in NR solar cells was ~150, larger than that observed in the NP DSSCs, resulting in an electron collection efficiency of nearly 100%. They presented two reasons for the longer recombination times observed in NWs relative to NPs. First, the diameter of the TiO_2 NWs is larger than both the diameter of typical TiO_2 NRs used for assembling DSSCs and the semiconductor's Debye length. This larger diameter and the cylindrical geometry allow the wires to support radial electric fields that keep the electrons away from the NW surface, thereby reducing surface electron densities and recombination. A second possibility is that the NWs and NPs have different spatial distributions of electron traps. In nanowire, a fraction of the traps may be at the internal grain boundaries, not exposed to the electrolyte.

Utilizing the TiO_2 NRs, several approaches to increase the specific surface area, and thereby improve the photoconversion efficiency, were tried by several researchers [44–46]. At first, the DSSC photoanode was composed of a composite of anatase TiO_2 NPs and single-crystalline anatase TiO_2 NWs [44]. The composite electrode showed a high surface area from NP aggregates and the fast electron transport rate and light scattering effect of NWs. At the optimum content (20 wt%) of TiO_2 NWs, the conversion efficiency ($\eta = 8.6\%$) was improved more than that ($\eta = 6.7\%$) of pure NP cells. In addition, they indirectly showed that the improved electron diffusion length of the composite electrode was caused by the continuous increase in the current density with increasing film thickness over the whole thickness range up to 17 μm. In addition, in the situation where the TiO_2 NWs baked

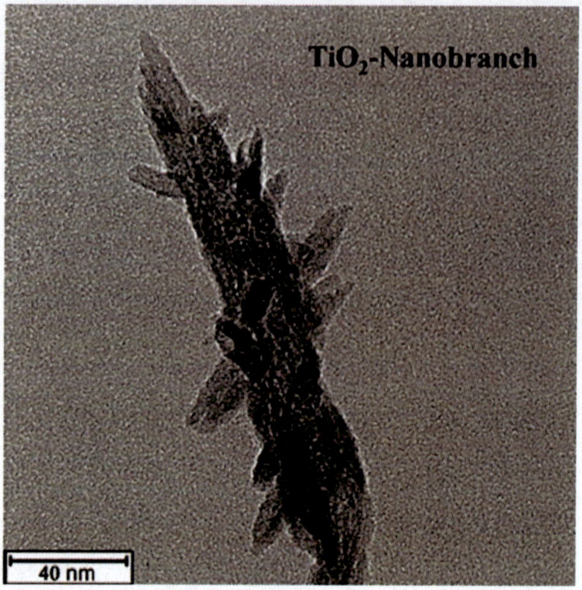

Fig. 8 FE-TEM image of the TiO_2 nanobranched structure (Reprinted with permission from Ref. [46]. Copyright 2010 American Chemical Society)

at 450–500°C showed a significant morphology change to NP, the niobium treatment of TiO_2 NWs acted to decrease the Na^+ content effectively and increase the thermal stability of NWs [45]. DSSCs composed of a mixture of TiO_2 NPs (P25) and the niobium-treated NWs (1:1 wt%) showed a photoconversion efficiency of 5.15% (V_{oc}: 0.78 V, J_{sc}: 11.2 mA/cm^2, FF: 0.59) compared to 4.73% (V_{oc} of 0.73, J_{sc}: 9.52 mA/cm2, FF: 0.68) of P25-based DSSCs. In addition, the branched TiO_2 NPs were synthesized by seeding with TiO_2 NWs, thereby transforming from NWs to a nanobranched shape [46]. Figure 8 shows the field-emission transmission electron microscopy (FE-TEM) image of the synthesized nanobranched structure.

Relative to a typical TiO_2 NW with few dye adsorption sites, the nanobranched TiO_2 showed the high roughness factor up to two times and thus the enhancement of charge harvesting efficiency. In the fabrication process of TiO_2 film, the high-temperature annealing did not affect the morphological modification of nanobranched TiO_2, so that the photocurrent and cell efficiency were enhanced from 6.25 mA/cm^2 (TiO_2 NW) to 12.18 mA/cm^2 (nanobranched TiO_2) and from 2.6 to 4.3%, respectively.

3 Preparation of Self-Ordered TiO_2 Nanorod/Nanotube Films for Dye-Sensitized Solar Cells

1-D TiO_2 NRs or NTs synthesized by chemical approach showed a randomly oriented, nanostructured electrode after the thin film formation for DSSCs. This randomly mixed film caused the degradation of the originally unique properties in the electron transport, similar to in the NP system [43]. The structural disorder at the contact between two crystalline NRs or NTs leads to enhanced scattering of free electrons, thus reducing electron mobility. Accordingly, to benefit from the improved electron transport in elongated nanostructures, it is necessary to assemble nanostructures directly on the surface of the electrode. More recently, ordered and strongly interconnected TiO_2 NTs on the Ti substrate has attracted attention as a fascinating material that offers a large internal surface area without damaging the geometrical and structural order. The precisely oriented nature of TiO_2 NT arrays makes them excellent electron percolation pathways for vertical charge transfer between the interfaces. Furthermore, TiO_2 NTs arrays show outstanding charge transport and carrier lifetime properties, and have been found to be suitable for a variety of applications, including Li-ion batteries, photo-electrolysis, and DSSCs [47, 48]. Furthermore, arrays of single-crystalline TiO_2 NWs on Ti foil were prepared by a novel alkali hydrothermal growth process. DSSCs using TiO_2 NWs arrays are promising prospects in research in this field. To minimize the light loss from back-side illumination, the TiO_2 NWs were prepared on an FTO substrate under hydrothermal conditions. This research is closely correlated with application to various types of solar cells, including DSSCs, heterojunction, hybrid, and organic solar cells [49–51]. The control of diverse experimental conditions such as film growth time, substrate, initial reaction concentration, acidity, titanium precursor, and surfactants or salts was reported for the DSSCs. As another synthesis tool, large oriented arrays and continuous films of TiO_2-based NTs were prepared by hydrothermal process on the Ti substrate. Several methodologies (electrochemical anodization and hydrothermal reactions) to form self-ordered TiO_2 NRs/NTs on the substrate are summarized in detail below.

3.1 Electrochemical Anodization of TiO_2 NTs on Ti for DSSCs

3.1.1 Electrochemical Anodization in Aqueous Electrolyte

In 1999, Zwilling et al. first reported titania films with porous surfaces formed by anodizing Ti metal in an F^- ion-based electrolyte [52]. Approximately 300 nm-thick TiO_2 NTs were formed on the Ti foil. The thin TiO_2 NTs were restricted to applications to other fields. Afterward, the tubular shape and length were optimized in the F^- ion-based electrolyte by tuning the pH, electrolyte concentration, and precursor species [53–55]. Schmuki et al. reported the fabrication of a 2.5 μm-thick TiO_2 NT array using NH_4F as a fluorine source [53] and Grimes et al. also succeeded in forming a longer TiO_2 NT (up to 4.4 μm) array using KF and NaF instead of HF as a precursor [54]. The key to achieve high-aspect-ratio growth is adjustment of the dissolution rate of TiO_2 by localized acidification at the pore bottom while a protective environment is maintained

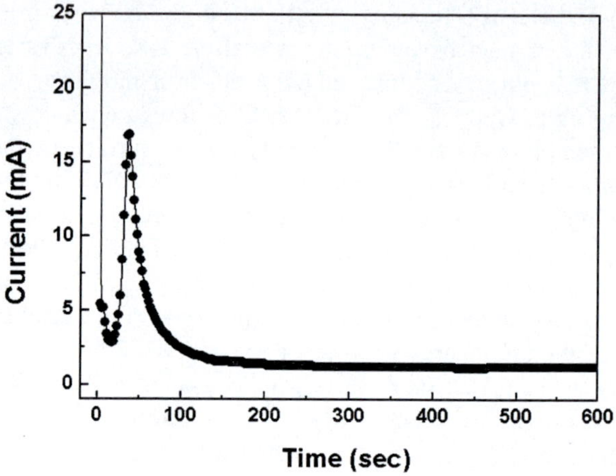

Fig. 9 Current-time response at a constant voltage of 20 V at pH 4.2 electrolyte using NaF precursor

along the pore walls and at the pore mouth [55]. Figure 9 shows the current density as a function of the anodization time in 10 min intervals.

In the initial stage of anodization (up to 30 s), an abrupt decrease of the current related to the formation of a barrier layer was followed by a sudden increase of current due to the active chemical dissolution reaction of the oxide layer formed by F^- ions from the electrolyte, leading to the formation of a porous structure. In the final stage, a relatively constant equilibrium state was maintained with increasing anodization time, while the current density slightly increased, showing an oscillation curve. In the equilibrium condition between the electrochemical formation of TiO_2 at the pore bottom and the chemical dissolution of this TiO_2 layer in the F^- ion-containing electrolyte, the TiO_2 NTs were grown continuously. The related mechanistic process for the formation of self-ordered titanium dioxide has been reported [56].

$$Ti^{4+} + 2H_2O \rightarrow TiO_2 + 4H^+ (\text{Hydrolysis of Ti}) \quad (1)$$

$$TiO_2 + 6HF \rightarrow [TiF_6]^{2-} + 2H_2O + 2H^+ (\text{Dissolution of } TiO_2) \quad (2)$$

The above reactions proceed alternatively in the overall reaction. Because the key to achieve longer TiO_2 NTs is control of the dissolution rate of TiO_2, which is highly dependent on the pH value, for the growth of high-aspect-ratio TiO_2 NTs. Figure 10 shows the field emission scanning electron microscope (FE-SEM), top- and cross-sectional views of TiO_2 NTs for various electrolyte pH values adjusted by a NaOH reagent.

The length of the TiO_2 NTs is dependent on the electrolyte's pH [57]. The length of the nanotubular layer gradually increases with increasing pH, being 1, 1.3, 1.7, and 2.6 μm at pH 1.5, 2.3, 3.6, and 4.2, respectively. In a strongly acidic electrolyte, there are enough H^+ ions to increase the chemical dissolution rate of the thin titanium dioxide layer formed on the pore bottom. If the chemical dissolution reaction is dominant, it limits the growth of the nanotubular TiO_2 layer

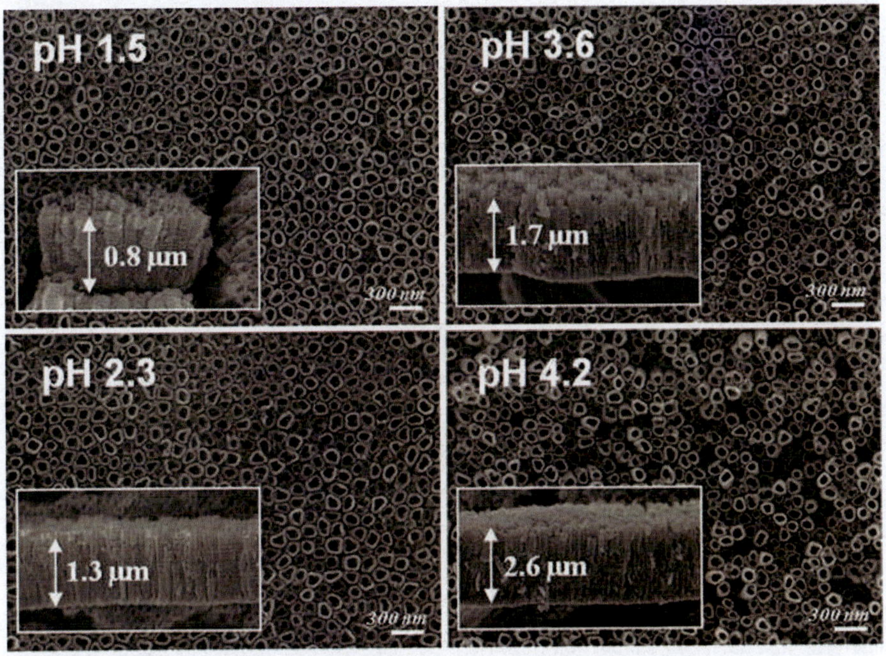

Fig. 10 FE-SEM top- and cross-sectional views of TiO_2 NTs for various electrolyte pH values adjusted by a NaOH reagent: **a** pH 1.5, **b** pH 2.3, **c** pH 3.6, and **d** pH 4.2

because the anodic electrolyte-containing F^- ions is directly connected to the bare Ti substrate, instead of the titanium dioxide layer, which terminates the anodic reaction. On the other hand, in a weakly acidic electrolyte, the chemical dissolution rate is retarded because the quantity of H^+ ions is decreased by Eq. 2, and is accompanied by a slight increase in the rate of the chemical oxidation reaction so that thicker anodic TiO_2 NTs are formed. In this condition, thicker TiO_2 NTs were grown. Above pH 5, thick TiO_2 layers with densely connected NPs were grown, instead of self-ordered TiO_2 NTs. In this case, the surface morphology exhibits a dendrite structure, which is formed due to the limited ion concentrations (not shown here). These are influenced by the limited ion concentration of the electrolyte and duration of anodization. The reduced number of H^+ ions at the pore bottom degrades the chemical dissolution rate and causes the formation of a thick and compact TiO_2 layer with dendrite-shaped surface morphology. All of the grown TiO_2 NTs were interconnected with a constant strain maintained among them to impede the formation of the exceptionally dominant grown TiO_2 NT that leans in the downward direction and limits the thickness of TiO_2 NTs.

Employing the TiO_2 NTs as a photoanode, the DSSCs were fabricated under the back-side illumination condition. Scheme 2 shows a representative diagram of a back-side illumination, NT-array dye solar cell [58]. Approximately 6 μm-thick TiO_2 NTs, grown in an electrolyte composed of 0.1 M KF, 1 M $NaHSO_4$, and 0.2 M

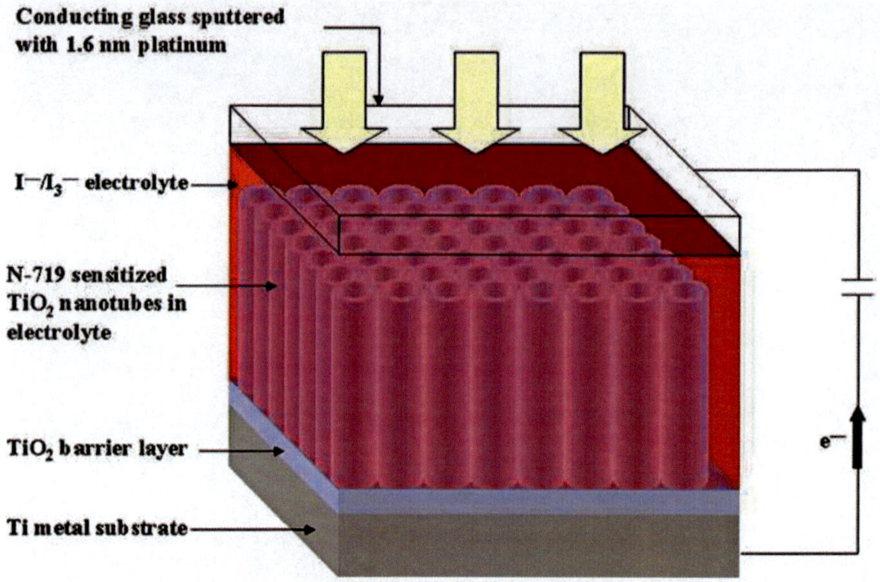

Scheme 2 Simple diagram describing a back-side illumination nanotube array dye-sensitized solar cell (Reprinted with permission from Ref. [58]. Copyright 2006 IOP Publishing Ltd.)

trisodium citrate with NaOH added to adjust the pH 5.0, were applied to the DSSCs with a resultant conversion efficiency of 4.24% (V_{oc}: 0.84 V, J_{sc}: 8.79 mA/cm^2, FF: 0.57) in the N719 dye dissolved in acetonitrile solvent. By adjusting the anodization conditions, such as the electrolyte composition and the anodization methods, different TiO_2 NTs with thicknesses of 1–2.5 μm (potential-sweeping) and 30 μm (rapid breakdown) were fabricated for the DSSCs [59]. The conversion efficiency was about 0.05% with 1–2.5 μm-thick TiO_2 NTs and 0.54% with 30 μm-thick TiO_2 NTs under back-side illumination. In addition, solid-state DSSCs with TiO_2 NTs electrodes were also made to use the easy penetration of viscous electrolyte along the large inner pores. At first, straight-stranded anatase TiO_2 NTs were produced by anodic oxidation on a pure titanium substrate in an aqueous solution containing a 0.45 wt% NaF electrolyte at a fixed pH of 4.3 [60]. The average length of the TiO_2 NTs was approximately 3 μm. To improve the conversion efficiency by retarding the charge recombination between the TiO_2 NT film and redox electrolyte, a thin ZnO shell was coated on the TiO_2 NT film, which increased V_{oc} (0.64–0.71 V) and J_{sc} (2.38–2.68 mA/cm^2), but similar FF (38–37%). Scheme 3 shows the interfacial charge-transfer process at the ZnO-coated TiO_2/dye/electrolyte of DSSC. To increase the FF, chemical etching was performed using hydrogen peroxide to reduce the compact TiO_2 layer formed between TiO_2 NT and the Ti foil, giving a final conversion efficiency of 0.906% (V_{oc}: 0.693 V, J_{sc}: 2.67 mA/cm^2, FF: 49%) with ZnO coating and chemical etching effect, compared to 0.578% for bare TiO_2 NT-based solid-state DSSCs.

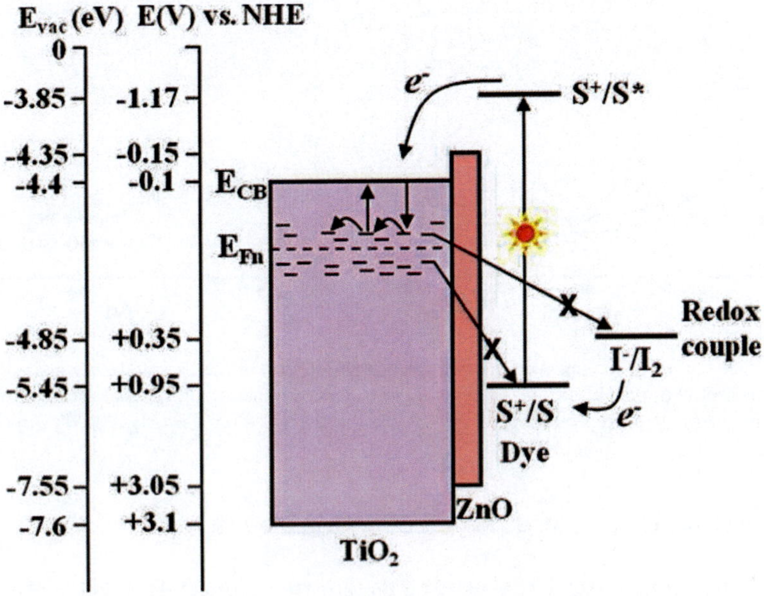

Scheme 3 Interfacial charge-transfer process at the ZnO-coated TiO_2/dye/electrolyte of DSSCs. Trapping/detrapping events of photoinjected electrons between the conduction band and trap sites of TiO_2 are shown in the region of E_{Fn} (Reprinted with permission from Ref. [60]. Copyright 2007 American Chemical Society)

Following from this result, the influence of ZnO coating in the charge-transfer process on the term of electron lifetime was investigated using open-circuit voltage decay (OCVD) [61]. Figure 11 shows the OCVD result from bare and ZnO-coated TiO_2 NTs. After 10 s of illumination, exponential decay of the photovoltage occurred immediately, followed by a steady decease. The ZnO-coated sample showed no significant change, while the bare TiO_2 film showed an irregular curve in the low photovoltage region. From the decay curve of the photo-voltage, the electron lifetime was calculated by applying Eq. 3, where k_BT is thermal energy, e is the electronic charge, and dV_{oc}/dt is the derivative of the open-circuit voltage transient.

$$\tau = -\frac{k_B T}{e}\left(\frac{dV_{oc}}{dt}\right)^{-1} \tag{3}$$

From this result, the ZnO-coated TiO_2 film showed a longer electron lifetime in the scanned potential range, indicating that more electrons were survived from the back-reaction, improving the photocurrent. On the other hand, the linear curve of the bare TiO_2 NT arrays showed some deviation in the low potential region, from 0.25 to 0.35 V, indicative of the formation of interfacial charge transfer from the trap/surface states to the cations of the redox electrolyte [62].

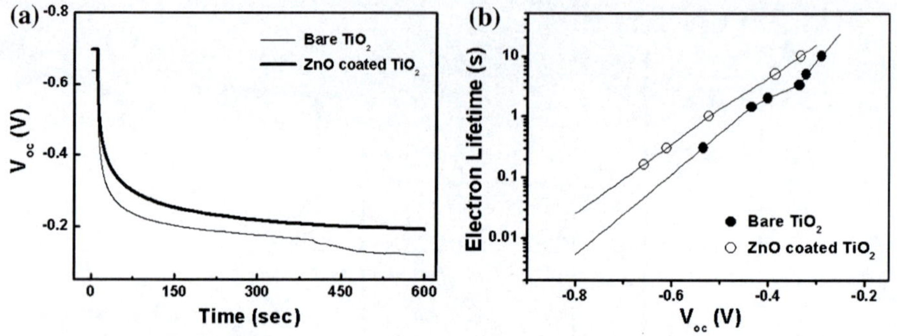

Fig. 11 The curves of the open-circuit voltage decay (OCVD) of the bare and ZnO-coated TiO_2 NTs: **a** the profile of the V_{oc} decay as a function of time, and **b** the log-plot of the electron lifetime as a function of V_{oc} (Reprinted with permission from Ref. [60]. Copyright 2007 American Chemical Society)

3.1.2 Electrochemical Anodization in the Organic Electrolyte

In the next step, the research focused on the growth of longer TiO_2 NT arrays in a polar organic-based electrolyte containing F^- ions. In the aqueous electrolyte, the thickness of the growing TiO_2 NTs was limited by a fast chemical dissolution process with the F^- ions and the concurrent chemical etching process on the top region of the tubular structure. In addition, the walls showed considerable disorder, like sidewall ripples and other etch effects. However, in the organic media containing the fluoride source, very long TiO_2 NTs with significantly improved tube wall roughness (smooth and high-aspect-ratio NTs) were formed on the Ti foil by controlling the high power potential and solvent species [63, 64]. In general, the anodization process occurred simultaneously in the three processes and was governed by (1) the field-assisted oxidation of Ti metal to form TiO_2, (2) the field-assisted dissolution of Ti metal ions in the electrolyte, and (3) the chemical dissolution of Ti and TiO_2 due to chemical etching by F^- ions [65].

Figure 12 shows the current transient as a function of anodization time in the ethylene glycol-based electrolyte containing 0.25 wt% NH_4F and 2 wt% water. In the initial stage (<100 s), gas evolution dominated from electronic conduction, which was attributed to the formation of a thinner oxide layer in the organic electrolyte. This is indicative of electronic conduction's dominance in the early part of the process. The current drops steeply thereafter due to the initial formation of an insulating oxide layer. In this region, electronic conduction decreases due to the blocking effect of the formed oxide layer and ionic conduction through the TiO_2 increases. Once the oxide layer is completely formed over the entire exposed surface of the anode, electronic conduction through the TiO_2 barrier layer becomes negligible and ionic conduction dominates the mechanistic behavior. The thinner oxide layer allows much greater ionic conduction than in aqueous electrolytes and faster movement of the Ti/TiO_2 interface into the Ti substrate, ultimately enabling substantial increases in NT length. Moreover, the higher anodization potential

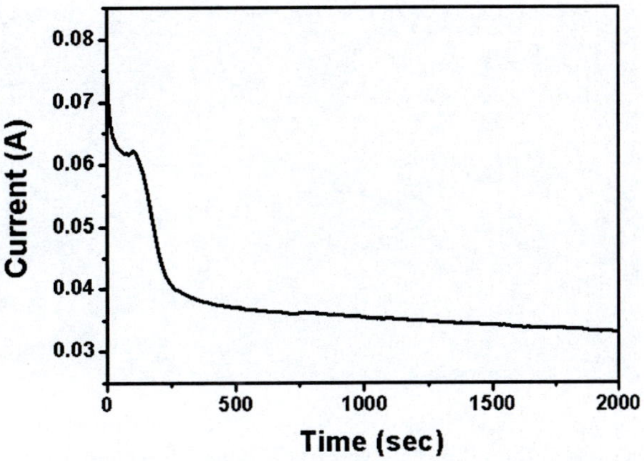

Fig. 12 Current-voltage profile during the anodization process in the organic electrolyte

exerts a greater driving force for both electronic and ionic conduction. As the anodization proceeds and the real surface area available for anodization is reduced, the current density is decreased in the final state.

Figure 13 displays the FE-SEM images of 10 μm-long TiO_2 NTs grown in ethylene–glycol-based electrolyte for 1 h anodization. Compared to TiO_2 NTs grown in aqueous electrolyte, well-aligned TiO_2 NTs with smooth wall were produced [66]. Depending on the anodization time, the length was controlled by a linear dependence. Furthermore, self-standing, 720 μm-thick TiO_2 NTs were fabricated using a double-sided electrochemical oxidation of Ti in an electrolyte comprised of water, NH_4F, and ethylene glycol [67]. From the effects of electrolyte composition, applied potential, and anodization duration on the length and diameter of the NTs, it was suggested that the reduced hydroxyl ion injected from the electrolyte, which enables faster high-field ionic conduction through the barrier layer, is responsible for the high NT growth rates.

Frank et al. studied the microstructure and dynamics of electron transport and recombination in DSSCs incorporating oriented TiO_2 NT arrays with a TiO_2 NP film as a control sample [68]. The arrays consisted of closely packed NTs several micrometers in length according to the anodization time, with typical wall thicknesses and intertube spacings of 8–10 nm and pore diameters of about 30 nm. The calcined material was fully crystalline with individual NTs consisting of approximately 30 nm-diameter crystallites. The transport and recombination properties of the NT and NP films used in DSSCs exhibited comparable transport times, while recombination was much slower in the NT films, indicating that the NT-based DSSCs have significantly higher charge-collection efficiencies (25%) than the NP-based films (Fig. 14).

This indicates that NT films can be made thicker than NP films for a given recombination loss. Besides, NT films exhibit higher light-harvesting efficiencies because of their stronger light-scattering effects. Accordingly, the conversion efficiencies of NT-based DSSCs were comparable or superior to those of NP film-based DSSCs of the same thickness.

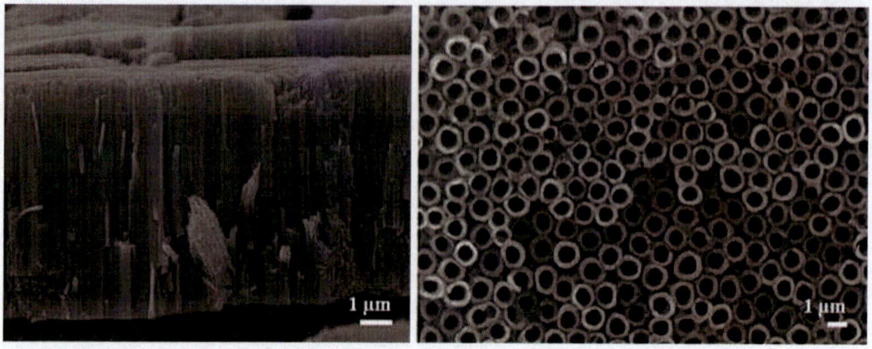

Fig. 13 FE-SEM images of TiO_2 NT arrays grown in ethylene–glycol-based electrolyte (Reprinted from Ref. [66]. Copyright 2008 with permission from Elsevier)

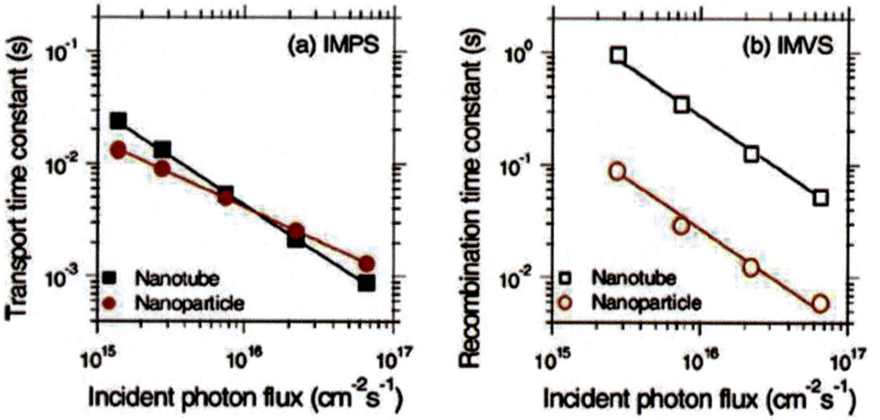

Fig. 14 Electron transport properties of TiO_2 NP and NT based DSSCs (Reprinted with permission from Ref. [68]. Copyright 2007 American Chemical Society)

However, with increasing anodization duration, the surface morphology became rough after an ultrasonic cleaning. Since the rough surface morphology affects light absorption and scattering under the illumination, uniform and self-ordered TiO_2 NTs, which look like anodic aluminum oxide, were formed by pretreatment of the Ti substrate [69]. Figure 15 shows the pretreated Ti substrate and the surface morphology after two-step anodization process. The TiO_2 NTs grown by two-step anodization showed a uniform surface morphology and thinner walls with a united wall structure between the pores like an aluminum oxide membrane, while one-step TiO_2 NTs showed a separated wall structure. The dependence of the electron behavior on the surface morphology of TiO_2 NTs in the DSSC was investigated. Anodically grown TiO_2 NTs, approximately 13 μm thick, were prepared on an ethylene glycol-based electrolyte. The TiO_2 NTs grown on the pretreated substrate showed uniform surface morphology with an

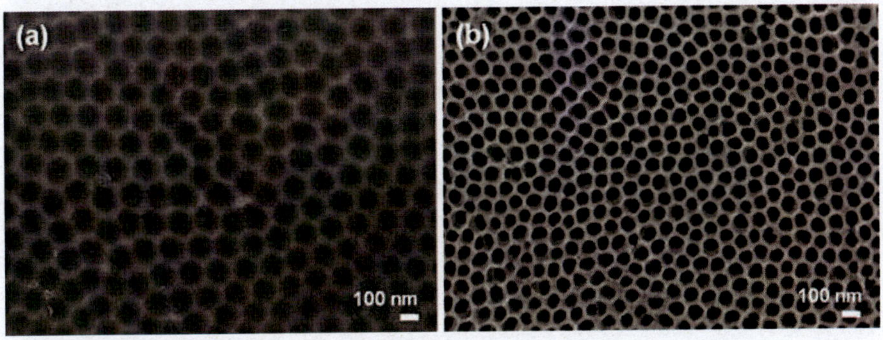

Fig. 15 FE-SEM images of pretreated Ti substrate and TiO_2 nanotube grown on pretreated Ti substrate (Reprinted with permission from Ref. [69]. Copyright 2009 IOP Publishing Ltd.)

interconnected nanotubular structure, while the surface morphology of the TiO_2 NTs formed on the bare substrate was quite rough. The photocurrent (8.4 mA/cm^2) of two-step TiO_2 NT-based DSSC was improved by 14% compared to that (7.2 mA/cm^2) of one-step TiO_2 NT-based DSSC. This improvement was attributed mainly to the increased light-harvesting efficiency (21%), that is, the light absorbance by a dye-sensitized TiO_2 NT film.

This was confirmed by a dye desorption experiment in NaOH after dye uptake. The discrepancy between the increasing light-harvesting yield (21%) and the overall photocurrent (14%) was attributed to the slow electron transport rate, a result of the large surface area and lateral movement along the 3-D network. Figure 16 shows the electron diffusion coefficient (D) and lifetimes in the DSSCs based on the one- and two-step TiO_2 NT electrodes with thicknesses of 7.5 ± 2.84 μm. The D values of the one-step TiO_2 NT-based DSSC were higher than those of the two-step TiO_2 NT-based DSSC over the light intensity range because the electrons in the two-step TiO_2 NT electrode were transported through a 3-D network instead of a 1-D conducting path. Generally, charge transport along a 3-D conducting path is slower than that along a 1-D conducting path because lateral movement reduces the transport rate significantly. The second explanation is that traps are located on or near the surface of the TiO_2 NPs, which are known as near-surface states. These reduce the electron transport rate with increasing internal surface area, which increases the number of traps sites. The mean surface area of the two-step TONT film was 21% higher than that of the one-step TiO_2 NT film. A third possible cause for inhibited electron diffusion is the wall thickness (electron pathway). The two-step TiO_2 had thinner walls and the jointed wall structure between the inner pores, while the pores of one-step TiO_2 NTs were divided separately. This means that the wall structure of the two-step TiO_2 NTs is unfavorable for electron transport.

On the other hand, the τ of the one-step TiO_2 NTs was lower than that of the two-step TiO_2 NT, possibly because of the distribution of traps near the surface region. In the case of the one-step TiO_2 NTs, the separation of each NT changes

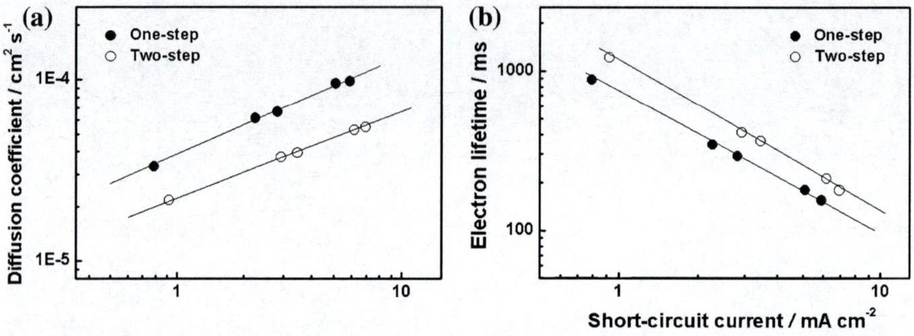

Fig. 16 **a** D and **b** τ in the DSSCs based on the one- and two-step, TONT electrode with a thickness of 7.5 ± 2.84 μm (Reprinted with permission from Ref. [69]. Copyright 2009 IOP Publishing Ltd.)

the transport mechanism to a mixture of one and three dimensions because each NT is bound strongly. The decrease in internal surface area is closely related to the decrease in trap site density. Photoinjected electrons were trapped preferentially on the energetically stable trap sites below the Fermi level. Electron transport to the contact involves multiple trapping events, which were modeled as the trapping and detrapping (thermal release) of the electrons along the distributed trap states. Unstable electrons in the conduction band leave after occupying the limited trap sites and can react easily with the cations from the redox electrolyte to maintain an energetically stable state. This might reduce the electron lifetimes on the one-step TiO_2 NT. Although the L value of two-step TiO_2 NT-based DSSCs is lower than that of one-step TiO_2 NT-based DSSCs, the L value is considerably higher than that of the NP-based DSSCs, suggesting that the nanotubular structure can support a thicker TiO_2 film for more efficient DSSCs. Therefore, the improved photocurrent of the two-step TiO_2 NT-based DSSC led to an enhancement (12.5%) of the overall power conversion efficiency.

Moreover, the diverse experimental conditions for electrochemical anodization affect the geometry and surface properties of the NT arrays [70–73]. Using appropriate alternating voltage switches between conditions, enabling tube growth and leading to a compact layer named the bamboo type structure [70]. By using different alternating voltage pulse durations, the distance between the bamboo rings can be adjusted. This structure resulted in an improved conversion efficiency of DSSCs due to the higher dye loading per unit volume by the bamboo ring, compared to normal NT structure where the carrier transport and recombination kinetics of bamboo and normal NT structure showed similar trends. The top surface morphology of the TONT was adjusted by several methods because this region is closely associated with electron transport and electrolyte penetration in DSSCs. To prevent the bundling and microcracks of oriented, NT film arrays, the supercritical CO_2 drying technique and the deposition of blocking layer composed of rutile-type oxide layers were widely used to grow the uniform self-ordered TiO_2 NT film [71]. Based on basic research on the TiO_2 NTs, the improvement of the photoconversion efficiency of the TiO_2 NT-based DSSC was also attractive because its maximum is about 4%. The main disadvantage of the TiO_2 NT electrode in DSSCs is the low surface area, especially in the case of TiO_2 NTs composed of large pores (>100 nm). Therefore, a $TiCl_4$ treatment of all areas of TiO_2 NTs was used to increase the dye adsorption area. Recently, the DSSC efficiency was doubled from 1.9 to 3.8% by such a treatment [72]. Then, 3 nm-sized TiO_2 nanocrystallites were uniformly coated on the surface region and annealed afterward. The amorphous TiO_2 NTs were transformed to anatase TiO_2 NTs by annealing at about 350°C, followed by the $TiCl_4$ treatment and, finally, annealing again at 450°C for 30 min. This research showed the potential for TiO_2 NT-based DSSCs to achieve a high efficiency. In addition, MgO was "coated" on the TiO_2 NT surface as an energy barrier to increase the efficiency by an order of magnitude higher than that of the MgO using bare TiO_2 NTs [73]. This was elucidated by the reduction of charge recombination between the TiO_2 and electrolyte and the improvement of dye adsorption due to the basicity of the MgO surface.

3.1.3 Electrochemical Anodization of Sputtered Grown Ti Metal on TCO Substrate for DSSCs

The micron-length TiO_2 NTs grown on opaque Ti foil should utilize back-side illumination from the counter electrode side to measure the DSSC efficiency. This structure decreases the total absorbing visible light because the light must pass through the counter-electrode and the light-absorbing electrolyte. Approximately 25% of the incident solar energy that is lost from the counter side can be recovered from the Ti metal film on the TCO substrate [74]. Grimes et al. reported highly ordered transparent TiO_2 NT arrays on TCO substrates in DSSCs [75]. The critical point to form transparent TiO_2 NTs on the FTO is the fabrication of a high quality Ti thin film grown at a high substrate temperature (500°C) and a slow sputtering rate during the sputtering process. After the electrochemical anodization at a constant potential of 12 V in an electrolyte of 0.5% HF and acetic acid mixed in a 7:1 ratio (constant temperature: 5°C), highly ordered NT arrays of 46 nm pore diameter, 17 nm wall thickness, and 360 nm length were grown perpendicular to a FTO-coated glass substrate. The $TiCl_4$ treatment on TiO_2 NTs enhanced the photocurrent in the commercial DSSC structure using ruthenium-based dyes. A photocurrent efficiency of 2.9% was attained with 360 nm-thick TiO_2 NTs. Furthermore, it was revealed that the highly ordered TiO_2 NT arrays, in comparison to nanoparticulate systems, have superior electron lifetimes and provide excellent pathways for electron percolation from voltage-decay measurements (Fig. 17). Furthermore, they succeeded in growing the highly ordered TiO_2 NTs (above 30 μm) on the FTO substrate 3 years later [76]. By employing several advantages of the TiO_2 NT structure, enhanced charge harvesting in the near-infrared region of the solar spectrum was thereby enabled. Three challenges remain to be overcome in the fabrication of transparent TiO_2 NT array films with extended lengths. The first is the formation

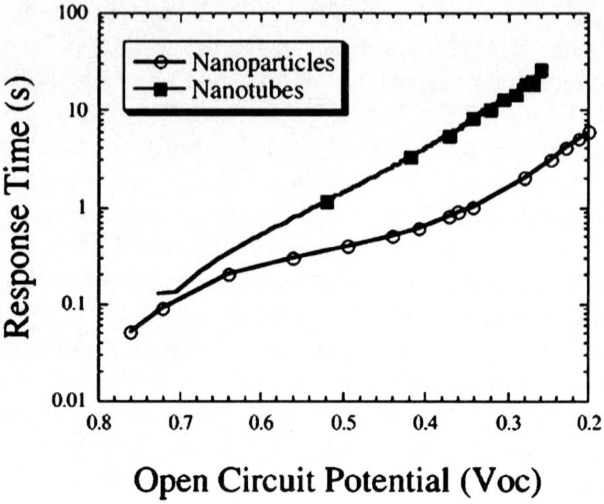

Fig. 17 Response time according to V_{oc} decay for a transparent TiO_2 NT array DSSC as well as the response time for a TiO_2 NP DSSC (Reprinted with permission from Ref. [75]. Copyright 2006 American Chemical Society)

of uniform, nonporous TiO_2 films with thicknesses of tens of micrometers and sufficient adhesion to the FTO glass layer. Second, the anodization of these thick TiO_2 films until uniform optical transparency is achieved. Finally, the third is the absence of any debris or clumping of the very long NTs after the anodization in the fluorine-containing, non-aqueous organic electrolytes. These challenges were overcome by the growth of a high quality Ti metal film on a FTO substrate and an appropriate electrolyte conductivity was selected to minimize interface etching and the surface debris problem, thus yielding long, uniform, non-clumped,and debris-free transparent NT array films. DSSCs were then fabricated using these TiO_2 NTs. Depending on the TiO_2 NT thickness, the efficiency ranged from 2.57% (1.2 μm) to 6.86% (20 μm). Even though the TiO_2 NT thickness was increased up to 33 μm, the efficiency remained low relative to that of NP DSSCs (10–11%). The major limitation of the power efficiency of NT DSSCs is their low FFs, which are about 25% lower than those of NP DSSCs. This was attributed to the degradation of the FTO film occurring during the extended, high-temperature treatment of the NTs for improving crystallinity and oxidizing trace metals under the NT film. It is expected that more efficient DSSCs will be fabricated after these problems are eliminated.

3.2 Hydrothermal Reaction for Growing TiO_2 NWs/NTs on Ti and TCO Substrates for DSSCS

Recently, several fascinating results related to TiO_2 NWs directly grown on the substrate were produced by hydrothermal process. The first is the growth of oriented, single-crystalline TiO_2 NW arrays on Ti foil [77]. After the formation of single crystal, sodium titanate NWs ($Na_2Ti_2O_5{\cdot}H_2O$), protonated bititanate NWs were formed by ion-exchange process without changing their morphology or crystal structure. The final calcination process led to the formation of the single-crystalline anatase TiO_2 NWs. This sequential process yielded 2–50 μm-long, single-crystalline NWs that oriented in the (100) direction and primarily normal to the Ti foil. The DSSCs fabricated using 12 μm-thick TiO_2 NW films on Ti foil showed a low-energy conversion efficiency ($\sim$1.4%), which was explained by five times lower roughness factor compared to the NP film and the formation of a thick, resistive TiO_2 layer between the NWs and Ti foil during the calcinations step. In a similar context, TiO_2 NWs were grown on a spiral-shaped titanium foil, synthesized by hydrothermal reactions at 230°C for 4 h to form NWs with diameters of 20–23 nm and lengths of 2–5 μm [78]. The TiO_2 NWs were randomly oriented and entangled to form a macroporous structured film. The TiO_2 NW film provided sufficient porosity for efficient dye adsorption and fast diffusion of redox couple. Accordingly, the DSSCs composed of TiO_2 NWs grown on the spiral-shaped titanium wire showed a conversion efficiency of 0.85% (V_{oc}: 0.616 V, J_{sc}: 2.3 mA/cm^2, FF: 0.61).

Furthermore, in a similar hydrothermal process using TiO_2 NPs as seeds, oriented TiO_2 NTs were formed on the Ti foil [79]. After the preparation of a dilute TiO_2 suspension by dispersing 1.0 g of Degussa P25 powder in deionized (DI) water, the TiO_2 NPs were deposited onto a Ti foil through dip coating in the suspension. The Ti foil containing the predeposited TiO_2 NPs reacted with an alkaline solution in a sealed Teflon® reactor containing 10 mL of 10 M NaOH solution. After reaction, the Ti foil, now covered with the newly formed film, was washed with DI water. The film showed an oriented texture and titanate structure composed of multilayered sheets.

Recently, single-crystalline rutile TiO_2 NRs or NWs were formed on FTO by adjusting parameters such as growth time, growth temperature, initial reactant concentration, acidity, and additives to control the diameters and lengths of the single-crystalline TiO_2 NRs [80]. The key factor in growing single- crystalline rutile TiO_2 NRs is a small lattice mismatch (epitaxial relation) between the FTO substrate and rutile TiO_2, driving the nucleation and growth of the nanorods. Figure 18 shows FE-SEM images of oriented, single crystal TiO_2 NWs on the FTO substrate. The entire surface of the FTO substrate is covered very uniformly with TiO_2 NRs with lengths of 90 ± 5 nm and diameters of 1.9 ± 0.1 μm. The NRs are nearly perpendicular to the FTO substrate. For application to DSSCs, an array of 4 μm-long TiO_2 NRs was used as the photoanode. To increase the dye adsorbing area, $TiCl_4$ treatment was carried out to give a photoconversion efficiency of 3% (V_{oc}: 0.71 V, J_{sc}: 6.05 mA/cm^2, FF: 0.7). However, the single-crystalline TiO_2 NWs didnot show any improvement in the electron dynamics relative to the TiO_2 NP systems due to the intrinsic properties of rutile TiO_2 [81] and a dependence on the slow and light intensity-dependent electron transport rate, indicating that trapping and detrapping, mostly in surface traps, still play an important role in electron transport [82]. Besides, a *c*-axis, highly oriented, sandwiched TiO_2 film was grown by hydrothermal process, in which one TiO_2 NP interlayer acted as the seed layer [83]. Two layers of TiO_2 NRs were grown toward both directions of the bulk solution and the substrate. Figure 19

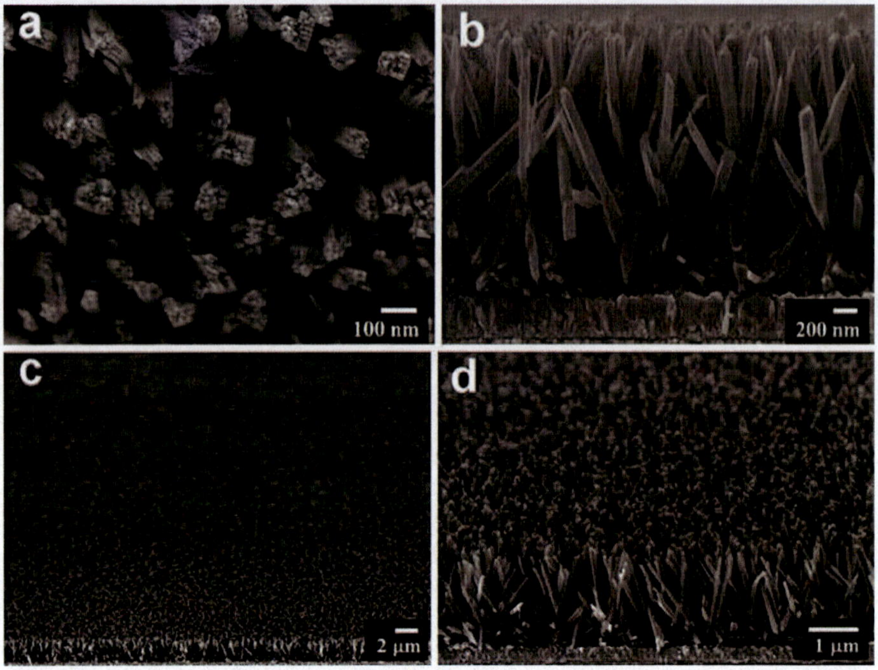

Fig. 18 FE-SEM images of oriented, rutile TiO_2 NR film grown on an FTO substrate at 150°C for 20 h: **a** top, **b** cross-sectional, and **c** and **d** tilted cross-sectional views (Reprinted with permission from Ref. [80]. Copyright 2009 American Chemical Society)

shows a schematic illustration of the formation mechanism of the sandwiched TiO_2 film. The high supersaturation degree of the solution leads to homogenous nucleation of TiO_2, which is precipitated on the FTO surface as seeds for further growth. From these seeds, subsequent crystal growth proceeds radially in the directions of the bulk solution and the substrate along with FTO etching. The density and length of the TiO_2 NRs can be tuned on both sides simply by controlling the reaction temperature and time.

Furthermore, Miyauchi et al. later reported a hydrothermal method for growing TiO_2 nanotubular arrays directly on the Ti substrate without the use of TiO_2 seeds [84]. The resultant nanotubular array structure exhibits superhydrophilic properties. Furthermore, the use of branched titanate NTs to grow a 3-D nanotubular network directly on the Ti substrate was reported [85]. The resultant 3-D nanotubular network exhibits a unique, three-dimensional, uniform, and porous structure. The 3-D nanotubular network structure was formed by the joining of branched NTs, as opposed to the previous vertical growth on the substrate. The inner and outer tubular diameters of the branched titanate NTs were approximately 6 and 12 nm, respectively. Branched titanate NTs were formed on the Ti substrate (Fig. 20). Therefore, a continuous seed formation-oriented, crystal growth mechanism was proposed for the branched titanate nanotubular network formation.

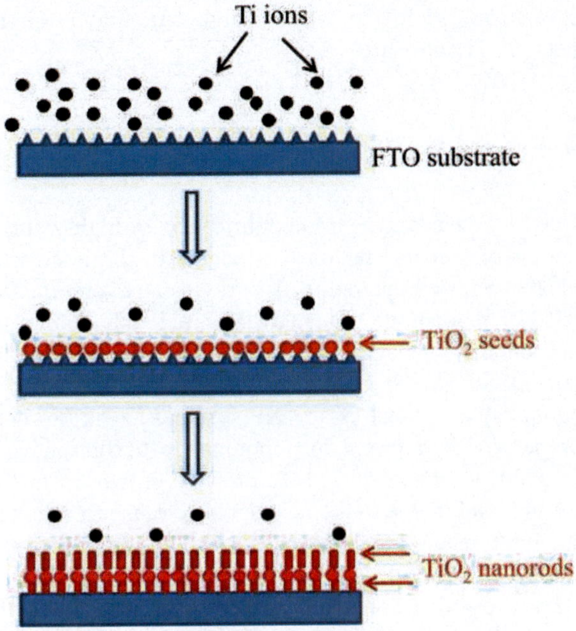

Fig. 19 Schematic illustration of the formation mechanism of sandwiched TiO_2 film: (I) formation of TiO_2 seeds, and (II) growth of TiO_2 NRs in both directions (Reprinted with permission from Ref. [83]. Copyright 2009 IOP Publishing Ltd.)

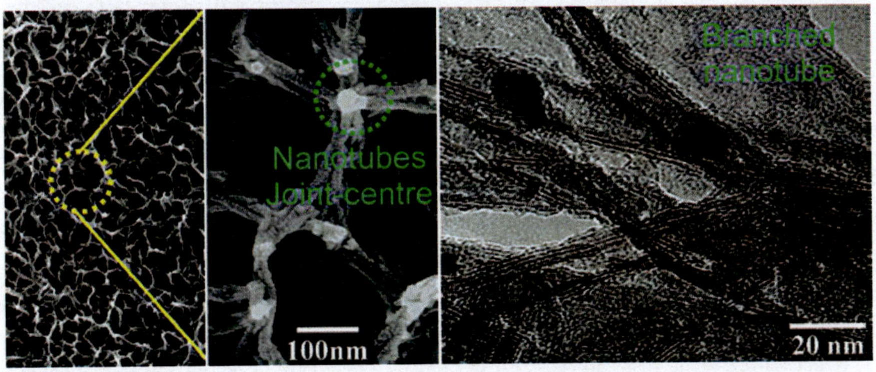

Fig. 20 FE-SEM and TEM images of hydrothermally prepared, nanobranched TiO_2 nanotubes (Reprinted with permission from Ref. [85]. Copyright 2010 American Chemical Society)

When implemented in a DSSC, this morphology produced a J_{sc} of 5.58 mA/cm^2, and V_{oc} of 0.74 V, and a FF of 60%. An overall photoconversion efficiency was 3.0%, compared to only 0.33% with a J_{sc} of 1.01 mA/cm^2, an V_{oc} of 0.65 V, and a FF of 0.42 for the DSSCs constructed using a TiO_2 porous film photoanode. From these results, the conversion efficiency of the 3-D titanate NT network photoanode cell was nine times higher than that of the TiO_2 porous film photoanode cell, which was attributed to the textural and structural properties of the 3-D nanotubular network architecture.

4 Future Outlook

One of the challenges in the fabrication of highly efficient DSSCs is the control of the charge recombination that occurs in the interfacial region between the TiO_2 photoanode and dye/redox electrolyte. To overcome this problem, the use of a 1-D NR/NT photoanode instead of the previously applied NP film has been researched to increase the electron transport properties. Despite active investigations of various DSSC photoanodes from NPs to 1-D NR/NT, the electron dynamics remain comparable to those of the NP system. To further enhance the electron transport properties, four promising photoanode designs have been proposed: (1) single-crystalline TiO_2 NRs or NTs directly grown on the substrate will hinder charge recombination due to the fast electron transport rate, (2) the development of a self-assembly method to deposit the 1-D TiO_2 nanomaterials, (3) the use of single wall carbon NT scaffolds to increase the charge separation and transport, and (4) the fabrication of a core-shell structure with a core material showing fast electron transport rate, such as ZnO, Cu, Sn, Ti, and doped TiO_2, and a shell material mainly consisting of TiO_2 due to the consideration of the electronic bandgap. The significant enhancement of electron transport ability in the DSSC photoanode that is expected to be achieved from these research efforts will contribute to the blocking of photogenerated electron loss, and hence the improvement of photoconversion efficiency in the DSSCs.

Acknowledgment This work was supported by the Research Center for Energy Conversion and Storage (Contract No. R11-2002-102-00000-0) and the WCU (World Class University) program (R31-10013) through the National Research Foundation of Korea funded by the Ministry of Education, Science and Technology.

References

1. Bennett JM, Pelletier E, Albrand G et al (1989) Comparison of the properties of titanium dioxide films prepared by various techniques. Appl Opt 28:3303–3317
2. Wang TM, Zheng SK, Hao WC et al (2002) Studies on photocatalytic activity and transmittance spectra of TiO_2 thin films prepared by r.f. magnetron sputtering method. Surf Coat Technol 155:141–145
3. Manna L, Scher EC, Li L-S, Alivisatos AP (2002) Epitaxial growth and photochemical annealing of graded CdS/ZnS shells on colloidal CdSe nanorods. J Am Chem Soc 124:7136–7145
4. Nelson J (1987) Organic photovoltaic films. Curr Opin Solid State Mater Sci 6:87–95
5. Law M, Greene LE, Yang PD et al (2005) Nanowire dye-sensitized solar cells. Nat Mater 4:455–459
6. Naoi K, Ohko Y, Tatsuma T (2004) TiO_2 films loaded with silver nanoparticles: control of multicolor photochromic behavior. J Am Chem Soc 125:3664–3668

7. Yang Z, Xu T, Ito Y, Welp U et al (2009) Enhanced electron transport in dye-sensitized solar cells using short ZnO nanotips on a rough metal anode. J Phys Chem C 113:20521–20526
8. O'Regan B, Grätzel M (1991) A low-cost, high-efficiency solar cell based on dye-sensitized colloidal TiO_2 films. Nature 353:737–740
9. Wang Z-S, Kawauchi H, Kashima T, Arakawa H (2004) Significant influence of TiO_2 photoelectrode morphology on the energy conversion efficiency of N719 dye-sensitized solar cell. Coord Chem Rev 248:1381–1389
10. Kopidakis N, Benkstein KD, van de Lagemaat J, Frank AJ (2003) Transport-limited recombination of photocarriers in dye-sensitized nanocrystalline TiO_2 solar cells. J Phys Chem B 107:11307–11315
11. Solbrand A, Henningsson A, Södergren S, Lindström H, Hagfeldt A, Lindquist S-E (1999) Charge transport properties in dye-sensitized nanostructured TiO_2 thin film electrodes studied by photoinduced current transients. J Phys Chem B 103:1078–1083
12. You M, Kim TG, Sung Y-M (2010) Synthesis of Cu-doped TiO_2 nanorods with various aspect ratios and dopant concentrations. Cryst Growth Des 10:983–987
13. Pradhan SK, Reucroft PJ, Yang F, Dozier A (2003) Growth of TiO_2 nanorods by metalorganic chemical vapor deposition. J Cryst Growth 256:83–88
14. Lei Y, Zhang LD, Meng GW et al (2001) Preparation and photoluminescence of highly ordered TiO_2 nanowire arrays. Appl Phys Lett 78:1125–1127
15. Wolcott A, Smith WA, Kuykendall TR et al (2009) Photoelectrochemical water splitting using dense and aligned TiO_2 nanorod arrays. Small 5:104–111
16. Polleux J, Pinna N, Niederberger M (2005) Ligand functionality as a versatile tool to control the assembly behavior of preformed titania nanocrystals. Chem Eur J 11:3541–3551
17. Cozzoli PD, Kornowski A, Weller H (2003) Low-temperature synthesis of soluble and processable organic-capped anatase TiO_2 nanorods. J Am Chem Soc 125:14539–14548
18. Schubert U, Tewinkel S, Lamber R (1996) Metal complexes in inorganic matrixes. 15. Coordination of metal ions by lysinate-modified titanium and zirconium alkoxides and the preparation of metal/titania and metal/zirconia nanocomposites. Chem Mater 8:2047–2055
19. Jiu J, Isoda S, Wang F, Adachi M (2006) Dye-sensitized solar cells based on a single-crystalline TiO_2 nanorod film. J Phys Chem B 110:2087–2092
20. Penn RL, Banfield JF (1998) Imperfect oriented attachment: dislocation generation in defect-free nanocrystals. Science 281:969–971
21. Penn RL, Banfield JF (1999) Morphology development and crystal growth in nanocrystalline aggregates under hydrothermal conditions: insights from titania. Geochim Cosmochim Acta 63:1549–1557
22. Penn RL (2004) Kinetics of oriented aggregation. J Phys Chem B 108:12707–12712
23. Han S, Choi S-H, Hyeon T et al (2005) Low-temperature synthesis of highly crystalline TiO_2 nanocrystals and their application to photocatalysis. Small 1:812–816
24. Kang SH, Choi S-H, Hyeon T, Sung Y-E (2008) Nanorod-based dye-sensitized solar cells with improved charge collection efficiency. Adv Mater 20:54–58
25. Nakade S, Kanzaki T, Yanagida S et al (2005) Stepped light-induced transient measurements of photocurrent and voltage in dye-sensitized solar cells: application for highly viscous electrolyte systems. Langmuir 21:10803–10807
26. Fisher AC, Peter LM, Ponomarev EA et al (2000) Intensity dependence of the back reaction and transport of electrons in dye-sensitized nanocrystalline TiO_2 solar cells. J Phys Chem B 104:949–958
27. Bisquert J, Zaban A, Salvador P (2002) Analysis of the mechanisms of electron recombination in nanoporous TiO_2 dye-sensitized solar cells. Nonequilibrium steady-state statistics and interfacial electron transfer via surface states. J Phys Chem B 106:8774–8782
28. Usami A, Ozaki H (2001) Computer simulations of charge transport in dye-sensitized nanocrystalline photovoltaic cells. J Phys Chem B 105:4577–4583
29. Adachi M, Murata Y et al (2004) Highly efficient dye-sensitized solar cells with a titania thin-film electrode composed of a network structure of single-crystal-like TiO_2nanowires made by the "oriented attachment" mechanism. J Am Chem Soc 126:14943–14949

30. Li S, Li Y, Wang H et al (2009) Peptization-hydrothermal method as a surfactant-free process toward nanorod-like anatase TiO_2 nanocrystals. Eur J Inorg Chem 2009:4078–4084
31. Brinker CJ, Schere GW (1989) Sol-gel science: the physics and chemistry of sol-gel processing. Elsevier Science, USA
32. Chemseddine A, Moritz T (1999) Nanostructuring titania: control over nanocrystal structure, size, shape, and organization. Eur J Inorg Chem 1999:235–245
33. Kasuga T, Hiramatsu M, Hoson A et al (1999) Titania nanotubes prepared by chemical processing. Adv Mater 11:1307–1311
34. Ohsaki Y, Masaki N, Yanagida S et al (2005) Dye-sensitized TiO_2 nanotube solar cells: fabrication and electronic characterization. Phys Chem Chem Phys 7:4157–4163
35. Koo B, Park J, Hyeon T et al (2006) Simultaneous phase- and size-controlled synthesis of TiO_2 nanorods via non-hydrolytic sol-gel reaction of syringe pump delivered precursors. J Phys Chem B 110:24318–24323
36. Cheng H, Ma J, Qi L et al (1995) Hydrothermal preparation of uniform nanosize rutile and anatase particles. Chem Mater 7:663–671
37. Matthews A (1976) The crystallisation of anatase and rutile from amorphous titanium dioxide under hydrothermal conditions. Am Mineral 61:419–424
38. Izumi F (1978) The polymorphic crystallization of Titanium(IV) oxide under hydro-thermal conditions II. The roles of inorganic anions in the nucleation of rutile and anatase from acid solutions. Bull Chem Soc Jpn 51:1771–1776
39. Matijevic E (1977) The role of chemical complexing in the formation and stability of colloidal dispersions. J Colloid Interface Sci 58:374–389
40. Wei M, Konishi Y, Zhou H, Arakawa H et al (2006) Utilization of titanate nanotubes as an electrode material in dye-sensitized solar cells. J Electrochem Soc 153:A1232–A1236
41. Chen A, Zhou W, Du G, Peng L-M (2002) Trititanate nanotubes made via a single alkali treatment. Adv Mater 14:1208–1211
42. Boercker JE, Enache-Pommer E, Adyil ES (2008) Growth mechanism of titanium dioxide nanowires for dye-sensitized solar cells. Nanotechnology 19:095604–095613
43. Enache-Pommer E, Boercker JE, Adyil ES (2007) Electron transport and recombination in polycrystalline TiO_2nanowire dye-sensitized solar cells. Appl Phys Lett 91:123116–123118
44. Tan B, Wu Y (2006) Dye-sensitized solar cells based on anatase TiO_2 nanoparticle/nanowire composites. J Phys Chem B 110:15932–15938
45. Beppu T, Yamaguchi S, Hayase S (2007) Improvement of heat resistant properties of TiO_2 nanowires and application to dye-sensitized solar cells. Jpn J Appl Phys 46:4307–4311
46. Oh J-K, Lee J-K, Park K-W (2010) TiO_2 branched nanostructure electrodes synthesized by seeding method for dye-sensitized solar cells. Chem Mater 22:1114–1118
47. Meekins BH, Kamat PV (2009) Got TiO_2 nanotubes lithium ion intercalation can boost their photoelectrochemical performance. ACS Nano 3:3437–3446
48. Kuang D, Brillet J, Chen P, Grätzel M et al (2008) Application of highly ordered TiO_2 nanotube arrays in flexible dye-sensitized solar cells. ACS Nano 2:1113–1116
49. Shankar K, Mor GK, Prakasam HE, Grimes CA (2007) Self-assembled hybrid polymer-TiO_2 nanotube array heterojunction solar cells. Langmuir 23:12445–12449
50. Yu B-Y, Tsai A, Tsai S-P, Wong K-T, Shyue J-J (2008) Efficient inverted solar cells using TiO_2 nanotube arrays. Nanotechnology 19:255202–255206
51. Wang J, Lin Z (2010) Dye-sensitized TiO_2 nanotube solar cells with markedly enhanced performance via rational surface engineering. Chem Mater 22:579–584
52. Zwilling V, Aucouturier M, Darque-Ceretti E (1999) Anodic oxidation of titanium and TA6 V alloy in chromic media. An electrochemical approach. Electrochim Acta 45:921–929
53. Macäk JM, Tsuchiya H, Schmuki P (2005) High-aspect-ratio TiO_2 nanotubes by anodization of titanium. Angew Chem Int Ed 44:2100–2102
54. Cai QPM, Varghese OK, Grimes CA et al (2005) The effect of electrolyte composition on the fabrication of self-organized titanium oxide nanotube arrays by anodic oxidation. J Mater Res 20:230–236

55. Taveira LV, Macäk JM, Tsuchiya H, Schmuki P et al (2005) Initiation and growth of self-organized TiO_2 nanotubes anodically formed in $NH_4F/(NH_4)_2SO_4$ electrolytes. J Electrochem Soc 152:B405–B410
56. Kang SH, Kim J-Y, Sung Y-E et al (2008) Formation and mechanistic study of self-ordered TiO_2 nanotubes on Ti substrate. J Ind Eng Chem 14:52–59
57. Bayoumi FM, Ateya BG (2006) Formation of self-organized titania nano-tubes by dealloying and anodic oxidation. Electrochem Comm 8:38–44
58. Paulose M, Shankar K, Varghese OK, Grimes CA et al (2006) Backside illuminated dye-sensitized solar cells based on titania nanotube array electrodes. Nanotechnology 17:1446–1448
59. Hahn R, Stergiopoulus T, Schmuki P et al (2007) Efficient solar energy conversion using TiO_2 nanotubes produced by rapid breakdown anodization—a comparison. Phys Stat Sol (RRL) 1:135–137
60. Kang SH, Kim J-Y, Sung Y-E et al (2007) Surface modification of stretched TiO_2 nanotubes for solid-state dye-sensitized solar cells. J Phys Chem C 111:9614–9623
61. Bisquert J, Zaban A, Greenshtein M, Mora-Seró I (2004) Determination of rate constants for charge transfer and the distribution of semiconductor and electrolyte electronic energy levels in dye-sensitized solar cells by open-circuit photovoltage decay method. J Am Chem Soc 126:13550–13559
62. Fabregat-Santiago F, García-Cañadas J (2004) The origin of slow electron recombination processes in dye-sensitized solar cells with alumina barrier coatings. J Appl Phys 96:6903–6907
63. Macak JM, Tsuchiya H, Schmuki P et al (2005) Smooth anodic TiO_2 nanotubes. Angew Chem Int Ed 44:7463–7465
64. Shankar K, Mor GK, Grimes CA et al (2007) Highly-ordered TiO_2 nanotube arrays up to 220 μm in length: use in water photoelectrolysis and dye-sensitized solar cells. Nanotechnology 18:065707–065717
65. Paulose M, Shankar K, Yoriya S, Grimes CA et al (2006) Anodic growth of highly ordered TiO_2 nanotube arrays to 134 μm in length. J Phys Chem B 110:16179–16184
66. Lee W, Kang SH, Sung Y-E, Han S-H et al (2008) Co-sensitization of vertically aligned TiO_2 nanotubes with two different sizes of CdSe quantum dots for broad spectrum. Electrochem Comm 10:1579–1582
67. Wang J, Lin Z (2008) Freestanding TiO_2 nanotube arrays with ultrahigh aspect ratio via electrochemical anodization. Chem Mater 20:1257–1261
68. Zhu K, Neale NR, Frank AJ et al (2007) Enhanced charge-collection efficiencies and light scattering in dye-sensitized solar cells using oriented TiO_2 nanotubes arrays. Nano Lett 7:69–74
69. Kang SH, Kim HS, Sung Y-E et al (2009) An investigation on electron behavior employing vertically-aligned TiO_2 nanotube electrodes for dye-sensitized solar cells. Nanotechnology 20:355307–355312
70. Kim D, Ghicov A, Schmuki P et al (2008) Bamboo-type TiO_2 nanotubes: improved conversion efficiency in dye-sensitized solar cells. J Am Chem Soc 130:16454–16455
71. Zhu K, Neale NR, Frank AJ et al (2007) Removing structural disorder from oriented TiO_2 nanotube arrays: reducing the dimensionality of transport and recombination in dye-sensitized solar cells. Nano Lett 7:3739–3746
72. Roy P, Kim D, Schmuki P et al (2009) Improved efficiency of TiO_2 nanotubes in dye sensitized solar cells by decoration with TiO_2 nanoparticles. Electrochem Comm 11:1001–1004
73. Park H, Yang D-J, Kim H-G et al (2009) Fabrication of MgO-coated TiO_2 nanotubes and application to dye-sensitized solar cells. J Electroceram 23:146–149
74. Ito S, Cevey Ha N-L, Grätzel M et al. (2006) High-efficiency (7.2%) flexible dye-sensitized solar cells with Ti-metal substrate for nanocrystalline-TiO_2 photoanode. Chem Comm 4004–4006

75. Mor GK, Shankar K, Paulose M, Grimes CA et al (2006) Use of highly-ordered TiO_2 nanotube arrays in dye-sensitized solar cells. Nano Lett 6:215–218
76. Varghese OK, Paulose M, Grimes CA (2009) Long vertically aligned titania nanotubes on transparent conducting oxide for highly efficient solar cells. Nat Mater 4:592–597
77. Liu B, Boercker JE, Aydil ES (2008) Oriented single crystalline titanium dioxide nanowires. Nanotechnology 19:505604–505610
78. Wang H, Liu Y, Li M et al (2009) Hydrothermal growth of large-scale macroporous TiO_2 nanowires and its application in 3D dye-sensitized solar cells. Appl Phys A 97:25–29
79. Tian ZR, Voigt JA, Xu H et al (2003) Large oriented arrays and continuous films of TiO_2-based nanotubes. J Am Chem Soc 125:12384–12385
80. Liu B, Adyil ES (2009) Growth of oriented single-crystalline rutile TiO_2 nanorods on transparent conducting substrates for dye-sensitized solar cells. J Am Chem Soc 131:3985–3990
81. Kang SH, Kang M-S, Sung Y-E et al (2008) Columnar rutile TiO_2 based dye-sensitized solar cells by radio-frequency magnetron sputtering. J Power Sources 184:331–335
82. Enache-Pommer E, Liu B, Adyil ES (2009) Electron transport and recombination in dye-sensitized solar cells made from single-crystal rutile TiO_2 nanowires. Phys Chem Chem Phys 11:9648–9652
83. Han Y, Wu G, Wang M, Chen H (2009) The growth of a *c*-axis highly oriented sandwiched TiO_2 film with superhydrophilic properties without UV irradiation on SnO:F substrate. Nanotechnology 20:235605–235611
84. Miyauchi M, Tokudome H (2007) Super-hydrophilic and transparent thin films of TiO_2 nanotube arrays by a hydrothermal reaction. J Mater Chem 17:2095–2100
85. Zhang H, Liu P, Zhao H et al (2010) Facile formation of branched titanate nanotubes to grow a three-dimensional nanotubular network directly on a solid substrate. Langmuir 26:1574–1578

Nanostructured Materials Formed by Molecular Layer Deposition for Enhanced Solar Energy Utilization with Optical Waveguides

Tetsuzo Yoshimura

Abstract We propose a waveguide-type sensitized photovoltaic device, where the porous semiconductors in a conventional dye-sensitized solar cell are replaced with flat thin-film semiconductors of high crystalline quality and high electron mobility. Although the light-absorbing layer on the semiconductor surface is very thin, sufficient light absorption is available in the guided light beam configurations. The light-absorbing layer is fabricated by molecular layer deposition (MLD), which achieves molecule-by-molecule polymer wire growth with designated molecular arrangements. MLD also forms self-assembled molecular structures utilizing electrostatic force. We demonstrated that MLD constructs polymer wires with multiple quantum dots, which are expected as efficient and robust light-absorbing layers. Measurements in the guided light beam configurations revealed that sensitized spectral regions of photocurrents are widened by self-assembled double layers of p-type and n-type dye molecules formed on ZnO thin films by liquid-phase MLD. Integration of the waveguide type devices into light beam collecting films will provide resource-saving flexible solar energy conversion systems.

1 Introduction

Nanostructured organic materials provide various kinds of optoelectronic devices. Molecular layer deposition (MLD) [1–8], which enables self-limiting monomolecular-step growth of organic materials, is a promising method for nanostructure construction. In MLD, different kinds of molecules are introduced onto a substrate

T. Yoshimura (✉)
Tokyo University of Technology, 1404-1 Katakura, Hachioji,
Tokyo 192-0982, Japan
e-mail: tetsu@cs.teu.ac.jp

L. Zang (ed.), *Energy Efficiency and Renewable Energy Through Nanotechnology*,
Green Energy and Technology, DOI: 10.1007/978-0-85729-638-2_10,

surface with designated sequences to grow organic materials with specific molecular arrangements. MLD can control electron wavefunction shapes in organic materials to optimize performance of electro-optic switches [5, 6, 9], electro-luminescent devices [5], photovoltaic devices [6], and so on.

In this chapter, the waveguide-type sensitized photovoltaic device [6, 7], which we proposed recently is presented. The nanostructured materials built by MLD are used as the light-absorbing layers in the photovoltaic device.

First, the structures of the waveguide-type sensitized photovoltaic devices, and the concept and experimental demonstrations of MLD are reviewed. Next, two examples of the light-absorbing layers fabricated by MLD are described; poly-azomethine (AM) wires with multiple quantum dots (MQDs), and self-assembled double layers consisting of p-type and n-type dye molecules. Since the MQDs can control the dimensionality of electron wavefunctions in the light-absorbing layers between 0-dimension and 1-dimension, they are useful to optimize spectral sensitization characteristics of the photovoltaic devices. Two-dye sensitization using the self-assembled double layers widens sensitized spectral regions in ZnO thin films. Then, preliminary experimental results of photocurrents enhanced by guided light beams in ZnO waveguides with the poly-AM wires and the self-assembled double layers are presented. Finally, film-based integrated solar energy conversion systems, which are suitable to integrate with waveguide type photovoltaic devices, are proposed.

2 Structures of Waveguide-Type Sensitized Photo-Voltaic Devices

In Fig. 1, the waveguide-type dye-sensitized photovoltaic device [6, 7] is compared with a conventional dye-sensitized solar cell. In the conventional device, in order to increase adsorbed dye molecule concentration on semiconductor surfaces, porous semiconductors are used. In this case, although sufficient light absorption is achieved for incident light normal to the semiconductor surfaces, the crystalline quality tends to be poor, suppressing electron mobility. Furthermore, the porous structure makes the carrier path in the semiconductors narrower, increasing the cell resistivity.

In the waveguide-type dye-sensitized photovoltaic device, on the other hand, thin-film semiconductors with flat surfaces, high crystalline quality, and high electron mobility, are used. The structure is suitable for reducing the internal resistivity. However, in this case, the adsorbed dye molecule concentration on the semiconductor surfaces is insufficient. Normally incident light onto the semiconductor surfaces passes through only the monolayer of dye molecules, resulting in very small light absorption. To solve this problem, instead of using normally incident light, we use guided light beams in the semiconductor thin films. In such configurations, light passes through many dye molecules to enhance light absorption. In addition, in the waveguide-type photovoltaic device, transparent electrodes can be replaced with metal electrodes, which reduce the series internal resistance of electrodes as well as the material costs. Thus, high-performance sensitized photovoltaic devices will be realized.

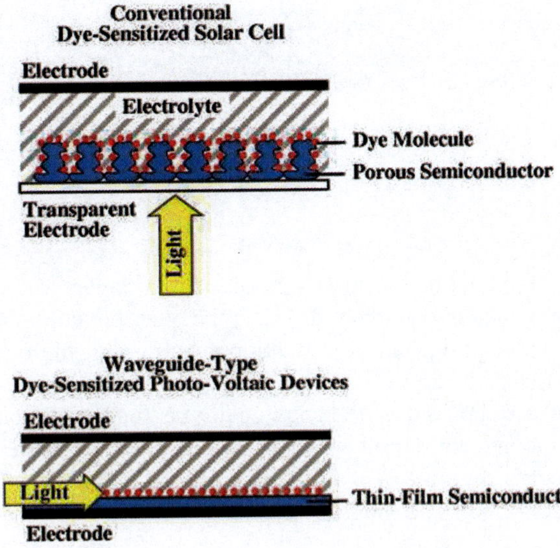

Fig. 1 The conventional dye-sensitized solar cell and the waveguide-type dye-sensitized photovoltaic device [7]

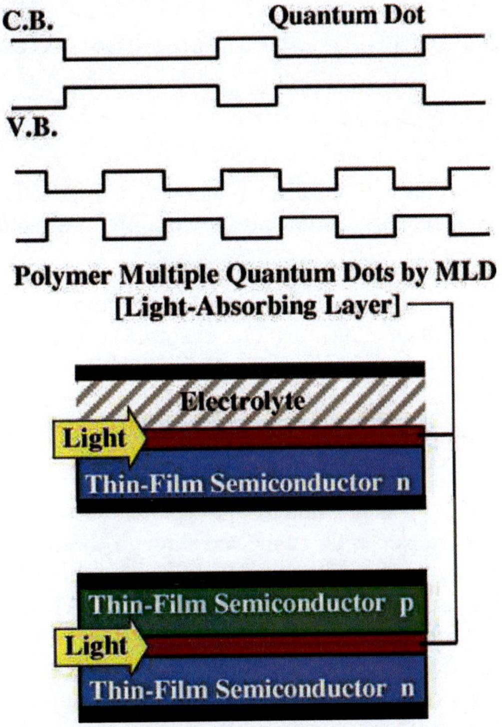

Fig. 2 Waveguide-type sensitized photovoltaic devices using polymer multiple quantum dots (MQD) for the light-absorbing layer [7]

Figure 2 shows a waveguide-type sensitized photovoltaic device with polymer MQDs [6, 7], defined as polymer wires having MQDs. In this device the dye molecules are replaced with the polymer MQDs, which are described more

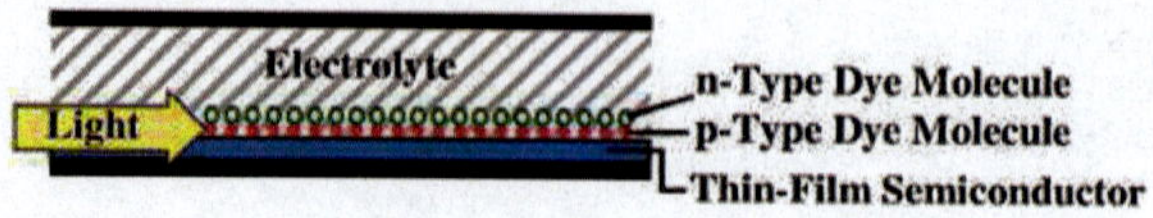

Fig. 3 Waveguide-type sensitized photovoltaic devices using self-assembled double layers of p-type and n-type dye molecules adsorbed on ZnO surfaces [7]

precisely in Sect. 4. The MQDs are fabricated by MLD. The polymer MQDs act as efficient and robust light-absorbing layers on the semiconductor surfaces. By controlling quantum dot size, the quantum confinement effects are optimized to provide appropriate sensitization spectra with high conversion efficiencies.

Figure 3 shows waveguide-type two-dye-sensitized photovoltaic devices using self-assembled double layers of p-type and n-type dye molecules adsorbed on the semiconductor surfaces. The two-dye sensitization can widen the sensitized spectral region to enhance the energy conversion efficiency [10, 11]. A detailed description is provided in Sect. 5.

3 Molecular Layer Deposition (MLD)

3.1 Concept

The concept of MLD is shown in Fig. 4 [1–8]. Different kinds of molecules, Molecules A, B, C, and D, were prepared under the following guidelines:

- A molecule has two or more reactive groups.
- Reactive groups in a molecule cannot react with reactive groups in the same kind of molecule. For example, reactive groups in Molecule A cannot react with reactive groups in Molecule A.
- Reactive groups in a molecule can react with reactive groups in other kinds of molecules. For example, reactive groups in Molecule A can react with reactive groups in Molecule B.

Then, same kind of molecules cannot be connected, and different kinds of molecules can be connected through the reactions between reactive groups. When Molecule A is deposited onto a substrate surface, a monolayer of Molecule A is formed. Once the surface is covered with Molecule A, the deposition of the molecules is automatically terminated due to the small reactivity between the same kinds of reactive groups, giving rise to thickness saturation. When Molecule B is provided to the surface, Molecule B connects to Molecule A to induce a rapid thickness increase followed by saturation, exhibiting monomolecular-step growth. This enables us to carry out sequential self-limiting reactions, as in atomic layer deposition (ALD) [12], by switching molecules. By repeating the switching with a designated sequence, polymer wires with a molecular sequence like ABCD, for example, can be obtained.

Fig. 4 The concept of molecular layer deposition (MLD) [6, 7]

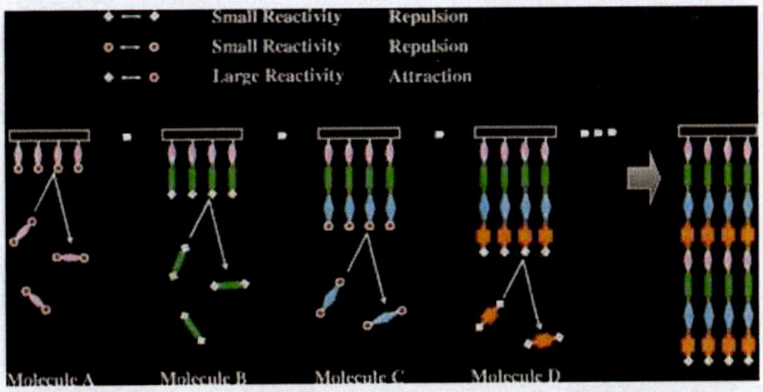

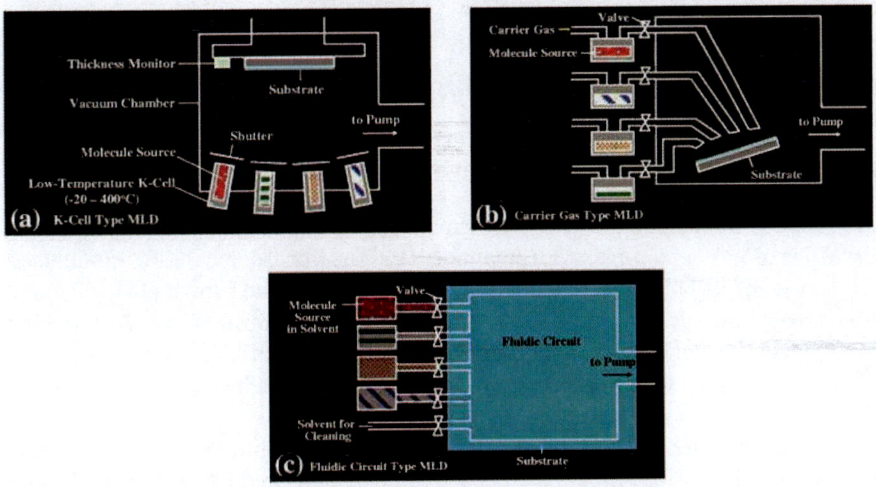

Fig. 5 Schematic illustration of MLD equipment: **a** K-cell type, **b** carrier gas type, and **c** fluidic circuit type

MLD can also be performed by using electrostatic forces. In this case, the monomolecular deposition and the self-limiting effect arises from attraction between different kinds of molecules and repulsion between the same kinds of molecules.

In MLD, source molecules are supplied to a substrate surface in either gas phase or liquid phase. Figure 5a shows schematic illustration of a K-cell type MLD. Molecular gases are provided to a surface in vacuum from low-temperature K-cells, in which source molecules are loaded. Molecular gas switching is performed by shutters or valves. In carrier gas-type MLD, as shown in Fig. 5b, molecular gases are provided with carrier gas onto a surface. Molecular gas switching is performed by valves. In fluidic circuit-type MLD, which is a kind of liquid-phase MLD [6, 13], source molecules are carried with solvent to be supplied to a surface, as shown in Fig. 5c. Fluidic processes using capillary circuits are effective to reduce the volume of solutions required for MLD as well as the solution switching duration. Figure 6 shows photographs of the K-cell-type MLD equipment and the carrier gas type MLD equipment in our laboratory.

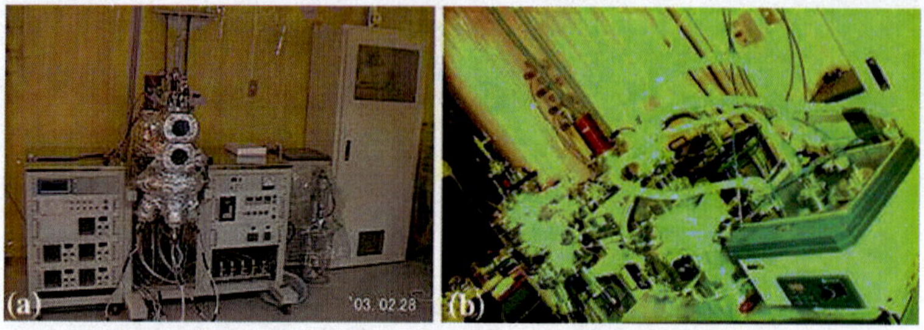

Fig. 6 Photograph of MLD equipment: **a** K-cell type and **b** carrier gas-type

3.2 Experimental Demonstrations of MLD

Experimental demonstrations of MLD [1–4, 6, 7] for polymer wire growth were carried out using K-cell MLD and carrier gas MLD. Figure 7 shows examples of reactions for conjugated polymer wire growth. By using terephthalaldehyde (TPA) for Molecule A and *p*-phenylenediamine (PPDA) for Molecule B, conjugated polymer wires of poly-azomethine (AM) [TPA/PPDA] were grown [14]. TPA and PPDA react through –CHO in TPA and $–NH_2$ in PPDA to form –C = N–, inducing π-conjugated systems. When oxalic acid (OA) was used with Molecule A and oxalic dihydrazide (ODH) with Molecule B, conjugated polymer wires of poly-oxadiazole (OXD) were grown [15].

We demonstrated MLD using TPA and PPDA [2]. Figure 8 shows the results obtained by K-cell MLD. When PPDA was deposited on TPA with a molecular gas pressure of $1–8 \times 10^{-2}$ Pa, step-like growth of ~0.7 nm, close to the molecular size, was observed at substrate temperature $T_S = 25°C$. After removing PPDA from the vacuum chamber, by depositing TPA on PPDA, step-like growth with saturation was observed again. These results indicate that monomolecular-step growth with self-limiting is achieved by MLD. It is also confirmed that the thickness is proportional to the molecular gas switching counts just like ALD.

If more kinds of source molecules are used, a variety of structures can be fabricated, as described in Fig. 4. This enables realization of molecule-by-molecule polymer wire growth with designated molecular sequences to control π-electron wavefunction shapes, optimizing nanostructured material performances.

MLD for polyamic acid, which is a precursor of polyimide, was demonstrated [1] using the reactions shown in Fig. 9 at molecular gas pressures ranging from 1 to 5×10^{-3} Pa. Polyamic acid [PMDA/DNB] or [PMDA/DDE] were obtained by using pyromellitic dianhydride (PMDA) for Molecule A and 2,4-diaminonitrobenzene (DNB) or 4,4′-diamino-diphenyl ether (DDE) for Molecule B [14]. Figure 10a shows the results of MLD for polyamic acid [PMDA/DNB]. When DNB was deposited on PMDA at $T_S = 50°C$, monomolecular-step growth with saturation was observed, and the thickness was preserved after removing the DNB

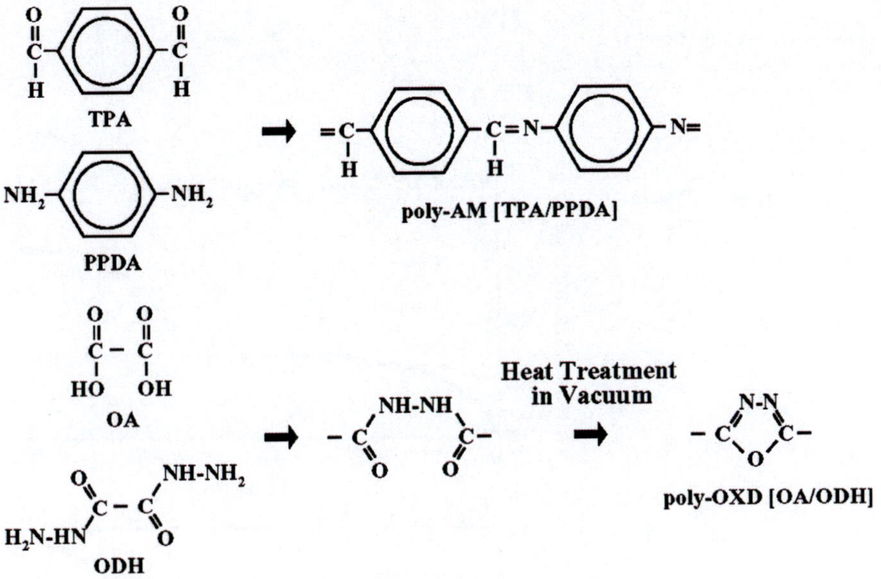

Fig. 7 Reactions for conjugated polymer wire growth

from the vacuum chamber. When T_S is increased to 65°C, a monomolecular-step thickness change was observed, but the thickness decreases after removing the DNB. This means that DNB re-evaporates from the PMDA before strong chemical bonds are formed. The average time τ_S for DNB to remain at PMDA surface corresponds to the time constant of the thickness change decay, about 30 s at $T_S = 65$°C.

When PMDA is deposited on DNB, as Fig. 10b shows, at $T_S = 50$°C an initial rapid increase in thickness is observed, but thereafter, the thickness continues to increase without saturation. This is indicative of PMDA molecules condensing on the surface to form a multi-molecular layer. When the substrate temperature is raised to 80°C, thickness saturation is observed, implying that PMDA condensation is avoided at this temperature and monomolecular-step growth of PMDA on DNB is achieved.

As described above, the optimal temperature for monomolecular-step growth of DNB on PMDA differs from that of PMDA on DNB. This might be due to the difference in vapor pressures between PMDA and DNB. To fabricate polyamic acid [PMDA/DNB] wires by MLD, the substrate temperature should be elevated synchronously with molecular gas switching.

As shown in Fig. 11, in the case of polyamic acid [PMDA/DDE], monomolecular-step growth is realized without elevating temperature. This might be attributed to small difference in vapor pressures between PMDA and DDE. A 10 nm growth is observed with 15 steps of molecular gas switching.

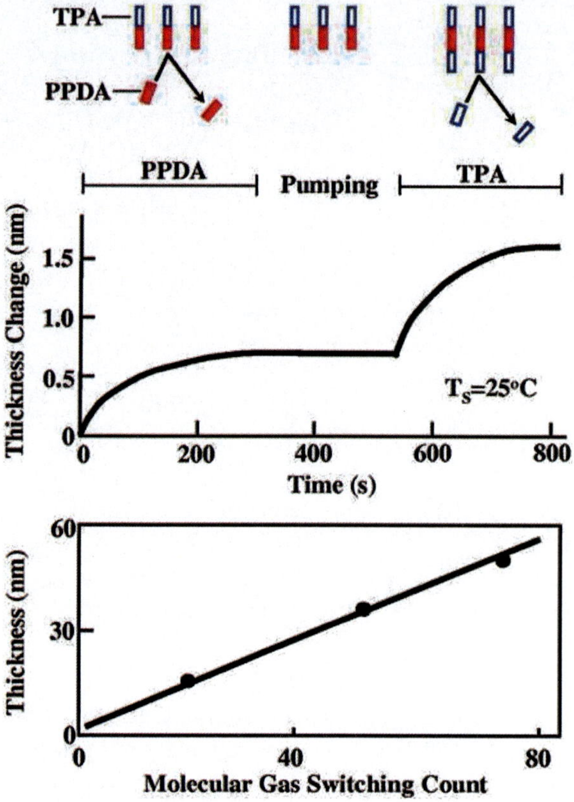

Fig. 8 Dependence of thickness changes on time, and thickness versus molecular gas switching count characteristics in the MLD of poly-AM [TPA/PPDA] [2]. Copyright 1992, American Institute of Physics

It should be noted that, in addition to the reactions shown in Figs. 7 and 9, MLD may be performed via other reactions using epoxy ring, –NCO, and so on [14].

3.3 MLD with Location/Orientation Control of Polymer Wire Growth

One of the future challenges in MLD is three-dimensional growth of polymer wires. The concept is shown in Fig. 12 [6, 8, 16]. On the top surfaces and/or sidewalls of the seed cores, a self-assembled monolayer (SAM) is formed selectively. The seed cores are used for controlling polymer wire growth locations and orientations. For vertical growth, the SAM is put on the top of the seed cores while for horizontal growth it is put on the sidewalls. The regions, where the SAM should not be formed, are covered with, for example, SiO_2 films. By distributing the seed cores with designed patterns, polymer wires are expected to be grown with designed configurations to construct a kind of polymer wire networks.

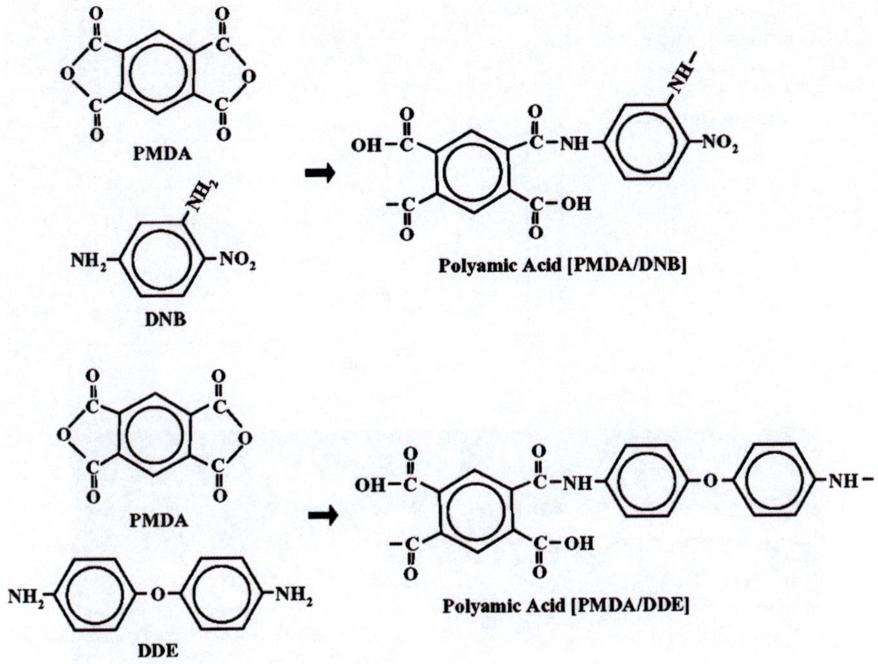

Fig. 9 Reactions for polyamic acid wire growth

Fig. 10 Dependence of thickness changes on time in the MLD of polyamic acid [PMDA/DNB] [1]. Copyright 1991, American Institute of Physics (**a**) DNB on PMBA. (**b**) PMDA on DNB

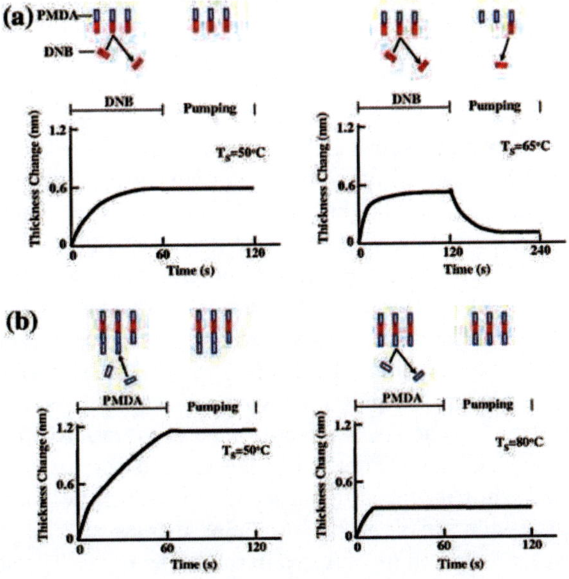

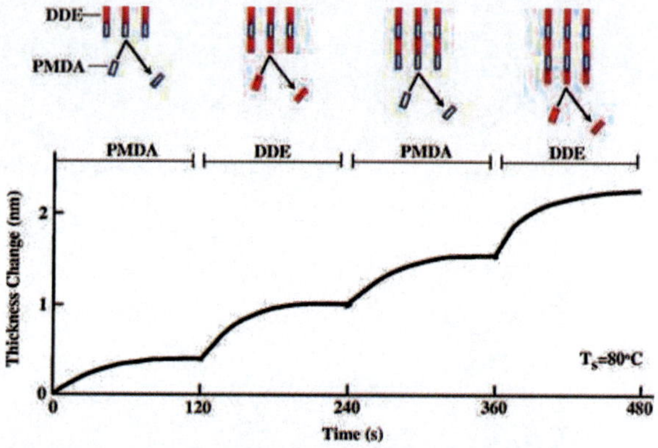

Fig. 11 Dependence of thickness changes on time in the MLD of polyamic acid [PMDA/DDE] [1]. Copyright (1991), American Institute of Physics

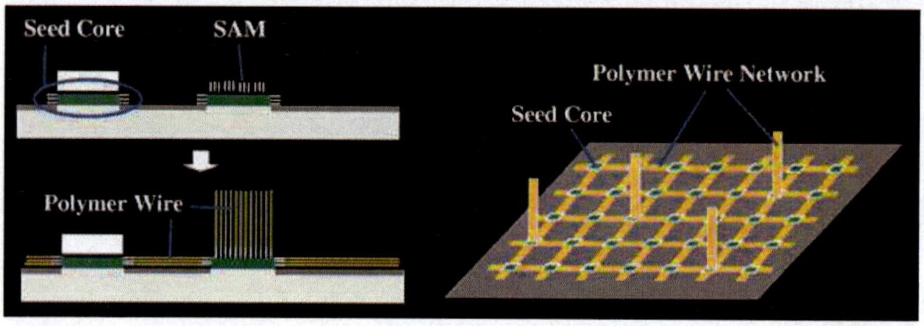

Fig. 12 Concept of polymer wire growth by MLD with location/orientation control

Figure 13 shows a possible process of polymer wire growth from seed cores for poly-AM [TPA/PPDA] using MLD [3, 4]. First, amino-alkanethiol SAM is formed on patterned Au films. Next, TPA is deposited onto the SAM surface to connect TPA to the SAM. Then, PPDA is supplied to the TPA surface, connecting the PPDA to the TPA. By sequential molecular gas switching to perform MLD, poly-AM wires were grown vertically from the Au film surfaces selectively.

Such location/orientation-controlled polymer wire growth can be done not only by MLD but also by other growth methods, such as the vacuum deposition polymerization [14] or organic chemical vapor deposition (CVD) [3].

Figure 14 shows the fourier transform infrared reflection absorption spectroscopy (FTIR-RAS) spectra of poly-AM grown by carrier gas-type MLD (step m = 20) [3]. Here, "SAM surface" indicates polymer wires grown on a surface with SAM and "non-SAM surface" describes polymer wires grown on a surface without SAM. It can be seen that both "SAM surface" and "non-SAM surface" have absorption peaks in wavenumber regions around 1,200–1,300 cm^{-1} and 850 cm^{-1}. The peaks are attributed to vibration in the in-plane direction (denoted "in-plane") and vibration in the surface-normal direction (denoted "surface-normal") in benzene rings, respectively. When polymer wires stand on the surface, absorption arising from "in-plane" increases. When polymer wires lie on the surface, absorption of "surface-normal" increases. The absorbance ratio of "in-plane"

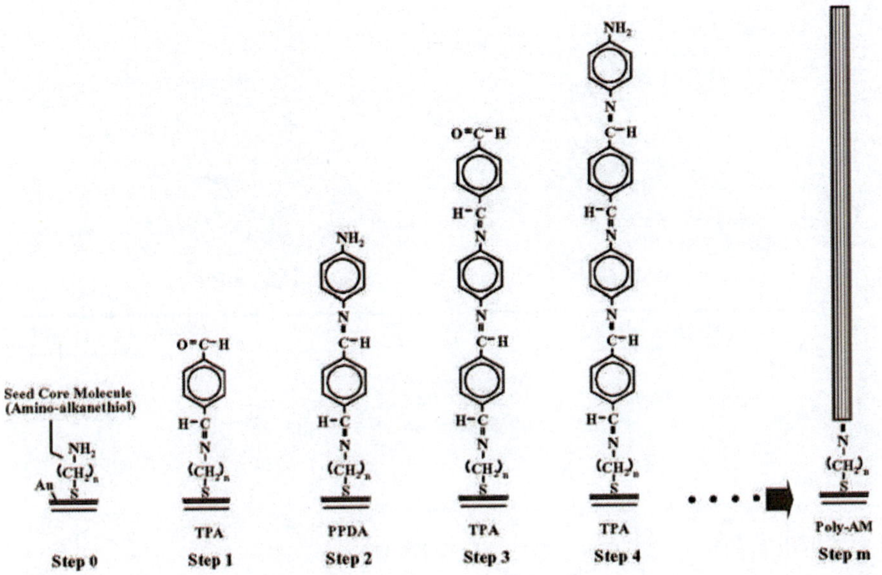

Fig. 13 The process of poly-AM wire growth using MLD from seed cores [3]. Copyright 2007, American Institute of Physics

to "surface-normal" I $_{\text{In-Plane}}$/I $_{\text{Surface-Normal}}$ is larger by factor of 2 in the "SAM surface" than in the "non-SAM surface." This suggests that polymer wires tend to grow along nearly-vertical directions in "SAM surface", as in the model shown in Fig. 13, while wires tend to grow along nearly-horizontal directions in "non-SAM surface".

The poly-AM wire growth direction described above was confirmed by the growth rate of poly-AM in carrier gas-type organic CVD. For the "non-SAM surface", the film thickness increases in the initial stage, followed by saturation. The contact angles of water droplets on the surface increase with growth time. For "SAM surface", film thickness increases linearly with growth time, without saturation.

As previously reported [17], hydrophilic TPA and PPDA molecules are adsorbed selectively on hydrophilic surfaces, resulting in the selective growth of poly-AM films on hydrophilic surfaces. As shown in the model illustrated in Fig. 15, on the "non-SAM surface", horizontal poly-AM wire growth occurs. Since the –CHO group and the $-NH_2$ group are hydrophilic, TPA and PPDA molecules are preferably adsorbed onto the –CHO and $-NH_2$ groups at the edge of the poly-AM wires while they are not adsorbed onto the body part of poly-AM wires, which are hydrophobic. Consequently, as poly-AM wires grow, the density of hydrophilic sites that exist at the edge of the poly-AM wires is reduced, i.e., the surface changes from hydrophilic to hydrophobic, resulting in the film growth's saturation. However, on the "SAM surface", poly-AM wires grow vertically, keeping the density of the hydrophilic sites constant. Therefore, poly-AM film does not exhibit growth saturation when grown on a "SAM surface".

It is clearly indicated from these results that nearly-vertical molecule-by-molecule growth of conjugated polymer wires can be achieved on substrates

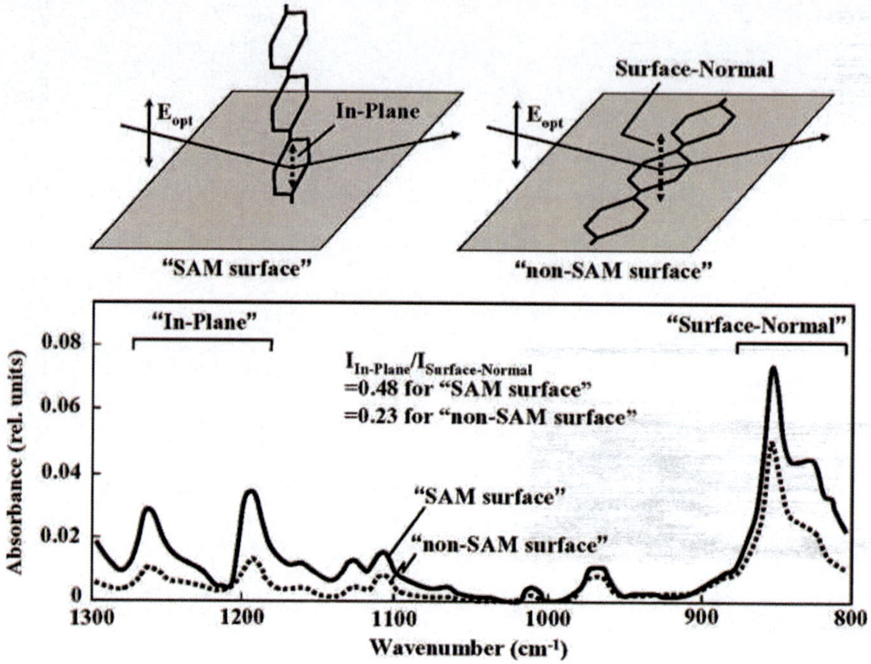

Fig. 14 FTIR-RAS spectra of poly-AM grown by carrier gas type MLD [3]. Copyright 2007, American Institute of Physics

Fig. 15 Models for poly-AM growth by carrier gas type organic CVD on a surface with SAM and a surface without SAM [3]. Copyright 2007, American Institute of Physics

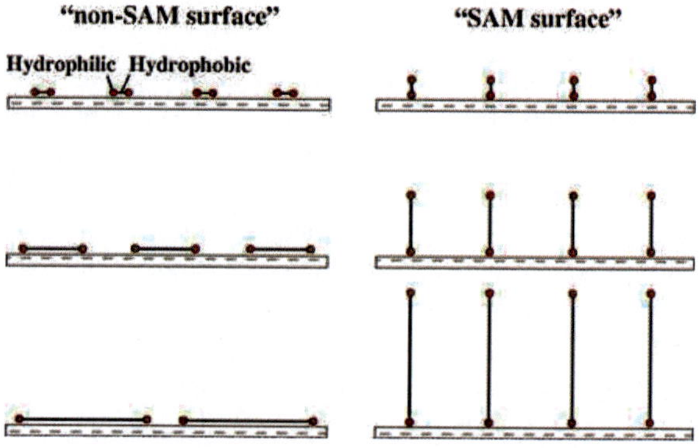

with SAMs by carrier gas-type MLD. This enables three-dimensional growth of polymer wires using seed cores, which can be ultra-small, e.g., nm-scale Au blocks with a SAM, or atomic-scale cores arranged by a scanning tunneling microscope.

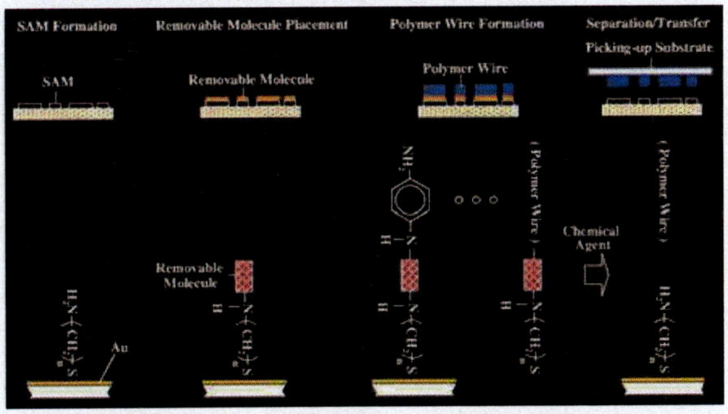

Fig. 16 Concept of molecular nanoduplication (MND) [16]

3.4 *Molecular Nanoduplication*

In order to reduce the fabrication cost of molecular nanosystems built by MLD, a mass production process is required. To do this, we proposed molecular nanoduplication (MND). Figure 16 shows the concept [8, 16]. On a growth substrate, surface treatments, such as SAM formation, are applied in nanoscale patterns. On the SAM, removable molecules are placed by MLD, followed by polymer wire growth. After attaching a picking-up substrate on the surface, the removable bonds are cut by a chemical agent to separate the polymer wires from the SAM. Thus, nanoscale patterns consisting of only polymer wires are transferred onto the picking-up substrate.

Since the growth substrate with patterned SAM is reusable, further nanoscale lithography is not required for nanoscale pattern duplication. By combining MND with the selectively occupied repeated transfer (SORT) [18], heterogeneous integration of nanoscale devices can be achieved. Although MND has not yet been demonstrated experimentally, we believe that this process will contribute to cost reduction of molecular nanosystems.

4 Multiple Quantum Dots in Polymer Wires by MLD

4.1 *Polymer MQD Made from Two Molecules*

First experimental demonstrations of polymer MQD construction were carried out using two kinds of molecules [2], shown in Fig. 17. The reaction between –CHO and $-NH_2$ is the same as described in Fig. 7 for poly-AM formation. Figure 18 diagrams the molecules and polymer wires with their corresponding electronic potential energy curves. Benzene rings in TPA, PPDA, and DDE are regarded as a short quantum dots ~0.5-nm long. In poly-AM [TPA/DDE], the π-conjugated part is regarded as a long quantum dot with barriers of –O–. The quantum dot contains three benzene rings and its length is ~2 nm. In poly-AM [TPA/PPDA], π-conjugation is spread along the wire to form a quantum wire.

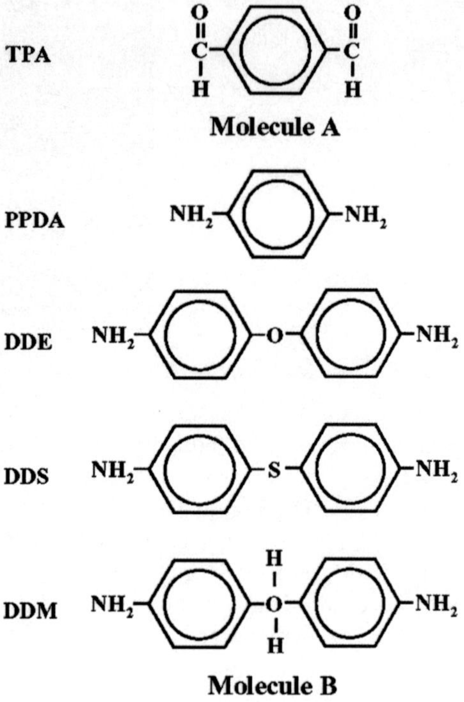

Fig. 17 Molecules for making quantum dot structures based on poly-AM [2]. Copyright 1992, American Institute of Physics

Figure 19 shows absorption spectra for TPA, PPDA, DDE, poly-AM [TPA/DDE], and poly-AM [TPA/PPDA]. In poly-AM [TPA/PPDA], a sharp peak of excitons are observed near 500 nm in wavelength. In poly-AM [TPA/DDE] the absorption band appears at a shorter wavelength than that in poly-AM [TPA/PPDA] due to electron confinement in the quantum dot structure. In TPA, PPDA, and DDE, the absorption band exhibits a further blue shift, indicating strong electron confinement in the short quantum dots.

In poly-AM [TPA/DDS] or poly-AM [TPA/DDM], –S– or $-CH_2-$ respectively make a barrier as –O– does in poly-AM [TPA/DDE]. In the order of –S–, –O–, and $-CH_2-$, the absorption peak shifts to the shorter wavelengths, indicating that the barrier height increases to produce strong electron confinement.

Energy of electrons confined in a quantum dot with infinite potential barrier height is expressed as $E_n = \hbar^2\pi^2n^2/(2m * m_0L^2)$, where, $n = 1, 2,\ldots$, m_0 is electron mass, m* is the effective mass in units of m_0, and L is the quantum dot length. Assuming the effective mass is 1 and the absorption peak energy for $L = \infty$ equals that of poly-AM [TPA/PPDA], the excitation energy for the lowest allowed transition is calculated as a function of L, as shown in Fig. 20. Experimental and calculated results agree fairly well, confirming the electron confinement in the quantum dots. These results demonstrate the possibility of controlling dimensionality to enhance the light-absorbing characteristics in polymer MQDs.

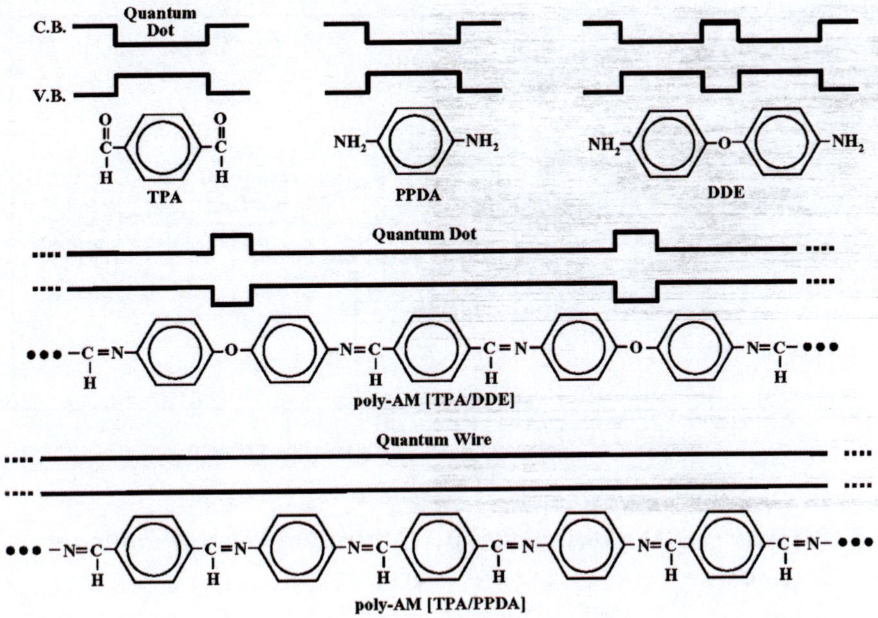

Fig. 18 Molecules and polymer wires, where quantum dots and quantum wires are formed, with corresponding electronic potential energy curves [2]. Copyright 1992, American Institute of Physics

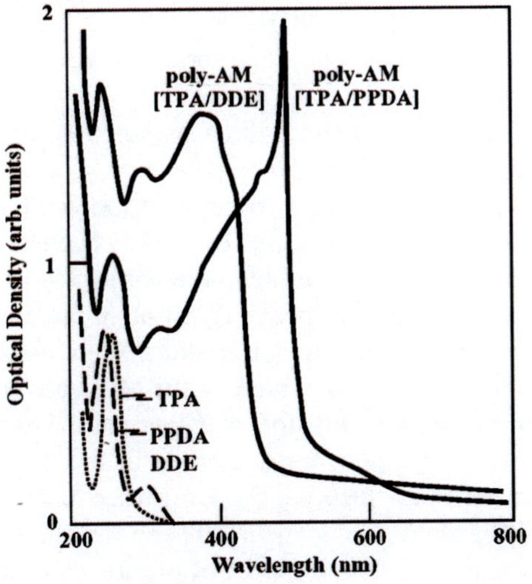

Fig. 19 Absorption spectra of TPA, PPDA, DDE, poly-AM [TPA/DDE], and poly-AM [TPA/PPDA] [2]. Copyright 1992, American Institute of Physics

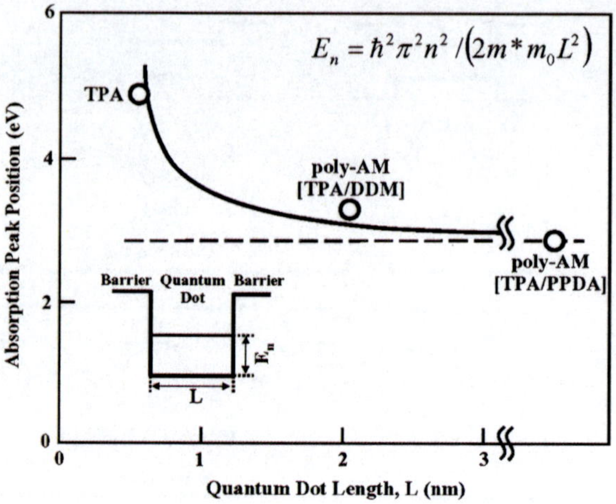

Fig. 20 The dependence of absorption peak positions on quantum dot length [2]. Copyright 1992, American Institute of Physics

4.2 Polymer MQD Constructed by Controlling Three-Molecule Sequences

Second, experimental demonstrations of polymer MQD construction were carried out by Three-Molecule MLD [6, 7], that is, by using three kinds of molecules, TPA, PPDA and ODH. TPA and PPDA were connected with a double bond, as shown in Fig. 7, allowing the wavefunctions of π-electrons to spread through the double bond to both molecules. TPA and ODH were connected with a single bond, as shown in Fig. 21, which cuts the wavefunctions at the bond, forming a barrier. By using these bond characteristics, quantum dots can be formed in polymer wires. Figure 22 shows a Three-Molecule MLD process for the construction of a structure called OTPT, where the molecules are connected sequentially as –ODH-TPA-PPDA-TPA–. The region between ODHs is regarded as a quantum dot since the wavefunctions of the π-electrons are spread within the region.

In order to control the wavefunctions and quantum dot lengths in polymer wires artificially, molecular sequence controls are essential. In Sect. 4.1, polymer MQD construction using two kinds of molecules is described. In that case, the polymer wire structures obtained by the MLD were not different from those obtained by the vacuum deposition polymerization [14] because the two molecules are automatically connected alternately in both cases.

When three kinds of molecules are used for polymer wire growth, the polymer wire structures obtained by the MLD are completely different from those obtained by vacuum deposition polymerization. In vacuum deposition polymerization, the three molecules are contained in the polymer wires with random arrangements. In MLD, on the other hand, the three molecules are contained in polymer wires with designated arrangements. Therefore, the essential advantage of the MLD appears when more than three kinds of molecules are used for polymer wire growth.

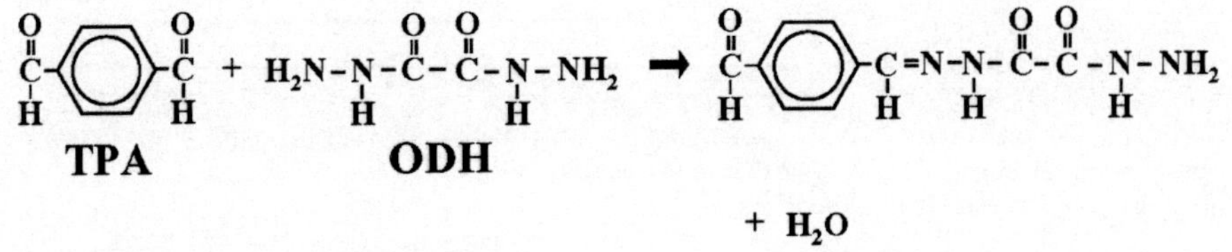

Fig. 21 The molecules and reaction for barrier formation in Three-Molecule MLD

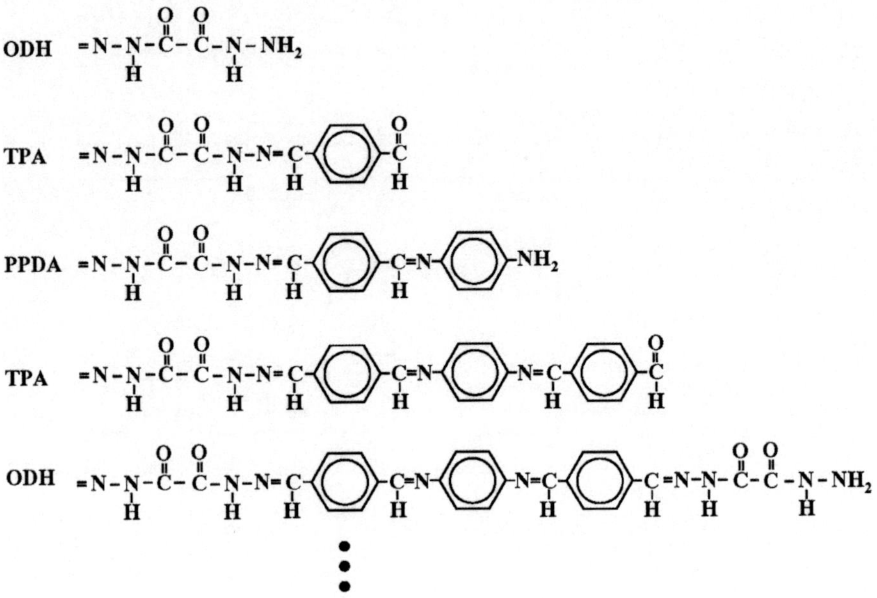

Fig. 22 Three-Molecule MLD process for OTPT formation [7]

Figure 23 shows structures of a poly-AM quantum wire and poly-AM wires with MQDs: OTPTPT, OTPT, and OT. In the poly-AM quantum wire, TPA and PPDA are alternately connected, so, the wavefunctions of π-electrons are spread through the polymer wire. In OTPTPT, molecules are connected in a sequence of —ODH-TPA-PPDA-TPA-PPDA-TPA—, resulting in a long quantum dot length of $\sim$3 nm. In OTPT, the quantum dot length is around 2 nm that is shorter than in OTPTPT. In OT, ODH and TPA are alternately connected, giving quantum dots with a very short length of less than 1 nm.

The poly-AM quantum wire, OTPTPT, OTPT, and OT were grown by the carrier gas-type MLD schematically shown in Fig. 24 [3]. A carrier gas of nitrogen (N_2) was employed to introduce molecular gases of TPA, PPDA, and ODH in temperature-controlled molecular cells onto a glass substrate surface with a flow rate of 4 NL/min. Excess gases were removed by a rotary pump. The molecular cell temperature was 50°C for PPDA and ODH, and 25°C for TPA. MLD operation was carried out by sequentially switching the three valves for

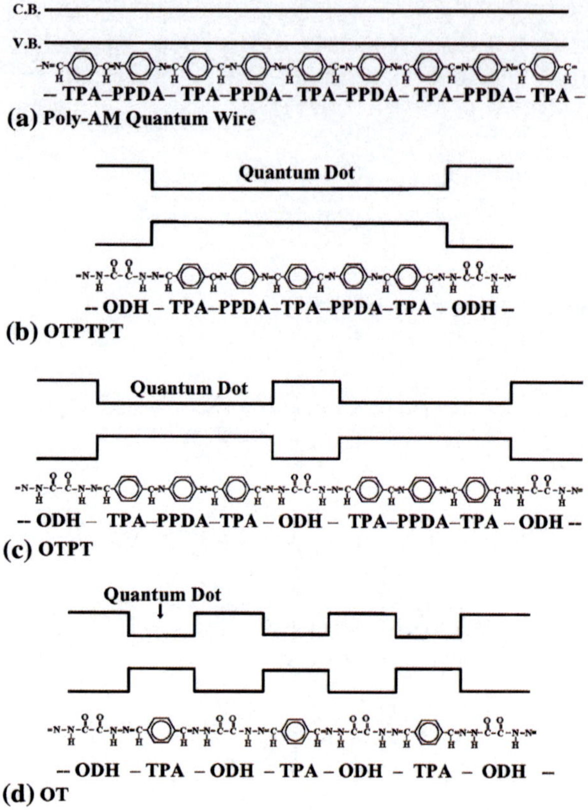

Fig. 23 A poly-AM quantum wire and poly-AM wires with MQDs (OTPTPT, OTPT, and OT) and their corresponding electronic potential energy curves [7]

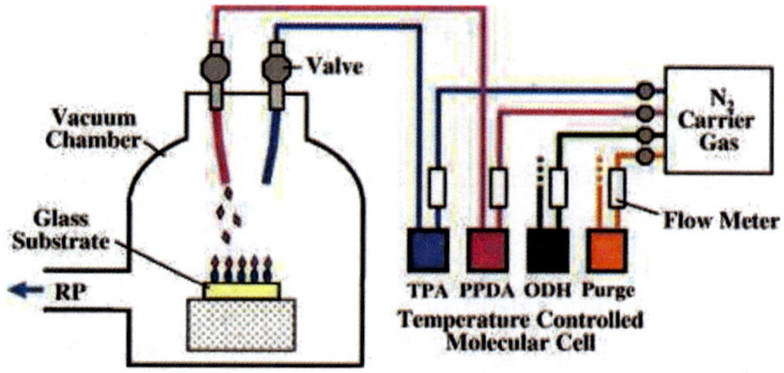

Fig. 24 Schematic illustration of the carrier gas type MLD [7]

TPA, PPDA, and ODH molecular gas injection. The substrate surface was exposed to molecular gas for 3–5 min in each molecular growth step. In order to switch molecular gases, after removing previous molecular gas by closing the valve for the gas and pumping for 5 min, the valve for the next molecular gas was opened. The switching step count was around 20, implying that three quantum dots were contained in a polymer wire for OTPTPT, four quantum dots for OTPTP, and ten quantum dots for OT.

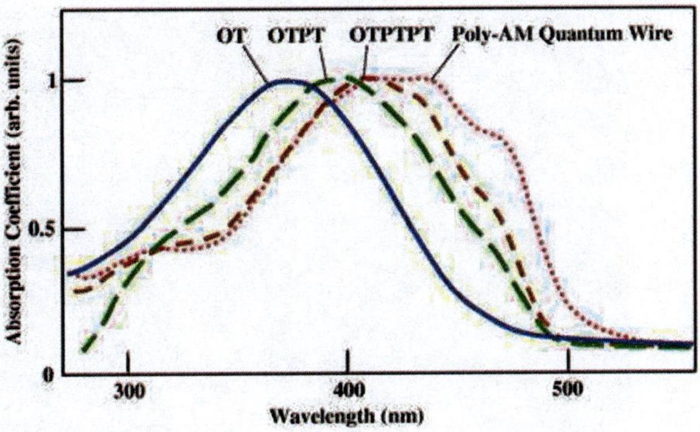

Fig. 25 Absorption spectra of a poly-AM quantum wire, OTPTP, OTPT, and OT [7]

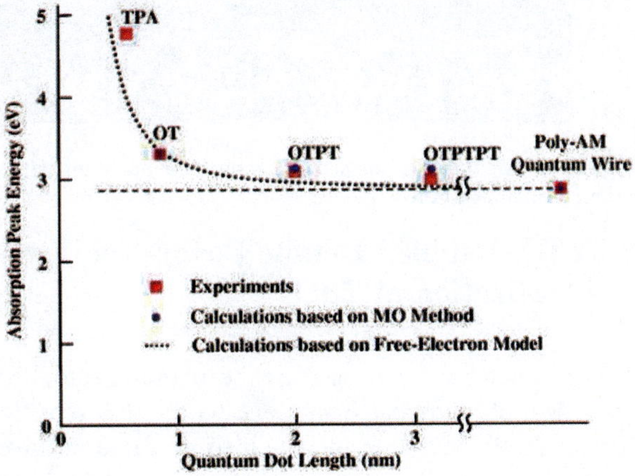

Fig. 26 The dependence of absorption peak positions on quantum dot length [7]

As Fig. 25 shows, in the order of the poly-AM quantum wire, OTPTPT, OTPT, and OT, with decreasing the quantum dot lengths, the absorption peak shifts to the shorter wavelength side. The shift is attributed to the quantum confinement of π-electrons in the quantum dots.

In Fig. 26, the absorption peak energy is plotted as a function of the quantum dot length. In this figure, the absorption peak energy calculated by the molecular orbital method is also plotted. The molecular structures for the calculation are shown in Fig. 27. The structures are obtained after structural optimization using WinMOPACK (Fujitsu). Since there is 0.3 eV offset between the calculated value and the experimental value in the absorption peak energy of the poly-AM quantum wire, the calculated absorption peak energy in Fig. 26 is plotted with sliding by 0.3 eV. In addition, the result derived from the free-electron model in a quantum dot, which is described in Sect. 4.1, is presented in Fig. 26. It was found that the experimental results and the calculated results reasonably agree, suggesting that the MQDs were successfully constructed in the polymer wires by MLD and present the quantum confinement effect.

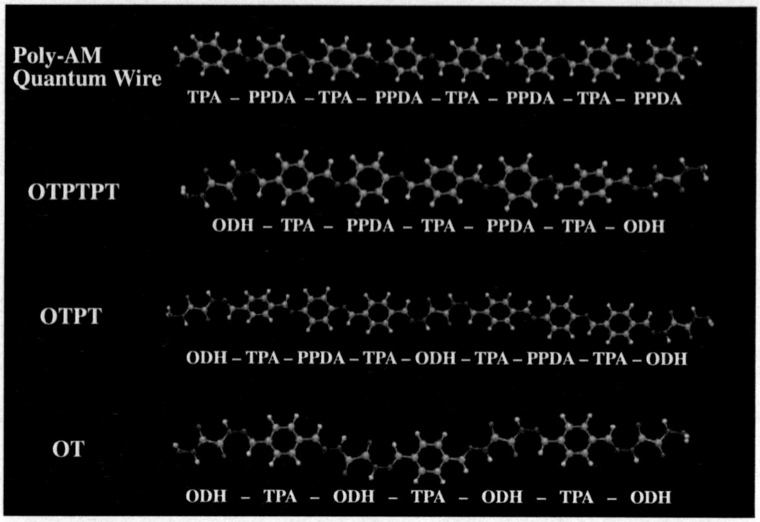

Fig. 27 Molecular structures of the poly-AM quantum wire, OTPTPT, OTPT, and OT after structural optimization

5 Self-Assembled Double Layers for Two-Dye Sensitization of ZnO

ZnO powder has been used as the photoreceptor for electrophotography, and a great deal of study has been done on the dye sensitization of ZnO. In the 1970s, Kiyota developed a technique to form a double layer of p-type and n-type dye molecules on ZnO surfaces and applied this self-assembled double layer to two-dye sensitization of ZnO-powder-based photoreceptors [19]. Recently, dyed ZnO has received much attention as the semiconductors in dye-sensitized solar cells. The sensitization mechanism of dyed ZnO in the solar cells is exactly the same as in the photoreceptors. Therefore, the self-assembled structure of p-type/n-type dye molecules on ZnO is expected to have applications in photovoltaic devices.

In this section, self-assembled double layers of p-type and n-type dye molecules and the two-dye sensitization of ZnO by the layers is reviewed [10, 11].

5.1 Self-Assembled Double Layers of p-Type and n-Type Dye Molecules

To analyze the self-assembled configuration of molecules on ZnO surfaces, surface potential measurements were carried out. The samples used for the measurement were prepared as follows. A ZnO powder layer was formed on a glass substrate with a SnO_2 electrode. An alcohol solution containing the first type of dye molecule was dropped on the ZnO layer and the solvent was removed. For adsorption of the second type of dye molecule, another alcohol solution containing the second

Fig. 28 Structures of dye molecule examples used in surface potential measurements

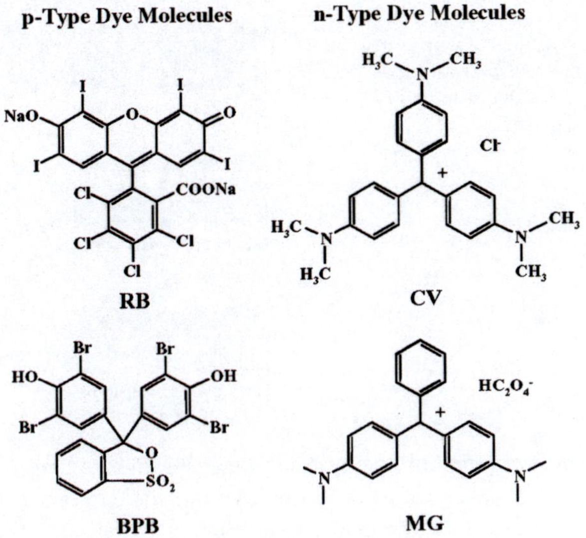

Fig. 29 Surface potential of ZnO adsorbing the first and second dye molecule layers, and the surface potential of ZnO co-adsorbing p-type dye molecules and n-type dye molecules simultaneously [11]

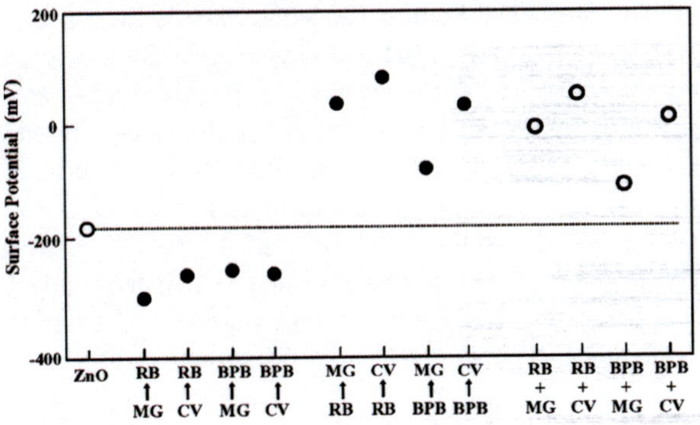

dye molecules was similarly dropped on top of the first dye molecule layer. Typically, rose bengal (RB) and bromphenol blue (BPB) are used for the p-type dye, while crystal violet (CV) and malachite green (MG) are used for the n-type dye. Structures of these dye molecules are illustrated in Fig. 28.

It was found that the surface potential becomes more negative than that of the undyed ZnO layer when a p-type dye layer is adsorbed on the ZnO layer. Conversely, when an n-type dye layer is adsorbed, the surface potential becomes less negative than that of the undyed ZnO layer. The results can be explained as follows. When a p-type dye layer is adsorbed on the ZnO layer, electrons in the ZnO layer are transferred to the dye molecule layer. When an n-type dye molecule layer is adsorbed on the ZnO layer, electrons in the n-type dye molecule layer are transferred to the ZnO layer.

In Fig. 29, the surface potential of ZnO adsorbing the first dye molecule layer and the second dye molecule layer is shown. Here, A→B indicates that the first dye molecule layer is A and the second dye molecule layer is B. Variations of the

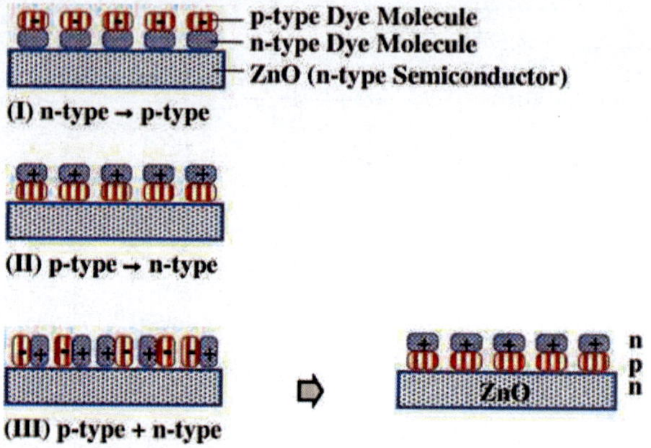

Fig. 30 Self-assembled structures of p-type dye molecules and n-type dye molecules adsorbed on ZnO surfaces

surface potential arising from the adsorption of two-dye double layers on the ZnO largely depend on the second dye molecule layer. This indicates that the surface potential is determined by differences in Fermi levels between the second dye molecule layer (the top layer) and ZnO.

The surface potentials of samples in which p-type and n-type dye molecules were co-adsorbed on the ZnO layer simultaneously are also shown in Fig. 29. The coadsorption is expressed by A + B. An alcohol solution containing both p-type dye molecules and n-type dye molecules at equal concentrations was dropped onto the ZnO. The surface potential of the ZnO after co-adsorbing the p-type and n-type dye molecules was very similar to that of the ZnO adsorbing p-type dye molecules as the first dye molecule layer and n-type dye molecules as the second dye molecule layer. This suggests that p-type dye molecule and n-type dye molecule are self-assembled to form an npn layered structure of ZnO (n-type)/p-type dye molecule/n-type dye molecule, as shown in Fig. 30.

The self-assembled structure arises from electron transfers from ZnO to p-type dye molecules and from n-type dye molecules to p-type dye molecules, resulting in strong connection between ZnO and p-type dye molecules, and n-type dye molecules and p-type dye molecules due to electrostatic force.

5.2 *Two-Dye Sensitization with Self-Assembled Double Layers*

The self-assembled double layered structure of p-type and n-type dye molecules adsorbed on the ZnO surface is desirable for spectral enhancement of dye sensitization. In this section, the two-dye sensitization of ZnO using the self-assembled structure is described precisely.

5.2.1 Experimental Demonstrations

A 20 μm-thick ZnO powder layer was formed on a glass substrate with a slit-type electrode (slit width: 10 mm, slit gap: 0.5 mm). Dye adsorption on the ZnO layer

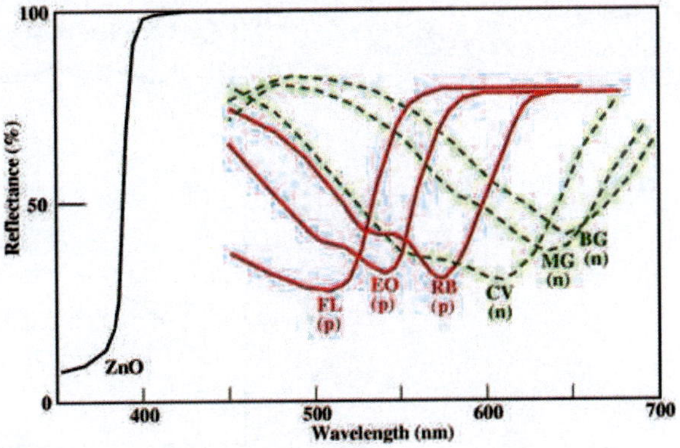

Fig. 31 Reflection spectra of dyed ZnO [10]

Table 1 Photocurrents in the dyed ZnO layers at the maximum absorption wavelengths of the dyes [10]

Dye	Type	Photocurrent (pA)
FL	p	120
EO	p	190
RB	p	180
CV	n	3
MG	n	0.02
BG	n	0.01

was carried out in the same fashion as described in Sect. 5.1. For photocurrent measurements, a 30 V potential was applied across the slit gap and the sample was illuminated by monochromatic light; the current was measured by an electrometer. The following six types of dye molecules were used: fluorescein (FL), eosine (EO), and RB for the p-type dye, and CV, MG, and brilliant green (BG) for the n-type dye. Figure 31 shows the reflection spectra of the dyed ZnO. N-type dyes have absorption bands in longer wavelength regions compared with p-type.

Photocurrents in the dyed ZnO layers at the maximum dye absorption wavelengths are listed in Table 1. It can be seen that the photocurrents are much larger for the p-type dyes than for the n-type dyes. This suggests, as Meier reported [20], that the p-type dye molecules inject photocarriers into the ZnO layer more efficiently than the n-type dye molecules.

A result of a typical two-dye sensitization is shown in Fig. 32. The photocurrents are for light intensity of 10 $\mu W/cm^2$. When RB and CV are co-adsorbed on ZnO, a self-assembled structure of ZnO/p-type dye molecule/n-type dye molecule is constructed, as shown in Fig. 30. As a result, anomalous sensitization of the photocurrent in the red regions is induced in the ZnO [RB + CV] sample. As it can be seen from Figs. 31 and 32, the profiles of the photocurrent spectra for ZnO [RB] and ZnO [CV] are respectively similar to those of the reflection spectra for ZnO [RB] and ZnO [CV] shown in Fig. 31. The photocurrent spectrum for ZnO [RB + CV] is obtained by an upward parallel shift of the spectrum for ZnO [CV] by two orders of magnitude keeping the spectral profile unchanged. Thus, in the 600–700 nm wavelength region, the photocurrent in ZnO [RB + CV] becomes much larger than that in either ZnO [RB] or ZnO [CV]. The same effect of two-dye sensitization was observed in ZnO

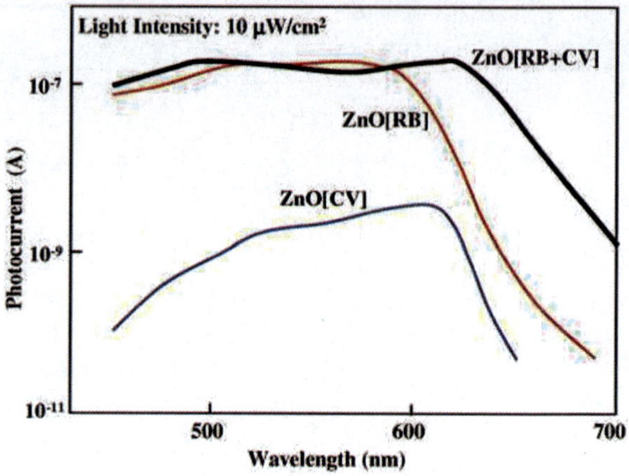

Fig. 32 Two-dye sensitization of ZnO (RB and CV) [10]

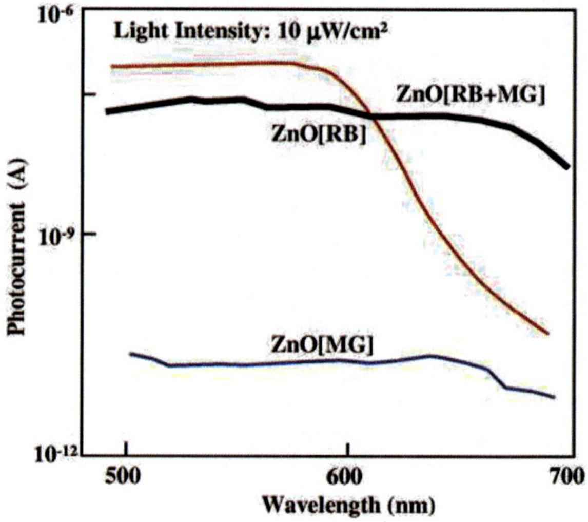

Fig. 33 Two-dye sensitization of ZnO (RB and MG) [10]

[RB + MG] (Fig. 33) and ZnO [EO + MG]. The spectral ranges of the dye sensitization are widened in the two-dye sensitization due to the superposition of the sensitization spectra of the p-type dye molecules and the n-type dye molecules.

5.2.2 Mechanism

Remarkable feature of the two-dye sensitization can be summarized as follows:

(1) Dramatic sensitization with two dyes arises only from the combination of a p-type dye and an n-type dye.
(2) When p-type and n-type dye molecules are co-adsorbed on ZnO, the p-type dye molecule layer is inserted between ZnO and the n-type dye molecule layer.

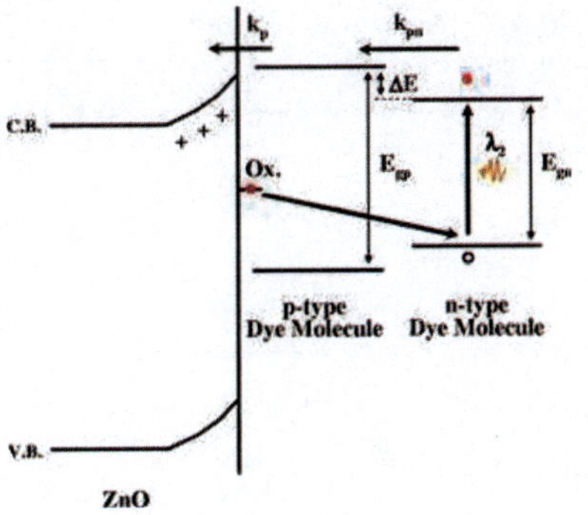

Fig. 34 The mechanism of two-dye sensitization [10]

(3) The n-type dye molecule layer absorbs almost all the red light illuminating the sample.

A schematic model of the two-dye sensitization is shown in Fig. 34. K_p is the efficiency with which the electrons are injected from the p-type dye molecule layer into ZnO. Similarly, K_pn is the efficiency with which the electrons excited in the n-type dye molecule layer are injected into the p-type dye molecule layer. Then the efficiency of the photocarriers generated in the n-type dye molecule layer being injected into ZnO is given by the following relationship:

$$\zeta = K_\mathrm{p}K_\mathrm{pn}$$

K_pn depends, to a large extent, on the combination of p-type and n-type dyes. Figure 35 shows the photocurrents caused by the various combinations of p-type dyes (FL, EO, and RB) and n-type dyes (CV, MG, and BG) at the maximum absorption wavelength of the n-type dye. The horizontal axis of this figure, ΔE_g, represents the difference between the energy gaps of the p-type dye and n-type dye combined that is, $\Delta E_\mathrm{g} = E_\mathrm{gp} - E_\mathrm{gn}$, where E_gp and E_gn are obtained from the absorption peaks of the dyes.

It was found from Fig. 35 that each point is approximately on a line with a slope of −24. If the p-type and n-type dye molecule layers form the energy structure shown in Fig. 34, the photocurrent may be proportional to $e^{-\Delta E/kT}$, where ΔE is the difference in energy of the excited states between the p-type and n-type dye molecules. If we assume K_p is constant, $\log I_p \propto -(\log e/kT)\,\Delta\mathrm{E}$, where I_p is the photocurrent. At room temperature, $-\log e/kT \approx -17$; the value is close to the slope of the line in Fig. 35. This implies that ΔE corresponds to ΔE_g in the 0th order approximation.

Since the efficiency of the electron injection into the ZnO is much higher for the p-type dye molecule than for the n-type dye molecule, the injection efficiency of the electrons from the n-type dye molecule into ZnO increases with the insertion of the p-type dye molecule layer. This is the reason why dramatic sensitization with two dyes occurs with the combination of p-type and n-type dyes.

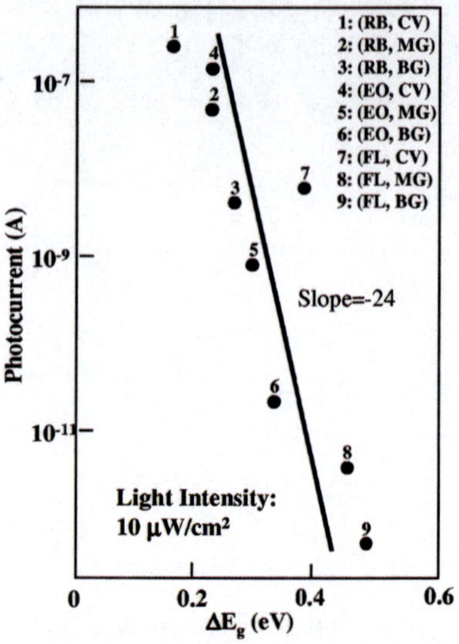

Fig. 35 Photocurrent versus ΔE_g characteristics in two-dye sensitization [10]

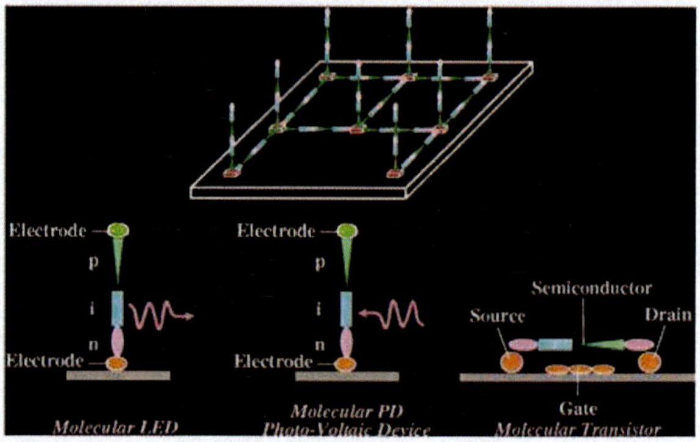

Fig. 36 Concept of molecular circuits consisting of molecular LEDs, PDs, photovoltaic devices, and transistors [16]

6 Molecular Circuits

Conjugated polymer wires grown by MLD are expected to be applied as molecular circuits, as schematically illustrated in Fig. 36 [8, 16], due to the following reasons:

- High carrier mobility characteristics of the π-conjugated systems.
- Small wire diameter, which is smaller than that of carbon nanotubes.
- Molecular sequence controllability in the wire with monomolecular precision, i.e., precise wavefunction shape controllability, which is not available in carbon nanotubes.

The circuits consist of molecular LEDs, PDs, photovoltaic devices, and transistors. Molecular LEDs, PDs, and photovoltaic devices are comprised of conjugated polymer wires with p-i-n junctions, sandwiched by a pair of conjugated polymer wire

Fig. 37 Tentative examples of p-i-n junctions in conjugated polymer wires

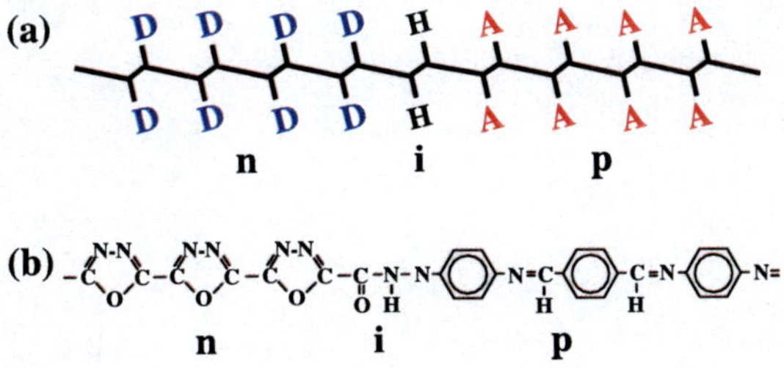

electrodes. In molecular transistors, a semiconductor conjugated polymer wire is aligned between source and drain electrodes, passing over a gate electrode.

Figure 37 shows proposed examples of conjugated polymer wires with p-i-n junctions. For the donor/acceptor substitution-type p-i-n junction, donors and acceptors are substituted into a common backbone. For the backbone-type p-i-n junction, an n-type backbone and a p-type backbone are connected. In proposed devices shown in Fig. 37, $-NH_2$ (D) and $-NO_2$ (A) are introduced into a polydiacetylene wire for the donor/acceptor substitution-type junction, and poly-AM and poly-OXD are directly connected for the backbone-type junction. Although p/n-type characterizations of poly-AM and poly-OXD are uncertain, if poly-AM and poly-OXD are formed within a polymer wire with molecular sequence arrangements like ABABABABABCDCDCDCDCD using the reactions shown in Figs. 7 and 21 with Molecule A: OA, Molecule B: ODH, Molecule C: TPA, and Molecule D: PPDA, some kind of "p-i-n-like" junctions may be formed in the poly-AM/poly-OXD interface region. For photovoltaic device operation, electrons and holes are generated by absorbing light beams in the i region to induce current/voltage. By using the p-i-n polymer wires in a form of flux or a film consisting of many p-i-n polymer wires, high-performance organic photovoltaic devices may become available in future.

7 Photocurrent Enhancement by Guided Light Beams in Waveguides

7.1 ZnO Thin-Film Waveguides

To confirm the photocurrent enhancement induced by guided light beams in waveguide-type photovoltaic devices, we measured photocurrents in a 600-nm thick ZnO thin film deposited by vacuum evaporation onto a glass substrate having a slit-type Al electrode with a 60-μm gap, as shown in Fig. 38 [6, 7]. The ZnO thin film was annealed in air at 400°C for 1 h after deposition. Light beams from a laser diode of 405 nm in wavelength were introduced into the ZnO thin film by the following two ways:

(1) by "normally incident light" onto the ZnO surface, and
(2) by "guided light" from the edge of the ZnO film.

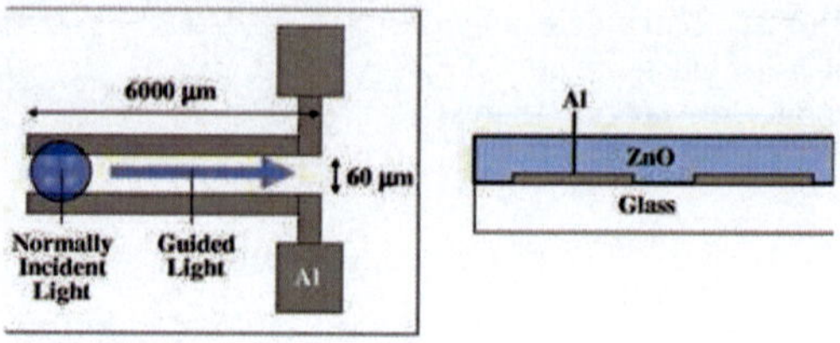

Fig. 38 Schematic illustration of a method for photocurrent measurements

Since the absorption coefficient of ZnO is low at 405 nm, small light absorption occurs for the "normally incident light". Therefore, photocurrent enhancement by "guided light" is expected to appear clearly.

Figure 39 shows guided light beams in the ZnO thin film and photocurrents induced by "normally incident light" and "guided light". It was found that the photocurrent was enhanced by one order of magnitude in the "guided light" case compared to "normally incident light". The photocurrent enhancement arises from an increase in light absorption in the "guided light" configuration.

7.2 ZnO Thin-Film Waveguides with Light-Absorbing Layers

We performed the same experiments for ZnO thin films with light-absorbing layers. We made a ZnO thin film, on which poly-AM wires were grown, and measured the photocurrents. Sensitization at 405 nm and 532 nm was observed [6].

This result indicates the possibility that the poly-AM-based polymer MQD can be used for the light-absorbing layer for sensitization in the photovoltaic device shown in Fig. 2. At 532 nm, considerable photocurrent enhancement was observed in the "guided light" configuration.

We also made ZnO thin films with dye adsorption. The dye layers on the ZnO thin films were formed by the liquid-phase MLD [6, 13]. In RB-adsorbed ZnO thin films, photocurrents were generated at 532 nm and the photocurrent enhancement by "guided light" was observed. Two-dye sensitization with a self-assembled double layer of RB and CV extended the sensitized spectra to red regions. The photocurrent enhancement was observed in the "guided light" configuration at 633 nm [6].

The next challenge is to combine the ZnO thin films with polymer MQDs or self-assembled double layers of dye molecules on the surface with an electrolyte layer to fabricate the photovoltaic devices shown in Figs. 2 and 3.

8 Film-Based Integrated Solar Energy Conversion Systems

In conventional solar cells, semiconductors are placed all over the modules, so the cost of the systems will increase with a shortage of semiconductor materials. In systems with Fresnel lens/mirror concentrators, semiconductor materials can be

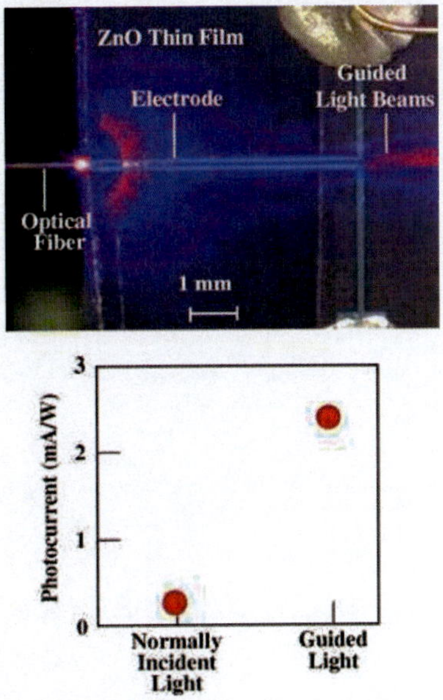

Fig. 39 Photocurrent enhancement by "guided light" in ZnO thin films [7]

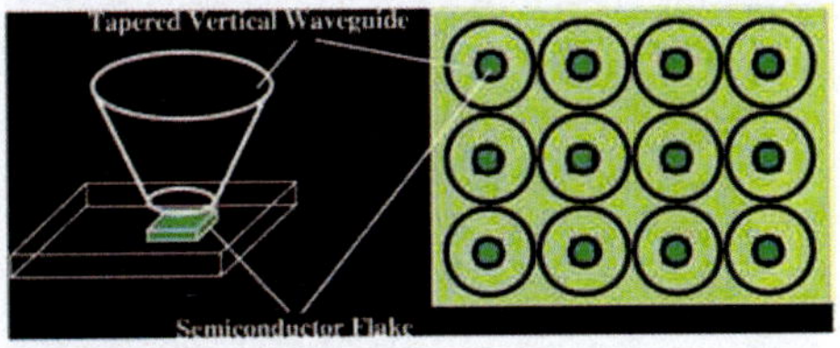

Fig. 40 Integrated solar energy conversion system with the tapered vertical waveguide-type light beam collecting film

saved because solar beams are focused on small semiconductor cells. However, the system size becomes very large and solar-beam tracking mechanisms are necessary. Recently, to reduce consumption of semiconductor materials, to provide wide-angle solar collecting capability to the systems, and to make systems flexible and compact, we proposed integrated solar energy conversion systems with waveguide-type light beam collecting films [21, 22–23]. In this section, the performance of the waveguide-type light-collecting film is predicted [24] using the beam propagation method (BPM) and the finite difference time domain (FDTD) method.

8.1 Proposed Structures

We consider three types of light beam collecting films. In the system with tapered vertical waveguide-type light beam collecting film (Fig. 40), semiconductor flakes for thin-film photovoltaic devices are embedded in the film. The incident

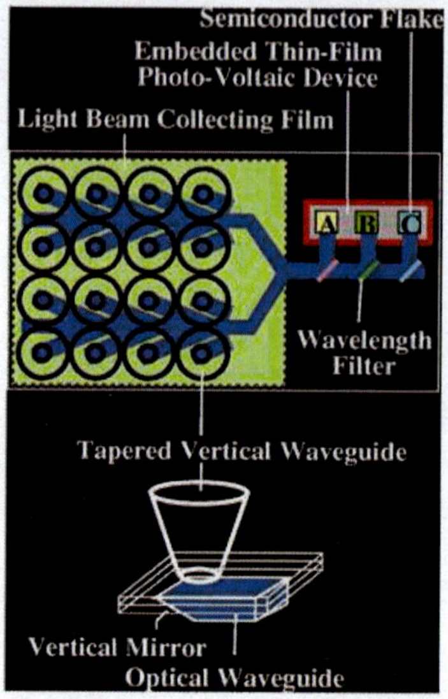

Fig. 41 An integrated solar energy conversion system with the tapered vertical/horizontal waveguide-type light beam collecting film [23]

light beams received by the tapered vertical waveguides are guided to the semiconductor flakes, whose areas are smaller than the areas of the light beam collecting aperture of the vertical waveguide. This enables a reduction of semiconductor consumption.

In the system with the tapered vertical/horizontal waveguide-type light beam collecting film (Fig. 41), a film with tapered vertical waveguides and a film with horizontal waveguides having vertical mirrors are stacked. The incident light beams are introduced into the horizontal waveguides through the vertical mirrors and guided to the semiconductor flakes embedded in the film. This enables further reduction of semiconductor consumption. By embedding wavelength filters into the optical waveguide films, solar beams can branch to semiconductors having appropriate absorbances. For example, in such a system, UV-blue beams are directed to ZnO, green–red beams to Si, and infrared beams to InP.

In the system with the multi-layer waveguide-type light beam collecting film (Fig. 42), many films with horizontal waveguides and vertical mirrors are stacked. The incident light beams are introduced through the vertical mirrors into the horizontal waveguides and guided to semiconductor flakes.

These film-based systems are suitable to integrate with waveguide-type photovoltaic devices because both the light beam collecting films and the photovoltaic devices have optical waveguide structures.

8.2 Simulation Procedure

The simulation procedure for predicting the light beam collecting performance is explained using a model example for the tapered vertical/horizontal waveguide-

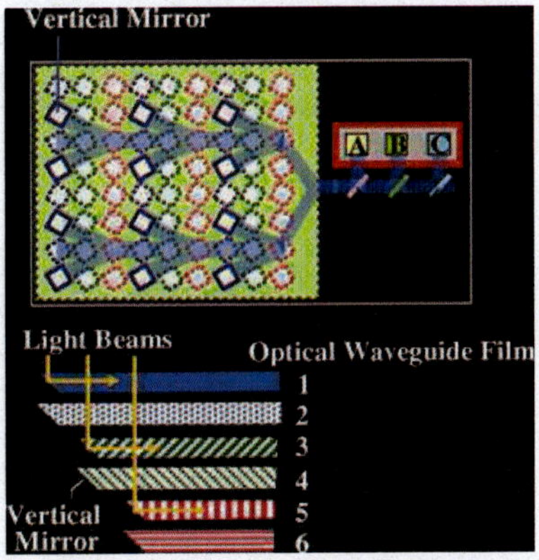

Fig. 42 Integrated solar energy conversion system with the multi-layer waveguide-type light beam collecting film

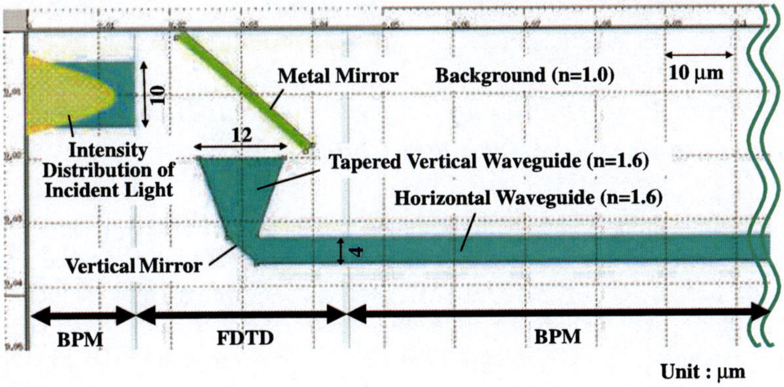

Fig. 43 An example of a simulation model for light beam collection in the waveguide-type light beam collecting films

type light beam collecting film shown in Fig. 43. Incident light beams from a metal mirror are introduced into a horizontal waveguide via a tapered vertical waveguide and a vertical mirror. The refractive indices of the optical waveguides are 1.6, and that of background is 1.0, assuming air. The width of the horizontal waveguide is 4 μm. An anti-reflective coating (ARC) with 100-nm thickness and refractive index of 1.3 is applied to the top and the bottom of the horizontal waveguide.

Incident light beams with a large width are preferable to reproduce the plane waves of sunlight. However, when the width is broadened, the area of the FDTD calculation increases and computation time becomes long. So in the present work, the simulations were performed using fundamental guided beams in an optical waveguide with a width of 10 μm for incident light beams. The incident angles of the light beams are adjusted by the metal mirror. The length of the model from the left boundary to the right boundary is 200 μm.

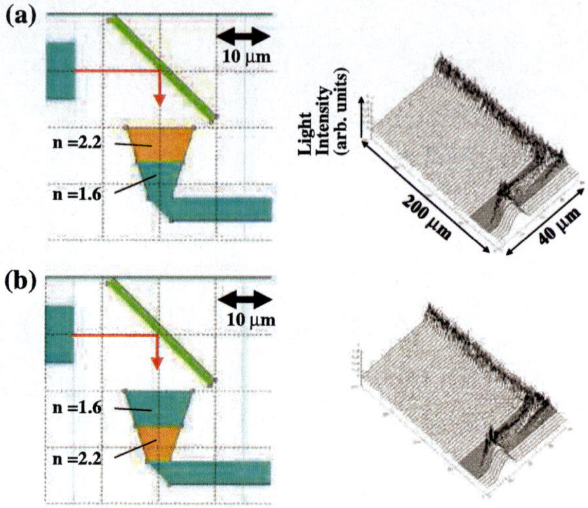

Fig. 44 Models of the tapered vertical/horizontal waveguide-type light collecting films and the light intensity distribution. **a** The *upper part* of the tapered vertical waveguide has a high refractive index (Model UH) and **b** the *lower part* of the tapered vertical waveguide has a high refractive index (Model LH)

8.3 Predicted Performance of Tapered Vertical/Horizontal Waveguide-Type Light Beam Collecting Films

Figure 44 shows models and the light intensity distribution of the tapered vertical waveguides with a high refractive index upper part (Model UH) and a high refractive index lower part (Model LH). θ_T is 17°. Wavelength of incident light beams is 600 nm. The light beam collecting efficiency was 61% in Model UH. The efficiency increased up to 77% in Model LH. Since stronger light beam confinement is necessary in the lower region of the tapered vertical waveguide, by putting a high refractive index material in the lower regions, effective light beam confinement can be achieved. For further enhancement of the collecting efficiency, the tapered angles and refractive index distribution profiles in the tapered vertical waveguides should be optimized.

8.4 Predicted Performance of Multi-Layer Waveguide-Type Light Beam Collecting Films

Figure 45 shows a detailed structure of a multi-layer waveguide-type light beam collecting film. Eighteen films are stacked. The numbers in the top view indicate vertical mirror apertures of the film. An example of optical circuits in a film is illustrated in Fig. 46. Merging waveguides transport light reflected off the mirrors to the main waveguide. By arranging the films regularly, the surface can be occupied by the light beam collecting apertures in high density, achieving an aperture occupation of 100% theoretically.

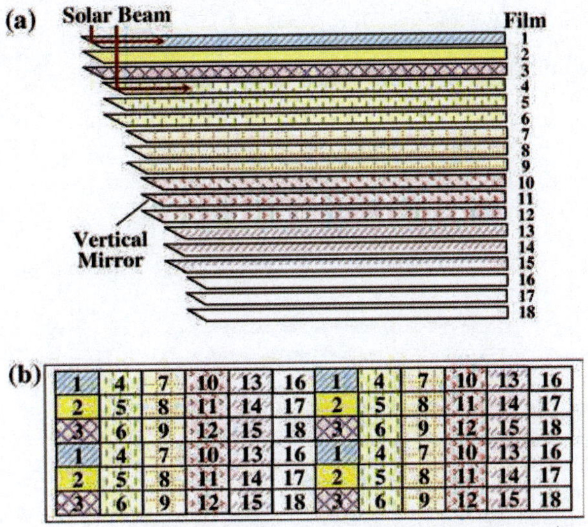

Fig. 45 A detailed structure of a multi-layer waveguide-type light collecting film [24]

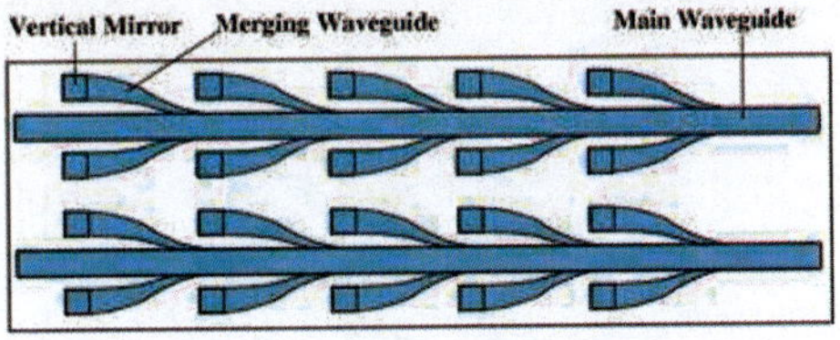

Fig. 46 An example of optical circuits with merging waveguides in a film

8.4.1 Effects of Incident Angles on Light Beam Collecting Efficiency

Figure 47 shows models with 45° mirrors and the light intensity distributions. The wavelength of the light beam is 600 nm. For an incident angle of 0°, the light beam is introduced to the optical waveguide through the 45° mirror, and then is guided through the optical waveguide smoothly, indicating the fundamental mode is dominant. For an incident angle of 45°, much of the light passes through the 45° mirror. At the same time considerably higher modes are excited in the guided beams.

Figure 48 shows models with 30° mirrors and the light intensity distributions. For an incident angle of 0°, the light beam guided in the optical waveguide contains considerably higher modes. For an incident angle of 45°, on the other hand, the guided beams propagate smoothly.

As summarized in Fig. 49, optical waveguides with 45° mirrors and optical waveguides with 30° mirrors are useful for incident light beams with vertical direction and with tilted directions, respectively.

8.4.2 Light Beam Collecting Efficiency of Multi-Layer Waveguides

We proposed multi-layer waveguide-type light beam collecting films consisting of optical waveguides with 45° and 30° mirrors. As illustrated in Fig. 50, the first layer is an optical waveguide with a 45° mirror, and the second and the third layers

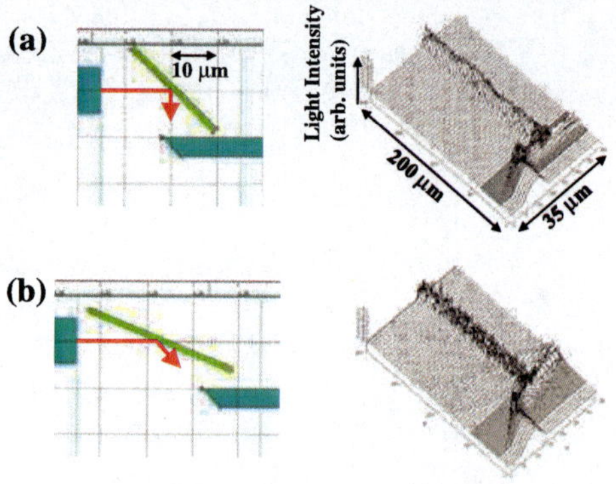

Fig. 47 Models with 45° mirrors and the light intensity distributions. **a** An incident angle of 0° and **b** an incident angle of 45°

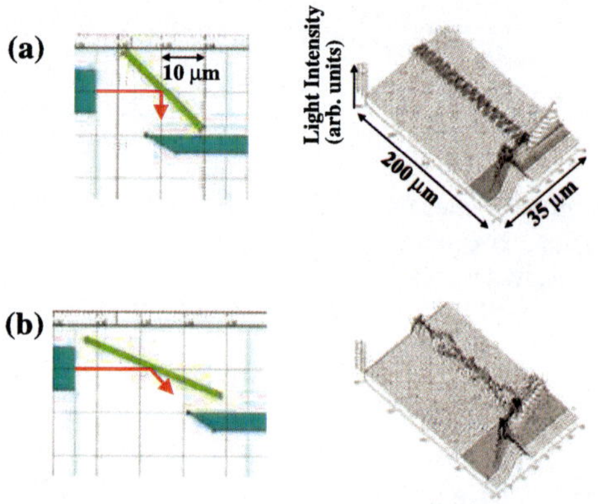

Fig. 48 Models with 30° mirrors and the light intensity distributions. **a** An incident angle of 0° and **b** an incident angle of 45°

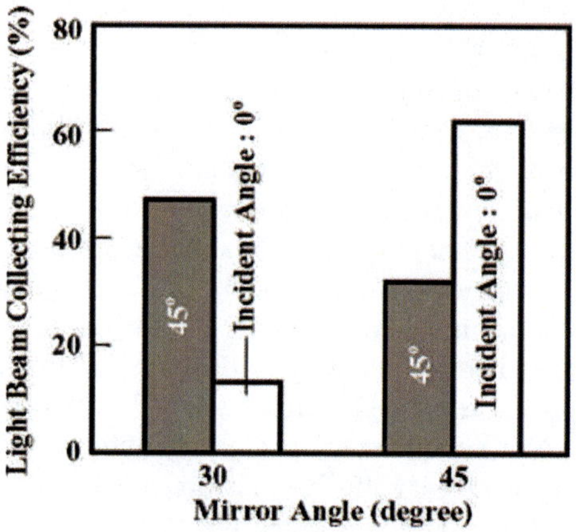

Fig. 49 Light beam collecting efficiency in models with 30° and 45° mirrors for incident angles of 0° and 45°

are optical waveguides with 30° mirrors. The two optical waveguides with 30° mirrors are arranged to direct light in opposing directions. The first, second, and third layers are for incident light beams from the vertical direction, from the left direction, and from the right direction. This structure is expected to provide wide-angle light beam collecting capability without tracking mechanisms.

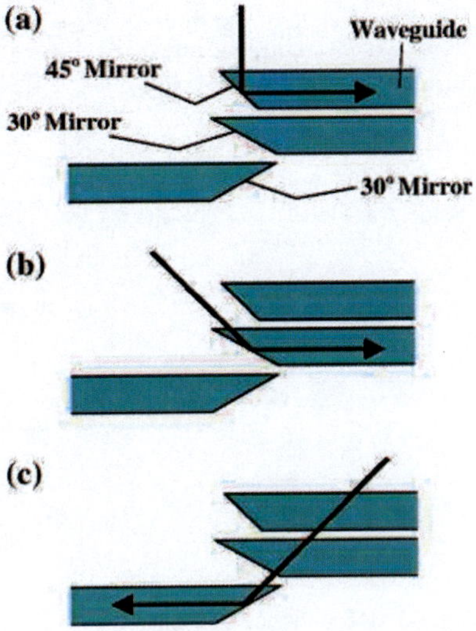

Fig. 50 An example of the three-layer structure with 45° and 30° mirrors. **a** Incident light beams with vertical directions, **b** incident light beams from the *left side* directions, and **c** incident light beams from the *right side* [24]

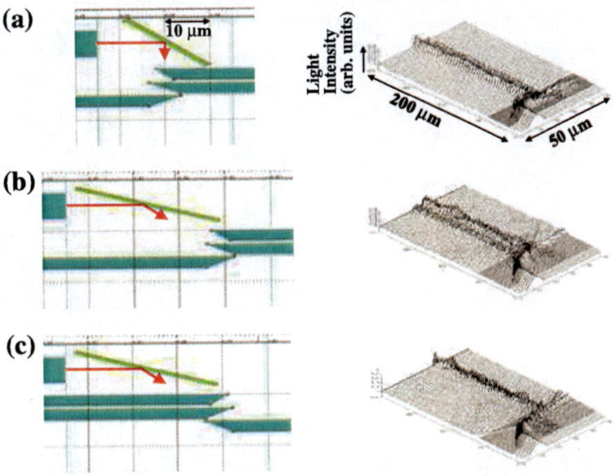

Fig. 51 Models of the three-layer structure with 45° and 30° mirrors and light intensity distributions. **a** An incident angle of 0°, **b** an incident angle of 45°, and **c** an incident angle of −45° [24]

Figure 51 shows models of the three-layer structure with 45° and 30° mirrors, and the light intensity distribution. Here, in the model of −45°, the optical waveguide direction is flipped horizontally, which effectively reproduces a configuration for −45° incidence with the same incident light beam direction as for the 45° incidence. The distance between optical waveguides is 1 μm. The wavelength of the incident light is 600 nm. It was found that the light was mainly collected by the first layer for vertical incidence, by the second layer for tilted incidence from the left, and by the third layer for tilted incidence from the right.

For comparison, the light beam collecting characteristics of the three-layer structure with only 45° mirrors were examined. As shown in Fig. 52, for tilted incidence, the collected light beams are reduced and propagate with considerably higher modes.

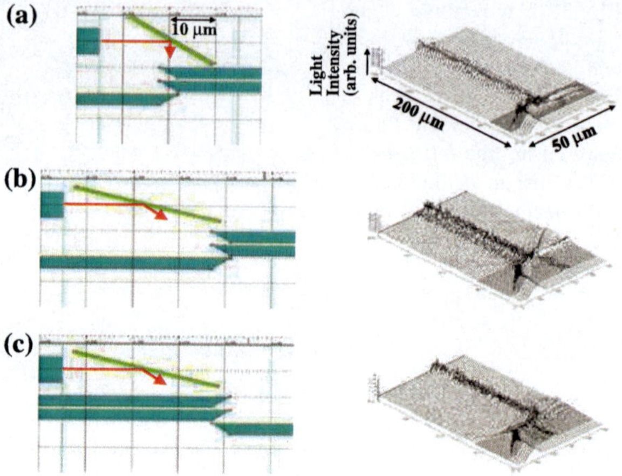

Fig. 52 Models of the three-layer structure with only 45° mirrors and the light intensity distributions. **a** An incident angle of 0°, **b** an incident angle of 45°, and **c** an incident angle of −45° [24]

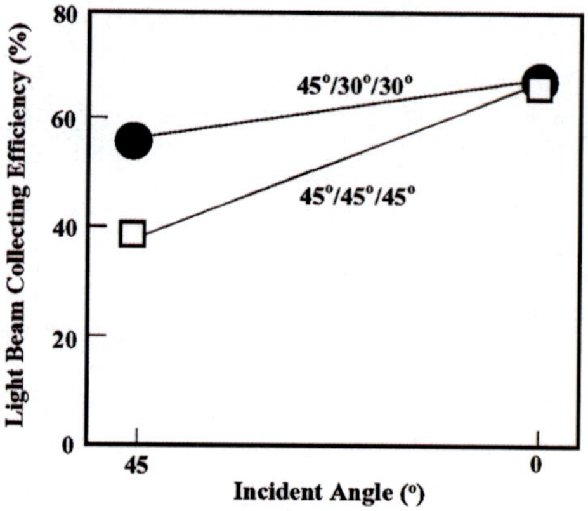

Fig. 53 Dependence of light collecting efficiency on incident angles for the three-layer structure having optical waveguides with 45° and 30° mirrors and the structure having optical waveguides with only 45° mirrors

Figure 53 shows dependence of the light beam collecting efficiency on incident angles for the three-layer structure having optical waveguides with 45° and 30° mirrors (denoted 45°/30°/30°) and the structure having optical waveguides with only 45° mirrors (denoted 45°/45°/45°). For an incident angle of 0°, almost the same efficiency is obtained in the 45°/30°/30° and 45°/45°/45° devices. For an incident angle of 45°, the efficiency is 50% larger in the 45°/30°/30° than in the 45°/45°/45° device. This indicates that the light beam collecting film consisting of mirrors of different angles provides wide-angle light beam collecting capability.

The light beam collecting efficiency was calculated to be 60%, 57%, and 65% for wavelengths of 400, 600, and 800 nm, respectively. Thus, it was found that the proposed model can operate in visible to near-infrared regions.

The models shown in Fig. 51 have 4 × 4 μm apertures. By stacking the 45°/30°/30° units three times and shifting the position by 4 μm to enlarge the effective aperture by a factor of three, the light beam collecting efficiency is expected to increase to 80% or more for incident angles of 45°, 0°, and −45°.

In the case of the model shown in Fig. 45, the layer count is eighteen for one angle of vertical mirrors. The eighteen-layer structure forms a module. To construct a multi-layer structure consisting of optical waveguides with 45° and 30° mirrors, one module unit for 45° mirrors and four module units for 30° mirrors are necessary to be stacked in order to collect light beams with vertical incident angles and tilted incident angles (from left side, right side, near side and far side). When the waveguide core thickness is 4 μm and the distance between optical waveguides is 1 μm, the module unit thickness is around 90 μm. Then, total thickness of the stack of five module units is 450 μm.

8.5 Overall Consideration

8.5.1 Overall Light Beam Collecting Efficiency

As mentioned in Sects. 8.3 and 8.4, the light beam collecting efficiency via vertical mirror for the tapered vertical/horizontal waveguide-type light beam collecting film (Model LH) and the efficiency via the multi-layer waveguide-type light beam collecting film are approximately 80%.

In addition to light beam collection through the vertical mirrors, light beam propagation via merging horizontal waveguides and couplings between optical waveguides and semiconductor flakes affect the overall light beam collecting efficiency. For the merging horizontal waveguides, BPM calculations reveal efficiency of ~70% for five-step merging.

For the optical transfer from optical waveguides (polymer, refractive index: 1.52) to semiconductor flakes (ZnO, refractive index: 2.1), the reflection loss at the interface is around 16%. The coupling efficiency may increase to nearly 100% by inserting an appropriate ARC between the optical waveguides and semiconductor flakes.

Overall light beam collecting efficiency η_{Overall} is given as follows:

η_{Overall} = [light beam collecting efficiency via vertical mirror] × [light beam propagation efficiency via merging waveguide] × [coupling efficiency between optical waveguide and semiconductor flake]

By substituting the light beam collecting efficiency of the vertical mirror (~80%), light beam propagation efficiency of the merging waveguide of (~70%), and the coupling efficiency between an optical waveguide and a semiconductor flake of ~100% into the above relation, the expected overall light collecting efficiency is calculated to be ~56% for five-step merging.

The estimated efficiency of the waveguide-type light beam collecting films is lower than that of Fresnel lens/mirror concentrators. So, at present these light collecting films do not seem suitable for large-scale solar energy conversion centers. However, they have an outstanding feature of flexible light-weight films, enabling us to carry them as sheets and put them on cars, robots, buildings, roofs of tents, surfaces of pavements/roads, walls, tables, various kinds of equipment, clothes, and anywhere else we desire. Since very high efficiency is not always necessary in these applications, these film-based integrated solar energy conversion systems may be useful yet.

8.5.2 "Efficiency/Material Consumption" Ratio

In the conventional solar cell, the overall light beam collecting efficiency is ~100%, and the reduction ratio of semiconductor consumption is 1. So, "efficiency/material consumption" ratio is 100.

For Model LH of the tapered vertical/horizontal waveguide-type light beam collecting film shown in Fig. 44b, the overall light beam collecting efficiency is ~56%. The light beam collecting aperture of the tapered vertical waveguides is 12 μm and that of the vertical mirror is 4 μm. So, a light collection factor of 3 will be achieved in this two-dimensional model. For three dimensions, the light collection factor is 9, i.e., the reduction ratio of semiconductor consumption is 1/9. In the case of five-step merging, since light beams from ten apertures are merged into one semiconductor flake, the reduction ratio of semiconductor consumption is expected to be (1/9) × (1/10) = 1/90 giving an "efficiency/material consumption" ratio of 56/(1/90) = 5040, fifty times larger than that of the conventional solar cell.

For the multi-layer waveguide-type light beam collecting film, the reduction ratio of semiconductor consumption is 1/10 for five-step merging, giving an "efficiency/material consumption" ratio of 56/(1/10) = 560, over five times that of the conventional solar cell. When collected light in different films are further merged and directed to a semiconductor flake, the "efficiency/material consumption" ratio will be increased.

8.5.3 Resource-Saving Heterogeneous Integration

In order to fabricate actual integrated solar energy conversion systems, it is important to provide a simple and low-cost process for embedding semiconductor flakes into a film. For this purpose, photolithographic packaging with SORT (PL-Pack with SORT) [18], which was developed for the optical interconnects within boxes, can be used. SORT enables us to place semiconductor flakes onto films at one time from a growth wafer, achieving semiconductor saving and low-cost device integration.

In addition, in the present design, we assumed 4-μm waveguide cores. In practice, larger cores with several tens-μm sizes can be used. For waveguide fabrication, the built-in mask method as well as imprinting and stamping can be used.

9 Summary

The waveguide-type sensitized photovoltaic devices consisting of ZnO thin films with flat surfaces and high-carrier mobility will enable us to improve the solar energy conversion efficiency. Polymer MQDs fabricated by MLD is promising for the light-absorbing layer in the sensitized photovoltaic devices. Sensitization using self-assembled double layers, which are fabricated by the liquid-phase MLD for p-type and n-type dye molecules on ZnO thin films, is also promising to widen the sensitized spectral regions. Integration with waveguide-type photovoltaic devices into light beam collecting films can provide flexible, light-weight solar energy conversion systems with wide-angle beam collecting capability.

References

1. Yoshimura T, Tatsuura S, Sotoyama W (1991) Polymer films formed with monolayer growth steps by molecular layer deposition. Appl Phys Lett 59:482–484
2. Yoshimura T, Tatsuura S, Sotoyama W, Matsuura A, Hayano T (1992) Quantum wire and dot formation by chemical vapor deposition and molecular layer deposition of one-dimensional conjugated polymer. Appl Phys Lett 60:268–270
3. Yoshimura T, Ito S, Nakayama T, Matsumoto K (2007) Orientation-controlled molecule-by-molecule polymer wire growth by the carrier-gas-type organic chemical vapor deposition and the molecular layer deposition. Appl Phys Lett 91:033103-1-3
4. Yoshimura T, Kudo Y (2009) Monomolecular-step polymer wire growth from seed core molecules by the carrier-gas type molecular layer deposition (MLD). Appl Phys Express 2:015502-1-4
5. Yoshimura T, Yano E, Tatsuura S, Sotoyama W (1995) Organic functional optical thin film, fabrication and use thereof. US Patent 5,444,811
6. Yoshimura T (2011) Thin-film organic photonics: molecular layer deposition and applications, CRC/Taylor and Francis, Boca Raton
7. Yoshimura T, Oshima A, Kim D, Morita Y (2009) Quantum dot formation in polymer wires by three-molecule molecular layer deposition (MLD) and applications to electro-optic/photovoltaic devices. ECS Transactions vol 25, no 4, Atomic Layer Deposition Applications 5. Austria Meeting, Vienna, pp 15–25
8. Yoshimura T (2007) Self-organized growth of polymer wire networks with designed molecular sequences for wavefunction-controlled nano systems (Ch. 5). In: Jost AR (ed) Trends in thin solid films research. Nova Science Publishers, New York
9. Yoshimura T (1989) Enhancing second-order nonlinear optical properties by controlling the wave function in one-dimensional conjugated molecules. Phys Rev B40:6292–6298
10. Yoshimura T, Kiyota K, Ueda H, Tanaka M (1981) Mechanism of spectral sensitization of ZnO coadsorbing p-type and n-type dyes. Jpn J Appl Phys 20:1671–1674
11. Yoshimura T, Kiyota K, Ueda H, Tanaka M (1979) Contact potential difference of ZnO layer adsorbing p-type dye and n-type dye. Jpn J Appl Phys 18:2315–2316
12. Pessa M, Makela R, Suntola T (1981) Characterization of surface exchange reactions used to grow compound films. Appl Phys Lett 38:131–133
13. Yoshimura T (1991) Liquid phase deposition. Japanese Patent, Tokukai Hei3-60487 (in Japanese)
14. Iijima M, Takahashi Y, Fukuda E (1989) Vacuum deposition polymerization. Nikkei New Materials, December 11:93–100 (in Japanese)
15. Murata H, Ukishima S, Hirano H, Yamanaka T (1996) A novel fabrication technique and new conjugated polymers for multilayer polymer light-emitting diodes. Polym Adv Technol 8:459–464
16. Yoshimura T, Suzuki Y, Shimoda N, Kofudo T, Okada K, Arai Y, Asama K (2006) Three-dimensional chip-scale optical interconnects and switches with self-organized wiring based on device-embedded waveguide films and molecular nanotechnologies. Proc SPIE 6126: 612609-1-15

17. Yoshimura T, Terasawa N, Kazama H, Naito Y, Suzuki Y, Asama K (2006) Selective growth of conjugated polymer thin films by the vapor deposition polymerization. Thin Solid Films 497:182–184
18. Yoshimura T, Ojima M, Arai Y, Asama K (2003) Three-dimensional self-organized micro optoelectronic systems for board-level reconfigurable optical interconnects–performance modeling and simulation. IEEE J Sel Top Quant Electron 9:492–511
19. Kiyota K, Yoshimura T, Tanaka M (1981) Electrophotographic behavior of ZnO sensitized by two dyes. Photogr Sci Eng 25:76–79
20. Meier HJ (1965) Sensitization of electrical effects in solids. Phys Chem 69:719–729
21. Yoshimura T (2009) Integrated photonic/electronic/chemical systems, solar energy conversion systems, light collectors, and optical waveguides. Japanese Patent, Tokukai 2009-4717 (in Japanese)
22. Yoshimura T, Asama K (2007) Integrated chemical systems and integrated light energy conversion systems. Japanese Patent, Tokukai 2007-107085 (in Japanese)
23. Yoshimura T (2007) Proposed applications of 3-D optical interconnect technologies to integrated chemical systems. 2007 Digest of the LEOS Summer Topical Meetings (Bio-Inspired Sensors and Application/Imprinting on Photonic Integrated Circuits), pp 129–130
24. Shioya R, Yoshimura T (2009) Design of solar beam collectors consisting of multi-layer optical waveguide films for integrated solar energy conversion systems. J Renew Sustain Energy 1:033106-1-15

Part II
Electricity Generation via Fuel Cells and Piezoelectric Materials

Nanomaterials for Proton Exchange Membrane Fuel Cells

Xin Wang and Shuangyin Wang

Abstract The development of fuel cell technology, especially the proton exchange membrane fuel cell, is progressed through the use of nanomaterials. This chapter introduces the recent development of nanostructured electrocatalysts for fuel cell applications. An overview of various synthesis methods for nanostructured electrocatalyts is first presented. The morphology control of the electrocatalysts is then discussed, including zero-dimensional, one-dimensional, multi-dimensional, and hollow structures. Multi-component electrocatalysts with improved durability and activity for various fuel cell reactions are also introduced. Finally, the advantage of using carbon nanotubes as electrocatalyst support and noncovalent functionalization of nanotube surface is discussed in detail.

1 Introduction of Proton Exchange Membrane Fuel Cell(PEMFC)

The fuel cell, a device that converts chemical energy directly into electrical energy, involves an electrochemical reaction, like the conventional battery [1–4]. The difference is that fuel cells have an external and continuous supply of fuels and therefore do not require a time-consuming charging process. Fuel cells have attracted much attention in the past few decades due to their advantages over conventional internal combustion engines, such as low pollutant emissions, high efficiency, and high theoretical power output density.

X. Wang (✉) · S. Wang
School of Chemical and Biomedical Engineering,
Nanyang Technological University,
Singapore, 637459, Singapore
e-mail: WangXin@ntu.edu.sg

L. Zang (ed.), *Energy Efficiency and Renewable Energy Through Nanotechnology*,
Green Energy and Technology, DOI: 10.1007/978-0-85729-638-2_11,

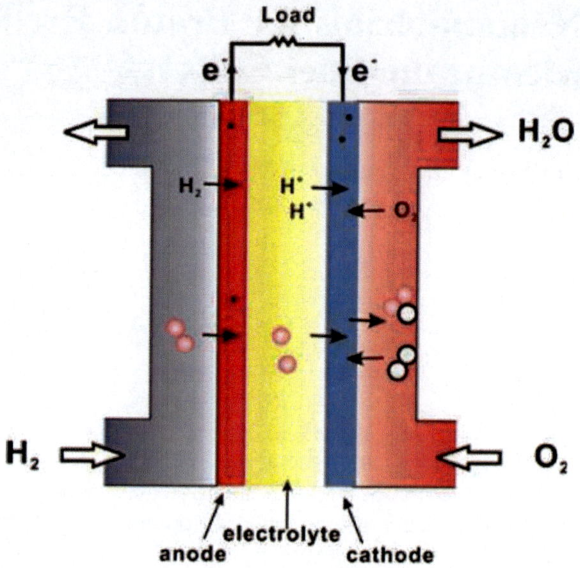

Fig. 1 Illustration of a proton exchange membrane fuel cell

An anode, electrolyte, and cathode are the main components of a fuel cell. Among the various members in the fuel cell family, PEMFCs are the most attractive candidate for automotive and portable applications [1, 4]. Figure 1 shows the main components of a PEMFC. The output voltage of a single cell is limited by the Nernst potential of the two half-cell reactions occurring on the two electrodes. Electrochemical oxidation of fuels such as hydrogen and small organic molecules (SOMs, such as methanol, ethanol, and formic acid) occurs on the anode and the electrochemical reduction of the oxidant takes place on the cathode. Ions generated from the electrochemical reduction or oxidation processes are transported through the electrolyte, which is ionically conductive but electronically insulating. During the reaction, the generated electrons will flow through the external circuit, producing electrical energy, while the ions will flow through the electrolyte from one side to the other side of the electrode, completing the overall fuel cell reactions. Depending on the fuels used, PEMFCs can be categorized as hydrogen/oxygen fuel cells, direct methanol fuel cells (DMFCs), and direct formic acid fuel cells (DFAFC) etc. [2, 5, 6].

In the various components of PEMFCs, nanotechnology is frequently employed to improve cell performance, increase durability, and reduce the cost. For example, the precious metal electrocatalysts used are reduced to the nanoscale, ensuring a high catalytic surface area and minimizing the amount of precious metal used to maintain high performance. Meanwhile, carbon nanomaterials are typically utilized as catalyst supports due to their high conductivity, good thermal and chemical stability, and excellent mechanical properties. Not limited to the catalyst, some nanosized materials such as carbon nanotube and silica oxide nanoparticles are also employed in the membrane to enhance its self-humidifying ability and mechanical properties. Therefore, nanomaterials have found extensive use in fuel cell system and will continue to be of research interest. Therefore, in this chapter, we will discuss the applications of nanomaterials in fuel cell systems. Specifically, we will focus on nanostructured electrocatalysts and their supports in PEMFC devices.

2 Synthesis Methods of Nanosized Electrocatalysts for PEMFC

Platinum-based metal nanoparticles supported on carbon are the most commonly used electrocatalysts for various reactions in PEMFCs. However, the high cost of this precious metal is a major barrier to a large-scale commercialization of PEMFCs. In order to reduce Pt usage without sacrificing cell performance, a great deal of effort must be devoted to improving the electrocatalyst utilization efficiency and enhancing the intrinsic catalytic activity of the electrocatalysts. One approach is to use high surface area carbon materials (e.g., carbon nanotube and carbon black) as supports to enhance the dispersion and distribution of metal nanoparticles and to improve their utilization efficiency [7]. The dispersion of metal electrocatalyst nanoparticles on carbon supports strongly depends on the interaction between them, and intrinsic properties of the supports. On the other hand, electrocatalytic activity is highly dependent on the morphology and particle size of metal electrocatalysts. Metal electrocatalysts exhibit shape, size, and dimension dependent catalytic properties, which have made controlling the morphology an active and prolific research area.

To date, various synthesis methods have been investigated to prepare supported metal electrocatalysts. These methods include impregnation, potential step electrodeposition, colloidal processing, and microwave-assisted polyol techniques [8]. The impregnation method, also called the electroless deposition method, is widely used to prepare highly-dispersed metal nanoparticles deposited on carbon supports. This approach involves the pre-mixing of carbon supports and metal precursors followed by the addition of reducing agents (such as sodium borohydride, ethylene glycol, and formic acid). The average particle size, dispersion, and distribution of the metal nanoparticles strongly depend on the properties of carbon supports and reducing agents. The synthesis and deposition of Pt nanoparticles by the electrodeposition is realized by pasting the carbon supports onto the working electrode and subsequent pulse deposition in an electrolyte containing metal precursors. However, such an electrodeposition method is not suitable for large-scale production. The colloidal route involves the pre-formation of metal nanoparticles followed by the deposition onto carbon supports. In this case, organic materials, such as polymers or surfactants, are used to stabilize metal nanoparticles to avoid aggregation in the colloidal dispersion. After the nanoparticles are deposited on the supports, these stabilizers should be removed from the metal surface. Otherwise, they can hinder the interaction between the reactant molecules and the active catalyst sites. Usually, a heat treatment or solvent rinse is used to remove the

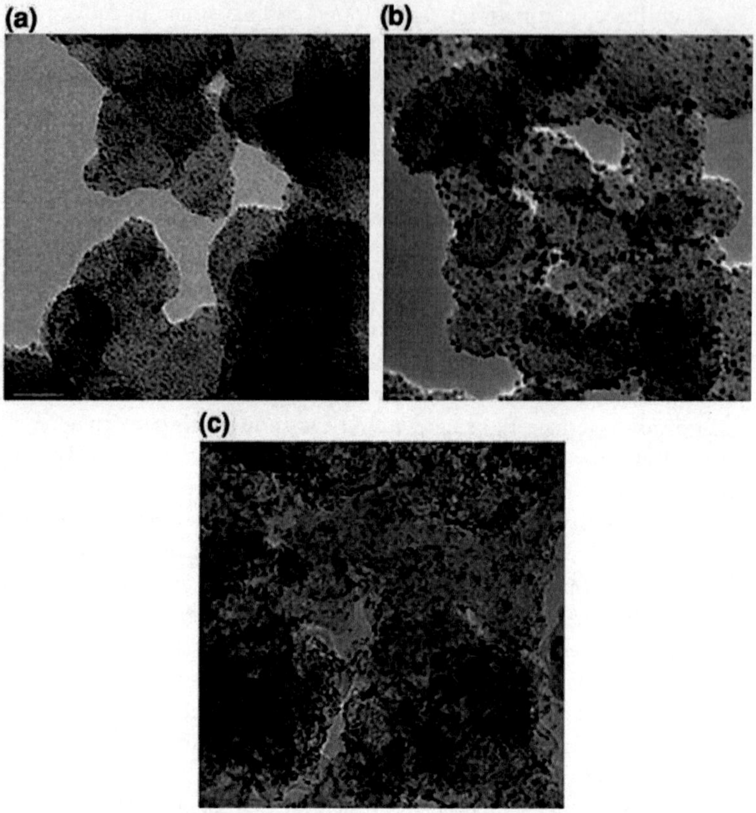

Fig. 2 TEM images of 6 wt% Pt and 3 wt% Ru catalysts deposited on Vulcan XC-72R. The PtRu catalysts were synthesized in ethylene glycol solutions with the following NaOH concentrations: 0.1 M (**a**), 0.075 M (**b**), and 0.068 M (**c**). The bars in (**a**) and (**b**) indicate a 20 nm scale, and the bar in (**c**) indicates a 50 nm scale. Reprinted with permission from Bock et al. [8] American Chemical Society

stabilizer molecules. However, it is difficult to completely remove them with a simple solvent rinse. On the other hand, the heat treatment method usually leads to the aggregation of metal nanoparticles, resulting in low utilization efficiency of the electrocatalysts and alteration of the surface composition of the catalysts if multi-component alloy catalysts are involved.

To date, the most commonly used chemical reduction method is the polyol reduction method. For this method, ethylene glycol is used as the solvent, reducing agent, and stabilizer. The formation of metal nanoparticles involves the reduction of metal precursor salts by ethylene glycol, which, in turn, is oxidized. It has been proposed that the oxidation of ethylene glycol mainly results in glycolic acid or glycolate anions, depending on the pH value of the solution. The stabilization of the metal nanoparticles by ethylene glycol system is realized by the electrostatic repulsion of the attached glycolate anions on the surface of the metal colloids.

The concentration of glycolate anions in the synthesis solution can be adjusted by controlling the pH of the solution. As the concentration of glycolate anions affects the stabilization behavior of the metal nanoparticles, the metal nanoparticle size can be easily tuned by controlling the pH of the synthesis solution. Bock et al. [8] performed such an investigation. As shown in Fig. 2, the average particle size and distribution of metal nanoparticles on carbon black can be adjusted by the concentration of NaOH. They concluded that the increase of the pH of the solution leads to smaller particle size. Such a strategy allows for the synthesis of highly-dispersed metal nanoparticles with tunable particle size on various substrates [8].

More recently, the microwave-assisted polyol reduction method has been developed to synthesize highly-dispersed metal nanoparticles deposited on carbon supports with much smaller particle size. Microwave irradiation facilitates the fast reduction of metal precursors to metal atoms and uniformly distributes heat in the reaction solution [9]. The creation of a high level of supersaturation within very short time range leads to fine particle size.

3 Morphology Control of Nanostructured Electrocatalysts

Extensive efforts have been directed toward the synthesis of morphology-controlled nanomaterials because of their extraordinary morphology-dependent properties and their wide application in various fields, such as catalysis, opto-electronics, biolabeling, chemical sensors, and fuel cells. To date, various forms of nanostructures have been developed, ranging from highly monodispersed spherical and cubic nanoparticles to various anisotropic nanostructures such as nanowires, nanorods, and nanotubes. Herein, we will review the recent progress of the morphology-controlled synthesis and its effect on the catalytic activities of metal nanomaterials [8, 10–14].

3.1 Zero-Dimensional Nano-Electrocatalysts

Fuel cell electrocatalysts are normally used in the form of zero-dimensional nanoparticles with small particle size, since small particles feature a high surface area to volume ratio, which is essential for surface catalysis applications [15]. Other than the surface area, the electrocatalytic activities are significantly affected by the surface properties of metal nanoparticles that could be tuned via size and shape-controlled synthesis. For example, Arenz et al. [16] investigated the size effect of Pt nanoparticles on the electrocatalytic activities for adlayer and bulk CO oxidation and found that the removal/oxidation of adsorbed CO on the Pt surface is sensitive to the surface properties of Pt nanoparticles. As shown in Fig. 3, the onset potential of adlayer CO oxidation is almost independent of the particle size, but the rate of CO_2 production decreases with the decrease of the particle size. They suggested that the removal/oxidation of CO is mainly controlled by the number of defects on the Pt surface. Large particles have "rougher" surfaces and therefore more defects than small particles, which have fairly smooth (111) facets. In contrast to CO adlayer oxidation, smaller particles are more active for bulk CO oxidation than larger particles because of the increased oxophilicity. Sun et al. [17] conducted shape-controlled synthesis of Pt nanoparticles. Polyhedral (3 nm), truncated cubic (5 nm), and cubic (7 nm) Pt nanoparticles were prepared, with dominating lattice fringes of Pt (111), Pt (100), and Pt (100) respectively, as shown in the TEM images in Fig. 4. The electrochemical investigation revealed that Pt (100)-dominated nanocubes are more active for the oxygen reduction reaction (ORR) in 0.5 M H_2SO_4 than the other two catalysts (polyhedral and truncated cubic nanoparticles). It is believed that such a shape-dependent ORR activity can be ascribed to the different adsorption properties of Pt surface. Therefore, these two examples demonstrate that the catalytic activities of the zero-dimensional nanoparticles could be efficiently tuned through the size and shape control.

3.2 One-Dimensional Nano-Electrocatalysts

In parallel, recently a lot of attention has been given to one-dimensional or branched metal nanostructures, such as nanorods, porous nanoparticles, multipods, nanowire networks, and dendritic nanoparticles. Zhuang et al. [18] successfully prepared one-dimensional Pd nanorods (Fig. 5) and zero-dimensional nanoparticles with electrochemical deposition technique by simply changing the precursor concentration. The electrochemical characterization indicated that Pd nanorods exhibited an unexpectedly higher area-specific activity toward the ORR than nanoparticles, which is close to that of bulk Pt. The electrochemical fingerprint experiment on CO stripping shows that the surface of Pd nanorods is dominated by Pd (110) facets, on which the adsorption of O is found to be exceptionally weak by subsequent density functional theory (DFT) calculations. It is believed that such a weak adsorption is the reason for the superior activity of Pd nanorods toward ORR. Another example of one-dimensional nanostructured electrocatalysts is the Pt nanotubes developed by Lu et al. [19]. The Pt nanotubes (Fig. 6) were synthesized via a redox replacement reaction in the presence of cetyltrimethylammonium bromide (CTAB). Improved catalytic activity toward methanol oxidation was observed compared to that of the nanowires, which is due to the high surface area and high preferential crystal facets of Pt nanotubes.

Using similar redox replacement reaction method, Yan et al. [20] successfully synthesized supportless Pt and PtPd alloy nanotubes (Fig. 7). As shown in Fig. 8, the as-synthesized PtPd alloy nanotubes not only exhibit significantly enhanced catalytic activities toward ORR, but also demonstrate an improved durability/stability, compared to monometallic Pt nanotubes, Pt black, and commercial Pt/C catalysts. The improved durability was explained by (1) the elimination of support

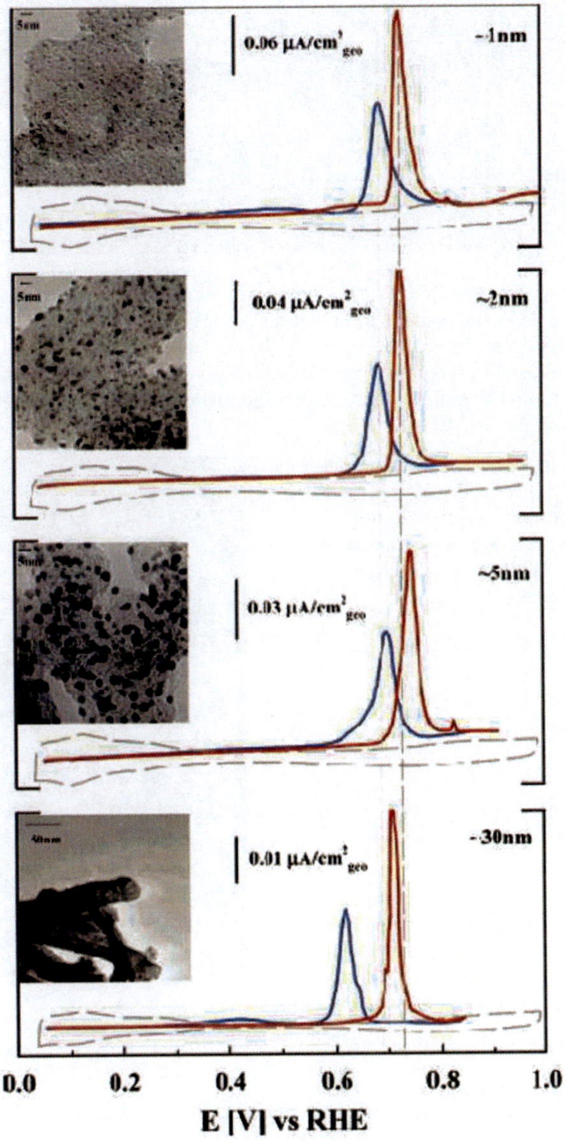

Fig. 3 Comparison of CO-stripping curves in 0.1 M HClO4 solution purged after CO adsorption with argon for 30 min at 0.05 V; scan rate 1 mV/s; dashed gray curves show the respective base voltammograms; the blue curves show the CO-stripping curves after oxide-annealing and the red curves after CO-annealing. Reprinted with permission from Arenz et al. [16] American Chemical Society

corrosion problem and (2) the one-dimensional nanostructure has a length on the order of microns, which makes it less vulnerable to dissolution, Ostwald ripening, and aggregation during fuel cell operation than the platinum nanoparticles. Additionally, the anisotropic morphology of the one-dimensional nanostructures also improves the mass transport properties and catalyst utilization efficiency for the electrocatalytic reaction. If properly assembled (e.g., cubic or hexagonal close-packing and vertical alignment on a Nafion® membrane), they can also lead to a

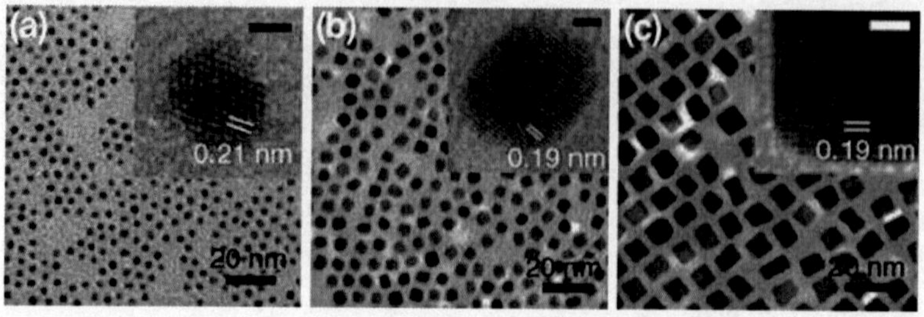

Fig. 4 TEM images of (**a**) the 3 nm polyhedral (**b**) the 5 nm truncated cubic, and (**c**) the 7 nm cubic Pt Nanoparticles. The insets are the representative HRTEM images of corresponding single particles, showing (**a**) Pt (111), (**b**) Pt (100), and (**c**) Pt (100) lattice fringes. All scale bars in the insets correspond to 1 nm. This is the pre-peer reviewed version of the following article: Wang et al. [17] Angewandte Chemie-International Edition 47:3588–3591

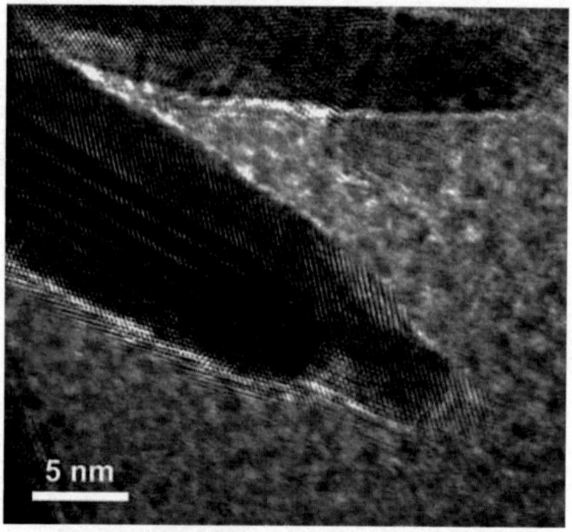

Fig. 5 A TEM image of Pd-NR deposited on carbon powder. Reprinted with permission from Xiao et al. [18] American Chemical Society

thin catalyst layer, further improving the mass-transfer characteristics within the catalyst layer [20]. On the basis of the above works, it can be concluded that, the electrocatalytic activity and durability of metal nanostructures can be controlled by changing the structure.

3.3 Two and Three-Dimensional Nano-Electrocatalysts

The increase in dimension of the metal nanostructures of electrocatalysts to two or three may produce other unexpected catalytic activities. Ding et al. [21] developed a simple method to fabricate a novel nanotubular mesoporous (NM) PdCu bimetallic catalyst with a nanoporous shell based on low-temperature dealloying in

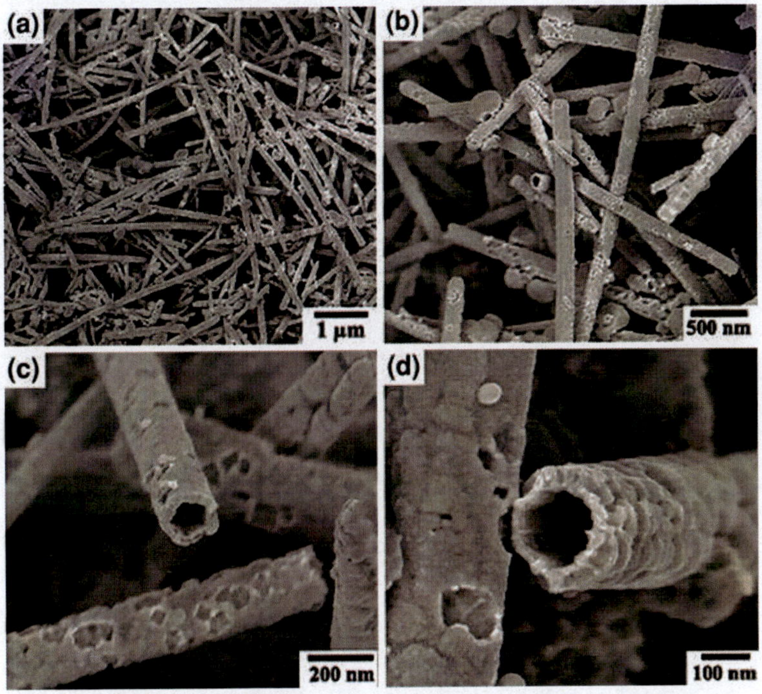

Fig. 6 **(a–d)** SEM and FE-SEM images of the platinum nanotubes. Reprinted with permission from Bi et al. [19] American Chemical Society

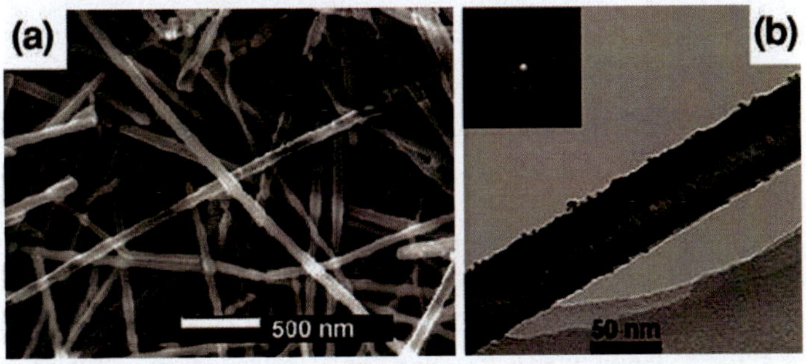

Fig. 7 **a** SEM image of PtPd nanotubes. **b** TEM image and electron diffraction pattern (*inset*) of PtPd nanotubes. This is the pre-peer reviewed version of the following article: Chen et al. [54]

an aqueous solution and a subsequent in situ galvanic replacement reaction. The as-prepared PdCu bimetallic catalyst exhibits a superior ORR activity, as shown in Fig. 9. The authors claimed that trimodal hollow bimetallic structure in the PdCu nanostructures play a crucial role in the enhancement of ORR activity. Another two-dimension example, Jiang et al. [22] fabricated Pd nanowire arrays (NWA) by the anodized aluminum oxide (AAO) templated electrodeposition

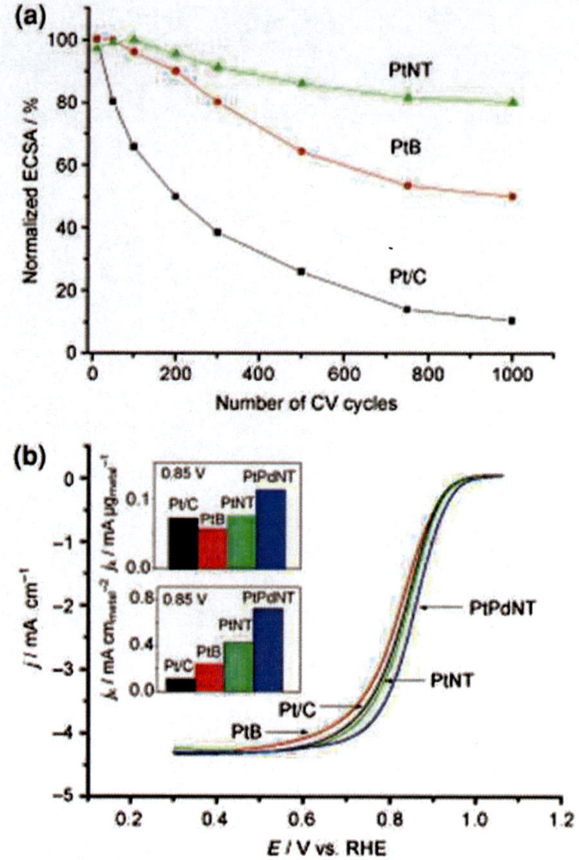

Fig. 8 **a** Loss of electrochemical surface area (ECSA) of Pt/C, platinum-black (PtB), and Pt nanotube catalysts with number of CV cycles in Ar-purged 0.5 M H_2SO_4 solution. **b** ORR curves in O_2-saturated 0.5 M H_2SO_4 solution. Inset: Mass activity (*top*) and specific activity (*bottom*) for the four catalysts at 0.85 V. This is the pre-peer reviewed version of the following article: Chen et al. [54]

method (Fig. 10). The Pd nanowire arrays have large amounts of electrochemically active surface area. The electrocatalytic activity and stability of the Pd nanowire array for ethanol electrooxidation are not only significantly higher than those of conventional Pd film electrodes, but also higher than those of commercial E-TEK PtRu/C electrocatalysts. The utilization efficiency and electrocatalytic activity of the Pd nanowire array electrode strongly depend on the length and structure of the Pd nanowires. The nanometer-sized gaps or pores between the nanowires in the array are straight and short, acting as effective transport channels to and from the electrode/electrolyte interface for the liquid fuel and products formed during the electrooxidation of ethanol [22].

Furthermore, Chen et al. [23], by use of a low-cost, one-step hydrothermal method, successfully prepared three-dimensional nanoporous PtRu networks (Fig. 11). The precursors Ru^{3+} and Pt^{4+} ions were reduced simultaneously by the reduction agent formaldehyde, resulting in the formation of PtRu alloys. The composition of the PtRu networks can be easily changed by varying the concentration of the Ru and Pt precursors. The surface characterization and

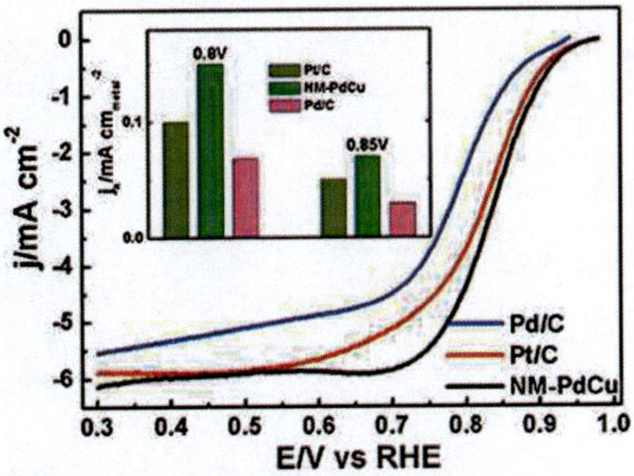

Fig. 9 Polarization curves for the ORR on NM-PdCu and Pd/C, Pt/C nanoparticles in an O_2-saturated 0.1 M $HClO_4$ at room temperature at 1600 rpm. Scan rate: 5 mV s^{-1}. The inset is the specific activity for NM-PdCu, Pt/C, and Pd/C catalysts at 0.8 and 0.85 V, respectively. Reprinted with permission from Xu et al. [21] American Chemical Society

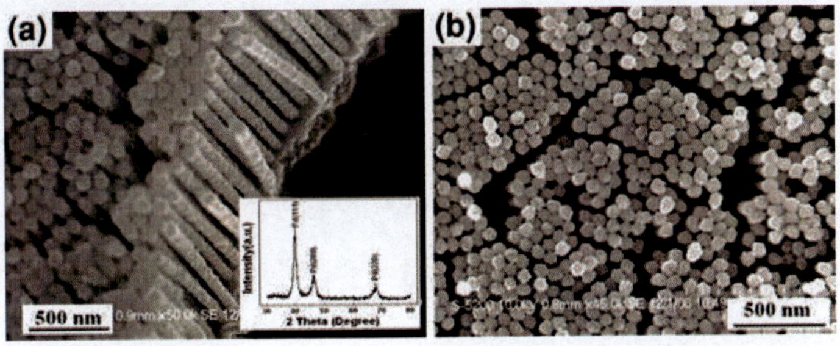

Fig. 10 SEM images of (**a**) cross section and (**b**) surface of Pd NWAs. Inset in (**a**): XRD pattern of Pd NWAs. This is the pre-peer reviewed version of the following article: Xu et al. [22]

electrochemical studies show that the active surface area of the synthesized novel nanoporous PtRu networks is much higher than that of the polycrystalline platinum electrode, and that the nanoporous PtRu networks have much higher activity for the electrochemical oxidation of methanol and CO.

In light of the potential benefits that multi-dimensional nanostructures possess, there is a pressing need to develop efficient approaches for the fabrication of multi-dimensional nanostructures with controllable morphology. A variety of approaches including chemical vapor deposition, laser assisted synthesis, electrochemical deposition, and hard or soft templating have been extensively attempted. The hard template method, e.g. using mesoporous silica, generally produces low metallic interconnectivity due to poor continuity of precursors in the mesoporous templates. Additionally, to remove the silica template, hydrofluoric acid, a serious environmental and safety hazard, is required. The template method based on the AAO is also limited by the channel diameter of the AAO available. On the other hand, low-temperature soft-template approaches based on wet-chemistry offer enormous process flexibility to assemble nanoscale building blocks into variety of nano-architectures. For example, metal nanowires can be synthesized using micelles, polymer-mediated nanoparticle assembly, and surfactant-stabilized water-in-oil microemulsions. Recently, a self-assembly soft-template method has been

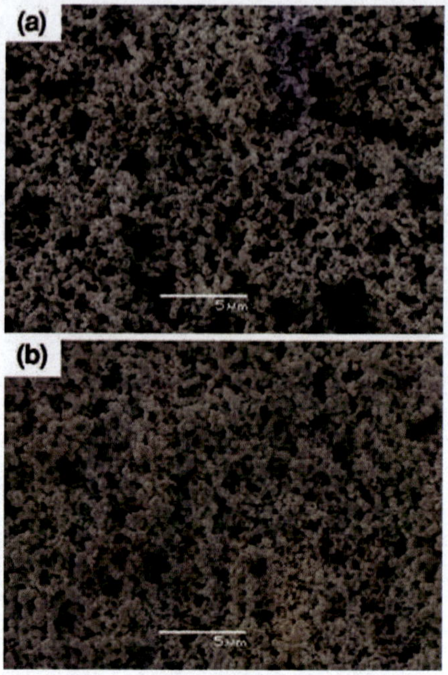

Fig. 11 SEM images of: (**a**) nanoporous Pt surface; and (**b**) typical nanoporous Pt-Ru electrode. Magnification: 5000×. This is the pre-peer reviewed version of the following article: Koczkur et al. [23]

developed to effectively synthesize Pd nanowire networks (NWNs) as electrocatalysts for formic acid oxidation [24]. The principle of the tunable synthesis of Pd nanostructures via the polysodium-p-styrenesulfonate (PSS) polyelectrolyte-mediated self-assembly process is shown in Fig. 12. Self-assembly occurs between positively charged Pd^{2+} ions and negatively charged SO_3 function groups attached to the pendent aromatic ring of PSS, forming anisotropic Pd^{2+}–PSS networks in solution. The reaction rate is controlled by pH. If the nucleation rate of Pd nanoparticles along the PSS molecular chains is fast, the crystallization and growth of Pd(0) would adopt the anisotropic Pd^{2+}-PSS network, forming Pd NWNs. The assembly and formation of Pd nanoparticles and NWNs occurs simultaneously. On the other hand, if the nucleation rate of Pd nanoparticles along the PSS polymeric chain is slow, the grain growth of Pd nanoparticles would dominate. The random nucleation and grain growth would disturb the anisotropic Pd^{2+}–PSS network in solution and form isolated Pd nanoparticles instead. This principle is confirmed by the TEM images shown in Fig. 13.

3.4 Hollow Metallic Electrocatalysts

The proper design of core–shell nanostructured electrocatalyst reduces the cost of precious metal electrocatalysts and improves their electrocatalytic activity. Meanwhile, hollow nanostructured electrocatalysts with empty cores have also

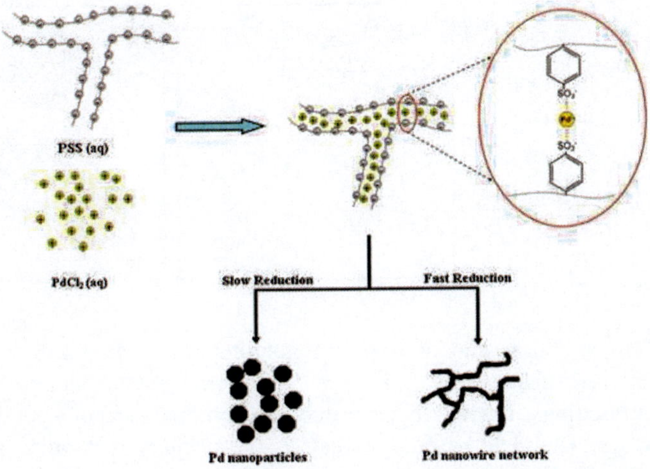

Fig. 12 A schematic illustration of the templateless and PSS-mediated self-assembly of Pd NWNs and nanoparticles. Reprinted with permission from Wang et al. [24] Copyright 2008 IOP Publishing Ltd

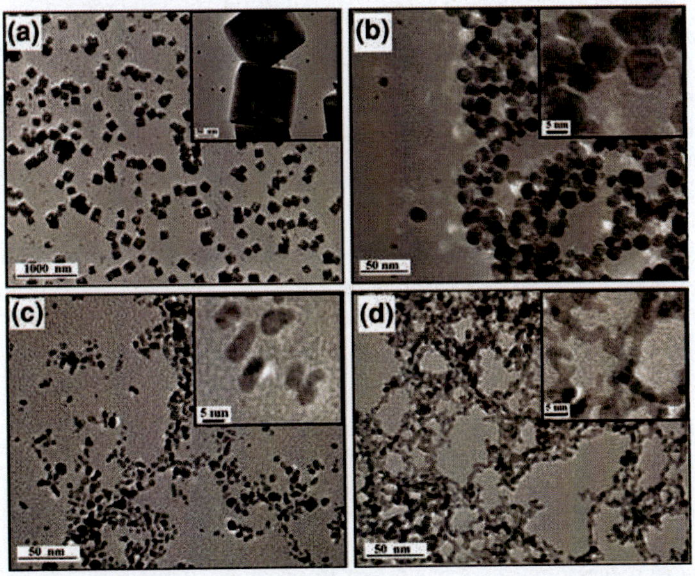

Fig. 13 TEM micrographs of Pd nanostructures synthesized in the presence of PSS polyelectrolyte. (**a**) Pd nanocubes synthesized at pH = 2, (**b**) Pd cuboctahedrons synthesized at pH = 6, (**c**) Pd nanorods synthesized at pH = 8, and (**d**) Pd nanowire networks synthesized at pH = 12. PSS/Pd ratio was 1:1 and ethanol was used as reducing agent. Inset is the high magnification TEM image. Reprinted with permission from Wang et al. [24] Copyright 2008 IOP Publishing Ltd

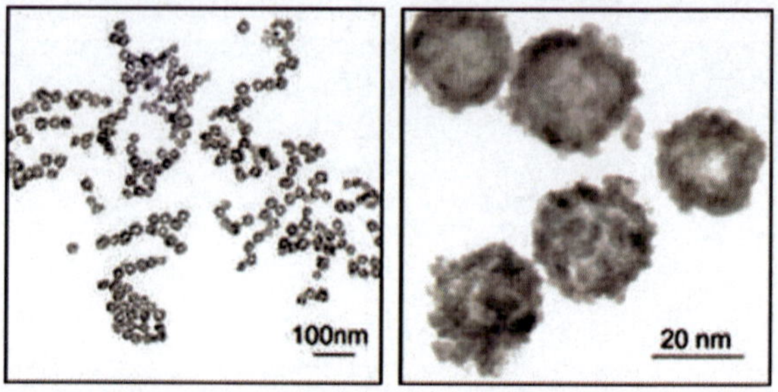

Fig. 14 TEM images of Pt hollow nanospheres (**a**) and (**b**). This is the pre-peer reviewed version of the following article: Liang et al. [27]

attracted much attention due to their unique properties. The hollow structure provides increased surface area and saves on material cost. It should be kept in mind that methodology is critically important to the preparation of a catalyst with a hollow structure. Template-based synthesis is the primary method to synthesize hollow metallic catalysts. Porous alumina, polystyrene spheres, silica spheres, and micelles have been employed as templates to prepare hollow metal nanostructures [25–28]. More recently, a galvanic replacement reaction-based template method was developed. This method uses a sacrificial Ag or Co nanoparticle as a template, which reduces aqueous metal ions, resulting in the formation of hollow metal nanostructures with various morphologies conformal to those of the templates. Bai and coworkers [27] developed a facile procedure for the synthesis of a Pt hollow nanosphere catalyst based on the replacement reaction between Co nanoparticles and H_2PtCl_6, as shown in Fig. 14. The as-prepared Pt hollow nanospheres had a very high surface area and therefore exhibited enhanced weight-normalized electrocatalytic performance. Based on this procedure, some bimetallic hollow nanospheres could also be synthesized. For example, Qiu [26] synthesized PtRu hollow nanospheres using Co as the sacrificial template and supported them on carbon nanotube support. The PtRu hollow nanospheres exhibited enhanced weight-normalized electrochemical performance for methanol oxidation compared with PtRu nanoparticles and commercial PtRu/C electrocatalysts due to the advantageous structural properties of the hollow structure.

4 Multi-component Electrocatalysts

Platinum is the best-known monometallic electrocatalyst for hydrogen oxidation and oxygen reduction reactions. However, at the anode side, Pt suffers from severe poisoning from fuel impurities, such as carbon monoxide (CO), and the electrocatalytic activity is quite low for other fuels like methanol or formic acid. Even for oxygen reduction at the cathode side, there exists a large overpotential. The enhancement of Pt electrocatalysts for fuel cell applications is usually realized by the appropriate design of multicomponent electrocatalysts with the introduction of another component, called the promoter.

4.1 Bimetallic Alloy Electrocatalysts

To overcome the low electrocatalytic activity of Pt toward fuels like methanol or CO containing hydrogen, the approach of incorporating other promoting metals has been investigated extensively, of which the most commonly used are ruthenium and tin. For example, it has been observed that the presence of Ru increases the rate of methanol adsorption and the maximum CO coverage of the surface is reached more rapidly at a PtRu surface than at a pure Pt catalyst at a given potential. A bifunctional mechanism originally proposed by Watanabe [29] is believed to be responsible for the activity improvement, where one metal (Pt) is available for the adsorption of the intermediate and the other metal on adjacent site acts as the adsorption site for oxygen-containing species. Since Ru can generate OH species from the water activation at lower potential than Pt [1, 12], the oxidation of the adsorbed CO species can be realized at lower potential.

PtSn bimetallic electrocatalysts have also been reported to show enhanced electrocatalytic activity for methanol oxidation in DMFCs [1]. It should be pointed out that the promotional role of a PtSn system is slightly different from that for PtRu electrocatalysts. The significant difference in electronegativity between Sn and Ru elements leads to different environments around the Pt sites in the two systems. Because of the small difference in electronegativity between Pt and Ru elements, the electronic effect of Ru on Pt is very weak and the bifunctional mechanism is responsible for the enhanced electrocatalytic performance of methanol oxidation on PtRu electrocatalysts. For PtSn electrocatalysts, Sn atoms donate electrons to Pt atoms due to Sn's smaller electronegativity. The charge transfer from Sn atoms to Pt atoms increases the electron density around Pt sites and downshifts their d-band center, which leads to weakened chemisorption energy with oxygen-containing species. Therefore, the enhancement of PtSn electrocatalysts is mainly due to the modification of electronic environment of Pt sites (electronic effect) [1].

At the cathode side, the ORR suffers from high polarization loss even with the use of the best catalyst Pt. There is an enormous impetus to find better and cheaper catalyst than Pt/C. Another consideration during the search of better electrocatalyst is its stability in highly corrosive and oxidizing environment in fuel cells. Throughout the effort to maintain the advantageous properties of Pt (i.e., relatively low overpotential, high stability) while simultaneously reducing the high Pt loading and searching for an improved ORR catalyst, a variety of binary and ternary Pt alloy systems have been investigated, with most of the alloyed metals being transition element metals, such as Co, Ni, Cu, Cr, Mo, and Fe. The main objective is to tune the ability of the metal to bond the key reaction intermediates in just the right way, either via the electronic effect or adjustment of the atomic distance between adjacent atoms. The appropriate chemisorption energy of O species with catalyst surface determines their electrocatalytic activity. An oxygen adsorption energy that is too high yields stable intermediates on the surface; if the interaction between oxygen and the surface leads is too weak, the adsorption is

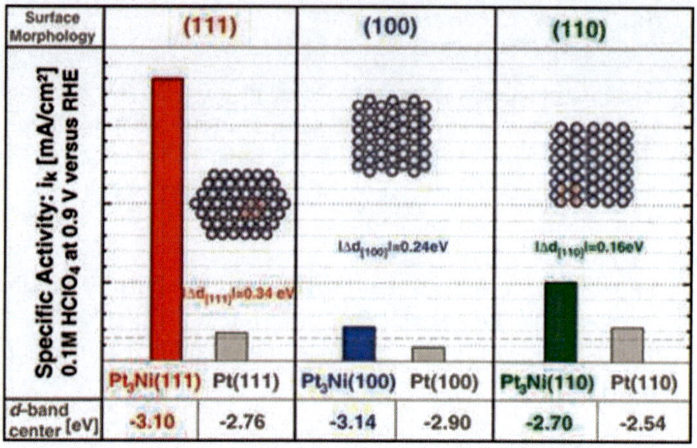

Fig. 15 Influence of the surface morphology and electronic surface properties on the kinetics of ORR. Specific activity is given as a kinetic current density i_k, measured at 0.9 V versus RHE. Values of d-band center position obtained from UPS spectra are listed for each surface morphology and compared between corresponding Pt_3Ni(hkl) and Pt(hkl) surfaces. Reprinted with permission from Stamenkovic et al. [30] Copyright 2007 Science Magazine

also weak, thereby limiting the activity of molecular oxygen. To date, the most extensively investigated cathodic bimetallic electrocatalysts are PtNi and PtCo. Stamenkovic et al. [30] demonstrated that the Pt_3Ni (111) surface is 10-times more active for oxygen reduction reaction (ORR) than the corresponding monometallic Pt (111) surface, as illustrated in Fig. 15. They claim that the Pt_3Ni (111) surface has an unusual electronic structure (d-band center shift) and the distribution of surface atoms in the near-surface region. The downshift of the d-band center of Pt on the surface was caused by the electronic modification by the underlying Ni, resulting in weakened chemisorption with –OH (nonreactive oxygenated species) on Pt surface, leading to increased number of active sites for O_2 adsorption and reaction [30].

4.2 Bimetallic Core–Shell Electrocatalysts

Although alloying Pt with other transition metal elements can significantly enhance the electrocatalytic activity of fuel cells, these materials tend to dissolve from the electrocatalyst during fuel cell operation because of the strong favorability of transition metals to leave their metal alloy crystal and form ions in acidic and oxidizing environment. The leaching out of transition metals forms a Pt skin (less than 1 nm) on the electrocatalyst nanoparticles and has been proposed to be the main reason for the improved activity with the use of bimetallic alloy catalysts [31]. In order to overcome this problem, incorporating these transition metals into the cores of Pt nanoparticles is a promising way to reduce the cost. Such a

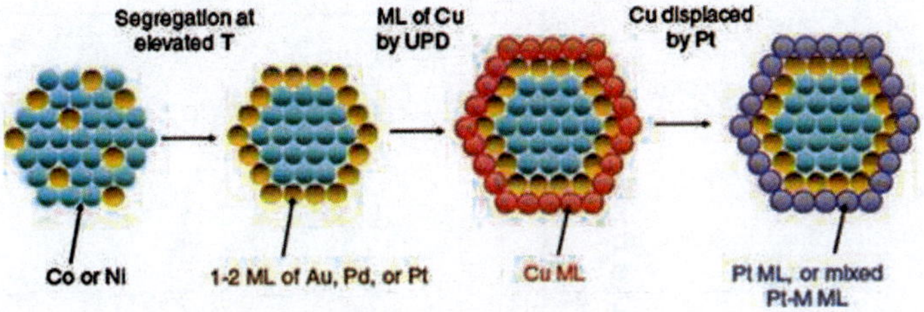

Fig. 16 Model for the synthesis of Pt monolayer catalysts on nonnoble metal-noble metal core–shell nanoparticles. Reprinted with permission from Zhang et al. [33] American Chemical Society

core–shell structure prevents the transition metals from dissolving out of the bimetallic structure and reduces the usage of precious Pt, since the catalytic reaction only occurs on the catalyst surface instead of inner core sites. Furthermore, the appropriate selection of inexpensive metal elements as the cores can modify the electronic structure of Pt shell, thereby improving the electrocatalytic activity of the Pt shell [32].

The first core–shell electrocatalyst for fuel cell reaction was reported by Adzic et al. [33]. They prepared a Pt monolayer on noble metal or nonnoble metal nanoparticles and examined them for O_2 reduction (see Fig. 16). These electrocatalysts showed a very high activity of O_2 reduction even though they contain only very small amounts of Pt. The Pt monolayer was prepared by galvanic displacement of a Cu monolayer deposited at underpotential condition. Geometric effects on the Pt monolayer and the effect of PtOH coverage, as revealed by electrochemical data, X-ray diffraction and X-ray absorption spectroscopy data, appear to be the sources of the enhanced catalytic activity. However, a full comparison to the common electrocatalyst Pt/C could not be obtained because of the large difference in the amount of active surface area. Moreover, from the viewpoint of mass production of the electrocatalyst, the electrochemical method used is less effective and less feasible compared to wet-chemistry approach [32].

More recently, Wang et al. [34, 35] developed an efficient wet-chemistry method to synthesize Au–Pt core–shell nanoparticles with controlled shell/core ratios for both methanol and formic acid oxidation. The synthesis process involves the pre-formation of Au seeds followed by the slow growth of Pt atoms on the Au cores in the presence of a weak reducing agent. Improved Pt utilization in terms of methanol oxidation is demonstrated with the core–shell structure. Furthermore, at very low Pt/Au molar ratios, incomplete Pt shells with decorated structures are formed. This catalyst not only significantly reduces the use of Pt via improved Pt dispersion, but also shows a high activity for formic acid oxidation as shown in Fig. 17. Compared to Pt/C, it undergoes a different reaction pathway by suppressing the formation of the CO intermediate where the Pt coverage on Au nanoparticles plays an important role. The origin behind this enhancement is proposed in Fig. 18. Formic acid oxidation on Pt generally follows a dual path mechanism. It is believed that the direct oxidation of formic acid on Pt surface (dehydrogenation) does not require the presence of continuous neighboring Pt

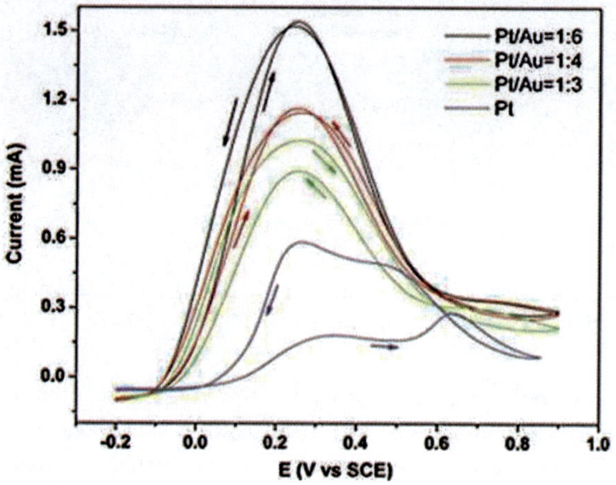

Fig. 17 CV of Pt/C and PtAu/C at 10 mV s21 in 0.5 M HCOOH + 0.5 M H_2SO_4. Reprinted with permission from Kristian et al. [35] Copyright 2007 The Royal Society of Chemistry

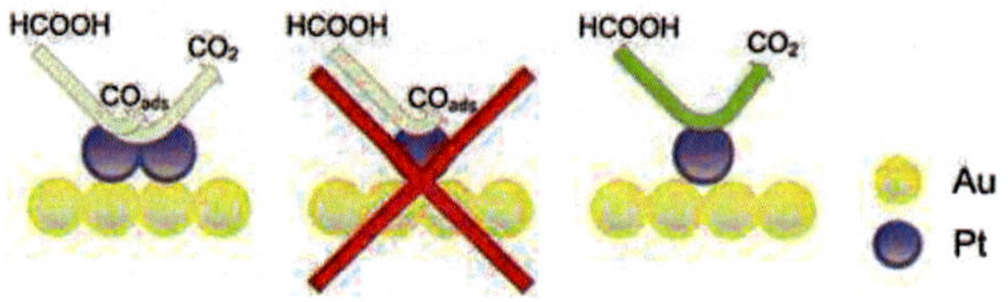

Fig. 18 Illustration of the catalytic reactions of formic acid oxidation (dissociative adsorption and dehydrogenation) on the Pt-decorated Au surface. Reprinted with permission from Kristian et al. [35] Copyright 2007 The Royal Society of Chemistry

sites, while the dissociative adsorption of formic acid to form CO requires at least two ensemble binding sites. A continuous decrease of Pt coverage on Au surface, i.e., decreasing the availability of Pt adjacent atoms, leads to a decrease in the dehydration rate and an eventual blocking of CO formation (the disappearance of the second oxidation peak at ~0.63 V), while the dehydrogenation reaction can still occur without being affected.

4.3 Metal Oxide as the Promotional Component of Electrocatalysts

The slow kinetics of the electrooxidation of methanol in PEMFCs is a large barrier to large-scale commercialization of fuel cells. As stated above, the most commonly used approach is to alloy Pt with other transition metals for use as catalysts. The detailed investigation on the electrocatalysis behavior of methanol oxidation on PtRu has shown that, in order to minimize the effect of the CO species, the cocatalyst, Ru should have a large fraction of oxide [36]. In such a case, CO tends to interact with the oxygenated species on oxide surface and converts to CO_2 at a relatively low potential via the classical bifunctional mechanism. On the basis of this concept, many researchers have been focusing on the promotional role of other metal oxides on the electrocatalytic reactions in fuel cells. Many metal oxides, such as TiO_2, CeO_2, SnO_2, ZrO_2, MgO, and MoO_2, have been extensively used as cocatalysts and found to play promotional role in alcohol oxidation or oxygen reduction either in acidic or alkaline media [20, 36–40]. Experimentally, researchers typically adopt two strategies to incorporate metal oxide into the catalyst system. One is to form a Pt-metal alloy, followed by the adjustment of the metal oxide component. Other researchers adopt a successive method; that is, loading metal oxide onto catalyst supports (such as carbon black and CNT/nanofiber), followed by the deposition of Pt nanoparticles onto the metal oxide/support composites.

Song et al. [36] prepared TiO_2 coated carbon nanotubes (TiO_2/CNTs) by a sol–gel method to act as Pt catalyst which support and enhance the electrooxidation of ethanol. The high-resolution TEM (HRTEM) images shown in Fig. 19 confirmed the successful coating of CNT and subsequent deposition of Pt nanoparticles with narrow particle size distribution. The electrochemical characterization for the ethanol oxidation and CO stripping demonstrates the promotional role of TiO_2/CNTs as Pt catalyst support. As shown in Fig. 20, the addition of TiO_2 significantly decreases the CO-stripping peak potential compared to Pt/CNTs and commercial Pt/C. This result shows that incorporating TiO_2 in the Pt/CNT catalyst system could enhance the electrochemical oxidation/removal of CO. It is believed that CO adsorption occurs only on the Pt surface and the bifunctional mechanism explains the promotional role of TiO_2. Plus, the addition of TiO_2 may enhance the durability of the catalysts because TiO_2 is very stable in acidic solution, which can hinder the corrosion of carbon nanotube supports. With a similar procedure, Peng et al. [41] synthesized MnO_2/CNT supported Pt and PtRu nanocatalysts for direct methanol fuel cells. It is concluded that the existence of MnO_2 on the surface of the CNTs effectively increases the proton conductivity of the catalyst, which remarkably enhances the electrochemical performance of the catalysts toward methanol oxidation.

In addition to mono-metal oxides as cocatalysts, multicomponent metal oxides are also used to enhance the electrochemical performance of ectrocatalysts in fuel cells. Qiu et al. [20] prepared Pt-$Ce_xZr_{1-x}O_2$/C catalysts with a co-precipitation method. This method involves the pre-mixing of zirconium nitrate and cerium nitrate with carbon black followed by the addition of an ammonia solution to prepare $Ce_xZr_{1-x}O_2$/C support composite. The deposition of Pt nanoparticles on the support composite completes the preparation of the catalysts. Figure 21 shows the typical CO-stripping voltammograms for the catalysts compared with Pt/C and PtRu/C. Catalysts with a metal oxide show much lower peak potential of CO stripping. The reason is that metal oxide favors the activation of water to produce the OH species on metal oxide surface, which subsequently interacts with poisonous CO adsorbed on Pt surface to generate CO_2 at lower potentials. The authors

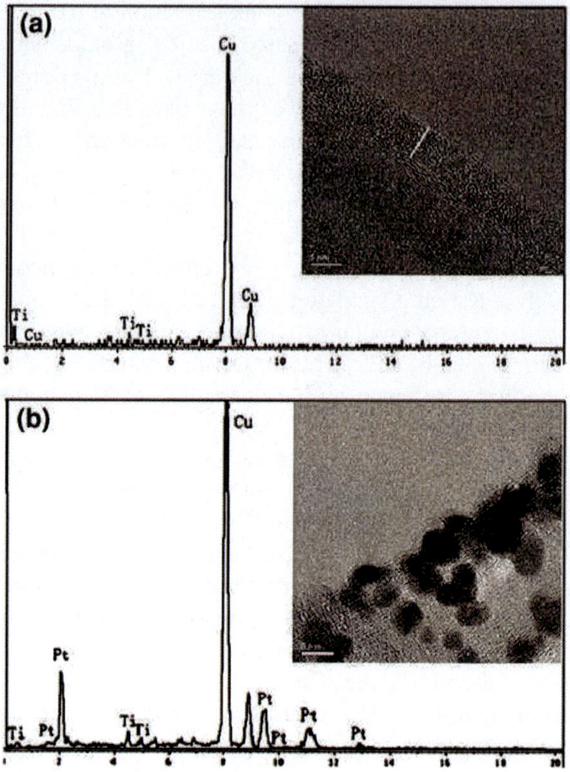

Fig. 19 HRTEM image and EDX spectrum of Pt–TiO_2/CNTs (1:1): (**a**) TiO_2/CNTs; (**b**) Pt–TiO_2/CNTs. The strong Cu-peak originates from the supporting HRTEM Cu grid. Reprinted from Song et al. [36] Copyright 2007 with permission from Elsevier

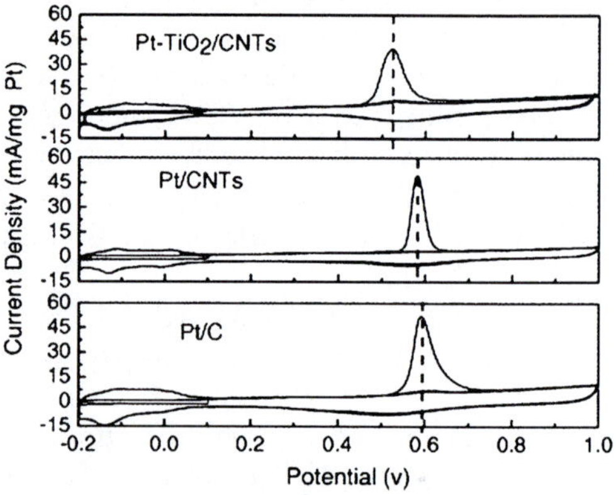

Fig. 20 CO stripping voltammograms in 1 M $HClO_4$ solutions with a scan rate of 10 mV s^{-1}. Reprinted from Song et al. [36] Copyright 2007 with permission from Elsevier

claimed that the peak potential of CO stripping depends on the content of ZrO_2 (1–x) in the catalysts and found that the peak potential increases with the increase of ZrO_2 content. The reason may be related to the different lattice structure of $Ce_xZr_{1-x}O_2/C$.

For the ethanol oxidation in direct ethanol fuel cells, the most serious barrier is the splitting of C–C bonds with the direct oxidation to CO_2. Recently, Adzic et al.

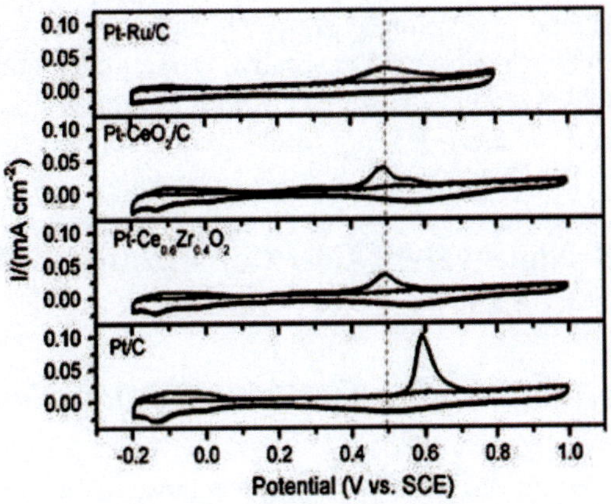

Fig. 21 The CO-stripping voltammograms recorded in 1 M $HClO_4$ solutions at 10 mV s^{-1} on Pt-CeO_2/C, Pt-$Ce_{0.6}Zr_{0.4}O_2$/C, PtRu/C, and Pt/C catalysts. Reprinted from Bai et al. [20] Copyright 2007 with permission from Elsevier

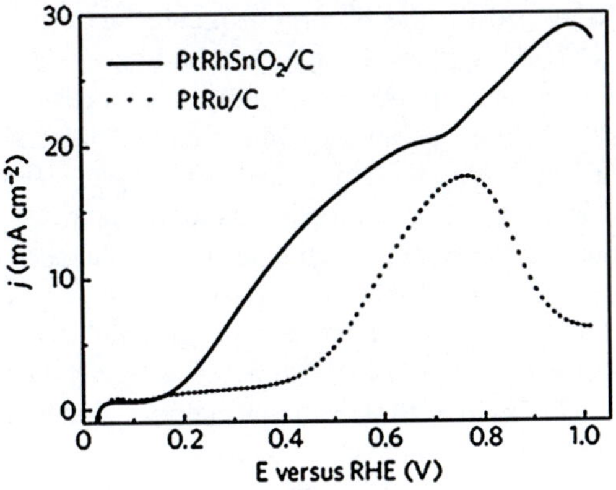

Fig. 22 Polarization curves of the oxidation of ethanol on PtRhSnO_2/C and PtRu/C (20% PtRu with 1:1 atomic ratio on C E-TEK). Reprinted with permission from Kowal et al. [42] Copyright 2009 Nature Publishing Group

[42] synthesized a ternary PtRhSnO_2/C electrocatalyst for ethanol electrooxidation by depositing platinum and rhodium atoms on carbon-supported tin dioxide nanoparticles. The as-synthesized electrocatalyst is capable of oxidizing ethanol with high efficiency and holds great promise for resolving the impediments to developing practical direct ethanol fuel cells. The electrochemical result shown in Fig. 22 demonstrates the significantly improved current density and lower onset potential of ethanol oxidation of PtRhSnO_2/C compared to the classical PtRu/C electrocatalysts. The PtRhSnO_2/C electrocatalyst effectively splits the C–C bond in ethanol at room temperature in acid solutions, facilitating its direct oxidation to CO_2 at low potentials, which has not been achieved with existing catalysts. It is concluded that the enhanced electrocatalytic activity is due to the specific property of each of the constituents, induced by their interactions. First, the addition of Rh effectively breaks C–C bonds based on interaction with Pt. Secondly, SnO_2, by strongly adsorbing water and interacting with the Pt and Rh, precludes the Rh and Pt sites from reacting with water to form M-OH, making them available for

ethanol oxidation. Also, SnO_2 activated with OH species facilitates CO removal from Rh active sites. Finally, Pt facilitates ethanol dehydrogenation and modifies the electronic structure of Rh to afford moderate bonding to ethanol, intermediates and products, leading to C–C bond splitting [42].

5 Noncovalent Functionalization of Carbon Nanotubes as Electrocatalyst Supports

5.1 Motivation for Using Carbon Nanotubes as Supports

Platinum-based electrocatalysts normally exist in the form of nano-scaled particles deposited on catalyst supports with a high surface area [43]. The catalyst supports not only to prevent the nano-scaled electrocatalyst particles from aggregation, but also to play a significant role in transporting the electrons generated from and consumed by the electrochemical reactions [12]. It has been demonstrated that catalyst supports have great influence on the performance and durability of electrocatalysts. An ideal catalyst support should meet these requirements: (a) high surface area to improve the catalyst dispersion; (b) low combustive reactivity; (c) high conductivity; and (d) high electrochemical and thermal stability. Currently, the most popular support material is porous carbon black (XC-72) with high surface area, which leads to well-dispersed electrocatalyst nanoparticles [43]. Recently, developing novel catalyst supports for PEMFC application has attracted much attention, a large portion of which has been devoted to the investigation of using carbon nanotubes (CNTs) as catalyst supports [43–45].

CNTs have attracted much attention due to their promising physical and chemical properties, since their discovery in 1990s by Ijima in Japan. Nowadays, the synthesis of carbon nanotubes can be realized by three main approaches: chemical vapor deposition, laser vaporization, and arc-discharge evaporation. CNTs are categorized as multi-walled carbon nanotubes (MWCNTs), double-walled carbon nanotubes (DWCNTs), and single-walled carbon nanotubes (SWCNTs) on the basis of the number of graphitic layers in the structure. CNTs have attracted increasing interest in both fundamental physics and a plethora of applications, including mechanically reinforced composites, field-effect transistors, chemical/electrochemical sensors, hydrogen storage, catalysis, and electronic nano-devices [12, 43–45].

CNTs are widely investigated as support materials for Pt and Pt alloy catalysts in fuel cells due to their high surface area, good electronic conductivity, and high chemical and thermal stability [3, 45, 46]. Compared with carbon black, CNTs allow for a higher utilization of Pt due to their smooth surfaces, as opposed to conventional carbon black supports where a big portion of Pt nanoparticles are trapped in the pores. An additional advantage of CNTs over carbon black is increased durability, owing to their inert nature. Furthermore, the high electrical conductivity of CNTs and their unique interaction with metal electrocatalysts (more specifically, the interaction of the π-electrons of carbon nanotubes and the d-electrons of platinum) leads to enhanced electrocatalytic activity [43]. The deposition, dispersion, and size of metal nanoparticles supported strongly depend on the surface properties of the CNTs [47]. The electrocatalytic activity of metal nanoparticles is significantly affected by the nature of their interaction with CNTs. The synthesis process of CNTs usually generates bundles, especially for single-walled CNTs, which limits their utility [43]. Furthermore, pristine CNTs are chemically inert; therefore, it is necessary to activate the graphitic surface of the nanotubes to disperse them in solution and efficiently deposit catalyst nanoparticles [48]. Thus, a pre-functionalization process is required for the use of CNTs as support material.

At the early development stage of CNT functionalization, the chemical oxidation of CNTs is most commonly used. Typically, this method involves the ultrasonication or refluxing CNTs in a harsh acid mixture, e.g., a mixture of sulfuric acid and nitric acid, resulting in the formation of oxygen-containing functional groups (i.e., carboxyl groups) on the ends and side-walls of CNTs [49]. The presence of carboxyl groups introduced by the chemical oxidation enables the further functionalization of CNTs via the covalent coupling. Various functional groups could be obtained on CNTs through the synthetic chemistry design [49]. However, such a vigorous functionalization process leads to the damage of the perfect CNT structure, namely, opening the nanotube caps, generating holes in the CNT side-walls or even cutting CNTs. In addition, the introduced functional groups prefer to locate at the defect sites of CNTs, and thus the distribution of functional groups is not uniform.

Alternatively, a gentler and more facile noncovalent functionalization method has been developed. Surfactant aggregating, bifunctional molecule stacking, macromolecule immobilization, and polymer wrapping on CNTs are the four main strategies to noncovalently functionalize CNTs. Surfactants are amphiphilic molecules, in which the hydrophobic ends attach to the hydrophobic CNT surface and the hydrophilic ends are directed to the solvent (aqueous solution), forming aggregates. A bifunctional molecule containing aromatic groups can be efficiently stacked on CNT surface via π–π interactions [50]. For macromolecules like proteins, the hydrophobic regions in the proteins could be responsible for the adsorption onto CNTs. CNTs can be effectively functionalized by most water-soluble polymers via a wrapping technique. These noncovalent functionalization methods effectively and uniformly introduce specific molecules/functional groups to the CNT surfaces without any damage to their structures. Compared to the covalent method, the noncovalent methods are more facile and effective for the functionalization of CNTs [12].

5.2 Noncovalent Functionalization of Carbon Nanotubes with Bifunctional Molecules

Recently, Wang et al. [9, 24, 51] developed a noncovalent functionalization method to modify the surface of CNTs with bifunctional molecules as electrocatalyst supports. As illustrated in Fig. 23, it involves the adsorption of

1-aminopyrene (1-AP) molecules onto the surface of MWCNTs. 1-AP is a bifunctional molecule with a pyrenyl group and an amino functional group. The pyrenyl group, being highly aromatic, is known to interact strongly with the basal plane of graphite via π-stacking. In a similar manner, the pyrenyl group of 1-AP also strongly interacts with the side-walls of MWCNTs, immobilizing the molecule. When the pH of the solution is controlled at a slightly acidic level (e.g., pH 6.5), the amino groups of 1-AP immobilized on the MWCNT surface are protonated and gain a weak positive charge. This leads to the electrostatic attraction of the negatively charged $PtCl_6^{2-}$, followed by the subsequent self-assembly of positively charged Ru^{3+} on the 1-AP-MWCNTs. The microwave-assisted polyol treatment in the presence of ethylene glycol reduces the PtRu precursors, forming PtRu nanoparticles on the MWCNTs. These surface groups also serve as anchoring sites for the direct deposition of reduced metal nanoparticles, which are negatively charged. Differing from the acid-oxidized MWCNTs, where the harsh chemical acids are used to produce carboxylic acid sites on the surface, the 1-AP-functionalization treatment preserves the integrity and the electronic structure of MWCNTs. As shown in Fig. 24, fine metal nanoparticles with a narrow size distribution were deposited on 1-AP-CNTs with a uniform distributions, a result of the evenly-distributed functional groups on the surface of CNTs. The average particle size is 2 nm and no aggregation occurs, even with high PtRu loading. However, on acid-oxidized CNTs, PtRu nanoparticles tend to form aggregates of large particle size due to the poor distribution of functional groups introduced by the harsh acid oxidation. The density of PtRu electrocatalysts can be controlled by adjusting the feeding concentration of the metal precursors. As a result, PtRu nanoparticles on 1-AP-MWCNTs have higher electrochemical surface area, much better activity, and enhanced stability for methanol electrochemical oxidation in acid solution than those on AO-MWCNTs [9].

5.3 Non-covalent Functionalization of Carbon Nanotubes by Polymer-Wrapping

Polymer-wrapping techniques are also used to functionalize carbon nanotubes as electrocatalyst supports. Wang et al. [24] used poly(diallyldimethylammonium chloride) (called PDDA) to wrap CNTs in an aqueous solution. PDDA is a water-soluble quaternary ammonium with a positive charge. During the functionalization process, sodium chloride salt is added to allow the PDDA chain to adopt a random configuration, leading to high coverage of PDDA chains on MWCNTs (PDDA-MWCNTs). This noncovalent functionalization not only leads to a high density and homogeneous dispersion of surface functional groups on MWCNTs, but also preserves the intrinsic properties of MWCNTs without damaging their perfect surface structures. As illustrated in Fig. 25, because of the positive surface charge on MWCNTs, a large amount of negatively charged Pt precursor can be anchored onto the MWCNT surfaces via electrostatic interaction. The subsequent reduction

Fig. 23 Schematic diagram of the synthesis of PtRu electrocatalysts on 1-AP-functionalized MWCNTs. Reprinted with permission from Wang et al. [9] American Chemical Society

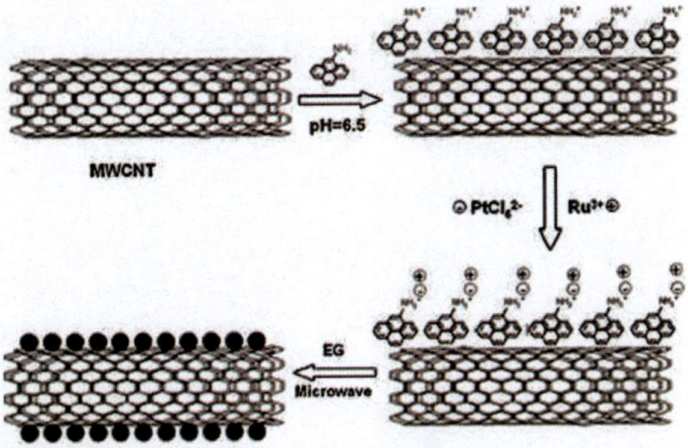

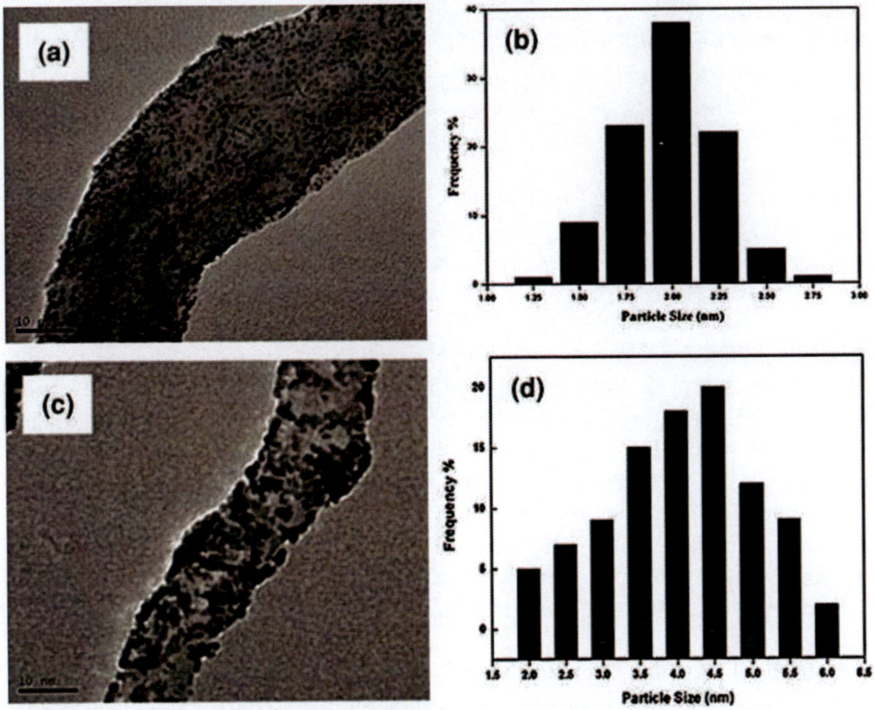

Fig. 24 TEM images and distribution histograms of PtRu nanoparticles on 1-AP-MWCNTs (**a** and **b**) and AO-MWCNTs (**c** and **d**). The PtRu loading was 40 wt%. Reprinted with permission from Wang et al. [9] American Chemical Society

by ethylene glycol (EG) yields Pt nanoparticles with uniform distribution and high density [12].

One of the most significant advantages of this noncovalent functionalization is the high density of functional groups introduced onto the CNT surface. Such high density allows for a high loading of metal nanoparticles. By combining the

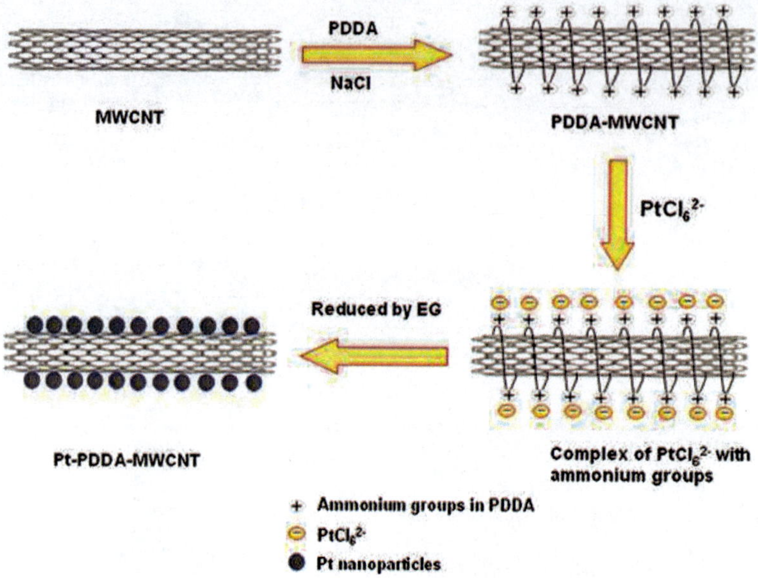

Fig. 25 Illustration of the synthesis of Pt nanoparticles on PDDA wrapped CNTs. Reprinted with permission from Wang et al. [24] Copyright 2008 IOP Publishing Ltd

noncovalent functionalization with polyelectrolyte and a seed-mediated growth method, Wang et al. successfully synthesized Pt/CNT electrocatalysts with various loadings ranging from 10–93 wt%, with the corresponding TEM images shown in Fig. 26. The correlation between the electrocatalytic activity and interconnectivity of Pt nanoparticles on CNT was further investigated. It was demonstrated for the first time that the electrocatalytic activities of Pt/CNTs catalysts are fundamentally correlated to the interconnectivity. The magnitude of the interconnectivity of Pt nanoparticles is a critical factor influencing their electrocatalytic activity, and the interconnected Pt nanoparticles are more active than the isolated Pt nanoparticles. The high electrocatalytic activity of highly interconnected Pt nanoparticles is considered to be related to the increased number of active grain boundaries, which promote the electrocatalytic activity of Pt nanoparticles. On the other hand, the interconnected Pt nanoparticles significantly weaken their chemical adsorption with oxygen-containing species (i.e., CO_{ad} and OH_{ad}), resulting in increased electrocatalytic activity for CO and methanol oxidation and oxygen reduction. The increase of interconnectivity of Pt nanoparticles also reduces the interface resistance among particles for electron transfer [12, 51].

Based on the same polymer-wrapping techniques, polyelectrolytes with various characteristic functional groups, to be used as interlinkers to catch Pt nanoparticles, were deployed to functionalize carbon nanotubes. The effect of

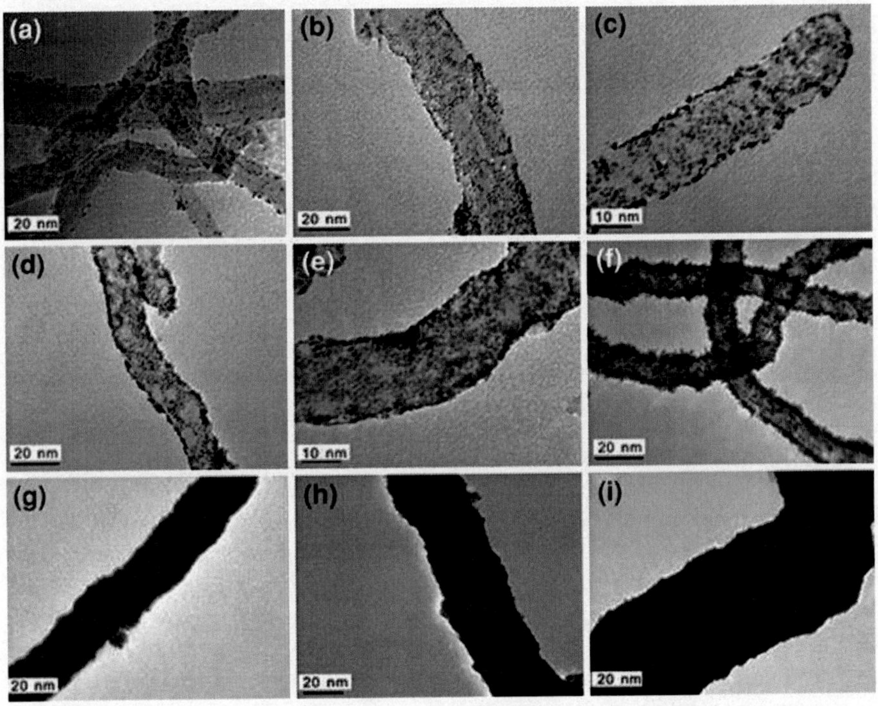

Fig. 26 TEM micrographs of Pt/MWCNTs catalysts with Pt loadings of (**a**) 10 wt%, (**b**) 20 wt%, (**c**) 30 wt%, (**d**) 40 wt%, (**e**) 50 wt%, (**f**) 69 wt%, (**g**) 81.6 wt%, (**h**) 86 wt%, and (**i**) 93 wt%. Reprinted with permission from Wanget al. [51] American Chemical Society

interlinkers between Pt nanoparticles and carbon nanotubes on the electrocatalytic activity for methanol oxidation was investigated by the same group. It was found that polyanions (e.g., poly(styrenesulfonic acid) (PSS) and poly(acrylic acid sodium) (PAA)) have a beneficial effect on methanol electrooxidation on Pt nanoparticles supported on CNTs. The effect is due to charge transfer from polyanions to Pt sites and supply of oxygen-containing species, as evidenced by X-ray photoelectron spectroscopy (XPS) results. The increased electron density around Pt sites via charge transfer causes partial filling of Pt 5*d*-bands, resulting in the downward shift of the *d*-band center and weaker chemisorption with oxygen-containing species (e.g., CO_{ad}). The weakened chemisorption of CO on Pt nanoparticles promotes methanol electrooxidation. However, polycations have a contrary effect on the electronic structure and chemisorption properties of Pt nanoparticles. On the other hand, the long-term stability testing shows that polycation-functionalized CNTs as Pt supports enhances their stability by the stronger interaction between Pt nanoparticles and CNTs contributed by the electrostatic attraction forces.

In addition to direct-stacking of bifunctional molecules and polymer wrapping, layer by layer self-assembly is another strategy to immobilize metal nanoparticles on the CNT surfaces. For example, Yang et al. [52] immobilized Pt nanocubes on CNT surface noncovalently using this technique. In this method, CNTs were first pre-functionalized with a negatively charged polyelectrolyte, PSS. Meanwhile, Pt nanocubes stabilized by a positively charged surfactant, CTAB, were synthesized separately. Mixing of the two batches leads to the deposition and immobilization of Pt nanocubes on CNT via electrostatic self-assembly. The Pt NC/CNT electrocatalyst shows relatively high catalytic activity toward the reduction of oxygen.

5.4 Noncovalent Functionalization of Carbon Nanotubes by Ionic Liquid

To develop another advanced strategy, Chen et al. [53] made use of an ionic liquid to functionalize CNT. This approach is based on free radical polymerization of the ionic-liquid monomer 3-ethyl-1-vinylimidazolium tetrafluoroborate via thermal-initiation to form ionic-liquid polymers (PIL) on the surfaces of the CNTs. This results in a large number of surface functional groups on the CNT with good dispersion. The introduction of the ionic species on the surfaces of CNTs provides positive charges to stabilize metal nanoparticles and prevents them from aggregating. Furthermore, the electrochemical performance of these catalysts was investigated by cyclic voltammetry in a nitrogen-saturated 0.5 M H_2SO_4 + 1.0 M CH_3OH solution, with the CV curves shown in Fig. 27. It was found that catalysts supported on ionic-liquid-functionalized CNTs show enhanced current density, reduced onset and peak potential, and improved CO tolerance, compared to those supported on conventional CNTs. This observation shows the advantage of ionic-liquid/CNT composites as electrocatalyst supports for fuel cell applications.

On the basis of the above examples, we conclude that noncovalent functionalization of CNTs effectively equips CNTs with specified functional groups with a high density and uniform distribution. These surface groups function as active sites to anchor metal precursors or metal nanoparticles, resulting in well-dispersed, narrowly-distributed, loading-controlled metal nanoparticles/CNT electrocatalysts. In addition to the physical role, these functional groups also interact with catalyst nanoparticles and affect their intrinsic electrochemical activities. To date, most strategies to tune the activity have been limited to the modification of Pt with other metals, while no specific attention has been paid to the anchoring groups connecting the metal nanoparticles and the support material, which may be equally important. This direction is certainly worthy further studying. Thus, the non-covalent functionalization strategy is an attractive method for the coating of CNTs with size and shape-controlled Pt-based nanoparticles and other nanocatalysts.

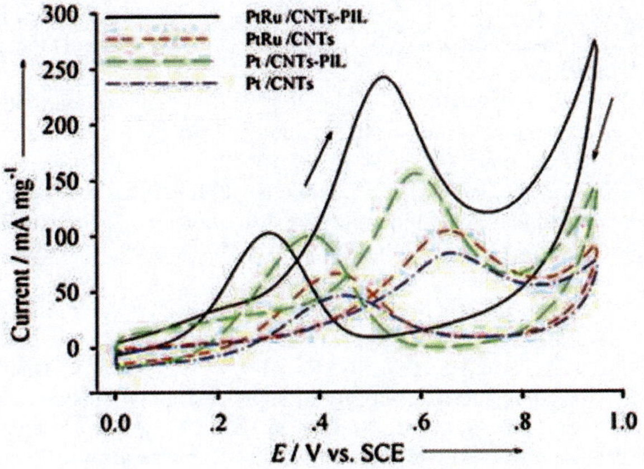

Fig. 27 Cyclic voltammograms of the different electrocatalysts in nitrogen-saturated 0.5 M H_2SO_4 +1.0 M CH_3OH at a scan rate of 50 mV s^{-1}. This is the pre-peer reviewed version of the following article: Wu et al. [53]

6 Conclusions

Nanoscale design of electrocatalysts provides a general approach to precisely control the composition, structure, and properties of electrocatalysts for fuel cells. It also furthers the fundamental understanding of electrocatalysis on various catalysts. For practical application in fuel cells, appropriate adoption of nanomaterials and nanotechnology has been shown to improve electrocatalytic activity, durability, and cost. Additionally, nanostructured materials are also being extensively used in other critical components of fuel cells.

References

1. Arico AS, Srinivasan S, Antonucci V (2001) DMFCs: DMFCs: from fundamental aspects to technology development. Fuel Cells 1:1–29
2. Borup R, Meyers J, Pivovar B, Kim YS, Mukundan R, Garland N, Myers D, Wilson M, Garzon F, Wood D, Zelenay P, More K, Stroh K, Zawodzinski T, Boncella J, Mcgrath JE, Inaba M, Miyatake K, Hori M, Ota K, Ogumi Z, Miyata S, Nishikata A, Zyun S, Uchimoto Y, Yasuda K, Kimijima K-i, Iwashita N (2007) Scientific aspects of polymer electrolyte fuel cell durability and degradation. Chem Rev 3904–3951
3. Carmo M, Paganin VA, Rosolen JM, Gonzalez ER (2005) Alternative supports for the preparation of catalysts for low-temperature fuel cells: the use of carbon nanotubes. J Power Sources 142:169–176
4. Du B, Guo QH, Pollard R, Rodriguez D, Smith C, Elter J (2006) PEM fuel cells: status and challenges for commercial stationary power applications. JOM 58:45–49
5. Smitha B, Sridhar S, Khan AA (2005) Solid polymer electrolyte membranes for fuel cell applications—a review. J Membr Sci 259:10–26
6. Zhu YM, Khan Z, Masel RI (2005) The behavior of palladium catalysts in direct formic acid fuel cells. J Power Sources 139:15–20
7. Ralph TR, Hards GA, Keating JE, Campbell SA, Wilkinson DP, Davis M, StPierre J, Johnson MC (1997) Low cost electrodes for proton exchange membrane fuel cells — performance in single cells and Ballard stacks. J Electrochem Soc 144:3845–3857

8. Bock C, Paquet C, Couillard M, Botton GA, MacDougall BR (2004) Size-selected synthesis of PtRu nano-catalysts: reaction and size control mechanism. J Am Chem Soc 126: 8028–8037
9. Wang SY, Wang X, Jiang SP (2008) PtRu nanoparticles supported on 1- aminopyrene-functionalized multiwalled carbon nanotubes and their electrocatalytic activity for methanol oxidation. Langmuir 24:10505–10512
10. Han MY, Gao XH, Su JZ, Nie S (2001) Quantum-dot-tagged microbeads for multiplexed optical coding of biomolecules. Nat Biotechnol 19:631–635
11. Kim F, Song JH, Yang PD (2002) Photochemical synthesis of gold nanorods. J Am Chem Soc 124:14316–14317
12. Wang S (2010) Nanostructured electrocatalysts for proton exchange membrane fuel cells (PEMFCs). PhD Thesis, Nanyang Technological University, Singapore
13. Xu Z, Xiao FS, Purnell SK, Alexeev O, Kawi S, Deutsch SE, Gates BC (1994) Size-dependent catalytic activity of supported metal clusters. Nature 372:346–348
14. Zhang ZB, Sun XZ, Dresselhaus MS, Ying JY, Heremans J (2000) Electronic transport properties of single-crystal bismuth nanowire arrays. Phys Rev B 61:4850–4861
15. Hsin YL, Hwang KC, Yeh CT (2007) Poly(vinylpyrrolidone)-modified graphite carbon nanofibers as promising supports for PtRu catalysts in direct methanol fuel cells. J Am Chem Soc 129:9999–10010
16. Arenz M, Mayrhofer KJJ, Stamenkovic V, Blizanac BB, Tomoyuki T, Ross PN, Markovic NM (2005) The effect of the particle size on the kinetics of CO electrooxidation on high surface area Pt catalysts. J Am Chem Soc 127:6819–6829
17. Wang C, Daimon H, Onodera T, Koda T, Sun SH (2008) A general approach to the size- and shape-controlled synthesis of platinum nanoparticles and their catalytic reduction of oxygen. Angewandte Chemie-International Edition 47:3588–3591
18. Xiao L, Zhuang L, Liu Y, Lu JT, Abruna HD (2009) Activating Pd by morphology tailoring for oxygen reduction. J Am Chem Soc 131:602–608
19. Bi YP, Lu GX (2008) Facile synthesis of platinum nanofiber/nanotube junction structures at room temperature. Chem Mater 20:1224–1226
20. Bai YX, Wu JJ, Qiu XP, Xi JY, Wang JS, Li JF, Zhu WT, Chen LQ (2007) Electrochemical characterization of Pt-CeO_2/C and Pt-CexZr1 − xO_2/C catalysts for ethanol electro-oxidation. Appl Catal B-Environ 73:144–149
21. Xu CX, Zhang Y, Wang LQ, Xu LQ, Bian XF, Ma HY, Ding Y (2009) Nanotubular mesoporous PdCu bimetallic electrocatalysts toward oxygen reduction reaction. Chem Mater 21:3110–3116
22. Xu CW, Wang H, Shen PK, Jiang SP (2007) Highly ordered Pd nanowire arrays as effective electrocatalysts for ethanol oxidation in direct alcohol fuel cells. Adv Mater 19:4256
23. Koczkur K, Yi QF, Chen AC (2007) Nanoporous Pt-Ru networks and their electrocatalytical properties. Adv Mater 19:2648
24. Wang SY, Wang X, Jiang SP (2008) Controllable self-assembly of Pd nanowire networks as highly active electrocatalysts for direct formic acid fuel cells. Nanotechnology 19:455602
25. Guo DJ, Cui SK (2009) Hollow PtCo nanospheres supported on multiwalled carbon nanotubes for methanol electrooxidation. J Colloid Interface Sci 340:53–57
26. Guo DJ, Zhao L, Qiu XP, Chen LQ, Zhu WT (2008) Novel hollow PtRu nanospheres supported on multi-walled carbon nanotube for methanol electrooxidation. J Power Sources 177:334–338
27. Liang HP, Zhang HM, Hu JS, Guo YG, Wan LJ, Bai CL (2004) Pt hollow nanospheres: facile synthesis and enhanced electrocatalysts. Angewandte Chemie-International Edition 43:1540–1543
28. Zhao J, Chen WX, Zheng YF, Li X (2006) Novel carbon supported hollow Pt nanospheres for methanol electrooxidation. J Power Sources 162:168–172
29. Kotobuki M, Watanabe A, Uchida H, Yamashita H, Watanabe M (2005) Reaction mechanism of preferential oxidation of carbon monoxide on Pt, Fe, and Pt-Fe/mordenite catalysts. J Catal 236:262–269

30. Stamenkovic VR, Fowler B, Mun BS, Wang GJ, Ross PN, Lucas CA, Markovic NM (2007) Improve oxygen reduction activity on Pt3Ni(111) via increased surface site availability. Science 315:493
31. Toda T, Igarashi H, Watanabe M (1999) Enhancement of the electrocatalytic O_2 reduction on Pt-Fe alloys. J Electroanal Chem 460:258–262
32. Kristian N (2010) Study of Pt-M (M=Au and Co) nano-catalysts with low Pt loading for PEMFC applications. PhD Thesis, Nanyang Technological University, Singapore
33. Zhang J, Lima FHB, Shao MH, Sasaki K, Wang JX, Hanson J, Adzic RR (2005) Platinum monolayer on nonnoble metal-noble metal core-shell nanoparticle electrocatalysts for O_2 reduction. J Phys Chem B 109:22701–22704
34. Kristian N, Wang X (2008) Pt-shell-Au-core/C electrocatalyst with a controlled shell thickness and improved Pt utilization for fuel cell reactions. Electrochem Commun 10:12–15
35. Kristian N, Yan Y, Wang X (2007) Highly efficient submonolayer Pt-decorated Au nano-catalysts for formic acid oxidation. Chem Commun 353–355
36. Song HQ, Qiu XP, Li FS, Zhu WT, Chen LQ (2007) Ethanol electro-oxidation on catalysts with TiO_2 coated carbon nanotubes as support. Electrochem Commun 9:1416–1421
37. Hou Z, Yi B, Yu H, Lin Z, Zhang H (2003) CO tolerance electrocatalyst of PtRu-$HxMeO_3$/C (Me = W, Mo) made by composite support method. J Power Sources 123:116–125
38. Jayaraman S, Jaramillo TF, Baeck SH, McFarland EW (2005) Synthesis and characterization of Pt-WO_3 as methanol oxidation catalysts for fuel cells. J Phys Chem B 109:22958–22966
39. Jiang L, Colmenares L, Jusys Z, Sun G, Behm RJ (2007) Ethanol electrooxidation on novel carbon supported Pt/SnOx/C catalysts with varied Pt:Sn ratio. Electrochim Acta 53:377
40. Waki K, Matsubara K, Ke K, Yamazaki Y (2005) Self-organized Pt/SnO_2 electrocatalysts on multiwalled carbon nanotubes. Electrochem Solid State Lett 8:A489–A491
41. Zhou CM, Wang HJ, Peng F, Liang JH, Yu H, Yang J (2009) MnO_2/CNT supported Pt and PtRu nanocatalysts for direct methanol fuel cells. Langmuir 25:7711–7717
42. Kowal A, Li M, Shao M, Sasaki K, Vukmirovic MB, Zhang J, Marinkovic NS, Liu P, Frenkel AI, Adzic RR (2009) Ternary Pt/Rh/SnO_2 electrocatalysts for oxidizing ethanol to CO_2. Nat Mater 8:325–330
43. Shao YY, Liu J, Wang Y, Lin YH (2009) Novel catalyst support materials for PEM fuel cells: current status and future prospects. J Mater Chem 19:46–59
44. Wang JJ, Yin GP, Shao YY, Wang ZB, Gao YZ (2008) Investigation of further improvement of platinum catalyst durability with highly graphitized carbon nanotubes support. J Phys Chem C 112:5784–5789
45. Wang X, Li WZ, Chen ZW, Waje M, Yan YS (2006) Durability inestigation of carbon nanotube as catalyst support for proton exchange membrane fuel cell. J Power Sources 158:154–159
46. Liu ZL, Lin XH, Lee JY, Zhang W, Han M, Gan LM (2002) Preparation and characterization of platinum-based electrocatalysts on multiwalled carbon nanotubes for proton exchange membrane fuel cells. Langmuir 18:4054–4060
47. Tian ZQ, Jiang SP, Liang YM, Shen PK (2006) Synthesis and characterization of platinum catalysts on muldwalled carbon nanotubes by intermittent microwave irradiation for fuel cell applications. J Phys Chem B 110:5343–5350
48. Balasubramanian K, Burghard M (2005) Chemically functionalized carbon nanotubes. Small 1:180–192
49. Yu RQ, Chen LW, Liu QP, Lin JY, Tan KL, Ng SC, Chan HSO, Xu GQ, Hor TSA (1998) Platinum deposition on carbon nanotubes via chemical modification. Chem Mater 10: 718–722
50. Ou YY, Huang MH (2006) High-density assembly of gold nanoparticles on multiwalled carbon nanotubes using 1-pyrenemethylamine as interlinker. J Phys Chem B 110:2031–2036
51. Wang SY, Jiang SP, White TJ, Guo J, Wang X (2009) Electrocatalytic activity and interconnectivity of Pt nanoparticles on multiwalled carbon nanotubes for fuel cells. J Phys Chem C 113:18935–18945

52. Yang W, Wang XL, Yang F, Yang C, Yang XR (2008) Carbon nanotubes decorated with Pt nanocubes by a noncovalent functionalization method and their role in oxygen reduction. Adv Mater 20:2579–2587
53. Wu BH, Hu D, Kuang YJ, Liu B, Zhang XH, Chen JH (2009) Functionalization of carbon nanotubes by an ionic-liquid polymer: dispersion of Pt and PtRu nanoparticles on carbon nanotubes and their electrocatalytic oxidation of methanol. Angewandte Chemie-International Edition 48:4751–4754
54. Chen ZW, Waje M, Li WZ, Yan YS (2007) Supportless Pt and PtPd nanotubes as electrocatalysts for oxygen-reduction reactions. Angewandte Chemie-International Edition 46:4060–4063

Energy Harvesting Based on PZT Nanofibers

Xi Chen, Nan Yao and Yong Shi

Abstract Energy harvesting technologies that are engineered to miniature sizes, while increasing the power delivered to wireless electronics [1, 2], portable devices, stretchable electronics [3] and implantable bio-sensors [4, 5] are strongly desired. Piezoelectric nanowire- and fiber-based generators have potential uses for powering such devices through conversion of mechanical energy into electrical energy [6]. However, the piezoelectric voltage constants of the semiconductor piezoelectric nanowires of the reported nanogenerators [7–12] are low. Here we introduce a piezoelectric nanogenerator based on lead zirconate titanate (PZT) nanofibers [13]. The PZT nanofibers, with diameters and lengths of approximately 60 nm and 500 μm, respectively, were aligned on interdigitated electrodes of platinum fine wires and packaged using a soft polymer on a silicon substrate. The measured output voltage and power under periodic stress applications to the soft polymer were 1.63 V and 0.03 μW, respectively.

1 Introduction

Recently, the piezoelectric properties of several nanowires, nanofibers and nanorods from zinc oxide [9], lead zirconate titanate (PZT) [14], cadmium sulphide [15], barium titanate [16] and gallium nitride [17], have been demonstrated. These

X. Chen · Y. Shi (✉)
Department of Mechanical Engineering, Stevens Institute of Technology, Castle Point on Hudson, Hoboken, NJ 07030, USA
e-mail: Yong.Shi@stevens.edu

N. Yao
Princeton Institute for the Science and Technology of Materials (PRISM), Princeton University, 70 Prospect Avenue, Princeton, NJ 08540, USA

L. Zang (ed.), *Energy Efficiency and Renewable Energy Through Nanotechnology*, Green Energy and Technology, DOI: 10.1007/978-0-85729-638-2_12,

one-dimensional piezoelectric nanostructures convert mechanical energy into electrical energy. As examples, various nanogenerators based on ZnO nanowires [9–12] and fine fibers [7, 8] proposed by Wang et al. have been demonstrated for potential applications in converting low frequency vibration and biomechanical energy into electrical energy.

However, the piezoelectric voltage constant of the piezoelectric nanomaterials, output voltage and output power of the nanogenerators still need to be improved further for practical applications. Furthermore, the fabrication method of the semiconductor piezoelectric nanomaterials may pose some drawbacks that can affect the performance of the nanogenerator. It is difficult to grow single crystal nanowires longer than 50 μm with diameters below 100 nm. The nanogenerator fabrication method and the output voltage of the nanogenerator could be significantly restricted by the short length of nanowires.

In order to overcome some of the drawbacks of the existing devices and demonstrate the possibility of energy harvesting using PZT nanomaterials, a highly efficient nanogenerator based on laterally aligned PZT nanofibers on interdigitated electrodes was created, and is reported herein. PZT is a widely used piezoelectric ceramic material with a high piezoelectric voltage constant and dielectric constant, which are ideal properties of the active materials for mechanical to electrical energy conversion. For a given volume under the same energy input, PZT can generate much higher voltage and power output than other semiconductor-types of piezoelectric materials.

As a ceramic material, bulk and thin film PZT structures are extremely fragile especially when subjected to alternating loads. Matters are made worse since thin film and micro fiber [18] structures are typically sensitive to high frequency vibration. However, unlike bulk, thin films or micro fibers, PZT nanofibers prepared by electrospinning process exhibit an extremely high piezoelectric voltage constant, high bending flexibility and high mechanical strength, which have been demonstrated in [14]. Therefore, utilizing PZT nanofibers in energy harvesting technology could provide a new way to make a portable, flexible, highly efficient device with low frequency vibration nature, since the nanofibers could be woven into fabrics and made into composites.

2 PZT Nanofiber Synthesis

PZT nanofibers were prepared by electrospinning [19]. The starting materials, PZT (52/48) sol-gel and poly vinyl pyrrolidone (PVP, Aldrich) were obtained from commercial sources. Alcohol was used as the solvent for PVP, while acetic acid was added to stabilize the solution and to control the hydrolysis reaction of the sol-gel precursor [20]. After being stirred vigorously for two hours at room temperature, the mixture was fed into a microscale metallic tube through a syringe pump. A droplet of the modified sol-gel solution was held at the orifice of the metallic tube due to the high surface tension. A high DC voltage (10 kV) between

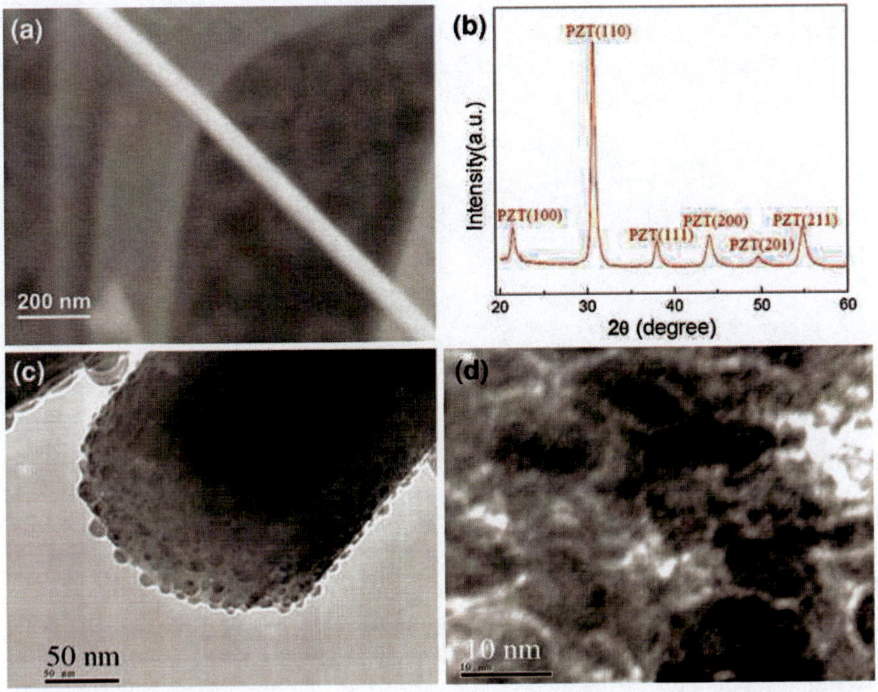

Fig. 1 **a** SEM image of single PZT nanofiber. **b** X-ray diffraction pattern for the PZT nanofibers. **c–d** TEM images of a PZT nanofiber. Reprinted with permission from [14]. Copyright 2009 American Institute of Physics

the tip of the micro metallic tube and the collecting substrate was applied to overcome the surface tension and generate an electrically charged jet of the modified solution. After electrospinning, the as-spun nanofibers were annealed at 650°C to obtain PZT nanofibers in the perovskite phase.

Figure 1 shows the morphology and X-ray diffraction pattern of the annealed PZT nanofibers. Figure 1a is a scanning electron microscopy (SEM) image of a selected nanofiber with a diameter of 70 nm. By varying the concentration of the polymer in the precursor, the average diameters of the nanofibers can be tuned from 52 ± 8 to 150 ± 12 nm. The X-ray diffraction (XRD) pattern of the annealed PZT nanofibers shown in Fig. 1b indicates that the PZT nanofibers annealed at 650°C were in the pure perovskite phase. The annealing temperature was lower than that used by most other methods. Transmission electron microscopy (TEM) images of the cross-section and surface of PZT nanofibers are shown in Fig. 1c–d. The grain size of the PZT nanofibers was about 10 nm. The structure of the nanofiber seemed porous, probably because of the polymer precursor solution of the sol-gel process [21]. By modifying the precursor and controlling the electrospinning parameters, a denser nanostructure can be obtained.

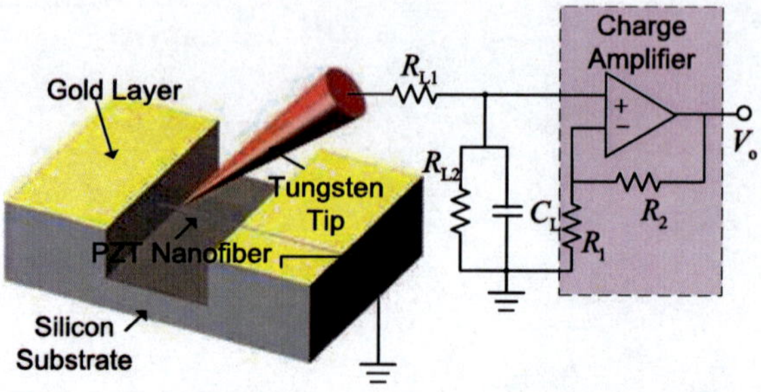

Fig. 2 Schematic diagram of the measurement system: R_{L1}, R_{L2} and C_L were load resistances and circuit capacitance respectively while V_o was the output voltage readout from the Keithley SPA. Reprinted with permission from [14]. Copyright 2009 American Institute of Physics

3 Piezoelectric Voltage Constant of a Single PZT Nanofiber

Piezoelectric response was measured by using a nanomanipulator under SEM. Two probes with tungsten tips were used for the measurements. The bending moment was applied to the nanofiber through one tungsten tip, while the other tip was used to form the ground by contacting the Au electrode directly near the PZT nanofiber. The output voltage from the PZT nanofiber was recorded in real time by a Keithley semiconductor parameter analyzer (SPA) with a charge amplifier; the equivalent circuit used for the measurement is shown in Fig. 2. To reduce the contact resistance between the tungsten tips and the PZT nanofiber, high currents were applied to burn the organic layer absorbed on the tungsten tips before the measurement. The resistance between the two tungsten tips, which were 1 μm apart, was around 30 Ω after the burning process. The current in the circuit shown in Fig. 2 was set to be at a picoampere level for signal readout using the software of the Keithley measurement system.

The PZT nanofiber across the trench was ruptured at one end using a tungsten tip to form a nanofiber cantilever, while the other end of the nanofiber remained anchored to the Au electrode. The tip of the nanofiber cantilever was firmly attached to the tungsten tip by electrostatic force when the two approached each other. The electrostatic potential, V_{nw}, of the nanofiber could be considered as an offset measured from the ground of the Keithley system when no force was applied on the nanofiber (Fig. 3a). When a concentrated load was applied to the tip of the nanofiber cantilever through the tungsten probe tip, the nanofiber was deformed, as shown in Fig. 3b. The measured voltage output is shown in Fig. 3c. The magnitude of the output voltage, V_o, generated from the single PZT nanofiber was 0.4 mV, while the discharging process due to the load resistor, R_L, of the measurement system lasted about 40 s. The higher noise level in the right part of the curve could be from the charge amplifier during the discharging process when the current in Keithley system was set to 0 nA. The results were reproducible and each voltage impulse lasted for the same period of time, as shown in Fig. 3d, when the nanofiber was bent repeatedly, as shown in Fig. 3a–b.

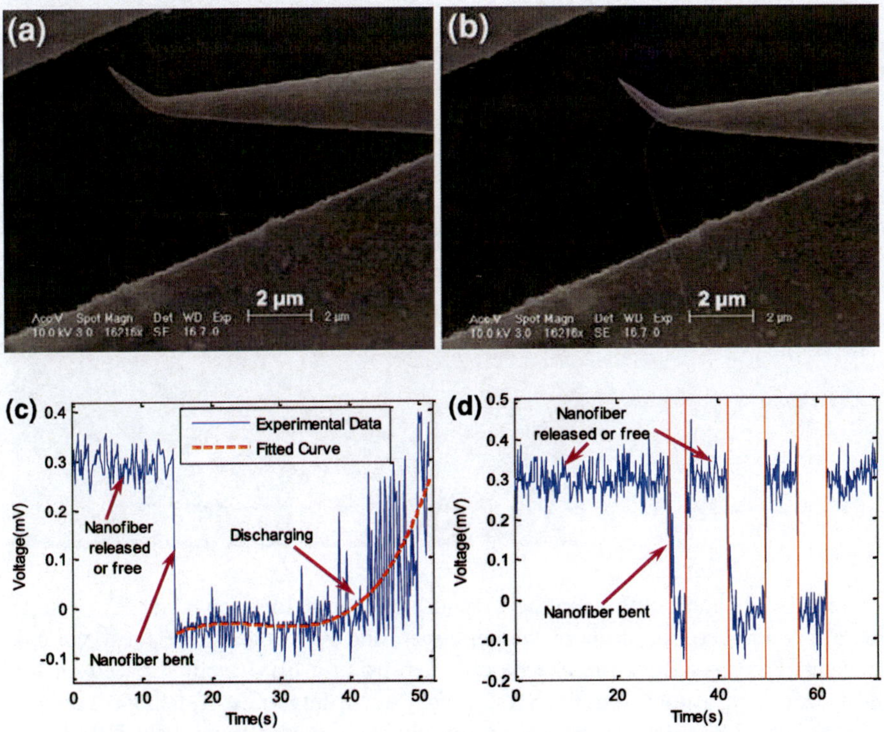

Fig. 3 **a–b** SEM images showing the free and bent single PZT nanofiber. **c** Voltage output and discharging from the single PZT nanofiber during the bending test. **d** Voltage output during the cycling tests. Reprinted with permission from [14]. Copyright 2009 American Institute of Physics

The measured voltage output due to the bending of the single PZT nanofiber was significantly higher than the noise level of the whole testing and measurement system, which was less than ~0.1 mV. When the electron beam of the SEM was switched between the on and off states, the differences in the measured voltage output results was less than ~0.05 mV. During the cycling test process, the nanofiber tip was attached to the tungsten probe perfectly during the test, which proves that the measured signal was not from the initial surface charges of the PZT nanofiber. The piezoelectric nanofiber also exhibited extremely high flexibility and strength during the test, which was very attractive in developing active micro and nanodevices or nanoenergy scavenging devices.

The voltage generation mechanism can be explained by a model similar to the one suggested in [22]. The voltage change on the PZT nanofiber surface during the bending test can be determined by:

$$\Delta V = \int_0^{\mathrm{r}} g_{33}\sigma(r)\mathrm{d}\mathrm{r} \tag{1}$$

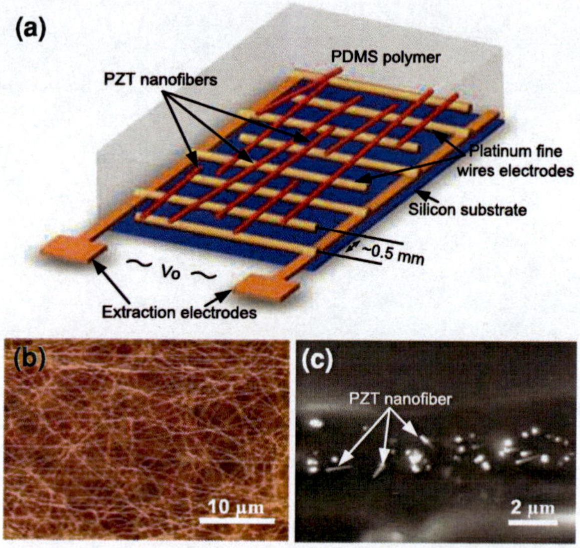

Fig. 4 Concept of the PZT nanofiber generator. **a** Schematic view of the PZT nanofiber generator. **b** SEM image of the PZT nanofiber mat across the interdigitated electrodes. **c** Cross-sectional SEM image of the PZT nanofibers in PDMS matrix. Reprinted in part with permission from Ref. [13]. Copyright 2010 American Chemical Society

where r is the radius of nanofiber, $\sigma(r)$is the stress function along the radius direction on the cross-section of the nanofiber; and g_{33} is the piezoelectric voltage constant. The stress function can be obtained using a FEM method based on the displacement measured from the SEM image. According to Eq. 1, using a Young's modulus of 42.99 GPa for the PZT nanofiber, as reported previously [19] and a voltage change of 0.403 mV, as measured in the test, the piezoelectric voltage constant, g_{33}, was determined to be about 0.079 Vm/N, which was larger than that found in bulk PZT. This was in agreement with the piezoelectric voltage constant measured from a bundle of PZT nanofibers during a dynamic bending test using a dynamic mechanical analyzer (DMA).

4 Fabrication Process for Nanogenerator

The nanogenerator device fabrication began with electrospinning the PZT nanofibers and depositing them on the pre-prepared interdigitated electrodes of platinum fine wire (diameter of 50 μm) arrays, which were assembled on a silicon substrate (Fig. 4a). The diameters of the PZT nanofibers were made to be around 60 nm (Fig. 4b) by varying the concentration of PVP in the modified sol-gel solution. The PZT nanofibers obtained were continuous while the distance between two adjacent electrodes was 500 μm as designed.

The pure perovskite phase was obtained by annealing at 650°C for about 25 min. Subsequently, a soft polymer (Polydimethylsiloxane, PDMS) was applied on top of the PZT nanofibers (Fig. 4c). The interdigitated electrodes of fine platinum wire were connected by extraction electrodes to transport harvested electrons to the external circuit. Finally, the PZT nanofibers were aligned by applying an electric

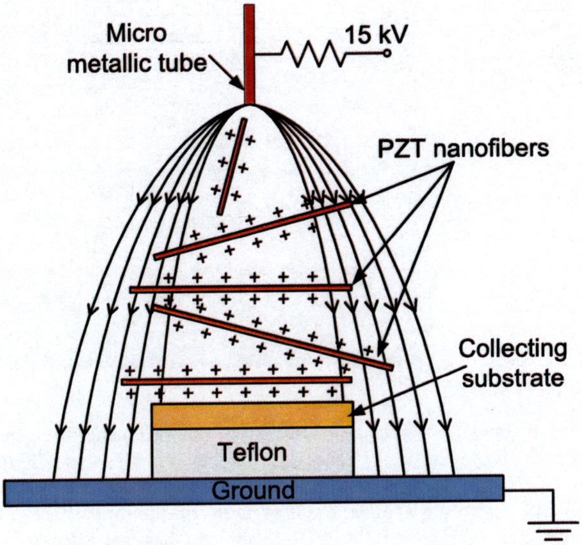

Fig. 5 Schematic of nanofiber alignment by controlling the electric field during the electrospinning process. Reprinted in part with permission from Ref. [13]. Copyright 2010 American Chemical Society

field of 4 V/μm across the electrodes at a temperature of above 140°C for about 24 h. The nanogenerator can be released from the silicon substrate or prepared on flexible substrates, depending on the requirements of the applications for energy harvesting.

By controlling the electric field distribution during the electrospinning process (Fig. 5), PZT nanofibers were laterally aligned on the interdigitated electrodes. PZT nanofibers, which were positively charged, were spun on the collection substrate under the electric field. PZT nanofibers were stretched by the controlled electric field in two directions and deposited across the substrate.

5 Working Principle

5.1 Power Generation Mechanism

The nanogenerator device and power generation mechanism are illustrated in Fig. 6a–b, where PZT nanofibers were operating in the longitudinal mode (an alternating pressure applied on the top surface of the nanogenerator). The applied pressure was transferred to the PZT nanofibers through the PDMS matrix and resulted in charge generation due to the combined tensile and bending stresses in the PZT nanofibers. A voltage difference between the two adjacent electrodes was thereby induced by this separation of charge. The interdigitated electrodes enhanced the power output of the nanogenarator. The piezoelectric nanofibers between each pair of adjacent electrodes served as unit cells and each cell was connected in parallel.

Electrons generated in the PZT nanofibers could transfer through the electrodes when the PZT nanofibers were subjected to external stresses. Compliant PDMS

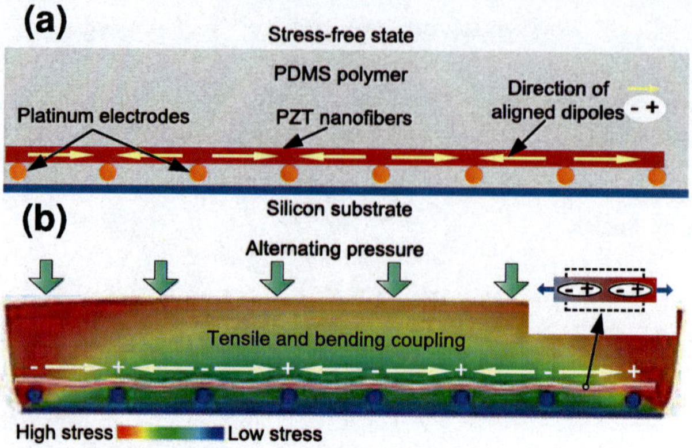

Fig. 6 Power generation mechanism of the PZT nanofiber generator. **a** Cross-sectional view of the PZT nanofiber in the generator. **b** Schematic view explaining the power output mechanism of the PZT nanofibers working in the longitudinal mode. The *color* presents the stress level in PDMS due to the application of pressure on the top surface. Reprinted in part with permission from Ref. [13]. Copyright 2010 American Chemical Society

covered the entire PZT nanofiber/electrode structure, due to the placement of the PZT nanofibers in a levitated position above the silicon substrate. The stress in the longitudinal direction, caused by Poisson's ratio of the composites, could be directly transferred to the PZT nanofibers when a stress was applied to the polymer matrix in the vertical direction. To avoid excessive stresses on the PZT nanofibers and to minimize the risk of damaging the electric connection of the electrodes, the silicon substrate was packaged with the nanogenerator as a rigid mechanical backing. This support could potentially be replaced by a flexible plastic backing for different applications. The final cured thickness of the PDMS polymer matrix was about 2 mm.

5.2 Mathematical Model

The potential generated from the PZT nanofibers between the interdigitated electrodes is given by:

$$\Delta V = \int_0^1 g_{33}\sigma(l)\mathrm{dl} \tag{2}$$

where l is the length of the nanofibers across two adjacent electrodes, $\sigma(l)$ is the stress function along the axial direction of the nanofiber and g_{33} is the piezoelectric voltage constant. Assuming perfect bonding between PZT nanofibers and PDMS polymer matrix, we have:

$$\varepsilon_\mathrm{p} = \varepsilon_\mathrm{m} = \varepsilon_\mathrm{c} \tag{3}$$

where ε_m and ε_c are the longitudinal strains in the PDMS matrix and the composite, respectively. Assuming both the PZT nanofibers and PDMS matrix are elastic, the nanogenerator can be considered as unidirectional, continuous fiber lamina, and we have [23]:

$$E_{11} = E_p v_p + E_m(1 - v_p) \tag{4}$$

$$v_p = A_p / A_c \tag{5}$$

where E_{11} is the longitudinal modulus, E_p is the PZT nanofiber modulus, E_m is the PDMS modulus, A_p is net cross-sectional area for the fibers and A_c is the cross-sectional area for the composite. The major Poisson's ratio can be calculated as:

$$\upsilon_{12} = \upsilon_{13} = \upsilon_p v_p + \upsilon_m(1 - v_p) \tag{6}$$

where $\upsilon_p$and υ_m are the Poisson's ratios of the PZT nanofibers and PDMS, respectively. The longitudinal strain can be given by:

$$\varepsilon_{xx} = \frac{\sigma_{xx}}{E_{11}} - \frac{\sigma_{yy}}{E_{11}}\upsilon_{12} - \frac{\sigma_{zz}}{E_{11}}\upsilon_{13} \tag{7}$$

where σ_{xx}, σ_{yy}, σ_{zz} are the stresses in three directions. By considering only the stress in the longitudinal direction:

$$\sigma(l) = E_p \cdot \left(\sigma_{xx}/E_{11} - \sigma_{yy}/E_{11} \cdot \upsilon - \sigma_{zz}/E_{11} \cdot \upsilon\right) \tag{8}$$

Thus, the output voltage can be written as:

$$\Delta V = \int_0^1 g_{33} \cdot E_p \cdot \left(\sigma_{xx}/E_{11} - \sigma_{yy}/E_{11} \cdot \upsilon - \sigma_{zz}/E_{11} \cdot \upsilon\right) \mathrm{dl} \tag{9}$$

For a given applied load or impact energy, the maximum output voltage is primarily determined by the piezoelectric voltage constant. From our previous study, the piezoelectric voltage constant of PZT nanofiber is roughly 0.079 Vm/N, which is much higher than that of bulk PZT (0.025 Vm/N) or the PZT microfiber composite (0.059 Vm/N) [24]. By inspection, the significantly larger g_{33} and *l*/*A* ratio of the nanofiber generator should result in much higher voltage output compared to that of the PZT microfibers [18] under the same loading condition. For the same reason, PZT nanofibers can also be used as an ultra-high sensitivity force/vibration sensor.

6 Characterization of Nanogenerator

The characteristics of the nanogenerator as a potential power supply were investigated by measuring generated voltage versus strain in the polymer matrix, under varying dynamic load frequencies, and power output versus load resistance, both using a DMA. The experimental setup is shown in Fig. 7.

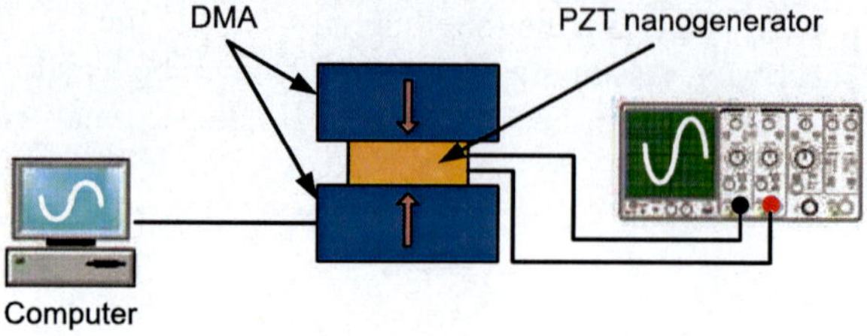

Fig. 7 The schematic of the experiment setup for the nanogenerator characterization test using the DMA. Reprinted in part with permission from Ref. [13]. Copyright 2010 American Chemical Society

The voltage generated by applying a harmonic force at a frequency of 250 rad/s (~39.8 Hz) and a specified maximum strain of 12% applied on the polymer matrix is shown in Fig. 8a. The positive and negative voltages were generated due to the sinusoidal load oscillations applied by the DMA (Fig. 8a). The peak to peak open circuit voltage, V_{p-p}, increased as the maximum strain applied increased. The maximum V_{p-p} was 1420 mV under a maximum applied strain along the PZT nanofibers of $\sim 7.5 \times 10^{-5}\%$ (established from mathematical and FEM models) at 250 rad/s, as shown in Fig. 8b.

The V_{p-p} versus various excitation frequencies under a maximum applied strain on the PDMS surface of 2.25% is illustrated in Fig. 8c. The highest output voltage of 62 mV occurred at a frequency of 220 rad/s (~35 Hz), corresponding to the lowest resonant frequency of the entire architecture. Voltage outputs were also recorded when varying the load resistances from 0.1 to 10 MΩ under a maximum specified strain of 10% applied to the PDMS surface and a harmonic load frequency of 250 rad/s (see Fig. 8d).

The power delivered to the load can be estimated from:

$$P_L = \frac{1}{T}\int \frac{V_o(t)^2}{R_L}\,\mathrm{d}t \tag{10}$$

where $V_o(t)$ is the real-time voltage, R_L is the load resistance and T is the period of load application. The maximum measured output power reached 0.03 μW with a load resistance of 6 MΩ, as shown in Fig. 8d.

7 Applications

Two applications of this nanogenerator are demonstrated. In the first application, the output voltage from the PZT nanofiber generator was measured when it underwent an impulsive loading, applied by tapping the top of the generator with a small Teflon® stack. As shown in Fig. 9a, the generated voltage, which was induced by piezopotential driven transient flow of electrons under the external load [25], reached 600 mV when a larger impact was applied on the nanogenerator by the periodic knocking. The higher the impact energy applied on the surface, the higher the output voltage generated by the device. The dampening effect of the soft

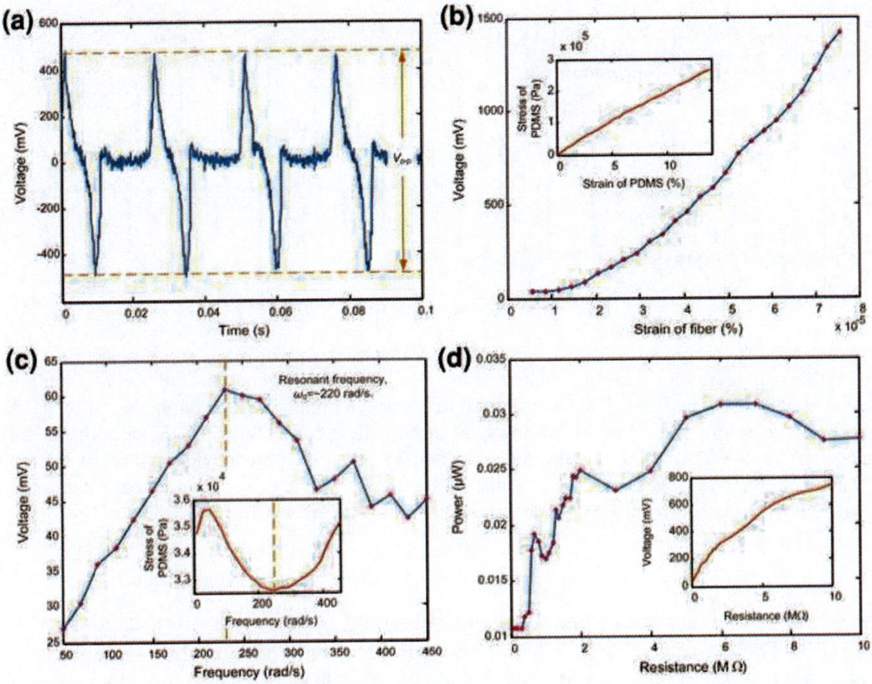

Fig. 8 The voltage generation properties of PZT nanofiber generator tested via DMA. **a** Voltage output when a harmonic force at the frequency of 250 rad/s (~39.8 Hz) and a maximum strain of 12% were applied on the PDMS surface. **b** The open circuit peak to peak voltage output versus strain of PZT nanofiber at the frequency of 250 rad/s (~39.8 Hz). The inset in b shows the stress of composites versus strain applied between the top and bottom PDMS surface. **c** The open circuit peak to peak voltage output versus frequencies of the harmonic forces at the maximum strain of 2.25% applied on PDMS. The inset in *c* shows the stress of composites versus frequency applied on the nanogenerator. **d** The power delivered to the load resistors versus the load resistance. The inset in d shows the voltage output versus the load resistance. Reprinted in part with permission from Ref. [13]. Copyright 2010 American Chemical Society

polymer matrix on the resonant frequency was also observed during the energy harvesting process.

In the second application, fingers were used to apply a periodic dynamic load to the top of the nanogenerator, during which positive and negative voltage outputs were observed (Fig. 9b). The negative voltage distribution was generated due to the carriers flowing in reverse when the external load was removed and the piezopotential vanished. The highest output voltage recorded during the test was 1.63 V. The amplitudes of the voltage outputs depended on how much pressure was applied on the nanogenerator surface.

In order to eliminate the influence of the bioelectric field of the human body and the electromagnetic interference from the testing equipment, a free vibration test using the PZT nanogenerator as a damper was conducted (Fig. 10a). The output voltage from the nanogenerator was measured when a Teflon® cantilever was

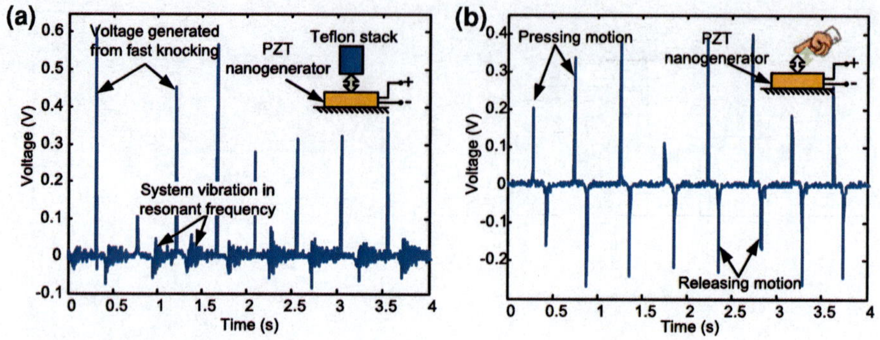

Fig. 9 Measurements of output voltage from PZT nanofiber generator. **a** Voltage output measured when a small Teflon® stack was used to impart an impulsive load on the top of the PZT nanofiber generator. The inset in a shows the schematic of a Teflon® stack tapping on the nanogenerator. **b** Voltage output measured when using a finger to apply a dynamic load on the top of the generator. The inset shows the schematic of a finger applying the dynamic load. Reprinted in part with permission from Ref. [13]. Copyright 2010 American Chemical Society

placed on top of the nanogenerator and subjected to free vibration, as shown in Fig. 10b. The damping ratio, ζ, and the natural frequency of this system were determined to be 0.064 and the 49.9 rad/s ($\sim$7.9 Hz), respectively. The output voltage from a dummy block without PZT nanofibers or any other active materials in it was also measured using the same setup. The measured result revealed that the amplitude of noise signal is only at about the 10 mV level. This confirmed that the power output from the PZT nanogenerator was in fact the energy harvested from mechanical vibration.

8 Conclusion

Energy harvesting technology that can increase operation time and decrease device size is urgently needed in wireless electronics, portable devices and implantable bio-sensors. This PZT nanofiber-based nanogenerator provides a promising solution for the challenges we are facing. This nanogenerator produces high voltage and power output, and has a high dielectric constant for a power source, which can be directly used for wireless sensors, personal electronics, implantable bio-sensors and bio-actuators. This low cost, flexible nanogenerator can also be packaged in bio-compatible polymers for embedding in shoes or clothes to harvest the kinetic energy of human body and charge batteries for devices such as iPods and cell phones on the go.

The peak output voltage from this nanogenerator was 1.63 V, and the output power was 0.03 μW with a load resistance of 6 MΩ. The piezoelectric voltage constant and dielectric constant of PZT nanofibers were much higher than those of

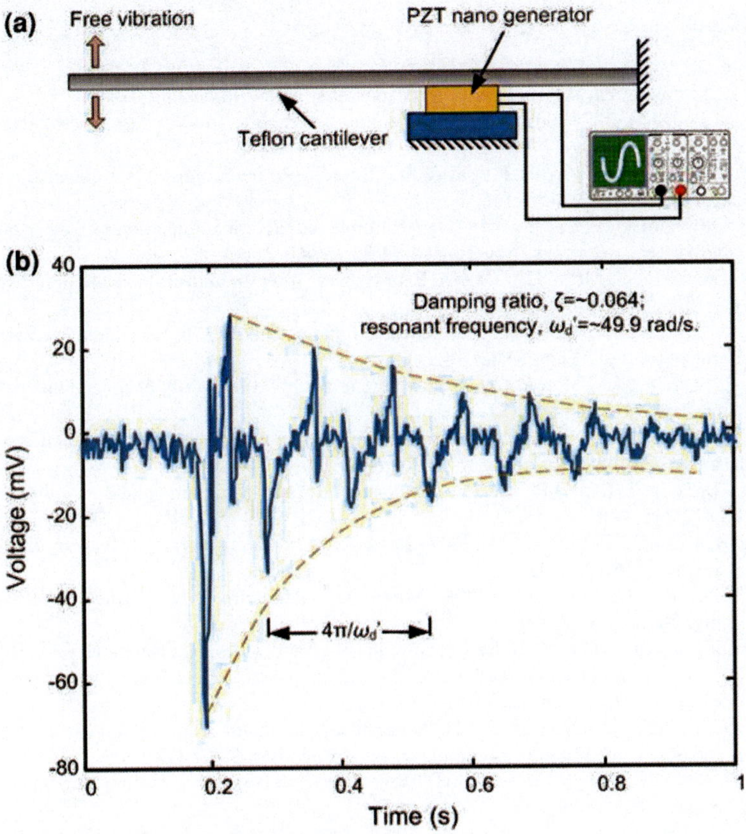

Fig. 10 Energy harvested from the free vibration of a Teflon® cantilever. **a** Schematic of the experimental setup. **b** The open circuit voltage output when the cantilever was under free vibration. Reprinted in part with permission from Ref. [13]. Copyright 2010 American Chemical Society

the semiconductor-type of piezoelectric nanowires and nanofibers, making this material ideal for nanogenerator or nanobattery applications. The flexible PZT nanofibers were embedded in soft PDMS polymer matrix, which helped to prevent the PZT nanofibers from being damaged, thereby extending the life cycle of the nanogenerator. The simple fabrication and assembly process would allow for the facile mass production of this type of nanogenerator.

Acknowledgment This work was supported in part by the National Science Foundation (Award No. CMMI-0826418 & No. ECCS-0802168), by the NSF MRSEC program through the Princeton Center for Complex Materials (grant DMR-0819860; N.Y.).

References

1. Lieber CM, Wang ZL (2007) Functional nanowires. MRS Bull 32:99–104
2. Paradiso JA, Starner T (2005) Energy scavenging for mobile and wireless electronics. IEEE Pervasive Comput 4:18–27
3. Lacour SP, Jones JE, Wagner S et al (2005) Stretchable interconnects for elastic electronic surfaces. Proc IEEE 93:1459–1467
4. Zheng GF, Patolsky F, Cui Y et al (2005) Multiplexed electrical detection of cancer markers with nanowire sensor arrays. Nat Biotechnol 23:1294
5. Torres EO, Rincón-Mora GA (2008) Energy-harvesting system-in-package microsystem. J Energy Eng 134:121–129
6. Wang ZL (2008) Towards self-powered nanosyatems: from nanogenerators to nanopiezotronics. Adv Func Mater 18:1–15
7. Yang RS, Qin Y, Dai LM et al (2009) Power generation with laterally packaged piezoelectric fine wires. Nat Nanotechnol 4:34–39
8. Yang RS, Qin Y, Li C et al (2009) Converting biomechanical energy into electricity by a muscle-movement-driven nanogenerator. Nano Lett 9:1201–1205
9. Wang ZL, Song JH (2006) Piezoelectric nanogenerators based on zinc oxide nanowire arrays. Science 312:242–246
10. Wang XD, Song JH, Liu J et al (2007) Direct-current nanogenerator driven by ultrasonic waves. Science 316:102–105
11. Qin Y, Wang XD, Wang ZL (2008) Microfiber—nanowire hybrid structure for energy scavenging. Nature 451:809–813
12. Xu S, Qing Y, Xu C et al (2010) Self-powered nanowire devices. Nat Nanotechnol 5:367–373
13. Chen X, Xu SY, Yao N et al (2010) 1.6 Volt nanogenerator for mechanical energy harvesting using PZT nanofibers. Nano Lett 10:2133–2137
14. Chen X, Xu SY, Yao N et al (2009) Potential measurement from a single lead zirconate titanate nanofiber using a nanomanipulator. Appl Phys Lett 94:253113
15. Lin YF, Song JH, Ding Y et al (2008) Piezoelectric nanogenerator using CdS nanowires. Appl Phys Lett 92:022105
16. Wang ZY, Hu J, Suryavanshi AP et al (2007) Voltage generation from individual BaTiO3 nanowires under periodic tensile mechanical load. Nano Lett 7:2966–2969
17. Su WS, Chen YF, Hsiao CL et al (2007) Generation of electricity in GaN nanorods induced by piezoelectric effect. Appl Phys Lett 90:063110
18. Mohammadi F, Khan A, Cass RB (2003) Power generation from piezoelectric lead zirconate titanate fiber composites. Proc Mat Res Soc Symp 736:D5.5.1–D5.5.6
19. Xu SY, Shi Y, Kim SG (2006) Fabrication and mechanical property of nano piezoelectric fibers. Nanotechnology 17:4497–4501
20. Li D, Xia Y (2003) Fabrication of tatania nanofibers by electrospinning. Nano Lett 3:555–560
21. Li D, Wang Y, Xia Y (2004) Electrospinning nanofibers as uniaxially aligned arrays and layer-by-layer stacked films. Adv Mater 16:361–366
22. Gao YF, Wang ZL (2007) Electrostatic potential in a bent piezoelectric nanowire. The fundamental theory of nanogenerator and nanopiezotronics. Nano Lett 7:2499–2505
23. Mallick PK (1993) Fiber-reinforced composites: materials, manufacturing and design. Marcel Dekker, New York
24. Swallow LM, Luo JK, Siores E et al (2008) A piezoelectric fibre composite based energy harvesting device for potential wearable applications. Smart Mater Struct 17:025017
25. Wang ZL (2010) Piezotronic and piezophototronic effects. J Phys Chem Lett 1:1388–1393

Part III
Clean Fuels Generation and Environmental Remediation by Sun Light

Nanostructured Materials for Photolytic Hydrogen Production

Jiefang Zhu, Dinko Chakarov and Michael Zäch

Abstract A hydrogen economy is often considered an attractive alternative to our current fossil fuel-based energy system. In order for such a hydrogen economy to become reality, several challenges associated with the production, storage, transportation and use of hydrogen must be solved. This chapter addresses the issue of hydrogen production. While the currently most widely used method to produce hydrogen is based on the conversion of fossil fuel resources and does not therefore fulfill the requirement of CO_2 neutrality, we discuss here the photolytic production of hydrogen via water splitting. This scheme is based on energy input from the most powerful and ultimately sustainable energy source mankind has at its disposal: the sun. Moreover, no carbon dioxide is released into the atmosphere, and the method has potential for cost-effective large-scale production.

1 Background and Introduction

Mankind is facing two challenges of hitherto unsurpassed importance and complexity, namely the "tera-Watt" and "tera-ton" challenges. The former refers to the issue of supplying a rapidly growing world population with larger and larger

J. Zhu (✉)
Department of Materials Chemistry, Uppsala University,
Lägerhyddsvägen 1,
751 21 Uppsala, Sweden
e-mail: jiefang.zhu@mkem.uu.se

D. Chakarov · M. Zäch
Department of Applied Physics, Chalmers University of Technology,
Fysikgränd 3, 412 96 Göteborg, Sweden
e-mail: michael.zach@chalmers.se

L. Zang (ed.), *Energy Efficiency and Renewable Energy Through Nanotechnology*,
Green Energy and Technology, DOI: 10.1007/978-0-85729-638-2_13,

amounts of energy [1], while the latter encompasses finding ways to deal with the huge amounts of carbon dioxide being emitted into the atmosphere as a consequence of our fossil fuel-based energy system [2]. The formidable task is to identify an alternative energy system, which at the same time is sustainable, virtually unlimited, safe and clean. The task becomes yet more challenging when also considering economy and security of supply aspects.

The hydrogen economy is one of many suggested scenarios that may be able to address these tera-challenges. The basic idea of the hydrogen economy is that hydrogen—a light, diatomic gas with a vast energy content by weight, but a small energy content per volume—would replace the fossil fuels which form the backbone of our current energy system. Hydrogen reacts cleanly with oxygen in a highly exothermic reaction, with pure water being the only exhaust product. The energy stored in the chemical bond of hydrogen can thus be released by burning it in a combustion engine, or more efficiently by oxidizing it in a fuel cell. Apart from zero-emissions, hydrogen has further attractive properties, including that it can be stored and transported, which are crucial requirements addressing the typically intermittent nature of renewable energy sources. In order for hydrogen to become an economically viable option, several hurdles must be surpassed. Besides providing efficient, inexpensive and durable fuel cells, the production of hydrogen and its storage constitute two other important cornerstones of the hydrogen economy. The former aspect will be discussed in this chapter, while the latter is covered separately in different chapters of this book.

1.1 Overview of Current and Potential Future Hydrogen Production Methods

While hydrogen (H) is abundant on earth, molecular hydrogen (H_2) is usually not found in large quantities in nature. Most of the hydrogen is bound to carbon, oxygen or both, meaning that H_2 must be generated by extracting it from other hydrogen sources via highly endothermic processes. The energy required to drive these extraction processes may come from essentially any energy source, including renewable energy sources, as illustrated in Fig. 1. A key aspect to bear in mind is that hydrogen is not an energy source but an energy vector.

A large fraction of today's hydrogen production is based on fossil fuel resources. The processes involved in the conversion of fossil fuels, such as the reforming of natural gas, coal gasification/liquefaction and coal electrolysis, require large amounts of energy, either in the form of heat or electricity, and all these methods are associated with the release of vast amounts of carbon dioxide. For the hydrogen economy to become a viable option of our energy future, we therefore have to access other feedstocks from which hydrogen can be produced, and we have to make sure that the energy required to transform these feedstocks to hydrogen is provided by a sustainable and carbon–neutral source. Alternative feedstocks under consideration include wood, other biomass, organic waste and

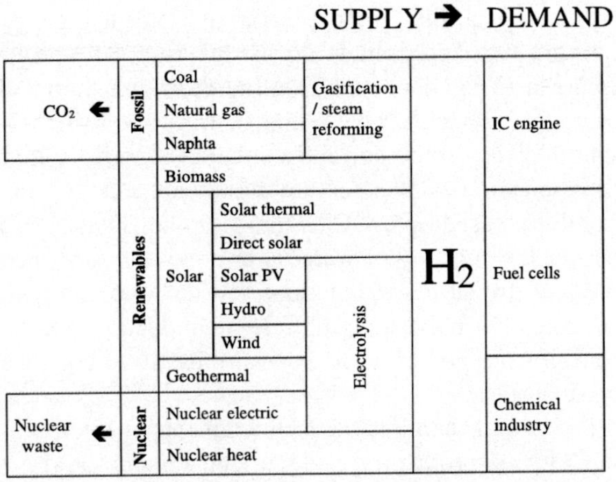

Fig. 1 Schematic illustration of various hydrogen production pathways using different primary energy sources and energy conversion schemes. The main application areas of hydrogen and unwanted emissions from selected processes are indicated on the right and left hand side, respectively

water, where the first three options are not truly carbon neutral. The last option constitutes an entirely sustainable energy system, where hydrogen is produced from water and later on recovered back into water upon conversion in a combustion engine or a fuel cell. This chapter therefore focuses on hydrogen production from water. In principle, water can be split into hydrogen and oxygen along several different pathways and utilizing various (sustainable) energy sources. The method used most commonly today is to dissociate water in an electrolysis cell, where the required electricity may be produced from nuclear, geothermal, gravitational or solar energy.[1] Solar energy is clean, quiet, sustainable, and its potential is enormous. The solar influx, i.e., the energy received by the planet from the sun (174 petawatts $= 1.7 \times 10^{17}$W), surpasses the current world energy consumption rate (ca. 15 TW $= 1.5 \times 10^{13}$W) by more than four orders of magnitude. In other words, less than one hour of sunshine would, in principle, be sufficient to provide us with energy for an entire year. Given this enormous potential, various forms of solar energy, including exploiting solar energy for heating and cooling purposes, converting solar energy into electricity using photovoltaic materials or solar thermal technologies and converting solar energy into storable chemical fuels, as discussed in this chapter, have recently received considerable attention.

Since electrolysis is a two-step process, the potential of achieving a highly efficient, simple, and at the same time cost-effective conversion is limited. A more appealing idea is to directly split water into its component gases using a single device without the prior production of electricity, i.e., with no external electron flow. High-temperature thermochemical cycles, i.e., a set of coupled, thermally-driven chemical reactions that sum to the decomposition of water into H_2 and O_2,

[1] Note that wind power and hydropower are considered variations of solar energy here, and that solar energy, and to a large extent also geothermal energy, actually are based on nuclear processes.

and photolysis (the direct splitting of water using light as the only energy input), represent two approaches in this direction. The primary heat sources, which are discussed in the context of high-temperature thermochemical cycles, are nuclear energy and solar energy. While thermochemical cycles potentially can be run at high efficiency, they currently either require very high temperatures (> 1500°C), where material challenges are significant, or they involve aggressive chemicals. Photolysis-based approaches, on the other hand, can run at ambient temperature because the splitting of water is not thermally induced but achieved via electron–hole pair driven redox reactions. To this end, energy-rich photons from the solar spectrum are harvested in a semiconductor material, where they may excite electron–hole pairs, which in turn may drive chemical reactions at the semiconductor surface.

The latter scheme is the subject of this chapter, which provides an overview of the CO_2-free production of hydrogen via photon-driven redox reactions on semiconductor materials and employing entirely renewable and abundant raw materials and energy sources, namely water and sunlight. Particular attention will be paid to the role of nanoscience and nanotechnology, which have attracted a continuously growing interest in this context, and which have shown a strong potential to improve the efficiency of various hydrogen production methods, not the least approaches based on photolysis.

Before diving into the details, we would like to note that the idea of using water as a raw material for energy production is not an entirely new concept. In 1874, Jules Verne, in his novel *The Mysterious Island*, painted a scenario where *water one day will be employed as fuel*, thereby replacing dwindling coal reserves and driving the machinery without which *there would be no railways, no steamers, no manufactories, nothing of that which is indispensable to modern civilization*. Today, almost 140 years later, Jules Verne's visionary ideas have neither been realized nor abandoned. In fact, the concept of a hydrogen economy still attracts a lot of interest, although the hurdles on the way sometimes seem insurmountable. The hope is that this chapter may provide the reader with an understanding of the state-of-the-art in the field of photolytic hydrogen production and identify the most burning scientific and technological issues that have prevented hydrogen from becoming our prime energy carrier.

The remainder of this chapter is organized as follows: after a general introduction, we briefly review the mechanism and challenges associated with photolytic water-splitting (Sect. 2). In Sect. 3, we present an overview of various semiconductor materials, which have been scrutinized for their ability to split water into hydrogen and oxygen under sunlight illumination. The advantages and shortcomings of these materials will be described. Next, we discuss various ways in which nanoscience and nanotechnology (N&N) can improve materials and their integration into water splitting devices (Sect. 4). Our emphasis is on N&N approaches: (i) to tune the optical and electronic properties of semiconductors, (ii) to manage and optimally utilize the incoming light, (iii) to improve the separation and extraction of photogenerated charge carriers via control of the photocatalyst structure and morphology, (iv) to tailor the transport of reactants and products, and

(v) to fabricate well-defined model systems. We conclude by outlining possible pathways toward sustainable hydrogen production schemes, which challenges need to be addressed on the way, and how N&N may contribute.

2 Mechanism of Photolytic Hydrogen Production

Photolytic water-splitting schemes integrate solar energy collection and water electrolysis into a single device, thereby transforming solar energy into chemical energy in the form of hydrogen. Several steps are required to achieve this transformation, which is initiated by the harvesting of solar photons (see Fig. 2). Since water is essentially transparent to the wavelengths constituting the solar spectrum, photolytic splitting of water requires the presence of a light harvesting structure, typically in the form of a semiconductor. Owing to their particular electronic band structure, semiconductors are able to interact with photons whose energy exceeds a threshold corresponding to the bandgap energy of the material, i.e., the energy separation between the valence and conduction band edges. If this condition is met, an electron may be promoted from the valence band to the conduction band, thereby leaving an empty state, normally referred to as a hole, behind. In order for such an excited electron–hole pair to drive water-splitting, it must have an energy that exceeds the standard potential, ΔE^{o}, for the decomposition of liquid water to form gaseous hydrogen and oxygen (-1.23 V, where the negative sign indicates that the water-splitting process is thermodynamically non-spontaneous). The excited electron–hole pairs are short lived, and the photogeneration of charge carriers must therefore be followed by efficient charge separation schemes in order for a significant fraction (defined by the quantum yield) of these electron–hole pairs to escape recombination and make their way to the semiconductor surface. If the photogenerated electrons and holes have sufficient energy to reduce protons and to oxidize water, respectively, they may be extracted at the solid/liquid interface and drive the photosplitting of water. Finally, hydrogen and oxygen gas need to be removed from solution and separated from one another in the case of photocatalytic water splitting.

Note that we distinguish here photocatalytic and photoelectrochemical water splitting. In the former case, the reaction takes place in a suspension or colloidal solution of semiconductor nanoparticles. Oxidation and reduction reactions occur in close proximity to one another, and a mixture of hydrogen and oxygen is thus evolved. In the latter case, the semiconductor is made into a photoanode, from which oxygen is evolved. Photogenerated electrons are extracted from the semiconductor by a charge collection layer and, through an external circuit, shuttled to a counter-electrode, where they recombine with protons to form hydrogen. Oxygen and hydrogen production sites are thus spatially separated.

There are several potential bottlenecks, which may hamper or even prevent photolytic water splitting. The most burning issues are summarized as follows: (i) mismatch between the semiconductor's absorption characteristics and the solar

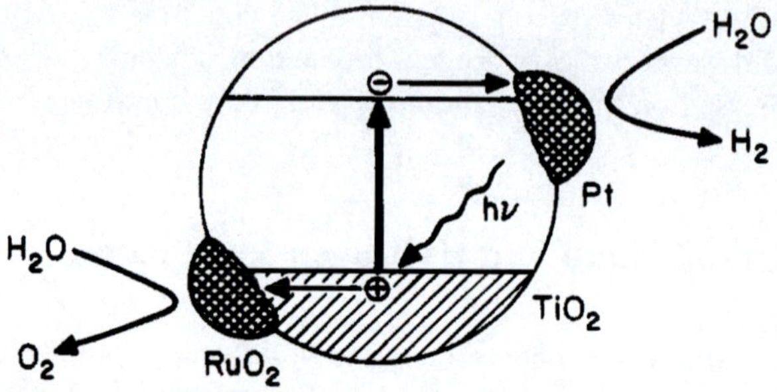

Fig. 2 Schematic illustration of the elementary reaction steps of photocatalytic hydrogen production via water splitting on semiconductor nanoparticles with additional co-catalysts for oxygen and hydrogen evolution. Reprinted with permission from [3]. Copyright 1995 American Chemical Society

spectrum (and other light sources), (ii) inappropriate semiconductor bandgap and/ or band edge positions, (iii) fast (bulk and/or surface) electron–hole recombination, (iv) short lifetime and thus limited diffusion length of photogenerated electrons and/or holes, (v) poor electron and/or hole transfer kinetics at the semiconductor-liquid interface (large over-potentials required), (vi) competing reactions, including photocorrosion and the back-reaction of hydrogen and oxygen to water in the case of photocatalytic water splitting

In the next section, different classes of materials and their properties will be discussed in light of these requirements and issues, followed by Sect. 4, which presents various nanoscience- and nanotechnology-based approaches towards improved photolytic water splitting.

3 Materials for Photolysis of Water

Materials play a key role in photolytic H_2 production, and their structure and electronic properties largely determine the efficiency of each step in the photolysis of water. The majority of photocatalysts and photoanodes are composed of metal oxides (Sects. 3.1–3.3), metal sulfides (Sect. 3.4), metal nitrides (Sect. 3.5), oxysulfides (Sect. 3.6), oxynitrides (Sect. 3.6) and composites thereof (Sect. 3.7). In these cases, metal cations with the highest oxidative states have d^0 (red area in Fig. 3) or d^{10} (green area) electronic configuration, while O, S and N (blue area) show their most negative states. The bottom of the conduction band consists of the d and sp orbitals of the metal cations, while the top of the valence band in metal oxides is composed of O 2p orbitals, which are normally located at ca. +3 V (vs NHE) or higher [4, 5]. The valence bands of metal oxysulfides and oxynitrides are formed by S 3p (and O 2p) and N 2p (and O 2p), respectively. Some alkali (Li, Na, K, Rb and Cs), alkaline earth (Mg, Ca, Sr and Ba) and transition metal

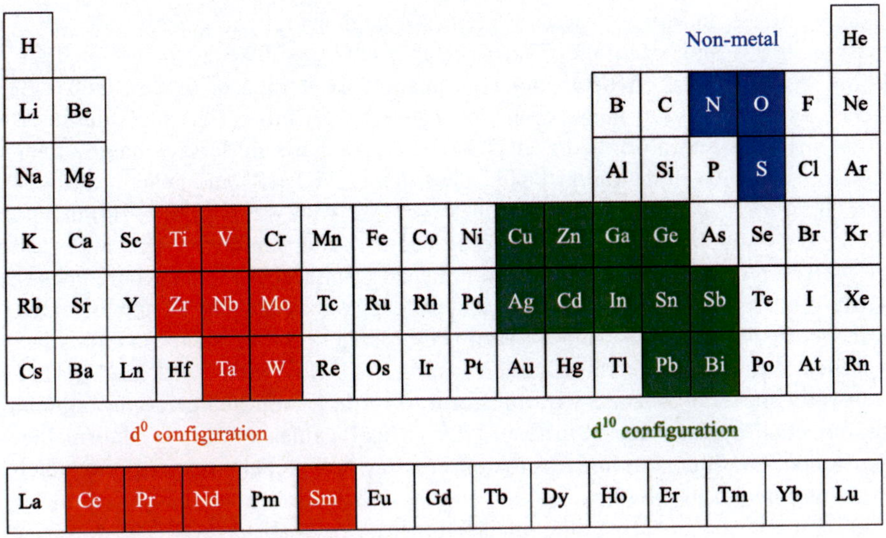

Fig. 3 Elements constructing photocatalysts for water splitting

(Y, La and Gd) ions can construct a crystal structure of layered perovskite and cubic pyrochlore compounds, but do not contribute to the energy band structure of these compounds [5]. In this section, the most widely used materials for photolytic H_2 production will be introduced. The focus is on material properties, which influence the photoconversion efficiency.

3.1 *Titanium Dioxide (TiO_2)*

Compared to other photoactive materials for H_2 production, TiO_2 has received much attention since the initial work of Fujishima and Honda [6] due to its low cost, abundance, stability and performance.

Despite a high rate of consumption, a shortage of TiO_2 seems impossible in the near future. Titanium is the ninth most abundant element in the world and constitutes about 0.63% in weight of the Earth's crust. Minerals, like rutile (TiO_2) and ilmenite ($FeTiO_3$), are the main ores of this element, and they are found in large deposits in Norway, Australia, China, Canada and many other countries [7]. The toxicity of TiO_2 is low, and it has been approved as a food colorant (E-171 in EU legislation). Actually, TiO_2 is widely used in many daily products such as toothpaste, pill coatings and chewing gum [8].

TiO_2 has four main crystal phases: anatase (tetragonal), rutile (tetragonal), brookite (orthorhombic) and TiO_2 (B) (monoclinic). All TiO_2 forms can be described as different arrangements of elongated TiO_6 octahedra connected by sharing corners and edges. These differences in lattice structure determine different mass densities and electronic band structures between different forms of TiO_2. Loose zigzag chains in octahedra coordination exist in anatase and TiO_2 (B) structures, compared to linear chains of octahedra in rutile. Therefore, the interstitial spaces between octahedra in TiO_2 (B) and anatase are larger, making them less compact than rutile and brookite (densities of TiO_2 (B) and anatase are 3.64 and 3.84 g cm^{-3}, respectively, compared to 4.26 g cm^{-3} for rutile and 4.17 g cm^{-3} for brookite) [9, 10]. Compared to anatase and rutile, brookite and TiO_2 (B) are uncommon, unstable and not easily synthesized in their pure phases. Therefore, they are seldom used as independent photocatalysts or photoanodes. Although rutile is the most stable crystal phase of TiO_2, the difference in Gibbs free energies of formation for rutile and anatase is small (less than 15 kJ mol^{-1}) [11]. Generally speaking, anatase will transform to rutile at high temperature, big size and/or small surface area. However, the critical values of this transformation depend on the pH conditions, adsorbates, pressure and preparation methods, which explains some of the conflicts in the literature. It has been shown that bicrystalline TiO_2 (mixture of anatase–rutile, anatase-brookite or anatase-TiO_2 (B)) normally has better photocatalytic and photoelectrochemical performance than its single-phase constituents. This synergistic effect is attributed to the formation of n–p junctions at the contact of the two crystal phases, which improves the efficiency of charge separation.

Due to the presence of a small amount of oxygen vacancies, which are compensated by the presence of Ti^{3+} centers, pure TiO_2 is an *n*-type semiconductor and is transparent to visible light. The top of the TiO_2 valence band is mainly formed by overlapping oxygen 2p orbitals (O $p\pi$ states), whereas the lower part of its conduction band is mainly constituted by the 3d orbitals (t_{2g} bands) of Ti^{4+} cations [12]. The band gaps are 3.2 eV (corresponding to an absorption edge of 380 nm) and 3.0 eV (400 nm) for anatase and rutile, respectively, which limits their ability to convert visible light. Another drawback of TiO_2 is its low photon efficiency. The band transitions of TiO_2 are indirect, and indirect-band semiconductors normally present less photon absorption compared to direct-band semiconductors. In addition, 90% of the photogenerated electron–hole pairs have short lifetimes and recombine in less than 10 ns; the photogenerated carriers available for surface reactions are very limited. Quantum yields for TiO_2 reactions in solution are typically below 1%, while they can exceed 25% for some gas-phase reactions. The values depend largely on the electronic transport at different interfaces and surfaces [13].

The conduction band energy E_{CB} of rutile coincides with the reversible hydrogen potential ($E^{H^+/H_2} = 0 V vs. NHE at pH 0$), whereas that for anatase is more negative by 0.2 eV. The valence band energy E_{VB} of both anatase and rutile is more than positive enough to oxidize water ($E_{O_2/H_2O} = 1.2$ V vs. NHE at pH 0), by about 1.8 eV. Note that these relationships are relatively fixed, since both E_{CB}, E_{VB}, and reduction and oxidation potentials simultaneously shift with solution pH (–59 mV per pH unit). Therefore, one of the main advantages of TiO_2 over other semiconductors is that its electronic structure allows both the reduction of protons and the oxidation of water, which are key processes for water splitting. TiO_2 is photo-corrosion resistant because water oxidation is thermodynamically more favored than the oxidation of oxide anions by photogenerated holes. This stability, which is maintained over a large pH range, is a crucial feature and explains the widespread use of TiO_2 in water splitting [13].

3.2 *Other Binary Metal Oxides*

In addition to TiO_2, there are some other traditional metal oxides, which have also been investigated extensively due to their specific advantages. Among them, ZnO, α-Fe_2O_3 and WO_3 are representative. However, they all have inherent drawbacks in photolytic H_2 production.

ZnO has a large application potential in optoelectronic devices, due to its diverse physical properties and finely tunable preparation. ZnO normally has a hexagonal (wurtzite) structure with the lattice parameters a = b = 3.25 Å and c = 5.12 Å. ZnO is intrinsically doped via oxygen and/or zinc interstitials, which act as *n*-type donors. However, it can be designed and doped to be either *n*- or *p*-type. ZnO is a direct-band gap semiconductor, and has a wide band gap of 3.2 eV with good charge carrier mobility. ZnO has similar conduction and valence positions as anatase TiO_2, as shown in Fig. 4, and therefore has been frequently considered an alternative to TiO_2 for photocatalytic applications. Although ZnO and TiO_2 have similar conduction band positions, the density of states is about one order of magnitude lower in ZnO. Also, the structure of the conduction bands differs. The conduction band of ZnO is mainly composed of Zn 4s and 4p orbitals, while that of TiO_2 consists of Ti 3d orbitals [14]. This may affect the transfer rate of photogenerated electrons from the valence band to the conduction band under irradiation. The electron mobility in single-crystalline ZnO is much higher than that of anatase TiO_2. It tends to decrease with doping due to electron scattering at impurities, and decreases substantially due to scattering and energy barriers at grain boundaries [14, 15].

The main drawback of ZnO in solution is its chemical instability. It is soluble in strong acids (Eq. 1) and its surface converts to $Zn(OH)_2$ in alkalis (Eq. 2), which limits its application at extreme pH settings as well as in the presence of Zn^{2+} chelating agents.

$$ZnO + 2H^+{}_{(aq)} \rightarrow Zn^{2+}{}_{(aq)} + H_2O \quad (1)$$

$$ZnO + H_2O \rightarrow Zn(OH)_2 \quad (2)$$

Besides its chemical instability, ZnO is easily photo-corroded, especially at low pH values. Under irradiation for a long time, ZnO suffers from photodecomposition due to the oxidation of O^{2-} in ZnO by photogenerated holes (Eq. 3).

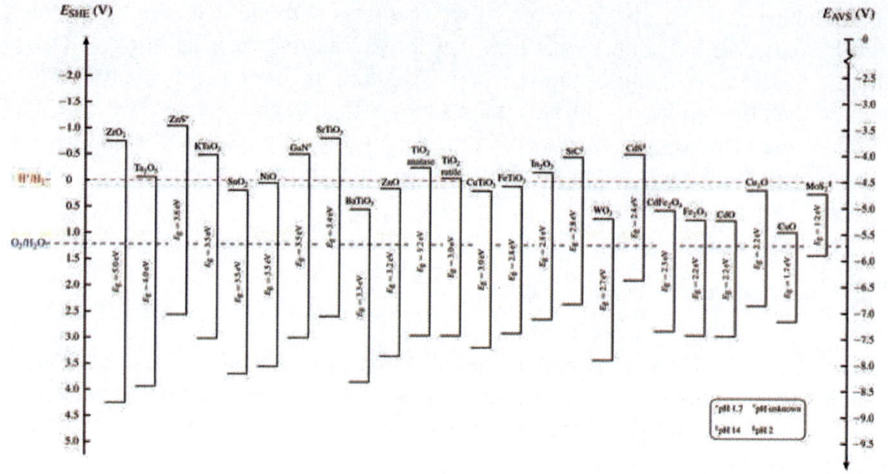

Fig. 4 Valence and conduction band positions of various semiconductors with respect to the standard hydrogen electrode (SHE) and the vacuum reference energy level E_{AVS}. All values are tested at pH 1, unless otherwise noted

$$2ZnO + 4h_{VB}^{+} \rightarrow 2Zn^{2+}{}_{(aq)} + O_{2(g)} \tag{3}$$

WO_3 is an *n*-type semiconductor with a relatively narrow band gap of 2.7 eV, which enables it to convert part of the visible solar light. Using WO_3 for solar energy conversion has been extensively investigated. WO_3, like TiO_2, is photo-corrosion resistant. The VB of WO_3 is close to that of TiO_2, exceeding the O_2/H_2O potential, which makes it a successful photocatalyst for O_2 evolution from water. However, the bottom of its CB lies below the redox potential of H^+/H_2, which means that the reduction of water to hydrogen is thermodynamically unfavorable for WO_3. Applying a bias can overcome the energy barrier for the ejection of photogenerated electrons from the conduction band to water. Coupling with other semiconductors (like TiO_2) and doping are alternative ways to achieve overall water splitting with WO_3 [16].

α-Fe_2O_3 is another attractive candidate for water splitting due to its narrow band gap (2.2 eV), good photo stability, chemical inertness and the abundance of iron mineral resources in the world. The band structure of α-Fe_2O_3 is similar to that of WO_3, with the VB top more negative, but still exceeding the O_2/H_2O potential, and the CB bottom lower than the redox potential of H^+/H_2. A variety of dopants have been employed to modify α-Fe_2O_3 and *p*-type semiconductor behavior has been obtained, in addition to its common *n*-type characteristics. The main bottlenecks of α-Fe_2O_3 are its fast e^--h^+ recombination rate, poor charge transport characteristics due to the trapping of electrons at defect sites and the poor mobility of holes. Attempts to circumvent these problems by using a bias, scavengers, coupling to other semiconductors, structure configurations and compositional tuning will contribute to further applications of this material.

There are many other binary metal oxides that have been used in photoelectrolysis of water, such as SnO_2, CeO_2, NiO, CdO, PdO, V_2O_5, MoO_3, Cr_2O_3, In_2O_3, Ga_2O_3, Cu_2O, CuO, Bi_2O_3, ZrO_2, Ta_2O_5 and so on. Although all of them have inherent issues, some of them are very good coupling partners to be used in conjunction with another semiconductor.

3.3 Ternary and Quaternary Metal Oxides

When TiO_2 is fused with other metal oxides (SrO, BaO, Ln_2O_3 (Ln = lanthanide)), metal titanates with the perovskite structure are formed. Perovskites have the general formula, ABX_3, with $SrTiO_3$ as an example. Their framework structure contains corner-sharing TiO_6 octahedral with the A cation in twelve-coordinate interstices. Several hundred oxides own this structure. In addition to the cubic structure exhibited by $SrTiO_3$, a series of distorted, non-cubic structures occurs, with the framework of TiO_6 octahedra being twisted. For example, $BaTiO_3$ is tetragonal. Both $SrTiO_3$ and $BaTiO_3$ with band gaps of 3.3 eV are interesting semiconductors for photoelectrolysis of water. Compared to $BaTiO_3$, $SrTiO_3$ has been more intensively studied.

Titanates with tunnel structures have been investigated for water splitting [17]. $BaTiO_9$ has a twin-type tunnel structure in which the TiO_6 octahedra are not parallel to one other, forming a pentagonal prism space. Alkaline metal hexatitanates ($M_2Ti_6O_{13}$; M = Na, K, Rb) have Wadsley-Andersson type structures in which TiO_6 octahedra share one edge at one level in linear groups of three, creating a tunnel structure with rectangular space. These materials have been used in powder form in suspensions, normally together with a co-catalyst.

There are more complex perovskites containing two different cations, which occupy either the A or B sites, and many of these have a layered structure. Two main classes of such oxides have been studied in water photolysis: the Dion-Jacobson series ($AM_{n-1}B_nO_{3n+1}$, e.g., $KCa_2Ti_3O_{10}$) and the Ruddlesden-Popper series ($A_2M_{n-1}B_nO_{3n+1}$, e.g., $K_2La_2Ti_3O_{10}$). Noble metal co-catalysts can be loaded onto these materials by photo deposition. Due to a negative charge (which is balanced by the alkali cations) of the oxide sheets, the noble metal salt anions will not be intercalated in the host lattice. Instead, the noble metal particles are formed on the external surfaces of the layered perovskites. In many cases, the H^+-exchanged layered oxides show higher activity for hydrogen production, due to the easy accessibility of the interlayer space to electron donor species [18, 19].

Another type of layered perovskite has the generic composition $A_nB_nO_{3n+2}$($n = 4, 5$; A = Ca, Sr, La; B = Nb, Ti). Unlike the (100)-oriented structures mentioned above, these perovskites have slabs parallel to (110) and are highly donor-doped. These structural and electronic characteristics are

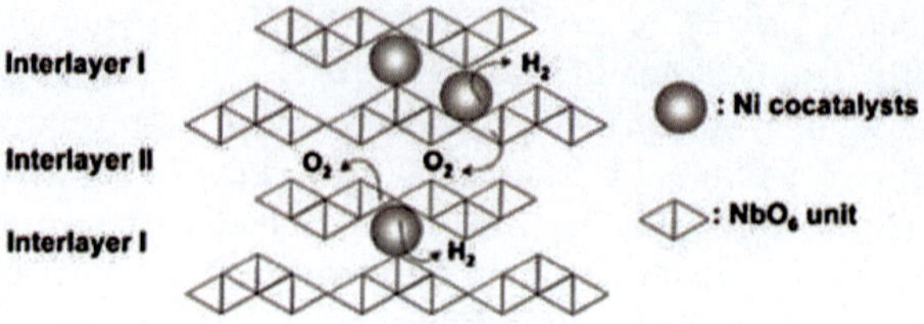

Fig. 5 Water splitting over $K_4Nb_6O_{17}$ photocatalyst with layered structure. From [5]. Reproduced by permission of The Royal Society of Chemistry (http://dx.doi.org/10.1039/b800489g)

thought to be responsible for the observed improved quantum yields for overall water splitting under UV irradiation. Among these compounds, $La_2Ti_2O_7$ (deduced from $La_4Ti_4O_{14}$) and $La_4CaTi_5O_{17}$ are representative titanates. Their band gaps are 3.2 and 3.8 eV, respectively, and they show quantum yields of 12% (< 360 nm) and 20% (< 320 nm), respectively [19, 20].

Tantalates and niobate oxides with corner-sharing octahedral MO_6 (M = Ta, Nb) structures have shown high photocatalytic activity for the cleavage of water. The high activity of these layered compounds can be assigned to the ease of migration and separation of photogenerated electron–hole pairs through the corner-shared framework of MO_6 units [21]. Tantalates, $MTaO_3$ (M = Li, Na, K) have been reported as effective photocatalysts for water splitting under UV irradiation. These oxides crystallize in perovskite structure, and their band gaps are 4.7 eV (Li), 4.0 eV (Na), and 3.7 eV (K) [22]. $NaTaO_3$ shows the highest photocatalytic activity among these $ATaO_3$ type photocatalysts when a NiO co-catalyst is loaded, which is due to the suitable conduction band level consisting of Ta 5d orbitals and energy delocalization caused by the small distortion of TaO_6 connections.

The layered oxides with ion-exchange characteristics have a net negative charge on the layered sheets. Thus, they can attract positively charged ions (such as K^+) in the interlamellar spaces. Interestingly, some of these materials (e.g., $K_4Nb_6O_{17}$) have two kinds of interlayers (I and II) that alternate, as shown in Fig. 5. H_2 is evolved from one interlayer, in which co-catalysts are introduced by ion-exchange or intercalation reactions, while O_2 production occurs in the other interlayer. In this case, the sites for H_2 and O_2 evolution are thus separated by the photocatalytic niobate sheet [5]. Furthermore, an electric field gradient originating from the uneven K^+ distribution on opposite sides of the niobate sheets assists electron–hole separation.

Generally speaking, oxides containing transition metal cations with d^0 electronic configuration like Ti^{4+}, Nb^{5+} or Ta^{5+} have wide band gaps (> 3.0 eV). Thus, these materials do not perform well under visible-light irradiation. The ternary and quaternary titanates, tantalates and niobates suffer from the same handicap. However, a series of H_2 production photocatalysts based on $K_4Ce_2M_{10}O_{30}$ (M = Ta, Nb) and their solid solution $K_4Ce_2Ta_{10-x}Nb_xO_{30}(x = 0 - 10)$ have appropriate band gaps of ca. 1.8–2.3 eV (corresponding to absorption edges of 540–690 nm)

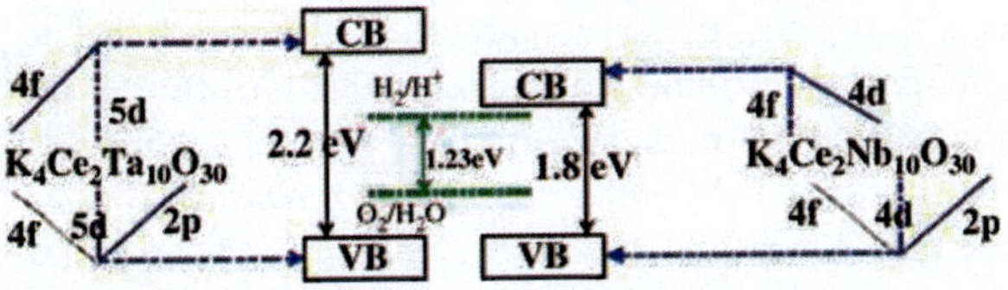

Fig. 6 Band structure of $K_4Ce_2M_{10}O_{30}$ (M = Ta, Nb) and comparison with redox couples for photocatalytic production of H_2 and O_2 from water. Reprinted from [23], Copyright 2007, with permission from, Elsevier (http://www.sciencedirect.com/science/journal/14686996)

[23, 24]. Density functional theory (DFT) calculations indicate that the valence bands of these photocatalysts are composed of hybridization with O 2p + Ta 5d (or Nb 4d) and occupied Ce 4f orbitals, while the conduction bands are mainly comprised of the Ta 5d (or Nb 4d) orbitals (Fig. 6). $K_4Ce_2M_{10}O_{30}$(M = Ta, Nb) has a parallelepiped (tunnel) surface structure. This is beneficial to the formation of "nano-nests", to which the co-catalysts, nanoparticles of Pt, RuO_2 and NiO_x, can be strongly associated, thus avoiding aggregation and improving photocatalytic H_2 generation greatly. This is an example of an effect of surface nanostructures on photocatalytic performance.

3.4 Metal Sulfides

Metal sulfides are considered attractive candidates for visible-light responsive photocatalysts. The valence bands of most metal sulfides consist of 3p orbitals of S, which results in a higher (more negative) valence band and narrower band gap compared to metal oxides.

Among the available metal sulfides, CdS with wurtzite structure is probably the most studied metal sulfide photocatalyst. Its narrow band gap (2.4 eV) means that it can absorb visible light below a wavelength of 510 nm. Compared to the VB of metal oxides being formed by O 2p orbitals, the VB of CdS is constituted by S 2p orbitals. The electronegativity of S is smaller than that of O, resulting in different VB positions. The CB potential of CdS is around -0.52 V (vs. NHE), high enough to reduce H_2O, and the top of its valence band is about 1.5 V (vs. NHE), suitable for the oxidation of H_2O. However, CdS is prone to photocorrosion (Eq. 4), which is a problem common to most metal sulfide photocatalysts.

$$2h_{VB}{}^{+} + CdS = Cd^{2+}{}_{(aq)} + S_{(s)} \tag{4}$$

Many methods have tried to overcome this photocorrosion, such as combining with other semiconductors or using scavengers (electron donors, such as cysteines, EDTA, sulfide or sulfite species) to attract or consume photogenerated holes. In particular, H_2 generation from an aqueous medium containing $S^{2-}/SO_3{}^{2-}$ is appealing. H_2S is produced in large quantities as an undesirable by-product in coal- and petroleum-related industries, and similarly, sulfite is a pollutant. Thus, the combination of photo-oxidation of H_2S and/or sulfite (to sulfate), and co-generation of H_2 using CdS has a value-added benefit.

ZnS (zinc blende) is another important metal sulfide for photochemical water splitting. It shows high activity without any assistance of co-catalysts. It has a wide band gap of 3.6 eV, which restricts light absorption to the UV (< 340 nm). Studies have been carried out to sensitize ZnS. Doping and the formation of solid solutions are promising ways to improve the optical absorption of ZnS in the visible. Similar to CdS, it undergoes photochemical decomposition into the constituents when irradiated in the absence of sacrificial electron donors [25] (Eq. 5).

$$2h^+_{VB} + ZnS = Zn^{2+}_{(aq)} + S_{(s)} \tag{5}$$

3.5 *Metal Nitrides*

We have seen that doping of nitrogen into the TiO_2 lattice has a favorable effect in terms of sensitizing it to the visible light range [26]. Reasonably, the band gap of Ta_2O_5 shrinks from ~4.0 to ~2.1 eV by nitriding it in a NH_3 atmosphere to yield Ta_3N_5. The shrinking of the band gap was attributed to a conduction band derived from Ta 5d orbitals and a higher-lying valence band derived from N 2p orbitals other than O 2p orbitals of Ta_2O_5. This material evolves H_2 and O_2 under visible irradiation (< 600 nm) in the presence of a sacrificial electron acceptor such as Ag^+ and a co-catalyst such as Pt [27].

Nitrides with d^{10} electronic configuration, such as Ge_3N_4 and GaN, are a new series of photocatalysts for the cleavage of water under UV-light irradiation. β-Ge_3N_4 exhibits the highest activity for the reaction among these nitrides. It is the first example of a non-oxide powdered photocatalyst for water splitting. The band gap of Ge_3N_4 is estimated to be 3.6–3.8 eV. The electronic structure of β-Ge_3N_4 was investigated using plane wave density functional theory (DFT) calculations. The density of states (DOS) indicate that the top of the valence band consists of N 2p orbitals, whereas the bottom of the conduction band is composed of hybridized Ge 4s4p orbitals. The band structure indicates that charge transfer under photo-excitation occurs from the N 2p orbitals to the hybridized Ge 4s4p orbitals. Because hybridized Ge 4s4p orbitals broadly expand, resulting from d^{10} electronic configuration, photogenerated electrons in such broad hybridized orbitals with large dispersion can be readily transferred to the co-catalyst, like RuO_2, on the surface without recombination, carrying out efficient overall water splitting [28]. The conduction band of GaN is 0.5 eV higher than the reduction potential of water, which is sufficient to overcome the over-potential for the release of hydrogen from a solution of methanol and Na_2S/Na_2SO_3 during the action of visible light in the absence of additional co-catalysts [29]. The results for Ge_3N_4 and GaN indicate that broad hybridized sp orbitals with large dispersion in nitrides with d^{10} electronic configuration have suitable band structures for overall water splitting.

3.6 *Oxynitrides and Oxysulfides*

Oxynitride photocatalysts consisting of metal cations of Ti^{4+}, Nb^{5+} and Ta^{5+} with d^0 configuration are active for H_2 or O_2 evolution in the presence of sacrificial reagents. TaON has been found to be active for water oxidation and reduction under irradiation between 420 and 500 nm [30, 31]. Two other d^0 metal oxynitrides, $LaTiO_2N$ and $Y_2Ta_2O_5N_2$, were reported effective for evolving H_2 and O_2 from H_2O under visible-light irradiation [32, 33].

In the development of oxynitrides with d^{10} electronic configuration, a solid solution of GaN and ZnO, $(Ga_{1-x}Zn_x)(N_{1-x}O_x)$ was first tested for decomposing water under visible-light illumination [34]. The solid solution of ZnO and $ZnGeN_2$, $(Zn_{1+x}Ge)(N_2O_x)$ is another active d^{10} metal oxynitride photocatalyst for pure water splitting under visible light [35].

In addition to N, sulfur is another non-metallic element that has sensitized TiO_2 to visible light. Thus, oxysulfides have also been studied as potential photocatalysts for water splitting. $Ln_2Ti_2S_2O_5$ (Ln = Pr, Nd, Sm, Gd, Tb, Dy, Ho and Er) with layered perovskite structure was reported for photocatalytic water splitting. The band gap of this series of oxysulfides varies from 1.94 eV (Er) to 2.13 eV (Sm). $Sm_2Ti_2S_2O_5$ was found to have the highest activity among all the homologues tested [21, 36].

3.7 *Other Materials for Photolytic H_2 Production*

Group III–V semiconductors have several positive features that make them attractive for water splitting. The combination of high charge carrier mobility, an optimal band gap (particularly for many of the alloys) and reasonable photoelectrochemical stability of the *p*-type materials for cathodic H_2 production, should inspire continuing scrutiny of Group III–V semiconductors. As with their chalcogenide semiconductor counterparts, Group III–V semiconductors, in *n*-type form, undergo photoanodic corrosion instead of evolving O_2 under illumination in aqueous media. Fortunately, their *p*-type form is relatively stable against cathodic photocorrosion, and the photogenerated electrons on these *p*-type semiconductor cathodes can be used to reduce water to H_2, particularly in the presence of a co-catalyst such as Pt or Ru [37]. p-InP photocathodes are capable of evolving H_2 from HCl or $HClO_4$ electrolytes with very high efficiency [38]. Photocathodes made from p-$GaInP_2$ (a solid solution of GaP and InP) biased with a GaAs p-n junction have also evolved H_2 with high efficiency [39].

Silicon has been studied for H_2 evolution under irradiation. One fundamental problem with Si from a water-splitting perspective is its low band gap (1.1 eV), which is too small for complete water splitting (1.23 eV). One way around this problem is to seek an alternative electron donor (e.g., ethanol, formic acid, HBr or HI), instead of water. For example, illumination of a Si powder photocatalyst whose anodic and cathodic surface were coated with polypyrrole and platinized Ag, respectively, caused hydrogen evolution from aqueous ethanol for longer than 150 h with a quantum efficiency of 2.1% at 550 nm [40]. Photocatalytic H_2 evolution from aqueous formic acid was successfully achieved by using platinized *n*-type silicon powder as a photocatalyst [41]. A tandem-type hydrogenated amorphous Si (a-Si) electrode having a [n-i-p-n-i-p] structure and a similar tandem a-Si electrode having [n-i-p-n-i-p] layers deposited on *p*-type crystalline Si showed cathodic photocurrents accompanied by H_2 evolution. These two electrodes, when connected to a RuO_2 counter-electrode (for O_2 evolution), caused sustained water splitting without external bias with solar-to-chemical conversion efficiencies of 1.98 and 2.93%, respectively [42].

CdSe nanoribbons show catalytic activity for photochemical hydrogen evolution from aqueous Na_2S/Na_2SO_3 solution under irradiation with ultraviolet and visible light [43]. This is one of few instances that displays the photocatalytic characteristics of CdSe, and the first evidence of the ability of CdSe to induce the reduction of water.

AgCl photoanodes combined with either p-$GaInP_2$ [44], or a single junction amorphous silicon solar cell connected to a platinum electrode [45] as the cathodic part of an electrochemical cell, split water. AgCl does not absorb light below the indirect-band gap transition, which is in the near-UV at about 3.3 eV (380 nm). The light sensitivity in the visible part of the spectrum is due to self-sensitization caused by reduced Ag species. The formation of these Ag clusters introduces new levels within the forbidden gap.

SiC was recently found to be able to reduce water upon exposure to visible light, even in the absence of electron-donating compounds [46].

4 Nanoscience and Nanotechnology Approaches Toward Improved Photolytic Water Splitting

Nature is a great source of inspiration, and the structure and function of both enzymes and the photosynthetic apparatus provide valuable guidelines for the design of optimized (photo) catalysts for water splitting. Key features of the biological analogues include: (i) judicious nanoscale design of reactant binding sites, (ii) ingenious arrangement of several components to form an electron transfer system and (iii) ultimate control of the reaction conditions by selectively operating transmembrane channels that connect the inside of a cell to the outside environment. Researchers have tried to mimic some of these concepts in artificial catalysts with varying degrees of success. We will review some selected approaches, including methods to control charge separation and transport, attempts to steer the flow of reactants and products and means to confine the reaction volume. The potential benefits of these approaches will be discussed from the viewpoint of the issues associated with photochemical water splitting outlined in Sect. 2. Additionally, we will discuss nano approaches aiming to tune the light absorption properties.

4.1 Tuning of Electronic and Optical Properties

As outlined in Sect. 2 above, a mismatch between the solar spectrum and a semiconductor's absorption properties may severely limit the photoconversion efficiency under solar illumination. Approaches to modify the semiconductor's electronic band structure and thus optical absorption are therefore urgently needed and discussed below. Another factor limiting conversion efficiency is the fast recombination of photoexcited charge carriers. Various ways to minimize such recombination losses are therefore covered in this section.

4.1.1 Quantum Size Effects

The most interesting feature of semiconductor nanoparticles is the remarkable blue shift in their optical absorption spectra due to size reduction, compared to the corresponding bulk material. Molecular orbital (MO) and the linear combination of atomic orbitals coupled with molecular orbitals (LCAO-MO) procedures can provide information about energy level diagrams for clusters of several molecules (left part of Fig. 7) up to bulk semiconductors (right part of Fig. 7). Increasing the number of molecules in a cluster requires the addition of filled and empty orbitals to the energy manifold. This decreases energy differences between the filled orbitals and the empty orbitals. The energy gap between the highest occupied molecular orbital (HOMO) and the lowest unoccupied molecular orbital (LUMO) is also decreased. For a bulk semiconductor, the filled and empty states form separated continuums, i.e., the valence and conduction bands, respectively. In the Q-size regime, the energy levels within the filled and empty states remain discrete, and the gap between the HOMO and the LUMO states increases, compared to the bulk material, as shown in Fig. 7 [47]. This leads to a blue shift of the absorption edge in nanoparticles.

Accompanying the bandgap broadening due to reduced particle size, electrons at the lower edge of the conduction band and holes at the upper edge of the valence band then possess more negative and positive potentials, respectively. Therefore, electrons and holes have stronger reduction and oxidation powers, respectively, in such nanoparticles.

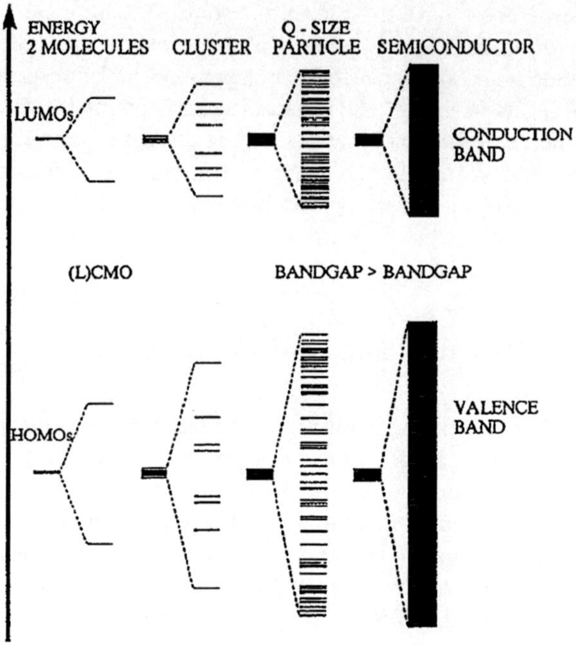

Fig. 7 Schematic diagram of the molecular orbital model for band structure. The number of energy levels increases with increasing number of atoms in the cluster. This results in a decrease in spacing within the occupied and unoccupied levels, and between the HOMO and LUMO levels. For bulk crystals the states within the occupied and unoccupied levels are so close that it is convenient to represent them as separated bands, i.e., the valence and conduction bands. The energy gap between these bands is the band gap, E_g, which increases in magnitude as the semiconductor particle radius decreases to the point where it becomes comparable to or smaller than the exciton radius. Reprinted from [47]. With kind permission from Springer Science + Business Media: Fig. 8

4.1.2 Modification by Doping

As mentioned above, semiconductors with a wide bandgap have stronger reduction and oxidation power. However, the main drawback of a large bandgap is the low utilization of solar light. For single semiconductor systems, dye sensitization has shown some encouraging results. It is, however, not easy to find a suitable and stable dye that can be illuminated over extended periods of time. Another way to sensitize semiconductors is by doping with ionic species. Although in many cases a red shift of the absorption edge into the visible region is observed in doped semiconductors, it should be noted that the photocatalytic activity may not necessarily increase under visible-light irradiation, since the measured absorption spectra of doped semiconductors result from several absorption transitions of different origins. Since the effects of doping depend on many factors, including the

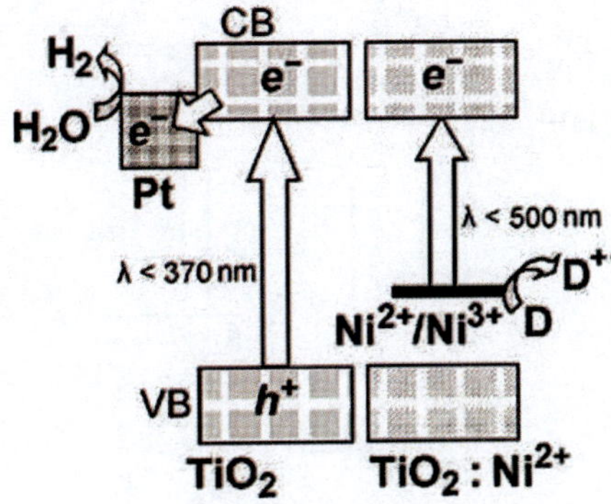

Fig. 8 Energy diagram of the operation of a photocatalytic system for the release of hydrogen from an aqueous solution containing an electron donor D for undoped titanium dioxide (LHS) and titanium dioxide doped with Ni^{2+} (TiO_2:Ni^{2+}, RHS). From [48]. With kind permission from Springer Science + Business Media, Fig. 3

doping method, the dopant species, the amount of dopant and the distribution of dopant in the semiconductor matrix, the role and behavior of dopants are still not completely clear and need to be investigated further.

The doping of wide-band gap semiconductors with transition metal ions creates local states (new energy levels) within the forbidden band. Visible-light absorption and photoactivity are induced by the interband, as shown in Fig. 8. Here, water is reduced by the photogenerated electrons, while the dopant metal ions get back electrons from a donor. This induced visible-light photoactivity is normally low, due to the small amount of visible-light absorption, as indicated by a small shoulder in the visible-light region, instead of a total red shift of the absorption edge. Increasing the dopant concentration in the semiconductor matrix can improve visible-light absorption, but excessive doping can easily disturb the original structure. Sometimes, doping can extend the lifetime of photogenerated charge carriers by trapping them transitorily and shallowly. On the other hand, dopants can also act as recombination centers for photogenerated electrons and holes and decrease photocatalytic activity.

Doping semiconductors with non-metal ions can also form intra-bandgap energy levels, as in the case of transition metal doping. For example, doping TiO_2 with F^- forms new energy states (color centers) close to the conduction band of TiO_2 (0.53 eV below the conduction band edge of TiO_2 for F center and 0.84 eV for F^+ center), resulting in an enhanced photocatalytic activity of TiO_2 under both UV and visible-light irradiation [49]. N-doping can locate N 2p states near the top of the valence band of TiO_2, which was predicted by theoretical calculations [50] and confirmed by photocatalytic experiments [51]. Interestingly, anionic doping of semiconductors with suitable N, C or S concentration can also narrow the bandgap due to mixing of the p orbitals of the dopant and oxygen, as shown in Fig. 9. The concentration should be higher than that required for the formation of discrete p

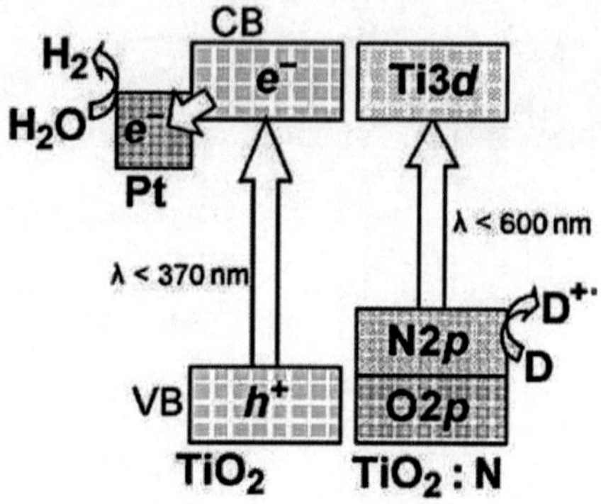

Fig. 9 Energy diagram of the operation of a photocatalytic system for the release of hydrogen from an aqueous solution containing an electron donor D for undoped titanium dioxide (LHS) and nitrogen-doped titanium dioxide (TiO_2:N, RHS). From [48]. With kind permission from Springer Science + Business Media, Fig. 4

states of dopants, which are close to the valence band. The dopant p orbitals then contribute to the top of the valence band, while the bottom of the conduction band is still formed by transition metal d orbitals without any change. Although some fruitful results have been achieved by this approach, it is difficult for a semiconductor to maintain its electronic balance and structural stability after incorporating large amounts of dopants. Heavily-doped oxides are non-stoichiometric and contain too many oxygen defects, due to different oxidation numbers of oxygen and dopants. These defects will reduce photocatalytic activity. Furthermore, excessive N (or S) doping can also change the structure of doped samples to stoichiometric oxynitrides (oxysulfides) or even nitrides (or sulfides), which, in many cases, are not very photoactive (there are also some exceptions, as mentioned in previous Sects. 3.4–3.6). The stability of dopants during long-term photocatalytic H_2 production should also be considered, since the bonds between transition metals and dopants are normally weaker than those between transition metals and oxygen, due to the smaller electronegativity of dopants as compared to oxygen.

4.1.3 Modification by Nanostructured Co-catalysts

Nanostructured co-catalysts, traditionally based on noble metals, have been widely investigated and used in photocatalytic systems for H_2 production. The function of these co-catalysts is to shuttle photogenerated electrons from the semiconductor to an acceptor (H^+ for H_2 production) in a photocatalytic process. Besides noble metals, co-catalyst chemistries also include nickel and its oxide, the mixed oxides of rhodium and chromium, tungsten carbide, MoS_2 and so on. The promoting mechanism of co-catalysts is illustrated in Fig. 10. Under irradiation by photons with enough energy (equal to or larger than the semiconductor bandgap), photogenerated electrons are promoted from the valence band to the conduction band of

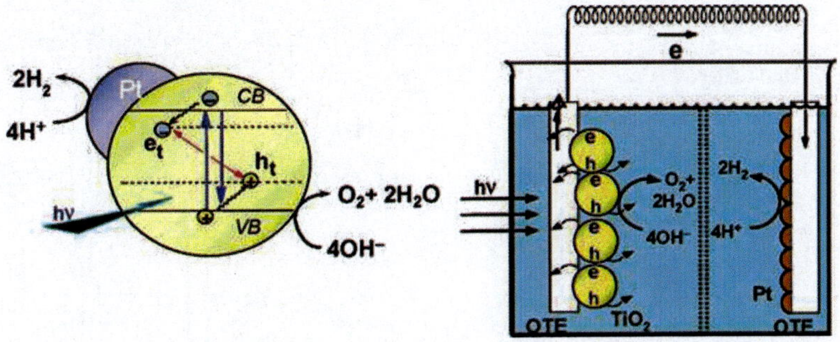

Fig. 10 Splitting of water by photocatalyst nanoparticles modified with Pt co-catalyst nanoparticles (LHS) and in a photoelectrochemical cell based on a nanostructured TiO_2 film anode and a nanostructured Pt cathode (RHS). Reprinted with permission from [52], Copyright 2007 American Chemical Society

the semiconductor, which raises the Fermi level of the semiconductor. Then, the energy difference at the semiconductor/metal interface drives electrons from the conduction band of the semiconductor to the noble-metal nanoparticle. The Fermi level of the noble-metal is thereby negatively shifted, which drives another electron transfer to an electron acceptor (H^+ in the case of H_2 production). This concept is also applicable to photoelectrochemical cells for H_2 production. Instead of a direct contact as in particulate photocatalysis, in photoelectrochemical water splitting, a noble metal counter-electrode (cathode) is connected to the semiconductor photoanode by an external circuit, which leads to spatial separation of H_2 and O_2 evolution sites and thus has the additional benefit of restraining their back-reaction to water (as shown on the RHS of Fig. 10).

One of the early studies indicated how different noble metals influence semiconductor-based photocatalytic H_2 production [53]. A direct correlation between the work function of the noble metals and the photocatalytic activity was also studied for metalized TiO_2 photocatalysis systems [54]. Besides the species of noble metal, the size of the noble-metal nanoparticle also largely affects the energetics and the electron transfer between the semiconductor and noble metal by shifting the Fermi level to more negative potential for smaller sizes. In the TiO_2-Au system, higher photocatalytic reduction efficiency was obtained with smaller Au nanoparticles [55]. Identifying and developing less expensive (noble) metals (like Ag) as co-catalysts for solar H_2 production is an important research topic.

4.1.4 Coupling to Another Nanostructured Semiconductor

Photoinduced charge separation can be realized by coupling one semiconductor with another. Matching their conduction and valence band levels can drive a vectorial transfer of photogenerated charge carriers from one semiconductor to the other, thus minimizing losses via electron–hole recombination. A case of special

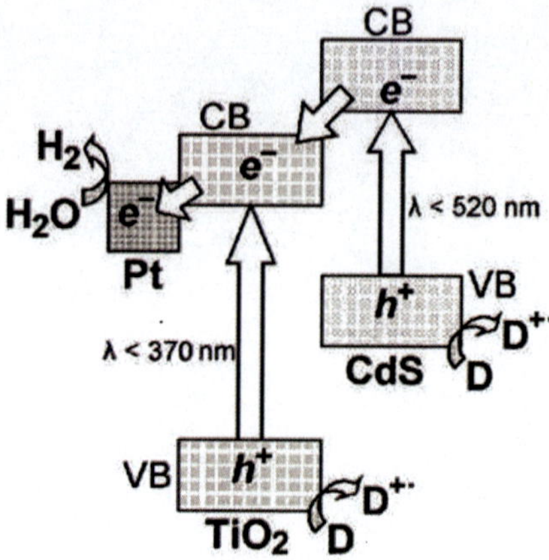

Fig. 11 Diagram illustrating spatial separation of photogenerated charge carriers in a CdS/TiO_2 heterostructure and the formation of hydrogen during the action of visible light. From [48]. With kind permission from Springer Science + Business Media, Fig. 2

interest is to select a narrow bandgap semiconductor as a coupling semiconductor in order to realize harvesting of visible light. Semiconductors, such as CdS, PbS, Bi_2S_3, CdSe and InP that can absorb visible light serve as sensitizers in semiconductor/semiconductor nanocomposites because they are able to transfer photogenerated electrons to wide-bandgap semiconductors, such as TiO_2, ZnO and SnO_2, as illustrated in Fig. 11. The band structure of two coupled semiconductors and the photogenerated charge transfer between them are determined by many factors, such as particle size, crystal structure, crystallinity, defect type and density and surface area; by tuning these factors, the flow of charge carriers can be adjusted and lead to enhanced photocatalytic and photoelectrochemical performance of such coupled-semiconductor systems. For example, the energetics can be controlled by varying the size of the semiconductor quantum dots. Increasing the band energies of these quantum dots was utilized to promote, suppress or rectify the electron transfer between two semiconductors [52].

In a traditional photocatalytic water-splitting system, semiconductor photocatalysts should possess both conduction band and valence band potentials that are suitable for H_2 and O_2 evolution, respectively, as shown in Fig. 12a. This requirement seriously limits photocatalyst selection and visible-light utilization. A Z-scheme photocatalytic water-splitting system, which involved two-step photoexcitation under visible-light irradiation, was developed by mimicking the natural photosynthesis of green plants [56]. The Z-scheme system consisted of a H_2-evolution photocatalyst (PS1 in Fig. 12b) an O_2–evolution photocatalyst (PS2 in Fig. 12b) and a reversible redox mediator (Ox/Red) that acted separately as the electron donor (for PS1) and acceptor (for PS2) for the respective half reactions.

Photocatalysts, which alone are only effective in one of the two half reactions in water splitting, are capable of running both half reactions when arranged in a Z-scheme. $SrTiO_3$, TaON, $CaTaO_2N$ and $BaTaO_2N$ can work as H_2 evolution photocatalysts, while WO_3, $BiVO_4$ and Bi_2MoO_6 can act as O_2 evolution photocatalysts. The IO_3^-/I^- and Fe^{3+}/Fe^{2+} redox couples normally serve as reversible electron mediators. An all-solid-state Z-scheme based on a CdS–Au–TiO_2

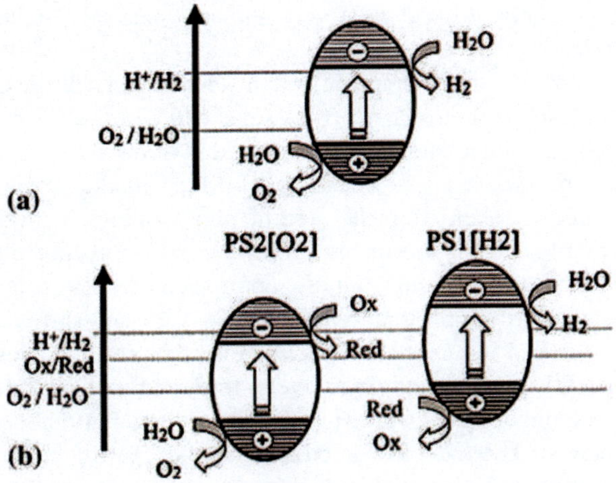

Fig. 12 Reaction mechanism of photocatalytic water splitting into H_2 and O_2 for: (**a**) a conventional one-step photo-excitation system, and (**b**) a system mimicking the Z-scheme of photosynthesis (two-step photo-excitation system). Reprinted from [56], Copyright 2002, with permission from Elsevier (http://www.sciencedirect.com/science/journal/10106030)

three-component nanojunction was recently reported, where PS1(CdS), PS2(TiO_2) and the electron-transfer system (Au) were spatially fixed [57]. The vectorial electron transfer of $TiO_2 \rightarrow Au \rightarrow CdS$ occurs as a result of excitation of both TiO_2 and CdS. The electron supply from TiO_2 to CdS via Au restricts the self-decomposition of CdS due to the oxidation of surface S^{2-} ions by the photo-generated holes in CdS. In a Z-scheme system, H_2 and O_2 are evolved separately by two different photocatalysts, which, to some extent, restrains their back-reaction to water. The key factors in the design of a Z-scheme system are to find a pair of photocatalysts for separate H_2 and O_2 production with high efficiency, and a reversible electron mediator, the redox potential of which can meet the requirements of being electron donor and acceptor in the respective half reactions. From what we have mentioned above, the energy levels of the components and charge transfer between them should be considered when designing both semiconductor–semiconductor nanocomposites and Z-scheme systems.

4.2 Micro- and Nanostructures for Light Management

An important issue in photovoltaic and photolytic energy conversion is a large mismatch between optical and electronic length scales for photon energies close to the semiconductor bandgap; while the absorption depth of light is on the order of (hundreds of) micrometers, the electronic diffusion length is at least an order of magnitude smaller. The optimal device thickness is thus a compromise between absorbing most of the incoming photons and reducing volume recombination of excited charge carriers. Ideally, one would like to construct optically thick but physically thin absorbers. This concept provides several advantages, including resource-efficient use of (in certain cases expensive and/or scarce) materials, the opportunity to use relatively impure materials and/or the potential to achieve higher device efficiency.

While various approaches to address this issue have been presented in the context of photovoltaic (PV) solar cells, our opinion is that their usefulness for photochemical energy conversion, and in particular photochemical water splitting, has not yet been exploited fully. Much of the work reviewed in this section is therefore taken from the area of photovoltaics, with a strong belief that the presented schemes are of high relevance also for photochemical water splitting.

Traditional light-trapping approaches in PV solar cells include the use of wavelength-scale textured substrates [58] and diffractive optical structures [59], which can increase the optically effective cell thickness by a factor of 4–5. More recently, light-scattering layers [60] and optical microcavities [61], including photonic crystals [62, 63] and whispering gallery modes, have attracted significant interest. The latter is named after the whispering gallery at St. Paul's Cathedral in London and occurs at particular resonant wavelengths of light for a given cavity size and shape when the light undergoes total internal reflection at the inner surface and becomes trapped within the void (or guide) for timescales of the order of nanoseconds [64]. Obviously, the increased photon path length increases the total absorption. An illustrative example of this effect is the observation of photolysis of water-soluble components inside cloud droplets by ultraviolet/visible radiation [65]. Most photonic dielectric cavities have traditionally been limited to sizes that are in the order of the wavelength of light. The lower bound on the effective mode volume (V_{eff}) arises from a mechanism of confinement based on interference effects and is therefore wavelength dependent. However, it has been shown [66] that by introducing dielectric discontinuities with sub-wavelength dimensions as a means of local field enhancement, the effective mode volume becomes wavelength independent. In this way cavities with large V_{eff} can be achieved, with a corresponding increase in the Purcell factor (a measure of the spontaneous emission rate enhancement for an emitter in a resonant cavity) of nearly two orders of magnitude relative to previously demonstrated high index photonic crystal cavities.

The most recent development in the area of light management for solar energy applications involves the utilization of plasmonically-active nanostructures. It is well-known that nanoparticles and perforated films of certain materials (e.g., silver, gold, platinum) show strongly enhanced optical absorption (i.e., increased cross-section for initial photon capture) due to the excitation of localized surface plasmon modes [67, 68]. Localized surface plasmon resonances (LSPRs) are collective oscillations of the conduction electrons, which may result in optical absorption cross-sections that exceed the geometric cross-section by several orders of magnitude. These plasmons are accompanied by a strongly enhanced electromagnetic near field. They may decay either radiatively (i.e., by re-emitting a photon), or non-radiatively into (quasi) particles such as electron–hole (e–h) pairs. For many metals, the resonance wavelength falls into the near-ultraviolet, visible or near-infrared regime for nanostructure sizes covering the range 20–200 nm. What is typical for the mentioned spectral range is that it covers (most of) the energetics of important chemical transformations, e.g., bond breaking and bond formation (0.5–6.5 eV). This should allow one to create conditions and propose schemes for enhanced (solar) light absorption in nanostructured materials and utilization of the deposited energy to run chemical transformations such as water splitting.

There are various pathways along which plasmonics may enhance solar-cell devices, as summarized in a recent review by Atwater and Polman [69]. In short, plasmonic structures can function as: (i) scattering structures that couple freely propagating waves from the sun into (waveguiding modes in) a thin semiconductor absorber layer (see [70, 71]), (ii) light-trapping structures, which effectively bend the incoming solar flux by 90° via the excitation of propagating surface plasmon polaritons (SPPs) [72, 73] and (iii) nanoantennas that increase the absorption in the adjacent semiconductor via their strongly enhanced plasmonic near-field [74, 75]. In the first two cases, the effective path length of the incident light in the semiconductor is increased, whereas in the final case, the incident light is concentrated into nanoscale volumes around the plasmonic nanoparticles. Experimental and theoretical work indicate that the active layer thickness of various types of solar cells, including organic solar cells, dye-sensitized solar cells and inorganic solar cells, may be decreased by 10–100-fold using such schemes with no significant impact on efficiency. While the three radiative pathways described above have shown great potential for PV devices, a potential fourth pathway, namely (iv) charge carrier generation via non-radiative plasmon decay has not been investigated to the best of our knowledge, and its usefulness for photovoltaic power generation thus remains to be shown.

Finally, we note that there are a number of recent design suggestions based on nanostructures that do not involve plasmonic effects. To mention a few, Kelzenberg and colleagues [76] demonstrated the use of microrod arrays in between which light is scattered and Zhu et al. [77] recently introduced nanodome solar cells.

In the case of plasmon-enhanced photochemistry, an area pioneered by Nitzan and Brus in the early eighties [78], both radiative and non-radiative enhancement pathways have been pursued, as illustrated in an extensive review article by Watanabe et al. [79]. To give two representative examples, the photocatalytic degradation of methylene blue was found to be significantly increased on nanocomposite photocatalysts consisting of silver nanoparticles embedded in titania. Radiative energy transfer from the Ag nanoparticles to the semiconductor was claimed to be the origin of the increased efficiency [80, 81]. Tian and Tasuma, on the other hand, reported that ethanol and methanol were photocatalytically oxidized by gold nanoparticle–nanoporous TiO_2 composites (at the expense of oxygen reduction) under visible-light illumination [82]. In this case, photoexcitation of the gold nanoparticles was followed by simultaneous charge transfer of electrons from the gold to the TiO_2 conduction band and from a donor in the solution to the gold nanoparticle, as illustrated in Fig. 13.

While the traditional light-trapping schemes presented at the beginning of this section seem to be primarily interesting for photoelectrochemical water splitting, the discussed plasmonic amplification schemes are applicable to both PEC and

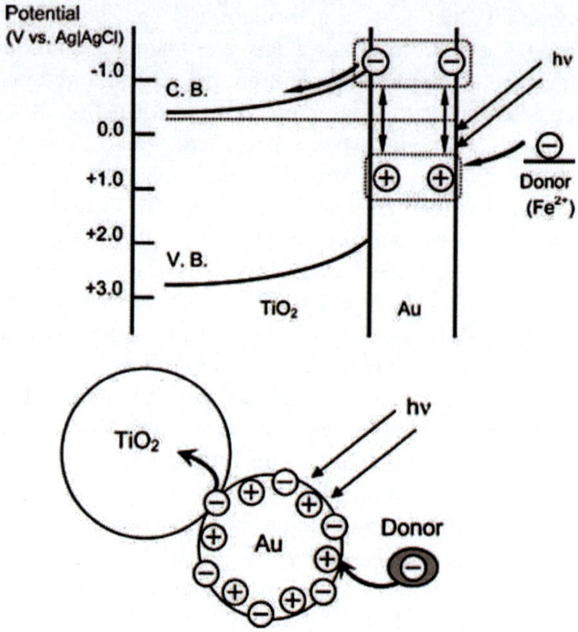

Fig. 13 Schematic illustration of charge separation at an illuminated gold nanoparticle–titania interface, enabling visible-light induced photocatalytic oxidation of alcohols. Reprinted with permission from [82]. Copyright 2005 American Chemical Society

photocatalytic water splitting. Despite a large potential, we are not aware of any actual realizations of plasmon-enhanced water splitting.

4.3 Control of Photocatalyst Structure and Morphology

Structure and morphology are important and hot issues in photocatalysis, since both affect the performance of a photocatalytic material. Small size and large surface area normally correspond to high photocatalytic activity. Low-dimensional nanostructures with high aspect ratio and porous nanostructures have attracted intense attention, since they favorably affect charge and mass transport, respectively.

4.3.1 Small Size and Big Surface Area: Catalytic and Photocatalytic Considerations

Heterogeneous catalysis is dependent on processes taking place at surfaces and interfaces. Since catalytic materials are often expensive, the goal has always been to fabricate catalysts with the highest possible surface-to-volume (S/V) ratio. In real-life catalysts, this is usually achieved by depositing small (< 5 nm), catalytically active nanoparticles onto a highly porous support material with very high surface area [83]. In parallel, it is known that nanostructures may possess catalytic power not found in their bulk counterparts. As an example, it was shown that tiny dots of gold on certain metal oxide supports exhibit extraordinary catalytic properties, despite the fact that bulk gold is chemically inert [84].

Similar to "normal" (dark) heterogeneous catalysis, the size of photocatalyst particles determines the surface area available for the adsorption and decomposition of reactants (e.g., water or H_2S in H_2 production), and for the harvesting of light. Besides providing more active sites due to large surface area compared to their bulk counterparts, nanosized photocatalysts can restrain bulk electron–hole recombination via charge-carrier trapping on their way to the surface. Photogenerated electrons and holes need to diffuse to the photocatalyst surface to react with electron and hole acceptors (reactants). If the dimensions of nanosized photocatalysts are small enough so that the transfer of photogenerated electrons and holes to the surface is faster than the recombination process, the photocatalytic efficiency will be largely enhanced. It should be noted, however, that the photocatalytic activity does not always increase with decreasing photocatalyst size. There are several potential disadvantages if the size becomes too small. First, too small a size cannot guarantee the best crystal structure for photocatalysis. Normally, very small particles (or quantum dots) are more prone to have amorphous structure. Secondly, very small particles easily aggregate. The size and morphology of these aggregates, or secondary particles, can affect the light-scattering properties of the photocatalyst, as well as the degree of photon penetration. The slow transport of reactants and products within the aggregates can also decrease the photocatalytic efficiency [85]. Finally, particles that are too small may result in increased surface electron–hole recombination, which offsets the benefits of high surface area. It is believed that there are respective optimal sizes at the nanoscale for different photocatalytic materials and reactions.

4.3.2 1-D Nanostructures: Nanowires, Nanorods, Nanotubes and Nanofibers

One-dimensional (1-D) nanostructured materials (nanowires, nanorods, nanotubes and nanofibers) have attracted more and more attention due to their specific properties, which differ from the bulk and other nanostructure (nanoparticle) counterparts. Compared to a nanoparticle with the same volume or weight, a 1-D nanostructure can provide higher surface area and faster interfacial charge transfer rate. As shown in Fig. 14, photoelectrodes composed of nanorods oriented perpendicular to the conductive substrate (see Fig. 14b) can shorten the transport distance for electrons to the back contact (the "electron expressway" concept [86]) and avoid recombination losses at grain boundaries between nanoparticles (see Fig. 14a) [87].

Furthermore, poor hole transport has been known to be one of the main factors that limit the conversion efficiency of Fe_2O_3 photo anodes. An elegant solution for this is to use high aspect-ratio nanowire electrodes. It was reported that the

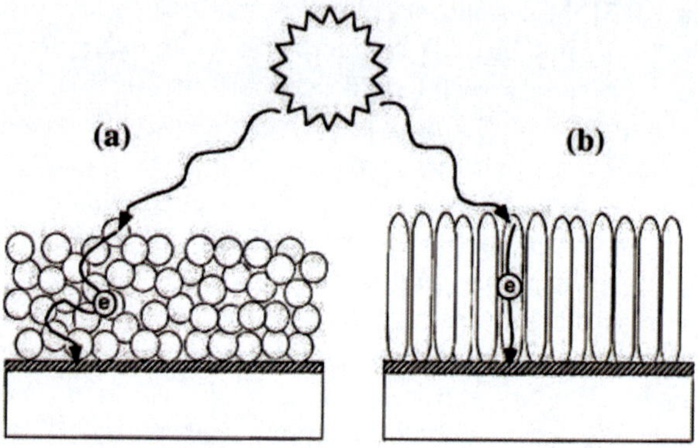

Fig. 14 Schematic representation of the electron transport through (**a**) spherical particles and (**b**) nanorods with preferential orientation. From [87]. Reproduced by permission of The Electrochemical Society

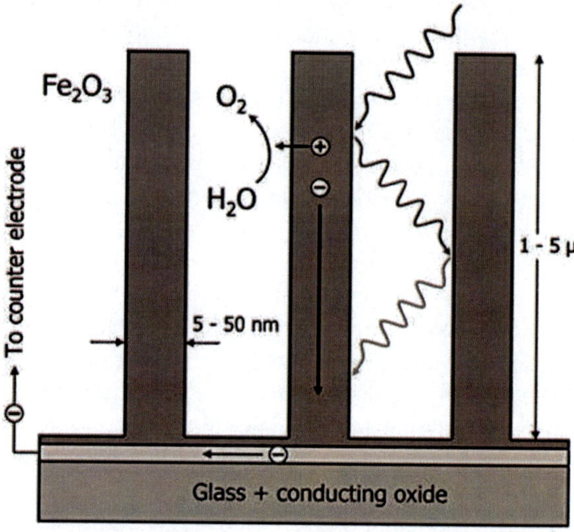

Fig. 15 Optimized morphology for an α-Fe_2O_3 photo-anode for water splitting. The small diameter of the nanowires ensures short hole diffusion path lengths. From [88]. Reproduced by permission of The Royal Society of Chemistry, http://dx.doi.org/10.1039/B718969A

transport distance for photogenerated holes to the Fe_2O_3–electrolyte interface was reduced in Fe_2O_3 nanowires with a small diameter, as illustrated in Fig. 15 [88]. One can imagine that the limitation with regard to hole transport can be overcome when the radius of the Fe_2O_3 nanowires is smaller than the hole diffusion length.

It should be noted that even with the same 1-D morphology of the basic building blocks, disorder, order and different orientations of the 1-D units affect the photoelectrochemical properties. As shown in Fig. 16, the photon-to-current efficiency is lower for the Fe_2O_3 electrode with nanorods parallel to the substrate. The IPCE at 360 nm is 3% compared to 5% with nanorods oriented perpendicularly. The electrons have a more straightforward pathway to the back contact with nanorods perpendicular to the substrate, which leads to a decrease in recombination losses (see Fig. 16b) and higher IPCE values [87].

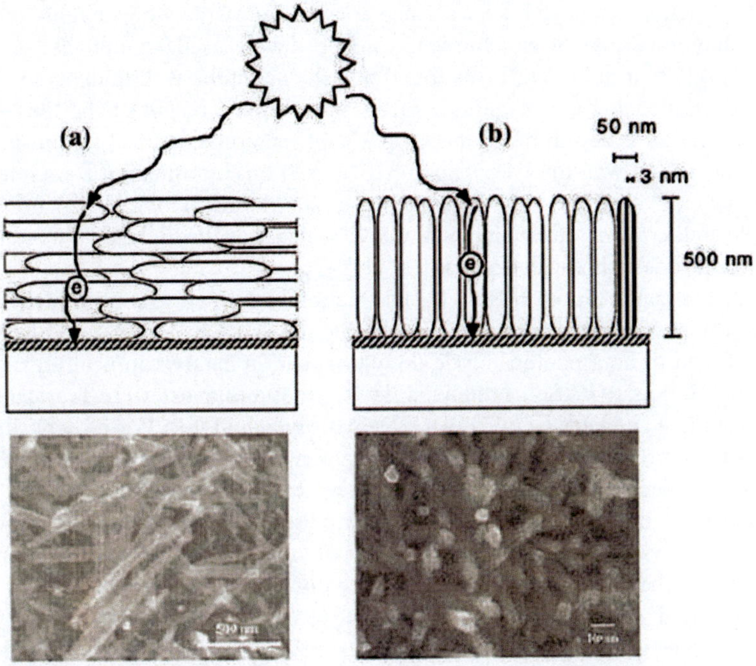

Fig. 16 Schematic illustration (*cross-section*) and scanning electron micrograph (*top view*) of hematite nanorods with (**a**) parallel and (**b**) perpendicular orientation to the conducting substrate. From [87]. Reproduced by permission of The Electrochemical Society

We additionally note that the basic concepts described above also hold for more complex nanostructures, such as the Fe_2O_3 "cauliflower" structure reported by Kay et al. [89].

Although the initial idea of using 1-D nanostructured (nanorod) electrodes for PEC water splitting was first reported for Fe_2O_3 [90], it has been widely investigated using other semiconductor electrodes, including novel materials such as VO_2 [91] and Ta_3N_5 [92], and, especially, TiO_2 (see [93] for a recent review). Thin film and nanowire electrodes of *n*-type titanium oxide (n-TiO_2) were fabricated and their photoresponse toward water-splitting was measured; a more than twofold increase in maximum photoconversion efficiency was observed when a single-layer thin film of n-TiO_2 was replaced by nanowires [94]. Photoelectrochemical water splitting using dense and aligned TiO_2 nanorod arrays with well-defined length was also reported; overall water splitting was observed with an applied overpotential of 1.0 V (versus Ag/AgCl) with a photon-to-hydrogen efficiency of 0.1%. The results suggest that these dense and aligned one-dimensional TiO_2 nanostructures are promising for hydrogen generation from water by PEC cells [95]. The concept of using 1-D nanostructures for H_2 production via PEC water splitting has also been extended to nanotubes. Compared to nanorods and nanowires, nanotubes have higher surface area for redox reactions. Highly ordered titania nanotube arrays of variable wall thickness were used to photocleave water under ultraviolet irradiation. It was found that the nanotube wall thickness is a key parameter influencing the magnitude of the photoanodic response and the overall efficiency of the water-splitting reaction. Using nanotubes with a 22 nm porediameter and 34 nm wall thickness, upon 320 – 400 nm illumination at an intensity of 100 mW/cm^2, hydrogen was generated at a rate of 960 μmol/h W (24 mL/h W) and an overall conversion efficiency of 6.8%, which was the highest value reported for a titania-based photoelectrochemical cell [96].

Besides the application in PEC H_2 production, 1-D nanostructures also play a very important role in photocatalytic H_2 production. TiO_2 nanotubes modified with Pt were found to be a photocatalytic dehydrogenation catalyst in neat ethanol for producing H_2 gas [97]. Self-organized TiO_2 nanotube-layers were fabricated by electrochemical anodization of Ti in a HF electrolyte, and then Pt was deposited on the TiO_2 nanotube layer by plasma sputtering. The Pt-TiO_2 nanotube photocatalyst generated H_2 successfully from an alkaline water solution [98]. Pt-ionized TiO_2 nanotubes were prepared for the stoichiometric production of H_2 and O_2 by water-splitting under visible light with hydrogen evolution rates of 14.6 and 2.3 μmol/h in aqueous methanol and pure water, respectively [99]. Nanostructured TiO_2 films with controlled morphology and thickness were synthesized for use in water splitting photocells and dye-sensitized solar cells. Two different morphologies were compared: a granular morphology and a highly crystalline columnar morphology. The columnar morphology outperformed the granular morphology for both applications, achieving a UV-light to hydrogen conversion efficiency of 11% for water splitting, and a visible light to electricity conversion efficiency of 6.0% for the dye-sensitized solar cell [100]. TiO_2 nanowires (TiO_2 NWs) were synthesized through a one-step hydrothermal process followed by post heat treatment. In this study, anatase TiO_2 NWs exhibited the highest photocatalytic H_2 evolution, which was also higher than that of the starting TiO_2 powder (Degussa P25) [101]. However, Lin et al. suggested that a bi-crystalline structure consisting of TiO_2 (B) nanotubes (or nanofibers) and anatase nanoparticles could act as an active, H_2-producing photocatalyst [102, 103]. A solvothermal method was applied to synthesize CdS nanorods [104] and nanowires [105], which have high photocatalytic activity for H_2 production.

We finally note that the concepts of 1-D nanostructured materials and semiconductor–semiconductor nano-composites, as discussed in Sect. 4.1.4, can be combined and lead to synergistic effects [106]. The same holds for the deposition of co-catalyst nanoparticles (Sect. 4.1.3) onto 1-D nanostructured materials, which leads to improved charge separation properties [107, 108]. Qu et al. demonstrated the fabrication of rationally designed 1-D nanostructures, which integrate several concepts in a single nanosystem. The combination of a charge-separating nanodiode structure, which was encased in a protective insulating shell and functionalized with two exposed metal co-catalysts, was both highly efficient and stable throughout the entire solar spectrum [109].

4.3.3 2-D Nanostructures: Nanosheets, Nanoscrolls and Layered Materials

From the microstructure point of view, nanosheets, nanoscrolls and layered materials belong to two-dimensionally (2-D) structured materials. For photocatalytic H_2 production, 2-D structured materials can maintain a high surface area, selectively expose certain photocatalyst facets with high photocatalytic activity, provide fast charge transfer and stack multicomponent structures by layer assembly.

Photocatalytic H_2 production over Pt/TiO_2 nanosheets with exposed (001) facets was reported recently [110]. The authors concluded that the exposed (001) facets contributed largely to the high photocatalytic activity in H_2 production, which was also proven by Lu et al. [111, 112]. In a dye-sensitized semiconductor photosystem, the semiconductor particle mediates electron transfer between the dye and a particle catalyst for hydrogen evolution (typically Pt or Rh). For the reaction to be efficient, the rate of electron transfer to this catalyst must be faster than that of back electron transfer to the oxidized dye or electron donor. Oxide nanosheets make particularly good electron transfer mediators. An external quantum yield of 20–25% was observed for hydrogen evolution from $EDTA^{2-}$ solutions with niobate nanoscrolls and nanosheets catalyzed by Pt and sensitized by a phosphonated $[Ru(bpy)_3]^{2+}$ derivative (Fig. 17). Taking into account light scattering, the low extinction coefficient of the dye, and the charge injection efficiency, this system has $> 50\%$ internal quantum yield for hydrogen evolution from photoinjected electrons. This high internal quantum yield shows that the single-crystal oxide nanoscrolls and nanosheets are good mediators of electron transfer between the dye molecules and Pt particles. Since the electron transfer from $EDTA^{2-}$ to Ru^{3+} was the slowest (limiting) step in the overall reaction, these results demonstrate that the nanosheet morphology was largely responsible for the enhanced overall photon conversion efficiency [113].

Layered structures have been widely used in photocatalytic H_2 production. $K_4Nb_6O_{17}$ has a layered structure with two kinds of interlayers. H_2 is evolved from one interlayer, in which co-catalysts are introduced by ion-exchange or interlayer reaction, while O_2 is produced in the other interlayer, as shown in Fig. 18. In this way, the sites for H_2 and O_2 evolution are separated by the photocatalytic niobate sheet [5]. Furthermore, an electric field gradient resulting from the uneven K^+ distribution on opposite sides of the niobate sheets assists electron–hole separation.

4.3.4 Porous Structures

Porous materials, behaving as spatially confined micro- and nanoreactors (see Sect. 4.4.2 below for further details), have attracted more and more attention in photocatalysis. They have high surface areas, and they can concentrate reactants with low concentrations, control the reaction environment and provide host–guest effects. Selective photocatalysis can be realized by adjusting the pore size. Co-catalysts can be easily deposited and dispersed onto these porous

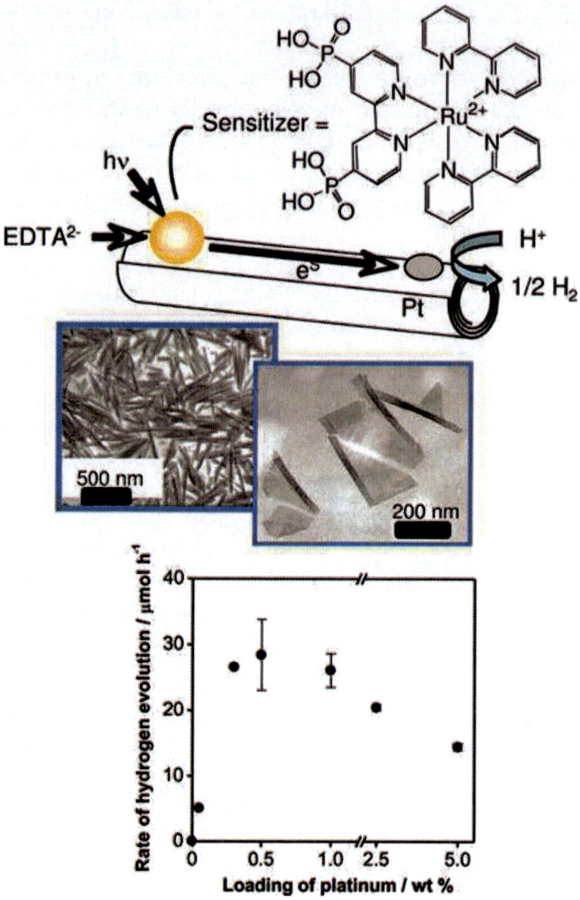

Fig. 17 (*Top*) Schematic representation of photo-induced electron transfer from a phosphonated $[Ru(bpy)_3]^{2+}$ sensitizer to Pt catalyst particles, mediated by $H_4Nb_6O_{17}$ nanoscrolls. (*Middle*) TEM images of individual nanosheets and of nanoscrolls precipitated from a suspension of exfoliated $H_4Nb_6O_{17}$. (*Bottom*) The dependence of hydrogen evolution rate on Pt loading and the observed high quantum yield establish that electron transfer from $EDTA^{2-}$ to Ru^{3+} is the slowest step in the overall process. Reprinted with permission from [113]. Copyright 2009 American Chemical Society

photocatalysts. Peng et al. reported that hydrothermally synthesized TiO_2 nanoparticles without calcination had a large specific surface area (438 m^2/g) with small crystallites (2.3 nm) dispersed among amorphous mesoporous domains, and exhibited much better photocatalytic activity for H_2 production compared to samples calcined at different temperatures and also the commercial photocatalyst P25 [114]. A novel synthesis was carried out using KCl electrolyte to control the electrostatic repulsive force between TiO_2 nanoparticles toward the formation of a mesoporous structure, which owned the highest photocatalytic activity for H_2 production, compared to nonporous colloidal-TiO_2, and commercial Degussa P25 and Hombikat UV-100 (HBK) samples [115]. The photocatalytic reduction of metal cations (M = Ni^{2+}, Co^{2+}, Cu^{2+}, Cd^{2+}, Zn^{2+}, Fe^{2+}, Ag^+, Pb^{2+}) on the surface of mesoporous TiO_2 (specific surface area 130–140 m^2/g, pore diameter 5–9 nm, and anatase content 70–90%) resulted in the formation of nanostructured metal–semiconductor composites (TiO_2/M). These metal–TiO_2 nanostructures showed a remarkable photocatalytic activity for hydrogen production from water–alcohol

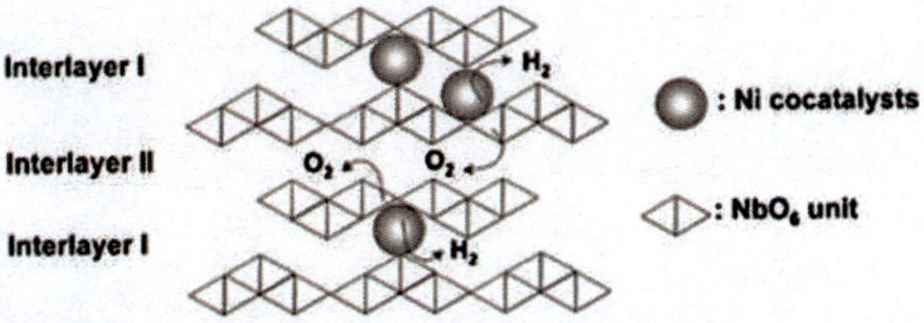

Fig. 18 Water splitting over $K_4Nb_6O_{17}$ photocatalyst with layered structure. From [5]. Reproduced with permission from The Royal Society of Chemistry (http://dx.doi.org/10.1039/b800489g)

mixtures, and the efficiency was 50–60% greater than that of the metal-containing nanocomposites based on Degussa P25. The anatase content and pore size proved to be the basic parameters determining the photoreaction rate [116].

Porous materials were also used as active supports for catalysts and photocatalysts. Tetrahedrally coordinated metal oxide (titanium, vanadium, chromium, and molybdenum oxides) moieties can be implanted and isolated in the silica matrices of microporous zeolite and mesoporous silica materials, referred to as "single-site photocatalysts". The single-site titanium oxide photocatalyst demonstrates a high reactivity and selectivity under UV-light irradiation, while the single-site chromium oxide operates as a visible-light sensitive photocatalyst [117]. The photocatalytic H_2 production by nanosized CdS was enhanced by immobilization of CdS on porous supports, such as aluminum-substituted mesoporous silica molecular sieve (Al-HMS) [118], microporous and mesoporous silicas [119], porous polyethylene terephthalate fibers (PET) [120], and ETS-4 zeolite [121].

4.4 Spatial Control of the Reaction Environment

Judicious control of the reaction environment, which includes tailoring the transport of reactants and products to and away from the reaction sites, respectively, and confining the reaction to nanoscale volumes, is an interesting option to fine-tune the kinetics of photolytic water splitting. It allows one to address issues related to the fast back-reaction of hydrogen and oxygen, photocorrosion and reaction selectivity.

4.4.1 Tailored Transport of Reactants and Products

Experience shows that noble metal co-catalysts, which are oftentimes used to assist water reduction, at the same time tend to act as good catalysts for water formation, i.e., the unwanted back-reaction of hydrogen and oxygen to water. While this is not an issue in photoelectrochemical water splitting due to hydrogen

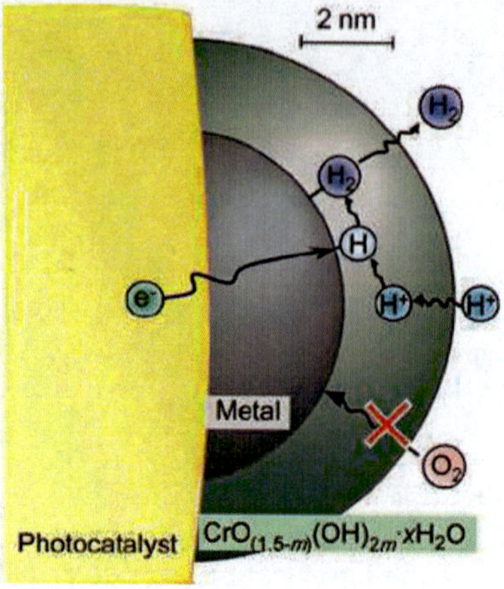

Fig. 19 Schematic illustrating the function of noble-metal/Cr_2O_3 core/shell nanoparticles as a co-catalyst for photocatalytic water splitting. Selective permeability of the chromia layer enables proton reduction and hydrogen evolution, while the back-reaction of hydrogen and oxygen to water is blocked. Reprinted with permission from [123]. Copyright 2009 American Chemical Society

and oxygen being evolved in separate compartments, it constitutes a major limiting factor for photocatalytic water splitting, where the two species are evolved in close proximity to one another. To alleviate this problem, Domen and co-workers pioneered a concept, where the transport of reactants to the co-catalyst sites is controlled judiciously by coating the noble metal co-catalyst nanoparticles with a very thin chromium oxide shell using a photodeposition method [122]. An oxynitride photocatalyst loaded with such core/shell co-catalyst nanoparticles showed a much higher water-splitting activity than the same photocatalyst with unmodified noble-metal co-catalyst nanoparticles. Recently, extensive work by the same group [123] demonstrated that the beneficial effect of the chromia shell is related to its relative permeability for protons, hydrogen and oxygen; while the Cr_2O_3 layer is permeable to protons and evolved hydrogen molecules and does not interfere with proton reduction and hydrogen evolution, it prevents oxygen from accessing the underlying noble-metal core and therefore blocks the back reaction (Fig. 19).

This modification method has also been shown to work in combination with an oxygen evolution co-catalyst on the same photocatalyst; the addition of a Mn_3O_4 oxygen evolution co-catalyst to a Rh/Cr_2O_3 core/shell loaded GaN:ZnO photocatalyst resulted roughly in a doubling of hydrogen and oxygen evolution rates [124].

Finally, we note that the approach of blocking the back reaction via the construction of core–shell nanoarchitectures is versatile and has been applied to different semiconductors and using other shell chemistries, such as SiO_2 [125].

Kale et al. [126] reported a strategy to protect CdS quantum dots from photocorrosion by carefully controlling the interaction between reactant and photocatalyst; by partially embedding the otherwise un-photostable CdS QDs in a glass matrix, photocatalytic decomposition of H_2S to hydrogen was possible with high activity over extended time periods, thus demonstrating an increase in photostability without compromising the catalytic function.

4.4.2 Nanoreactors to Confine the Reaction Volume

Traditionally, catalytic systems are described in terms of interactions between the reacting species and the (open) catalytic surface. Architectures in which the interaction between the catalyst and the reactant(s) is confined to nano-liter volumes, i.e., nanoreactors, have recently attracted a great deal of attention. Various methods of confining the reaction volume have been identified and include diverse porous inorganic structures (mesoporous materials, zeolites, etc.), approaches based on molecular self-assembly (oil-in-water emulsions, liquid foams, polyelectrolyte capsules [127], co-polymer vesicles [128], lipid vesicles [129] and lipid nanotube-vesicle networks [130], etc.,) and lithographically-defined nanocavities [131].

While the most evident way in which confinement can affect a reactive system is to alter the thermodynamic properties, there are a number of other effects that can influence chemical reactions in nanospace, such as geometrical constraints in pores with a size comparable to the molecular sizes, selective adsorption of reacting molecules, and changes to the potential energy surface.

A nanoreactor may stabilize a transition state of a chemical reaction and thereby change the activation parameters of the reaction. Both enthalpic stabilization, i.e., non-covalent interactions between the transition state and the surrounding, and entropic stabilization must be considered. The latter contribution may be appreciable since the translational and rotational degrees of freedom of reactants are reduced inside a confined nanospace, leading to a pre-organization of reactants toward the transition state. Even in cases where the nanoreactor does not change the activation parameters, the kinetics of the reaction may be modified due to a locally increased concentration of reactants. This mechanism is important for low (bulk) reactant concentrations.

Stabilization of reaction intermediates, or reaction products for that matter, has the added benefit of providing new pieces of information regarding the reaction mechanism, which would not have been accessible otherwise owing to the reactive and/or labile nature of the intermediates and products; encapsulation traps intermediates and/or final products for spectroscopic analysis. Such confinement also means that several successive reactive events can be induced in a very small confined space of molecular dimensions. For example, a first photon can generate an electron that causes one bond to break and thereby creates some reaction intermediates that are trapped at the interface. A second photon can then later generate another electron that induces a second type of reactive event where the intermediates from the first event participate, and so on. This creates significant probabilities for reaction paths and products that would be vanishingly small if the reactants and intermediates were not confined. We would like to point out that this reaction mechanism, as presented, is a batch process. In order to continue, it is necessary to take the products out and to regenerate the active sites to make this a true catalytic process. There are several possible ways to achieve this, such as photo-excitation, or further reaction with other molecules. If such a process could

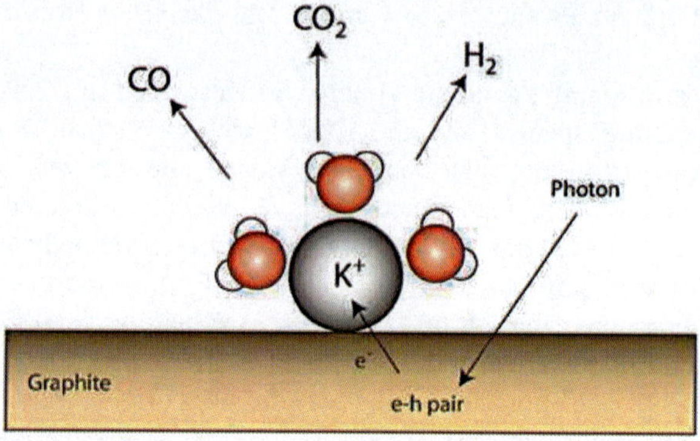

Fig. 20 Schematic illustrating the mechanism of photodissociation of water on potassium-functionalized graphite. Reprinted from [144]

be found, the system would thereafter work as a molecular energy conversion cycle: a true nanoscale chemical reactor.

The combined result of these effects is that confinement may increase the activity and/or selectivity of a photocatalytic reaction. Additionally, the large surface-area-to-volume ratio, associated with small dimensions and small volumes, short material transport and high attempt frequency for the reaction in nanoreactors make it possible to conduct unique processes not possible for ordinary heterogeneous schemes [132–135].

While the nanoreactor concept has been applied to a diversity of chemical (photo) processes [136–139], it has not yet found widespread attention in the photocatalytic water-splitting community. Early experimental efforts report successful incorporation of photocatalyst nanoparticles into the cavities of lipid and surfactant vesicles and the ability of these to generate hydrogen under visible-light illumination. The focus of these studies was, however, mainly on electron relays for charge separation across the membrane rather than confinement effects [140–143]. Some more recent and encouraging results for photocatalytic conversion in porous media are presented in Sect. 4.3.4 above.

A recent example from our own research [144] further illustrates the applicability of the nanoreactor concept to photocatalytic water splitting. Our experimental system consists of a graphite substrate onto which water and potassium are co-adsorbed. Upon photoexcitation of the graphite substrate, energetic charge carriers are generated and drive photo-dissociation reactions of water molecules at the graphite/K^+/water interface, leading to the formation of H_2, CO and CO_2 (Fig. 20). Covering the graphite substrate with ice layers of varying thickness and with different morphologies drastically changes the product composition due to (partial) confinement effects imposed by the ice layer, the product permeability of which depends on thickness and morphology.

Although the limited body of results generated to date is perhaps mainly of academic interest, we believe that further research efforts will demonstrate the usefulness of the concept for large-scale industrial schemes.

4.5 *Fabrication of Supported Model Systems*

Enormous progress in the area of nanoscience and nanotechnology has resulted in a plethora of nanofabrication and characterization methods becoming available today. While the nano-architectures in most of the examples given in the previous sections were realized via bottom-up (often wet-chemical) approaches, the role of lithographic, top-down methods will be discussed in this section. Lithographic nanofabrication schemes such as electron beam lithography, nanosphere lithography [145] and (hole-mask) colloidal lithography [146] allow one to manufacture photocatalysts and photoelectrodes with an extremely high degree of control over structural parameters such as the size, shape and mutual arrangement of nanostructures. Given the high cost, limited speed and restrictions regarding the maximum area that can be patterned typically associated with lithographic nanofabrication approaches, we note that these methods are perhaps not expected to be of great importance for large-scale applications. They are, however, an ideal playground to fabricate supported model systems, the nano-architecture of which can be controlled and tuned accurately and over a large parameter space. By systematically varying certain key parameters, such model systems allow one to construct structure–activity maps, providing a detailed understanding of the physical and chemical processes underlying water splitting, and predictions for improved photocatalyst architectures may also be derived. An additional attractive feature of lithographically fabricated, well-defined model systems is that they are easily modeled, which is often not so for wet-chemically synthesized systems.

The concept of supported model systems has entered several fields related to energy and environment, including catalysis [147], solar cells [75], hydrogen storage materials [148] and fuel cells [149–151]. It has, to our surprise, not seen a breakthrough in the photolytic water-splitting area however. Considering the enormous, demonstrated power of this approach, we believe that it is only a matter of time until suitable model systems will be utilized to study the intriguing details of photolytic water splitting.

5 Conclusions and Outlook

There is no doubt that the world energy consumption will continue to increase rapidly, and man-made global warming is a fact, which is closely associated with our current, fossil fuel-based energy system. In order to avoid serious political unrest, energy shortages and catastrophic climate change, mankind must tap into renewable, CO_2-neutral energy sources as soon as possible. Hydrogen might be part of the answer to this immense challenge. Yet, for hydrogen to become a viable option numerous fundamental problems associated with both the production, the storage, the distribution and the use of hydrogen must still be solved. Without cost-competitive production schemes, which neither emit greenhouse gases nor have any other negative environmental impacts, the production of hydrogen will be the bottleneck of a future hydrogen economy.

Currently, commercially available hydrogen production methods are limited to fossil fuel-based procedures and electrolysis (as driven by a variety of electricity sources). Initial efforts in the transition to the hydrogen economy may be based on these methods, with improved efficiency, lower costs and minimized carbon dioxide footprint being important prerequisites. On a longer time scale, only hydrogen which is derived from carbon-neutral sources and produced using renewable energy sources fulfills all demands posed on a sustainable energy carrier. Research and development in all fields of hydrogen production conforming to these basic criteria should be intensified, so that a wide variety of processes will become available and constitute a flexible energy system, which can be adapted to varying local conditions and requirements.

Among the renewable energy sources, we believe that solar energy will be the dominant resource for hydrogen production. Of the various pathways to harvest solar energy and use it to produce hydrogen, photolytic schemes seem to be particularly interesting, owing to their potential to achieve an attractive trade-off between high efficiency and low costs. Considering that "traditional" metal oxides, such as TiO_2, WO_3 and Fe_2O_3, have, despite several decades of investigation, not reached the target of 10% solar-to-hydrogen conversion efficiency, it is clear that further research into materials that are inexpensive, durable, abundant and environmentally-benign is needed; materials with good absorption and carrier-transport properties need to be developed, and novel interfaces need to be designed with energetic and kinetic properties favoring the water-splitting reactions while inhibiting corrosion reactions. In the case of PEC hydrogen production, the materials challenge is not only limited to the photoactive material, but also concerns the materials found in the transparent conductive layer used for charge collection. Experimental efforts in the area will be complemented by computational catalysis.

Further down the road, we foresee a significant engineering challenge to integrate improved materials and interfaces into integrated devices that can be manufactured on a commercial and large-scale basis at low cost.

But what is the role of Nanoscience and Nanotechnology (N&N) in this process? We believe that N&N will have several roles to play and that they will continue to push the field forward. Many of the improvements we have seen over the years have, in fact, been made possible thanks to N&N, albeit with a lack of consciousness and direction. This has changed over time, however, due to analytical tools to analyze very small structures becoming widely available as a consequence of developments in N&N. Today N&N are being used to systematically investigate and improve water-splitting schemes. The design of nanocomposites, which feature a genuine combination of light-harvesting and charge-separation units, is perhaps the most prominent research topic at present.

Considering the rapidly growing number of important N&N contributions to other related fields, there seems to be an immense backlog demand for such schemes to be implemented in the area of photolysis. In particular, light-harvesting and light-management schemes, which have been investigated in conjunction with solar photovoltaics, are expected to make their way into the water-splitting community in the near future. On a somewhat longer term, we expect nanoreactor concepts to gain momentum and become an area of intense research. Throughout this development, N&N will enable scientists to achieve a better understanding of the basic physics and chemistry involved in photolytic water splitting, owing to the ability to manufacture and characterize well-controlled model systems. Based on such an improved understanding, more efficient and durable systems can be constructed.

We conclude by noting that alternatives to the hydrogen economy are being and should be considered. One such example is the concept of the methanol economy [152], where methanol rather than hydrogen is used as an energy storage medium. The potential advantages of such a scheme include convenient and safe storage and handling options, and readily available transport and distribution infrastructure. Additionally, methanol may replace oil and gas resources to serve as a feedstock for the synthesis of hydrocarbons.

The most attractive scheme of producing methanol is via the photocatalytic conversion of carbon dioxide and water, i.e., using abundant and clean energy from the sun and offering ways to chemically recycle CO_2. This process has many similarities to photocatalytic hydrogen production as described in this chapter. The oxidation of water is in fact a prerequisite for both reactions, and there are thus obvious synergies. Many of the N&N concepts described here for water splitting are thus expected to be readily applicable also to photocatalytic CO_2 conversion.

Acknowledgments We acknowledge financial support by the Foundation for Strategic Environmental Research (Mistra, Dnr 2004-118), Ångpanneföreningen's Foundation for Research and Development (09-370), the Environmental Foundation of the Swedish Association of Graduate Engineers and N-INNER through the Solar Hydrogen project (P30938-1 Solväte).

References

1. Smalley RE (2005) Future global energy prosperity: the Terawatt challenge. MRS Bull 30:412–417
2. Mikkelsen M, Jørgensen M, Krebs FC (2010) The teraton challenge, a review of fixation and transformation of carbon dioxide. Energy Environ Sci 3:43–81
3. Linsebigler AL, Lu G, Yates JT Jr (1995) Photocatalysis on TiO_2 surfaces: principles, mechanisms, and selected results. Chem Rev 95:735–758
4. Maeda K, Domen K (2007) New non-oxide photocatalysts designed for overall water splitting under visible light. J Phys Chem C 111:7851–7861
5. Kudo A, Miseki Y (2009) Heterogeneous photocatalyst materials for water splitting. Chem Soc Rev 38:253–278
6. Fujishima A, Honda K (1972) Electrochemical photolysis of water at a semiconductor electrode. Nature 238:37–38
7. Carp O, Huisman CL, Reller A (2007) Photoinduced reactivity of titanium dioxide. Prog Solid State Chem 32:33–177

8. Salazar K, Kimball SM (2009) Mineral commodities summaries 2009, US Geological Survey. http://minerals.usgs.gov. Accessed 31 Aug 2010
9. Zhu J, Zhang J, Chen F et al (2005) High photocatalytic activity TiO_2 prepared by a modified sol-gel method: characterization and their photocatalytic activity for degradation of XRG and X-GL. Top Catal 35:261–268
10. Zhu J, Zhang J, Chen F et al (2005) Preparation of high photocatalytic activity TiO_2 with a bicrystalline phase containing anatase and TiO_2 (B). Mater Lett 59:3378–3381
11. Zhang HZ, Banfield JF (2000) Understanding polymorphic phase transformation behavior during growth of nanocrystalline aggregates: insights from TiO_2. J Phys Chem B 104:3481–3487
12. Chen X, Mao SS (2007) Titanium dioxide nanomaterials: synthesis, properties, modifications, and applications. Chem Rev 107:2891–2959
13. Hernández-Alonso MD, Fresno F, Suárez S et al (2009) Development of alternative photocatalysts to TiO_2: challenges and opportunities. Energy Environ Sci 2:1231–1257
14. Boschloo G, Edvinsson T, Hagfeldt A (2006) Dye-sensitized nanostructured ZnO electrodes for solar cell application. In: Tetsuo S (ed) Nanostructured materials for solar energy conversion. Elsevier, Amsterdam
15. Özgür Ü, Alivov YI, Liu C et al (2005) A comprehensive review of ZnO materials and devices. J Appl Phys 98:041301
16. Zhang H, Chen G, Bahnemann DW (2009) Photoelectrocatalytic materials for environmental applications. J Mater Chem 19:5089–5121
17. Anpo M, Chapter 10, pp 175–185; Inoue Y, Chapter 15, pp 249–261 (2002) In: Kaneko M, Okura I (eds) Photocatalysis: science and technology, Springer, New York
18. Domen K, Chapter 16, pp 261–278 (2002) In: Kaneko M, Okura I (eds) Photocatalysis: science and technology, Springer, New York
19. Rajeshwar K (2008) Hydrogen generation from irradiated semiconductor-liquid interfaces. In: Rajeshwar K, McConnell R, Licht S (eds) Solar hydrogen generation. Springer, New York
20. Kim HG, Hwang DW, Kim J et al (1999) Highly donor-doped (110) layered perovskite materials as novel photocatalysts for overall water splitting. Chem Commun 1999:1077–1078
21. Ishikawa A, Takata T, Kondo JN et al (2002) Oxysulfide $Sm_2Ti_2S_2O_5$ as a stable photocatalyst for water oxidation and reduction under visible light irradiation ($\lambda \leq 650$ nm). J Am Chem Soc 124:13547–13553
22. Kato H, Kudo A (2001) Water splitting into H_2 and O_2 on alkali tantalate photocatalysts $ATaO_3$ (A = Li, Na, and K). J Phys Chem B 105:4285–4292
23. Shangguan WF (2007) Hydrogen evolution from water splitting on nanocomposite photocatalysts. Sci Tech Adv Mater 8:76–81
24. Tian MK, Shangguan WF, Yuan J et al (2007) Promotion effect of nanosized Pt, RuO_2 and NiO_x loading on visible light-driven photocatalysts $K_4Ce_2M_{10}O_{30}$ (M = Ta, Nb) for hydrogen evolution from water decomposition. Sci Tech Adv Mater 8:82–88
25. Osterloh FE (2008) Inorganic materials as catalysts for photochemical splitting of water. Chem Mater 20:35–54
26. Asahi R, Morikawa T, Ohwaki T et al (2001) Visible-light photocatalysis in nitrogen-doped titanium oxides. Science 293:269–271
27. Hitoki G, Ishikawa A, Takata T et al (2002) Ta_3N_5 as a novel visible light-driven photocatalyst ($\lambda < 600$ nm). Chem Lett 33:736–737
28. Sato J, Saito N, Yamada Y et al (2005) RuO_2-loaded β-Ge_3N_4 as a non-oxide photocatalyst for overall water splitting. J Am Chem Soc 127:4150–4151
29. Kida T, Minami Y, Guan G et al (2006) Photocatalytic activity of gallium nitride for producing hydrogen from water under light irradiation. J Mater Sci 41:3527–3534
30. Hara M, Takata T, Kondo JN et al (2004) Photocatalytic reduction of water by TaON under visible light irradiation. Catal Today 90:313–317

31. Yamasita D, Takata T, Hara M et al (2004) Recent progress of visible-light-driven heterogeneous photocatalysts for overall water splitting. Solid State Ionics 172:591–595
32. Kasahara A, Nukumizu K, Takata T et al (2003) $LaTiO_2N$ as a visible-light ($\leq$600 nm)-driven photocatalyst (2). J Phys Chem B 107:791–797
33. Liu M, You W, Lei Z et al. (2004) Water reduction and oxidation on Pt–Ru/$Y_2Ta_2O_5N_2$ catalyst under visible light irradiation. Chem Commun 2004: 2192–2193
34. Maeda K, Teramura K, Lu DL et al (2006) Photocatalyst releasing hydrogen from water-enhancing catalytic performance holds promise for hydrogen production by water splitting in sunlight. Nature 440:295
35. Lee Y, Terashima H, Shimodaira Y et al (2007) Zinc germanium oxynitride as a photocatalyst for overall water splitting under visible light. J Phys Chem C 111:1042–1048
36. Ishikawa A, Takata T, Matsumura T et al (2004) Oxysulfides $Ln_2Ti_2S_2O_5$ as stable photocatalysts for water oxidation and reduction under visible-light irradiation. J Phys Chem B 108:2637–2642
37. Finklea HO (1988) Semiconductor electrodes. Elsevier, Amsterdam
38. Heller A (1984) Hydrogen-evolving solar cells. Science 223:1141–1148
39. Khaselev O, Turner JA (1998) A monolithic photovoltaic-photoelectrochemical device for hydrogen production via water splitting. Science 280:425–427
40. Taniguchi Y, Yoneyama H, Tamura H (1983) Hydrogen evolution on surface-modified silicon powder photocatalysts in aqueous ethanol solutions. Chem Lett 12:269–272
41. Yoneyama H, Matsumoto N, Tamura H (1986) Photocatalytic decomposition of formic acid on platinized n-type silicon powder in aqueous solution. Bull Chem Soc Jpn 59:3302–3304
42. Sakai Y, Sugahara S, Matsumura M et al (1988) Photoelectrochemical water splitting by tandem type and heterojunction amorphous silicon electrodes. Can J Chem 66:1853–1856
43. Frame FA, Carroll EC, Larsen DS et al (2008) First demonstration of CdSe as a photocatalyst for hydrogen evolution from water under UV and visible light. Chem Commun 19:2206–2208
44. Schürch D, Currao A, Sarkar S et al (2002) The silver chloride photoanode in photoelectrochemical water splitting. J Phys Chem B 106:12764–12775
45. Currao A, Reddy VR, van Veen MK et al (2004) Water splitting with silver chloride photoanodes and amorphous silicon solar cells. Photochem Photobiol Sci 3:1017–1025
46. Gao Y, Wang Y, Wang Y (2007) Photocatalytic hydrogen evolution from water on SiC under visible light irradiation. React Kinet Catal Lett 91:13–19
47. Levy B (1997) Photochemistry of nanostructured materials for energy applications. J Electroceramics 1:239–272
48. Stroyuk AL, Kryukov AI, Kuchmii SY et al (2009) Semiconductor photocatalytic systems for the production of hydrogen by the action of visible light. Theor Exp Chem 45: 209–233
49. Li D, Haneda H, Hishita S et al (2005) Fluorine-doped TiO_2 powders prepared by spray pyrolysis and their improved photocatalytic activity for decomposition of gas-phase acetaldehyde. J Fluor Chem 126:69–77
50. Lin ZS, Orlov A, Lambert RM et al (2005) New insights into the origin of visible light photocatalytic activity of nitrogen-doped and oxygen-deficient anatase TiO_2. J Phys Chem B 109:20948–20952
51. Nakamura R, Tanaka T, Nakato Y (2004) Mechanism for visible light responses in anodic photocurrents at n-doped TiO_2 film electrodes. J Phys Chem B 108:10617–10620
52. Kamat PV (2007) Meeting the clean energy demand: nanostructure architectures for solar energy conversion. J Phys Chem C 111:2834–2860
53. Baba R, Nakabayashi S, Fujishima A et al (1985) Investigation of the mechanism of hydrogen evolution during photocatalytic water decomposition on metal-loaded semiconductor powders. J Phys Chem 89:1902–1905
54. Nosaka Y, Norimatsu K, Miyama H (1984) The function of metals in metal-compounded semiconductor photocatalysts. Chem Phys Lett 106:128–131

55. Subramanian V, Wolf EE, Kamat PV (2004) Catalysis with TiO_2/Au nanocomposites effect of metal particle size on the Fermi level equilibration. J Am Chem Soc 126:4943–4950
56. Sayama K, Mukasa K, Abe R et al (2002) A new photocatalytic water splitting system under visible light irradiation mimicking a Z-scheme mechanism in photosynthesis. J Photochem Photobio A: Chem 148:71–77
57. Tada H, Mitsui T, Kiyonaga T et al (2006) All-solid-state Z-scheme in CdS–Au–TiO_2 three-component nanojunction system. Nature Mater 5:782–786
58. Campbell P, Green MA (1987) Light trapping properties of pyramidally textured surfaces. J Appl Phys 62:243–249
59. Heine C, Morf RH (1995) Submicrometer gratings for solar energy applications. Appl Optics 34:2476–2482
60. Ito S, Murakami TN, Comte P et al (2008) Fabrication of thin film dye sensitized solar cells with solar to electric power conversion efficiency over 10%. Thin Solid Films 516:4613–4619
61. Vahala KJ (2003) Optical microcavities. Nature 424:839–846
62. Halaoui LI, Abrams NM, Mallouk TE (2005) Increasing the conversion efficiency of dye-sensitized TiO_2 photoelectrochemical cells by coupling to photonic crystals. J Phys Chem B 109:6334–6342
63. Bermel P, Luo C, Zeng L et al (2007) Improving thin-film crystalline silicon solar cell efficencies with photonic crystals. Opt Express 15:16986–17000
64. Wiersma DS, Sapienza R, Mujumdar S et al (2005) Optics of nanostructured dielectrics. J Opt A: Pure Appl Opt 7:S190–S197
65. Mayer B, Madronich S (2004) Actinic flux and photolysis in water droplets: Mie calculations and geometrical optics limit. Atmos Chem Phys 4:2241–2250
66. Robinson JT, Manolatou C, Chen L et al (2005) Ultrasmall mode volumes in dielectric optical microcavities. Phys Rev Lett 95:143901
67. Kreibig U, Vollmer M (1995) Optical properties of metal clusters. Springer, Berlin
68. Coyle S, Netti MC, Baumberg JJ et al (2001) Confined plasmons in metallic nanocavities. Phys Rev Lett 87:176801
69. Atwater HA, Polman A (2010) Plasmonics for improved photovoltaic devices. Nat Mater 9:205–213
70. Tsai F-J, Wang J-Y, Huang J-J et al (2010) Absorption enhancement of an amorphous Si solar cell through surface plasmon-induced scattering with metal nanoparticles. Opt Express 18:A207–A220
71. Hägglund C, Zäch M, Petersson G et al (2008) Electromagnetic coupling of light into a silicon solar cell by nanodisk plasmons. Appl Phys Lett 92:053110
72. Ferry VE, Sweatlock LA, Pacifici D et al (2008) Plasmonic nanostructure design for efficient light coupling into solar cells. Nano Lett 8:4391–4397
73. Bai W, Gan Q, Bartoli F et al (2009) Design of plasmonic back structures for efficiency enhancement of thin-film amorphous Si solar cells. Optics Lett 34:3725–3727
74. Kirkengen M, Bergli J, Galperin YM (2007) Direct generation of charge carriers in c-Si solar cells due to embedded nanoparticles. J Appl Phys 102:093713
75. Hägglund C, Zäch M, Kasemo B (2008) Enhanced charge carrier generation in dye sensitized solar cells by nanoparticle plasmons. Appl Phys Lett 92:013113
76. Kelzenberg MD, Boettcher SW, Petykiewicz JA et al (2010) Enhanced absorption and carrier collection in Si wire arrays for photovoltaic applications. Nat Mater 9:239–244
77. Zhu J, Hsu C-M, Yu Z et al (2010) Nanodome solar cells with efficient light management and self-cleaning. Nano Lett 10:1979–1984
78. Brus L (2008) Noble metal nanocrystals: plasmon electron transfer photochemistry and single-molecule Raman spectroscopy. Acc Chem Res 41:1742–1749
79. Watanabe K, Menzel D, Nilius N et al (2006) Photochemistry on metal nanoparticles. Chem Rev 106:4301–4320
80. Awazu K, Fujimaki M, Rockstuhl C et al (2008) A plasmonic photocatalyst consisting of silver nanoparticles embedded in titanium dioxide. J Am Chem Soc 130:1676–1680

81. Christopher P, Ingram DB, Linic S (2010) Enhancing photochemical activity of semiconductor nanoparticles with optically active Ag nanostructures: photochemistry mediated by Ag surface plasmons. J Phys Chem C 114:9173–9177
82. Tian Y, Tatsuma T (2005) Mechanisms and applications of plasmon-induced charge separation at TiO_2 films loaded with gold nanoparticles. J Am Chem Soc 127:7632–7637
83. Ertl G, Knözinger H, Weitkamp J (1997) Handbook of heterogeneous catalysis. Wiley, Weinheim
84. Haruta M, Kobayashi T, Sano H et al (1987) Novel gold catalysts for the oxidation of carbon-monoxide at a temperature far below 0°C. Chem Lett 2:405–406
85. Maira AJ, Yeung KL, Lee CY et al (2000) Size effects in gas-phase photo-oxidation of trichloroethylene using nanometer-sized TiO_2 catalysts. J Catal 192:185–196
86. Suzuki Y, Ngamsinlapasathian S, Yoshida R et al (2006) Partially nanowire-structured TiO_2 electrode for dye-sensitized solar cells. Cent Eur J Chem 4:476–488
87. Beermann N, Vayssieres L, Lindquist S-E et al (2000) Photoeletrochemical studies of oriented nanorod thin films of hematite. J Electrochem Soc 147:2456–2461
88. van de Krol R, Liang Y, Schoonman J (2008) Solar hydrogen production with nanostructured metal oxides. J Mater Chem 2008(18):2311–2320
89. Kay A, Cesar I, Grätzel M (2006) New benchmark for water photooxidation by nanostructured α-Fe_2O_3 films. J. Am. Chem Soc 128:15714–15721
90. Lindgren T, Wang H, Beermann N et al (2002) Aqueous photoelectrochemistry of hematite nanorod array. Sol Energy Mater Sol Cells 71:231–243
91. Wang Y, Zhang Z, Zhu Y et al (2008) Nanostructured VO_2 photocatalysts for hydrogen production. ACS Nano 2:1492–1496
92. Feng X, LaTempa TJ, Basham JI et al (2010) Ta_3N_5 nanotube arrays for visible light water photoelectrolysis. Nano Lett 10:948–952
93. Shankar K, Basham JI, Allam NK et al (2009) Recent advances in the use of TiO_2 nanotube and nanowire arrays for oxidative photoelectrochemistry. J Phys Chem C 113:6327–6359
94. Khan SUM, Sultana T (2003) Photoresponse of n-TiO_2 thin film and nanowire electrodes. Sol Energy Mater Sol Cells 76:211–221
95. Wolcott A, Smith WA, Kuykendall TR et al (2009) Photoelectrochemical water splitting using dense and aligned TiO_2 nanorod arrays. Small 5:104–111
96. Mor GK, Shankar K, Paulose M et al (2005) Enhanced photocleavage of water using titania nanotube arrays. Nano Lett 5:191–195
97. Lin CH, Lee CH, Chao JH et al (2004) Photocatalytic generation of H_2 gas from neat ethanol over Pt/TiO_2 nanotube catalysts. Catal Lett 98:61–66
98. Nam W, Han GY (2007) Preparation and characterization of anodized Pt-TiO_2 nanotube arrays for water splitting. J Chem Eng Jpn 40:266–269
99. Khan MA, Akhtar MS, Woo SI et al (2008) Enhanced photoresponse under visible light in Pt ionized TiO_2 nanotube for the photocatalytic splitting of water. Catal Commun 10:1–5
100. Thimsen E, Rastgar N, Biswas P (2008) Nanostructured TiO_2 films with controlled morphology synthesized in a single step process: performance of dye-sensitized solar cells and photo water splitting. J Phys Chem C 112:4134–4140
101. Jitputti J, Suzuki Y, Yoshikawa S (2008) Synthesis of TiO_2 nanowires and their photocatalytic activity for hydrogen evolution. Catal Commun 9:1265–1271
102. Kuo HL, Kuo CY, Liu CH et al (2007) A highly active bi-crystalline photocatalyst consisting of TiO_2 (B) nanotube and anatase particle for producing H_2 gas from neat ethanol. Catal Lett 113:7–12
103. Lin CH, Chao JH, Liu CH et al (2008) Effect of calcination temperature on the structure of a Pt/TiO_2 (B) nanofiber and its photocatalytic activity in generating H_2. Langmuir 24:9907–9915
104. Janet CM, Viswanath RP (2006) Large scale synthesis of CdS nanorods and its utilization in photo-catalytic H_2 production. Nanotechnology 17:5271–5277
105. Jang JS, Joshi UA, Lee JS (2007) Solvothermal synthesis of CdS nanowires for photocatalytic hydrogen and electricity production. J. Phys Chem C 111:13280–13287

106. Wang G, Yang X, Qian F et al (2010) Double-sided CdS and CdSe quantum dots co-sensitized ZnO nanowire arrays for photoelectrochemical hydrogen generation. Nano Lett 10:1088–1092
107. Costi R, Saunders AE, Elmalem E et al (2008) Visible light-induced charge retention and photocatalysis with hybrid CdSe-Au nanodumbbells. Nano Lett 8:637–641
108. Amirav L, Alivisatos AP (2010) Photocatalytic hydrogen production with tunable nanorod heterostructures. J Phys Chem Lett 1:1051–1054
109. Qu Y, Liao L, Cheng R et al (2010) Rational design and synthesis of freestanding photoelectric nanodevices as highly efficient photocatalysts. Nano Lett 10:1941–1949
110. Yu JG, Qi LF, Jaroniec M (2010) Hydrogen Production by Photocatalytic Water Splitting over Pt/TiO_2 nanosheets with exposed (001) facets. J Phys Chem C 114:13118–13125
111. Liu G, Yang HG, Wang XW et al (2009) Visible light responsive nitrogen doped anatase TiO_2 sheets with dominant 001 facets derived from TiN. J Am Chem Soc 131:12868–12869
112. Liu G, Yang HG, Wang XW et al (2009) Enhanced photoactivity of oxygen-deficient anatase TiO_2 sheets with dominant 001 facets. J Phys Chem C 113:21784–21788
113. Youngblood WJ, Lee SHA, Maeda K et al (2009) Visible light water splitting using dye-sensitized oxide semiconductors. Acc Chem Res 42:1966–1973
114. Yi H, Peng T, Ke D et al (2008) Photocatalytic H_2 production from methanol aqueous solution over titania nanoparticles with mesostructures. Int J Hydrogen Energy 33:672–678
115. Lakshminarasimhan N, Bae E, Choi W (2007) Enhanced photocatalytic production of H_2 on mesoporous TiO_2 prepared by template-free method: role of interparticle charge transfer. J Phys Chem C 111:15244–15250
116. Korzhak A, Ermokhina N, Stroyuk A et al (2008) Photocatalytic hydrogen evolution over mesoporous TiO_2 /metal nanocomposites. J Photochem Photobio A: Chem 198:126–134
117. Yamashita H, Mori K (2007) Applications of single-site photocatalysts implanted within the silica matrixes of zeolite and mesoporous silica. Chem Lett 36:348–353
118. Zhang YJ, Zhang L (2008) Synthesis of composite material CdS/Al-HMS and hydrogen production by photocatalytic pollutant degradation under visible light irradiation. J Inorg Mater 23:66–70
119. Ryu SY, Balcerski W, Lee TK et al (2007) Photocatalytic production of hydrogen from water with visible light using hybrid catalysts of CdS attached to microporous and mesoporous silicas. J Phys Chem C 111:18195–18203
120. Lunawat PS, Senapati S, Kumar R et al (2007) Visible light-induced splitting of water using CdS nanocrystallites immobilized over water-repellant polymeric surface Int. J Hydrogen Energy 32:2784–2790
121. Guan GQ, Kida T, Kusakabe K et al (2004) Photocatalytic H_2 evolution under visible light irradiation on CdS/ETS-4 composite. Chem Phys Lett 385:319–322
122. Maeda K, Teramura K, Lu D et al (2006) Noble-metal/Cr_2O_3 core/shell nanoparticles as a co-catalyst for photocatalytic overall water splitting. Angew Chem Int Ed 45: 7806–7809
123. Yoshida M, Takanabe K, Maeda K et al (2009) Role and function of noble-metal/Cr-layer core/shell structure co-catalysts for photocatalytic overall water splitting studied by model electrodes. J Phys Chem C 113:10151–10157
124. Maeda K, Xiong A, Yoshinaga T et al (2010) Photocatalytic overall water splitting promoted by two co-catalysts for hydrogen and oxygen evolution under visible light. Angew Chem Int Ed. 49:4096–4099
125. Ikeda S, Hirao K, Ishino S et al (2006) Preparation of platinized strontium titanate covered with hollow silica and its activity for overall water splitting in a novel phase-boundary photocatalytic system. Catal Today 117:343–349
126. Kale BB, Baeg J-O, Apte SK et al (2007) Confinement of nano CdS in designated glass: a novel functionality of quantum dot-glass nanosystems in solar hydrogen production. J Mater Chem 17:4297–4303

127. Dähne L, Leporatti S, Donath E et al (2001) Fabrication of micro reaction cages with tailored properties. J Am Chem Soc 123:5431–5436
128. Nardin C, Thoeni S, Widmer J et al (2000) Nanoreactors based on (polymerized) ABA-triblock copolymer vesicles. Chem Commun 2000:1433–1434
129. Ruysschaert T, Germain M, da Silva Gomes JFP et al (2004) Liposome-based nanocapsules. IEEE Trans Nanobiosci 3:49–55
130. Karlsson M, Davidson M, Karlsson R et al (2004) Biomimetic nanoscale reactors and networks. Ann Rev Phys Chem 55:613–649
131. Doshi DA, Huesing N, Lu M et al (2000) Optically defined multifunctional patterning of photosensitive thin-film silica mesophases. Science 290:107–111
132. Shtykov SN (2002) Chemical analysis in nanoreactors: main concepts and applications. J Anal Chem 57:859–868
133. Herrmann J-M (2005) Heterogeneous photocatalysis: state of the art and present applications. Top Catal 34:49–65
134. Santiso EE, George AM, Turner CH et al (2005) Adsorption and catalysis: the effect of confinement on chemical reactions. Appl Surf Sci 252:766–777
135. Shchukin DG, Sviridov DV (2006) Photocatalytic processes in spatially confined micro- and nanoreactors. J Photochem Photobiol C 7:23–39
136. Koblenz TS, Wassenaar J, Reek JNH (2008) Reactivity within a confined self-assembled nanospace. Chem Soc Rev 37:247–262
137. Li J, Zeng HC (2005) Size tuning, functionalization, and reactivation of Au in TiO_2 nanoreactors. Angew Chem Int Ed 44:4342–4345
138. Yen CW, Mahmoud MA, El-Sayed MA (2009) Photocatalysis in gold nanocage nanoreactors. J Phys Chem A 113:4340–4345
139. Harris C, Kamat PV (2009) Photocatalysis with CdSe nanoparticles in confined media: mapping charge transfer events in the subpicosecond to second timescales. ACS Nano 3:682–690
140. Parmon VN, Lymar SV, Tsvetkov IM et al (1983) Development of microheterogeneous systems based on lipid vesicles for photocatalytic charge separation in molecular converters of solar energy. J Mol Catal 21:353–363
141. Khramov MI, Parmon VN (1993) Synthesis of ultrafine particles of transition metal sulphides in the cavities of lipid vesicles and the light-stimulated transmembrane electron transfer catalysed by these particles. J Photochem Photobiol A: Chem 71:279–284
142. Efimova EV, Lymar SV, Parmon VN (1994) 1,4-bis(1, 2, 6-triphenyl-4-pyridyl)benzene as a novel hydrophobic eletron relay for dihydrogen evolution in photocatalytic systems based on lipid vesicles. J. Photochem Photobiol A: Chem 83:153–159
143. Tricot Y-M, Emeren Å, Fendler JH (1985) In situ generation of catalyst-coated CdS particles in polymerized and unpolymerized surfactant vesicles and their utilization for efficient visible-light induced hydrogen production. J Phys Chem 89:4721–4726
144. Bergeld J, Kasemo B, Chakarov D (2008) Photocatalytic reactions at the graphite/ice interface. Phys Chem Chem Phys 10:1151–1155
145. Hulteen JC, van Dyne RP (1994) Nanosphere lithography: a materials general fabrication process for periodic particle array surfaces. J Vac Sci Technol A 13:1553–1558
146. Fredriksson H, Alaverdyan Y, Dmitriev A et al (2007) Hole-mask colloidal lithography. Adv Mater 19:4297–4302
147. Johanek V, Laurin M, Grant AW et al (2004) Fluctuations and bistabilities on catalyst nanoparticles. Science 304:1639–1644
148. Langhammer C, Zoric I, Kasemo B et al (2007) Hydrogen storage in Pd nanodisks characterized with a novel nanoplasmonic sensing scheme. Nano Lett 7:3122–3127
149. Komanicky V, Iddir H, Chang KC et al (2009) Shape-dependent activity of platinum array catalyst. J Am Chem Soc 131:5732–5733

150. Brown EC, Wilke SK, Boyd DA et al (2010) Polymer sphere lithography for solid oxide fuel cells: a route to functional, well-defined electrode structures. J Mater Chem 20:2190–2196
151. Seidel YE, Scheider A, Jusys Z et al (2010) Transport effects in the electrooxidation of methanol studied on nanostructured Pt/glassy carbon electrodes. Langmuir 26:3569–3578
152. Olah GA, Goeppert A, Prakash GKS (2009) Beyond oil and gas: the methanol economy, 2nd edn. Wiley-VCH, Weinheim

(Oxy)nitrides and Oxysulfides as Visible-Light-Driven Photocatalysts for Overall Water Splitting

Kazuhiko Maeda, Tsuyoshi Takata and Kazunari Domen

Abstract Overall water splitting to form hydrogen and oxygen using a particulate photocatalyst with solar energy is a promising process for clean hydrogen production in large-scale. In recent years, numerous attempts have been made for the development of photocatalysts that work under visible-light irradiation to efficiently utilize solar energy. This chapter presents recent research progress in the development of visible-light-driven photocatalysts, focusing on the refinement of non-oxide type photocatalysts such as (oxy)nitrides and oxysulfides. These materials harvest visible photons (450–700 nm), and work as stable photocatalysts for water reduction and oxidation under visible-light.

1 Introduction

Catalytic splitting of pure water into hydrogen and oxygen in the presence of semiconductor powders using visible-light is a promising approach for storing solar energy as chemical energy. The reaction (Eq. 1) is a typical "uphill reaction", having a large positive change in the Gibbs free energy ($\Delta G^0 = 238\ \text{kJ} \cdot \text{mol}^{-1}$).

$$2H_2O \rightarrow 2H_2 + O_2 \tag{1}$$

The half-reactions are described as follows:

$$2H^+ + 2e^- \rightarrow H_2 \tag{2}$$

$$2H_2O + 4h^+ \rightarrow O_2 + 4H^+ \tag{3}$$

K. Maeda · T. Takata · K. Domen (✉)
Department of Chemical System Engineering, The University of Tokyo,
7-3-1 Hongo, Bunkyo-ku, Tokyo 113-8656, Japan
e-mail: domen@chemsys.t.u-tokyo.ac.jp

L. Zang (ed.), *Energy Efficiency and Renewable Energy Through Nanotechnology*, Green Energy and Technology, DOI: 10.1007/978-0-85729-638-2_14,

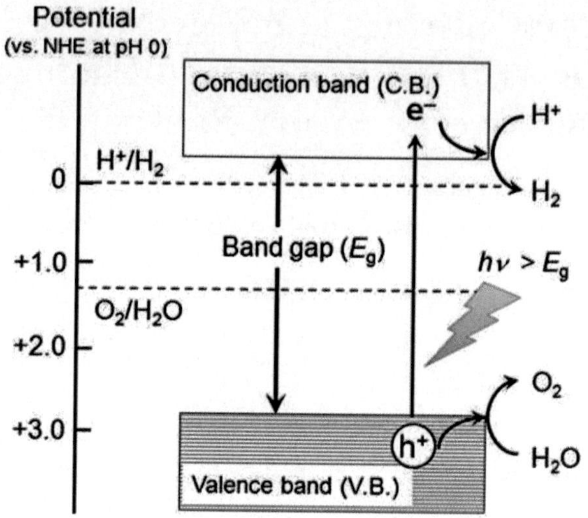

Fig. 1 Basic principle of water splitting on a particulate photocatalyst

Figure 1 shows a schematic illustration of the basic principle of overall water splitting on a particulate photocatalyst. Under irradiation at an energy equal or greater than the band-gap of the semiconductor photocatalyst, electrons in the valence band are excited into the conduction band, leaving holes in the valence band. These photogenerated electrons and holes cause reduction and oxidation reactions, respectively. To achieve overall water splitting, the bottoms of the conduction bands must be located at a more negative potential than the reduction potential of H^+ to H_2 (0 V vs. NHE at pH 0), while the tops of the valence bands must be positioned more positively than the oxidation potential of H_2O to O_2 (1.23 V vs. NHE). Therefore, the minimum photon energy thermodynamically required to drive the reaction is 1.23 eV, corresponding to a wavelength of ca. 1,000 nm, in the near-infrared region. Accordingly, it would appear to be possible to utilize the entire spectral range of visible-light. However, there is an activation barrier in the charge transfer process between photocatalysts and water molecules, necessitating a photon energy greater than the band-gap of the photocatalyst to drive the overall water-splitting reaction at reasonable reaction rates. In addition, the backward reaction, that is, water formation from H_2 to O_2, must be strictly inhibited, and the photocatalysts themselves must be stable in the reaction.

Research in this field was initially triggered by the demonstration of photoelectrochemical water splitting using a single-crystal titanium dioxide (rutile) photoanode and a platinum cathode with an external bias [1]. Over the past decades, while many semiconducting materials have been reported to be utile in this reaction, none had demonstrated a stable, reproducible water-splitting capability under visible-light irradiation. The major obstacle to progress in this field has been the lack of a suitable compound that meets the following three requirements: (1) a band-gap narrower than 3 eV, (2) band-edge potentials suitable for overall water

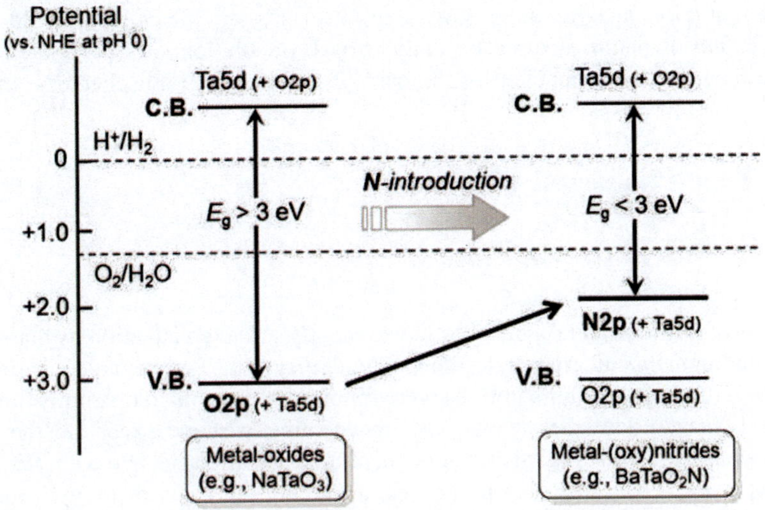

Fig. 2 Schematic band structures of metal oxides and metal (oxy)nitrides

splitting, and (3) stability in the photocatalytic reaction. Specifically, the band-gap must be sufficiently small to allow the absorption of visible-light (<3 eV), and the band edges must be located in a position that allows for the reduction and oxidation of water by photoexcited electrons and holes, as mentioned earlier. In general, efficient photocatalytic materials contain either transition-metal cations with a d^0 electronic configuration (e.g., Ti^{4+} and Ta^{5+}) [2–14] or typical-metal cations with a d^{10} electronic configuration (e.g., In^{3+} and Sn^{4+}) [15–20] as principal cation components. The empty d or sp orbitals form the bottom of the conduction band. The top of the valence band of a metal-oxide photocatalyst with d^0- or d^{10}-metal cations usually consists of O2p orbitals, which are located at about +3 eV or higher versus NHE, thereby producing a band-gap too wide to absorb visible-light [21]. Because some non-oxide compounds such as CdS and CdSe have band edge potentials suitable for overall water splitting under visible-light, these compounds were carefully examined for visible-light utilization. However, no successful photocatalytic systems have been established because of a lack of oxygen production due to the instability of the materials [22, 23].

The authors' group has studied such materials, in particular (oxy)nitrides and oxysulfides, as photocatalysts for overall water splitting under visible-light. Note that these types of materials are different from those doped with nitrogen or sulfur. Figure 2 shows the schematic band structures of metal oxides and (oxy)nitrides. The tops of the valence band (HOMO) for metal oxides are composed of O2p orbitals. When N atoms are partially or fully substituted for O atoms in a metal oxide, the HOMO of the material must be shifted to higher level compared to corresponding metal oxide without affecting the bottoms of the conduction bands (LUMO) level. In this chapter, (oxy)nitride and oxysulfide photocatalysts for overall water splitting under visible-light are presented. Although our research has shown that photocatalytic activity of (oxy)nitrides is significantly enhanced upon modification of nanostructured cocatalysts as H_2 evolution sites [24], we would like to focus on the refinement of the base photocatalyst in this chapter.

2 d^0-Type Photocatalysts

2.1 *Ta^{5+}-Based (Oxy)nitrides*

Some metal nitrides and oxynitrides have recently attracted attention as non-toxic inorganic pigments alternating to the CdS–CdSe system. For example, a series of $LaTaON_2$–$CaTaO_2N$ solid solutions were reported to vary in color from yellow to red [25]. In addition, these materials are chemically stable except in the very strong acids, e.g., aqua regia and hot concentrated sulfuric acid. We were attracted to oxynitrides because of such properties, triggering our study of (oxy)nitrides as novel photocatalysts for visible-light-induced water cleavage.

For the utilization of narrow gap materials, band-gap position and stability are essential factors for achieving water splitting. In the preliminary stage of our study, various transition-metal (oxy)nitrides were surveyed to judge the effect of applying visible-light to induce photocatalysis. Various transition-metal (oxy)nitrides were synthesized and characterized, and subsequently their abilities for photocatalytic H_2 and O_2 evolution were investigated.

First, several tantalum-based (oxy)nitrides were examined. Many tantalum-based d^0 transition-metal oxides have been reported as photocatalysts for overall water splitting [11–14]. Therefore, it is natural to expect that colored oxynitrides based on Ta^{5+} with d^0 electronic configuration would exhibit photocatalytic activity under visible-light. In Fig. 3, the schematic structures of examined Ta^{5+}-based (oxy)nitrides are depicted. The crystal structures of $LaTaON_2$ and $AETaO_2N$ (AE = Ca, Sr, Ba) are perovskite, generally formulated as ABX_3 [26, 27]. There are two distinct cationic sites in the perovskite structure. The coordination number of A- and B-site cations are 12 and 6 with respect to anions located at X-site, respectively. A-site is generally occupied by La^{3+} or AE^{2+} with relatively large ionic radius, whereas smaller Ta^{5+} ions are located at B-site. For the occupancy of X-site with O^{2-} and N^{3-}, the total anion number in a unit is 3, and the O/N ratio here is determined by the total positive charge of the constituent cations, thereby maintaining charge balance. TaON has a baddeleyite structure, which is isostructural to the monoclinic ZrO_2 [28]. Each Ta^{5+} ion is coordinated with total seven O^{2-} and N^{3-} anions, and the occupancies of O^{2-} and N^{3-} in the lattice are in 8 ordered arrangements. Ta_3N_5 has an anosovite structure consisting of corner- and edge- sharing TaN_6 octahedra [29].

Transition-metal (oxy)nitrides were prepared by thermal ammonolysis of precursor oxides in our experiments. For the synthesis of Ta_3N_5 and TaON, commercially available Ta_2O_5 powder consisting of crystalline particles with several hundred nanometers in diameter was the precursor. The corresponding

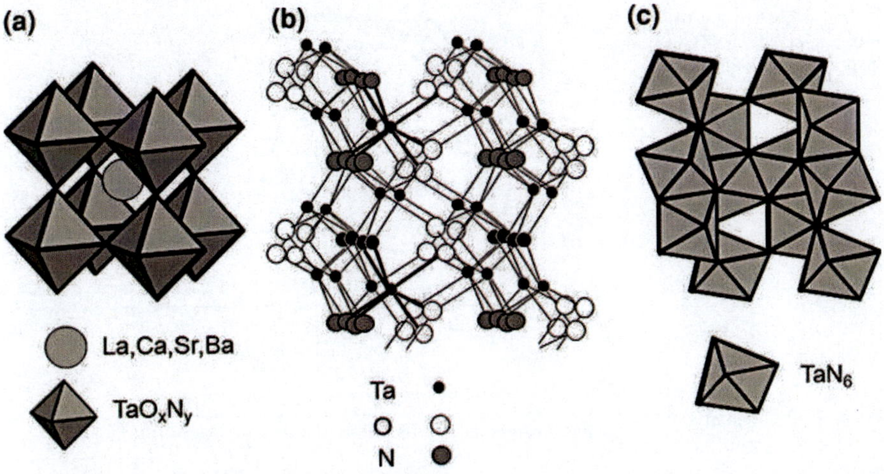

Fig. 3 Schematic crystal structures of Ta^{5+}-based (oxy)nitrides. **a** $LaTaON_2$ and $AETaO_2N$ (AE = Ca, Sr, Ba), **b** TaON, and **c** Ta_3N_5

mixed-oxide precursors for perovskite oxynitrides were prepared via a molecular route called polymerized complex method [30]. In this method, the component were intimately mixed. Otherwise, Ta_3N_5 was formed in some cases as a result of direct nitridation of Ta_2O_5 without forming perovskite oxynitride structure due to insufficient mixing. Typically, $TaCl_5$ and anhydrous citric acid (CA) were dissolved in ethylene glycol (EG); then, a methanol solution of $La(NO)_3{\cdot}6H_2O$ or ACO_3 was added. The molar ratio of La or AE/Ta/CA/EG was 1/1/30/120. The mixture was heated at 403 K with stirring until a transparent gel was formed. Then the polymer was carbonized at 623 K followed by calcination in air at 923 K for the removal of carbon species. The precursor was wrapped with quartz wool, and was inserted in the center of a horizontal alumina tube-furnace, and heated under the flow of dry NH_3 at 1,223 K for 10–15 h. Flow rates were typically 1 L min^{-1} except for the preparation of TaON to employ a flow rate of 20 mL min^{-1}. The products were characterized by powder X-ray diffraction.

Figure 4 shows the UV-visible diffuse reflectance spectra of various tantalum-based (oxy)nitrides. Each (oxy)nitride absorbs photons in visible region, while the precursor oxides (e.g., $LaTaO_4$, shown for reference) absorb only UV light. Most of the examined oxynitrides have absorption band edges in between 500 and 650 nm. $BaTaO_2N$ has a characteristic absorption band in longer wavelength region, up to 700 nm. This is attributable to the d-d transition based on reduced tantalum species (e.g., Ta^{4+} and Ta^{3+}) created during nitridation. Band-gap energies of various Ta^{5+}-oxynitrides, as estimated from their absorption band edges, are summarized in Table 1. For $AETaO_2N$ series, the band-gap values decrease in the order of Ca- > Sr- > Ba-form. In the case of some perovskite type oxides (e.g., $ATaO_3$ (A = Li, Na, K)), the band-gap values decrease in the order of Li > Na > K-form [13]. Band-gap energy seems to decrease with increasing the

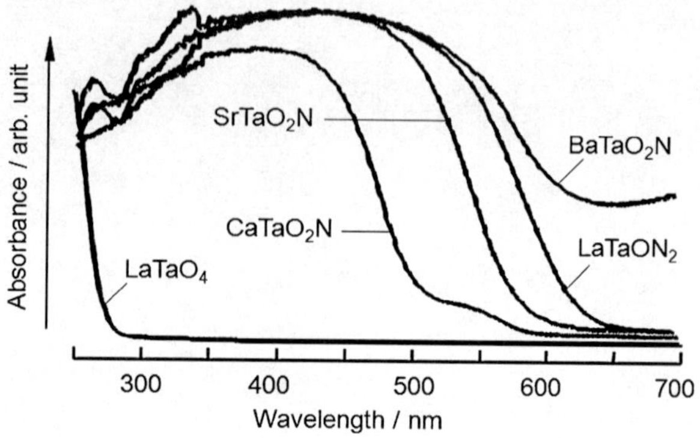

Fig. 4 UV-visible diffuse reflectance spectra for some Ta^{5+}-based oxynitrides

Table 1 Photocatalytic activities of some Ta^{5+}-based (oxy)nitrides for H_2 or O_2 evolution in the presence of sacrificial reagents under visible-light ($\lambda > 420$ nm)

Photocatalyst	Band-gap[a]/eV	Activity/μmol h^{-1}	
		H_2[b]	O_2[c]
TaON	2.5	15	220
Ta_3N_5	2.1	6	46
$LaTaON_2$	2.0	20	0
$CaTaO_2N$	2.4	23	0
$SrTaO_2N$	2.1	20	0
$BaTaO_2N$	1.9	12	0

Reaction conditions: 0.2–0.4 g of catalyst, 200 ml aqueous solution containing sacrificial reagents, 300 W xenon lamp light source, Pyrex top irradiation-type reaction vessel with cutoff filter

[a] Estimated from onset wavelength of diffuse reflectance spectra

[b] Loaded with nanoparticulate Pt as a cocatalyst, reacted in the presence of methanol (10 vol%), sacrificial reagent

[c] Sacrificial reagent: silver nitrate (0.01 M)

radius of A-site cations, which is related to distorted linkage of TaO_6 octahedra and affects the extent of delocalization of the electronic bands. This tendency also holds true for the $AETaO_2N$ series. The linkage of Ta–O/N bonds is the origin of visible-light absorption and the distortion of the octahedra would also affect the band-gap energy. Apparently, band-gap narrowing to sufficiently absorb visible-light up to ca. 500–650 nm is possible for these (oxy)nitrides by the introduction of nitrogen to change the valence band's position. These results show that band-gap energy varies depending on the constituent cations and O/N ratio in the material, indicating the tunable band-gap energy and photocatalysis, as well.

The photocatalytic performances of H_2 and O_2 evolution for various Ta^{5+}-based oxynitrides were examined. H_2 and O_2 evolution was separately evaluated in the presence of sacrificial reagents to examine the abilities of the oxynitrides in each reaction. The water-splitting process is often inhibited for various but unclear reasons, even though a given photocatalyst satisfies the thermodynamic requirement to split water into stoichiometric H_2 and O_2. Therefore, H_2 and O_2 evolutions were separately carried out in the presence of following sacrificial reagents for tentative evaluation of photocatalytic capabilities. One is H_2 evolution from aqueous methanol solution, and the other is O_2 evolution from aqueous $AgNO_3$ solution as denoted in the Eqs. 4 and 5, respectively [24].

$$H_2O + CH_3OH \rightarrow 3H_2 + CO_2 \quad (4)$$

$$2H_2O + 4AgNO_3 \rightarrow O_2 + 4Ag + 4HNO_3 \quad (5)$$

In the former case, oxidation of methanol by valence-band holes is facilitated, which allows accumulation of electrons in the conduction band and leads to accelerated H^+ reduction to H_2. In the latter case, a conduction band electron promptly reduces Ag^+ adsorbed on the surface of photocatalyst instead of H^+ depositing Ag on the surface, while positive holes in the valence band oxidize H_2O to O_2. The results of both reactions are essential for a given photocatalyst to be regarded as a candidate for visible-light-induced water splitting.

Typical experimental procedures of photocatalytic reactions are as follows. The photocatalytic reaction was carried out in a closed gas circulation system under visible-light irradiation using a Xe lamp (300 W) through a cutoff ($\lambda > 420$ nm) filter. Residual air both in gas phase and the reactant solution was completely removed by evacuation prior to photo-irradiation. The catalyst (0.2–0.4 g) was suspended in an aqueous solution (200 mL) by magnetic stirring. An aqueous methanol solution (H_2O 180 mL + MeOH 20 mL) was employed as the sacrificial electron donor for H_2 evolution, and an aqueous silver nitrate solution (0.01 M) was employed as the sacrificial electron acceptor for O_2 evolution. For H_2 evolution, Pt was loaded by the impregnation method from an aqueous $[Pt(NH_3)_4]Cl_2$ solution followed by H_2 reduction for 2 h at 573 K. In the case of O_2 evolution, La_2O_3 (0.2 g) was added to the reaction solution to maintain the pH at ca. 8.5. Evolved gases were analyzed by gas chromatograph (Ar carrier, TCD, MS-5A column) connected directly to the closed gas circulation system. A schematic illustration of the reaction setup is shown in Fig. 5.

Photocatalytic activities of various Ta^{5+}-based (oxy)nitrides were examined [31]. As summarized in Table 1, the activities of H_2 evolution are in a similar order (a few tens of $\mu mol{\cdot}h^{-1}$) in every case. For O_2 evolution, TaON and Ta_3N_5 were active while others did not show any activity. Despite many efforts to improve the activity of O_2 evolution on $LaTaON_2$ and $AETaO_2N$, no success has been achieved thus far. In the case of oxide photocatalysts, there are many examples of H_2 and O_2 evolution over compounds containing La^{3+} or alkaline earth metal cations. Therefore, the presence of these cationic species should not be the cause of O_2 evolution inhibition. One conceivable reason for the lack of O_2 evolution is that the tops of the valence band of $LaTaON_2$ and $AETaO_2N$ lie at potentials more negative than water oxidation potential. On the other hand, both TaON and Ta_3N_5 have the ability to evolve H_2 and O_2 under band-gap excitation, suggesting that these materials are candidates for overall water splitting under visible-light. In this context, structural, physicochemical and photocatalytic properties of Ta_3N_5 and TaON were examined in detail [32].

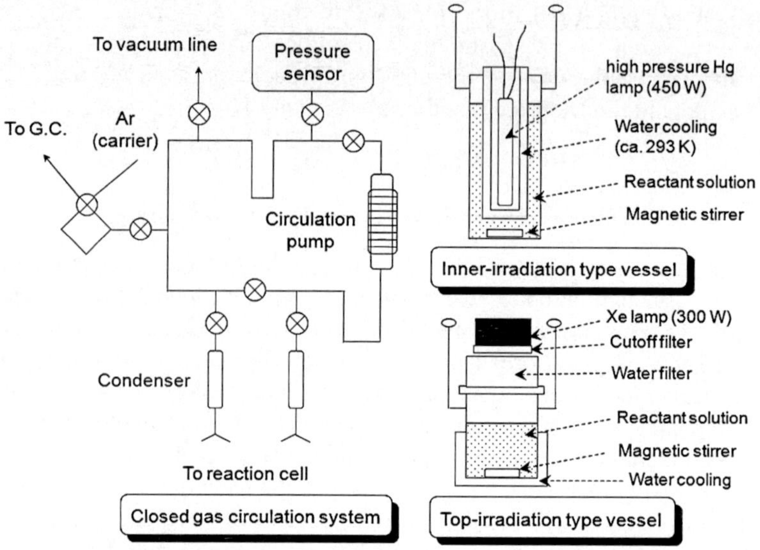

Fig. 5 A schematic illustration of the reaction setup for photocatalytic water splitting

Typically, TaON and Ta_3N_5 can be obtained by nitriding Ta_2O_5 for 15 h under NH_3 flow of 20 mL•min^{-1} and 1 L•min^{-1}, respectively. A mixture of TaON and Ta_3N_5 phases is produced in intermediate nitriding conditions. The detailed phase transition from Ta_2O_5 to TaON (and to Ta_3N_5) was investigated by slowing the progress of nitridation. Figure 6 shows the XRD patterns of nitrided Ta_2O_5 at various time intervals under 20 mL•min^{-1} of NH_3 flow. The XRD pattern of Ta_2O_5 heated for 1 h under this nitriding condition is almost the same as that of the original Ta_2O_5. Heating for 5–10 h under NH_3 flow resulted in the formation of TaON as a major phase, with small amounts of unreacted Ta_2O_5. TaON as a single phase was obtained with 15 h of nitridation. Further nitridation led to the formation of a mixture of Ta_3N_5 and TaON. The TaON phase completely disappeared after 100 h of nitridation, resulting in the single phase of Ta_3N_5.

It was also clearly seen from the structural analysis by XRD that Ta_3N_5 was formed from Ta_2O_5 via the formation of TaON. This means that TaON is an intermediate and metastable phase produced during the phase transition from Ta_2O_5 to Ta_3N_5. Other crystalline phases of tantalum oxynitrides, TaO_xN_y ($0 < x < 2.5$, $x + 3/2y = 2.5$), were not detected. The O/N ratio changes discontinuously due to transformation of the crystal structure. As two N^{3-} ions are substituted for three O^{2-} ions in order to maintain the charge balance, the number of atoms coordinating to one Ta atom inevitably changes upon nitridation, with a corresponding change in the crystal structure. Therefore, O/N exchange proceeds intermittently rather than continuously, following the transformation of the crystal structure from Ta_2O_5 to Ta_3N_5 via TaON. Although the XRD patterns of the samples nitrided for 15 and 100 h corresponded to those of the single phases of TaON and Ta_3N_5, respectively, the compositions of obtained samples were

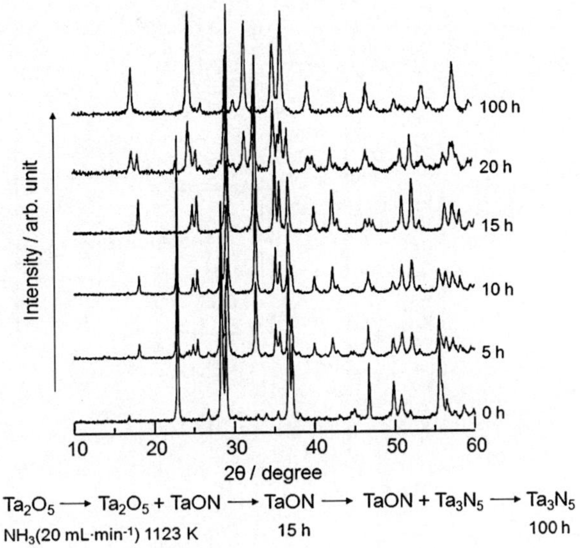

Fig. 6 XRD patterns of Ta_2O_5 subjected to various durations of nitridation

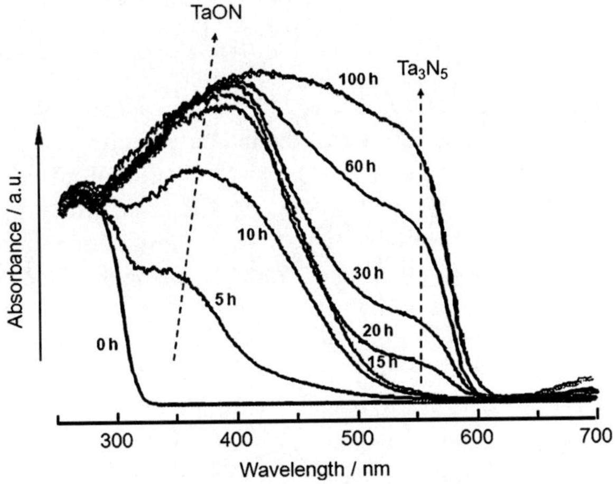

Fig. 7 UV-visible diffuse reflectance spectra of Ta_2O_5 at various time intervals during nitridation

somewhat different from those expected from their stoichiometry. This means that the obtained samples are non-stoichiometric and defective compounds.

The UV-visible diffuse reflectance spectra for nitrided Ta_2O_5 subjected to various durations of nitridation are shown in Fig. 7. Although pure Ta_2O_5 only absorbs light at wavelengths shorter than 320 nm, the absorption band extended to 500 nm after nitridation for 1 h. A broad absorption band up to around 500 nm increased with nitridation duration up to 10 h, after which another absorption band at around 500–600 nm appeared and increased upon further nitridation. Correlation of these results with the XRD results suggests that these two absorption peaks, up to ca. 500 and 600 nm, are attributable to TaON and Ta_3N_5, respectively. From the absorption band edges, the band-gap energies of Ta_2O_5, TaON and Ta_3N_5 are

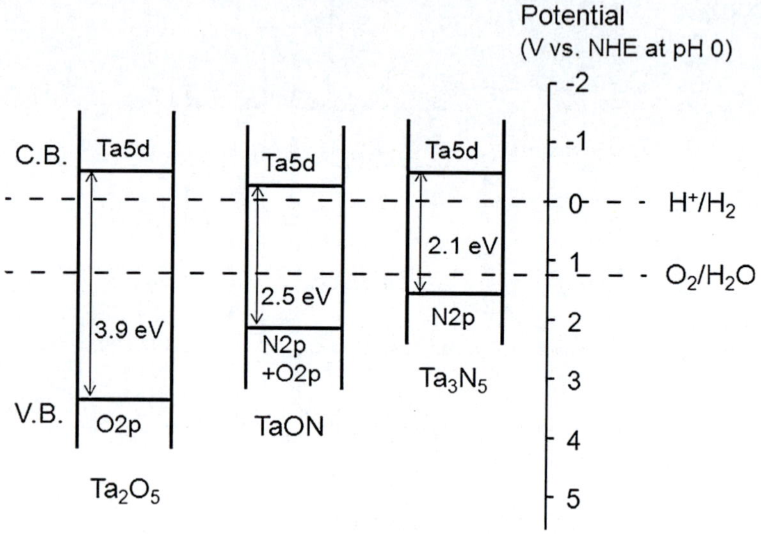

Fig. 8 Band-gap positions of Ta_2O_5, TaON and Ta_3N_5

estimated to be 3.9, 2.5 and 2.1 eV, respectively. The observed band-gap narrowing occurs as a result of the increase in nitrogen content through changes in the electronic structure due to the incorporation of nitrogen. Since the conduction and valence bands of tantalum oxynitride consist primarily of Ta5d orbitals and hybrid O2p and N2p orbitals, respectively, the incorporation of nitrogen is expected to result mainly in electronic structural changes of the valence-band states. Density functional theory (DFT) calculations [33] show that the upper part of the valence band is dominated by N2p orbitals on account of the higher potential energy of the N2p orbital compared to the O2p orbital. However, DFT calculations were unable to reveal the absolute band positions or determine whether the change in the absorption band is due to a shift of the valence band's maximum or the conduction band's minimum. The absolute band positions of Ta_2O_5, TaON and Ta_3N_5 were measured by electrochemical and photoelectron spectroscopic methods [34], as illustrated in Fig. 8. Major changes due to the incorporation of nitrogen appear at the valence-band level. The observed band-gap narrowing due to nitridation is therefore attributable to the negative shift of the valence band's maximum.

H_2 evolution was performed in an aqueous methanol solution as a sacrificial electron donor with the use of a platinized catalyst. For O_2 evolution, an aqueous $AgNO_3$ solution was employed as a sacrificial electron acceptor. The dependences of the activities of nitrided Ta_2O_5 for H_2 and O_2 evolution on the nitridation time are shown in Fig. 9. The activities of H_2 and O_2 evolution increased monotonically with nitridation time and reached maximum for 15 h nitridation in both cases. Further nitridation led to a gradual suppression of H_2 evolution activity and decay in O_2 evolution activity.

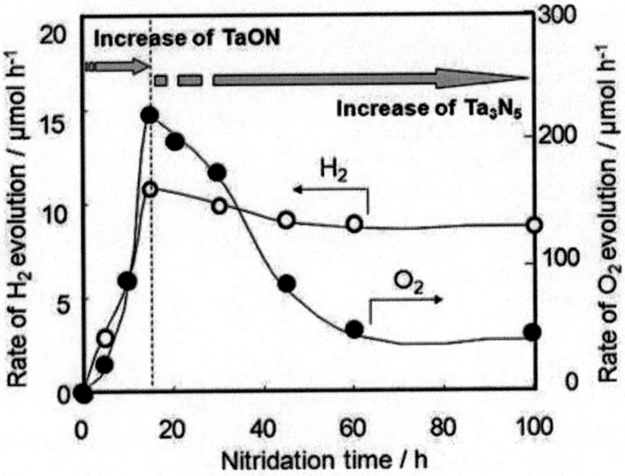

Fig. 9 Dependence of the photocatalytic activity of nitrided Ta_2O_5 for H_2 and O_2 evolution on nitridation time. Reaction conditions: 0.4 g of catalyst, 200 mL aqueous solution containing sacrificial reagents, 300 W xenon lamp light source, Pyrex top irradiation-type reaction vessel with cutoff filter ($\lambda > 420$ nm)

Taking into consideration the structural transformation with nitridation time, the increase in the activities for H_2 and O_2 evolution during the first 15 h of nitridation is attributed to the broadening of the absorption band into the visible region through the increase of TaON content. The decrease in activity associated with nitridation for more than 15 h is considered to be due to the replacement of the TaON phase with the Ta_3N_5 phase. The highest activities were obtained after nitridation for 15 h, corresponding to an almost pure TaON phase, in the case of both H_2 and O_2 evolution. Therefore, TaON appears to have higher activity for water decomposition than Ta_3N_5 under visible-light. The decrease in activity for O_2 evolution due to the increase in the Ta_3N_5 phase was more prominent than for H_2 evolution. The characteristics of TaON and Ta_3N_5 in terms of photocatalytic activity can be differentiated as follows: the capacity for O_2 evolution on TaON is significantly greater than that on Ta_3N_5, whereas the activity of H_2 evolution on TaON is moderately higher compared to Ta_3N_5. These characteristics can presumably be explained from the band positions. The top of the valence-band position shifts negatively as nitridation proceeds, whereas the position of the conduction band bottom remains largely unchanged as mentioned above. It is considered that the difference between the position of the valence band maximum in TaON and Ta_3N_5 determines the activity of O_2 evolution. The sufficiently positive potential of TaON with respect to water oxidation may be beneficial for O_2 evolution. On the other hand, the bottom of the conduction band did not change significantly upon nitridation, which may be a possible reason for the similar H_2 evolution activity of TaON and Ta_3N_5.

Figure 10 displays a typical time course of H_2 evolution on Ta_3N_5 in aqueous methanol solution under visible-light irradiation. The gas phase was evacuated at every 24 h. H_2 evolved under visible-light irradiation, accompanied by a small amount of N_2 evolution in the early stage of the reaction. N_2 evolution was suppressed almost immediately, and no N_2 evolution was observed after the first iteration. The evolution of N_2 is attributed to the oxidation of nitrogen species in the catalyst via valence-band holes. Less than 0.5% of the nitrogen present was

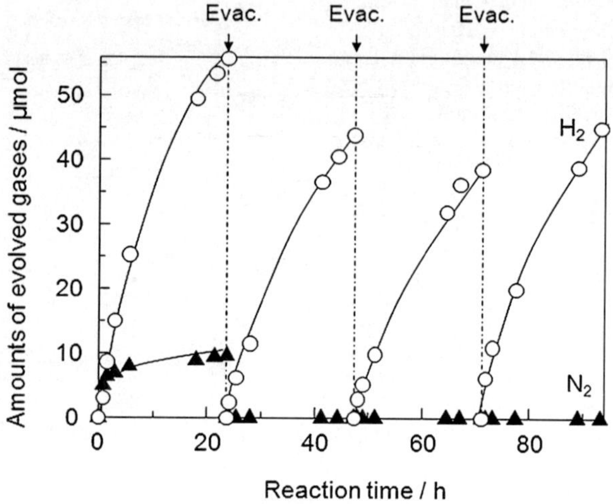

Fig. 10 Time course of photocatalytic H_2 evolution on Pt-loaded Ta_3N_5. Reaction conditions: 0.4 g of catalyst, 200 mL aqueous solution containing methanol (10 vol.%), 300 W xenon lamp light source, Pyrex top irradiation-type reaction vessel with cutoff filter ($\lambda > 420$ nm)

oxidized during H_2 evolution in both cases. TaON showed similar and somewhat superior photocatalytic performance for H_2 evolution to that of Ta_3N_5. Thus, Ta_3N_5 and TaON can be regarded as essentially stable photocatalysts for H_2 evolution.

Figure 11 shows the time courses of O_2 evolution over Ta_3N_5 in an aqueous $AgNO_3$ solution under different conditions. When O_2 evolution was performed without the addition of La_2O_3 powder, the activity of O_2 evolution was quite low and was exceeded by N_2 evolution. The relatively high N_2 evolution is attributed to the anodic dissociation of oxynitrides according to the following reaction:

$$2N^{3-} + 6h^{+} \rightarrow N_2 \tag{6}$$

The oxidation of N^{3-} via valence-band holes competes with the oxidation of water (Eq. 3). However, the addition of La_2O_3 powder resulted in a remarkable enhancement of O_2 evolution and suppression of N_2 evolution. The pH of the reactant solution decreases with the progress of this reaction because H^+ is released (Eq. 5). The La_2O_3 powder maintained the pH of the reaction solution at about 8.5, balancing the consumption of OH^- ions that occurs with the progress of O_2 evolution in the presence of Ag^+. Thus, La_2O_3 functions as a buffer to maintain the pH at about 8.5. Any other buffers to maintain the pH above 8.5 are not available due to the precipitation of Ag^+ by OH^-. It is clear that the activity of O_2 evolution is strongly affected by the pH of the reaction solution, and alkaline conditions appear to be favorable for enhancing O_2 evolution and suppressing N_2 evolution. The addition of La_2O_3 is an effective means of maintaining the pH of the reaction solution. The dominant reaction is therefore controllable by changing the pH of the reaction solution. Similarly, TaON showed better activity of O_2 evolution in alkaline condition compared with the case in acidic condition. However, N_2 evolution on TaON was inhibited in both acidic and basic conditions.

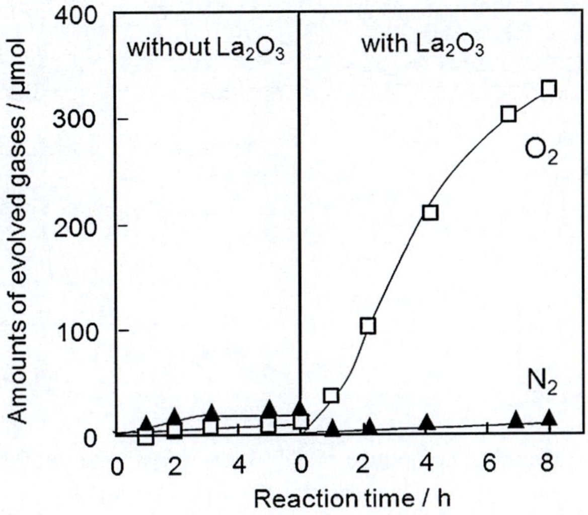

Fig. 11 Time course of photocatalytic H_2 evolution on Ta_3N_5 with and without La_2O_3 addition as a pH buffer. Reaction conditions: 0.4 g of catalyst, 200 mL aqueous solution containing silver nitrate (0.01 M), 300 W xenon lamp light source, Pyrex top irradiation-type reaction vessel with cutoff filter ($\lambda > 420$ nm)

As such, both Ta_3N_5 and TaON proved to be potential photocatalysts for the splitting of H_2O into H_2 and O_2 under visible-light irradiation. Although other narrow-band-gap photocatalysts such as sulfides or phosphides have valence bands consisting primarily of S3p or P3p orbitals, such non-oxide photocatalysts are generally unstable against water oxidation due to photo-anodic dissociation [22, 23, 35]. The stability of the (oxy)nitrides presented here under alkaline conditions is a distinctive and advantageous feature, differentiating these materials from previously examined non-oxide photocatalysts.

2.2 Ti^{4+}-Based Oxynitrides

Oxynitrides based on Ti^{4+} with d^0 electronic configuration were also examined as another possible material class for visible-light photocatalysis. Several titanium-based oxides, such as TiO_2 and $SrTiO_3$, are known to be active photocatalysts for overall water splitting [3–5, 7–10]. Generally, the band-gap energies of Ti^{4+}-based compounds are smaller than those of Ta^{5+}-based compounds. This is mainly due to the lower potential of conduction-band minimum of Ti^{4+}-based compounds. This minimum consists mainly of Ti3d orbitals, rather than the Ta5d orbitals found in the conduction-band minimum of Ta^{5+}-based materials. However, most Ti^{4+}-based oxides are known to have ability to evolve both H_2 and O_2. Therefore, it is expected that Ti^{4+}-based oxynitrides to decompose water under visible-light. In addition, it should be noted that Ti^{4+}-based materials are more cost-effective than Ta^{5+}-based materials because Ti is a cheaper and more abundant. Therefore, Ti^{4+}-based oxynitrides are important materials in terms of very large-scale solar energy utilization.

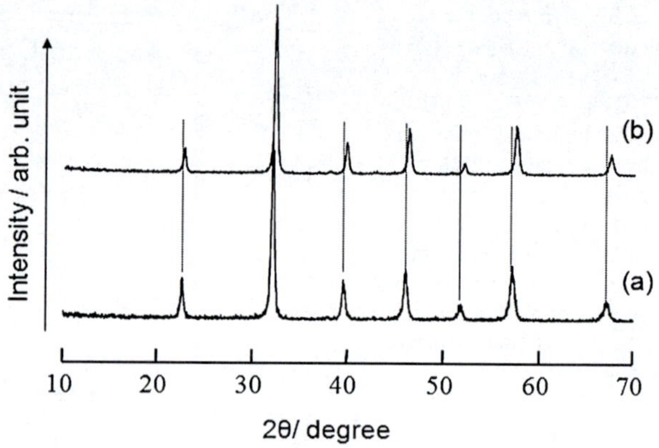

Fig. 12 XRD patterns of (**a**) $LaTiO_2N$ and (**b**) $Ca_{0.25}La_{0.75}TiO_{2.25}N_{0.25}$

A typical compound of Ti^{4+}-based oxynitride with d^0 electronic configuration is $LaTiO_2N$ [36]. This material has a perovskite structure that is isostructural to that of $LaTaON_2$. The detail of the crystallographic analysis has already been reported. When a binary oxide of TiO_2 was nitrided by dry NH_3 in a similar condition to the synthesis of Ta_3N_5, TiN was produced instead of Ti_3N_4. In this case, original valence of Ti^{4+} was reduced to Ti^{3+} during nitridation because NH_3 provides strongly reductive atmosphere at high temperature. When the nitridation of TiO_2 was carried out in low temperature, very small part of oxygen in the lattice was replaced by nitrogen, namely to form N-doped TiO_2 [37]. However, the formation of any clear crystalline phase consisting of Ti^{4+} and O^{2-}/N^{3-} has never been recognized. Since the absorption coefficient of N-doped TiO_2 in visible region is quite low on account of the small nitrogen content, this type of materials are not suitable for an application to huge scale photon energy conversion. Therefore, the study of photocatalysis in the d^0-Ti^{4+}-based oxynitrides was focused on $LaTiO_2N$ and its variant with modified composition [38, 39]. A partial substitution of Ca^{2+} for La^{3+} in $LaTiO_2N$ creates single negative charges, which is compensated by creating single positive charges with the substitution of O^{2-} for N^{3-} to perform an isostructural substitution. This system can namely be regarded as a solid solution of $CaTiO_3$ and $LaTiO_2N$, generally formulated as $Ca_xLa_{1-x}TiO_{2+x}N_{1-x}$.

The Ti^{4+}-based oxynitrides were prepared in a similar manner to that of above-introduced perovskite-type Ta^{5+}-based (oxy)nitrides. Figure 12 shows XRD patterns of $LaTiO_2N$ and $Ca_{0.25}La_{0.75}TiO_{2.25}N_{0.75}$ prepared by thermal ammonolysis at 1,223 K. The diffraction pattern of $LaTiO_2N$ agrees with the reference data, which is similar to that of $Ca_{0.25}La_{0.75}TiO_{2.25}N_{0.75}$. This indicates that isostructural replacement was successfully performed for the Ca^{2+}-doped variant. Although the diffraction patterns were similar for the two samples, each diffraction peak of Ca^{2+}-doped sample was located at a slightly higher diffraction angle compared to the corresponding peaks in $LaTiO_2N$, indicative of lattice shrinkage in Ca^{2+}-substituted sample. The ionic radii of La^{3+} and Ca^{2+} are 150 pm and 148 pm [40], respectively, which are close enough not to cause lattice shrinkage. The change in anionic composition accompanied by La/Ca-substitution, i.e., the change of N/O

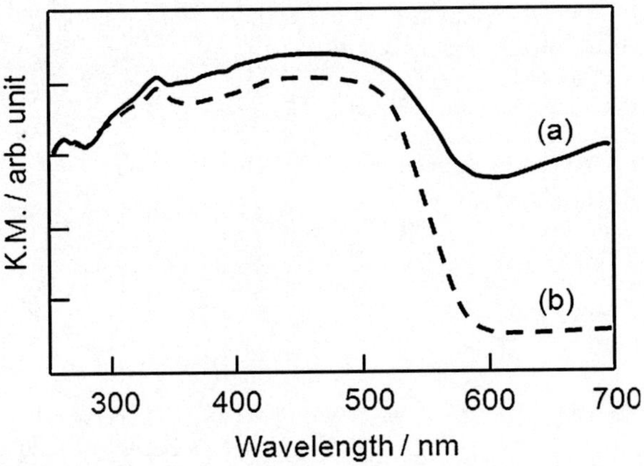

Fig. 13 UV-visible diffuse reflectance spectra of $LaTiO_2N$: **a** as-prepared **b** heated in air at 573 K

contents, leads to the decrease of lattice constant due to the difference in the radii between O^{2-} (126 pm) and N^{3-} (157 pm).

Figure 13 shows the UV-visible diffuse reflectance spectra of $LaTiO_2N$ samples. The spectrum of as-prepared $LaTiO_2N$ consists of two broad bands. These absorption bands are attributed to the band-gap and d-d transitions of the reduced Ti^{3+} species in shorter and longer wavelength regions, respectively. Heating the sample in air at 573 K for 24 h caused no change in the XRD pattern, but weakened the band in longer wavelength region. The absorption edge of charge transfer band in $LaTiO_2N$ was found to be about 600 nm. Thus, the band-gap energy of $LaTiO_2N$ is approximately 2.1 eV.

Figure 14 shows the evolution of H_2 from an aqueous methanol solution. No reaction took place in the dark, but H_2 evolution started upon irradiation. The rate of H_2 evolution in the second and subsequent runs with intermittent evacuation was somewhat slower than that of the first run, but continued to proceed steadily. Up to 1.6 μmol of N_2 was detected in the first run, but was not detectable by gas chromatography after the second run. Therefore, it is concluded that the oxynitride was stable during the reaction.

Figure 15a shows the time course of O_2 evolution under visible-light irradiation ($\lambda > 420$ nm). In this case, La_2O_3 was also added to the reaction solution for pH control as in the case of Ta_3N_5; otherwise the activity of O_2 evolution was poor and experienced photo-anodic dissociation. $LaTiO_2N$, prepared at 1,223 K, was used in the following photoreactions because of increased activity for O_2 and H_2 evolution under visible-light irradiation. No reaction took place in the dark, and O_2 evolution began with the irradiation. The total amount of O_2 evolved over 43 h is estimated to be 440 μmol. This amount corresponds to ca. 60% of the lattice oxygen in the catalyst (0.2 g of $LaTiO_2N$). There was no noticeable difference in the XRD pattern of the sample before and after the reaction, except for the emergence of a diffraction peak attributable to metallic Ag. The evolution of O_2 is therefore attributable to the oxidation of water on $LaTiO_2N$ under visible-light irradiation.

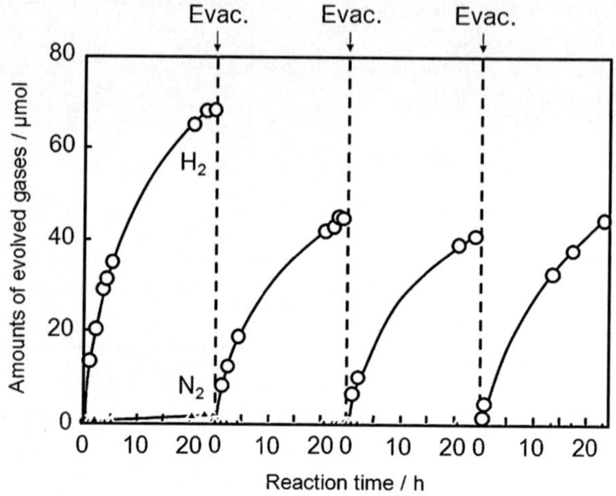

Fig. 14 Time course of photocatalytic H_2 evolution on Pt-loaded $LaTiO_2N$. Reaction conditions: 0.2 g of catalyst, 200 mL aqueous solution containing methanol (20 vol.%), 300 W xenon lamp light source, Pyrex top irradiation-type reaction vessel with cutoff filter ($\lambda > 420$ nm)

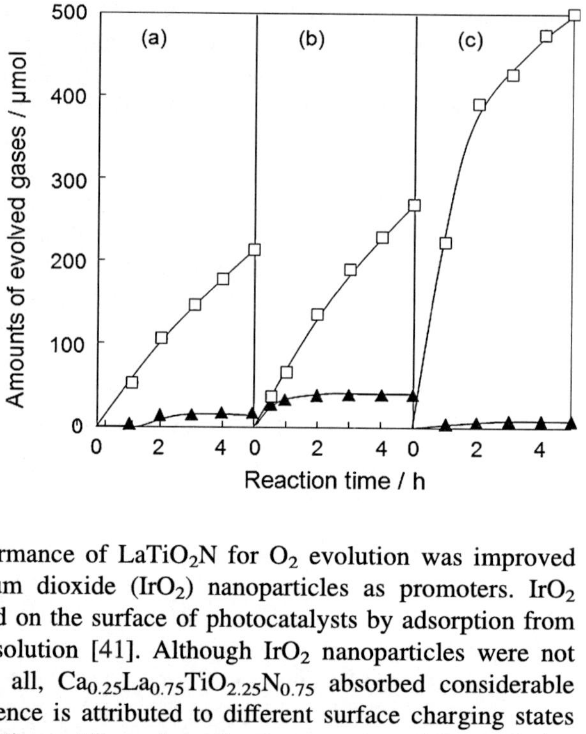

Fig. 15 Time courses of photocatalytic O_2 evolution on $LaTiO_2N$-based materials: **a** $LaTiO_2N$, **b** $Ca_{0.25}La_{0.75}TiO_{2.25}N_{0.25}$ (c) IrO_2/$Ca_{0.25}La_{0.75}TiO_{2.25}N_{0.25}$. Reaction conditions: 0.2 g of catalyst (0.2 g of La_2O_3), 200 mL aqueous solution containing silver nitrate (0.01 M), 300 W xenon lamp light source, Pyrex top irradiation-type reaction vessel with cutoff filter ($\lambda > 420$ nm)

The photocatalytic performance of $LaTiO_2N$ for O_2 evolution was improved by modification with iridium dioxide (IrO_2) nanoparticles as promoters. IrO_2 nanoparticles were deposited on the surface of photocatalysts by adsorption from an aqueous colloidal IrO_2 solution [41]. Although IrO_2 nanoparticles were not adsorbed onto $LaTiO_2N$ at all, $Ca_{0.25}La_{0.75}TiO_{2.25}N_{0.75}$ absorbed considerable amount of IrO_2. This difference is attributed to different surface charging states between the two materials. Figure 15b, c show the time courses of O_2 evolution under visible-light irradiation on bare and IrO_2(2 wt%)-loaded $Ca_{0.25}La_{0.75}$-$TiO_{2.25}N_{0.75}$, respectively. $Ca_{0.25}La_{0.75}TiO_{2.25}N_{0.75}$ was prepared at 1,223 K. There was no noticeable difference in the activity of O_2 evolution between $LaTiO_2N$ and $Ca_{0.25}La_{0.75}TiO_{2.25}N_{0.75}$. However, as shown $Ca_{0.25}La_{0.75}$-$TiO_{2.25}N_{0.75}$ loaded with IrO_2-nanoparticles showed higher activity for O_2

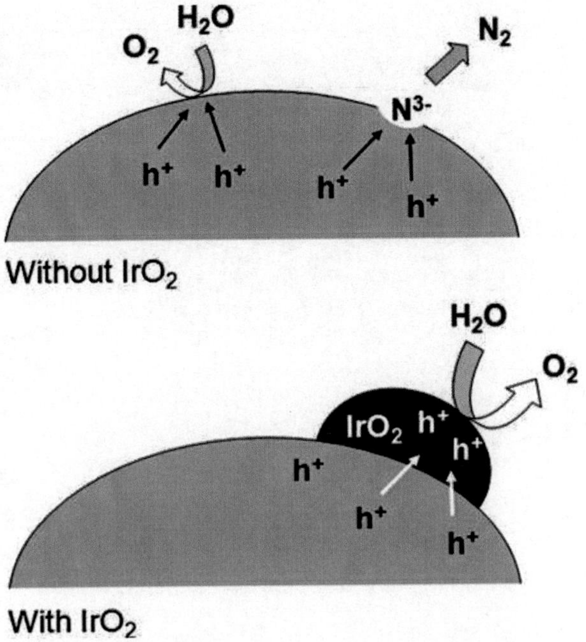

Fig. 16 Effect of IrO_2-loading on the improvement of photocatalytic O_2 evolution by an (oxy)nitride

evolution and lower activity for N_2 evolution than those with unmodified $Ca_{0.25}La_{0.75}TiO_{2.25}N_{0.75}$ and $LaTiO_2N$. With the evolution of 500 μmol of O_2 (5 h irradiation), only 4 μmol of N_2 was produced, i.e., 99% of holes were used for water oxidation. This result clearly demonstrates that it is possible to stabilize this system via kinetic control with respect to the degradation of the oxynitride with the appropriate modifications, as schematically depicted in Fig. 16.

$LaTiO_2N$ and its variant of $Ca_{0.25}La_{0.75}TiO_{2.25}N_{0.75}$ were shown to evolve H_2 and O_2 in the presence of sacrificial electron donor and acceptor, respectively. Both the activities of H_2 and O_2 were moderately high for $LaTiO_2N$. O_2 evolution on $LaTiO_2N$ was remarkably improved by the combination of IrO_2 as a cocatalyst to promote O_2 formation, and the Ca^{2+}-substitution is necessary to make the IrO_2 nanoparticles adsorbed on the surface. However, simultaneous H_2 and O_2 evolution from water in the absence of sacrificial reagent have yet to be successful.

2.3 *Oxysulfide*

Sulfur-containing materials are also an interesting research area. Some metal sulfides, such as CdS and CdSe, have suitable band-gap positions for absorption of visible-light and decomposition of water. The energy level of S3p state is more negative than that of O2p state, resulting in a shallower valence band maximum and narrower band-gap in sulfides compared to oxides. However, in practice, these metal chalcogenides are not stable for water oxidation to O_2 because surface

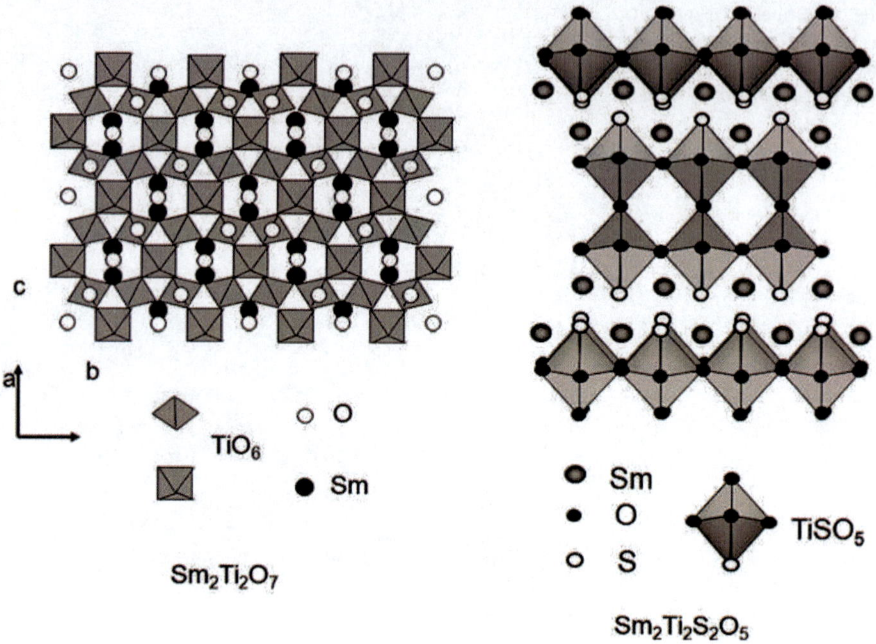

Fig. 17 Schematic structures of $Sm_2Ti_2O_7$ and $Sm_2Ti_2S_2O_5$

S^{2-} and Se^{2-} are more susceptible to oxidation than water, i.e., the CdS or CdSe itself is oxidized [22, 23]. Oxysulfides, on the other hand, are expected to be somewhat durable against self-oxidation by stabilizing surface S^{2-} via hybridization of S3p and O2p orbitals.

Next, we will introduce the photocatalytic and photoelectrochemical properties of an oxysulfide, $Sm_2Ti_2S_2O_5$ [42, 43]. $Sm_2Ti_2S_2O_5$ can be synthesized by the following two methods. The first is a solid-state reaction carried out by heating a mixture of Sm_2S_3, Sm_2O_3 and TiO_2 (molar ratio, $Sm_2S_3/Sm_2O_3/TiO_2 = 2/1/6$) in an evacuated quartz tube at 1,273 K for 1 week [42]. The other method is the sulfurization of a corresponding oxide under H_2S flow at 1,223 K. Here, the results obtained from a sample synthesized by the former method are presented [43].

Figure 17 shows the structures of $Sm_2Ti_2O_7$ [44] and $Sm_2Ti_2S_2O_5$ [45]. $Sm_2Ti_2S_2O_5$ belongs to I4/mmm crystallographic group and has a similar structure to the Ruddlesden-Popper type layered perovskite oxide, represented by $(AX)(ABX_3)_2$ (A, B: metal cations, X: anions), and the layers are composed of S-(TiO_2)-O-(TiO_2)-S double octahedra. The oxide with corresponding cationic composition is $Sm_2Ti_2O_7$, which has a pyrochlore structure. Although S^{2-} was substituted for equimolar amount of O^{2-}, structural transformation occurs because of the large difference in ionic radius between S^{2-} and O^{2-}.

Figure 18 shows UV-visible diffuse reflectance spectra for $Sm_2Ti_2O_7$ and $Sm_2Ti_2S_2O_5$. The absorption band edge of $Sm_2Ti_2S_2O_5$ lies at around 600 nm with a tail that extends up to 650 nm. The absorption band edge of $Sm_2Ti_2S_2O_5$ is

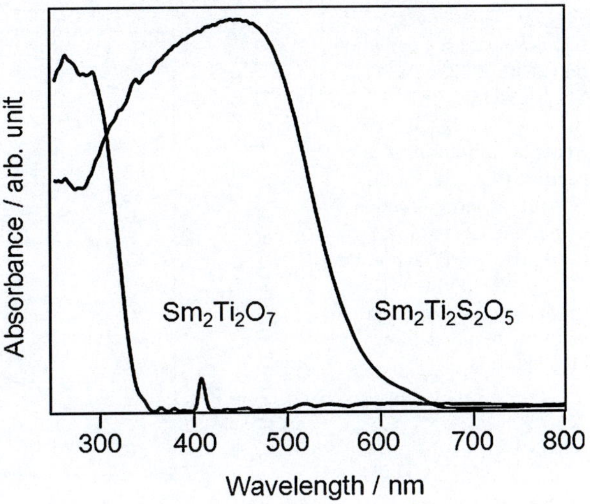

Fig. 18 UV-visible diffuse reflectance spectra for $Sm_2Ti_2O_7$ and $Sm_2Ti_2S_2O_5$

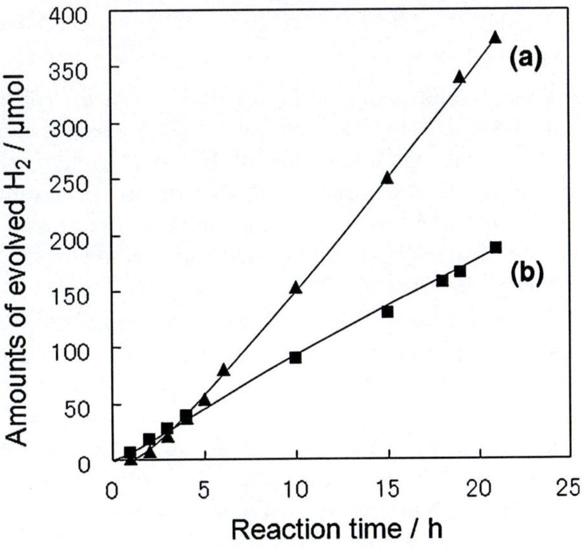

Fig. 19 Time courses of photocatalytic H_2 evolution on Pt-loaded $Sm_2Ti_2S_2O_5$. **a** From aqueous $Na_2S + Na_2SO_3$ solution (0.01 M) using a Pt-photodeposited catalyst, and **b** from aqueous methanol solution (10 vol.%) using a Pt-impregnated catalyst. Reaction conditions: 0.2 g of catalyst, 200 mL aqueous solution (20 vol.%), 300 W xenon lamp light source, Pyrex top irradiation-type reaction vessel with cutoff filter ($\lambda \geq 440$ nm)

shifted about 250 nm from that of $Sm_2Ti_2O_7$. The band-gap energy of $Sm_2Ti_2S_2O_5$ is estimated to be 2.1 eV. The observed band-gap narrowing is likely due to the contribution of S3p orbitals to the upper region of valence band.

Figure 19 shows the time courses of H_2 evolution on 1.0 wt% Pt-$Sm_2Ti_2S_2O_5$ in the presence of Na_2S-Na_2SO_3 or methanol under visible-light irradiation. In the case of the Na_2S-Na_2SO_3 system, a faster rate of H_2 evolution was obtained by loading Pt of 1.0 wt% via in situ photodeposition. In the early stage of the reaction (~3 h), 1.0 μmol of H_2PtCl_6 is reduced to Pt^0 as an H_2 evolution promoter on the $Sm_2Ti_2S_2O_5$ surface. After the induction period, H_2 evolved steadily, indicating that excited electrons in the oxysulfide can reduce H^+ to H_2. XRD and XPS

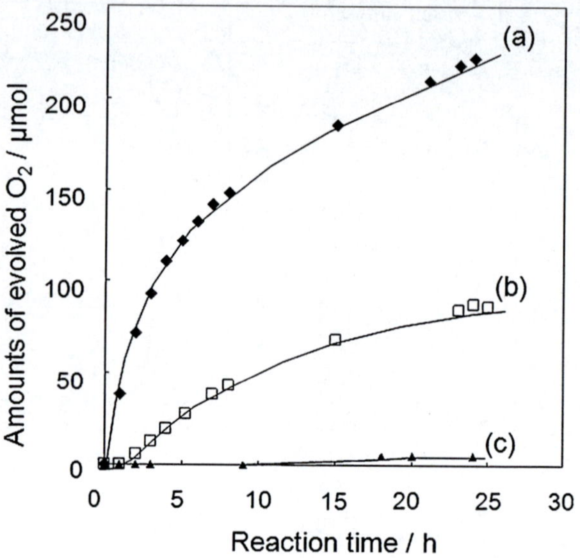

Fig. 20 Time courses of photocatalytic O_2 evolution on (**a**) IrO_2(0.25wt%)/$Sm_2Ti_2S_2O_5$, (**b**) $Sm_2Ti_2S_2O_5$, and (**c**) CdS. Reaction conditions: 0.2 g of catalyst (0.2 g of La_2O_3), 200 mL aqueous solution containing silver nitrate (0.01 M), 300 W xenon lamp light source, Pyrex top irradiation-type reaction vessel with cutoff filter ($\lambda \geq 440$ nm)

analyses confirmed that H_2 evolution proceeded without any degradation of $Sm_2Ti_2S_2O_5$. This indicates that $Sm_2Ti_2S_2O_5$ also functions as a stable photocatalyst for the photoreduction of H^+ to H_2 under visible-light irradiation. This photocatalyst also showed steady activity of H_2 evolution under visible-light irradiation in the presence of methanol as a sacrificial electron donor. In this case, Pt was loaded via the impregnation method from $[Pt(NH_3)_4]Cl_2$, followed by reduction in H_2 at 573 K.

Time courses of O_2 evolution on $Sm_2Ti_2S_2O_5$ in $AgNO_3$ solution under visible-light irradiation are shown in Fig. 20. In evaluating the as-prepared $Sm_2Ti_2S_2O_5$ samples, O_2 was evolved with visible-light irradiation after short induction period (ca. 1 h). The XRD patterns of $Sm_2Ti_2S_2O_5$ samples before and after the reaction were essentially identical except for the appearance of diffraction peaks assigned to metallic Ag in the latter sample. O_2 evolution was examined in a basic solution (ca. pH 8.5) buffered by La_2O_3 as described previously. When the reaction was carried out in the absence of La_2O_3, the rate of O_2 evolution was suppressed, as in oxynitrides. This implies that alkaline conditions are favorable for O_2 evolution on $Sm_2Ti_2S_2O_5$. For comparison, time course on CdS under the same experimental condition is also shown in Fig. 20. CdS in the $AgNO_3$ solution turned black immediately because of dissociation of CdS by the formation of Ag_2S, and only a trace amount of O_2 evolution was observed.

Modification of $Sm_2Ti_2S_2O_5$ with IrO_2 was also examined to promote O_2 evolution. Although adsorption of colloidal IrO_2 onto $Sm_2Ti_2S_2O_5$ was not achieved, a prior treatment of $Sm_2Ti_2S_2O_5$ with $Ca(OH)_2$ made adsorption possible. As shown in Fig. 20, the photocatalytic activity of O_2 evolution was tripled upon modification with IrO_2. These results demonstrate that IrO_2, in nanoparticle form, functions as an effective O_2 evolution promoter for $Sm_2Ti_2S_2O_5$.

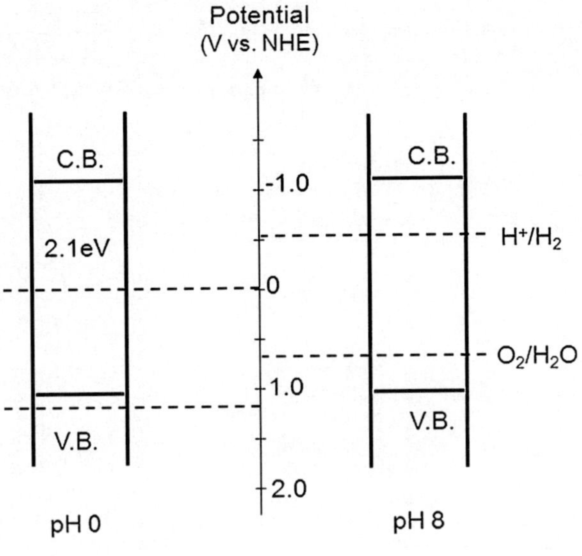

Fig. 21 Band-gap position of $Sm_2Ti_2S_2O_5$ at pHs 0 and 8

The band-gap position of $Sm_2Ti_2S_2O_5$ was measured by electrochemical methods. The estimated band structure of $Sm_2Ti_2S_2O_5$ is illustrated in Fig. 21. The conduction and valence bands have appropriate potentials for reduction of H^+ to H_2 and oxidation of water to O_2 at a pH of 8 under irradiation, consistent with the results of photocatalytic reactions. At a pH of 0, the tops of the valence bands of Ti^{4+}-based metal oxides are located at ca. 3 V, whereas that for $Sm_2Ti_2S_2O_5$ is more negative than 3 V, resulting in a smaller band-gap energy. This also suggests that S3p orbitals contribute to the valence band of $Sm_2Ti_2S_2O_5$. Interestingly, the flat band potential of $Sm_2Ti_2S_2O_5$ is almost independent of pH. In metal-oxide photocatalysts such as TiO_2, there is a flat band potential shift to more negative potentials with increasing pH because of dissociation of surface hydroxyl groups. Although $Sm_2Ti_2S_2O_5$ has surface hydroxyl groups as shown by XPS analysis, the flat band potential of $Sm_2Ti_2S_2O_5$ is not sensitive to pH. A possible explanation for this is the small acid dissociation constant of hydroxyl groups on $Sm_2Ti_2S_2O_5$. S^{2-} in S–(TiO_2)–O–(TiO_2)–S or Sm^{3+} might affect this feature of the hydroxyl groups.

$Sm_2Ti_2S_2O_5$, a Ti^{4+}-based oxysulfide, was confirmed to be a stable, visible-light-driven photocatalyst for the oxidation of water to O_2 and reduction of H^+ to H_2. The bottom of the conduction band and the top of the valence band are sufficiently positive and negative to allow for water reduction and oxidation, respectively. Therefore, this oxysulfide performs as a stable photocatalyst for H_2 or O_2 evolution from aqueous solution containing a sacrificial electron donor or acceptor under visible-light. The deposition of IrO_2 nanoparticles on $Sm_2Ti_2S_2O_5$ as an O_2 evolution promoter enhanced the photooxidation activity by 3 times, demonstrating that such surface modification can also improve the photocatalytic activity of $Sm_2Ti_2S_2O_5$.

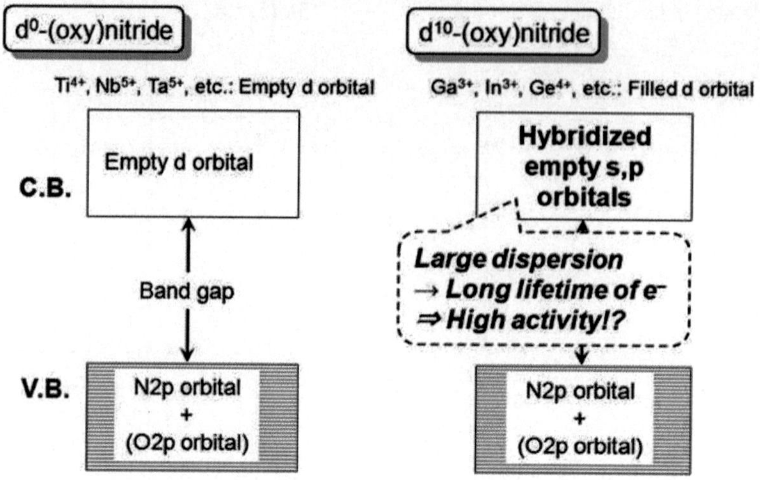

Fig. 22 A schematic illustration of the difference in electronic band structure between d^0- and d^{10}-based compounds

3 d^{10}-Type Photocatalysts

3.1 *Difference in Band Structure Between d^0- and d^{10}-Compound*

As described in the previous sections, compounds including early transition-metal cations with d^0 electronic configuration, such as Ta_3N_5, $LaTiO_2N$ and $Sm_2Ti_2S_2O_5$ are potential candidates for overall water splitting under visible-light. From the viewpoint of the electronic band structure, however, d^{10}-based semiconducting materials are advantageous over the d^0 configurations as photocatalysts. In metal-oxides, for example, the top of the valence band consists of O2p orbitals and the bottom of the conduction band is composed of hybridized s,p orbitals (Fig. 22) [15–20]. Hybridized s,p orbitals possess large dispersion, leading to increased mobility of photogenerated electrons in the conduction band and increased photocatalytic activity. This sparked study of (oxy)nitrides or oxysulfides with the d^{10} electronic configuration as photocatalysts for overall water splitting.

3.2 *β-Ge_3N_4*

As a typical-metal-nitride with d^{10} electronic configuration, β-Ge_3N_4 has been shown to act as a photocatalyst for stoichiometric water splitting into H_2 and O_2 when modified with ruthenium dioxide (RuO_2) nanoparticles as promoters [46, 47]. Germanium nitride is a well-known polymorphic compound, exhibiting

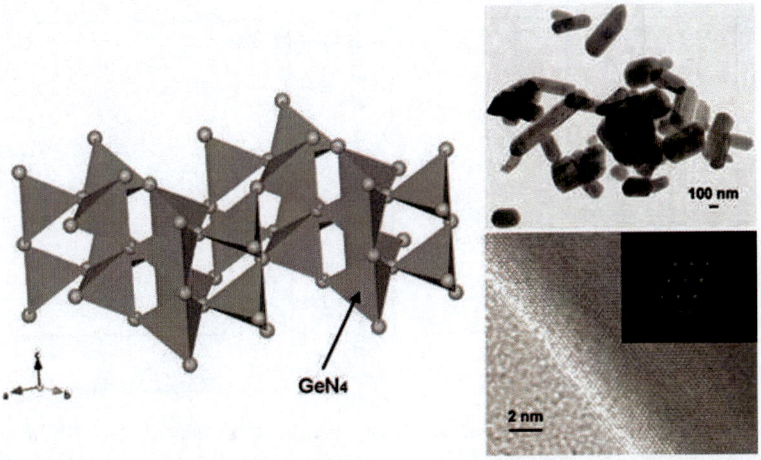

Fig. 23 Crystal structure of β-Ge_3N_4 (*left*) and HR-TEM images and an electron diffraction pattern (*right*)

α, β, γ and δ phases with various crystal structures [48–50]. The synthesis of Ge_3N_4 was pioneered by Johnson in 1930 [51]. This material can be readily obtained by heating Ge or GeO_2 powder under a flow of ammonia; the crystal phase of the product is dependent upon the process conditions, such as temperature, duration time, flow rate and so on. Our comparative experiments have shown that the β-phase exhibits the highest activity. Additionally, γ and δ phases were difficult to obtain under the present synthetic conditions. In a typical synthesis, β-Ge_3N_4 can be synthesized from GeO_2 powder by nitridation under a flow of NH_3 at 1,123–1,173 K for ~20 h [52]. The compound consists of GeN_4 tetrahedrons that connect to share each corner, as schematically shown in Fig. 23 (left). Figure 23 (right) shows typical HR-TEM images and an electron diffraction pattern of the as-prepared β-Ge_3N_4. As can be seen, the material consists of primary well-crystallized rod-like particles with a hexagonal crystal system. The photocatalytic activity of the as-synthesized β-Ge_3N_4 for overall water splitting was negligible. However, when modified with RuO_2 nanoparticles, the material became photocatalytically active for the stoichiometric evolution of H_2 and O_2 from pure water. This demonstration of the photodecomposition of water is the first involving a non-oxide-based photocatalyst.

Effects of reaction conditions and post-treatment on activity of RuO_2/β-Ge_3N_4 were examined. The photocatalytic performance of RuO_2-loaded β-Ge_3N_4 was strongly dependent on the pH of the aqueous solution, with the activity increasing with decreasing pH from neutral with the addition of H_2SO_4, passing through a maximum at a pH of 0 and then decreasing [47]. Figure 24 shows the time courses of overall water splitting on the RuO_2-loaded β-Ge_3N_4 under UV irradiation ($\lambda > 200$ nm) at pH's of 7 and 0. Both H_2 and O_2 evolved, but the rates of both H_2 and O_2 evolution decreased with reaction time. The reason for this deactivation will be explained later. In pure water (pH $\approx$ 7), the rate of O_2 evolution was lower

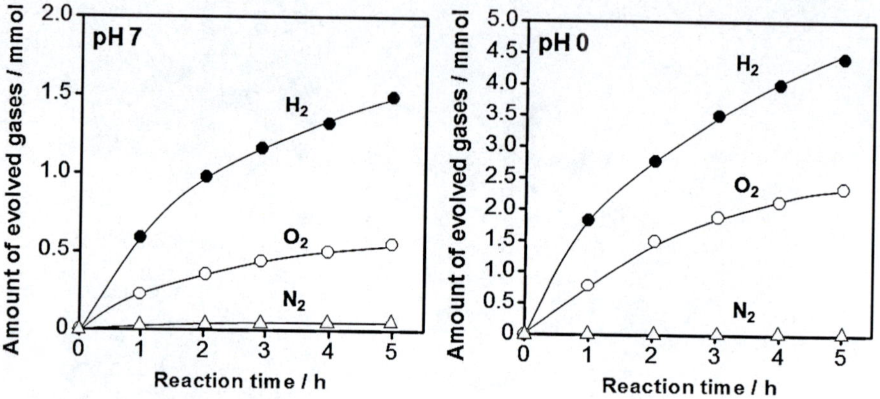

Fig. 24 Time courses of overall water splitting on the RuO_2-loaded β-Ge_3N_4 under UV irradiation ($\lambda > 200$ nm) at pH's of 7 and 0. Reaction conditions: 0.5 g of catalyst (1 wt% RuO_2-loaded), 390 mL pure water or aqueous H_2SO_4 solution, 450 W high-pressure mercury lamp light source, Quartz inner irradiation-type reaction vessel

than that expected from the stoichiometry. Furthermore, a relatively high level of N_2 evolution was observed continuously in the reaction, indicating that the material is gradually decomposed by photogenerated holes. This decomposition consumes photogenerated holes that would otherwise be consumed in the oxidation of water, as described earlier (Eq. 6). However, as the concentration of H_2SO_4 was increased (up to 1 M, pH 0), the rate of H_2 and O_2 evolution became stoichiometric and increased. At H_2SO_4 concentrations exceeding 1 M, the evolution rates began to degrade. N_2 evolution, indicative of partial decomposition of the nitride photocatalyst, was also increasingly suppressed at higher H_2SO_4 concentrations. This characteristic pH dependence of the RuO_2-loaded β-Ge_3N_4 photocatalyst deviates significantly from the general character of transition-metal-oxide-based photocatalysts. For example, NiO-loaded $NaTaO_3$ [13] favors alkaline solutions for overall water splitting. It appears that the enhanced activity of RuO_2-loaded β-Ge_3N_4 with increasing H_2SO_4 concentration is associated with the suppression of hydrolysis. In general, typical-metal (oxy)nitrides are inherently unstable in basic media due to hydrolysis, but are stable in acidic media. Another possible explanation for this improvement is the promotion of water reduction (Eq. 2) due to an increase in H^+ concentration with the addition of H_2SO_4. Increasing the H_2SO_4 concentration above 1 M reduces activity, possibly a result of the detrimental effect of high H^+ concentrations in the water oxidation reaction (Eq. 3). It was also found that a high-pressure treatment of β-Ge_3N_4 with NH_3 gas results in a four-fold increase in activity by reducing the density of defects [53].

For non-oxide photocatalysts, it is important to discuss the effect of self-decomposition by photogenerated holes in the valence band of the material, since this process competes with the water photo-oxidation reaction. An interesting feature is that β-Ge_3N_4, having lower crystallinity, released N_2 continuously with reaction time, while well-crystallized material did not, even under neutral reaction conditions [52]. The observed N_2 evolution during the reaction occurs near the

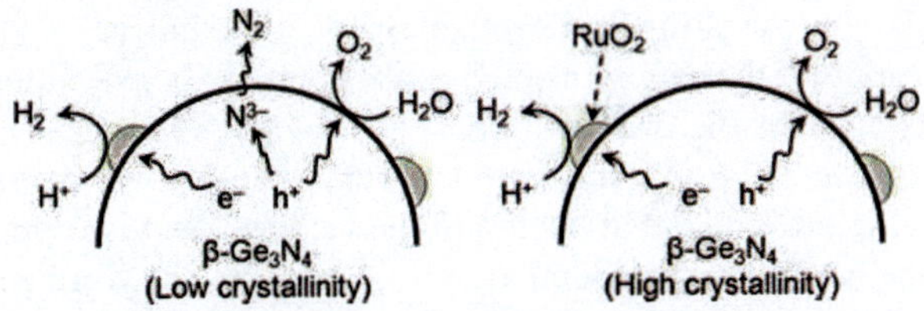

Fig. 25 A schematic illustration of overall water splitting using β-Ge_3N_4 with different degree of crystallinity

catalyst surface, while O_2 evolution takes place directly on the Ge_3N_4 surface. Therefore, photogenerated holes contributing to O_2 evolution must migrate over a longer distance compared to those involved in N_2 evolution. This situation is schematically illustrated in Fig. 25. As the crystallinity of Ge_3N_4 improves, photogenerated electrons and holes can more readily migrate longer distances, resulting in a higher likelihood that the electrons and holes will reach the surface reaction sites without recombining. As mentioned above, a decrease in the pH of the reactant solution suppresses N_2 evolution in overall water splitting using RuO_2-loaded β-Ge_3N_4 by inhibiting hydrolysis-induced deactivation in neutral or basic media, leading to enhanced activity. In addition to such a kinetic control, N_2 evolution can be effectively suppressed by improving the crystallinity of β-Ge_3N_4. In metal-oxide photocatalysts, high crystallinity has a positive effect on the rates of H_2 and O_2 evolution in overall water splitting, since the density of defects, which act as recombination centers between photogenerated carriers, decreases with increasing crystallinity [7, 13, 16–20]. In metal-nitride photocatalysts, however, high crystallinity is a crucially important factor affecting the stability of the material in addition to promoting H_2 and O_2 evolution in overall water splitting.

Even under the optimal reaction conditions, the rates of H_2 and O_2 evolution via water splitting on RuO_2-loaded β-Ge_3N_4 decreased with reaction time. After 24 h of reaction, however, the total production of H_2 and O_2 reached 11.2 mmol, substantially greater than the amount of RuO_2-loaded β-Ge_3N_4 used (0.5 g; 1.8 mmol of β-Ge_3N_4). This result is indicative of a catalytic cycle in the reaction. In order to examine the reason(s) for the decreased activity, catalysts, after reacting for 24 h in an aqueous 1 M H_2SO_4 solution, were investigated by a variety of techniques. XRD and XPS revealed that the crystal structure and the valence state of the catalyst are almost unchanged by the photocatalytic reaction. However, inductively coupled plasma (ICP) mass spectrometry revealed that the solution contained approximately 7.5 mg of Ge cations and no Ru cations after the reaction. Since GeO_2 is slightly soluble in aqueous media [54], Ge–O species on the β-Ge_3N_4 surface are presumably eluted from the photocatalyst during the reaction. Such elusion of the catalyst component affected the interface between β-Ge_3N_4 and loaded RuO_2 nanoparticles before and after the reaction. Before the reaction, the surface of β-Ge_3N_4 was smooth, and β-Ge_3N_4 and the loaded RuO_2 nanoparticle were in contact with each other. However, the surface structure became rougher after the reaction, which is attributed to elution of the surface Ge cations. This situation would cause a partial collapse and relaxation of the interfacial structure between β-Ge_3N_4 and the RuO_2 nanoparticles. This degredation of the structure hinders prompt electron migration from the β-Ge_3N_4 to the RuO_2, thereby resulting in deactivation of the catalyst. Another reason for the deactivation of the catalytic activity was found to be photoreduction of O_2; however, this is a reversible deactivation.

The electronic band structure of β-Ge_3N_4 was examined by DFT calculations, which revealed that the top of the valence band is formed by N2p orbitals, whereas the bottom of the conduction band consists of hybridized 4s,4p orbitals of Ge. This result indicates that photoexcitation under irradiation occurs from the N2p orbitals to the Ge4s,4p hybridized orbitals. Water oxidation to produce O_2 on conventional metal-oxide photocatalysts takes place as a result of contributions from photogenerated holes in the valence band consisting of O2p orbitals. Conventional non-oxide photocatalysts, such as CdS and CdSe, are unable to achieve overall water splitting, as mentioned earlier. It is therefore noteworthy that the N2p orbitals in the valence band both generate photogenerated holes and contribute to the photocatalytic water oxidation in the overall water-splitting reaction. Unfortunately, β-Ge_3N_4 has a band-gap of ca. 3.8 eV and is thus only active under UV irradiation. Nevertheless, the discovery of such a non-oxide photocatalyst (β-Ge_3N_4) achieving the same function as metal-oxide photocatalysts provides strong encouragement for research on non-oxide materials for visible-light-driven photocatalysis.

3.3 GaN–ZnO Solid Solution

To devise a new (oxy)nitride with d^{10} electronic configuration that can decompose water under visible-light, the solid solution of GaN and ZnO, $(Ga_{1-x}Zn_x)(N_{1-x}O_x)$, was examined. As both GaN and ZnO have wurtzite structures with similar lattice parameters [55, 56], a solid solution can easily be formed between the two. Taking into account the large band-gap energies of GaN and ZnO (>3 eV), the resulting composite was expected to have a band-gap exceeding 3 eV. However, for II–VI semiconductors, it has been shown that p–d repulsion (e.g., O2p–Zn3d) shifts the valence-band maximum upward without affecting the conduction-band minimum [57]. Similarly, it was hypothesized that p–d repulsion in the GaN–ZnO solid solution (i.e., N2p–Zn3d repulsion) may cause the top of the valence band formed by N2p atomic orbitals to shift to higher potential energy, resulting in a narrower band-gap for GaN. The expected band structure of GaN–ZnO solid solution is illustrated in Fig. 26.

Preparation of the solid solution was first attempted by nitriding a mixture of β-Ga_2O_3 and ZnO under a flow of ammonia at ~1,223 K. The product was a yellow powder. The yellow color clearly indicates visible-light absorption by this material, consistent with the initial hypothesis. Elemental analysis by inductive coupled plasma optical emission spectroscopy (ICP-OES) revealed that the ratios of Ga to N and Zn to O were close to 1, and that the nitrogen and oxygen concentrations increased with increasing gallium and zinc concentrations [58, 59]. X-ray diffraction (XRD) analysis showed that the prepared material had a wurtzite structure similar to GaN and ZnO, and that the diffraction peaks were between

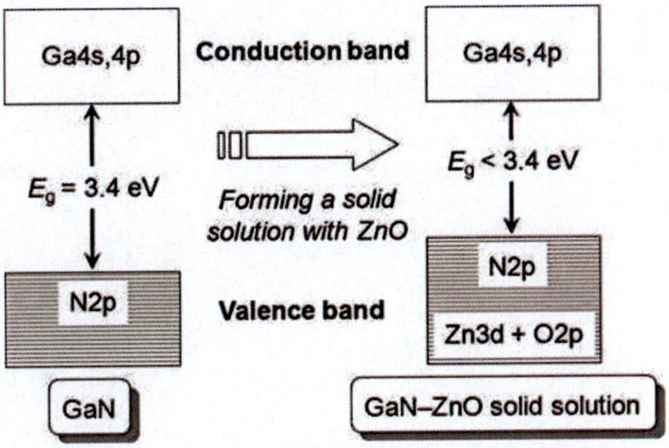

Fig. 26 A schematic illustration of the formation of GaN–ZnO solid solution. Shown are schematic band structures

those of GaN and ZnO (as will be discussed below in more detail). Figure 27a shows typical XRD patterns of samples with various compositions. All samples exhibit single-phase diffraction patterns indicative of the wurtzite structure similar to GaN and ZnO. The positions of the diffraction peaks are shifted successively to lower angles ($2\ \theta$) with increasing Zn and O concentrations, indicating that the samples are not physical mixtures of GaN and ZnO phases but rather solid solutions of GaN and ZnO. This peak shift is reasonable, as the lattice parameters of GaN are larger than those of ZnO (GaN: $a = b = 0.319$, $c = 0.519$ nm. ZnO: $a = b = 0.325$, $c = 0.521$ nm) [55, 56]. This same trend was confirmed by Rietveld analysis using the computer program RIETAN-2000 [59] and neutron powder diffraction analysis has confirmed that the oxygen atoms are substituted directly for the nitrogen atom in $(Ga_{1-x}Zn_x)(N_{1-x}O_x)$, without causing interstitial sites or large amounts of disorder in the material [60].

While GaN and ZnO both have band-gap energies greater than 3 eV and therefore do not absorb visible-light, the $(Ga_{1-x}Zn_x)(N_{1-x}O_x)$ solid solution has absorption edges in the visible region. Figure 27B shows the UV-visible diffuse reflectance spectra for several samples. The absorption edge shifts to longer wavelengths with increasing Zn and O concentration (x) in $(Ga_{1-x}Zn_x)(N_{1-x}O_x)$. The band-gap energies of the solid solutions are estimated to range between 2.4 and 2.8 eV based on the diffuse reflectance spectra. To elucidate the origin of the visible-light absorption of the material, photoluminescence spectroscopy measurements were conducted. The photoluminescence and photoluminescence excitation spectra of $(Ga_{1-x}Zn_x)(N_{1-x}O_x)$ with compositions of $x = 0.05$–0.20 (i.e., GaN-rich) measured at 10 K suggested that the visible-light absorption of these materials occur via Zn-related acceptor levels [61]. The proposed band structure model is illustrated in Fig. 28. This idea, however, contradicts our previous DFT calculations, which indicates that the bottom of the conduction band for $(Ga_{1-x}Zn_x)(N_{1-x}O_x)$ is mainly composed of 4s and 4p orbitals of Ga, while the top of the valence band consists of N2p orbitals followed by Zn 3d and O2p orbitals [62]. This idea remains valid for the solid solution with stoichiometric unit cells. For a dilute solid solution, however, it is necessary to consider the local inhomogeneity

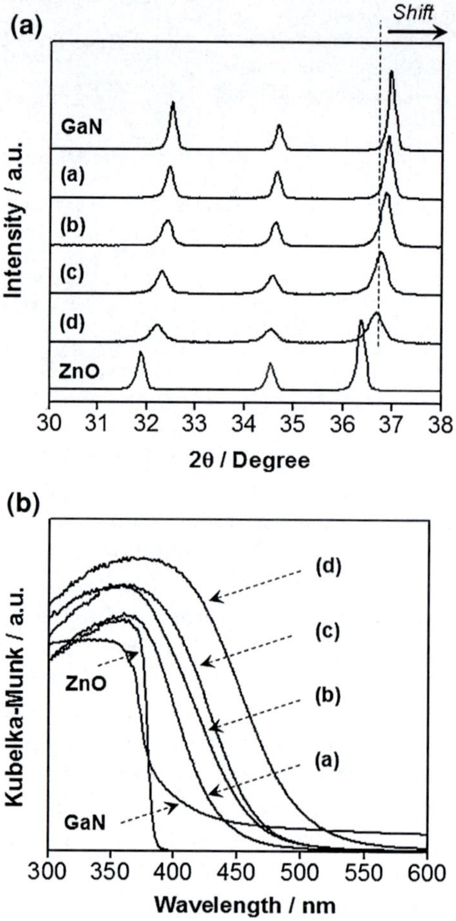

Fig. 27 **a** Powder XRD patterns and **b** UV-visible diffuse reflectance spectra for various $(Ga_{1-x}Zn_x)(N_{1-x}O_x)$ solid solutions with $x =$ (a) 0.05, (b) 0.12, (c) 0.23, and (d) 0.42

of the Zn and O atom densities. In such circumstances, empty impurity levels can appear just above the valence band or filled impurity levels may appear just below the conduction band. These impurity levels may explain the present experimental results, and were characterized by our DFT calculations with nonstoichiometric unit cells. Muckerman et al. and Huang et al. have studied this material using theoretical calculations [63, 64]. Their results are essentially identical to those obtained in our study.

On the basis of the characterization above, it is concluded that the samples prepared from the Ga_2O_3 and ZnO mixture under ammonia flow are solid solutions of GaN and ZnO. Therefore, the ideal composition of the present samples can be expressed as $(Ga_{1-x}Zn_x)(N_{1-x}O_x)$, although the real compositions deviated somewhat from the ideal one. A similar type of solid solution is that between Si_3N_4 and Al_2O_3, $(Si_{3-x}Al_x)(N_{4-x}O_x)$, which is known as "SiAlON" [65].

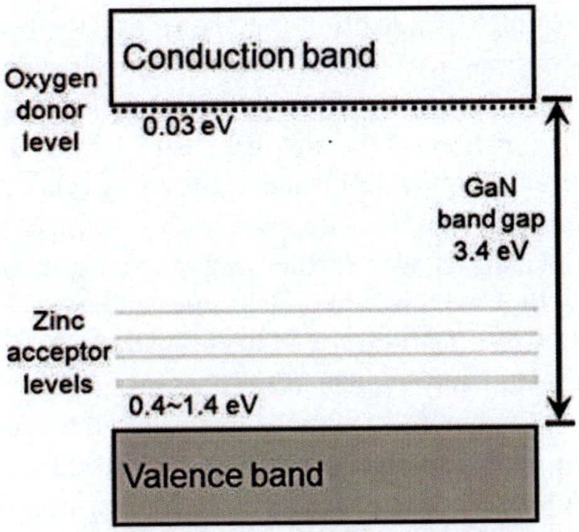

Fig. 28 A proposed energy level diagram for GaN-rich $(Ga_{1-x}Zn_x)(N_{1-x}O_x)$ by means of photoluminescence spectroscopy

The as-prepared $(Ga_{1-x}Zn_x)(N_{1-x}O_x)$ exhibits little photocatalytic activity for water decomposition even under UV irradiation. However, modification by RuO_2 nanoparticles as H_2 evolution sites resulted in H_2 and O_2 evolution. The photocatalytic activity increased remarkably with increasing RuO_2 content to a maximum at 5 wt% RuO_2, with the activity dropping gradually at higher RuO_2 loadings. It was determined by scanning electron microscopy (SEM), XPS and X-ray absorption fine structure (XAFS) spectroscopy that the enhancement of photocatalytic activity by RuO_2 loading is related to the formation of crystalline RuO_2 nanoparticles with optimal particle size and coverage [66]. The photocatalytic performance of the RuO_2-loaded $(Ga_{1-x}Zn_x)(N_{1-x}O_x)$ was also dependent on both the pH of the aqueous solution [58] and the crystallinity and composition of $(Ga_{1-x}Zn_x)(N_{1-x}O_x)$ [59]. The activity increased with decreasing pH from neutral and reaching maximum activity at a pH of 3. Below 3, where the activity began to decrease, the surface of the catalyst is no longer completely stable due to corrosion of the surface Zn species. This tendency is similar to that observed for RuO_2-loaded β-Ge_3N_4 [47]. The occurrence of peak activity in an acidic medium is consistent with the general characteristics of d^{10}-type (oxy)nitride materials, which are inherently unstable in basic environments but stable in acids. Although it is known that O_2 evolution occurs on ZnO electrode as a result of degradation when employed as a photoanode for water oxidation in a photoelectrochemical cell [67], it has been confirmed through ^{18}O-isotopic H_2O cleavage experiments that O_2 evolution on $(Ga_{1-x}Zn_x)(N_{1-x}O_x)$ is due to water oxidation, with the XRD pattern of the sample remaining unchanged after the reaction [58]. These results indicate that $(Ga_{1-x}Zn_x)(N_{1-x}O_x)$ functions as a stable, visible-light-driven photocatalyst for overall water splitting. To the best of our knowledge, this is the first successful and reproducible example of overall water splitting via one-step photoexcitation using a particulate photocatalyst having a band-gap in the visible region.

The photocatalytic activity of a given material is dependent on the cocatalyst, which plays the essential roles of inducing the photogeneration of carriers, providing reaction sites, and reducing the activation energy required for gas evolution.

A range of transition-metal oxides have been examined as potential cocatalysts for promotion of overall water splitting using $(Ga_{1-x}Zn_x)(N_{1-x}O_x)$. The cocatalysts were loaded by impregnation with the transition-metal precursors followed by calcination in air. Ni-, Ru-, Rh-, Ir- and Pt-oxides were found to be effective for promoting overall water splitting by $(Ga_{1-x}Zn_x)(N_{1-x}O_x)$ under UV irradiation. The activity of $(Ga_{1-x}Zn_x)(N_{1-x}O_x)$ modified with various transition-metal oxides as cocatalysts was further enhanced by coloading Cr-oxide, although modification with Cr-oxide alone was not effective for promoting activity in this reaction [68, 69]. The activity of the coloaded catalyst was found to be strongly dependent on the amount of Cr, which varies according to the paired transition-metal. The improvement in activity is attributed to the formation of suitable reaction sites by intimate interaction between Cr and the paired metal component. Among the various materials examined, the largest improvement in activity was obtained when $(Ga_{1-x}Zn_x)(N_{1-x}O_x)$ was loaded with a mixed oxide of Rh and Cr. Analyses by SEM, HR-TEM, EDX, XPS and XAFS confirmed that the loaded Rh and Cr species interact to form mixed-oxide ($Rh_{2-y}Cr_yO_3$) nanoparticles of 10–30 nm in size with a trivalent electronic state [70]. The photocatalytic performance of the $Rh_{2-y}Cr_yO_3$-loaded $(Ga_{1-x}Zn_x)(N_{1-x}O_x)$ was strongly dependent on the pH of the reactant solution [71], as in the case of RuO_2-loaded $(Ga_{1-x}Zn_x)(N_{1-x}O_x)$. The photocatalyst exhibits stable and high photocatalytic activity in an aqueous solution at a pH of 4.5 for as long as 3 days. The photocatalytic performance at pHs of 3.0 and 6.2 was much lower, attributable to corrosion of the cocatalyst and hydrolysis of the catalyst. It has also been confirmed by XRD, XPS and XAFS that the crystal structure of the catalyst and the valence state of both the surface and bulk material do not change even after reaction for 3 days at the optimal pH (4.5). As shown in Fig. 29, at each optimal condition, the activity of the $Rh_{2-y}Cr_yO_3$-loaded $(Ga_{1-x}Zn_x)(N_{1-x}O_x)$ catalyst for overall water splitting was at least 5 times greater than that of the RuO_2-loaded catalyst. The improvement in activity achieved by loading the $(Ga_{1-x}Zn_x)(N_{1-x}O_x)$ catalyst with $Rh_{2-y}Cr_yO_3$ is attributable to promotion of the H_2 evolution reaction, which is the rate-determining step of overall water splitting, and inhibition of O_2 photoreduction [71]. This $Rh_{2-y}Cr_yO_3$ cocatalyst was also applicable to other photocatalytic systems, such as TiO_2 and GaN, regardless of electronic configuration or the type of photocatalyst [69].

From the viewpoint of practical application, the ability to produce hydrogen from natural seawater, the most abundant water resource, would be highly desirable. Interestingly, $Rh_{2-y}Cr_yO_3$-loaded $(Ga_{1-x}Zn_x)(N_{1-x}O_x)$ catalyst produced H_2 and O_2 simultaneously from artificially-prepared seawater, although the activity was approximately half of that observed in pure water. Experimental results suggest that the main culprit of the decreased activity in seawater is Cl^-, which suppresses water oxidation and may induce other reactions, thereby contributing to a decrease in overall water-splitting efficiency. Nevertheless, the results demonstrate the possibility of a seawater-based reaction system for photocatalytic overall water splitting under visible-light [72].

In a typical synthesis, $(Ga_{1-x}Zn_x)(N_{1-x}O_x)$ can be obtained by nitriding a mixture of β-Ga_2O_3 and ZnO at temperatures higher than 1,123 K under a flow of

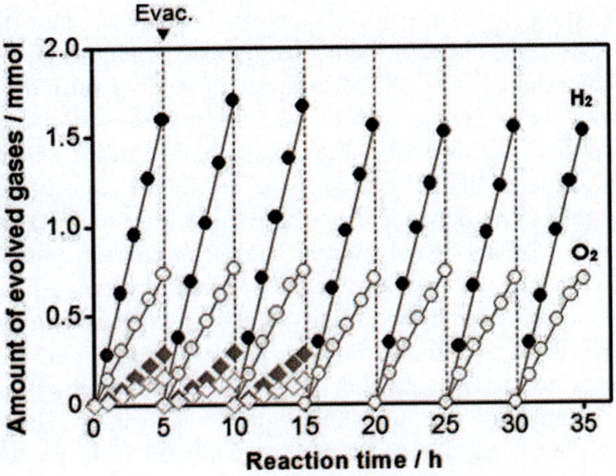

Fig. 29 Time courses of overall water splitting using $(Ga_{1-x}Zn_x)(N_{1-x}O_x)$ with various cocatalysts under visible-light irradiation ($\lambda > 400$ nm). Circles denote Rh-Cr mixed-oxide cocatalyst, diamonds denote RuO_2, solid symbols denote H_2 production, open symbols denote O_2. Reaction conditions: 0.3 g of catalyst, an aqueous H_2SO_4 solution adjusted at pH 3.0 for RuO_2-loaded sample and at pH 4.5 for $Rh_{2-y}Cr_yO_3$-loaded sample (370 mL), 450 W high-pressure mercury lamp light source, Pyrex inner irradiation-type reaction vessel made of Pyrex with an aqueous $NaNO_2$ solution (2 M) filter

NH_3, as described above. In order to establish a guideline to prepare a highly-active photocatalyst, the effects of the nitridation conditions were examined in detail. It was found that there is a trade-off between activity and nitridation conditions; that is, the activity increased up to a certain point, as nitridation progressed, but further nitridation resulted in decreased activity. Structural analyses by means of XRD, SEM, EDX and XPS revealed that promoting nitridation results in crystallization of $(Ga_{1-x}Zn_x)(N_{1-x}O_x)$, with volatilization of Zn species from the surface [59]. Crystallization of a photocatalyst leads to reducing the number of defects acting as recombination centers between photogenerated electrons and holes, thereby improving activity [24]. A decrease in the ZnO concentration of $(Ga_{1-x}Zn_x)(N_{1-x}O_x)$, on the other hand, suggests that the density of surface defects related to zinc and/or oxygen species are increased at such high temperatures or during long time nitridation. This is supported by the fact that post-calcination of the as-prepared $(Ga_{1-x}Zn_x)(N_{1-x}O_x)$ in air at $\sim$873 K enhances the water-splitting activity, and the degree of enhancement by post-calcination decreases with increasing ZnO composition [73]. It also appears that the activity is nearly independent of the specific surface area, unlike the behavior generally observed in metal-oxide photocatalysts [59]. These trends were seen regardless of cocatalysts employed. To summarize, photocatalytic activity of $(Ga_{1-x}Zn_x)(N_{1-x}O_x)$ for overall water splitting is primarily dependent on the crystallinity and the composition of the material.

It is generally known that the precursor material affects the physicochemical properties (e.g., crystallinity, surface structure and composition) of the final product and plays a crucial role in determining the activity. The influence of starting materials on the physicochemical and photocatalytic properties of $(Ga_{1-x}Zn_x)(N_{1-x}O_x)$ were investigated in an attempt to optimize the preparation conditions [74, 75]. The catalyst was successfully prepared by nitriding a starting mixture of ZnO and Ga_2O_3. A mixture of metallic zinc and GaN, however, did not produce the desired compound. The crystallinity, surface area, composition and absorption characteristics of the resultant $(Ga_{1-x}Zn_x)(N_{1-x}O_x)$ solid solution were found to be dependent on the morphology of ZnO but largely insensitive to the choice of Ga_2O_3 polymorph. The use of coarser-grained ZnO resulted in a coarser-grained catalyst with elevated zinc and oxygen content and reduced uniformity in composition and crystallinity. The results demonstrate the importance of selecting appropriate ZnO and Ga_2O_3 starting materials for maximizing the photocatalytic activity of $(Ga_{1-x}Zn_x)(N_{1-x}O_x)$ for overall water splitting under visible-light.

$ZnGa_2O_4$ was also found to be a useful starting material for $(Ga_{1-x}Zn_x)(N_{1-x}O_x)$, but the final product exhibited little photocatalytic activity [76]. The addition of ZnO was essential for preparing a high-activity catalyst. XRD analysis and SEM observation revealed that the sample prepared from $ZnGa_2O_4$ had relatively low crystallinity and many structural imperfections, compared to an analogous sample prepared with an addition of ZnO. This was the probable cause of its negligible activity, and strongly suggests the importance of ZnO in the preparation of $(Ga_{1-x}Zn_x)(N_{1-x}O_x)$. The importance of ZnO was further indicated when nitridation of a mixture of metallic Zn powder and β-Ga_2O_3 in a similar manner did not lead to the formation of $(Ga_{1-x}Zn_x)(N_{1-x}O_x)$. XRD analysis showed that $ZnGa_2O_4$ and some other impurity phases were in the nitridation products in addition to desired phase, and the intensity of the peaks assigned to the solid solution was much lower than in the sample prepared using ZnO. This result indicates that the changes in physicochemical properties of the resulting $(Ga_{1-x}Zn_x)(N_{1-x}O_x)$ are derived not from metallic Zn but from ZnO, suggesting that ZnO in the starting material facilitated the formation of $(Ga_{1-x}Zn_x)(N_{1-x}O_x)$.

As described in the previous sections, certain (oxy)nitrides containing d^0 transition-metal cations are presented to be photocatalytically active for H_2 and O_2 evolution in the presence of sacrificial reagents (i.e., half-reactions) under visible-light. However, while most of the (oxy)nitride materials examined to date exhibit visible-light photocatalytic activity for such half-reactions in the presence of a sacrificial reagent, there are only two (oxy)nitrides that are active for the full overall water-splitting reaction under visible-light: $(Ga_{1-x}Zn_x)(N_{1-x}O_x)$ and $(Zn_{1+x}Ge)(N_2O_x)$, the latter of which will be introduced below. As half-reactions are carried out as test reactions for overall water splitting, comparison of the activities for the half-reactions with those for overall water splitting is of interest and is expected to provide useful information for the development of efficient catalysts. Such a comparison has been made for a number of ultraviolet-driven metal-oxide photocatalysts, and the relationship between the structural characteristics of the material and photocatalytic activity has been reported. However, no comparative study had been made for visible-light-responsive (oxy)nitride photocatalysts.

Therefore, the photocatalytic activity of $(Ga_{1-x}Zn_x)(N_{1-x}O_x)$ for H_2 and O_2 evolution in the presence of methanol and silver nitrate as sacrificial reagents under visible-light was investigated in detail [77]. $(Ga_{1-x}Zn_x)(N_{1-x}O_x)$ evolved H_2 from an aqueous methanol solution when loaded with nanoparticulate $Rh_{2-y}Cr_yO_3$ as a cocatalyst, and evolved O_2 from an aqueous silver nitrate solution without a cocatalyst. Structural analyses indicated that the H_2 evolution activity is strongly dependent on the crystallinity and composition of the catalyst, while the rate of O_2 evolution is proportionally related to the specific surface area. The activity for H_2 evolution from methanol solution was of the same order as for overall water splitting, but was an order of magnitude lower than that for O_2 evolution from silver nitrate solution. The results of photocatalytic reactions and photoelectrochemical measurements suggested that the rate-determining step for overall water splitting using $(Ga_{1-x}Zn_x)(N_{1-x}O_x)$ is the H_2 evolution process.

To further investigate aspects of the water-splitting mechanism on $(Ga_{1-x}Zn_x)(N_{1-x}O_x)$ photocatalyst powder, the water-splitting rate was investigated on the basis of the effects of cocatalyst loading, light intensity, hydrogen/deuterium isotopes and reaction temperature on photocatalytic activity [78]. The loading amount of $Rh_{2-y}Cr_yO_3$ cocatalyst significantly affected the photocatalytic behavior with varying light intensity. When the loading amount of $Rh_{2-y}Cr_yO_3$ as a H_2 evolution cocatalyst was insufficient, the reaction yield was lower than unity. This is because photoexcited electrons accumulated in the photocatalyst and recombined with photoexcited holes more frequently than they contributed to the water-splitting reaction. Conversely, when a sufficient amount of cocatalyst was loaded accumulation of photoexcited electrons was suppressed and the water-splitting rate increased monotonically with light intensity. At a light intensity equivalent to solar irradiation, AM1.5, the water-splitting rate using modified $(Ga_{1-x}Zn_x)(N_{1-x}O_x)$ remained below the light-intensity-proportional regime. The isotope effect of H_2O/D_2O on overall water splitting using $Rh_{2-y}Cr_yO_3$-loaded $(Ga_{1-x}Zn_x)(N_{1-x}O_x)$ was smaller than previously reported for electrochemical H_2 evolution reactions. Moreover, the apparent activation energy of the overall water splitting was as small as 8 kJ mol^{-1}, and independent of the reactants. These results arise from the limited number of photoexcited carriers present in the $Rh_{2-y}Cr_yO_3$-loaded $(Ga_{1-x}Zn_x)(N_{1-x}O_x)$, which should become available for redox reactions on the surface of the photocatalyst. A kinetic model of photocatalytic water splitting proposed in this work suggests that the reaction probability of photoexcited holes for O_2 evolution versus recombination with intrinsic electrons of the photocatalyst determines the photocatalytic activity of the $Rh_{2-y}Cr_yO_3$-loaded $(Ga_{1-x}Zn_x)(N_{1-x}O_x)$.

3.4 $ZnGeN_2$-ZnO Solid Solution

A solid solution of $ZnGeN_2$ and ZnO, $(Zn_{1+x}Ge)(N_2O_x)$ was also found to be an active and stable photocatalyst for overall water splitting under visible-light [79, 80]. This material has several similarities with $(Ga_{1-x}Zn_x)(N_{1-x}O_x)$. $ZnGeN_2$ forms a hexagonal monoclinic phase with similar lattice parameters to ZnO, constituting a hexagonal wurtzite structure [81]. Both $ZnGeN_2$ and ZnO are

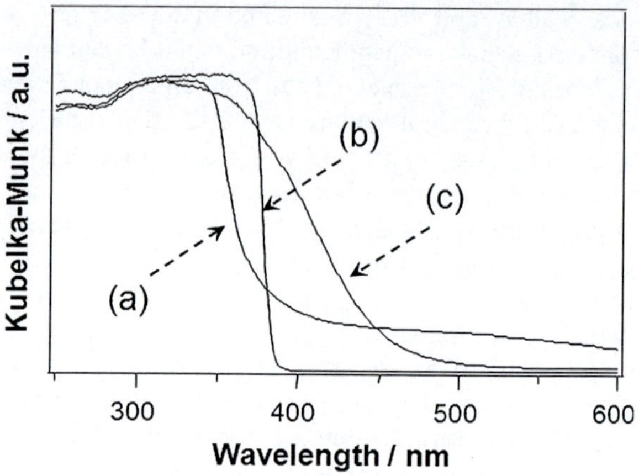

Fig. 30 UV-visible diffuse reflectance spectra for **a** $ZnGeN_2$, **b** ZnO, and **c** $(Zn_{1.44}Ge)$ $(N_{2.08}O_{0.38})$.

wide-gap semiconductors with band-gaps larger than 3 eV, although the band-gap of $ZnGeN_2$ is dependent on the crystal structure and composition [82]. Nitridation of ZnO and GeO_2 (Zn/Ge = 5 by mole) at 1,123 K for 15 h under a flow of NH_3 (20 mL min^{-1}) resulted in a pale yellow powder with an atomic composition of $(Zn_{1.44}Ge)(N_{2.08}O_{0.38})$. The crystal structure of the material was confirmed by Rietveld analysis and neutron powder diffraction to be wurtzite with space group $P6_3mc$. In the solid solution between $ZnGeN_2$ and ZnO, oxygen atoms are replaced with nitrogen. Figure 30 shows the diffuse reflectance spectrum for $(Zn_{1.44}Ge)$ $(N_{2.08}O_{0.38})$, along with $ZnGeN_2$ and ZnO references. The solid solution exhibits an absorption band at around 460 nm, while the absorption edges of $ZnGeN_2$ and ZnO are both located in UV light region. The band-gap of $(Zn_{1.44}Ge)(N_{2.08}O_{0.38})$ is estimated to be ca. 2.7 eV based on the absorption edges, which is smaller than the band-gaps of β-Ge_3N_4 (ca. 3.8 eV), $ZnGeN_2$ (ca. 3.3 eV), and ZnO (ca. 3.2 eV). DFT calculations indicated that the visible-light-response of the material originates from the wide valence bands consisting of N2p, O2p and Zn3d atomic orbitals and p–d repulsion between Zn3d and N2p + O2p electrons in the upper part of the valence bands. This implies that the band-gap of $(Zn_{1+x}Ge)(N_2O_x)$ solid solutions decreases with increasing Zn content in the compounds, consistent with the experimental results [80]. Although both $(Ga_{1-x}Zn_x)(N_{1-x}O_x)$ and $(Zn_{1+x}Ge)(N_2O_x)$ solid solutions are structurally similar to each other, the mechanism of the band-gap reduction appears different. Photoluminescence spectroscopy for $(Zn_{1+x}Ge)$ (N_2O_x) samples revealed that the materials excited by 420 nm light exhibit an emission centered at ca. 650 nm, similar to $(Ga_{1-x}Zn_x)(N_{1-x}O_x)$. However, the excitation spectrum of $(Zn_{1+x}Ge)(N_2O_x)$ was similar to the absorption profile, which suggests that the visible-light absorption of this material occurs via band-gap transition from the valence band formed by hybridized N2p, O2p and Zn3d atomic orbitals to the conduction band consisting of hybridized Ge4s,4p orbitals. A schematic illustration of the band structure of $(Zn_{1+x}Ge)(N_2O_x)$ is depicted in Fig. 31.

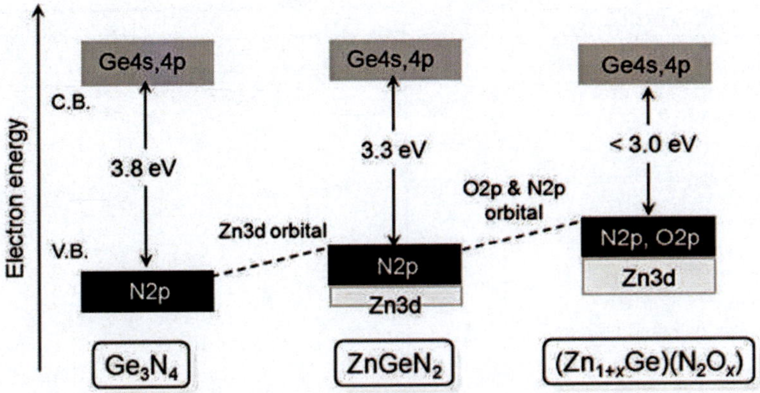

Fig. 31 A schematic of the band structure of $(Zn_{1+x}Ge)(N_2O_x)$

Neither $ZnGeN_2$ nor ZnO alone exhibited photocatalytic activity for overall water splitting under UV irradiation. However, the solid solution $(Zn_{1.44}Ge)(N_{2.08}O_{0.38})$ became photocatalytically active under UV and visible irradiation when loaded with nanoparticulate RuO_2. Overall water splitting on RuO_2-loaded $(Zn_{1.44}Ge)(N_{2.08}O_{0.38})$ proceeds by band-gap photoexcitation from the valence band formed by N2p, O2p and Zn3d atomic orbitals to the conduction band consisting of Ge4s,4p hybridized atomic orbitals, without noticeable degradation. As demonstrated in the $(Ga_{1-x}Zn_x)(N_{1-x}O_x)$ section earlier, the photocatalytic activity of the transition-metal-loaded catalysts was improved markedly by co-loading Cr-oxide. The largest improvement in activity was obtained by loading the base catalysts with both 3 wt% Rh and 0.2 wt% Cr. It was confirmed that catalytic gas evolution of $Rh_{2-y}Cr_yO_3$-loaded $(Zn_{1+x}Ge)(N_2O_x)$ are stable as long as 60 h.

In addition to cocatalyst-loading, preparation conditions of $(Zn_{1+x}Ge)(N_2O_x)$ and the kind of precursors had a significant impact on the photocatalytic activity. As mentioned earlier, $(Zn_{1+x}Ge)(N_2O_x)$ are typically prepared by reaction of GeO_2 and ZnO under a NH_3 flow at 1,123 K. With increasing nitridation time, the zinc and oxygen concentrations decreased due to reduction of ZnO and volatilization of zinc, and the crystallinity and band-gap energy of the product increased. The highest activity for overall water splitting was obtained from the sample subjected to nitridation for 15 h. Structural analyses revealed that the photocatalytic activity of $(Zn_{1+x}Ge)(N_2O_x)$ for overall water splitting depends heavily on the crystallinity and composition of the material. The preparation route has a strong influence on photocatalytic performance of $(Zn_{1+x}Ge)(N_2O_x)$ [83]. Improving the homogeneity of the $(Zn_{1+x}Ge)(N_2O_x)$ powder, reducing the number of superficial defects or increasing the crystallinity, shows a direct positive impact on the photocatalytic activity for overall water splitting. However, photoreduction of water, a half-reaction of overall water splitting, occurred with low Zn/Ge ratios and water oxidation requires a high crystallinity. Moreover, overall water splitting was achieved only if the crystal phase is active enough for photoreduction. It was also

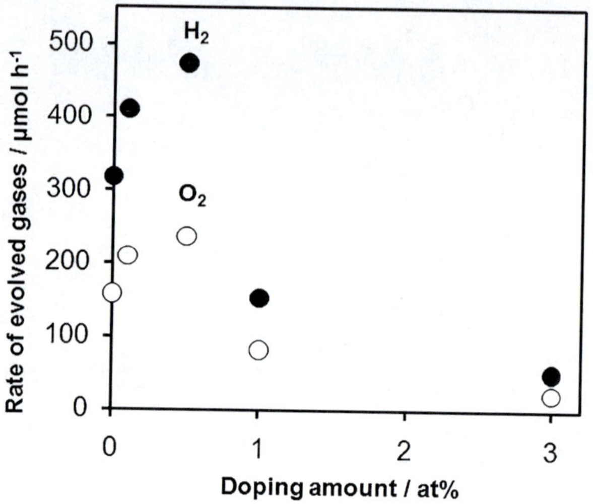

Fig. 32 Effects of Mn-doping on activity of $(Zn_{1+x}Ge)(N_2O_x)$ loaded with $Rh_{2-y}Cr_yO_3$ for overall water splitting under visible-light ($\lambda > 400$ nm). Reaction conditions: 0.2 g of catalyst (3 wt% Rh and 0.2 wt% Cr), pure water (390 mL), 450 W high-pressure mercury lamp light source, Pyrex inner irradiation-type reaction vessel made of Pyrex with an aqueous $NaNO_2$ solution (2 M) filter. The same tendency was observed in Cu-doped $(Zn_{1+x}Ge)(N_2O_x)$

found that post-calcination of $(Zn_{1+x}Ge)(N_2O_x)$ with N_2 or air at moderate temperatures (~673 K) is effective for enhancing photocatalytic activity for overall water splitting [84, 85]. These tendencies are very similar to those observed in the $(Ga_{1-x}Zn_x)(N_{1-x}O_x)$ system, as introduced above.

Another interesting feature of this material, which has not been explored in $(Ga_{1-x}Zn_x)(N_{1-x}O_x)$, is that doping with certain metal cations such as Mn or Cu into $(Zn_{1+x}Ge)(N_2O_x)$ improves activity [85]. As shown in Fig. 32, the activity increases with dopant concentration to a maximum at 0.5 wt%, above which it began to drop. This tendency suggests that there are at least two conflicting factors affecting the activity. The activity-decrease of catalysts with higher doping concentration is ascribed to the excess dopants that can increase the probability of recombination between electrons and holes, as believed in other doping-type photocatalysts. On the other hand, it is speculated that the promotional effect observed in doped-$(Zn_{1+x}Ge)(N_2O_x)$ can originate from the possibility that the doped species act as cocatalysts to promote surface redox reactions and/or the intrinsic properties of the material are positively changed by doping. The former possibility was investigated by a series of test reactions using Cu-doped $(Zn_{1+x}Ge)(N_2O_x)$ samples. As listed in Table 2, Cu-doped $(Zn_{1+x}Ge)(N_2O_x)$ without cocatalyst exhibited little activity for H_2 evolution, even in the presence of methanol as an electron donor. In addition, O_2 evolution activity from aqueous $AgNO_3$ solution decreased upon doping of Cu. It indicates that doped Cu species do not provide active sites for H_2 evolution and have a negative effect on water oxidation process. With modification by $Rh_{2-y}Cr_yO_3$ cocatalyst, however, Cu-doped $(Zn_{1+x}Ge)(N_2O_x)$ gave higher activity for H_2 evolution from aqueous methanol solution than a similarly-modified undoped sample. On the other hand, the O_2 evolution activity of the doped $(Zn_{1+x}Ge)(N_2O_x)$ modified with $Rh_{2-y}Cr_yO_3$ was still lower, compared to the undoped analogue. The results clearly indicate that Cu-doping promotes the H_2 evolution but decreases the O_2 evolution on the

Table 2 H_2 and O_2 production rates for the photocatalytic reactions with sacrificial reagents using $(Zn_{1+x}Ge)(N_2O_x)$-based photocatalysts under visible-light ($\lambda > 400$ nm)

Photocatalyst	Cocatalyst	Activity/μmol h^{-1}	
		H_2^b	O_2^c
$(Zn_{1+x}Ge)(N_2O_x)$	None	n.d.	349
$(Zn_{1+x}Ge)(N_2O_x)$	None	n.d.	318
Cu-doped $(Zn_{1+x}Ge)(N_2O_x)$	$Rh_{2-y}Cr_yO_3$[a]	204	247
Cu-doped $(Zn_{1+x}Ge)(N_2O_x)$	$Rh_{2-y}Cr_yO_3$[a]	468	127

Reaction conditions: 0.05 g of catalyst, 390 mL aqueous solution containing sacrificial reagents, 3,450 W high-pressure mercury lamp light source, Pyrex inner irradiation-type reaction vessel made of Pyrex with an aqueous $NaNO_2$ solution (2 M) filter

[a] 3 wt% Rh and 0.2 wt% Cr

[b] In the presence of methanol (40 mL methanol, 350 mL H_2O)

[c] In the presence of silver nitrate (0.01 M)

photocatalyst. Considering the importance of promoting H_2 evolution [71, 78, 80], the effect with respect to H_2 evolution by Cu-doping would enhance the activity for overall water splitting. It was therefore considered that the improvement arises from the altered band structure by forming the impurity levels that allow for more visible-light absorption, increasing the number of photons available for photocatalytic reaction, or from the suppression of recombination [86]. UV-visible spectroscopy showed that no additional absorption band formed by dopants is observed in Cu-doped $(Zn_{1+x}Ge)(N_2O_x)$, precluding the former possibility. Interestingly, $(Zn_{1+x}Ge)(N_2O_x)$ alone exhibited a photoluminescence band centered at ca. 650 nm upon excitation at 77 K, which might be produced by recombination. The intensity of the luminescence band decreased upon doping, suggesting that the density of such recombination sites in $(Zn_{1+x}Ge)(N_2O_x)$ is effectively reduced as a result. It should be noted that the same relationship between activity and photoluminescence was observed in $(Zn_{1+x}Ge)(N_2O_x)$ post-calcinated with N_2 or air [84], but post-calcination for the doped samples did not result in any activity-increase. Presumably, the effects of doping and post-calcination give identical changes, which would have a positive effect on activity. Although further investigation on the nature of defects in $(Zn_{1+x}Ge)(N_2O_x)$ is required, it is considered that the improvement of photocatalytic activity is attributed mainly to the reduction in defects in the photocatalyst.

3.5 *A Mixed Oxynitride of Gallium, Zinc and Indium*

As presented in above sections, both $(Ga_{1-x}Zn_x)(N_{1-x}O_x)$ and $(Zn_{1+x}Ge)(N_2O_x)$ work as stable water-splitting photocatalysts under irradiation of light with wavelength shorter than 500 nm. For more efficient utilization of solar spectrum, it is highly desirable to develop a photocatalyst having a more extended visible-light absorption, ideally up to 600 nm [24]. A straightforward way to reduce band-gaps of solid

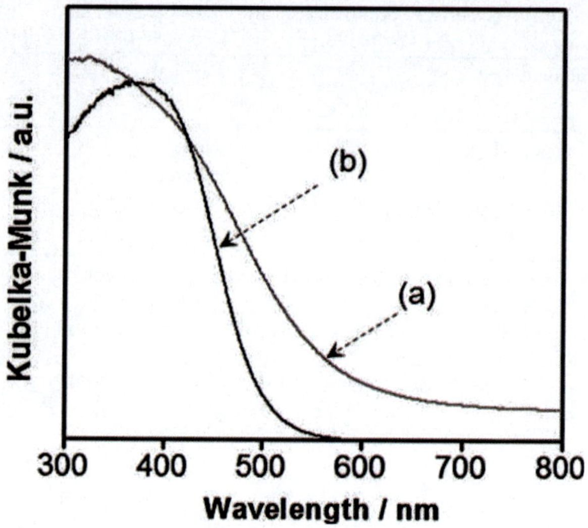

Fig. 33 UV-visible diffuse reflectance spectra for (**a**) Ga–Zn–In–O–N and (**b**) $(Ga_{1-x}Zn_x)(N_{1-x}O_x)$ with $x = 0.42$

solutions having wurtzite structure is to make a solid solution with wurtzite indium nitride (InN), which has a much narrower band-gap (~1 eV) [87] than $(Ga_{1-x}Zn_x)(N_{1-x}O_x)$ and $(Zn_{1+x}Ge)(N_2O_x)$ (~2.5 eV). Although it is easy to draw such an ideal picture, there exists a major challenge in how InN is incorporated due to its relatively low thermal stability [88]. Nevertheless, our preliminary attempt resulted in a material consisting of gallium, zinc, indium, oxygen and nitrogen as principal elements and exhibits a visible-light absorption band around 600 nm [89].

This new material is brown powder, and can be prepared by nitriding an oxide precursor containing gallium, zinc and indium, which is prepared by a low-temperature precursor route, under a flow of NH_3 at 873 K for 15 h. Controlling the nitridation temperature is very important because the desired material was not produced at lower or higher temperatures than 873 K. Figure 33 shows DRS spectrum for the as-prepared material (Ga–Zn–In–O–N), along with a reference of $(Ga_{1-x}Zn_x)(N_{1-x}O_x)$ having the same Zn/Ga concentration as Ga–Zn–In–O–N. It is clear that Ga–Zn–In–O–N has a wider absorption band than $(Ga_{1-x}Zn_x)(N_{1-x}O_x)$, attributed to the contribution from indium content in the material. Structural characterization by means of XRD and TEM revealed that Ga–Zn–In–O–N is incompletely-crystallized aggregated nanoparticles with ~10 nm in size. XAFS analyses also showed that the valence states of Ga, Zn and In are somewhat close to those in the corresponding metal oxides, but not completely consistent. This is attributed to the lower nitrogen concentration than that expected from an ideal solid solution among GaN, ZnO and InN.

Photocatalytic activity of this sample was evaluated using methanol and silver nitrate as sacrificial electron donors and acceptors, respectively, to test whether the sample can reduce and oxidize water under visible-light. When Rh metal and cobalt oxide were loaded as cocatalysts for water reduction and oxidation onto Ga–Zn–In–O–N, both H_2 and O_2 evolution were achieved upon visible-light irradiation. In the case of O_2 evolution, the Co-oxide worked as a cocatalyst to improve O_2 evolution rate while suppressing N_2. It was also confirmed that O_2 evolution is observed up to 560 nm irradiation, which corresponds to the absorption band of the material. However, the same experiment for H_2 evolution was difficult because of its lower activity. Although the activity of this material is at least an order of magnitude lower than that of $(Ga_{1-x}Zn_x)(N_{1-x}O_x)$ at the present stage, the results of photocatalytic reactions indicate that Ga–Zn–In–O–N has the abilities to reduce and oxidize water with visible-light. To enhance the performance of this sample, improvement of the crystallinity is considered to be a straightforward approach, judging from experiences with $(Ga_{1-x}Zn_x)(N_{1-x}O_x)$ and $(Zn_{1+x}Ge)(N_2O_x)$.

4 Summary and Outlook

A range of non-oxide materials for photocatalytic overall water splitting have been developed by the authors' group. (Oxy)nitrides and oxysulfides have been found to function as stable photocatalysts for water reduction and oxidation under visible irradiation, and the $(Ga_{1-x}Zn_x)(N_{1-x}O_x)$ and $(Zn_{1+x}Ge)(N_2O_x)$ solid solutions have been shown to achieve overall water splitting under visible-light without noticeable degradation. A tentative goal for research on hydrogen production via solar-driven overall water splitting using a particulate photocatalyst is to develop a stable material that can achieve quantum efficiency of 30% at 600 nm. The state of the art in this field is, at most, quantum efficiency of a few percent at wavelengths as long as 500 nm. Therefore, more efficient photocatalytic materials with a band-gap as narrow as 2 eV (corresponding to 600 nm) must be developed.

References

1. Fujishima A, Honda K (1972) Electrochemical photolysis of water at a semiconductor electrode. Nature 238:37
2. Domen K, Naito S, Soma M, Onishi T, Tamaru K (1980) Photocatalytic decomposition of water vapour on an NiO-$SrTiO_3$ catalyst. J Chem Soc, Chem Commun 543
3. Sato S, White JM (1980) Photodecomposition of water over Pt/TiO_2 catalysts. Chem Phys Lett 72:83
4. Lehn JM, Sauvage JP, Ziessel R (1980) Photochemical water splitting continuous generation of hydrogen and oxygen by irradiation of aqueous suspensions of metal loaded strontium titanate. Nouv J Chim 4:623
5. Domen K, Kudo A, Shinozaki A, Tanaka A, Maruya K, Onishi T (1986) Photodecomposition of water and hydrogen evolution from aqueous methanol solution over novel niobate photocatalysts. J Chem Soc, Chem Commun 356
6. Kudo A, Tanaka A, Domen K, Maruya K, Aika K, Onishi T (1988) Photocatalytic decomposition of water over NiO-$K_4Nb_6O_{17}$ catalyst. J Catal 111:67
7. Kudo A, Tanaka A, Domen K, Onishi T (1988) Nickel-loaded $K_4Nb_6O_{17}$ photocatalyst in the decomposition of H_2O into H_2 and O_2: structure and reaction mechanism. J Catal 111:296

8. Inoue Y, Kubokawa T, Sato K (1990) Photocatalytic activity of sodium hexatitanate, $Na_2Ti_6O_{13}$, with a tunnel structure for decomposition of water. J Chem Soc, Chem Commun 1298
9. Takata T, Furumi Y, Shinohara K, Tanaka A, Hara M, Kondo JN, Domen K (1997) Photocatalytic decomposition of water on spontaneously hydrated layered perovskites. Chem Mater 9:1063
10. Ikeda S, Hara M, Kondo JN, Domen K, Takahashi H, Okubo T, Kakihana M (1998) Preparation of $K_2La_2Ti_3O_{10}$ by polymerized complex method and photocatalytic decomposition of water. J Mater Res 13:852
11. Kudo A, Kato H (1997) Photocatalytic decomposition of water into H_2 and O_2 over novel photocatalyst $K_3Ta_3Si_2O_{13}$ with pillered structure consisting of three TaO_6 chains. Chem Lett 26:867
12. Kudo A, Kato H, Nakagawa S (2000) Water splitting into H_2 and O_2 on new $Sr_2M_2O_7$ (M=Nb and Ta) photocatalysts with layered perovskite structures: factors affecting the photocatalytic activity. J Phys Chem B 104:571
13. Kato H, Kudo A (2001) Water splitting into H_2 and O_2 on alkali tantalate photocatalysts $ATaO_3$ (A=Li, Na, and K). J Phys Chem B 105:4285
14. Kato H, Asakura K, Kudo A (2003) Highly efficient water splitting into H_2 and O_2 over lanthanum-doped $NaTaO_3$ photocatalysts with high crystallinity and surface nanostructure. J Am Chem Soc 125:3082
15. Sato J, Saito N, Nishiyama H, Inoue Y (2001) New photocatalyst group for water decomposition of RuO_2-loaded p-block metal (In, Sn, and Sb) oxides with d^{10}configuration. J Phys Chem B 105:6061
16. Ikarashi K, Sato J, Kobayashi H, Saito N, Nishiyama H, Inoue Y (2002) Photocatalysis for water decomposition by RuO_2-dispersed $ZnGa_2O_4$ with d^{10} configuration. J Phys Chem B 106:9048
17. Sato J, Saito N, Nishiyama H, Inoue Y (2002) Photocatalytic water decomposition by RuO_2-loaded antimonates, $M_2Sb_2O_7$ (M = Ca, Sr), $CaSb_2O_6$ and $NaSbO_3$, with d^{10} configuration. J Photochem Photobiol A 148:85
18. Sato J, Kobayashi H, Saito N, Nishiyama H, Inoue Y (2003) Photocatalytic activities for water decomposition of RuO_2-loaded $AInO_2$ (A = Li, Na) with d^{10} configuration. J Photochem Photobiol A 158:139
19. Sato J, Saito N, Nishiyama H, Inoue Y (2003) Photocatalytic activity for water decomposition of indates with octahedrally coordinated d^{10} configuration. I. Influences of preparation conditions on activity. J Phys Chem B 107:7965
20. Sato J, Kobayashi H, Ikarashi K, Saito N, Nishiyama H, Inoue Y (2004) Photocatalytic activity for water decomposition of RuO_2-dispersed Zn_2GeO_4 with d^{10} configuration. J Phys Chem B 108:4369
21. Scaife DE (1980) Oxide semiconductors in photoelectrochemical conversion of solar energy. Sol Energy 25:41
22. Williams R (1960) Becquerel photovoltaic effect in binary compounds. J Chem Phys 32:1505
23. Ellis AB, Kaiser SW, Bolts JM, Wrighton MS (1997) Study of n-type semiconductoring cadmium chalcogenide-based photoelectrochemical cells employing polychalcogenide electrolytes. J Am Chem Soc 99:2839
24. Maeda K, Domen K (2007) New non-oxide photocatalysts designed for overall water splitting under visible light. J Phys Chem C 111:7851
25. Jansen M, Letschert HP (2000) Inorganic yellow-red pigments without toxic metals. Nature 404:980
26. Gunter EU, Hagenmayer R, Jansen MZ (2000) Structural investigations on the oxidenitrides $SrTaO_2N$, $CaTaO_2N$ and $LaTaON_2$ by neutron and x-ray powder diffraction. Anorg Allg Chem 626:1519
27. Pors F, Marchand R, Laurent Y, Bacher P, Roult G (1988) Neutron diffraction structural study of the strontium tantalum oxynitride ($SrTaO_2N$) perovskite and of the barium strontium tantalum oxynitride (Ba_1-xSrxTaO_2N) solid solution. Mater Res Bull 23:1447

28. Armytage B, Fender BEF (1974) Anion ordering in tantalum oxynitride. Powder neutron-diffraction investigation. Acta Crystallogr B 30:809
29. Brese NE (1991) Structure of tantalum nitride (Ta_3N_5) at 16 K by time-of-flight neutron diffraction. Acta Crystallogr C 47:2291
30. Kakihana M (1996) Sol-gel preparation of high temperature superconducting oxides. J Sol-Gel Sci 5:7
31. Hitoki G, Takata T, Kondo JN, Hara M, Kobayashi H, Domen K (2002) (Oxy)nitrides as new photocatalysts for water splitting under visible light irradiation. Electrochem 70:463
32. Takata T, Hitoki G, Kondo JN, Hara M, Kobayashi H, Domen K (2007) Visible-light-driven photocatalytic behavior of tantalum-oxynitride and nitride. Res Chem Intermed 33:13
33. Fang CM, Orhan E, Wijs GA, Hintzen HT, Groot RA, Marchand R, Saillard JY, With G (2001) The electronic structure of tantalum (oxy)nitrides TaON and Ta_3N_5. J Mater Chem 11:1248
34. Chun WA, Ishikawa A, Fujisawa H, Takata T, Kondo JN, Hara M, Kawai M, Matsumoto Y, Domen K (2003) Conduction and valence band positions of Ta_2O_5, TaON, and Ta_3N_5 by ups and electrochemical methods. J Phys Chem B 107:1798
35. Bard AJ, Wrighton MS (1977) Thermodynamic potential for the anodic dissolution of n-type semiconductors. A crucial factor controlling durability and efficiency in photoelectrochemical cells and an important criterion in the selection of new electrode/electrolyte systems. J Electrochem Soc 124:1706
36. Clarke SJ, Guinot BP, Michie CW, Calmont MJC, Rosseinsky MJ (2002) Oxynitride perovskites: synthesis and structures of $LaZrO_2N$, $NdTiO_2N$, and $LaTiO_2N$ and comparison with oxide perovskites. Chem Mater 14:288
37. Asahi R, Morikawa T, Ohwaki T, Aoki K, Taga Y (2001) Visible-light photocatalysis in nitrogen-doped titanium oxides. Science 293:269
38. Kasahara A, Nukumizu K, Hitoki G, Takata T, Kondo JN, Hara M, Kobayashi H, Domen K (2002) Photoreactions on $LaTiO_2N$ under visible light irradiation. J Phys Chem A 106:6750
39. Kasahara A, Nukumizu K, Takata T, Kondo JN, Hara M, Kobayashi H, Domen K (2003) $LaTiO_2N$ as a visible-light ($\leq$600 nm)— driven photocatalyst. J Phys Chem B 107(2):791
40. Shannon RD (1976) Revised effective ionic radii and systematic studies of interatomie distances in halides and chaleogenides. Acta Crystallogr Sect A 32:751
41. Harriman A, Pickering IJ, Thomas JM, Christensen PA (1988) Redox reactions with colloidal metal oxides. Comparison of radiation-generated and chemically generated $RuO_2{\cdot}2H_2O$ and MnO_2 colloids. J Chem Soc Faraday Trans 1 84:2795
42. Ishikawa A, Takata T, Kondo JN, Hara M, Kobayashi H, Domen K (2002) Oxysulfide $Sm_2Ti_2S_2O_5$ as a stable photocatalyst for water oxidation and reduction under visible light irradiation ($\lambda \leq$650 nm). J Am Chem Soc 124:13547
43. Ishikawa A, Yamada Y, Takata T, Kondo JN, Hara M, Kobayashi H, Domen K (2003) Novel synthesis and photocatalytic activity of oxysulfide $Sm_2Ti_2S_2O_5$. Chem Mater 15:4442
44. Tabira Y, Withers RL, Minervini L, Grimes RW (2000) Systematic structural change in selected rare earth oxide pyrochlores as determined by wide-angle CBED and a comparison with the results of atomistic computer simulation. J Solid State Chem 153:16
45. Goga M, Seshadri R, Ksenofontov V, Gütlich P, Tremel W (1999) $Ln_2Ti_2S_2O_5$ (Ln=Nd, Pr, Sm): a novel series of defective Ruddlesden–Popper phases. Chem Commun 979
46. Sato J, Saito N, Yamada Y, Maeda K, Takata T, Kondo JN, Hara M, Kobayashi H, Domen K, Inoue Y (2005) RuO_2-loaded β-Ge_3N_4 as a non-oxide photocatalyst for overall water splitting. J Am Chem Soc 127:4150
47. Maeda K, Saito N, Lu D, Inoue Y, Domen K (2007) Photocatalytic properties of RuO_2-loaded β-Ge_3N_4 for overall water splitting. J Phys Chem C 111:4749
48. Dong J, Sankey OF, Deb SK, Wolf G, McMillan PF (2000) Theoretical study of β-Ge_3N_4 and its high-pressure spinel γ phase. Phys Rev B 61:11979
49. Deb SK, Dong J, Hubertc H, McMillana PF, Sankey OF (2000) The Raman spectra of the hexagonal and cubic (spinel) forms of Ge_3N_4: an experimental and theoretical study. Solid State Commun 114:137

50. Soignard E, McMillan PF, Hejny C, Leinenweberd K (2004) Pressure-induced transformations in α- and β-Ge_3N_4: in situ studies by synchrotron x-ray diffraction. J Solid State Chem 177:299
51. Johnson WC (1930) Nitrogen compounds of germanium. I. The preparation and properties of germanic nitride. J Am Chem Soc 52:5160
52. Maeda K, Saito N, Inoue Y, Domen K (2007) Dependence of activity and stability of germanium nitride powder for photocatalytic overall water splitting on structural properties. Chem Mater 19:4092
53. Lee Y, Watanabe T, Takata T, Hara M, Yoshimura M, Domen K (2006) Effect of high-pressure ammonia treatment on the activity of Ge_3N_4 photocatalyst for overall water splitting. J Phys Chem B 110:17563
54. Linke WF (1958) Solubilities: inorganic and metal-organic compounds, (A-Ir), vol I, 4th edn. D. Van Nostrand, Princeton
55. Suhulz H, Thiemann KH (1977) Crystal structure refinement of AlN and GaN. Solid State Commun 23:815
56. Garcia-Martinez O, Rojas RM, Vila E, Martin de Vidales JL (1997) Microstructural characterization of nanocrystals of ZnO and CuO obtained from basic salts. Solid State Ionics 63:442
57. Wei SH, Zunger A (1988) Role of metal d states in II-VI semiconductors. Phys Rev B 37:8958
58. Maeda K, Takata T, Hara M, Saito N, Inoue Y, Kobayashi H, Domen K (2005) GaN:ZnO solid solution as a photocatalyst for visible-light-driven overall water splitting. J Am Chem Soc 127:8286
59. Maeda K, Teramura K, Takata T, Hara M, Saito N, Toda K, Inoue Y, Kobayashi H, Domen K (2005) Overall water splitting on $(Ga_{1-x}Zn_x)(N_{1-x}O_x)$ solid solution photocatalyst: relationship between physical properties and photocatalytic activity. J Phys Chem B 109:20504
60. Yashima M, Maeda K, Teramura K, Takata T, Domen K (2005) Crystal structure and optical properties of $(Ga_{1-x}Zn_x)(N_{1-x}O_x)$ oxynitride photocatalyst ($x = 0.13$). Chem Phys Lett 416:225
61. Hirai T, Maeda K, Yoshida M, Kubota J, Ikeda S, Matsumura M, Domen K (2007) Origin of visible light absorption in GaN-rich $(Ga_{1-x}Zn_x)(N_{1-x}O_x)$ photocatalysts. J Phys Chem C 111:18853
62. Maeda K, Teramura K, Saito N, Inoue Y, Kobayashi H, Domen K (2006) Overall water splitting using (oxy)nitride photocatalysts. Pure Appl Chem 78:2267
63. Jensen LL, Muckerman JT, Newton MD (2008) First-principles studies of the structural and electronic properties of the $(Ga_{1-x}Zn_x)(N_{1-x}O_x)$ solid solution photocatalyst. J Phys Chem C 112:3439
64. Wei W, Dai Y, Yang K, Guo M, Huang B (2008) Origin of the visible light absorption of GaN-rich $(Ga_{1-x}Zn_x)(N_xO_x)$ ($x = 0.125$) solid solution. J Phys Chem C 112:15915
65. Izhevskiy VA, Genova LA, Bressiani JC, Aldinger F (2000) Progress in SiAlON. J Eur Ceram Soc 20:2275
66. Teramura K, Maeda K, Saito T, Takata T, Saito N, Inoue Y, Domen K (2005) Characterization of ruthenium oxide nanocluster as a cocatalyst with $(Ga_{1-x}Zn_x)(N_{1-x}O_x)$ for photocatalytic overall water splitting. J Phys Chem B 109:21915
67. Gerischer H (1966) Electrochemical behaviour of semiconductors under illumination. J Electrochem Soc 113:1174
68. Maeda K, Teramura K, Lu D, Takata T, Saito N, Inoue Y, Domen K (2006) Photocatalyst releasing hydrogen from water. Enhancing catalytic performance holds promise for hydrogen production by water splitting in sunlight. Nature 440:295
69. Maeda K, Teramura K, Saito N, Inoue Y, Domen K (2006) Improvement of photocatalytic activity of $(Ga_{1-x}Zn_x)(N_{1-x}O_{1-x})$ solid solution for overall water splitting by co-loading Cr and another transition metal. J Catal 243:303
70. Maeda K, Teramura K, Lu D, Takata T, Saito N, Inoue Y, Domen K (2006) Characterization of Rh-Cr mixed-oxide nanoparticles dispersed on $(Ga_{1-x}Zn_x)(N_{1-x}O_x)$ as a cocatalyst for visible-light-driven overall water splitting. J Phys Chem B 110:13753

71. Maeda K, Teramura K, Masuda H, Takata T, Saito N, Inoue Y, Domen K (2006) Efficient overall water splitting under visible-light irradiation on $(Ga_{1-x}Zn_x)(N_{1-x}O_x)$ dispersed with rh-cr mixed-oxide nanoparticles: effect of reaction conditions on photocatalytic activity. J Phys Chem B 110:13107
72. Maeda K, Masuda H, Domen K (2009) Effect of electrolyte addition on activity of $(Ga_{1-x}Zn_x)(N_{1-x}O_x)$ photocatalyst for overall water splitting under visible light. Catal Today 147:173
73. Maeda K, Teramura K, Domen K (2008) Effect of post-calcination on photocatalytic activity of $(Ga_{1-x}Zn_x)(N_{1-x}O_x)$ solid solution for overall water splitting under visible light. J Catal 254:198
74. Hisatomi H, Maeda K, Lu D, Domen K (2009) The effects of starting materials in the synthesis of $(Ga_{1-x}Zn_x)(N_{1-x}O_x)$ solid solution on its photocatalytic activity for overall water splitting under visible light. Chem Sus Chem 2:336
75. Maeda K, Domen K (2010) Solid solution of GaN and ZnO as a stable photocatalyst for overall water splitting under visible light. Chem Mater 22:612
76. Sun X, Maeda K, Le Faucheur M, Teramura K, Domen K (2007) Preparation of $(Ga_{1-x}Zn_x)(N_{1-x}O_x)$ solid-solution from ZnGa2O4 and ZnO as a photo-catalyst for overall water splitting under visible light. Appl Catal A: Gen 327:114
77. Maeda K, Hashiguchi H, Masuda H, Abe R, Domen K (2008) Photocatalytic activity of $(Ga_{1-x}Zn_x)(N_{1-x}O_x)$ for visible-light-driven H2 and O2 evolution in the presence of sacrificial reagents. J Phys Chem C 112:3447
78. Hisatomi T, Maeda K, Takanabe K, Kubota J, Domen K (2009) Aspects of the water splitting mechanism on $(Ga_{1-x}Zn_x)(N_{1-x}O_x)$ photocatalyst modified with $Rh_{2-y}Cr_yO_3$ cocatalyst. J Phys Chem C 113:21458
79. Lee Y, Terashima H, Shimodaira Y, Teramura K, Hara M, Kobayashi H, Domen K, Yashima M (2007) Zinc germanium oxynitride as a photocatalyst for overall water splitting under visible light. J Phys Chem C 111:1042
80. Lee Y, Teramura K, Hara M, Domen K (2007) Modification of $(Zn_{1+x}Ge)(N_2O_x)$ solid solution as a visible light driven photocatalyst for overall water splitting. Chem Mater 19:2120
81. Larson WL, Maruska HP, Stevenson DA (1974) Synthesis and properties of zinc germanium nitride ($ZnGeN_2$). J Electrochem Soc 121:1673
82. Misaki T, Wu X, Wakahara A, Yoshida A (2000) Theoretical analysis of multinary nitride semiconductors by density functional theory. In: Proc. Int. Workshop nitride semiconductors IPAP Conf. Series 1 685
83. Tessier F, Maillard P, Lee Y, Bleugat C, Domen K (2009) Zinc germanium oxynitride: influence of the preparation method on the photocatalytic properties for overall water splitting. J Phys Chem C 113:8526
84. Wang X, Maeda K, Lee Y, Domen K (2008) Enhancement of photocatalytic activity of $(Zn_{1+x}Ge)(N_2O_x)$ for visible-light-driven overall water splitting by calcination under nitrogen. Chem Phys Lett 457:134
85. Takanabe K, Uzawa T, Wang X, Maeda K, Katayama M, Kubota J, Kudo A, Domen K (2009) Enhancement of photocatalytic activity of zinc-germanium oxynitride solid solution for overall water splitting under visible irradiation. Dalton Trans 10055
86. Kudo A, Niishiro R, Iwase A, Kato H (2007) Effects of doping of metal cations on morphology, activity, and visible light response of photocatalysts. Chem Phys 339:104
87. Bechstedt F, Furthmuller J (2002) Do we know the fundamental energy gap of InN? J Cryst Growth 246:315
88. Gao L, Zhang Q, Li J (2003) Preparation of ultrafine InN powder by the nitridation of In_2O_3 or $In(OH)_3$ and its thermal stability. J Mater Chem 13:154
89. Kamata K, Maeda K, Lu D, Kako Y, Domen K (2009) Synthesis and photocatalytic activity of gallium–zinc–indium mixed oxynitride for hydrogen and oxygen evolution under visible light. Chem Phys Lett 470:90

Heterogeneous Photocatalytic Conversion of Carbon Dioxide

Hisao Yoshida

Abstract Carbon dioxide is a greenhouse gas, which may contribute to the global warming. The conversion of carbon dioxide to more useful chemicals is not an easy task because of a high thermodynamic barrier, which requires much energy consumption. However, we should not use energy from fossil resources to convert the carbon dioxide because the use of them produces carbon dioxide; therefore, it is desirable to use natural energy for this purpose. Photocatalysis, which can utilize solar energy and break the thermodynamic limitation, is a possible green technology available for the carbon dioxide conversion and many studies have been carried out. In this chapter, after a description of the importance of the photocatalytic system, the physical and chemical basis for carbon dioxide conversion, and the basis for photocatalysis and photocatalysts, we will review a brief history about heterogeneous photocatalytic conversion of carbon dioxide to other compounds, such as methane, methanol and carbon monoxide, by using reducing reagents such as water, hydrogen and methane. The perspectives related to the field of nanotechnology will also be described.

1 Introduction

Since the global warming issue has become worldwide concern, the methodology for the mitigation of carbon dioxide as a greenhouse gas has been investigated. There are several strategies to reduce carbon dioxide emissions, such as energy-

H. Yoshida (✉)
Department of Applied Chemistry, Graduate School of Engineering,
Nagoya University, Furo-cho, Chikusa-ku, Nagoya 464-8603, Japan
e-mail: yoshidah@apchem.nagoya-u.ac.jp

L. Zang (ed.), *Energy Efficiency and Renewable Energy Through Nanotechnology*, Green Energy and Technology, DOI: 10.1007/978-0-85729-638-2_15,

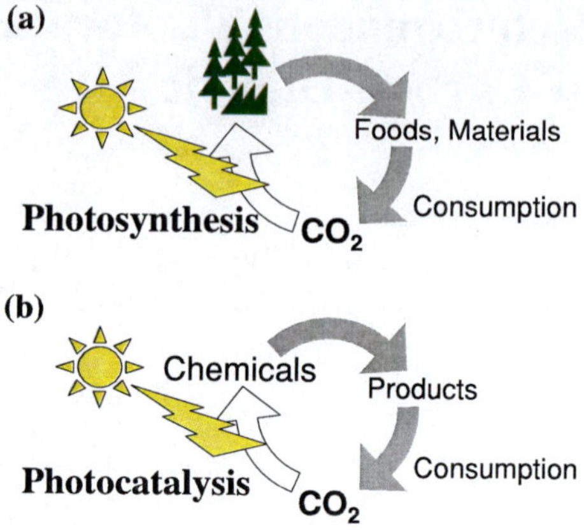

Fig. 1 Carbon neutral cycles through (**a**) agriculture and forestry with photosynthesis, and (**b**) artificial route with photocatalysis

conservation, carbon dioxide capture and storage, enhanced oil recovery and so on [1]. One such method, artificial conversion of carbon dioxide to other useful chemicals, has attracted much attention.

In the ecological system in nature, photosynthesis (Eq. 1) takes place in plants to convert carbon dioxide and water to sugars and oxygen by using solar energy [2].

$$6\ CO_2 + 6\ H_2O\ +\ \text{photoenergy} \rightarrow C_6H_{12}O_6 + 6\ O_2 \qquad (1)$$

Plant photosynthesis demonstrates the viability of converting carbon dioxide to chemicals and also converting sunlight into chemical potential. Since carbon dioxide is one of the most stable molecules, as mentioned later, the conversion of carbon dioxide to other chemicals must involve energy injection. This means that plants can store the energy from the incident solar irradiation in the form of chemical bonds. The carbon neutral cycle is driven by solar energy (Fig. 1a).

However, the efficiencies of energy-conversion in natural photosynthesis are not enough to satisfy our demands. We need an artificial photosynthetic method. Photocatalysis with solar light is one potential approach as an artificial photosynthesis (Fig. 1b).

Clearly, utilization of the diluted carbon dioxide in atmosphere is difficult because of its large entropy and therefore requires much cost for collection, separation and purification [3]. Diluted carbon dioxide is released from stationary sources, such as electric power stations and blast furnaces in industry, as well as from the mobile sources such as cars, aircrafts, trains and ships [3, 4]. On the other hand, concentrated carbon dioxide is more preferred as a carbon dioxide source. There are many stationary sources emitting the concentrated carbon dioxide, including several kinds of industries for processing of natural gas, production of ethanol by fermentation, and production of bulk chemicals such as ammonia and ethylene oxide, as well as inorganic industries for production of cement, limestone and soda ash [3, 4]. Thus, we should just diminish the diluted emission from both the stationary and mobile sources, and should effectively utilize the concentrated carbon dioxide before diffusion.

Currently, carbon dioxide is used in various ways: for beverage carbonation, or fire extinguishers, also as a refrigerant for food preservation (as dry ice), a supercritical solvent, a pressurizing agent and a neutralizing agent. In addition, carbon dioxide is regarded as an important green chemical reagent as a mild oxidant or as a weak acid for many synthetic reactions [3, 4]. Catalytic reactions involving carbon dioxide and producing valuable products have been well reviewed, e.g., by Xiaoding and Moulijn [4], Song [3], Sakakura et al. [5] and Aresta and Dibenedetto [6]. Although establishing a chemical industry based on a renewable feedstock like carbon dioxide is still challenging, some reactions have already been industrialized or are one step away from commercialization [5]. The fixation of carbon dioxide as compounds is valuable for reducing the carbon dioxide in the atmosphere until the products will finally be disposed and transformed to carbon dioxide again. However, we must also consider that the amount of fixed carbon dioxide converted to chemical compounds will be much smaller than that of carbon dioxide emission through combustion of fossil fuels on the global scale. Moreover, fixation reactions require activation energy, which is usually produced in conjunction with more carbon dioxide emissions. Thus, these reactions using carbon dioxide as a raw material may not necessarily contribute to reduce the greenhouse effect as much as desired. Currently, less than 1% of carbon dioxide emissions are used in chemical reactions [4]. Of course, even if it is only 1%, it is still very important. However, we must also provide and develop other methods that are more environmentally benign. The photocatalytic conversion of carbon dioxide described in this chapter has a potential to become a useful method [1, 7] because these methods require no chemical reagents except for water, hydrogen and methane.

2 Properties of Carbon Dioxide in Chemical and Thermodynamical View

2.1 *Acidic Properties of Carbon Dioxide*

Carbon dioxide is one of the most stable molecules among carbon compounds, and is, therefore, the final product in the oxidation/combustion of organic and carbonaceous compounds. The molecule consists of a linear connection of a carbon and two oxygen atoms (O=C=O) and it does not have a dipole moment. However, as a chemical reagent, it is not an inert molecule.

This molecule is often recognized as an acidic gas. Since it can be adsorbed by the surface base sites of heterogeneous catalysts, it can be used as a probe molecule for these sites in measurements of FT-IR spectra and temperature programmed desorption (TPD) profiles [8]. In this adsorption, the carbon atom of carbon dioxide molecule accepts the electron lone pair of the surface oxygen ion at the metal oxide surface, and the oxygen atom having a lone pair in the carbon

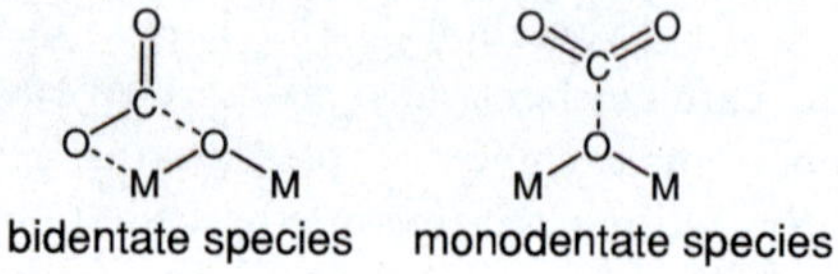

Fig. 2 Adsorbed forms of carbon dioxide

Table 1 Gibbs free energy (ΔG_f^0/kJ mol^{-1}) for each compound

Gas phase		Liquid phase		Solid phase	
C_6H_6	+130	C_6H_6	+124		
C_2H_4	+68			C (graphite)	0
H_2, O_2	0			Ni, Mg	0
NH_3	−16	C_6H_{14}	−4		
CH_4	−51	HCHO	−102		
CO	−137	CH_3OH	−166		
H_2O	−237	H_2O	−228	NiO	−212
CO_2	−394	HCOOH	−361		
		CO_2	−368		
		CO_3^{2-}	−528	MgO	−569
		HCO_3^-	−586	CaO	−603
		$C_2O_4^{2-}$	−671	$NiCO_3$	−613
				$MgCO_3$	−1012
				$CaCO_3$	−1130

The values are from [3]

dioxide molecule can donate these electrons to the surface acidic sites, to form various kinds of adsorbed species, such as carbonate-like species (Fig. 2).

The acidic property of carbon dioxide promotes the formation of carbonate with alkaline and alkaline earth cations to become more stable. In the Table 1 the values of Gibbs free energy for the representative compounds are given [3]. We can recognize the thermodynamically stable properties of carbon dioxide, carbonate anion and metal carbonates. Thus, carbon dioxide is the last product in combustion of compounds containing carbon, and metal carbonates are found as stable natural resources.

2.2 *Chemical Reactions of Carbon Dioxide*

On the other hand, carbon dioxide can function as a mild oxidant since carbon dioxide cannot be oxidized any further. For example, it is well-known that the reaction between methane and carbon dioxide can produce syngas (carbon monoxide and hydrogen), which is called a dry reforming of methane (DRM) or CO_2 reforming of methane (Table 2, entry 7) [9]. In this reaction, carbon dioxide oxidizes methane to yield carbon monoxide, while carbon dioxide itself is reduced by methane to become carbon monoxide. The DRM is a thermodynamically unfavorable reaction and can only proceed at high temperature, typically more than 1,000 K, even by using catalysts [10]. There are many other reactions using

Table 2 Change of Gibbs free energy at 298 K for various reactions related to the photocatalytic reduction of carbon dioxide

Entry	Reductant	Product	Chemical equation			ΔG^0/kJ mol^{-1}
1	H_2	Carbon monoxide	$CO_2 + H_2$	→	$CO + H_2O$	29
2		Formic acid	$CO_2 + H_2$	→	HCOOH	43
3		Formaldehyde	$CO_2 + 2\ H_2$	→	$HCHO + H_2O$	63
4		Methanol	$CO_2 + 3\ H_2$	→	$CH_3OH + H_2O$	4
5		Methane	$CO_2 + 4\ H_2$	→	$CH_4 + 2\ H_2O$	−114
6		Carbon	$CO_2 + 2\ H_2$	→	$C + 2\ H_2O$	−63
7	CH_4	Carbon monoxide	$CO_2 + CH_4$	→	$2\ CO + 2\ H_2$	171
8		Acetone	$CO_2 + 2\ CH_4$	→	$CH_3COCH_3 + H_2O$	115
9		Acetic acid	$CO_2 + CH_4$	→	CH_3COOH	71
10		Formaldehyde	$CO_2 + CH_4$	→	2 HCHO	240
11		Carbon	$CO_2 + CH_4$	→	$2\ C + 2\ H_2O$	−12
12	H_2O	Formic acid	$CO_2 + H_2O$	→	$HCOOH + 1/2\ O_2$	272
13		Carbon monoxide	$CO_2 + H_2O$	→	$CO + H_2 + O_2$	486
14		Formaldehyde	$CO_2 + H_2O$	→	$HCHO + O_2$	521
15		Methanol	$CO_2 + 2\ H_2O$	→	$CH_3OH + 3/2\ O_2$	690
16		Methane	$CO_2 + 2\ H_2O$	→	$CH_4 + 2\ O_2$	801
17		Carbon monoxide	CO_2	→	$CO + 1/2\ O_2$	257
18			$H_2O(g)$	→	$H_2 + 1/2\ O_2$	229
19			$H_2O(l)$	→	$H_2 + 1/2\ O_2$	237

Calculated from the data of gaseous states in most cases [12, 13]

carbon dioxide as an oxidant, where carbon dioxide is reduced instead, as reviewed in the literature [4–6].

Table 2 shows the representative reactions of carbon dioxide with simple small molecules, i.e., hydrogen, methane and water, with the values of the change of Gibbs free energy at 298 K for each reaction. In these reactions, these molecules reduce the carbon dioxide as reductant to yield products such as carbon monoxide, formic acid, acetic acid, formaldehyde, acetaldehyde, methanol, methane and carbon. A reaction can proceed preferably when the change of Gibbs free energy is negative. However, most of reactions listed here show positive values, meaning that they are thermodynamically unfavorable reactions.

When we reduce the carbon dioxide with hydrogen, the products will be the reduced C1 compounds or carbon (Table 2, entries 1–6). In the case of the reactions to form carbon monoxide, formic acid and formaldehyde, the changes of Gibbs free energy are positive values in range from 29 to 63 kJ mol^{-1} (Table 2, entries 1–3). This means that these reaction systems require the energy and store it as chemical potential in the products. This type of the reaction is called an "uphill reaction" (Fig. 3a) [11]. On the other hand, in the reactions with hydrogen producing methane and carbon, the changes of Gibbs free energy are negative values such as −114 and

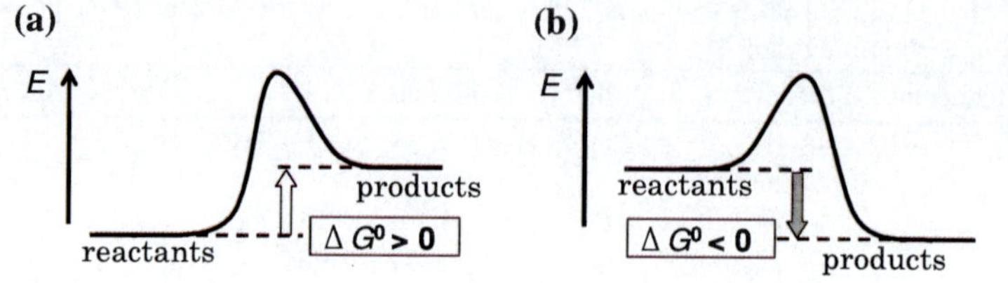

Fig. 3 Energy diagrams for (**a**) up-hill reaction and (**b**) downhill reaction

−63 kJ mol^{-1} (Table 2, entries 5 and 6). This type of the reaction is called a "downhill reaction" (Fig. 3b). The reduction of carbon dioxide by hydrogen to produce methane (Table 2, entry 5) is not an attractive reaction since the system must lose the chemical potential by consuming hydrogen of high chemical potential, and in the economical sense, hydrogen is more expensive than methane [4]. However, if the hydrogen could be obtained easily through utilizing solar energy, e.g., through the photocatalytic water splitting, rather than through consumption of fossil fuels, this reaction may become appealing.

When we reduce carbon dioxide with methane, the products vary more than the case of the reduction by hydrogen. In Table 2, entries 7–11, the representative products are listed. Among them, most of the reactions are uphill. Thus, when we try the photocatalytic reduction of carbon dioxide by methane, the yields of these products will be valuable.

The reduction of carbon dioxide with water seems extremely thermodynamically-unfavorable reaction, since the changes of Gibbs free energy are largely positive values greater than 272 kJ mol^{-1}. These values are larger than the value for the water splitting (Table 2, entries 18 and 19), especially the value is as high as 801 kJ mol^{-1} for the methane formation (Table 2, entry 16); this value is too high to proceed through usual catalysis. These reactions are the reverse reactions for the complete oxidation of carbon compounds. However, there are many studies reporting the photocatalytic reduction of carbon dioxide by water as a reductant with a heterogeneous photocatalyst to yield carbon monoxide, methanol, methane and so on, as mentioned later.

There are many reports on the utilization of carbon dioxide through the organic synthetic reactions [4–6]. In these cases, carbon dioxide is regarded as a chemical feedstock that is required for the specific reaction. However, it is not clear whether these reactions can contribute to the mitigation of carbon dioxide. In these reactions, we use the chemically active reagents having high chemical potentials such as epoxide, acetylene, diyne, diene, allene and amine, to yield oxygen-containing stable molecules such as carbonate, carbamate carboxylic acid, ester and lactone [5], where the changes of Gibbs free energy are usually negative and thus the reaction can proceed more favorable, but always with energy consumption. When we discuss the conversion of carbon dioxide in the viewpoint of environmental issue, it is very important that we avoid using the fossil energy. If we use the energy obtained from fossil fuels or the active chemicals derived from the fossil sources for the carbon dioxide conversion, it may be almost meaningless. Most desirable is a direct utilization of the solar energy, that is, photoenergy.

Table 3 Some electrochemical reactions with their corresponding reduction potential (E^0_{redox} vs. NHE at pH 7)

Entry	Product	Reaction			E^0_{redox}/V vs. NHE
1	Carbonate anion radical	$CO_2 + e^-$	→	$\bullet CO_2^-$	−1.90
2	Formic acid	$CO_2 + H^+ + 2\ e^-$	→	HCO_2^-	−0.49
3	Carbon monoxide	$CO_2 + 2\ H^+ + 2\ e^-$	→	$CO + H_2O$	−0.53
4	Formaldehyde	$CO_2 + 4\ H^+ + 4\ e^-$	→	$HCHO + H_2O$	−0.48
5	Hydrogen	$2\ H^+ + 2\ e^-$	→	H_2	−0.41
6	Methanol	$CO_2 + 6\ H^+ + 6\ e^-$	→	$CH_3OH + H_2O$	−0.38
7	Methane	$CO_2 + 8\ H^+ + 8\ e^-$	→	$CH_4 + 2\ H_2O$	−0.24
8	Oxygen	H_2O	→	$1/2\ O_2 + 2\ H^+ + 2\ e^-$	0.82

The values are from [1]

Direct decomposition to produce carbon monoxide and oxygen is thermodynamically difficult with a large positive change of the Gibbs free energy (Table 2, entry 17). Photoexcitation by a part of solar light can dissociate carbon dioxide into excited states of carbon monoxide and oxygen atom. However, the required light for the photoabsorption and photodissociation of carbon dioxide are VUV light. The threshold wavelength is around 90 nm (corresponding to 1,330 kJ mol^{-1}) [14]; neither UV nor visible light can excite carbon dioxide. Thus, the solar energy reaching the surface of the earth is not enough for photochemical dissociation. Therefore, in order to utilize available sunlight, we must apply suitable photocatalysis to activate the carbon dioxide.

2.3 *Redox of Carbon Dioxide*

When carbon dioxide is activated in a redox system, such as electrochemical, photochemical and photocatalytical systems, it can be reduced by receiving one or several electrons to become the corresponding reduced forms (Table 3) [1]. For example, a carbon dioxide molecule that receives one electron becomes a carbonate anion radical ($CO_2^{\bullet-}$) (Table 3, entry 1). It is notable that this reaction is highly endergonic because of the negative adiabatic electron affinity of CO_2 [1, 15]. This step may be the first step for the photoreduction of carbon dioxide. This suggests difficulty in the photoreduction of carbon dioxide, which is even more difficult than the hydrogen production from water (Table 3, entry 5). However, the experimental data revealed that adsorbed carbon dioxide was photocatalytically reduced to form the carbonate anion radical (CO_2^-), as mentioned later. Thus, it is suggested that the special interaction between the photocatalyst surface and carbon dioxide is very important. This suggests that the development of the structure and surface state of the semiconductor photocatalyst on the nanoscale will enable enhancement of their photocatalytic activity.

When the anion radical is further reduced by receiving multiple electrons with protons (or hydrogen radicals), a formate anion (HCO_2^-), carbon monoxide (CO), formaldehyde (HCHO), methanol (CH_3OH) and methane (CH_4) are obtained. We usually consider that carbon dioxide is reduced gradually in this order, since the production of each product requires two, four, six or eight electrons and protons. Thus, we often assume that formic acid, carbon monoxide and formaldehyde are the early products, while methanol and methane are subsequently obtained by deep reduction of these intermediates. However, in the sense of potential energy, the reactions producing methanol and methane can more easily proceed than the formation of other compounds and hydrogen. It has not yet been determined which proceeds more easily: consecutive photoreduction or complete reduction to methane.

To reduce carbon dioxide, we can use the electrochemical method by using electricity produced by solar cells. However, the use of the photoenergy for the chemical conversion by using photocatalysis is a more direct and effective way. Thus, in the following chapter, we focus on the photocatalytic conversion.

3 Photocatalytic Reactions and Photocatalysts

Here, introduced are the basis of photocatalysis in view of thermodynamics and two types of heterogeneous photocatalysts.

3.1 Photocatalytic Reactions

Photocatalytic reactions can promote thermodynamically unfavorable and difficult reactions at low temperature (e.g., room temperature). This brings some advantages, such as reduced energy consumption, less catalyst deactivation and more safety and stability of the reactor. Usually, the photoenergy that can be used for photocatalytic reactions, typically UV light, is larger than the activation energy for many kinds of chemical reactions. We can determine the energy of one photon in a certain wavelength light using Planck's constant. For example, photons with 300 nm wavelengths (UV light) have energies of 399 kJ mol^{-1}. Even though this energy is too low to directly break the C=O bond in carbon dioxide (estimated to be 799 kJ mol^{-1}) [16], it is expected that the use of a suitable photocatalyst would enable activation of even stable molecules. As mentioned above, the molecules adsorbed by the surface of a heterogeneous catalyst are in a certain stable state as shown in Fig. 2, which may provide an easier route for the reaction.

For example, the changes of Gibbs free energy for the reduction of carbon dioxide with water to produce methane is a largely positive value (801 kJ mol^{-1}), as mentioned above (Table 2, entry 16) and this value corresponds to the energy of the VUV light of about 150 nm in wavelength. Normally, in order to promote a reaction, activation energy (ΔE_a) is required and it must be larger than the change

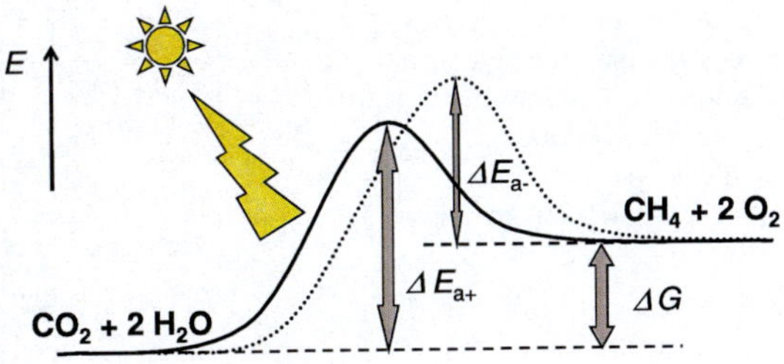

Fig. 4 Energetic scheme of photocatalytic reduction of carbon dioxide, where the photoenergy is partly converted to a higher chemical potential of products. ΔG is the change of Gibbs free energy, ΔE_{a+} is the activation energy for the forward reaction and ΔE_{a-} is that for the backward reaction

in Gibbs free energy (ΔG), as shown in Fig. 4. Thus, one may think that the direct promotion of this reaction should be impossible. However, in actuality, this reaction is reported to photocatalytically proceed on a suitable photocatalysts upon irradiation by UV light or even visible light, as discussed later. Thus, we believe that photocatalysts with the multiple photons of high energy can enable the thermodynamically unfavorable reactions ($\Delta G > 0$) to proceed photocatalytically under mild conditions (Table 3, entry 7).

In principle, a photocatalyst uses the photoenergy for the activation energy and the increment of the Gibbs free energy (ΔG^0). When the ΔG^0 is positive (uphill reaction), the photoenergy is converted to the higher chemical potential of products as shown in Fig. 3. Solar energy cannot be stored as it is. Thus, it is very valuable that we can store the photoenergy in chemicals, such as hydrogen, which can be stored and transferred easily. It opens the possibility to utilize the abundant solar energy conveniently and efficiently.

In the case of the thermal catalytic reaction, the reaction is finally controlled by thermodynamic equilibrium, where the forward and backward reactions can equally proceed. The thermodynamic values of the molecules and the reaction temperature determine the equilibrium composition. Since a usual catalyst promotes both the forward and backward reactions, the system will reach the equilibrium composition at the temperature. On the other hand, some photocatalytic reaction systems can break the thermodynamic equilibrium [9], although there are some exceptions. In many cases, the forward and backward reactions could be considered two different reactions in photocatalysis. The excited molecules and the excitation light wavelength are usually different from each other: e.g., the excited molecules are carbon dioxide or water in the forward reaction and methane or oxygen in the backward reaction. In addition, the reaction mechanism and the excitation state are also different (Fig. 4). Thus, it is possible to avoid the progression of the undesired backward reaction in some cases, meaning it is possible for the photocatalytic reaction to break the thermodynamic equilibrium.

All of these matters mentioned above are the benefits of the photocatalytic reactions. However, unfortunately, the yield obtained from the photocatalytic

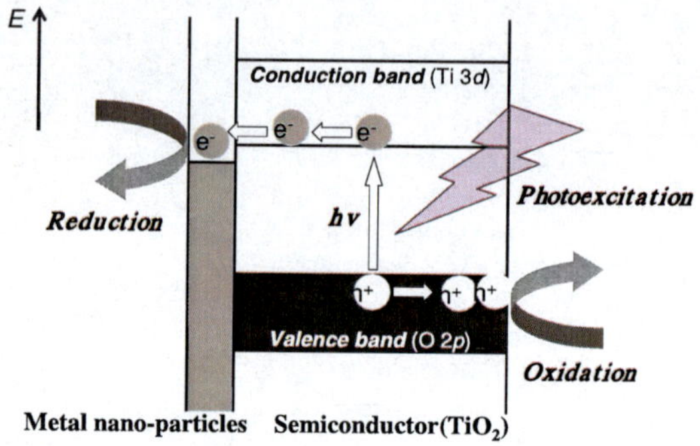

Fig. 5 Photoexcitation and successive photocatalytic reactions on a titanium dioxide semiconductor photocatalyst with a co-catalyst of metal nano-particles

reactions is still very low in many cases. Therefore, development of photocatalytic system is required at present.

3.2 *Types of Heterogeneous Photocatalysts*

The heterogeneous photocatalysts introduced in this chapter can be roughly divided into two groups: semiconductor photocatalysts and highly-dispersed photocatalysts.

Examples of the semiconductor photocatalysts include metal oxide, sulfide, nitride and oxynitride [11]; the most famous is TiO_2. Metal additives, such as Pt nanoparticles, are often employed to enhance the photocatalytic activity. When the TiO_2 photocatalyst absorbs incident light of larger energy than its band gap, photoexcitation occurs as electrons transfer from the valence band (O2p) to the conduction band (Ti 3d), as shown in Fig. 5, and the holes and electrons in the valence and conduction bands, respectively, migrate to the surface. When the metal nanoparticles are deposited on the surface, the excited electrons migrate to them. Holes and electrons promote oxidative and reductive reactions, respectively. This type of photocatalyst has a large and intense band in the UV spectrum. For example, the spectrum of TiO_2 is shown in Fig. 6a [17].

On the other hand, the highly-dispersed photocatalyst has unique structures and excitation states [18]. An example is monomers of titanium oxide supported by insulating materials, such as silica, where one titanium cation is coordinated by oxygen anions on the surface of the support. The photoexcitation takes place at the isolated titanium sites, which is often expressed as follows:

$$Ti^{4+} - O^{2-} + \text{photon} \rightarrow Ti^{3+} - O^{-} \qquad (2)$$

Since the species exist on insulating materials, the excitation state is localized on the moiety. Compared to the semiconductor photocatalysts, this type photocatalyst shows a narrow and less intense absorption band in UV spectra. Figure 6b shows an example, the spectrum of silica-supported titanium oxide photocatalyst containing 0.3 mol% of Ti, Ti(0.3)/SiO_2 [17]. In the following chapter, another

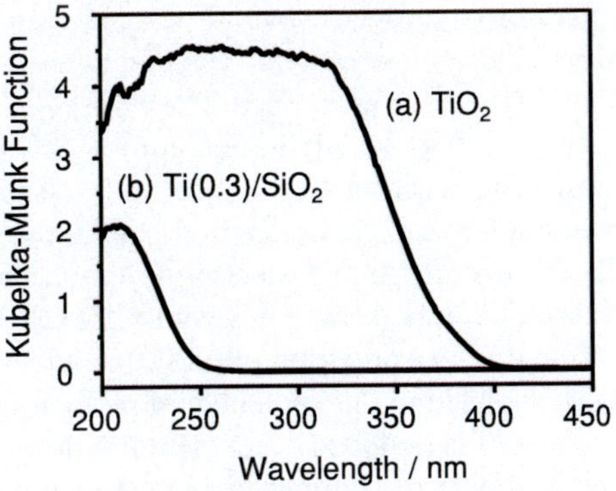

Fig. 6 Diffuse reflectance UV-visible spectra of (**a**) TiO_2 as a semiconductor photocatalyst and (**b**) Ti(0.3)/SiO_2 as a highly-dispersed photocatalyst

type of photoactive site, the surface carbonaceous species derived from adsorbed carbon dioxide, will be also described.

4 Photocatalytic Conversion of Carbon Dioxide

Photocatalytic conversion of carbon dioxide will be a most desirable method because this can mitigate the carbon dioxide as a greenhouse gas, utilize the carbon dioxide as a carbon source, and convert the solar energy to storable chemical energy. In addition, the processes consume no fossil energy in principal and are free from carbon dioxide emission. The photoenergy is high enough to activate even a stable molecule such as carbon dioxide on the surface, and the reactions have a potential to proceed with breaking the thermodynamics barrier. Thus, the photocatalytic conversion of carbon dioxide can proceed if the condition of the system including the photocatalyst is satisfied. So far, many researchers have examined many photocatalytic reaction systems in both homogeneous and heterogeneous systems. In this chapter, we will survey only the heterogeneous ones, and describe the brief history and the current situation of the photocatalytic conversion of carbon dioxide.

In this chapter, the photocatalytic reductions of carbon dioxide are classified into three groups by means of the reductant; i.e., water, hydrogen and methane as shown in Table 2.

4.1 Photocatalytic Reduction of Carbon Dioxide by Water

4.1.1 Semiconductor Photocatalysts

Since the discovery of the electrochemical photocatalysis on TiO_2 for water splitting by Fujishima and Honda in 1972 [19], this system has received much attention. This led to the revelation that the photocatalytic system can convert solar energy to hydrogen. Several years later, in 1978, although the reduction of carbon dioxide is a more difficult reaction as mentioned above, Hemminger et al. reported that methane was formed from gaseous water and carbon dioxide (ca. 4 kPa) through a photocatalytic reaction on $SrTiO_3$ single crystal in contact with Pt foil [20]. The reaction took place on a clean and reduced surface of $SrTiO_3$ (111) without any externally applied potential between the foil and the substrate, and in the absence of liquid electrolyte, upon the photoirradiation with light of greater energy than the band gap (from a 500 W high-pressure mercury lamp through an infrared filter consisting of $NiSO_4$ solution). The initial rate of methane formation corresponded to one molecule of methane for each 10^4 incident photons. Although the reaction stopped after 10 min due to poisoning, the total amount of the methane was 5–10 times larger than the number of the surface sites on the $SrTiO_3$ crystal, indicating the photocatalytic reaction proceeded. Other products were not observed. In the same year, Halmann reported that carbon dioxide in water was reduced to formic acid, formaldehyde and methanol on a p-type, Zn-doped GaP photocathode in an electrochemical photocell using a high-pressure mercury lamp [21]. The maximal solar energy conversion efficiencies for formaldehyde and methanol in this system were estimated to be 0.97 and 0.61%, respectively.

In 1979, Inoue et al. showed that carbon dioxide was photocatalytically converted to organic compounds, such as formic acid, formaldehyde, methanol and methane, in the presence of various semiconductor powders, like TiO_2, ZnO, CdS, GaP and SiC, suspended in water as photocatalysts [22]. Reduced compounds, such as formaldehyde and methanol, were mostly obtained on SiC, which has a conduction band with a highly negative potential. A correlation between the conduction band energy and the yield of methanol was found: the yields of methanol increased as the conduction band became more negative with respect to the redox potential of H_2CO_3/CH_3OH (Table 3, entry 6), while methanol was not produced on WO_3 (with a more positive conduction band than the redox potential of H_2CO_3/CH_3OH). The estimated quantum yields of production for photocatalytic reduction of carbon dioxide were ca. 5.0×10^{-4} for formaldehyde and 1.9×10^{-4} for methanol with a TiO_2 photocatalyst, and 4.5×10^{-3} for methanol on SiC, measured against the absorbable incident photons. These photocatalytic systems employing semiconductor photocatalysts have also been tested under sunlight irradiation for utilization of solar energy, and the formation of formic acid, formaldehyde, methanol, methane, acetaldehyde and ethanol were confirmed [23–25].

Metal-loaded semiconductors, such as Pt/TiO_2, function as micro-photoelectrochemical cells [26]. For example, Pt-loaded TiO_2 was found to decompose water to hydrogen and oxygen photocatalytically, in 1980 [27]. Then, the Pt/TiO_2 photocatalysts were examined as one of the typical heterogeneous photocatalysts for the reduction of carbon dioxide or carbonate. In 1983, Chandrasekaran and Thomas reported that photoirradiation of an aqueous deaerated sodium carbonate solution containing Pt/TiO_2 gave rise to a small yield of oxygen and formaldehyde in the solution [28]. However, extended irradiation led to the loss of formaldehyde. In this work, phthalocyanine-coated TiO_2 was also tested and gave the similar results, where phthalocyanine absorbed light and transferred an electron to the conduction band of TiO_2. In 1984, Halmann et al. found that RuO_2-loaded TiO_2 enhanced the photocatalytic reduction of carbon dioxide to formic acid, formaldehyde and methanol, where formic acid was the major product [29]. Tennakone examined several kinds of metals (Pt, Au, Ag, Co, Pb, Hg) as the additive on TiO_2 [30] and found that Hg/TiO_2 provided the highest yield of formaldehyde from the water containing CO_2 upon photoirradiation from a 200 W medium-pressure Hg lamp. Raphael and Malati reported that Pt/TiO_2 in aqueous sodium carbonate gave methanol, carbon, formaldehyde and formic acid [31]. Hirano et al. [32] reported that the coexistence of suspended TiO_2 and Cu metal powders promoted the photocatalytic reduction of carbon dioxide to yield methanol, formaldehyde and a trace amount of formic acid and carbon monoxide. Since Cu(II) ions were also observed in this system, the copper metal particles do not only offer the reaction sites for photoreduction of carbon dioxide with the excited electrons but also act as a sacrificial reagent to react with positive holes in the valence band. In this work, they mentioned that the presence of $KHCO_3$ in the solution enhanced the photocatalytic yield of methanol. In 1993, Ishitani et al. investigated the activity of various metal-deposited TiO_2 photocatalysts [33]. Metal particles of 2 wt%, such as Pd, Rh, Pt, Au, Cu and Ru, were loaded on TiO_2 by the photochemical deposition method and examined for the photocatalytic reduction of carbon dioxide in water. Although only a small quantity of methane was produced by using non-deposited TiO_2, the deposition of Pd, Rh, Pt or Au on TiO_2 greatly enhanced the methane formation. Among them, Pd/TiO_2 exhibited the highest activity. The second one was Rh/TiO_2. They also confirmed that when $^{13}CO_2$ was employed instead of $^{12}CO_2$, $^{13}CH_4$ was obtained, while $^{13}CH_4$ was not detected under $^{12}CO_2$, which clearly evidenced the photocatalytic reduction of carbon dioxide to methane. Solymosi and Tombácz reported that Rh-loading TiO_2 and Rh-loading W-doped TiO_2 photocatalysts provided formic acid, formaldehyde and methanol [34]. Reductive pretreatment for the latter photocatalyst offered high conversion and high selectivity to methanol.

Cu/TiO_2 has been investigated by several groups. In 1994, Yamashita et al. showed that anatase TiO_2, which has a large band gap and numerous hydroxyl groups, showed high efficiency for photocatalytic formation of methane while Cu-loaded TiO_2 showed additional formation of methanol [35]. Further, they observed photogenerated Ti^{3+}, H and CH_3 radicals as reactive intermediates by ESR at 77 K. Tseng et al. also reported that methanol was obtained on Cu/TiO_2, where TiO_2 was prepared by a sol-gel method [36, 37]. They found that the addition of NaOH increased the methanol yield, suggesting that the basic solution dissolved a larger amount of carbon dioxide than pure water and the hydroxy anions in the solution react with holes to accelerate the photocatalytic reaction. Wu et al. showed that the most active Cu species on TiO_2 surface were Cu_2O clusters and they played an important role for the formation of methanol in a study using an optical fiber photoreactor [38]. Nguyen and Wu reported that Fe and Cu co-loaded TiO_2 produced ethylene and methane and trace amount of ethane and methanol [39]. This catalyst exhibited higher activity for ethylene production than the catalysts loaded with either Cu or Fe, indicating a synergistic effect for photoreduction of carbon dioxide with water to ethylene. In this condition, Cu/TiO_2 produced methane more favorably than ethylene.

In many of the reports mentioned above, the formation of molecular oxygen was scarcely mentioned in the photocatalytic reduction of carbon dioxide with water, although oxygen should be formed, as indicated in Table 2, entries 12–16. In some papers, the formation of molecular oxygen was confirmed. For example, an oxygen sensor detected the molecular oxygen dissolved in the solution, which was produced over the Cu/TiO_2 photocatalyst, though the amount was less than expected from the stoichiometric ratio [36]. In the case of water splitting, non-stoichiometric evolution of oxygen has also been frequently observed and ascribed to the photoadsorption of O_2 or O_2^- onto semiconductor particles [40–42] or the formation of surface O_2^{2-} peroxide species [40, 41]. Photoadsorption of O_2 on highly hydrated TiO_2 surface was also directly observed [43, 44]. A H_2/O_2 ratio greater than stoichiometric is observed only when the photocatalytic reaction is slow, and this is attributed to the adsorption of O_2 on TiO_2; if the reaction is fast, the reaction stoichiometry is not influenced by the photoadsorption [42]. In the system of the photocatalytic reduction of carbon dioxide with water, the photoadsorption of oxygen was mentioned as a possible reason for the difficulty in detecting oxygen [35], implying that the reaction rate was low in these reaction systems. The decreasing rate for products evolution might be due to this photoadsorption.

In 1993, Sayama and Arakawa discovered that the photocatalytic decomposition of pure water proceeded over non-loaded ZrO_2 powder, without any loaded metals upon photoirradiation in a closed gas-circulating system with an inner irradiation quarts reactor equipped with a 400 W high-pressure Hg lamp, at an almost constant rate for long period ($\sim$20 h) [45]. Moreover, in the presence of $NaHCO_3$ in water, H_2, O_2 and CO were constantly produced at rates of 309, 167 and 3 μmol/h, respectively, over the ZrO_2 photocatalyst, and 19, 11 and 2.5 μmol/h, respectively, over Cu-loaded ZrO_2. These results suggest that photocatalytic water decomposition (Table 2, entry 19, $2H_2O(l) \rightarrow 2H_2 + O_2, \Delta G^0 = 474\,\text{kJ mol}^{-1}$) and photocatalytic reduction of carbon dioxide (in water HCO_3^-) to form CO, H_2 and O_2 (Table 2, entry 13, $CO_2 + H_2O(l) \rightarrow CO + H_2 + O_2, \Delta G^0 = 494\,\text{kJ mol}^{-1}$) can proceed, although photocatalytically produced hydrogen might reduce the carbon dioxide, yielding CO (Table 2, entry 1, $CO_2 + H_2 \rightarrow CO + H_2O(l)$, $\Delta G^0 = 20\,\text{kJ mol}^{-1}$). Since the ZrO_2 semiconductor has a highly negative flat-band potential and a wide band gap, it can promote both photocatalytic water decomposition and the photoreduction of carbon dioxide without loaded metals. In this study, some kinds of loaded metal and metal oxides, such as Pt, RuO_2, Au, and Cu, were examined, and it was found that the loaded co-catalyst change the reaction selectivity.

To reveal the fundamental aspects, several studies have been carried out on TiO_2 photocatalysts with and without loaded metals. High-pressure CO_2 (1 MPa) was effective for the production of methanol as a major product on a TiO_2

photocatalyst in pure water, and higher pressure (2.5 MPa) was effective for gaseous minor products, such as methane, ethane and ethene on Cu-loaded TiO_2 and pure TiO_2. While in the NaOH solution, the higher pressure gave high yields of acetaldehyde, ethanol and methanol [46, 47]. Under a high-pressure of carbon dioxide liquid (13.8 MPa), bare TiO_2 homogeneously dispersed on Nafion® films produced formic acid and methanol as major products and acetic acid as a minor product [48], while Ag-loaded TiO_2 onto Nafion® films produced methanol as a major product and formic acid as a minor one [49]. The effect of the reaction temperature was varied between 298 and 473 K on TiO_2 and the reaction rate increased with increasing temperature [50], where the activation energy (E_a) for the formation of methane was calculated to be ca. 2 kJ mol^{-1} around room temperature and almost zero at 473 K. The product desorption was suggested to be the rate limitation step. In supercritical carbon dioxide (8 MPa, 308–323 K) with a small amount of water, TiO_2, and a mixture of TiO_2 and metal Cu photocatalytically converted carbon dioxide to formic acid and methanol [51]. During photoirradiation of TiO_2 in liquid-phase carbon dioxide without any proton source was found to produce carbonate anion radicals, and after pressure reduction the addition of water resulted in the formation of formic acid, suggesting that the photoreduction of carbon dioxide with water to produce formic acid proceed in two steps, the photoreduction of carbon dioxide to form the radical intermediates and the successive protonation [52]. Yahaya et al. reported the effect of the photoirradiation time by using high-intensity light of 355 nm wavelength from a Nd:YAG laser [53]. The amount of methanol produced from carbon dioxide increased with the irradiation time on TiO_2 and ZnO and then decreased, while hydrogen was continuously produced. From these results, they discussed reaction schemes including consecutive oxidation of methanol to formaldehyde and carbon monoxide with formation of hydrogen.

The effect of the particle size also has been studied. Kočí et al. investigated the effect of the particle size of anatase TiO_2 in the range from 4.5 to 29 nm crystallites, determined by X-ray diffraction [54]. Hydrogen, methane, methanol and CO were obtained, and the optimum particle size for the production of both methane and methanol was 14 nm, which would result from some competing effects such as specific surface area, photoexcited carrier dynamics and light absorption efficiency. Yang et al. studied that the size effect by using the TiO_2 nanocrystallites supported on SAB-15 mesoporous silica [55], where the size was changed from 1.4 to 8.3 nm by changing the loading amount. The optimum amount of TiO_2 loading for methanol production was 45 wt%, corresponding to the 5.5 nm crystallites of TiO_2. Quantum size effects of the metal sulfide photocatalysts in the photocatalytic reduction of carbon dioxide using sacrificial electron donors were reviewed by Yoneyama [56].

The structure of the TiO_2 is also important. A study on the single crystals of TiO_2(100) and TiO_2(110) showed that the former was more active for the formation of methane and methanol than the latter [57]. The photocatalytic activity of TiO_2 nanotubes (20 nm in diameter × 300 nm in length × 5 nm in thickness) was compared to that of TiO_2 nanoparticles (P-25, Degussa, 20–50 nm in size

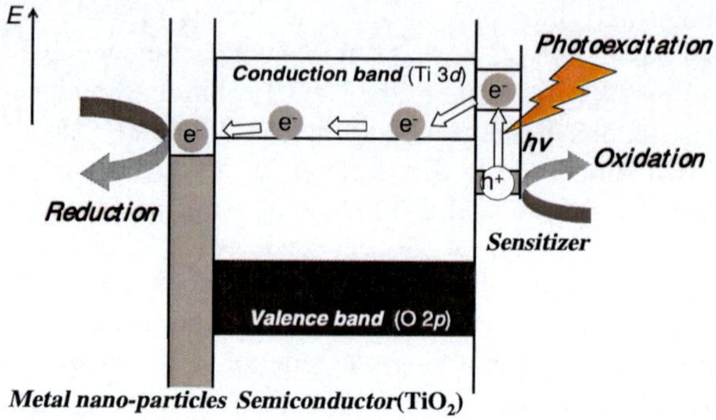

Fig. 7 The mechanism of photosensitization with dye or quantum dot upon visible light irradiation

determined by HRTEM), both of which were loaded with Pt. For the reaction between gaseous carbon dioxide and water vapor [58], the Pt-loaded TiO_2 nanotubes exhibited a higher activity in forming methane than the nanoparticles, especially under a high H_2O/CO_2 ratio.

In recent years, many researchers have developed photocatalysts that can function upon irradiation of visible light, since it is the main component of solar light. One strategy is the application of the dye-sensitized semiconductor photocatalysts, which is conceptually depicted in Fig. 7. For example, the addition of tris(2,2'-bipyridyl) Ru(II) chloride, perylene diimide derivatives [59] and Ru(II) (2,2'-bipyridyl-4,4'-dicarboxylate)$_2$-(NCS)$_2$ (N3 dye) [60] as visible light sensitizers on metal loaded TiO_2, such as Pt/TiO_2 and Cu-Fe/TiO_2, offered the activity to yield methane in visible light. In these systems, the photoexcited electron generated on the sensitizer is transferred to the conduction band of the semiconductor and then to the metal co-catalyst. The electron leads the reduction of carbon dioxide on the metal. In these examples, carbon dioxide was reduced to form methane. However, at the same time, the hole remains on the sensitizer. When the holes cannot be consumed, the photocatalytic reaction stops and the dye sensitizer is oxidized. This means that some sacrificial reagent is required to consume the holes. Like dyes, semiconductors can also serve this function. Recently, Wang et al. applied CdSe quantum dots of 2.5 nm in size as a photosensitizer to Pt/TiO_2 photocatalyst [61]. Although the conduction band of bulk CdSe is only slightly above that of TiO_2, quantum confinement shifts the conduction band of CdSe quantum dots to higher energies, which enables charge injection into the TiO_2. Then, the electron on the Pt/TiO_2 reduces carbon dioxide. On this CdSe/Pt/TiO_2 heterostructured photocatalyst, carbon dioxide was reduced to give methane as a major product and methanol, hydrogen and CO as minor products. However, the consumption of hole generated on the quantum dots was still a problem, which led to the oxidation of the quantum dots, decreasing the photocatalytic activity during the photoreaction.

As another strategy to design visible light photocatalysts, many materials have been examined and many viable semiconductor photocatalysts have been found, especially in the field of photocatalytic water splitting [11]. Also in the field of

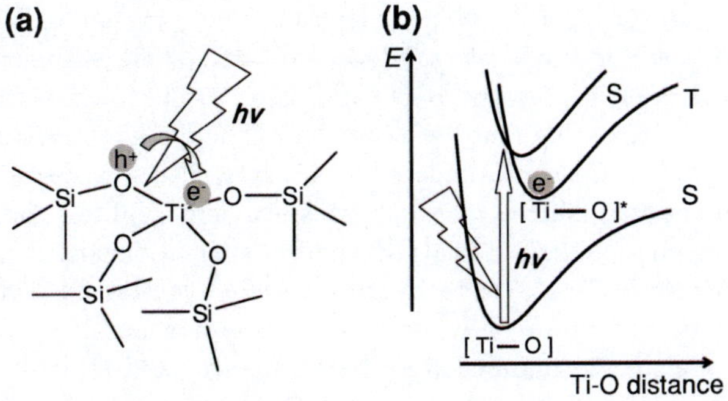

Fig. 8 Photoexcitation of highly-dispersed titanium species in silica matrix: **a** a model of the photoactive tetrahedral sites on silica, and **b** a schematic model for localized photoexcitation at Ti-O moiety

photocatalytic reduction of carbon dioxide, many photocatalysts have been found. $CaFe_2O_4$, a p-type semiconductor with a small band gap (1.9 eV) was found to produce formaldehyde and methanol under UV light, but the activity was not enough to perform the photoreduction under visible light irradiation without a sacrificial reducing agent [62]. In 2007, Pan and Chen reported that $NiO/InTaO_4$ photocatalysts produce methanol upon visible light irradiation [63]. Jia et al. reported that C and Fe co-doped $LaCoO_3$ exhibited photocatalytic activity to yield formaldehyde and formic acid [64]. Kudo and co-workers developed photocatalysts, such as $Ag/NaTaO_3$:Ba and $Ag/BaLa_4Ti_4O_{15}$, which produced H_2, O_2 and CO from an aqueous medium with a carbon dioxide flow system [65, 66]. These photocatalysts exhibited a high production rate for CO (38 and 25 μmol/h, respectively). These semiconductors were previously reported as active photocatalysts for water splitting [67, 68] and the Ag co-catalyst was found to be suitable co-catalyst for these photocatalysts to reduce carbon dioxide. Liu et al. found that a $BiVO_4$ photocatalyst reduced carbon dioxide with water to yield ethanol and O_2 under visible light irradiation from a 300 W xenon lamp [69]. When the irradiated light was extended to the UV region, the rate of ethanol production increased on the monoclinic $BiVO_4$ photocatalyst. The intensity of the light plays an important role in ethanol formation; both methanol and ethanol were formed when a 36 W fluorescent lamp was used.

4.1.2 Highly-Dispersed Photocatalysts

For the photocatalytic reduction of carbon dioxide with water, highly-dispersed photocatalysts (Fig. 8) have been also developed. In 1992, Anpo and Chiba prepared highly-dispersed titanium oxide anchored onto porous Vycor® glass through a facile reaction between the surface hydroxy groups and $TiCl_4$ [70]. This anchored catalyst showed photocatalytic activity for the reduction of carbon dioxide with water to yield methane, methanol and carbon monoxide. A small amount of O_2 was also detected. An ESR study revealed the formation of C radicals and H atoms, as well as Ti^{3+} ions. Thus, the highly-dispersed titanium oxide species was photoexcited to a charge-transfer excited state, a trapped hole center (O^-) paired with an electron center (Ti^{3+}), i.e., [Ti^{3+}–O^-]*, which activated the carbon dioxide to produce radicals. When the reaction system involved CO_2 and D_2O in place of H_2O, the major composition of photoformed methane was CD_4, indicating that the water functioned as the reductant.

When titanium oxide species are well-dispersed on silica materials, such as amorphous silica, mesoporous silica and zeolites, the highly-dispersed species become tetrahedral [18]. Mesoporous silica materials containing Ti, Ti-MCM-41 and Ti-MCM-48 were hydrothermally synthesized and applied to the photocatalytic reduction of carbon dioxide with water [71]. Although it is difficult to compare the photocatalytic performance of the highly-dispersed titanium oxide catalysts with bulk powdered TiO_2, the product selectivity was clearly different, i.e., while the bulk material produced predominantly methane, the highly dispersed one produced methanol in addition to methane. The Ti-MCM-41 showed similar activity to TS-1, which is a famous MFI-type zeolite containing Ti, while the Ti-MCM-48 showed much higher activity than either TS-1 or Ti-MCM-41, and similar selectivity to them. TiO_2-loaded zeolites and Ti ion-exchanged zeolites were prepared by the impregnation and ion-exchange methods, respectively, and examined for the photoreduction of carbon dioxide with water [72]. The sample containing 10 wt% TiO_2 showed high selectivity to methane, similar to bulk anatase, while the sample of 1 wt% TiO_2 gave methanol as minor product in addition to methane. The Ti ion-exchanged MFI-zeolite produced methanol and methane with comparable rates. These results suggest that the highly-dispersed titanium oxide species are more suitable for methanol formation than bulk TiO_2. Furthermore, the addition of Pt to the Ti ion-exchanged zeolite resulted in a high methane yield with high selectivity. Although the structure and size of the platinum species and the junction between Pt species and Ti species were still unclear, the Pt species would promote the charge separation, increasing the methane yield.

Ikeue et al. prepared two types of Ti-β zeolites by a hydrothermal synthesis method under different conditions using OH^- and F^- as anions of the structure-directing agents, referred to as Ti-β(OH) and Ti-β(F), respectively [73]. The titanium species in both samples were in a highly-dispersed, tetrahedral coordination state in the zeolite frameworks. The Ti-β(OH) was hydrophilic, while the Ti-β(F) was hydrophobic. It was confirmed by XANES that the highly-dispersed Ti species in Ti-β(OH) interacted with water molecules. Both samples exhibited photocatalytic activity in the photocatalytic reduction of carbon dioxide with water to yield methane, methanol, and trace amounts of CO, C_2H_4 and O_2. The Ti-β(OH) exhibited a much higher activity and higher selectivity to methane than TS-1, and Ti-β(F) exhibited a lower activity and higher selectivity to methanol than TS-1. This means that hydrophilic behavior is important to enhancing photocatalytic activity in the reduction of carbon dioxide with water to methane, while hydrophobic behavior is suitable for methanol production. This is consistent with another report: among other type of Ti-containing porous silica films prepared by solvent evaporation, the more hydrophobic sample with small amount of surface hydroxy groups exhibited higher methanol selectivity [74].

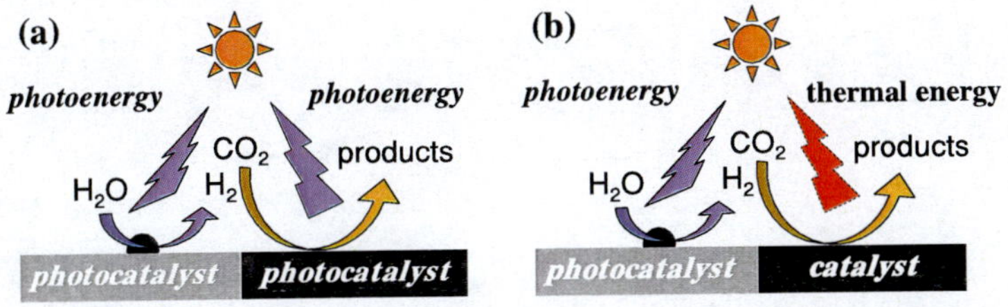

Fig. 9 The reduction of carbon dioxide with water by hybrid photocatalysts in which a photocatalyst for water splitting is combined with (**a**) a photocatalyst or (**b**) a catalyst for the reduction of carbon dioxide with hydrogen

On the other hand, Frei et al. found the formation of CO and O_2 in the photocatalytic reduction of carbon dioxide with water over Ti-MCM-41 upon photoirradiation by 266 nm light from a Nd:YAG laser by using in situ FT-IR spectroscopy and mass spectrometric analysis [75]. They concluded that a single UV photon induced the splitting of CO_2 with H_2O to CO and a surface OH radical, i.e., $CO_2 + H_2O \rightarrow CO + 2\ OH \rightarrow CO + 1/2\ O_2 + H_2O$. They also observed that this type of photoreduction of CO_2 could be promoted by a metal to metal charge transfer (MMCT) at Zr–O-Cu(I) moiety formed on MCM-41, Zr(IV)-O-Cu(I) →[Zr(III)-O-Cu(II)]* [76], although a MMCT on Ti–O-Cu(I) moiety could not induce the reaction.

4.2 Photocatalytic Reduction of Carbon Dioxide by Hydrogen

The reduction of carbon dioxide using hydrogen seems unattractive because of economics, i.e., hydrogen is relatively expensive. However, if we can use hydrogen as a reductant, the carbon dioxide can be converted to other products more easily, since the reduction with hydrogen is less unfavorable thermodynamically than the reduction with water, as shown in Table 2. Since many types of highly active photocatalysts have been developed for water splitting to produce solar hydrogen [11], we can use the solar hydrogen as the reductant for the reduction of carbon dioxide (Fig. 9a). Thus, we should also consider studying the photocatalytic reduction of carbon dioxide with hydrogen.

In 1982, Thampi et al. [77] found that methanation of carbon dioxide, the reduction of carbon dioxide by hydrogen to produce methane and water (Table 2, entry 5), was promoted by highly-dispersed Ru/RuOx loaded on TiO_2 at room temperature even in the dark, and furthermore the reaction rate was greatly enhanced by photoirradiation. The reaction proceeded photocatalytically and the selectivity was very high (>99%). They proposed that (i) the carbon dioxide was first reduced by 4 electrons on the Ru species to form Ru–C and active oxygen species, such as O^{2-}, (ii) hydrogen was oxidized by holes to form protons and react with the active oxygen species to form water, and (iii) the Ru–C species also

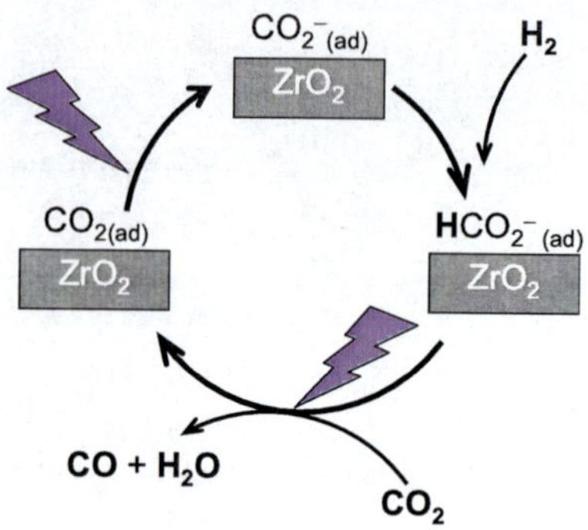

Fig. 10 Proposed mechanism for photocatalytic reduction of carbon dioxide by hydrogen over ZrO_2

reacts with hydrogen to yield methane and regenerate the Ru. Since this reaction is not an uphill reaction (Table 2, entry 5), the suitable catalyst can promote the reaction catalytically, even at room temperature in the dark. However, combined with water splitting, considering utilization of the solar hydrogen (Table 2, entry 18), the whole reaction becomes thermodynamically unfavorable (Table 2, entry 16). In 1999, Kohno et al. reported that the reduction of CO_2 with hydrogen to form CO or CH_4 proceeded on Rh-loaded TiO_2 in the dark, but was enhanced upon photoirradiation [78]. In this system, the metallic Rh species showed low activity but high selectivity to CH_4, while the mixture of Rh in metallic and oxidized states showed high activity and high selectivity to CO. Furthermore, it was found that the photocatalytic activity decreased with a decrease of the particle size of the Rh species during the photoreaction [79].

As mentioned above, the ZrO_2 photocatalyst without any co-catalyst promoted the photoreduction of carbon dioxide with water, where the carbon dioxide was actually aqueous carbonate [45]. The products were hydrogen, oxygen and carbon monoxide. Since the product distribution was not consistent with the equation shown in Table 2, entry 13, it was clear that water splitting also occurred. Thus, in this system, there is a possibility that the carbon oxide is reduced photocatalytically by hydrogen or hydrogen radical species in water produced through photocatalytic water splitting. In 1997, Kohno et al. [80] reported that gaseous carbon dioxide was reduced to carbon monoxide by hydrogen on a ZrO_2 photocatalyst without loading a co-catalyst in a closed static system with light of sub-300 nm wavelength from a 500 W ultrahigh-pressure Hg lamp. Among several samples, including TiO_2, ZrO_2, V_2O_5, Nb_2O_5, Ta_2O_5, WO_3 and ZnO, only ZrO_2 was active for this reaction. FT-IR spectroscopy clarified that the surface formate species existed as surface intermediates [81]. On ZrO_2, it was revealed that the photocatalytic reaction consisted of the following three steps, as shown in Fig. 10; (i) adsorbed carbon dioxide or carbonate was photoexcited and received an electron from ZrO_2 to become an adsorbed $\bullet CO_2^-$ anion radical [82], (ii) the anion radical reacted with hydrogen to form an adsorbed formate, even in the dark, and (iii) upon photoirradiation the formate species reduced the carbon dioxide molecule to yield carbon monoxide and water, where the surface formate species functioned as

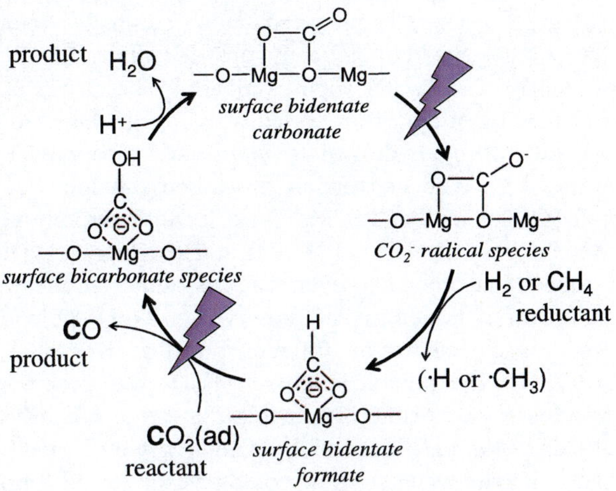

Fig. 11 A main mechanism of photocatalytic reduction of carbon dioxide on MgO by hydrogen or methane as a reductant

a reductant to become adsorbed carbon dioxide. For the photocatalytic reduction of carbon dioxide, at least two photons were required at the first and the third steps. ZrO_2 is a semiconductor with a wide band gap (5.0 eV) and can be excited by the light with wavelength shorter than 250 nm. However, the reaction proceeded even under light with wavelengths longer than 290 nm and the surface formate was formed even under the light with wavelengths longer than 330 nm. The photoluminescence spectroscopy revealed that adsorbed carbon dioxide species (or carbonate) can absorb the light above 300 nm in wavelength [81]. Therefore, the photoactive species is not the bulk ZrO_2, but the surface photoactive species. This means that a new type of photocatalytic reduction system was found: the adsorbed carbon dioxide species act as the photocatalytic active species [83].

Since it is known that the CO_2^- anion radical can form on MgO [84], the photocatalytic reduction of carbon dioxide by hydrogen was examined and discovered on MgO [85]. Teramura et al. revealed that the reaction proceeded by a similar mechanism to that on ZrO_2 (Fig. 11) [86]. On the surface of MgO, the carbon dioxide was adsorbed in the form of the surface bidentate carbonate and can be photoactivated to form a surface CO_2^- anion radical. This can be reduced by hydrogen to form the surface bidentate formate species, which can reduce another carbon dioxide to carbon monoxide as an actual reductant upon photoirradiation. These surface species can act as photocatalytically active species. These studies confirmed the new type of photocatalytic reduction of carbon dioxide that occurred in the manner of the surface photochemistry on the metal oxide of an insulator, rather than a semiconductor. Hydrogen is used as a reductant of the surface species in the photocatalytic cycle. As mentioned later, methane can similarly act as a reductant of the surface species on both MgO [86] and ZrO_2 [83, 87, 88].

Lo et al. [89] confirmed that the photoreduction of carbon dioxide over TiO_2 was enhanced using a gaseous mixture of hydrogen and water vapor to yield methane as a major product, compared with the using solely hydrogen or water.

On the other hand, the photoreduction of carbon dioxide over ZrO_2 with hydrogen gave a higher yield of carbon monoxide compared with using water or the mixture of hydrogen and water vapor, supporting the results presented above.

Teramura et al. [90] reported that Ga_2O_3 promoted the photoreduction of carbon dioxide by hydrogen to yield carbon monoxide selectively around room temperature. The conversion of carbon dioxide was approximately 3% using β-Ga_2O_3 calcined at 1,073 K for 3 h. Both carbon dioxide and hydrogen can be chemisorbed on Ga_2O_3 [91, 92] and the highest yield was obtained when the amount of adsorbed hydrogen reached saturation. Since the effect of the introduced amount of hydrogen was not observed on MgO, the photoreduction mechanism on Ga_2O_3 is expected to be different from that over basic metal oxide such as MgO and ZrO_2. The photoreaction on Ga_2O_3 was proposed to proceed between the adsorbed substrates according to the Langmuir–Hinshelwood mechanism.

Guan et al. [93] successfully demonstrated that the hydrogen produced through photocatalytic water splitting could be used for the reduction of carbon dioxide by a hybrid catalyst, in which a Pt-loaded $K_2Ti_6O_{13}$ photocatalyst (Pt/$K_2Ti_6O_{13}$) was combined with an Fe-based catalyst supported on a dealuminated Y-type zeolite (Fe-Cu-K/DAY) under concentrated sunlight in water. Upon photoirradiation from an Hg lamp around room temperature, the Pt/$K_2Ti_6O_{13}$ photocatalyst promoted the water splitting to produce hydrogen and somewhat reduced carbon dioxide to organic compounds though the Fe-Cu-K/DAY catalyst did not work. On the other hand, in concentrated sunlight, since the temperature of the system reached about 600 K, the hybrid catalyst produced hydrogen, formic acid, methanol, ethanol and methane, where the Pt/$K_2Ti_6O_{13}$ photocatalyst promoted the water splitting and the Fe-Cu-K/DAY catalyzed the reduction of carbon dioxide with produced hydrogen. The sunlight provided both the excitation light for the photocatalyst to produce hydrogen and the thermal energy to enhance the catalysis for the reduction of carbon dioxide (Fig. 9b). This study shows a new strategy to utilize the sunlight as both photoenergy and thermal energy.

4.3 Photocatalytic Reduction of Carbon Dioxide by Methane

Methane might be a good candidate as a reductant for the conversion of carbon dioxide [9]. Methane is known as an abundant underground resource (a fossil fuel), and also a renewable resource that can be naturally generated from ecosystems and artificially produced from biomass. Moreover, since methane can act as a greenhouse gas with an even higher global warming potential (GWP) than carbon dioxide [94], it is desirable to reduce its release into the environment and to utilize and convert it to more useful compounds. The reaction between methane and carbon dioxide can be catalyzed at high temperature to form syngas (carbon monoxide and hydrogen). This reaction is called a "dry reforming of methane" (DRM) or "CO_2 reforming of methane", as shown in Table 2, entry 7. A short history of the DRM can be found in literature [95]. The DRM is a thermodynamically unfavorable reaction and only can proceed at a high temperature, typically over 1,000 K, even when using catalysts. However, the photocatalytic reaction between methane and carbon dioxide can proceed at room temperature or mild temperatures.

Photocatalytic reduction of carbon dioxide with methane was found to proceed over ZrO_2 [83, 87, 88] and MgO [86] at room temperature. After the photoreaction, carbon monoxide and hydrogen were detected, although the amount of hydrogen was very low compared to the amount of carbon monoxide. It was confirmed that carbon monoxide was produced from carbon dioxide, not from methane. Methane acted as a reductant of the surface carbonate anion radical species to form the surface bidentate formate, as shown in Fig. 11. In these systems, the photoexcitation of adsorbed carbon dioxide forms a surface carbonate anion radical, followed by the reduction with methane to form the surface bidentate formate, and another carbon dioxide is photocatalytically reduced by the surface bidentate formate to yield carbon monoxide.

A Cu/CdS-TiO_2/SiO_2 photocatalyst was reported to yield acetone as the major product (Table 2, entry 8) with a high selectivity in the photocatalytic conversion of carbon dioxide with methane under UV irradiation at 393 K [96], where the selectivity for acetone was 92% at a 0.74% conversion of carbon dioxide. Other products, such as ethane and CO were also detected with selectivities of 3.1% and 4.6%, respectively. At a lower temperature (353 K) only ethane and CO were produced with selectivities of 47% and 53%, respectively. However, the conversion of methane and carbon dioxide ceased after about 2 h due to the photocorrosion of CdS by the TiO_2.

Recently, we found the photocatalytic reduction of carbon dioxide by methane to produce CO and hydrogen, i.e., photocatalytic DRM (Table 2, entry 7) by using Ga_2O_3 under photoirradiation at 473 K in a closed reactor [97]. As a side reaction, photocatalytic non-oxidative coupling of methane [98] to yield hydrocarbons (mainly ethane) was also observed. Since only a trace amount of carbon monoxide was obtained at room temperature, it was clear that thermal energy is required to yield the desired products. The production of carbon monoxide increased with increasing reaction temperature. The thermal activation energy (E_a) for carbon monoxide formation was ca. 10 kJ mol^{-1} at 473–673 K and ca. 60 kJ mol^{-1} at 673–873 K, which were much lower than that obtained from the catalytic thermal DRM on this Ga_2O_3 sample without UV irradiation ($E_a = 110$ kJ mol^{-1} at 773–973 K). This mild thermal activation energy at 473–673 K can be used for a mild activation step in the photocatalytic reaction mechanism, such as product desorption or electron migration from the bulk to the surface of Ga_2O_3. As mentioned above, sunlight can provide both the photoenergy and thermal energy and, therefore, the thermally assisted photocatalytic reaction at a moderate or high temperature is also possible. The methane conversion at 473 K was 0.27% after 3 h, which was superior to the equilibrium conversion (0.21%). This means that the photocatalysis can promote the difficult reaction under mild condition beyond the thermodynamic equilibrium conversion.

In this system, both carbon dioxide and hydrogen existed as reactant and product, respectively, and a Ga_2O_3 photocatalyst was reported to promote the photocatalytic reduction of carbon dioxide by hydrogen to yield carbon monoxide [90]. Quite recently, we prepared a series of Ga_2O_3 photocatalysts by homogeneous precipitation and examined the photocatalytic reduction of carbon dioxide with methane at 523 K [99]. We obtained Ga_2O_3 samples consisting of homogeneous granules with a short, rod-like morphology with similar size (300–500 nm in diameter and 1–2 μm in length), and calcination at various temperatures provided two polymorphs with various specific surface areas and crystallites sizes. We found that the Ga_2O_3 photocatalysts of low specific surface area selectively promoted reduction of carbon dioxide by methane at moderate temperature to yield carbon monoxide and hydrogen, while those of high specific surface area additionally promoted consecutive photocatalytic reduction of carbon dioxide by produced hydrogen to provide carbon monoxide selectively. The surface sites of Ga_2O_3, such as low coordination sites or surface hydroxy groups, would preferably catalyze the photoreduction of carbon dioxide by hydrogen. This demonstrates that the structure of the semiconductor photocatalysts, including crystallites size (or the specific surface area) and the surface sites, is very important for controlling the photocatalytic reactions. Further improvement is expected with the aid of recent developments in the nanotechnology.

5 Conclusion and Perspective

In this chapter, we reviewed the bases and history of the photocatalytic reduction of carbon dioxide by heterogeneous photocatalysts with three kinds of reductants: water, hydrogen and methane. The most preferable way to reduce carbon dioxide uses only water and sunlight. The investigations of using of hydrogen or methane accelerate the development of photocatalytic systems for carbon dioxide conversion. The idea of using the hydrogen generated by photocatalytic water splitting for the reduction of carbon dioxide is one of the realistic ways. The utilization of biomethane or other biomass is also meaningful.

Now, we note that, among the investigations introduced in this chapter, many photocatalysts have been prepared by conventional methods. On the other hand, we are now developing nanomaterial syntheses that can produce finely designed materials. As shown, the structure of the photocatalysts is very important, e.g., the crystallites size, specific surface area, surface structure and electronic band structure of the semiconductors. The structure and physicochemical properties of the co-catalysts and the local structure of the dispersed photoactive sites are also important. It is expected that the nanotechnology will enable us to design the structure and function of these photocatalysts with precise control, which will provide effective photocatalysts. The conversion of carbon dioxide using solar energy is fundamentally important for our sustainable future. We hope that the development of the photocatalysis with nanomanufacturing realizes the reduction of carbon dioxide with solar energy in the near future.

References

1. Indrakanti VP, Kubicki JD, Schobert HH (2009) Photoinduced activation of CO_2 on Ti-based heterogeneous catalysts: current state, chemical physics-based insights and outlook. Energy Environ Sci 2:745–758
2. Armor JN (2007) Addressing the CO_2 dilemma. Catal Lett 114:115–121
3. Song C (2006) Global challenges and strategies for control, conversion and utilization of CO_2 for sustainable development involving energy, catalysis, adsorption and chemical processing. Catal Today 115:2–32
4. Xiaoding X, Moulijn JA (1996) Mitigation of CO_2 by chemical conversion: plausible chemical reactions and promising products. Energy Fuels 10:305–325
5. Sakakura T, Choi JC, Yasuda H (2007) Transformation of carbon dioxide. Chem Rev 107:2365–2387
6. Aresta M, Dibenedetto A (2007) Utilisation of CO_2 as a chemical feedstock: opportunities and challenges. Dalton Trans 2975–2992
7. Usubharatana P, McMartin D, Veawab A, Tontiwachwuthikul P (2006) Photocatalytic process for CO_2 emission reduction from industrial flue gas streams. Ind Eng Chem Res 45:2558–2568
8. Hattori H (1995) Heterogeneous Basic Catalysis. Chem Rev 95:537–558
9. Yuliati L, Yoshida H (2008) Photocatalytic conversion of methane. Chem Soc Rev 37: 1592–1602
10. Roh H-S, Potdar HS, Jun K-W (2004) Catal Today 93–95:39–44
11. Kudo A, Miseki Y (2009) Heterogeneous photocatalyst materials for water splitting. Chem Soc Rev 38:253–278
12. The Chemical Society of Japan (1984) Kagakubinran kisohen, 3rd edn. Maruzen, Tokyo
13. Atkins PW (1998) Physical chemistry, 6th edn. Oxford University Press, Oxford
14. Shaw DA, Holland DMP, Hayes MA et al (1995) A study of the absolute photoabsorption, photoionisation and photodissociation cross sections and the photoionisation quantum efficiency of carbon dioxide from the ionisation threshold to 345 Å. Chem Phys 198:381–396
15. Freund HJ, Roberts MW (1996) Surface chemistry of carbon dioxide. Surf Sci Rep 25: 225–273
16. The CODATA Task Group on key values for thermodynamics (1978) CODATA recommended key values for thermodynamics, 1977. J Chem Thermodyn 10:903–906
17. Murata C, Yoshida H, Kumagai J, Hattori T (2003) Active sites and active oxygen species for photocatalytic epoxidation of propene by molecular oxygen over TiO_2–SiO_2 binary oxides. J Phys Chem B 107:4364–4373
18. Yoshida H (2003) Silica-based quantum photocatalysts for selective reactions. Curr Opin Solid State Mater Sci 7:435–442
19. Fujishima A, Honda K (1972) Electrochemical photolysis of water at a semiconductor electrode. Nature 238:37–38
20. Hemminger JC, Carr R, Somorjai GA (1978) The photoassisted reaction of gaseous water and carbon dioxide adsorbed on the $SrTiO_3$ (111) crystal face to form methane. Chem Phys Lett 57:100–104
21. Halmann M (1978) Photoelectrochemical reduction of aqueous carbon dioxide on p-type gallium phosphide in liquid junction solar cells. Nature 275:115–116
22. Inoue T, Fujishima A, Konishi S, Honda K (1979) Photoelectrocatalytic reduction of carbon dioxide in aqueous suspensions of semiconductor powders. Nature 277:637–638

23. Aurian-Blajeni B, Halmann M, Manassen J (1980) Photoreduction of carbon dioxide and water into formaldehyde and methanol on semiconductor materials. Sol Energy 25:165–170
24. Halmann M, Ulman M, Aurian-Blajeni B (1983) Photochemical solar collector for the photoassisted reduction of aqueous carbon dioxide. Sol Energy 31:429–431
25. Irvine JTS, Eggins BR, Grimshaw J (1990) Solar energy fixation of carbon dioxide via cadmium sulphide and other semiconductor photocatalysts. Sol Energy 45:27–33
26. Bard AJ (1980) Photoelectrochemistry. Science 207:139–144
27. Sato S, White JM (1980) Photodecomposition of water over Pt/TiO2 catalysts. Chem Phys Lett 72:83–86
28. Chandrasekaran K, Thomas JK (1983) Photochemical reduction of carbonate to formaldehyde on TiO_2 powder. Chem Phys Lett 99:7–10
29. Halmann M, Katzir V, Borgarello E, Kiwi J (1984) Photoassisted carbon dioxide reduction on aqueous suspensions of titanium dioxide. Solar Energy Mater 10:85–91
30. Tennakone K (1984) Photoreduction of carbonic acid by mercury coated n-titanium dioxide. Solar Energy Mater 10:235–238
31. Raphael MW, Malati MA (1989) The photocatalysed reduction of aqueous sodium carbonate using platinized titania. J Photochem Photobiol A 46:367–377
32. Hirano K, Inoue K, Yatsu T (1992) Photocatalysed reduction of CO_2 in aqueous TiO_2 suspension mixed with copper powder. J Photochem Photobiol A 64:255–258
33. Ishitani O, Inoue C, Suzuki Y, Ibusuki T (1993) Photocatalytic reduction of carbon dioxide to methane and acetic acid by an aqueous suspension of metal-deposited TiO_2. J Photochem Photobiol A 72:269–271
34. Solymosi F, Tombácz I (1994) Photocatalytic reaction of H_2O+CO_2 over pure and doped Rh/TiO_2. Catal Lett 27:61–65
35. Yamashita H, Nishiguchi H, Kamada N et al (1994) Photocatalytic reduction of CO_2 with H_2O on TiO_2 and Cu/TiO_2 catalysts. Res Chem Intermed 20:815–823
36. Tseng IH, Chang WC, Wu JCS (2002) Photoreduction of CO_2 using sol-gel derived titania and titania-supported copper catalysts. Appl Catal B 37:37–48
37. Tseng IH, Wu JCS, Chou HY (2004) Effects of sol-gel procedures on the photocatalysis of Cu/TiO_2 in CO_2 photoreduction. J Catal 221:432–440
38. Wu JCS, Lin HM, Lai CL (2005) Photo reduction of CO_2 to methanol using optical-fiber photoreactor. Appl Catal A 296:194–200
39. Nguyen TV, Wu JCS (2008) Photoreduction of CO_2 in an optical-fiber photoreactor: effects of metals addition and catalyst carrier. Appl Catal A 335:112–120
40. Kiwi J, Grätzel M (1984) Optimization of conditions for photochemical water cleavage. Aqueous Pt/TiO_2 (anatase) dispersions under ultraviolet light. J Phys Chem 88:1302–1307
41. Kiwi J, Morrison C (1984) Heterogeneous photocatalysis. Dynamics of charge transfer in lithium-doped anatase-based catalyst powders with enhanced water photocleavage under ultraviolet irradiation. J Phys Chem 88:6146–6152
42. Yamaguti K, Sato S (1985) Pressure dependence of the rate and stoichiometry of water photolysis over platinized TiO_2 catalysts. J Phys Chem 89:5510–5513
43. Munuera G, Rives-Arnau V, Saucedo A (1979) Photo-adsorption and photo-desorption of oxygen on highly hydroxylated TiO_2 surfaces. Part 1.–Role of hydroxyl groups in photoadsorption. J Chem Soc, Faraday Trans 1 75:736–747
44. Gonzalez-Elipe A, Munuera G, Soria J (1979) Photo-adsorption and photo-desorption of oxygen on highly hydroxylated TiO_2 surfaces. Part 2.–Study of radical intermediates by electron paramagnetic resonance. J Chem Soc, Faraday Trans 1 75:748–761
45. Sayama K, Arakawa H (1993) Photocatalytic decomposition of water and photocatalytic reduction of carbon dioxide over ZrO_2 catalyst. J Phys Chem 97:531–533
46. Adachi K, Ohta K, Mizuno T (1994) Photocatalytic reduction of carbon dioxide to hydrocarbon using copper-loaded titanium dioxide. Sol Energy 53:187–190
47. Mizuno T, Adachi K, Ohta K, Saji A (1996) Effect of CO_2 pressure on photocatalytic reduction of CO_2 using TiO_2 in aqueous solutions. J Photochem Photobiol A 98:87–90

48. Pathak P, Meziani MJ, Li Y et al (2004) Improving photoreduction of CO_2 with homogeneously dispersed nanoscale TiO_2 catalysts. Chem Commun 1234–1235
49. Pathak P, Meziani MJ, Castillo L, Sun YP (2005) Metal-coated nanoscale TiO_2 catalysts for enhanced CO_2 photoreduction. Green Chem 7:667–670
50. Saladin F, Alxneit I (1997) Temperature dependence of the photochemical reduction of CO_2 in the presence of H_2O at the solid/gas interface of TiO_2. J Chem Soc, Faraday Trans 93:4159–4163
51. Mizuno T, Tsutsumi H, Ohta K, et al. (1994) Photocatalytic reduction of CO_2 with dispersed TiO_2/Cu powder mixtures in supercritical CO_2. Chem Lett 1533–1536
52. Kaneco S, Kurimoto H, Ohta K et al (1997) Photocatalytic reduction of CO_2 using TiO_2 powders in liauid CO_2 medium. J Photochem Photobiol A 109:59–63
53. Yahaya AH, Gondal MA, Hameed A (2004) Selective laser enhanced photocatalytic conversion of CO_2 into methanol. Chem Phys Lett 400:206–212
54. Kočí K, Obalová L, Matějová L et al (2009) Effect of TiO_2 particle size on the photocatalytic reduction of CO_2. Appl Catal B 89:494–502
55. Yang HC, Lin HY, Chien YS et al (2009) Mesoporous TiO_2/SBA-15, and Cu/TiO_2/SBA-15 composite photocatalysts for photoreduction of CO_2 to methanol. Catal Lett 131:381–387
56. Yoneyama H (1997) Photoreduction of carbon dioxide on quantized semiconductor nanoparticles in solution. Catal Today 39:169–175
57. Yamashita H, Kamada N, He H et al (1994) Reduction of CO_2 with H_2O on TiO_2(100) and TiO_2(110) single crystals under UV-irradiation. Chem Lett 23:855–858
58. Zhang QH, Han WD, Hong YJ, Yu JG (2009) Photocatalytic reduction of CO_2 with H_2O on Pt-loaded TiO_2 catalyst. Catal Today 148:335–340
59. Ozcan O, Yukruk F, Akkaya EU, Uner D (2007) Dye sensitized CO_2 reduction over pure and platinized TiO_2. Top Catal 44:523–528
60. Nguyen TV, Wu JCS, Chiou CH (2008) Photoreduction of CO_2 over ruthenium dye-sensitized TiO_2-based catalysts under concentrated natural sunlight. Catal Commun 9: 2073–2076
61. Wang C, Thompson RL, Baltrus J, Matranga C (2010) Visible light photoreduction of CO_2 using CdSe/Pt/TiO_2 heterostructured catalysts. J Phys Chem Lett 1:48–53
62. Matsumoto Y, Obata M, Hombo J (1994) Photocatalytic reduction of carbon dioxide on p-type $CaFe_2O_4$ powder. J Phys Chem 98:2950–2951
63. Pan PW, Chen YW (2007) Photocatalytic reduction of carbon dioxide on NiO/$InTaO_4$ under visible light irradiation. Catal Commun 8:1546–1549
64. Jia L, Li J, Fang W (2009) Enhanced visible-light active C and Fe co-doped $LaCoO_3$ for reduction of carbon dioxide. Catal Commun 11:87–90
65. Miseki Y, Iizuka K, Saito K et al (2009) Water splitting and CO_2 reduction over $ALa_4Ti_4O_{15}$ (A=Ca, Sr, Ba) photocatalysts layered perovskite structure. Catal Catal 51:84–86
66. Iizuka K, Kojima Y, Kudo A (2009) CO_2 reduction using heterogeneous photocatalysts aiming at artificial photosynthesis. Catal Catal 51:228–233
67. Iwase A, Kato H, Okutomi H, Kudo A (2004) Formation of surface nano-step structures and improvement of photocatalytic activities of $NaTaO_3$ by doping of alkaline earth metal ions. Chem Lett 33:1260–1261
68. Miseki Y, Kato H, Kudo A (2009) Water splitting into H_2 and O_2 over niobate and titanate photocatalysts with (111) plane-type layered perovskite structure. Energy Environ Sci 2:306–314
69. Liu Y, Huang B, Dai Y et al (2009) Selective ethanol formation from photocatalytic reduction of carbon dioxide in water with $BiVO_4$ photocatalyst. Catal Commun 11:210–213
70. Anpo M, Chiba K (1992) Photocatalytic reduction of CO_2 on anchored titanium oxide catalysts. J Mol Catal 74:207–212
71. Zhang SG, Fujii Y, Yamashita H et al (1997) Photocatalytic reduction of CO_2 with H_2O on Ti-MCM-41 and Ti-MCM-48 mesoporous zeolites at 328 K. Chem Lett 26:659–660

72. Anpo M, Yamashita H, Ichihashi Y et al (1997) Photocatalytic reduction of CO_2 with H_2O on titanium oxide anchored within micropores of zeolites: Effects of the structure of the active sites and the addition of Pt. J Phys Chem B 101:2632–2636
73. Ikeue K, Yamashita H, Anpo M, Takewaki T (2001) Photocatalytic reduction of CO_2 with H_2O on Ti-β zeolite photocatalysts: effect of the hydrophobic and hydrophilic properties. J Phys Chem B 105:8350–8355
74. Ikeue K, Nozaki S, Ogawa M, Anpo M (2002) Characterization of self-standing Ti-containing porous silica thin films and their reactivity for the photocatalytic reduction of CO_2 with H_2O. Catal Today 74:241–248
75. Lin W, Han H, Frei H (2004) CO_2 splitting by H_2O to CO and O_2 under UV light in TiMCM-41 silicate sieve. J Phys Chem B 108:18269–18273
76. Lin W, Frei H (2006) Bimetallic redox sites for photochemical CO_2 splitting in mesoporous silicate sieve. C R Chimie 9:207–213
77. Thampi KR, Kiwi J, Grätzel M (1987) Methanation and photo-methanation of carbon dioxide at room temperature and atmospheric pressure. Nature 327:506–508
78. Kohno Y, Hayashi H, Takenaka S et al (1999) Photo-enhanced reduction of carbon dioxide with hydrogen over Rh/TiO_2. J Photochem Photobiol A 126:117–123
79. Kohno Y, Yamamoto T, Tanaka T, Funabiki T (2001) Photoenhanced reduction of CO_2 by H_2 over Rh/TiO_2. Characterization of supported Rh species by means of infrared and X-ray absorption spectroscopy. J Mol Catal A 175:173–178
80. Kohno Y, Tanaka T, Funabiki T, Yoshida S (1997) Photoreduction of carbon dioxide with hydrogen over ZrO_2. Chem Commun 841–842
81. Kohno Y, Tanaka T, Funabiki T, Yoshida S (1998) Identification and reactivity of a surface intermediate in the photoreduction of CO_2 with H_2 over ZrO_2. J Chem Soc, Faraday Trans 94:1875–1880
82. Kohno Y, Tanaka T, Funabiki T, Yoshida S (2000) Photoreduction of CO_2 with H_2 over ZrO_2. A study on interaction of hydrogen with photoexcited CO_2. Phys Chem Chem Phys 2:2635–2639
83. Yoshida S, Kohno Y (2000) A new type of photocatalysis initiated by photoexcitation of adsorbed carbon dioxide on ZrO_2. Catal Surv Jpn 4:107–114
84. Lunsford JH, Jayne JP (1965) Formation of CO_2^- radical ions when CO_2 is adsorbed on irradiated magnesium oxide. J Phys Chem 69:2182–2184
85. Kohno Y, Ishikawa H, Tanaka T, Funabiki T, Yoshida S (2001) Photoreduction of carbon dioxide by hydrogen over magnesium oxide. Phys Chem Chem Phys 3:1108–1113
86. Teramura K, Tanaka T, Ishikawa H et al (2004) Photocatalytic reduction of CO_2 to CO in the presence of H_2 or CH_4 as a reductant over MgO. J Phys Chem B 108:346–354
87. Kohno Y, Tanaka T, Funabiki T, Yoshida S (1997) Photoreduction of carbon dioxide with methane over ZrO_2. Chem Lett 993–994
88. Kohno Y, Tanaka T, Funabiki T, Yoshida S (2000) Reaction mechanism in the photoreduction of CO_2 with CH_4 over ZrO_2. Phys Chem Chem Phys 2:5302–5307
89. Lo CC, Hung CH, Yuan CS, Wu JF (2007) Photoreduction of carbon dioxide with H_2 and H_2O over TiO_2 and ZrO_2 in a circulated photocatalytic reactor. Solar Energy Mater Solar Cells 91:1765–1774
90. Teramura K, Tsuneoka H, Shishido T, Tanaka T (2008) Effect of H_2 gas as a reductant on photoreduction of CO_2 over a Ga_2O_3 photocatalyst. Chem Phys Lett 467:191–194
91. Collins SE, Baltanás MA, Bonivardi AL (2005) Hydrogen chemisorption on gallium oxide polymorphs. Langmuir 21:962–970
92. Collins SE, Baltanás MA, Bonivardi AL (2006) Infrared spectroscopic study of the carbon dioxide adsorption on the surface of Ga_2O_3 polymorphs. J Phys Chem B 110:5498–5507
93. Guan G, Kida T, Yoshida A (2003) Reduction of carbon dioxide with water under concentrated sunlight using photocatalyst combined with Fe-based catalyst. Appl Catal B 41:387–396
94. Forster P, Ramaswamy V, Artaxo P et al (2007) Changes in atmospheric constituents and radiative forcing. In: Solomon S, Qin D, Manning M et al (eds) Climate change 2007: the

physical science basis. Contribution of Working Group I to the fourth assessment report of the intergovernmental panel on climate change. Cambridge University Press, Cambridge, pp 129–234
95. Hu YH, Ruckenstein E (2002) Binary MgO-based solid solution catalysts for methane conversion to syngas. Catal Rev 44:423–453 and references therein
96. Shi D, Feng Y, Zhong S (2004) Photocatalytic conversion of CH_4 and CO_2 to oxygenated compounds over Cu/CdS–TiO_2/SiO_2 catalyst. Catal Today 98:505–509
97. Yuliati L, Itoh H, Yoshida H (2008) Photocatalytic conversion of methane and carbon dioxide over gallium oxide. Chem Phys Lett 452:178–182
98. Yuliati L, Hattori T, Itoh H, Yoshida H (2008) Photocatalytic nonoxidative coupling of methane on gallium oxide and silica-supported gallium oxide. J Catal 257:396–402
99. Yoshida H, Maeda K (2010) Preparation of gallium oxide photocatalysts for reduction of carbon dioxide. Stud Surf Sci Catal 175:351–354

Nanostructured Electrodes and Devices for Converting Carbon Dioxide Back to Fuels: Advances and Perspectives

Gabriele Centi and Siglinda Perathoner

Abstract The production of solar fuels from water and CO_2 is an efficient solution to store and use solar energy and reduce the negative effects associated with large volumes of CO_2 emissions. In this vision, the liquid fuels produced by recycling CO_2 using solar energy are an ideal energy source: easy to store/transport and providing full integration into the existing energy infrastructure. After discussing the possible option to reach this objective, the use of a novel photoelectrocatalytic (PEC) device is analyzed in a more detail. New characteristics of the photoanode and electrocatalyst are required. Some aspects related to the characteristics of nanostructured (a) TiO_2 thin films (based on an ordered array of titania nanotubes) and their performances as photoanodes and (b) carbon nanotube-based electrodes for the gas phase reduction of CO_2 to liquid fuels (mainly isopropanol) together with their application for the design of a novel photoelectrocatalytic (PEC) approach for the synthesis of solar fuels will be presented.

1 Introduction

Rising concerns over carbon dioxide emissions, shown at the recent United Nations Climate Change Conference (Copenhagen, December 2009) has accelerated the R&D activities on carbon dioxide sequestration and storage (CSS) [1–3].

G. Centi · S. Perathoner (✉)
Dipartimento di Chimica Industriale ed Ingegneria dei Materiali and
CASPE (INSTM Laboratory of Catalysis for Sustainable Production and Energy),
Università di Messina, Salita Sperone 31, 98166 Messina, Italy
e-mail: perathon@unime.it

G. Centi
e-mail: centi@unime.it

L. Zang (ed.), *Energy Efficiency and Renewable Energy Through Nanotechnology*,
Green Energy and Technology, DOI: 10.1007/978-0-85729-638-2_16,

However, at the same time the interest in considering CO_2 as a value instead of a waste increased [4–6]. The increasing amounts of CO_2 available from current and planned CSS plants will make carbon dioxide a feedstock of zero (or even negative!) cost for conversion to fuels and chemicals. Furthermore, incentives exist, such as goodwill for companies adopting policies of CO_2 emissions reduction.

The other factor stimulating the interest in CO_2 chemical recycling is the presence of many emissions for which the CSS option is unsuitable: distance from safe sequestration sites, diluted concentration of CO_2 in the emission, small-medium size sources, and uncertainty on the impact on environment. Considering that at least 5–10% of the total CO_2 emissions (about 30 Gt worldwide in 2009 [7]) could be suitable for production of fuels and chemicals, e.g., at least one order of magnitude higher than the actual use of CO_2 in industry, there are large opportunities to develop new approaches for recycling [8–14]. The utilization of CO_2 as a raw material in the synthesis of chemicals and liquid energy carriers offers a way to mitigate the increasing CO_2 build-up in conjunction with economic opportunities.

The volume of CO_2 emissions is so large with respect to the amount which may be converted to chemicals that recycling it back to fuel using renewable sources of energy is the necessary and sustainable option to consider [15–20]. Worldwide fuel consumption is two orders of magnitude higher than that of chemicals. Therefore, the fuel market has the largest potential to utilize the products of CO_2 conversion. In addition, valorization of carbon dioxide emissions could be one important part of the general strategy for reducing CO_2 emissions and pushing the chemical and energy companies toward a more sustainable use of the resources [21, 22].

From a different perspective, it should be pointed out that in the near-term, we could forecast an increasing use of solar energy, allowing it to become predominant in a longer-term energy scenario. A massive use of solar energy would require converting light energy to forms that can be easily stored to be used when and where required. In fact, photovoltaic (PV) cells for converting solar to electrical energy can have a mismatch between the time and location of production and use. Actual methods for storing electrical energy (mainly based on potential energy) are not very efficient. The storage of electrical energy by devices such as batteries and capacitors, notwithstanding the significant developments in nanostructured materials for them [23, 24], still remains an issue for large energy amounts. Therefore, there is a strong need to develop efficient chemical energy vectors to store and easy transport energy produced from solar light.

H_2 is a clean energy vector that has received a large amount of attention, part of the idea to implement the "hydrogen economy" [25–27]. However, the technical limits in efficient and cost-effective hydrogen storage put serious constraints on the possibility of its use as the primary energy vector. Other alternatives, such as ammonia, have relevant limits in terms of safety and toxicity. Furthermore, huge investments are necessary to rebuild the energy infrastructure using H_2 or other alternative energy vectors. Liquid products such as methanol or better longer-chain alcohols or hydrocarbons are preferable options as energy vectors in terms of energy density, low toxicity, easy and safe storage/transport, and, especially, capability of integration into the existing energy infrastructure for both mobile and stationary applications. We have thus proposed these *C-based energy vectors* as the preferable and more sustainable options toward solar fuels [28–30].

1.1 Options for Converting Carbon Dioxide Back to Fuels

In principle, there are different possibilities in converting CO_2 back to fuels. The most investigated area is the hydrogenation of CO_2 to form oxygenates and/or hydrocarbons. Methanol synthesis from CO_2 and H_2 has been investigated up to pilot-plant stage with promising results [16]. An alternate possibility is the production of dimethyl ether (DME), a clean-burning fuel that is a potential diesel substitute. Ethanol formation, either directly or via methanol homologation, or the conversion of CO_2 to formic acid are also potentially interesting routes. Methanol, ethanol, and formic acid can also be used as feedstocks in fuel cells, providing a route to store energy using CO_2 and subsequently produce electricity. Alcohols are, in principle, preferable over hydrocarbons because their synthesis requires less hydrogen per unit of product. In fact, the key problem in this route is the availability of H_2. If the latter is produced from hydrocarbons (the main current route is by steam reforming of methane) there are no real advantages in converting CO_2. H_2 must be derived from renewable sources. The possible options are the following:

- Water electrolysis, coupled with a renewable source of electrical energy (photovoltaic cells, wind or waves, etc.). This technology is already available, but the need of multiple steps, the overpotential in the electrolyzer, and other issues limit the overall efficiency. The technology is mature with a limited degree of further possible improvements.
- Biomass conversion, preferably using waste materials and in conditions which require low energy consumption. An example is the catalytic production of H_2 directly in liquid phase from aqueous solutions (ethanol waste streams, for example). This option could be a way for the valorization of side waste streams from a biorefinery, but it is not an efficient method if considered alone. In fact, if we consider the whole life cycle from growing the plant, harvesting, fermentation, etc. and finally H_2 production (from bioethanol, for example), the overall energy consumption (and thus amount of CO_2 produced) is higher than the advantage in hydrogenating CO_2 back to fuels.
- Production of H_2 via biogas produced from anaerobic fermentation of biomass. Also in this case, it could be a valuable option using waste biomass, but is a quite complex process considering the whole production chain. There are also problems with the purification of biogas.
- Production of H_2 using cyanobacteria or green-algae. This is an interesting option, but with low productivity and still under development.

- Direct H_2 production by water photoelectrolysis, which suffers low productivity and, in some cases, of the need for further separation/recovery of hydrogen.

Renewable H_2, after the eventual compression and heating to the requisite reaction temperature, may then be used for the hydrogenation of CO_2. The first step is the reversible reaction of reverse water gas shift (RWGS), a mildly endothermic reaction with enthalpy and free energy changes of $\Delta H_{298K} = 41.2$ kJ/mol and $\Delta G_{298K} = 28.6$ kJ/mol:

$$CO_2 + H_2 \leftrightharpoons CO + H_2O \quad (1)$$

Carbon monoxide and hydrogen, also called syngas, may then be converted to methanol and/or DME, or Fischer–Tropsch products (hydrocarbons, mainly) by known catalytic processes. These processes may also be combined into a single process with the RWGS reaction, but the formation of water in the latter is an issue in syngas transformation.

If we consider H_2 produced at 10% efficiency by a combination of PV + electrolyzer cell, the need of three H_2 molecules for converting CO_2 to methanol is shown by:

$$CO_2 + 3H_2 \leftrightharpoons CH_3OH + H_2O \quad (2)$$

and the energy associated with the whole process chain (including the separation, etc.) in a complete life cycle assessment (LCA), it is possible to calculate the effective energy efficiency (EEE) of the process (the ratio between energy content of the product (methanol) and the total energy input necessary for the whole chain) and the CO_2 recycle effectiveness (CRE) [the ratio between CO_2 recycled and CO_2 emitted due energy consumption (due to the use of fossil fuels in producing and running the various devices)].

A few LCA studies exist on the use of CO_2, H_2 production, PV cells, methanol production, etc. [31–38] and they are not specific to the above case or its analogous paths. It is therefore not possible to make conclusions, but it can be estimated that EEE is significant below 1% and the CRE is only slightly negative. The routes using H_2 produced via biomass have significantly lower EEEs and positive CRE values. In other words, specific studies are necessary to evaluate the effective sustainability of CO_2 hydrogenation routes. This technology is nearly viable, although some further improvement is necessary. The overall complexity (multi-stage process) makes the overall efficiency low, with a limited degree of possible improvement. Therefore, it is not a long-term, sustainable option for converting CO_2 back to fuels.

Dry reformation of methane with CO_2 is an alternative possibility to produce syngas:

$$CH_4 + CO_2 \leftrightharpoons 2CO + 2H_2 \quad \Delta H_{298K} = 247 \text{ kJ/mol} \quad (3)$$

This is a strong endothermic reaction that suffers from fast deactivation due to carbon formation. The coupling of the reaction of steam reforming of methane and partial methane combustion (the tri-reforming process) reduces the issue of deactivation and allows authothermic operations. The process is interesting and has been developed to pilot-scale operations. However, the CO_2 recycle effectiveness is probably low, even if specific LCA studies are not present in the literature.

There are other routes and options in converting CO_2 to fuels, but it is not our aim to discuss them in detail here. More specific aspects are reported elsewhere [1, 15–17, 39–46]. The above discussion highlights the concept that when complex and multi-stage processes are necessary, the final effectiveness is low, as is the effective potential in reducing CO_2 emissions. It is necessary to explore direct routes for producing fuels from carbon oxides using solar energy. Although all of these routes are still at an early stage of development, they have the potential to be more effective and more economical. They are therefore the necessary direction for research, although in a medium to long-term perspective.

1.2 Solar Fuels Based on Carbon Dioxide Recycle

There are essentially three main routes for solar fuels based on carbon dioxide recycling:

- Bio-route, based on plants, algae, bacteria, etc.
- Concentrated solar energy, where the solar radiation is concentrated using mirrors, providing the required heat to perform endothermic reactions or suitable thermodynamic cycles
- Low-temperature conversions using semiconductors, where the light absorption is used to generate electrons and holes by charge separation, which subsequently react with water and CO_2.

Photosynthesis in green plants is the best known example of solar energy used for the conversion of CO_2, but its efficiency is relative low (around 1%) and the CO_2 is converted (in a slow process) to carbohydrates, lipids, etc. Producing fuels from biomass is possible and the research interest is growing, especially on biomass conversion to liquid fuels (the BtL process) [47, 48]. Many steps are necessary, from plant growth to harvesting, treatment, transport, and finally conversion in typically multi-step processes before producing biofuels. This is only a feasible short-term solution but, in our opinion, not in the long-term perspective of sustainable energy.

Using algae, the efficiency in using solar light is higher (around 10%) and CO_2 (from power plants) can be even fed directly to the photobioreactors or open ponds [49, 50]. However, the process of producing biofuels from (micro)algae is quite complex and critical problems are the controlled growth of the algae and cost. Biofuels from microalgae are considered third generation biofuels, after the second generation processes based on biofuels produced from lignocellulosic materials [50]. However, from a conceptual point of view, direct routes of solar energy use for conversion of CO_2 to fuels that do not pass through the phase of complex molecules (cellulose, hemicellulose, lignine, starch, lipides, oils, etc.) are preferable. A life cycle assessment of biodiesel production confirms the necessity to decrease the energy and fertilizer consumed by the process [51].

Some algae and cyanobacteria can use light directly to produce fuels [45, 52, 53], but only hydrogen can be produced. It could be possible to genetically manipulate enzymes, fungi, or bacteria to convert CO_2 directly to biofuels, but it is difficult to predict if this will ever be realized. Bio-routes for the production of solar fuels by recycling CO_2 suffer the intrinsic limitation to go "uphill" in energy by creating complex molecules that are then decomposed ("downhill" in energy) to produce fuels. It is evident that it is preferable to preserve this complexity to make chemicals instead of fuels and that a direct route to use solar energy that directly (or at least in fewer steps) converts CO_2 and H_2O to liquid fuels and O_2 is the energetically preferable route. In addition, biomaterials (algae, cyanobacteria, etc.) are typically quite sensitive to the environment and it is necessary to have sophisticated control of the reaction conditions, which in turn demand higher costs.

Concentrated solar radiation (in solar high-temperature furnaces) can be used to produce H_2 (and O_2) from water or CO (and O_2) from CO_2 [1, 43, 54–56]. Thermochemical cycles are necessary to lower the temperatures required. An example is the use of a metal oxide that spontaneously reduces at high temperature and is then reoxidized by interaction with H_2O to form H_2 or alternatively with CO_2 to produce CO. Nevertheless, temperatures above 1200–1400°C are necessary. This creates a number of issues in terms of materials, stability, cost-effectiveness, and output. The syngas (CO/H_2) should then be catalytically upgraded to fuels (e.g., methanol and FT hydrocarbons). This approach is essentially suitable for solar plants, while it may be difficult to adapt to a delocalized production of solar fuels. In addition, scaling problems are likely.

A variation of the concept is to use this solar concentrators and reactors to drive CO_2 reforming with methane (solar dry reforming of methane) to produce syngas, which can be then converted to methanol and Fischer–Tropsch products. The advantage is the possibility of continuous operations, instead of cyclic, while the disadvantage is the need of a methane feed. Solar illumination provides the heat necessary for the endothermic, dry reforming of methane with CO_2. Also in this case, there are issues related to the difficulty of maintaining a homogeneous temperature in the monolith, materials stability, and carbon formation. Scaling of the solar system to larger production of solar fuels remains an issue.

Possible alternatives are the solar, wet reforming of methane (e.g., feeding water instead of CO_2) and the solar decomposition of methane to carbon and hydrogen. All of these reactions are endothermic and the sun provides the heat of reaction. However, in the wet reforming of methane, the water to methane ratio should be greater than three in order to avoid excessive build-up of carbon over the catalyst. In the conventional catalytic process, with an external heat supply, this dilution is not critical, but in solar reformers, where the uniformly heated zone is limited, the productivity and cost-effectiveness is significantly affected. The reaction is controlled by the heat transfer rate and the parameters to improve the reaction rate are therefore very limited. In solar decomposition of methane, the production of a solid (carbon) is a clear issue, particularly for larger-scale applications. Solar dry reforming of methane with CO_2 appears the preferable choice between these three options, although in all cases the problem of catalyst stability (due to inhomogeneities in the heating) is a big issue.

Therefore, although the potential gain in energy efficiency of solar reformers over the conventional catalytic process to generate syngas or H_2 is attractive, the low productivity and limited scale of economy are the main issues, not to forget the problem of the materials. The feasibility of concentrated solar power (CSP) for producing solar fuels was proven, but not its stability of operations or its economic value. The expansion of the market for CSP (mainly to produce electrical energy) provides incentive to further explore this technology.

Delocalization of energy production is also an important aspect for better integration, reduction of eco-impact, and of the fuel transport costs. Low-temperature approaches in producing solar fuels present advantages in this respect.

In the low-temperature approach, solar energy is used by a suitable semiconductor to generate electrons and holes by charge separations, which react with water and CO_2. The reduction of the latter can be a two-step approach (e.g., generation of electricity and then using the electrons to reduce electrochemically/catalytically CO_2 in a physically separate cell) [57]. Alternatively, a one-step approach is possible by coupling the two processes in a single unit, a photoelectrochemical/catalytic approach [18, 28, 29, 58–60]. The physical separation of the two reactions of water oxidation and CO_2 reduction, in a photoanode and electrocathode, respectively, is necessary to increase the efficiency of the two reactions and limit charge recombination. The same device can also be used for the production of physically separated flows of H_2 and O_2 during water photoelectrolysis.

Many studies have been dedicated to water splitting on semiconductor catalysts under solar irradiation. Recent developments have been reviewed in detail by various research groups [61–63]. Remarkable progress has been made since the pioneering work by Fujishima and Honda in 1972, but the development of photocatalysts with improved efficiencies for hydrogen production from water using solar energy still faces major challenges. Most of the recent efforts focus in the search for active and efficient photocatalysts, for example through new materials and synthesis methods. While good quantum efficiencies (>50%) have been obtained with ultraviolet light, the use of visible light still poses major problems. While in principle the same materials could be used also for the photocatalytic reduction of CO_2 to fuels, the progresses in this field has been much more limited.

There are several basic problems in the photocatalytic reduction of CO_2 to fuels. The addition of one electron to CO_2 leads to the formation of the corresponding anion radical $^{\bullet}CO_2^{-}$, which is easily transformed to acids (formic and oxalic acids) in water solution. These are the main products of reaction, but their recovery from the water solution is energetically very expensive. While they could be used in fuel cells in principle, the concentration that can be achieved by the photocatalytic reduction of CO_2 is too small to be used without a further treatment. Moreover, another relevant problem is the low solubility of CO_2 in water and the abundance of quenching reactions present, which lower the quantum efficiency.

Although the photocatalytic reduction of CO_2 to gaseous fuels such as CH_4 has been reported [64, 65], the amounts produced are very limited. In addition, O_2 forms simultaneously and must be separated from methane. Also, potential safety issues are created. Alcohols such as methanol can also be obtained by photocatalytic reduction of CO_2 on Ag/TiO_2 [66]. A maximum methanol yield of 4.12 μmole/g_{cat} h was observed, while using TiO_2–SiO_2 doped with Cu and Fe, hydrocarbons (methane and ethylene) were formed. It is possible to form alcohols or hydrocarbons by photocatalytic reduction of CO_2 using different doped photomaterials, but the problems mentioned above are not solved. When liquid products form, their recovery from solution is energetically too expensive. When gaseous hydrocarbons form, they are not separated from the oxygen derived from the simultaneous reaction of water oxidation. In addition, these gaseous products have exactly the same problems of storage/transport discussed previously for H_2.

Although the direct photocatalytic reduction of CO_2 to fuels is potentially attractive, we suggest that it will be not practically feasible and the approach should be based on the physical separation of the water photo-oxidation reaction from the electrocatalytic reduction of CO_2, i.e., the photoelectrocatalytic (PEC) approach.

2 PEC Approach

The photoelectrocatalytic or photoelectrochemical (PEC) approach involves the separation of the oxidation and reduction processes into two half-cell reactions [59]. Three approaches are possible: the use of (i) photovoltaic (PV) cells, (ii) semiconductor-liquid junctions (SCLJ), or (iii) a combination of the two (PV/SCLJ). The approach based on solid-state photovoltaics is to couple a PV cell and electrolyzer into a single system. Semiconductor layers are connected in series to form a single monolithic device capable of generating the potential needed to split water. These tandem cells or multi-junction cells are modified with, or connected to, H_2 and O_2 producing electrodes, like Pt and RuO_2-modified Pt acting as cathode and anode, respectively.

An example of recent achievement in this area is a flexible, thin film Cu(In,Ga)Se_2 solar cell deposited on a titanium foil which was combined with a TiO_2 photocatalyst layer and modified by a niobium-doped titanium oxide front electrode to function as a photoelectrochemical tandem cell/membrane to facilitate the direct, light-driven hydrogen evolution from aqueous solution [67]. Under illumination with UV/visible light, the system produced up to 0.052 μL_{H2}/s cm^2 (i.e., the hydrogen formation rate was about 7250 μmol/h g relative to the amount of TiO_2 used). Several aspects of the operating principles of the photoelectrochemical devices, the materials requirements, main bottlenecks, and the various device concepts (in relation to H_2 production) were discussed in a chapter of a recent book on sustainable energy technologies [68]. However, this approach may be adapted with difficulty to the reduction of CO_2, and to date, the attempts in this direction have been quite limited.

In the semiconductor—liquid junction approach, the water splitting occurs at the semiconductor—liquid interface. In 1972, Fujishima and Honda first reported the sunlight-assisted electrolysis of water using crystalline TiO_2 photoelectrodes [69]. The photoelectrochemical cell consisted of TiO_2 (rutile) as a photoanode and platinum as a cathode. Illumination of the TiO_2 electrode led to O_2 evolution on the anode and H_2 evolution on the cathode. The quantum efficiency increased with an increase in alkalinity in the TiO_2 photoanode compartment and in the acidity of the Pt cathode compartment. This means that the thermodynamic potential of 1.23 V required for water splitting was substantially decreased due to the presence of a large pH gradient between the compartments (ΔpH $\sim$ 13, chemical bias $\sim$0.77 V).

A photoelectrolysis cell based on two illuminated semiconductor—liquid junctions is another possibility. An *n*-type semiconductor is used for the evolution of O_2 and a *p*-type semiconductor for the evolution of H_2. By separating the oxidation and reduction processes into half-cell reactions, the optimization of each reaction is possible, in addition to reducing recombination. Besides, two semiconductors with smaller band gaps can be utilized since each needs only to provide part of the water-splitting potential. These smaller band gaps mean more absorption in the visible region of the solar spectrum where the sun has a greater photon flux. As a result, the maximum theoretical efficiency is considerably higher.

In the PV/SCLJ approach for overall water splitting, a PV cell is directly combined with a semiconductor that is in direct contact with the electrolyte to form a single integrated unit. The PV cell can be combined either with a reduction (photocathode) or an oxidation (photoanode) photocatalyst.

Using these approaches, good efficiencies in separate H_2 and O_2 production are possible even with visible light, although in several cases stability was a major issue, in addition to complexity and cost. These systems may be adapted with difficulty for the solar conversion of CO_2 to fuels. In general, there is a need to develop robust and cost-effective systems that can be scaled and are suitable for small to medium-sized installations. The key issue for applicability is an approach that intrinsically solves the problem of recovering the fuels produced from CO_2 and at the same time allows the production of liquid fuels to be used directly in the energy infrastructure.

The feasibility of CO_2 conversion depends on the ability to form liquid fuels under solventless conditions; such liquid fuels include long-chain hydrocarbons and/or alcohols, which can be easily collected without the need to distill from liquid solutions (a very energy intensive process). There are very few studies based on this novel approach, which will be discussed in more detail in the following section. The need to recycle carbon dioxide to liquid fuels (methanol and dimethyl ether) has been addressed in the perspective paper of Nobel Laureate Olah and co-workers [20], which summarizes the more extensive discussion made in the recent 2nd edition of their famous book on the methanol economy [70].

2.1 Novel PEC Solar Cells

For practical use of PEC solar cells, the design of the cells should be quite different from what is used commonly in the literature: devices based essentially on slightly modified conventional electrochemical cells operating in the liquid phase.

The cathode and anode in the PEC device should take the form of thin films separated by a proton-conducting membrane (Nafion® for example, but other membranes could be used) and deposited over a porous conductive substrate that allows efficient collection/transport of the electrons over the entire film as well the diffusion of protons to/from the membrane. It is also necessary to allow an efficient evolution of the gas. The reactor geometry and the spatial relation between the reactor and light source are also important, as is the efficient control of the temperature to avoid overheating during operations. It is necessary to have different types of nanostructured electrodes (both at the photoanode and at the cathode).

On the cathode side, gas phase operation is necessary for CO_2 reduction to avoid the problem of forming a gas cap over the electrode, CO_2's limited solubility in solution and change the type of products formed. Note, the electrocatalytic reduction of CO_2 in gas phase is essentially a new area of development; most of studies focus on liquid phase CO_2 reduction. This would also require new approaches in the type of electrodes to be used.

Electrochemical utilization of CO_2 (in liquid phase) has been studied for many years, as recently reviewed [71, 72]. There are two main approaches, depending whether the conversion of CO_2 is studied in aqueous or non-aqueous solutions. Formic acid is the main reaction product in the electrolysis of aqueous solutions of CO_2 because the carbon dioxide anion radical forms easily by addition of one electron to CO_2 on the electrode surface. Desorption of this product is faster than its consecutive reduction. In addition, a problem in the utilization of CO_2 in aqueous solution derives from its low solubility in water at standard temperature and pressure. Higher pressures are necessary to increase the CO_2 concentration in the liquid phase, but electrode stability in these conditions is limited [73].

Solvents with high solubility for CO_2 are used in the non-aqueous electrochemical reduction. However, high CO_2 solubility requires larger current density, but low electrolytic conductivity leads to high Ohmic losses. High current densities are also necessary to maximize the formation of hydrocarbons. Also, a fast deactivation can occur under these conditions [74].

A solventless electrocatalytic reduction of CO_2 would overcome these problems and can form more valuable products (liquid fuels) as it will be discussed later. A novel design for the PEC solar cell is necessary for this scope. An example is presented in Fig. 1. The apparatus for the photocatalytic experiments consists of a solar illuminator source, a photo-reactor and gas chromatographs for on-line analysis. The lamp housing is furnished with a Xe-arc lamp (ORIEL, 300 W), a set of lenses for light collection and focusing, and a water filter, to eliminate the infrared radiation.

The PEC solar cell consists of three layers: a nanostructured TiO_2 thin film supported on Ti prepared by anodization, a Nafion® membrane for electrical

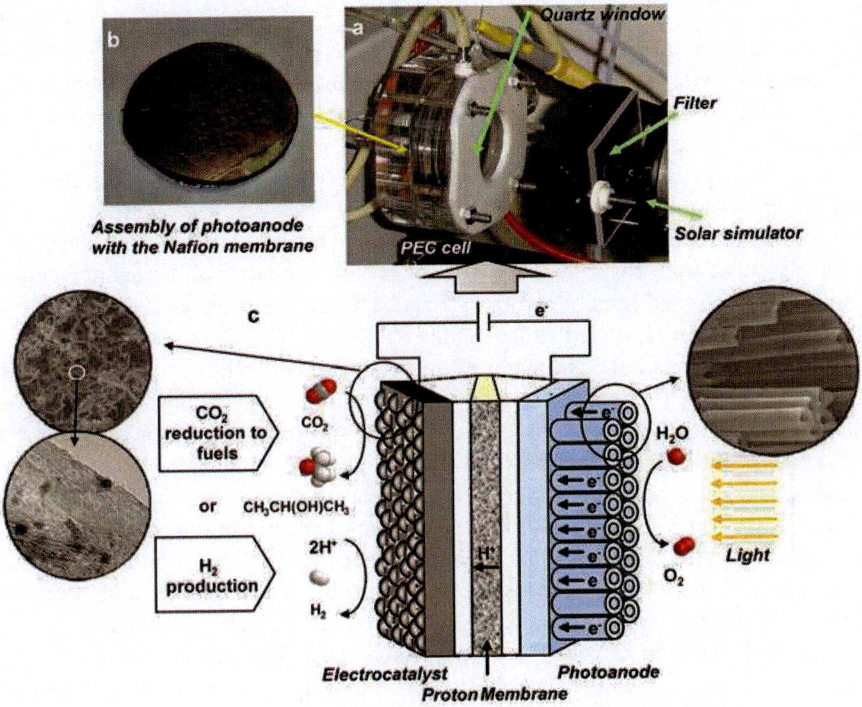

Fig. 1 **a** View of the lab-scale PEC device. **b** Image of the assembly of the photoanode with the Nafion® membrane and the electrocatalyst (*on the back*). **c** Scheme of the PEC device for CO_2 reduction to fuels and H_2 production. In the inset, SEM images of the photoanode and the electrocatalyst. Adapted from Ampelli et al. [29]

insulation and proton transport, and an electrocatalyst (Fig. 1). This PEC device can be used either for water photo-splitting/-electrolysis with separate evolution of O_2 and H_2, or for the CO_2 reduction to liquid fuels using solar energy. In the H_2 photoproduction configuration, the electrocatalyst is a carbon cloth (E-tek®) with a high platinum dispersion (20%). For CO_2 reduction, the electrocatalyst is the electrode prepared by the deposition of Fe/CNT or Pt/CNT (carbon nanotubes) on carbon cloth, described later. The assembly of the disc was performed by hot-pressing at 120°C under a pressure of 20 kg/cm^2.

2.2 Photoanode and Role of Titania Thin-Film Nanostructure

The photoanode in a PEC solar cell, as introduced before, should be in the form of a porous thin film allowing good light harvesting, fast transport of the protons and electrons produced during the water oxidation, and good contact with both the electron-collector substrate and the proton-conductor membrane. There is a need to have a specific nanostructure in the photoanode.

The use of an array of 1D aligned nanostructures (nanorods, nanotubes, etc.) improves light harvesting and limits charge recombination at the grain boundaries with respect to an assembly of nanoparticles, while maintaining a high surface area necessary to improve the photoresponse [23, 75–77]. An optimal contact/interface with the H^+-membrane is also necessary.

A further general issue is the need of a preparation method to produce the photoanode that (i) is cost-effective, (ii) allows a good control of the nanostructure in terms of 1D-array characteristics (vertical alignment, density, and thickness), and (iii) can be easily scaled to large films (at least 10×10 cm). The choice of suitable preparation methods having all of these characteristics is quite restricted [75, 76]. We have therefore focused our attention on the anodic oxidation of titanium thin foils [76, 78, 79] to synthesize 1D nanostructured titania arrays to be used as photoanode in PEC solar cells.

These TiO_2 materials show excellent properties in the preparation of solar cells and photoelectrodes [80–82]. We limit discussion here to not doped titania films, but their reactivity properties, which can be further promoted by doping or creating of heterostructures. Such structures promote visible-light absorption and effective transfer of electrons from the valence to the conduction band of titania.

A critical question is whether clear proofs exist that the use of a specific nanostructure has advantages in terms of photo-performances. In fact, notwithstanding the fast-growing literature on the subject (discussed in detail in the cited reviews and books [76, 78, 79]), often a clear comparison with conventional titania films (under comparable conditions) is not made.

An example of such a comparison is shown in Fig. 2 [29], which reports the performances in ethanol photoreforming for H_2 production of three different titania films: (a) produced by pressing commercial P25 Degussa TiO_2, (b) prepared by anodic oxidation of Ti foils, and (c) prepared by sol–gel dip-coating method. For comparison, the results obtained with P25 Degussa using a slurry-type well-mixed reactor are also reported. All of these titania samples were loaded with 0.5 wt% Pt by wet impregnation because without Pt all the samples show very low activity in photoreforming [83]. The data in Fig. 2 demonstrate that the specific photoreforming activity (per amount of titania) considerably depends on both the reactor geometry and the nanostructure of the titania thin film.

P25 TiO_2 Degussa is a classical reference material in photocatalysis. Under the tested reaction condition this material shows about twice the specific activity when used in the form of a thin film than as a suspended powder (slurry reactor), due to reduced light scattering. The sample prepared by sol–gel dip-coating, also a conventional method to prepare titania thin films, shows performances quite similar to those of the thin film prepared by compacting the P25 TiO_2 nanoparticles. On the contrary, the thin film in the form of an ordered array of vertically aligned TiO_2 nanotubes (produced by anodic oxidation) shows higher activity than the other samples. It is evident that the nanostructure of TiO_2 has a large effect. There are several reasons for this improvement: better light harvesting, lower rate

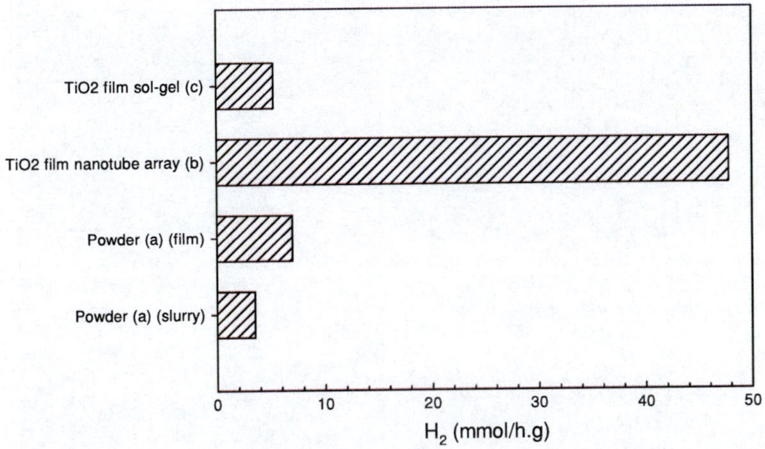

Fig. 2 Hydrogen evolution rate for different kinds of TiO_2 samples during photoreforming of 10% ethanol in water. **a** Degussa P-25 TiO_2 powder; **b** ordered array of TiO_2 nanotubes prepared by anodization on Ti foil; **c** TiO_2 dense film prepared by sol–gel dip-coating. In all samples 0.5 wt% Pt was added by wet impregnation. Adapted from Ampelli et al. [29]

of recombination at the grain boundaries, reduced effects related to formation of a H_2 cap, faster electron transport and charge separation, and a nanostructure that increases the absorption of visible-light components. The result of these effects is that the specific activity of the TiO_2 nanotube array thin film has significant higher activity in H_2 photoreforming.

Similar observations have been made in water photoelectrolysis (splitting), although the activity in hydrogen production is lower than in the photoreforming of ethanol [29]. A parallel trend for the photocurrent generated during the reaction and the amount of H_2 formed was also demonstrated [29]. This is a general observation that is in accordance with the indication that the reaction of H_2 production from water on semiconductors such as titania is effectively a photoelectrolysis rather than a photocatalytic splitting as often described. In other words, the semiconductor generates an electric current by adsorbing solar light and these promoted electrons are responsible for the electrolysis of water at short-circuited nanoelectrodes. The latter are associated with the metal nanoparticles (Pt, for example).

The process of anodic anodization used to prepare the nanostructured titania thin films can be summarized, by analogy with the formation of porous alumina by anodization [84], as follows. When the Ti foil, after the preliminary cleaning treatment, is immersed in the electrolyte for the anodization process, a fast surface oxidation occurs with formation of a thin TiO_2 layer. This process can be monitored by a fast decrease of the current, since the TiO_2 layer is not conductive. Due to the presence of an aqueous solution of HF as electrolyte, the solubilization of Ti^{4+} ions and/or of small TiO_2 particles starts simultaneously with the formation of the oxide layer. These processes lead to the formation of holes,

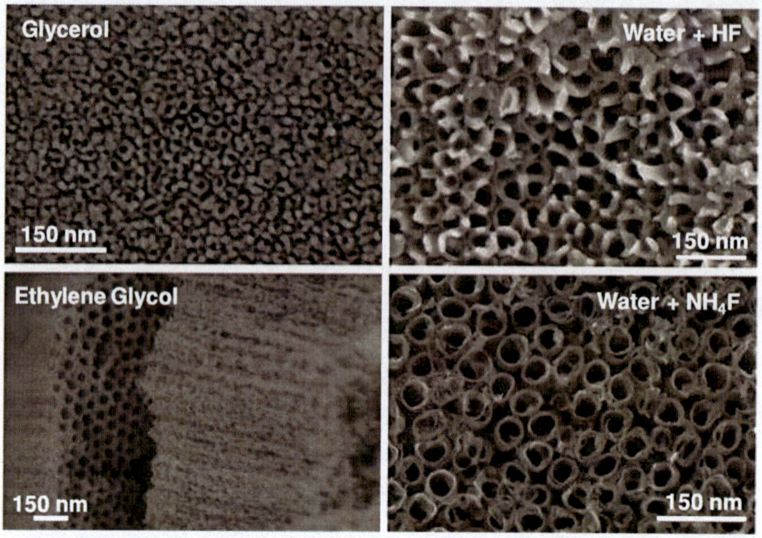

Fig. 3 Role of the electrolyte on the nanostructure of titania films produced a by anodic oxidation. SEM images (*top view*, apart cross-section in the case of ethylene glycol) of titania nanotube arrays prepared by anodic oxidation under different conditions: (i) in glycerol + 0.5 wt% NH_4F at pH 6 applying a 20 V potential for 45 min. (ii) in water + 0.5 wt% HF at pH 0 applying a 15 V potential for 45 min. (iii) in ethylene glycol + 0.3 wt% NH_4F and 2 vol% H_2O applying a 50 V potential for 6 h. (iv) in water + 0.5 wt% HF at pH 4 (adjusted with NH_4OH) applying a 20 V potential for 45 min

which locally modify the electric field and induce from one-side the acceleration of the process of dissolution, due to field-enhanced effects, and from the other side the oriented growing of a 1D structure (nanotubes, nanorods, etc.). All of these processes are strongly influenced by the reaction conditions, which, in turn, determine the characteristics of the different nanostructures.

The electrolyte and conditions of anodization influence the film characteristics (thickness, type of nanostructure, specific features on 1D nano-objects, etc.). An example is given in Fig. 3 which reports the scanning electron microscopy (SEM) images for two cases using organic electrolytes and two different anodization conditions in an aqueous medium in the presence of fluoride ions. All these titania nanostructured films give very different performances in terms of photocurrent generation upon irradiation and photocatalytic activity.

The film thickness increases with the time of anodization with a consequent increase of the photocurrent. While minor differences are observed using either water or glycerol as the solvent, a very large difference is observed using ethylene glycol as the solvent. In this case, a photocurrent density of about an order of magnitude higher is observed, particularly for the longer anodization times. Using ethylene glycol, very regular straight TiO_2 nanotubes with thick walls are obtained (see Fig. 3). At longer times of anodization, the presence of an amorphous titania deposit on the surface is observed, but this debris can be removed from the surface by sonication. The photocurrent generated during irradiation could be measured in situ also during the catalytic reaction using the apparatus schematized in Fig. 1.

2.3 Nanostructured Electrodes for CO_2 Gas-Phase Reduction

In the liquid phase, electrochemical reduction of carbon dioxide Cu-based foils as electrode materials gives the best results to form hydrocarbons [71, 72]. High current densities are necessary to maximize the formation of hydrocarbons. Hori et al. [74], using immobilized CuCl on a Cu-mesh electrode, reported a Faradaic efficiency of about 70% to C_2H_4, although the electrode quickly deactivates. In addition, corrosive media (high pressure, metal halides) are used because they are necessary to promote a high surface concentration of CO_2 at the electrode. Also of note, Cu is the only metal, which gives appreciable amounts of C2 hydrocarbons. Several critical aspects determine the performance, such as (a) the gas evolution in the electrochemical cells, which reduces electrolyte conductivity and increases Ohmic resistance, (b) the pH and reaction temperature, which negatively influence the solubility of CO_2 and selectivity to C2 products [85], and (c) the porosity of the electrodes. Using TiO_2 nanotube composite electrodes for the electrochemical reduction of CO_2 to methanol [86], current efficiencies over 60% were reported.

The reaction network in the CO_2 electrocatalytic reduction on Cu-electrodes involves an initial stage of formation of carbon dioxide anion radicals $^{\bullet}CO_2^{-}$, which shows why metal halides are necessary to promote the reaction. The next step leads to the formation of adsorbed carbon monoxide (CO) which can react with protons and electrons (in the presence hydroxide anions) to give water and chemisorbed methylene (:CH_2). The latter may be further hydrogenated to CH_4 or react with another methylene intermediate following a Fischer–Tropsch like chain growth mechanism.

Alkanes and alkenes up to C6 hydrocarbons were recently reported by Shibata et al. [87] using a Cu-electrode that was not pre-treated by electropolishing. The product distribution follows the Schultz–Flory distribution and, depending on the applied potential, the chain growth probability (α) ranges from 0.23 to 0.31, values lower than those obtained in Fischer–Tropsch synthesis over heterogeneous Co- or Fe-based catalysts. When the same electrode material was pre-treated by electropolishing it behaved like a pure Cu-electrode, yielding mainly methane and ethene.

These results show that the electrode's specific nanostructure affects the performances and type of products formed. In these tests, flat-type electrodes have been used, as typically made for electrochemical studies. However, as discussed before, the feasibility of CO_2 conversion to fuels depends on the formation of long-chain hydrocarbons and/or alcohols and the use of solventless reaction conditions. This suggests using different types of electrodes.

The PEC solar cell reported in Fig. 1 is analogous to proton-exchange membrane (PEM) fuel cells, where the reaction of O_2 reduction is also made in the gas phase. The commercial electrodes for PEM fuel cells typically use a tissue of carbon macrofibers (carbon cloth—CC) as a substrate, which has the function of electron transport and allows a homogeneous dispersion of the electrocatalyst and good diffusion of the gases. On CC, the electrocatalyst is deposited on the side in contact with the proton-conducting membrane (Nafion® 112, for example).

On the anode side, Pt (20 wt%) deposited on carbon black (for example, Vulcan XC-72 carbon black) is used as the electrocatalyst. The typical size of noble metal particles is around 2 nm. On the cathode side (O_2 reduction) Pt-Ru bimetallic electrocatalysts also deposited on carbon black are used. Moreover, on the sides of the CC, which are in contact with the gas phase, a hydrophobic porous layer is created (for example, by deposition of a Teflon solution), in order to avoid the dehydration of the proton-conducting membrane. This structure is called the gas diffusion electrode (GDE). The anode and cathode GDE are then hot pressed with the Nafion to realize the final composite known as membrane-electrode assembly (MEA) to be used in the PEMFC.

For the electrocatalytic reduction of CO_2 commercial-type electrodes for PEM fuel cells were used [87, 88] (e.g., a carbon cloth/Pt on carbon black/Nafion assembled electrode). The electrocatalyst is Pt supported on carbon black, which is then deposited on a conductive carbon cloth to allow the electrical contact and the diffusion of gas-phase CO_2 to the electrocatalyst. The Pt particles are in contact with Nafion, through which protons diffuse.

On the Pt nanoparticles, gas-phase CO_2 reacts with electrons and protons to be reduced to long-chain hydrocarbons and alcohols, with relative distributions dependant upon the reaction temperature. At room temperature, long-chain hydrocarbons (up to C9) were formed, but with low productivities [88, 89]. A 60°C and atmospheric pressure, acetone was found to be the major product of conversion.

Flat-type carbon electrodes (i.e., glassy carbon) were not effective. Due to the active carbon's high porosity, the Pt nanoparticles were mainly located inside nanopores. To realize an efficient reduction of CO_2, it is necessary to optimize the three-phase reaction at the catalyst surface between the protons diffusing through the Nafion membrane, the electrons necessary for the carbon dioxide reduction, and the CO_2 from the gas phase. Furthermore, the formation of C–C bonds should be enhanced when CO_2 (and its products of conversion) remains confined over the electrocatalyst surface. Thus, nanoporosity (nanoconfinement concept) should have a role in forming products with longer C-chains from CO_2.

However, it is equally important to guarantee fast diffusion of the protons and electrons to the electrocatalytic nanoparticles. The latter should be in good contact with a conductive (graphitic) carbon support. For this reason, carbon nanotubes should be used as the substrate instead of active carbon because they combine good electron conductivity with a specific nanostructure, thereby optimizing the nanoconfinement.

Using a similar GDE configuration, but with carbon nanotubes as the substrate for the electrocatalyst nanoparticles, it is possible to form isopropanol as the main reaction product [90, 91] (Fig. 4). This figure also shows that the electrocatalyst

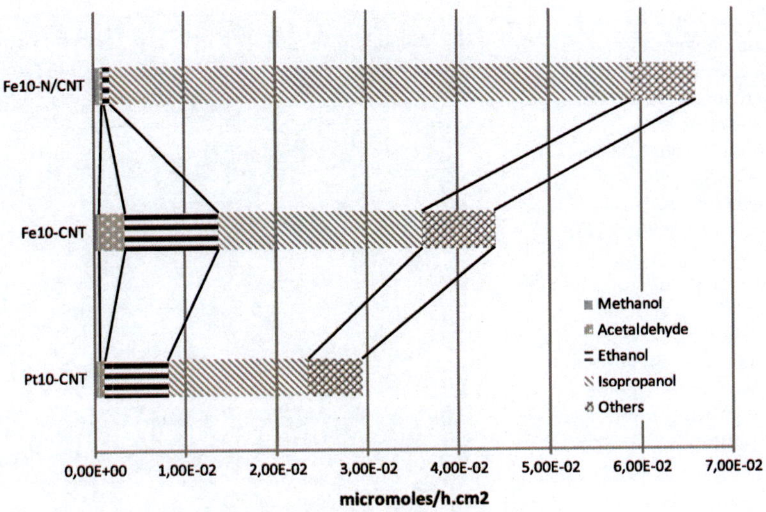

Fig. 4 Products distribution at 60°C in the electrocatalytic reduction of carbon dioxide in gas phase over Nafion 117/(Pt or Fe(10%)/CNT)20%/carbon cloth GDM (gas diffusion membrane) electrode. Tests in a semi-batch cell, using a 0.5 $KHCO_3$ electrolyte on the anode side and operating the cathode in the gas phase with a continuous feed of 50% CO_2 in humidified nitrogen. Adapted from Centi and Perathoner [90] and Gangeri et al. [91]

can be based on iron nanoparticles instead of a noble metal. In any case, the Pt-based electrodes show better stability on carbon nanotubes. In addition, it is shown that the use of N-doped carbon nanotubes (N/CNT) allows a further improvement in the productivity of isopropanol. This is related to two concepts. The first is a better dispersion and stabilization of the metal particles, due to the creation of nitrogen moieties, which anchor the nanoparticles. The nature and amount of these nitrogen species (which depends on the modalities adopted to dope the carbon nanotubes with nitrogen sources, and the thermal treatment after this doping procedure) influences the performances. The second is a modification of the hydrophilic properties of the carbon nanotubes which influences the contact interface with the Nafion® and the diffusivity of the protons.

In general, the doping and creation of surface defect sites in carbon nanotubes, increasing the number of surface functional groups on each CNT, influences various aspects: (i) the efficiency of the three-phase boundary and the transport of protons to or from the active metal particles, (ii) the resistance of electron transfer, and (iii) the intrinsic reactivity properties of the metal nanoparticles. Many aspects determine the final properties.

There are other aspects to consider in the optimization of the performance of nanostructured electrodes for CO_2 gas-phase reduction, such as the presence of a suitable 3D organization to maximize the contact with the Nafion, promotion of an efficient diffusion of CO_2 and back-diffusion of the reaction products, and limited dehydration of Nafion, which negatively affects the transport properties of the

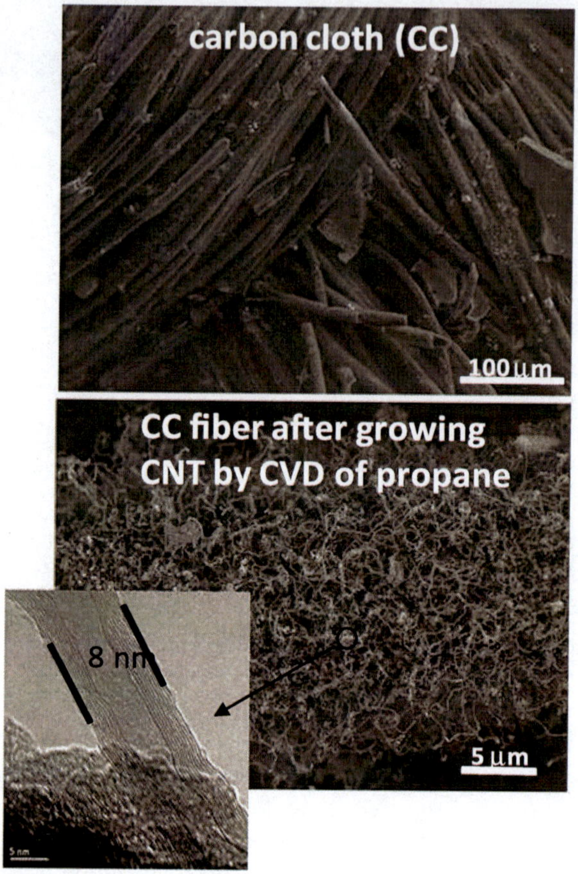

Fig. 5 Top image: electron microscopy images of carbon cloth used as substrate to create the necessary flexible grid to allow electron transport and good diffusion of the gases. *Bottom image*: carbon nanotubes covering a carbon macrofiber; the CNTs were obtained by propane CVD after deposition on the carbon cloth of a suitable catalyst (Fe, Co nanoparticles); in the inset a higher resolution image of one CNT growing from the catalyst [90]

membrane and proton diffusion. For this objective, the use of a hierarchically organized structures by growing carbon nanotubes over carbon maro fibres (CC) has been proposed [90, 92].

Figure 5 shows an example of these materials prepared by deposition of a Co-Fe/SBA-15 catalyst onto carbon cloth and the use of these catalysts to grow CNT by the propane CVD method [90]. It was observed that the CNTs uniformly cover the carbon macrofibres. This allows for the improvement of the surface area of carbon while avoiding the need to use the carbon black as the support for Pt. The use of hierarchically organized structures provides a better 3D geometry for contact with the Nafion. However, the properties of these materials are not optimal in terms of wettability by the Nafion and transport of protons. Thus modification is necessary to optimize the performances. Also tuning the conductivity properties by annealing in an inert atmosphere is necessary. There are therefore many aspects that must be controlled and optimized to prepare the optimal electrocatalysts for the CO_2 conversion to fuels.

3 Conclusions

The interest in the use of solar energy is quickly growing, as is the need to find efficient solutions to reduce the negative effects associated with large volumes of CO_2 emissions. We have discussed here an approach to combine these two aspects by producing solar fuels from water and CO_2. In this vision, liquid fuels produced by recycling CO_2 using solar energy are ideal energy vectors, easy to store/transport and fully integratable with the existing energy infrastructure.

The approach discussed to reach this objective is based on a novel photo-electrocatalytic (PEC) device, which solves some of the potential issues in the photoelectrochemical or photocatalytic approaches to convert carbon dioxide. This device needs some specific new characteristics in the photoanode and electrocatalyst, which were briefly discussed. In particular, some aspects related to the characteristics of nanostructured (a) TiO_2 thin films (based on an ordered array of titania nanotubes) and their performances as photoanodes, and (b) carbon-nanotube-based electrodes for the gas phase reduction of CO_2 to liquid fuels (mainly isopropanol) was presented. Finally, their application to the design of a novel photoelectrocatalytic (PEC) approach for the synthesis of solar fuels has been analyzed.

This is an area currently under fast development worldwide and significant progress is expected in the near future. We feel that the PEC reactor design discussed here is a good attempt toward the practical implementation of solar fuels, even if the passage from the proof-of-the-concept (actual state) to the practical implementation still requires significant research.

Acknowledgments This paper derives from many discussions we had in the frame of various EU projects (NATAMA, ELCAT, IDECAT) dedicated to this topic, and in the frame of the activities of the European Laboratory of Catalysis and Surface Science (ELCASS) which partners (FHI-MPG, and LMSPC-ECPM-ULP) are gratefully acknowledged. In particular, we are grateful for useful discussions with Prof. R. Schlögl and Dr. D.S. Su (Fritz Haber Institute of Max Plank Gesellschaft, Berlin, Germany) and Prof. R. Lambert (University of Cambridge, U.K.).

References

1. Figuero JD, Fout T, Plasynski S et al (2008) Advances in CO_2 capture technology-The U.S. department of energy's carbon sequestration program. Int J Greenhouse Gas Control 2:9–20
2. Choi S, Drese JH, Jones CW (2009) Adsorbent materials for carbon dioxide capture from large anthropogenic point sources. ChemSusChem 2(9):796–854
3. Yu KM, Curcic I, Gabriel J, Tsang SC (2008) Recent advances in CO_2 capture and utilization. ChemSusChem 1(11):893–899
4. Mikkelsen M, Jørgensen M, Krebs FC (2010) The teraton challenge. A review of fixation and transformation of carbon dioxide. Energy Env Sci 3(1):43–81
5. Peters M, Mueller T, Leitner W (2009) CO_2: from waste to value. Tce 813:46–47
6. Ritter SK (2007) What can we do with carbon dioxide? Chem Eng News 85(18):11
7. International energy agency—IEA (2009) World energy outlook 2009, Paris

8. Aresta M, Dibenedetto A (2007) Utilisation of CO_2 as a chemical feedstock: opportunities and challenges. Dalton Trans (28): 2975–2992
9. Sakakura T, Choi J-C, Yasuda H (2007) Transformation of carbon dioxide. Chem Rev 107(6):2365–2387
10. Omae I (2006) Aspects of carbon dioxide utilization. Catal Today 115(1–4):33–52
11. Song C (2006) Global challenges and strategies for control, conversion and utilization of CO_2 for sustainable development involving energy, catalysis, adsorption and chemical processing. Catal Today 115(1–4):2–32
12. Sun Y (2004) Carbon dioxide utilization for global sustainability. Stud Surf Sci Catal 153:9–16
13. Aresta M, Dibenedetto A (2004) The contribution of the utilization option to reducing the CO_2 atmospheric loading: Research needed to overcome existing barriers for a full exploitation of the potential of the CO_2 use. Catal Today 98(4):455–462
14. Arakawa H, Aresta M, Armor JN et al (2001) Catalysis research of relevance to carbon management: progress, challenges, and opportunities. Chem Rev 101(4):953–996
15. Graham-Rowe D (2008) Turning CO_2 back into hydrocarbons. New Scientist 2645 (March): 32
16. Centi G, Perathoner S (2009) Opportunities and prospects in the chemical recycling of carbon dioxide to fuels. Catal Today 148(3–4):191–205
17. Centi G, Perathoner S (2004) Carbon dioxide utilization for global sustainability. Stud Surf Sci Catal 153:1–8
18. Centi G, Perathoner S, Rak ZS (2003) Reduction of greenhouse gas emissions by catalytic processes. Appl Catal B: Env 41(1–2):143–155
19. Bell AT, Gates BC, Ray D (2007) Basic research needs: catalysis for energy (PNNL-17214). U.S. Department of Energy, Washington, DC
20. Olah GA, Goeppert A, Prakash GKS (2009) Chemical recycling of carbon dioxide to methanol and dimethyl ether: from greenhouse gas to renewable, environmentally carbon neutral fuels and synthetic hydrocarbons. J Org Chem 74(2):487–498
21. Centi G, van Santen RA (2007) Catalysis for renewables. Wiley-VCH, Weinheim (Germany)
22. Cavani F, Centi G, Perathoner S et al (2009) Sustainable industrial chemistry—principles, tools and industrial examples. Wiley-VCH, Weinheim (Germany)
23. Centi G, Perathoner S (2009) The role of nanostructure in improving the performance of electrodes for energy storage and conversion. Eur J Inorg Chem 26:3851–3878
24. Serrano E, Rus G, García-Martínez J (2009) Nanotechnology for sustainable energy. Renew Sustain Energy Rev 13(9):2373–2384
25. Muradov NZ, Veziroğlu TN (2008) "Green" path from fossil-based to hydrogen economy: an overview of carbon-neutral technologies. Int J Hydrogen Energy 33(23):6804–6839
26. Sahaym U, Norton MG (2008) Advances in the application of nanotechnology in enabling a 'hydrogen economy'. J Mater Sci 43(16):5395–5429
27. Züttel A, Borgschulte A, Schlapbach L (2008) Hydrogen as a future energy carrier. Wiley-VCH, Weinheim (Germany)
28. Centi G, Perathoner S (2010) Towards solar fuels from water and CO_2. ChemSusChem 3:195–208
29. Ampelli C, Centi G, Passalacqua R, Perathoner S (2010) Synthesis of solar fuels by a novel photoelectrocatalytic approach. Energy Env Sci 3(3):292–301
30. Centi G, Perathoner S, Passalacqua R, Ampelli C (2010) Solar production of fuels from water and CO_2. In: Veziroglu N, Muradov N (eds) Carbon neutral fuels and energy carriers: science and technology. Taylor & Francis, London
31. Higo M, Dowaki K (2010) A life cycle analysis on a Bio-DME production system considering the species of biomass feedstock in Japan and Papua New Guinea. Appl Energy 87(1):58–67
32. Azzopardi B, Mutale J (2010) Life cycle analysis for future photovoltaic systems using hybrid solar cells. Renew Sustain Energy Rev 14(3):1130–1134

33. Dufour J, Gálvez JL, Serrano DP et al (2009) Life cycle assessment of hydrogen production by methane decomposition using carbonaceous catalysts. Int J Hydrogen Energy 35(3):1205–1212
34. Lee J-Y, Yoo M, Cha K et al (2009) Life cycle cost analysis to examine the economical feasibility of hydrogen as an alternative fuel. Int J Hydrogen Energy 34(10):4243–4255
35. Li Z, Gao D, Chang L, Liu P et al (2010) Coal-derived methanol for hydrogen vehicles in China: energy, environment, and economic analysis for distributed reforming. Chem Eng Res Des 88(1):73–80
36. Xiao J, Shen L, Zhang Y et al (2009) Integrated analysis of energy, economic, and environmental performance of biomethanol from rice straw in China. Ind Eng Chem Research 48(22):9999–10007
37. Dowaki K, Genchi Y (2009) Life cycle inventory analysis on Bio-DME and/or Bio-MeOH products through BLUE tower process. Int J Life Cycle Assess 14(7):611–620
38. Dowaki K, Eguchi T, Ohkubo R et al (2009) A life cycle assessment on a fuel production through distributed biomass gasification process. IEEJ Trans Electron, Inform Syst 128(2):168–175
39. Wu JCS (2009) Photocatalytic reduction of greenhouse gas CO_2 to fuel. Catal Surv Jpn 13(1):30–40
40. Armor JN (2007) Addressing the CO_2 dilemma. Catal Letters 114(3–4):115–121
41. Centi G, Cum G (2008) Conversion of carbon dioxide to fuels and chemicals. In: Centi G, Cum G, Fierro JLG et al. (eds), Direct conversion of methane, ethane and carbon dioxide to fuels and chemicals, CAP Report, The catalyst group resources. Spring House, Pa (US)
42. Dey GR (2007) Chemical Reduction of CO_2 to different products during photo catalytic reaction on TiO_2 under diverse conditions: an overview. J Natural Gas Chem 16(3):217–226
43. Kodama T (2003) High-temperature solar chemistry for converting solar heat to chemical fuels. Prog Energy Combust Sci 29(6):567–597
44. Morris AJ, Meyer GJ, Fujita E (2009) Molecular approaches to the photocatalytic reduction of carbon dioxide for solar fuels. Acc Chem Res 42(12):1983–1994
45. Magnuson A, Anderlund M, Johansson O et al (2009) Biomimetic and microbial approaches to solar fuel generation. Acc Chem Res 42(12):1899–1909
46. Jessop PG, Joó F, Tai C-C (2004) Recent advances in the homogeneous hydrogenation of carbon dioxide. Coord Chem Rev 248(21–24):2425–2442
47. Stöcker M (2008) Biofuels and biomass-to-liquid fuels in the biorefinery: Catalytic conversion of lignocellulosic biomass using porous materials. Angew Chemie–Int Ed 47(48):9200–9211
48. van Steen E, Claeys M (2008) Fischer-Tropsch catalysts for the biomass-to-liquid process. Chem Eng Techn 31(5):655–666
49. Mata TM, Martins AA, Caetano NS (2010) Microalgae for biodiesel production and other applications: a review. Renew Sustain Energy Rev 14(1):217–232
50. Brennan L, Owende P (2010) Biofuels from microalgae-A review of technologies for production, processing, and extractions of biofuels and co-products. Renew Sustain Energy Rev 14(2):557–577
51. Lardon L, Hélias A, Sialve B et al (2009) Life-cycle assessment of biodiesel production from microalgae. Env Sci Technol 43(17):6475–6481
52. Allakhverdiev SI, Kreslavski VD, Thavasi V et al (2009) Hydrogen photoproduction by use of photosynthetic organisms and biomimetic systems. Photochem Photobiolo Sci 8(2):148–156
53. Angermayr SA, Hellingwerf KJ, Lindblad P et al (2009) Energy biotechnology with cyanobacteria. Current Opinion Biotechnol 20(3):257–263
54. Kodama T, Gokon N (2007) Thermochemical cycles for high-temperature solar hydrogen production. Chem Rev 107(10):4048–4077
55. Licht S (2005) Thermochemical solar hydrogen generation. Chemical Comm 37(7):4635–4646

56. N'Tsoukpoe KE, Liu H, Le Pierrès N et al (2009) A review on long-term sorption solar energy storage. Renew Sustain Energy Rev 13(9):2385–2396
57. Tributsch H (2008) Photovoltaic hydrogen generation. Int J Hydrogen Energy 33(21):5911–5930
58. Kaneko M, Nemoto J, Ueno H et al (2006) Photoelectrochemical reaction of biomass and bio-related compounds with nanoporous TiO_2 film photoanode and O_2-reducing cathode. Electrochem Comm 8(2):336–340
59. Currao A (2007) Photoelectrochemical water splitting. Chimia 61(12):815–819
60. Barton EE, Rampulla DM, Bocarsly AB (2008) Selective solar-driven reduction of CO_2 to methanol using a catalyzed p-GaP-based photoelectrochemical cell. J Am Chem Soc 130(20):6342–6344
61. Navarro Yerga RM, Alvarez Galván MC, del Valle F et al (2009) Water splitting on semiconductor catalysts under visible-light irradiation. ChemSusChem 2(6):471–485
62. Kudo A, Miseki Y (2009) Heterogeneous photocatalyst materials for water splitting. Chem Soc Rev 38(1):253–278
63. Minero C, Maurino V (2007) Chapter 16 Solar photocatalysis for hydrogen production and CO_2 conversion. In: Centi G, van Santen R (eds) Catalysis for renewables. Wiley-VCH, Weinheim (Germany), pp 351–385
64. Dey GR, Belapurkarb AD, Kishore K (2004) Photo-catalytic reduction of carbon dioxide to methane using TiO_2 as suspension in water. J Photochem Photobiol: Chem 163(3):503–508
65. Tana SS, Zoub L, Hu E (2006) Photocatalytic reduction of carbon dioxide into gaseous hydrocarbon using TiO_2 pellets. Catal Today 115(1–4):269–273
66. Wu JCS (2009) Photocatalytic reduction of greenhouse gas CO_2 to fuel. Catal Surv Jpn 13(1):30–40
67. Neumann B, Bogdanoff P, Tributsch H (2009) TiO2-protected photoelectrochemical tandem Cu(In, Ga)Se2 thin film membrane for light-induced water splitting and hydrogen evolution. J Phys Chem C 113(49):20980–20989
68. Van de Krol R, Schoonman J (2008) Photo-electrochemical production of hydrogen. In: Hanjali K, Van de Krol R, Lekić A (eds) Sustainable energy technologies. Springer, Netherlands, pp 121–142
69. Fujishima A, Honda K (1972) Electrochemical photolysis of water at a semiconductor electrode. Nature 238:37–38
70. Olah GA, Goeppert A, Surya Prakash GK (2009) Beyond oil and gas: the methanol economy. Wiley-VCH, (Germany)
71. DuBois DL (2006) Chapter 6.2 Electrochemical reactions of carbon dioxide. In: DuBois DL (ed) Encyclopedia of Electrochemistry, vol 7. Wiley-VCH, (Germany), pp 202–225
72. Gattrell M, Gupta N, Co A (2006) A review of the aqueous electrochemical reduction of CO_2 to hydrocarbons at copper. J Electroanal Chem 594(1):1–19
73. Hara K, Kudo A, Sakata T (1995) Electrochemical reduction of carbon dioxide under high pressure on various electrodes in an aqueous electrolyte. J Electroanalytical Chem 391(1–2):141–147
74. Hori Y, Konishi H, Futamura T et al (2005) Deactivation of copper electrode in electrochemical reduction of CO_2. Electrochim Acta 50(27):5354–5369
75. Centi G, Perathoner S (2007) Nano-architecture and reactivity of titania catalytic materials. Quasi-1D nanostructures. In: Spivey JJ (ed) Catalysis (specialist periodical reports). Royal society of chemistry, vol 20. Cambridge, UK, pp 367–394
76. Centi G, Perathoner S (2009) Nano-architecture and reactivity of titania catalytic materials bidimensional nanostructured films. In: Spivey JJ, Dooley KM (eds) Catalysis (specialist periodical reports). Royal society of chemistry, vol 21. Cambridge, UK, pp 82–130
77. Centi G, Passalacqua R, Perathoner S et al (2007) Oxide thin films based on ordered arrays of 1D nanostructure. A possible approach toward bridging material gap in catalysis. Phys Chem Chem Phys 9:4930–4938
78. Grimes CA, Mor GK (2009) TiO_2 nanotube arrays: synthesis properties and applications. Springer, (Germany)

79. Schmuki P (2009) Self-organized oxide nanotube layers on titanium and other transition metals. In: Schmuki P, Virtanen S (eds) Electrochemistry at the nanoscale. Springer Science, (New York, US), pp 435–466
80. Mor GK, Shankar K, Paulose M et al (2006) Use of highly-ordered TiO_2 nanotube arrays in dye-sensitized solar cells. Nano Lett 6(2):215–218
81. Kuang D, Brillet J, Chen P et al (2008) Application of highly ordered TiO_2 nanotube arrays in flexible dye-sensitized solar cells. ACS Nano 2(6):1113–1116
82. Shankar K, Basham JI, Allam NK et al (2009) Recent advances in the use of TiO_2 nanotube and nanowire arrays for oxidative photoelectrochemistry. J Phys Chem C 113(16):6327–6359
83. Antoniadou M, Kondarides DI, Lianos P (2009) Photooxidation products of ethanol during photoelectrochemical operation using a nanocrystalline titania anode and a two compartment chemically biased cell. Catal Letters 129(3–4):344–349
84. Patermarakis G (1998) Development of a theory for the determination of the composition of the anodizing solution inside the pores during the growth of porous anodic Al_2O_3 films on aluminium by a transport phenomenon analysis. J Elec Anal Chem 447(1–2):25–41
85. Salimon J, Kalaji M (2003) Electrochemical reduction of CO_2 at polycrystalline copper in aqueous phosphate buffered solution: pH and temperature dependence. Malaysian J Chem 5(1):23–29
86. Qu J, Zhang X, Wang Y et al (2005) Electrochemical reduction of CO_2 on RuO_2/TiO_2 nanotubes composite modified Pt electrode. Electrochim Acta 50(16–17):3576–3580
87. Shibata H, Moulijn JA, Mul G (2008) Enabling electrocatalytic Fischer–Tropsch synthesis from carbon dioxide over copper-based electrodes. Catal Lett 123(3–4):186–192
88. Centi G, Perathoner S, Wine G et al (2006) Converting CO2 to fuel: a dream or a challenge? Prepr Symp–Am Chem Soc Div Fuel Chem 51:745–746
89. Centi G, Perathoner S, Wine G et al (2007) Electrocatalytic conversion of CO_2 to long carbon-chain hydrocarbons. Green Chem 9(6):671–678
90. Centi G, Perathoner S (2009) Catalysis: role and challenges for a sustainable energy. Top Catal 52(8):948–961
91. Gangeri M, Perathoner S, Caudo S et al (2009) Fe and Pt carbon nanotubes for the electrocatalytic conversion of carbon dioxide to oxygenates. Catal Today 143(1–2):57–63
92. Perathoner S, Gangeri M, Lanzafame P et al (2007) Nanostructured electrocatalytic Pt-carbon materials for fuel cells and CO_2 conversion. Kinet Catal 48:877–883

Nitrogen Photofixation at Nanostructured Iron Titanate Films

Horst Kisch

Abstract A nanostructured iron titanate thin film on glass is prepared by a sol-gel method from iron(III) chloride and titanium tetraisopropylate. Energy dispersive X-ray analysis, Mößbauer spectroscopy, and wavelength dependent measurements of the quasi-Fermi level suggest the presence of an $Fe_2Ti_2O_7$ phase having *n*-type semiconductor characteristics. In the presence of ethanol or humic acids and traces of oxygen, the film photocatalyzes the fixation of dinitrogen to ammonia and nitrate under ultraviolet or visible light irradiation. In the first observable reaction step, hydrazine is produced, which then undergoes further photoreduction to ammonia. Oxidation of the latter by oxygen yields nitrate as the final product. Since the reaction also occurs in air and with visible light ($\lambda > 455$ nm), and since the iron titanate phase can be formed by weathering of ilmenite minerals, the system may be a model for a previously unknown non-enzymatic nitrogen fixation in nature.

1 Introduction

Nitrogen fixation is the most important chemical process in nature after photosynthesis. The mild reaction conditions of the enzymatic reaction compared to the Haber–Bosch synthesis stimulated a large variety of investigations on the synthesis and reactivity of N_2 transition metal complexes under thermal reaction conditions. Relevant examples are the reductive protonation to ammonia under concomitant oxidation of the central metal, oxidative alkylation with alkyl halogenides

H. Kisch (✉)
Department of Chemistry and Pharmacy, Institute of Inorganic Chemistry,
Friedrich-Alexander-Universität Erlangen-Nürnberg,
Egerlandstraße 1, 91058 Erlangen, Germany
e-mail: Horst.Kisch@chemie.uni-erlangen.de

L. Zang (ed.), *Energy Efficiency and Renewable Energy Through Nanotechnology*,
Green Energy and Technology, DOI: 10.1007/978-0-85729-638-2_17,

affording alkyldiazenido complexes [1] and successive addition of methyl lithium and trimethyloxonium tetrafluoroborate leading to a 1, 2-dimethyldiazene complex [2]. Comparably little work is known for photofixation, especially with respect to a sunlight-induced non-enzymatic nitrogen fixation at a simple inorganic photocatalyst. In 1977, Schrauzer and Guth reported for the first time that powdered, humidified, Fe-doped TiO_2 photoreduces molecular nitrogen to ammonia with the concomitant production of molecular oxygen. It was proposed that nitrogen is reduced at photochemically generated, reduced Ti-centers that react with adsorbed H_2O to yield ammonia via diazene and hydrazine intermediates. Oxygen is formed by the reaction of surface Ti-OH groups with photochemically-generated valence band holes. The most active photocatalyst activity was obtained from powdered anatase doped with 0.2% Fe_2O_3 and subjected to a three-hour heat treatment at 1000°C. This treatment caused a partial conversion to rutile, producing crystals of diameters from 0.1 to 0.3 microns. Increasing the iron content produced fewer active photocatalysts because Fe accelerates the anatase–rutile conversion and speeds up the growth of the newly formed rutile crystals with attendant diminution of the active surface.

No ammonia was formed when nitrogen was bubbled through an aqueous suspension of this powder. Higher iron contents resulted in inactive materials [3]. Subsequent work of other authors confirmed these results, although the nature of the reducing agent was unknown in most cases since oxygen was only rarely identified [4–20]. In general, ammonia concentrations were in the range of 1–10 μM, and excitation by UV light was necessary. Very recently, in a short communication, it was reported that an electrochemically-formed titania layer is also active without iron doping [21, 22].

These partly contradictory results induced adverse discussions, particularly by Edwards et al., and culminated in the conclusion that all the previously reported ammonia yields originate from traces of this gas as always present in a laboratory [23, 24]. However, this possibility can be easily excluded using corresponding precautions. Furthermore, it is well-known that the photocatalytic properties of semiconductor powders are strongly influenced by crystal phase composition, crystal size, and the presence of impurities. Therefore the contrary results most likely stem from difficulties in preparing the photocatalyst by calcination procedures. We therefore checked alternative approaches and prepared nanostructured thin films of iron titanates by a sol-gel method. Contrary to the previously employed titania powders containing only 0.2% of iron, the films have an Fe:Ti ratio of 1:1 and are able to fix nitrogen also with visible light.

2 Film Preparation and Characterization

The thin films were prepared from an alcoholic solution of anhydrous iron(III) chloride and titanium tetraisopropylate in ratios of Fe:Ti = 1:1 or 1:2 by dip-coating glass slides, followed by hydrolysis in humid air and annealing at 600°C.

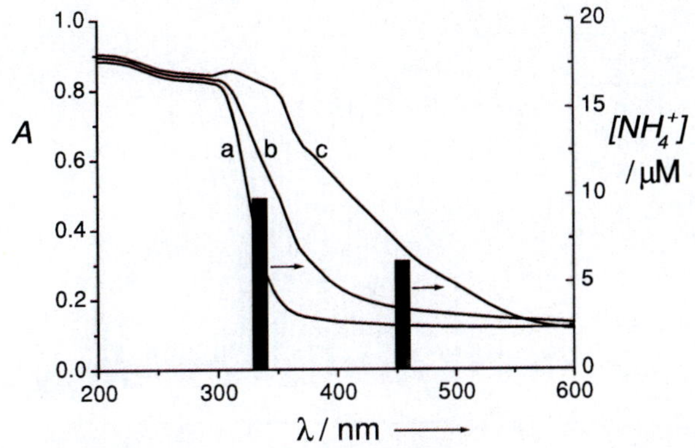

Fig. 1 UV/V is absorption spectra of various films and wavelength dependence of ammonia formation (75 vol% EtOH; 90 min of irradiation time). **a** TiO_2, **b** Fe:Ti = 1:2, **c** Fe:Ti = 1:1. The *two vertical bars* indicate ammonia concentrations when cut-off filters were employed ($\lambda \geq 335$ and 455 nm). From Ref. [29], Copyright 2003 Wiley-VCH Verlag GmbH & Co. KGaA, reproduced with permission (http://dx.doi.org/10.1039/B608396J)

At 500 and 700°C only inactive films were produced. Iron-free titania and titania-free iron oxide films were prepared analogously. Electron microscopy of the iron titanate film as obtained from a 1:1 metal ratio indicated the presence of a nanostructured matrix of about 300 nm thickness. It contains 15–20 vol.% of cubic crystals with an average diameter of 150 nm. The ratio of Fe:Ti:O was determined by energy dispersive X-ray analysis (EDX) as 1:1:3.5 for both the matrix and the crystals. This composition suggests that the compound $Fe_2Ti_2O_7$ is present. It was previously obtained only as an intermediary phase by heating ilmenite minerals ($FeTiO_3$) in oxygen atmosphere at 700°C [25]. This assignment is corroborated by the good agreement between the published and measured XRD spectra, although they also suggest the presence of traces of pseudobrookite (Fe_2TiO_5) and anatase [26]. The iron-free titania film exhibited only anatase peaks.

In the Mößbauer spectrum of the 1:1 film, the doublet at δ (relative to α-Fe) = 0.462 mm s^{-1} ($\Delta E_Q = 0.910$ mm s^{-1}, line widths of 0.294 mm s^{-1}) points to the presence of hexa-coordinated Fe^{III}. Although these values are almost identical with those of pseudobrookite, they cannot originate from this phase since the Fe:Ti ratio of the film is 1:1. There is no indication of the presence of an iron(III) oxide phase which should have a similar isomer shift but a much smaller quadrupole splitting (0.24 mm s^{-1}) [27].

2.1 Absorption and Photocurrent Action Spectra

The UV/V is spectrum of titania (Fig. 1, curve a) is red shifted down to 600 nm upon increasing the Fe:Ti ratio of the starting metal compounds from 0:1 to 1:2 and 1:1 (Fig. 1, curves b and c, respectively).

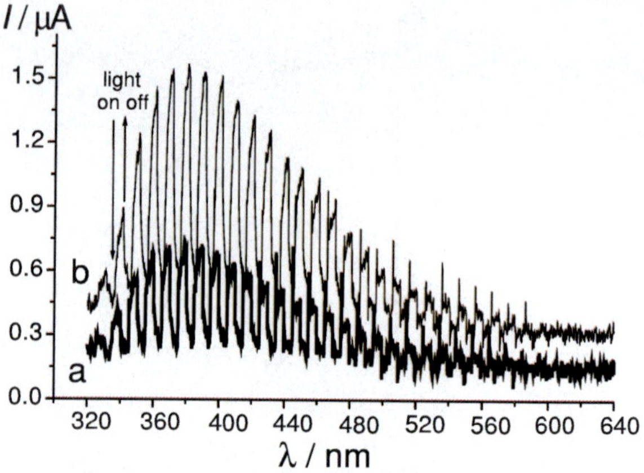

Fig. 2 Photocurrent action spectrum of the $Fe_2Ti_2O_7$/ITO electrode (**a**) before and (**b**) after the addition of methanol. Reprinted with permission from Ref. [34]. Copyright 2005 American Chemical Society

Absorbance starts at about 650 nm and increases sharply around 400 nm. To evaluate the bandgap energy, the square and square root of the absorption coefficient, α, should be plotted against the light energy for direct and indirect semiconductors, respectively [28]. It turns out that the use of the square root is necessary to obtain the required linear relationship. Surprisingly two linear regimes are found, resulting in bandgap energies of about 2.0 and 3.0 eV for the 1:1 film. Since this $Fe_2Ti_2O_7$ film upon UV-Vis irradiation produced the highest ammonia amounts (Fig. 6), all results discussed in the following refer to this material.

The anodic photocurrent observed upon irradiation of an $Fe_2Ti_2O_7$/ITO (ITO is indium tin oxide, a conducting glass) electrode at wavelengths between 320 and 640 nm suggests the presence of an n-type semiconductor (Fig. 2, curve a). Addition of methanol to the electrolyte results in a photocurrent increase by a factor of up to ~3 (at 380 nm, Fig. 2, curve b), an effect known in photoelectrochemistry as *current amplification*. This effect is based on electron injection of the CH_2OH radical, formed by primary hole-oxidation, into the semiconductor conduction band. When the film was calcined at 700, 500°C, or lower temperatures, the resulting electrode did not exhibit current amplification by methanol and was inactive in nitrogen photofixation.

2.2 *Measurement of Quasi-Fermi Levels*

Photoelectrochemical measurements of the quasi-Fermi level, i.e., the Fermi level observed under irradiation, were performed by recording the photovoltage of the iron titanate film on glass as a function of pH value in the presence of methylviologen.

This method was originally reported for titania powders [29, 30] but can also be applied to thin films on glass. At the inflection point (pH_0) of the resulting titration

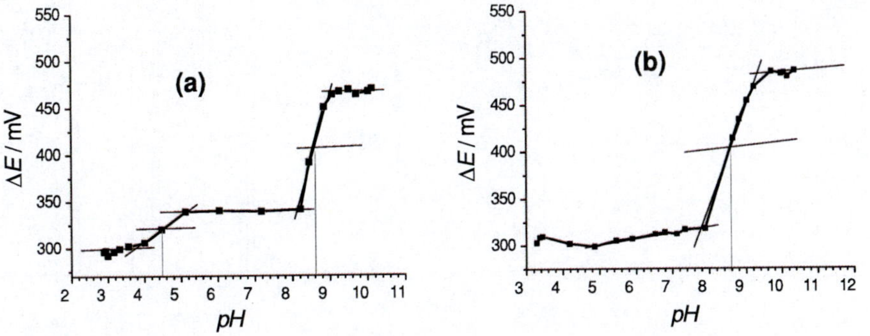

Fig. 3 Photovoltage versus pH value recorded for the $Fe_2Ti_2O_7$ film in 0.1 M KNO_3 in the presence of 1 mM methylviologen dichloride. Irradiation with full light (**a**) and through a 455 nm cut-off filter (**b**). Reprinted with permission from Ref. [34]. Copyright 2005 American Chemical Society

curve, the photovoltage is identical to the quasi-Fermi level of electrons ($_nE_F$*) in the semiconductor [30–33]. The plot of photovoltage versus pH value surprisingly exhibited two inflection points: $pH_0(1)$ and $pH_0(2)$ at pH 4.5 ($\pm$ 0.2) and pH 8.6 ($\pm$ 0.2), respectively (Fig. 3a). Whereas at the first point, where no color change was observed, the deep blue color of the methylviologen radical cation appeared at the second inflection point $pH_0(2)$. The first and second inflection points correspond to quasi-Fermi levels of −0.60 and −0.32 V (at pH 7), respectively. In agreement with the XRD analysis, the first level is assigned to traces of anatase. When a cut-off filter of 455 nm was inserted into the exciting light beam, the first inflection point disappeared whereas the second was still present at the same pH_0 value of 8.6 (Fig. 3b). Under this irradiation condition, the titania component cannot be excited and therefore the photovoltage must originate from the $Fe_2Ti_2O_7$ phase, exhibiting a quasi-Fermi level at 0.32 V. It is noted, that such a wavelength dependent photovoltage measurement offers an easy method for testing the composition of a semiconducting thin film or powder. These conclusions are also supported by spectroelectrochemical experiments [34].

3 Nitrogen Photofixation

3.1 *Ammonia Formation*

Irradiations ($\lambda \geq 320$ nm) of the iron-titania films in $EtOH/H_2O$ solutions were performed under N_2-bubbling or in a closed system after saturation with N_2.

The background concentration of ammonia, as measured before each experiment, was in the range up to 2 μM. Depending on the solvent composition, ammonia concentrations of 3–17 μM were observed (Fig. 6). It is noted that no significant amounts of ammonia formed under argon bubbling. To exclude the

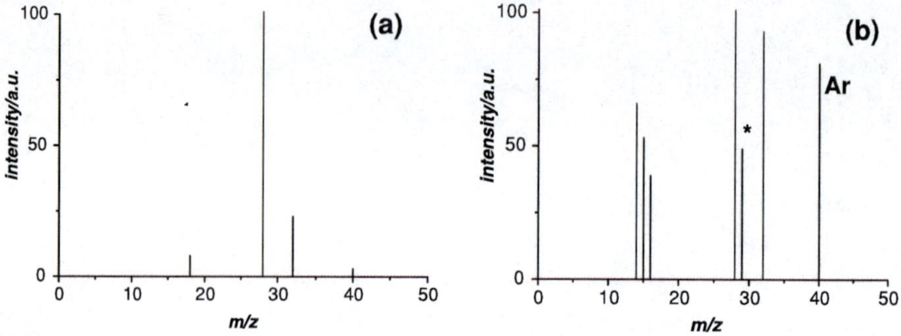

Fig. 4 Mass spectra of gas samples obtained from hypobromite oxidation of the reaction solution after 4 h irradiation time. Gas phase in Schlenk tube (**a**) before and (**b**) after filling with product gas (in EtOH (75 vol.%)). Reprinted with permission from Ref. [35]. Copyright 2006 the Royal Society of Chemistry (RSC)

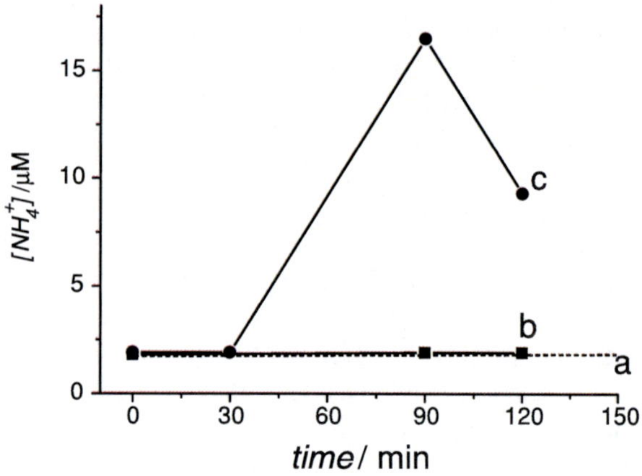

Fig. 5 Ammonia formation versus irradiation time in EtOH (75 vol.%): **a** background concentration of ammonia; **b** under bubbling a mixture of N_2:CO = 10:1; and **c** under N_2 after washing the same film. From Ref. [29], Copyright 2003 Wiley-VCH Verlag GmbH & Co. KGaA, reproduced with permission (http://dx.doi.org/10.1039/B608396J)

possibility that ammonia may originate not from N_2 but rather from other nitrogen contaminants, the experiments were performed with $^{15,15}N_2$. After acidification of the reacted solution, addition of excess $(^{14}NH_4)_2SO_4$, and oxidation with sodium hypobromite, $^{29}N_2$ was generated, as indicated by the intense peak at m/z = 29 (Fig. 4). The same result was obtained when ethanol was replaced by humic acid as the reducing agent [35].

Since it is known that carbon monoxide inhibits thermal nitrogen fixation [36], a mutual influence on the photofixation was tested. Upon bubbling with a mixture of N_2:CO = 10:1, complete inhibition was observed (Fig. 5, line b). The effect is reversible; the same film induced ammonia formation after it had been washed with water (Fig. 5, curve c).

The iron titanate films produced largest amounts of ammonia when the alcohol content was 75% (v/v) (Fig. 6). While the 1:1 film was inactive both in pure water

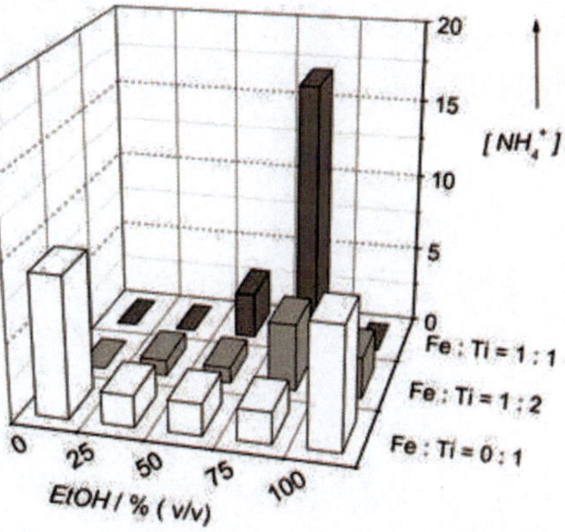

Fig. 6 Dependence of ammonia concentration on alcohol content of water and on the Fe/Ti ratio employed in the thin film preparation; $\lambda \geq 320$ nm, 90 min of irradiation time (standard irradiation conditions). From Ref. [29], Copyright 2003 Wiley-VCH Verlag GmbH & Co. KGaA, reproduced with permission

and pure ethanol, the 1:2 film afforded small ammonia amounts in pure ethanol. This differs significantly from the solvent dependence of the iron-free titania phase, which was active both in pure water and pure alcohol, and did not exhibit a maximum ammonia yield at 75% alcohol content. From these differences, we conclude that the 1:2 film is a mixture of the $Fe_2Ti_2O_7$ phase and anatase, and that traces of pristine titania are not responsible for nitrogen reduction. The results further suggest that in this latter case, water is the reducing agent, as also observed for the rutile/anatase powder containing 0.2% of iron [3], whereas in the case of the 1:1, film ethanol is the reductant. Why significant activity is observed only at ethanol contents of 50–75% but not in pure alcohol remains an open question. It is further noted that ammonia is also produced upon visible light excitation ($\lambda \geq 455$ nm, Fig. 1) of the 1:1 film, excluding the possibility that the traces of anatase present in this film may be responsible for nitrogen photofixation under polychromatic irradiation. Finally, it is mentioned that an analogously prepared titania-free iron oxide film did not produce ammonia.

Since the 1:1 film afforded the highest yield of ammonia in 75% EtOH, all of the following experiments were performed with this reaction system unless otherwise noted. Under these experimental conditions, acetaldehyde was detected after 90 min of irradiation and reached a concentration of 13 mM after 24 h. Ethanol may also be replaced by the natural reducing agent humic acid (see Fig. 7, curve e).

Figure 7 displays the time dependence of ammonia concentration under various reaction conditions. Line a represents the values measured for the solution containing the immersed film before irradiation and before N_2-bubbling. Line b corresponds to the values observed upon purging this system with nitrogen in the dark, and line c represents the system under irradiation and purging with argon. In all these blank experiments, the ammonia concentrations never exceeded values of 2 μM. However, when the irradiation was performed under N_2-bubbling, formation of ammonia started after an induction period of 30 min and passed through a maximum at 90 min of irradiation time (Fig. 7, curve d). Using air instead of nitrogen purging, the ammonia concentration decreased by about 60%.

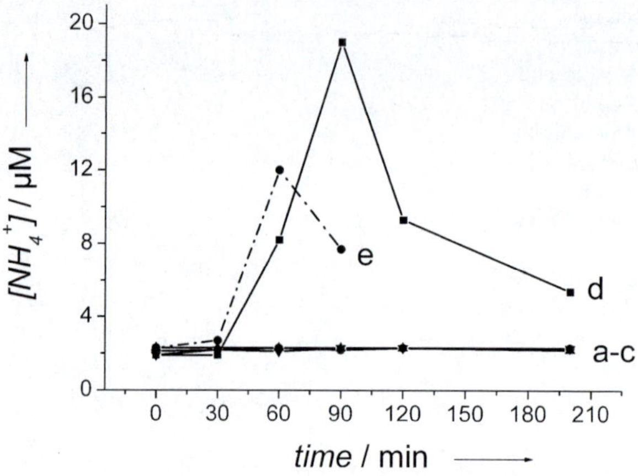

Fig. 7 Ammonia formation as function of irradiation time in EtOH (75 vol.%): **a** solution with immersed film prior to nitrogen bubbling and irradiation, **b** subsequent nitrogen purging in the dark, **c** irradiation under argon bubbling; **d** irradiation under N_2; and **e** irradiation under N_2 in the presence of 10^{-2} g L^{-1} of aqueous humic acid in ethanol-free water. From Ref. [29], Copyright 2003 Wiley-VCH Verlag GmbH & Co. KGaA, reproduced with permission (http://dx.doi.org/10.1039/B608396J)

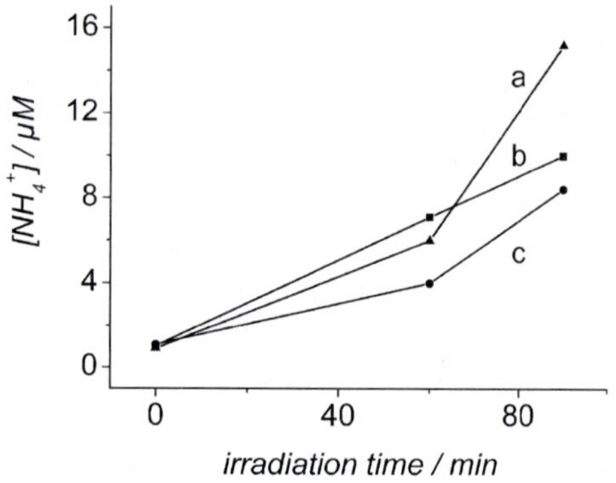

Fig. 8 Photofixation of dinitrogen to ammonia at thin films prepared from **a** $FeCl_3$, **b** $FeBr_3$, and **c** $Fe(acac)_3$. Reprinted with permission from Ref. [35]. Copyright 2006 the Royal Society of Chemistry (RSC)

When instead of iron(III) chloride the corresponding anhydrous bromide or acetylacetonate was employed for the preparation of the 1:1 film, slightly less active materials were obtained when irradiating under standard conditions (Fig. 8).

3.2 Hydrazine Formation

As evidenced by Fig. 9 (curve a), the photoreduction has a rather long induction period of about 30 min, suggesting formation of the true photocatalyst or of a reaction intermediate like hydrazine. In fact, small amounts of hydrazine are

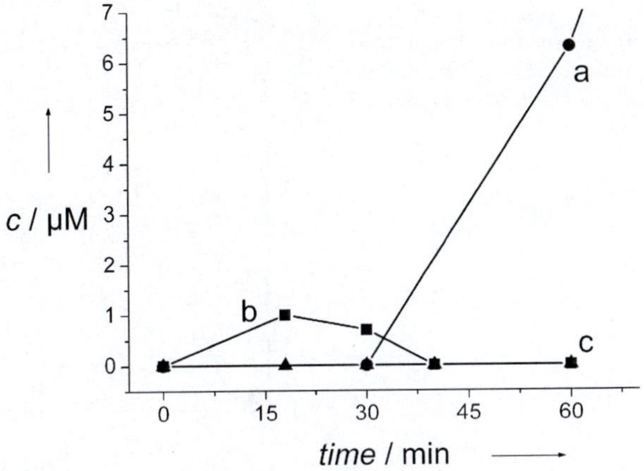

Fig. 9 Concentrations of (**a**) ammonia, (**b**) and hydrazine under N_2 and (**c**) under Ar. From Ref. [29], Copyright 2003 Wiley-VCH Verlag GmbH & Co. KGaA, reproduced with permission

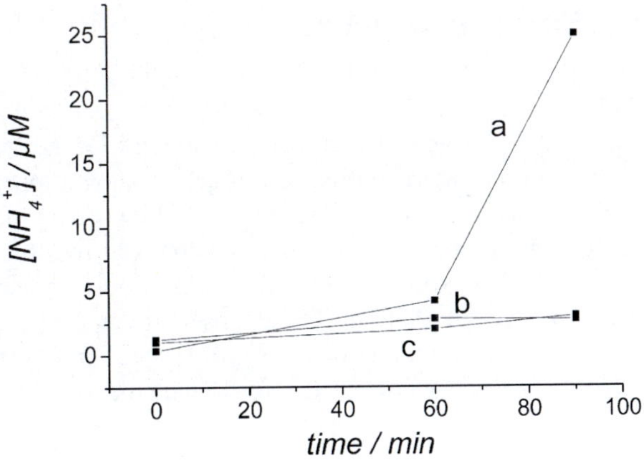

Fig. 10 Formation of ammonia from hydrazine in the presence of the iron titanate film (**a**) upon irradiation or (**b**) stirring in the dark at 50°C and (**c**) at room temperature. Reprinted with permission from Ref. [35]. Copyright 2006 the Royal Society of Chemistry (RSC)

produced during the first 15–30 min with a concentration maximum at about 20 min (Fig. 9, curve b). This result clearly suggests that in the induction period, nitrogen is photoreduced to hydrazine. No hydrazine was detectable when nitrogen was replaced by argon.

To find out if the further reduction of hydrazine occurs thermally or photochemically, a 15 μM solution of hydrazine in 75% EtOH in the presence of the 1:1 film was either stirred in the dark or subjected to the standard irradiation procedure. Whereas the photoreaction after 90 min produced a 25 μM solution of ammonia, the thermal reactions, conducted at 50°C and room temperature, did not exceed the blank value of about 2 μM (Fig. 10).

Although this result shows that hydrazine is reduced to ammonia only photochemically, it does not unequivocally establish that this is also the case during the actual nitrogen photofixation. If so, two photochemical steps would be required for ammonia formation and the reaction should exhibit a corresponding intensity dependence (vide infra).

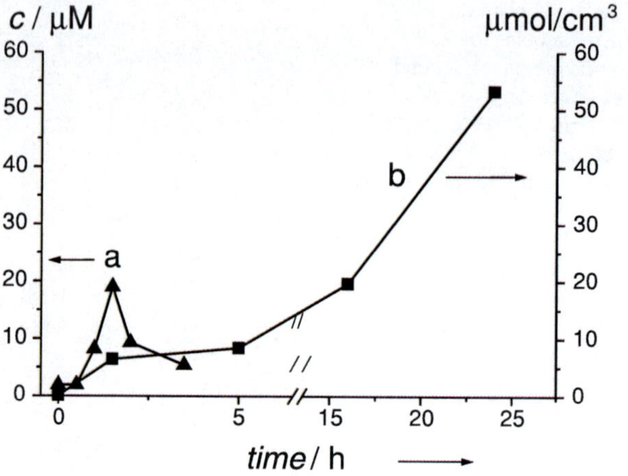

Fig. 11 Concentration of (**a**) ammonia in solution and (**b**) total nitrate concentration as a function of irradiation time

3.3 Nitrate Formation

The disappointing decrease of ammonia concentration after a 90 min irradiation time (Fig. 7, curve d) suggests a deactivation of the iron titanate film or a consecutive reaction like oxidation to nitrate, as known to occur upon UV excitation of titania [37–42]. Since after repeated washing of the film with water re-irradiation afforded ammonia in a concentration lowered by only 15%, photocatalyst deactivation is rather unlikely. Furthermore, no iron ions were detectable in the solution after 24 h of irradiation. These observations suggested that ammonia may be oxidized to nitrite/nitrate by traces of oxygen introduced during nitrogen-bubbling. Whereas only traces of nitrite were detectable, the nitrate concentration in the film reached 45 and 7 µM in solution (Fig. 11). When air was substituted for nitrogen, the total concentration of nitrate was 30 µM. Nitrate was formed in appreciable amounts only when ammonia reached its maximum concentration. This suggests that nitrate is formed via intermediary ammonia and not by direct oxidation of nitrogen. In accordance with this postulate, no nitrite/nitrate formed in the absence of EtOH.

To find out if the oxidation of the initially-produced ammonia is a photochemical or thermal process, an ammonium chloride solution of comparable concentration was stirred in the dark in the presence of the iron titanate film. Surprisingly, it turned out that nitrate formation is a thermal process occurring only in the presence of the film. Whereas at room temperature the oxidation is slow, it is much faster at 50°C, a temperature that could be easily reached during the prolonged irradiation experiment since the photoreactor was cooled only by ambient air.

The unexpected finding that ammonia oxidation is a thermal catalytic process prompted us to investigate whether or not the related semiconducting films of titania and iron(III) oxide, prepared analogously, would also be active. Whereas titania is active as the $Fe_2Ti_2O_7$ phase, iron oxide induced only about one-third of the activity.

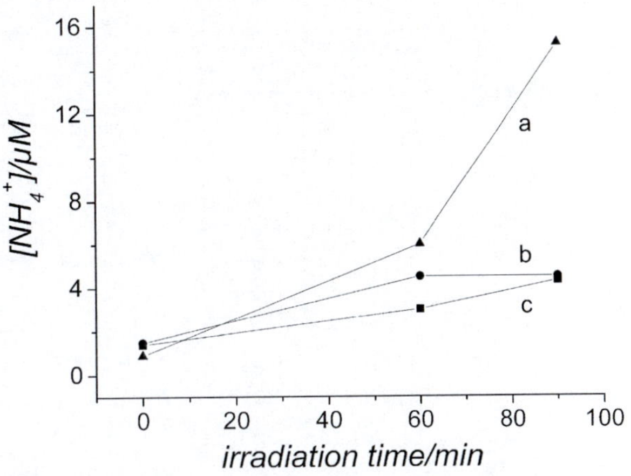

Fig. 12 Dependence of photochemical ammonia formation on the presence of chloride ions. **a** unwashed film, **b** washed film in the presence of 20 μM sodium chloride solution, **c** washed film. Reprinted with permission from Ref. [35]. Copyright 2006 the Royal Society of Chemistry (RSC)

3.4 Role of Chloride Ions

During ion chromatographic nitrate measurements varying amounts of chloride ions were detected. This suggested that hydrolysis of the employed anhydrous iron trichloride was not completed under the experimental conditions of film preparation. It turned out that a solution of up to 23 μM chloride was obtained upon prolonged irradiation of the film or even stirring in the dark. Since this is accompanied by a decrease of the pH-value from 7.2 to 6.3, it suggests that hydrochloric acid is formed. The resulting film induced formation of only traces of ammonia, even upon addition of 20 or 40 μM of sodium chloride solution (Fig. 12). From these results it follows that chloride is desorbed from the film surface probably by hydrolysis of remaining Fe–Cl bonds and therefore cannot be re-adsorbed at the original surface sites responsible for nitrogen photofixation.

In agreement with this observation is the finding that the iron titanate film produced from iron chloride hydrate is inactive.

From this dominating role of chloride ions we propose that in the oxidative primary step not ethanol but chloride is oxidized (Eq. 1). The resulting adsorbed chlorine atom then oxidizes ethanol to the hydroxyethyl radical (Eq. 2) followed by re-formation of the reactive surface iron chloride species (Eq. 3). This unprecedented role of surface Fe–Cl bonds differs from titania photocatalyzed oxidation reactions in the presence of dissolved chloride ions. In this case, both inhibiting and accelerating effects are known [43, 44]:

$$h_r^+ + [\mathrm{Fe(III)Cl}]_s \rightarrow [\mathrm{Fe(III)}]_s^+ + \mathrm{Cl}^{\bullet}_{ads} \quad (1)$$

$$\mathrm{Cl}^{\bullet}_{ads} + \mathrm{CH_3CH_2OH} \rightarrow \mathrm{CH_3^{\bullet}CHOH} + \mathrm{Cl}^- + \mathrm{H}^+ \quad (2)$$

$$[\mathrm{Fe(III)}]_s^+ + \mathrm{Cl}^- \rightarrow [\mathrm{Fe(III)Cl}]_s \quad (3)$$

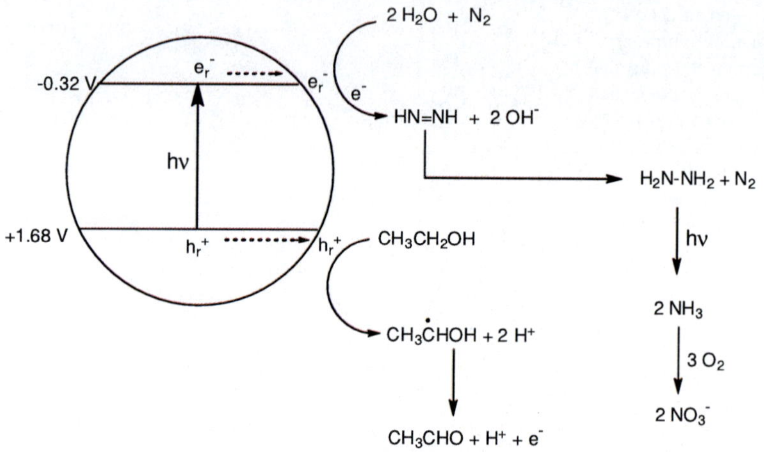

Scheme 1 Simplified mechanistic scheme of the visible light photofixation of nitrogen to nitrate. The upper and lower horizontal lines represent conduction and valence band edges, respectively. Reprinted with permission from Ref. [35]. Copyright 2006 the Royal Society of Chemistry (RSC) (http://dx.doi.org/10.1039/B608396J)

When instead of iron(III) chloride the corresponding bromide is used as starting material, the resulting film is a little less active, in agreement with the smaller oxidation potential of a bromine atom (Fig. 8, curves a and b). About the same activity is also obtained when the film is prepared from iron(IIII) acetylacetonate, suggesting that ethanol oxidation can proceed also in the absence of an Fe-X bond (X = Cl, Br) (Fig. 8, curve c), either via an initially formed acetylacetonyl radical or through direct reaction with the reactive hole.

3.5 Mechanism

From the experimental results presented above, a semiconductor photocatalysis mechanism is proposed for this unique nitrogen photofixation process, as schematically depicted in Scheme 1. Absorption of a photon by the $Fe_2Ti_2O_7$ phase generates a reactive electron-hole pair trapped at iron and oxygen surface centers. The hole oxidizes ethanol to the hydroxyethyl radical directly or via Eqs. 1–3. Subsequently the hydroxyethyl radical injects an electron into the conduction band affording acetaldehyde as oxidation product, analogous to the current amplification effect (Fig. 2). It is recalled that films that do not exhibit this effect also do not induce ammonia formation.

Thus, absorption of one quantum of light yields the two electrons necessary for the crucial first reaction step: the proton-coupled one photon/two-electron reduction of N_2 to diazene, known as the primary reduction product in homogeneous model systems of biological nitrogen fixation [3, 45]. Disproportionation of this short lived intermediate produces N_2 and hydrazine [36, 46–48], the latter of which is subsequently photoreduced to ammonia. The following oxidation of ammonia to nitrate by traces of oxygen is the thermally catalyzed final reaction step.

Instead of ethanol, humic acid can also act as a reductant, as shown in Fig. 7. Since humic acids are ubiquitous in nature and $Fe_2Ti_2O_7$ phases may be formed through solar oxidative weathering of ilmenite, this novel nitrogen photofixation may be an example for a light-driven non-enzymatic nitrogen fixation under natural conditions. It is known that solar nitrogen reduction occurs on rutile-containing desert sands [49].

Acknowledgments This work was supported by Deutsche Forschungsgemeinschaft. Helpful assistance by Dr. P. Zabek is highly appreciated.

References

1. Hidai M, Mizobe Y (1995) Recent advances in the chemistry of dinitrogen complexes. Chem Rev (Washington DC) 95(4):1115–1133
2. Sellmann D, Weiss W (1978) Consecutive nucleophilic and electrophilic attack on nitrogen ligands: synthesis of azomethane from molecular nitrogen. Angewandte Chemie 90(4): 295–296
3. Schrauzer GN, Guth TD (1977) Photocatalytic reactions. 1. Photolysis of water and photoreduction of nitrogen on titanium dioxide. J Am Chem Soc 99(22):7189–7193
4. Radford PP, Francis CG (1983) Photoreduction of nitrogen by metal-doped titanium dioxide powders: A novel use for metal vapor techniques. J Chem Soc Chem Commun 24:1520–1521
5. Endoh E, Bard AJ (1987) Heterogeneous photoreduction of nitrogen to ammonia on catalyst-loaded titanium(IV) oxide powders. New J Chem 11(3):217–219
6. Endoh E, Leland JK, Bard AJ (1986) Heterogeneous photoreduction of nitrogen to ammonia on tungsten oxide. J Phys Chem 90(23):6223–6226
7. Lichtin NN, Vijayakumar KM (1986) Photoassisted solid-catalyzed reduction of molecular nitrogen by water. Evidence for a photostationary state and for catalytic activity of many oxides. J Indian Chem Soc 63(1):29–34
8. Augugliaro V, Lauricella A, Rizzuti L, Schiavello M, Sclafani A (1982) Conversion of solar energy to chemical energy by photoassisted processes—I. Preliminary results on ammonia production over doped titanium dioxide catalysts in a fluidized-bed reactor. Int J Hydrogen Energy 7(11):845–849
9. Augugliaro V, D'Alba F, Rizzuti L, Schiavello M, Sclafani A (1982) Conversion of solar energy to chemical energy by photoassisted processes–II. Influence of the iron content on the activity of doped titanium dioxide catalysts for ammonia photoproduction. Int J Hydrogen Energy 7(11):851–855
10. Khader MM, Lichtin NN, Vurens GH, Salmeron M, Somorjai GA (1987) Photoassisted catalytic dissociation of water and reduction of nitrogen to ammonia on partially reduced ferric oxide. Langmuir 3(2):303–304
11. Miyama H, Fujii N, Nagae Y (1980) Heterogeneous photocatalytic synthesis of ammonia from water and nitrogen. Chem Phys Lett 74(3):523–524
12. Rao NN, Dube S, Natarajan P, Manjubala (1994) Photocatalytic reduction of nitrogen over (Fe, Ru or Os)/Tio2 catalysts. Appl Catal B 5(1–2):33–42
13. Litter MI, Navío JA (1996) Photocatalytic properties of iron-doped titania semiconductors. J Photochem Photobiol A: Chem 98:171–181
14. Yue PL, Khan F, Rizzuti L (1983) Photocatalytic ammonia synthesis in a fluidized bed reactor. Chem Eng Sci 38(11):1893–1900

15. Khan MMT, Chatterjee D, Bala M (1992) Photocatalytic reduction of nitrogen to ammonia sensitized by the [ruthenium(III)-ethylenediaminetetraacetate-2, 2′-bipyridyl]-complex in a platinum-titania semiconductor particulate system. J Photochem Photobiol A 67(3):349–352
16. Soria J, Conesa JC, Augugliaro V, Palmisano L, Schiavello M, Sclafani A (1991) Dinitrogen photoreduction to ammonia over titanium dioxide powders doped with ferric ions. J Phys Chem 95(1):274–282
17. Sclafani A, Palmisano L, Schiavello M (1992) Photoreduction of dinitrogen and photooxidation of phenol and nitrophenol isomers as examples of heterogeneous photocatalytic reactions. Res Chem Intermed 18(2–3):211–226
18. Palmisano L, Augugliaro V, Sclafani A, Schiavello M (1988) Activity of chromium-ion-doped titania for the dinitrogen photoreduction to ammonia and for the phenol photodegradation. J Phys Chem 92(23):6710–6713
19. Augugliaro V, Soria J (1993) Concerning "an opinion on the heterogeneous photoreduction of nitrogen with water". Angew Chem 105(4):579 (See also Angew Chem, Int Ed Engl, 1993, 1932(1994), 1550)
20. Palmisano L, Schiavello M, Sclafani A (1993) Concerning "an opinion on the heterogeneous photoreduction of nitrogen with water". Angew Chem 105 (4):580 (See also Angew Chem, Int Ed Engl, 1993, 1932(1994), 1551)
21. Hoshino K, Inui M, Kitamura T, Kokado H (2000) Fixation of dinitrogen to a mesoscale solid salt using a titanium oxide/conducting polymer system. Angew Chem Int Ed 39(14):2509–2512
22. Hoshino K, Kuchii R, Ogawa T (2008) Dinitrogen photofixation properties of different titanium oxides in conducting polymer/titanium oxide hybrid systems. Appl Catal B Appl Catal B: Environ 79(1):81–88
23. Boucher DL, Davies JA, Edwards JG, Mennad A (1995) An investigation of the putative photosynthesis of ammonia on iron-doped titania and other metal oxides. J Photochem Photobiol A 88(1):53–64
24. Edwards JG, Davies JA, Boucher DL, Mennad A (1992) Comments on heterogeneous photoreaction of nitrogen with water. Angew Chem 104 (4):489–491 (See also Angew Chem Int Ed Engl, 1992, 1931(1994), 1480-1992)
25. Gupta SK, Rajakumar V, Grieveson P (1991) Phase transformations during heating of ilmenite concentrates. Metall Trans B 22B(5):711–716
26. Joint commitee on powder diffraction standards 1970–1989, powder diffraction file(Swarthmore, Pennsylvania: International Center for Diffraction Data)
27. Goldanski VI, Herber RH (1968) Chemical applications of mössbauer spectroscopy. Academic Press, New York
28. Tauc J, Grigorovici R, Vancu A (1966) Optical properties and electronic structure of amorphous germanium. J Phys Soc Jpn Suppl 21:123–126
29. Rusina O, Linnik O, Eremenko A, Kisch H (2003) Nitrogen photofixation on nanostructured iron titanate films. Chem Eur J 9(2):561–565
30. Roy AM, De GC, Sasmal N, Bhattacharyya SS (1995) Determination of the flatband potential of semiconductor particles in suspension by photovoltage measurement. Int J Hydrogen Energy 20(8):627–630
31. Gerischer H (1966) Electrochemical behavior of semiconductors under illumination. J Electrochem Soc 113 (11):1174–1181 (discussion 1181–1172)
32. Ward MD, White JR, Bard AJ (1983) Electrochemical investigation of the energetics of particulate titanium dioxide photocatalysts. The methyl viologen-acetate system. J Am Chem Soc 105(1):27–31
33. White JR, Bard AJ (1985) Electrochemical investigation of photocatalysis at cadmium sulfide suspensions in the presence of methylviologen. J Phys Chem 89(10):1947–1954
34. Rusina O, Macyk W, Kisch H (2005) Photoelectrochemical properties of a dinitrogen-fixing iron titanate thin film. J Phys Chem B 109(21):10858–10862
35. Linnik O, Kisch H (2006) On the mechanism of nitrogen photofixation at nanostructured iron titanate films. Photochem Photobiol Sci 5(10):938–942

36. Zones SI, Palmer MR, Palmer JG, Doemeny JM, Schrauzer GN (1978) Hydrogen evolving systems. 3. Further observations on the reduction of molecular nitrogen and of other substrates in the vanadium(II) hydroxide-magnesium hydroxide system. J Am Chem Soc 100(7):2113–2121
37. Ileperuma OA, Kiridena WCB, Dissanayake WDDP (1991) Photoreduction of nitrogen and water on montmorillonite clays loaded with hydrous ferric oxide. J Photochem Photobiol A 59(2):191–197
38. Schrauzer GN, Guth TD, Salehi J, Strampach N, Hui LN, Palmer MR (1986) Photoreduction and -oxidation of molecular nitrogen on titanium dioxide and titanium containing minerals. NATO ASI Ser, Ser C (Homogeneous Heterog Photocatal) 174:509–520
39. McLean WR, Ritchie M (1965) Reactions on titanium dioxide photooxidation of ammonia. J Appl Chem 15(10):452–460
40. Mozzanega H, Herrmann JM, Pichat P (1979) Ammonia oxidation over UV-irradiated titanium dioxide at room temperature. J Phys Chem 83(17):2251–2255
41. Pichat P, Herrmann JM, Courbon H, Disdier J, Mozzanega MN (1982) Photocatalytic oxidation of various compounds over titanium dioxide and other semiconductor oxides; mechanistic considerations. Can J Chem Eng 60(1):27–32
42. Pollema CH, Milosavljevic EB, Hendrix JL, Solujic L, Nelson JH (1992) Photocatalytic oxidation of aqueous ammonia (ammonium ion) to nitrite or nitrate at titanium dioxide particles. Monatsh Chem 123(4):333–339
43. Calza P, Pelizzetti E (2001) Photocatalytic transformation of organic compounds in the presence of inorganic ions. Pure Appl Chem 73(12):1839–1848
44. Lewandowski M, Ollis DF (2003) Halide acid pretreatments of photocatalysts for oxidation of aromatic air contaminants: rate enhancement, rate inhibition, and a thermodynamic rationale. J Catal 217(1):38–46
45. Schrauzer GN, Palmer JG (2003) The chemical evolution of a nitrogenase model, XXIV. Correlational analysis of effects of organic acids on in vitro MoFe-protein substrate reduction activities. [erratum to document cited in ca136:305895]. Z Naturforsch B: Chem Sci 58(8):820
46. Zones SI, Vickrey TM, Palmer JG, Schrauzer GN (1976) Hydrogen evolving systems. 2. The reduction of molecular nitrogen, organic substrates, and protons by vanadium(II). J Am Chem Soc 98(23):7289–7295
47. Schrauzer GN, Palmer MR (1981) Hydrogen-evolving systems. 4. The reduction of molecular nitrogen and of other substrates in the vanadium(II)-pyrocatechol system. J Am Chem Soc 103(10):2659–2667
48. Schrauzer GN, Strampach N, Palmer MR, Zones SI (1981) Reduction of acetylene and nitrogen in the vanadium(II)-magnesium hydroxide system: concerning reaction stoichiometries and mechanisms of substrate reduction. Nouv J Chim 5(1):5–10
49. Schrauzer GN, Strampach N, Hui LN, Palmer MR, Salehi J (1983) Nitrogen photoreduction on desert sands under sterile conditions. Proc Natl Acad Sci USA 80(12):3873–3876

Photoreduction of Nitrogen on TiO_2 and TiO_2-Containing Minerals

Gerhard N. Schrauzer

Abstract A report appearing in 1975 demonstrated that incompletely out-gassed, powdered titanium dioxide photo-reduces acetylene to ethylene and prompted attempts to achieve the photoreduction of molecular nitrogen under similar conditions. Whereas pure, powdered anatase was only slightly active, a substantial enhancement of its photocatalytic activity was achieved by impregnating it with a small percentage of iron, causing its partial conversion to rutile by a heat-treatment at 1000°C. After humidification, this TiO_2 substratum, on exposure to light in an argon atmosphere, exhibited activity in the water-splitting reaction and yielded NH_3 in the presence of molecular nitrogen. Subsequent studies demonstrated that this N_2 photoreduction occurs in a stepwise fashion via diazene and hydrazine as the intermediates. These reactions were shown to take place under simulated terrestrial conditions on the surface of rutile-containing minerals on exposure to sunlight. Evidence for the secondary photoxidation of NH_3 on rutile-containing minerals was also obtained. In view of the wide distribution of titanium minerals on the Earth's crust, these abiological, light-driven reactions contribute to the nitrogen ecology in semiarid regions of the Earth and have been suggested to occur on the surface of Mars.

G. N. Schrauzer (✉)
Department of Chemistry and Biochemistry, University of California San Diego, Revelle College, La Jolla, CA 92037, USA
e-mail: gschrauzer@ucsd.edu

L. Zang (ed.), *Energy Efficiency and Renewable Energy Through Nanotechnology*, Green Energy and Technology, DOI: 10.1007/978-0-85729-638-2_18,

1 Introduction

Following the description of the first TiO_2-based hydrogen-producing solar cells by Fujishima and Honda in 1971 [1], the photoreduction of nitrogen to ammonia was deemed theoretically possible: since TiO_2 is known to absorb molecular nitrogen, it would only be necessary to divert the electrons generated by the photodecomposition of water to the absorbed nitrogen molecule. Interest in this reaction was initially primarily theoretical, although the possibility that the process could provide fertilizer ammonia in remote areas was considered; the idea that nitrogen could be fixed on titanium-containing minerals on Earth or perhaps on other planets, also seemed persuasive. The observation that N_2 inhibited the TiO_2-sensitized photolytic production of H_2 and the photoreduction of C_2H_2 led to the discovery that N_2 was photoreduced on Fe-doped TiO_2 [2] as well as on desert sands containing titanium minerals [3–5]. In the following chapter, the photoreduction of N_2 on TiO_2 substrata will be described from a general, methodological and historical point of view. Inasmuch as the general aspects of the reaction are concerned, the emphasis is on energetics and the mechanism. Methodological details are given primarily to make it easier for other investigators to enter the field. Since this review focuses on the photoreduction of N_2, the discussion of other substrates has been limited to those that are relevant to this reaction. This limitation was deemed necessary in view of the many substrates whose photocatalytic reactions on TiO_2 have already been investigated (for recent reviews see Mills and Le Hunter [6], Anandan et al. [7]). Historical details are given to show that the concept of N_2 photoreduction is not entirely new—the reaction was actually postulated to occur already more than half a century ago but remained obscure until it was independently rediscovered by the present author in the mid-1970s.

2 Studies Leading to the Discovery of Nitrogen Photoreduction

2.1 Water Photocleavage on Platinized TiO_2 Substrata

The photosplitting of water into hydrogen and oxygen on TiO_2 will be discussed first because it provides the energetic and chemical basis for nitrogen photoreduction; the coverage of the literature in this subject is not exhaustive since it was recently reviewed [6–9]. The reaction is possible energetically because the band gap of TiO_2 is 2.9–3.2 eV (290–335 kJ mol^{-1}), larger than the ΔG^0 of Reaction 1:

$$H_2O(l) \rightarrow H_2(g) + \tfrac{1}{2}\,O_2(g)\ (\Delta G^0 \text{ at } 298\text{ K} = 286\text{ kJ mol}^{-1}) \qquad (1)$$

For the reaction to occur, near-UV radiation of wavelengths 390–420 nm is required. The reaction is believed to be initiated by the formation of hole-electron exciton pairs in TiO_2, followed by the photochemical cleavage of surface or sub-surface Ti–OH groups, yielding OH radicals and titanium(III) centers; these products have been detected under specified experimental conditions on light-exposed TiO_2 [9–11]. The formation of H_2 requires the subsequent reaction of protons with reducing sites. On pure TiO_2 surfaces such sites are rare and the overpotential for the release of hydrogen is high, making it a poor substratum for H_2O photosplitting. To lower this overpotential, platinized TiO_2 electrodes are used. The oxygen is believed to have formed from OH radicals at the TiO_2 photoanode, which are generated by the reaction of H_2O with photochemical vacancies at the TiO_2 surface. RuO_2 at the anode was found to reduce the O_2 overvoltage, but at 0.78 V, according to recent Density Function Theory (DFT) calculations [12] is still too high and must be compensated by the application of an external potential, making oxygen production the most difficult step in the photochemical splitting of water.

2.2 *Water Photocleavage on Powdered TiO_2 Substrata*

The efficiency of water photosplitting is not only reduced in the absence of a supporting potential—the presence of excessive amounts of water is also undesirable because it prevents desorption of the reaction products from TiO_2. The photochemical water-splitting reaction was accordingly explored with powdered, humidified TiO_2 in gas–solid systems in the absence of excess water [13]. Under these conditions, the production of H_2 and O_2 from water in the molar ratio of 2:1 was observed. The preparation of the TiO_2 substrata requires special attention. In the studies conducted in the author's laboratory, the substrata with the highest photoreactivity were obtained using powdered anatase after it was partially converted to rutile by a heat treatment and rehumidified in an argon atmosphere. The starting material was a commercial-type powdered, essentially rutile-free anatase, which had been prepared by hydrolysis of titanyl sulfate [2], with an average particle size of 2 μm. The absence of rutile in the anatase is emphasized because any rutile accelerates the rate of the anatase → rutile transformation during the thermal pretreatment, causing the rapid growth of rutile crystals with diminution of active surface and photocatalytic activity. After heating to 1000°C for 4 h in the presence of air, the size of the newly-formed rutile crystals was only about 0.1 μm, affording a high active surface. After rehumidification, this substratum produced H_2 and O_2 at the molar ratio of 2:1 when exposed to UV light in an argon atmosphere at 50°C. Substrata with significantly higher photocatalytic activity were obtained by doping the anatase with small amounts of iron, added prior to the heat-pretreatment, usually in the form of ferric sulfate. As it will be discussed further in Sects. 3.6 and 3.7, iron accelerates the rate of the anatase → rutile conversion. Accordingly, the thermal pretreatment time was reduced from 4 to 1 h at 1000°C. In this manner, a substratum with a rutile content of 23% and an average particle size of approximately 0.1 μm was obtained, which proved to be significantly more photoactive than any previously examined TiO_2 substrata. Substrata with the same rutile content and the same small particle size can be reproducibly obtained only if the heat treatment is conducted under rigorously

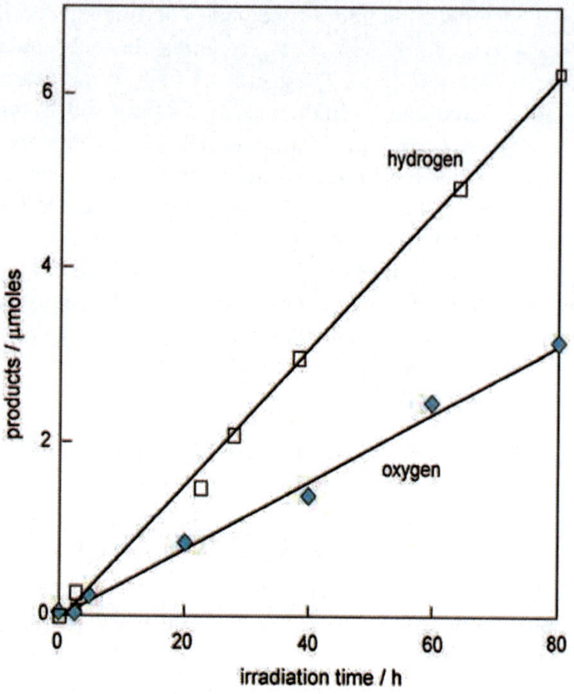

Fig. 1 Time dependence of H_2 and O_2 evolution from chemisorbed H_2O on Fe-doped TiO_2 (0.2 wt% Fe_2O_3) upon exposure to light from a 500 W Hanovia Hg-arc lamp at 80°C. Sample weight: 0.1 g, reactor volume: 38 ml. Adapted from Salehi and Schrauzer [6]

identical conditions with respect to temperature and time of heating. The presence of air or oxygen in surrounding gas-atmosphere during thermal pretreatment is also important as oxygen slows the rate of rutile formation. Under conditions of insufficient oxygen supply, the rate of the anatase → rutile conversion may be too fast to control rutile crystallite growth. The optimal duration of the thermal pretreatment must be determined empirically for each batch of anatase.

The substrata with the highest relative photocatalytic activity contain 0.2 wt% Fe [2,3.5]. The increased photocatalytic activity may be attributed primarily to the high surface area of the newly-formed rutile. The iron present in the TiO_2 lattice was suggested by Bickley et al. [14, 15] to introduce defect sites and enhance photocatalytic activity by improving the charge separation on exposure to UV light. The strong dependence of photocatalytic activity of the substrata on the pretreatment conditions caused some investigators to work with inactive substrata. For example, unable to observe significant yields of H_2 and O_2 in their experiments, Van Damme and Hall questioned the catalytic nature of the water-splitting reaction and suggested that it was a photoassisted stoichiometric decomposition of the hydroxylated TiO_2 surface instead [16]. An important variable that must be considered within this context is the reaction temperature: it must be high enough to favor product desorption. Figure 1 shows an experiment with an appropriately prepared and pretreated Fe-doped TiO_2, which produced H_2 and O_2 from H_2O continuously and at the molecular ratio of 2.00 for at least 80 h when irradiated with UV light in a gas–solid reactor at 80°C [5].

2.3 The Photoreduction of Acetylene and Ethylene on Fe-Doped TiO_2

The photoreduction of acetylene (C_2H_2) is mentioned here because acetylene is widely used as a model substrate in nitrogen-reducing systems and was also used in the initial experiments to demonstrate the feasibility of photoreducing N_2 on TiO_2. Exposure of powdered TiO_2 to UV light in the presence of acetylene was first shown by Boonstra and Mutsaers [17] to yield a mixture of methane (CH_4), ethylene (C_2H_4), ethane (C_2H_6), and C_3 hydrocarbons, provided that the TiO_2 was incompletely outgassed, i.e., surface Ti–OH groups were present. Ethylene under similar conditions was converted to a mixture of CH_4, C_2H_6, propane (C_3H_8), and n-butane (C_4H_{10}). With TiO_2 rendered anhydrous by vacuum outgassing at 500°C, C_2H_2 was phototrimerized to benzene (C_6H_6) instead. In the presence of N_2 in the gas phase, the photoreduction of C_2H_2 to C_2H_4 was inhibited and NH_3 was formed, indicating that N_2 was photoreduced at the C_2H_2-binding sites [2]. At the optimal temperature for N_2 photoreduction (40°C), nitrogen in the gas phase had little effect on CH_4 formation, but almost completely suppressed C_2H_4 production, indicating that N_2 was reduced at two-electron reducing sites. Diazene, N_2H_2, is the primary product of N_2 photoreduction.

3 The Photoreduction of Nitrogen on Fe-Doped TiO_2

3.1 Demonstration of the Reaction Stoichiometry

In the presence of increasing amounts of N_2 in the gas phase, proportionately more electrons are diverted to reduce nitrogen, causing a decline of the yields of H_2 from the photodecomposition of water, but not of O_2, while the yields of NH_3 and N_2H_4 increase (Fig. 2). These experiments prove the stoichiometry of N_2 photoreduction, as given in Eq. 2:

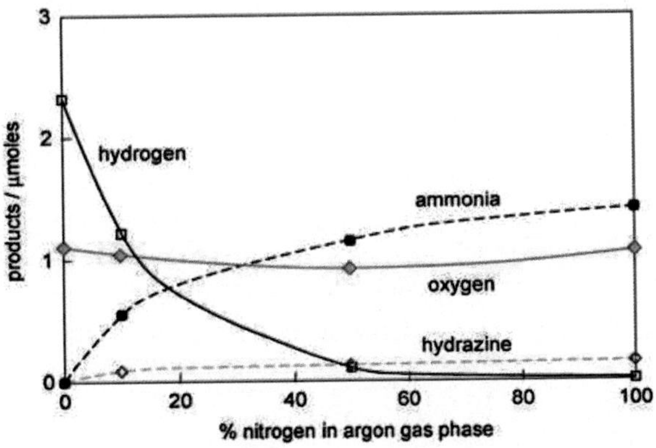

Fig. 2 Yields of H_2, O_2, NH_3 and N_2H_4 as a function of the percentage of N_2 in the gas phase; 100% N_2 = 1 atm at 40°C. Reaction time 3 h, TiO_2 photocatalyst and light source, as in Fig. 1; photocatalyst mass 0.2 g, water content 0.3 wt%, estimated from weight loss on heating for 2 h at 1000°C. Data from Schrauzer and Guth [2]

$$N_2 + 3H_2O + nh\nu \rightarrow 2NH_3 + 1^1/_2\ O_2 \quad (2)$$

Quantitatively, the observed NH_3/O_2 ratio at 100% N_2 in the gas phase was 1.32, close to the theoretical value of 1.33, according to Eq. 2. If the O_2 yields are corrected for the observed formation of N_2H_4 and of H_2, the NH_3/O_2 ratio increases to 1.54, indicating an apparent oxygen deficit of 14%. The ratio of the yield of NH_3 at 100% N_2 vs. the yield of H_2 generated in argon, 1.68, was also somewhat larger than the theoretical ratio of 1.50. However, in argon, slightly more O_2 was produced than in N_2 (1.10 *vs.* 1.05 μM), and if this difference is taken into account, the H_2/NH_3 ratio approaches the theoretical ratio of 1.50. The observed H_2/O_2 ratio from the water-photolysis experiment in argon was 2.11, 6% above the theoretical ratio of 2.00. This difference, as well as the deviation from the theoretical NH_3/O_2 ratio at 100% N_2 are most likely caused by the incomplete desorption of O_2, which is less complete at 40°C, the temperature at which these experiments were run, than at 80°C, at which temperature the observed H_2/O_2 ratio was 2.00 (see Fig. 2).

3.2 Energetics of Nitrogen Photoreduction

The question now is whether N_2 is reduced to ammonia and hydrazine in one step or sequentially. In view of the large ΔG^0 of Reaction Eq. 3:

$$N_2(g) + 3H_2O(l) \rightarrow 2NH_3(g) + 1^1/_2\ O_2(g)(\Delta G^0 \text{ at } 298\text{ K} = 766\text{ kJ mol}^{-1}) \quad (3)$$

the photoreduction of N_2 to NH_3 on the surface of TiO_2 cannot proceed in one step. The same is true for the direct photoreduction of N_2 to N_2H_4 according to Eq. 4:

$$N_2(g) + 2\ H_2O(l) \rightarrow N_2H_4(g) + O_2(g)\ (\Delta G^0 \text{ at } 298\text{ K} = 626\text{ kJ mol}^{-1}) \quad (4)$$

It thus remains to consider diazene, N_2H_2, the two-electron reduction product of nitrogen, as the intermediate:

$$N_2(g) + H_2O(g) \rightarrow N_2H_2(g) + {}^1/_2O_2(g)\ (\Delta G^0 \text{ at } 298\text{ K} = 440 - 482\text{ kJ mol}^{-1}) \quad (5)$$

Although the ΔG^0 in Eq. 5 is still larger than the bandgap energy of TiO_2, it applies to N_2H_2 generated in the gas phase. However, since under the actual conditions of the experiment, N_2H_2 would be chemisorbed on the surface of TiO_2 rather than in the gaseous state, this could diminish the value of ΔG^0 by as much as 100–140 kJ mol^{-1}, sufficiently to bring it down to the accessible range of between 290 and 335 kJ mol^{-1} [18].

3.3 Diazene as the Intermediate of Nitrogen Reduction

For diazene to be formed from N_2, two electrons must be transferred into the lowest antibonding π^* molecular orbital of N_2, which lies at –7.0 eV [19]. This is energetically possible on the surface of TiO_2 since its band gap ranges from –4.25 to –7.8 eV [9]. With sufficiently strong reducing agents, nitrogen can also be reduced to diazene under non-photochemical conditions, as in the case of the nitrogen-reducing systems first described by Shilov and co-workers [20]. In one such system, vanadium(II) hydroxide co-precipitated with magnesium hydroxide is the reducing agent; in another, the reducing agent is a V(II)-pyrocatechol complex. Since alkynes, RC≡CR, are reduced to olefins, RCH=CHR, the V(II) species in these systems acts as two-electron reducing agents and nitrogen is reduced to hydrazine and ammonia. From the four- and six-electron reduction products of N_2, it follows that the reduction of nitrogen occurs in a stepwise fashion via diazene, N_2H_2, as the intermediate. From the stereochemical course of the reduction of the alkynes, which is exclusively *cis*, the reduction of N_2 was postulated to proceed by way of a side-on interaction of N_2 with V(II), producing a vanadium diazenido intermediate from which N_2H_2 is released on protolysis [21–24]. The subsequent fate of the generated N_2H_2 depends on the reaction conditions: as it will be shown in the next section, diazene either disproportionates to N_2 and N_2H_4 or decomposes into the elements. This stepwise mechanism of N_2-reduction was shown also to operate in other protic nitrogen-reducing systems [25–27].

The reaction of N_2 with the photochemically excited Ti(II) sites at the surface of TiO_2 is envisaged to proceed analogously by way of a 'side-on' Ti-diazenido intermediate, see Eq. 6, from which diazene is released by the subsequent reaction with two protons from the adjacent Ti–OH groups:

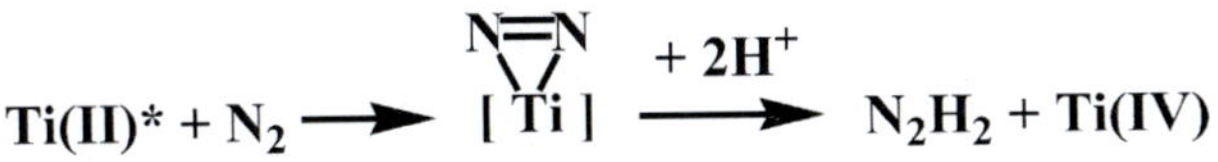

(6)

The Ti(II) centers can form on TiO_2 from photochemically generated Ti(III) ions by trapping an additional electron or from two photochemically generated Ti(III) ions by a disproportionation reaction. The involvement of Ti(II) species has also been suggested in the light-enhanced reduction of N_2 on *p*-GaP phosphide in the presence of Ti(IV) isopropoxide [28].

3.4 Formation of N_2H_4 and NH_3

N_2H_2 is a short-lived, endothermic compound. Since it is a strong reducing agent, it is not reduced further, but either disproportionates into N_2H_4 + N_2, or decomposes spontaneously into N_2 and H_2 [29]. The disproportionation of diazene is a unimolecular reaction and occurs quantitatively at lower concentrations; it is

Scheme 1 Reduction of N_2 via diazene (N_2H_2)

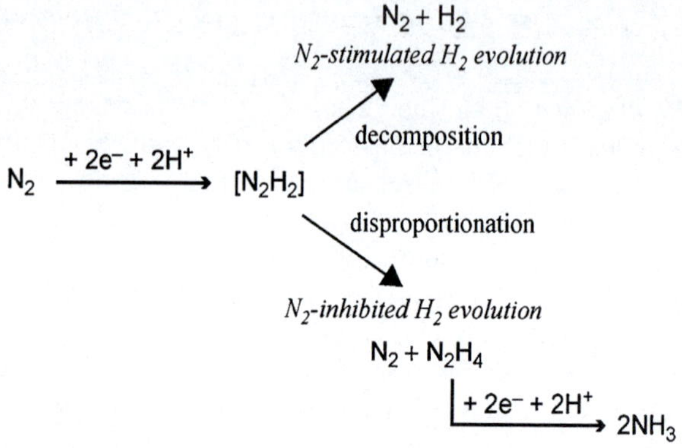

also base-catalyzed and favored under alkaline conditions. For N_2H_2 to disproportionate into N_2H_4, it must be generated in sufficiently high concentrations and preferably in a neutral or acidic environment. Humidified TiO_2 fulfills the latter condition since about half of its surface OH groups are acidic, with a pKa of about 2.9 [30]. The former condition, however, is fulfilled only by the most active TiO_2 substrata because only on these N_2H_2 is generated at sufficiently high stationary concentrations for its disproportionation to occur. Although formed in detectable amounts, the absolute yields of N_2H_4 are generally low since it is rapidly reduced further to NH_3 (see Sect. 3.5). With less active TiO_2 substrata, the N_2H_2 generated decomposes into N_2 and H_2, giving rise to a characteristic N_2-stimulated evolution of H_2 (see Scheme 1). For similar observations in several chemical N_2-reducing systems, see [21–27].

3.5 *Reaction Parameters*

The photoreduction of N_2 requires the use of specially prepared TiO_2 substrata with surface Ti–OH groups and proceeds optimally in *gas–solid systems* with humidified, powdered substrata; no reaction was observed with completely outgassed TiO_2 photocatalysts. The ammonia yields were also low when the TiO_2 substrata were exposed to light in aqueous slurries in the presence of N_2 in the gas phase. The fact that N_2 photoreduction experiments with these substrata should not be conducted in aqueous slurries was ignored by later investigators and resulted in several erroneous claims in the literature (see Sect. 3.9).

Maximum yields of NH_3 were observed with bulk TiO_2 substrata that were rehumidified after the thermal pretreatment (see Sect. 3.6) by storage in an H_2O-saturated Ar or N_2 atmosphere. The H_2O content of active substrata was, on average, 0.30 wt% [2]. Radiation in the near-UV region (390–420 nm) as emitted by a 500 W Hg arc lamp or by the Sun was photo-chemically active, as in other TiO_2-sensitized reactions; light from a tungsten filament lamp whose maximum intensity ranged from 520 to 575 nm was ineffective [2]. Although some of the

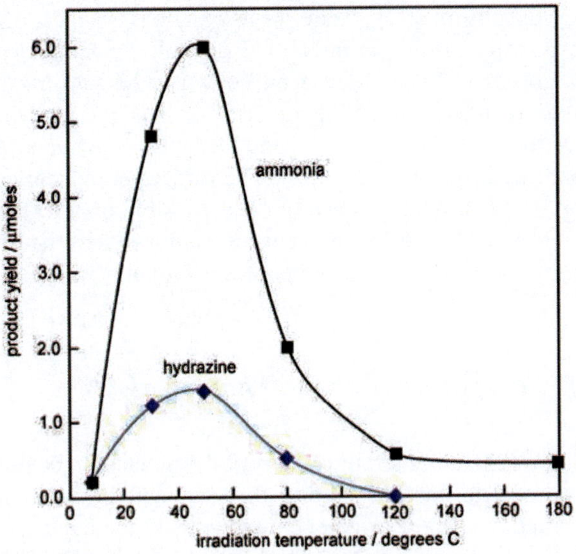

Fig. 3 Yields of NH_3 and of N_2H_4 from N_2 (1 atm) as a function of the 'irradiation temperature', i.e., the temperature of the reaction vessel during UV irradiation. Other conditions as indicated in Fig. 2. Adapted from Schrauzer and Guth [2]

reactions were conducted in quartz vessels, in most experiments Pyrex glass reactors were used. This is possible because Pyrex is transparent to light in the 300–400 nm wavelength range. The temperature at which the irradiation is conducted is also an important parameter. The yields of NH_3 and N_2H_4 are highest at irradiation temperatures of 30–50°C; only traces of N_2 reduction products are detected at irradiation temperatures below 10°C or above 80°C (Fig. 3).

Hydrazine yields also increased with N_2 pressure, but only up to about 2 atm. As a function of time at the irradiation temperature of 40°C, yields of NH_3 increased for 6 h of irradiation, mostly during the first 3 h. The cause(s) of plateauing of the yields of NH_3 on prolonged irradiation remains to be discovered. Possible causes include product inhibition (chemisorbed NH_3 blocking active reducing sites or the diminishing availability of acidic surface Ti–OH groups due to their reaction with NH_3) or NH_3 photooxidation.

The formation of NH_4^+ ions causes some of the product NH_3 to be relatively firmly held; only approximately two-thirds of the NH_3 formed can be removed by vacuum degassing at about 250°C. Repeated irradiation-desorption cycles resulted in gradual declines of photocatalytic activity, presumably due to surface-aging phenomena, and/or the occupation of lattice vacancies by oxygen. The latter view is supported by the fact that inactivation of the substrata is also observed on prolonged storage in O_2, but not on storage in N_2, argon or H_2 atmospheres [2].

3.6 Photooxidation of NH_3

Lower yields of NH_3 are observed if N_2 photoreduction experiments are conducted in the presence of O_2. In air, for example, the yields of NH_3 reach about 40% of the yields observed at 1 atm of N_2 in the absence of O_2. Prolonged irradiation, especially in the presence of O_2, results in larger losses of NH_3. The photooxidation of NH_3 has long been known to be sensitized by TiO_2 [31]; more recent studies showed that NH_2 chemisorbed on TiO_2 is photooxidized in the presence of O_2 to N_2, NO_2, N_2O, NO, and NH_2OH, as evidenced by selective cold-trapping and mass spectrometry [32]. The direct photochemical oxidation of chemisorbed N_2 to NO would be energetically possible on TiO_2 and was reported to occur in 1974 [10]. However, later studies demonstrated that the oxidized nitrogen species were the products of NH_3 photooxidation [18, 33].

3.7 Preparation of the Photocatalysts

The preparation of the photocatalysts usually begins with the selection of the appropriate anatase starting material and its impregnation with a suitable iron compound. In the original experiments [2], anatase with an average particle size of 2 μm, prepared by the hydrolysis of titanium(IV) sulfate, was impregnated with Fe(III) sulfate and dried. The Fe-impregnated anatase was subsequently subjected to a heat treatment to affect the dissolution of the iron and the partial transformation of the anatase into rutile. As it was already discussed in Sect. 2.2, the emphasis is on 'partial' because the newly-formed rutile crystals must be small so that a large surface area is obtained for maximum chemisorption of N_2. The rate of the anatase → rutile transformation is dependent on the type of anatase used, the concentration and composition of the iron dopant, the geometry of the furnace, and the atmosphere in contact with the sample during the heat treatment [34, 35]. The heat treatment should be conducted in a furnace that allows air to access the sample since oxygen slows down the anatase → rutile conversion. The metal dopant should preferably be Fe(III) sulfate because the sulfate ion also slows down the anatase → rutile conversion and its decomposition upon heating into SO_3 presumably produces micropores and defect sites in the TiO_2 lattice that increases H_2O and N_2 chemisorption. The highest photocatalytic activity was observed with 0.2% iron-doped anatase containing 20–30% rutile, which required a heating period at 1000°C for 1 h (Fig. 4).

The photocatalytic activity of the Fe-doped substrata was found to depend primarily on the presence of the optimal amount of dissolved iron in the rutile lattice and the small size of the newly-formed rutile crystals, as evidenced by SEM studies [3]. Increasing the iron content further accelerates the anatase → rutile transformation and growth of the rutile crystals, resulting in lower photocatalytic activities. Heating for longer periods also resulted in significant declines in photocatalytic activity. The optimal duration of the heat treatment must be determined empirically for each batch of TiO_2, also because of its dependence on the type of furnace used (see below). The pretreatment temperature, usually 1000°C, should be maintained as accurately as possible to obtain substrata with reproducible photoactivity. At pretreatment temperatures of 920–960°C, the rate of the anatase → rutile transformation was slow, yielding substrata containing only a small percentage of microcrystalline rutile.

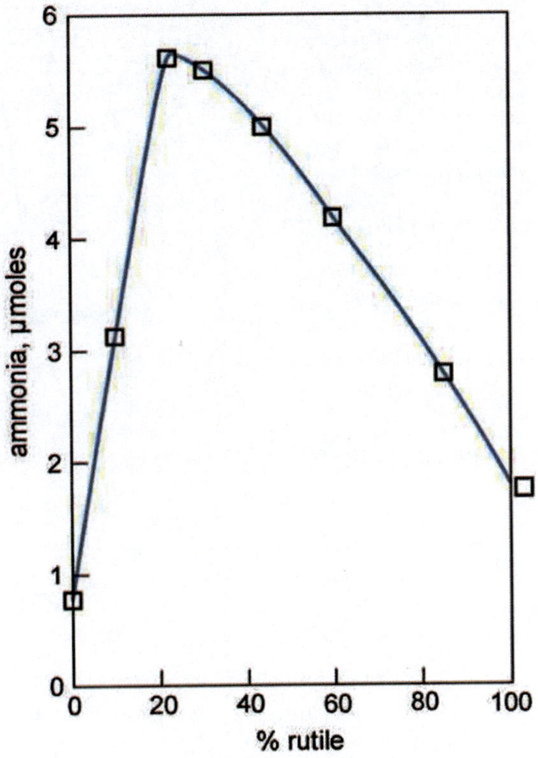

Fig. 4 Yields of ammonia as a function of the percentage rutile content in 2% Fe-doped TiO_2. Irradiation temperature: 40°C. Other conditions as in Fig. 2. Adapted from Schrauzer and Guth [2]

These still produced H_2 and O_2 from H_2O on UV irradiation under argon but did not yield NH_3 under N_2. Instead, N_2-stimulated H_2 evolution was observed (Fig. 5). An N_2-stimulated evolution of H_2 is typically observed in reactions in which N_2H_2 is formed at low stationary concentrations favoring its decomposition into N_2 and H_2. At pretreatment temperatures above 1000°C, rutile is formed too rapidly and crystal growth is accelerated so much that active substrata are usually not obtained. As to the furnace, a vertical heat crucible furnace was used (maximum temperature 1200°C) with an opening in its ceramic lid for air access. It is generally not recommended to perform the heat treatment in a vacuum, or in the presence of inert or flowing gases. As furnace dimensions may vary, the duration of the heat treatment must be determined empirically and chosen so that the anatase → rutile conversion of a 0.2% Fe-doped sample is complete in about 5 h at 1000°C. Substrata with photocatalytic activity are obtained after a heating period of 1–2 h at 1000°C. All prospective substrata must be characterized by X-ray diffraction analysis before and after the heat treatment; analysis by SEM or surface area determinations is also recommended. The 'apparent quantum yields' or 'turnover numbers' as defined by Schiavello et al. [36] can also be used for photocatalyst characterization.

Not all commercially available samples of anatase are suitable for photocatalyst preparation. Some of the newer products are produced by vapor-phase reactions,

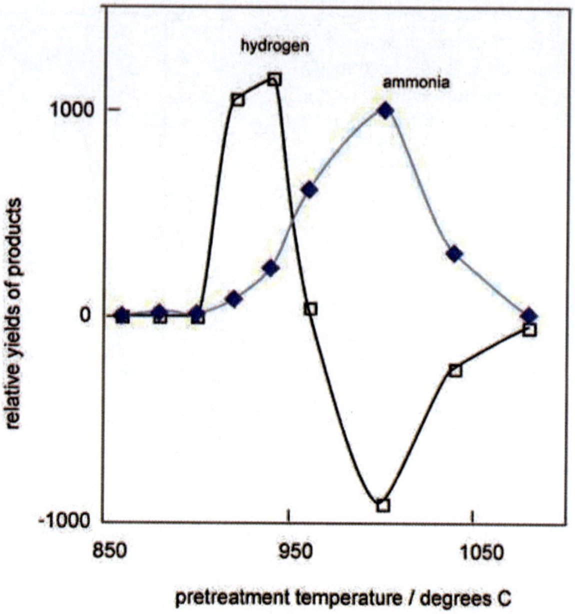

Fig. 5 Yields of H_2 in the presence of N_2 relative to Ar in the gas phase (both at 1 atm), at irradiation temperature of 80°C. Note stimulation and inhibition of H_2 production with TiO_2 substrata pretreated at 920–940 and 960–1040°C, respectively, by N_2. Maximum inhibition of H_2 production by N_2 coincides with the maximum of NH_3 formation

such as the combustion of $TiCl_4$ in oxygen. These failed to produce active substrata due to rapid rutile crystal growth under standard thermal pretreatment conditions. Other commercial TiO_2 products were found to be unsuitable for photocatalyst studies due to the undisclosed presence of proprietary brighteners and stabilizers. Several authors accordingly preferred to prepare their photocatalysts themselves, either by hydrolyzing the sulfate or the hydrolytic decomposition of $TiCl_3$, a method known to produce amorphous anatase. The hydrolytic methods allow co-precipitation of titanium oxide with metal dopants and were preferred by some workers in this field (Cordischi et al. [37], Bickley et al. [38]). Hydrazine yields also increased with N_2 pressure, but only up to about 2 atm. As a function of time at 40°C, yields of NH_3 increased for 6 h of irradiation, mostly during the first 3 h. The cause(s) of plateauing of the yields of NH_3 on prolonged irradiation remain to be discovered. Possible causes include product inhibition (chemisorbed NH_3 blocking active reducing sites or the diminishing availability of acidic surface Ti–OH groups due to reaction with NH_3) or NH_3 photooxidation.

3.8 The Role of the Metal Dopants

The function of iron and other dopants in the photocatalysts is not fully understood. In the presence of iron, the conversion of anatase to rutile and the grain growth of TiO_2 are accelerated during the thermal pretreatment; other metals, e.g., Co, Mo, and Ni, have the same effect [34] and have produced active substrata; the roles of the metals were therefore judged to be nonspecific [2]. It is possible that the thermal treatment of TiO_2 in the presence of these metals introduces oxygen vacancies into the rutile lattice, which then facilitates N_2 chemisorption. A number of other metals, including Pd, Pt Ag, Au, Pb, and Cu, did not produce significantly active photocatalysts. However, most of these metals also do not accelerate the anatase → rutile trans-formation under the pretreatment conditions chosen. In the active photocatalysts, Fe^{3+} ions are dispersed in the rutile lattice and no other phases are observed. At high Fe concentrations (>5%) the solids become multiphasic, indicating the presence of Fe_2TiO_5 [4, 5]. Such substrata generally showed low photo-catalytic activity. For additional studies on the preparation properties of Fe-doped TiO_2 photocatalysts, see Navio et al. [39], Martin et al. [40], Palmisano et al. [41], Litter and Navio [42] and Ranjit and Viswanathan [43].

3.9 Newer Studies

In the 1980s to early 1990s N_2 photoreduction reactions on TiO_2 substrata was further investigated by Augugliaro and his associates [44–51]. These investigators conducted the reaction in gas–solid fluidized bed reactors and used Fe-doped TiO_2 prepared in a manner similar to those originally employed by Schrauzer and Guth, except that some were supported on Al_2O_3 to improve photocatalyst fluidizability. Active substrata were also obtained by co-precipitating titanium dioxide with iron hydroxide and subsequent heat treatments. Control experiments revealed little or no N_2 photoreducing activity with undoped anatase, rutile, Fe_2TiO_5, or alumina. Several authors subsequently studied the photoreduction of nitrogen with Fe-doped TiO_2 substrata with various preparation methods or with chemical compounds that were altogether different. In some of these studies, it is difficult to assess whether the ammonia was formed by a genuine photocatalytic process or a photoassisted reaction, as pointed out by Schrauzer et al. [18]. For example, Radford and Francis [52] reported in 1983 that ammonia was formed upon irradiating an aqueous slurry of anatase impregnated with 1–2% iron by thermally decomposing an organoiron compound. As the NH_3 may have been formed by the reaction of N_2 with finely divided elemental iron that could have been formed on light irradiation of the organoiron compound, this system would more likely have to be regarded as photoassisted rather than photocatalytic. The same is true for the reported photoreduction of N_2 with co-precipitated hydrous oxides of samarium(III) [53] or Fe(III) [54] and vanadium(III) in which the V(III) hydroxide was considered to be the sacrificial reducing agent. Another system operating in the presence of a sacrificial electron donor (ethanol) and utilizes nanostructured iron titanate films as the photocatalysts was described by Linnik and Kisch [55]; for details see Chap. 17 of this volume.

Failures to achieve photoreduction of N_2 on TiO_2 substrata were reported in the early 1990s by Edwards and Boucher et al. [56, 57]. One problem of the studies of these authors was that their experiments were conducted with poorly characterized and probably inactive substrata; their conclusions were rejected in subsequent rebuttal communications [58, 59].

4 Nitrogen Photoreduction on Desert Sands and Titanium Minerals

4.1 The 1979 Study and Earlier Work by Dhar

The wide distribution of titanium minerals on the surface of the earth suggested that nitrogen photoreduction should also take place in nature on titanium-containing soil minerals and contribute to the nitrogen cycle, especially in semi-arid regions. In 1979, the first results were published [3] that demonstrated that sand from a Californian desert photoreduced N_2 on exposure to sunlight. Sterilized samples of this sand (10 g) produced 140 nmol of NH_3 when exposed to La Jolla, California July sunlight for 5 days at 1 atm in a $^{15}N_2/O_2$ mixture (78:22%). The same sand produced 600 nmol of $^{15}NH_3$ when irradiated in 1 atm of $^{15}N_2$ with light from a 360 W Hanovia mercury arc lamp at a distance of about 20 cm at 28°C. Virtually all the photocatalytic activity was concentrated in the heavy, rutile-containing fraction. Encouraged by this result, and with the support of a grant from the U.S. Department of Agriculture, sands from deserts in different countries were investigated for photoreducing activity. Among the sands tested, only those containing substantial amounts of rutile were found to be active [4]. While this study was in preparation, an extensive literature search revealed that the concept of nitrogen photoreduction was not entirely new—it was actually first proposed by the Indian soil scientist Dhar in the 1940s. Dhar and co-workers also conducted experiments that showed, or at least suggested, that a reduction of N_2 to NH_3 takes place photochemically on the surface of TiO_2 minerals. Dhar believed that the reducing equivalents needed for the reduction of nitrogen were generated by the photodecomposition of water on TiO_2 into hydrogen atoms and hydroxyl radicals, but also considered that organic substances could act as sacrificial electron donors [60–62]. Although some of Dhar's observations were later confirmed in 1955 by Eschena [63], his hypothesis was rejected and fell into oblivion until 1983, when a sample of sand was specifically obtained from the same location Dhar had been using for his experiments shown with $^{15}N_2$ as the tracer in the present author's laboratory to be photoactive (see below).

4.2 The 1983 Study

A total of 12 samples of sand were obtained from the locations given in Table 1 and were tested for N_2-photoreducing and -photooxidizing activity under

Table 1 Sites of sand samples used in the 1983 N_2 photoreduction study

No.	Site
1	Imperial Sand Dunes, Imperial Valley (California)
II	Death Valley Dune, base material (California)
III	Kuwait desert (Kuwait)
IV	Jumna River near Allahabad (India)
V	Desert near Cairo (Egypt)
VI	Tengger Desert (China)
VII	Panamint Spring area, Death Valley (California)
VIII	Death Valley, top of dune (California)
IX	Desert, gray sand (Saudi Arabia)
X	Entrance to Windrose Canyon, Death Valley (California)
XI	Desert (Saudi Arabia)
XII	Nemo Canyon, Death Valley (California)

sterile conditions [4]. For comparative purposes, the sands were also tested for C_2H_2-photoreducing activity. Six of the samples were from California deserts, two from deserts in Saudi Arabia and one each from Kuwait, Egypt, China, and India. To identify the photocatalytically active minerals present, samples of each sand were subjected to a density fractionation. For this, weighed amounts of sand were suspended in C_2Br_4, yielding a heavy ($\rho > 3$ g mL^{-1}) and a light ($\rho < 3$ g mL^{-1}). fraction. The latter consisted mainly of SiO_2, $CaCO_3$ and other common light minerals, and contained only traces of titanium, showing little or no photocatalytic activity. The heavy fractions were collected and, after removal of ferromagnetic minerals (mainly magnetite) by means of a magnet, were analyzed by X-ray powder diffractometry. The data were processed with a General Nova 4X minicomputer using ICPDS cards from 1981 for mineral identification. The compositions of the heavy nonmagnetic fractions, as determined by X-ray analysis, are given in Table 2. The detection limits for minerals are dependent on the total number of other minerals present; for rutile it was 2%. In addition, the samples were analyzed by atomic absorption spectroscopy for Ti, Fe, and 16 other elements. The results for Ti and Fe and the percentage rutile contents are given in Table 3.

The sands were tested for photocatalytic activity in their native states without grinding or other treatments affecting the surface of the minerals present. To eliminate any possible measurement bias, the samples were coded randomly. Each sample was washed with deionized water and heat-sterilized at 100°C for 4 h. Aliquots of the samples (6 or 10 g, as indicated) were placed into Pyrex glass test tubes (10 cm long, 1 cm diameter).

A calculated volume of air was withdrawn and replaced by the same volume of $^{15}N_2$ to enrich the N_2 to 25% $^{15}N_2$. The test tubes were subsequently exposed to sunlight, or, in other experiments, to UV light from a 400 W Hanovia mercury arc lamp. In all cases, simultaneous dark controls were used. These were prepared using the same experimental samples except that they were covered with aluminium foil or black cloth during the exposure to light.

Table 2 Composition of heavy nonmagnetic fractions

No.	Heavy minerals identified in nonmagnetic heavy fraction[a]	heavy fraction (%)
I	41% hornblende, 13% augite, 14% haematite, 6–9% rutile	0.57
II	60% hornblende, 17% augite, 17% Fe_2TiO_5, 2–3% rutile	1.20
III	47% diopsite, 6% banalsite, 16% haematite, 17% Fe_2TiO_5, 3% rutile	0.35
IV	53% a-quartz, 17% kyonite, 4% aphrosiderite, 10% rutile	0.19
V	26% chantalite, 22% bakerite, 15% ilmenite, 12% rutile	0.25
VI	30% paragasite, 23% haematite, 18% tourmaline, 5% sphene	0.19
VII	35% diopsite, 16% diopsite-Hegenbergite, 10% Al-diopsite 21% augite, 10% high albite	3.33
VIII	42% hornblende, 28% diopsite, 9% sphene, 6% pseudorutile, $Fe_2Ti_3O_9$	1.05
IX	30% augite, 21% piedmontite, 19% a-quartz, 6% haematite	0.31
X	41% piedmontite, 34% anorthoclase, 9% kryzanovskite	0.65
XI	77% dolomite, 22% goethite	6.97
XII	43% goethite, 20% diopsite-Hedenbergite, 12% grenalite, 9% biotite, 8% aphrosiderite	0.43

[a] $\rho > 3.0$ g ml^{-1}

Table 3 Results of N_2 and C_2H_2 photoreduction and photooxidation experiments and analytical data for samples I–XII

	Product yield (nmol)			Weight(%)		Ti/Fe (wt/wt)	Rutile in sample[a]%[α]
Sample	NH_3	N(ox)	C_2H_4	Ti	Fe		
I	59.2	20.0	27.0	0.08	0.47	0.178	0.0513
II	38.4	3.2	19.0	0.27	2.13	0.126	0.036
III	28.4	5.2	18.0	0.05	0.28	0.178	0.011
IV	25.2	0.11	14.0	0.22	1.62	0.135	0.019
V	24.0	9.6	16.0	0.03	0.38	0.079	0.030
VI	23.4	12.3	13.0	0.16	0.03	0.132	0.010[b]
VII	20.0	8.0	11.0	0.15	1.25	0.12	0.016[c]
VIII	14.4	0.12	13.5	0.20	1.86	0.108	0.010[b]
IX	10.8	2.0	9.5	0.06	0.54	0.111	0.007
X	5.7	8.8	6.6	0.18	2.57	0.07	~0.001
XI	5.2	0.0	13.0	0.05	0.05	0.091	~0.001
XII	1.4	7.6	3.6	0.43	5.26	0.082	~0.001

From Schrauzer et al. [4]. Samples of 6 g each were exposed to $^{15}N_2$-enriched air and daylight with intermittent sunshine (42%) for 14 days. C_2H_4 was from C_2H_2 reduction experiments at 1 atm, after 3 h of UV irradiation at 60°C

[a] Previously unreported data calculated from X-ray diffraction data or as indicated

[b] Estimated; photocatalytic species is probably sphene ($CaTiSi_3O_9$)

[c] Rutile below X-ray diffraction detection limit, estimated from Ti contents and Ti/Fe ratios

For analysis, the exposed sand samples were extracted twice with HCl, followed by water. The combined extracts were made alkaline and distilled into a measured amount of dilute HCl. Since it was expected that some of the ammonia formed would have been oxidized to nitrite or nitrate, the residual solutions in the Kjeldahl flasks were reduced to NH_3 with Devarda's alloy and collected by Kjeldahl distillation as outlined above. No attempt was made to distinguish between NO_2^- and NO_3^-. The NH_3 was oxidized to N_2 with NaOBr, collected and subjected to mass spectrographic analysis. All yields were corrected for background from parallel experiments with the same sands kept under the same conditions in the dark.

Acetylene photoreduction experiments were conducted with heat-sterilized sand at 1 atm (1.013×10^5 Pa) of C_2H_2. The bottles were irradiated at 60°C with a 400 W Hanovia mercury arc lamp for 3 h and yields of C_2H_4 were determined by gas–liquid-phase chromatography.

4.3 Results

N_2 and C_2H_2 photoreduction experiments with native sands. The results of the N_2 photoreduction and photooxidation field experiments and the C_2H_2 laboratory photoreduction experiments are shown in Table 3 in sequence of declining photocatalytic activity, together with the weight percentage of Ti and Fe, the Ti/Fe ratio and the rutile content.

Figure 6 demonstrates that the relative N_2- and C_2H_2 photoreducing activities of the 12 sand samples are directly correlated, suggesting that the two substrates are photoreduced by the same photoactive minerals present, Fig. 7 reveals the N_2-photoreducing activities of the sands to be proportional to their native rutile contents. This is also in accord with the results of intercorrelation calculations shown in Table 4. The observed yields of NH_3 from N_2 and of C_2H_4 from C_2H_2 are most significantly correlated with the rutile contents.

The highest yields of NH_3 were obtained in sample I, consisting of sand from the Imperial Sand Dunes, located 100 miles east of San Diego, California, which also contained the largest amount of rutile (0.0513%). Samples II and VIII were samples from the base and the top of a dune in Death Valley, California. The base material (sample II) contained much more rutile than the sand from the top of the dune, suggesting that gravity separation had occurred. Sample IV, the fourth most active, was from Jumna River near Allahabad, India. This sample was specifically investigated because sand from this location was previously used by Dhar in his experiments. Its heavy fraction was isolated and found to contain 10% rutile. Samples IX—XII showed negligible photocatalytic activities and contained very little or no rutile.

Experiments with isolated 'light' and 'heavy' sand fractions. Independent photoreduction experiments with the isolated 'light' and 'heavy' mineral fractions showed that virtually all the photocatalytic activity was concentrated in the heavy fraction ($\rho > 3.0$ g mL^{-1}), which contained all the rutile ($\rho = 4.25$ g mL^{-1}). Rutile emerged as the most active, and perhaps the only active photocatalytic mineral in this and the 1979 study; ilmenite, $FeTiO_3$ and magnetite, Fe_3O_4, and 30 other minerals present in the sands showed no photoactivity, with the possible exception of sphene, $CaTiSiO_5$ ($\rho = 3.53$ g mL^{-1}), which was detected in samples VI and VIII (see Table 2). However, the photocatalytic species in these cases

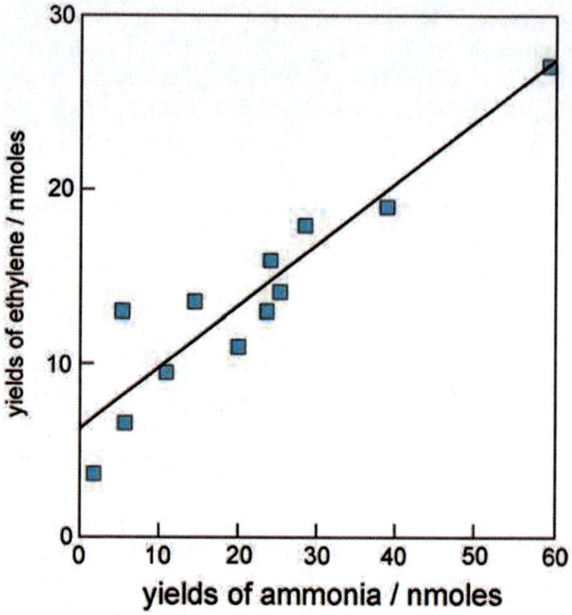

Fig. 6 Plot of photochemically produced C_2H_4 from C_2H_2 vs. NH_3 from N_2. Experiments with sand samples I –XII; data from Table 3 (from Schrauzer et al. 4)

were probably also rutile, since sphene is known [64] to lose calcium ions to leaching, resulting in a surface layer of TiO_2. Since reliable estimates of the amount of TiO_2 present could not be made, the linear correlation coefficients in the last column of Table 4 were calculated with and without samples VI and VIII; inclusion of the data by assuming sphene to be as photocatalytically active as rutile did not significantly alter absolute values of the correlation coefficients.

Photooxidation reactions. The desert sands exhibiting N_2-photoreducing activity also produced some ^{15}N-containing nitrate and nitrite, designated N(ox) in Table 3. The amounts of $^{15}N(ox)$ were, in most cases, significantly lower than those of $^{15}NH_3$. The degree to which the photooxidation of ammonia is inhibited by the possible concurrent oxidation of organic substances remains to be determined. However, the observed correlations of N(ox) with the percentage rutile contents and with the yields of NH_3 were both direct and significant with $P < 0.05$, indicating that the photooxidation of NH_3 occurs predominantly on the rutile minerals.

5 Conclusion and Perspectives

The photoreduction of N_2 is one of the most interesting and challenging reactions known to occur on TiO_2. Future investigations should focus on the characterization of the photoactive sites on TiO_2, the roles of the metal dopants, lattice defects, and of the surface Ti–OH groups on photoactivity. The reaction is not only of academic interest; it also has practical significance since it can take place

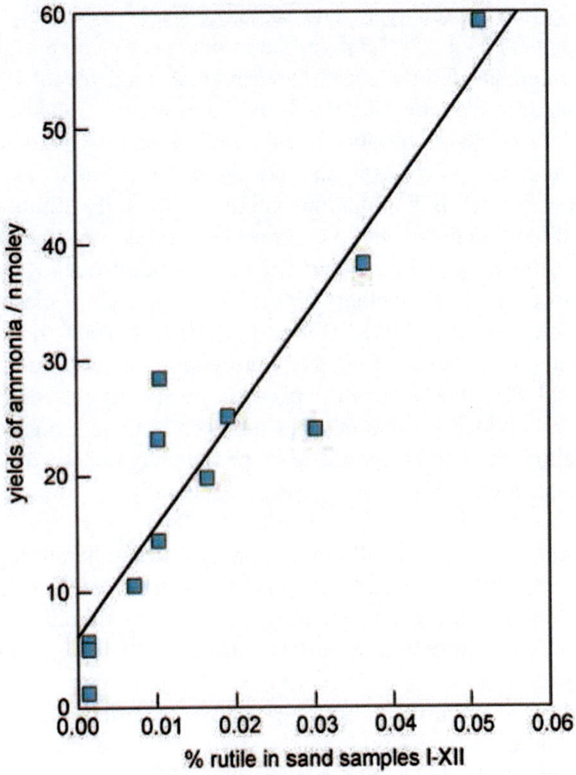

Fig. 7 Plot of ammonia yields against the rutile contents of sands I–XII

Table 4 Linear correlation coefficients between pairs of data in Table 3[a]

	NH_3	C_2H_4	N(ox)	Ti	Fe	Ti/Fe	% rutile[b]
NH_3		0.93[c]	0.55[d]	–0.25	–0.44	0.78[e]	0.94[c]
C_2H_4			0.375	–0.47	–0.623[d]	0.71[e]	0.87[c]
N(ox)				–0.38	–0.33	0.28	0.59[d]
Ti					0.94	0.01	–0.145
Fe						–0.19	–0.33
Ti/Fe							0.54[d]

[a] P (=probability) values are from one-tailed t-tests
[b] Previously unreported data (see text)
[c] $P < 0.001$
[d] $P < 0.05$
[e] $P < 0.01$

on naturally occurring TiO_2-containing minerals exposed to sunlight and thus could play a role in semi-arid regions where conditions for biological nitrogen fixation are unfavorable. Based on the average TiO_2 concentration of 0.25% of terrestrial sands or sandstones [65], the photoreduction of N_2 was estimated to produce 10^7 tons of NH_3 per year on the 19×10^6 square miles of deserts on the Earth [4]. This corresponds to about one-third of the N_2 annually oxidized by lightning discharges, and to about 10% of the N_2 reduced biologically, or to about 10 kg of NH_3 per acre per year—enough to support the sparse plant growth in deserts. Field experiments could reveal if the amount of bioavailable nitrogen can be increased by addition of TiO_2-rich sands to soils. A recent Chinese study produced encouraging results by demonstrating that nano-structured TiO_2 increases the photoreduction activities of the spinach chloroplast membrane [66]. Titanium minerals in soils stimulating nitrogen fixation was already claimed by Japanese researchers in the 1930s, but the effect was attributed to the development of the nodules of legumious plants by Ti(IV) [67]. According to another report [68], spraying plants with titanium ascorbate increases crop yields. Since this effect appears to be unrelated to the ability of TiO_2 to promote nitrogen photoreduction, the effect of Ti on plant growth thus could be studied from a broader perspective. Within this context, the catalytic photooxidation of ammonia nitrogen and of biomass on the surface of Ti-containing minerals should also be investigated to assess the degree to which it contributes to nitrogen losses, especially in tropical regions where sunlight is intense. On the other hand, the TiO_2–promoted nitrogen photoreduction is expected to be of little or importance in temperate regions with abundant rainfall or in aquatic environments. Last but not least, there is a question of whether or not ammonia can form on the surface of titanium minerals on planets such as Mars, as suggested as early as in 1977 [2], and in 2004 tentatively confirmed by spectroscopic data collected by the Planetary Fourier Spectrometer (PFS) on board the European Space Agency's orbiting Mars Express craft. Initially considered to be of biological origin [69], the presence of ammonia on Mars, if confirmed, could have a more plausible, if less spectacular, explanation.

Acknowledgments The author wishes to express his thanks to T. D. Guth, Liu Nan Hui, J. Palmer, M. R. Palmer, J. Salehi and N. Strampach for their dedicated collaboration. Support of this research by the U.S. National Science Foundation and the U.S. Department of Agriculture is gratefully acknowledged.

References

1. Fujishima A, Honda K (1971) Electrochemical evidence for the mechanism of the primary stage of photosynthesis. Bull Chem Soc Japan 44:1148–1150
2. Schrauzer GN, Guth TD (1977) Photolysis of water and photoreduction of nitrogen on titanium dioxide. J Am Chem Soc 99:7189–7193
3. Schrauzer GN, Guth TD, Palmer MR, Salehi J (1979) Nitrogen-reducing solar cells. In: Hautala RR, King RB, Kutal C (eds) Solar energy: chemical conversion and storage. Humana Press, Clifton, NJ, pp 261–269
4. Schrauzer GN, Strampach N, Liu N-H, Palmer MR, Salehi J (1983) Nitrogen photoreduction on desert sands under sterile conditions. Proc Nat Acad Sci 80:3873–3876
5. Schrauzer GN and Salehi J (1986) Unpublished, in: Salehi J, Nitrogen photo-chemistry on titanium semiconductors, doctoral thesis, University of California, San Diego
6. Mills A, Le Hunte S (1997) An overview of semiconductor photocatalysis. J Photochem Photobiol A: Chem 108:1–35

7. Anandan S, Ikuma Y, Niva K (2010) An overview of semi-conductor photocatalysis: modification of TiO_2 nanomaterials. Solid State Phenom 162:239–260
8. Kiwi J, Grätzel M (1984) Optimization of conditions for photochemical water cleavage. Aqueous Pt/TiO_2 (anatase) dispersions under ultraviolet light. J Phys Chem 88:1302–1307
9. Linsebigler AL, G-Qu Lu, Yates JT (1995) Photocatalysis on TiO_2 surfaces: principles, mechanisms, and selected results. Chem Rev 95:735–758
10. Bickley RI, Jayanty RKM (1974) Photoadsorption and photocatalysis on titanium dioxide surfaces. Disc Farad Soc 58:194–204
11. Anpo M, Shima T and Kubokawa Y (1985) Electron-spin-resonance and photo-luminescence evidence for the photocatalytic formation of hydroxyl radicals on small TiO_2 particles. Chem Lett 1799–1802
12. Valdes A, Qu Z-W, Kroes G-J, Rossmeisl J, Nørskov K (2008) Oxidation and photo-oxidation of water on TiO_2 surface. J Phys Chem C 112:9872–9879
13. Yamaguti K, Sato S (1985) Photolysis of water over metallized powdered titanium dioxide. J Chem Soc. Faraday Trans 1:1237–1246
14. Bickley RI, Gonzalez-Carreno T, Palmisano L (1991) A study of the inter-action between iron(III) oxide and titanium(IV) oxide at elevated temperatures. Mat Chem Phys 29:475–487
15. Bickley RI, Lees JS, Tilley RJD, Palmisano L, Schiavello M (1992) Characterization of iron/ titanium oxide photocatalysts. Part 1. Structural and magnetic studies. J Chem Soc Faraday Trans 88:377–383
16. Van Damme H, Hall WK (1979) On the photoassisted decomposition of water at the gas-solid interface on TiO_2. J Am Chem Soc 101:4373–4374
17. Boonstra AH, Mutsaers CAHA (1975) Photohydrogenation of ethyne and ethene on the surface of titanium dioxide. J Phys Chem 79:2025–2027
18. Schrauzer GN, Guth TD, Salehi J, Strampach N, Liu N-H and Palmer MR (1986) Photo-reduction and -oxidation of molecular nitrogen on titanium dioxide and titanium-containing minerals. In: Pelizzetti E and Serpone N (eds) Homogeneous and heterogeneous photo-catalysis, Reidel D Publishing Co. pp 509–520
19. Chatt J and Richards RJ (1971) In: Postgate JR (ed) The chemistry and biochemistry of nitrogen fixation. Plenum Press, London and New York, p 75
20. Shilov AE (1974) Nitrogen fixation in solutions in the presence of transition metal complexes. Usp. Khim 43:863–902
21. Zones SI, Vickrey TM, Palmer JG, Schrauzer GN (1976) Hydrogen-evolving systems. II. The reduction of molecular nitrogen in the $V(OH)_2/Mg(OH)_2$ system. J Am Chem Soc 98:7289–7295
22. Zones SI, Palmer MR, Palmer JG, Schrauzer GN (1978) Hydrogen-evolving systems. 3. Further observations on the reduction of molecular nitrogen and of other substrates in the $V(OH)_2/Mg(OH)_2$ system. J Am Chem Soc 100:2113–2121
23. Schrauzer GN, Palmer MR (1981) 'Hydrogen-evolving systems. 4. The reduction of molecular nitrogen and of other substrates in the vanadium(II)-pyro-catechol system. J Am Chem Soc 103:2659–2667
24. Schrauzer GN, Strampach N, Palmer MR, Zones SI (1981) Reduction of acetylene and of molecular nitrogen in the vanadium(II)–magnesium hydroxide system: concerning reaction stoichiometries and mechanism of substrate reduction. Nouveau J Chim 5:5–8
25. Liu N-H, Strampach N, Palmer JG, Schrauzer GN (1984) Reduction of molecular nitrogen on molybdenum(III–V) hydroxide/titanium(III) hydroxide systems. Inorg Chem 23:2772–2777
26. Schrauzer GN, Kiefer GW, Tano K, Doemeny PA (1974) The chemical evolution of a nitrogenase model VII. The reduction of nitrogen. J Am Chem Soc 96:641–665
27. Schrauzer B, Strampach N, Hughes LA (1982) Hydrogen evolving systems. 5. Nitrogen reduction in the V(OH)2/Mg(OH)s and V(OH)2/ZeO2.H2O systems. Factors influencing the yields of hydrazine and ammonia production. Inorg Chem 21:2184–2188
28. Dickson CR, Nozik AJ (1975) Photoelectrolysis of water using semi-conduct-ing TiO_2 crystals. Nature 257:383–385
29. Willis C, Back RA (1973) Di-imide: some physical and chemical properties, and the kinetics and stoichiometry of the gas-phase decomposition. Can J Chem 51:3605–3619

30. Herrmann M, Boehm HP (1969) Über die Chemie der Oberfläche des Titandioxids II. Saure Hydroxylgruppen auf der Oberfläche. Z Anorg Allg Chem 368:73–89
31. Rao G (1939) Photosensibilisierung durch feste Stoffe: Titandioxid Photo-sensibilis-ierte Oxidation von Ammoniak in wässeriger Lösung, Teil I. Z Physik Chem 184(A):377–384
32. Bickley RI, Navio-Santos JA and Vishwanathan V (1986) Routes to the photo-oxidative fixation of dinitrogen on TiO_2 (rutile) surfaces. In: Proceedings 6th international conference photochemical conversion and storage of solar energy, Paris, pp C 5–6, 21–25 July 1986
33. Bickley RI, Vishwanathan V (1979) Photocatalytically induced fixation of molecular nitrogen by near UV radiation. Nature 280:306–308
34. Iida Y, Ozaki S (1961) Grain growth and phase transformation of titanium oxide during calcinations. J Am Ceramic Soc 44:120–127
35. Shannon RD, Pask J (1965) Kinetics of the anatase–rutile transformation. J Am Ceramic Soc 48:391–398
36. Schiavello M, Augugliari V, Palmisano L (1991) An experimental method for the determination of the photon flow reflected and absorbed by aqueous dispersions containing polycrystalline solids in heterogeneous photocatalysts. J Catal 172:332–341
37. Cordischi D, Burriesci N, D'Alba F, Petrera M, Pelizotti G, Schiavello M (1985) Structural characterization of Fe/Ti oxide photocatalysts by X-ray, ESR, and Mössbauer methods. Solid-State Chem 54:182–190
38. Bickley RI, Gonzalez-Carreno T and Palmisano L (1987) Preparation and characterization of some ternary titanium oxide photocatalysts. In: Preparation of Catalysts IV, Elsevier, Amsterdam, pp 297–308
39. Navio JA, Macias M, Gonzalez-Catalan M, Justo A (1992) Bulk and surface characterization of powdered iron-doped titania photocatalysts. J Mar Sci 27:3036–3042
40. Martin C, Martin I, Rives V, Palmisano L, Schiavello M (1992) Structural and surface characterization of the polycrystalline system $Cr_xO_y.TiO_2$ employed for photoreduction of dinitrogen and photo-degradation of phenol. J Catal 134:434–444
41. Palmisano L, Schiavello M, Sclafani A, Martin C, Martin I, Rives V (1994) Surface properties of iron-titania photocatalysts employed for 4-nitrophenol photodegradation in aqueous TiO_2 dispersion. Catal Lett 24:303–315
42. Litter MI, Navio JA (1996) Photocatalytic properties of iron-doped titania semi-conductors. J Photochem Photobiol A: Chem 98:171–181
43. Ranjit KT, Viswanathan BJ (1997) Synthesis, characterization and photo-catalytic properties of iron-doped TiO_2 catalysts. Photochem Photobiol A: Chemistry 108:79–84
44. Augugliaro V, D'Alba F, Rizutti M, Schiavello M, Sclafani A (1982) Conversion of solar energy to chemical energy by photoassisted processes I. Preliminary results on ammonia production over doped titanium dioxide catalysts in a fluidized bed reactor. Int J Hydrogen Energy 7:845–850
45. Augugliaro V, D'Alba F, Rizutti M, Schiavello M, Sclafani A (1982) Conversion of solar energy to chemical energy by photoassisted processes II. Influence of the iron content on the activity of doped titanium dioxide catalysts for ammonia photoproduction. Int J Hydrogen Energy 7:851–855
46. Augugliaro V and Palmisano L (1988) Photoreduction of dinitrogen to ammonia in irradiated heterogeneous systems. In: Photocatalysis and Environment, Schiavello M, ed., Kluwer, Dordrecht, pp. 425–444, and references therein
47. Schiavello M and Sclafani A (1985) Photocatalytic reactions: and overview on the water splitting and on the dinitrogen reduction. In: Schiavello M (ed) NATO ASI Ser, Ser C. 146: Photochemistry, photocatalysis and photoreactors. Fundamentals and developments. D. Reidel, Dordrecht, pp 503–519 and references therein
48. Soria J, Conesa JC, Augugliaro V, Palmisano L, Schiavello M, Sclafani A (1991) Dinitrogen photoreduction to ammonia over titanium dioxide powders doped with ferric ions. J Phys Chem 95:274–282
49. Yue P, Khan F and Rizzuti L (1983) Photocatalytic ammonia synthesis in a fluidised bed reactor, Chem Eng Sci 1893–1900

50. Cordischi D, Burriesci N, D'Alba F, Petrera M, Pelizotti G, Schiavello M (1985) Structural characterization of Fe/Ti oxide photocatalysts by X-ray, ESR, and Mössbauer methods. Solid State Chem 54:182–190
51. Sclafani A, Palmisano L, Schiavello M (1992) N_2 photoreduction and phenol and nitrophenol isomers photooxidation as examples of heterogeneous photo-catalytic reactions. Res Chem Intermed 18:211–226
52. Radford PP and Francis CG (1983) Photoreduction of nitrogen by metal-doped titanium dioxide powders: a novel use for metal vapour technique. J Chem Soc Chem Commun pp 1520–1521
53. Tennakone K, Thaminimulla CTK, Kiridena WCB (1993) Nitrogen photoreducion by coprecipitated hydrous oxides of Samarium(III) and Vanadium(III). Laangmuir 9:723–726
54. Tennakone K, Thaminimulla CTK, Bandara JMS (1992) Nitrogen photoreduction by vanadium(III)-substituted hydrous ferric oxide. J Photochem Photobiol A: Chem 66:131–135
55. Linnik O, Kisch H (2006) On the mechanism of nitrogen photofixation at nanostructurediron titanate films. Photochem Photobiol Sci 5:938–942
56. Edwards JG, Davies JA, Boucher DL, Mennad A (1992) An opinion on the heterogeneous photoreductions of N_2 with H_2O. Angew Chem Int Ed 31:480–481
57. Boucher DL, Davies JA, Edwards JG, Mennad A (1995) An investigation of the putative photosynthesis of ammonia on iron-doped titania and other metal oxides. J Photochem Photobiol 88:53–64
58. Augugliaro V, Soria J (1993) Concerning "An opinion on the heterogeneous photoreduction of N_2 with H_2O", First letter. Angew Chem Int Ed 32:550
59. Palmisano L, Schiavello M, Sclafani A (1993) Concerning "An opinion on the heterogeneous photoreduction of N_2 with H_2O". Second letter. Angew Chem Int Ed 32:551
60. Dhar NR, Shchacharyulu EV, Mukerji SK (1941) Influence de la lumière sur la fixation de l'azote. Ann Agronom 11:83–86
61. Dhar NR, Francis AM (1951) Fixation of atmospheric nitrogen under sterile conditions in sand, using different energy materials. Proc Nat Acad Sci India 20(A):112–117
62. Dhar NR and Chowdhry S (1968) Fixation of nitrogen in presence of pure oxide surfaces by the slow oxidation of energy materials. Proc Nat Acad Sci India 38 (A), III & IV: 485–490
63. Eschena T (1955) La fissazione chimico-fisicas dell'azoto nel terreno. Ann Sper Agrar Nuova Serie 9:1399–1415
64. Bancroft GM, Metson JB, Kanetkar SM, Brown JD (1982) Surface studies on leached sphene glass. Nature 299:708–710
65. Clarke FW (1908) The data of geochemistry. US Geol Surv Bull No 330:20
66. Yang F, Liu C, Gao F, Su M, Wu X, Zheng L, Hog F, Yang P (2007) The improvement of spinach growth by nano-anatase TiO2 treatment is related to nitrogen photoreduction. Biol Trace El. Res 119:77–88
67. Konishi K and Tsuge T (1936) Mineral matters of certain leguminous crops. II. Nodule formation and titanium supply. Mem. Coll. Agr., Kyoto Imp. Univ. 37: (Chem Ser No. 20) 24–35
68. Fehér M, Papp K and Pais I (1985) New results in the application of titanium ascorbate as a plant-conditioning agent. In: Pais I (ed) Proceedings Symposium on New Results in the Research on Hardly-Known Trace Elements, University of Horticulture, Budapest pp 197–211
69. Whitehouse D (2004) Ammonia on Mars could mean life, BBC News, 15 July 2004

Photocatalytic Degradation of Water Pollutants Using Nano-TiO_2

R. Vinu and Giridhar Madras

Abstract This review discusses the utilization of photocatalysis for the degradation of water pollutants. Emphasis is placed on TiO_2 nanoparticles as a benchmark photocatalyst for the destruction of microorganisms and the degradation of a wide variety of organic compounds like phenolics, dyes, pesticides and pharmaceuticals. The mechanism of photocatalytic degradation is elucidated, underlining the importance of reaction kinetics for the efficient design of the processes. The effects of different reaction parameters on photocatalytic degradation are discussed. Surface modification of TiO_2 for visible light response by doping and heterostructuring is outlined. Finally, the challenges in the implementation of this technology for "real" waste water systems are summarized with a few attainable goals.

1 Introduction

The industrial revolution has led to the pollution of the natural resources in massive proportions. Environmental pollution caused by anthropogenic sources is a day-to-day problem faced by both developing and developed countries. Among the different types of pollution, air and water pollution from the point sources contribute a major share of the overall imbalance of the ecosystem. Common pollutants are usually toxic organic compounds like chlorinated and non-chlorinated aliphatic and aromatic compounds, dyes, detergents and surfactants, agro

---

R. Vinu · G. Madras (✉)
Department of Chemical Engineering, Indian Institute of Science,
Bangalore 560 012, India
e-mail: giridhar@chemeng.iisc.ernet.in

L. Zang (ed.), *Energy Efficiency and Renewable Energy Through Nanotechnology*,
Green Energy and Technology, DOI: 10.1007/978-0-85729-638-2_19,

wastes like insecticides, pesticides and herbicides, disinfection byproducts, volatile organic compounds, plastics, inorganic compounds (e.g., heavy metals), noxious gases (e.g., NO_x, SO_x, CO and NH_3) and pathogens (e.g., bacteria, fungi and viruses). Hence, strict environmental regulations on the use of these recalcitrant pollutants and their safe disposal drives the research community to develop clean and green processes to degrade the pollutants before they are discharged into the atmosphere and water bodies. This review examines "photocatalysis" as an advanced oxidation process for the degradation of different kinds of organic pollutants for environmental remediation.

Photolysis involves the use of ultraviolet (UV) or visible light radiation to effect chemical transformations. Photocatalysis uses a catalyst, usually, a semiconducting oxide to accelerate the photoreaction. The pioneering work of Fujishima and Honda in 1972 [1] on the photosplitting of water (i.e., simultaneous oxidation and reduction of water to oxygen and hydrogen, respectively) on a TiO_2 single crystal electrode led to the onset of photo induced redox reactions on semiconductor surfaces. This remarkable discovery attracted worldwide attention because it showed the possibility of generating hydrogen as a clean source of energy from water. It was soon realized that such redox processes could be utilized for environmental cleanup applications by Frank and Bard in 1977 [2], when they showed the photocatalytic oxidation of CN^- and SO_3^- using different semiconductor materials like TiO_2, ZnO, CdS, Fe_2O_3 and WO_3. This was followed by the demonstration of the TiO_2-catalyzed photodegradation of chlorinated organic compounds by Ollis [3, 4], and the Pt-loaded-TiO_2-catalyzed photochemical sterilization of microorganisms by Matsunaga et al. in the early 1980s [5]. Later, Grätzel, in 1991 [6], showed the first high-efficiency solar cell based on dye-sensitized colloidal TiO_2 films for photovoltaic power generation. Today, the principles of photocatalysis, especially using TiO_2 and its variants are widely employed in water and air purification, self-cleaning surfaces, self-sterilizing surfaces, antifogging surfaces, anticorrosive surface treatments, lithography, photochromic materials, microchemical systems and the selective and green synthesis of organic compounds.

Photocatalysis, as a research area, has witnessed tremendous development over the past two decades, with progress in the synthesis of novel materials and nanostructures, and the design of efficient processes for degradation of organic pollutants. Several classic review articles are dedicated to the principles and mechanism of photocatalysis [7–21], with special emphasis on the electron transfer processes, lattice and electronic structure of TiO_2, the surface chemistry of semiconductor oxides, generation of reactive radicals, chemisorption of small and large molecules, surface modification by doping and photooxidation of organic and inorganic substrates. Hence, photocatalysis can be regarded a well-understood field; yet, immense challenges and opportunities exist in realizing this technology on a large enough scale for pollution abatement.

Some of the traditional methods of water purification include neutralization of acidic and alkaline liquors, ultrafiltration, air stripping, flocculation, extraction, activated carbon adsorption, biological and peroxide/ozone treatment. It is important to note that most of the above methods are non-destructive, i.e., they only transfer the pollutant to another phase and do not ensure its destruction. The peroxide/ozone (H_2O_2/O_3) treatment is an economical way to oxidize pollutants, but O_3 causes air pollution. Biological degradation methods are effective in reducing the biological oxygen demand of the effluent. However, achieving complete mineralization (i.e., complete conversion to CO_2 and H_2O) of the parent organic compound using the above methods is difficult. The potential of photocatalysis to accelerate the rate of degradation reactions is attributed to the generation of reactive hydroxyl radicals ($OH^{\bullet}$), which are the precursors of degradation of any organic or inorganic compound and microorganisms. These reactive species possess a higher oxidation potential (2.80 V) compared to other common oxidants like atomic oxygen (2.42 V), O_3 (2.07 V), H_2O_2 (1.78 V), hydroperoxy radicals (1.70 V) and chlorine dioxide (1.57 V) [15]. Moreover, the thrust to utilize solar radiation has made photocatalysis, the technology of choice for the detoxification of water pollutants.

This review article begins with an explanation of the basic principles of UV photocatalysis and the different semiconductor oxides, which are used as photocatalysts. Emphasis is on TiO_2 as a "benchmark photocatalyst", and the different synthesis procedures and properties of TiO_2 are briefly discussed. The photocatalytic degradation of various organic compounds like aliphatic and aromatic compounds, dyes, pesticides, pharmaceutical compounds and the destruction of microorganisms are extensively discussed in terms of their pathway of degradation. The effect of different reaction parameters like temperature, pH, solvents, UV light intensity, catalyst loading, substrate concentration, photocatalyst and the presence of anions and cations are discussed. The possibility of extending the absorption spectrum of UV photocatalysts to the visible region by appropriate surface modification is examined. The various mechanistic kinetic models proposed for the degradation of organic compounds are evaluated. The industrial applicability of photocatalysis is analyzed in the chemical engineering point of view, in terms of the different photoreactors that are currently being developed. Finally, the future research perspectives and the possibilities for commercializing this technology for the degradation of "real" industrial effluents are probed.

2 Mechanism of UV Photocatalysis

The mechanism of UV photocatalysis is well-documented [8, 10, 17, 18], and the following reactions represent the key steps in the generation of charge-carriers and hydroxyl radicals, which mediate photocatalytic reactions. A pictorial representation of the mechanism of TiO_2 photocatalysis is shown in Fig. 1.

Charge-carrier generation:

$$TiO_2 + hv \xrightarrow{hv > E_{BG}} h^+_{VB} + e^-_{CB} \tag{1}$$

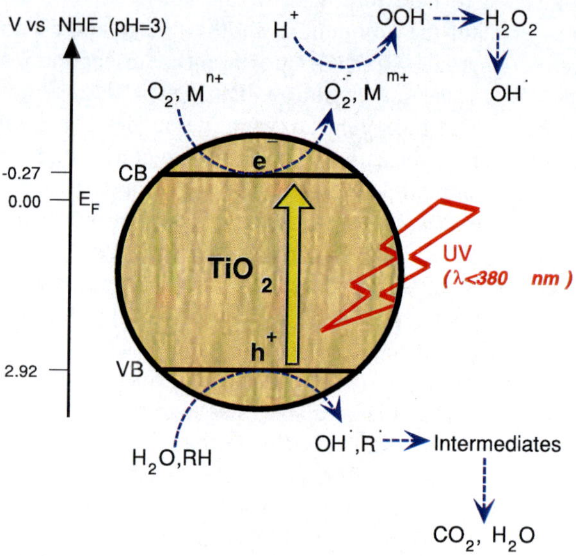

Fig. 1 Mechanism of photocatalysis on the surface of TiO_2 in presence of UV radiation

Charge-carrier trapping:

$$\mathrm{Ti^{IV}OH^-} + h^+_{\mathrm{VB}} \longrightarrow \mathrm{Ti^{IV} - OH^\bullet} \quad (2)$$

$$\mathrm{Ti^{IV}OH^-} + e^-_{\mathrm{CB}} \rightleftharpoons \mathrm{Ti^{III} - OH^-} \quad (3)$$

$$\mathrm{Ti^{IV}} + e^-_{\mathrm{CB}} \longrightarrow \mathrm{Ti^{III}} \quad (4)$$

Electron–hole recombination:

$$\mathrm{Ti^{IV} - OH^\bullet} + e^-_{\mathrm{CB}} \longrightarrow \mathrm{Ti^{IV}OH^-} \quad (5)$$

$$\mathrm{Ti^{III}} + h^+_{\mathrm{VB}} \longrightarrow \mathrm{Ti^{IV}} \quad (6)$$

$$h^+_{\mathrm{VB}} + e^-_{\mathrm{CB}} \longrightarrow \mathrm{heat} \quad (7)$$

Reactions in aqueous medium:

$$\mathrm{Ti^{IV} - H_2O} + h^+_{\mathrm{VB}} \longrightarrow \mathrm{Ti^{IV} - OH^\bullet + H^+} \quad (8)$$

$$\mathrm{Ti^{III} + O_2} \rightleftharpoons \mathrm{Ti^{III} - O_2^{-\bullet}} \quad (9)$$

$$\mathrm{Ti^{IV} - O_2^{-\bullet} + 2H^+} + e^-_{\mathrm{CB}} \rightleftharpoons \mathrm{Ti^{IV} - H_2O_2} \quad (10)$$

$$\mathrm{Ti^{IV} - O_2^{-\bullet} + H^+} \rightleftharpoons \mathrm{Ti^{IV} - HO_2^\bullet} \quad (11)$$

$$\mathrm{Ti^{IV} - 2HO_2^\bullet} \longrightarrow \mathrm{Ti^{IV} - H_2O_2 + O_2} \quad (12)$$

$$\mathrm{Ti^{IV}-H_2O_2+e^-_{CB}\longrightarrow Ti^{IV}-OH^\bullet+OH^-} \quad (13)$$

$$\mathrm{Ti^{IV}-H_2O_2+O_2^{-\bullet}\longrightarrow Ti^{IV}-OH^\bullet+OH^-+O_2} \quad (14)$$

Adsorption–desorption of a reductant (e.g., any organic substrate *S*):

$$\mathrm{Ti^{IV}}+S\rightleftharpoons \mathrm{Ti^{IV}}-S \quad (15)$$

$$\mathrm{Ti^{IV}-OH^\bullet}+S\rightleftharpoons \mathrm{Ti}^{IV}-\mathrm{OH^\bullet}-S \quad (16)$$

Adsorption–desorption of an oxidant (e.g., metal ion)

$$\mathrm{Ti^{IV}}+M^{n+}\rightleftharpoons \mathrm{Ti^{IV}}-M^{n+} \quad (17)$$

Photooxidation of a reductant:

$$\mathrm{Ti^{IV}}-S+\mathrm{Ti^{IV}-OH^\bullet}\longrightarrow \mathrm{Ti^{IV}}+\mathrm{Ti^{IV}}-I\longrightarrow \mathrm{CO_2+H_2O} \quad (18)$$

$$\mathrm{Ti^{IV}}-S+\mathrm{OH^\bullet}\longrightarrow \mathrm{Ti^{IV}}-I\longrightarrow \mathrm{CO_2+H_2O} \quad (19)$$

$$\mathrm{Ti^{IV}-OH^\bullet}-S\longrightarrow \mathrm{Ti^{IV}}-I\longrightarrow \mathrm{CO_2+H_2O} \quad (20)$$

$$\mathrm{Ti^{IV}}-S+\mathrm{Ti^{IV}}(h^+_{VB})\longrightarrow \mathrm{Ti^{IV}}+\mathrm{Ti^{IV}}-I\longrightarrow \mathrm{CO_2+H_2O} \quad (21)$$

Adsorption–desorption of the organic intermediate:

$$\mathrm{Ti^{IV}}+I\rightleftharpoons \mathrm{Ti^{IV}}-I \quad (22)$$

Photoreduction of a metal ion:

$$\mathrm{Ti^{IV}}-M^{n+}+\mathrm{me^-}\longrightarrow \mathrm{Ti^{IV}}-M^{(n-m)+} \quad (23)$$

When TiO_2 or any semiconductor oxide is irradiated with UV light of energy greater than or equal to its bandgap (E_{BG}), valence band holes (h^+_{VB}) and conduction band electrons (e^-_{CB}) are produced (reaction 1). Hoffmann et al. [8] have found by laser flash photolysis studies that the characteristic time scale for this charge-carrier generation reaction is of the order of femtoseconds (fs). The charge carriers generated get trapped on the TiO_2 surface, which occurs over a time scale of tens of nano seconds (ns). Reaction 2 represents the trapping of the holes by the surface hydroxyl groups present in TiO_2, and reactions 3–4 represent the reversible trapping of the electrons on the surface of TiO_2 (shallow traps), and the irreversible trapping or relaxation of the electrons in the bottom of the conduction band (CB) (deep traps), respectively. Ti^{IV} denotes the four coordinated surface functionality of TiO_2 or the "active site". An important reaction that competes with the charge-carrier trapping is the electron–hole recombination reaction. As shown by reactions 5–7, electron–hole recombination can occur either at the surface states of the TiO_2, or in the bulk medium due to the delocalization of the electrons and holes from the surface. This is one of the most detrimental reactions in photocatalysis as this affects the interfacial charge transfer processes and hence the quantum efficiency of the photoprocess. Hence, several techniques have been adopted to modify the semiconductor surface to increase the lifetime of the charge carriers by introducing surface trapping sites by the deposition of noble metals or transition metals. Due to the contact between the semiconductor and the metal, electrons flow from the CB of the semiconductor to the metal, until the Fermi levels of the two materials are aligned. This results in enhanced charge-carrier separation.

In an aqueous medium, valence band (VB) holes can react with the surface adsorbed water molecules to from hydroxyl species (reaction 8), and the trapped CB electrons can react with the dissolved oxygen in the system to form superoxide radicals (reaction 9). These superoxide radicals then undergo a series of reactions 10–14 with the solvated protons and the CB electrons to form hydrogen peroxide, hydroperoxy radicals, hydroxyl anions and hydroxyl radicals. Thus, hydroxyl radicals are generated by both the hole and electron pathways of photocatalysis. When the reactions are carried out in a non-aqueous (organic) medium, the surface bound hydroxyl species present in the semiconductor plays a major role (reaction 2), and the contribution of reactions 8–14 for the overall oxidation of the substrate is negligible.

The next reaction in the sequence is the adsorption of the target material (oxidant or reductant) onto the active sites of the photocatalyst. This is one of the crucial steps in heterogeneous photocatalysis as the reactions take place on the surface of the semiconductor photocatalyst. This is followed by the oxidation of the reductant and the concomitant reduction of the oxidant by the attack of hydroxyl radicals and CB electrons, respectively. The characteristic time scales for the above two processes are of the order of 100 ns and ms, respectively [8, 13]. This shows that the oxidizing powers of VB holes and hydroxyl radicals are always higher than that of the reducing power of the CB electrons. Moreover, these interfacial electron transfer steps compete with the electron–hole recombination reaction (10 ns), and hence, the practical efficiency or quantum yield is always lower than that of the theoretical yield. Reactions 18–21 denote the different possibilities by which the surface bound or solvated hydroxyl radicals and the VB holes can attack and degrade organic materials. Initially, organic intermediates are formed, which, on longer exposure to UV radiation, undergo mineralization to form CO_2 and H_2O. Once the reaction is complete, the unreacted reactants, intermediates and the products desorb from the surface of the catalyst (reactions 15–17, 22). Similarly, when metal ions are present in the system, they are reduced to their thermodynamically stable oxidation states by the CB electrons (reaction 23). This opens up a possibility to reduce the metal ions from their toxic to non-toxic states (e.g., Cr^{6+}–Cr^{3+}) by photocatalysis. Therefore, the overall photocatalysis reaction can be represented by reaction 24, wherein, the oxidants are reduced and the reductants are oxidized by the action of UV radiation on the semiconductor photocatalyst.

$$(\mathrm{Ox}_1)_{\mathrm{ads}} + (\mathrm{Red}_2)_{\mathrm{ads}} \xrightarrow{\mathrm{hv/TiO_2}} \mathrm{Red}_1 + \mathrm{Ox}_2 \quad (24)$$

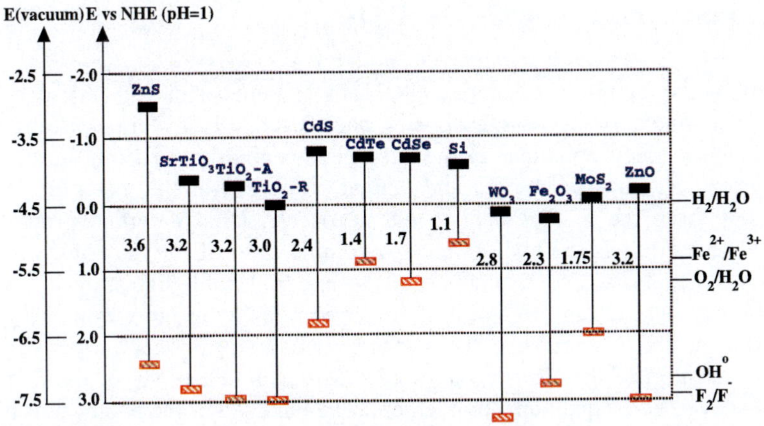

Fig. 2 Band gap energy and band edge positions of different semiconductor oxides and chalcogenides, along with selected redox potentials [12, 17, 22]

3 Semiconductor Photocatalytic Materials

Several metal oxides (TiO_2, ZnO, MoO_3, CeO_2, ZrO_2, WO_3, α-Fe_2O_3, SnO_2, $SrTiO_3$) and metal chalcogenides (ZnS, CdS, CdSe, WS_2, MoS_2) can be used as photocatalysts [10, 12, 17, 18]. However, energetics dictates that, for a semiconductor photocatalyst to be active, the redox potential of photogenerated VB holes must be sufficiently positive to generate hydroxyl radicals and that of CB electrons must be sufficiently negative to generate superoxide radicals. Figure 2 depicts the band structure diagram of the different materials, which have been tested for the photooxidation of organic compounds. It is clear that, TiO_2, ZnO, $SrTiO_3$ and CdS exhibit favorable bandgap positions compared to the other materials. The material selection is also based on the stability of the material toward photocorrosion. For example, ZnO and CdS have only one stable oxidation state (+2), and are prone to decomposition by VB holes. Furthermore, ZnO undergoes incongruous dissolution, yielding $Zn(OH)_2$ on the surface, thereby leading to the deactivation of the material over a period of time [18]. However, Ti in TiO_2 is capable of reversibly changing its oxidation state from +4 to +3; hence, TiO_2 is more favorable compared to the other materials. Between the two common crystal structures of TiO_2, viz., anatase and rutile—anatase phase TiO_2 is widely used in photocatalysis while rutile phase TiO_2 is used in pigments. Although rutile phase TiO_2 possesses a smaller band-gap energy (3.0 eV) compared to anatase phase (3.2 eV), indicating the possibility of the absorption of longer wavelength radiation, the former exhibits a higher photoactivity because of the position of the CB edge, which is more negative. This shows that TiO_2-anatase possesses a higher reduction power compared to that of TiO_2-rutile. In addition to the above advantages, other factors like its non-toxic nature (environmentally benign), low cost and the ease of synthesis makes TiO_2 the "photocatalyst of choice" for photocatalytic degradation reactions.

4 Synthesis and Properties of TiO_2

Different methodologies are being adopted for the synthesis of TiO_2. Many studies have concentrated on the synthesis of “nano-sized” TiO_2 for applications in photocatalysis. The importance of the nanometer regime stems from the fact that the physical, chemical, electrical and optical properties of the materials change immensely from the macroscopic to the nanoscale. One important property of nanoscale materials is the high surface to volume ratio of the material, which is beneficial to catalysis. This results in catalytic materials with high surface area and high porosity, which can promote high reaction rates owing to the high level of interaction of the reactants on the surface. Another important property is the efficient transport of the electrons and holes across the band gap of the material, which is governed by quantum confinement. The bandgap of the photocatalyst also dictates the wavelength at which the material can be excited. Hence, lower bandgap energies are beneficial for visible light absorption.

TiO_2 photocatalysts have been synthesized in different shapes and morphologies, which include, nanoparticles, nanorods, nanotubes, nanopillars and nanowire arrays, nanobowls, nanowhiskers, aerogels, nanosheets, opal and inversed opals [12, 14]. The synthesis routes can be broadly classified as solution phase and gas phase techniques. Solution phase synthesis is the most preferred technique for the preparation of TiO_2 in the form of powders and thin films. Some of the solution phase techniques are the precipitation method, hydrothermal synthesis, solvothermal synthesis, sol–gel method, sol method (nonhydrolytic sol–gel), micelle and inverse micelle method, combustion synthesis, electrochemical synthesis, sonochemical synthesis and microwave synthesis methodologies. Gas phase technique is widely employed for the synthesis of thin film samples. Some of the gas phase techniques are chemical vapor deposition (CVD), physical vapor deposition and spray pyrolysis deposition. Each of the above synthesis methods possesses a unique advantage over the other techniques, and the characteristics of the final product vary from one method to another. The review article by Chen and Mao [14] discusses each of the above synthesis techniques in detail, with particular emphasis on the morphology, size, shape and properties of the final TiO_2 product obtained in each method. Table 1 [23–30] presents a representative list of the different techniques adopted for the synthesis of TiO_2. These works elucidate that the final properties of TiO_2 nanoparticles like the phase composition (anatase : rutile), particle size, porosity, surface area, band gap and surface hydroxyl content can be tailored by varying the reaction conditions like: the precursor compound ($TiCl_4$, titanium(IV) isopropoxide, etc.), fuel (glycine, oxalyl dihydrazide, hexamethylene tetramine in the case of solution combustion synthesis), hydrolyzing agent (in the case of sol–gel synthesis), molar composition of the reactants, reaction temperature, reaction time (ageing time), calcination temperature and presence of gas atmosphere (air, Ar, NH_3).

One of the synthesis methodologies that has been proven to yield nano-TiO_2 with enhanced characteristics is the solution combustion method (CS TiO_2),

Table 1 Physicochemical properties of nano-TiO_2 by various synthesis techniques

Sl. No.	Synthesis method	Highlights/properties of TiO_2	Reference
1	Thermal decomposition of ammonium titanyl sulfate	Properties of TiO_2 tuned by varying the gas atmosphere and reaction temperature; mesoporous texture with mean pore dia. = 15 nm; particle size = 20–30 nm; surface area c.a. 64 $m^2\ g^{-1}$	[23]
2	Precipitation of a mixture of $Ti(OPr)_4$, stearic acid and 1-propanol, followed by calcination at different temperatures	Pore size tuned by adjusting the molar composition of stearic acid; pore dia. = 5–15 nm; surface area = 92–130 $m^2\ g^{-1}$	[24]
3	Four different synthesis routes involving the hydrolysis of $Ti(OPr)_4$ or $TiCl_4$ followed by calcination at different temperatures	Samples made from $TiCl_4$ exhibited the highest photoactivity; 100% anatase phase TiO_2 was obtained with crystallite size = 7–30 nm; surface area = 100 $m^2\ g^{-1}$; pore size = 7–14 nm	[25]
4	Combustion of aqueous titanyl nitrate with stoichiometric amounts of glycine at 350°C; precursor—$Ti(OPr)_4$	100% anatase phase TiO_2 was obtained; particle size = 8 ± 2 nm; band gap = 2.21 and 2.85 eV; surface area = 240 $m^2\ g^{-1}$; TGA wt. loss = 15.5%; high surface acidity	[26]
5	Hydrothermal synthesis using $TiCl_4$ using cationic surfactants like CTAB and CPB	Crystallite size = 10–18 nm; morphology change—nano-spheres to cotton fibrils; surface area = 240–418 $m^2\ g^{-1}$; pore size = 2–4 nm; mixture of anatase and rutile phase was obtained	[27]
6	Sol–gel synthesis using $Ti(OPr)_4$	Max. surface area = 125 $m^2\ g^{-1}$; min. crystallite size = 6 nm; band gap = 3.2±0.1 eV; mixture of anatase and rutile phase was obtained	[28]
7	One-pot hydrothermal synthesis using tetramethylammonium hydroxide (TMAOH)	Anatase phase TiO_2 nano-pillar arrays; c.a. 250 nm in width and c.a. 700 nm in length with a tetrahedral bipyramidal tip; optimum conditions—1M TMAOH, 200°C, 8 h	[29]
8	Combustion of aqueous titanyl nitrate with stoichiometric amounts of glycine at 350°C; precursor – $Ti(OBu)_4$	100% anatase phase TiO_2 was obtained; crystallite size = 4–6 nm; surface area = 257 $m^2\ g^{-1}$; band gap = 2.92 eV	[30]

$Ti(OPr)_4$—titanium(IV)isopropoxide; $Ti(OPr)_4$—titanium(IV)isobutoxide; CTAB—cetyltrimethylammonium bromide; CPB—cetylpyridinium bromide

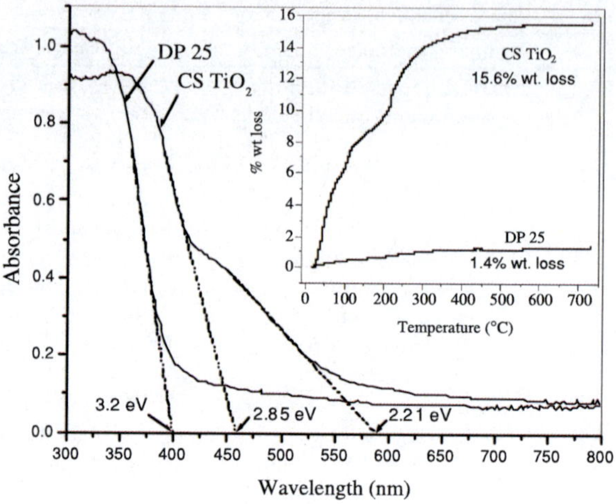

Fig. 3 UV/visible absorption spectra of CS TiO_2 and DP-25. The band gap values corresponding to the absorption threshold are shown. Inset: TGA thermograms [26]

developed by Nagaveni et al. [26]. CS TiO_2 exists in anatase phase, possesses a smaller particle size, higher surface area, lower band gap, higher amount of surface hydroxyl species and a highly acidic surface compared to the commercial P-25 TiO_2 (DP-25) produced by Degussa corporation. Figure 3 shows the UV/visible spectra and thermogravimetric analysis of CS TiO_2 and DP-25. It is clear that CS TiO_2 exhibits strong absorption in the visible region due to the carbide ion substitution for oxide ion, of the form $TiO_{2-2x}C_xV_x$, where 'V' denotes the oxide ion vacancy. Moreover, CS TiO_2 exhibits a higher weight loss compared to DP-25, indicating the presence of higher amount of surface hydroxyl groups. Table 2 [26, 31–35] compares the characteristics of CS TiO_2 with the other commercially available TiO_2 samples. It is evident that the activity of the photocatalyst cannot be standardized based on any single property, but by striking a balance between the different physicochemical properties. In a series of publications it was shown that the photoactivity of CS TiO_2 toward the photooxidation of different class of dyes [36–38], phenolic compounds like chlorophenols, nitrophenols and multiple substituted phenols [39, 40], substituted nitrobenzenes [41], water soluble polymers and plastics [42, 43] and the reduction of metal ions [44], is much higher compared to that of DP-25. More detailed discussions on the degradation of the organic compounds are available in later sections.

5 UV Photocatalytic Degradation of Organic Compounds

Chlorinated organic compounds like chloroalkanes were the first organic compounds to be degraded by photocatalysis [3, 4, 45]. Besides being widely used as solvents, pesticides, refrigerants and in plasticizers and plastics, these organochlorine compounds are toxic and accumulate in the environment. Hsiao et al. [3], and Pruden and Ollis [4] demonstrated complete mineralization of dichloromethane

Table 2 Comparison of the physicochemical properties of combustion synthesized TiO_2 (CS TiO_2) with commercially available TiO_2 samples

TiO_2	BET surface area, $m^2\ g^{-1}$	Crystallite size, nm	pH_{zpc}	Surface hydroxyl content	Reference
CS TiO_2	150	8±2 nm	2.4	15.6% total wt. loss (TGA)	[26]
DP-25 (80% A: 20% R)	50	A—37; R—90	6.3	1.4% (TGA); 163 (FTIR)	[26, 31]
Hombicat UV 100	290	5	6.0	843 (FT-IR)	[32, 33]
Junsei	9.7	34	4.4	52 (FT-IR)	[32]
Aldrich Anatase	9.2	37	4.2	54 (FT-IR)	[32]
Millenium PC 10	11	75	–	–	[34]
Millenium PC 50	50	25	–	–	[34]
Millenium PC 100	89.6	21	5.9	505 (FT-IR)	[32, 34]
Millenium PC 500	287	5–10	6.2	–	[34]
Tronox A-K-1	90	20	–	–	[35]
Ishihara ST-01	340	11	5.8	719 (FT-IR)	[32]

A—anatase; R—rutile; unless otherwise mentioned all the TiO_2 samples are 100% A

(CH_2Cl_2), chloroform ($CHCl_3$), carbon tetrachloride (CCl_4) and trichloroethylene in the presence of TiO_2, to HCl and CO_2. They observed that the chloride ions generated during the reaction act as inhibitors for the conversion of chloroalkanes. A Langmuir–Hinshelwood kinetic model was proposed to find the adsorption and degradation rate coefficients of the chloroalkanes and the chloride ions. The relative order of degradation of the chloromethanes is: $CHCl_3 > CH_2Cl_2 > CCl_4$.

A general stoichiometric reaction for the photocatalytic degradation of any organochlorine compound can be written as [8]:

$$C_xH_yCl_z + \left(x + \frac{y-z}{4}\right)O_2 \xrightarrow{hv,TiO_2} xCO_2 + zHCl + \left(\frac{y-z}{2}\right)H_2O \quad (25)$$

Many studies have been devoted to the photocatalytic degradation of chlorinated aromatic compounds like chlorophenols and chlorobenzenes, chlorinated pesticides like DDT, hexachlorobenzene, atrazine and parathion, surfactants like sodium dodecyl benzene sulfonate and trimethyl phosphate, aliphatic and olefinic compounds, dyes, nitrogeneous compounds like nitrophenols and nitrobenzenes, carboxylic acids, alcohols and heteroatom compounds. The review articles by Hoffmann et al. [8], Mills and Hunte [10], Legrini et al. [15], Bhatkhande et al. [17] and Blake [46] have discussed the different studies on the photocatalytic degradation of different classes of organic compounds.

Although many organic compounds have been shown to photocatalytically degrade in presence of TiO_2, the mechanism of degradation of phenolic compounds like phenol, chlorophenols, nitrophenols and other substituted phenols, which are used as intermediates in the manufacturing of pesticides and herbicides,

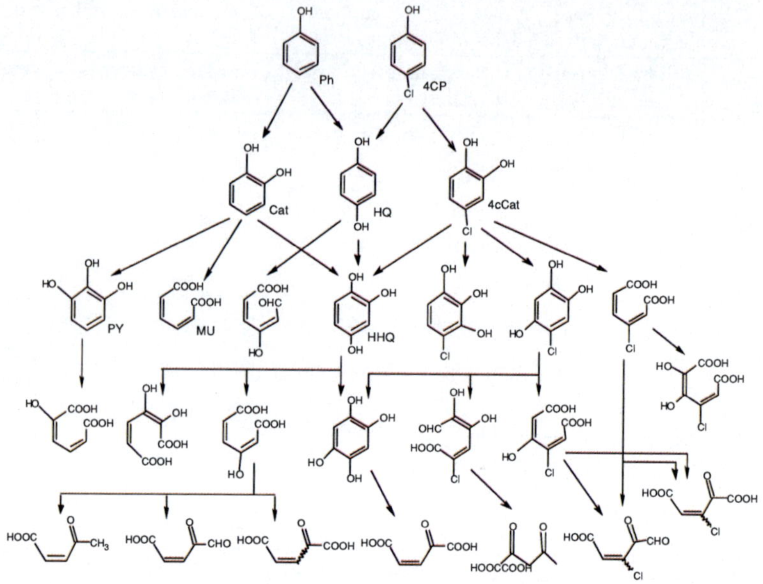

Fig. 4 Detailed pathway of degradation of phenol and 4-chlorophenol adapted from different studies. Some of the primary and secondary intermediates are identified as Ph—phenol, 4CP—4-chlorophenol, Cat—catechol, HQ—hydroquinone, 4cCat—4-chlorocatechol, PY—pyrogallol, MU—muconic acid and HHQ—hydroxyl hydroquinone [39, 47]

is well understood and documented. The degradation proceeds primarily by the attack of the hydroxyl radicals, which results in the hydroxylation of the parent compound at the ortho and para-positions, due to the ortho and para-directing nature of the phenolic –OH moiety. These are the primary intermediates, which, upon further exposure, results in the formation of a fully hydroxylated secondary intermediate species. The next step is the fragmentation of the benzene ring to form C-6 and C-5 aliphatic carboxylic acids and aldehydes.

In the case of chloro or nitro-substituted phenols, the hydroxyl group replaces the substituent group before the ring fragmentation. The longer (C-6, C-5, C-4) chain organic acids and aldehydes on longer exposure periods yield C-3, C-2 and C-1 organic acids. Finally, these shorter chain compounds mineralize to form CO_2 and H_2O. Li et al. [47] elucidated the mechanism of formation of the intermediates during the photocatalytic degradation of 4-chlorophenol and 4-chlorocatechol. Figure 4 shows the degradation pathway of phenol and 4-chlorophenol, adapted from different studies [39, 40, 47]. The photocatalytic degradation of phenol and 4-chlorophenol yields catechol, hydroquinone and chlorocatechol as the primary hydroxylated intermediates. Further hydroxylation steps result in the formation of pyrogallol, hydroxy hydroquinone and 4-chloro dihydroxy phenols. This step is followed by the fragmentation of the dihydroxy phenols to form muconic acid and muconic aldehydes with chloro and hydroxy-substitutions. A detailed mechanism of scission of the benzene ring is depicted in Figure 5. Muconic acid further

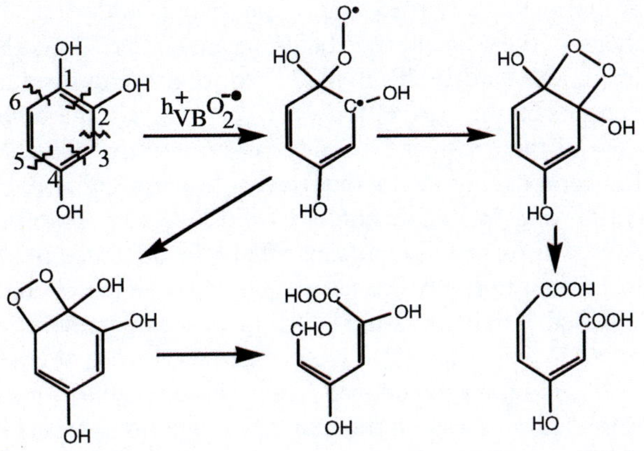

Fig. 5 Mechanism of cleavage of benzene ring in hydroxy hydroquinone, resulting in the formation of carboxylic acids and aldehydes. The possible sites of ring cleavage are represented by broken lines [47]

degrades to smaller organic acids like oxalic acid, maleic acid, succinic acid, fumaric acid, acrylic acid, etc. Finally, these acids mineralize to CO_2 and H_2O after long exposure periods.

Sivalingam et al. [39] studied the effects of different substituents like chloro group, nitro group and methyl group on the kinetics of photocatalytic degradation of phenol, and concluded that the order of degradation is: pentachlorophenol > trichlorophenol > dichlorophenol > 4-chlorophenol ≈ 2-chlorophenol > 2-methyl phenol ≈ 3-methyl phenol > phenol. Similarly, Priya and Madras [40] conducted a thorough study on the effect of multiple substitutions of the above groups on the degradation of phenol. Their results indicate that chloro-methylphenols (cresols) degraded much faster compared to chloro-nitrophenols, which was attributed to the ring deactivating nature of the nitro group for the hydroxyl radicals to react with the phenolic compound. Another interesting conclusion of their study shows that the degradation is independent of the position of the substituents, but depends on the nature of the substituent group. In another study [41], it was found that chloro and hydroxy-substitution in nitrobenzene accelerated the degradation rate, while nitro-substitution resulted in a reduced degradation rate of nitrobenzene.

Other important organic compounds, which have attracted interest in the past decade, include dyes, pesticides and pharmaceutical compounds. The forthcoming sections will examine the various studies on the photocatalytic degradation of these pollutants.

5.1 Dyes

Dyes constitute a major class of organic compounds, which find a multitude of applications in our daily life in clothing, food, paper, leather, cosmetics, plastics, drugs, electronics and printing. The modern textile industry consumes about 80% of all synthetic dyes produced, with 700,000 tons of dyes applied to 40 million tons of fabric annually [48]. One of the major bottlenecks in the textile industry is dye fixation (i.e., spent dye baths, residual dye liquors and water from washing operations contain dye in the hydrolyzed and unfixed form). Although dye fixation depends on the class of the dye, type of fabric and other dyeing parameters, nearly 10% of the dyes used are discharged into the effluent as a result of this process. Hence, reducing the toxicity levels to permissible limits before admitting the dye waste water to aquatic bodies is a critical issue. In this regard, photocatalysis, as an AOP, has played a significant role in the degradation of textile dye effluents. Many reviews exclusively discuss the degradation of different classes of dyes under UV and visible light exposure [48–50], various parameters affecting the degradation of dyes [51], and modified TiO_2 materials for the degradation of dyes [52].

Based on the functional group that constitutes the dye molecules, dyes can be classified as azoic, anthraquinonic, heteropolyaromatic, aryl methane, xanthene, indigo, acridine, nitro, nitroso, cyanine and stilbene. Among all the above dye functionalities, photodegradation of azo dyes has been investigated extensively [50], as these dyes contribute 50% of the commercial dyes. The degradation of a dye can be characterized in two ways: percent decolorization and percent mineralization. Decolorization refers to the reduction in concentration of the parent dye molecule under consideration at its characteristic wavelength, but does not refer to the complete removal of the organic carbon content. This is due to the formation of colored dye intermediates, which might absorb at a different wavelength. Hence, complete degradation or mineralization occurs when all the organic carbon is converted to CO_2. Therefore, analyzing the mineralization of the dyes in terms of the total organic carbon (TOC) content assumes importance. The mechanism of CO_2 evolution from an organic dye intermediate is well-documented and it follows the photo-Kolbe decarboxylation mechanism [53]. The reaction is given by:

$$\Phi - COO^- + h^+_{VB} \longrightarrow \Phi^\bullet + CO_2 \tag{26}$$

In Eq. 26, Φ denotes the organic component of the dye or the dye intermediate. The radical $\Phi^\bullet$ formed as a result of this reaction can undergo further transformation to yield other intermediates with smaller sizes.

Epling and Lin [54] degraded 15 dyes belonging to different classes of functionality in the presence of visible light. Their results show that the presence of electron withdrawing groups retards the degradation of the dye. The order of degradation among the different dyes follows this order: indigo $\approx$ phenanthrene $>$ triphenyl methane $>$ azo $\approx$ quinoline $>$ xanthenes $\approx$ thiazine $>$ anthraquinone. Moreover, the order of degradation of the dyes in the presence of different light sources follows: natural sunlight $\gg$90 W halogen flood light $>$150 W spotlight. They have attributed the degradation of the dye to both photosensitized oxidation and reduction mechanisms.

Vinu et al. [38] studied the degradation of 5 anionic, 8 cationic and 3 solvent dyes, each belonging to a different class by molecular structure, with CS TiO_2 and DP-25 in the presence of UV irradiation. It was observed that the order of degradation of the dyes with CS TiO_2 and DP-25 was different, and this was correlated to the surface charge of the catalysts. By monitoring the shift in the characteristic wavelength of the dyes during degradation, it was concluded that the formation and consumption rates of the intermediates were high in the presence of highly acidic surface of CS TiO_2, while significant amounts of intermediates were observed in the presence of DP-25. Overall, anionic dyes exhibited higher decolorization rates compared to the cationic dyes, and the solvent dyes exhibited adsorption dependent degradation.

Silva et al. [55] studied the degradation of mono, di and tri-azo dyes and found that the decolorization followed the order: Solophenyl Green BLE (tri-azo) > Erionyl Red B (di-azo) > Chromotrop 2R (mono-azo). However, the mineralization of the dyes followed the opposite trend. All the above studies suggest that a correlation does not exist between the degradation rate of the dye with the dye structure or the functional groups that characterize the dye. Therefore, it is important to analyze the pathway of degradation of the dyes in order to assess the stability of the different intermediates.

A representative list of the studies devoted to the analysis of the organic intermediates during the degradation of the dyes is presented in Table 3 [53, 56–63]. Rajeshwar et al. [48] provided an extensive list of the studies on the degradation pathway of azo and non-azo dyes. A generalized conclusion is that alkyl substituted (methyl or ethyl) dyes like triphenyl methane or rhodamine dyes degrade by the N-de-alkylation mechanism, which involves the abstraction of the alkyl groups by the hydroxyl radicals resulting in the formation of the corresponding aldehyde as a side product. Once complete dealkylation occurs, fragmentation of the molecule occurs, which results in the formation of different substituted benzene compounds. At long exposure periods, the cleavage of benzene ring results in the formation of low molecular weight organic acids. The degradation of azo dyes involves the cleavage in the vicinity of the azo bond and results in the formation of naphthol and benzoic acid intermediates. Likewise, the degradation pathway of anthraquinone dyes result in the formation of phthalic acid intermediates.

The dye molecules also contain other hetero atoms like N, S, Br and Cl in their structure, and hence, an assessment of the fate of these elements during mineralization is also critical. It has been well established that sulfonate groups ($-SO_3^-$) in anionic dyes are transformed to innocuous sulfate ions (SO_4^{2-}), primary ($-NH_2$) and secondary (–NH–) amino groups are converted into ammonium ions (NH_4^+), azo nitrogen (–N=N–) is converted to N_2 and halogen atoms are released into the solution as their respective anions. The following reactions describe the mechanism by which the above transformations take place [53]:

Formation of sulfate ions:

$$\Phi - SO_3^- + OH^\bullet \rightarrow \Phi - OH + SO_3^{\bullet -} \tag{27}$$

$$SO_3^{\bullet -} + OH^- \rightarrow H^\bullet + SO_4^{2-} \tag{28}$$

Table 3 Survey of the organic intermediates formed during the photocatalytic degradation of different classes of dyes

Sl. No.	Dye/dye category	Organic intermediates detected	Reference
1	Acid Orange 7/mono-azo	22 intermediates were detected; coumarin, 2-maphthol, naphthoquinone derivatives, phthalic acid, phthalic acid, phthalimide, benzoic acid derivatives, phenol, succinic acid, maleic acid, malonic acid, etc.	[56]
2	Metanil Yellow/Mono-azo	Benzenesulfonic acid, hydroxylated diphenyl amine, diphenyl amine, benzene, aniline, phenol, hydroquinone	[57]
3	Procion Red MX-5B/ Triazine+mono-azo	12 intermediates were detected; *p*-hydroxy-phenyl-3-hydroxy propanedioic acid, 2-hydroxy benzoic acid, *p*-hydroxy cinnamic acid, phthalic acid, malic acid, oxalic acid, etc.	[58]
4	Remazol Brilliant Blue R/Anthraquinonic	Amino and hydroxyl substituted anthraquinone, phthalic acid, amino substituted phthalic acid	[59]
5	Indigo and Indigo Carmine/Indigo	2-Nitro benzaldehyde, 2,3-dihydroxy indoline, anthranilic acid, tartaric acid, malic acid, amino-fumaric acid, pyruvic acid, malonic acid, glycolic acid, oxalic acid, acrylic acid, acetic acid, etc.	[60]
6	Methylene Blue/ Heteropolyaromatic	(3-dimethyl amino) aniline, benzene sulfonic acid, phenol, hydroxylation of amino and sulfoxide groups was observed	[53]
7	Rhodamine B/Xanthene fluorene	Mono-, di-, tri- and tetra-*N*-de-ethylated rhodamine species, rhodamine 110	[61, 62]
8	Methyl Green/Triphenyl methane	33 intermediates were identified and characterized; major intermediates—colorless carbinol base and crystal violet dye; degradation proceeded by *N*-de-methylation and *N*-de-alkylation pathway; methylated and ethylated derivatives of aminophenol and benzophenone were observed	[63]

Formation of ammonium ion:

$$\Phi - NH - \Phi' + H^{\bullet} \rightarrow \Phi^{\bullet} + \Phi' - NH_2 \tag{29}$$

$$\Phi' - NH_2 + H^{\bullet} \rightarrow \Phi'^{\bullet} + NH_3 \tag{30}$$

$$NH_3 + H^{+} \rightarrow NH_4^{+} \tag{31}$$

Formation of N_2:

$$\Phi - N = N - \Phi' + H^{\bullet} \rightarrow \Phi - N = N^{\bullet} + \Phi' - H \tag{32}$$

$$\Phi - N = N^{\bullet} \rightarrow \Phi^{\bullet} + N \equiv N \tag{33}$$

Formation of chloride/bromide ions:

$$\Phi - Cl/Br + OH^{\bullet} \rightarrow \Phi - OH + Cl^{-}/Br^{-} \quad (34)$$

5.2 *Pesticides*

Pesticides encompass a class of organic compounds used for preventing, destroying or controlling pests. These are widely applied on agricultural crops to protect them from insects, pathogens, weeds, moths, microbes, etc. However, the continuous usage of pesticides results in poor soil quality, reduced nitrogen fixation for plants and poisoned food sources. Also serious is the transfer of these compounds to water bodies by leaching, percolation, spillage or runoff, which poses ill-effects to human health and other forms of life. Based on the chemical structure, pesticides can be classified as organochlorine compounds, organophosphorous compounds and carbamate compounds, which contain nitrogen, phosphorous, sulfur, chlorine and heterocyclic nitrogen atoms. Therefore, the mineralization of pesticides should result in the conversion of these hazardous compounds to innocuous products of N, S and P, along with CO_2 and H_2O.

A variety of pesticides like atrazine (s-triazine) [64], propanil (acetanilide) [64], molinate (thiocarbamate) [64], pyridaben [65], methamidophos [66], methyl parathion [67], cyproconazole [68], hexachloro benzene [69], dicofol [69], pyrimethanil (fungicide) [70], primicarb (insecticide) [70], propyzamide (herbicide) [70], etc., have been photocatalytically degraded using TiO_2. Many studies have proposed a detailed pathway of degradation of the pesticides by monitoring the formation of intermediates by HPLC/MS/MS and GC/MS techniques. It was found that, during mineralization, phosphorous is converted to orthophosphoric acid (H_3PO_4), nitrogen to NO_3^- and NH_4^+ ions and sulphur to SO_4^{2-} ions. Figure 6 depicts the pathway of degradation of triazophos, an organophosphorous pesticide [71].

5.3 *Pharmaceutical Compounds*

Recently, photocatalytic degradation of pharmaceutical compounds and drugs has received attention, as these compounds are found in the effluents in levels ranging from ng L^{-1}–μg L^{-1}. These compounds are admitted into waste water and aquatic bodies by manufacturing operations, spillage, human and animal excretion and hospital wastes. Besides leading to serious hazards like genotoxicity, endocrine disruption and aquatic toxicity, these xenobiotic substances increase the resistance of the pathogenic bacteria in waste water. These compounds are resistant toward biological degradation and other common oxidation techniques, and thus, build up their concentration in the ecosystem. In fact, clofibric acid, which is used as a

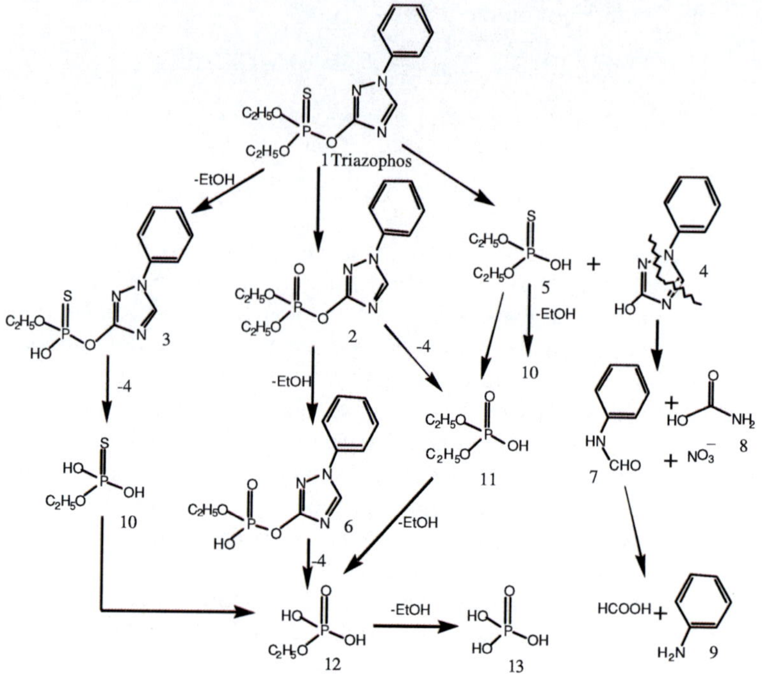

Fig. 6 Pathway of degradation of the pesticide, triazophos, established using LC/MS/MS and GC/MS/MS [71]

regulator of lipids in the blood, is still detected in lakes and rivers, even after 21 years of persistence in the environment [72]. Therefore, photocatalysis has emerged as an alternative for the effective degradation and mineralization of these persistent organic pollutants.

The different pharmaceutical compounds that were subjected to photocatalytic degradation using TiO_2 are listed in Table 4 [72–79]. The degradation of a wide class of compounds has been studied, and many studies have optimized the reaction parameters like pH, O_2 concentration and TiO_2 loading for maximum conversion of the parent drug compound. Most of these studies have also established the pathway of degradation of the pharmaceutical compounds, by following the evolution of the intermediates via mass spectroscopy. A generalized conclusion in each of the above studies is that the time taken for the complete mineralization of the pharmaceutical compound was longer compared to the removal of the parent compound. Calza et al. [80] studied the photocatalytic degradation and ecotoxicity of amiloride, a diuretic agent, and concluded that although amiloride is non-toxic, the degradation intermediates were toxic compounds. Similar results were also observed for salbutamol [79], indicating the importance of complete mineralization of the drug to CO_2, H_2O and inorganic ions. Figure 7 depicts the pathway of degradation of diclofenac, a non-steroidal anti-inflammatory drug [78].

Table 4 Listing of recent studies on the photocatalytic degradation of different pharmaceutical compounds

Sl. No.	Compound class	Pharmaceutical compounds
1	Antibiotic	Chloramphenicol [73], sulfamethoxazole [74], ciprofloxacin [75], ofloxacin [72]
2	Analgesic and antipyretic	Paracetamol [76], phenazone [72]
3	Non-steroidal anti-inflammatory drug (NSAID)	Diclofenac [77, 78], naproxen [77], ibuprofen [77]
4	Beta blocker	Atenolol [76]
5	Anti convulsant	Carbamzepine [72]
6	Anti cholestric	Clofibric acid [72]
7	Histamine H_2—receptor antagonist	Ranitidine hydrochloride [72]
8	Loop diuretic	Furosemide [72]
9	Beta 2-agonist	Salbutamol [79]

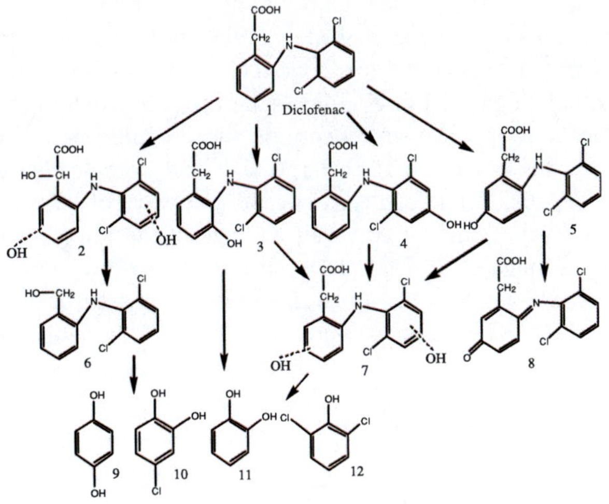

Fig. 7 Degradation pathway of the non-steroidal anti-inflammatory drug, diclofenac, established using ESI–MS. The dotted lines in the compounds 2 and 7 shows that the hydroxyl radicals can attack either of the benzene rings to form the intermediate [78]

6 Photocatalytic Destruction of Microorganisms

Microorganisms like pathogenic bacteria, fungi, viruses and protozoa in drinking water and air pose a severe threat to human and animal life; they are the primary carriers of infectious diseases like gastroenteritis, cholera, tuberculosis, anthrax, malaria and yellow fever. Conventional disinfection technologies employ chlorine, chlorine dioxide (ClO_2), hypochlorite (OCl^-) or ozone as disinfectants, owing to their low operating cost and applicability in a wide pH range. However, the byproducts of the above techniques involve the formation of halomethanes and halo acetic acids, which are carcinogenic even at low concentrations. Hence, photolytic (UV-C radiation at 254 nm) and photocatalytic disinfection

(UV-A at 365 nm or visible radiation) of microorganisms, using TiO_2 as the photocatalyst has been researched extensively for the past two decades. Table 5 [81–90] shows the different studies devoted to the deactivation of various kinds of microorganisms using different semiconductor materials. Among the different microorganisms, *Escherichia coli* (*E. coli*), a Gram negative rod shaped bacterium, owing to its presence in the intestines of humans, animals and birds, serves as a biological indicator to test environmental samples for fecal contamination. Unlike organic compounds, the concentration of microorganisms is usually expressed in terms of the number of colony forming units per mL of the solution (cfu mL^{-1}). CFU refers to the number of viable microbial colonies, which are measured by counting the cell aggregates on an agar plate.

Cho et al. [91] evaluated the inactivation of *E. coli* under different light intensities, TiO_2 concentration and pH, and have observed a linear correlation of the inactivation time (for 2-log reduction) with the hydroxyl radical concentration in the system. This study suggests that hydroxyl radicals are more important than the reactive oxygen species, and that they are 1000–10000 times as effective as common oxidants like chlorine, chlorine dioxide or ozone for the disinfection of *E. coli*. The mechanism of photocatalytic disinfection of microorganisms is different from that of the degradation and mineralization of organic compounds, although the hydroxyl radicals are the precursors of degradation in both the cases. The primary reason concerning the complex nature of inactivation of microorganisms is due to their large size and complex structure compared to organic molecules. This results in several pathways of cell rupture and regeneration.

Many modes of cell wall damage of the microorganisms have been proposed. Matsunaga et al. [92] observed a decrease in Coenzyme A (CoA) content in the cell and a concomitant increase in concentration of the dimeric form of CoA, when *E. coli* was irradiated in presence of TiO_2. Since the electron transfer between the cell and the TiO_2 is mediated by CoA, the reduction in its concentration was attributed to cell damage. Saito et al. [93] observed a "rapid" leakage of potassium ions and the "slow" release of protein and RNA during the lysis of *Streptococcus sobrinus*, in presence of UV/TiO_2. Thus, they concluded that the inactivation is due to the loss of permeability of the cell wall of the bacteria. Maness et al. [94] have observed an exponential increase in the concentration of malondialdehyde during the inactivation of *E. coli*, suggesting that the active species like hydroxyl, hydroperoxy and superoxide radicals attack the polyunsaturated phospholipids in *E. coli*, which results in lipid peroxidation and hence the loss of respiratory activity of the cell.

Thus, it is unequivocally accepted that the cell wall damage induced by the active species in presence of UV/TiO_2 is the primary cause for the death of the microorganisms. However, it must be ensured that complete disinfection (99.999%) takes place and microorganisms are indeed in their inactive state, even after the irradiation and/or TiO_2 is removed from the system. This is because, the presence of active bacteria may result in their reproduction, as organic compounds like aldehydes, ketones and carboxylic acids, which are usually the products of cell damage, serve as nutrients for further growth. It has widely been observed that the

Table 5 Photocatalytic disinfection of microorganisms using TiO_2 in presence of UV and visible light radiation

Sl. No.	Microorganism	Concentration	Catalyst	Loading	Light source	Reference
	UV radiation					
1	*Legionella pneumophila*	10^7 cfu mL^{-1}	TiO_2	1 g L^{-1}	$\lambda = 365$ nm; $I = 108$ μW cm^{-2}	[81]
2	*E. coli, Pseudomonas aeruginosa, Salmonella typhimurium, Enterobacter cloacae*	10^6–10^7 cfu mL^{-1}	DP-25 TiO_2	0.1 g L^{-1}	$\lambda = 365$ nm; $I = 5.5,1.4$ mW cm^{-2}	[82]
3	*Bacillus anthracis*	700–1900 cfu mL^{-1}	TiO_2	10–100 mg	$\lambda = 320$–400 nm; $I = 0.42$–2.43 mW cm^{-2}	[83]
4	*E. coli, Lactobacillus helveticus*	10^8 cfu m L^{-1}	TiO_2 and ZnO	1,2 g L^{-1}	$\lambda = 365$ nm; $I = 20$ W m^{-2}	[84]
5	*E. coli*	10^9 cfu mL^{-1}	Ag impregnated TiO_2	2 g L^{-1}	$\lambda = 280$ nm; $I = 0.5$ W m^{-2}	[85]
6	*E. coli, Staphylococcus aureus*	5×10^7 cfu mL^{-1}	AgI/TiO_2	0.2 g L^{-1}	350 W XL; $I = 2.8$ mW cm^{-2}	[86]
7	*E. coli*	1 – 4×10^8 cfu mL^{-1}	Montmorillonite supported Ag/TiO_2	0.4 g L^{-1}	40 W FL	[87]
8	*E. coli, Pseudomonas aeruginosa, Staphylococcus aureus*	10^7 cfu mL^{-1}	PdO and N doped TiO_2	1 g L^{-1}	$I = 0.4$–1.6 mW cm^{-2}	[88]
9	*Micrococcus lylae*	3×10^7 cfu mL^{-1}	S doped TiO_2	0.2 g L^{-1}	100 W WHL; $I = 47$ mW cm^{-2}	[89]
10	*Microcystin-LR*	1 mg L^{-1}	N–F-codoped TiO_2		Two 15 W FL; $I = 78$ μW cm^{-2}	[90]

XL—xenon arc lamp; FL—fluorescent lamp; WHL—tungsten halogen lamp inactivation rate of microorganisms in pure, deionized water is higher compared to that in the presence of tap water or water contaminated with ionic species or humic substances [95]. This is due to the osmotic stress in deionized water, which results in the easy leakage of potassium, calcium and magnesium ions through the cell membrane, thereby leading to the loss of permeability. Moreover, humic substances compete for the reactive radicals, which results in the inhibition of the disinfection rate.

From Table 5, it is clear that TiO_2 and its modified forms have shown superior photocatalytic activity for the disinfection of a wide class of Gram-positive and Gram-negative bacteria and other microorganisms in water, both in presence of UV and visible light irradiation. The initial concentration of the microorganisms that is widely used in the recent works corresponds to 10^3–10^9 cfu mL^{-1}. It is also interesting to note that the optimum catalyst concentration varies widely across the different studies, suggesting the strong dependence on the intensity of irradiation, with high intensities requiring lesser catalyst loading. Pal et al. [96] recently evaluated the disinfection of air borne *E. coli* in a continuous annular reactor in presence of TiO_2. Complete inactivation of *E. coli* was observed using UV-A irradiation of 3.4 mW cm^{-2} intensity, 85% relative humidity and 1516 mg m^{-2} of TiO_2, within a residence time of 1.1 min. Recently, Ag is incorporated into TiO_2, either on the surface or in the lattice, to enhance the biocidal activity of bare TiO_2 [85–87]. The advantages of having Ag as a component in the photocatalyst are three fold: (i) Ag^+ ions possess an inherent antimicrobial activity, which complements the disinfection ability of TiO_2, (ii) Ag, when impregnated on the surface of TiO_2, acts as electron trapping sites, which retards the unwanted charge-carrier recombination, (iii) nano-sized Ag promotes the formation of electrons in presence of visible radiation, due to the surface-plasmon excitation. These electrons are transferred to the conduction band of TiO_2, thereby leading to the formation of reactive superoxide radicals. However, it has been observed in a recent study that an optimum loading of 1% Ag onto TiO_2 yielded a maximum rate of disinfection of *E. coli* [85]. The synthesis of Ag nanoparticles and their incorporation onto different substrates by different techniques, and the anti-microbial activity are discussed elsewhere [97].

7 Effect of Reaction Conditions

7.1 Temperature

Minor incremental changes in temperature do not significantly affect the rate of photocatalytic reactions. However, higher temperatures might result in the reduction of dissolved oxygen in the solution. Dissolved oxygen is one of the key elements in photocatalysis, as it aids in scavenging CB electrons, thereby leading to the formation of hydroxyl radicals by the electron pathway. Moreover, higher temperatures result in desorption of the organic compounds from the surface of the photocatalyst. Because photocatalytic degradation is a surface mediated phenomenon, desorption of the organic compound before the reaction results in a decrease in the reaction rate. Naeem et al. [98] studied TiO_2 photocatalyzed degradation of phenol, 4-chlorophenol and 4-nitrophenol in the temperature range 25–45°C and found that increasing the temperature does not significantly enhance the degradation rates. The apparent activation energy of photocatalytic degradation ranges from 9.68 to 21.44 kJ mol^{-1}, which is much lower than that of thermal degradation reactions (e.g., 66.5 kJ mol^{-1} for the thermal degradation of poly(α-methyl styrene) in solution [99]).

7.2 pH

Photocatalytic degradation reactions are highly sensitive to pH of the reaction mixture. This is due to the fact that the adsorption of the organic compound onto the photocatalyst surface, and, therefore, the degradation reaction depends on the surface charge of the photocatalyst and also on the ionization state of the organic compound. TiO_2 exists in the following protonated and deprotonated forms in acidic and basic medium, respectively [48].

Acidic medium:

$$> \mathrm{Ti-OH} + \mathrm{H}^+ \rightleftharpoons\ > \mathrm{Ti-OH}_2^+ \tag{35}$$

Basic medium:

$$> \mathrm{Ti-OH} + \mathrm{OH}^- \rightleftharpoons\ > \mathrm{TiO}^- + \mathrm{H_2O} \tag{36}$$

Hence, the adsorption of positively charged organic compounds is facilitated at a basic pH, while that of negatively charged species is favorable at an acidic pH of the TiO_2 surface. In this regard, pH_{zpc}, the zero-point charge pH or the isoelectric pH of the TiO_2 photocatalyst plays a crucial role. At a $pH < pH_{zpc}$, the TiO_2 surface becomes positively charged (acidic), while at a $pH > pH_{zpc}$, it becomes negatively charged (basic). The effect of pH on the photocatalytic degradation of dyes was extensively reviewed by Konstantinous and Albanis [50], and Akpan and Hameed [51]. Sivalingam et al. [39] observed that dyes like Orange G (anionic) and Methyl Red (neutral) degrade at higher rates at low pH (4.1) values, while Congo Red (anionic) and Alizarin Red S (anionic) degrade at high pH (9.1) values. Devi et al. [100] investigated the degradation of anionic and cationic dyes in presence of Mo^{6+} ion-substituted TiO_2 under different pH's, and correlated the degradation behavior of the dyes with the adsorption tendency of the dye at different pH levels. They observed that the rate coefficients for the degradation of anionic dyes were higher at an acidic pH (5.6), while that of cationic dyes were higher at a basic pH (8.0). Therefore, it can be concluded that pH is a very complex parameter, which depends both on the surface charge of the photocatalyst and the nature of the organic compound, and hence, it is difficult to standardize the pH conditions for the degradation of a specific class of organic compounds. However, for the inactivation of microorganisms, it was found that the disinfection rate is nearly the same in the pH range of 4–9. This was attributed to the acid-induced proteins in *E. coli* and *Salmonella typhimurium*, which impart acid tolerance to these cells [95, 101].

7.3 Wavelength and Intensity of the Light Source

The properties of the UV light source, viz., the wavelength and intensity, play a key role in the excitation of semiconductor photocatalyst and the generation of reactive radicals. Matthews and McEvoy [102] degraded phenol and salicylic acid in the presence of 254 nm (UV-C; germicidal lamp) and 350 nm (UV-A; black-light fluorescent lamp), and concluded that the photocatalytic degradation rate is higher in presence of 254 nm radiation. This is because of the high energy of the 254 nm photons (112.6 kcal Einstein^{-1}) compared to that of the 350 nm photons (81.7 kcal Einstein^{-1}). However, the thrust to realize photocatalytic degradation in the presence of sunlight or visible radiation has resulted in light sources operating at wavelengths 365 nm and above, preferable to the 254 nm high-energy radiation, which is filtered by the atmosphere.

The intensity of radiation is related to the photon energy flux and quantum yield of the photoprocess. Photon energy flux is defined as the number (or energy) of photons per second per unit area incident on the reaction mixture. It is expressed in terms of the intensity of the incident radiation (I_0), volume of the reaction mixture (V) and the area of the exposed solution (A) as $I_0 V N_A E_\lambda / A$, where N_A is Avagadro's number of photons per mole, and E_λ is the energy of the incident photon at wavelength λ. Photon flux is expressed in terms of W m^{-2} and is usually determined by chemical actinometric techniques [103]. Quantum yield, $\Phi(\lambda)$, is defined as the ratio of number of molecules undergoing chemical transformation to the number of photons absorbed in that period of time. As the number of photons absorbed depends on the input energy density of the photons, the intensity of the light source has a positive effect on the photocatalytic degradation rate of an organic compound. Ollis et al. [45] classified the variation of the reaction rate with intensity into three regimes. At low I_0, the degradation rate is linearly proportional to I_0; at intermediate I_0, the degradation rate is proportional to $I_0^{1/2}$, and at high I_0, the degradation rate is independent of I_0, due to mass transfer limitations. Meng et al. [104], by assuming that the total number of H_2O/OH^- in the system is insufficient to capture the holes, showed that, in the intensity range from 0.86 to 6.8 mW cm^{-2}, the adsorption and degradation rate constants signifying the degradation of p-chlorobenzoate vary with the intensity as $I^{-1/2}$ and $I^{1/2}$, respectively. Wu and Chern [105] used a first-order correlation of the form, $[hv] = I_0 \exp(-k[TiO_2])$, to account for the screening effect of the TiO_2 particles at high intensities of the UV radiation. The nonlinear dependence of the rate coefficients on the UV intensity was predicted by their model. Thus, the effect of light intensity on the photocatalytic degradation rate is not straightforward and is dependent on the reactor configuration and the concentration of the TiO_2 photocatalyst.

7.4 Catalyst Loading

The concentration of the photocatalyst is one of the critical parameters that determines the degradation rate of any organic compound. This is dependent on the volume of the solution being treated and the initial concentration of the organic compound. It has been widely observed that the degradation rate increases with an increase in catalyst loading. This is due to the availability of more active catalyst sites at higher concentrations, which results in the generation of more hydroxyl species. However, when loading is increased beyond an optimum value, there is no appreciable increase in the degradation rate. This is because the catalyst particles scatter light, reducing the effective light intensity reaching the bulk reaction solution. In the literature, 0.4–3.5 g L^{-1} of catalyst has been used for the photocatalytic degradation of different organic compounds. Sivalingam et al. [36] observed that the optimum loading of CS TiO_2 for the degradation of dyes is 1 g L^{-1}.

The above argument also holds good when the catalyst particles are immobilized on a substrate like fiberglass, glass slides, glass tubes, etc. In such cases the loading is characterized by the surface coverage of the catalyst, denoted as mass per unit area of the substrate. Hence, the number of coating cycles is a critical parameter, which has to be tuned for the maximum degradation rate. Lim et al. [106] have compared the different dip-coating procedures for the immobilization of TiO_2, 5 coating cycles on woven fiberglass exhibited the highest photocatalytic degradation rate for Methylene Blue. However, in another study, it was found that a direct correlation of optimum catalyst loading for the degradation of organic compounds with the deactivation of microorganisms does not exist. The optimum loading in the case of microorganisms is always lesser than that required for the degradation of organic compounds. For the TiO_2 coated glass sheets, the optimum loading was 0.5 mg cm^{-2} for the deactivation of *E. coli*, while it was 1.1 mg cm^{-2} for the degradation of atrazine and formic acid, under the same conditions of illumination [101]. Therefore, owing to the economics of the catalyst, an optimum concentration is always favorable to study the photocatalytic degradation of organic compounds.

7.5 Presence of Organic Solvents

Organic solvents pose a deleterious effect on the degradation rate of an organic compound. Epling and Lin [107] observed that the degradation rate of Methylene Blue decreased significantly, when 1% of acetonitrile was added to the reaction mixture. Aarthi and Madras [37] also observed a decrease in the degradation rate of Rhodamine B when the composition of acetonitrile or ethanol was increased in the aqueous medium. This observation is attributed to the fact that the solvation of the CB electrons is reduced in the presence of solvent, which results in the electron–hole recombination. Another possible reason is that the organic solvents scavenge the VB holes, thereby reducing the possibility of generation of hydroxyl radicals. However, organic solvents are beneficial for the partial oxidation of organic compounds to achieve high selectivities for a particular product. This finds applications in the synthesis of fine chemicals by photocatalysis.

7.6 Pollutant Concentration

It is well-known that the rate of degradation is higher at higher concentrations of the organic compound. Especially, for the degradation of colored organic compounds like dyes, the degradation rate exhibits an initial increase with dye concentration, but decreases after a certain critical concentration. This is attributed to the screening of UV radiation by the dye molecules before reaching the catalyst surface. However, the catalyst concentration can be tuned based on the concentration of the organic compound, so that the organic compound can be significantly adsorbed onto the catalyst surface and degraded. Most of the degradation studies have employed a concentration of the organic compounds in the range of 10–200 mg L^{-1}, which agrees well with the concentration of the common pollutants in real wastewater.

7.7 Presence of Anions and Oxidizing Agents

Industrial effluents are composed of a large number of organic compounds, humic substances and inorganic substances, such as anions and metal ions. Hence, an evaluation of the degradability of an organic compound under the influence of such externally added ions is important. It is generally observed that anions like carbonate, bicarbonate, chloride and sulfate ions retard the degradation of organic compounds, in accordance with the following reactions [48–51]:

$$CO_3^{2-} + OH^{\bullet} \longrightarrow CO_3^{\bullet -} + OH^{\bullet -} \quad (37)$$

$$HCO_3^{-} + OH^{\bullet} \longrightarrow CO_3^{\bullet} + H_2O \quad (38)$$

$$Cl^{-} + OH^{\bullet} \longrightarrow Cl^{\bullet} + OH^{-} \quad (39)$$

$$SO_4^{2-} + OH^{\bullet} \longrightarrow SO_4^{\bullet -} + OH^{-} \quad (40)$$

$$SO_4^{\bullet -} + H_2O \longrightarrow SO_4^{2-} + OH^{\bullet} + H^{+} \quad (41)$$

It is clear that all these anions scavenge the hydroxyl radicals to form the respective anion radicals. In the case of the sulfate anion, a sulfate anion radical can result in the formation of hydroxyl radicals according to reaction 41. However, these radicals are again scavenged according to reaction 40, and hence there is a net reduction in the rate of photocatalytic degradation of the organic compound. Azevedo et al. [108] evaluated the effect of saline media on the photocatalytic degradation of phenol and concluded that, at a low concentration of NaCl (2 g L^{-1}), the rate was unaffected, while at a high concentration of NaCl (50 g L^{-1}), there was a drastic reduction in the degradation rate of phenol. However, complete removal and mineralization of phenol was observed. They modeled the reaction using extended-lumped saline kinetics and predicted the time evolution profiles of phenol, intermediates and CO_2. Devi et al. [100] conducted extensive experiments to evaluate the effect of anions on the degradation of anionic and cationic dyes, and found that the presence of sulfate anions result in the highest reduction in the degradation rate of the dyes, while the presence of oxalate anions show the lowest reduction in rate. Thus the reduction in rate due to the presence of different anions for the degradation of Rhodamine B follows the order: $SO_4^{2-} > Cl^- > HCO_3^- > NO_3^- > CO_3^{2-} > C_2O_4^{2-}$. A similar retardation effect was also observed for the inactivation of microorganisms in presence of TiO_2. According to the results of Rincón and Pulgarin [95], the order of retardation of *E. coli* in presence of different anions is as follows: $Cl^- > NO_3^- > SO_4^{2-} \gg HCO_3^- > HPO_4^{2-}$, where the chloride and phosphate ions exhibit the minimum and maximum inhibiting effect, respectively.

The presence of oxidizing agents such as persulfate and H_2O_2 has a positive effect on the degradation rate of the organic compound, which is attributed to the generation of hydroxyl radicals by these oxidizing agents, according to the following reactions [48–51]:

$$S_2O_8^{2-} + e_{CB}^- \longrightarrow SO_4^{2-} + SO_4^{\bullet -} \tag{42}$$

$$H_2O_2 + e_{CB}^- \longrightarrow OH^\bullet + OH^- \tag{43}$$

$$H_2O_2 + O_2^{\bullet -} \longrightarrow OH^\bullet + OH^- + O_2 \tag{44}$$

The sulfate anion radical generated by reaction 42 reacts with water to form a hydroxyl radical by reaction (41). Reactions (43)–(44) are the key reactions for the generation of hydroxyl radicals in presence of H_2O_2. It was previously shown that these two reactions are responsible for the enhanced photocatalytic inactivation of *E. coli* in presence of TiO_2, even at very low concentrations of H_2O_2 (0.3 mM). However, when the concentration of H_2O_2 is higher than the optimum, the following reactions occur, which result in the quenching of the hydroxyl radicals, thereby resulting in a reduction of the degradation rate [48–51]:

$$H_2O_2 + 2h_{VB}^+ \longrightarrow O_2 + 2H^+ \tag{45}$$

$$H_2O_2 + OH^\bullet \longrightarrow H_2O + HOO^\bullet \tag{46}$$

$$HOO^{\bullet} + OH^{\bullet} \longrightarrow H_2O + O_2 \qquad (47)$$

Therefore, careful optimization of the concentration of the oxidants is a must for achieving maximum photocatalytic degradation rates.

7.8 Presence of Metal Ions

In the recent past, extensive research has been conducted on the effects of metal ions on the degradation of organic compounds. It is imperative to note that metal ions like Ag^+, Hg^{2+}, Cu^{2+}, Pb^{2+}, Cd^{2+}, Ni^{2+} and Cr^{6+} are toxic and have infinite lifetimes. Hence, their accumulation in the environment leads to biomagnification. Reduction of metal ions by photocatalysis is an important subject and many studies have dealt with this aspect [44, 109]. A number of studies have also focused on the simultaneous degradation of organic compounds and reduction of metal ions using TiO_2. Prairie et al. [110] observed that the rate constant for the reduction of Cr^{6+}–Cr^{3+} is strongly dependent on the concentration of salicylic acid, and the rate constant exhibits a sudden drop at salicylic acid concentrations higher than the optimum value. Moreover, the highest degradation rate of salicylic acid was observed in presence of Au^{3+}. An important result of their work shows that only those metal ions whose half-reaction standard reduction potential is greater than 0.3 V (vs NHE) can be treated by photocatalysis.

A careful evaluation of the works of Chen et al. [111], Aarthi and Madras [37], and Kyung et al. [112] shows that the rate of degradation of Rhodamine B decreases in the presence of Cu^{2+}, Fe^{3+}, Zn^{2+} and Al^{3+}, while increasing with the presence of Cr^{6+} and Ag^+ ions. This suggests that the electronic state of the metal ions also plays a major role in the degradability of the organic compound. Wang et al. [113] studied the photocatalytic degradation of phenol in presence of Cu^{2+} and F^- ions using DP-25 TiO_2. The I-order degradation rate coefficient of phenol in presence of different ions exhibited the following trend: $k(Cu^{2+}+F^-) > k(F^-) > k(Cu^{2+}) > k(\text{without } Cu^{2+} \text{ or } F^-)$. The observed effect was attributed to the enhanced charge-carrier separation induced by Cu^{2+} and F^-.

A common feature in most of the works reporting synergistic degradation of organic compounds in the presence of metal ions is that the pH of the reaction mixture was in the acidic regime. For example, the reduction of Cr^{6+}–Cr^{3+} occurs in the presence of protons according to the following reaction [114]:

$$2Cr_2O_7^{2-} + 16H^+ \rightarrow 4Cr^{3+} + 8H_2O + 3O_2 \qquad (48)$$

Hence, an acidic medium can provide excess protons for the reduction of Cr^{6+} compared to a neutral or basic medium. Moreover, metal ions like Cr^{6+} and Cu^{2+} precipitate as hydroxides in the basic pH regime, which prevents them from getting adsorbed onto the TiO_2 surface. In a recent study, Vinu and Madras [114] observed that the presence of phenol accelerated the reduction of Cu^{2+}–Cu^+, while this was not possible in the absence of phenol (Fig. 8). The presence of Cu^{2+} ions

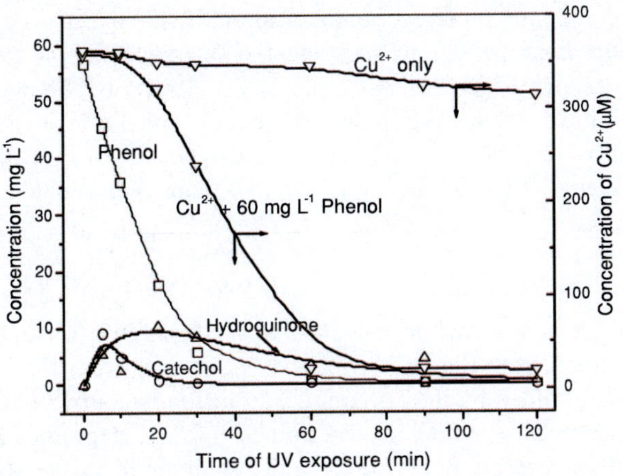

Fig. 8 Concentration profiles of phenol and Cu^{2+} during the simultaneous oxidation–reduction of phenol+Cu^{2+} using CS TiO_2. The phenolic intermediates, viz., catechol and hydroquinone were observed only in the presence of Cu^{2+} [114]

did not significantly alter the degradation rate of phenol, while the presence of Cr^{6+} enhanced the degradation of phenol.

Sun et al. [115] showed that, among the different metal ions like Cu^{2+}, Fe^{3+}, Mn^{4+}, Ce^{4+}, V^{5+} and Cr^{6+}, only Cr^{6+} accelerated the photooxidation of 4-chlorophenol in presence of visible light. Therefore, it can be summarized that the simultaneous degradation-reduction of such two-component systems containing an organic compound and a metal ion is dependent on the pH of the solution, concentration of the metal ions and the organic compound, adsorption of the metal ion and the electronic states of the metal ions.

8 Mechanistic Models for Photocatalytic Degradation

8.1 Models for the Degradation of Organic Compounds

It is well reported that the rate of photocatalytic degradation of the organic compounds follows the classical Langmuir–Hinshelwood (L–H) kinetics [116]. Sivalingam et al. [36] derived the rate equation for the photocatalytic degradation of dyes by incorporating the reactions including charge-carrier generation, recombination, adsorption–desorption of the substrate, generation of hydroxyl radicals and the oxidation of the adsorbed organic compound by hydroxyl radicals and holes. The rate equation, which resembles the L–H equation, is given by:

$$-r_{\mathrm{A}} = \frac{K_0((k_{0\mathrm{h}}/K_0) + k_0)C_{\mathrm{A}}}{1 + K_0C_{\mathrm{A}}} \tag{49}$$

where, r_A denotes the rate of consumption of the organic compound, C_A is the concentration of the organic compound, and K_0 is the equilibrium adsorption rate coefficient, which includes the adsorption of the organic compounds and hydroxyl radicals. The rate coefficients k_{0h} and k_0 signify the oxidation of the organic compound by direct hole attack and by the hydroxyl radicals, respectively. Hence, the compound rate coefficient $k_r = (k_{0h}/K_0) + k_0$ denotes the oxidation of the organic compound. A simple way to determine the rate coefficients K_0 and k_r is the initial rate method. In this method, the equation is inverted and rearranged so that a plot of $1/C_{A0}$ versus $1/r_{A0}$ yields the slope and intercept, from which K_0 (in L mg^{-1}) and k_r (in mg L^{-1} min^{-1}) can be evaluated. The equation is given by

$$-\frac{1}{r_{A0}} = \frac{1}{k_r K_0} \frac{1}{C_{A0}} + \frac{1}{k_r} \tag{50}$$

It is worthwhile to note that the equilibrium adsorption coefficient, K_0 determined from the L–H equation is different from the adsorption coefficient that is determined by the Langmuir adsorption isotherm. This is because the equilibrium adsorption coefficient determined by the Langmuir isotherm denotes the adsorption–desorption of the organic compound in the dark, while K_0 signifies the adsorption of the organic compounds onto the catalyst surface and the desorption of the products from the surface during the photoreaction. Hence, it is imperative to note that K_0 is dependent on other reaction parameters including the intensity of UV irradiation, concentration of oxygen and the catalyst concentration.

As most of the organic compounds are present at trace levels in the atmosphere or effluent stream, the term $K_0 C_A \ll 1$ in the denominator of Eq. 49. Hence, the rate equation can be simplified to a first-order equation with 'k' as the first-order rate constant signifying the degradation of the organic compound [39].

$$-r_A = -\frac{dC_A}{dt} = kC_A \tag{51}$$

The solution of the above equation is given by $C_A = C_{A0} \exp(-kt)$. The rate coefficient, k, can be determined by a linear plot of $\ln(C_{A0}/C_A)$ versus t. The value of the rate coefficient reported in the literature for the degradation of organic compounds varies over a wide range and, therefore, it is difficult to establish a specific range of variability. This is because the rate coefficient is dependent upon the type of the catalyst (anatase/rutile TiO_2), intensity of the UV radiation, catalyst loading and other reaction parameters like pH, and the presence of anions and cations.

Li et al. [117] developed a kinetic model for the rate of degradation of Rhodamine B dye in presence of a TiO_2-coated, activated carbon catalyst by including the fractional surface coverage of the hydroxyl radicals and the dye, on the catalyst surface. The L–H parameters were found to depend on the intensity (I) of the UV radiation according to the following equations:

$$k_r^{-1} = \alpha I^{-1/2} + \beta \tag{52}$$

$$K_0 = \eta I^{-1/2} + \varpi \tag{53}$$

where, α, β, η and ω are lumped constants, expressed in terms of the rate coefficients and signifying the individual steps. They have observed good agreement of the model with the experimental data at different light intensities.

Wu and Chern [105] adopted a cyclic network reduction technique to derive the rate expression for the degradation of Methylene Blue. The rate of decomposition of Methylene Blue was found to depend on the initial concentration of the dye, light intensity, TiO_2 loading and dissolved oxygen concentration. The rate expression is given by:

$$-r_A = \frac{k_a' C_A}{1 + k_b' C_A} \tag{54}$$

where, $k_a' = \frac{k_4[h\nu][TiO_2]}{1+k_1[h\nu]}$ and $k_b' = \frac{k_2+k_3[h\nu]}{1+k_1[h\nu]}$. The concentration profiles were simulated by integrating Eq. 54 in the limits $C_A|_{t=0} = C_{A0}$ and $C_A|_{t=t'} = C_{At}$. The rate coefficients were determined by nonlinear curve fitting of the following expression with the experimental data:

$$\ln\left(\frac{C_{At}}{C_{A0}}\right) + k_b'(C_{At} - C_{A0}) = -k_a' t \tag{55}$$

Almquist and Biswas [118] investigated the effect of dissolved oxygen on the kinetics of degradation of phenol in presence of DP-25. Their results show that dissolved oxygen has an inhibiting effect on the degradation of phenol when the concentration of phenol is of the order of the solubility of oxygen in water. They developed a three-parameter model that describes the dependence of the degradation rate on the concentration of phenol and dissolved oxygen. The rate equation is given by:

$$-\frac{dC_A}{dt} = \frac{bC_{O_2}C_A\left\{\sqrt{1 + a/C_{O_2}} - 1\right\}}{1 + cC_A} \tag{56}$$

The model parameter 'a' depends on the light intensity profile inside the reaction slurry, light absorption and the efficiency of electron-hole formation and recombination. The parameters 'b' and 'c' depend mainly on the type of TiO_2. The parameters were determined by nonlinear curve fitting. Therefore, the differing kinetic equations suggest that, although L–H and the first-order model satisfactorily describes the overall kinetics of degradation of organic compounds, more thorough models must be used to unravel the effects of different reaction parameters like the UV light intensity, catalyst loading, oxygen concentration and the mode of reactor operation.

8.2 Modeling of Microbial Inactivation

The kinetics of photocatalytic disinfection of microorganisms is quite different from that of the organics because the inactivation profiles are characterized by three distinct regimes. In the first regime, a "shoulder" or a slow deactivation of the microorganisms occurs due to the self-defense and auto-repair mechanisms of the bacterial cell membrane against the hydroxyl radicals [119]. Once the hydroxyl radical concentration builds up in the system and a certain number of critical molecules are denatured, perforation of the bacterial membrane occurs, as the anti-stress enzymes are unable to prevent the membrane from oxidation. Thus, in the second regime, a log-linear reduction in concentration of the microorganism is observed. The last part of the deactivation process is characterized by a slow reduction in concentration or tailing of the concentration profile, due to the release of organic compounds into the medium, which competes with the inactivation of the microorganism.

Several empirical models have been proposed to account for the different regimes of deactivation of microorganisms [120]. These include a delayed Chick-Watson, modified Chick-Watson, Hom and modified Hom equations. The modified Hom equation, which fits well with the bacterial concentration profile has three adjustable parameters and is expressed as:

$$\log \frac{C_0}{C} = k_1[1 - \exp(-k_2 t)]^{k_3} \tag{57}$$

where, C_0 and C are the initial concentration and concentration at time 't' of the microorganism in the system. The parameters k_1, k_2 and k_3 in the above model have no physical significance. Therefore, rigorous mechanistic models are necessary to describe the kinetics of the deactivation of microorganisms. Initially, the mechanism of rupturing the cell wall of the microorganism, thereby inactivating the organism, was represented as a series of consecutive events of different damage levels [120, 121]:

$$C_0 \xrightarrow{k_1} C_1 \xrightarrow{k_2} C_2 \xrightarrow{k_3} C_3 \cdots \xrightarrow{k_n} C_n \tag{58}$$

where, k_i denotes the kinetic rate coefficient for the damage from level $i - 1$ to i. Labas et al. [121] coupled the radiation transport equation with this model and observed that the threshold limit for bacterial damage corresponds to two levels ($n = 2$). Other parameters in their model correspond to the inactivation rate coefficient k, and the reaction order with respect to the local volumetric rate of photonic absorption of the bacteria.

Marugán et al. [120] recently proposed a modified mechanism for the deactivation of *E. coli*, represented as:

$$C_{\text{undamaged}} \xrightarrow{k_1} C_{\text{damaged}} \xrightarrow{k_2} C_{\text{inactivated}} \xrightarrow{k_3} C_{\text{product}} \tag{59}$$

where, $C_{\text{undamaged}}$ denotes the undamaged population of the microorganism, C_{damaged} denotes the lumped population at all intermediate levels of damage, $C_{\text{inactivated}}$ denotes the microorganisms in the inactive state and C_{product} denotes the products of microbial cell lysis, which are released into the reaction medium. The L–H-like model for the above mechanism is given by:

$$\frac{dC_{\text{undamaged}}}{dt} = -k \frac{KC^{n}_{\text{undamaged}}}{1 + KC^{n}_{\text{undamaged}} + KC^{n}_{\text{damaged}}} \tag{60}$$

$$\frac{dC_{damaged}}{dt} = k\frac{KC^{n}_{undamaged} - KC^{n}_{damaged}}{1 + KC^{n}_{undamaged} + KC^{n}_{damaged}} \quad (61)$$

where, the rate coefficients k, K and n correspond to the interaction of the hydroxyl species with the microorganism, microorganism with TiO_2, and the products of bacterial lysis with the hydroxyl species, respectively. The above expression is based on the assumption that the rate coefficients for the intrinsic reaction steps for the undamaged and the damaged cells are the same. Thus, the rate coefficients K, k and n signifying the pseudo-adsorption, inactivation and inhibition, represent the initial lag, log-linear decrease and final lag phases of the bacterial disinfection profiles, respectively. This model has been validated for the deactivation of *E. coli* with different catalysts of varying concentrations, and in presence of anions like chloride, bicarbonate and phosphate, and humic acids [122].

8.3 Modeling of Multicomponent Systems

Industrial effluents are composed of a mixture of different organic compounds (dyes, phenolics, chlorinated compounds) and metal ions. The rate of degradation of an organic compound in the presence of another will be different from individual degradation rates. Therefore, kinetic modeling of multicomponent systems yields useful information on the interaction between the different species in the system, competition of the reactants and products for the active sites, and the mechanism of degradation. Priya and Madras [123] studied the photocatalytic degradation of mixtures of 4-chlorophenol and 4-nitrophenol to simulate the industrial effluent. They modeled the rate of degradation of chlorophenol (cp) and nitrophenol (np) using the following rate equations with individual L–H rate parameters (k_{cc}, K_{cp}, k_{nn} and K_{nn}) and interaction parameters (k_{cn} and k_{nc}):

$$-r_{cp} = \frac{(k_{cc} - k_{cn}C_{np})C_{cp}}{1 + K_{cp}C_{cp} + K_{np}C_{np}}; \quad -r_{np} = \frac{(k_{nn} - k_{nc}C_{cp})C_{np}}{1 + K_{cp}C_{cp} + K_{np}C_{np}} \quad (62)$$

The above model accounts for the competitive inhibition of the rate of degradation of cp by np, and vice-versa. The denominator in the above equations accounts for the competition of cp and np for the active site of TiO_2. By linearizing the above expressions, the rate coefficients were determined. Based on the values of k_{cn} and k_{nc}, it was found that the degradation of chlorophenol was unaffected by the presence of nitrophenol, while the degradation of the nitrophenol was significantly affected by the presence of the chlorophenol.

Similarly, when a metal ion is present in the system along with an organic compound, simultaneous oxidation and reduction reactions occur, and the kinetics of the reaction is affected by the complex interaction of the organic compound and metal ion with the TiO_2 surface. A detailed discussion on this topic is available in Sect. 7.8. The retardation of the rate of degradation of dyes in presence of metal ions was modeled by Aarthi et al. [37], and their kinetic model relates the rate of degradation of the dye (r_D) with the concentration of the dye $[D]$ and the metal ion $[M^{n+}]$. The equation is given by:

$$-\frac{1}{r_D} = \left(\frac{1}{[D]} + K_2\right)\frac{(1 + K_6[M^{n+}])}{(k_0 + K_6 k_7 [M^{n+}])} \tag{63}$$

The lumped rate coefficients in the above model, viz. K_2, K_6, k_0 and k_7, were determined by nonlinear regression, and the model was validated for the degradation of different dyes, including Rhodamine B [37], Rhodamine 6G [37], Sulforhodamine B [37] and Azure B [124], in presence of Cu^{2+}. One shortcoming of this model is that it does not account for the concomitant reduction of metal ions during the degradation of the organic compound. With this in mind, Vinu and Madras [114] modeled the simultaneous and synergistic degradation of phenolic compounds (phenol and 4-nitrophenol), and the reduction of metal ions (Cu^{2+} and Cr^{6+}), by assuming competitive adsorption of metal ions and the phenolic compounds onto the TiO_2 surface. The steady-state rates through the individual phenol degradation and metal ion reduction pathways, derived using the network reduction technique, are given by:

$$\left(-\frac{1}{r_{Ph,0}} - K_2\right) = \frac{1}{[Ph]_0}\left(\frac{1}{K_1} + \frac{[M^{n+}]}{K_3}\right) \tag{64}$$

$$\left(-\frac{1}{r_{M^{n+},0}} - K_5\right) = \frac{1}{[M^{n+}]_0}\left(\frac{1}{K_4} + \frac{[Ph]}{K_6}\right) \tag{65}$$

where, $r_{Ph,0}$ and $r_{Mn+,0}$ denote the initial rate of degradation of phenolic compound and the reduction of metal ions, respectively. This model accounts for the contribution of the metal ions for the rate of degradation of the phenolic compound and vice versa. The lumped rate coefficients K_2 and K_5 were determined by individually degrading phenolic compounds and metal ions, and the other rate coefficients were evaluated by conducting multiple experiments at different phenol and metal ion concentrations. Using this model, the rate coefficients can be determined uniquely by a simple linear regression.

8.4 Modeling of Degradation Intermediates

The intermediates formed during the degradation of organic compounds can have a deleterious effect on the overall degradation rate, since these compounds compete for the active sites of the catalyst. Hence, by knowing the formation and consumption rate of the different intermediates, the relative importance of each of them on the overall mineralization of the parent compound can be assessed. Previous studies on the degradation of phenolic compounds, dyes and pesticides have monitored the concentration of the various intermediates during degradation.

For example, the predominant intermediates observed during the degradation of phenol include catechol, hydroquinone, hydroxy hydroquionone, benzoquinone and oxalic acid. The kinetics of formation and consumption of these intermediates was modeled by considering a simple series reaction mechanism with first-order reactions [39, 40], represented as:

$$\mathrm{A} \xrightarrow{k_f} \text{Intermediate} \xrightarrow{k_c} \text{Product} \tag{66}$$

where, k_f and k_c are the formation and consumption rate coefficients of the intermediates. By writing rate equations for A, intermediate and the product, an expression for the concentration of the intermediate can be derived as [125]:

$$\frac{C_{\text{int}}}{C_{A0}} = \frac{k_f}{k_c - k_f}(\exp(-k_f t) - \exp(-k_c t)) \tag{67}$$

The concentration profiles of the primary hydroxylated species show an initial increase and subsequent decrease with time due to the formation of secondary hydroxylated species. Hence, there is a time (t_{max}) at which the concentration of the intermediate is maximum (C_{max}). Therefore, by differentiating the above equation with respect to t and C_{int}, expressions for t_{max} and $C_{\text{int,max}}$ were derived as [125]:

$$t_{\text{max}} = \frac{\ln(k_f/k_c)}{k_c(k_f/k_c - 1)} \tag{68}$$

$$\frac{C_{\text{int,max}}}{C_{A0}} = \left(\frac{k_f}{k_c}\right)^{1/(1-k_f/k_c)} \tag{69}$$

Thus, by solving these two equations with t_{max} and $C_{\text{int,max}}$ from the experimental data, k_c and k_f can be determined. More importantly, the ratio k_c/k_f, which signifies the relative rate at which the intermediate is consumed, can be used as a quantifying factor to assess the stability of the different intermediates and the effective mineralization of the parent compound. The formation of catechol and hydroquinone during the photocatalytic degradation of phenol in presence of Cu^{2+} ions and CS TiO_2 is depicted in Fig. 8. The curves for catechol and hydroquinone represent the model fits.

9 Visible Light Degradation of Organic Compounds

9.1 *Second Generation TiO_2 Photocatalysts*

The first generation TiO_2 or pristine TiO_2 proves to be an efficient photocatalyst in the UV region. However, the wide band-gap (3.2 eV) of anatase TiO_2, corresponding to an absorption threshold of 390 nm, limits its use in the visible range

Table 6 Effect of different metal ion substitutions in TiO_2 for the photocatalytic degradation of organic compounds

Sl. No.	Organic compound	Substituted metal ion in TiO_2	Results	Reference
1	CCl_4 and $CHCl_3$	Fe^{3+}, Mo^{5+}, Ru^{3+}, Os^{3+}, Re^{5+}, V^{4+}, Rh^{3+}, Co^{3+}, Al^{3+}	Co^{3+} and Al^{3+} doping reduces the photoactivity, while all other metal ions in the concentration range from 0.1 to 0.5 at. % enhance the photoactivity	[126]
2	Oxalic acid	Cr^{3+}, Fe^{3+}, V^{5+}	The photoactivity of all the samples was lower compared to bare TiO_2	[127]
3	2-Chlorophenol	Nd^{3+}, Pd^{2+}, Pt^{4+}, Fe^{3+}	Order of photoactivity: $Nd^{3+} > Pd^{2+} > Pt^{4+} \approx$ undoped $> Fe^{3+}$	[128]
4	5,5-Dimethyl-1-pyrroline N-oxide (DMPO spin trap)	Cr^{3+}, Mn^{2+}, Co^{2+}	All the metal ion-doped samples exhibit a lower photocatalytic activity for the generation of DMPO-OH and DMPO-O_2^- compared to undoped DP-25	[129]
5	4-Nitrophenol	Cu^{2+}, Fe^{3+}, Ce^{4+}, Zr^{4+}, V^{5+}, W^{6+}	Order of photoactivity: undoped CS $TiO_2 > Fe/TiO_2 > W/TiO_2 > Ce/TiO_2 > Zr/TiO_2 > V/TiO_2 \approx Cu/TiO_2$	[130]
6	Orange II	La^{3+}, Ce^{4+}, Pr^{3+}, Nd^{3+}, Sm^{3+}, Eu^{3+}, Dy^{3+}, Gd^{3+}	High visible light photoactivity compared to undoped TiO_2 in the concentration range of 0.5–1 wt%	[131]

(400–800 nm) in practical applications. Moreover, solar radiation is comprised of c.a. 50% visible radiation and less than 5% UV radiation. Hence, extending the spectral absorption of TiO_2 to the visible region is vital for the development of energy efficient degradation processes.

Second generation TiO_2 catalysts encompass a wide variety of the cationic and anionic substituents in TiO_2. The cationic substituents mainly include metal ions, while the anionic substituents include C, N and S. Numerous studies have examined the substitution of lower valent (+1, +2, +3), iso valent (+4) and higher valent (+5, +6) metal ions in the lattice of TiO_2 for the photocatalytic degradation of organic compounds. This includes noble metals, transition metals, lanthanide metals and alkaline metals. Table 6 [126–131] presents a list of the various studies conducted on the photocatalytic activity of metal ion-doped TiO_2. Although there is a general consensus that metal ion doping extends the absorption spectrum of TiO_2 to the visible region, the photocatalytic activity of the metal ion-doped TiO_2 compared to the undoped TiO_2 varies across different studies. From the table, it is evident that doping of some metal ions enhances the photoactivity, while others result in a reduction of photoactivity.

Choi et al. [126], by studying the transient absorption decay of the trapped electrons, showed that the dopant energy levels serve as trap sites for the electrons and holes, apart from the surface trap sites. The low activity exhibited by metal ion-doped TiO_2 is due to these states acting as recombination centers. Hence, the photoactivity of metal ion-doped TiO_2 strongly depends on the dopant concentration, energy level of the dopant within the TiO_2 lattice, *d*-electronic configuration, distribution of dopant, interfacial charge transfer and light intensity. Nagaveni et al. [130] conducted a thorough photoluminescence study of the different metal ion-doped TiO_2 ($M_xTi_{1-x}O_{2\pm\delta}$), synthesized by the solution combustion technique, and observed a reduction in the emission intensity of the metal ion-doped samples. The metal ions form inter-band energy levels above the valence band or below the conduction band, which result in lower band gap of the doped TiO_2 materials. Serpone et al. [127] showed that doping of Cr^{3+}, Fe^{3+} and V^{5+} in the lattice of TiO_2 results in a lower photoactivity for the oxidation of oxalic acid, but an enhanced activity for the photoreduction of water to H_2. Similarly, Pd^{2+} ion substitution in CS TiO_2 was found to be beneficial for gas phase CO oxidation, NO reduction and NO decomposition [132], while it exhibited a negative effect for liquid phase degradation of organic compounds [133]. This shows that a generalization of the activity of metal doped TiO_2 compared to the undoped TiO_2 is not possible for a wide class of reactions.

Another method by which metal ions can be incorporated in TiO_2 is by impregnation on the surface. Paola et al. [134] investigated the effect of different transition metals impregnated TiO_2 for the photocatalytic degradation of aliphatic and aromatic compounds and found that the highest mineralization efficiency was obtained with bare TiO_2. In addition to the above result, Vinu and Madras [133, 135] showed that TiO_2 impregnated with Pd or Ag exhibits higher photoactivity compared to doped-TiO_2. The higher activity exhibited by the metal ion-impregnated TiO_2 is attributed to the formation of a Schottky barrier, which results in the scavenging of electrons and holes, thereby preventing unfavorable recombination reaction.

While metal ions are substituted for the Ti atoms in the lattice, anions are substituted for oxygen in the oxide. Hence, these are represented as $TiO_{2-x}D_x$, where D is usually N, C or S. The first study on anion-substituted TiO_2, $TiO_{2-x}N_x$, was carried out by Asahi et al. [136] for the photocatalytic degradation of Methylene Blue and gaseous acetaldehyde in presence of visible radiation. Based on X-ray photoelectron spectroscopic analysis, they observed an optimum concentration of N to be 0.25 atomic%. Khan et al. [137] incorporated C in TiO_2 by flame pyrolysis, which shows a band gap of 2.32 eV, and high activity for photosplitting of water (photoconversion efficiency = 8.35%). Unlike the cation-doped TiO_2, anion-doped TiO_2 exhibits a high photoactivity in the visible region compared to undoped and commercial DP-25 due to narrowing of the band-gap. In fact, the band gap narrowing of N doped TiO_2 was attributed to the mixing of N 2p and O 2p energy levels. This requires the elevation of valence band maximum or the lowering of conduction band minimum by the introduction of new energy level of the anion dopant, with a more homogeneous distribution and without a loss of the crystal structure of the host TiO_2. The rules for any non-metal to be substituted for oxygen in TiO_2 to elevate the valence band, are as follows [138]: (i) the electronegativity of the non-metal dopant should be lesser than that of oxygen, and

(ii) the radius of the dopant should be comparable to that of oxygen for a more uniform distribution.

Serpone [139] demonstrated that the visible light activity of the anion-doped TiO_2 is not due to the narrowing of the band-gap, but due to the defects associated with the oxide ion vacancy, which results in the formation of color centers. Color centers are essentially a single or a pair of electrons associated with an oxygen vacancy. For MgO, it has been shown that the ground state of the color centers lie above the O 2p valence band. Table 7 [140–144] shows the different studies on the anion-doped TiO_2 for the visible light degradation of organic compounds.

9.2 *Heterostructuring of TiO_2*

Heterostructuring refers to the modification of the surface of pristine TiO_2 by employing (i) narrow band-gap semiconductor dopants (like CdS, PbS, CdSe, Bi_2S_3), (ii) dyes as sensitizers and (iii) co-catalysts [138]. Different schemes of charge-carrier transfer have been proposed for heterostructured TiO_2 materials, viz. traditional charge-carrier transfer, sensitization, indirect Z-scheme, direct Z-scheme, vectorial electron transfer and co-catalyst coupling. The main idea of these heterostructuring procedures is to isolate the electrons and holes at two different reaction sites in order to prevent recombination. This enhances the efficiency of simultaneous oxidation and reduction reactions. Moreover, the incorporation of small band-gap semiconductors, dyes and co-catalysts, increases the probability of absorption in the visible range. Liu et al. [138] reviewed such schemes in terms of their mechanisms, materials and the key issues involved in their implementation. Figure 9 depicts the two different modes of charge transfer in presence of a sensitizing dye and a narrow band-gap semiconductor. Some important design considerations for the heterostructured systems, for effective interfacial charge transfer, are as follows:

- The main semiconductor (TiO_2) and the dopant semiconductor need to have suitable electronic structure, i.e., the dopant should have a higher conduction band minimum and valence band maximum compared to TiO_2, for the smooth injection of electrons downhill from the conduction band of the dopant, and the transfer of holes uphill to the valence band of the dopant.
- The above condition is also applicable for sensitizers, although there is no transfer of holes to the HOMO. Moreover, high surface area of TiO_2 is necessary for the enhanced adsorption of the sensitizer.
- Intimate contact between the two different phases (e.g., a p–n junction in the case of traditional transfer or Ohm/Schottky contact in the case of co-catalysts) is necessary.

Table 8 [145–154] presents the different studies on the sensitized degradation of organic compounds using modified TiO_2 materials. It is clear that a wide variety of sensitizers like narrow band-gap semiconductors, organic dyes and conjugated polymers are being coupled to TiO_2 to enhance the visible light response.

Table 7 A survey of the recent studies on the photocatalytic degradation of organic compounds using anion-doped TiO_2

Sl. No.	Organic compound	Photocatalyst/loading	Light source	Initial concentration	% degradation	Time taken	Reference
1	Orange G, RBBR, Methylene Blue	CS TiO_2; 1 g L^{-1}	Sunlight; 753 W m^{-2}	25, 100, 100 ppm	100, 80, 90	2, 1.5, 4 h	[140]
2	Methylene Blue	C-doped TiO_2; 1 g L^{-1}	Sunlight; 21.28 W m^{-2}	10 ppm	100	1 h	[141]
3	2,4-Dichlorophenol, Acid Orange 7	C deposited TiO_2; 1 g L^{-1}	1,000 W HL	50, 20 ppm	60, 100	5, 4 h	[142]
4	Reactive Brilliant Red X-3B	C, N, S—tridoped mesoporous TiO_2; 1.5 g L^{-1}	250 W HL	100 ppm	70	2 h	[143]
5	Methylene Blue	$Ti_{1-x}Ce_xO_{1-y}N_y$; x = 0.007; 1 g L^{-1}	30 W FL	15 ppm	100	4 h	[144]

RBBR—Remazol brilliant Blue R; HL—halogen lamp; FL—fluorescent lamp

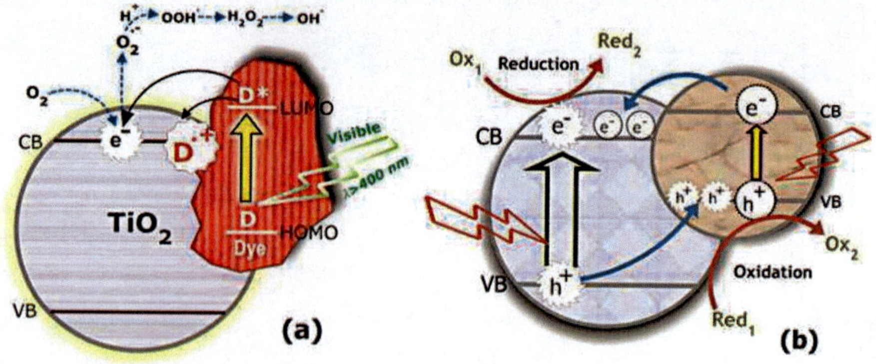

Fig. 9 Mechanism of charge carrier transfer in **a** dye sensitized TiO_2, and **b** narrow band gap semiconductor coupled TiO_2

Moreover, the time taken for the complete degradation of the organic compounds varies from 1 to 10 h, which shows the importance of optimizing the various reaction conditions like the initial concentration of the organic compound, catalyst loading and the intensity of the light source. Hence, a fair comparison of the photoactivity of the different modified TiO_2 materials requires the evaluation of the kinetic rate coefficients of degradation.

10 Photocatalysis in Practice

10.1 Immobilization of TiO_2

The photocatalytic degradation reactions considered thus far have been conducted in small volume (0.1–10 L) batch reactors, usually in immersion well photoreactors [36], in which the catalyst particles are suspended and continuously stirred in the reaction mixture. While this configuration is beneficial for the high mass transfer of the reactants and products on the catalyst surface, which results in high degradation rates of organic compounds, this mode of operation is not suitable for large-scale applications involving large volumes of effluents, due to the inherent problem associated with the separation of the catalyst particles after the completion of the reaction. This escalates the costs related to the operation of the unit due to the complex downstream operations involved.

A solution to the above scale up issue is to immobilize the catalyst on a suitable support material. Many different supports have been explored for the immobilization of TiO_2, viz. glass beads [155], flexible fiber glass cloth [156], glass tubes [157], cotton [158], stainless steel [159], ZnO tetrapods [160], silicone rubber film [161], PTFE mesh sheets [162], polystyrene beads [163], low density support (perlite) [164] and porous lavas (pumice stone) [165]. A good support is

Table 8 Survey of the recent studies on sensitized degradation of organic compounds using modified TiO_2

Sl. No.	Organic compound	Sensitizer in TiO_2	Light source	Initial concentration	% degradation	Time taken	Reference
1	Methyl Orange, 4-Chlorophenol	MoS_2 and WS_2	300 W WHL	20 μM		30–120 min	[145]
2	Methyl Orange	Poly(3-hexyl thiophene)	300 W IWL	10 ppm	90	10 h	[146]
3	4-Chlorophenol	Al tetracarboxy phthalocyanine	500 W HL	0.23 mM	90	8 h	[147]
4	2,3-Dichlorophenol	Poly(thiophene)	150 W HPSL	20 ppm	50	7 h	[148]
5	2,4-Dichlorophenol	Xanthene dyes	500 W HL	16.3 mg L^{-1}		5 h	[149]
6	Phenol	Metallo phthalo-cyanines	100 W HL	100 ppm	90	1 h	[150]
7	Terbutyl azine	Rose Bengal (10 ppm)	500 W XL	5 ppm	50	2 h	[151]
8	Atrazine	Tetra(4-carboxy phenyl) porphyrin	XL	20 ppm	80	1 h	[152]
9	Trichloroacetate and CCl_4	$Pt/TiO_2/Ru^{II}L_3$ (10 μM)	450 W XL	1 mM		2–3 h	[153]
10	Methylene Blue and 4-Chlorophenol	CdS quantum dot	300 W WHL	100 ppm, 100 μM	95, 70	3, 4 h	[154]

WHL—Tungsten halogen lamp; IWL—iodine tungsten lamp; HL—halogen lamp; XL—xenon arc lamp; HPSL—high pressure sodium lamp characterized by its inertness to UV radiation, non-toxic nature, corrosion and erosion resistance at highly acidic and basic conditions and good adhesion to the catalyst particles. The different immobilization procedures include dip-coating, pasting, spray coating from suspension, sol–gel technique, CVD, sputtering and electrophoretic deposition. Usually, catalyst particles are coated many times on the supports to achieve better photocatalytic activity. However, excessive layers results in the detachment of the outer catalyst layers and cracking of the surface. Lim et al. [106] compared four different dip-coating methods for the mineralization of Methylene Blue and concluded that a hybrid method with 5 coating cycles and a calcination period of 1 h at 500°C yielded the best results. Moreover, the suitability of the substrates for immobilization followed the order: woven fiberglass $\approx$ fiberglass $\gg$ Al plate $\geq$ glass plate. However, the optimum number of coating cycles for the effective degradation of an organic compound is different from that required for the degradation of a microorganism. This was demonstrated in wall and fixed-bed reactors, where 2 dip-coating cycles exhibited the highest deactivation of *E. coli*, whereas, 3 dip-coating cycles were necessary for the effective degradation of Methylene Blue [166].

The degradation of organic compounds and microorganisms is, however, always lesser in an immobilized catalyst system compared to the suspended catalyst system. This is because of the reduction in the surface area of the catalyst, which results in the lesser adsorption of the reactants. Moreover, there is a significant reduction in activity of the immobilized catalyst with time due to the eventual loss of the catalyst particles from the surface, and the fouling of the catalyst, caused by the adsorption of the degradation products on the catalyst surface. Although the adsorbed products can be eliminated by calcination of the immobilized catalyst, this might affect the surface characteristics of the catalyst. Rao et al. [167] evaluated the reduction in activity of TiO_2 on different supports for the degradation of Acid Orange 7 and observed that the reductions in efficiency were a factor of four on TiO_2/polymer film, five on TiO_2/cellulosic fibers and ten on TiO_2/pumice stone after a period of 4 weeks. In a different study, Nakashima et al. [162] investigated the degradation of endocrine disrupting compounds like 17β-estradiol, bisphenol-A and 2,4-dichlorophenol using TiO_2 immobilized on PTFE mesh sheets. They observed 4.5 times enhancement in the mass transfer rate of these compounds when the PTFE sheets were rotated at 60 rpm compared to conventional circulation of the reaction mixture. In spite of the shortcomings, the recent research in the development of novel supports and immobilization protocols makes immobilized systems the preferred configuration for large-scale utilization of photocatalysis for the degradation of water pollutants.

10.2 Photocatalytic Reactor Modeling

Small scale batch reactors are widely used for the evaluation of the catalytic activity of novel photocatalytic materials, and for establishing the mechanism and kinetics of photocatalytic degradation reactions. Nevertheless, industrial decontamination of effluents requires large volumes to be processed. Hence, continuous operation is the preferred method, with catalyst particles immobilized on suitable supports. Different reactor configurations have been found suitable for photocatalytic degradation reactions, which include, thin film, flat plate, fluidized bed, packed bed, rotating disc, fountain type slurry, tubular, membrane, fiber, monolithic, bubble column, air-lift loop, Taylor-vortex flow and concentrating and non-concentrating solar reactors. A detailed description of the various photocatalytic reactor configurations is available elsewhere [168]. Given a wide range of reactor configurations, it is important to critically evaluate them based on the mass transfer limitations, spectral energy distribution of the light source, axial and radial dispersion of the reactants, catalyst layer thickness and energy efficiency, in order to find the best configuration for a particular class of pollutant and reaction conditions. Therefore, in depth modeling of these reactors based on the first principles chemical engineering approach is important.

The key component that distinguishes photocatalytic reactors from conventional reactors is the dependence of the kinetics of the reaction on the irradiance of the light source. Hence, models for photocatalytic reactors incorporate local volumetric rate of energy absorption (LVREA) in the rate expression. This is represented as [169–171]:

$$r = k_{\mathrm{R}}[e_{\lambda}^{\mathrm{a},\nu}(\underline{x})]^{n}\frac{K_{\mathrm{R}}C_{\mathrm{R}}}{1+K_{\mathrm{R}}C_{\mathrm{R}}} \tag{70}$$

where, r denotes the rate of the reaction per unit reactor volume, k_{R} denotes the rate coefficient for the degradation of the pollutant R, which is independent of light intensity, K_{R} denotes the equilibrium adsorption coefficient, C_{R} denotes the concentration of the pollutant, and $e_{\lambda}^{\mathrm{a},\nu}$ signifies the LVREA, which is dependent on the position vector $\underline{x}$ in the reaction space, and the exponent n signifies the dependence of the rate on LVREA. The selection of n is based on the light intensity and was discussed in Sect. 7.3. An expression for LVREA is [171]:

$$e_{\lambda}^{\mathrm{a},\nu}(\underline{x}) = \kappa_{\lambda}(\underline{x})G_{\lambda}(\underline{x}) \tag{71}$$

where, κ_{λ} is the spectral volumetric absorption coefficient, which is a function of the concentration of the absorbing species, and G_{λ} is the spectral local incident radiation [171].

Equation 70 shows that the rate follows L–H kinetics, but any appropriate rate form (like first-order or multicomponent L–H) can be used, based on the influence of the different reactants. Thus, by incorporating the rate of the reaction in the mass balance or continuity equations, and solving the equation with appropriate boundary conditions at steady-state, yields the concentration profiles of the reactant as a function of space inside the reactor. Similarly, by solving the momentum balance equation, the velocity profiles can be obtained. Jarandehei and Visscher [172] carried out a computational fluid dynamics simulation of a flat plate photocatalytic reactor with a serpentine geometry for the photocatalytic degradation of trichloroethylene. They found that the kinetics of the reaction is better captured by the L–H model than the first-order model. By analyzing the velocity and concentration profiles inside the reactor they observed that the laminar flow becomes unstable at a Reynolds number around 900, and mixing was efficient in the 180° sharp turns inside the reactor due to the formation of vortices.

Imoberdorf et al. [171] determined the LVREA inside a fluidized bed photoreactor using a Monte Carlo approach to track the photons. They have considered the shadowing effect of the spheres, absorption of radiation by the TiO_2-coated particles and reflection by the spheres. It was found that operating the bed at low expansions resulted in the effective absorption of the radiation. Chen et al. [173] evaluated the internal and external mass transfer resistances for the immobilized TiO_2 photoreactor, used for the degradation of benzoic acid. Their results indicate that, although the external mass transfer resistance can be minimized by increasing the velocity of the flow field (i.e., the Reynolds number), the internal mass transfer was limited by the TiO_2 film thickness. They observed the optimum catalyst layer thickness to be 5 μm, above which there was no effect of mass transfer on the rate of reaction. This correlates to the discussion in the previous section, where the number of coating cycles was found to be detrimental to the degradation of the pollutants in immobilized catalyst systems. Dijkstra et al. [174] evaluated the effects of different parameters like radial dispersion, external mass transfer, flow rate, catalyst loading and oxygen concentration on the degradation of formic acid in different immobilized reactor configurations. Their results suggest that membrane and packed bed reactors show better performance compared to fiber and tubular reactors. Imoberdorf et al. [175] employed the kinetic information obtained from an 81 cm^2 flat plate lab scale photoreactor to model a 5209 cm^2 multi-annular, series flow, pilot-scale photoreactor, for the degradation of perchloroethylene. The model, based on the mass balance and radiation transport equations, without any adjustable parameters, was able to predict the experimental conversion of perchloroethylene within an error limit of 5.6%. Thus, modeling of photocatalytic reactors is important for performance assessment, optimal design and scaling up, in large-scale industrial applications.

11 Concluding Remarks

In this comprehensive review, we have covered the fundamental aspects of photocatalysis as applied to the degradation of water pollutants. Over the past two decades, nano-sized TiO_2 has emerged as a leading photocatalyst with the potential to catalyze the degradation and mineralization of a wide range of toxic organic compounds, such as chlorinated organics, phenolics, dyes, pesticides and herbicides, surfactants, pharmaceutical compounds and drugs and microorganisms. The current trend in photocatalysis research is aimed at extending the spectral absorption of pristine or bare TiO_2 to the visible region, better harvesting solar radiation. Toward realizing this goal, many modified forms of TiO_2, such as anion-substituted TiO_2, dye-sensitized TiO_2 and heterostructured-TiO_2, have produced encouraging results.

One of the difficulties faced by the photocatalysis community is that TiO_2 synthesized by different protocols exhibit different physicochemical properties and substrate specific photoactivities and, hence, a single photocatalytic test with a specific material is not representative of its behavior for a wide class of organic compounds. Ryu and Choi [32] conducted the photocatalytic degradation of 19 test substances belonging to different classes of organic compounds using eight commercial TiO_2 samples, and concluded that each TiO_2 sample exhibits better photocatalytic activity compared to others for each class of organic compounds. Only DP-25 TiO_2 exhibited a better photoactivity for most of the organic compounds and can therefore be regarded as a yardstick for the assessment of the activity of novel photocatalytic materials. Therefore, owing to the ever increasing number of publications in this research area, the goal is to standardize the procedures, test substrates and photocatalytic activity, in order to have a rational comparison of the results of the different studies. Moreover, testing the degradation behavior of a single substrate is not representative of the "real" waste water or an effluent from a chemical plant. Hence, a degradation study of the multicomponent systems is a prospective research area. Although it is well-documented that the presence of humic substances, organic solvents, and chloride, sulfate and phosphate species retard the rate of degradation of organic compounds, evaluation of the degradation of systems with two or more organic compounds, metal ions or microorganisms is a relatively less explored topic.

The industrial viability of photocatalysis is driving research in the development of immobilized catalysts and photocatalytic reactors. In this venture, kinetic modeling of the reaction and reactor assumes paramount importance in the evaluation of the various rate limiting factors, optimum reaction conditions and scaling of reactors for the degradation of organic compounds. Currently, the large-scale applicability of photocatalysis for the degradation of organic pollutants is hindered by the costs involved in the operation of the treatment plant. Hence, more research in the selection of materials for the design of equipments and cheaper ways of catalyst synthesis and immobilization procedures is the key for further development. The authors do believe that exciting research opportunities exist in this field in the years to come.

References

1. Fujishima A, Honda K (1972) Electrochemical photolysis of water at a semiconductor electrode. Nature 238:37–38
2. Frank SN, Bard AJ (1977) Heterogeneous photocatalytic oxidation of cyanide and sulfite in aqueous solutions at semiconductor powders. J Phys Chem 81:1484–1488
3. Hsiao C-Y, Lee C-L, Ollis DF (1983) Heterogeneous photocatalysis: degradation of dilute solutions of dichloromethane (CH_2Cl_2), chloroform ($CHCl_3$) and carbon tetrachloride (CCl_4) with illuminated TiO_2 photocatalyst. J Catal 82:418–423

4. Pruden AL, Ollis DF (1983) Heterogeneous photocatalysis: the degradation of trichloroethylene in water. J Catal 82:404–417
5. Matsunaga T, Tomato R, Nakajima T, Wake H (1985) Photoelectrochemical sterilization of microbial cells by semiconductor powders. FEMS Microbiol Lett 29:211–214
6. O'Regan B, Grätzel M (1991) A low-cost, high efficiency solar cell based on dye-sensitized colloidal TiO_2 films. Nature 353:737–740
7. Fox MA, Dulay MT (1993) Heterogeneous photocatalysis. Chem Rev 93:341–357
8. Hoffmann MR, Martin ST, Choi W, Bahnemann DW (1995) Environmental applications of semiconductor photocatalysis. Chem Rev 95:69–96
9. Linsebigler AL, Lu G, Yates JT (1995) Photocatalysis on TiO_2 surfaces: principles, mechanisms and selected results. Chem Rev 95:735–758
10. Mills A, Hunte SL (1997) An overview of semiconductor photocatalysis. J Photochem Photobiol A: Chem 108:1–35
11. Fujishima A, Rao TN, Tryk DA (2000) Titanium dioxide photocatalysis. J Photochem Photobiol C: Photochem Rev 1:1–21
12. Carp O, Huisman CL, Reller A (2004) Photoinduced reactivity of titanium dioxide. Prog Solid State Chem 32:33–177
13. Fujishima A, Zhang X, Tryk DA (2008) TiO_2 photocatalysis and surface related phenomena. Surf Sci Rep 63:515–582
14. Chen X, Mao SS (2007) Titanium dioxide nanomaterials: synthesis, applications, modifications and applications. Chem Rev 107:2891–2959
15. Legrini O, Oliveros E, Braun AM (1993) Photochemical processes for water treatment. Chem Rev 93:671–698
16. Herrmann J-M (1999) Heterogeneous photocatalysis: fundamentals and applications to the removal of various types of aqueous pollutants. Catal Today 53:115–129
17. Bhatkhande DS, Pangarkar VG, Beenackers (2001) AACM Photocatalytic degradation for environmental applications—a review. J Chem Technol Biotechnol 77:102–116
18. Kabra K, Chaudhary R, Sawhney RL (2004) Treatment of hazardous organic and inorganic compounds through aqueous-phase photocatalysis: a review. Ind Eng Chem Res 43: 7683–7696
19. Zhao J, Chen C, Ma W (2005) Photocatalytic degradation of organic pollutants under visible light irradiation. Top Catal 35:269–278
20. Gaya UI, Abdullah AH (2008) Heterogeneous photocatalytic degradation of organic contaminants over titanium dioxide: a review of fundamentals, progress and problems. J Photochem Photobiol C: Photochem Rev 9:1–12
21. Malato S, Fernández-Ibáñez P, Maldonado MI, Blanco J, Gernjak W (2009) Decontamination and disinfection of water by solar photocatalysis: recent overview and trends. Catal Today 147:1–59
22. Wada Y, Yin H, Yanagida S (2002) Environmental remediation using catalysis driven under electromagnetic irradiation. Catal Sur Japan 5:127–138
23. Zhang Q, Gao L (2006) One-step preparation of size-defined aggregates of TiO_2 nanocrystals with tuning of their phase and composition. J Eur Ceram Soc 26:1535–1545
24. Sato S, Oimatsu S, Takahashi R, Sodesawa T, Nozaki F (1997) Pore size regulation of TiO_2 by use of a complex of titanium tetraisopropoxide and stearic acid. Chem Commun 22:2219-2220
25. Addamo M, Augugliaro V, Paola AD, García-López E, Loddo V, Marcí G, Molinari R, Palmisano L, Schiavello M (2004) Preparation, characterization and photoactivity of polycrystalline nanostructured TiO_2 catalysts. J Phys Chem B 108:3303–3310
26. Nagaveni K, Hegde MS, Ravishankar N, Subbanna GN, Madras G (2004) Synthesis and structure of nanocrystalline TiO_2 with lower band gap showing high photocatalytic activity. Langmuir 20:2900–2907
27. Mohamed MM, Bayoumy WA, Khairy M, Mousa MA (2007) Synthesis of micro-mesoporous TiO_2 materials assembled via cationic surfactants: morphology, thermal stability and surface acidity characteristics. Micropor Mesopor Mater 103:174–183

28. Venkatachalam N, Palanichamy M, Murugesan V (2007) Sol-gel preparation and characterization of nanosize TiO_2: its photocatalytic performance. Mater Chem Phys 104:454–459
29. Dong X, Tao J, Li Y, Zhu H (2010) Oriented single crystalline TiO_2 nano-pillars directly grown on titanium substrate in tetramethylammonium hydroxide solution. Appl Surf Sci 256:2532–2538
30. Cheng Y, Sun H, Jin W, Xu N (2007) Photocatalytic degradation of 4-chlorophenol with combustion synthesized TiO_2 under visible light irradiation. Chem Eng J 128:127–133
31. Zhang Z, Wang C-C, Zakaria R, Ying JY (1998) Role of particle size in nanocrystalline TiO_2-based photocatalysts. J Phys Chem B 102:10871–10878
32. Ryu J, Choi W (2008) Substrate-specific photocatalytic activities of TiO_2 and multiactivity test for water treatment application. Environ Sci Technol 42:294–300
33. Hidalgo MC, Colón G, Navío JA (2002) Modification of the physicochemical properties of commercial TiO_2 samples by soft mechanical activation. J Photochem Photobiol A: Chem 148:341–348
34. Hathway T, Jenks WS (2008) Effects of sintering of TiO_2 particles on the mechanisms of photocatalytic degradation of organic molecules in water. J Photochem Photobiol A: Chem 200:216–224
35. Kritikos DE, Xekoukoulotakis NP, Psillakis E, Mantzavinos D (2007) Photocatalytic degradation of reactive black 5 in aqueous solutions: effect of operating conditions and coupling with ultrasound irradiation. Water Res 41:2236–2246
36. Sivalingam G, Nagaveni K, Hegde MS, Madras G (2003) Photocatalytic degradation of various dyes by combustion synthesized nano anatase TiO_2. Appl Catal B: Environ 45:23–38
37. Aarthi T, Madras G (2007) Photocatalytic degradation of rhodamine dyes with nano-TiO_2. Ind Eng Chem Res 46:7–14
38. Vinu R, Akki SU, Madras G (2010) Investigation of dye functional group on the photocatalytic degradation of dyes by nano-TiO_2. J Hazard Mater 176:765–773
39. Sivalingam G, Priya MH, Madras G (2004) Kinetics of photodegradation of substituted phenols by solution combustion synthesized TiO_2. Appl Catal B: Environ 51:67–76
40. Priya MH, Madras G (2006) Kinetics of photocatalytic degradation of phenols with multiple substituent groups. J Photochem Photobiol A: Chem 179:256–262
41. Priya MH, Madras G (2006) Photocatalytic degradation of nitrobenzenes with combustion synthesized nano-TiO_2. J Photochem Photobiol A: Chem 178:1–7
42. Vijayalakshmi SP, Madras G (2006) Photocatalytic degradation of poly(ethylene oxide) and polyacrylamide. J Appl Polym Sci 100:3997–4003
43. Sivalingam G, Madras G (2004) Photocatalytic degradation of poly(bisphenol-A-carbonate) in solution over combustion-synthesized TiO_2: mechanism and kinetics. Appl Catal A: Gen 269:81–90
44. Aarthi P, Madras G (2008) Photocatalytic reduction of metals in presence of combustion synthesized nano-TiO_2. Catal Commun 9:630–634
45. Ollis DF, Pelizzetti E, Serpone N (1991) Photocatalyzed destruction of water contaminants. Environ Sci Technol 25:1522–1529
46. Blake DM (2001) Bibliography of work on the heterogeneous photocatalytic removal of hazardous compounds from water and air. NREL/TP-510-31319, National Renewable Energy Laboratory, Golden
47. Li X, Cubbage JW, Jenks WS (1999) Photocatalytic degradation of 4-chlorophenol. 2. The 4-chlorocatechol pathway. J Org Chem 64:8525–8536
48. Rajeshwar K, Osugi ME, Chanmanee W, Chenthamarakshan CR, Zanoni MVB, Kajitvichyanukul P, Krishnan-Ayer R (2008) Heterogeneous photocatalytic treatment of organic dyes in air and aqueous media. J Photochem Photobiol C: Photochem Rev 9:171–192
49. Rauf MA, Ashraf SS (2009) Fundamental principles and application of heterogeneous photocatalytic degradation of dyes in solution. Chem Eng J 151:10–18

50. Konstantinou IK, Albanis TA (2004) TiO_2-assisted photocatalytic degradation of azo dyes in aqueous solution: kinetic and mechanistic investigations: a review. Appl Catal B: Environ 49:1–14
51. Akpan UG, Hameed BH (2009) Parameters affecting the photocatalytic degradation of dyes using TiO_2-based photocatalysts: a review. J Hazard Mater 170:520–529
52. Han F, Kambala VSR, Srinivasan M, Rajarathnam D, Naidu R (2009) Tailored titanium dioxide photocatalysts for the degradation of organic dyes in wastewater treatment: a review. Appl Catal A: Gen 359:25–40
53. Houas A, Lachheb H, Ksibi M, Elaloui E, Guillard C, Herrmann J-M (2001) Photocatalytic degradation pathway of methylene blue in water. Appl Catal B: Environ 31:145–157
54. Epling GA, Lin C (2002) Photoassisted bleaching of dyes utilizing TiO_2 and visible light. Chemosphere 46:561–570
55. Silva CG, Wang W, Faria JL (2006) Photocatalytic and photochemical degradation of mono-, di- and tri-azo dyes in aqueous solution under UV irradiation. J Photochem Photobiol A: Chem 181:314–324
56. Stylidi M, Kondarides DI, Verykios XE (2003) Pathways of solar light-induced photocatalytic degradation of azo dyes in aqueous TiO_2 suspensions. Appl Catal B: Environ 40:271–286
57. Sleiman M, Vildozo D, Ferronato C, Chovelon J-M (2007) Photocatalytic degradation of azo dye Metanil Yellow: optimization and kinetic modeling using a chemometric approach. Appl Catal B: Environ 77:1–11
58. Hu C, Yu JC, Hao Z, Wong PK (2003) Photocatalytic degradation of triazine-containing azo dyes in aqueous TiO_2 suspensions. Appl Catal B: Environ 42:47–55
59. Saquib M, Muneer M (2002) Semiconductor mediated photocatalyzed degradation of an anthraquinone dye, Remazol Brilliant Blue R under sunlight and artificial light source. Dyes Pigments 53:237–249
60. Vautier M, Guillard C, Herrmann J-M (2001) Photocatalytic degradation of dyes in water: case study of indigo and of indigo carmine. J Catal 201:46–59
61. Wu T, Liu G, Zhao J, Hidaka H, Serpone N (1998) Photoassisted degradation of dye pollutants. V. Self-photosensitized oxidative transformation of Rhodamine B under visible light irradiation in aqueous TiO_2 dispersions. J Phys Chem B 102:5845–5851
62. Park H, Choi W (2005) Photocatalytic reactivities of nafion-coated TiO_2 for the degradation of charged organic compounds under UV or visible light. J Phys Chem B 109:11667–11674
63. Chen C-C, Lu C-S (2007) Mechanistic studies of the photocatalytic degradation of Methyl Green: an investigation of products of the decomposition processes. Environ Sci Technol 41:4389–4396
64. Konstantinou IK, Zarkadis AK, Albanis TA (2001) Photodegradation of selected herbicides in various natural waters and soils under environmental conditions. J Environ Qual 30:121–130
65. Zhu X, Yuan C, Bao Y, Yang J, Wu Y (2005) Photocatalytic degradation of pesticide pyridaben on TiO_2 particles. J Mol Catal A: Chem 229:95–105
66. Wei L, Shifu C, Wei Z, Sujuan Z (2009) Titanium dioxide mediated photocatalytic degradation of methamidophos in aqueous phase. J Hazard Mater 164:154–160
67. Moctezuma E, Leyva E, Palestino G, de Lasa H (2007) Photocatalytic degradation of methyl parathion: reaction pathways and intermediate reaction products. J Photochem Photobiol A: Chem 186:71–84
68. Lhomme L, Brosillon S, Wolbert D (2007) Photocatalytic degradation of a triazole pesticide, cyproconazole, in water. J Photochem Photobiol A: Chem 188:34–42
69. Yu B, Zeng J, Gong L, Zhang M, Zhang L, Chen X (2007) Investigation of the photocatalytic degradation of organochlorine pesticides on a nano-TiO_2 coated film. Talanta 72:1667–1674
70. Navarro S, Fenoll J, Vela N, Ruiz E, Navarro G (2009) Photocatalytic degradation of eight pesticides in leaching water by use of ZnO under natural sunlight. J Hazard Mater 172:1303–1310

71. Aungpradit T, Sutthivaiyakit P, Martens D, Sutthivaiyakit S, Kettrup AAF (2007) Photocatalytic degradation of triazophos in aqueous titanium dioxide suspension: identification of intermediates and degradation pathways. J Hazard Mater 146:204–213
72. Molinari R, Pirillo F, Loddo V, Palmisano L (2006) Heterogeneous photocatalytic degradation of pharmaceuticals in water by using polycrystalline TiO_2 and a nanofiltration membrane reactor. Catal Today 118:205–213
73. Chatzitakis A, Berberidou C, Paspaltsis I, Kyriakou G, Sklaviadis T, Poulios I (2008) Photocatalytic degradation and drug activity reduction of chloramphenicol. Water Res 42:386–394
74. Abellán MN, Bayarri B, Giménez J, Costa J (2007) Photocatalytic degradation of sulfamethoxazole in aqueous suspension of TiO_2. Appl Catal B: Environ 74:233–241
75. An T, Yang H, Li G, Song W, Cooper WJ, Nie X (2010) Kinetics and mechanism of advanced oxidation processes (AOPs) in degradation of ciprofloxacin in water. Appl Catal B: Environ 94:288–294
76. Radjenović J, Sirtori C, Petrović M, Barceló D, Malato S (2009) Solar photocatalytic degradation of persistent pharmaceuticals at pilot-scale: kinetics and characterization of major intermediate products. Appl Catal B: Environ 89:255–264
77. Méndez-Arriaga F, Esplugas S, Giménez J (2008) Photocatalytic degradation of non-steroidal anti-inflammatory drugs with TiO_2 and simulated solar irradiation. Water Res 42:585–594
78. Calza P, Sakkas VA, Medana C, Baiocchi C, Dimou A, Pelizzetti E, Albanis T (2006) Photocatalytic degradation study of diclofenac over aqueous TiO_2 suspensions. Appl Catal B: Environ 67:197–205
79. Sakkas VA, Calza P, Medana C, Villioti AE, Baiocchi C, Pelizzetti E, Albanis T (2007) Heterogeneous photocatalytic degradation of the pharmaceutical agent salbutamol in aqueous titanium dioxide suspensions. Appl Catal B: Environ 77:135–144
80. Calza P, Massolino C, Monaco G, Medana C, Baiocchi C (2008) Study of the photolytic and photocatalytic transformation of amiloride in water. J Pharm Biomed Anal 48:315–320
81. Cheng YW, Chan RCY, Wong (2007) Disinfection of *Legionella pneumophila* by photocatalytic oxidation. Water Res 41:842–852
82. Ibáñez JA, Litter AI, Pizarro RA (2003) Photocatalytic bactericidal effect of TiO_2 on *Enterobacter cloacae*: comparative study with other gram (−) bacteria. J Photochem Photobiol A:Chem 157:81–85
83. Prasad GK, Agarwal GS, Singh B, Rai GP, Vijayaraghavan R (2009) Photocatalytic inactivation of *Bacillus anthracis* by titania nanomaterials. J Hazard Mater 165:506–510
84. Liu H-L, Yang TC-K (2003) Photocatalytic inactivation of *Escherichia coli* and *Lactobacillus helveticus* by ZnO and TiO_2 activated with ultraviolet light. Process Biochem 39:475–481
85. Kubacka A, Ferrer M, Martínez-Arias A, Fernández-García M (2008) Ag promotion of TiO_2-anatase disinfection capability: study of *Escherichia coli* inactivation. Appl Catal B: Environ 84:87–93
86. Hu C, Guo J, Qu J, Hu X (2007) Photocatalytic degradation of pathogenic bacteria with AgI/TiO_2 under visible light irradiation. Langmuir 23:4982–4987
87. Wu T-S, Wang K-X, Li G-D, Sun S-Y, Sun J, Chen J-S (2010) Montmorillonite-supported Ag/TiO_2 nanoparticles: an efficient visible-light bacteria photodegradation material. ACS Appl Mater Interf 2:544–550
88. Wu P, Xie R, Imlay JA, Shang JK (2009) Visible-light-induced photocatalytic inactivation of bacteria by composite photocatalysts of palladium oxide and nitrogen-doped titanium oxide. Appl Catal B: Environ 88:576–581
89. Yu JC, Ho W, Yu J, Yip H, Wong PK, Zhao J (2005) Efficient visible-light-induced photocatalytic disinfection on sulfur-doped nanocrystalline titania. Environ Sci Technol 39:1175–1179

90. Pelaez M, de la Cruz AA, Stathatos E, Falaras P, Dionysiou DD (2009) Visible light-activated N–F-codoped TiO_2 nanoparticles for the photocatalytic degradation of microcystin-LR in water. Catal Today 144:19–25
91. Cho M, Chung H, Choi W, Yoon J (2004) Linear correlation between inactivation of *E. coli* and OH radical concentration in TiO_2 photocatalytic disinfection. Water Res 34:1069–1077
92. Matsunaga T, Tomoda R, Nakajima T, Nakamura N, Komine T (1988) Continuous-sterilization system that uses photosemiconductor powders. Appl Environ Microbiol 54:1330–1333
93. Saito T, Iwase T, Horie J, Morioka T (1992) Mode of photocatalytic bactericidal action of powdered semiconductor TiO_2 on mutans streptococci. J Photochem Photobiol B: Biol 14:369–379
94. Maness P-C, Smolinski S, Blake DM, Hyang Z, Wolfrum EJ, Jacoby WA (1999) Bactericidal activity of photocatalytic TiO_2 reaction: toward an understanding of its killing mechanism. Appl Environ Microbiol 65:4094–4098
95. Rincón A-G, Pulgarin C (2004) Effect of pH, inorganic ions, organic matter and H_2O_2 on *E. coli* K12 photocatalytic inactivation by TiO_2: implications in solar water ds infection. Appl Catal B: Environ 51:283–302
96. Pal A, Pehkonen SO, Yu LE, Ray MB (2008) Photocatalytic inactivation of airborne bacteria in a continuous-flow reactor. Ind Eng Chem Res 47:7580–7585
97. Sharma VK, Yngard RA, Lin Y (2009) Silver nanoparticles: green synthesis and their antimicrobial activities. Adv Coll Interf Sci 145:83–96
98. Naeem K, Weiqian P, Ouyang F (2010) Thermodynamic parameters of activation for photodegradation of phenolics. Chem Eng J 156:505–509
99. Madras G, Smith JM, McCoy BJ (1996) Thermal degradation of poly(α-methyl styrene) in solution. Polym Degrad Stab 52:349–358
100. Devi LG, Murthy BN, Kumar SG (2009) Heterogeneous photo catalytic degradation of anionic and cationic dyes over TiO_2 and TiO_2 doped with Mo^{6+} ions under solar light: correlation of dye structure and its adsorptive tendency on the degradation rate. Chemosphere 76:1163–1166
101. Alrousan DMA, Dunlop PSM, McMurray TA, Byrne JA (2009) Photocatalytic inactivation of *E. coli* in surface water using immobilized nanoparticle TiO_2 films. Water Res 43:47–54
102. Matthews RW, McEvoy SR (1992) A comparison of 254 and 350 nm excitation of TiO_2 in simple photocatalytic reactors. J Photochem Photobiol A: Chem 66:355–366
103. Kuhn HJ, Braslavsky SE, Schmidt R (2004) Chemical actinometry. Pure Appl Chem 76:2105–2146
104. Meng Y, Huang X, Wu Y, Wang X, Qian Y (2002) Kinetic study and modeling on photocatalytic degradation of para-chlorobenzoate at different light intensities. Environ Poll 117:307–313
105. Wu C-H, Chern J-M (2006) Kinetics of photocatalytic decomposition of methylene blue. Ind Eng Chem Res 45:6450–6457
106. Lim LLP, Lynch RJ, In S-I (2009) Comparison of simple and economical photocatalyst immobilization procedures. Appl Catal A: Gen 365:214–221
107. Epling GA, Lin C (2002) Investigation of retardation effects on the titanium dioxide photodegradation system. Chemosphere 46:937–944
108. Azevedo EB, Neto FRA, Dezotti M (2004) TiO_2-photocatalyzed degradation of phenol in saline media: lumped kinetics, intermediates and acute toxicity. Appl Catal B: Environ 54:165–173
109. Chen D, Ray AK (2001) Removal of toxic metal ions from wastewater by semiconductor photocatalysis. Chem Eng Sci 56:1561–1570
110. Prairie MR, Evans LR, Stange BM, Martinez SL (1993) An investigation of TiO_2 photocatalysis for the treatment of water contaminated with metals and organic chemicals. Environ Sci Technol 27:1776–1782

111. Chen C, Li X, Ma W, Zhao J, Hidaka H, Serpone N (2002) Effect of transition metal ions on the TiO_2-assisted photodegradation of dyes under visible irradiation: a probe for interfacial electron transfer process and reaction mechanism. J Phys Chem B 106:318–324
112. Kyung H, Lee J, Choi W (2005) Simultaneous and synergistic conversion of dyes and heavy metal ions in aqueous TiO_2 suspensions under visible-light illumination. Environ Sci Technol 39:2376–2382
113. Wang N, Chen Z, Zhu L, Jiang X, Lv B, Tang H (2007) Synergistic effects of cupric and fluoride ions on photocatalytic degradation of phenol. J Photochem Photobiol A: Chem 191:193–200
114. Vinu R, Madras G (2008) Kinetics of simultaneous photocatalytic degradation of phenolic compounds and reduction of metal ions with nano-TiO_2. Environ Sci Technol 42:913–919
115. Sun B, Reddy EP, Smirniotis PG (2005) Visible light Cr(VI) reduction and organic chemical oxidation by TiO_2 photocatalysis. Environ Sci Technol 39:6251–6259
116. Turchi CS, Ollis DF (1990) Photocatalytic degradation of organic water contaminants: mechanisms involving hydroxyl radical attack. J Catal 122:178–192
117. Li Y, Sun S, Ma M, Ouyang Y, Yan W (2008) Kinetic study and model of the photocatalytic degradation of rhodamine B(RhB) by a TiO_2-coated activated carbon catalyst: effects of initial RhB content, light intensity and TiO_2 content in the catalyst. Chem Eng J 142:147–155
118. Almquist CB, Biswas P (2001) A mechanistic approach to modeling the effect of dissolved oxygen in photo-oxidation reactions on titanium dioxide in aqueous systems. Chem Eng Sci 56:3421–3430
119. Benabbou AK, Derriche Z, Felix C, Lejeune P, Guillard C (2007) Photocatalytic inactivation of *Escherischia coli*: effect of concentration of TiO_2 and microorganism, nature and intensity of UV irradiation. Appl Catal B: Environ 76:257–263
120. Marugán J, van Grieken R, Sordo C, Cruz C (2008) Kinetics of photocatalytic disinfection of *Escherichia coli* suspensions. Appl Catal B: Environ 82:27–36
121. Labas MD, Brandi RJ, Martín CA, Cassano AE (2006) Kinetics of bacteria inactivation employing UV radiation under clear water conditions. Chem Eng J 121:135–145
122. Marugán J, van Griekan R, Pablos C, Sordo C (2010) Anologies and differences between photocatalytic oxidation of chemicals and photocatalytic inactivation of microorganisms. Water Res 44:789–796
123. Priya MH, Madras G (2006) Kinetics of photocatalytic degradation of chlorophenol, nitrophenol and their mixtures. Ind Eng Chem Res 45:482–486
124. Aarthi P, Narahari P, Madras G (2007) Photocatalytic degradation of azure and sudan dyes using nano-TiO_2. J Hazard Mater 149:725–734
125. Levenspiel O (1999) Chemical reaction engineering. John Wiley, Singapore
126. Choi W, Termin A, Hoffmann MR (1994) The role of metal ion dopants in quantum-sized TiO_2: correlation between photoreactivity and charge carrier recombination dynamics. J Phys Chem 98:13669–13679
127. Serpone N, Lawless D, Disdier J, Herrmann J-M (1994) Spectroscopic, photoconductivity, and photocatalytic studies of TiO_2 colloids: naked and with the lattice doped with Cr^{3+}, Fe^{3+} and V^{5+} cations. Langmuir 10:643–652
128. Shah SI, Li W, Huang C-P, Jung O, Ni C (2002) Study of Nd^{3+}, Pd^{2+}, Pt^{4+} and Fe^{3+} dopant effect on photoreactivity of TiO_2 nanoparticles. Proc Nat Acad Sci 99:6482–6486
129. Dvoranová D, Brezová V, Mazúr M, Malati MA (2002) Investigations of metal-doped titanium dioxide photocatalysts. Appl Catal B: Environ 37:91–105
130. Nagaveni K, Hegde MS, Madras G (2004) Structure and photocatalytic activity of $Ti_{1-x}M_xO_{2\pm\delta}$ (M = W, V, Ce, Zr, Fe, and Cu) synthesized by solution combustion method. J Phys Chem B 108:20204–20212
131. Štengl V, Bakardjieva S, Murafa N (2009) Preparation and photocatalytic activity of rare earth doped TiO_2 nanoparticles. Mater Chem Phys 114:217–226

132. Roy S, Hegde MS, Ravishankar N, Madras G (2007) Creation of redox adsorption sites by Pd^{2+} ion substitution in nano TiO_2 for high photocatalytic activity of CO oxidation, NO reduction and NO decomposition. J Phys Chem C 111:8153–8160
133. Vinu R, Madras G (2008) Synthesis and photoactivity of Pd substituted nano-TiO_2. J Mol Catal A: Chem 291:5–11
134. Paola AD, García-López E, Ikeda S, Marcí G, Ohtani B, Palmisano L (2002) Photocatalytic degradation of organic compounds in aqueous systems by transition metal doped polycrystalline TiO_2. Catal Today 75:87–93
135. Vinu R, Madras G (2009) Photocatalytic activity of Ag-substituted and impregnated nano-TiO_2. Appl Catal A: Gen 366:130–140
136. Asahi R, Morikawa T, Ohwaki T, Aoki K, Taga Y (2002) Visible-light photocatalysis in nitrogen-doped titanium oxides. Science 293:269–271
137. Khan SUM, Al-Shahry M, Ingler WB Jr (2002) Efficient photochemical water splitting by a chemically modified n-TiO_2. Science 297:2243–2245
138. Liu G, Wang L, Yang HG, Cheng H-M, Lu GQM (2010) Titania-based photocatalysts–crystal growth, doping and heterostructuring. J Mater Chem 20:831–843
139. Serpone N (2006) Is the band gap of pristine TiO_2 narrowed by anion- and cation-doping of titanium dioxide in second-generation photocatalysts? J Phys Chem B 110:24287–24293
140. Nagaveni K, Sivalingam G, Hegde MS, Madras G (2004) Solar photocatalytic degradation of dyes: high activity of combustion synthesized nano TiO_2. Appl Catal B: Environ 48:83–93
141. Xiao Q, Zhang J, Xiao C, Si Z, Tan X (2008) Solar photocatalytic degradation of methylene blue in carbon-doped TiO_2 nanoparticles suspension. Sol Energy 82:706–713
142. Zhong J, Chen F, Zhang J (2010) Carbon-deposited TiO_2: synthesis, characterization and visible photocatalytic performance. J Phys Chem C 114:933–939
143. Ao Y, Xu J, Fu D, Yuan C (2009) Synthesis of C, N, S-tridoped mesoporous titania with enhanced visible light-induced photocatalytic activity. Micropor Mesopor Mater 122:1–6
144. Yu T, Tan X, Zhao L, Yin Y, Chen P, Wei J (2009) Characterization, activity and kinetics of a visible light driven photocatalyst: cerium and nitrogen co-doped TiO_2 particles. Chem Eng J. doi: 10.1016/j.cej.2009.10.051
145. Ho W, Yu JC, Lin J, Yu J, Li P (2004) Preparation and photocatalytic behavior of MoS_2 and WS_2 nanocluster sensitized TiO_2. Langmuir 20:5865–5869
146. Wang D, Zhang J, Luo Q, Li X, Duan Y, An J (2009) Characterization and photocatalytic activity of poly(3-hexyl thiophene)-modified TiO_2 for degradation of methyl orange under visible light. J Hazard Mater 169:546–550
147. Sun Q, Xu Y (2009) Sensitization of TiO_2 with aluminium phthalocyanine: factors influencing the efficiency for chlorophenol degradation in water under visible light. J Phys Chem C 113:12387–12394
148. Liang H-C, Li X-Z (2009) Visible-induced photocatalytic reactivity of polymer sensitized titania nanotube films. Appl Catal B: Environ 86:8–17
149. Xiangzhong L, Wei Z, Jincai Z (2002) Visible light-sensitized semiconductor photocatalytic degradation of 2, 4-dichlorophenol. Sci China Ser B 45:421–425
150. Gilma GO, Carlos APM, Fernando MO, Edgar AP-M (2005) Photocatalytic degradation of phenol on TiO_2 and TiO_2/Pt sensitized with matallophthalocyanines. Catal Today 107–108:589–594
151. Ross H, Bendig J, Hecht S (1994) Sensitized photocatalytical oxidation of terbutylazine. Solar Energy Mater Solar Cells 33:475–481
152. Granados-Oliveros G, Páez-Mozo EA, Ortega FM, Ferronato C, Chovelon J-M (2009) Degradation of atrazine using metalloporphyrins supported on TiO_2 under visible light irradiation. Appl Catal B: Environ 89:448–454
153. Bae E, Choi W (2003) Highly enhanced photoreductive degradation of perchlorinated compounds on dye-sensitized metal/TiO_2 under visible light. Environ Sci Technol 37:147–152

154. Li G-S, Zhang D-Q, Yu JC (2009) A new visible-light photocatalyst: CdS quantum dots embedded mesoporous TiO_2. Environ Sci Technol 43:7079–7085
155. Chiou C-S, Shie J-L, Chang C-Y, Liu C-C, Chang C-T (2006) Degradation of di-*n*-butyl phthalate using photoreactor packed with TiO_2 immobilized on glass beads. J Hazard Mater B 137:1123–1129
156. Horikoshi S, Watanabe N, Onishi H, Hidaka H, Serpone N (2002) Photodecomposition of a nonylphenol polyethoxylate surfactant in a cylindrical photoreactor with TiO_2 immobilized fiberglass cloth. Appl Catal B: Environ 37:117–129
157. Lee J-M, Kim M-S, Kim B-W (2004) Photodegradation of bisphenol-A with TiO_2immobilized on the glass tubes including the UV light lamps. Water Res 38:3605–3613
158. Tryba B (2008) Immobilization of TiO_2 and Fe-C-TiO_2 photocatalysts on the cotton material for application in a flow photocatalytic reactor for decomposition of phenol in water. J Hazard Mater 151:623–627
159. Chen Y, Dionysiou DD (2007) A comparative study on physicochemical properties and photocatalytic behavior of macroporous TiO_2–P25 composite films and macroporous TiO_2 films coated on stainless steel substrate. Appl Catal A: Gen 317:129–137
160. Zhang Q, Fan W, Gao L (2007) Anatase TiO_2 nanoparticles immobilized on ZnO tetrapods as a highly efficient and easily recyclable photocatalyst. Appl Catal B: Environ 76:168–173
161. Gao Y, Liu H (2005) Preparation and catalytic property study of a novel kind of suspended photocatalyst of TiO_2-activated carbon immobilized on silicone rubber film. Mater Chem Phys 92:604–608
162. Nakashima T, Ohko Y, Tryk DA, Fujishima A (2002) Decomposition of endocrine-disrupting chemicals in water by use of TiO_2 photocatalysts immobilized on polytetrafluoroethylene mesh sheets. J Photochem Photobiol A: Chem 151:207–212
163. Magalhães F, Lago RM (2009) Floating photocatalysts based on TiO_2 grafted on expanded polystyrene beads for the solar degradation of dyes. Sol Energy 83:1521–1526
164. Faramarzpour M, Vossoughi M, Borghei M (2009) Photocatalytic degradation of furfural by titania nanoparticles in a floating-bed photoreactor. Chem Eng J 146:79–85
165. Rachel A, Lavedrine B, Subrahmanyam M, Boule P (2002) Use of porous lavas as supports of photocatalysis. Catal Commun 3:165–171
166. Grieken RV, Marugán J, Sordo C, Pablos C (2009) Comparison of the photocatalytic disinfection of *E. coli* suspensions in slurry, wall and fixed-bed reactors. Catal Today 144:48–54
167. Rao KVS, Subrahmanyam M, Boule P (2004) Immobilized TiO_2 photocatalyst during long-tem use: decrease of activity. Appl Catal B: Environ 49:239–249
168. Lasa H, Serrano B, Salaices M (2005) Photocatalytic reaction engineering. Springer, New York
169. Puma GL, Yue PL (2001) A novel fountain photocatalytic reactor: model development and experimental validation. Chem Eng Sci 56:2733–2744
170. Denny F, Scott J, Pareek V, Peng GD, Amal R (2009) CFD modeling for a TiO_2-coated glass-bead photoreactor irradiated by optical fibres: photocatalytic degradation of oxalic acid. Chem Eng Sci 64:1695–1706
171. Imoberdorf GE, Taghipour F, Keshmiri M, Mohseni M (2008) Predictive radiative field modeling for fluidized bed photocatalytic reactors. Chem Eng Sci 63:4228–4238
172. Jarandehei A, Visscher AD (2009) Three-dimensional CFD model for a flat plate photocatalytic reactor: degradation of TCE in a serpentine flow field. AIChE J 55:312–320
173. Chen D, Li F, Ray AK (2000) Effect of mass transfer and catalyst layer thickness on photocatalytic reaction. AIChE J 46:1034–1045
174. Dijkstra MFJ, Hoerts ECB, Beenackers AACM, Wesselingh JA (2003) Performance of immobilized photocatalytic reactors in continuous mode. AIChE J 49:734–744
175. Imoberdorf GE, Irazoqui HA, Alfano OM, Cassano AE (2007) Scaling-up from first principles of a photocatalytic reactor for air pollution remediation. Chem Eng Sci 62: 793–804

Advanced Photocatalytic Nanomaterials for Degrading Pollutants and Generating Fuels by Sunlight

Dieqing Zhang, Guisheng Li and Jimmy C. Yu

Abstract This chapter focuses on the recent development of sunlight-driven heterogeneous photocatalysts with different chemical compositions and nanostructures. Various photocatalytic nanomaterials, including metal oxides, heterojunction nanocomposites, oxynitrides, oxysulfides, and graphitic carbon nitride, are described. Their preparation methods as well as the mechanisms involved are introduced. These materials can be used to degrade pollutants and generate fuels. Photocatalytic evolution of H_2 from water and conversion of CO_2 to fuels are discussed in detail. The development of advanced photocatalytic technology involving novel nanomaterials may allow the construction of clean and facile systems for solving the global energy and environmental problems.

1 Introduction

Environment and energy are two of the world's most challenging issues. For a sustainable society, it is absolutely necessary to develop efficient pollution treatment technologies and harvest clean energy. Photocatalysis has a role to play in both aspects. Photocatalysis is a natural phenomenon that promotes chemical reactions on the surface of an irradiated semiconductor. The essence of the photocatalysis is attributed to the property of photo-excited carriers

D. Zhang · J. C. Yu (✉)
Department of Chemistry, Environmental Science Programme,
and Institute of Environment, Energy and Sustainability,
The Chinese University of Hong Kong, Shatin, New Territories,
Hong Kong, China
e-mail: jimyu@cuhk.edu.hk

D. Zhang · G. Li
Department of Chemistry, Shanghai Normal University, 200234 Shanghai, China
e-mail: lgscuhk@yahoo.com.cn

L. Zang (ed.), *Energy Efficiency and Renewable Energy Through Nanotechnology*,
Green Energy and Technology, DOI: 10.1007/978-0-85729-638-2_20,

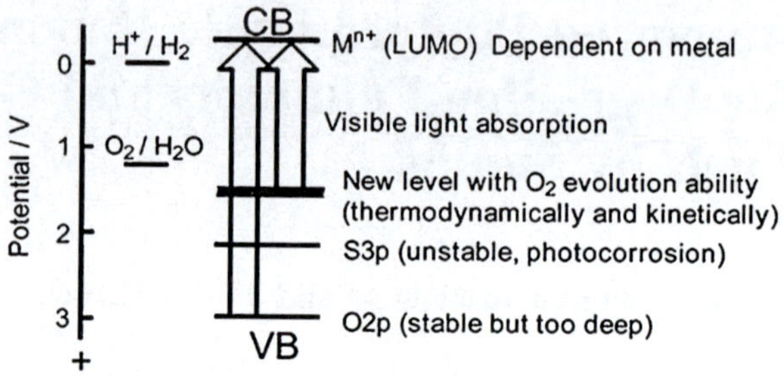

Fig. 1 Band structure control to develop visible-light-driven photocatalysts for water splitting. Reprinted with permission from Ref. [2]. Copyright 2009 American Chemical Society

(electrons and holes) with strong oxidization and reduction power. The redox reactions contribute to the decomposition of hazardous pollutants, conversion of carbon dioxide to valuable hydrocarbons, and the decomposition of water to hydrogen and oxygen.

Photocatalysts are materials that can realize the photocatalysis process. Much attention has been paid to these materials for the development of environment-friendly technology [1–9]. Titanium dioxide (TiO_2) is the most widely used photocatalyst. However, TiO_2 displays a high activity only when it is irradiated by UV light, where the light wavelength is shorter than 400 nm. It is therefore not efficient under sunlight irradiation. Suitable band engineering is required to develop new photocatalysts for solar applications (shown in Fig. 1) [2]. In this chapter, we discuss the fundamental issues that govern the design of visible-light responsive photocatalysts. We also describe some of their applications in environmental and energy aspects.

2 Solar-Light-Driven Photocatalysts for Degrading Pollutants

2.1 *Doped TiO_2 Photocatalysts*

The band gap of bulk TiO_2 lies in the UV regime (3.0 eV for the rutile phase and 3.2 eV for the anatase phase) [10]. Solar application of TiO_2 materials is limited by its wide band gap because pure TiO_2 can only absorb a small fraction of the sun's energy (<10%). To improve the efficiency, doping TiO_2 with metal/nonmetal atoms has proven an efficient route to broadening the photoresponse of TiO_2 to include the visible-light region. Recently, Chen et al. reviewed the modification of TiO_2 with metal/non-metal atoms in detail [10]. Herein, we will focus on the latest reported work related to doping metal/non-metal atoms into the TiO_2 framework.

2.1.1 Metal Doped TiO_2

A number of metal atoms have been doped into the framework of TiO_2 nanomaterials [11–17]. Besides the traditional methods, such as wet chemistry, high-temperature treatment, and ion implantation, various novel routes have been developed for doping metal atoms into the TiO_2 nanomaterials. These include hydro-alcohol thermal, electrospinning, and flame spray pyrolysis (FSP) techniques. Yu et al. fabricated Fe-doped TiO_2 (Fe-TiO_2) nanorods with an impregnating-calcination method using a hydrothermally-prepared titanate nanotube as a precursor and $Fe(NO_3)_3$ as the dopant. Fe-doping greatly enhanced the visible-light photocatalytic activity of mesoporous TiO_2 nanorods, and when the atomic ratio of Fe/Ti (R-Fe) was in the range of 0.1–1.0%, the photocatalytic activity of the samples was higher than that of Degussa P25 and pure TiO_2 nanorods. At R-Fe = 0.5%, the photocatalytic activity of Fe-TiO_2 nanorods exceeded that of Degussa P25 by a factor of more than two [11]. Wang et al. prepared mesoporous W^{6+}-doped TiO_2 thin film photocatalysts by electrospinning and sol–gel chemistry through employing a triblock copolymer as a structure-directing agent. 3% W^{6+} was found to be the most suitable doping concentration, at which the recombination of photoinduced electrons and holes were effectively inhibited [12]. Li et al. fabricated V-doped TiO_2 (V-TiO_2) nanoparticles using a simple one-step FSP technique. Under visible-light irradiation, the degradation rate of 2, 4-dichlorophenol over 1% V-TiO_2 was two times higher than that of undoped TiO_2 [13]. Li et al. further utilized the one-step FSP technique to fabricate Cr-doped TiO_2 nanoparticles. The optimal Cr^{3+} concentration was found to be 1% [14]. Lorret et al. prepared nanocrystalline tungsten-doped titanium dioxide powders using a sol–gel method based on the hydrolysis of $TiCl_4$ in aqueous solution. Introducing tungsten into the TiO_2 framework could effectively extend light absorption of the TiO_2-based photocatalysts toward the visible-light range [15]. Dai et al. used a hydro-alcohol thermal method to fabricate Fe-doped titanium dioxide (TiO_2) microspheres with special core–shell structures. The concentration of Fe^{3+} played a key role in the photocatalytic degradation of phenol. Moreover, the 0.5 mol% Fe^{3+} doping was an optimal amount [16]. Yang et al. found that doping ruthenium, by an ion-exchange method, on the hydrothermally synthesized titania nanotube (Ti-NT) greatly enhanced the photocatalytic activity for degrading methylene blue (MB) dye under visible-light irradiation [17].

2.1.2 Non-Metal Doped TiO_2

Different nonmetal elements, such as B, C, N, F, and S, have been utilized recently to modify TiO_2 nanomaterials [18–26]. Xu et al. fabricated B-doped titania hollow spheres. They found that doping boron atoms effectively enhanced the photocatalytic activity of the hollow titania spheres in the degradation of Reactive Brilliant Red dye X-3B (C.I. Reactive Red 2) under-visible light irradiation [18]. Choi et al. reported a carbon-doped TiO_2 (C-TiO_2) photocatalyst prepared from a conventional sol–gel synthesis without using external carbon precursors. The carbon atoms from the titanium alkoxide precursor were incorporated into the lattice of TiO_2, creating mid-bandgap electronic states through controlled calcination [19]. Lu et al. demonstrated a facile route for the one-pot synthesis of visible-light responsive nitrogen-doped anatase TiO_2 sheets with dominant facets of TiN.

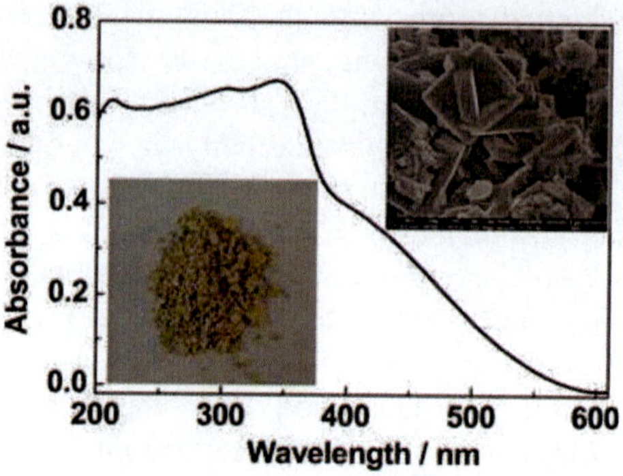

Fig. 2 UV–visible absorption spectra of (**a**) pure anatase TiO_2 sheets and (**b**) nitrogen-doped anatase TiO_2 sheets; the *insets* in the *upper right* and *lower left* corners are the plot of transformed Kubelka–Munk function vs the energy of light and optical photo of nitrogen-doped anatase TiO_2 sheets. Reprinted with permission from Ref. [27]. Copyright 2009 American Chemical Society

The as-synthesized anatase TiO_2 sheets showed a strong and stable ability to generate hydroxyl radicals [27].

The UV–visible absorption spectra of the anatase TiO_2 sheets (Fig. 2) shows an additional high visible-light absorption band from 400 nm to ca. 570 nm, consistent with the yellow color of the sample (see inset of Fig. 2). Derived from the plot of the Kubelka–Munk function versus the energy of the light absorbed, assuming titania is an indirect semiconductor, the bandgap of the obtained anatase TiO_2 sheets is extrapolated to be 3.11 eV, which is nearly identical to that of pure bulk anatase TiO_2. However, such nitrogen-doped {001}-dominant anatase TiO_2 sheets show a significantly enhanced visible-light absorption [27]. Yu et al. used a one-step low-temperature hydrothermal approach to fabricate hierarchical porous F-doped TiO_2 microspheres as shown in Fig. 3. These hierarchical porous microspheres exhibited high activity in the photocatalytic degradation of 4-chlorophenol under visible-light illumination [28].

Yu et al. proposed a one-step low-temperature hydrothermal route to synthesize S-doped TiO_2 photocatalysts from TiS_2 and HCl. Sulfur atoms could be efficiently doped into the anatase lattice under the mild hydrothermal conditions. The S-doped TiO_2 prepared by this hydrothermal approach exhibited much higher photocatalytic activity than that obtained by the traditional high-temperature thermal annealing method for the degradation of 4-chlorophenol under visible-light irradiation [29]. Li et al. prepared a S-doped TiO_2 by treating a TiO_2 xerogel under supercritical conditions in CS_2/ethanol fluid. The TiO_2 was modified through forming S–Ti–O bonds rather than adsorbing CS_2. During liquid-phase photocatalytic degradation of MB under visible-light irradiation, the S-doped TiO_2 exhibited higher activity than that of the undoped TiO_2 and even the N-doped TiO_2 obtained via supercritical treatment. A maximum activity of nearly eight times higher than that of commercially available Degussa P25 was obtained at a 1.8% S/Ti molar ratio [22].

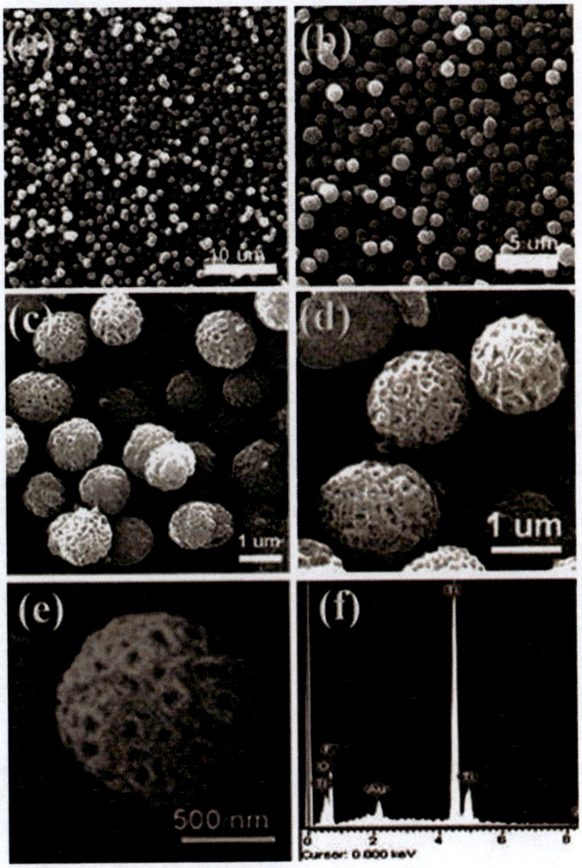

Fig. 3 SEM images of (**a–d**) porous F-doped TiO_2 microspheres with different magnifications; (**e**) an individual single microsphere showing detailed texture and porosity (ca. 1 μm in diameter); (**f**) EDX microanalysis spectrum of porous F-doped TiO_2 microspheres. Ref. [28]—reproduced by permission of The Royal Society of Chemistry

2.2 *Nano-heterojunction (TiO_2-Based) Photocatalytic Materials*

Modification of the TiO_2 band gap by doping [30–34] and development of new semiconductor materials capable of absorbing visible light [35] are the major strategies for developing visible-light photocatalysts. However, the low quantum efficiency, owing to the fast recombination of photo-generated electron–hole pairs, is still a challenge. This can be partially overcome by the construction of a heterojunction interface between semiconductors with matching band potentials. This allows electric-field-assisted charge transport from one particle to the other [36]. To date, the reported heterojunction semiconductors mainly fall into two categories: TiO_2-based photocatalysts [37, 38] and a small number of non-TiO_2-based systems [39, 40].

Recently, Yu et al. reported a cadmium sulfide quantum dots (QDs) sensitized mesoporous TiO_2 heterojunction photocatalyst [41]. It was prepared by preplanting cadmium oxide as crystal seeds into the framework of ordered mesoporous titanium dioxide and then converting CdO to CdS QDs through ion-exchange (as shown in Fig. 4).

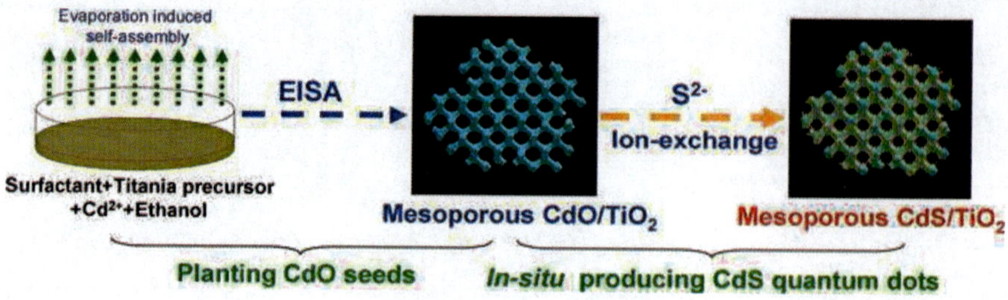

Fig. 4 Schematic synthesis route to ordered mesoporous CdS/TiO_2. Reprinted with permission from Ref. [41]. Copyright 2009 American Chemical Society

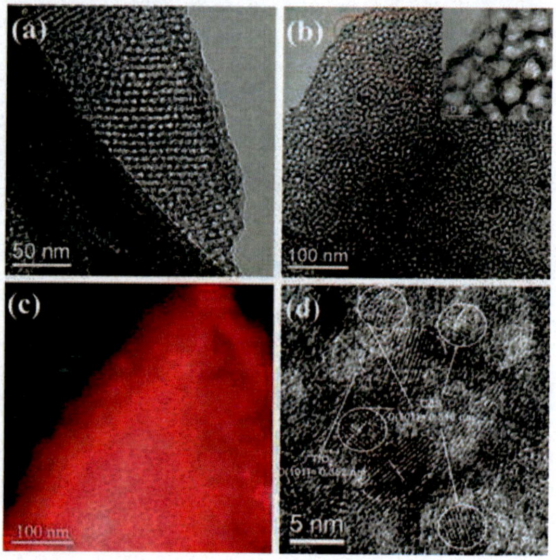

Fig. 5 **a** Standard TEM of CdO/TiO_2, **b** TEM image of CdS/TiO_2, **c** The chemical map of CdS/TiO_2 (*red areas* correspond to the S distribution) and **d** HRTEM image of CdS/TiO_2. Reprinted with permission from Ref. [41]. Copyright 2009 American Chemical Society

The presence of CdS QDs in the TiO_2 framework extended its photoresponse to the visible-light region by accelerating the photogenerated electron transfer from the inorganic sensitizer to TiO_2. The new photocatalyst showed excellent photocatalytic efficiency for both the oxidation of NO gas in air and the degradation of organic compounds (MB and 4-chlorophenol) in aqueous solution under visible-light irradiation [41]. As shown in Fig. 5a, the CdO/TiO_2 sample owned a long-range order structure. The ordered structure could be well maintained even after ion-exchange with S^{2-} (Fig. 5b), though distortions of the pore channels were observed, owing to the in situ transformation of CdO to CdS. The red areas in Fig. 5c represent the S distribution, and the black areas correspond to the pores of the mesoporous CdS/TiO_2. As illustrated in the map, virtually all CdS QDs are highly dispersed on the pore walls of the mesoporous TiO_2. This confirms that CdS QDs are well-integrated into the TiO_2 mesoporous network. The nanocrystalline nature of hexagonal CdS (solid ellipses) and anatase TiO_2 (dot ellipses) are well-

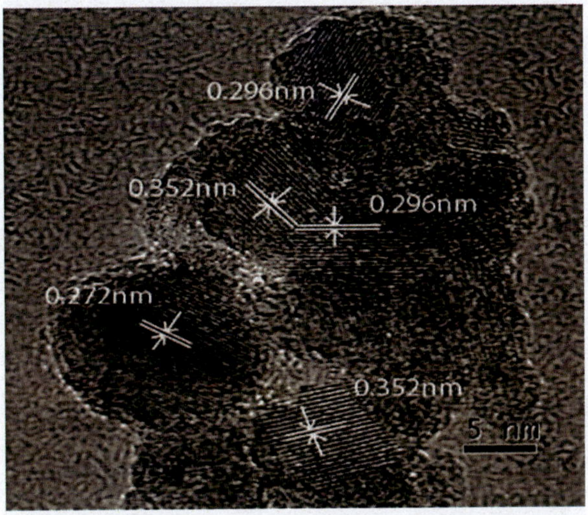

Fig. 6 High-resolution TEM image of $LaVO_4/TiO_2$ nanocomposite. Reprinted with permission from Ref. [43]. Copyright 2009 American Chemical Society

defined in the HRTEM image of CdS/TiO_2 as shown in Fig. 5d. These indicate that the heterojunction between CdS and TiO_2 were formed, owing to the intimate contact between CdS and TiO_2. These CdS/TiO_2 heterojunction will lead to a more efficient inter-electron transfer between the two components and improve the charge separation and, therefore, the photocatalytic activity [41, 42].

Li et al. fabricated $LaVO_4/TiO_2$ nanocomposite material with interconnected nanocrystal heterojunction by using a simple coupled method [43]. As shown in Fig. 6. The fringes of $d = 0.352$ nm matched the (101) crystallographic planes of anatase TiO_2, while the fringes of $d = 0.296$ nm and $d = 0.272$ nm matched the (012) and (202) crystallographic planes of monoclinic $LaVO_4$ nanoparticles. Meanwhile, interconnected fine nanoparticulate morphologies that confirmed the formation of $LaVO_4/TiO_2$ nanocrystal heterojunctions in the composite photocatalyst were observed [43].

This new type of heterojunction $LaVO_4/TiO_2$ nanocomposite exhibited very strong photocatalytic activity for decomposition of benzene under visible-light irradiation ($450 < \lambda < 900$ nm) with high photocatalytic stability. As shown in Fig. 7, the photocatalytic activities of T500, T400, P25, and $LaVO_4$ were very low under visible-light irradiation. Nevertheless, the $LaVO_4/TiO_2$ nanocomposite catalyst showed notably high visible-light photocatalytic activity [43]. Such enhanced photocatalytic performance of $LaVO_4/TiO_2$ can be attributed to the matched band potentials and the interconnected nanocrystal heterojunction of $LaVO_4$ and TiO_2. Figure 8 demonstrates a possible photocatalysis process for the degradation of benzene under visible-light irradiation. It includes four steps: (1) Upon visible-light irradiation, electrons and holes generated by $LaVO_4$ are separated. (2) Some electrons are injected into TiO_2 nanoparticles quickly because the conduction band (CB) of $LaVO_4$ is more negative than that of TiO_2. The formed nanostructure heterojunction on $LaVO_4/TiO_2$ composite can also lead to a more

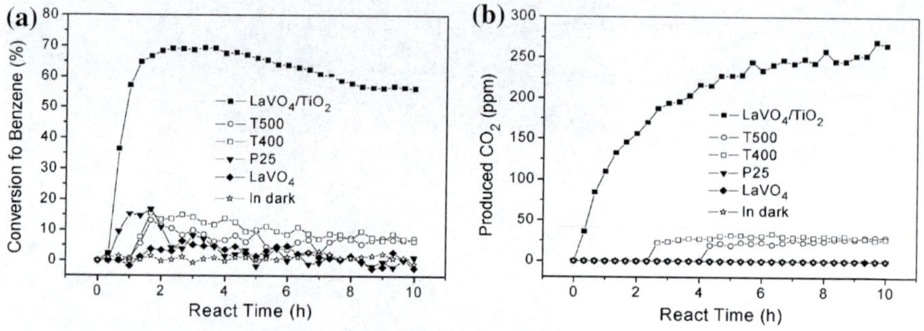

Fig. 7 Conversion of C_6H_6 (**a**) and the amount of produced CO_2 (**b**) on $LaVO_4/TiO_2$, T500, T400, P25, and $LaVO_4$ under visible-light irradiation and on $LaVO_4/TiO_2$ in the dark. Reprinted with permission from Ref. [43]. Copyright 2009 American Chemical Society

efficient inter-electron transfer between the two components [42]. (3) The photogenerated electrons are then captured by O_2 to yield O_2•- and H_2O_2, and then the OHo can be formed by reacting O_2•- with H_2O_2 [32]. The OH• owns a high ability to attack any organic molecules. (4) The photogenerated hole in $LaVO_4$ also may serve as oxidants to activate some unsaturated organic pollutants (e.g., benzene), leading to subsequent decomposition [43]. Since then, Li et al. reported a $Pb(Zr_{0.52}Ti_{0.48})O_3/TiO_2$ (PZT/TiO_2) composite photocatalyst with nanostructured heterojunction prepared by a simple sol–gel method. The as-prepared PZT/TiO_2 photocatalyst with large special surface area exhibited enhanced visible-light absorption and high efficient photocatalytic activity for decomposition of ethylene under visible-light irradiation with high photochemical stability [44].

A $FeTiO_3/TiO_2$ heterojunction structure containing a $FeTiO_3$ nanodisc and Degussa P25 was prepared by using maleic acid as an organic linker [45]. The $FeTiO_3$ nanodisc was a single-crystalline ilmenite phase with its face oriented in (001) plane and grown to the (110) direction. The 5/95 $FeTiO_3/TiO_2$ exhibited the optimized photocatalytic activity in removing 2-propanol and evolving CO_2 in the gas phase under visible-light irradiation. Its degradation constant (k) for removing 2-propanol was 25 times that of Degussa P25. The remarkably enhanced photocatalytic activity of $FeTiO_3/TiO_2$ was attributed to the intersemiconductor hole-transfer mechanism due to the unique relative band positions of these two semiconductors [45]. As shown in Fig. 9, the $FeTiO_3/TiO_2$ system is an example of the type-B heterojunction. The photocatalytic reaction takes place based on inter-semiconductor hole-transfer . The valence band (VB) position of $FeTiO_3$ is very close to that of TiO_2, while its CB is much lower than that of TiO_2 ($\sim$0.5 V lower). The VB of $FeTiO_3$ is rendered partially vacant by band gap excitation under visible-light irradiation. The electrons in the VB of TiO_2 can be transferred to that of $FeTiO_3$. Thus, the holes generated in VB of TiO_2 have a sufficient lifetime to initiate the various photocatalytic oxidation reactions [45].

$FeOOH/TiO_2$, a heterojunction structure between FeOOH and TiO_2, was prepared by covering the surface of the similar to 100-nm-sized FeOOH particles with

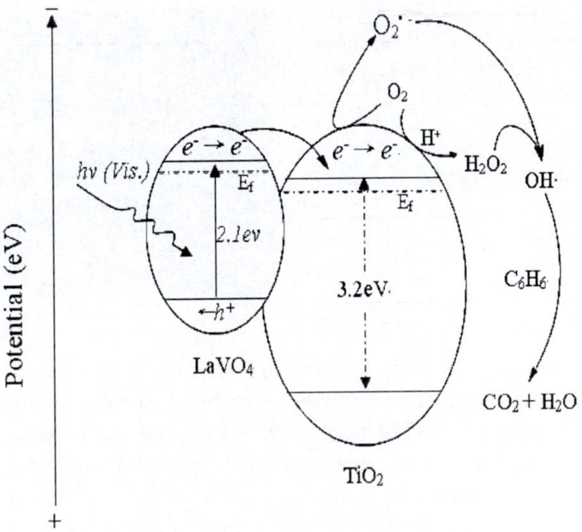

Fig. 8 Proposed mechanism for the visible-light photodegradation of benzene on $LaVO_4/TiO_2$ nanocomposite. Reprinted with permission from Ref. [43]. Copyright 2009 American Chemical Society

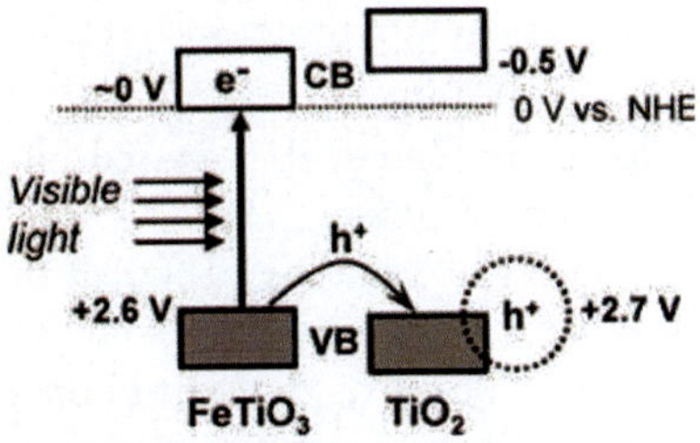

Fig. 9 Energy-band diagram illustrating the type-B heterojunction of TiO_2 and $FeTiO_3$ with visible-light irradiation. Reprinted with permission from Ref. [45]. Copyright 2009 American Chemical Society

Degussa P25 by applying maleic acid as ail organic linker [46]. Under visible-light irradiation, such $FeOOH/TiO_2$ heterojunction structure showed notable photocatalytic activity for the removal of gaseous 2-propanol and evolution of CO_2.

2.3 Non-TiO_2 Photocatalytic Materials

2.3.1 Bi_2WO_6

Semiconducting materials of the Aurivillius oxides $Bi_2A_{n-1}B_nO_{3n+3}$ (A = Ca, Sr, Ba, Pb, Na, K, and B = Ti, Nb, Ta, Mo, W, Fe) have been extensively studied because of their layer structure and unique properties [47, 48]. Among these compounds, Bi_2WO_6, as the simplest member of the Aurivillius family of layered perovskites, has been extensively utilized as an excellent photocatalyst for water splitting and photodegradation of organic compounds under visible-light irradiation [49–51]. Kudo et al. found that Bi_2WO_6 had photocatalytic activity for O_2 evolution [52] and Zou et al. revealed that Bi_2WO_6 could degrade organic compounds under visible-light irradiation [53]. Wang et al. fabricated flower-like

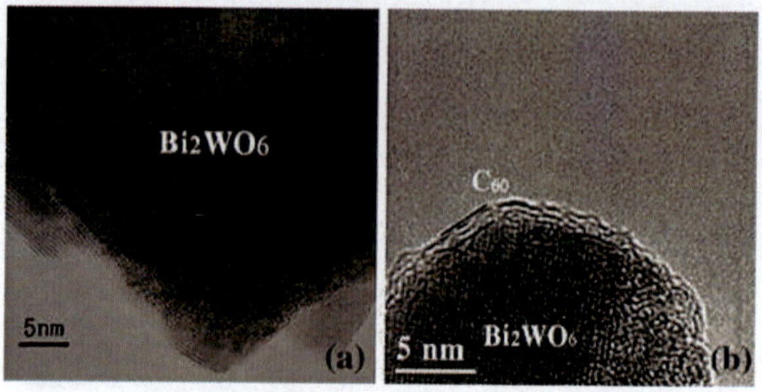

Fig. 10 HRTEM images of (**a**) Bi_2WO_6 and (**b**) the C_{60}-modified Bi_2WO_6 sample. Reprinted with permission from Ref. [57]. Copyright 2007 American Chemical Society

structured Bi_2WO_6 through the hydrothermal route without using any surfactants or templates. The products exhibited strong visible-light-driven photocatalytic performance for the treatment of RhB due to the novel hierarchical transport pores of the flower-like superstructures [54]. Subsequently, Xie et al. used the hydrothermal method to synthesize a Bi_2WO_6 hierarchical nest-like structure with the assistance of PVP [55]. Since then, new types of Bi_2WO_6 with complex morphologies, namely, flower-like, tyre- and helix-like, and plate-like shapes, were selectively synthesized via a hydrothermal process with P123 as a template by Wang et al. [56].

Recently, Zhu et al. prepared fullerene (C_{60}) modified Bi_2WO_6 photocatalyst by an absorbing process [57]. As shown in Fig. 10, the lattice structure of Bi_2WO_6 was observed from the center to the boundary (Fig. 10a). The outer boundary of Bi_2WO_6, modified by C_{60}, was distinctly different (Fig. 10b). An outer layer with an amorphous structure surrounded the surface of the Bi_2WO_6 nanosheet. The thickness of the layer was estimated to be about 1 nm, very close to the diameter of C_{60}. Therefore, it was concluded that C_{60} was dispersed on the surface of Bi_2WO_6 with a monolayer structure [57].

Such new composite photocatalyst exhibited a high efficiency for the degradation of nonbiodegradable azodyes MB and rhodamine B (RhB) under visible-light ($\lambda > 420$ nm) and simulated solar light ($\lambda > 290$ nm). The enhanced photocatalytic activity for the C_{60}-modified Bi_2WO_6 could come from the high migration efficiency of the photo-induced electrons on the interface of the C_{60} and Bi_2WO_6. The delocalized conjugated π structure of C_{60} made the transfer of photoinduced electrons easier [58]. The schematic of photocatalytic mechanism is shown in Fig. 11 [57].

More recently, Zhu et al. used a two-step process to synthesize F-substituted Bi_2WO_6 ($Bi_2WO_{6-x}F_{2x}$) photocatalysts with high activity. F-substitution changed the original coordination around the W and Bi atoms. Compared with Bi_2WO_6, the photocatalytic activity of $Bi_2WO_{6-x}F_{2x}$ increased about two times for the degradation of MB under visible light irradiation. Density functional calculations revealed that $Bi_2WO_{6-x}F_{2x}$ has a wider valence bandwidth and lower VB position. The high activities of $Bi_2WO_{6-x}F_{2x}$ photocatalysts come from its VB, which increase the mobility of photo-excited charge carriers and possess a stronger oxidation power [59].

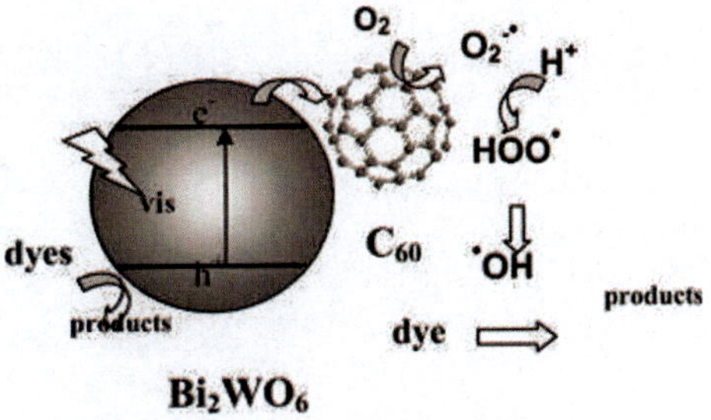

Fig. 11 Possible pathway of the photoelectron transfer excited by visible-light irradiation including photocatalytic process for C60-modified Bi_2WO_6. Reprinted with permission from Ref. [57]. Copyright 2007 American Chemical Society

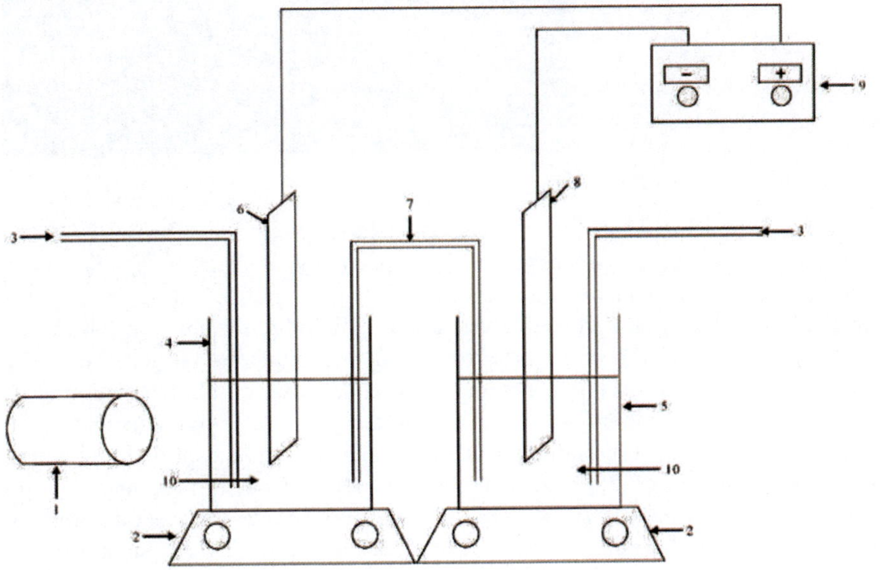

Fig. 12 Schematic illustration of a PEC system: (1) visible-light lamp; (2) magnetic stirrer; (3) air compressor; (4) anodic cell; (5) cathodal cell; (6) Bi_2WO_6/ITO electrode; (7) KCl bridge; (8) Pt cathode; (9) CHI-600A potentiostat; (10) RhB solutions. Reprinted with permission from Ref. [60]. Copyright 2007 American Chemical Society

Lin et al. introduced the photoelectrochemical (PEC) concept to the photocatalytic oxidation application of Bi_2WO_6 [60]. Hydrothermal combined with a spin coating technique was utilized to fabricate a Bi_2WO_6 nanoplate film electrode. As shown in Fig. 12, PCE experiments were performed in the anodic cell, using Bi_2WO_6/ITO electrode with the area of 3 cm^2 in 0.005 mol L^{-1} of Na_2SO_4 electrolyte solution under visible-light illumination. The voltage applied in the EC and PEC systems was 1.2 V. During the entire experiment, the solutions in the anodic and cathodal cells were magnetically stirred. During photocatalytic oxidation process, the Bi_2WO_6/ITO electrode only worked as a photocatalyst without an applied bias [60]. The PEC system based on Bi_2WO_6 nanoplate film electrode degraded 87.2% of RhB with concentration of 5 mg L^{-1} in 120 min, operated at low voltage and under visible-light irradiation, whereas only 36.8 and 39.5%

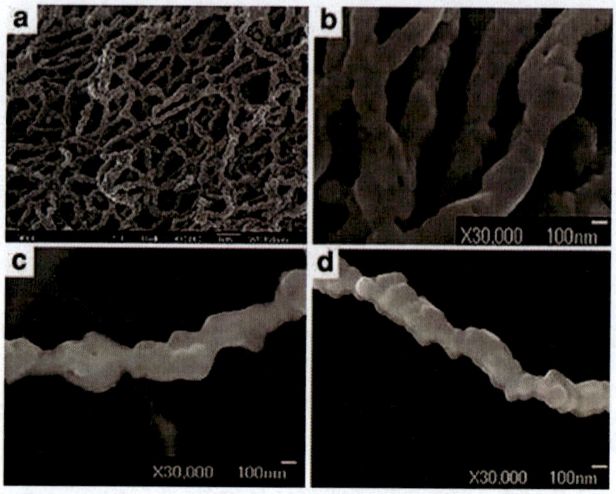

Fig. 13 The SEM images of Bi_2WO_6 nanofibers with (**a**) $R = 0.5$, (**b**) $R = 1$, (**c**) $R = 1.5$, and (**d**) $R = 2$. Ref. [61]—reproduced by permission of The Royal Society of Chemistry

degradation of RhB were observed for the electro-oxidation process (EC) and photocatalytic oxidation processes (PC), respectively, operated under the same condition. These results revealed a significant synergetic effect on degrading RhB via electro-oxidation and photocatalysis under visible-light irradiation [60].

Wang et al. developed an electrospinning technique to fabricate Bi_2WO_6 nanofibrous mat with excellent photoactivity under visible-light irradiation [61]. The as-prepared nanofibers are made of single-crystalline Bi_2WO_6 nanoparticles about 100 nm in size. The diameters of the nanofibers can be controlled by tuning the weight ratio (R) of Bi_2WO_6 to poly(vinyl pyrrolidone) (PVP). As shown in Fig. 13, when the R is 0.5 (Fig. 13a), it was unfavorable for the formation of uniform nanofibers, presumably due to the low content of Bi_2WO_6 nanoparticles that could not be connected during the calcination process. The average diameter of the Bi_2WO_6 nanofibers was about 450 nm. By increasing R from 1 to 2, the average diameter of the fibers decreased from 300 to 170 nm (Fig. 13b–d) [61]. In addition to the favorable recycling characteristics, the mat with $R = 2$ exhibited higher photocatalytic activity in the decomposition of acetaldehyde (CH_3CHO) and aqueous ammonia than that of the sample prepared by the solid-state reaction (SSR) and the nanoparticles. Electrospinning is favorable for the effective improvement of the photocatalytic activity of catalysts. It also can provide a solution to the separation problem in conventional catalysts that are small in size. Thus, it is worth considering for the preparation of other photocatalysts [61].

2.3.2 $BiVO_4$

$BiVO_4$, with a narrower band gap of 2.4 eV, is an important visible-light responsive photocatalyst, widely used in the evolution of photocatalytic O_2 and the photocatalytic degradation of organic pollutants [62, 63]. There are three naturally occurring crystal forms of $BiVO_4$, namely tetragonal zircon, monoclinic scheelite,

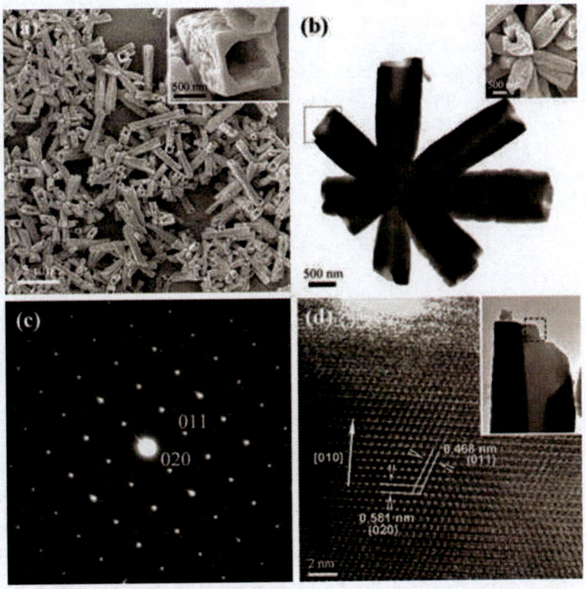

Fig. 14 $BiVO_4$ microtubes synthesized at 80°C for 6 h: **a** low-magnification SEM image of the product and high-magnification SEM image (*inset*) for a single $BiVO_4$ microtube, showing the hollow structure and wall thickness; **b** TEM and SEM image (*inset*) of an individual $BiVO_4$ microflower; **c** the corresponding SAED pattern taken from the rectangular part of the microtube in Fig. 14b; **d** HRTEM image near the nozzle of a single microtube (*inset*). Reprinted with permission from Ref. [68]. Copyright 2007 American Chemical Society

and tetragonal scheelite. Monoclinic scheelite presents the best photocatalytic performance under visible-light irradiation [64]. To date, various methods of synthesizing monoclinic scheelite $BiVO_4$ have been developed, including aqueous [62], solid-state [65], and hydrothermal processes [66], as well as organometallic decomposition [67].

Single-crystalline $BiVO_4$ microtubes, with novel square cross-sections and a flower-like morphology, were synthesized by a reflux method at 80°C [68]. The as-synthesized products were almost entirely microtubes with lengths of 2–5 μm. The individual tubes had well-defined square cross-sections with side lengths of ca. 800 nm and wall thicknesses of ca. 100 nm. The clear lattice fringe indicated the high-crystallinity and single-crystalline nature of the microtubes. The d spacings were found to be 0.581 and 0.468 nm, which agreed well with the lattice spacings of (020) and (011) of monoclinic $BiVO_4$ (Fig. 14c, d) [68]. The as-synthesized microtubes exhibited higher photocatalytic activity under visible-light radiation than that of the reference sample prepared by a solid-state reaction. This was ascribed to the special single-crystalline tubular structure and/or flower-like morphology [68].

Yu et al. used a nanocasting route to fabricate ordered mesoporous bismuth vanadate ($BiVO_4$) crystals using bismuth nitrate hydrate and ammonia meta-vanadate as bismuth and vanadium sources and silica (KIT-6) as a template [69]. Monoclinic scheelite $BiVO_4$ crystals were formed inside the mesopores of silica through a mild thermal process, and $BiVO_4$ was obtained after the removal of the hard template (silica) by NaOH treatment. Such product exhibited a superior photocatalytic performance in the photochemical degradation of MB and photo-catalytic oxidation of NO gas in air under visible-light irradiation compared to

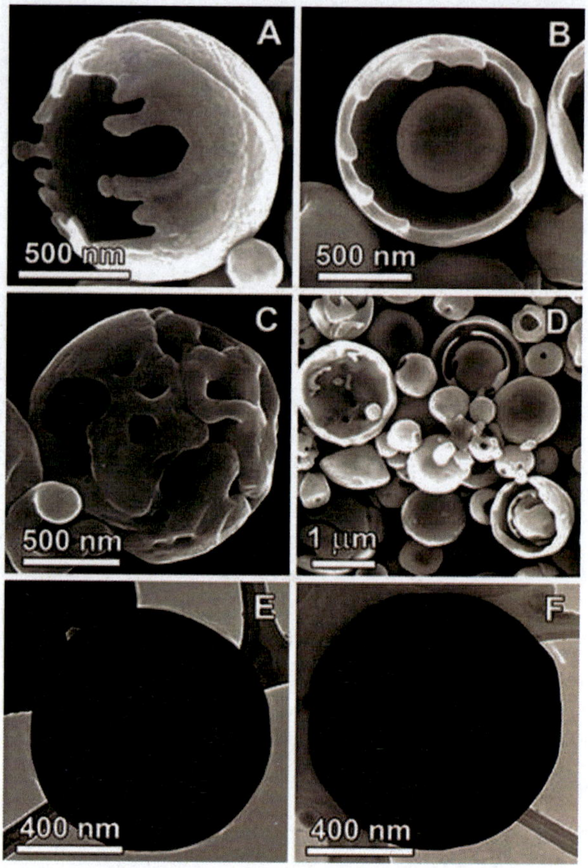

Fig. 15 **a–d** SEM and **e** and **f** TEM micrographs illustrating typical particle morphologies obtained during USP synthesis of $BiVO_4$. Reprinted with permission from Ref. [71]. Copyright 2009 American Chemical Society

conventional $BiVO_4$ [69]. Wang et al. used an aqueous strategy to synthesize two novel types of $BiVO_4$ mesocrystals with flower-like and dendrite morphology. The primary building units perfectly aligned in three-dimensions and built up well-defined mesocrystals with sharp facets and edges. The pH value and reaction temperature had great influences on the formation of these unique mesocrystals [70]. Recently, $BiVO_4$ powders with unique particle architectures were synthesized using ultrasonic spray pyrolysis (USP) [71]. These nanostructured $BiVO_4$ with particles range from thin, hollow, and porous shells to ball-in-ball type structures, as shown in Fig. 15. The $BiVO_4$ powders were utilized as an oxygen evolving photocatalyst and the kinetics of O_2 formation was studied in a $AgNO_3$ solution irradiated with $\lambda > 400$ nm light. USP prepared $BiVO_4$ was found to have superior photocatalytic activity compared to commercial $BiVO_4$ and WO_3, likely due to the differences in particle morphology [71].

Li et al. prepared a series of Cu-loaded $BiVO_4$ (Cu-$BiVO_4$) photocatalysts by the impregnation method [72]. The Cu-$BiVO_4$ series catalysts had significant optical absorption in the visible region between 550 and 800 nm and the

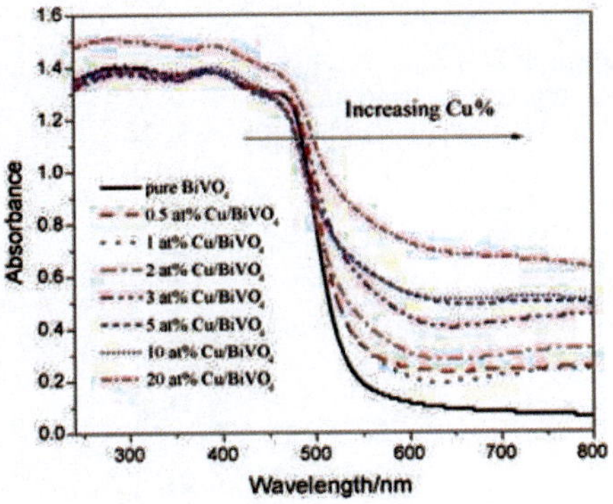

Fig. 16 UV–Vis diffuse-reflectance spectra of pure and Cu–$BiVO_4$ series catalysts. Reprinted from Xu et al. [72], copyright 2008 with permission from Elsevier

absorption intensity increased with the enhancement of Cu content as shown in Fig. 16. The photocatalytic activities of Cu-$BiVO_4$ catalysts for the degradation of MB were found to depend largely on the Cu content and the calcination temperature. The optimum Cu loading and calcination temperature were found to be 5% and 300°C [72]. Novel Pt/$BiVO_4$ composite photocatalysts with visible-light activities were prepared by the impregnation method [73]. Monoclinic scheelite structures of $BiVO_4$ were retained after the Pt species was doped. The visible-light absorption of the $BiVO_4$ photocatalysts was obviously enhanced upon modifying with Pt species (Fig. 17). The platinum doped in the composite photocatalyst was present in the form of platinum chloride ($PtCl_4$). The Pt species doping effectively enhanced the photoactivities of $BiVO_4$ samples in decomposition of methyl orange under visible-light irradiation [73].

2.3.3 Graphitic Carbon Nitrides (g-C_3N_4) Polymeric Photocatalyst

Among the different allotropes of carbon nitrides, the graphitic phase is usually regarded as the most stable under ambient conditions. Recently, graphitic carbon nitride (g-C_3N_4), as a conducting polymer semiconductor, has been proven an efficient photocatalyst for water splitting, even in the absence of noble metals [74]. Thermal polycondensation of common organic monomers was utilized to synthesize graphitic carbon nitrides (g-C_3N_4) with various architectures [75, 76]. As shown in Fig. 18, the graphitic planes are constructed from tri-s-triazine units connected by planar amino groups (Fig. 18a). Upon condensation at 823 K, an in-plane repeat period of 0.681 nm (for example, the distance between nitride pores) in the crystal is evident from the X-ray powder diffraction (XRD) pattern (Fig. 18b). It is smaller than one tri-s-triazine unit (0.713 nm), presumably owing to the presence of a small tilt angularity in the structure. The strongest XRD peak at 27.4, corresponding to 0.326 nm, is due to the stacking of the conjugated aromatic system. The band gap of the gold–yellow condensed graphitic carbon nitride is estimated to be 2.7 eV from its ultraviolet–visible spectrum (Fig. 18c). It implies an intrinsic semiconductor-like absorption in the blue region of the

Fig. 17 UV–Vis diffuse absorption spectra of different photocatalysts. Reprinted from Ge [73], copyright 2008 with permission from Elsevier

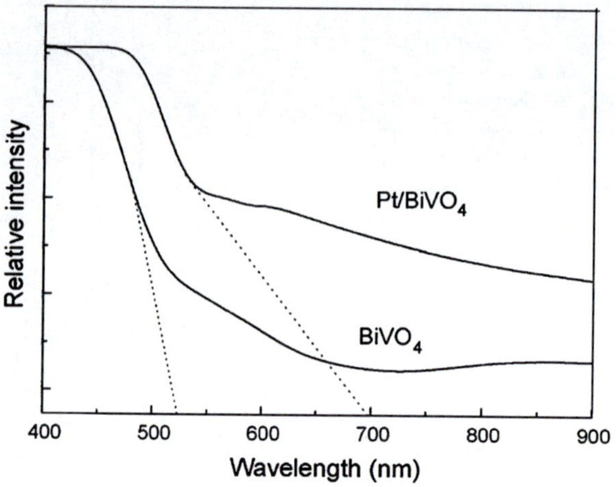

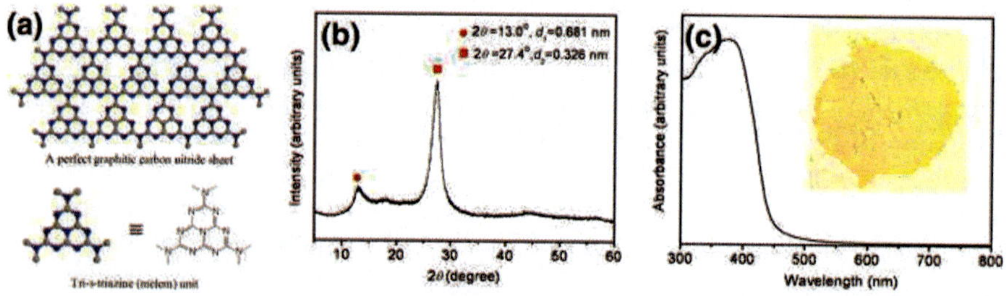

Fig. 18 Crystal structure and optical properties of graphitic carbon nitride. **a** Schematic diagram of a perfect graphitic carbon nitride sheet constructed from melem units. **b** Experimental XRD pattern of the polymeric carbon nitride, revealing a graphitic structure with an interplanar stacking distance of aromatic units of 0.326 nm. **c** Ultraviolet–visible diffuse-reflectance spectrum of the polymeric carbon nitride. *Inset*: Photograph of the photocatalyst. Reprinted by permission from Macmillan Publishers Ltd: (Nature Materials) (Ref. [74]), copyright 2009

visible spectrum. Such bandgap is sufficiently large to overcome the endothermic character of the water-splitting reaction (requiring 1.23 eV theoretically) [74].

Recently, the electronic and optical functions of polymeric g-C_3N_4 were further modified by the inclusion of metal species, such as Fe^{3+}, in its structure by a simple soft-chemical method without destroying the graphitic structure of the host. The metal components strongly affected the electronic properties of g-C_3N_4 and provided the material with additional new functionalities such as mimicking metalloenzymes in H_2O_2 activation. The metal species can significantly lower the bandgap and expand the light absorption of the material further into the visible region of the electromagnetic spectrum, while keeping a sufficient overpotential for carrying out oxidation reactions [77]. As shown in Fig. 19, the optical band gap energy gradually shifts to lower energies with increasing Fe content in the Fe/g-C_3N_4 hybrid materials. This suggests a host–guest interaction between

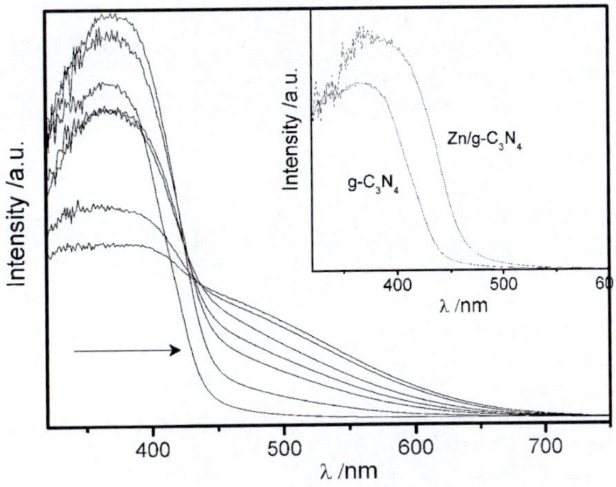

Fig. 19 Optical absorption spectra of Fe/g-C_3N_4 complexes and g-C_3N_4; the bandgap is shifted toward lower energies in the hybrid materials. Arrow direction: g-C_3N_4, 1%-, 3%-, 5%-, 10%-, 15%-, and 20%-Fe/g-C_3N_4. The *inset* is the optical spectrum of 10%-Zn/g-C_3N_4, which also demonstrates narrowing of the g-C_3N_4 bandgap by the metal inclusion. Wang et al. [77], copyright Wiley–VCH Verlag GmbH & Co. KGaA, reproduced with permission

g-C_3N_4 and the metal. A change in the optical absorption was also observed for Zn/g-C_3N_4 and is probably caused by the d–p repulsion of the Zn 3d and N2p orbitals [77, 78].

Photocatalytic experimental results confirmed that various organic dyes (e.g., RhB, MB, methyl orange, and p-hydroxyazobenzene) were be degraded by using H_2O_2 and Fe/g-C_3N_4. The photocatalyst can also be recovered and reused [77]. Figure 20 shows the results from the RhB-oxidation and control experiments. The overall efficiency of the process can be enhanced by photoillumination ($\lambda > 420$ nm). Various intermediates, such as hydroxylated and de-ethylated, were observed during the oxidation of RhB by H_2O_2 activated with Fe/g-C_3N_4. These can be further decomposed to smaller organic molecules, eventually being mineralized to CO_2 [77].

More recently Fe/g-C_3N_4 was found to be capable of activating hydrogen peroxide for the direct oxidation of benzene to phenol in mild conditions. Fe-g-C_3N_4 is active for the direct oxidation of benzene to phenol using hydrogen peroxide. By taking advantage of the photocatalytic functions of g-C_3N_4, the yield of the phenol synthesis can be markedly improved [79]. Figure 21a shows the catalytic properties of graphitic carbon nitride. Figure 21b implies that it is able to adsorb and activate benzene chemically (Fig. 21b) and, thus, catalyzed not only Friedel–Crafts reactions of benzene but also phenol synthesis using benzene and CO_2 [75, 80].

2.3.4 Heterojunction of Non-TiO_2 Semiconductors

Much progress has been achieved in the development of TiO_2-based heterojunction photocatalytic systems. However, there is still need to develop highly active photocatalysts working under visible light in order to put this technology into practical applications. Recently, heterojunctions of non-TiO_2 semiconductors has become a hot topic in the photocatalytic research field.

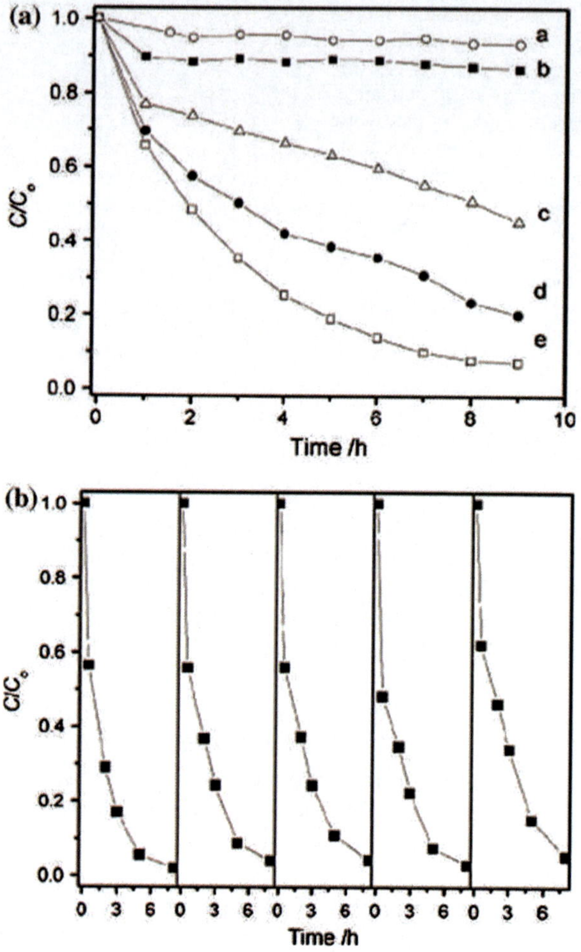

Fig. 20 **a** Concentration changes of RhB (10 mM) as a function of reaction time under different conditions: *a* H_2O_2 (0.01 M); *b* Fe/g-C_3N_4(40 mg); *c* Fe_2O_3(40 mg)/H_2O_2(0.01 M); *d* Fe/g-C_3N_4(40 mg)/H_2O_2(0.01 M); *e* Fe/g-C_3N_4(40 mg)/H_2O_2(0.01 M) under visible-light irradiation. **b** Cyclic runs of RhB (10 mM) degradation by H_2O_2 (0.05 M) activated by the Fe/g-C_3N_4 catalyst (40 mg). All reactions were carried out at neutral pH using the 10%-Fe/g-C_3N_4 catalyst. C: concentration; C_0: initial concentration. Wang et al. [77], copyright Wiley–VCH Verlag GmbH & Co. KGaA, reproduced with permission

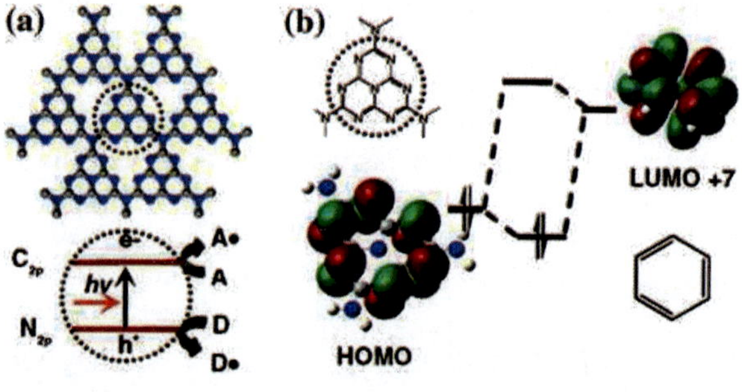

Fig. 21 **a** Stacked g-C_3N_4 sheets function as an all-organic solid-state photocatalyst promoting redox reactions with visible light. **b** Chemical interaction of benzene and defective g-C_3N_4 via HOMO–LUMO hybridization of melem and benzene. Reprinted with permission from Ref. [79]. Copyright 2009 American Chemical Society

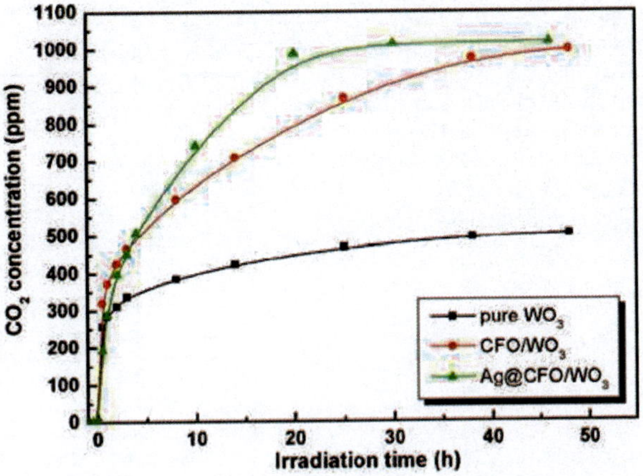

Fig. 22 CO_2 photogeneration profile at a long time visible irradiation. The initial acetaldehyde concentration is 500 ppm. The visible light density is 80 mW/cm^2. Reprinted with permission from Ref. [81]. Copyright 2009 American Chemical Society

Efficient visible-light active $CaFe_2O_4$ (CFO)/WO_3 composite photocatalysts were prepared by Miyauchi et al. [81]. The composite containing 5 wt% CFO showed optimized performance. Surface modification was made by heating the CFO/WO_3 composite or by coating the particle surface with Ag or ITO. The photocatalytic activity was greatly improved by coating the CFO particles with a Ag or ITO layer. The CO_2 generation profile over pure WO_3, 5 wt% CFO/WO_3, Ag@CFO/WO_3 photocatalysts were presented in Fig. 22. The CO_2 generation almost stops after 40 h of irradiation over pure WO_3. Pure WO_3 cannot completely decompose acetaldehyde to CO_2 even after more than 100 h under the present experimental conditions. However, acetaldehyde can be completely decomposed to CO_2 over a 5 wt% CFO/WO_3 composite in 48 h. It is impressive that the Ag modified composite Ag@CFO/WO_3 can reach complete decomposition of acetaldehyde in 20 h, which is more than twice as fast as the CFO/WO_3 photocatalysts [81].

A network-structured SnO_2/ZnO heterojunction nanocatalyst with high photocatalytic activity was synthesized through a simple two-step solvothermal method [82]. As shown in Fig. 23, the UV–vis absorption edges of the as-synthesized SnO_2 semicrystals and ZnO nanorods are located at about 305 and 380 nm, respectively. There are two prominent absorption bands for the SnO_2/ZnO sample. The former is assigned to the absorption of SnO_2 semicrystals and the latter is attributed to the characteristic absorption of ZnO nanocrystals. The absorption edges of SnO_2 and ZnO nanocrystals in the SnO_2/ZnO sample slightly shift toward blue. This indicates that the sizes of SnO_2 and ZnO in the SnO_2/ZnO sample are smaller than the corresponding values of pure SnO_2 or ZnO [82].

The photocatalytic activity of SnO_2/ZnO heterojunction nanocatalysts for the degradation of methyl orange is much higher than those of solvothermally synthesized SnO_2 and ZnO samples. Figure 24 shows the proposed band structure of the as-synthesized SnO_2/ZnO heterojunction nanocatalyst [82, 83]. Upon formation of the heterojunction between SnO_2 and ZnO, the different work functions will induce the negatively charged carriers to move from SnO_2 (the material with low

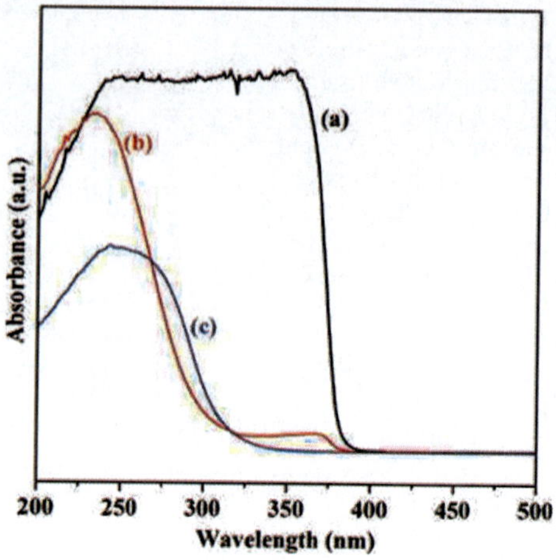

Fig. 23 UV–vis diffuse-reflectance spectra of the as-synthesized samples: (a) ZnO, (b) SnO_2/ZnO, and (c) SnO_2. Reprinted with permission from Ref. [82]. Copyright 2009 American Chemical Society

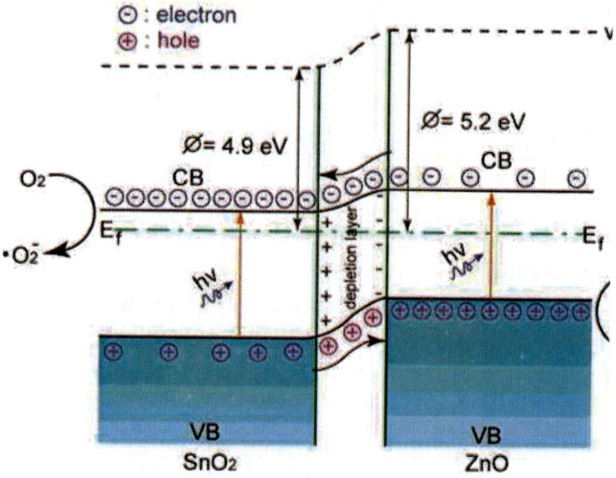

Fig. 24 Energy-band diagram and photocatalytic mechanism of the as-synthesized SnO_2/ZnO heterojunction nanocatalyst, where vac is the vacuum level, E_f is the Fermi level, CB is the conduction band, and VB is the valence band. Reprinted with permission from Ref. [82]. Copyright 2009 American Chemical Society

work function) to ZnO (the one with high work function) until their Fermi levels align (i.e., the system reaches thermal equilibrium). Thus, an electrostatic field is created at the interface. At thermal equilibrium, the CBs and VBs of SnO_2 and ZnO bend, and a depletion layer forms around the interface, too. Under UV light irradiation, electrons (e−) in the VB can be excited to the CB while simultaneously generating the same amount of holes (h+) in the VB. The photogenerated electrons and holes can be separated under the influence of the electrostatic field induced by different work functions. Therefore, electrons move to the SnO_2 side and holes to the ZnO side. The photogenerated electrons and holes in the SnO_2/ZnO heterojunction nanocatalyst can be injected into a reaction medium and participate in chemical reactions [82].

Bi_2O_3 is a good n-type semiconductor and $BaTiO_3$ is a high dielectric and ferroelectric material, where some atoms in the lattice are movable. Based on this, Huang et al. used a milling-annealing technique to prepare a heterojunction

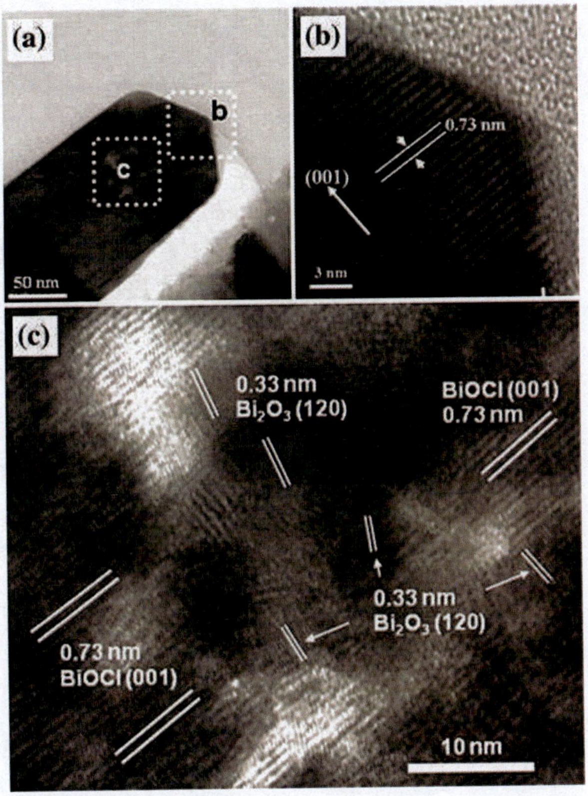

Fig. 25 TEM images for an 85/15 $BiOCl/Bi_2O_3$ particle. Typical TEM image (**a**), and HRTEM images for the outer (**b**) and inner (**c**) parts of the sample. Reprinted from Chai [35], copyright 2009 with permission from Elsevier

photocatalyst, $Bi_2O_3/BaTiO_3$ [36]. This technique has an advantage over the direct mixing method because it can construct a tight chemically bonded interface between the coupled materials. The heterojunction semiconductors $Bi_2O_3/BaTiO_3$ showed better photocatalytic activities than single-phase $BaTiO_3$ or Bi_2O_3 for degrading methyl orange and MB. The remarkable enhancement in the photocatalytic performance of $Bi_2O_3/BaTiO_3$ was ascribed mainly to the electric field-driven electron–hole separation at the interface and in the two semiconductors. Besides, the fair mobility for electron and hole transportation in Bi_2O_3 and $BaTiO_3$, respectively, were also favorable for the high photocatalytic property [36].

Lee et al. also reported a novel $BiOCl/Bi_2O_3$ heterojunction-type photocatalyst [35]. The TEM image in Fig. 25a reveals that the 85/15 $BiOCl/Bi_2O_3$ is a well-defined longish particle of $\sim$200 nm width. As shown in the high-resolution TEM image of Fig. 25b, the outer part of the $BiOCl/Bi_2O_3$ particle reveals a clear image, suggesting the presence of crystalline BiOCl. The uniform fringe, with an interval of 0.73 nm, is in good agreement with the (001) lattice plane of the tetragonal BiOCl. On the other hand, as shown in Fig. 25c, the TEM image for the core of the $BiOCl/Bi_2O_3$ particle is not as clear as that of outer part and several sets of mixed fringes are found, indicating the presence of mixed phases of BiOCl and Bi_2O_3 that is, the interlayer distance of 0.73 nm corresponds to the (001) lattice plane of BiOCl, and that of 0.33 nm is consistent with the (120) plane of α-Bi_2O_3. These observations indicate that the nano-sized Bi_2O_3 grains are embedded here and there inside the BiOCl matrix [35]. Though both the individual BiOCl and Bi_2O_3 show very low photocatalytic efficiency under visible-light irradiation, their heterojunctions provide unexpectedly high efficiency in decomposing organic compounds. The BiOCl/Bi_2O_3 can induce complete mineralization without formation of intermediate species by utilizing the holes generated in the VB of BiOCl. Compared to the Degussa P25, it demonstrates 5.7 times the efficiency in evolving CO_2 from gaseous 2-propanol (IP) and 10.5 times the efficiency in removing aqueous 1, 4-terephthalic acid (TA) under visible-light irradiation. In this BiOCl/Bi_2O_3 system, the BiOCl seems to work as the main photocatalyst, while the role of Bi_2O_3 is a sensitizer, absorbing visible light [35].

3 Solar-Light-Driven Photocatalysts for Generating Fuels

3.1 Solar-Light-Driven Photocatalysts for H_2 Evolution

The concern for the depletion of fossil fuels and the environmental problems accompanying their use fostered the research for viable alternatives. Many research efforts have been devoted to the generation of hydrogen since it is the fuel with the highest energy capacity per unit mass. Hydrogen is also a clean energy carrier because it produces neither CO_2 nor pollutants. Many reviews on photocatalytic water splitting have been published [2, 9, 84–89]. In the following sections, we focus on visible-light-driven heterogeneous photocatalytic materials, such as metal oxides, metal oxynitrides, metal oxysulfides, metal sulfides, and polymers for H_2 evolution.

3.1.1 Metal Oxide Photocatalysts

To obtain photocatalytic activity under visible-light irradiation, it is essential to control the interdependence between the electronic, microstructural, and surface properties of photocatalysts by means of a careful design of both bulk and surface properties. The strategies can be classified in five categories: (a) developing new single-phase photocatalysts; (b) tuning the band gap energy with ion doping; (c) surface modification by depositing co-catalysts; (d) sensitization; and (e) controlling the defects, size, and morphology. This section briefly reviews the recent developments in oxide photocatalysts (Table 1) that show activity under visible light.

3.1.2 Oxynitride and Oxysulfide Photocatalysts

Domen and co-workers have done extensive studies on oxynitride and oxysulfide visible-light-driven photocatalysts [87, 88, 90–99]. The hydrogen generation application for these catalysts has been reviewed in previous reports [100, 101].

Table 1 Overview of recently developed metal oxide photocatalysts for H_2 generation under visible-light illumination

Photocatalsyt	Co-catalyst	Sacrificial reagent	H_2 evolution/ μmol h^{-1}	Ref.
$K_4Nb_6O_{17}$	tris(2,2′-bipyridyl)ruthenium(II) chloride	EDTA	3.6	[127]
$SrTiO_3$:Cr/Ta	Pt	CH_3OH	140	[128]
$SrTiO_3$:Cr/Sb	Pt	CH_3OH	156	[129]
$SrTiO_3$:Ni/Ta	Pt	CH_3OH	2.4	[130]
$SrTiO_3$:Rh	Pt	CH_3OH	117	[131]
$CaTiO_3$:Rh	Pt	CH_3OH	8.5	[132]
$La_2Ti_2O_7$:Cr	Pt	CH_3OH	30	[133]
$La_2Ti_2O_7$:Fe	Pt	CH_3OH	20	[133]
$Sm_2Ti_2S_2O_5$	Pt	CH_3OH	40	[134]
$PbTiO_3$	Pt	CH_3OH	13.6	[135]
$In_{0.9}Ni_{0.1}TaO_4$	NiO, RuO_2	–	16.6	[136, 137]
$K_{0.5}La_{0.25}Bi_{0.25}Ca_{0.75}Pb_{0.75}Nb_3O_{10}$	Pt	CH_3OH	168	[135]
$PbBi_2Nb_2O_9$	Pt	CH_3OH	7.6	[138]
$In_2O_3/Cr:In_2O_3$	NiO, Pt	CH_3OH	0.36	[139]
$Ba_2In_2O_5$	NiO, Pt	CH_3OH	3.2	[139]
$RbPb_2Nb_3O_{10}$	Pt	CH_3OH	4	[140]
$SnNb_2O_6$	Pt	CH_3OH	14.4	[141–143]
$AgNbO_3$	Pt	CH_3OH	8.2	[144]
$Sn^{2+}/K_4Nb_6O_{17}$	Pt	CH_3OH	23	[145]
$Sn^{2+}/KTiNbO_5$	Pt	CH_3OH	54	[145]
$Sn^{2+}/CsTi_2NbO_7$	Pt	CH_3OH	18	[145]
$Sn^{2+}/K_2Ti_4O_9$	Pt	CH_3OH	23	[145]
$Sn^{2+}/K_2Ti_4O_9$	Pt	CH_3OH	5	[145]
$Sn^{2+}/Cs_2Ti_6O_{13}$	Pt	CH_3OH	7	[145]

Herein, we summarize the more recent publications related to the oxynitride and oxysulfide photocatalysts.

Takanabe et al. studied the photocatalytic water-splitting reactions over the $(Zn_{1+x}Ge)(N_2O_x)$ photocatalyst [102]. The photocatalyst showed high rates for overall water splitting under visible irradiation. Their results showed negligible changes in the structure and composition of the photocatalyst after the photocatalytic reaction. The photocatalytic activity was improved by metal doping in the oxynitride formulation and post-calcination after nitridation. The improvement of photocatalytic activity was ascribed to the reduction of the number of defects in the photocatalyst materials.

TaON nanotube arrays (shown in Fig. 26) were synthesized via sono-electrochemical anodization followed by nitridation [103]. They exhibited efficient performance for photoelectrochemical generation of hydrogen from water.

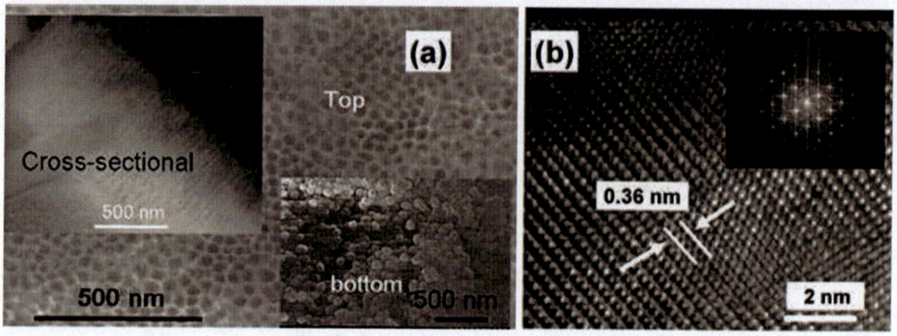

Fig. 26 **a** FESEM images of TaON nanotube arrays on Ta foil. The *insets* show the cross sectional image of Ta_2O_5 NT arrays. **b** HRTEM and FFT pattern of TaON NTs. Ref. [103]—reproduced by permission of The Royal Society of Chemistry

The photocatalytic activity of $(Ga_{1-x}Zn_x)(N_{1-x}O_x)$, a solid solution of GaN and ZnO, for H_2 evolution in the presence of methanol as a sacrificial reagent under visible light was investigated in detail [104]. $(Ga_{1-x}Zn_x)(N_{1-x}O_x)$ evolved H_2 from an aqueous methanol solution when loaded with nanoparticulate $Rh_{2-y}Cr_yO_3$ as a cocatalyst. The H_2 evolution activity was strongly dependent on the crystallinity and composition of the catalyst. The quantum efficiency for overall water splitting increased to 2.5% at 420–440 nm [105]. This represented a tenfold increase in efficiency over the highest efficiency previously obtained using nanoparticulate RuO_2 as a cocatalyst. Besides, the dispersion and size of cocatalyst nanoparticles were identified as important factors affecting the degree of enhancement for stoichiometric water splitting. The results of photocatalytic reactions and photoelectrochemical measurements suggested that the rate-determining step for overall water splitting using $(Ga_{1-x}Zn_x)(N_{1-x}O_x)$ was the H_2 evolution process [105].

Lee et al. developed a zinc germanium oxynitride, a solid solution between ZnO and $ZnGeN_2$, through a reaction of GeO_2 and ZnO under an NH_3 flow [106]. The samples nitrided for 5–15 h under these conditions exhibited a single phase of wurtzitic $(Zn_{1+x}Ge)(N_2O_x)$ and were responsive to visible light with a band gap of ca. 2.7–2.8 eV. Nitridation for 15 h afforded $(Zn_{1+x}Ge)(N_2O_x)$ with the highest photocatalytic activity for overall water splitting. A variety of cocatalysts were also examined, and $Rh_{2-x}Cr_xO_3$ was identified as the most effective cocatalyst for $(Zn_{1+x}Ge)(N_2O_x)$, which caused an increase in the activity for hydrogen evolution. Modification of the optimized $(Zn_{1.44}Ge)(N_{2.08}O_{0.38})$ sample by loading with $Rh_{2-x}Cr_xO_3$ (3.0 wt% Rh, 0.2 wt% Cr) resulted in an effective photocatalyst for overall water decomposition with a quantum efficiency of ca. 0.20% at 420 nm.

A rose-red color $Nb_2Zr_6O_{17-x}N_x$ oxynitride photocatalyst was synthesized by thermal ammonolysis of $Nb_2Zr_6O_{17}$ at 1073 K [107]. TEM images of the $Nb_2Zr_6O_{17-x}N_x$ sample showed prismatic pseudo orthorhombic shaped particles with clear edges and an average particle size in the range of 80–90 nm. The oxynitride $Nb_2Zr_6O_{17-x}N_x$ gave a quantum yield of 13.5% in the production of

Table 2 Sulfide photocatalysts for H_2 evolution from aqueous solutions in the presence of sacrificial reagents

Photocatalyst	BG/eV	Incident light/nm	Light source	Reactant solution	H_2 evolution/ μmol h^{-1}	QY (%)	Ref.
$Mn_{0.9}Cd_{0.1}S$	2.2	>420	500 W Xe	$Na_2S + Na_2SO_3$	71	7	[146]
CuS(5.9%)-$Zn_{0.65}Cd_{0.35}S$:Pt	2.5	>420	300 W Xe	$Na_2S + Na_2SO_3$	740	19	[147]
PdS(0.13%)/CdS:Pt	2.3	>420	300 W Xe	$Na_2S + Na_2SO_3$	8770	93	[148]
$ZnS_{-1x-0.5y}O_x(OH)_y$(1:1)	2.2	>420	400 W metal halide lamp	$Na_2S + Na_2SO_3$	14	3	[149]
$Cd_{0.7}Zn_{0.3}S$	2.6		300 W arc lamp	$Na_2S + Na_2SO_3$	350	–	[150]
$Cd_{0.1}Zn_{0.9}S:Ni^{2+}$	2.4	>420	350 W Xe	$Na_2S + Na_2SO_3$	585	16	[151]
$AgGa_{0.9}In_{0.1}S_2$:Pt	2.4	>420	450 W Hg	$Na_2S + Na_2SO_3$	350	–	[152]
$Cd_{0.8}Zn_{0.2}S$	2.5	>430	350 W Xe	$Na_2S + Na_2SO_3$	190	10	[153]
ZnS:Cu	–	>400	550 W Xe	$Na_2S + Na_2SO_3$	650	–	[154]
CdS:Pt	2.3	>420	300 W Xe	$Na_2S + Na_2SO_3$	4100	60	[155]
$AgIn_5S_8$:Pt	1.8	>420	400 W Xe	$Na_2S + K_2SO_3$	60	5.3	[156]
$ZnS:Ni^{2+}$		>400	300 W Xe	$Na_2S + K_2SO_3$	18	2.1	[157]

hydrogen from the decomposition of hydrogen sulfide under visible-light irradiation. A d^0–d^{10} complex photocatalyst, zinc, and titanium spinel oxynitride ($Zn_xTiO_yN_z$) reduced H^+ to H_2 in the presence of a sacrificial electron donor under visible-light irradiation [108]. Ogisu et al. reported a lanthanum–indium oxysulfide visible-light-driven ($420 < \lambda < 480$ nm) photocatalyst for water splitting. Loading with Pt is effective for promoting H_2 evolution [99].

3.1.3 Metal Sulfide Photocatalysts

Metal sulfides photocatalysts have been widely studied due to their outstanding performance in hydrogen generation via photocatalysis. The VB usually consists of S 3p orbitals the level of which is more negative than O 2p. The photocorrosion problem is usually solved by adding sacrificial reagents such as S^{2-} and ${SO_3}^{2-}$ into the water-splitting system. Many visible-light-driven metal sulfide photocatalysts have been summarized in a previous report [2]. Table 2 lists the new photocatalysts reported in the last 3 years.

3.1.4 Polymeric Photocatalysts

Synthetic polymer semiconductors such as polyparaphenylene have also been used for hydrogen production under UV illumination [109]. Recently, a metal-free polymeric, visible-light-driven photocatalyst for hydrogen production was

Fig. 27 **a** TEM image of mpg-C_3N_4, showing a 3D porous framework constructed from tri(s)triazine units. The stacking distance of 0.332 nm is evident by the intense electron diffraction ring(*inset*), providing high partial crystallinity of the wall. **b** TEM image of bulk g-C_3N_4. Reprinted with permission from Ref. [110]. Copyright 2009 American Chemical Society

reported [74]. It was a graphitic carbon nitride (g-C_3N_4) synthesized via a thermal polycondensation of cyanamide. The bandgap of g-C_3N_4 was estimated to be 2.7 eV from its ultraviolet–visible spectrum, showing an intrinsic semiconductor-like absorption in the blue region. Different thermal condensation enabled the finer adjustment of the electronic and optical properties. The photocatalyst produced H_2 from water containing triethanolamine as a sacrificial electron donor upon light illumination ($\lambda > 420$ nm) in the absence of noble metal catalysts such as Pt. No N_2 evolution was observed for this catalyst, even after a very long irradiation time, indicating excellent stability due to the strong binding of N in the covalent carbon nitride. This is the first polymeric photocatalyst that is cheap and commonly available. It will open new avenues for organic semiconductors as energy transducers.

However, the quantum yield of the above system (0.1% at 420–460 nm) must be improved. The efficiency of hydrogen production over g-C_3N_4 can be improved by tailoring its nanostructure. A mesoporous structure can enhance the light harvesting ability and mass transfer due to its large surface and multiple scattering effects. Wang et al. advanced g-C_3N_4 by generating a nanoporous structure into the polymeric matrix to improve its structural and electronic functions for solar energy conversion [110]. The photocatalyst mpg-C_3N_4 has a 3D porous framework (shown in Fig. 27), exhibiting an improved efficiency by an order of magnitude. The improved catalytic efficiency was due to the large surface area, which was a basic requirement for a heterogeneous (photo)catalyst to be chemically productive. This example shows excellent artificial photosynthesis over mesoporous polymer semiconductors.

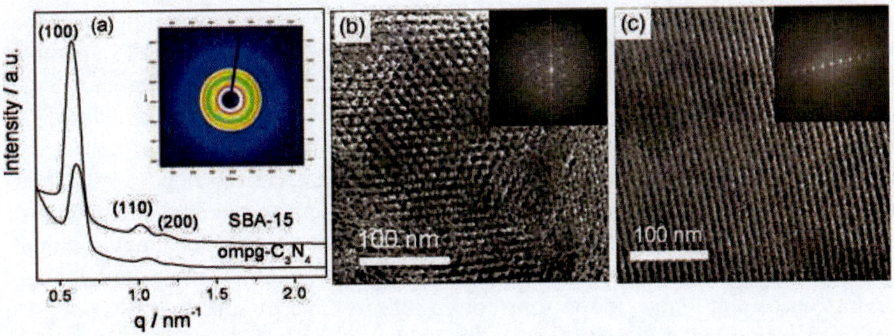

Fig. 28 **a** SAXS patterns of ompg-C_3N_4 and SBA-15 template. The *inset* shows the corresponding 2D SAXS image of ompg-C_3N_4. **b**, **c** Typical TEM images of ompg-C_3N_4. The *insets* show the corresponding fast Fourier transforms of the patterns. Reprinted with permission from Ref. [113]. Copyright 2009 American Chemical Society

The ordered mesostructure permits the structural orientation of guest molecules in the periodic nanopores, which enhances the selectivity and activity in photocatalysis [69, 111, 112]. Very recently, highly ordered porous g-C_3N_4 materials (shown as ompg-C_3N_4 in Fig. 28) were synthesized via a SBA-15 template route [113]. The photocatalytic activity was evaluated by photochemical reduction of water in the presence of an electron donor with visible light [113]. The total evolution of H_2 reached 2.1 mmol during the course of 25 h visible-light irradiation. The H_2 evolution on the ordered mesoporous C_3N_4 was about five times higher than that of bulk g-C_3N_4 [110]. Such a structure is promising as a host semiconductor scaffold for the design of hybrid visible-light photocatalyst. Furthermore, the photocatalyst surface can be functionalized easily via surface reaction or deposition. Various cocatalysts such as chromophoric antenna molecules can be coassembled into the ordered mesoporus carbon nitride, generating new biomimetic photocatalyst systems.

3.2 *Solar-Light-Driven Photocatalysts for Valuable Hydrocarbon Evolution from CO_2*

In recent years, carbon dioxide emissions from the burning of fossil fuels have grown to $\sim 2.5 \times 10^{10}$ metric tons per annum. This presents a significant environmental challenge for the twenty-first century. In nature, CO_2 is removed from the environment by photosynthesis. The energy obtained from sunlight is ultimately used to convert CO_2 into glucose, a sugar molecule that stores solar energy in the form of chemical energy. However, the efficiency of energy transformation is low. Even under the optimal artificial conditions (microalgae in full sunlight), the energy efficiency is only about 7% [8].

Carbon sequestration is considered a promising interim solution to global warming. It involves the capture and storage of fossil fuel-derived CO_2 emissions to prevent their release into the atmosphere. The captured CO_2 is stored in the oceans or in depleted gas and oil fields. The main drawback of the technology is the temporary nature of the storage. CO_2 stored in the ocean, for example, will inevitably return to the atmosphere in periods estimated from hundreds to thousands of years. Another drawback of ocean storage is the acidic characteristic of dissolved CO_2 and the effects of pH change in seawater on the local environment. Storage of large amounts of non-converted, concentrated CO_2 in oil and gas reservoirs needs continuous monitoring for an infinite time. A sudden release of CO_2 could be lethal, as demonstrated in the 1986 Lake Nyos disaster in Cameroon [114]. Carbon capture and storage will therefore not be commercially available until the cost and safety issues are satisfactorily addressed.

The conversion of CO_2 to useful fuels by physiochemical means not only reduces CO_2 in the atmosphere, but also eases our dependence on oil. The conventional approach involves the thermal hydrogenation of CO_2 into hydrocarbons under relatively high temperatures and pressures [115]. The major problem with the catalytic reduction of CO_2 is that huge amounts of H_2 are required as the reducing agent and in addition fossil fuels are consumed to provide the heat needed for the reaction to proceed.

An attractive alternative to thermal hydrogenation is photocatalytic reduction, especially if this utilizes sunlight. In this approach, CO_2 from industrial waste gases is converted to valuable fuels, such as methane and methanol. These products can be easily transported, stored and used in industry or, in the case of methanol, as a gasoline-additive for automobiles. Moreover, they can be transformed into other useful chemicals by using conventional technologies. This is a perfect solution to both the global warming and energy shortage problems. This section reviews the use of photocatalysts to produce valuable fuels from the virtually free resources of carbon dioxide, water and sunlight. The potential products are methane, methanol, or even longer chain hydrocarbons via a Fisher-Tropsch type condensation. From the viewpoints of both energy and environment, the conversion of CO_2 to fuels by solar energy is an ideal solution to the current global warming and energy crises.

For solar-driven catalytic conversion of CO_2 to fuels to be practical, highly efficient photocatalysts are required. Titania (TiO_2) has been considered the most appropriate candidate due to its powerful oxidizing nature, superior charge transport properties, and corrosion resistance. Earlier studies, however, could only achieve low CO_2 conversion rates in spite of using UV illumination for band gap excitation. Anpo et al. carried out a series of studies on Ti-zeolites and Ti-mesoporous materials [116, 117]. Powdered TiO_2 was also used by Adachi and co-workers as a photocatalyst for the reduction of CO_2 with H_2O. A total hydrocarbon (methane, ethene and ethane) generation rate of about 1.7 μL/(h g) was achieved under xenon lamp illumination when copper-loaded titania nanoparticles were dispersed in CO_2-pressurized water [118]. Tan et al., using titania pellets, obtained a maximum rate of about 0.25 μmol/h of methane from the irradiation of moist carbon dioxide by monochromatic ultraviolet light (253.7 nm wavelength) [119, 120]. Using UV irradiation of a hydrogen (90%), water, carbon dioxide combination, a rate of 4.1 μmol/(h g) was obtained by Lo and co-workers [121].

Recently, numerous studies on the preparation of solar-light-driven photocatalysts for hydrocarbon formation were reported. They can be classified into two categories: TiO_2-based photocatalysts and composite photocatalysts.

3.2.1 TiO_2-Based Photocatalysts

Metal doped TiO_2 catalyst sensitized with N3 dye was employed to photoreduce CO_2 with H_2O under concentrated natural sunlight to fuels in an optical-fiber photoreactor [122]. A methane production rate of 0.617 μmol/(g h) was achieved on N3-dye-Cu(0.5 wt%)-e(0.5 wt%)/TiO_2 coated onto optical fibers under an average solar light intensity of 20 mW/cm^2. The N3 dye substantially improved the photoactivity of Cu(0.5 wt%)-Fe(0.5 wt%)/TiO_2 catalyst toward methane production under concentrated natural sunlight due to its full visible-light adsorption. The photocatalyst was stable up to 6 h.

Recently, N-doped TiO_2 nanotubes with copper and platinum nanoparticles loaded onto the surfaces were developed to realize efficient solar conversion of carbon dioxide and water vapor to methane and other hydrocarbons [123]. The experiments were conducted in outdoor sunlight at University Park, PA (shown in Fig. 29). Using outdoor global AM 1.5 sunlight, 100 mW/cm^2, a hydrocarbon production rate of 111 ppm cm^{-2} h^{-1}, or ~160 μL/(g h), was obtained when the nanotube array samples were loaded with both 52% Cu and 48% Pt nanoparticles. The authors pointed out that the efficiency of the catalyst was still quite low, but were optimistic that further work could improve it.

3.2.2 Composite Photocatalysts

Carbon dioxide can be reduced with water to organic compounds over a hybrid catalyst under concentrated sunlight [124]. The catalyst was Pt-loaded $K_2Ti_6O_{13}$ coupled with an Fe-based catalyst supported on a dealuminated Y-type zeolite (Fe-Cu-K/DAY) [124]. The Pt/$K_2Ti_6O_{13}$ catalyst decomposed water to produce H_2 and the Fe-Cu-K/DAY catalyst reduced CO_2, with resulting organic compounds of CH_4, HCOOH, HCHO, CH_3OH, and C_2H_5OH. The Pt/$K_2Ti_6O_{13}$ catalysts can be combined with another CO_2 hydrogenation catalyst of Cu/ZnO [125]. The generation of CH_3OH over this composite photocatalyst under concentrated sunlight means successful photocatalytic conversion carbon dioxide to fuels. These studies suggest that sunlight-driven photocatalytic processes have potential for organic compound evolution from CO_2 and water.

Recently, a NiO/$InTaO_4$ photocatalyst with a band gap of 2.6 eV was developed by Chen's Group [126]. The NiO cocatalyst was loaded by incipient-wetness impregnation with an aqueous solution of $Ni(NO_3)_2$. The product was calcined at

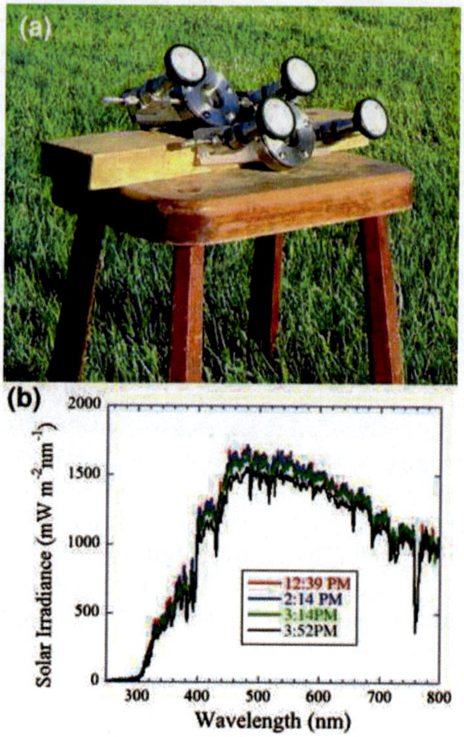

Fig. 29 **a** Digital photograph of the reaction chambers kept under natural sunlight for photocatalytic CO_2 conversion. **b** Spectral irradiance recorded from 12:39 p.m. to 3:52 p.m. for an experiment conducted on September 1, 2008 at University Park, PA. Reprinted with permission from Ref. [123]. Copyright 2009 American Chemical Society

350°C for 1 h in air, and then pretreated by H_2 reduction at 500°C for 2 h and subsequent O_2 oxidation at 200°C for 1 h. This catalyst was able to reduce CO_2 to methanol under visible-light illumination. A 1.0 wt% NiO-$InTaO_4$ photocatalyst in 0.2 M $KHCO_3$ gave the highest activity (1.394 μmol/(h g)) (shown in Fig. 30). The reduction–oxidation pretreatment had a positive effect on the activity of the catalyst.

In summary, the ultimate goal is to design advanced catalysts with high photon efficiencies. For the most promising catalysts the rate limiting steps in the conversions of water and CO_2 need to be determined. Relationships between the photocatalytic efficiency and the characteristics of the catalyst such as morphology, pore structure, surface area, surface electronic states and band gap must be investigated.

4 Summary and Future Prospects

Various photocatalytic semiconductor nanomaterials with great potential for energy and environmental applications have been prepared. Despite the tremendous research efforts on the synthesis and modification of photocatalysts, many

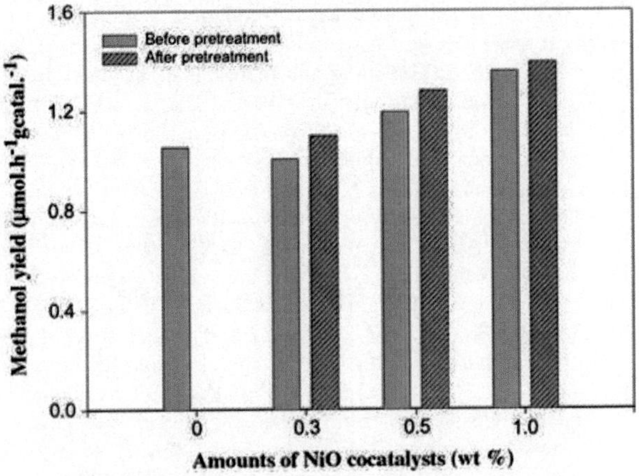

Fig. 30 Methanol yield on $NiO-InTaO_4$ in 0.2 M $KHCO_3$ aqueous solution under visible-light irradiation. Reprinted from Pan and Chen [126], copyright 2007 with permission from Elsevier

problems still exist. This is particularly true with regard to low photocatalytic efficiencies and the lack of understanding of the intrinsic mechanisms for the systems. Thus, the design of novel photocatalytic materials with higher efficiency is a perennial subject of interest in the field of photocatalysis. As more and more attention is paid to energy issues, the development of practical systems for H_2 evolution from water and the photo-reduction of carbon dioxide to fuels cannot be overemphasized. Advanced photocatalytic nanomaterials may be the key to a sustainable future.

Acknowledgments We would like to thank supports from the Research Grants Council of Hong Kong (General Research Fund CUHK404810), the Program for Professor of Special Appointment (Eastern Scholar) at Shanghai Institutions of Higher Learning, and the National Natural Science Foundation of China (21007040).

References

1. Ravelli D, Dondi D, Fagnoni M et al (2009) Photocatalysis. A multi-faceted concept for green chemistry. Chem Soc Rev 38:1999–2011
2. Kudo A, Miseki Y (2009) Heterogeneous photocatalyst materials for water splitting. Chem Soc Rev 38:253–278
3. Han F, Kambala VSR, Srinivasan M et al (2009) Tailored titanium dioxide photocatalysts for the degradation of organic dyes in wastewater treatment: a review. Appl Catal A Gen 359:25–40
4. Hoffmann MR, Martin ST, Choi WY et al (1995) Environmental applications of semiconductor photocatalysis. Chem Rev 95:69–96
5. Yu HT, Quan X (2009) Nano-heterojunction photocatalytic materials in environmental pollution controlling. Prog Chem 21:406–419
6. Fujishima A, Zhang XT, Tryk DA (2008) TiO_2 photocatalysis and related surface phenomena. Surf Sci Rep 63:515–582
7. Hernandez-Alonso MD, Fresno F, Suarez S et al (2009) Development of alternative photocatalysts to TiO_2: challenges and opportunities. Energy Environ Sci 2:1231–1257

8. Wu JCS (2009) Photocatalytic reduction of greenhouse gas CO_2 to fuel. Catal Surv Asia 13:30–40
9. Osterloh FE (2008) Inorganic materials as catalysts for photochemical splitting of water. Chem Mater 20:35–54
10. Chen X, Mao SS (2007) Titanium dioxide nanomaterials: synthesis, properties, modifications, and applications. Chem Rev 107:2891–2959
11. Yu JG, Xiang QJ, Zhou MH (2009) Preparation, characterization and visible-light-driven photocatalytic activity of Fe-doped titania nanorods and first-principles study for electronic structures. Appl Catal B Environ 90:595–602
12. Yang Y, Wang HY, Li X et al (2009) Electrospun mesoporous w^{6+}-doped TiO_2 thin films for efficient visible-light photocatalysis. Mater Lett 63:331–333
13. Tian BZ, Li CZ, Gu F et al (2009) Flame sprayed V-doped TiO_2 nanoparticles with enhanced photocatalytic activity under visible light irradiation. Chem Eng J 151:220–227
14. Tian BZ, Li CZ, Gu F et al (2009) Visible-light photocatalytic activity of Cr-doped TiO_2 nanoparticles synthesized by flame spray pyrolysis. J Inorg Mater 24:661–665
15. Lorret O, Francova D, Waldner G et al (2009) W-doped titania nanoparticles for UV and visible-light photocatalytic reactions. Appl Catal B Environ 91:39–46
16. Li JX, Xu JH, Dai WL et al (2009) Direct hydro-alcohol thermal synthesis of special core-shell structured Fe-doped titania microspheres with extended visible light response and enhanced photoactivity. Appl Catal B Environ 85:162–170
17. Khan MA, Han DH, Yang OB (2009) Enhanced photoresponse towards visible light in Ru doped titania nanotube. Appl Surf Sci 255:3687–3690
18. Xu JJ, Ao YH, Chen MD (2009) Preparation of B-doped titania hollow sphere and its photocatalytic activity under visible light. Mater Lett 63:2442–2444
19. Park Y, Kim W, Park H et al (2009) Carbon-doped TiO_2 photocatalyst synthesized without using an external carbon precursor and the visible light activity. Appl Catal B Environ 91:355–361
20. Tafen DN, Wang J, Wu NQ et al (2009) Visible light photocatalytic activity in nitrogen-doped TiO_2 nanobelts. Appl Phys Lett 94:093101.1–093101.3
21. Lv YY, Yu LS, Huang HY et al (2009) Preparation of F-doped titania nanoparticles with a highly thermally stable anatase phase by alcoholysis of $TiCl_4$. Appl Surf Sci 255:9548–9552
22. Li HX, Zhang XY, Huo YN et al (2007) Supercritical preparation of a highly active S-doped TiO_2 photocatalyst for methylene blue mineralization. Environ Sci Technol 41:4410–4414
23. Tian GH, Pan K, Fu HG et al (2009) Enhanced photocatalytic activity of S-doped TiO_2-ZrO_2 nanoparticles under visible-light irradiation. J Hazard Mater 166:939–944
24. Xu JJ, Ao YH, Fu DG et al (2008) Low-temperature preparation of F-doped TiO_2 film and its photocatalytic activity under solar light. Appl Surf Sci 254:3033–3038
25. Wu GS, Chen A (2008) Direct growth of F-doped TiO_2 particulate thin films with high photocatalytic activity for environmental applications. J Photoch Photobio A Chem 195:47–53
26. Cui Y, Du H, Wen LS (2009) Origin of visible-light-induced photocatalytic properties of S-doped anatase TiO_2 by first-principles investigation. Solid State Commun 149:634–637
27. Liu G, Yang HG, Wang XW et al (2009) Visible light responsive nitrogen doped anatase TiO_2 sheets with dominant 001 facets derived from tin. J Am Chem Soc 131:12868–12869
28. Ho W, Yu JC, Lee S (2006) Synthesis of hierarchical nanoporous F-doped TiO_2 spheres with visible light photocatalytic activity. Chem Commun 1115–1117
29. Ho WK, Yu JC, Lee SC (2006) Low-temperature hydrothermal synthesis of S-doped TiO_2 with visible light photocatalytic activity. J Solid State Chem 179:1171–1176
30. Asahi R, Morikawa T, Ohwaki T et al (2001) Visible-light photocatalysis in nitrogen-doped titanium oxides. Science 293:269–271
31. Yamashita H, Harada M, Misaka J et al (2002) Degradation of propanol diluted in water under visible light irradiation using metal ion-implanted titanium dioxide photocatalysts. J Photochem Photobiol A Chem 148:257–261

32. Sakthivel S, Kisch H (2003) Daylight photocatalysis by carbon-modified titanium dioxide. Angew Chem Int Ed 42:4908–4911
33. Yamashita H, Harada M, Misaka J et al (2003) Photocatalytic degradation of organic compounds diluted in water using visible light-responsive metal ion-implanted TiO_2 catalysts: Fe ion-implanted TiO_2. Catal Today 84:191–196
34. Hong XT, Wang ZP, Cai WM et al (2005) Visible-light-activated nanoparticle photocatalyst of iodine-doped titanium dioxide. Chem Mater 17:1548–1552
35. Chai SY, Kim YJ, Jung MH et al (2009) Heterojunctioned $BiOCl/Bi_2O_3$, a new visible light photocatalyst. J Catal 262:144–149
36. Lin XP, Xing JC, Wang WD et al (2007) Photocatalytic activities of heterojunction semiconductors $Bi_2O_3/BaTiO_3$: a strategy for the design of efficient combined photocatalysts. J Phys Chem C 111:18288–18293
37. Pal B, Hata T, Goto K et al (2001) Photocatalytic degradation of o-cresol sensitized by iron-titania binary photocatalysts. J Mol Catal A Chem 169:147–155
38. Bessekhouad Y, Robert D, Weber J (2004) Bi_2S_3/TiO_2 and CdS/TiO_2 heterojunctions as an available configuration for photocatalytic degradation of organic pollutant. J Photochem Photobiol A Chem 163:569–580
39. Zhang HT, Ouyanga SX, Lia ZS et al (2006) Preparation, characterization and photocatalytic activity of polycrystalline $Bi_2O_3/SrTiO_3$ composite powders. J Phys Chem Solids 67:2501–2505
40. Zhang ML, An TC, Hu XH et al (2004) Preparation and photocatalytic properties of a nanometer $ZnO\text{-}SnO_2$ coupled oxide. Appl Catal A Gen 260:215–222
41. Li GS, Zhang DQ, Yu JC (2009) A new visible-light photocatalyst: CdS quantum dots embedded mesoporous TiO_2. Environ Sci Technol 43:7079–7085
42. Zong X, Yan HJ, Wu GP et al (2008) Enhancement of photocatalytic H_2 evolution on CdS by loading MoS_2 as cocatalyst under visible light irradiation. J Am Chem Soc 130:7176–7177
43. Huang HJ, Li DZ, Lin Q et al (2009) Efficient degradation of benzene over $LaVO_4/TiO_2$ nanocrystalline heterojunction photocatalyst under visible light irradiation. Environ Sci Technol 43:4164–4168
44. Huang HJ, Li DZ, Lin Q et al (2009) Efficient photocatalytic activity of PZT/TiO_2 heterojunction under visible light irradiation. J Phys Chem C 113:14264–14269
45. Kim YJ, Gao B, Han SY et al (2009) Heterojunction of $FeTiO_3$ nanodisc and TiO_2 nanoparticle for a novel visible light photocatalyst. J Phys Chem C 113:19179–19184
46. Rawal SB, Chakraborty AK, Lee WI (2009) Heterojunction of FeOOH and TiO_2 for the formation of visible light photocatalyst. Bull Korean Chem Soc 30:2613–2616
47. Tsunoda Y, Sugimoto W, Sugahara Y (2003) Intercalation behavior of n-alkylamines into a protonated form of a layered perovskite derived from aurivillius phase $Bi_2SrTa_2O_9$. Chem Mater 15:632–635
48. Kim JY, Chung I, Choy JH et al (2001) Macromolecular nanoplatelet of Aurivillius-type layered perovskite oxide, $Bi_4Ti_3O_{12}$. Chem Mater 13:2759–2761
49. Zhang C, Zhu YF (2005) Synthesis of square Bi_2WO_6 nanoplates as high-activity visible-light-driven photocatalysts. Chem Mater 17:3537–3545
50. Yu JG, Xiong JF, Cheng B et al (2005) Hydrothermal preparation and visible-light photocatalytic activity of Bi_2WO_6 powders. J Solid State Chem 178:1968–1972
51. Fu HB, Pan CS, Yao WQ et al (2005) Visible-light-induced degradation of rhodamine B by nanosized Bi_2WO_6. J Phys Chem B 109:22432–22439
52. Kudo A, Hijii S (1999) H_2 or O_2 evolution from aqueous solutions on layered oxide photocatalysts consisting of Bi^{3+} with $6s^2$ configuration and d^0 transition metal ions. Chem Lett 1103–1104
53. Tang JW, Zou ZG, Ye JH (2004) Photocatalytic decomposition of organic contaminants by Bi_2WO_6 under visible light irradiation. Catal Lett 92:53–56
54. Zhang LH, Wang WZ, Chen ZG et al (2007) Fabrication of flower-like Bi_2WO_6 superstructures as high performance visible-light driven photocatalysts. J Mater Chem 17:2526–2532

55. Wu J, Duan F, Zheng Y et al (2007) Synthesis of Bi_2WO_6 nanoplate-built hierarchical nest-like structures with visible-light-induced photocatalytic activity. J Phys Chem C 111:12866–12871
56. Zhang LS, Wang WZ, Zhou L et al (2007) Bi_2WO_6 nano- and microstructures: shape control and associated visible-light-driven photocatalytic activities. Small 3:1618–1625
57. Zhu SB, Xu TG, Fu HB et al (2007) Synergetic effect of Bi_2WO_6 photocatalyst with C_{60} and enhanced photoactivity under visible irradiation. Environ Sci Technol 41:6234–6239
58. Hasobe T, Imahori H, Fukuzumi S et al (2003) Light energy conversion using mixed molecular nanoclusters. Porphyrin and C_{60} cluster films for efficient photocurrent generation. J Phys Chem B 107:12105–12112
59. Shi R, Huang GL, Lin J et al (2009) Photocatalytic activity enhancement for Bi_2WO_6 by fluorine substitution. J Phys Chem C 113:19633–19638
60. Li JP, Zhang X, Ai ZH et al (2007) Efficient visible light degradation of rhodamine B by a photo-electrochemical process based on a Bi_2WO_6 nanoplate film electrode. J Phys Chem C 111:6832–6836
61. Shang M, Wang WZ, Ren J et al (2009) A practical visible-light-driven Bi_2WO_6 nanofibrous mat prepared by electrospinning. J Mater Chem 19:6213–6218
62. Kudo A, Omori K, Kato H (1999) A novel aqueous process for preparation of crystal form-controlled and highly crystalline $BiVO_4$ powder from layered vanadates at room temperature and its photocatalytic and photophysical properties. J Am Chem Soc 121:11459–11467
63. Kohtani S, Koshiko M, Kudo A et al (2003) Photodegradation of 4-alkylphenols using $BiVO_4$ photocatalyst under irradiation with visible light from a solar simulator. Appl Catal B Environ 46:573–586
64. Tokunaga S, Kato H, Kudo A (2001) Selective preparation of monoclinic and tetragonal $BiVO_4$ with scheelite structure and their photocatalytic properties. Chem Mater 13:4624–4628
65. Lim AR, Choh SH, Jang MS (1995) Prominent ferroelastic domain-walls in $BiVO_4$ crystal. J Phys Condens Mat 7:7309–7323
66. Liu JB, Wang H, Wang S et al (2003) Hydrothermal preparation of $BiVO_4$ powders. Mater Sci Eng B 104:36–39
67. Sayama K, Nomura A, Zou ZG et al (2003) Photoelectrochemical decomposition of water on nanocrystalline $BiVO_4$ film electrodes under visible light. Chem Commun 2908–2909
68. Zhou L, Wang WZ, Zhang L et al (2007) Single-crystalline $BiVO_4$ microtubes with square cross-sections: microstructure, growth mechanism, and photocatalytic property. J Phys Chem C 111:13659–13664
69. Li GS, Zhang DQ, Yu JC (2008) Ordered mesoporous $BiVO_4$ through nanocasting: a superior visible light-driven photocatalyst. Chem Mater 20:3983–3992
70. Zhou L, Wang WZ, Xu HL (2008) Controllable synthesis of three-dimensional well-defined $BiVO_4$ mesocrystals via a facile additive-free aqueous strategy. Cryst Growth Des 8:728–733
71. Dunkle SS, Helmich RJ, Suslick KS (2009) $BiVO_4$ as a visible-light photocatalyst prepared by ultrasonic spray pyrolysis. J Phys Chem C 113:11980–11983
72. Xu H, Li HM, Wu CD et al (2008) Preparation, characterization and photocatalytic properties of Cu-loaded $BiVO_4$. J Hazard Mater 153:877–884
73. Ge L (2008) Novel visible-light-driven Pt/$BiVO_4$ photocatalyst for efficient degradation of methyl orange. J Mol Catal A Chem 282:62–66
74. Wang XC, Maeda K, Thomas A et al (2009) A metal-free polymeric photocatalyst for hydrogen production from water under visible light. Nat Mater 8:76–80
75. Goettmann F, Fischer A, Antonietti M et al (2006) Chemical synthesis of mesoporous carbon nitrides using hard templates and their use as a metal-free catalyst for Friedel-Crafts reaction of benzene. Angew Chem Int Ed 45:4467–4471
76. Groenewolt M, Antonietti M (2005) Synthesis of g-C_3N_4 nanoparticles in mesoporous silica host matrices. Adv Mater 17:1789–1792

77. Wang XC, Chen XF, Thomas A et al (2009) Metal-containing carbon nitride compounds: a new functional organic-metal hybrid material. Adv Mater 21:1609–1612
78. Yamamoto S, Andersson K, Bluhm H et al (2007) Hydroxyl-induced wetting of metals by water at near-ambient conditions. J Phys Chem C 111:7848–7850
79. Chen XF, Zhang JS, Fu XZ et al (2009) Fe-g-C_3N_4-catalyzed oxidation of benzene to phenol using hydrogen peroxide and visible light. J Am Chem Soc 131:11658–11659
80. Goettmann F, Thomas A, Antonietti M (2007) Metal-free activation CO_2 by mesoporous graphitic carbon nitride. Angew Chem Int Ed 46:2717–2720
81. Liu ZF, Zhao ZG, Miyauchi M (2009) Efficient visible light active $CaFe_2O_4/WO_3$ based composite photocatalysts: effect of interfacial modification. J Phys Chem C 113:17132–17137
82. Zheng LR, Zheng YH, Chen CQ et al (2009) Network structured SnO_2/ZnO heterojunction nanocatalyst with high photocatalytic activity. Inorg Chem 48:1819–1825
83. Zheng YH, Chen CQ, Zhan YY et al (2008) Photocatalytic activity of Ag/ZnO heterostructure nanocatalyst: correlation between structure and property. J Phys Chem C 112:10773–10777
84. Eisenberg R (2009) Rethinking water splitting. Science 324:44–45
85. Maeda K, Teramura K, Lu DL et al (2006) Photocatalyst releasing hydrogen from water—enhancing catalytic performance holds promise for hydrogen production by water splitting in sunlight. Nature 440:295
86. Kudo A, Kato H, Tsuji I (2004) Strategies for the development of visible-light-driven photocatalysts for water splitting. Chem Lett 33:1534–1539
87. Hitoki G, Takata T, Kondo JN et al (2002) (Oxy)nitrides as new photocatalysts for water splitting under visible light irradiation. Electrochemistry 70:463–465
88. Yamasita D, Takata T, Hara M et al (2004) Recent progress of visible-light-driven heterogeneous photocatalysts for overall water splitting. Solid State Ionics 172:591–595
89. Yerga RMN, Galvan MCA, del Valle F et al (2009) Water splitting on semiconductor catalysts under visible-light irradiation. ChemSusChem 2:471–485
90. Kasahara A, Nukumizu K, Hitoki G et al (2002) Photoreactions on $LaTiO_2N$ under visible light irradiation. J Phys Chem A 106:6750–6753
91. Hitoki G, Takata T, Kondo JN et al (2002) An oxynitride, TaON, as an efficient water oxidation photocatalyst under visible light irradiation (lambda $\leq$ 500 nm). Chem Commun 1698–1699
92. Hara M, Nunoshige J, Takata T et al (2003) Unusual enhancement of H_2 evolution by Ru on TaON photocatalyst under visible light irradiation. Chem Commun 3000–3001
93. Hara M, Takata T, Kondo JN et al (2004) Photocatalytic reduction of water by TaON under visible light irradiation. Catal Today 90:313–317
94. Takata T, Hitoki G, Kondo JN et al (2007) Visible-light-driven photocatalytic behavior of tantalum-oxynitride and nitride. Res Chem Intermed 33:13–25
95. Liu MY, You WS, Lei ZB et al (2004) Water reduction and oxidation on Pt-Ru/$Y_2Ta_2O_5N_2$ catalyst under visible light irradiation. Chem Commun 2192–2193
96. Nukumizu K, Nunoshige J, Takata T et al (2003) $TiN_xO_yF_z$ as a stable photocatalyst for water oxidation in visible light (<570 nm). Chem Lett 32:196–197
97. Maeda K, Shimodaira Y, Lee B et al (2007) Studies on $TiN_xO_yF_z$ as a visible-light-responsive photocatalyst. J Phys Chem C 111:18264–18270
98. Ishikawa A, Yamada Y, Takata T et al (2003) Novel synthesis and photocatalytic activity of oxysulfide $Sm_2Ti_2S_2O_5$. Chem Mater 15:4442–4446
99. Ogisu K, Ishikawa A, Teramura K et al (2007) Lanthanum–indium oxysulfide as a visible light driven photocatalyst for water splitting. Chem Lett 36:854–855
100. Maeda K, Teramura K, Saito N et al (2006) Overall water splitting using (oxy)nitride photocatalysts. Pure Appl Chem 78:2267–2276
101. Maeda K, Domen K (2007) New non-oxide photocatalysts designed for overall water splitting under visible light. J Phys Chem C 111:7851–7861

102. Takanabe K, Uzawa T, Wang XC et al (2009) Enhancement of photocatalytic activity of zinc-germanium oxynitride solid solution for overall water splitting under visible irradiation. Dalton Trans 10055–10062
103. Banerjee S, Mohapatra SK, Misra M (2009) Synthesis of TaON nanotube arrays by sonoelectrochemical anodization followed by nitridation: a novel catalyst for photoelectrochemical hydrogen generation from water. Chem Commun 7137–7139
104. Maeda K, Hashiguchi H, Masuda H et al (2008) Photocatalytic activity of $(Ga_{1-x}Zn_x)(N_{1-x}O_x)$ for visible-light-driven H_2 and O_2 evolution in the presence of sacrificial reagents. J Phys Chem C 112:3447–3452
105. Maeda K, Teramura K, Domen K (2007) Development of cocatalysts for photocatalytic overall water splitting on $(Ga_{1-x}Zn_x)(N_{1-x}O_x)$ solid solution. Catal Surv Asia 11:145–157
106. Lee YG, Teramura K, Hara M et al (2007) Modification of $(Zn_{1+x}Ge)(N_2O_x)$ solid solution as a visible light driven photocatalyst for overall water splitting. Chem Mater 19:2120–2127
107. Kanade KG, Baeg JO, Kale BB et al (2007) Rose-red color oxynitride $Nb_2Zr_6O_{17-x}N_x$: a visible light photocatalyst to hydrogen production. Int J Hydrogen Energ 32:4678–4684
108. Hisatomi T, Hasegawa K, Teramura K et al (2007) Zinc and titanium spinel oxynitride $(Zn_xTiO_yN_z)$ as a d^0–d^{10} complex photocatalyst with visible light activity. Chem Lett 36:558–559
109. Yanagida S, Kabumoto A, Mizumoto K et al (1985) Poly(para-phenylene)-catalyzed photoreduction of water to hydrogen. J Chem Soc Chem Commun 474–475
110. Wang XC, Maeda K, Chen XF et al (2009) Polymer semiconductors for artificial photosynthesis: hydrogen evolution by mesoporous graphitic carbon nitride with visible light. J Am Chem Soc 131:1680–1681
111. Yu JC, Wang XC, Fu XZ (2004) Pore-wall chemistry and photocatalytic activity of mesoporous titania molecular sieve films. Chem Mater 16:1523–1530
112. Jun YS, Hong WH, Antonietti M et al (2009) Mesoporous, 2D hexagonal carbon nitride and titanium nitride/carbon composites. Adv Mater 21:4270–4274
113. Chen XF, Jun YS, Takanabe K et al (2009) Ordered mesoporous SBA-15 type graphitic carbon nitride: a semiconductor host structure for photocatalytic hydrogen evolution with visible light. Chem Mater 21:4093–4095
114. Kling GW, Clark MA, Compton HR et al (1987) The 1986 Lake Nyos gas disaster in Cameroon, West-Africa. Science 236:169–175
115. Freund HJ, Roberts MW (1996) Surface chemistry of carbon dioxide. Surf Sci Rep 25:225–273
116. Anpo M, Yamashita H, Ichihashi Y et al (1997) Photocatalytic reduction of CO_2 with H_2O on titanium oxides anchored within micropores of zeolites: effects of the structure of the active sites and the addition of Pt. J Phys Chem B 101:2632–2636
117. Ikeue K, Yamashita H, Anpo M et al (2001) Photocatalytic reduction of CO_2 with H_2O on Ti-beta zeolite photocatalysts: effect of the hydrophobic and hydrophilic properties. J Phys Chem B 105:8350–8355
118. Adachi K, Ohta K, Mizuno T (1994) Photocatalytic reduction of carbon-dioxide to hydrocarbon using copper-loaded titanium-dioxide. Sol Energy 53:187–190
119. Tan SS, Zou L, Hu E (2007) Photosynthesis of hydrogen and methane as key components for clean energy system. Sci Technol Adv Mater 8:89–92
120. Tan SS, Zou L, Hu E (2006) Photocatalytic reduction of carbon dioxide into gaseous hydrocarbon using TiO_2 pellets. Catal Today 115:269–273
121. Lo CC, Hung CH, Yuan CS et al (2007) Photoreduction of carbon dioxide with H_2 and H_2O over TiO_2 and ZrO_2 in a circulated photocatalytic reactor. Sol Energy Mater Sol Cells 91:1765–1774
122. Nguyen TV, Wu JCS, Chiou CH (2008) Photoreduction of CO_2 over ruthenium dye-sensitized TiO_2-based catalysts under concentrated natural sunlight. Catal Commun 9:2073–2076
123. Varghese OK, Paulose M, LaTempa TJ et al (2009) High-rate solar photocatalytic conversion of CO_2 and water vapor to hydrocarbon fuels. Nano Lett 9:731–737

124. Guan GQ, Kida T, Yoshida A (2003) Reduction of carbon dioxide with water under concentrated sunlight using photocatalyst combined with Fe-based catalyst. Appl Catal B Environ 41:387–396
125. Guan GQ, Kida T, Harada T et al (2003) Photoreduction of carbon dioxide with water over $K_2Ti_6O_{13}$ photocatalyst combined with Cu/ZnO catalyst under concentrated sunlight. Appl Catal A Gen 249:11–18
126. Pan PW, Chen YW (2007) Photocatalytic reduction of carbon dioxide on NiO/$InTaO_4$ under visible light irradiation. Catal Commun 8:1546–1549
127. Maeda K, Eguchi M, Youngblood WJ et al (2008) Niobium oxide nanoscrolls as building blocks for dye-sensitized hydrogen production from water under visible light irradiation. Chem Mater 20:6770–6778
128. Ishii T, Kato H, Kudo A (2004) H_2 evolution from an aqueous methanol solution on $SrTiO_3$ photocatalysts codoped with chromium and tantalum ions under visible light irradiation. J Photochem Photobiol A Chem 163:181–186
129. Kato H, Kudo A (2002) Visible-light-response and photocatalytic activities of TiO_2 and $SrTiO_3$ photocatalysts codoped with antimony and chromium. J Phys Chem B 106:5029–5034
130. Niishiro R, Kato H, Kudo A (2005) Nickel and either tantalum or niobium-codoped TiO_2 and $SrTiO_3$ photocatalysts with visible-light response for H_2 or O_2 evolution from aqueous solutions. Phys Chem Chem Phys 7:2241–2245
131. Konta R, Ishii T, Kato H et al (2004) Photocatalytic activities of noble metal ion doped $SrTiO_3$ under visible light irradiation. J Phys Chem B 108:8992–8995
132. Nishimoto S, Matsuda M, Miyake M (2006) Photocatalytic activities of Rh-doped $CaTiO_3$ under visible light irradiation. Chem Lett 35:308–309
133. Hwang DW, Kirn HG, Lee JS et al (2005) Photocatalytic hydrogen production from water over M-doped $La_2Ti_2O_7$ (M = Cr, Fe) under visible light irradiation (lambda > 420 nm). J Phys Chem B 109:2093–2102
134. Ishikawa A, Takata T, Kondo JN et al (2002) Oxysulfide $Sm_2Ti_2S_2O_5$ as a stable photocatalyst for water oxidation and reduction under visible light irradiation (lambda <= 650 nm). J Am Chem Soc 124:13547–13553
135. Kim HG, Becker OS, Jang JS et al (2006) A generic method of visible light sensitization for perovskite-related layered oxides: substitution effect of lead. J Solid State Chem 179:1214–1218
136. Zou ZG, Arakawa H (2003) Direct water splitting into H_2 and O_2 under visible light irradiation with a new series of mixed oxide semiconductor photocatalysts. J Photochem Photobiol A Chem 158:145–162
137. Zou ZG, Ye JH, Sayama K et al (2001) Direct splitting of water under visible light irradiation with an oxide semiconductor photocatalyst. Nature 414:625–627
138. Kim HG, Hwang DW, Lee JS (2004) An undoped, single-phase oxide photocatalyst working under visible light. J Am Chem Soc 126:8912–8913
139. Wang DF, Zou ZG, Ye JH (2005) Photocatalytic water splitting with the Cr-doped $Ba_2In_2O_5/In_2O_3$ composite oxide semiconductors. Chem Mater 17:3255–3261
140. Yoshimura J, Ebina Y, Kondo J et al (1993) Visible-light induced photocatalytic behavior of a layered perovskite type niobate, $RbPb_2Nb_3O_{10}$. J Phys Chem 97:1970–1973
141. Hosogi Y, Shimodaira Y, Kato H et al (2008) Role of Sn^{2+} in the band structure of SnM_2O_6 and $Sn_2M_2O_7$ (M = Nb and Ta) and their photocatalytic properties. Chem Mater 20:1299–1307
142. Hosogi Y, Tanabe K, Kato H et al (2004) Energy structure and photocatalytic activity of niobates and tantalates containing Sn(||) with a $5s^2$ electron configuration. Chem Lett 33:28–29
143. Hosogi Y, Kato H, Kudo A (2006) Synthesis of $SnNb_2O_6$ nanoplates and their photocatalytic properties. Chem Lett 35:578–579
144. Kato H, Kobayashi H, Kudo A (2002) Role of Ag^+ in the band structures and photocatalytic properties of $AgMO_3$ (M: Ta and Nb) with the perovskite structure. J Phys Chem B 106:12441–12447

145. Hosogi Y, Kato H, Kudo A (2008) Photocatalytic activities of layered titanates and niobates ion-exchanged with sn^{2+} under visible light irradiation. J Phys Chem C 112:17678–17682
146. Ikeue K, Shiiba S, Machida M (2010) Novel visible-light-driven photocatalyst based on Mn-Cd-S for efficient H_2 evolution. Chem Mater 22:743–745
147. Zhang W, Xu R (2009) Surface engineered active photocatalysts without noble metals: CuS-$Zn_xCd_{1-x}S$ nanospheres by one-step synthesis. Int J Hydrogen Energ 34:8495–8503
148. Yan HJ, Yang JH, Ma GJ et al (2009) Visible-light-driven hydrogen production with extremely high quantum efficiency on Pt-PdS/CdS photocatalyst. J Catal 266:165–168
149. Li YX, Ma GF, Peng SQ et al (2009) Photocatalytic H_2 evolution over basic zincoxysulfide ($ZnS_{1-x-0.5y}O_x(OH)(_y)$) under visible light irradiation. Appl Catal A Gen 363:180–187
150. del Valle F, Ishikawa A, Domen K et al (2009) Influence of Zn concentration in the activity of $Cd_{1-x}Zn_xS$ solid solutions for water splitting under visible light. Catal Today 143:51–56
151. Zhang XH, Jing DW, Liu MC et al (2008) Efficient photocatalytic H_2 production under visible light irradiation over Ni doped $Cd_{1-x}Zn_xS$ microsphere photocatalysts. Catal Commun 9:1720–1724
152. Jang JS, Borse PH, Lee JS et al (2008) Indium induced band gap tailoring in $AgGa_{1-x}In_xS_2$ chalcopyrite structure for visible light photocatalysis. J Chem Phys 128:154717-1–154717-6
153. Zhang K, Jing DW, Xing CJ et al (2007) Significantly improved photocatalytic hydrogen production activity over $Cd_{1-x}Zn_xS$ photocatalysts prepared by a novel thermal sulfuration method. Int J Hydrogen Energ 32:4685–4691
154. Arai T, Senda SI, Sato Y et al (2008) Cu-doped ZnS hollow particle with high activity for hydrogen generation from alkaline sulfide solution under visible light. Chem Mater 20:1997–2000
155. Bao NZ, Shen LM, Takata T et al (2008) Self-templated synthesis of nanoporous CdS nanostructures for highly efficient photocatalytic hydrogen production under visible. Chem Mater 20:110–117
156. Chen D, Ye JH (2007) Photocatalytic H_2 evolution under visible light irradiation on $AgIn_5S_8$ photocatalyst. J Phys Chem Solids 68:2317–2320
157. Bang JH, Hehnich RJ, Suslick KS (2008) Nanostructured ZnS: Ni^{2+} photocatalysts prepared by ultrasonic spray pyrolysis. Adv Mater 20:2599–2603

Part IV
Energy Storage

Lithium-Based Batteries for Efficient Energy Storage: Nanotechnology and Its Implications

Jiajia Tan and Ashutosh Tiwari

Abstract Demand for energy overshadows all other problems mankind will face during the next half century. With more countries achieving higher economic development, this demand will continue rising dramatically. Plaguing the energy demand is our persistent dependence on fossil fuels and its related environmental effects. Renewable sources of energy are the most possible option to reduce this dependence. Since many green energy sources cannot provide consistent power at all times, efficient storage and transmission of generated energy is necessary. For this we require high-energy density devices that are rechargeable and cost-efficient. Although the development of new energy storage technology should continue, current high- energy density lithium-ion batteries should also be researched in great detail to improve their performance and widespread use. Toward this aim, we review the historic and recent development of cathode, anode and electrolyte materials in detail. We also review the mechanisms of charge transport and phase stability in these compounds. The use of nanotechnology has already found great influence in modifying these materials towards higher energy density and greater reliability. Nanotechnologies will go on to provide breakthroughs not only in better materials, but also better battery design for energy storage, such as in thin film and lithium-air batteries.

1 Introduction

Lithium-ion(Li-ion) batteries have changed our lifestyles tremendously. In the 1980s lithium-ion batteries (LIBs) began finding applications as power sources in watches, calculators and implantable medical devices. In today's age of

J. Tan · A. Tiwari (✉)
Department of Materials Science and Engineering,
University of Utah, Salt Lake City, UT 84112, USA
e-mail: tiwari@eng.utah.edu

L. Zang (ed.), *Energy Efficiency and Renewable Energy Through Nanotechnology*,
Green Energy and Technology, DOI: 10.1007/978-0-85729-638-2_21,

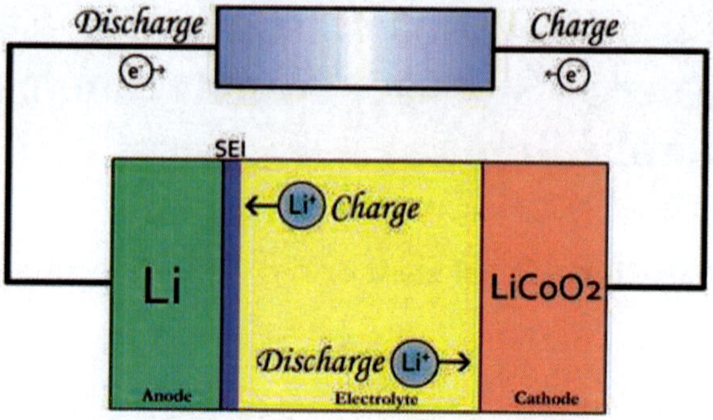

Fig. 1 Operating process of lithium-ionbatteries

information and speed, LIBs have become integral components in portable computing, entertainment, telecommunication equipment and even automobiles. Since conventional automobiles are one of the major contributors of greenhouse gases, there is significant effort to reduce or replace fossil fuel demand by developing hybrid and electric vehicles. LIBs will provide a breakthrough in electric vehicle acceptance and widespread use. In this section, we will give a brief introduction of the basic parts and working principles of LIBs, and their advantages and challenges.

A primary battery is an energy transformation device that converts the chemical energy stored in its active materials into electric energy through electron and lithium-ion transfer caused by an oxidation–reduction reaction. For Li-ion battery systems, the reaction is reversible when supplied with external electric power source. Normally, a battery is composed of one or more cells (basic electrochemical unit) that are electrically connected in series or parallel to provide the required operating voltage and capacity. In our discussion, we generally refer to LIB as a single cell.

Each cell consists of positive and negative electrodes, an electrolyte and a separator. The positive electrode has a higher chemical potential than the negative electrode. The operating process is demonstrated in Fig. 1. The positive electrode provides lithium ions and becomes oxidized as an anode when charged. When discharging, lithium ions are inserted back into positive electrode, which is reduced as a cathode. In contrast, the negative electrode is the opposite electrode during cycling. Since the lithium ions can be inserted and removed from the electrodes, they are also called insertion electrode materials. In general, the positive electrode is referred to as the cathode and negative electrode as the anode, which is correct only when the cell is discharged. The ionic conducting electrolyte provides the medium for transfer of ions between the anode and cathode. The electrolyte is typically a liquid composed of solvents, aqueous or non-aqueous and dissolved lithium salts, acids or alkali to enable ionic conductivity. Some batteries use solid electrolytes, which require high operating temperatures (>80 °C) in order to realize acceptable ionic conductivity of electrolyte. The anode and cathode electrodes should be electronically separated to prevent internal short-circuiting. Thus the electrolyte should not be electronically conductive, and a mesoporous polymer separator is designed to isolate electrodes while allowing ions to move through it.

The main parameters that determine the electrochemical performances of a LIB are cell potential (V), specific energy capacity, rate capability and cycling ability. Energy capacity describes the ability of hosting lithium ions, with units of Wh/kg or Wh/L, representing the energy stored per mass or volume. High-energy capacity is critical for the application of light small power sources in portable electronic devices and hybrid vehicles. The specific rate, with units of A/kg or A/L, is crucial when large current is required, as in hybrid vehicles and large power devices. The good rate capability is mainly determined by the electric conductivity within the electrode and the ionic conductivity within the electrodes, electrolyte and through the electrode solid interface (SEI). The operating current is often expressed as 1C, 2C or 0.1C, where C stands for the current (A/Kg) that will charge an electrode to its theoretical capacity in 1 h. If an operating current is xC, then the theoretical specific capacity is obtained in $1/x$ hour. Rate capability is considered good only if the experimentally obtained capacity is close to theoretical capacity, no matter what operating current is applied. The cycling ability reflects the stability of the cell after several hundred times of charging/discharging and is determined by factors such as volume expansion/contraction, structure-maintaining ability and thermal stability.

Li-based batteries have outperformed other battery systems and account for much more than a half of worldwide sales in portable batteries due to several advantageous properties against which no other battery system can compete.

First, LIBs have high operating voltages and high-energy densities. A single cell has an average operating potential of approximately 3.6–4 V, much higher than that of Ni-Cd, Ni-MH and Pb-acid batteries. Lithium is the lightest and the most electropositive metal in the periodic table; hence the high energy density is predicted. The specific capacity is more than 1.5 times that of Ni-Cd batteries. Figure 2 [1] shows the power density and energy density of various battery systems. In this picture, 4G/5G refers to devices that provide users with IP telephony, ultra-broadband Internet access, gaming services and streamed multimedia. EV and HEV are abbreviation for electric vehicles and hybrid electric vehicles. It can be seen that the Li-ion batteries have the highest power and energy densities, and come to the closest in meeting projected requirements for further application, though there are still improvements to be made.

The second advantage of LIBs is the impressively high charging and discharging rate. Up to 3C of the discharging rate and 1C of charging rate are attainable. The operating temperature range is large, from −20 to +60 °C. In addition to the advantages described above, LIBs have superior cycle life, exceeding 500 cycles; low self-discharge (8–12% per month), and long shelf lives. There is also no memory-effect and thus can be recharged at any time.

With these advantages in mind, there are still some problems that limit the usage of LIBs. Some problems proposed several decades ago still need solutions, such as safety issues and restricted assembly conditions. Moreover, energy capacity and cycling life requirements will continue to increase, as shown in Fig. 2.

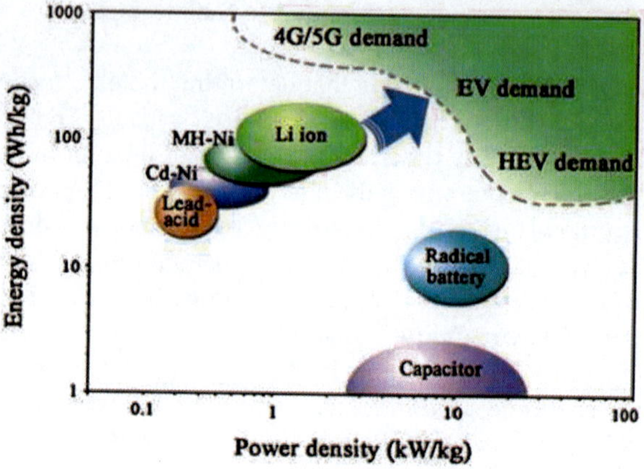

Fig. 2 Schematic distribution of power density and energy density for different batteries and capacitors (from Tarascon (2001), with permission)

One problem associated with metallic Li is the formation of dendrites, leading to short-circuiting, then thermal runaway and possible explosion. Moreover, metallic lithium is highly reactive with oxygen, nitrogen and moisture; thus glove boxes or dry-rooms become necessary for assembly. Battery assembly is much simpler if insertion electrodes composed of Li compounds are used as both anode and cathode, since they are stable in the ambient atmosphere. Therefore, several insertion anode materials have been explored, including graphite, titanium oxide, silicon, carbon nanofibers, etc. Research on anode materials is very popular today, since new anode materials have much higher specific capacity than the commonly used graphite. Their developmental status and properties will be discussed in detail in the anode section.

Another big problem with lithium–ion batteries is that all Li ions that have participated in a secondary reaction cannot be reversibly removed; this phenomenon is described as small Coulombic efficiency and poor reversibility. The capacity will also fade due to processes such as electrolyte decomposition, phase changes in the insertion/desertion of lithium ions and passive film formation on the electrodes. The passive films are called solid electrolyte interface (SEI), and are formed by decomposition of electrolyte solvents and dissolved lithium salts. On one hand, the SEI helps protect against further corrosion of lithium. On the other, if the SEI becomes very thick due to unnecessary side reactions between electrolyte and electrodes, it can cause a significant loss in capacity and will become highly resistive to ionic conduction. This is also a critical factor when choosing suitable electrode materials and electrolyte for LIBs.

Many efforts have been made to improve the performance of the LIBs by developing novel cathode materials. For example, substitution of Mn in $LiMn_2O_4$ with Ni and Cr is known to increase the operating voltage (up to 5 V) and reduce capacity fading. The main problem with many cathode materials is low ionic conductivity, which significantly limits the rate capability and reversibility. Nanotechnologies can help mitigate this problem by reducing the particle size in a variety of ways. Nanosized or nanostructured materials exhibit faster rate and higher specific capacity since the Li^+ diffusion length is shortened, the reaction sites are modified and the electrolyte–electrode contact area is increased. However, too large a contact area between electrode and electrolyte is not preferable, since significant amount of side reactions will take place, resulting in low Coulombic efficiency. More advanced mesoporous materials with ordered or disordered micro pores can be developed, which have many advantages and disadvantages of their own. These will be discussed in the cathode section.

Electrolytes have also attracted attention because they will lower the risk of danger when using a low vapor pressure or even solid electrolyte. Novel electrolytes are essential when making thin film batteries and lithium-oxygen batteries. These issues are covered in the last three sections of this chapter.

2 Positive Electrodes

2.1 History of Cathode Materials

Li metal was first utilized in the 1970s to assemble primary Li cells. This kind of cell found applications in implantable medical devices, after hydrogen gas production was avoided in the cell [2]. Simultaneously, researchers revealed the intercalation ability of alkali metal ions into dichalcogenides. From then on, LIBs have become one of the most highlighted research areas. Over last 40 years, various kinds of cathode materials have been investigated in order to achieve the synthesis of high capacity, high-electrochemical potential, fast rate, low cycling fading, low cost and environment friendly material. An ideal cathode material should satisfy all these criteria:

(1) The cathode should have a high redox potential;
(2) The potential should not change much, normally within 0.4 V, for different degrees of lithium insertion/extraction;
(3) The ability to host lithium ions should be high to give satisfactory gravimetric and volumetric capacity;
(4) The ionic and electric conductivities should be high to guarantee small polarization during charge and large operation current;
(5) The structure of cathode material should be thermally stable and with no disordered ions pinning in lithium sites;
(6) The volume should not expand/contract too much with lithium insertion/de-insertion to ensure cycling stability.

In this subsection we will describe the development of several kinds of cathode materials. The research can be separated roughly by two stages: before and after 1990.

2.1.1 The Birth of lithium-ion Batteries

Whittingham concluded that the use of TiS_2 as the positive electrode, embarked by Exxon in 1972, brought in the era of rechargeable lithium batteries [3]. TiS_2 was the best intercalation or host compound among all the layered dichalcogenides for the application of cathode materials [4]. Titanium disulfide attracted so much

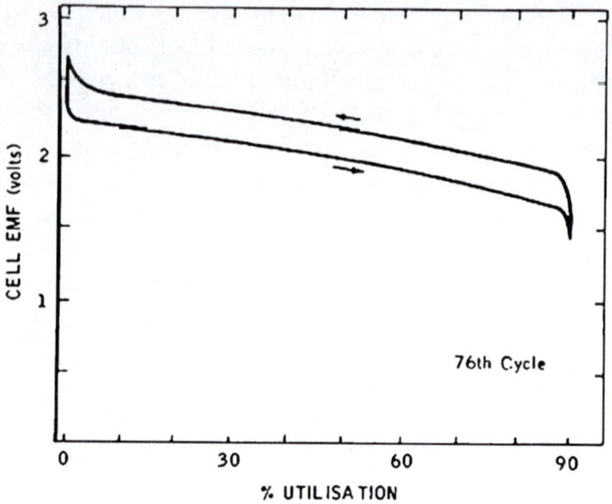

Fig. 3 Charge/discharge curve of Li_xTiS_2 (from Whittingham [121], with permission from Elsevier)

attention due to its light weight, good electronic conductivity (semimetal) and free from phase transition in the entire range of Li_xTiS_2 for $0 \leq x \leq 1$. For detailed explanations, please refer to [3]. Figure 3 shows the typical charge/discharge cycling of TiS_2 at a rate of 10 mA/cm^2. It is visible that the operating voltage is around 2 V and there is no phase transition in the entire lithium intercalation range. With titanium disulfide, button cells were successfully applied as power sources for watches in 1977–1979.

2.1.2 The Concept of Positive Materials Shifts to Layered Oxides

Significant advances in intercalation materials were achieved due to interest in transferring from chalcogenides to oxides, resulting in higher voltage and capacities [5]. Vanadium pentoxide, V_2O_5, has been investigated for over 30 years. It possesses a layered structure with weak vanadium-oxide bonds between layers. The intercalation process of lithium into vanadium pentoxide is complex, forming multiple phases for different intercalation degrees. From α-phase to ε-phase and to δ-phase at high voltages above 3 V, and later to γ and then ω-phase at voltages close to 2 V, the multi-phase transitions of vanadium pentoxide prohibits its practical application as a power source [6], since a practical cell can only allow a small change of potential plateau within 0.5 V. Beside materials like crystalline vanadium pentoxide and V_6O_{13}, double sheets materials like xerogel and aerogel vanadium oxide, have also been explored in depth [7–9]. However, they still cannot diminish the potential change.

Later, Goodenough recognized that $LiCoO_2$ has a layered structure similar to that of vanadium pentoxide and dichalcogenides [10]. And it has a flat charge/discharge voltage around 4 V, making it a promising cathode material for practical applications [11]. Its lattice structure is displayed in Fig. 4, where a layer sequence of O–Li–O–Co–O– is formed along the c axis. Lithium can be extracted and inserted with only mild distortion of the crystal structure. A typical intercalation/de-intercalation curve

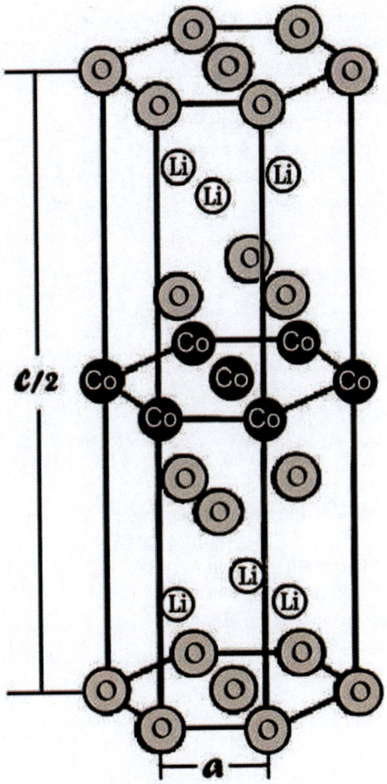

Fig. 4 Layered structure for $LiCoO_2$ crystal

of $LiCoO_2$ is shown in Fig. 5. Combined with a graphite anode, $LiCoO_2$ was the first cathode material that enabled commercially successful Li-ion batteries. However, $LiCoO_2$ has its own limitations, such as low theoretical capacity and high price. Those factors are not a concern when $LiCoO_2$ batteries are applied in small devices, including cell phones, computers and cameras. But their applications are impeded in large-scale systems such as hybrid vehicles and clean energy storage where the advanced requirements for high capacity and low cost cannot be satisfied.

2.2 *Diverse Positive Materials Under Investigation Later than 1990s*

2.2.1 Spinel $LiMn_2O_4$

Spinel $LiMn_2O_4$ has attracted so much attention as an insertion material because of its low cost and high voltage. The two-stage charge/discharge curve of pure $LiMn_2O_4$ is well known, as shown in Fig. 6. It shows two plateaus of voltage at around 4 V and 3 V. The plateau around 4 V is due to lithium insertion into the tetragonal sites of spinel Fd3m structure, while the plateau at

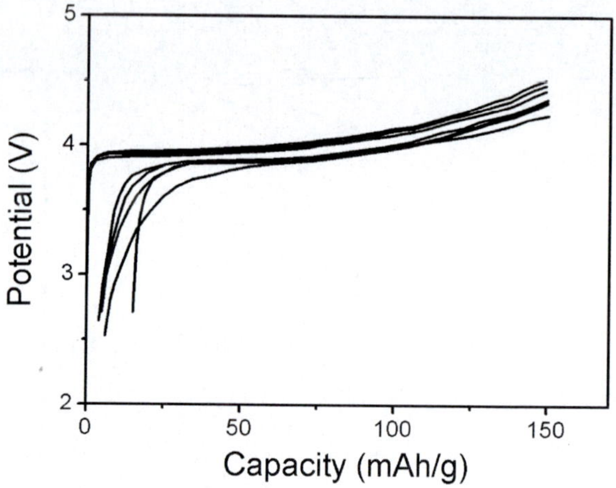

Fig. 5 Charge and discharge curves for $LiCoO_2$ at a rate of 0.17 mA/cm^2 in 1 M $LiClO_4$/PC solution at 30 °C (from Ohzuku and Ueda [11], reproduced by permission of The Electrochemistry Society)

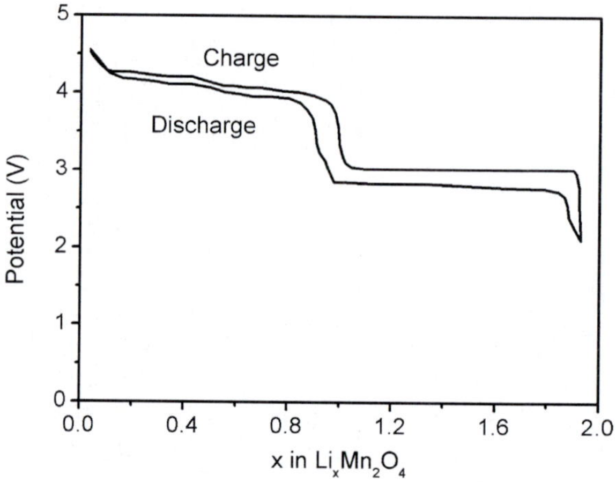

Fig. 6 Charge and discharge curve for $Li_xMn_2O_4$ at a rate of 0.2 mA/cm^2 (from Thackeray [12], with permission from Elsevier)

3 V indicates the lithium insertion into the octahedral sites. [12] People have proved that, during cycling, the higher voltage stage is more stable than the lower one, having less capacity fading after each cycle. [13] Thus the higher voltage range is practically in use, especially when coupled with high-voltage anode, such as Titanates. The electrochemical performance of spinel $LiMn_2O_4$ has been explored in several ways, including thin film techniques using electron-beam evaporation or radio-frequency sputtering. $LiMn_2O_4$ thin films of around 0.5 μm showed that lithium could be inserted into the electrode at about 3 V, the same with that of bulk $LiMn_2O_4$. By contrast, the extraction of lithium from thin film $LiMn_2O_4$ occured at approximately 4.5 V, 0.5 V higher than the bulk $LiMn_2O_4$. [14] Together with thin film electrolyte, flexible thin film LIBs were assembled. Further, researchers have tried substitution of manganese with metals like nickel and chromium to stabilize the spinel structure and resulted in flatter insertion voltage. [15, 16].

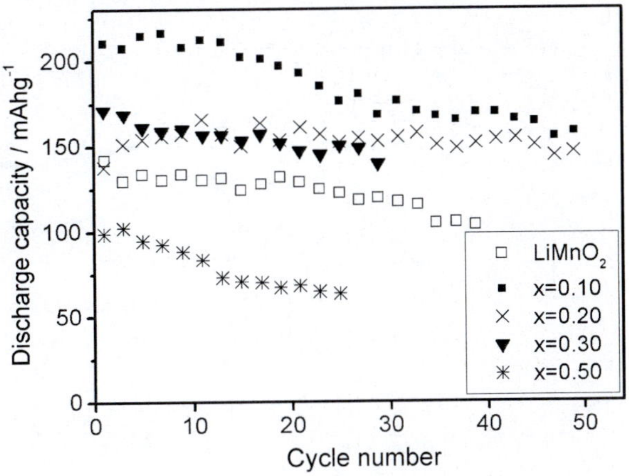

Fig. 7 Discharge capacity vs. cycle number for $LiMnO_2$ and $Li_xMn_{1-x}Co_xO_2$ at rate of 0.1 mA/cm^2 and a potential window between 2.6 and 4.8 V (from Armstrong et al. [21], with permission from Elsevier)

2.2.2 Metal-Doped (Co, Cr, Ni, Ti and Mg) Layered $LiMnO_2$

Similar to $LiCoO_2$, $LiMO_2$ (M = Ti, Ni, Mn, Cr or Fe) has been a topic of continued interest for a long period of time [3, 17, 18]. Among them, research on $LiMnO_2$ has received the most attention and pragmatic value. Since layered $LiMnO_2$ is of lower cost, high capacity and is environmentally benign, it is promising for large-scale usage [19]. The only problem with layered $LiMnO_2$ is the instability of its structure. The layered phase converts to the spinel-like $Li_xMn_2O_4$ upon cycling, even at ambient temperature [20]. Unfortunately, the application of LIBs on hybrid vehicles requires stability at room temperature and higher due to the heat released by engine operation. This problem leads to research for efficient solutions, such as the partial substitution of Mn by Ni, Cr, Mg, Co, etc.

The idea to substitute manganese in $LiMnO_2$ by cobalt (forming $LiMn_{1-y}Co_yO_2$) was stimulated by the desire to make a $LiCoO_2$-like material. $LiCoO_2$ can sustain many cycles; thus the introduction of Mn is supposed to reduce the cost and toxicity of $LiCoO_2$. Bruce et al. did a thorough investigation of Co substitution in the range of $0 \leq y \leq 0.5$ [21]. In their report, a 10% substitution of Mn by Co was found to be sufficient to suppress the distortion of layered $Li_{0.9}Mn_{0.9}Co_{0.1}O_2$. They also drew a conclusion that higher replacement of Mn by Co resulted in somewhat decreased capacity and better cycle stability, as indicated in Fig. 7. In this case, the electrochemically active material is Mn ion, while Co ion plays a role as a structure stabilizer.

$Li(NiM_n)O_2$ have attracted much attention since 2001, after Ohzuku et al. It is a promising cathode material due to its high capacity and stability. The compound $LiNi_{0.5}Mn_{0.5}O_2$, called 550 (0.5 Ni, 0.5 Mn, 0.0 Co), showed a very good capacity around 200 mAh/g at the rate of 0.12 mA/cm^2, as well as good cycling stability [22]. Figure 8 shows the charge/discharge curves of 550 for 30 cycles. Theoretical calculations confirmed the behavior of Ni and Mn during insertion and deinsertion of the compound. The Mn ion has a 4^+ valence independent of the lithium

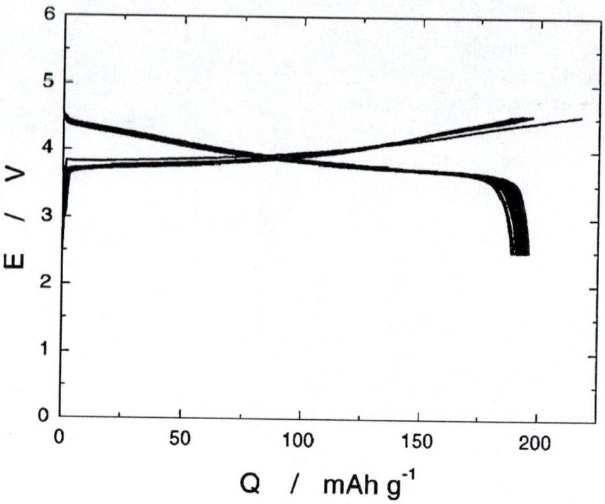

Fig. 8 Charge/discharge of $LiNi_{0.5}Mn_{0.5}O_2$ at 0.17 mA/g for 30 cycles (from Makimura and Ohzuku [22], with permission from Elsevier)

concentration in the material, so it works as a structure frame, while the valence of nickel ion changes from 2^+ to 4^+ upon lithium removal. Therefore, the Ni ion is the electrochemically active element giving the capacity [23]. However, substitution by nickel still cannot raise the electronic conductivity of doped $LiMnO_2$ as much as that achieved via substitution by cobalt or iron.

Although the compound of $LiNi_{0.5}Mn_{0.5}O_2$ works well, there is still over 5% nickel pinning in the lithium layer, causing problems such as slow rates and loss of capacity. The addition of Co into the compound of $Li(Mn_{1-y}Ni_y)O_2$ can reduce the amount of transition metal in lithium layer to 2.4% for $LiMn_{0.2}Ni_{0.5}Co_{0.3}O_2$ [24]. The role of cobalt was verified by earlier research on Co-doped $LiNiO_2$. Another concern of reducing the nickel disorder is the synthesis temperature which has been optimized in between 800 and 900 °C. Among different substitution ratios and degrees, the $LiMn_{0.4}Ni_{0.4}Co_{0.2}O_2$ (442) material was reported to show the highest capacity, thermal stability and retention ability upon cycling. Besides $LiMn_{0.4}Ni_{0.4}Co_{0.2}O_2$, $LiMn_{0.33}Ni_{0.33}Co_{0.33}O_2$ also attracted much interest. In both situations, the transition metal in the lithium layers is only nickel, and the introduction of cobalt into the compound increases the conductivity by some degree; however, the improved performance over cobalt-free compound is mainly caused by the decrease of nickel in the lithium layer [25] rather than conductivity.

Chromium-substituted $LiMn_{1-x}Cr_xO_2$ also received considerable attention [26]. Hwang et al. reported that replacement of Mn by Cr induces a decrease in crystallite size. This smaller crystal size results from shortening of Mn–O bonds. Also the presence of chromium ions hinders manganese migration to the lithium layer, leading to stabilization of the cathode material during cycling. $LiMn_{0.9}Cr_{0.1}O_2$ shows a capacity of 190 mAh/g with little capacity fading from the second cycle onward. In Guo's work, $LiMn_{0.7}Cr_{0.3}O_2$ showed a better performance than that of many other metal-doped $LiMnO_2$, as shown in Fig. 9 [27]. Unfortunately, practical usage cannot be realized since Cr is both expensive and toxic.

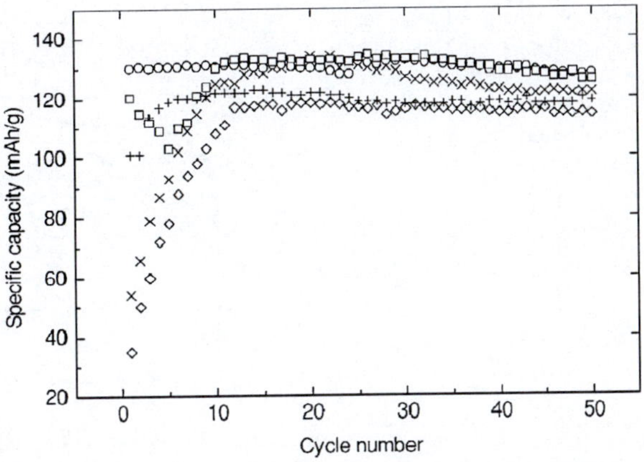

Fig. 9 Cycling data for (X) $LiMnO_2$ and $LiMn_{0.7}M_{0.3}O_2$ cathodes: (*open circle*) M = Cr; (+) M = V; (*open square*) M = Zn; (*lozenge*) M = Mo (from Guo et al. [27], with permission from Elsevier)

There has also been research on other compounds, like $LiNi_{1-x}Co_xO_2$, $Li[Co_{0.17}Li_{0.28}Mn_{0.55}]O_2$, $LiNi_{1-x}Ti_{x/2}Mg_{x/2}O_2$, etc. [28–30]. Generally speaking, those substitution metals are combined for two reasons: first, to reduce toxicity and cost; second, to stabilize the lattice structure.

2.2.3 Polyanion Olivine-Structured Materials

It was recently found that olivine-structured oxyanion scaffolded materials provide interesting possibilities. Those materials consist of corner-sharing MO_6 (M = Fe, Ti, V or Nb) octahedra and XO_4 (X = S, P, As, Mo or W) tetrahedral anions [31]. The redox potentials of Fe^{3+}/Fe^{2+} and V^{4+}/V^{3+} couples lie at a higher level in the phosphate polyanions PO_4^{3-} than in their oxidized forms [32]. The creative work of Goodenough and co-workers motivated the use of $LiFePO_4$ as a positive cathode material for LIBs. Figure 10 demonstrates charge-discharge curves for secondary lithium batteries using $LiFePO_4$ cathode material [33]. It was found that a high theoretical capacity of 165 mAh/g and a flat discharge voltage at 3.4 V can be achieved from $LiFePO_4$-assembled batteries. This kind of battery also has high thermal and chemical stability [34], and possesses the advantages of low cost [35] and small environmental impact [34]. However, it has the drawback of less good rate capability due to its poor electrical conductivity and slow ionic diffusivity [36, 37]. To overcome these barriers, much effort has been made, such as size reduction, uniformity improvement, metal dispersion and carbon coating.

As it is well-known, the electrochemical behavior of $LiFePO_4$ is strongly influenced by its preparation methods. $LiFePO_4$ has conventionally been synthesized by a solid-state reaction that involves successive steps of grinding and annealing at high temperatures for long periods of time. The main steps are as follows: stoichiometric starting materials are mixed by the ball milling method; then the mixture is heated at around 300–400 °C to decompose any organic compound involved; it is subsequently annealed at around 500–800 °C to form and fully crystallize the desired phase [38]. The solid-state method has several

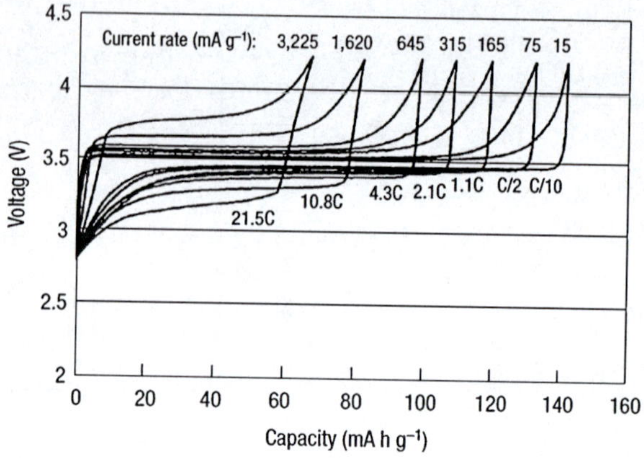

Fig. 10 Charge–discharge curves for $LiFePO_4$ cathode materials at different rates (from Chung et al. [33], with permission)

disadvantages, such as large particle size, broad particle size distribution and the low conductivity of the as-synthesized products. Thus, carbon is added to increase the conductivity of the cathode. The post synthesis addition of carbon leads to its inhomogeneous distribution surrounding the $LiFePO_4$ particles.

Recently, researchers developed some excellent solution-based techniques for the synthesis of $LiFePO_4$/C composites with carbon homogeneously distributed around the $LiFePO_4$ particles [39, 40]. Typical solution-based techniques, such as hydrothermal processing, sol–gel processing, precipitation, emulsion-drying and spray pyrolysis, are effective methods for the synthesis of olivine-structured materials [41]. The sol–gel process is incredibly versatile, allowing for a wide selection of precursor materials and carbon sources. This technique also allows for the synthesis of particles with excellent morphology and uniform size distribution of small particles, while offering a high purity and homogeneity at low-processing temperatures. Figure 11 shows the microstructures of $LiFePO_4$ synthesized by sol–gel technique. The carbon incorporation during the synthesis process helps reduce the particle size and increase the electric conductivity [42]. One concern about the heat treatment procedure is the temperature range. According to Kepler's report, the preparation process should be below a temperature of around 600 °C, or errant Fe ions will exist on the Li sites and reduce the reactivity and diffusion of the lithium ions [43].

2.3 The Contributions of Nanotechnology and Meso-Porous Methods

Tuning the particle size and microstructures of positive materials will significantly influence their electrochemical performance, such as the intercalation rate and capacity. The advantages of nanostructured materials include: (i) a faster rate of lithium insertion/removal caused by shorter lithium diffusion length within the particles; (ii) improvement of electron transport velocity within the particles; (iii) a higher lithium–ion flux resulting from the high surface area between electrode and electrolyte and (iv) longer cycling life due to the buffering of the volume change of

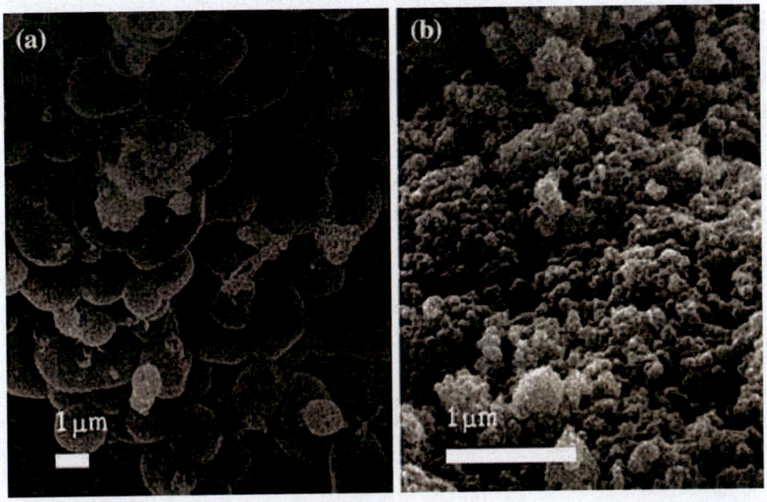

Fig. 11 **a** Without carbon, **b** with the oxidized carbon (from Huang et al. [42], reproduced by permission of The Electrochemistry Society)

lithium. There are also associated disadvantages, such as the difficulty of size control, more side reactions due to the higher surface area, poor contact between particles and a lower density of materials [44]. Besides nanoparticles, nano-scaled tubes and wires are of interest to some degree. Another nanostructure that attracts much attention today is the meso-porous positive materials. They are discussed in detail in the following text.

2.3.1 Nanoparticles

Inherent to the nature of its particle size, small diffusion length and high surface area of nanoparticles, materials with low electronic and ionic conductivities are the most suitable to be synthesized into nanoscale sizes. As mentioned above, the main drawbacks of olivine-structured $LiFePO_4$ are of poor electric and ionic conductivities caused by pure iron valence. Reducing the particle sizes enables $LiFePO_4$ a good enough rate capability for practical value [42]. Thus, nanoparticles of $LiFePO_4$ have achieved success in the commercial market. However, there is also a risk of large capacity fading caused by the thick and large area of SEI from significant decomposition of electrolyte materials.

2.3.2 Ordered and Disordered Porous Cathode Materials

Compared to nanosized particles, tubes or wires, the meso-porous positive electrodes, especially when ordered, have profound advantages. Those mesoporous solids are composed of micro-sized particles, thus having better electronic contact between each particle and an almost similar packing density as conventional micro

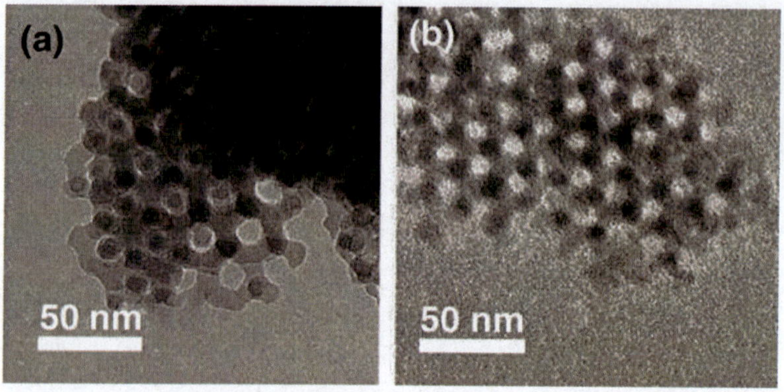

Fig. 12 TEM figures of mesoporous Co_3O_4: **a** Co_3O_4-500 and **b** Co_3O_4-600 (Co_3O_4-500 means the precursor materials were calcinated at 500 °C.) (Reprinted from [122], with permission from Elsevier)

powders. And within those particles, ordered pores of 2–50 nm in diameter are present and separated by walls of several nanometers thick.

The ordered porous cathode materials generally have better rate capability, faster lithium-ion insertion/removal and smaller capacity fading [45]. The internal pores can be filled with electrolyte, leading to high surface area and large lithium flux between the electrode and electrolyte interfaces. Those ordered pores ensure even distribution of electrolyte and reduce the speed of decomposition process in high concentration electrolyte [44]. The high rate capability can be achieved by shortening diffusion lengths of lithium ion within the thin walls between pores.

There are two basic template methods to fabricate ordered meso-porous materials: soft and hard. For the soft template technique, the materials often lack the thermal stability and crystallinity required by many applications. For the hard template method, where silica is employed as the template to help form the mesoporous structure, the materials are well crystallized and the pores are ordered well. However, the precursors that result in alkali composite cannot be used together with silica, thus limiting the usage of silica for production of positive materials for LIBs. Jiao et al. reported in their work that the obtained transition metal oxide Co_3O_4 can then react with a lithium source, forming mesoporous lithium-doped positive materials [45]. The detailed nanostructure of ordered mesoporous Co_3O_4 is shown in Fig. 12.

As for the disordered mesoporous positive materials, they still enjoy continued research interest. Aerogel V_2O_5 and MnO_2 have been reported to possess tremendously large specific capacities compared to graphite. Dong et al. found that the electrochemical capacity of mesoporous V_2O_5 is about 100% greater than in normal polycrystalline powders [46]. In addition, the preparation of disordered meso-porous materials is more straightforward than that for ordered mesopous materials. The problems with those conductive areogels are poor packing density and the possibility of residual moisture after preparation.

2.4 *Summary and Conclusions*

Having reviewed nearly 40 years of scientific adventure on positive rechargeable lithium battery materials, we now hope to develop materials that can provide faster rate capability and higher capacity, as required by the hybrid vehicle industry and advanced portable electrical devices. There is progress within the last 10 years, mainly on modifying the nanostructure of positive materials. We also look forward to proposals of novel positive electrode materials that are conceptually new, based on the development of inorganic and organic electrochemistry, physics and surface sciences.

3 Negative Electrodes

Lithium-ion batteries emerged in the 1970s, employing lithium metal, the most powerful reducing element, as the anode. When coupled with $LiCoO_2$ or V_2O_5, the lithium metal battery gives high voltage and high-energy capacity. However, the well-known formation of dendritic lithium during cycling causes safety issues and a short practical cycling life. These problems require research into alternative materials to replace lithium metal and obtain safe anode materials.

3.1 *Graphite and Highly Disordered Carbonaceous Anode Materials*

The smart solution of lithium metal alternating with carbon graphite enabled the commercialization of secondary lithium batteries by Sony in 1991. Graphite is a lithium host material that can absorb and release lithium ion reversibly. The potential of graphite vs. lithium is around 300 mV; thus, the voltage loss compared to lithium metal is very little while avoiding the risk of forming lithium dendrite. Here we briefly state the method of measuring the performances of anode materials and the characterization parameters. In the measurement of the charge–discharge ability, the anode materials and a pure lithium metal foil are assembled together in a coin cell. During discharging, lithium ions migrate from the lithium metal and enter the anode carbon, and will return to lithium metal during charging. As stated in the introduction of this chapter, the cathode has a higher electrochemical potential; thus, graphite works as the cathode when coupled with lithium metal. However, unlike a normal practical cell, the lithium source here is the lithium metal anode.

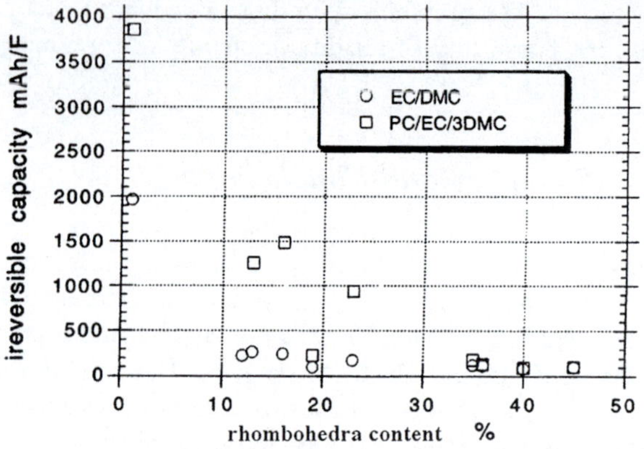

Fig. 13 The faradaic losses per unit of double-layer capacitance for graphite samples of various rhombohedra contents (from Flandrois and Simon [47], with permission from Elsevier)

3.1.1 Initial Graphite Materials

Armand et al. first published the use of graphite as an intercalation/deintercalation material; then Sony Corp. realized the commercial production of LIBs with graphite as the anode and $LiCoO_2$ as cathode [47]. At first, their graphite anode suffered from significant reaction with electrolyte. This was improved by Sony using a large variety of high crystallinity carbon materials, which finally brought about commercial success. The concern of the crystallinity or morphologies of electrode materials then became important to the search for materials that are relatively stable in contact with electrolyte and during lithium insertion/removal.

At the beginning, people used naturally existing ores to synthesize highly crystallized graphite as anode materials. The most common graphite possesses the hexagonal Bernal structure [48]. The hexagonal graphite layers are arranged by …ABAB… sequences. It was found that another crystalline phase can exist in a rhombohedra structure with stacking sequences of …ABC… Those two phases often exist together in the graphite anode materials. However, the rhombohedra phase is not as thermally stable as the hexagonal structure since it disappears when heat treated at 2000 °C [47].

The content of rhombohedra graphite plays an important role in minimizing the first irreversible capacity of the total anode electrode, though is not very critical in long-term cycling reversibility. Thus, the faradaic losses depend much on the rhombohedral content. Figure 13 shows that the faradaic losses per unit double-layer capacitance are reduced with increasing rhombohedral phase content in two different electrolytes. It shows that higher rhombohedral content gives higher faradiac retention than pure hexagonal graphite. This is because of the higher content of grain boundaries and dislocations in the impure phase that makes it more difficult to open interlayer graphite for large solvent ions, which will cause formation of SEI and even exfoliation.

Another important characteristic of graphite intercalation compounds is the multiphase transitions during charge/discharge. Within a voltage window from 0 to

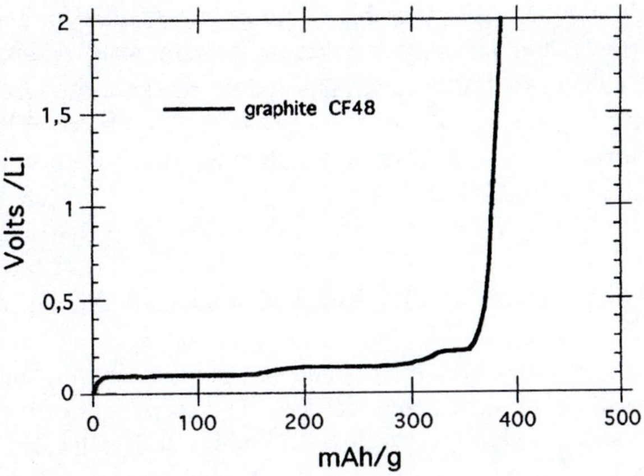

Fig. 14 Cell voltage as a function of capacity during Li removal (from Flandrois and Simon [47], with permission from Elsevier)

0.4 V, four plateaus are shown, indicating phase transitions at each stage. The plateau voltages of 0.09, 0.12, 0.14 and 0.2 V represent the corresponding transitions: stage-1 (LiC_6) to stage-2 (LiC_{12}), stage-2 to stage-2L (LiC_{18}) and then to stage-3 and stage-4. Those details are demonstrated in Fig. 14 [47].

3.1.2 Soft and Hard Carbons

Soft carbon can gradually transfer to a graphite structure while heated at elevated heat treatment temperatures (HTTs), up to 3000 °C. They are comprised of disoriented crystallites that grow larger and more ordered during heating. When heated at temperatures higher than 2200 °C, the sizes of the crystallites become hundreds of angstroms and the interlayer spacing approaches that of graphite, as large as 3.354 Å.

Flandrois and co-workers concluded that a higher HTT results in better anode materials with less faradaic loss and higher energy capacity within the range of 1300–2400 °C [47]. Inspecting the structures formed at higher HTTs, they find that the crystallization is better and total surface is smaller since the content of disordered crystallites is reduced. The smaller faradaic loss or irreversible capacity for higher HTTs could be caused by smaller surface area contacting electrolyte while the higher energy capacity results from better crystallinity.

Hard carbons (HTT 500–1000 °C), also called low-temperature or non-graphitable carbons, are synthesized from the carbonization of various kinds of organic compounds at around 500 °C. The cross-linking groups from the network of the organic compounds hinder the formation of columns and coalescence. Therefore, the structures of hard carbon are less crystallized. The hard carbons are also found to include large amount of micro or nano-porosity. These features will affect the electrochemical performance of the hard carbons. First, the discrepancy between charge and discharge voltage plateaus is large; second, the reversible capacity is high, much larger than graphite; and third, there is high irreversible capacity in the first cycle [49]. These were explained by many mechanisms [50, 51]. Zhang and co-workers found a linear relationship between the capacity and edge length of carbon layer. Thus, larger amount of edge carbons resulting from organic compounds will provide more sites to incorporate lithium ions [52].

3.1.3 Conclusion of Initial Carbonaceous Anode Materials

Research has been performed on many different kinds of carbon materials, from graphite to disordered carbon. The most promising material is still graphite, especially highly crystallized graphite with some amount of rhombohedra phase, due to its good electrochemical performances: high reversible capacity, small capacity fading and excellent rate capability. As for soft carbon synthesized at high temperature, the capacity is low. For carbons synthesized at lower temperatures, the energy capacity is high; however, the capacity fading and the hysteresis of the voltage cannot be avoided efficiently [47]. In recent years, carbonaceous materials have attracted much interest in the field pursuing novel structured, mesoporous and metal dispersed carbons.

3.2 *Novel Nanostructured Carbon Materials*

3.2.1 Carbon Nanotubes as Lithium Storage Materials

Studies on lithium insertion into carbon nanotubes have been widely carried out. Carbon nanotubes (CNTs) are developed for energy usage in the field of hydrogen production and LIBs. CNTs are supposed to exhibit excellent performance as anode materials in secondary lithium batteries due to their special structural features. CNTs can host large amount of Li+ owing to the abundant nanosized pores that may provide reaction sites for lithium-ion storage, thus, the energy capacity can be improved. Another advantage of CNTs is the short diffusion distance for the lithium ions through the tube walls, which can increase the rate capability by some degree.

Based on those expectations, many groups have performed studies on the insertion of lithium into CNTs. Che and co-workers reported for the first time that a high intercalation capacity of 490 mAh/g was achieved from CNT lithium batteries [53]. And later, Frackowiak et al. studied the annealing temperature effects on the capacity of CVD synthesized nanotubes. They observed a large insertion capacity of 952 mAh/g of CNTs for the first cycle; however, only 447 mAh/g was reversible at the second time. They also found that the capacity decreased with higher HTTs [54]. Other than pure CNTs, SnO-doped, Sn-doped, $SnSb_{0.5}$ and Cu-doped CNTs have also been studied [55–57]. Compared to pure CNTs, all doped CNTs showed less capacity fading, thus better cycling ability.

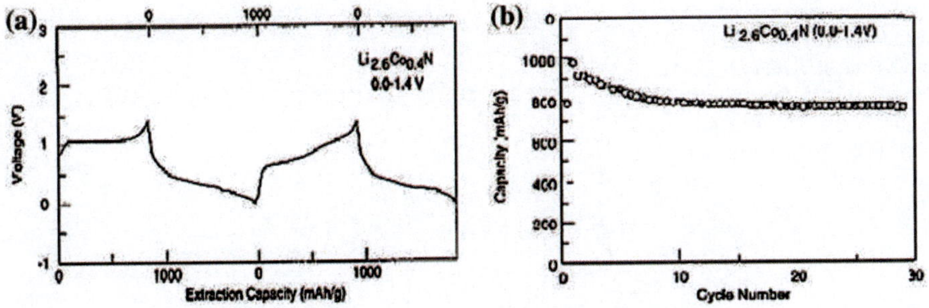

Fig. 15 Charge–discharge curve (**a**) and cycling performance (**b**) for $Li_{2.6}Co_{0.4}N$ at the rate of 0.5 mA/g (from Shodai et al. [58], with permission from Elsevier)

This was explained by two main reasons: first, the doped small particles may release the stress caused by lithium insertion/removal; second, these doped particles can reduce the interface between the electrolyte and highly porous CNTs, thus forming thinner SET and retaining more capacity.

Nevertheless, the practical use of chemically or physically modified CNTs as the anode material for secondary lithium batteries is not realistic till now. It has its own drawbacks, in addition to the high cost of synthesis. The general concern is the high surface area; although it can be reduced by doping, it is still large enough to cause large amount of irreversibility.

In parallel, research on materials for carbon alternatives is underway. Slightly higher insertion potential is preferred when considering lithium plating at low potentials of carbon. And larger capacities are desired to trigger the fast growth of portable electric devices and hybrid vehicles. Studies on those materials are discussed in the following section.

3.3 Li Transition-Metal Nitrides

Lithium transition-metal nitrides have attracted researchers' attention in some degree due to large capacity and good cyclability. Shodai et al. did a breakthrough work by exploring $Li_{2.6}Co_{0.4}N$ and brought in a new class of lithium insertion materials—Li-M-N (M = Co, Ni or Cu). Figure 15a shows the charge–discharge curve of $Li_{2.6}Co_{0.4}N$ whose charging voltage is around 1 V. This potential is much higher than the lithium electroplating (Li+ reduction) potential and can help avoid the safety problems associated with a thin layer of electroplated lithium. At the same time, they show the high capacity of 760 mAh/g achieved even after 30 cycles, as displayed in Fig. 15b [58]. However, the high calculated capacity compared to graphite may be contributed a lot by the higher cycling voltage vs. lithium. When coupling with cathode materials such as $LiCoO_2$ or $LiFePO_4$, the cell potential is strongly reduced. Furthermore, the use of $Li_{3-x}Co_xN$ is constrained by the restrictive manufacturing requirements for handling such moisture-sensitive negative electrodes.

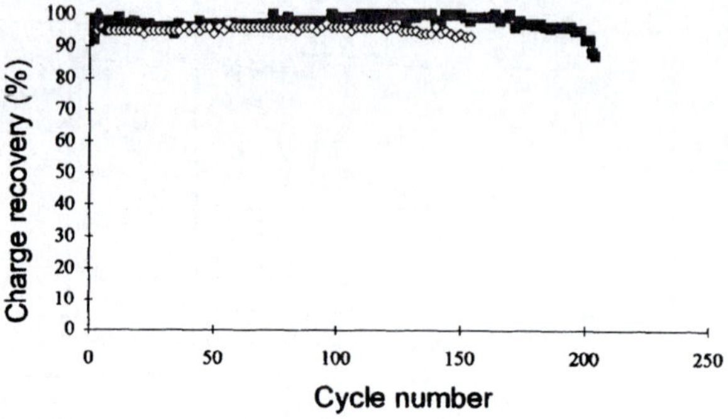

Fig. 16 Energy recovery of (*filled square*) Sn0.72Sb0.28 and (*open diamond*) $Sn_{0.91}Ag_{0.09}$ at the rate of 0.25 mA/cm^2 (from Yang et al. [62], with permission from Elsevier)

3.4 Lithium Alloy

In order to avoid dentrite formation linked to lithium metal, alternative metals (M), such as aluminum, as anode materials have been studied since the 1970s [59]. Studies on other metals (M = Sn, Zn, Bi, In, Cd and Sb) have been performed extensively [32, 60]. Unfortunately, none of them showed good cycling performances with increased depth of discharge, though their gravimetric capacities were satisfactory. When alloying with lithium, these metals suffer from large volume swings (up to 200%); hence the structures deteriorate after lithium removal and the electrical contacts between particles get lost. Some groups tried to reduce the alloy particle sizes to benefit the cycling stability based on the nature of small particles' tolerance to stress cracking, but what they obtained still could not satisfy the practical requirements [61]. However, research is still in progress to find possible solutions to structural deterioration of lithium alloys during cycling.

In Yang and co-worker's report, an impressive improvement of cycling performance of lithium alloy anode was claimed to be achieved by using multi-phase metallic host materials like $Sn/SnSb_n$ or $Sn/SnAg_n$ [62]. They observed a little capacity fading even after 150 cycles for one compound $Sn_{0.72}Sb_{0.28}$, as shown in Fig. 16. The insertion of lithium during the test was limited to 1.7 Li per formula. This was explained to be assisted by the two-phase alloy composite. The volume change of the more reactive phase was compensated or buffered by the matrix of the inactive material. Another example of ternary alloy was examined by Thackeray et al. [43]. They discovered that an intermetallic compound of Cu_6Sn_5 can accept lithium in a two-phase reaction to yield $Li_xCu_6Sn_5$ ($x = 13$, corresponding to capacity of 358 mAh/g, close to the theoretical capacity of graphite Li_6C 372 mAh/g). The reaction occurs at 0.4 V vs. lithium. Using the density of 8.28 g/ml, the volumetric capacity of this $Li_xCu_6Sn_5$ is up to 1656 mAh/ml, much larger than that of the graphite alloy (850 mAh/ml). Another advantage for this alloy is its relatively low irreversibility. Since the inactive phases, like Cu in Cu_6Sn_5, do not react significantly with the electrolyte, the energy loss should be lower than the case of tin oxides, where Li_2O is formed irreversibly when lithiated. However, the cycling performances are still not as good as the graphite.

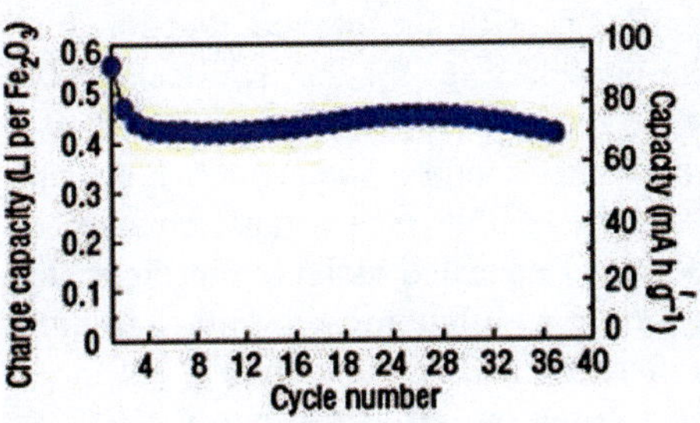

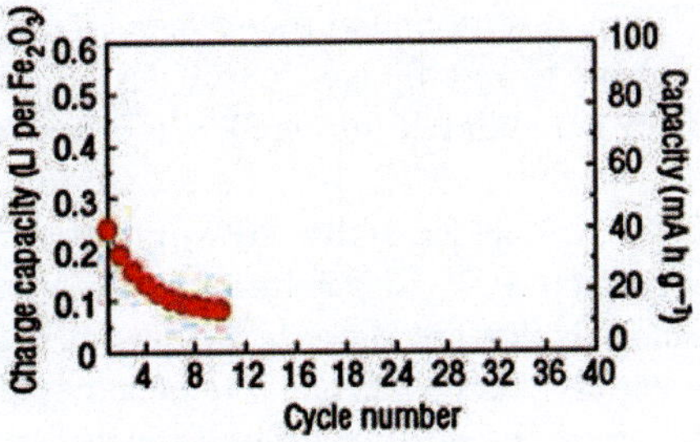

Fig. 17 Electrochemical behavior of bulk (*right*) and nanostructured (*left*) (–Fe_2O_3 (from Larcher [66], reproduced by permission of The Electrochemistry Society)

A similar approach to improve the cycling ability by intermetallic phases is the Sn-Fe-C system [63]. According to Dahn and coworkers' design and experiments, Sn reacts with lithium and the embedded inert iron works as the buffer matrix in the Sn_2Fe alloy. They obtained high capacity of 800 mAh/g and volumetric capacity of 4500 mAh/ml for the first time. Similar to other alloys, this material does not behave well after extensive cycling. In general, the basic concept behind these ternary alloys is the idea of a non-reactive "buffer matrix" to compensate for the volume expansion of the reactive phases. However, the concept still has not worked very well for achieving improved reversibility.

3.5 *Transition Metal Oxides*

There are two classes of metal oxides for LIBs. One is classic lithium insertion/de-insertion materials, like titanates. By contrast, the other displays lithium capacity through the reaction of metal oxides with lithium and forms metal dispersed in a matrix of Li_2O during charging. Later on, the charge/discharge is realized by the formation and decomposition of Li-M in the Li_2O matrix. Internally nanostructured metal oxides, M-O (M = Co, Ni, Fe, Cu, Sn, Mn or Mo), began to be taken into serious consideration only from the realization of their non-traditional reaction mechanism with lithium. For this reaction mechanism, these metal oxides exhibited 100% capacity retention after 100 cycles. This high reversibility may also be supported by their internal nanostructures [64]. Fuji announced the use of amorphous tin composite oxide as the anode material in 1997. The capacity was measured to be twice that of graphite. A Li_2O phase was verified by in situ X-ray diffraction studies [65]. Despite good results, they are far from commercialization, owing to poor long-term cyclability, large voltage hysteresis and the huge irreversible capacity during the first cycle.

Besides titanates, Larcher et al. observed large lithium insertion into a nanoscale transition metal oxide that can hold only a small amount of lithium when in bulk or microsize particles. It was also found that through nanoscaling of inorganic materials, the electrochemical potentials could be tuned. They compared the large lithium insertion ability of nano-sized (α-Fe_2O_3 with the limited reaction that happens to micro-sized particles [66]. The particle sizes for the nanoscaled iron oxide is around 20 nm, and is 1–2 μm for the micro-sized ones. From Fig. 17, we can tell that the lithium inserted into the nanosized metal oxide can reach 0.6 per Fe_2O_3, while the bulk Fe_2O_3 undergoes an irreversible phase transformation at only about 0.05 Li per formula. This experiment educated us on exploring many materials that are supposedly not suitable for use as lithium insertion host in bulk materials. As the particle size becomes smaller, the modification not only enhances the ability of hosting lithium ions in large amount but also deep into the lattice. Besides nanoparticles, nanotubes for (α-Fe_2O_3 were also examined by Chen et al. They measured 1400 mAh/g for the first discharge; however, the capacity fading is fast and large [67].

The reaction mechanism is the same for transition metal oxides and tin oxide. A group of transition metal oxides, such as CoO, CuO, NiO, Co_3O_4 and MnO, have been thoroughly studied [64]. This concept has expanded to areas of metal nitrides, sulfides, chlorides and fluorides. This is based on the fact that weaker M-X (metal-anion) bonding strength gives a higher potential [64] which is directly linked to the change of the metal oxidation state during cycling. Thus, the combination of different metals and their oxidation states, as well as the type of anions, will give potentials ranging from 0 to 3.5 V [68]. However, the synthesis of those rapid (high electrical conductivity) and reversible nanocomposites is not an easy task.

3.6 Titanates and Titanium Oxide Polymorphs

One issue associated with the nanoscaling of primary anode particles is the risk of possible side-reactions with the electrolyte, leading to capacity fading and safety problems. For titanates or titanium oxide, the lithium insertion potential is not within the range of electrolyte decomposition. This is essential for high-power cells. Therefore, since the SEI formation is suppressed, the capacity loss is largely reduced and the advantages of nanoparticles, such as deeper insertion and more reaction sites are realized. One such material is spinel $Li_{x+4}Ti_5O_{12}$ ($0 < x < 3$), which is inherently safe. The lithium insertion (charging potential) is around 1.55–1.6 V vs. Li^+/Li. It has three basic advantages over graphite or any other anode material mentioned above: insertion safety, minimal volume and structural changes, and 100% Coulombic efficiency with no side reactions [69]. Thus, $Li_{x+4}Ti_5O_{12}$ is of serious consideration as an anode for high energy, fast rate batteries. The electrode has been coupled with $LiMn_2O_4$ and $LiFePO_4$, forming low cost and safe cells with around 2.5 V [3]. It would be preferred if a higher-voltage cathode is coupled to produce at least 3.5 V.

The advantage of inert safety due to high redox potential is valid for both lithium titanate ($Li_4Ti_5O_{12}$) and titanium oxides polymorphs. Compared to lithium titanate, titanium oxide possesses a capacity up to 335 mAh/g, much higher than

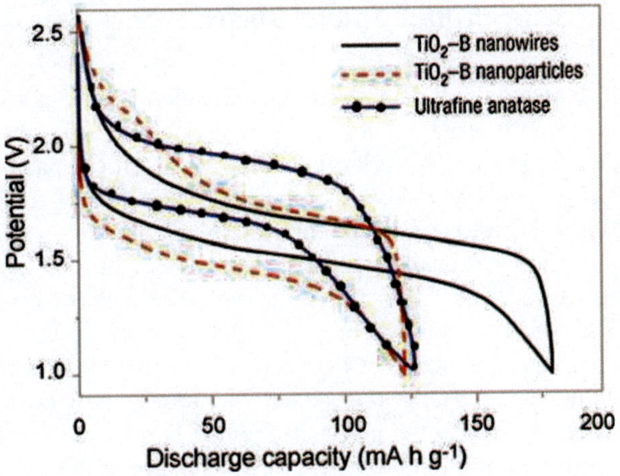

Fig. 18 Comparison of charge–discharge curves for TiO_2-B nanowires, particles and anatase TiO_2 (from Yang et al. [69], with permission from Elsevier)

175 mAh/g for titanate. The only problem with titanium oxide is poor electronic conductivity. And we know nanosized particles tend to have higher electronic and ionic conductivity. Therefore, TiO_2 nanotubes and nanowires, as well as nanoparticles, have been extensively synthesized and characterized. TiO_2 can exist in different lattice structures: rutile, anatase, brookite and bronze. Detailed information on those four different structures is provided below. (i) Rutile TiO_2 has negligible Li insertion ability (<0.1 Li per TiO_2) at room temperature and the lithium diffusion through the structure is anisotropic. Despite these drawbacks, nanotechnology greatly increases its electrochemical performances. For example, the rutile nano-rods were irreversibly transformed into cubic rocksalt $LiTiO_2$ nanorods, and thereafter the structure remained stable [70]. Also, up to 0.8 mol Li-insertion into nanostructured rutile TiO_2 (10 nm × 40 nm) was reported at room temperature [71]. (ii) Anatase TiO_2 suffers from a structure transformation from a body-centered space group to an orthorhombic group and the volume change is 4%. However, nanotubes grown by hydrothermal approaches increase the capacity significantly. (iii) Bronze TiO_2, called TiO_2-B is the most promising titanium oxide. Although there is volume change during lithium insertion, the bronze structure is more open than the first two types, allowing easier and increased lithium intercalation. A recent study on TiO_2-B nanotubes and wires gave 305 mAh/g with excellent cycling ability at 1.5–1.6 V vs. lithium. And the capacity and voltage hysteresis for bronze TiO_2 nanowires are better than TiO_2-B nanoparticles and TiO_2 anatase particles (see Fig. 18) [69]. (iv) Brookite nanoparticles of TiO_2 exhibited over 170 mAh/g for the first 40 cycles and quickly dropped to around 60 mAg/g for the 50th cycle [72], whereas TiO_2 nanotubes showed 160 mAh/g after 50 cycles, as expected [69].

3.7 Silicon and Silicon-Based Alloys as Anode Materials

This section will discuss silicon and silicon-based alloys as anode materials, which is currently very popular in the research field of LIBs. Silicon has received much attention, owing to its large theoretical specific capacity for the fully inserted phase $Li_{4.4}Si$ of 4200 mAh/g, compared to 372 mAh/g for graphite. The problem with silicon is similar to that of lithium metal alloys: big volume expansion and a phase transition associated with large amounts of lithium accommodated into the lattice structure. This large change of volume leads to cracking and crumbing of the electrode and this will not only decrease the reversible capacity but also induce poor electronic contact between silicon particles. In this case, silicon has limited volumetric capacity: the capacity per unit volume does not change much from LiSi to $Li_{4.4}Si$ due to the compensation of large volume expansion.

There are basically three ways to improve the cycling performance of silicon. The foremost improvement is to reduce the size of the silicon particles. A second method is to maintain electronic contact between particles by mixing with conductive additives, such as carbon graphite or nano-structured carbon black. The third is to synthesize silicon-metal alloys, dispersing silicon in an inert or active matrix. These three methods can be used together, which means the silicon particles are first reduced to the nanoscale and then dispersed into either conductive additives or a metal buffer matrix. Moreover, limiting lithium insertion to a certain depth will reduce volume expansion and lengthen the cycle life. For example, the cycling number of above 99% charge retention can be increased from 25 to 90 cycles, when the capacity is reduced from 900 to 500 mAh/g [73].

3.7.1 Si/C Composite in Powder Phase

Several methods have been attempted to improve the electrochemical performance of silicon-based anodes. One of them is to mix silicon powders with carbon or graphite powders. There are various synthesis processes to mix these powders: pyrolysis of polymers containing silicon and carbon, chemical vapor deposition (CVD), ball milling of carbon and silicon powders, chemical reaction of gels and so on [74–77]. The ratio of carbon and silicon and their structures play an important role in determining the capacity, Coulombic efficiency and cycling behavior.

When synthesized by pyrolysis, the obtained powders often contain oxygen, along with silicon and carbon, forming S–C–O glass. Carbon works as the conduction pathways for electrons and lithium ions; thus a higher carbon ratio will bring about a fast reaction and ensure good capacity. Conversely, introduction of silicon and oxygen will also increase the irreversible capacity; and no matter what the ratio is, the large irreversible capacities cannot be avoided [78]. Later, it was observed that the high irreversible capacity of the pyrolyzed anodes is caused by the presence of oxygen in the Si/C composite. In order to get rid of the oxygen, materials like $SiCl_4$ (SiH_4) and $(CH_3)_2Cl_2Si$ have been used as carbon and silicon source materials, respectively [79, 80]. In this case, vapor-phase deposition was employed to disperse nano-silicon onto graphite. Experiments gave good reversible capacity but were still plagued by high irreversible capacities, caused by the large surface area of graphite. However, the cycling ability was greatly increased to over 100 cycles.

This early research made Si/C composites more attractive for further investigation. However, they still show high irreversible capacity which may be caused by high surface area of the carbon phase. To overcome it, carbon was deposited on silicon powders to suppress the SEI formation. Dimov et al. deposited carbon on homogenously mixed graphite and silicon powders with a 1:1 weight ratio [81]. However, they measured a 1000 mAh/g irreversible capacity as a result of a dead area where lithium could not be extracted for further cycling. This is often caused by non-uniform distribution of silicon in the anode.

Besides carbon additives on silicon, an inert material, specifically copper, was also introduced to form Si-Cu alloys [82]. The Si-Cu-Cu_3Si/C composites showed advanced cycling stability because highly conductive Cu_3Si worked as an electric conductor and binder, revealed by EPMA analysis; and the substitution of carbon by Cu_3Si reduced the surface area in contact with the electrolyte. Although Si/C composites, whether carbon coated by Si or Si coated by carbon, have been helpful in maintaining capacity during cycling, very high cycling performance was not achieved.

Using pyrolysis and CVD, the structures and particle sizes are difficult to control in the Si/C composites. Ball milling, in contrast, can maintain a homogenous mixture and specific composition, structures and particle sizes. Unfortunately, though ball milled Si/C composites improved the cycling ability; they suffered from high irreversibility on the first cycle which should be eliminated in practical secondary cells. More work must be done to modify the micro or nanostructures of silicon-based anode materials.

3.7.2 Advanced Binder for Si/C Composite

As required for high rate capacity, the adhesion of active silicon particles to electric conductor and collector is essential. This can be significantly influenced by the type and amount of binder. Usually, polyvinylideneflouride (PVDF) is used as the binder for anode and cathode materials. However, in the case of silicon-based anodes with large volume expansion, PVDF may not be a suitable choice due to its limited flexibility as a thermoplastic material. Therefore, studies on novel binders have been conducted, such as polyethylene oxide PEO–$LiClO_4$, cross-linked polyethylene glycol PEG—$LiClO_4$ and styrene butadiene rubber (SBR).

The $LiClO_4$ was added to both PEO and PEG to improve their ionic conductivities. The anode including cross-linked PEG–$LiClO_4$ showed higher elasticity and mechanical strength compared to PEO–$LiClO_4$. Thus, the electrochemical performance of the PEG-supported anode showed a higher energy capacity and

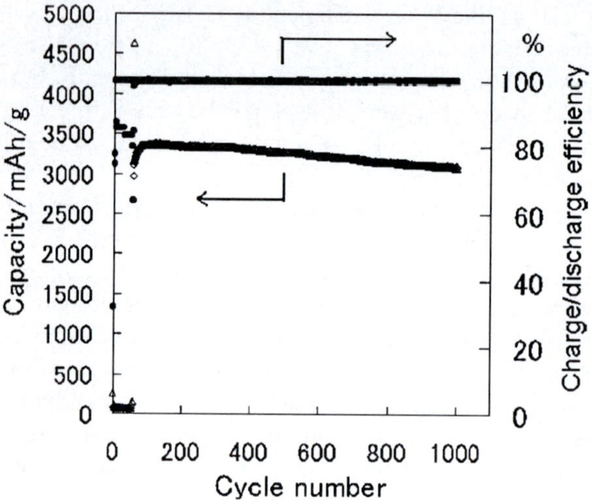

Fig. 19 Cycling performance of a 50 nm-thick film operated at 12C (from Ohara et al. [85], with permission from Elsevier)

capacity retention. The reason for the improvement of the capacity is the mechanical strength enhanced by PEG cross-linking. However, the cycling performance of this kind of anode was worse compared to anodes mixed with PVDF [83]. Other polymers have been investigated as well. One promising material is the highly elastic SBR. Improvement in the electrochemical performance of anodes with SBR binders was ascribed to a smaller modulus that allows the anode to easily expand and contract and a higher adhesive strength between the particles and current collector [84].

Since silicon has the highest theoretical capacity, it has a potential to outperform any other anode material. However, severe capacity fade in the first cycle results in low Coulombic efficiency, which is an energy waste process. Although nanosized silicon, as well as silicon dispersed onto a buffer matrix, was proved to be somewhat helpful, the improvement is not sufficient for practical use. Anyhow, this research paved the way for research in thin film silicon anodes with nanoparticles and amorphous silicon.

3.7.3 Thin Film Si/C Composite

Development in thin film silicon or silicon alloy anodes has impact on both research and practice for thin film secondary lithium batteries. Much effort has been made on pure silicon thin films deposited on different kinds of metals that are also electronic collectors. It was found that the thickness of the thin film will influence the cyclability and lithium insertion capacity in a strong way. This is reasonable, since thinner films will allow for larger volume changes and accommodate more lithium without severe structure cracking. One of the most impressive works was carried out by Takamura and co-workers, where they

Table 1 Overall electrochemical performances of the seven main kinds of anode materials

	Specific capacity (mAh/g)	Cycling number (85%)	Initial irreversible capacity (%)	Discharge potential (V)	Rate capacity
Original graphite [123]	300	≥200	20	0.3	C/7
Nanostructure carbon [124]	260	28	4	0.55	C/5
Li–M–N [58]	900	10	–	0.4	0.5 mA/cm^2
Li alloy [62]	100 ($Sn_{0.72}Sb_{0.28}$)	170	–	0.7	0.25 mA/cm^2
Metal oxide [125]	1200 (Fe_2O_3)	200	5	0.8	0.2 mA/cm^2
Titanium oxide [126]	160 (anatase)	100	20	1.5	4C
Silicon powder [77]	1500	≥50	25	0.3	C/10
Silicon thin film [85]	3200	≥1000	1.5	0.3	12C

deposited silicon thin films of 200–1500 Å thickness on a nickel surface in vacuum [85]. Among all the films deposited, the 50 nm thick film exhibited a stable reversible capacity of 3500 mAh/g for the first 200 cycles. The capacity retention is 97% for the 200th cycle. Furthermore, the charge-discharge rate is 12C, indicating a fast rate and good electronic conduction. The results are shown in Fig. 19.

Research on active Si–M (M = Sn, Ag, Zn, Mg, V, or O) thin films have been carefully explored by many groups [86–91], as well as inert Si–M (Cr, Fe, Mn, Ni, Co, Zr, and TiN) alloy thin films (refer to the review in [92]). Si–M thin film anodes have also been coupled with thin film electrolytes and cathodes. Though the reversible capacities of silicon-based thin film anodes are several times higher than those of graphite and the cycle lives are also much longer than in graphite, the high cost of thin film deposition makes them unsuitable for cheap and largely demanded portable power sources.

In summary, seven kinds of the most researched anode materials have been reviewed.Each of them have their own advantages and disadvantages, as summarized in Table 1. This table is a rough summary of the most frequently reported data and there should be some exceptions. From this table, we can see that the most impressive anode material is the silicon thin film, despite its ordinary volumetric capacity, not included in the table. The metal oxide materials seem to be suitable for large-scale and commercial applications due to their large capacities, long cyclabilities and fast operation rates. However, their voltage is not stable for different degrees of lithium insertion. Graphite still looks very practical and cheap. Titanium oxide anode materials can find unique use when the safety requirement is quit stringent. Notes: The values in the column of “cycling stability” are calculated to be the number of cycles that can retain 85% of the initial maximum capacity; the rate capacity column shows the operating rates that were most frequently reported for specific anode materials; some of the rates are in the unit of mA/cm^2, this is because the theoretical capacity (related to C) is not fixed or the anode is too thin to give an accurate weight.

4 Electrolytes

Along with the anode and cathode, the electrolyte as the ionic transport media between them has been explored and analyzed extensively. So far, four main types of electrolytes have been under careful inspection: conventional liquid, ionic liquid, solid polymer and inorganic solid electrolytes. They each have preferences for coupling with different electrodes and operation conditions. A good combination of electrodes and electrolytes affords excellent electrochemical performance and guarantees safety. Accordingly, it is crucial to choose the most suitable electrolyte when designing a new LIB. The related considerations and handling precautions are discussed below.

4.1 Conventional Liquid Electrolyte

In the earlier times (1970s), lithium perchlorate in propylene carbonate-dimethoxyethane mixtures was used as the electrolyte and coupled with lithium metal to assemble secondary lithium batteries [93]. It took almost 10 years for the emergence of ethylene carbonate-added (EC) electrolyte. Soon it gained attention, owing to a sustainable improvement of charge-discharge performance [94]. As it is well-known, EC works as an active material that decomposes on the anode surface to form a solid electrolyte layer (SEI) when charged to around 0.8 V versus Li^+/Li during the first half cycle. The SEI is composed of mainly two layers. The inner layer closer to electrode is more densified and comprises of lithium salts like LiF, Li_2O, and lithium carbonate, while the outer layer is polymorphous and is comprised of lithium polymers, such as $Li_2(OCO_2(CH_2)_2OCO_2)_2$ and $(CH_2OCO_2Li)_2$ [95]. In most of today’s liquid electrolyte LIBs in ambient temperature, EC, whose melting point is 36.8 °C, is used in mixture with one or two low melting point carbonates. These carbonates are aprotic, polar and of high dielectric constant; hence they are able to solvate lithium salts to more than 1 M. The lithium salts widely used in current research and commercial LIBs are $LiPF_6$ and sometimes $LiClO_4$. However, realized in this century, novel lithium bis-oxalatoborate,

LiBOB, shows better electrochemical performance than $LiPF_6$ dissolved electrolyte [96]. Anyway, the improvement is not large.

The liquid electrolytes of carbonates with dissolved lithium salts are electrochemically stable under 5.5 V and, thus, can satisfy most of the high potential electrodes [97]. Their high ionic conductivity of larger than 10^{-3} S/cm also improves their popularity. Even though they are used extensively for commercial and household applications, they have problems concerning safety and the environment. Since the solvents are flammable, the battery may catch fire or even explode if it shorts accidentally. The solvents are also toxic and should be dealt with cautiously. Additionally, the limited operating temperature range (−20 to 50 °C) and voltage range fall short when required in severe natural conditions.

4.2 Ionic Liquid Electrolyte

Starting in the 1980s, interest in room temperature ionic liquid (RTIL) electrolyte persists. Room temperature ionic liquid is also called room temperature molten salt, with melting points below room temperature. Countless types of ionic liquids have been evaluated as liquid electrolytes, especially in the last 10 years. To make it simple and clear, a molten salt is represented by $[A^+][X^-]$, the cation and anion. The most commonly used RTIL are quaternary ammonium salts, such as tetraalkylammonium $[R_4N^+]$. The counter anions can be organic or inorganic, such as $[BF_4^-]$, $[PF_6^-]$, $[AsF_6^-]$, triflate $[CF_3SO_3^-]$, imide $[N(CF_3SO_2)_2^-]$ and $[N(F_2SO_2)_2^-]$ [98]. When they are used as the electrolyte in the cell, lithium salt $[Li^+][X^-]$ is added, forming $[Li^+][A^+][X^-]$ ionic liquid. The lithium salt is often LiTFSI instead of $LiPF_6$ in carbonate solvents. LiTFSI became popular recently, since it can greatly improve the ionic conductivity of solid polymer electrolytes (as discussed in the next section). Their features have been thoroughly studied and the relationships between those characteristics and electrochemical behavior inside the LIBs are addressed in this section.

One of the most promising RTILs is PP13–TFSI (N-Methyl-N-propylpiperidinium bis(trifluoromethanesulfonyl)imide, reported by Sakaebe and co-workers. They prepared a RTIL composed of quaternary ammonium cation and imide anion. This novel salt was tested in $Li/LiCoO_2$ cell and showed very good performance of more than 97% Coulombic efficiency at the C/10 cycling rate and 85% of the capacity at C/2 [99]. This high Coulombic efficiency was attained, indicating less decomposition of electrolyte and a thinner SEI. The formation of SEI is the basic and most important factor determining the quality of electrolyte. Generally speaking, however, the cycling ability of $Li/RTIL/LiCoO_2$ is not as good as that of $Li/LiPF_6$ in $EC:DMC/LiCoO_2$. The addition of molecular additives to ionic liquid has been proved to be the most effective method, since they form a protective coating on the surface of electrodes.

A controversial issue in RTILs is the viscosity. The ionic liquid has much higher viscosity than that of water, normally in the range of 30–50 cP. This is the result of strong Coulombic interaction between ions. Accordingly, the vapor pressure is always low for ionic liquids, which is very good because low vapor pressures reduce flammability. But the highly viscous liquid makes it difficult to fill the gaps between electrode particles. Thus, special treatment is required to improve the contact between the electrolyte and electrodes, for instance, setting them together for long time under low ambient pressure [100]. Though there are many problems yet to be solved, research interest in RTIL never drops due to the importance of safety issues.

4.3 Solid Polymer or Gel Electrolyte

Solid polymer electrolyte is most attractive when considering thin film batteries. These thin film cells can be packed together, forming a high voltage or high rate battery while being flexible. Besides, the cost and reliability of solid polymer electrolytes are appealing. In contrast to various liquid electrolytes, the high-molecular weight polymers for solid electrolytes are limited to polyethylene oxide (PEO). Because of the low ionic conductivity and solid state, they are suitable for application in thin film batteries where the limited thickness increases their conductance.

A high temperature (60–80 °C) is needed to keep the PEO from crystallizing; thus, the ionic conductivity of amorphous PEO can reach a useful value of about 10^{-4} S/cm. Hence, effort has been placed on improving the ionic conductivity of the polymer electrolyte at room temperature. One of the most effective methods is to add liquid plasticizers, but this promotes problems such as deterioration of the mechanical strength and risk of reaction with the lithium metal anode. Scrosati et al. reported in 1998 that nanosized ceramic powders of TiO_2 and Al_2O_3 can be added in as solid plasticizers to lower the operation temperature of PEO [101]. And they achieved ionic conductivity of 10^{-5} at 30 °C, which is more than three orders of magnitude higher than plasticizer-free PEO. They opened a new area in raising the ionic conductivity of the solid electrolyte; however, no other techniques for new and better plasticizers were reported to improve the performance. Therefore, solid electrolyte LIBs are only favorable when the operation temperature is above 60 °C. Other approaches about polymer development have been investigated as well. Hammond and Delongchamp designed a polymer comprised of PEO/PAA in a layer-by-layer configuration [102]. The molecular weight of the integrated film can be very high due to the low PEO/PAA cross-link density. Even though they attained an ionic conductivity of 10^{-4} S/cm, the requirement of pre-exposure to humidity does not seem suitable in humidity-sensitive LIB applications.

Although large amount of efforts have been focused on enhancing the ionic conductivity of polymer electrolytes, the progress has not been sufficient to allow practical operation at room temperature. Despite this, polymer electrolytes continue to stimulate researchers' thoughts regarding polymer design and additive materials. The reason is that in terms of their cost and safety, they outperform all liquid electrolytes.

4.4 Inorganic Solid Electrolyte

Inorganic solid electrolytes are pointing to fast ion conducting lithium glasses and ceramics, where lithium ions move within a naturally static framework. The fast ionic conduction mechanism describes the lithium ion movement by rotational disorder and vacancies existing in the crystal structures of electrolyte. As the same in polymer solid electrolytes, inorganic solid electrolytes are supposed to be safe (no leakage or pollution), economical and suitable for thin film batteries. The problem associated with this type of electrolyte is low ionic conductivity at room temperature, which greatly limits their practical application. There are various kinds of fast ionic conductors for not only lithium, but also monovalent protons, sodium, silver, potassium, copper, and fluoride ions [103]. But herein, we only discuss the lithium ionic conductors and their applications.

The lithium ionic conductors can be divided into three groups. The first is perovskite-type—$RE_{2/3-x}Li_{3x}TiO_3$ (RE: rare earth ion). Much work has been done on lithium lanthanum titanates since 1993 when Inaguma et al. reported a very high lithium conductivity of 10^{-3} S/cm for $La_{0.5}Li_{0.34}TiO_{2.94}$ at room temperature [104]. This is a very impressive ionic conducting electrolyte; however, its performance is reduced when used in batteries since grain boundaries exist for polycrystalline electrolytes. The main factor limiting its usage is the large amount of lithium intercalation that occurs when the operation potential is around 1.2 V. This will also cause the coexistence of Ti^{4+} and Ti^{3+}, leading to increased electronic conductivity [105]. The second kind of lithium ionic conducting solid electrolytes is of three-dimensional structure (LISICON) similar to the well-known Na super-ionic conductor (NASICON). The composition is set to be $Li_{1+x}M_{2-x}M'_xP_3O_{12}$, where M = Ti, Ge or Hf and M' = Al or In. They have shown a good ionic conductivity of 2–8 × 10^{-4} S/cm at around 25 °C [106]. But in practice, the grain boundaries increase the difficulty of ion transport by promoting scattering sites, inducing a drop in conductivity to 10^{-5} S/cm. The third kind of lithium ionic conductor electrolyte is garnet-structured glasses. They are reported to have two main compositions: $Li_5La_3M_2O_{12}$ (M = Nb, Ta) and $Li_6ALa_2M_2O_{12}$ (A = Ca, Sr, Ba; M = Nb, Ta). Among all these types, $Li_6BaLa_2Ta_2O_{12}$ is the most attractive, due to its high bulk ion conductivity of 4 × 10^{-5} S/cm and high electrochemical stability of 6 V versus. Li^+/Li at room temperature. The problem is its reactivity with the cathode materials [107]. In conclusion, all of perovskite-type, NASICON type and garnet-structured fast ion conducting electrolyte need much improvement.

Because of the limited ionic conductivities of all inorganic solid electrolytes, they are used in thin film batteries with minimal inter-electrode distances. The solid electrolyte mostly used today is none of the above, but rather lithium phosphorus oxynitride (LiPON). This is mainly because of its large stable voltage range of 0–5.5 V versus. Li^+/Li and thin film-compatible deposition techniques [108]. The deposition of LiPON can be carried out by reactive sputtering of Li_3PO_4 in nitrogen atmosphere or nitrogen ion beam-assisted deposition (IBAD) of thermally evaporated Li_3PO_4 [108, 109]. The typical ionic conductivity of LiPON is 10^{-6} S/cm, which is acceptable. In 2009, Orimo et al. announced a new super ionic conductor: halide-stabilized $LiBH_4$ [110]. $LiBH_4$ showed an ionic conductivity of 10^{-2} S/cm at about the super ionic transition temperature, 115 °C, and the halide helps stabilize the super ionic phase, even at around 35 °C; however, the ionic conductivity is on the order of 10^{-5} S/cm and there is no other report related to this kind of material.

Finally, we draw a conclusion on these different kinds of liquid and solid electrolytes. For conventional coin cells or prismatic cells, liquid electrolyte of $LiPF_6$ in EC-DMC takes more than 90% weight. And the lithium salt of LiTFSI was recently shown to exhibit stability toward electrodes. Thus, attempts have been made to add it both in carbonate solvents and ionic liquids to form a novel liquid electrolyte. As for the progress of solid polymer and inorganic electrolytes, their use in thin film batteries becomes more and more widespread. However, more types of polymers should be included and an understanding of the additives should also be improved. As well, fast ion conductors are developing very fast right now. We hope to see more reliable and powerful all-solid-state thin film batteries in the future.

5 Thin Film Batteries

Thin film batteries are classified by small thickness, ranging from several microns to several millimeters. Their areas can be minimized to $100 \times 100\ \mu m^2$ or enlarged to around 20 cm^2. The concept of designing thin film batteries is the deposition of thin films of anode, electrolyte and cathode sequentially on a substrate. The typical features of this kind of battery are very unique, compared to conventional coin, cylindrical and prismatic cells. Additionally, they are reported to have higher volumetric and gravimetric energy densities, power capabilities and superior design flexibility. We will discuss the similarities and differences of thin film and conventional secondary lithium batteries.

Similar to conventional lithium-ion cells, thin film batteries suffer from problems such as volume expansion during charge-discharge cycles, side reactions between nanoscale electrodes and electrolyte and low ionic conductivities of cathode and electrolyte. Differing from other lithium-ion cells, thin film batteries have issues of high cost from thin film deposition techniques, but may provide solutions to many problems via modification of the nanostructured materials. Therefore, it is very useful to investigate thin film techniques and their impact on the structures and electrochemical performances of battery materials. Normally, the lithium metal is grown by vacuum thermal vapor deposition and the solid cathode and electrolyte are deposited by various techniques: radio-frequency (RF) sputtering, RF magnetic sputtering, CVD, e-beam evaporation deposition, electrostatic spray deposition, pulsed laser deposition or sol–gel coating, etc.

Fig. 20 Micrograph of a five cell serial battery (from West et al. [113], with permission)

Many types of cathodes, anodes and electrolytes have been investigated in thin film batteries. Most of them use lithium metal as the anode, LiPON as electrolyte and lithium transition metal oxide as cathode. For instance, Wang et al. presented their work on the deposition of lithium metal by thermal vaporization and $LiCoO_2$, LIPON by radio-frequency sputtering [111]. The obtained films have thicknesses of 0.05–0.5 μm, 3 μm and 1–2 μm for $LiCoO_2$, lithium and LIPON, respectively. They obtained very impressive results from the cell comprised of 0.05 μm thick $LiCoO_2$. Capacity fading of 0.0001% per cycle was achieved. And the 0.5 μm thick layer showed capacity fading of 0.002% per cycle for the same discharge depth. The better cyclability of thinner films results from released strain. With less strain, the structure can be maintained while cycling, thus the resistance caused by structure deformation can be suppressed. Therefore, thin film batteries benefit from the small thickness of electrodes.

A very attractive area of thin film batteries is the microscaling of battery arrays, which are called smart cards. They are widely used and demanded for fingerprint scanners, face recognition, hand geometry, eye scanners, signature verification and voice recognition [112]. Ratnakumar et al. reported all-solid-state microscale thin film batteries in 2002 [113]. They prepared batteries as small as $50 \times 50\ \mu m^2$ by photolithography. The batteries were arranged in parallel and serial to yield higher voltage or capacity. Figure 20 shows the series arrangement of the micro-batteries. The big squares are metal contacts and the small dark areas are active batteries. The obtained voltage for each cell is 3.9 V and current is 10 $\mu A/cm^2$. However, they observed a short cycle life of tens of cycles due to contamination in the electrodes and electrolyte during etching. The electrolyte used in thin film batteries can also be a liquid electrolyte. But the transition from liquid to solid electrolyte solves the problem of leakage and ensures higher mechanical strength. And even within the solid electrolyte area, fast ion conducting inorganic electrolyte is more favorable than polymer electrolyte due to the high temperature needed for polymer electrolyte.

Briefly speaking, thin film batteries take advantage of small thicknesses, which benefit the less ion conductive electrode and electrolyte materials. It also has problems similar to other LIBs, including large volume change, cycling instability, etc. The developments of anode materials as carbon and lithium alternatives are popular focal points, since they possess much higher capacity than graphite and are safer than lithium. Since the weight and volume of thin film batteries are small, the ratio of pre-packed material over a packed cell should be as large as possible. So, it requires a rational distribution of anode, cathode, electrolyte, collectors and outer shell. Besides, the mechanical strength is important, since it relates to the tolerance to be bent or pressed. Further, the condensed serial package of thin films can give a high voltage, thus, the design of thin film cells is carefully considered.

6 Lithium-Air Batteries

Metal-air batteries have an intrinsic advantage in using oxygen in air as the cathode, increasing the specific capacity in the package by sharing the reduction material in air. People started working on metal-air batteries before 1970. There are varieties of metal-air batteries: zinc-, aluminum-, magnesium-, calcium-, iron-, lithium-air cells and so on. Among them, the lithium-air cell shows the highest theoretical discharge voltage of 3.45 V and largest specific capacity of 13.3 kWh/kg [114]. The research on lithium-air batteries began very late because of its fast reaction with oxygen and moisture in air, which leads to a very short shelf life. Consequently, people developed new electrolytes, where the amount of oxygen and water dissolved is largely reduced and opened the door for fast-growing research on lithium-air batteries [115]. We will discuss the key factors for improvement of the electrochemical performances of lithium-air batteries. Some of the issues will also be helpful for other metal/oxygen systems.

A lithium-air battery is composed of a lithium metal anode, graphite cathode and electrolyte in between. Oxygen diffuses into the pores of the graphite and forms lithium oxides. To achieve better cycle life, specific capacity and shelf life, the electrolyte of ionic liquid is very fundamental. The same as in LIBs, the electrolytes available are lithium salt dissolved carbonates or ionic liquids, and polymer solid electrolytes, which were mentioned in the fourth section of this chapter. First, the lithium salt dissolved carbonates and lithium salts co-synthesized liquid electrolyte have been shown to be responsible for the discharge capacity, rate capability and rechargeability [116]. The electrolyte formula of PC:THF:$LiPF_6$ showed the best specific capacity, 1198 mAh/g at a rate of 0.2 mA/cm^2. This is related to its high solubility of oxygen. However, this kind of electrolyte has leakage and vaporization problems, which shorten the testing time to several hours.

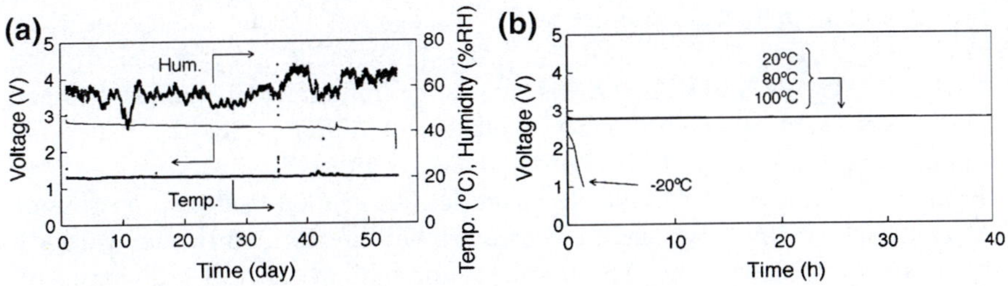

Fig. 21 The effect of humidity and time (**a**) and temperature (**b**) on open circuit voltages of lithium air batteries (from Kuboki et al. [117], with permission from Elsevier)

Ionic liquids were examined carefully by Kuboki and co-workers [117]. They promoted the use of a hydrophobic ionic liquid as electrolyte: 1-Ethyl-3-methyl-imidazolium bis (trifluoromethylsulfonyl) amide (EMITSI). It showed high conductivity, low evaporation and low solubility of water, thus preventing the hydrolysis of lithium. The cell made of Li/EMITSI/graphite worked for 56 days in air and kept a constant discharge capacity under different humidity conditions. Besides, the working temperature can vary form 20–100 °C. These features can be seen in Fig. 21. Compared to organic liquid electrolytes, the hydrophobic ionic liquid is more promising for lithium air batteries.

Another type of lithium air battery uses a polymer as its electrolyte. Early in 1996, Abraham and Jiang assembled a Li/polymer/carbon cell using a mixture of PAN, EC, PC, and $LlPF_6$ as the starting polymer materials. It is a hybrid electrolyte, combining liquid organic and solid polymer electrolytes. After synthesis, there is still solvent within the polymer; however, it was fully absorbed by the polymer, making the electrolyte a mechanically strong, freestanding film [114]. Although featuring good mechanical properties, they were unable to perform well. Recently, Xu and Ye proposed a novel polymer electrolyte: oligomeric polyether/cross-linked PMMA [118]. Their electrolyte had excellent mechanical integrity and strength, and exhibited an ionic conductivity of 4.3×10^{-4} S/cm at room temperature. Still they did not show the capacity; and the cyclic voltammetry revealed a large difference between the cathodic and anodic peaks caused by large resistance in the 0.8 mm thick polymer electrolyte.

Another important part of cell, the cathode is a very intrinsic factor on determining the cell's performance. The most used cathode is conductive graphite. Its mesoporic volume, not its surface area, has been reported repeatedly to affect the discharge capacity to a great extent [116, 117]. It is explained that a larger amount of lithium can be deposited into the interior of graphite through the mesopores.

Beyond lithium-air batteries, Sammes et al. described a lithium water–air cell with an even higher open-circuit voltage of 3.7 V for the reaction:

$$2\text{Li} + \frac{1}{2}\text{O}_2 + \text{H}_2\text{O} = 2\text{LiOH}. \tag{1}$$

The lithium is protected from the water by a layer of lithium conducting glass ceramic: $Li_{1+x+y}Ti_{2-x}Al_xSi_yP_{3-y}O_{12}$ (LTAP) [119]. The polymer electrolyte $PEO_{18}Li(CF_3SO_2)_2N$ ($PEO_{18}LiTFSI$) separates the lithium metal and LTAP, since LTAP is not stable when in direct contact with lithium. So, the battery is assembled like Li/PEO_{18}LiTFSI/LTAP/aqueous LiCl/Pt [120].

From this brief review of lithium-air batteries, we can tell that the properties of electrolytes, especially ionic liquid electrolytes, will greatly improve the capacity and cyclability of these cells. The problems for lithium-air cells are caused by exposure to ambient air and difficult packing. Hence, we propose a thin film lithium-air battery that combines the advantages of thin film and lithium-air batteries. Inorganic lithium ion-conducting solid electrolytes have not been used in lithium-air battery systems, to our knowledge. The reason behind this should relate to the poor contact of lithium, so thin film techniques may provide a solution to this problem by shortening electrolyte thickness. Packing techniques for thin film batteries will also benefit lithium-air batteries by providing a flexible shape and size.

Acknowledgments Thanks for help from Yizhao Lang and Makarand Karmarkar.

References

1. Gao XP, Yang HX (2010) Multi-electron reaction materials for high energy density batteries. Energy Environ Sci 3:174–189
2. Greatbalch W, Lee JH, Schneider AA et al (1971) Solid-state lithium battery. A new improved chemical power source for implantable cardiac pacemakers. IEEE Trans Biomed Eng 18:317–324
3. Whittingham MS (2004) Lithium batteries and cathode materials. Chem Rev 104: 4271–4301
4. Whittingham MS (1976) Electrical energy storage and intercalation. Chem Sci 192: 1126–1127
5. Broadhead J, DiSalvo FJ, Trumbore FA, Non-aqueous battery using chalcogenide electrode. US Patent 3864167
6. Delmas C, Doumerc JP et al (1994) The $Li_xV_2O_5$ system: An overview of the structure modifications induced by the lithium intercalation. Solid State Ionics 69:257–264
7. Howing J, Gustafsson T, Thomas JO (2003) Low-temperature structure of V_6O_{13}. Acta Crystallogr B59:747–752
8. Chandrappa GT, Steunou N, Livage J (2002) Materials chemistry: macroporous crystalline vanadium oxide foam. Nature 416:702
9. Le DB, Passerini S, Guo J, Ressler J, Owens BB, Smyrl WH (1996) Aerogels and xerogels of V2O5 as intercalation hosts. J Electrochem Soc 143:L102–L103
10. Mitzushima K, Jones PC, Wiseman PJ, Goodenough JB (1980) Li_xCoO_2 ($0 < x < -1$): a new cathode material for batteries of high energy density. Mater Res Bull 15:783–789
11. Ohzuku T, Ueda A (1994) Solid-state redox reactions of LiCoO (R3m) for 4 volt secondary lithium cells. J Electrochem Soc 141:2972–2977
12. Thackeray MM, David WIF, Bruce PG, Goodenough JB (1983) Lithium insertion into manganese spinels. Mater Res Bull 18:461–472
13. Tarascon JM, Wang E, Shokoohi FK (1991) The spinel phase of $LiMn_20_4$ as a cathode in secondary lithium cells. J Electrochem Soc 138:2859–2864

14. Bates JB, Lubben D, Dudney NJ, Hart FX (1995) 5 volt plateau in $LiMn_2O_4$ thin films. J Electrochem Soc 142:L149–L151
15. Zhong Q, Bonakdarpour A, Zhang M, Gao Y, Dahn JR (1997) Synthesis and electrochemistry of $LiNi_xMn_{2-x}O_4$. J Electrochem Soc 144:205–213
16. Robertson AD, Lu SH, Averill WF, Howard WF (1997) M^{3+}-modified $LiMn_2O_4$ spinel intercalation cathodes: I. Admetal effects on morphology and electrochemical performance. J Electrochem Soc 144:3500–3505
17. Hewston TA, Chamberland BL (1987) A survey of first-row ternary oxides $LiMO_2$ (M = Sc-Cu). J Electrochem Soc 48:97–108
18. Obrovac MN, Mao O, Dahn JR (1998) Structure and electrochemistry of $LiMO_2$ (M = Ti, Mn, Fe, Co, Ni) prepared by mechanochemical synthesis. Solid State Ionics 112:9–19
19. Armstrong AR, Bruce PG (1996) Synthesis of layered $LiMnO_2$ as an electrode for rechargeable lithium batteries. Nature 381:499–500
20. Choi S, Manthiram A (2002) Factors influencing the layered to spinel-like phase transition in layered oxide cathodes. J Electrochem Soc 149:A1157–A1163
21. Armstrong AR, Robertson AD, Gitzendanner R, Bruce PG (1999) The layered intercalation compounds $Li(Mn_{1-y}Co_y)O_2$: positive electrode materials for lithium–ion batteries. J. Solid State Chem 145:549–556
22. Makimura Y, Ohzuku T (2003) Lithium insertion material of $LiNi_{1/2}Mn_{1/2}O_2$ for advanced lithium–ion batteries. J Power Sources 119:156–160
23. Reed J, Ceder G (2002) Charge, potential, and phase stability of layered $Li(Ni_{0.5}Mn_{0.5})O_2$. Electrochem. Solid State Lett 5:A145–A148
24. Ohzuku T, Makimura Y (2001) Layered lithium insertion material of $LiCo_{1/3}Ni_{1/3}Mn_{1/3}O_2$ for lithium–ion batteries. Chem Lett 30:642–643
25. Sun Y, Ouyang C, Wang Z, Huang X, Chen L (2004) Effect of Co content on rate performance of $LiMn_{0.5-x}Co_{2x}Ni_{0.5-x}O_2$ cathode materials for lithium–ion batteries. J Electrochem Soc 151:A504–A508
26. Hwang SJ, Park HS, Choy JH (2000) Effects of chromium substitution on the chemical bonding nature and electrochemical performance of layered lithium manganese oxide. J Phys Chem B 104:7612–7618
27. Guo ZP, Zhong S, Wang GX, Liu HK, Dou SX (2003) Structure and electrochemical characteristics of $LiMn_{0.7}M_{0.3}O_2$ (M = Ti, V, Zn, Mo, Co, Mg, Cr). J Alloys Compd 348: 231–235
28. Li D, Peng Z, Ren H, Guo W, Zhou Y (2008) Synthesis and characterization of $LiNi_{1-x}O_2$ for lithium batteries by a novel method. Mater Chem Phys 107:171–176
29. Park YJ, Hong YS, Wu X, Kim MG, Ryu KS, Chang SH (2004) Cyclic properties of $Li[Co_{0.17}Li_{0.28}Mn_{0.55}]O_2$ cathode material. Bull Korean Chem Soc 25:511–516
30. Gao Y, Yakovleva MV, Ebner WB (1998) Novel $LiNi_{1-x}Ti_{x/2}Mg_{x/2}O_2$ compounds as cathode materials for safer lithium–ion batteries. Electrochem Solid State Lett 1:117–119
31. Padhi AK, Nanjundaswamy KS, Masquelier C, Okada S, Goodenough JB (1997) Effect of structure on the Fe^{3+}/Fe^{2+} redox couple in iron phosphates. J Electrochem Soc 144: 1609–1613
32. Tarascon JM, Armand M (2001) Issues and challenges facing rechargeable lithium batteries. Nature 414:359–367
33. Chung SY, Bloking JT, Chiang YM (2002) Electronically conductive phospho-olivines as lithium storage electrodes. Nat Mater 1:123–128
34. Sauvage F, Baudrin E, Gengembre L, Tarascon JM (2005) Effect of texture on the electrochemical properties of $LiFePO_4$ thin films. Solid State Ionics 176:1869–1876
35. Ritchie A, Howard W (2006) Recent developments and likely advances in lithium–ion batteries. J Power Sources 162:809–812
36. Kim JK, Choi JW, Chauhan GS, Ahn JH, Hwang GC, Choi JB, Ahn HJ (2008) Enhancement of electrochemical performance of lithium iron phosphate by controlled sol–gel synthesis. Electrochim Acta 53:8258–8264

37. Hsu KF, Tsay SY, Hwang BJ (2005) Physical and electrochemical properties of $LiFePO_4$/carbon composite synthesized at various pyrolysis period. J Power Sources 146:529–533
38. Koltypin M, Aurbach D, Nazar L, Ellis B (2007) More on the performance of $LiFePO_4$ electrodes—the effect of synthesis route, solution composition, aging, and temperature. J Power Sources 174:1241–1250
39. Choi D, Kumta PN (2007) Surfactant based sol–gel approach to nanostructured $LiFePO_4$ for high rate Li-ion batteries. J Power Sources 163:1064–1069
40. Lin Y, Pan H, Gao M, Liu Y (2007) Effects of reductive conditions on the microstructure and electrochemical properties of sol–gel derived $LiFePO_4$/C. J Electrochem Soc 154: A1124–A1128
41. Jugović D, Uskoković D (2009) A review of recent developments in the synthesis procedures of lithium iron phosphate powders. J Power Sources 190:538–544
42. Huang H, Yin SC, Nazar LF (2001) Approaching theoretical capacity of $LiFePO_4$ at room temperature at high rates. Electrochem Solid State Lett 4:A170–A172
43. Kepler KD, Vaughey JT, Thackeray MM (1999) $LixCu_6Sn_5$ ($0 < x < 13$): an intermetallic insertion electrode for rechargeable lithium batteries. Electrochem Solid State Lett 2: 307–358
44. Bruce PG, Scrosati B, Tarascon JM (2008) Nanomaterials for rechargeable lithium batteries. Angew Chem Int Ed 47:2930–2946
45. Jiao F, Shaju KM, Bruce PG (2005) Synthesis of nanowire and mesoporous low-temperature $LiCoO_2$ by a post-templating reaction. Angew Chem Int Ed 44:6550–6553
46. Dong W, Rolison DR, Dunn B (2000) Electrochemical properties of high surface area vanadium oxides aerogels. Electrochem Solid State Lett 3:457–459
47. Flandrois S, Simon B (1999) Carbon materials for lithium–ion rechargeable batteries. Carbon 37:165–180
48. Bernal JD (1924) The structure of graphite. Proc R Soc Lond A 106:749–773
49. Matsumura Y, Wang S, Mondori J (1995) Interactions between disordered carbon and lithium in lithium ion rechargeable batteries. Carbon 33:1457–1462
50. Yata S, Kinoshita H, Komori M, Ando N, Kashiwamura T, Harada T, Tanaka K, Yamabe T (1994) Structure and properties of deeply Li-doped polyacenic semiconductor materials beyond C_6Li stage. Synth Met 62:153–158
51. Mori Y, Iriyama T, Hashimoto T, Yamazaki S, Kawakami F, Shiroki H, Yamabe T (1995) Lithium doping/undoping in disordered coke carbons. J Power Sources 56:205–208
52. Matsumura Y, Wang S, Mondori J (1995) Mechanism leading to irreversible capacity loss in Li ion rechargeable batteries. J Electrochem Soc 142:2914–2918
53. Che GL, Lakshmi BB, Fisher ER, Martin CR (1998) Carbon nanotubule membranes for electrochemical energy storage and production. Nature 393:346–349
54. Frackowiak E, Gautier S, Gaucher H, Bonnamy S, Beguin F (1999) Electrochemical storage of lithium in multiwalled carbon nanotubes. Carbon 37:61–69
55. Chen MH, Huang ZC, Wu GT, Zhu GM, You JK, Liu ZG (2003) Synthesis and characterization of SnO-carbon nanotube composite as anode material for lithium–ion batteries. MRS Bull 38:831–836
56. Chen WX, Lee JY, Liu ZL (2003) The nanocomposites of carbon nanotube with Sb and $SnSb_{0.5}$ as Li-ion battery anodes. Carbon 41:959–966
57. Yang ZH, Li ZF, Wu HQ, Simard B (2003) Effects of doped copper on electrochemical performance of the raw carbon nanotube anode. Mater Lett 57:3160–3166
58. Shodai T, Okada S, Tobishima S, Yamaki J (1996) Study of $Li_{3-x}M_xN$ (M = Co, Ni or Cu) system for use as anode material in lithium rechargeable cells. Solid State Ionics 86: 785–789
59. Rao BML, Francis RW, Christopher HA (1977) Lithium–aluminum electrode. J Electrochem Soc 124:1490–1492
60. Wang X, Nishina T, Uchida I (2002) Lithium alloy formation at bismuth thin layer electrode and its kinetics in propylene carbonate electrolyte. J Power Sources 104:90–96

61. Winter M, Besenhard JO (1999) Electrochemical lithiation of tin and tin-based intermetallics and composites. Electrochem Acta 45:31–50
62. Yang J, Winter M, Besenhard JO (1996) Small particle size multiphase Li-alloy anodes for lithium–ion batteries. Solid State Ionics 90:281–287
63. Mao O, Dunlap RA, Dahna JR (1999) Mechanically alloyed Sn–Fe (–C) powders as anode materials for li-ion batteries: I. The Sn_2Fe–C system. J Electrochem Soc 146:405–413
64. Poizot P, Laruelle S, Grugeon S, Dupont L, Tarascon JM (2000) Nano-sized transition-metal oxides as negative-electrode materials for lithium–ion batteries. Nature 407:496–499
65. Chouvin J, Vicente CP, Fourcade JO, Jumas JC, Simon B, Biensan P (2004) Deeper insight on the lithium reaction mechanism with amorphous tin composite oxides. Solid State Sci 6:39–46
66. Larcher D (2003) Effect of particle size on lithium intercalation into (-Fe_2O_3. J Electrochem Soc 150:A133–A139
67. Chen J, Xu L, Li W, Gou X (2005) (-Fe_2O_3 nanotubes in gas sensor and lithium–ion battery applications. Adv Mater 17:582–586
68. Aricòl AS, Bruce P, Scrosati B, Tarascon JM, Schalkwijk WV (2005) Nanostructured materials for advanced energy conversion and storage devices. Nat Mater 4:366–377
69. Yang Z, Choi D, Kerisit S, Rosso KM, Wang D, Zhang J, Graffa G, Liu J (2009) Nanostructures and lithium electrochemical reactivity of lithium titanites and titanium oxides: a review. J Power Sources 192:588–598
70. Wang D, Choi D, Yang Z, Viswanathan VV, Nie Z, Wang C, Song Y, Zhang J, Liu J (2008) Synthesis and Li-ion insertion properties of highly crystalline mesoporous rutile TiO_2. Chem Mater 20:3435–3442
71. Hu YS, Lorenz K, Guo YG, Maier J (2006) High lithium electroactivity of nanometer-sized rutile TiO_2. Adv Mater 18:1421–1426
72. Reddy MA, Pralong V, Varadaraju UV, Raveau B (2008) Crystallite size constraints on lithium insertion into brookite TiO_2. Electrochem Solid State Lett 11:A132–A134
73. Dimov N, Kugino S, Yoshio M (2004) Mixed silicon–graphite composites as anode material for lithium ion batteries—influence of preparation conditions on the properties of the material. J Power Sources 136:108–114
74. Wilson AM, Reimers JN, Fuller EW, Dahn JR (1994) Lithium insertion in pyrolyzed siloxane polymers. Solid State Ionics 74:249–254
75. Xie J, Cao GS, Zhao XB (2004) Electrochemical performances of Si-coated MCMB as anode material in lithium–ion cells. Mater Chem Phys 88:295–299
76. Wang CS, Wu GT, Zhang XB, Qi ZF, Li WZ (1998) Lithium insertion in carbon–silicon composite materials produced by mechanical milling. J Electrochem Soc 145:2751–2758
77. Wang GX, Ahn JH, Yao J, Bewlay S, Liu HK (2004) Nanostructured Si–C composite anodes for lithium–ion batteries. Electrochem Commun 6:689–692
78. Xue JS, Myrtle K, Dahn JR (1995) Epoxy-silane approach to prepare anode materials for rechargeable lithium ion batteries. J Electrochem Soc 142:2927–2935
79. Wilson AM, Way BM, Dahn JR, Van Buuren T (1995) Nanodispersed silicon in pregraphitic carbons. J Appl Phys 77:2363–2369
80. Holzapfel M, Buqa H, Scheifele W, Novák P, Petrat FM (2005) A new type of nano-sized silicon/carbon composite electrode for reversible lithium insertion. Chem Commun 12:1566–1568
81. Yang XQ, McBreen J, Yoon WS, Yoshio M, Wang H, Fukuda K, Umeno T (2002) Structural studies of the new carbon-coated silicon anode materials using synchrotron-based in situ XRD. Electrochem Commun 4:893–897
82. Kim JH, Kim H, Sohn HJ (2005) Addition of Cu for carbon coated Si-based composites as anode materials for lithium–ion batteries. Electrochem Commun 7:557–561
83. Kasavajjula US, Wang C (2005) Nano Si/G composite anode in Li ion battery for aerospace applications. Indian J Chem A 44:975–982
84. Liu CL (2005) Using mutual information for adaptive item comparison and student assessment. Educ Technol Soc 8:100–119

85. Ohara S, Suzuki J, Sekine K, Takamura T (2004) A thin film silicon anode for Li-ion batteries having a very large specific capacity and long cycle life. J Power Sources 136:303–306
86. Hatchard TD, Dahn JR (2004) Study of the electrochemical performance of sputtered $Si_{1-x}Sn_x$ films. J Electrochem Soc 151:A1628–A1635
87. Hatchard TD, Dahn JR (2005) Electrochemical reaction of the SiAg binary system with Li. J Electrochem Soc 152:A1445–A1451
88. Hatchard TD, Obrovac MN, Dahn JR (2005) Electrochemical reaction of the $Si_{1-x}Zn_x$ binary system with Li. J Electrochem Soc 152:A2335–A2344
89. Song SW, Striebel KA, Reade RP, Roberts GA, Cairns EJ (2003) Electrochemical studies of nanocrystalline Mg_2Si thin film electrodes prepared by pulsed laser deposition. J Electrochem Soc 150:A121–A127
90. Lee SJ, Baik HK, Lee SM (2003) An all-solid-state thin film battery using LISIPON electrolyte and Si-V negative electrode films. Electrochem Commun 5:32–35
91. Miyachi M, Yamamoto H, Kawai H, Ohta T, Shirakata M (2005) Analysis of SiO anodes for lithium–ion batteries. J Electrochem Soc 152:A2089–A2091
92. Kasavajjula U, Wang C, Appleby AJ (2007) Nano- and bulk-silicon-based insertion anodes for lithium–ion secondary cells. J Power Sources 163:1003–1039
93. Caïola A, Guy H, Sohm JCH (1970) Etude des systemes $Li/LiCH_3CO_2$, $Cu/Cu(CH_3CO_2)_2$ ET $Hg/Hg_2(CH_3CO_2)_2$ en milieu organique. Electrochim Acta 15:1733–1746
94. Tobishima S, Yamaji A (1983) Electrolytic characteristics of mixed solvent electrolytes for lithium secondary batteries. Electrochim Acta 28:1067–1072
95. Aurbach D (2000) Review of selected electrode–solution interactions which determine the performance of Li and Li ion batteries. J Power Sources 89:206–218
96. Xu K, Zhang S, Jow TR, Xu W, Angell CA (2002) LiBOB as salt for lithium–ion batteries. A possible solution for high temperature operation. Electrochem Solid State Lett 5: A26–A29
97. Dominey LA (1994) Current state of the art on lithium battery electrolytes. Ind Chem Libr 5:137–165
98. Galiński M, Lewandowski A, Stepniak I (2006) Ionic liquids as electrolytes. Electrochim Acta 51:5567–5580
99. Sakaebe H, Matsumoto H (2003) N-Methyl-N-propylpiperidinium bis (trifluoromethanesulfonyl) imide (PP13–TFSI)—novel electrolyte base for Li battery. Electrochem Commun 5:594–598
100. Hayashi K, Nemoto Y, Akuto K, Sakurai Y (2005) Alkylated imidazolium salt electrolyte for lithium cells. J Power Sources 146:689–692
101. Croce F, Appetecchi GB, Persi L, Scrosati B (1998) Nanocomposite polymer electrolytes for lithium batteries. Nature 394:456–458
102. DeLongchamp DM, Hammond PT (2004) Highly ion conductive poly (ethylene oxide)-based solid polymer electrolytes from hydrogen bonding layer-by-layer assembly. Langmuir 20:5403–5411
103. Thangadurai V, Weppner W (2006) Recent progress in solid oxide and lithium ion conducting electrolytes research. Ionics 12:81–92
104. Inaguma Y, Liquan C, Itoh M, Nakamura T, Uchida T, Ikuto H, Wakihara M (1993) High ionic conductivity in lithium lanthanum titanate. Solid State Commun 86:689–693
105. Shan YJ, Chen L, Inaguma Y, Itoh M, Nakamura T (1995) Oxide cathode with perovskite structure for rechargeable lithium batteries. J Power Sources 54:397–402
106. Xu X, Wen Z, Gu Z, Xu X, Lin Z (2004) Lithium ion conductive glass ceramics in the system $Li_{1.4}Al_{0.4}(Ge_{1-x}Ti_x)_{1.6}(PO_4)_3$ ($x = 0$–1.0). Solid State Ionics 171:207–213
107. Thangadurai V, Weppner W (2005) Investigations on electrical conductivity and chemical compatibility between fast lithium ion conducting garnet-like $Li_6BaLa_2Ta_2O_{12}$ and lithium battery cathodes. J Power Sources 142:339–344
108. Vereda F, Goldner RB, Haas TE (2002) Rapidly grown IBAD LiPON films with high Li-ion conductivity and electrochemical stability. Electrochem Solid State Lett 5:A239–A241

109. Bates JB, Dudney NJ, Gruzalzki GR, Zuhr RA, Choudhury A, Luck CF, Robertson JD (1992) Electrical properties of amorphous lithium electrolyte thin films. Solid State Ionics 53:647–654
110. Maekawa H, Matsuo M, Takamura H, Ando M, Noda Y, Karahashi T, Orimo S (2009) Halide-stabilized $LiBH_4$, a room-temperature lithium fast-ion conductor. J Am Chem Soc 131:894–895
111. Wang B, Bates JB, Hart FX, Sales BC, Zuhr RA, Robertson JD (1996) Characterization of thin-film rechargeable lithium batteries with lithium cobalt oxide cathodes. J Electrochem Soc 143:3203–3213
112. Patil A, Patil V, Shin DW, Choi JW, Paik DS, Yoon SJ (2008) Issue and challenges facing rechargeable thin film lithium batteries. Mater Res Bull 43:1913–1942
113. West WC, Whitacre JF, White V, Ratnakumar BV (2002) Fabrication and testing of all solid-state microscale lithium batteries for microspacecraft applications. J Micromech Microeng 12:58–62
114. Qingfeng L, Bjerrum NJ (2002) Aluminum as anode for energy storage and conversion: a review. J Power Sources 110:1–10
115. Abraham KM, Jiang Z (1996) A polymer electrolyte-based rechargeable lithium/oxygen battery. J Electrochem Soc 143:1–5
116. Read J (2002) Characterization of the lithium/oxygen organic electrolyte battery. J Electrochem Soc 149:A1190–A1195
117. Kuboki T, Okuyama T, Ohsaki T, Takami N (2005) Lithium-air batteries using hydrophobic room temperature ionic liquid electrolyte. J Power Sources 146:766–769
118. Xu JJ, Ye H (2005) Polymer gel electrolytes based on oligomeric polyether/cross-linked PMMA blends prepared via in situ polymerization. Electrochem Commun 7:829–835
119. Zhang T, Imanishi N, Hasegawa S, Hirano A, Xie J, Takeda Y, Yamamoto O, Sammes N (2008) Li/polymer electrolyte/water stable lithium-conducting glass ceramics composite for lithium–air secondary batteries with an aqueous electrolyte. J Electrochem Soc 155: A965–A969
120. Zhang T, Imanishi N, Shimonishi Y, Hirano A, Xie J, Takeda Y, Yamamoto O, Sammes N (2010) Stability of a water-stable lithium metal anode for a lithium–air battery with acetic acid–water solutions. J Electrochem Soc 157:A214–A218
121. Whittingham MS (1978) Chemistry of intercalation compounds: metal guests in chalcogenide hosts. Prog Solid State Chem 12:41–99
122. Ren Y (2009) Tailoring the pore size/wall thickness of mesoporous transition metal oxides. Microporous Mesoporous Mater 121:90–94
123. Joho F, Rykart B, Imhof R, Novák P, Spahr ME, Monnier A (1999) Key factors for the cycling stability of graphite intercalation electrodes for lithium–ion batteries. J Power Sources 81:243–247
124. Li JX, Wu CX, Guan LH (2009) Lithium insertion/extraction properties of nanocarbon materials. J Phys Chem C 113:18431–18435
125. Wang PC, Ding HP, Bark T, Chena CH (2007) Nanosized (-Fe_2O_3 and Li–Fe composite oxide electrodes for lithium–ion batteries. Electrochim Acta 52:6650–6655
126. Gao XP, Lan Y, Zhu HY, Liu JW, Ge YP, Wu F, Song DY (2004) Electrochemical performance of anatase nanotubes converted from protonated titanate hydrate nanotubes. Electrochem Solid State Lett 8:A26–A29

Computational Nanostructure Design for Hydrogen Storage

Jianjun Liu, James Tyrrell and Qingfeng Ge

Abstract Developing an efficient and affordable hydrogen-storage technology for on-board vehicular applications is a grand challenge to the success of a hydrogen economy. This challenge provides great opportunities for nanoscience and nano-engineering. Novel synthesis and characterization methods allow for an unprecedented degree of manipulation and tracking of the atomic structure in nanoassemblies. Furthermore, computational tools based on density functional theory, which combine fundamental predictive power with atomic resolution, provide a complementary and powerful means for the study and characterization of existing materials and prediction of new compounds and structural motifs, including those for hydrogen storage. In this chapter, we review the development of density functional theory-based computational studies of nanostructure design for hydrogen storage. Our emphasis is on complex metal hydrides. We also discuss the new developments in high surface area materials, including carbon-based materials, and metal- and covalent organic framework-based materials.

1 Introduction and Scope

In his lecture entitled "Our Energy Challenge" given at Southern Illinois University on April 3, 2005, the late Professor Richard E. Smalley identified "Energy" as No.1 on his list of Humanity's top ten problems for the next 50 years. Fossil fuels in the form of coal, oil, and natural gas have been the main source of energy since the industrial revolution. World energy consumption has been

J. Liu · J. Tyrrell · Q. Ge (✉)
Department of Chemistry and Biochemistry, Southern Illinois University,
Carbondale, IL 62901, USA
e-mail: qge@chem.siu.edu

L. Zang (ed.), *Energy Efficiency and Renewable Energy Through Nanotechnology*,
Green Energy and Technology, DOI: 10.1007/978-0-85729-638-2_22,

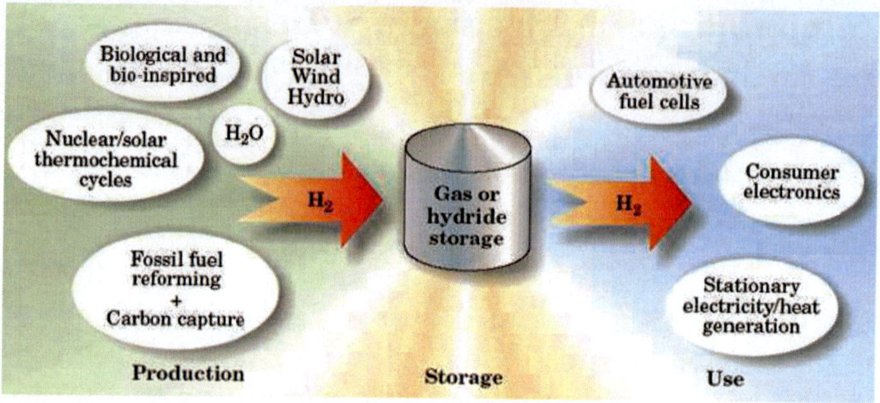

Fig. 1 The hydrogen economy as a network of primary energy sources linked to multiple end users through hydrogen as an energy carrier. Reprinted with permission from [3]. Copyright 2004, American Institute of Physics

constantly increasing since then. However, the supply of fossil fuel is limited. Furthermore, the use of fossil fuel puts human health at risk through chemical and particulate pollutants and affects the global climate through CO_2 and other greenhouse gas emissions. Professor Smalley suggested that *nanotechnology* "holds the answer to most of our pressing material needs", including hydrogen storage, a view shared by many scientists and engineers [1].

Hydrogen is a promising alternative to fossil fuel as an energy carrier. Similar to electricity, hydrogen needs to be produced using a different energy source, such as solar or nuclear. Once hydrogen is produced, a hydrogen-based energy system with a competitive fuel cell technology only produces water and, therefore, is pollution-free [2]. Crabtree et al. described the hydrogen economy as a network composed of three functional steps: production, storage, and use, as depicted in Fig. 1 [3]. These authors indicated that storing hydrogen in a high-energy–density form is a key element of the hydrogen economy, especially for on-board vehicular applications.

For on-board applications, the hydrogen storage systems are required to provide the needed quantity of hydrogen with acceptable volume, weight, cost, and safety risk compared to the current gasoline-driven combustion engine [4]. The available hydrogen storage modalities – including compressed gaseous hydrogen, cryogenic liquid hydrogen, and metal hydrides – each meet some, but not all, of the requirements for critical applications, as shown in Fig. 2 [5]. Therefore, developing a compact and efficient hydrogen-storage technology is the most technically challenging aspect of achieving a hydrogen economy.

Solid-state hydrogen storage, using nanoscience and nanotechnology, offers perhaps the best opportunity for meeting the requirements of on-board applications. Nanoscience and nanotechnology involve studying and working with matter on a nanometer scale. Nanomaterials are categorized as those that have structural components with at least one dimension smaller than 100 nm. Nanomaterials can

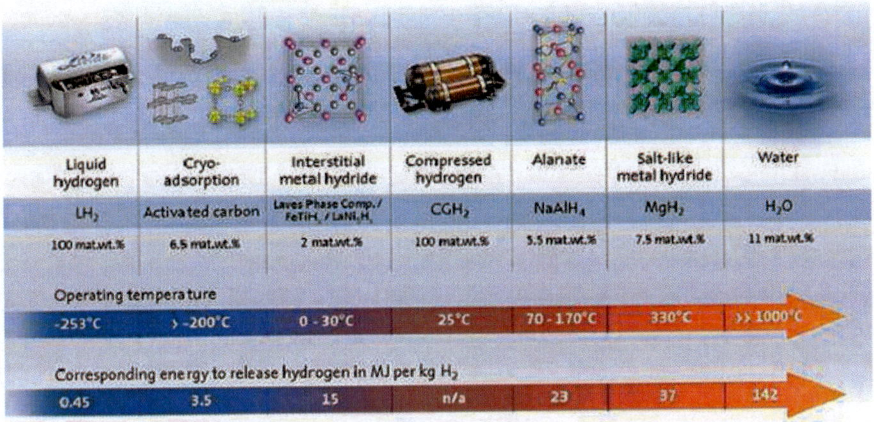

Fig. 2 Available hydrogen-storage technology and the corresponding operating conditions. Reprinted from [5], Copyright 2007, with permission from Elsevier

be in the form of thin films or surface coatings, nanowires and nanotubes, or nanoparticles. The unique properties of nanomaterials originate from the increased surface area and quantum effects, both of which relate to the small size. Consequently, the reactivity, strength and electrical characteristics of nanomaterials may be significantly different from those of conventional materials. For example, a particle of size 30 nm has only 5% of its atoms on its surface. When the size is reduced to 3 nm, the particle has 50% of its atoms exposed on the surface. This gives nanoparticles a much greater surface area per unit mass than larger particles. Both surface area and exposed atoms on the surface could be useful for storing hydrogen. Therefore, nanotechnology is expected to play a key role, as suggested by Smalley, in designing high-capacity solid-state hydrogen storage materials. In fact, the concept of nanoscience and nanoengineering has been actively exploited in improving existing hydrogen storage materials and searching for new hydrogen storage candidates. For example, nanocatalysts have been used to improve the kinetics of hydrogen uptake and release [6] and to improve hydrogen storage capacity through spillover [7]. Novel building blocks have been proposed to maximize the hydrogen capacity and optimize the strength of hydrogen binding [8]. Large-surface area nanomaterials that offer more host atoms/sites for hydrogen and allow easy access to these sites have been synthesized in materials such as metal-organic frameworks (MOFs) [9] and covalent organic frameworks (COFs) [10], and show promise as hydrogen storage media.

Depending on the nature of the hydrogen interaction with the storage media, solid-state hydrogen storage materials can be categorized into two groups: atomically bound hydrogen in either hydrides or other compounds (amines or imides) and molecularly adsorbed hydrogen in highly porous materials. The two groups can also be combined, e.g., using the high surface area material as a support/scaffold for hydrides or other hydrogen-containing compounds.

For all hydrogen storage materials, structural information is critical to understand their properties and performance. However, experimentally determining the structure of materials at the nanometer scale, especially for those materials involving hydrogen, is extremely challenging. Tracking the chemistry of hydrogen is even more difficult. On the other hand, computational chemistry and molecular simulation are advantageous for studies of well-defined systems, including those containing hydrogen. First-principles approaches developed on the basis of density functional theory (DFT) [11, 12] and implemented for both clusters and periodic slabs can handle systems significantly larger than those using traditional ab initio methods. The combined predictive power and atomic resolution provide a quantitative characterization of new materials. They have been applied to a wide range of problems of practical interest, including materials and structures related to hydrogen storage. Theory and computation can be used not only to understand experimental results, but also to guide the search for, and design of, new hydrogen-storage materials.

In the past few years, there have been a number of reviews on hydrogen storage materials with different foci [13–17]. In this chapter, we offer our perspective on the progress of research in hydrogen storage. We do not intend to provide a complete review of the literature, but rather to focus on nanostructure design, with emphasis on computation-based studies. The research is categorized into three parts: (i) nanosized effects on hydrogen storage; (ii) nanostructured materials for hydrogen storage; and (iii) nanobuilding-blocks for novel hydrogen storage materials, as detailed in the following sections.

2 Nanosized Effects in Hydrogen Storage

Metal hydrides, in particular complex hydrides, have been actively studied as hydrogen storage media. However, these hydrides suffer from the fact that hydrogen is held atomically through strong ionic or covalent bonds, making release of hydrogen occur only at high temperatures. Nanoengineering can be used to modify the thermodynamics and kinetics involved in hydrogen release and uptake. For example, ball-milling has been often used to mix the additives and hydrides and to reduce the size of the hydride particles, thereby, increasing the surface area and reducing the diffusion path. We will start by discussing complex metal hydrides and move on to other hydrides.

2.1 *Complex Metal Hydrides*

The alkali metal aluminohydrides and borohydrides are compounds belonging to a larger class of complex hydrides. In the past, these hydrides were known to liberate copious amounts of hydrogen either by direct thermal decomposition or hydrolysis [18]. The bonding characteristics of these complex hydrides determine that their dehydriding and hydriding are unfavorable either thermodynamically or kinetically under moderate conditions. As such, the processes were generally considered irreversible and not useful as reversible hydrogen storage materials. This bleak outlook persisted until Bogdanović and Schwickardi demonstrated that $NaAlH_4$ reversibly releases and re-adsorbs hydrogen under relatively mild conditions when doped with Ti-containing compounds [19]. The sodium alanate releases/uptakes hydrogen through a series of reversible decomposition/recombination reactions:

$$NaAlH_4 \rightleftarrows 1/3Na_3AlH_6 + 2/3Al + H_2 \quad (3.7\text{ wt }\%) \tag{R1}$$

$$1/3Na_3AlH_6 \rightleftarrows NaH + 1/3Al + 1/2H_2 \quad (1.9\text{ wt }\%) \tag{R2}$$

$$NaH \rightleftarrows Na + 1/2H_2 \quad (1.9\text{ wt }\%) \tag{R3}$$

The first two combined reactions give a theoretical hydrogen capacity of 5.6 wt% at low to medium temperatures (<250°C), which is considered practically reversible. The remaining 1.9 wt% of hydrogen released in the third step through the decomposition of NaH occurs at temperatures above 400°C, which is considered too high for most technical applications [20]. Clearly, the process couples a series of chemical reactions with solid-state phase transitions that require all elements to be mobile during the reactions. This mechanism is significantly different from that of hydrogen stored in conventional metal hydrides, in which hydrogen atoms are the only mobile species with the metal lattice providing the framework.

The structural change and energetics associated with the creation of the surfaces are expected to have strong effects on the thermodynamics of the complex hydrides and the kinetics of hydrogen release and uptake from either the bulk crystalline phase or nanosized particles. As such, the relative stability of various crystalline surfaces is critical to the size and shape of a particle. Furthermore, the stability of a surface under varying conditions is based on thermodynamics arguments. In the following, we will use $NaAlH_4$ to illustrate the thermodynamic analysis used to predict the shape of a particle. We also relate the surface stability to the ambient condition, including temperature and gas phase compositions.

2.1.1 Thermodynamic Analysis of Complex Hydrides

Considering $NaAlH_4$, a complete dehydrogenation to form Na and Al takes place through Na_3AlH_6 and NaH as reaction intermediates. Conceptually, the alanate consist of three regions: bulk alanates, a gas phase that is in contact with the solid, and an interfacial region with chemical composition $Na_xAl_yH_z$. To prevent $NaAlH_4$ from decomposing, the hydrogen partial pressure and temperature must be kept in a range where the alanate is thermodynamically stable, i.e., the hydrogen partial pressure is sufficiently high and the temperature sufficiently low. At thermodynamic equilibrium of the solid–gas interface, the chemical potentials of all species are equal in each region. For the Na and Al atoms in the system, the large region of alanate determines their chemical potential. The chemical potential of hydrogen is given by its value in the gas phase of pressure, P, and temperature, T. Therefore, the free energy, Ω, of an alanate surface under a hydrogen atmosphere can be written as:

$$\Omega = G - N_{Na}\mu_{Na} - N_{Al}\mu_{Al} - N_{H}\mu_{H}, \tag{1}$$

where G is the Gibbs free energy of the surface region, $N_{Na}\mu_{Na}$ and $N_{Al}\mu_{Al}$ are the Gibbs free energies of all Na and Al atoms in the surface region with the chemical potentials of Na and Al equal to those in bulk alanate, and $N_{H}\mu_{H}$ is the total Gibbs free energy of all hydrogen atoms in the surface region with a chemical potential equal to that in the gas phase. At a given temperature and pressure, the thermodynamically most stable system minimizes the surface free energy by adapting the stoichiometry of the surface region, i.e., by varying N_{Na}, N_{Al} and N_{H}. Conceptually, this is accomplished by an exchange of Na and Al atoms from the bulk and surface regions of the alanate and by exchanging H atoms between the solid and the gas phases.

For $P_{H} = 0$ and $T = 0$ with a slab model for the surface, $G \cong E$, Eq. 1 leads to the familiar expression for surface energies, namely:

$$\gamma = \frac{1}{2A}(E_{slab} - E_{bulk}) \tag{2}$$

where γ is the surface energy, E_{slab} is the total energy of the slab, E_{bulk} is the total energy of the bulk system with the same number of atoms as in the slab, and A is the surface area of the unit cell.

The relationship between Eqs. 1 and 2 can be established by the following steps: (i) approximating the Gibbs free energy, $G = E_{el} + E_{vib} + E_{other,internal} + PV - TS$, with the electronic energy, E_{el}, which is often the dominant term; (ii) replacing the surface region by a single slab with the stoichiometry of the bulk phase. The Gibbs free energy of the surface region is then approximately $G \cong E_{el,slab}$; with the stoichiometry given above and the constraint that the chemical potential of each species is equal to that of the species in the bulk, the term $N_{Na}\mu_{Na} + N_{Al}\mu_{Al} + N_{H}\mu_{H} = G_{bulk} \cong E_{el,bulk}$. Normalizing to the surface area, A, and taking into account the fact that a slab has two surfaces, Eq. 1 becomes Eq. 2.

The Gibbs free energy of the surface region is given by:

$$G = E + PV - TS \tag{3}$$

with E being the internal energy. It is assumed that the terms PV and TS are similar for different surface terminations and thus will cancel out. Furthermore, the internal energy is approximated by the total electronic energy of the surface region, as obtained from DFT calculations. This assumption implies that the phonon density of states of the solid does not depend strongly on the surface structure and composition. In particular, the zero-point energies of the various systems are assumed similar and thus can be ignored. Alternatively, these approximations can be corrected to the first-order using linear-response theory [21].

For $NaAlH_4$, the chemical potentials of Na, Al and H, μ_{Na}, μ_{Al} and μ_{H}, are linked through bulk alanate, which serves as a reservoir of Na and Al atoms:

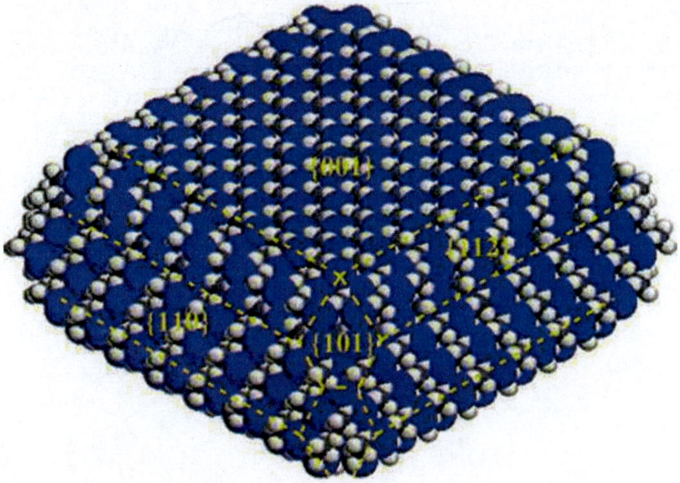

Fig. 3 The predicted equilibrium crystal shape of $NaAlH_4$ determined from the Wulff construction [22]. Reproduced by permission of the PCCP Owner Societies

$$\mu_{Na} + \mu_{Al} + \mu_H = \mu_{NaAlH_4}^{bulk}, \tag{4}$$

where $\mu_{NaAlH_4}^{bulk}$ is the chemical potential of bulk $NaAlH_4$. This term is approximated by the total electronic energy per formula unit of bulk $NaAlH_4$. With these assumptions and approximations, we obtain the gas-surface interfacial free energy:

$$\Omega = E_{slab} - (N_{Na} + N_{Al})\mu_{NaAlH_4}^{bulk} + (4N_{Na} - N_H)\mu_H \tag{5}$$

By normalizing to the surface area, we get the surface free energy:

$$\gamma = \frac{1}{2A}\Omega \tag{6}$$

Using the calculated surface energies for surfaces with different indices, Vegge [22] applied the Wulff construction [23] to predict the shape of a $NaAlH_4$ particle, as shown in Fig. 3. Similar methods were used by Kim et al. to predict the equilibrium shapes of particles for Sc, Ti and their hydrides [24].

The gas phase above the alanate surface contains molecular hydrogen, as well as other hydrogen-containing species. If the pressure is sufficiently low and the temperature sufficiently high, one can neglect intermolecular interactions and treat the pressure dependence using the ideal gas model. To describe the thermodynamic equilibrium of any dehydriding reaction step or the overall reaction, we use the van't Hoff equation to relate hydrogen partial pressure and the enthalpy of the reaction:

$$\ln\frac{P}{P_0} = -\frac{\Delta G^0}{RT} = -\frac{\Delta H^0}{RT} + \frac{\Delta S^0}{R} \tag{7}$$

For a dehydriding reaction that produces a gaseous product, ΔS is always positive, and was estimated to lie in a small range of 100–130 J $K^{-1}(mol\ H_2)^{-1}$ [25].

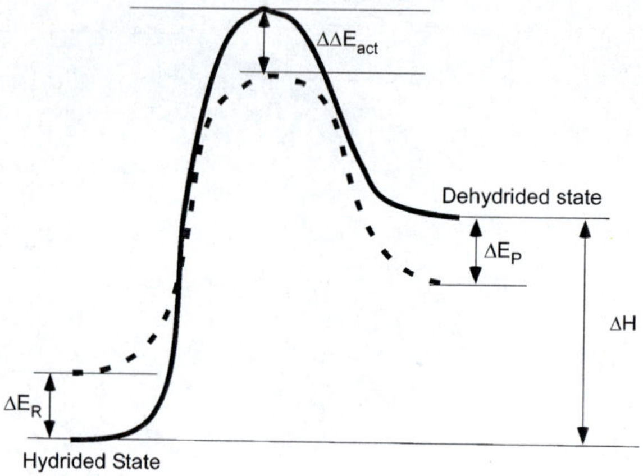

Fig. 4 Schematic potential energy profile for a hydrogen release reaction. The overall reaction enthalpy and activation energy can be manipulated by destabilizing/stabilizing the reactant and product states or adding catalysts

As a result, the equilibrium hydrogen pressure will depend on the enthalpy change of the dehydriding reactions. If the dehydriding reaction is exothermic, we would expect $P > P^0$ all the time, i.e., thermodynamically favoring hydrogen release. On the other hand, dehydriding is endothermic in most cases. As such, a relatively narrow window of 30–78 kJ (mol H_2)$^{-1}$ for ΔH^0 is expected for potential hydrogen storage materials to operate in the temperature range of 300 to 600 K [25]. The challenge is then to identify or design materials that will have a dehydriding/dehydrogenation enthalpy in the range of 30–78 kJ (mol H_2)$^{-1}$. The reaction enthalpy can be modified by using additives to change the hydrided/hydrogenated or dehydrided/dehydrogenated states involved in hydrogen release/uptake reactions. In the following sections, we will use examples to illustrate some of the strategies.

2.1.2 Modifying Sodium Aluminohydrides

The decomposition reaction steps R1 and R2 for releasing hydrogen are significantly endothermic, with the heats of reaction being 37 and 15.7 kJ, respectively [20]. Generally, dehydriding reactions are endothermic and follow a schematic potential energy profile corresponding to the solid line shown in Fig. 4. The hydride releases its hydrogen by going through a transition state with a substantial barrier. There are three states involved in the reaction that can be modified to promote hydrogen release. Based on the potential energy profile, destabilizing the reactant state and/or stabilizing the product state will make the reaction less endothermic, i.e., reducing ΔH, and thereby energetically favor the dehydrided state. The reduction of endothermicity will shift the reaction toward the product state. On the other hand, to improve the kinetics of the reaction, one can introduce catalysts to the system or reduce the size of the hydride particles to lower the barrier of the reaction, as we will discuss in the following sections.

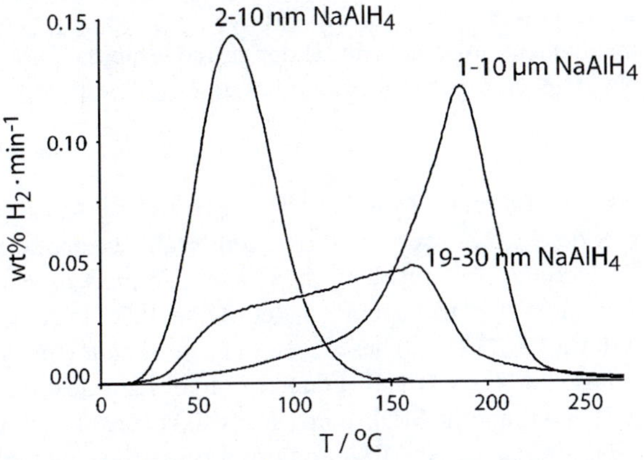

Fig. 5 Temperature programmed desorption profile of H_2 for $NaAlH_4$ supported on carbon nanofiber. Reprinted with permission from [36]. Copyright 2008, American Chemical Society

One or a combination of these approaches may be applied to many hydrides to reduce the hydrogen release temperature and improve the kinetics of the reactions.

For complex metal hydrides, attempts have been made to reduce ΔH by introducing a second cation. Chemically, the decomposition enthalpy of hydrides can be reduced by substituting a cation with a metal of larger ionization potential (IP). The two cations mixed in one hydride are expected to function synergistically to maintain reasonable stability, and at the same time provide a favorable decomposition enthalpy. Extensive DFT calculations showed that bialkali hexahydrides, such as K_2LiAlH_6, K_2NaAlH_6, KNa_2AlH_6, and $LiNa_2AlH_6$, are stable compared to the pure alanates [26, 27]. In fact, $LiNa_2AlH_6$ was synthesized experimentally [28, 29]. Mixed aluminohydrides such as $LiMg(AlH_4)_3$ and $LiMgAlH_6$ were also predicted based on DFT studies, and synthesized and characterized experimentally [30, 31]. Although their overall hydrogen storage performance was not fully examined, some of these compounds exhibit favorable decomposition temperatures.

An alternative approach to modify sodium alanate is to reduce the particle size to the low nanometer regime and maintain the size during hydrogen cycling. In fact, nanosizing effects have been widely exploited in metal hydrides to improve their performances [32–34]. This approach was recently applied to $NaAlH_4$ by supporting $NaAlH_4$ on carbon nanofibers with controlled size ranges: 1–10 μm, 19–30 nm, and 2–10 nm [35, 36]. The activation energies for hydrogen desorption decreased with decreasing particle size, from 116 kJ mol^{-1}, to 80 kJ mol^{-1} and to 58 kJ mol^{-1}, respectively [36]. Figure 5 shows a clear decrease in hydrogen desorption temperature as the size of the particles is decreased. The authors attributed the decrease in activation energies for hydrogen desorption to the expected higher fraction of more open planes, as well as corners and edges exposed on a nanoparticle. The authors also suggested that the rate-limiting step for the nano-$NaAlH_4$ is hydrogen desorption from the particle, in contrast to the case of the bulk samples. Recent studies from the same group showed that the decomposition of the nano-confined $NaAlH_4$ does not go through the Na_3AlH_6

intermediate state [37, 38]. These latest results confirm the Na_3AlH_6 formation mechanism suggested by Balema and Balema [39]. These authors proposed that Na_3AlH_6 is a product of a solid-state reaction:

$$NaAlH_4 + 2NaH \rightleftarrows Na_3AlH_6 \quad (R4)$$

instead of directly from $NaAlH_4$ by liberating Al and H_2. The fact that no Na_3AlH_6 was detected suggests that reaction R4 is not operative in nano-confined $NaAlH_4$. This could be a result of rapid consumption of Na by other reactions, such as oxidation. Other support materials include mesoporous silica [40] and MOFs (HKUST-1) [41]. These materials were shown to promote the formation of nanoscale $NaAlH_4$ and reduce the hydrogen desorption temperature. In fact, $NaAlH_4$ confined in MOF pores starts to release H_2 at a temperature as low as 70°C [41], similar to $NaAlH_4$ supported on carbon nanofibers [36]. This may suggest that the particles in the two systems have similar sizes.

2.1.3 Other Complex Hydrides

$LiBH_4$ is considered as an attractive candidate for hydrogen storage due to its intrinsically high gravimetric and volumetric hydrogen capacities (18.2 wt%, 121 kg/m^3) and has attracted much interest [42–49]. Unfortunately, the B–H bond in the pure $LiBH_4$ material is extremely strong and only liberates 2% of the hydrogen around the melting point (541–559 K) [42, 43]. Starting from $LiBH_4$, the partial decomposition to $LiH(s) + B(s) + 3/2H_2(g)$ has the standard enthalpy of 100.3 kJ/mol. The highly endothermic decomposition reaction indicates hydrogen release from $LiBH_4$ must occur at elevated temperatures. The experimental results of Züttel et. al showed a significant hydrogen desorption peak started at 673 K and reached its maximum value around 773 K [42, 43]. Nickels et al. attempted to tune the hydrogen desorption temperature by introducing K to $LiBH_4$ and successfully synthesized mixed alkali metal borohydride ($KLi(BH_4)_2$). The new compound has a similar structure to $LiBH_4$ but decomposes at a lower temperature [50].

Another complex hydride similar to $NaAlH_4$ but having an even higher intrinsic hydrogen capacity is $LiAlH_4$. The decomposition of $LiAlH_4$ is believed to undergo similar steps to $NaAlH_4$. The first decomposition step from tetrahedral $LiAlH_4$ to octahedral Li_3AlH_6 is weakly endothermic [51, 52]. The second decomposition reaction from octahedral Li_3AlH_6 to LiH and Al phase was found to be endothermic with ΔH of 25 kJ/mol of H_2. Its dehydriding was observed to occur at 228–282°C, likely due to kinetic limiting steps. Apparently, the decomposition temperature is too high for practical purposes. The decomposition of $LiAlH_4$ is very slow without a catalyst. Adding catalysts results in an enhanced kinetics [53–56].

By introducing the cation of a metal with a high ionization potential, the hydride/hydrogenated state will destabilize, making the enthalpy of the hydrogen release reaction favorable. Another way to tune the overall reaction enthalpy is achieved by incorporating a second species into the reaction to stabilize the product state, which will also make the overall reaction enthalpy shift toward hydrogen release. Extensive DFT computations have been performed to assess a large number of possible destabilized metal hydrides [25, 57–60]. By assessing the enthalpies of all possible reactions, more than 300 destabilization reactions are predicted to have favorable reaction enthalpies [25]. Wolverton et al. proposed several guidelines to thermodynamically destabilize the metal hydrides in order to design novel hydrogen storage materials [61]. Basically, the enthalpy of the proposed destabilized reaction must be less than the decomposition enthalpies of the individual reactant phases. In addition, if the proposed reaction involves a reactant that can absorb hydrogen, the formation enthalpy of the corresponding hydride cannot be greater in magnitude than the enthalpy of the destabilized reaction.

Vajo et al. examined this strategy by altering the thermodynamics and kinetics of several metal hydrides [62]. The equilibrium hydrogen pressure and reaction enthalpies change with additives that form new alloys or compound phases upon dehydriding. The formation of new phases lowers the energy of the dehydrided state and efficiently destabilizes the component hydrides. A series of experimental explorations have been performed to destabilize the reaction products of $LiBH_4$ and reduce the dehydriding temperatures [49, 62–72].

2.1.4 Structure and Effect of Doped Transition Metal in $NaAlH_4$

In heterogeneous catalysis, nanoparticles have been used in many industrially and environmentally important reactions. As such, they represent one of the earlier applications of nanotechnology before nanoscience and nanotechnology were formally established [73]. Catalysts have been widely exploited in hydrogen storage to improve the kinetics of hydrogen release and uptake in complex hydrides and metal hydrides, following the pioneering work of Bogdanović and Schwickardi [19]. A great number of studies have been devoted to experimentally characterizing alanates, as well as developing new dopants as catalysts to improve the kinetics and cyclability [15, 20, 40, 55, 74–112]. Attempts have been made to elucidate the effect of Ti-based dopants on the structure and hydrogen-storage characteristics. Early results showed that the rates of hydrogen absorption and desorption strongly depend on the level of catalyst doping; the total reversible hydrogen capacity decreases as the Ti-halide doping level is increased. Many types of dopants result in enhanced hydrogen release/uptake kinetics in alanates. Although great progress has been made in characterizing the state of Ti in dehydrided and hydrided materials, an overall understanding of the role of Ti additives in hydrogen release and uptake has still not been achieved.

Many studies indicate that a surface-localized species consisting of a nascent binary phase Ti–Al alloy formed during cyclic dehydriding and rehydriding processes [113–115]. The alloy was investigated in amorphous form by an X-ray diffraction and absorption study [113], as shown Fig. 6a. Combined TEM-EDX and XAFS measurement determined that Ti was atomically dispersed on an Al phase and forms an Al–Ti alloy on the surface of the Al particle, separating from

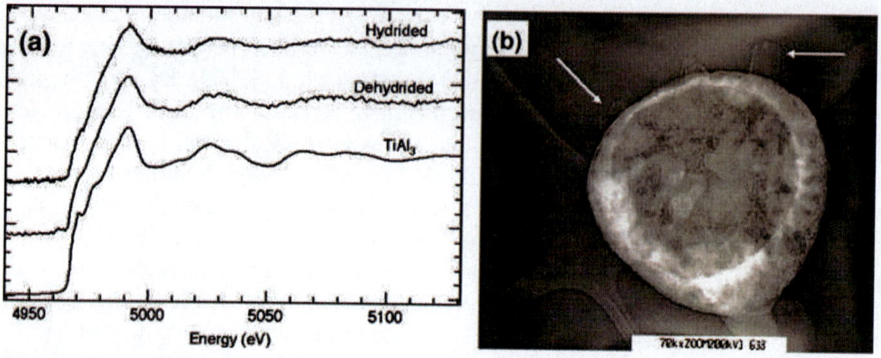

Fig. 6 **a** Ti K edges from $TiAl_3$ and 2 mol % Ti-doped sodium alanate in hydrided and dehydrided states. Reprinted with permission from [113]. Copyright 2004, American Institute of Physics; **b** TEM image of a dehydrogenated titanium doped $NaAlH_4$ sample [114]. Reproduced by permission of The Royal Society of Chemistry

the NaH phase, as shown in Fig. 6b [114, 115]. Based on the thermodynamic stability, $TiAl_3$ is the most likely form after dehydriding Ti-doped $NaAlH_4$. The local structure of these species had Ti–Al and Ti–Ti bond distances of 2.79 and 3.88 Å, respectively. However, directly doped $TiAl_3$ alloy in $NaAlH_4$ was found to be substantially less effective than the Ti halides [116].

Many experimental measurements established that highly dispersed Ti in a predominantly Al phase plays an important role in hydrogen uptake and release processes [78, 81, 94, 117]. On the other hand, there is limited experimental data to support the suggestion that the Ti hydrides formed in the process will catalytically enhance dehydriding and rehydriding processes [39, 83, 86, 118]. Although extensive experimental studies have been performed on the Ti-doped $NaAlH_4$ system, the mechanism by which the $NaAlH_4$ system is activated in the presence of a small amount of Ti is still not well understood.

First of all, the Ti is unlikely to simply act as a classic catalyst assisting the conversion between hydrogen atoms and molecules at the solid-state surface. There is a strong thermodynamic driving force for Ti to form a hydride in the presence of hydrogen. Therefore, Ti was proposed to play a role in improving the transport of mobile species, such as AlH_3, to the catalytic site and further make AlH_3 dissociate into Al and H_2 [119, 120], in addition to the enhancement of bond-breaking and bond-forming processes. On the other hand, both hydriding and dehydriding involve complex phase transitions of NaH and Al $\leftrightarrow$ Na_3AlH_6 $\leftrightarrow$ $NaAlH_4$. Different local structures and/or chemical states of Ti may accompany the transformation process. However to accurately determine the structures and location of Ti in such a complicated system and follow it during the hydriding/dehydriding process is extremely challenging to the available experimental techniques. In this aspect, DFT-based first-principles methods are advantageous.

Several theoretical studies have been performed to focus on substitution of Ti for Al and Na atoms in Ti-doped $NaAlH_4$ bulk and surfaces. Substitution of Ti for Al was shown theoretically to be the preferred location in bulk $NaAlH_4$ [121–123]. Íñiguez et al. studied the structure, energetics, and dynamics of pure and Ti-doped sodium alanate ($NaAlH_4$), focusing on the possibility of substitutional Ti doping in the bulk [124]. These authors optimized several possible structures for the substitution of Al and Na by Ti, accompanied by neighboring vacancies using the DFT plane-wave method implemented in CASTEP. Their results showed that doped Ti prefers to substitute for Na and further attract surrounding hydrogen atoms, softening and/or breaking the Al–H bonds. The same group of authors extended their study to determine the location of Ti [125]. These later results showed that Ti prefers to be on the surface, substituting for Na, and attracting a large number of H atoms to its vicinity. They predicted that a $TiAl_n$ ($n > 1$) structure might form on the surface of the sodium alanate. On the other hand, this process would cost an energy of 1.7 eV and is, therefore, not likely to happen. However, Løvvik et al. also suggested that substitution of Ti in bulk $NaAlH_4$ is less favorable than that near surface or defect positions. On the $NaAlH_4$ (001) surface, DFT calculations by Yildirim and Íñiguez showed substitution of Ti for Na is the preferred site [125] whereas Løvvik and Opalka found substitution of Ti for Al is more favorable [122, 123]. The difference was attributed to the different reference states used in energy calculations. Although thermodynamic stability of local structure can determine the possible sites of Ti, the improvement of kinetics remains unclear.

Doping with $TiCl_3$ was believed to result in vacancies in $NaAlH_4$ and was simulated by generating Na-vacancies next to Ti [126, 127]. The calculated results show that Na-vacancies play a larger role in reducing hydrogen desorption energies. Na-vacancies lead to a lower hydrogen desorption energy (0.4 eV/H_2) than 0.9 eV/H_2 in the substitutional model. In addition, Vegge's calculations revealed that Na-vacancy formation energies were significantly reduced when Ti is doped in $NaAlH_4$ surfaces [22].

We approached the problem based on a surface model of $NaAlH_4$ and proposed an alternative mechanism based on the surface interstitial complex structure that we identified in Ti-doped $NaAlH_4$ (001) and (100) surfaces [128–131]. This is different from many other computational studies that adopted bulk $NaAlH_4$ models and focused primarily on substitutional modes of doping. The resulting local structures after doping the surface interstitial sites in both surfaces are very similar [128, 129], as shown in Fig. 7. Our results show that the hydrogen desorption energies from many positions of $TiAl_3H_x$ are reduced considerably compared with that from the corresponding clean, undoped $NaAlH_4$ surfaces. The almost invariant local environment surrounding Ti during dehydriding makes the $TiAl_3H_x$ complex a precursor state for the formation of experimentally observed $TiAl_3$ after dehydriding Ti-doped $NaAlH_4$ [113]. Furthermore, we showed that the $TiAl_3H_{12}$ complex has an extended effect beyond locally reducing the hydrogen desorption energy. It also facilitates hydrogen desorption at a reduced desorption energy by either transferring the hydrogen to $TiAl_3H_x$ or by reducing the hydrogen

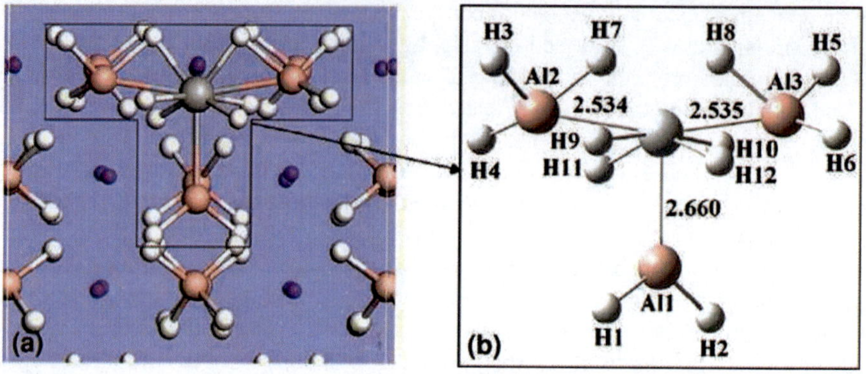

Fig. 7 **a** DFT-GGA relaxed structure of Ti-doped $NaAlH_4(001)$ with Ti in the surface interstitial site. **b** Detailed local structure of the $TiAl_3H_{12}$ complex shown in (**a**). Reproduced from [128]

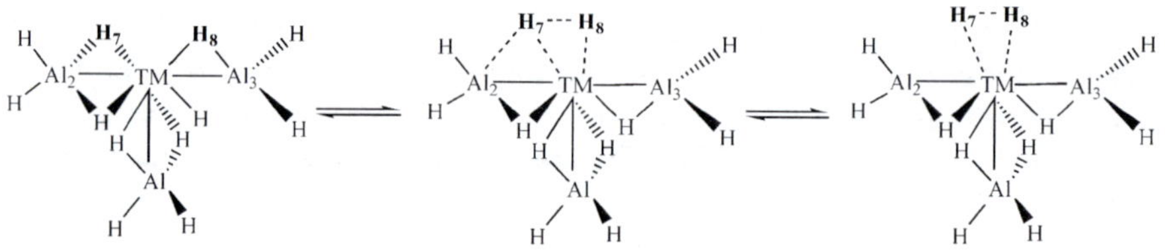

Fig. 8 A proposed ligand exchange mechanism involved in dehydriding and hydriding of transition metal doped $NaAlH_4$. Reproduced from [131]

desorption energy in neighboring AlH_4^- by linking these AlH_4^- units with the complex structure. Our predicted interstitial $TiAl_3H_x$ structure was supported by a recent combined Ti K-edge EXAFS, Ti K-edge XANES, and XRD study of $TiCl_3$-doped $NaAlH_4$ by Baldé et al. [78]. These authors observed that the interstitial structure accounts for more than 70% of all Ti doped in $NaAlH_4$.

The general consensus of the DFT studies of Ti-doped $NaAlH_4$ is that Ti attracts surrounding H atoms, accompanied by breaking the Al–H bonds and forming Ti–H bonds [125, 128, 129]. During the recharging state, transition metals should assist in breaking the molecular hydrogen bond but not prevent the hydrogen atoms from migrating to Al atoms. Based on extensive computational studies, we proposed that a dihydrogen complex is formed in both dehydriding and hydriding processes. In fact, both dehydriding and hydriding can be viewed as ligand exchange between H–H and Al–H, as shown in the scheme in Fig. 8 [131]. Based on stability and calculated hydrogen desorption energy from the $TMAl_3H_{12}$ structures, we concluded that the 18-electron rule is applicable in understanding the variation of binding strength between active site and ligands (H–H and Al–H bonds). Our analysis of electronic structure revealed that electron transfer from d-orbitals of TM to σ^* antibonding bonds plays an important role in hydrogen release and uptake of $NaAlH_4$ [131]. This interaction mechanism between TM and

$NaAlH_4$ can be ascribed to Kubas-type interaction [132], i.e., metal coordinating with σ-bonds to form complexes. In fact, electron transfer from d-orbitals to σ^* bonds of H_2 was also found in hydriding of Ti-doped Al surface structures according to this interaction mechanism [133, 134]. Very recently, Ljubić and Clary confirmed this model by DFT and CCSD(T) calculations for Ti(Sc)-2Na-AlH_4 clusters [135]. They showed that the barrier (10–20 kJ/mol) for H_2 release is reduced considerably with respect to that ($\sim$120 kJ/mol) of the pure $NaAlH_4$.

Based on the above analysis, we can predict the effective catalysts that will improve the performance of complex metal hydrides. In the case of $NaAlH_4$, the early TMs may be more effective than the late TMs in promoting hydrogen release and uptake. In fact, many transition metals have been assessed for their activity in dehydriding and rehydriding of $NaAlH_4$, and among them, Sc, Ti, Zr, Ce, and Pr were shown to be effective [84, 117, 136]. Furthermore, doped transition metal cations reduce to a lower oxidation state, which is favorable for an electron donation from the occupied σ orbitals of Al–H and H–H to transition metal d-orbitals. A transition metal in a higher oxidation state has a higher formal charge and favors the electrostatic interaction between the metal and σ bonding molecules. In experiments, many different precursors, $TiCl_3$, $Ti(OBu^n)_4$, $TiCl_4$, $TiBr_4$, and TiF_3, have been used to examine the catalytic effect [98]. These extensive studies have identified that zero-valence Ti particles are dispersed in the Al phase after hydrogen cycling. Thus, reduction of Ti is expected to occur during the initial doping process such as ball-milling.

In contrast, a different interaction mechanism was found to be operative in Ti-doped $LiBH_4$ surfaces [47]. Although the most stable structures in Ti-doped $LiBH_4$ are similar to those in Ti-doped $NaAlH_4$, the bond-activation by doping Ti into $LiBH_4$ is much less dramatic than in $NaAlH_4$. Our analysis showed that the bond-activation difference can be attributed to the inability of back-donation of electrons from the d-orbitals of Ti to antibonding orbitals of B–H bond. Based on the analysis of structural stability, the symmetry-adapted occupied orbital overlap between d-orbitals and B–H bonding orbitals was found to play an important role in stabilizing structures. This structure promotes hydrogen desorption locally but does not show an extended effect, as in the case of Ti-doped $NaAlH_4$. Consequently, the amount of Ti needed to achieve a significant reduction of hydrogen desorption energy in $LiBH_4$ would be significantly higher than in $NaAlH_4$, consistent with the experimental observations [66, 68, 137].

In summary, the understanding of the interaction mechanism between catalyst and host material will help screen the catalysts that improve performance of the complex hydrides as hydrogen storage media. Clearly, a complicated mechanism is involved in the heterogeneous processes because of phase transition and mass transport. Both experimental measurements and theoretical studies on the catalytic effect of Ti involved in these dynamic processes have been challenging. A recent hydrogen–deuterium scrambling experiments showed that the mass transport of heavy species (NaH or AlH_3) may be rate-determining steps in dehydriding $NaAlH_4$ [85]. Using first-principles DFT simulations, Gunaydin et al. identified that the Al mass transport mediated by AlH_3 vacancy has the lowest barrier of mass transport, but still is higher than that for the bond-breaking steps [138]. However, the effect of the doped transition metal in mass transport and phase transition remains unclear. A first-principles dynamics simulation may be needed to address the questions involving bond-breaking and making, mass transport, and phase transition.

2.2 Magnesium Hydrides

Magnesium is an attractive and promising material for hydrogen storage applications due to its high hydrogen capacity (7.7 wt%), low cost, and being rich in natural abundance. However, three major problems hamper its commercial usage as a hydrogen storage material. First, the decomposition energy is high, at 75 kJ/mol H_2 [139, 140], which corresponds theoretically to an equilibrium temperature with 1 bar H_2 pressure of 288°C. Kinetic limitations push the desorption temperature to at least 300°C to reach 1 bar H_2 pressure. Second, the hydriding and dehydriding rates are very slow at room temperature, mainly due to the slow diffusion of hydrogen atoms through the hydrides. Third, the hydrogen molecule does not readily dissociate on the Mg surface to generate hydrides for recharging. In fact, these phenomena are common for the main group metals, as reflected by the fact that the enthalpies of formation of the hydrides are usually very large. Catalysts can be added to assist breaking the H–H bond for the recharging phase. The hydrogen atoms generated at the catalytic center will be transferred to Mg sites through hydrogen spillover. Both theoretical and experimental studies have been performed to explore the effectiveness of hydrogen spillover to recharge Mg using a palladium-based catalyst [141–144].

During the past two decades, many experimental methods have been developed to lower the desorption temperature and increase the rate of sorption/desorption. An example is the formation of the complex hydride of Mg_2NiH_4 [145], which resulted in a lowered sorption temperature of ~200°C. However, the hydrogen capacity is significantly reduced, to 3.6 wt%, through this alloying process.

Nanostructured materials have been exploited to improve hydrogen sorption properties from magnesium hydrides. Hydrogen sorption properties of magnesium were greatly improved by combining the material with nanostructures through chemical/physical synthesis [146–149] or high-energy ball-milling with additives [150–156]. The improved sorption kinetics can be ascribed to the increase of the specific surface area, the decrease of diffusion path lengths, and change of electronic structure properties. Despite many improvements, hydrogen desorption temperature remains too high for practical applications.

The size effect of nanoparticles on the hydrogen interaction in MgH_2 was analyzed by Wagemans et al. using ab initio Hartree–Fock (HF) and DFT calculations [157]. These authors found that both magnesium and magnesium hydride become less stable with decreasing cluster size, notably for clusters smaller than 20 magnesium atoms. However, magnesium hydride destabilizes more strongly

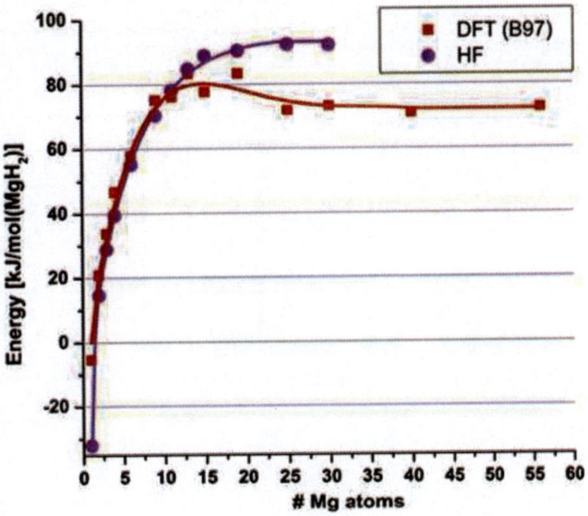

Fig. 9 Calculated average hydrogen desorption energy from Mg_nH_{2n} for clusters with different number of Mg atoms. Desorption energy was normalized to MgH_2. Reprinted with permission from [157]. Copyright 2005 American Chemical Society

than magnesium. As a result, the hydrogen desorption energy decreases significantly when the crystal grain size becomes smaller than 1.3 nm, as shown in Fig. 9. For instance, an MgH_2 crystallite size of 0.9 nm corresponds to a desorption temperature of only 200°C. This predicted decrease of the hydrogen desorption temperature is an important step toward the application of Mg as a hydrogen storage material. Cheung et al. simulated the hydrogen desorption/sorption processes using a reactive force field method and focused on the size dependency of MgH_2 and Mg nanoparticles [158]. They predicted that the heat of MgH_2 formation is very close to that of the bulk structure for clusters larger than 2 nm. It is expected that an ionic bond exists between Mg and H atoms in MgH_2 structures due to the large difference (0.7) of their electronegativities. A MgH_2 nanostructured particle exposes its highly reactive edges and corners on its surface. Therefore, fabricating and maintaining a small size nanoparticle (less than 2 nm) in hydrogen storage applications using MgH_2 is very challenging. Mechanical milling of MgH_2 leads to particles with sizes of 20–300 nm [159]. Consequently, the MgH_2 particles produced by ball-milling are not expected to show any significant difference in thermodynamic properties from the bulk crystalline particles. Small size (2–5 nm or smaller) of the Mg crystallites have been synthesized in nanoporous carbon by infiltration [147].

A DFT calculation of the electronic structure and total energies of MgH_2 nanoclusters that are interacting with transition metal dopants has been carried out to understand the catalytic effect of transition metals (Ti, V, Fe, and Ni) in dehydriding of MgH_2 nanoclusters by Larsson et al. [160]. The authors reported both the effect of doping transition metal on average and single-site desorption energies. Their results showed that the doped transition metal did not change the

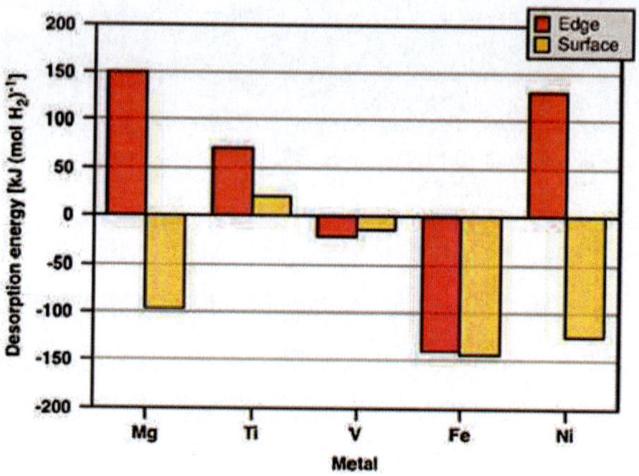

Fig. 10 Calculated single-site hydrogen desorption energies for $Mg_{30}MH_{62}$ clusters at two different surface sites. Reprinted with permission from [160]. Copyright 2008, National Academy of Sciences, U.S.A

average desorption significantly but has a dramatic effect on the single-site desorption energy, as shown in Fig. 10, which reveals the local effects of the transition metal catalysts. In MgH_2 nanoclusters, the removal of hydrogen bound to edge-site Mg atoms costs energy, whereas the process is exothermic for the surface Mg atoms. This is consistent with previous studies that nanostructuring can by itself lower the temperature required for the onset of hydrogen release [157, 158, 161–163], even without the addition of catalysts. Most strikingly, doping Fe or Ni led to the negative removal energies of H atoms from both of the surface sites. The authors argued that this kind of exothermic dissociation might lower the minimum temperature needed to initiate hydrogen desorption, especially because an application of the Hammond–Leffler postulate suggests that the corresponding activation energies will be lower as well. They propose that the transition metal, in particular, Fe, is mobile and remains at the Mg/MgH_2 interface to catalyze de-hydriding. On the other hand, searching for a method to stabilize the nanosized magnesium hydrides particles as well as highly active catalysts remains a challenge for implementing magnesium hydrides in practical hydrogen storage.

3 Nanostructured Materials for Hydrogen Storage

High surface porous materials have been used as gas separation and storage materials for a long time. Carbon-based nanostructures, including nanotubes and amorphous carbon have been widely studied for their hydrogen storage properties [14, 164–166]. Metal–organic frameworks (MOFs) are a novel class of nanoporous materials that are built with metal ion clusters linked by rigid organic linkers. The potential of using such materials as hydrogen storage media was first demonstrated by Rosi et al. [9]. Recently, a new family of three-dimensional covalent organic frameworks (COFs) were synthesized and explored for hydrogen storage [167].

Both MOFs and COFs have shown exceptionally high hydrogen capacities because of their large surface areas and porous structures [168, 169]. In COFs, the organic building units are held together by strong covalent bonds (C–C, C–O, B–O, and Si–C), which replace the heavy metal ions in MOFs materials. Therefore, COFs-based materials have not only a larger gravimetric capacity of hydrogen, but also show a high thermal and architectural stability. Hydrogen stored in this type of material remains molecular. According to the thermodynamic analysis by Bathia and Myers, a heat of adsorption of 15.1 kJ/mol is required for an adsorbent to remain attractive to hydrogen at ambient temperature [170]. On the other hand, MOFs and COFs have a weak binding toward H_2 (4–7 kJ/mol) [167, 171, 172]. Hydrogen storage in these materials with significant capacity can only occur at cryogenic temperatures ($\sim$77 K).

High surface area and pore volume form the basis of MOFs and COFs materials as hydrogen storage media. The working temperature is limited by their weak interaction with dihydrogen. On the other hand, the flexibility and diversity of MOFs' and COFs' structures allow some improvements for hydrogen storage at ambient conditions by increasing the affinity of dihydrogen toward host materials. In elastic neutron scattering experiments and grand canonical Monte Carlo (GCMC) simulations suggest that the open metal clusters within MOF are the preferred binding sites at low pressures while organic linkers play a more crucial role at higher pressures [9, 173]. This was attributed to the fact that, although the absorbent sites in the organic linker have lower binding energies, a much greater number of such sites produce an additive effect and resulted in increased H_2 loading [174]. Consequently, many studies have been reported to enhance H_2 binding strength with MOFs and COFs by modifying the metal ions or organic linkers.

The nature of dihydrogen interaction with the metal ions in MOF materials can be understood by examining similar metal-modified carbon-based materials. In the latter case, the metal cation interacts with H_2 through electrostatic interaction and possibly electron transfer as described in the Dewar, Chatt, and Duncanson (DCD) model [175]. For early transition metals, electron transfer is expected to be the main contributor to H_2 interaction with host sites [176]. The main group metals and late transition metals rely on electrostatic interaction to bind H_2 [177]. The metal ions in MOFs are expected to behave similarly to the metal ions in the metal-modified carbon-based materials. However, the local environment of metal ions may lead to a different situation from the metal-modified carbon-based materials. Monte Carlo simulations were performed for hydrogen interaction in the MOF structure of $[In_3O(C_{16}N_2O_8H_6)_{1.5}]NO_3$ [178]. In this case, hydrogen interacts with the MOF through three principle attractive forces: van der Waals, electrostatic, and induction. The DFT study of Zhou and Yildirim on the binding of H_2 on Mn cation in Mn_4Cl-MOF structure (Fig. 11) showed that the major contribution to the overall binding arises from electrostatic effects [179]. The interaction strength was also found to depend on the local environment, including the spin-state of the metal ion and H_2 orientation. Obviously, a single system cannot represent all possible hydrogen interaction mechanisms in MOFs.

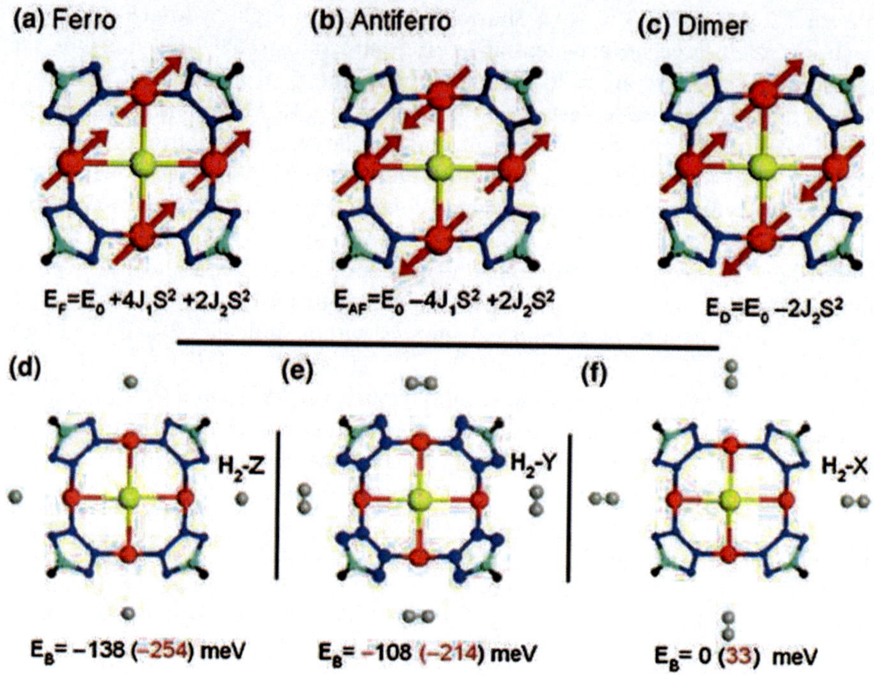

Fig. 11 *Top* view of the Mn_4Cl-MOF cluster with three magnetic configurations and their energies in terms of nearest (J_1) and next-nearest (J_2) exchange interactions. The bottom panel shows H_2—Mn_4Cl—MOF cluster for three different H_2 orientations. The H_2 binding energies (E_B) and pure Coulomb contributions (in parentheses) are also given. Reprinted with permission from [179]. Copyright 2008, American Chemical Society

In general, terminal ligands are bound to metal centers by a Lewis acid/base interaction. The open metal site may be formed by removing these terminal ligands without destruction of the framework. Mostly, these exposed metals are divalent cations, electron deficient, and expected to form dihydrogen σ complexes. By generating frameworks bearing open metal coordination sites, it is possible to increase the H_2 affinity of the surface, giving rise to a higher storage capacity at room temperature [172, 180, 181]. For example, it was reported that the exposed Mn^{2+} coordination sites within $Mn_3[(Mn_4Cl)_3(BTT)_8]_2$ contribute to its record uptake of 1.49 wt% and 12.1 g/L at 298 K and 90 bar [172]. By examining various metal ions in MOF structure, Zhou et al. provided some insight into the increased binding strength of H_2 in MOFs [182]. These authors performed DFT calculations on a series of isostructural MOFs, M_2(dhtp) (M = Mg, Mn, Co, Ni, Zn, and dhtp = 2,5-dihydroxyterrphthalate). They reported that the H_2 binding strength has the trend, Zn^{2+}(4.43 kJ/mol) < Mn^{2+}(4.70 kJ/mol) < Mg^{2+}(5.99 kJ/mol) < Co^{2+}(6.28 kJ/mol) < Ni^{2+}(7.33 kJ/mol). These results suggest that the relative strength of the metal center may be empirically correlated with the ionic radius of cations in the same coordination environment. Sun et al. also studied the

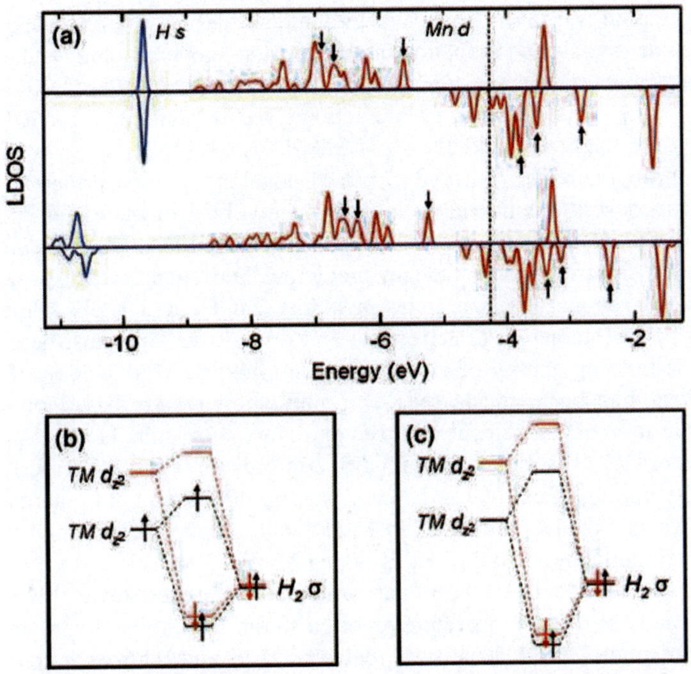

Fig. 12 **a** Local spin density of states on Mn d and H s before (*upper*) and after (*lower*) H_2 adsorption. **b** Schematic diagram illustrating the orbital interactions between TM dz^2 and H_2 σ in the Mn – and Cr – MOFs. Reprinted with permission from [183]. Copyright 2007, American Chemical Society

binding of H_2 in MOFs using a different set of transition metals [183]. Based on the analysis of electronic structure change before and after H_2 adsorption on the Mn site of Mn-doped model MOFs (Fig. 12), these authors predicted that doping with an early transition metal would lead to stronger H_2 binding. They further calculated the binding energies of 8.4, 10.4, 21.9, 34.6, and 46.5 kJ/mol for Mn, Sc, Ti, and V doped MOFs. Clearly, early transition metals in MOFs have a larger binding energy than late ones. The authors proposed a special Kubas-type interaction and attributed the orbital interactions between a TM center and an approaching dihydrogen to four key factors: (1) the separation of the H_2 σ level and the TM d levels, which decreases when the atomic number increases; (2) the splitting of the spin-up and spin-down d-levels which is reflected by the magnetic moment of the TM center; (3) the position of the most responsive d-levels to the approaching H_2 which is determined by the crystal field splitting of the d-orbitals according to the local symmetry of the TM center; and (4) the occupancy of the responsive d levels which is determined by the number of valence electrons and the oxidation states.

Both experimental and theoretical studies revealed that the aromatic rings on the organic linkers play an important role in H_2 adsorption, particularly at a higher hydrogen pressures [184]. However, the interaction between H_2 and organic linkers, van der Waals and induction interactions, are generally too weak to have a significant impact on increasing the binding energy of H_2 molecule to host materials [185]. The dispersive interactions are a challenge for DFT-based approaches. On the other hand, the size of MOFs is too big for the wave-function-based ab initio methods. Han and Goodard developed a first-principles-derived force field to describe nonbonding interaction involved in hydrogen interactions in MOFs [186]. This new potential represents a significant improvement from Lenard-Jones empirical force field in predicting hydrogen storage properties in high surface area porous materials, including MOFs and COFs. However, a stronger type of interaction (electrostatic and orbital overlap) must be evoked to improve the binding energy of H_2 with organic linkers. Modification of the aromatic linkers has been considered. The simulation results based on the new potential are in good agreement with the experimental results [184, 187].

A series of GCMC simulations with different organic linkers were performed to examine the doping effect of Li cations based on this force field potential [188].

As shown in Fig. 13, doping Li in MOFs increases hydrogen uptake at room temperature significantly. At 300 K, Li-doped MOF-C30 was found to absorb 3.89 wt% at 20 bar and 4.56 wt% at 50 bar, which is the highest reversible hydrogen storage capacity at room temperature reported so far. In contrast, hydrogen storage capacities of pure MOF-C30 are calculated as 0.25 and 0.56 wt% at the corresponding pressures and temperature. Other organic linkers doped with Li also showed increased hydrogen capacities over unmodified MOFs. The increased hydrogen capacity can be attributed to increased electrostatic interactions. In Li-doped MOF structures, Li is attracted to the high electron affinity of aromatic linkers and becomes a positive Li cation. The cooperative electrostatic interaction to H_2 from metal ion clusters and doped Li ions leads to a strong binding toward H_2 molecules. Indeed, the Li-doped MOF-C30 exhibited a high binding energy of 16.7 kJ/mol, which is significantly attractive to H_2 at ambient temperature. Therefore, these computational studies established that doping Li in MOFs is a very effective way to improve the binding strength of H_2 within MOFs. The prediction was confirmed by Mulfort and Hupp, who used chemical reduction methods to obtain Li-doped MOF and found that the hydrogen adsorption capacity nearly doubled [189]. Such improvement of binding strength between H_2 and Li-doped MOFs was also confirmed by other ab initio calculations and GCMC simulations [190, 191]. Han et al. further predicted that this doping effect also can increase the binding energies of H_2 with COF material to achieve the hydrogen storage target at room temperature [10].

Klontzas et al. explored another modification for organic linkers by substituting a terminal hydrogen atom with a Li atom and forming a modified Li alkoxide group [192], as shown in Fig. 14. Results from ab initio calculations showed interaction energies between hydrogen molecules and the modified group are up to three times larger than H_2 with unmodified MOFs. They obtained a very promising result, 4.5 wt% hydrogen storage at room temperature [192].

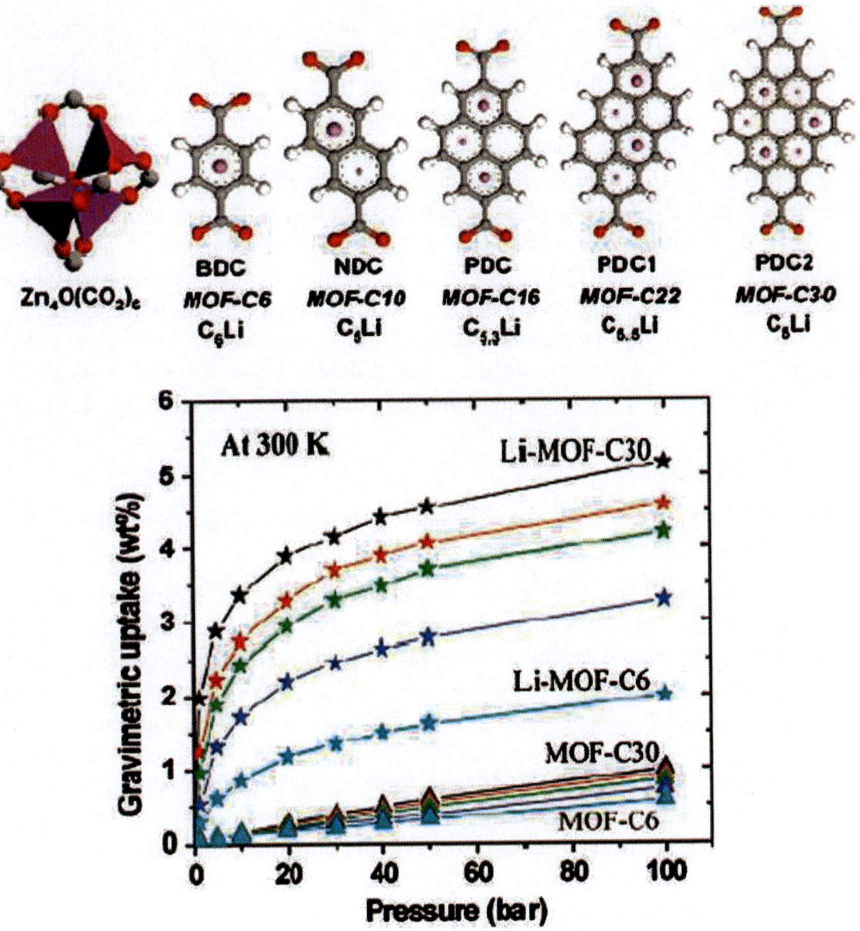

Fig. 13 The building blocks (metal oxide center and Li-doped organic linkers) for MOF structures (*upperpanel*) and predicted H_2 adsorption isotherms at 300 K for assembled MOFs (*lower panel*). Reprinted with permission from [188]. Copyright 2007, American Chemical Society

Similarly, Li-doped in covalent organic frameworks also improves their hydrogen capacity at room temperature [193, 194]. First-principles calculations showed that doping Li in the COF materials led to a higher gravimetric adsorption capacity at T = 298 K and p = 100 bar, 6.84 and 6.73 wt% for Li-doped COF-105 and COF-108. A multiscale theoretical method combined with the first-principles calculation and GCMC simulation was performed to investigate the hydrogen adsorption properties in undoped and Li-doped COF-202. The GCMC simulation predicted that the total gravimetric and volumetric uptakes of hydrogen in the Li-doped COF-202 could reach 4.39 wt% and 25.86 g/L at T = 298 K and p = 100 bar. In contrast, the maximum H_2 gravimetric and volumetric uptake of the undoped COF-202 are only 1.52 wt% and 8.08 g/L at $T = 298$ K and $p = 100$ bar.

Experimentally, significant effort has been made to understand and improve binding strength. Many MOFs are found to have a large space between their links. For example, the distance between phenylene faces in MOF-5 is measured to be 15 Å in diameter [9]. There would be an unused void volume at the center of these pores. In order to increase the attractive interactions between the wall and hydrogen, one would reduce the size of the pore. A smaller pore facilitates the overlapping of potentials from two or more linkers, thereby creating a synergy toward binding the hydrogen molecule. Dinca et al. observed H_2 binding energies of up to 9.5 kJ/mol by increasing van der Waals contact area associated with a very small pore size [172]. Four-fold catenated IRMOFs also have been reported to have a larger hydrogen capacity at room temperature and 48 bar [195, 196].

4 Nanobuilding-Blocks for Novel Hydrogen Storage Materials

Inspired by the early work on the use of carbon nanotubes for hydrogen storage at room temperature [164], carbon nanotube materials have been extensively investigated as hydrogen storage media [14]. Some studies showed that carbon-based materials can store a large amount of hydrogen molecules since they contain defect structures that act as trapping sites [197]. Very different hydrogen storage values between 0.25 and 56% were reported under various experimental conditions for carbon nanotube materials [164, 198–201]. Later, it was found that hydrogen storage of > 1.0 wt% was practically impossible to achieve and higher values were possibly the result of measurement errors due to contamination of the analysis gas with water [202, 203]. Theoretical studies also confirmed that high hydrogen content in the pure carbon nanotubes cannot be achieved through physical sorption [204]. On the other hand, Kim et al. studied the hydrogen storage properties of multiwalled carbon nanotubes (MWCNTs) with Ni nanoparticles at moderate conditions of 340–520 K and 4 MPa [205]. Hydrogen uptake up to 2.8 wt% was achieved under these conditions. In fact, other carbon-based materials (activated carbon, and graphite nanofibers) modified by some metals showed increased hydrogen storage capacity [205–208].

For pure carbon-based materials, a low hydrogen storage capacity actually originates from the small binding energy between hydrogen molecules and carbon porous materials. The interaction with carbon nanotubes or carbon fullerenes is interpreted as instantaneous transition dipole or dispersion interaction, which is very weak, on the order of a few kJ/mol. Modification by metals results in increased hydrogen storage capacity, which may be attributed to stronger binding, resulting from electrostatic interaction between the hydrogen molecule and metals. The binding energies between hydrogen and metal are really dependent on reactivity and structural properties of metal clusters.

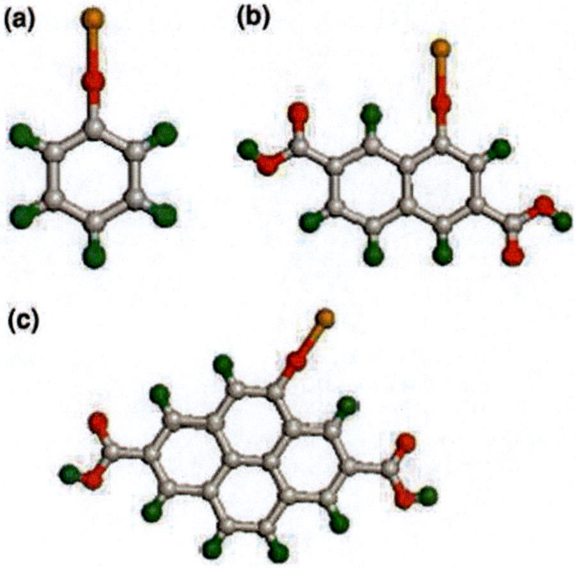

Fig. 14 The modified orgianic linkers. (**a**) Li-alkoxide-benzene, (**b**) Li-alkoxide-anphalene-dicarboxylate (IRMOF-8 linker), and (**c**) Li-alkoxide-pryene-dicarboxylate (IRMOF-14 linker). Reprinted with permission from [192]. Copyright 2008, American Chemical Society

In order to make use of porous carbon materials with high surface areas and improve the weak binding situation, chemists considered metal/transition metal coating of carbon nanotubes and fullerenes. In the following section, we will discuss computational studies on metal/transition metal-coated carbon-based materials, focusing on two important topics: hydrogen capacity and binding energy between hydrogen and metal.

In order to increase hydrogen storage capacity and binding energy of hydrogen, many metals, including alkali metals, alkali earth metals, and transition metals, were used to coat carbon-based materials. Yildirium and Ciraci performed the DFT calculations for a single Ti coated on a single-walled nanotube (SWNT) which could bind up to four hydrogen molecules [209]. The first H_2 adsorption is dissociative with no energy barrier whereas the other hydrogen molecules are adsorbed physically with slightly elongated H–H bonds. The calculated binding energy was 0.13 eV/H_2, which is 4–5 times stronger than dispersion force between hydrogen molecules and SWNT. At a high Ti coverage, SWNT can strongly adsorb up to 8 wt% hydrogen. Later, they further reported computational work on 3d transition metal-decorated C_{60} for hydrogen storage [210]. For early transition metals, the average binding energy of H_2 molecule is 0.3–0.5 eV/H_2 and the maximum hydrogen storage was predicted to be 9.0 wt% [176]. Late transition metals such Fe, Co, and Ni, do not bond to the C_{60} cluster. However, transition metals tend to form clusters on C_{60} instead of uniformly distributing because the binding energy of the nanoparticle is stronger than between TM and C_{60} [211]. In the case of Ti (Fig. 15), the binding energy of nanoparticle (2.8 eV/Ti) is 0.6 eV/Ti stronger than between Ti and C_{60} (2.2 eV) [210]. It is obvious that the clustering effect significantly reduces the gravimetric hydrogen storage capacity.

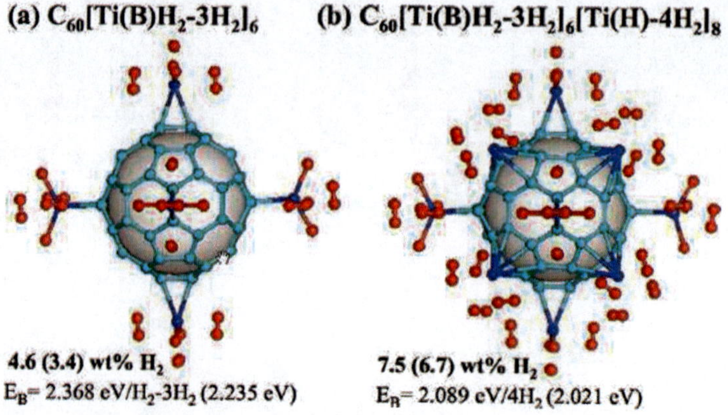

Fig. 15 Ti-coated C_{60} systems with high-density H coverage. Reprinted with permission from [210]. Copyright 2005, American Physical Society

Further, Sun et al. proposed lighter metals, such as alkali and alkali earth metals, coating C_{60} fullerenes to enhance hydrogen capacity [8, 212]. $Li_{12}C_{60}$ was found to have high gravimetric (9 wt%) and volumetric densities (70 g/L) [8]. Although Li metal tends to uniformly coat on the surface of C_{60}, the binding energies between $Li_{12}C_{60}$ and hydrogen molecules are very weak, with a maximum value of 0.075 eV. On the other hand, the weak binding energy (averaging 1.78 eV/Li) between Li and C_{60} may lead to instability of the structure at ambient temperature. Very recently, they also investigated Ca-coated C_{60} as hydrogen storage materials [212]. Ca tends to coat on the surface of C_{60} as single atoms rather than forming clusters. The calculated average binding energy increased to 0.45 eV/H_2 and the gravimetric density of this material can reach 6.2 wt%.

In fact, hydrogen storage capacity also can be increased by applying lighter-weight organic molecules or nanostructures, for example, boron nitride and boron fullerenes, as the supporter of metals [176, 213–215]. Li et al. studied Ca-coated boron fullerenes and nanotubes as hydrogen storage materials. They found that the B_{80} fullerene with 12 Ca can attract up to 60 H_2 molecules with binding energy 0.12–0.40 eV/H_2, corresponding to a 8.2 wt% capacity [216]. In addition, one must consider how many hydrogen molecules can be attracted by one metal atom. Very recently, Gagliardi and Pyykkö suggested the maximum number of 12 hydrogen atoms can be bound to one metal atom (Cr, Mo, W, V, and Ti) [217]. Therefore, the selectivity of a suitable metal plays an important role in increasing hydrogen storage capacity.

In addition, understanding the fundamental role of the carbon fullerene in hydrogen storage, as a support for metal or attractive sites for hydrogen molecules, is very important to improve hydrogen content. Yoon et al. studied charged fullerenes as hydrogen storage materials [218]. They found that the binding strength of molecular hydrogen on either positively or negatively charged fullerenes can be dramatically enhanced to 0.18–0.32 eV, which ensures hydrogen storage at ambient conditions. At full hydrogen coverage, a charged fullerene can achieve storage capacities of more than ~8.0 wt%. The enhanced binding between charged fullerenes and hydrogen is attributed to the polarization of hydrogen molecules by the high electronic field generated near the surface of the charged fullerene. It suggests that both charged carbon fullerene and metals can be hosts for hydrogen molecules.

The hydrogen storage capacity in novel organometallic fullerenes and nanotubes is dependent on metal nanosize and structural properties of nanoparticles. The binding energy between hydrogen molecules and host plays an important role in validating hydrogen storage materials. Therefore, it is worth further discussing the interaction nature of host and hydrogen molecules. Niu et al. studied the binding of transition metal Ni and its ion with hydrogen molecules [177]. They found that only one hydrogen molecule can chemisorb dissociatively with a neutral Ni atom, forming a linear structure. The interaction between the Ni and hydrogen molecule is attributed to a consequence of the Pauli Exclusion Principle. In contrast, Ni^+ can bind several hydrogen molecules by electrostatic interaction. However, very recent computational studies showed that transition metals (Cr, Mo, W, Ti, and V) can not only break up the hydrogen molecule and form metal hydrides, but also that several hydrogen molecules can be adsorbed on the metal atom. The computational results showed that the more hydrogen molecules are adsorbed, the more stable metal hydrides became.

The interaction with hydrogen is enhanced due to the presence of metal, attributed to interaction between the charged metal and the induced dipole of the hydrogen molecule. However, this type of interaction is still not strong enough to allow hydrogen storage at ambient temperature. The molecular orbital analysis showed that there is a large hybridization among Ti-d, H-σ^*, and C-p orbitals, as shown in Fig. 16 [209]. The mechanism of the bonding can be explained by the DCD donation-backdonation model. According to this model, electrons of Ti-d orbital are simultaneously donated to the π^* antibonding orbital of C6 ring of SWNTN and σ^* antibonding orbital of H–H. This weakens the H–H bond or causes H–H bond to break. Therefore, a combination of forces, including electron transfer and electrostatic interaction contribute to the overall effect. For transition metals, electron transfer interaction actually is the dominant force between H_2 and TM. In fact, electron transfer interaction is weakened because electrons in metals transferred to the low-energy π^* in carbon system. With more hydrogen molecules being attracted to the TM site, electrostatic interaction will become more important. As a result, hydrogen adsorption takes the form of a hydrogen molecule after electrostatic interaction becomes dominant.

In summary, preventing the clustering of the coated metals and enhancing electrostatic interactions between the metals and hydrogen play an important role in designing novel hydrogen storage materials based on metal-coated SWNT and fullerenes.

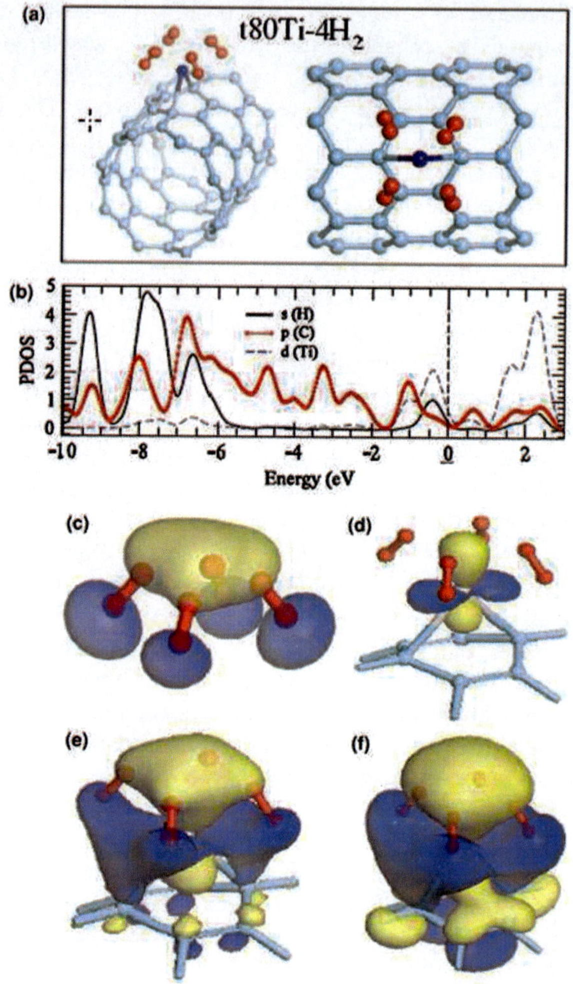

Fig. 16 Geometrical and electronic structures of Ti-coated carbon nanotube. **a** Two different views of the optimized structure of t80Ti-4H_2, **b** PDOS of Ti, C, and H atoms, **c** The σ^* antibonding orbital of four H_2 complex, **d–f** isosurface of the state just below E_f at three different values (0.08, 0.04, and 0.2). Reprinted with permission from [209]. Copyright 2005, American Physical Society

5 Concluding Remarks

Hydrogen storage has been a remarkably active field for research and development over the past decade. Nanoscience and nanotechnology have made a significant impact on hydrogen storage research. The new characterization techniques have started to provide unprecedented details of the system, including atomic and electronic structures. The new synthesis methods developed over the same period provide new opportunities to control the size, shape, and local composition on the nanometer scale. Developments in density functional theory-based computational methods led to a remarkable degree of reliability in structural analyses for systems relevant to hydrogen storage. Theoretical studies help to screen materials that are potentially useful for hydrogen storage. Theory also helps to search for and design systems to have favorable properties for hydrogen storage. Nanosize effects have been studied both theoretically and experimentally. In this aspect, theoretical studies clearly demonstrated the improvement in both thermodynamic and kinetics for hydrogen storage. The structural and electronic analysis of the transition metal-doped hydride helps us understand the change in the hydrogen interaction induced by the transition metal and to select more effective catalysts and design new catalytic structures. Theoretical studies allow us to explore novel structures that have not yet been synthesized and predict their properties. The novel structures may be synthesized with the aid of the newly developed nanofabrication technology. We hope that the issues preventing hydrogen storage from becoming a reality will be overcome with the use of nanotechnology, as predicted by Professor Smalley, so that the hydrogen economy becomes a reality.

Acknowledgments We acknowledge support by U. S. Department of Energy, Basic Energy Science grant DE-FG02-05ER46231.

References

1. Baxter J, Bian ZX, Chen G, Danielson D, Dresselhaus MS, Fedorov AG, Fisher TS, Jones CW, Maginn E, Kortshagen U, Manthiram A, Nozik A, Rolison DR, Sands T, Shi L, Sholl D, Wu YY (2009) Nanoscale design to enable the revolution in renewable energy. Energy Environ Sci 2:559–588
2. Schlapbach L, Züttel A (2001) Hydrogen-storage materials for mobile applications. Nature 414:353–358
3. Crabtree GW, Dresselhaus MS, Buchanan MV (2004) The hydrogen economy. Phys Today 57(12):39–44
4. US DOE (2010) Hydrogen Storage http://hydrogenenergygov/storagehtml
5. von Helmolt R, Eberle U (2007) Fuel cell vehicles: status 2007. J Power Sour 165:833–843
6. Zaluska A, Zaluski L, Ström-Olsen JO (2001) Structure, catalysis and atomic reactions on the nano-scale: a systematic approach to metal hydrides for hydrogen storage. Appl Phys A: Mater Sci Process 72:157–165
7. Li YW, Yang RT (2006) Significantly enhanced hydrogen storage in metal-organic frameworks via spillover. J Am Chem Soc 128:726–727
8. Sun Q, Jena P, Wang Q, Marquez M (2006) First-principles study of hydrogen storage on $Li_{12}C_{60}$. J Am Chem Soc 128:9741–9745
9. Rosi NL, Eckert J, Eddaoudi M, Vodak DT, Kim J, O'Keeffe M, Yaghi OM (2003) Hydrogen storage in microporous metal-organic frameworks. Science 300:1127–1129
10. Han SS, Furukawa H, Yaghi OM, Goddard WA (2008) Covalent organic frameworks as exceptional hydrogen storage materials. J Am Chem Soc 130:11580
11. Hohenberg P, Kohn W (1964) Inhomogeneous electron gas. Phys Rev 136:B864
12. Kohn W, Sham LJ (1965) Self-consistent equations including exchange and correlation effects. Phys Rev 140:A1133
13. Graetz J (2009) New approaches to hydrogen storage. Chem Soc Rev 38:73–82
14. Wang LF, Yang RT (2008) New sorbents for hydrogen storage by hydrogen spillover - a review. Energy Environ Sci 1:268–279
15. Orimo SI, Nakamori Y, Eliseo JR, Züttel A, Jensen CM (2007) Complex hydrides for hydrogen storage. Chem Rev 107:4111–4132

16. van den Berg AWC and Arean CO (2008) Materials for hydrogen storage: current research trends and perspectives. Chem Commun. 668–681
17. Yang J, Sudik A, Wolverton C, Siegel DJ (2010) High capacity hydrogen storage materials: attributes for automotive applications and techniques for materials discovery. Chem Soc Rev 39:656–675
18. Finholt AE, Bond AC, Schlesinger HI (1947) Lithium aluminum hydride, aluminum hydride and lithium gallium hydride, and some of their applications in organic and inorganic chemistry. J Am Chem Soc 69:1199–1203
19. Bogdanović B, Schwickardi M (1997) Ti-doped alkali metal aluminium hydrides as potential novel reversible hydrogen storage materials. J Alloys Compd 253:1–9
20. Bogdanović B, Brand RA, Marjanović A, Schwickardi M, Tölle J (2000) Metal-doped sodium aluminium hydrides as potential new hydrogen storage materials. J Alloys Compd 302:36–58
21. Baroni S, de Gironcoli S, Dal Corso A, Giannozzi P (2001) Phonons and related crystal properties from density-functional perturbation theory. Rev Mod Phys 73:515–562
22. Vegge T (2006) Equilibrium structure and Ti-catalyzed H_2 desorption in $NaAlH_4$ nanoparticles from density functional theory. Phys Chem Chem Phys 8:4853–4861
23. Johnson CA, Chakerian GD (1965) On the proof and uniqueness of Wulff's construction of the shape of minimum surface free energy. J Math Phys 6:1403–1404
24. Kim KC, Dai B, Johnson JK, Sholl DS (2009) Assessing nanoparticle size effects on metal hydride thermodynamics using the Wulff construction. Nanotechnology 20:204001
25. Alapati SV, Johnson JK, Sholl DS (2007) Using first principles calculations to identify new destabilized metal hydride reactions for reversible hydrogen storage. Phys Chem Chem Phys 9:1438–1452
26. Løvvik OM, Swang O, Opalka SM (2005) Modeling alkali alanates for hydrogen storage by density-functional band-structure calculations. J Mater Res 20:3199–3213
27. Opalka SM, Løvvik OM, Brinks HW, Saxe PW, Hauback BC (2007) Integrated experimental-theoretical investigation of the Na-Li-Al-H system. Inorg Chem 46: 1401–1409
28. Huot J, Boily S, Güther V, Schulz R (1999) Synthesis of Na_3AlH_6 and Na_2LiAlH_6 by mechanical alloying. J Alloys Compd 283:304–306
29. Brinks HW, Hauback BC, Jensen CM, Zidan R (2005) Synthesis and crystal structure of Na_2LiAlD_6. J Alloys Compd 392:27–30
30. Tang X, Opalka SM, Laube BL, Wu FJ, Strickler JR, Anton DL (2007) Hydrogen storage properties of Na-Li-Mg-Al-H complex hydrides. J Alloys Compd 446:228–231
31. Grove H, Brinks HW, Heyn RH, Wu FJ, Opalka SM, Tang X, Laube BL, Hauback BC (2008) The structure of $LiMg(AlD_4)_3$. J Alloys Compd 455:249–254
32. Yamauchi M, Kobayashi H, Kitagawa H (2009) Hydrogen storage mediated by Pd and Pt nanoparticles. ChemPhysChem 10:2566–2576
33. Pundt A (2004) Hydrogen in nano-sized metals. Adv Eng Mater 6:11–21
34. Pundt A, Kirchheim R (2006) Hydrogen in metals: microstructural aspects. Ann Rev Mater Res 36:555–608
35. Baldé CP, Hereijgers BPC, Bitter JH, de Jong KP (2006) Facilitated hydrogen storage in $NaAlH_4$ supported on carbon nanoribers. Angew Chem-Int Edit 45:3501–3503
36. Baldé CP, Hereijgers BPC, Bitter JH, de Jong KP (2008) Sodium alanate nanoparticles - Linking size to hydrogen storage properties. J Am Chem Soc 130:6761–6765
37. Verkuijlen MHW, Gao J, Adelhelm P, van Bentum PJM, de Jongh PE, Kentgens APM (2010) Solid-state NMR studies of the local structure of $NaAlH_4$/C nanocomposites at different stages of hydrogen desorption and rehydrogenation. J Phys Chem C 114: 4683–4692
38. Gao J, Adelhelm P, Verkuijlen MHW, Rongeat C, Herrich M, van Bentum PJM, Gutfleisch O, Kentgens APM, de Jong KP, de Jongh PE (2010) Confinement of $NaAlH_4$ in nanoporous carbon: impact on H_2 release, reversibility, and thermodynamics. J Phys Chem C 114: 4675–4682

39. Balema VP, Balema L (2005) Missing pieces of the puzzle or about some unresolved issues in solid state chemistry of alkali metal aluminohydrides. Phys Chem Chem Phys 7: 1310–1314
40. Zheng SY, Fang F, Zhou GY, Chen GR, Ouyang LZ, Zhu M, Sun DL (2008) Hydrogen storage properties of space-confined $NaAlH_4$ nanoparticles in ordered mesoporous silica. Chem Mat 20:3954–3958
41. Bhakta RK, Herberg JL, Jacobs B, Highley A, Behrens R, Ockwig NW, Greathouse JA, Allendorf MD (2009) Metal-organic frameworks as templates for nanoscale $NaAlH_4$. J Am Chem Soc 131:13198
42. Züttel A, Rentsch S, Fischer P, Wenger P, Sudan P, Mauron P, Emmenegger C (2003) Hydrogen storage properties of $LiBH_4$. J Alloys Compd 356:515–520
43. Züttel A, Wenger P, Rentsch S, Sudan P, Mauron P, Emmenegger C (2003) $LiBH_4$ a new hydrogen storage material. J Power Sour 118:1–7
44. Lodziana Z, Vegge T (2004) Structural stability of complex hydrides: $LiBH_4$ revisited. Phys Rev Lett 93:145501
45. Orima S, Nakamori Y, Kitahara G, Miwa K, Ohba N, Towata S, Züttel A (2005) Dehydriding and rehydriding reactions of $LiBH_4$. J Alloys Compd 404:427–430
46. Ge Q (2004) Structure and energetics of $LiBH_4$ and its surfaces: a first-principles study. J Phys Chem A 108:8682–8690
47. Liu J, Ge Q (2009) Hydrogen interaction in Ti-doped $LiBH_4$ for hydrogen storage: a density functional analysis. J Chem Theory Comput 5:3079–3087
48. Kang XD, Wang P, Ma LP, Cheng HM (2007) Reversible hydrogen storage in $LiBH_4$ destabilized by milling with Al. Appl Phys A 89:963–966
49. Kostka J, Lohstroh W, Fichtner M, Hahn H (2007) Diborane release from $LiBH_4$/Silica-Gel mixtures and the effect of additives. J Phys Chem C 111:14026–14029
50. Nickels EA, Jones Martin O, David William IF, Johnson Simon R, Lowton Rebecca L, Sommariva M, Edwards Peter P (2008) Tuning the decomposition temperature in complex hydrides: synthesis of a mixed alkali metal Borohydride13. Angew Chem-Int Edit 47: 2817–2819
51. Claudy P, Bonnetot B, Letoffe JM, Turck G (1978) Determination des constantes thermodynamiques des hydrures simples et complexes de l'aluminium. IV. Enthalpie de formation de $LiAIH_2$ et Li_3AIH_6. Thermochimica Acta 27:213–221
52. Dymova TN, Aleksandrov DP, Konoplev VN, Silina TA, Sizareva AS (1994) Spontaneous and thermal-decomposition of Lithium Tetrahydroaluminate $LiAlH_4$ - the promoting effect of mechanochemical action on the process. Koord Khimiya 20:279–285
53. Balema VP, Dennis KW and Pecharsky VK (2000) Rapid solid-state transformation of tetrahedral AlH_4 (-) into octahedral AlH_6 (3-) in lithium aluminohydride. Chem Commun. 1665–1666
54. Balema VP, Pecharsky VK, Dennis KW (2000) Solid state phase transformations in $LiAlH_4$ during high-energy ball-milling. J Alloys Compd 313:69–74
55. Balema VP, Wiench JW, Dennis KW, Pruski M, Pecharsky VK (2001) Titanium catalyzed solid-state transformations in $LiAlH_4$ during high-energy ball-milling. J Alloys Compd 329:108–114
56. Chen J, Kuriyama N, Xu Q, Takeshita HT, Sakai T (2001) Reversible hydroen storage via titanium-catalyzed $LiAlH_4$ and Li_3AlH_6. J Phys Chem B 105:11214–11220
57. Alapati SV, Johnson JK, Sholl DS (2006) Identification of destabilized metal hydrides for hydrogen storage using first principles calculations. J Phys Chem B 110:8769–8776
58. Alapati SV, Johnson JK, Sholl DS (2007) Predicting reaction equilibria for destabilized metal hydride decomposition reactions for reversible hydrogen storage. J Phys Chem C 111:1584–1591
59. Yu XB, Grant DM and Walker GS (2006) A new dehydrogenation mechanism for reversible multicomponent borohydride systems - The role of Li-Mg alloys. Chem Commun. 3906–3908

60. Wolverton C, Siegel DJ, Akbarzadeh AR, Ozoliņš V (2008) Discovery of novel hydrogen storage materials: an atomic scale computational approach. J Phys-Condes Matter 20:14
61. Siegel DJ, Wolverton C, Ozoliņš V (2007) Thermodynamic guidelines for the prediction of hydrogen storage reactions and their application to destabilized hydride mixtures. Phys Rev B 76:134102
62. Vajo JJ, Olson GL (2007) Hydrogen storage in destabilized chemical systems. Scr Mater 56:829–834
63. Fang ZZ, Kang XD, Dai HB, Zhang MJ, Wang P, Cheng HM (2008) Reversible dehydrogenation of $LiBH_4$ catalyzed by as-prepared single-walled carbon nanotubes. Scr Mater 58:922–925
64. Vajo JJ, Skeith SL, Mertens F (2005) Reversible storage of hydrogen in destabilized $LiBH_4$. J Phys Chem B 109:3719–3722
65. Yu XB, Grant DM, Walker GS (2009) Dehydrogenation of $LiBH_4$ destabilized with various oxides. J Phys Chem C 113:17945–17949
66. Au M, Jurgensen A, Zeigler K (2006) Modified lithium borohydrides for reversible hydrogen storage (2). J Phys Chem B 110:26482–26487
67. Zhang Y, Zhang WS, Wang AQ, Sun LX, Fan MQ, Chu HL, Sun JC, Zhang T (2007) $LiBH_4$ nanoparticles supported by disordered mesoporous carbon: hydrogen storage performances and destabilization mechanisms. Int J Hydrogen Energy 32:3976–3980
68. Au M, Jurgensen AR, Spencer WA, Anton DL, Pinkerton FE, Hwang SJ, Kim C, Bowman RC (2008) Stability and reversibility of lithium borohydrides doped by metal halides and hydrides. J Phys Chem C 112:18661–18671
69. Gross AF, Vajo JJ, Van Atta SL, Olson GL (2008) Enhanced hydrogen storage kinetics of $LiBH_4$ in nanoporous carbon scaffolds. J Phys Chem C 112:5651–5657
70. Walker GS, Grant DM, Price TC, Yu XB, Legrand V (2009) High capacity multicomponent hydrogen storage materials: investigation of the effect of stoichiometry and decomposition conditions on the cycling behaviour of $LiBH_4$-MgH_2. J Power Sources 194:1128–1134
71. Opalka SM, Tang X, Laube BL, Vanderspurt TH (2009) Experimental and theoretical screening of nanoscale oxide reactivity with $LiBH_4$. Nanotechnology 20:204024
72. Oguchi H, Matsuo M, Hummelshoj JS, Vegge T, Norskov JK, Sato T, Miura Y, Takamura H, Maekawa H, Orimo S (2009) Experimental and computational studies on structural transitions in the $LiBH_4$-LiI pseudobinary system. Appl Phys Lett 94:141912
73. Somorjai GA, Borodko YG (2001) Research in Nanosciences–Great opportunity for Catalysis Science. Catal Lett 76:1–5
74. Berseth PA, Harter AG, Zidan R, Blomqvist A, Araújo CM, Scheicher RH, Ahuja R, Jena P (2009) Carbon nanomaterials as catalysts for hydrogen uptake and release in $NaAlH_4$. Nano Lett 9:1501–1505
75. Singh S, Eijt SWH, Huot J, Kockelmann WA, Wagemaker M, Mulder FM (2007) The $TiCl_3$ catalyst in $NaAlH_4$ for hydrogen storage induces grain refinement and impacts on hydrogen vacancy formation. Acta Mater 55:5549–5557
76. Lee GJ, Shim JH, Cho YW, Lee KS (2007) Reversible hydrogen storage in $NaAlH_4$ catalyzed with lanthanide oxides. Int J Hydrogen Energy 32:1911–1915
77. Felderhoff M, Weidenthaler C, von Helmolt R, Eberle U (2007) Hydrogen storage: the remaining scientific and technological challenges. Phys Chem Chem Phys 9:2643–2653
78. Baldé CP, Stil HA, van der Eerden AMJ, de Jong KP, Bitter JH (2007) Active Ti species in $TiCl_3$-doped $NaAlH_4$. mechanism for catalyst deactivation. J Phys Chem C 111:2797–2802
79. Léon A, Schild D, Fichtner M (2006) Chemical state of Ti in sodium alanate doped with $TiCl_3$ using X-ray photoelectron spectroscopy (vol 404, pg 766, 2005). J Alloys Compd 407:340–340
80. Léon A, Kircher O, Rösner H, Décamps B, Leroy E, Fichtner M, Percheron-Guégan A (2006) SEM and TEM characterization of sodium alanate doped with $TiCl_3$ or small Ti clusters (Ti_{13} * 6THF). J Alloys Compd 414:190–203

81. Léon A, Kircher O, Fichtner M, Rothe J, Schild D (2006) Evolution of the local structure around Ti atoms in $NaAlH_4$ doped with $TiCl_3$ or Ti_{13} center dot 6THF by ball milling using X-ray absorption and X-ray photoelectron spectroscopy. J Phys Chem B 110:1192–1200
82. Canton P, Fichtner M, Frommen C, Léon A (2006) Synchrotron X-ray studies of Ti-doped $NaAlH_4$. J Phys Chem B 110:3051–3054
83. Brinks HW, Sulic M, Jensen CM, Hauback BC (2006) $TiCl_3$-enhanced $NaAlH_4$: impact of excess Al and development of the Al1-gamma Ti gamma phase during cycling. J Phys Chem B 110:2740–2745
84. Bogdanović B, Felderhoff M, Pommerin A, Schüth T, Spielkamp N (2006) Advanced hydrogen-storage materials based on Sc-, Ce-, and Pr-doped $NaAlH_4$. Adv Mater 18:1198
85. Bellosta von Colbe JM, Schmidt W, Felderhoff M, Bogdanović B, Schüth F (2006) Hydrogen-isotope scrambling on doped sodium alanate. Angew Chem-Int Edit 45: 3663–3665
86. Wang P, Kang XD, Cheng HM (2005) Exploration of the nature of active Ti species in metallic Ti-doped $NaAlH_4$. J Phys Chem B 109:20131–20136
87. Wang J, Ebner AD, Zidan R, Ritter JA (2005) Synergistic effects of co-dopants on the dehydrogenation kinetics of sodium aluminum hydride. J Alloys Compd 391:245–255
88. von Colbe JMB, Felderhoff M, Bogdanović B, Schüth F and Weidenthaler C (2005) One-step direct synthesis of a Ti-doped sodium alanate hydrogen storage material. Chem Commun. 4732–4734
89. Resan M, Hampton MD, Lomness JK, Slattery DK (2005) Effect of TixAly catalysts on hydrogen storage properties of $LiAlH_4$ and $NaAlH_4$. Int J Hydrogen Energy 30:1417–1421
90. Majer G, Stanik E, Banuet LEV, Grinberg F, Kircher O, Fichtner M (2005) Effects of catalysts on the dehydriding of alanates monitored by proton NMR. J Alloys Compd 404:738–742
91. Léon A, Schild D, Fichtner M (2005) Chemical state of Ti in sodium alanate doped with $TiCl_3$ using X-ray photoelectron spectroscopy. J Alloys Compd 404:766–770
92. Isobe S, Ichikawa T, Hanada N, Leng HY, Fichtner M, Fuhr O, Fujii H (2005) Effect of Ti catalyst with different chemical form on Li-N-H hydrogen storage properties. J Alloys Compd 404:439–442
93. Gomes S, Renaudin G, Hagemann H, Yvon K, Sulic MP, Jensen CM (2005) Effects of milling, doping and cycling of $NaAlH_4$ studied by vibrational spectroscopy and X-ray diffraction. J Alloys Compd 390:305–313
94. Haiduc AG, Stil HA, Schwarz MA, Paulus P, Geerlings JJC (2005) On the fate of the Ti catalyst during hydrogen cycling of sodium alanate. J Alloys Compd 393:252–263
95. Bellosta von Colbe JM, Felderhoff M, Bogdanović B, Schüth F and Weidenthaler C (2005) One-step direct synthesis of a Ti-doped sodium alanate hydrogen storage material. Chem Commun: 4732–4734
96. Wang P, Jensen CM (2004) Preparation of Ti-doped sodium aluminum hydride from mechanical milling of NaH/Al with off-the-shelf Ti powder. J Phys Chem B 108: 15827–15829
97. Wang P, Jensen CM (2004) Method for preparing Ti-doped $NaAlH_4$ using Ti powder: observation of an unusual reversible dehydrogenation behavior. J Alloys Compd 379: 99–102
98. Srinivasan SS, Brinks HW, Hauback BC, Sun DL, Jensen CM (2004) Long term cycling behavior of titanium doped $NaAlH_4$ prepared through solvent mediated milling of NaH and Al with titanium dopant precursors. J Alloys Compd 377:283–289
99. Bogdanović B, Felderhoff M, Kaskel S, Pommerin A, Schlichte K, Schüth F (2003) Improved hydrogen storage properties of Ti-doped sodium alanate using titanium nanoparticles as doping agents. Adv Mater 15:1012
100. Bogdanović B, Felderhoff M, Germann M, Hartel M, Pommerin A, Schüth F, Weidenthaler C, Zibrowius B (2003) Investigation of hydrogen discharging and recharging processes of Ti-doped $NaAlH_4$ by X-ray diffraction analysis (XRD) and solid-state NMR spectroscopy. J Alloys Compd 350:246–255

101. Thomas GJ, Gross KJ, Yang NYC, Jensen C (2002) Microstructural characterization of catalyzed $NaAlH_4$. J Alloys Compd 330:702–707
102. Sandrock G, Gross K, Thomas G (2002) Effect of Ti-catalyst content on the reversible hydrogen storage properties of the sodium alanates. J Alloys Compd 339:299–308
103. Bogdanović B, Schwickardi M (2001) Ti-doped $NaAlH_4$ as a hydrogen-storage material - preparation by Ti-catalyzed hydrogenation of aluminum powder in conjunction with sodium hydride. Appl Phys A 72:221–223
104. Zaluska A, Zaluski L, Ström-Olsen JO (2000) Sodium alanates for reversible hydrogen storage. J Alloys Compd 298:125–134
105. Zidan RA, Takara S, Hee AG, Jensen CM (1999) Hydrogen cycling behavior of zirconium and titanium-zirconium-doped sodium aluminum hydride. J Alloys Compd 285:119–122
106. Jensen CM, Zidan R, Mariels N, Hee A, Hagen C (1999) Advanced titanium doping of sodium aluminum hydride: segue to a practical hydrogen storage material? Int J Hydrogen Energy 24:461–465
107. Stephens RD, Gross AF, Van Atta SL, Vajo JJ, Pinkerton FE (2009) The kinetic enhancement of hydrogen cycling in $NaAlH_4$ by melt infusion into nanoporous carbon aerogel. Nanotechnology 20:204018
108. Adelhelm P, de Jong KP and de Jongh PE (2009) How intimate contact with nanoporous carbon benefits the reversible hydrogen desorption from NaH and $NaAlH_4$. Chem Commun. 6261–6263
109. Shi Q, Yu X, Feidenhans'l R, Vegge T (2008) Destabilized $LiBH_4$-$NaAlH_4$ Mixtures Doped with Titanium Based Catalysts. J Phys Chem C 112:18244–18248
110. Majzoub EH, Ozolinš V (2008) Prototype electrostatic ground state approach to predicting crystal structures of ionic compounds: application to hydrogen storage materials. Phys Rev B 77:104115
111. Yin LC, Wang P, Kang XD, Sun CH, Cheng HM (2007) Functional anion concept: effect of fluorine anion on hydrogen storage of sodium alanate. Phys Chem Chem Phys 9:1499–1502
112. Wang J, Ebner AD, Ritter JA (2007) Synthesis of metal complex hydrides for hydrogen storage. J Phys Chem C 111:14917–14924
113. Graetz J, Reilly JJ, Johnson J, Ignatov AY, Tyson TA (2004) X-ray absorption study of Ti-activated sodium aluminum hydride. Appl Phys Lett 85:500–502
114. Schüth F, Bogdanović B, Felderhoff M (2004) Light metal hydrides and complex hydrides for hydrogen storage. Chem Commun: 2249–2258
115. Felderhoff M, Klementiev K, Grunert W, Spliethoff B, Tesche B, Bellosta von Colbe JM, Bogdanović B, Hartel M, Pommerin A, Schüth F, Weidenthaler C (2004) Combined TEM-EDX and XAFS studies of Ti-doped sodium alanate. Phys Chem Chem Phys 6:4369–4374
116. Kang XD, Wang P, Song XP, Yao XD, Lu GQ, Cheng HM (2006) Catalytic effect of Al_3Ti on the reversible dehydrogenation of $NaAlH_4$. J Alloys Compd 424:365–369
117. Weidenthaler C, Pommerin A, Felderhoff M, Bogdanović B, Schüth F (2003) On the state of the titanium and zirconium in Ti- or Zr-doped $NaAlH_4$ hydrogen storage material. Phys Chem Chem Phys 5:5149–5153
118. Gross KJ, Majzoub EH, Spangler SW (2003) The effects of titanium precursors on hydriding properties of alanates. J Alloys Compd 356:423–428
119. Gross KJ, Guthrie S, Takara S, Thomas G (2000) In situ X-ray diffraction study of the decomposition of $NaAlH_4$. J Alloys Compd 297:270–281
120. Gross KJ, Sandrock G, Thomas GJ (2002) Dynamic in situ X-ray diffraction of catalyzed alanates. J Alloys Compd 330:691–695
121. Voss J, Shi Q, Jacobsen HS, Zamponi M, Lefmann K, Vegge T (2007) Hydrogen dynamics in Na_3AlH_6: a combined density functional theory and quasielastic neutron scattering study. J Phys Chem B 111:3886–3892
122. Løvvik OM, Opalka SM (2006) Stability of Ti in $NaAlH_4$. Appl Phys Lett 88:161917
123. Løvvik OM, Opalka SA (2005) Density functional calculations of Ti-enhanced $NaAlH_4$. Phys Rev B 71:054103

124. Íñiguez J, Yildirim T, Udovic TJ, Sulic M, Jensen CM (2004) Structure and hydrogen dynamics of pure and Ti-doped sodium alanate. Phys Rev B 70:060101
125. Íñiguez J, Yildirim T (2005) First-principles study of Ti-doped sodium alanate surfaces. Appl Phys Lett 86:103109
126. Araújo CM, Li S, Ahuja R, Jena P (2005) Vacancy-mediated hydrogen desorption in $NaAlH_4$. Phys Rev B 72:165101
127. Du AJ, Smith SC, Lu GQ (2007) Vacancy mediated desorption of hydrogen from a sodium alanate surface: an ab initio spin-polarized study. Appl Phys Lett 90:143119
128. Liu J and Ge Q (2006) A precursor state for formation of $TiAl_3$ complex in reversible hydrogen desorption/adsorption from Ti-doped $NaAlH_4$. Chem Commun. 1822–1824
129. Liu J, Ge Q (2006) A first-principles analysis of hydrogen interaction in Ti-doped $NaAlH_4$ surfaces: structure and energetics. J Phys Chem B 110:25863–25868
130. Liu J, Ge Q (2007) A first-principles study of Sc-doped $NaAlH_4$ for reversible hydrogen storage. J Alloys Compd 446–447:267–270
131. Liu J, Han Y, Ge Q (2009) Effect of doped transition metal on reversible hydrogen release/uptake from $NaAlH_4$. Chem Eur J 15:1685–1695
132. Kubas GJ (2009) Hydrogen activation on organometallic complexes and H_2 production, utilization, and storage for future energy. J Organomet Chem 694:2648–2653
133. Chaudhuri S, Rangan S, Veyan JF, Muckerman JT, Chabal YJ (2008) Formation and bonding of alane clusters on Al(111) surfaces studied by infrared absorption spectroscopy and theoretical modeling. J Am Chem Soc 130:10576–10587
134. Chaudhuri S, Muckerman JT (2005) First-principles study of Ti-catalyzed hydrogen chemisorption on an Al surface: a critical first step for reversible hydrogen storage in $NaAlH_4$. J Phys Chem B 109:6952–6957
135. Ljubić I, Clary DC (2010) Towards understanding a mechanism for reversible hydrogen storage: theoretical study of transition metal catalysed dehydrogenation of sodium alanate. Phys Chem Chem Phys 12:4012–4023
136. Anton DL (2003) Hydrogen desorption kinetics in transition metal modified $NaAlH_4$. J Alloys Compd 356:400–404
137. Au M, Jurgensen A (2006) Modified Lithium Borohydrides for Reversible Hydrogen Storage. J Phys Chem B 110:7062–7067
138. Gunaydin H, Houk KN, Ozoliņš V (2008) Vacancy-mediated dehydrogenation of sodium alanate. Proc Natl Acad Sci 105:3673–3677
139. Grochala W, Edwards PP (2004) Thermal decomposition of the non-interstitial hydrides for the storage and production of hydrogen. Chem Rev 104:1283–1315
140. Bogdanović B, Ritter A, Spliethoff B (1990) Active MgH_2-Mg systems for reversible cehmical energy storage. Angew Chem Int Edit 29:223–234
141. Du AJ, Smith SC, Yao XD, Lu GQ (2007) Hydrogen spillover mechanism on a Pd-doped Mg surface as revealed by ab initio density functional calculation. J Am Chem Soc 129:10201–10204
142. Yoshimura K, Yamada Y, Okada M (2004) Hydrogenation of Pd capped Mg thin films at room temperature. Surf Sci 566:751–754
143. Shalaan E, Schmitt H (2006) Mg nanoparticle switchable mirror films with improved absorption-desorption kinetics. Surf Sci 600:3650–3653
144. Berlouis LEA, Honnor P, Hall PJ, Morris S, Dodd SB (2006) An investigation of the effect of Ti, Pd and Zr on the dehydriding kinetics of MgH_2. J Mater Sci 41:6403–6408
145. Saita I, Li LQ, Saito K, Akiyama T (2003) Hydriding combustion synthesis of Mg_2NiH_4. J Alloys Compd 356:490–493
146. Aguey-Zinsou KF, Ares-Fernandez JR (2008) Synthesis of colloidal magnesium: a near room temperature store for hydrogen. Chem Mat 20:376–378
147. de Jongh PE, Wagemans RWP, Eggenhuisen TM, Dauvillier BS, Radstake PB, Meeldijk JD, Geus JW, de Jong KP (2007) The preparation of carbon-supported magnesium nanoparticles using melt infiltration. Chem Mat 19:6052–6057

148. Li WY, Li CS, Ma H, Chen J (2007) Magnesium nanowires: enhanced kinetics for hydrogen absorption and desorption. J Am Chem Soc 129:6710
149. Kooi BJ, Palasantzas G, De Hosson JTM (2006) Gas-phase synthesis of magnesium nanoparticles: a high-resolution transmission electron microscopy study. Appl Phys Lett 89:161914
150. Schimmel HG, Huot J, Chapon LC, Tichelaar FD, Mulder FM (2005) Hydrogen cycling of niobium and vanadium catalyzed nanostructured magnesium. J Am Chem Soc 127: 14348–14354
151. Bystrzycki J, Plociński T, Zieliński W, Wiśniewski Z, Polanski M, Mróz W, Bojar Z, Kurzdlowski KJ (2009) Nano-engineering of magnesium hydride for hydrogen storage. Microelectron Eng 86:889–891
152. Huot J, Tremblay ML, Schulz R (2003) Synthesis of nanocrystalline hydrogen storage materials. J Alloys Compd 356:603–607
153. de Castro JFR, Yavari AR, LeMoulec A, Ishikawa TT, Botta WJ (2005) Improving H-sorption in MgH_2 powders by addition of nanoparticles of transition metal fluoride catalysts and mechanical alloying. J Alloys Compd 389:270–274
154. Dehouche Z, Peretti HA, Hamoudi S, Yoo Y, Belkacemi K (2008) Effect of activated alloys on hydrogen discharge kinetics of MgH_2 nanocrystals. J Alloys Compd 455:432–439
155. Lu HB, Poh CK, Zhang LC, Guo ZP, Yu XB, Liu HK (2009) Dehydrogenation characteristics of Ti- and Ni/Ti-catalyzed Mg hydrides. J Alloys Compd 481:152–155
156. Tanaka K, Miwa T, Sasaki K, Kuroda K (2009) TEM studies of nanostructure in melt-spun Mg-Ni-La alloy manifesting enhanced hydrogen desorbing kinetics. J Alloys Compd 478:308–316
157. Wagemans RWP, van Lenthe JH, de Jongh PE, van Dillen AJ, de Jong KP (2005) Hydrogen storage in magnesium clusters: quantum chemical study. J Am Chem Soc 127:16675–16680
158. Cheung S, Deng WQ, van Duin ACT, Goddard WA (2005) ReaxFF(MgH) reactive force field for magnesium hydride systems. J Phys Chem A 109:851–859
159. Aguey-Zinsou KF, Fernandez JRA, Klassen T, Bormann R (2007) Effect of Nb_2O_5 on MgH_2 properties during mechanical milling. Int J Hydrogen Energy 32:2400–2407
160. Larsson P, Araújo CM, Larsson JA, Jena P, Ahuja R (2008) Role of catalysts in dehydrogenation of MgH_2 nanoclusters. Proc Natl Acad Sci 105:8227–8231
161. Hanada N, Ichikawa T, Isobe S, Nakagawa T, Tokoyoda K, Honma T, Fujii H, Kojima Y (2009) X-ray absorption spectroscopic study on valence state and local atomic structure of transition metal oxides doped in MgH_2. J Phys Chem C 113:13450–13455
162. Hanada N, Ichikawa T, Fujii H (2005) Catalytic effect of nanoparticle 3D-transition metals on hydrogen storage properties in magnesium hydride MgH_2 prepared by mechanical milling. J Phys Chem B 109:7188–7194
163. Hanada N, Ichikawa I, Fujii H (2005) Catalytic effect of Ni nano-particle and Nb oxide on H-desorption properties in MgH_2 prepared by ball milling. J Alloys Compd 404:716–719
164. Dillon AC, Jones KM, Bekkedahl TA, Kiang CH, Bethune DS, Heben MJ (1997) Storage of hydrogen in single-walled carbon nanotubes. Nature 386:377–379
165. Cheng HS, Chen L, Cooper AC, Sha XW, Pez GP (2008) Hydrogen spillover in the context of hydrogen storage using solid-state materials. Energy Environ Sci 1:338–354
166. Züttel A, Orimo S (2002) Hydrogen in nanostructured, carbon-related, and metallic materials. MRS Bull 27:705–711
167. Wong-Foy AG, Matzger AJ, Yaghi OM (2006) Exceptional H_2 saturation uptake in microporous metal-organic frameworks. J Am Chem Soc 128:3494–3495
168. Zhao D, Yuan DQ, Zhou HC (2008) The current status of hydrogen storage in metal-organic frameworks. Energy Environ Sci 1:222–235
169. Li JR, Kuppler RJ, Zhou HC (2009) Selective gas adsorption and separation in metal-organic frameworks. Chem Soc Rev 38:1477–1504
170. Bhatia SK, Myers AL (2006) Optimum conditions for adsorptive storage. Langmuir 22: 1688–1700

171. Dinca M, Yu AF, Long JR (2006) Microporous metal-organic frameworks incorporating 1, 4-benzeneditetrazolate: syntheses, structures, and hydrogen storage properties. J Am Chem Soc 128:8904–8913
172. Dinca M, Dailly A, Liu Y, Brown CM, Neumann DA, Long JR (2006) Hydrogen storage in a microporous metal-organic framework with exposed Mn^{2+} coordination sites. J Am Chem Soc 128:16876–16883
173. Rowsell JLC, Millward AR, Park KS, Yaghi OM (2004) Hydrogen sorption in functionalized metal-organic frameworks. J Am Chem Soc 126:5666–5667
174. Rowsell JLC, Eckert J, Yaghi OM (2005) Characterization of H_2 binding sites in prototypical metal-organic frameworks by inelastic neutron scattering. J Am Chem Soc 127:14904–14910
175. Chatt J, Duncanson LA (1953) Olefin co-ordination compounds. Part III. Infra-red spectra and structure: attempted preparation of acetylene complexes. J Chem Soc 2939–2947
176. Zhao YF, Kim YH, Dillon AC, Heben MJ, Zhang SB (2005) Hydrogen storage in novel organometallic buckyballs. Phys Rev Lett 94:155504
177. Niu J, Rao BK and Jena P (1992) Binding of hydrogen molecules by a transition-metal ion. Phys Rev Lett 68:2277–2280
178. Belof JL, Stern AC, Eddaoudi M, Space B (2007) On the mechanism of hydrogen storage in a metal-organic framework material. J Am Chem Soc 129:15202–15210
179. Zhou W, Yildirim T (2008) Nature and tunability of enhanced hydrogen binding in metal-organic frameworks with exposed transition metal sites. J Phys Chem C 112:8132–8135
180. Kaye SS, Dailly A, Yaghi OM, Long JR (2007) Impact of preparation and handling on the hydrogen storage properties of Zn_4O(1, 4-benzenedicarboxylate)(3) (MOF-5). J Am Chem Soc 129:14176
181. Chen BL, Ockwig NW, Millward AR, Contreras DS, Yaghi OM (2005) High H_2 adsorption in a microporous metal-organic framework with open metal sites. Angew Chem-Int Edit 44:4745–4749
182. Zhou W, Wu H, Yildirim T (2008) Enhanced H_2 adsorption in isostructural metal-organic frameworks with open metal sites: strong dependence of the binding strength on metal ions. J Am Chem Soc 130:15268
183. Sun YY, Kim YH, Zhang SB (2007) Effect of spin state on the dihydrogen binding strength to transition metal centers in metal-organic frameworks. J Am Chem Soc 129:12606
184. Frost H, Duren T, Snurr RQ (2006) Effects of surface area, free volume, and heat of adsorption on hydrogen uptake in metal-organic frameworks. J Phys Chem B 110: 9565–9570
185. Lochan RC, Head-Gordon M (2006) Computational studies of molecular hydrogen binding affinities: the role of dispersion forces, electrostatics, and orbital interactions. Phys Chem Chem Phys 8:1357–1370
186. Han SS, Deng WQ, Goddard WA (2007) Improved designs of metal-organic frameworks for hydrogen storage. Angew Chem-Int Edit 46:6289–6292
187. Frost H, Snurr RQ (2007) Design requirements for metal-organic frameworks as hydrogen storage materials. J Phys Chem C 111:18794–18803
188. Han SS, Goddard WA (2007) Lithium-doped metal-organic frameworks for reversible H_2 storage at ambient temperature. J Am Chem Soc 129:8422
189. Mulfort KL, Hupp JT (2007) Chemical reduction of metal-organic framework materials as a method to enhance gas uptake and binding. J Am Chem Soc 129:9604
190. Blomqvist A, Araújo CM, Srepusharawoot P, Ahuja R (2007) Li-decorated metal-organic framework 5: a route to achieving a suitable hydrogen storage medium. Proc Natl Acad Sci 104:20173–20176
191. Mavrandonakis A, Tylianakis E, Stubos AK, Froudakis GE (2008) Why Li doping in MOFs enhances H_2 storage capacity? A multi-scale theoretical study. J Phys Chem C 112: 7290–7294

192. Klontzas E, Mavrandonakis A, Tylianakis E, Froudakis GE (2008) Improving hydrogen storage capacity of MOF bv functionalization of the organic linker with lithium atoms. Nano Lett 8:1572–1576
193. Lan J, Cao D, Wang W (2010) Li-doped and nondoped covalent organic borosilicate framework for hydrogen storage. J Phys Chem C 114:3108–3114
194. Cao D, Lan J, Wang W, Smit B (2009) Lithium-doped 3D covalent organic frameworks: high-capacity hydrogen storage materials. Angew Chem Int Edit 48:4730–4733
195. Kesanli B, Cui Y, Smith MR, Bittner EW, Bockrath BC, Lin WB (2005) Highly interpenetrated metal-organic frameworks for hydrogen storage. Angew Chem-Int Edit 44:72–75
196. Rowsell JLC, Yaghi OM (2006) Effects of functionalization, catenation, and variation of the metal oxide and organic linking units on the low-pressure hydrogen adsorption properties of metal-organic frameworks. J Am Chem Soc 128:1304–1315
197. Orimo S, Züttel A, Schlapbach L, Majer G, Fukunaga T, Fujii H (2003) Hydrogen interaction with carbon nanostructures: current situation and future prospects. J Alloys Compd 356:716–719
198. Wu XB, Chen P, Lin J, Tan KL (2000) Hydrogen uptake by carbon nanotubes. Int J Hydrogen Energy 25:261–265
199. Chambers A, Park C, Baker RTK, Rodriguez NM (1998) Hydrogen storage in graphite nanofibers. J Phys Chem B 102:4253–4256
200. Ye Y, Ahn CC, Witham C, Fultz B, Liu J, Rinzler AG, Colbert D, Smith KA, Smalley RE (1999) Hydrogen adsorption and cohesive energy of single-walled carbon nanotubes. Appl Phys Lett 74:2307–2309
201. Darkrim FL, Malbrunot P, Tartaglia GP (2002) Review of hydrogen storage by adsorption in carbon nanotubes. Int J Hydrogen Energy 27:193–202
202. Shiraishi M, Takenobu T, Ata M (2003) Gas-solid interactions in the hydrogen/single-walled carbon nanotube system. Chem Phys Lett 367:633–636
203. Kajiura H, Tsutsui S, Kadono K, Kakuta M, Ata M, Murakami Y (2003) Hydrogen storage capacity of commercially available carbon materials at room temperature. Appl Phys Lett 82:1105–1107
204. Dag S, Ozturk Y, Ciraci S, Yildirim T (2005) Adsorption and dissociation of hydrogen molecules on bare and functionalized carbon nanotubes. Phys Rev B 72:155404
205. Kim BJ, Lee YS, Park SJ (2008) A study on the hydrogen storage capacity of Ni-plated porous carbon nanofibers. Int J Hydrogen Energy 33:4112–4115
206. Kim HS, Lee H, Han KS, Kim JH, Song MS, Park MS, Lee JY, Kang JK (2005) Hydrogen storage in Ni nanoparticle-dispersed multiwalled carbon nanotubes. J Phys Chem B 109:8983–8986
207. Liu C, Chen Y, Wu CZ, Xu ST, Cheng HM (2010) Hydrogen storage in carbon nanotubes revisited. Carbon 48:452–455
208. Jordá-Beneyto M, Suárez-García F, Lozano-Castelló D, Cazorla-Amorós D, Linares-Solano A (2007) Hydrogen storage on chemically activated carbons and carbon nanomaterials at high pressures. Carbon 45:293–303
209. Yildirim T, Ciraci S (2005) Titanium-decorated carbon nanotubes as a potential high-capacity hydrogen storage medium. Phys Rev Lett 94:175501
210. Yildirim T, Íñiguez J, Ciraci S (2005) Molecular and dissociative adsorption of multiple hydrogen molecules on transition metal decorated C_{60}. Phys Rev B 72:153403
211. Sun Q, Wang Q, Jena P, Kawazoe Y (2005) Clustering of Ti on a C_{60} surface and its effect on hydrogen storage. J Am Chem Soc 127:14582–14583
212. Wang Q, Sun Q, Jena P, Kawazoe Y (2009) Theoretical study of hydrogen storage in Ca-Coated fullerenes. J Chem Theory Comput 5:374–379
213. Kiran B, Kandalam AK, Jena P (2006) Hydrogen storage and the 18-electron rule. J Chem Phys 124:224703
214. Shevlin SA, Guo ZX (2006) Transition-metal-doping-enhanced hydrogen storage in boron nitride systems. Appl Phys Lett 89:153104

215. Shevlin SA, Guo ZX (2008) High-capacity room-temperature hydrogen storage in carbon nanotubes via defect-modulated Titanium doping. J Phys Chem C 112:17456–17464
216. Li M, Li Y, Zhou Z, Shen P, Chen Z (2009) Ca-Coated boron fullerenes and nanotubes as superior hydrogen storage materials. Nano Lett 9:1944–1948
217. Gagliardi L, Pyykkö P (2004) How many hydrogen atoms can be bound to a metal? Predicted MH_{12} species. J Am Chem Soc 126:15014–15015
218. Yoon M, Yang SY, Wang E, Zhang ZY (2007) Charged fullerenes as high-capacity hydrogen storage media. Nano Lett 7:2578–2583

Part V
Energy Efficiency and Saving

Use of Nanostructures for High Brightness Light-Emitting Diodes

G. B. Stringfellow

Abstract Light-emitting diodes or LEDs are expected to play a major role in efforts to utilize less energy for lighting applications due to their high efficiency, long operating life, and other "green" characteristics. The history of LEDs began in the 1960s. Since that time, the performance has increased exponentially while the cost has decreased dramatically. LEDs dominate the market for monochromatic displays and indicators, and are slated to provide an increasing share of the white light market. During the last decade, advances in efficiency have been obtained partly as a result of the use of nanotechnology. LEDs and lasers provided some of the first applications for quantum-well structures with nm dimensions. Future advances will almost certainly be linked to advances in the use of quantum wire and quantum dot structures. They appear to offer attractive new alternatives for single-junction white light generation. The use of self-assembled structures also offers the promise of allowing the fabrication of high efficiency devices in highly defected materials, such as those grown on less expensive substrates. This chapter reviews the basic aspects of LED devices and materials, with a focus on the AlGaInP system for red and yellow emitters and AlGaInN for blue, green, and white emitters, all grown by the organometallic vapor phase epitaxial technique. The focus is on the present and future use of nanotechnology for lighting applications.

G. B. Stringfellow (✉)
Department of Materials Science and Engineering, University of Utah,
Salt Lake City, UT, USA
e-mail: stringfellow@eng.utah.edu

L. Zang (ed.), *Energy Efficiency and Renewable Energy Through Nanotechnology*,
Green Energy and Technology, DOI: 10.1007/978-0-85729-638-2_23,

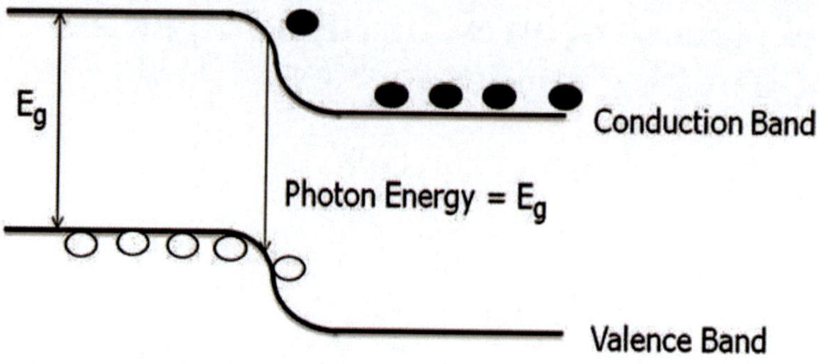

Fig. 1 Schematic diagram of a LED structure showing the electron–hole recombination giving a photon with energy equal to the bandgap energy

1 Introduction

Light emitting diodes, or LEDs, are p/n junction semiconductor devices that convert electrons flowing through the device into nearly monochromatic photons, as shown schematically in Fig. 1, with an energy (color) determined by the bandgap energy of the semiconductor. Using two III/V alloy systems, AlInGaP and AlInGaN, the bandgaps can be tuned to give highly efficient generation of any color of the visible spectrum, as well as IR or UV light, as shown in Figs. 2 and 3. LEDs have much longer operating lifetimes than any other conventional light source. They can operate continuously for as long as 10 years (100,000 h) before the light output declines significantly and normally do not fail catastrophically. Because the LEDs are typically small, about the size of a grain of salt, the direction of light output is easily tailored to suit particular applications. They turn on and off in less than a microsecond, so there is no irritating delay, as for fluorescent lights, and they can be used with dimmers. Their rapid modulation capability also allows the use of LEDs and lasers in high-bandwidth, optical communications applications. Furthermore, LEDs do not contain harmful substances, such as mercury, so they are much more environmental friendly.

Light emitting diodes are sometimes referred to as the "ultimate light bulb" because they approach the theoretical maximum efficiency for turning electrical current into light. External quantum efficiencies (photons out per electron in) exceeding 50% have been achieved in both the red and blue regions of the spectrum. The efficiency is plotted as a function of photon wavelength in Fig. 4 [36–38, 51, 57, 58, 95]. It is seen that the efficiency is not as high in the mid-spectral range, resulting in the so-called "green gap".

LEDs are inherently monochromatic; thus, they are more than 20 times more efficient than filtered incandescent lights for applications requiring a particular color, such as the red brake lights on cars and the red, green, and amber lights in traffic semaphores. This, plus their very long operating lives, has led to their complete dominance for these applications. Even for white lights, LEDs are more efficient by over a factor of five compared to incandescent sources. They are

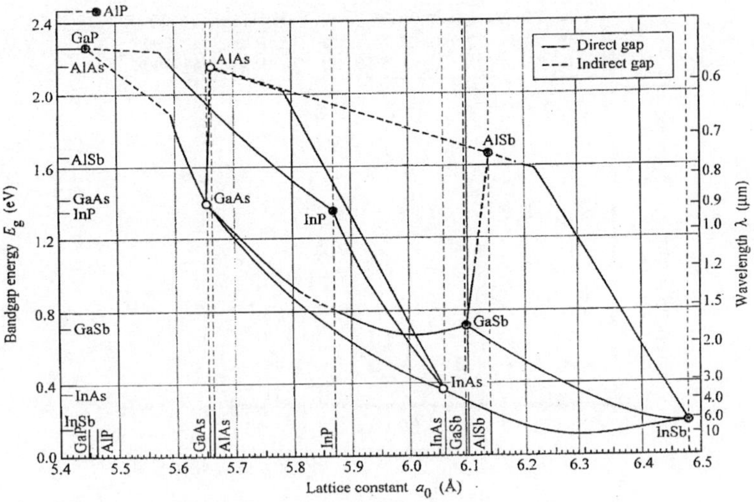

Fig. 2 Energy gap vs lattice constant for common semiconductor elements, compounds, and alloys not containing nitrogen (after P. K. Tien, 1988, Unpublished work from AT&T Bell Laboratories with permission)

Fig. 3 Bandgap versus lattice constant for AlInGaN alloys (after Stringfellow [85], with permission)

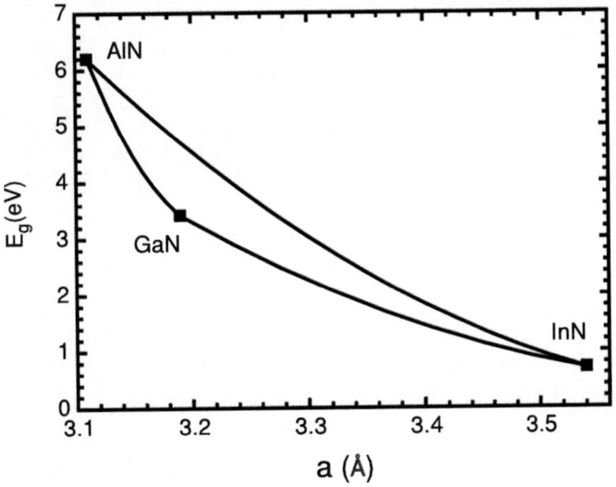

significantly more expensive than conventional light bulbs, but their efficiency and long operating life make them economically preferable to conventional light bulbs after less than 2 years of operation. Since 50% of residential lighting is provided by 60 Watt incandescent bulbs, replacement by white LED light bulbs offers a potentially staggering savings in energy usage. If all incandescent lamps in the USA were replaced by LEDs, the energy savings would be approximately 1000 TW-h/year (estimated to be equivalent to the energy used to light 15 million homes), representing cost savings of $100 B/year. This would also lead to a decrease in the annual amount of carbon emissions into the atmosphere of 200 M

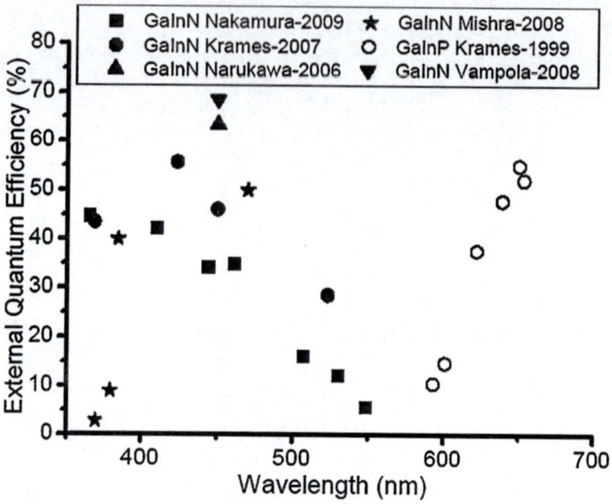

Fig. 4 External quantum efficiency of LEDs versus emission wavelength comparing data for AlInGaN (solid data points) with AlInGaP (open data points)

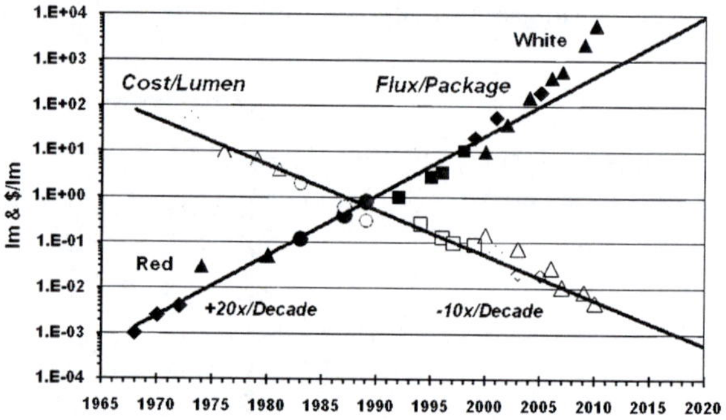

Fig. 5 Time evolution of LED luminous flux and purchase price per lumen (R. Haitz, 2010, private communication)

tons [38]. In the near future, the efficiencies of white lights will increase somewhat and the cost will decrease dramatically, as is common with virtually all semiconductor devices; as the volume increases the unit cost decreases. Both trends are shown in Fig. 5 (Haitz 2010). The sales volume of white LED lights has soared recently due to applications where LEDs dominate on the basis of factors other than cost, such as headlamps for automobiles and the light sources for backlit liquid–crystal display (LCD) television screens. The competing plasma TV screens are so energy inefficient that they have been banned in California after 2011. It is estimated that the switch to LCD/LED TVs will save $8.1 billion in energy costs according to the California Energy Commission, as reported by NewsFactor on November 20, 2009.

The other light-emitting device that will be discussed in this chapter is the injection laser. It operates on the same principle as the conventional LED. In fact, the laser operates as an LED until a certain threshold current is exceeded. For higher currents, stimulated emission dominates spontaneous emission and the device emits a high-intensity laser beam. For a laser, the p/n junction is enclosed in an optical cavity to allow a build-up of photon intensity to promote stimulated emission.

Lasers in the visible region of the spectrum are used in display and projection applications. They are also the key components of optical memories, including CD and DVD drives. Lasers are necessary components of high speed optical communication systems. The optimum wavelengths for high frequency, long-distance communications is in the IR, at either 1.3 or 1.55 μm. These lasers are typically fabricated in the InGaAsP quaternary alloy system [34, 83, Chap. 10]. Operation in the red region of the spectrum is suitable for short distance communication via plastic fibers. Si detectors have their maximum photon detection efficiency in the red. Red semiconductor lasers are also being considered for quantum cryptography applications, where single photon sources are required [72]. A particularly important application is for high speed optical interchip communications. This requires integration of III/V lasers with the Si ICs. Thus, the growth of high quality III/V semiconductors on Si is an area of intense research activity.

The performance of LEDs has evolved exponentially over the last 30 years with order-of-magnitude increases in efficiency coupled with order-of-magnitude decreases in cost, as seen in Fig. 5. The threshold current densities of lasers have decreased by orders of magnitude during this time. These advances have been based on an increased understanding and control of the materials properties as well as the device and package design. The importance of the formation and control of nanostructures has played a central role in these advances. Future efforts to reach higher performance at lower device costs appear to be dependent on developing techniques for the low-cost use of other nanostructures, including quantum wells, quantum wires, and quantum dots (QDs), nanostructures with reduced dimensionality in 1–3 dimensions, respectively.

Some benefits of nanostructures for laser performance are similar to those seen for LEDs; however, there are certain aspects of laser devices where the use of nanostructures is even more important. For example, the use of low-dimensional structures produces a density of states that is optimal for low threshold current densities [100, p. 213], as discussed in Sect. 4. In addition, the temperature dependence of the threshold current density is much superior for lasers fabricated in low-dimensional structures. The temperature dependence completely disappears for ideal QD lasers [100, p. 212]. The large band offsets in InP/AlInGaP QD structures is considered to be important for the fabrication of red-emitting lasers for quantum information applications [72]. In addition, the use of quantum wire structures to mitigate lattice-mismatch effects is a promising approach to the fabrication of high-performance lasers on Si for optical interconnect applications [92]. Finally, the concept of self-assembled laser structures in nanowires offers tremendous promise [3], especially since, in some ways, the performance of nanowire structures is superior to typical quantum-well device structures.

State-of-the-art LED and laser materials, structures, and devices are complex. Since all commercial LEDs and lasers are composed of III/V semiconductors and their alloys, this chapter will be limited to a discussion of these materials.

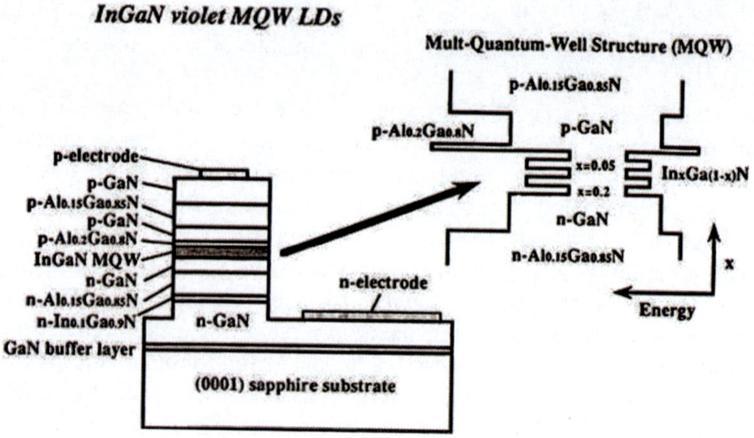

Fig. 6 The structure of InGaN MQW laser diode (after Nakamura [56], with permission)

An overview of the materials used can be obtained by examining the plots of bandgap energy versus lattice constant, Figs. 2 and 3. The bandgap energy is depicted as a unique function of solid composition in these plots, which is true only for random alloys. In other words, the effects of micro-and nano-structure have been neglected. This will be rectified in later sections. The two main materials for LEDs are the AlInGaP system, for red, orange, and yellow LEDs, and the AlInGaN system, for blue, green, and white LEDs. These two materials will be the focus of this chapter.

A high-performance LED or laser typically consists of several tens of layers of different semiconductor materials grown epitaxially on a particular substrate, as illustrated in Fig. 6. In order to emit light efficiently, the semiconductor materials must be direct bandgap and single crystalline, having a low density of mechanical defects, such as dislocations, and point defects including impurities, vacancies, interstitials, etc. This means that the materials and structures used in LED and laser devices must be grown by epitaxial techniques on single-crystalline substrates. Several epitaxial techniques have been used. Liquid phase epitaxy (LPE) and hydride vapor phase epitaxy (HVPE) were used for early LED and laser devices, but in the last two decades molecular beam epitaxy (MBE) and, particularly, organometallic vapor phase epitaxy (OMVPE—sometimes written MOVPE or MOCVD) have become predominant.

For the wavelengths required for specific applications, the simultaneous achievement of both increasing performance (efficiency of converting electrons into photons escaping from the device) and lower cost represents the basic challenge of LEDs and lasers in the coming decade. Reductions in cost are often obtained at the expense of creating more defects. One way of producing efficient light production in defected materials is to localize the recombination of electrons and holes to produce photons in small regions of highly perfect material using nanostructures, as discussed in Sect. 4.

Before examining the details of the semiconductor materials and structures used in light-emitting devices, with an emphasis on the role of nanostructures, it is worthwhile to enumerate the overarching concerns.

The first requirement is the ability to produce perfect and pure single-crystalline layers capable of high efficiency conversion of electrons crossing the p/n junction into photons of a particular wavelength (bandgap energy), determined by the application. Modern device structures require that multiple layers of several different materials be stacked upon one another via epitaxial growth.

A high conversion efficiency for the production of photons of a specified wavelength requires two factors: a high radiative recombination efficiency, or a short radiative lifetime, which, in turn, requires a direct bandgap and a large overlap of electron and hole wavefunctions and a long non-radiative lifetime, i.e., a slow rate of wasteful recombination that does not produce photons.

The ultimate goal is to have a high conversion efficiency even in a material with a high density of defects, such as dislocations. This would allow the growth of the desired layers on a substrate having a dissimilar lattice constant. Particularly important would be the growth of III/V structures on Si substrates. The ability to allow the various layers to have different lattice parameters is also highly desirable.

The photons produced at the p/n junction must be extracted efficiently, i.e., a large fraction must exit the device, rather than being absorbed.

The high conversion efficiency must be maintained at very high current densities while maintaining a long operating life. This allows for the efficient use of the expensive LED structures, i.e., more photons are extracted per unit area of the device. A problem observed for LEDs fabricated in the InGaN system is the so-called "droop", where the photon intensity becomes nearly constant (i.e., the efficiency decreases) for high current densities [38].

All of the above must be accomplished at a reasonable cost, since LED lighting competes directly with the very well-developed and low-cost alternatives of incandescent and fluorescent lighting.

In this introduction, these factors will be discussed briefly with special emphasis on the role of nanotechnology in producing the desired results. Later, in Sect. 4, the use of nanostructures will be described in more detail, followed by descriptions of the use of nanostructures to improve the performance of LEDs in several specific III/V materials systems.

This chapter will emphasize the use of single-crystalline III/V alloys for the production of visible photons. Organic LEDs are also used, especially for low-cost devices, but will not be discussed here. Other compound semiconductors can also be used for crystalline semiconductor LEDs, but all commercial LEDs are produced in III/V materials. Thus, this chapter will discuss only single-crystalline III/V semiconductor systems with emphasis on the two quaternary alloys AlInGaP and AlInGaN. The commercial production of LEDs in both of these materials systems uses the technique of OMVPE. This technique is versatile, so it can produce all of the materials and structures, including quantum wells, quantum wires, and QDs, required for LEDs and lasers. Thus, this chapter will focus on the

OMVPE technique, although results from layers grown by other techniques will be included where they add to our understanding of the effects of the nanostructure on materials properties and device performance.

The choice of bandgap energy nominally determines the color of the emitted light. Every color is needed between red (Eg = 1.9 eV) and blue (Eg = 2.5 eV) for various applications. This determines the materials from which the LEDs are fabricated, as indicated in Figs. 2 and 3. Devices emitting in the IR and UV spectral regions are also important for non-lighting applications.

The three primary colors, red, green, and blue, are required for white light and full color displays. The initial approach to making an LED of a particular color was based on the idea that the selection of solid composition uniquely determines the bandgap energy or the color of the light emitted from a particular material. For example, the lines shown in Figs. 2 and 3 are meant to indicate the bandgap energy of particular alloy compositions in the various bulk III–V materials (i.e., with no quantum confinement). However, more recently it has become clear that the solid composition does not uniquely determine the materials properties; the microstructure (more appropriately, the nanostructure) is a major determinant of the properties of alloy semiconductors and the devices made from them. The judicious control of nanostructure is an invaluable tool in the arsenal of modern materials and device designers. For example, the use of low-dimensional structures having dimensions smaller than the de Broglie wavelength of an electron, i.e., of dimensions of 1–30 nm, in 1, 2, or 3 directions allows tailoring of the emission energy (LED color) due to the quantum size effect. Confinement of electrons and holes together in these small regions can be used to improve the radiative recombination efficiency. Nanostructures can also have a positive effect on the light extraction efficiency as described in Sect. 4.1.

The use of quantum wells, where the electron and hole wave functions are confined in 1 direction, forming a two-dimensional (2D) system, was one of the first practical applications of nanotechnology for any material. A seminal paper by Dingle et al. [15] demonstrated that the bandgap of GaAs could be increased when a thin layer was sandwiched between higher bandgap AlGaAs layers, with the entire quantum-well structure grown by MBE. Today, quantum wells are used in essentially every commercial LED and laser device as a means of improving device performance. Subsequently, quantum wires and QDs, giving 1D and 0D systems, respectively, have been found to give additional enhancements in the performance of both LEDs and lasers as described in Sect. 4.1. The use of such nanostructures promises to improve green, and perhaps yellow and red, AlInGaN LEDs. The efficiency of green LEDs is much lower than for red and blue LEDs, as seen in Fig. 4. Nanostructures are being actively explored for the production of improved green laser and LED devices in the AlInGaN materials system. A particularly important example currently under intensive investigation is the use of QDs to allow the production of single-junction LEDs that the human eye perceives as white. Of course, white LEDs must emit at several wavelengths which, when added together, produce white light. This can be achieved by combining individual monochromatic LEDs of the three primary colors, red, green,

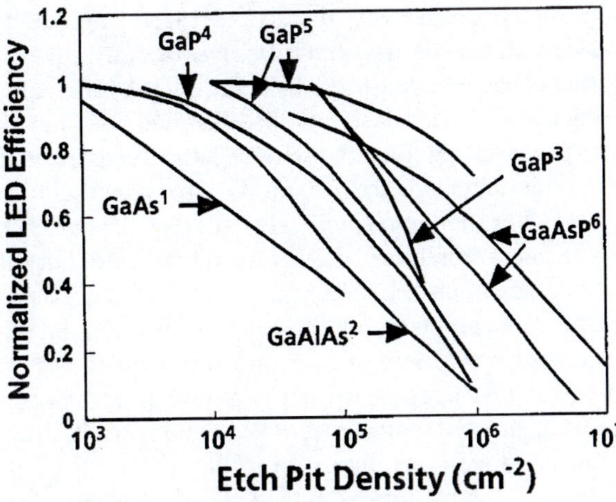

Fig. 7 LED efficiency versus dislocations density for conventional III/V materials (reprinted with permission from Lester et al. [47]. Copyright (1995), American Institute of Physics)

and blue (the RGB approach). In the future, the goal is to produce a single LED that simultaneously produces 2 or more colors that combine to produce light that appears white.

For the III/V semiconductors used in LEDs, the choice of the substrate material is vital. When the crystal structure and lattice constant of the epitaxial layer match those of the substrate, virtually perfect epitaxial layers can be produced. This severely limits the choice of materials for LEDs. As seen in Fig. 2, one choice for a so-called "lattice matched" system would be the growth of an InGaP epitaxial layer on a GaAs substrate; however, only one InGaP alloy satisfies this constraint, that with an InP mole fraction of approximately 0.5. When this condition is not satisfied, the lattice parameter mismatch gives rise to dislocations, generated to relieve the strain, which are formed in the interface plane and propagate up through the epitaxial layer. These dislocations have been known for decades to severely degrade device performance, as shown in Fig. 7 [47]. For high LED performance, dislocation densities of 10^4 cm^{-2} and less are needed for conventional III/V materials. The nitrides behave differently, which appears to be related to a spontaneously generated nanostructure, as discussed in detail in Sect. 4.3.

One approach to improve lattice-mismatched systems for LEDs is to develop a so-called "conformal" substrate. Such substrates allow the strain from the mismatch to be confined to the substrate, i.e., the substrate deforms to allow the growth of a dislocation-free epitaxial layer. All of the strain and any dislocations generated will be entirely confined to the substrate, where they will have no deleterious effect on device performance. This has been a topic of research for more than a decade, with less than satisfactory results for the large substrates required in order to contain the cost per device [9]. It will not be discussed further here, although there remains some hope of ultimate success.

A more promising approach is to form the light-emitting regions of the device in nano-wires or -rods that extend in a direction perpendicular to the substrate. For these very thin, wire-shaped single-crystalline materials, a change in lattice constant is accommodated elastically, by lateral relaxation of the wire, without the generation of dislocations [8]. This approach offers promise for the production of III/V lasers on Si substrates for inter chip IC communications [92]. Elastic deformation of the wire allows structures with a change in lattice constant along the length of the wire to be grown without the generation of misfit dislocations. For the "core/shell" structure, where the change in lattice constant occurs in the lateral direction, the layers are strained, but dislocations are not generated.

Another approach that uses nanotechnology is to somehow "passivate" the misfit dislocations produced in the epitaxial layer, i.e., to allow the efficient generation of photons even in a material with a very high dislocation density. This idea was first suggested for GaAs grown on Si substrates [23, 99]. It is an area where nanotechnology is believed to play a vital role for LEDs fabricated in the AlInGaN system, as discussed in Sect. 4.

Finally, the photons produced at the p/n junction must be extracted efficiently in order to produce a high efficiency LED. More than a decade ago, techniques were developed to minimize the absorption of photons within the semiconductor material [35], which often involves the removal of the absorbing substrate and wafer bonding to a new transparent "substrate". Use of nanowire growth to remove the lattice-match constraint might allow this expensive step to be eliminated.

In addition, techniques have been developed to reduce the change in refractive index experienced by the photons as they exit the high refractive index LED materials. This may involve coating the LEDs with high index, transparent materials. Shaping of the individual diodes is also used to enhance light extraction [36]. More recently, the use of nanotechnology has been developed to produce surfaces structures to enhance the light extraction efficiency [28].

Important aspects of controlling the cost of LEDs include: increasing yield in large-scale production techniques giving excellent uniformity of thickness and solid composition for the multiple thin layers of which the LED is composed; decreasing the cost of the substrate by using inexpensive materials or by developing processes to allow the re-use of the substrate; and producing materials that emit light with high external efficiencies at very high current densities for very long times, which allows a minimum of material to be used for the LED. Of course, the cost of the LED will scale with the device area for a given optical output. Currently, the efficiency of light output of blue and green GaN LEDs begins to "droop" at high current densities. The cause has not been unambiguously identified, but the use of nanotechnology is being explored to solve the problem.

In the remainder of this chapter, we will first review the fundamental aspects of LED performance, including the benefits of using nanotechnology to improve both cost and performance, the generation of light at a p/n junction and the extraction of the photons from the device. This will be followed by a review of the OMVPE growth technique used for the production of LED materials and structures. Finally, this background will be used as the basis of a discussion of the practical application of nanotechnology for LEDs and lasers in several specific systems, with an emphasis on LEDs emitting in the red-yellow spectral region, made from AlInGaP, and the green–blue spectral region, made from AlInGaN, including the use of this material for white LEDs.

2 LED Devices

The two major factors that must be optimized to produce an efficient LED are the recombination efficiency and the optical extraction efficiency. Both factors can be significantly enhanced via the judicious use of nanostructures.

Consider first the internal quantum efficiency, the fraction of electrons traversing the p/n junction that produce photons. The rate of the radiative process which produces the desired visible photons is the product of the np product and the recombination coefficient B_r [89, Chap. 13],

$$R_r = npB_r. \tag{1}$$

The former is determined by the majority carrier concentration and the deviation from equilibrium, i.e., the injected minority carrier concentration, and the latter by the fundamental properties of the semiconductor material, principally the band structure. Direct bandgap materials, where the conduction band minimum and the valence band maximum occur at the same place in k-space, allow rapid recombination of electrons and holes, with conservation of energy allowed by the creation of a photon with an energy equal to the difference in energy between the electron and hole in their initial states, as shown in Fig. 1. In indirect bandgap materials, a third particle, a phonon, is required to allow conservation of momentum. Because it is indirect, Si is a poor material for light emission. The high-bandgap energy III/V semiconductors may also be indirect, as shown in Fig. 2, where direct bandgap materials are indicated using solid lines and indirect materials by broken lines. Indirect bandgap materials can be used for LEDs [71, Chap. 12]; however, all commercial devices are made from direct materials. Thus, the LED and laser materials described in this chapter will be restricted to those with direct bandgaps.

At this point it is worth mentioning that when an indirect bandgap material is produced as low-bandgap QDs dispersed in a higher bandgap semiconductor, the bandgap may become direct [91], thus increasing the radiative recombination coefficient. In the simplest case, conservation of momentum becomes a simple selection rule, e.g., $\Delta n = 0$ [100, p. 65], where Δn is the difference in quantum numbers for the confined electron and hole states. However, in real systems, the converse, direct to indirect transitions, may occur for small QDs [49].

The electrons in the conduction band and holes in the valence band can also recombine non-radiatively. This normally occurs via defects either in the bulk (point defects or dislocations) or on the surface. Many III/V semiconductors have very high surface recombination rates. In general the non-radiative recombination rate is given by the equation [89, Chap. 13]:

$$R_{nr} = N_t \sigma v_{th} \Delta p, \tag{2}$$

where N_t is the concentration of defects, σ the capture cross-section, v_{th} the thermal velocity, and Δp the injected hole concentration, assuming the semiconductor to be n-type. The internal quantum efficiency is the ratio of the rate of radiative recombination to the total recombination rate, written

$$\eta_{int} = R_r/(R_r + R_{nr}). \tag{3}$$

High internal quantum efficiencies require both high values of B_r and low defect densities. Both of these factors can be enhanced in non-classical ways by using nanostructures. As the dimensionality of the recombination volume decreases, the electron and hole wavefunctions are forced to overlap, which increases the quantum mechanical matrix element for radiative recombination and, consequently, B_r. In addition, the exciton binding energy increases making excitonic recombination dominant, even at room temperature, which also increases the radiative recombination rate [100, p. 64].

An additional factor that affects B_r in (0001) quantum wells in the AlInGaN system is the quantum confined Stark effect (QCSE), which *decreases* the value of B_r. This is a serious issue for blue and green LEDs. One approach to circumventing this problem involves growing the quantum-well structures on crystallographic directions other than (0001), where the piezoelectric effect does not occur [57]. The use of very small QDs also appears to offer promise for reducing the QCSE by forcing an increase in wavefunction overlap, as discussed in Sect. 4.1 [54].

In addition to increasing the radiative recombination rate, nanostructures can also be used to suppress non-radiative recombination simply by insuring that the minority carriers are not able to diffuse to non-radiative defects [69, 85]. The diffusion of minority carriers can be reduced by producing fluctuations in the conduction and valence bands due to the controlled formation of compositional fluctuations. One way of producing spontaneous compositional fluctuations is to take advantage of the natural phenomenon of phase separation (PS) or spinodal decomposition [85]. This is generally believed to enhance the internal efficiency in InGaN LEDs. PS may also be a mechanism to explain the very high recombination efficiency observed in InGaN quantum wires, as compared with epitaxial layers [41], as discussed in Sect. 4.3. The effects of nanostructures on LED performance will be considered in more detail in Sect. 4.1.

The second major factor affecting LED performance is the photon escape efficiency, i.e., the probability of escape from the LED for photons generated by electron–hole recombination at the p/n junction. A major loss mechanism is due to the substrate, itself. When the substrate has a smaller bandgap than the photon energy, all photons propagating toward the substrate are absorbed. As an example, this is a major loss mechanism for InGaP LEDs grown lattice matched to GaAs substrates. Due to its small bandgap energy, GaAs absorbs all visible photons. Since photon emission in the GaInP is isotropic, the use of a GaAs substrate guarantees that approximately half of the emitted photons will be absorbed. The actual fraction absorbed is even greater due to total internal reflection of many photons as they attempt to exit the very high refractive index semiconductor [35, 71]. There are two ways to overcome this problem: 1. Grow on high-bandgap substrates that do not absorb the photons, which is often not practical due to the lattice parameter mismatch to such substrates, which results in the production of dislocations where non-radiative recombination occurs, and 2. Remove the growth substrate and wafer-bond the epitaxial layer structure to a transparent "substrate", typically GaP, for mechanical support [35]. The latter is the technique commonly used for InGaP LEDs. It results in very high efficiency red and yellow LEDs; however, the substrate removal and wafer-bonding steps add significantly to the cost of the finished LED.

The use of nanotechnology suggests a novel, alternative approach—namely the growth of quantum wires on a transparent substrate. Dislocations due to the lattice parameter mismatch do not propagate into the quantum wires, as discussed in Sect. 1. The strain is relieved elastically since the wires are able to relax in the two directions orthogonal to the growth axis. Thus, the LED structure can be grown directly on a high-bandgap substrate such as GaP. The use of nanowires for both LED and laser devices is an active research area that will be described in Sect. 4.1.

As mentioned above, the other factor leading to the absorption of photons is total internal reflection due to the large refractive index of III/V semiconductors. Only the photons with directions within a cone, having an angle given by Snell's Law, are able to escape from the semiconductor [71, Chap. 5]. Photons striking the surface at all other angles are reflected back into the semiconductor, where they can be absorbed by the LED structure, itself, or the substrate. Decades ago, device designers were able to increase the solid angle for photon escape simply by encapsulating the LED in a transparent, high index material such as silicone or plastic. The LED can also be shaped into structures, such as truncated pyramids, that increase the fraction of photons that can escape from the LED [36].

More recently, techniques based on nanotechnology have been developed to improve the probability of photon escape. Simply roughening the surface produces surface angles that allow more of the photons to escape total internal reflection [71, Chap. 9]. A more controlled approach is to produce a periodic arrangement of surface structures, a "photonic crystal", at the surface that prevents total internal reflection. For example, an array of GaN nano-pyramids produced by etching [28, 57] produces a significant increase in light extraction efficiency.

As seen in Fig. 4, the combination of increasing the internal efficiency, using quantum wells as the light-emitting region, and using various techniques to increase photon escape, has produced red and blue LEDs having amazingly high external efficiencies of 55% in the red and well-above 60% in the blue. Figure 6 shows the structure used for low threshold current density blue lasers. LED structures are very similar. It will be noted that the InGaN is grown on a transparent sapphire substrate. This results in a very high defect level in the active layers. However, the sapphire is transparent, so photons can escape without absorption in the substrate. Since sapphire is an insulator, the laser and LED structures have both the n- and p-contacts on the top surface. This necessitates the growth of a high-conductivity buried n-type layer between the sapphire and the p/n junction to avoid a high series resistance which results in low overall power efficiencies and non-uniform photon generation across the p/n junction. Typically, the LED devices are mounted upside-down so that the light emerges through the sapphire substrate. In some devices, the sapphire substrate is removed and the LED structure is bonded to a mirror mounted on an inexpensive, absorbing substrate, such as Si [57]. In this case, the p-contact is produced on the top of the device and the n-contact on the substrate. For this structure, which closely resembles the structure for InGaP LEDs, all of the photons are emitted from the top of the device.

3 Materials Issues for LEDs

The only materials to be discussed in this chapter are the III/V semiconductor alloys shown in Figs. 2 and 3. As mentioned above, the solid lines connecting the points for the individual compounds represent random ternary alloys, where the bandgap energy and lattice constant are uniquely specified by the solid composition. The areas between lines represent quaternary alloys, where the bandgap energy and lattice constant are independent parameters, both specified by the two solid composition parameters. These alloys were considered to be random or ideal during the early development of our understanding of these materials. The zinc-blende and wurtzite crystal structures in which the III/V materials crystallize can be thought of as two interpenetrating sublattices, with the group III atoms on one sublattice and the group V atoms on the other. A random ternary alloy consists of random mixing on one of the sublattices. Thermodynamically, this corresponds to an ideal mixture. As our understanding has matured, we have come to realize that non-random mixing, described in terms of the microstructure and, indeed, the nanostructure, can have a major role in determining the properties of the solid, particularly the bandgap energy, for an alloy having a specific composition. Thus, the bandgap energy is not uniquely specified by the solid composition, so the lines shown in Figs. 2 and 3 must be specified as being for random alloys.

Essentially all of the materials and structures described in this chapter are grown by the OMVPE technique. Early LED and laser devices were fabricated in the GaAs/AlGaAs system using the LPE technique. Problems with the growth of the phosphide and nitride alloys containing Al [79] as well as the difficulty of using LPE for the large-scale, high-yield processes required for inexpensive device manufacture have prevented it from becoming an important production method. Early LEDs in the GaAsP/GaAs system were made by HVPE, however, the same problems with the growth of Al-containing phosphide and nitride alloys has limited the application of this technique for LED and laser structures [79], although this process is used for the production of thick, compound semiconductor (typically GaP or GaN) layers for either window layers (GaP) or pseudo-substrate materials (GaN). MBE growth is a much more important technique. As mentioned above, it was the first to demonstrate the formation and advantages of nano-structures, particularly quantum wells, especially for laser devices. However, it has proven to be a more expensive and less flexible production technique than OMVPE, which now dominates the commercial production of both AlInGaP and AlInGaN for LED and laser applications. Thus, only the OMVPE technique will be described in what follows.

OMVPE is a cold-wall technique where the source molecules, organometallic group III precursors, such as trimethyl-Ga, -In, and -Al, TMGa, TMIn, and TMAl, respectively, are combined with either hydride or organometallic group V molecules. Examples of the group V precursors are the hydrides, AsH_3, PH_3, and NH_3, and organometallic precursors, tertiarybutylarsine (tBAs), tertiarybutylphospine (tBP), and dimethylhydrazine $(CH_3)\ _2H_2N_2$. The precursors are introduced into the gas stream flowing over the substrate either by metering them from high-pressure cylinders, in which they may be diluted in hydrogen (typical for the hydrides), or by bubbling hydrogen through temperature-controlled reservoirs (typical for the low vapor pressure organometallic precursors). The precursors are present at pressures less than their room temperature vapor pressures, so they do not condense on the room temperature reactor walls. The only hot part of the reactor is the susceptor, on which the substrates are placed. When the gasses reach the heated substrate, they decompose to produce the desired elements on the surface of the single-crystalline substrate. They diffuse rapidly on the surface, reaching low-energy sites, typically at step edges [83].

The growth temperature is chosen to be sufficiently high to pyrolyze the precursors and to allow adequate surface mobilities to allow the growth of a highly perfect, single-crystalline solid. Normally, very high temperatures are avoided to minimize the generation of defects (due to entropy considerations) and incorporation of undesirable background impurities. Also, the growth rate decreases at high temperature for these highly exothermic processes for thermodynamic reasons [83, Chap. 1].

The growth process is optimized with respect to the substrate temperature and orientation, the total flow rate of the group III precursors, which determines the growth rate, and the ratio of group V to group III elements introduced into the vapor stream, the so-called "V/III ratio". OMVPE is a growth process that occurs with a very high supersaturation. This means that the free energy of the solid semiconductor produced is very much lower than that of the precursors at the growth temperature [83, Chap. 2]. Thus, OMVPE is sometimes referred to as a "non-equilibrium" process. However, in fact, a near-equilibrium state is typically reached between the solid and the vapor adjacent to the substrate during growth [83, Chap. 2]. Thus, thermodynamics normally determines many of the aspects of the growth process, itself, and the properties of the resulting solid.

One advantage of OMVPE is the versatility: essentially all III/V compounds and ternary and quaternary alloys as well as low-dimensional structures can be produced. MBE and chemical beam epitaxy (CBE) offer similar versatility, but the large-scale of OMVPE and the need for UHV equipment for MBE and CBE gives significant economic advantages to OMVPE. Thus, it is virtually the only technique used for the commercial production of LEDs and lasers, as well as other devices such as high performance, multijunction solar cells.

Some results obtained by MBE are included in this Chapter. MBE is a process that also operates at very high supersaturation. In this case, the elements are introduced into the vapor, an ultra-high vacuum in this case, by simple evaporation. Thus, the pure elements reach the surface by ballistic transport, where they react to form the solid [94].

During OMVPE growth, thermodynamics can be used to understand the solid composition in terms of the composition of the vapor and the growth temperature. For alloys with mixing on the group III sublattice, such as AlInGaP and AlInGaN, grown at relatively low temperatures with an excess of the group V element in the vapor phase, the group III distribution coefficients are unity [83, Chap. 2]. This means that the ratios of the group III elements in the solid are the same as those in the input vapor. This is an extremely important factor that allows the growth of these alloys with good control of solid composition [83, Chap. 2]. Thus, the alloys we are interested in, with Al, Ga, and In mixed on the group III sublattice, can easily be grown across the entire compositional range by OMVPE, but not by other techniques such as LPE or HVPE [79]. This is the key factor that has led to the dominant position of OMVPE for the growth of LED materials. MBE is the other technique capable of producing these alloys with good control of solid composition [11].

At the high growth temperatures typically used for the OMVPE growth of GaN (>1000°C), the volatility of metallic In means that the In distribution coefficient is less than unity and decreases with increasing temperature [4, 56, 83, Chap. 2]. The distribution coefficients for mixing on the group V sublattice can also be predicted from thermodynamics and these distribution coefficients are often far from unity and dependent on the growth parameters [83, Chap. 2].

One other factor is vital for the determination of the composition of the solid, independent of the growth technique. The equilibrium composition of the solid is determined by minimization of the free energy of the system. Since the strain energy of an epitaxial layer can be large, when the lattice parameter of the thin epitaxial layer or island differs from that of the substrate, a strain energy term must be included in the free energy. This gives rise to the so-called “lattice pulling” effect. It was first observed in InGaP layers grown on GaAs substrates by LPE [77]. The equilibrium solid composition of the coherent InGaP epitaxial layer was found to differ from the composition of tall, thin, rod-shaped crystallites formed around the edges of the epitaxial layer. The composition of the epi-layer was always pulled toward that giving a lattice parameter match with the substrate, a value of approximately $x_{\mathrm{In}} = 0.5$. As discussed below, this is an important factor in determination of the composition of low-dimensional structures.

For thin epitaxial layers having a different equilibrium lattice constant than the substrate this difference is accommodated elastically. However, as the thickness of the epitaxial layer increases, a point is reached where the system begins to generate dislocation at the epilayer/substrate interface [71, Chap. 7]. The dislocations take the form of a cross-hatched array contained entirely in the interface plane. However, these dislocations interact, resulting in the propagation of so-called “threading dislocations” into the epitaxial layer. Dislocations are generally harmful to device

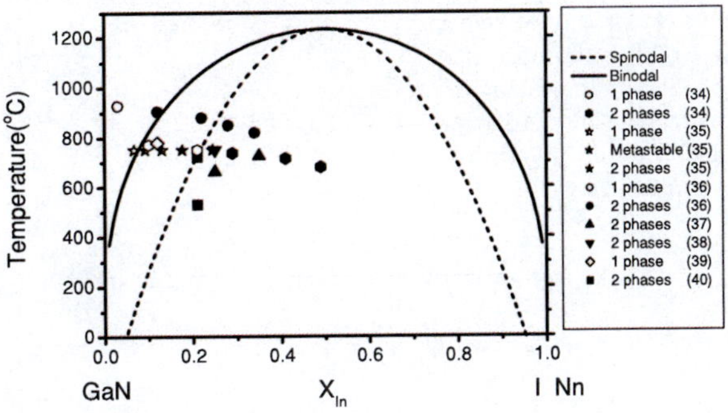

Fig. 8 Calculated bimodal (*solid*) and spinodal (*dashed*) curves [26] compared to experimental data [16, 18, 32, 63–65, 68] for InGaN. Open data points = single phase, half-filled data points = metastable, filled data points = 2 phase mixtures (after Stringfellow [85], with permission)

performance. They act as non-radiative recombination centers, and so reduce the internal quantum efficiency of LEDs, as shown in Fig. 7 and discussed in Sect. 1. They also result in an increase in the threshold current density of lasers. As described in Sect. 1, the exploration of methods for eliminating the misfit dislocations has been an important area of research, but no solutions have been discovered. Thus, the need for lattice matching, i.e., the selection of alloy composition to provide lattice matching with the substrate, has been a major constraint on the choice of materials to be used in LED structures. The use of quantum wires has recently emerged as a possible solution to this problem, as discussed in the next section.

The thermodynamics of semiconductor alloys was invoked above to describe the determination of solid composition and the effect of strain energy on the composition of strained epitaxial layers. An understanding of the thermodynamics of mixing of these alloys is also important for understanding the micro- or nano-structure produced spontaneously during growth. A number of models have been developed to understand the thermodynamics of III/V alloys. All of the models predict that the enthalpy of mixing is positive and is mainly determined by the size difference between the elements on the sublattice where mixing occurs [85]. The simplest model is the DLP model, where the enthalpy of mixing is proportional to the square of the difference in lattice constants between the two end components in a ternary alloy [78]. This correctly predicts that AlGaAs alloys are ideal solid solutions, i.e., the enthalpy of mixing (ΔH^M) is equal to 0, and that alloys, such as GaPN, where the lattice constants of GaP and GaN are very different, will have very large positive values of ΔH^M, leading to miscibility gaps [78]. The very important InGaN alloys are predicted to have a miscibility gap, as seen in Fig. 8 [26, 85]. This has enormous consequences for the micro- and nano-structures of these alloys [85].

Basic thermodynamics predicts that an AC–BC alloy with a positive enthalpy of mixing will spontaneously form clusters of like (AA or BB) atoms, while a negative enthalpy of mixing leads to ordering, where the number of AB pairs is larger than for a random alloy [88]. Thus, it was a surprise to discover that III/V alloys have a tendency to form *both* clusters and ordered structures. As described in detail below, clustering and PS are driven by the thermodynamics of the bulk for

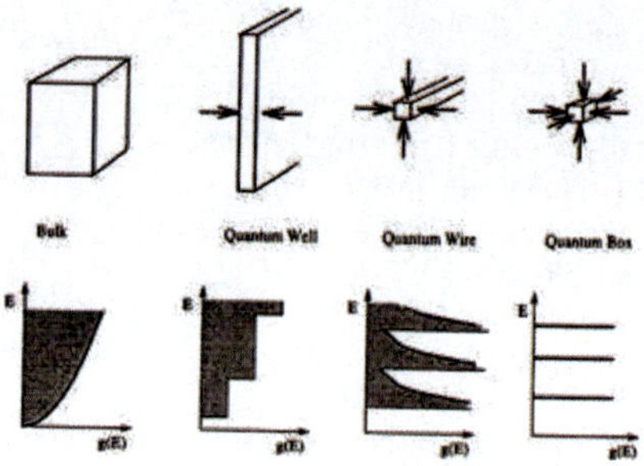

Fig. 9 Density of states, $g(E)$, as the number of confining dimensions is increased (after Adams and O'Reilly [2], with permission)

both InGaP and InGaN. However, when considering the surface thermodynamics, certain types of surface reconstruction lead to the formation of ordered structures, which may lead to significant reductions in the bandgap energy. Ordering is particularly important for the InGaP alloys used in red and yellow LEDs.

4 Natural and Artificial Nanostructures

Nanostructures will be defined here as structures with decreased dimensionality in either 1, 2, or all 3 directions, as shown in Fig. 8, with feature sizes of less than 30 nm. In the bulk, the zincblende or wurtzite structures are isotropic that is, the properties are independent of crystallographic direction and the wavefunctions for electrons (and holes) are Block functions, traveling waves in all 3 directions. The energy is nearly continuously variable. When the dimensionality is reduced in 1 direction (z), forming a 2D system, the wavefunctions in the x and y directions are still travelling waves. However, in the z direction, the wavefunctions are standing waves. This gives quantization of the energy in this direction. This is termed a quantum well when the well thickness is less than a de Broglie wavelength, approximately 30 nm. One may also count ordered structures as a part of this group of nanostructures since they consist of layered structures along specific crystallographic directions, forming a 2D system, although the periodicity is less than 1 nm. When the lattice periodicity is interrupted in two directions, the resulting structures are 1D quantum wires when the size of the structure is less than 30 nm.

When the lattice periodicity is interrupted in all 3 directions, forming a 0-D system, and the size is less than 30 nm, the structure is termed a QD. In this case, energy is quantized in all 3 directions. This gives a density of states for electrons like that of an atom, as shown in Fig. 9, consisting of discrete states, as discussed below. As the dimensionality of the system decreases the gain spectrum for lasing increases and becomes narrower [5], giving rise to a decrease in the threshold current density, J_{th}, as shown in Fig. 10. These nanostructures may form naturally during epitaxial growth. The spontaneous formation of both ordered structures and QDs will be discussed below. Additionally, quantum wells, wires, and dots can be

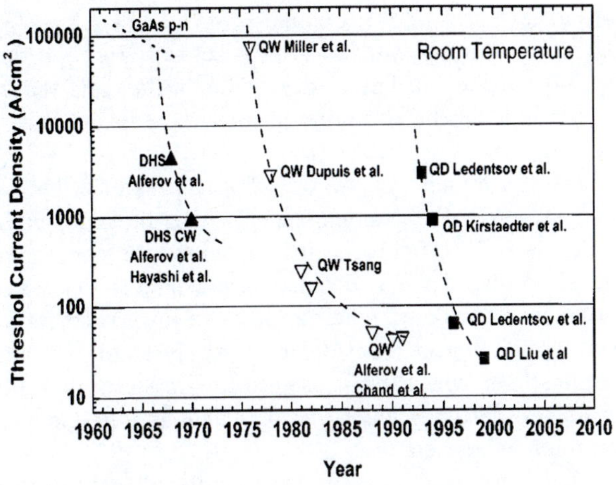

Fig. 10 Reduction of threshold current density from homojunction laser to double heterostructure, quantum-well, and quantum-dot lasers (reprinted with permission from Ledentsov et al. [44]. Copyright (2000), IEEE)

formed artificially, by switching the composition during growth, the use of masks, and other techniques. These will also be discussed in Sect. 4.1.

The first spontaneously-formed nanostructure to be discussed is atomic-scale ordering. In the 1980s, transmission electron microscopy (TEM) studies led to the surprising discovery that alloys such as AlGaAs, InGaP, and GaAsSb consisted of alternating monolayers (MLs) having different compositions along a particular crystallographic direction [82]. This was a phenomenon observed many years earlier in metal alloy systems; hence the nomenclature used refers to metallic systems: Cu-Pt (with ordering of {111} planes) and Cu-Au (with ordering of {100} planes). However, in metallic systems, the ordering process is driven by the thermodynamics of the bulk alloys. For systems with negative values of the ΔH^M, ordering is predicted to occur on the basis of simple thermodynamic calculations [83, 88, Chap. 2]. However, as discussed above, the enthalpy of mixing in semiconductor alloys is always greater than or equal to zero. Eventually, this apparent paradox was explained using surface thermodynamics. Real, low-index surfaces in semiconductor systems reconstruct [83, Chap. 3]. For growth by OMVPE where the V/III ratio is much greater than unity, the surface is typically terminated by group V atoms. The formation of dimer bonds between the group V atoms produces a (2×4) type of reconstruction composed of (110) dimer rows. The strain in the layers beneath these dimer rows provides a thermodynamic driving force for the formation of the Cu-Pt ordered structure [83, Chap. 3]. Thus, ordering is an example of a self-assembly process driven by surface thermodynamics.

Today, the degree of order has been firmly, empirically linked to the surface reconstruction [84]. An understanding of the thermodynamics and kinetics of growth has led to the ability to control the degree of order in epitaxial layers using parameters such as the growth temperature, V/III ratio, growth rate, and substrate misorientation [84]. Perhaps most interesting is the use of surfactants to control ordering. It is found that elements such as Sb and Bi added to the system in small quantities during epitaxial growth act to suppress ordering. This is believed to be mainly due to the larger size of the surfactants that accumulate on the surface, which reduces the strain in the subsurface layers and, hence, the driving force for ordering [84].

For the InGaP system, ordering has major effects on the materials properties. For example, the bandgap energy can be reduced by 160 meV due to Cu-Pt ordering [87]. Of course, this has enormous consequences for LEDs. Those produced in disordered InGaP lattice matched to GaAs emit red light, whereas those produced in ordered material produce infrared radiation. Thus, ordering must be controlled. Similar considerations apply to the InGaP used in tandem solar cell applications, where, Sb is used to suppress ordering. In the important InGaN alloy system, ordering is much less prevalent and no one has yet demonstrated an effect on the bandgap energy.

Artificial structures similar to ordered structures are intentionally produced during epitaxial growth to produce superlattices and multiple quantum-well structures. For the production of the low-dimensional structures required for modern LED and laser devices, the growth technique must allow rapid changes in solid composition. The change in solid composition must occur over a distance of a single atomic layer for the growth of very thin quantum wells. MBE was the first technique to demonstrate such exquisitely fine control. However, extensive development efforts eventually allowed these fine structures to be controlled using OMVPE. Abrupt changes in solid composition require several factors: rapid gas switching, laminar flow (eddies provide for virtual memory effects), little or no memory due to adsorption/desorption on reactor walls, and atomically flat surfaces during growth [83, Chap. 9].

A novel type of quantum-well construction involves the growth of InGaN structures on patterned substrates. The In content and quantum-well thickness are dependent on the facet orientation [20, 21]. This gives the simultaneous emission of several different colors from a single epitaxial layer, which is potentially important for single-junction white LEDs.

Techniques similar to those described above were used in the initial attempts to produce quantum wire structures. In the GaAs/AlGaAs system, grown by MBE, Petroff et al. [62] produced quantum wire structures in the growth plane by using deposition at step edges on intentionally misoriented substrates. Selective growth techniques have also been used on patterned substrates to produce lateral quantum wires [31, 97]. Today, most quantum wire structures are produced in a vertical direction by using selective growth on the surface catalyzed by various techniques, including the addition of tiny Au balls to the surface (see, for example, Siefert et al. [74], Lauhon et al. [43], and Quian et al. [67]. In this case, the liquid Au catalyzes the low-temperature OMVPE growth, resulting in the production of a high density of nearly vertical wires each topped by a Au drop. The diameter of the wire is determined by the diameter of the Au droplets on the surface. Techniques have also been developed for the growth of nanowires without the use of Au droplets. As discussed below, quantum wires have properties that can be quite different than the properties of thin epitaxial layers of the same solid composition, so are potentially interesting for both LED and laser devices.

The other nanostructure that yields desirable properties for light-emitting devices is the QD. Early research used masking techniques to produce 0D quantum structures [100, Chap. 6]. However, such techniques are unlikely to produce the desired results in an economical fashion. Thus, more recent research has focused on the self-assembly of collections of QDs. It is desirable to produce a high density of mono-dispersed QDs using rapid, simple growth techniques. The most promising technique is to take advantage of the Stranski–Krastanov (S–K) growth of strained epitaxial layers.

Strained epitaxial layers in semiconductor systems are normally found to grow by the S–K mechanism, involving the formation of a high density of small islands as well as a thin wetting layer [40, 96]. The S–K process for growth of a lattice-mismatched epitaxial layer produces a series of islands because the strain energy is lower than for a uniform layer; thus, there is a thermodynamic driving force for island formation [40, 73]. In S–K growth, a thin 1–2 ML uniform "wetting" layer is first formed by 2-D growth on the substrate. As growth continues, an array of three-dimensional (3D) islands forms spontaneously. The transition from 2-D to 3-D growth at a certain critical layer thickness has been observed for many semiconductor materials, grown by many techniques.

The phenomenon of S–K growth has been widely studied because the islands formed may constitute, when properly controlled, an array of QDs (QDs). This self-assembly process is a candidate as a practical technique for the fabrication of QDs for commercial injection laser devices, as discussed below. The reduced dimensionality of QDs gives rise to markedly superior device performance [45].

Among the first systems studied was the growth of Ge QDs on Si substrates. Strain-induced self-assembly of coherent islands was observed for Ge layers nominally a few MLs in thickness (see, for example, Ross et al. [70] and references therein). Of the III/V semiconductors, the growth of InAs and InGaAs QDs on GaAs has been perhaps the most widely studied. For example, Leonard et al. [46] found that $In_{0.5}Ga_{0.5}As$ coherently strained islands were spontaneously formed on (001) GaAs substrates by MBE after deposition of 4 ML of InGaAs. Moison et al. [53] studied the MBE deposition of InAs on GaAs. They found the transition from 2D to 3D growth to occur at a coverage of 1.75 ML. AFM measurements showed that islands 3 nm in height and 24 nm across were formed. The value of this self-assembly technique was shown by the remarkably uniform size of the islands formed. This is, of course, of vital importance for achieving the desired device performance. The OMVPE technique has been used to produce multi-layered QD structures in this system that have been used to produce injection laser devices with very low threshold current densities [45]. Interestingly, an early use of these InAs QD structures was to localize recombination in GaAs layers grown on Si substrates, where the high dislocation density kills the radiative recombination efficiency in homogeneous GaAs layers. When the density of the QDs exceeds the dislocation density, the InAs QDs collect the minority carriers for radiative recombination before they can reach a dislocation [23, 100].

4.1 Advantages of Nanotechnology

The major advantage of the use of low-dimensional structures with lengths of <30 nm was first recognized when quantum wells were found to allow GaAs, which has a bandgap energy giving IR photons, to emit visible photons [15]. This is based on the well-known quantum mechanical concept of a “particle in a box”. For a 1-D box with a potential of 0 within the well and with infinite barriers, the energy levels of electrons are quantized

$$E_{\mathrm{n}} = \hbar^2/2m\,(n\pi/L_z)^2 \tag{4}$$

where L_z is the dimension of the 1D box, m is the effective mass, and n is the quantum number. This means that the lowest energy level of an electron in the conduction band of the well layer increases as L_z decreases. The same is true of the holes in the valence band. Thus, as the dimension L_z decreases the effective bandgap energy increases. This is termed the “quantum size effect” or QSE. For a finite well, the QSE is less than for the infinite well and the magnitude plateaus for very thin wells. Nevertheless, the QSE for electrons and holes results in photon energies considerably larger than the bandgap of the bulk materials. This adds an important degree of freedom to the design of the optimum material for an LED producing a specific wavelength of light. However, the advantages of low-dimensional nanostructures do not end here. By confining the electrons and holes to a small space, the wavefunction overlap increases, which leads to an increase in the matrix element for radiative recombination, in other words, the electrons and holes are squashed together which makes recombination to produce a photon more rapid. An additional advantage of low-dimensional structures for light-emitting devices is the increase in exciton binding energy [100]. This leads to exciton recombination at room temperature, which further significantly increases the rate of radiative recombination. The localization of recombination can also provide significant benefits in materials with high densities of dislocations and other defects by insuring that the electrons and holes are captured before they can reach the defects, where non-radiative recombination occurs. Thus, localization of recombination increases the radiative recombination rate and decreases the non-radiative recombination rate. Both effects lead to an increase in the internal quantum efficiency.

There is a potentially negative effect of electron and hole localization. The QCSE leads to a decrease in internal quantum efficiency for LEDs in (0001) quantum wells in the nitride systems. The nitrides crystallize in the hexagonal wurtzite structure, where strain in the (0001) direction gives rise to an electric field due to the piezoelectric effect. This field forces electrons and holes apart in the quantum well, leading to a reduction in the matrix element for radiative recombination [57]. However, this effect is diminished in very small QDs, where the electrons and holes are forced to occupy the same space in spite of the QCSE [54]. Thus, in addition to the benefits described above, the use of very small QD structures has the benefit of reducing the QCSE in the InGaN alloy system.

This offers the hope of a future improvement of green, yellow, and even red LEDs in this alloy system.

The use of coupled quantum wells opens the door to another type of laser, particularly suited to emission at very long wavelengths. In quantum-cascade lasers, the light-emitting transition occurs between the confined electron states in the quantum well [89, Chap. 12]. Thus, in this case, the photon energy can be much smaller than the bandgap energy. These lasers have been designed to operate at wavelengths as long as 70 μm.

The low-dimensional nanostructures produce a fundamental alteration of the band structure for semiconductor materials. In bulk material the density of states (number of states/unit of energy) increases as $E^{1/2}$ at the bottom of the band. For a quantum well, the density of states becomes independent of energy, with a much larger density of states at the bottom of the band. This leads directly to a decrease in threshold current density of lasers. Progressively more favorable density of states profiles in the conduction and valence bands are produced as the dimensionality of the nanostructures decreases, as shown in Fig. 9. This leads to increases in the gain at a given current density, and consequently progressive reductions in threshold current density for quantum wires and QDs, as seen in Fig. 10. The temperature dependence of the threshold current density is also dramatically improved as the dimensionality of the nanostructures decreases [100]. However, it is important to keep the volume of material where the emission actually occurs reasonably large. This requires a large number of QDs, often produced in several layers [45]. In addition, the size of the QDs, for example, must be uniform to avoid inhomogeneous broadening of the gain spectrum. Finally, to make the processes economically feasible, the QDs must be formed by a self-assembly process. This is normally accomplished by the production of the QDs by the S–K growth of strained materials.

Initially, quantum wires generated research interest due to the theoretically expected high electron mobilities [100, Sect. 26]. The geometry of wire structures provides the additional advantage for chemical and biological sensors of having a large surface area to volume ratio [59]. However, this may be a disadvantage for light-emitting devices, since surface recombination provides a non-radiative pathway. Thus, efficient LED operation may require use of core/shell nanostructures, where the high- bandgap shell prevents minority carriers in the core from reaching the surface.

Recent interest in quantum wires has centered on the use of vertical quantum wire structures as a self-assembly technique for the fabrication of tiny lasers [3]. This led to the unexpected discovery that the photoluminescence (PL) and LED efficiency of wires do not fall off as the amount of In in InGaN alloys increases. As mentioned above, and indicated in Fig. 4, for the MQW structures used for LED and laser devices, the radiative recombination efficiency falls dramatically as the In content increases above 15%. This extraordinary quantum wire behavior may partially be due to the much reduced dislocation density due to lattice mismatch in quantum wires, where lateral relaxation obviates the need for dislocation generation. However, this is unlikely to be the entire story. Compositional fluctuations due to spinodal decomposition are more likely to occur in these unstrained structures, as discussed in Sect. 4.3, which may retard non-radiative recombination. Prcliminary experimental results support this conclusion [27, 67]. Nevertheless, it is not entirely clear how PS is beneficial in structures having no dislocations. A third possibility is that PS occurs during growth (X. Niu et al., 2010, Controlled composition profiles of semiconductor alloy quantum dots and nanowires by selecting the growth mode, unpublished results) to spontaneously produce core/shell structures that retard surface recombination. A fourth possibility is that the removal of strain from the structure reduces the QCSE.

Nanostructured surfaces provide a further benefit for LEDs. They lead to less wave guiding in the high refractive index LED materials, resulting in an increased probability of photon escape. The nanostructured surfaces may consist of a periodic array of nano-sized features created on the surface by etching [57] or by selective-area, self-limiting growth. The roughened surface leads directly to an increase in the ratio of the number of photons escaping from the LED to the number produced at the p/n junction and, hence, the external quantum efficiency.

The use of low-dimensional nanostructures has become indispensable for LED and laser devices. It allows flexibility in the choice of the energy of the emitted photons, improves both the internal quantum efficiency and the photon extraction efficiency, and reduces costs by allowing the production of high-performance devices in highly defected materials, including those grown in lattice-mismatched to dissimilar substrates. The use of nanostructures, such as QDs and quantum wires, also offers promise for improvement of green, yellow, and even red LEDs in the InGaN alloy system.

The first system where the efficacy of nanostructures was demonstrated is the combination of GaAs and AlGaAs. Heterostructures in these materials yielded the first low threshold current density laser diodes and were intensively studied. One advantage of this materials system is that the lattice parameter is independent of the ratio of Al to Ga in the solid. Thus, high quality heterostructures can be grown epitaxially on GaAs substrates without the formation of misfit dislocations. As mentioned above, the bandgap of GaAs can be increased by sandwiching nm thick layers between layers of the higher bandgap AlGaAs. This was the beginning of "band gap engineering", the engineering of the properties of materials at the nanometer scale.

The need for lasers operating in the 1.3 and 1.55 μm wavelength ranges for long-distance, fiber-optic communication systems motivated early research. This application required bandgap energies lower than those of GaAs, leading to the development of GaInAsP, which can be grown lattice matched to InP substrates. In this materials system, a difficulty associated with limited solid miscibility was encountered that is associated with mixing elements having different covalent radii on the same sublattice, as discussed in Sect. 3. Miscibility gaps are more extensive in quaternary alloy systems, such as GaInAsP [83, Chap. 2]. Solid immiscibility led to difficulty in the growth of some alloys and adverse effects on materials properties, such as the electron mobility, where fluctuations in solid composition from spinodal decomposition produce scattering of free carriers [6].

Another system of interest is the simpler ternary GaInAs/GaAs system. However, the difference in covalent radii of Ga and In leads to misfit dislocations in thick layers with a significant In content grown on GaAs substrates. It was discovered that this problem can, again, potentially be overcome by the use of nanostructures. For very thin GaInAs layers grown on GaAs, the energy of strained GaInAs with no misfit dislocations is lower than that of GaInAs with misfit dislocations at the interface [50, 71, Chap. 7]. Thus, quantum-well structures can be fabricated with the desired 1.3 and 1.55 micron emission wavelengths. Interestingly, the strain inherent in such quantum-well structures gives a reduction in the hole effective mass, leading to reduced values of threshold current density [2]. It is, of course, vital to keep the well thickness below the critical layer thickness for dislocation generation. This system also became a vehicle for early investigations of even lower dimensional structures, such as quantum wires and QDs.

Since the beginnings of research on compound semiconductors, there has been a desire to grow epitaxial layers of these materials on Si substrates because Si substrates are much less expensive than GaAs or InP substrates and the integrated circuit business is based on Si. This is the third item of the overarching concerns articulated in Sect. 1. As an example, lasers grown on Si could be used for interchip communications, a current area of intensive research. There are a number of problems associated with GaAs growth on Si, but a major difficulty is with the generation of misfit dislocations due to the lattice parameter mismatch. The use of InGaAs QDs presents an ingenious solution to this problem. A high density of GaInAs QDs on a GaAs underlayer, generated by self-assembly of S–K islands, can mitigate the effects of dislocations in the GaAs generated by the lattice mismatch with the Si substrate [23, 99]. The minority carriers injected into the GaAs layer are captured by the smaller bandgap InGaAs regions before they can reach the dislocations where they generate the longer wavelength photons characteristic of the InGaAs, modified, of course, by the quantum size effect.

InGaAs/GaAs QD lasers have very low threshold current densities. The performance of lasers with QDs formed by the S–K process [45] have the lowest values of J_{th} reported due to the narrow gain spectrum, discussed in Sect. 2. However, the performance is still not optimal due to the dispersion in size of the self-assembled InGaAs QDs, giving rise to inhomogeneous broadening of the gain spectrum. Recent work has used the growth of QDs on the tops of pyramids formed using selective-area growth [102] to improve the control of QD size; however, this is necessarily an expensive process.

4.2 Importance of Nanostructures for AlInGaP Alloys

Early materials systems used for visible LEDs suffered from the presence of dislocations running through the p/n junction due to the lattice parameter mismatch with the substrate. For example, GaAsP was grown on GaAs substrates for early LEDs [14, 81]. The dislocation density could not be reduced below approximately 10^6 cm^{-2}, even through the use of grading techniques to minimize dislocations at the GaAs/GaAsP interface [86]. As indicated in Fig. 7, this results in low quantum efficiencies of the order of 10^{-3} [81]. This is shown as the data point at the year 1968 on the plot of external quantum efficiency vs time in Fig. 5. The need for lattice-matched structures led to the development of $Ga_{0.5}In_{0.5}P$ grown on GaAs substrates [12, 14, 35, 81]. It was realized that the bandgap of this lattice matched structure could be increased by substituting Al for part of the Ga, forming the AlInGaP quaternary. In this system, the bandgap can be increased until, at high Al concentrations, the bandgap becomes indirect. As discussed above, indirect materials are very poor for light-emitting devices.

The push into AlInGaP alloys was a defining time in the development of materials for LEDs, since this alloy can only be grown by "non-equilibrium" techniques such as OMVPE [79] or MBE [11]. As discussed above, the high Al distribution coefficient makes growth by LPE or HVPE, the techniques used for AlGaAs and GaAsP, virtually impossible. This led to the domination of the LED industry by the OMVPE technique.

The development of OMVPE for the growth of AlInGaP for LEDs and lasers involved overcoming obstacles with unintentional incorporation of oxygen, which is known to kill the internal quantum efficiency [83, Chap. 8]. This is particularly problematic for alloys with high Al concentrations, since Al has such a high affinity for oxygen. P-type doping is also a problem, especially for the highest bandgap materials. In addition, ordering must be avoided in order to obtain the highest bandgaps. It is harmful to LED performance in this materials system. However, another nanostructure, namely QDs, is potentially beneficial, since QDs can have large direct bandgaps. This might be potentially useful for pushing the AlInGaP alloys to higher photon energies, beyond the direct–indirect transition for bulk alloys. Of course, the situation is complicated by the need for higher bandgap barrier materials.

4.2.1 Spontaneous Ordering

Early in the development of semiconductors for LED and laser applications, the materials properties of an alloy were considered to be uniquely specified by the solid composition. Over the last two decades a number of critically important micro- and nano-structural details, with dramatic impact on materials properties, have emerged for III/V alloys, including both AlInGaP and AlInGaN. Perhaps the simplest to analyze is atomic-scale ordering. The occurrence of ordering was unanticipated, since, as discussed in Sect. 3, the enthalpy of mixing for III/V alloys is always greater than or equal to zero: it is proportional to the difference in lattice constant squared [78]. Thus, for the widely studied InGaP alloys, used for red and yellow LEDs, the cation distribution is expected to be non-random; clustering and PS are predicted. Surprisingly, the opposite—atomic-scale ordering—is observed [83, Chap. 2, 84]. The basic thermodynamics of mixing predicts ordering only for alloys with a negative enthalpy of mixing [88]. For (001) InGaP alloys, the ordering normally occurs in the form of a {111} ML superlattice, termed Cu-Pt ordering. This is a nanostructure with a periodicity of less than 1 nm. Formation of this structure is driven by surface thermodynamics [83, Chap. 3]. The driving force is the local, periodic strain associated with the surface reconstruction. Other ordered structures have also been observed, depending on the alloy system, the substrate orientation, the growth parameters, and the addition of surfactants during OMVPE growth [84, 87]. The order parameter has a major effect on the materials properties. Cu-Pt ordering is found to decrease the bandgap energy in $Ga_{0.5}In_{0.5}P$ by as much as 160 meV [87]. Thus, control of Cu-Pt ordering is important for LED and laser applications [84]. The use of ordering in InGaP with higher Ga concentrations has recently been suggested as a means of pushing the emission energies deeper into the green region of the spectrum [7]. It might similarly be used for AlInGaP alloys. Unlike the InGaP alloys, ordering does not appear to be a major factor determining the properties, in particular the bandgap energy, of InGaN, although it may adversely affect LED performance [54].

4.2.2 Artificial Nanostructures

Quantum wells are used in all high-performance AlInGaP LEDs and lasers. The use of quantum wells allows the production of high photon energies, in the orange and yellow regions of the spectrum, in materials with low Al concentrations, due to the QSE. This avoids the problems encountered at high Al concentrations, discussed above. In addition, quantum wells increase the radiative recombination rate.

Structures with lower dimensionality are also being explored in this materials system. As an example, small InP QDs dispersed in an AlInGaP matrix can be made to emit visible light [17] due to the QSE, even though InP has a smaller bandgap energy than GaAs, as seen in Fig. 2. The large band offsets in this system allow room temperature operation of lasers in the red region of the spectrum for data storage and medical applications. Room temperature InP QD lasers emitting at 740 nm have been reported [39]. They are also being considered for single photon sources for quantum cryptography emitting in the red, near the maximum sensitivity of Si photodetectors [72].

Quantum wire structures have also been fabricated using OMVPE with GaAs cores and InGaP shells [75]. The GaAs cores were grown using the standard vapor/liquid/solid (VLS) process at a low temperature of 450°C and the InGaP shells at higher temperature where growth can occur directly on the sides of the wires. The higher bandgap cladding layer inhibits surface recombination and so increases the GaAs PL intensity by a factor of 100–1000. The elastic strain from the cladding layer, when the Ga/In ratio is greater than unity, can be used to blue shift the emission from the core by as much as 240 meV. Thus, the quantum wire structures represent a method of tuning the emission wavelength. A similar increase in PL intensity was reported for GaAsP/GaP core–shell nanowires [52]. Significantly, in this case the spontaneous formation of a P-rich core was reported for GaAsP nanowires grown by the VLS mechanism. This presents a potentially useful opportunity for the self-assembly of core–shell nanowires ((X. Niu et al., 2010, Controlled composition profiles of semiconductor alloy quantum dots and nanowires by selecting the growth mode, unpublished results)).

The use of nanostructured surfaces has been explored to increase the photon escape efficiency of AlInGaP LEDs. However, the effect is nearly eliminated when the LED chip is encapsulated in the polymer packaging material [71, p. 154].

4.3 Importance of Nanostructures for AlInGaN Alloys

InGaN and AlInGaN alloys have assumed increasing importance during the last decade. This is because the bandgap of AlInGaN can be tuned over the entire near IR to deep UV range, from 0.7 (bandgap of InN) to 6.2 eV (bandgap of AlN), as seen in Fig. 3. These alloys are essential for the commercial fabrication of blue, green, and white LEDs [4, 56]. However, the performance of yellow and, especially, red LEDs in this system is poor, solely because of a degradation of material quality at high In concentrations. These materials are also used for detectors, high power FETs and solar cells.

The AlInGaN alloys have proven to be most difficult to understand and control. The basic properties of the alloys are often masked by the large defect density induced by the lack of a native substrate. This typically means that they are grown on sapphire (or SiC) substrates having a dissimilar crystal structure, lattice spacing, and thermal expansion coefficient [4, 56]. This results in epitaxial AlInGaN layers having very high dislocation densities of 10^{8}–10^{10} cm^{-2}, in addition to stacking faults, twins, and other defects near the sapphire–epilayer interface [1]. GaN and InGaN alloys for device applications have been grown by MBE and HVPE techniques, but are nearly always grown by OMVPE for commercial LEDs and lasers. Typically, a thin GaN or AlN buffer layer is grown first at a low temperature followed by the high temperature (>1000°C) growth of a thick GaN or AlGaN layer. Then the InGaN is grown at a lower temperature, ranging from 700 to 900°C [55]. Typical LED structures consist of multiple quantum wells with well thicknesses of 2–3 nm and In contents ranging from 15% (blue) to 20% (green) InN, and even higher for yellow and red LEDs. The performance of these LEDs is stunningly good, considering the high density of threading dislocations. As discussed above (and shown in Fig. 7), in other III/V semiconductor materials systems a dislocation density of 10^{8} cm^{-2} would be sufficient to kill the radiative recombination efficiency.

For LED performance, a number of materials issues turn out to be of critical importance. Issues relating to OMVPE growth, defect generation, and p-type doping, all major materials issues, will not be dealt with here. Instead, this section will concentrate on issues related to nanostructures in AlInGaN, or more frequently InGaN alloys.

The first issue to be considered is the control of alloy composition. In common with all III/V alloys, before the complexities of the micro- and nanostructures of real AlInGaN alloys were fully appreciated, the sole parameter thought to control materials properties was the alloy composition. Thus, the bandgap energies of the ternaries included in the AlInGaN system are represented as solid lines in Fig. 3. The actual color of an LED with an InGaN active layer is, in fact, a function of the nominal alloy composition, but with additional effects from strain and the nanostructure.

For InGaP alloys, ordering is the most important naturally occurring nanostructure. For the InGaN alloys, a much more critical nanostructural phenomenon is the spontaneous formation of non-uniform alloys due to the occurrence of PS during growth. PS will be defined here as the occurrence of a non-uniform alloy composition on a nanometer scale. PS can be caused by a number of different effects. Even though random alloys have some degree of inhomogeneity, they are, by definition, a single phase.

First, we will consider the basic thermodynamics of InGaN alloys. Early valence force field calculations of Ho and Stringfellow [26] indicated that the phase diagram for bulk, unstrained, zincblende InGaN has a large region of solid immiscibility at typical InGaN growth temperatures. In Fig. 8, the region within the solid, binodal line defines the alloys that are thermodynamically unstable. Between the binodes the broken line indicates the spinode. For alloys having compositions between the spinodes at a given temperature, there is no energy barrier to PS, i.e., the solid is unstable to any compositional fluctuations. The calculated phase diagram gives a solubility of InN in GaN at 750°C of $<5\%$. Later, more detailed first principles calculations have indicated that the equilibrium solubility may be even smaller, perhaps as small as 2% [22]. Even allowing for uncertainties in the calculations, the alloys used for blue and green MQW LEDs ($x = 0.15$ and $x = 0.20$, respectively) are well within the calculated region of solid immiscibility.

At equilibrium, these alloys could not be grown. In fact, the non-equilibrium growth of immiscible III/V alloys by OMVPE has been shown to be possible for many alloys [83, Chap. 2]. For thick InGaN epilayers, the beginnings of PS by spinodal decomposition occur during OMVPE growth for alloys within the miscibility gap as seen in Fig. 8. This gives rise to experimentally observed compositional fluctuations within the epitaxial layer. As seen in Fig. 8, the experimental data are in general agreement with the calculated phase diagram. Uniform layers were observed for In concentrations of less than the solid solubility (binode). For compositions between the spinode and the binode, the alloys are metastable. In these regions, an energy barrier must be surmounted for PS to occur. This suggests that PS may occur only near lattice disruptions, such as dislocations and other imperfections. Ponce et al. [64] reported that In-rich regions were formed only at elastically distorted regions of the layers, namely at dislocations, for these metastable alloys. As seen in Fig. 8, PS in this region is found to vary between the various experimental investigations. With larger In concentrations, giving alloys between the spinodes, spontaneous PS was observed throughout the layer in all experimental studies, as indicated in Fig. 8.

Analysis of the thermodynamics of alloy formation for InGaN in the elaborate structures used for MQW LEDs and lasers is more complex than the calculations of [26], which are for incoherent, bulk systems. First, the strain energy associated with coherent spinodal decomposition (the formation of In-rich regions that are coherent, i.e., having no mismatch dislocations at the interfaces where x changes) will suppress PS. This is because coherent In-rich regions are under significant compressive strain in a more Ga-rich matrix, which has a high energy cost for the system. This coherency strain is predicted to completely suppress spinodal decomposition in most III/V alloys [80]. Second, biaxial strain due to lattice parameter mismatch between the substrate and underlying layers on the InGaN layer under consideration also significantly affects spinodal decomposition. InGaN coherently grown on GaN is under compressive strain. This mismatch strain, itself, suppresses the formation of In-rich clusters, since they have a larger equilibrium lattice constant than the matrix. Karpov [33] calculated the InGaN phase diagram, taking this strain energy into account. The results indicate that for $In_xGa_{1-x}N$, with $x = 0.15$ or 0.2, grown coherently on GaN, spinodal decomposition will be suppressed for typical growth temperatures.

Any understanding of these phenomena, as applied to the real InGaN materials used in the quantum wells of LED and laser devices, requires analysis of the mechanisms of epitaxial growth at the microscopic level. For example, strained epitaxial layers in semiconductor systems are normally found to grow by the S–K mechanism, described above, involving the formation of a high density of nanometer-sized islands as well as a thin wetting layer [40, 96]. This mechanism is expected to dominate the growth process for the InGaN quantum-well structures used in blue and green LEDs, due to the large lattice parameter mismatch with GaN. Thus, the good agreement between the experimental results and the phase diagram, as shown in Fig. 8, is perhaps surprising. Especially interesting are the results of Potin et al. [65] and Tran et al. [93] for QW structures.

S–K growth is expected for InGaN grown on GaN due to the large lattice mismatch of 11% between InN and GaN. A 2D to 3D transition is experimentally found to occur after a critical thickness of approximately three MLs [24]. Moustakis et al. [54] reported S–K growth of InGaN QDs on AlN by rf plasma assisted MBE. A transition from 2D to 3D growth occurred after the deposition of several MLs of InGaN. After 12 ML coverage, the average diameter and height of the InGaN islands, with $x = 0.43$, were 30 nm and 3 nm, respectively. Yamaguchi et al. [101] reported the S–K growth of InGaN islands on GaN by both MBE and OMVPE. RHEED studies during MBE clearly showed a 2D to 3D transition as the InGaN layer thickness increased. For samples grown by OMVPE, the 2D to 3D transition occurred for layer thicknesses of 4–5 nm, with an In concentration of 0.22. The aspect ratio for the faceted, entirely coherent islands was found to be approximately 2. Pristovsek et al. [66] used in situ ellipsometry to study the thickness at the 2D–3D transition in InGaN layers grown by OMVPE as a function of the In composition. They found a value of approximately 1.5 nm for $x_{In} = 0.20$ and a value of approximately 2 nm for $x_{In} = 0.15$. InGaN QDs 1–2 nm in height and 40–50 nm in diameter have been observed to form by the S–K growth mode during OMVPE growth by Jung et al. [30]. Self-assembled InGaN QDs were also produced by atmospheric pressure OMVPE by Tachibana et al. [90].

S–K growth was specifically demonstrated to occur for InGaN layers in MQW structures grown by OMVPE. In single quantum-well structures used for LEDs, Florescu et al. [19] reported, for the 3 nm wells containing approximately 15% InN, 3D, i.e., S–K, growth, giving much enhanced 300 K PL intensities. The effects of an inhomogeneous distribution of In on optical properties and device performance will be discussed below.

Efforts have been made to control spinodal decomposition during quantum-well growth, since it may be an important factor affecting device performance. One approach has been the use of anti-surfactants (Si in particular) in an effort to enhance QD formation during the S–K growth of InGaN on AlGaN by OMVPE. For growth at 800°C, Hirayama et al. [25] produced InGaN QDs in an AlGaN matrix for layers 3 nm thick. The diameter and height of the QDs were reported to be 10 and 5 nm, respectively. Intense room temperature PL was reported. Other authors have inserted buried layers to control the strain in the InGaN quantum-well layer(s). This will affect both the mismatch strain in the quantum-well layer(s) and the In incorporation, via the lattice pulling effect. Both will affect the formation of S–K islands. For example, Huang et al. [29] found enhanced emission efficiency for green InGaN/GaN QW LEDs when a prestrained buried layer was added. Park et al. [60] used surface roughening to enhance PS in thin InGaN layers grown on GaN. The layer thicknesses were less than the critical thickness for strain relaxation, t_c. However, intentional surface roughening was believed to produce unstrained, incoherent S–K islands. In the layers with enhanced spinodal decomposition, they observed intense PL emission from the In-rich regions at significantly lower energies than the PL of the material grown on flat GaN layers.

Lai et al. [42] used a five period InGaN/GaN MQW structure, grown by OMVPE at 780–880°C, with 2 nm InGaN wells, sandwiched by GaN to produce materials emitting at wavelengths from blue into the yellow region of the spectrum. They reported PS to occur due to spinodal decomposition during S–K growth of the InGaN, resulting in the formation of In-rich clusters, coherent with the matrix, having very uniform diameters. As discussed below, the control of PS to produce yellow emission is important for the fabrication of single junction, white LEDs.

In terms of the basic thermodynamics, discussed in Sect. 3, several phenomena can be expected to lead to PS for the S–K growth of *thin*, coherent InGaN islands on GaN. First, strain will suppress spinodal decomposition in coherent thin layers, but this constraint will be relaxed non-uniformly in S–K islands. In addition, the elastic relaxation will give more In incorporation in the more strain relaxed parts of the islands, due to the lattice pulling effect. Both phenomena will result in PS, i.e., the formation of local In-rich regions during growth [85].

The experimental evidence for the occurrence of PS in thin layers has been reviewed recently [85] and is overwhelming. EXAFS studies show the formation of non-random In distributions, characteristic of In clustering. Careful, high resolution TEM clearly shows a very high density of 2–4 nm In-rich regions. The CL and PL splitting as well as the Stokes shift are consistent with the formation of In-rich regions in a more Ga-rich matrix.

The spontaneous formation of QD structures during growth due to PS is of more than academic interest. Formation of In-rich clusters is found to have profound effects on the performance of laser and LED devices. PS will produce an emission wavelength in an LED or laser that is longer than expected for a random alloy because carriers collect in low-bandgap regions, so the emission energy is controlled by the local In composition. In addition, there is a great deal of evidence that the formation of In-rich nano-clusters produces a remarkable increase in radiative recombination efficiency in these highly dislocated materials. Chichibu et al. [13] were the first to explain this mystery in terms of localization of recombination at the In-rich clusters. Essentially, minority carriers are collected in the small bandgap regions at the In-rich clusters before they can recombine at dislocations. This yields an enormous increase in radiative recombination efficiency if the distance between the In-rich regions is smaller than the dislocation spacing. The S–K island density has been measured to be from 10^{10} to 10^{12} cm^{-2}, depending on growth conditions, which is orders of magnitude higher than the dislocation density in the high quality material used for high efficiency LEDs. Amazingly high values of external efficiency, well exceeding 60%, as seen in Fig. 4, have been reported for blue MQW LEDs in materials with extremely high dislocation densities, exceeding 10^7 cm^{-2}. A dramatic increase in the external quantum efficiency as In is added to GaN in MQW LEDs, which increases the wavelength, is seen in Fig. 4. A small concentration of In gives a dramatic boost in LED efficiency. Indeed, it is impossible to produce high efficiency LEDs without the addition of In. The onset of this beneficial effect appears to coincide roughly with the onset of PS.

A major effect that may be related to the reduced radiative recombination in MQW LED structures with high In content is the QCSE. Wurtzite AlInGaN alloys are piezoelectric. Thus, stress along the c-axis of InGaN in quantum wells will give rise to a piezoelectric field that decreases the overlap between electron and hole wavefunctions. This results in a decrease in the quantum efficiency for the production of visible photons. As the In concentration increases, this QCSE increases. Thus, the efficiency of green LEDs would be expected to be lower than for blue LEDs. This is consistent with the well-known "green gap" discussed above. At the higher In concentrations required for yellow and red LEDs, the effect is even larger because of the increased piezoelectric field. This may, in part, explain their very low efficiencies. A beneficial effect of the QDs formed by PS may be that the deleterious effect of the QCSE is reduced by the formation of very small QDs, 1–2 nm in size. Of course, the wave function overlap will increase as the size of the QD decreases because the electrons and holes are squeezed together. This presents a promising approach for the development of longer wavelength InGaN LEDs. Moustakis and co-workers used MBE to produce small QDs that were used for green LEDs [1, 54]. This mechanism may also explain the improved yellow emission intensity obtained by Lai et al. [42] for InGaN grown by OMVPE, which is found to spinodally decompose to produce small-sized clusters behaving as quasi-QDs. Park et al. [61] used the S–K process for the self-assembly of InGaN QDs to produce a five layer LED structure. They specifically reported improved LED performance as the size of the QDs was decreased.

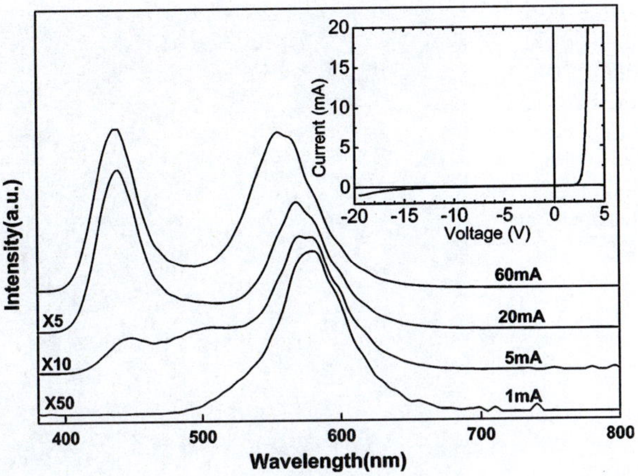

Fig. 11 Electroleuminescence spectra of InGaN layers with In-rich clusters at several currents (reprinted with permission from Wang et al. [98]. Copyright (2007), American Institute of Physics)

Recently, several strategies have been developed for the use of PS in InGaN for the production of white LEDs. The basic idea is to generate yellow photons in the InN-rich regions and blue photons in the matrix. Together, the light produced will appear as white. This is a very promising approach for white LEDs because the two competing techniques use either a blue LED and an overlayer of phosphor to down-convert some blue photons to yellow or the use of three (red, green, and blue) or sometimes four (adding yellow) individual LEDs packaged together. Both approaches have significant disadvantages. A white LED having a single p/n junction and no phosphor is very attractive.

Several approaches have been used to foster and control PS in the InGaN for white LEDs, as described above. Park et al. [60] used roughening of the surface to facilitate PS. The surface roughening is believed to produce incoherent In-rich InGaN islands. Other groups have used the control of strain in the epitaxial layers, using partially or completely relaxed buried layers, to control PS, as described above. These approaches are novel and promising. Wang et al. [98] used spinodal decomposition to form InN-rich QDs. The 220 nm thick underlying layer, containing 5% In, is relaxed, since the thickness is much greater than t_c. This is followed by a GaN layer and then the MQW structure. Four, 3 nm thick InGaN layers, containing 17% In form the well layers. Thus, the InGaN QW layers are partially relaxed, enhancing PS due to spinodal decomposition. The EL spectra reported by Wang et al. [98] are reproduced in Fig. 11. The yellow emission from the In-rich QDs and the blue emission from the more Ga-rich matrix are clearly resolved. The combined emission at high current densities looks white. These authors also demonstrated the production of In-rich regions using TEM. The results are reproduced in Fig. 12. InN rich QDs are clearly distinguished.

Lu et al. [48] also used an InGaN under-layer to reduce strain in the QW layers. They reported enhanced PS, as evidenced by TEM studies. Atomic level images

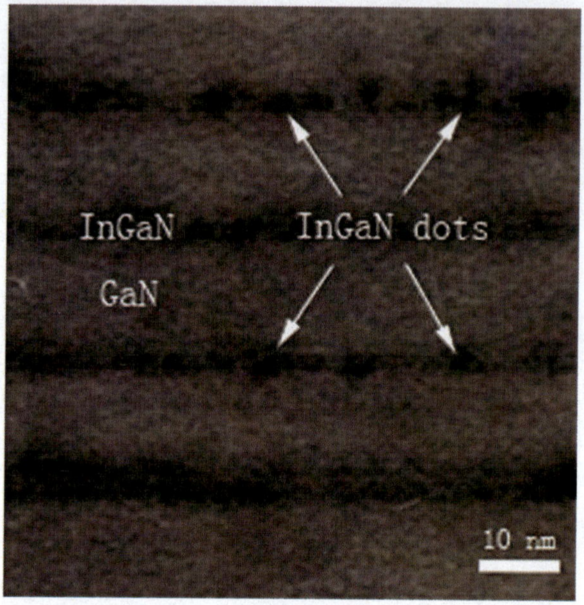

Fig. 12 Cross-section TEM image of the In-rich InGaN quantum dots in the LED structure giving the EL spectra shown in Fig. 11 (reprinted with permission from Wang et al. [98]. Copyright (2007), American Institute of Physics)

using strain state analysis yielded the In variation from QD to matrix in the light-emitting InGaN layers. They used changes in the thickness of the underlying, strain-relieving layer to produce values of the variation of In content from the In-rich region to the matrix of 14–18%, 15–25%, 17–37% and 18–49% as the pre-strained layer thickness was decreased. For five period MQW structures with 3 nm InGaN well layers, the best results were obtained with a 5 nm thick underlying layer having 7% In. This is less than the critical layer thickness, so the layers are coherent. Soh et al. [76] also made white LEDs using In-rich QDs to produce the yellow emission. They used dual stacked MQWs, using a complex procedure that includes the use of In as an antisurfactant.

Another approach to produce materials for improved green, yellow, and red LEDs in the AlInGaN alloys is the use of quantum wires. As mentioned above, the internal recombination efficiency of AlInGaN MQW structures drops dramatically as the In content of the InGaN is increased. There is evidence that quantum wires do not suffer from this behavior. Recent research [41] has shown that quantum wires grown by HVPE with bandgap energies across the entire visible spectrum emit PL with a near-constant efficiency. Work by Lieber and his group at Harvard has produced entire LED devices in triangular, core/multishell, radial InGaN nanostructures [67] that efficiently emit light from the purple through the yellow region of the spectrum (365–600 nm), with In concentrations of 1–35%. High quantum efficiencies of 5.8% at 440 nm and 3.9% at 540 nm were observed. They report the TEM observation of formation of localized In-rich clusters in the InGaN layers.

The reasons for the increased recombination efficiency in the longer wavelength InGaN wires have not been fully explained. One factor may be the absence of dislocations in the quantum wires, as described above. However, this does not appear to be the sole factor. Hong et al. [27] recently reported intense green PL from nanorod arrays grown by plasma assisted MBE. These authors report localized recombination in the quantum rods, with an increase in the localization depth as compared with 2D epitaxial layers. This is attributed to an increase in PS in the rods. This confirms the results of Quian et al. [67], discussed above. It is also consistent with the results for MQW samples that localization of the recombination is beneficial. The removal of elastic strain will increase the occurrence of spinodal decomposition, as described above. In this case the effect cannot be due to preventing minority carriers from reaching dislocations. Perhaps, the localization of recombination prevents recombination at other defects or at surfaces. This may tie in with the results described in Sect. 4.2.2 for GaAsP/GaP nanowires, where there was a natural segregation of the constituents in a lateral direction. This was confirmed for InGaN quantum wires grown by chemical vapor deposition using elemental sources and ammonia, where a natural segregation of Into the core of the wire was observed for Si substrates [10].

Another possible explanation is the decrease in the QCSE in rods, where the strain is relaxed elastically, thus removing the field due to the piezoelectric effect. Hong et al. [27] also report an increase in optical extraction efficiency for the samples made from quantum rods, as expected. A combination of these effects results in the PL intensity for the nanorod samples 15 times greater than that for InGaN layers with nearly the same composition.

5 Summary

The use of nanostructures, in the form of quantum wells, quantum wires, and QDs, has led to major advances in both the performance and affordability of visible LEDs, as summarized in Table 1. All commercial LED and laser devices across the entire spectrum from infrared to ultraviolet are fabricated using the OMVPE technique, using quantum wells to enhance performance. The quantum size effect leads to increased photon energies. This allows materials with bandgaps in the IR spectrum, such as GaAs and InP, to emit visible photons. In the commercially important AlInGaP and AlInGaN alloys, the use of quantum wells allows the desired color to be obtained while minimizing the amount of Al, a particularly difficult element due to C and O contamination, in the active region. The use of quantum wells also increases the radiative recombination rate and decreases the rate of non-radiative recombination. Both effects enhance the conversion efficiency of electrons into photons. The use of QDs yields even greater increases in LED performance as well as the lowest threshold current densities for laser devices. Current research is heavily directed to the use of QDs for increased performance of LEDs and lasers in the green, yellow, and red regions of the spectrum, as well as

Table 1 Advantages of nanostructures for light-emitting devices

Quantum wells	Increased bandgap energy as L_z decreases
	B_r increases as L_z decreases
	More favorable density of states for lasers and LEDs
Quantum wires	Increased free carrier mobility
	Increased bandgap energy as L_z, L_y decrease
	B_r increases as L_z, L_y decrease
	More favorable density of states for lasers and LEDs
	Relaxation of strain due to lattice mismatch, without dislocations
	Spontaneous phase separation during growth
Quantum dots	Increased bandgap energy as L_z, L_y, L_x decrease
	Stronger exciton binding energy
	B_r increases as L_z, L_y, L_x decrease
	Decrease in recombination at defect (including dislocation) sites
	Most favorable density of states (delta function) for lasers and LEDs
	Narrow electroluminescence and gain spectra
	Lowest threshold current densities for lasers
	No temperature dependence of threshold current density
	Reduced QCSE in InGaN
	Increased light extraction for photonic crystals on the surface

white LEDs, using AlInGaN alloys. The use of quantum wires also appears promising for such applications. Nanostructures on the surface have been demonstrated to increase the efficiency of extraction of photons generated within the LED. It is anticipated that future efforts to increase performance and decrease cost will center on the increased use of nanostructures.

References

1. Abell J, Moustakis TD (2008) The role of dislocations as nonradiative recombination centers in InGaN quantum wells. Appl Phys Lett 92:091901
2. Adams AR, O'Reilly EP (1996) Semiconductor band structure and related properties. In: Quillec M (ed) Materials for optoelectronics. Kluwer Academic, Boston, p 61
3. Agarwal R, Lieber CM (2006) Semiconductor nanowires: optics and optoelectronics. Appl Phys A 85:209
4. Akasaki I, Amano H (1997) OMVPE of GaN for high-brightness blue LEDs. In: Stringfellow GB, Craford MG (eds) High brightness LEDs. Academic Press, New York, Ch. 7
5. Asada M, Miyamoto Y, Suematsu Y (1996) Gain and threshold of three-dimensional quantum-box lasers. IEEE J Quantum Electron QE-22:1915
6. Benchimol JL, Quillec M, Slempkes S (1983) Improved mobility in InGaAsP alloys using high temperature LPE. J Cryst Growth 64:96
7. Bhusal L, Fluegel B, Steiner MA, Mascarenhas A (2009) Ordering induced direct–indirect transformation in unstrained GaInP for $0.76 < x < 0.78$. J Appl Phys 106:114909
8. Bjork MT et al (2002) One-dimensional heterostructures in semiconductor nanowhiskers. Appl Phys Lett 80:1058
9. Brown AS, Doolittle WA (2000) The status and promise of compliant substrate technology. Appl Surf Sci 166:392

10. Cai XM et al (2006) Straight and helical InGaN core-shell nanowires with a high In core content. Nanotechnology 17:2330
11. Casey HC, Panish MB (1978) Heterostructure lasers. Academic Press, New York
12. Chen CH, Stockman SA, Peansky MJ, Kuo CP (1997) OMVPE growth of AlGaInP for high-efficiency visible LEDs in high-brightness LEDs. In: Stringfellow GB, Craford MG (eds) High brightness LEDs. Academic Press, New York, Chap. 4
13. Chichibu S, Azuhata T, Sota T, Nakamura S (1996) Spontaneous emission of localized excitons in InGaN single and MQW structures. Appl Phys Lett 69:4188
14. Craford MG (1997) Overview of device issues in high-brightness LEDs. In: Stringfellow GB, Craford MG (eds) High brightness LEDs. Academic Press, New York, Chap. 2
15. Dingle R, Wiegmann W, Henry CH (1974) Quantum states of confined carriers in very thin $Al_xGa_{1-x}As$-GaAs-$Al_xGa_{1-x}As$ heterostructures. Phys Rev Lett 33:827
16. Doppalapudi D, Basu SN, Ludwig KF, Moustakas TD (1998) Phase separation and ordering in InGaN alloys grown by molecular beam epitaxy. J Appl Phys 84:1389
17. Eichfelder M et al (2009) Room-temperature lasing of electrically pumped red-emitting InP/AlGaInP quantum dots embedded in a vertical microcavity. Appl Phys Lett 95:131107
18. Faleev N et al (2009) Correlation of crystalline defects with photoluminescence of InGaN layers. Appl Phys Lett 95:051915
19. Florescu DI et al (2004) AFM and temperature-dependent photoluminescence studies of the degree of localization induced by quantum-dot like states in InGaN single quantum well light emitting diodes grown by MOCVD on (0 0 0 1) sapphire. J Cryst Growth 272:449
20. Funato M et al (2006) Tailored emission color synthesis using microfacet quantum wells consisting of nitride semiconductors without phosphors. Appl Phys Lett 88:261920
21. Funato M et al (2008) Emission color tunable LEDs composed of InGaN multifacet quantum wells. Appl Phys Lett 93:021126
22. Gan CK, Feng YP, Srolovitz DJ (2006) First-principles calculation of the thermodynamics of InGaN alloys. Phys Rev B 73:235214
23. Gerard JM, Cabrol O, Sermage B (1996) InAs quantum boxes: Highly efficient radiative traps for light emitting devices on Si. Appl Phys Lett 68:3123
24. Grandjean N, Ilegems M (2007) Visible InGaN/GaN quantum-dot materials and devices. Proc IEEE 95:1854
25. Hirayama H, Tanaka S, Ramvall P, Aoyagi Y (1998) Intense photoluminescence from self-assembling InGaN quantum dots artificially fabricated on AlGaN surfaces. Appl Phys Lett 72:1736
26. Ho IH, Stringfellow GB (1996) Solid phase immiscibility in GaInN. Appl Phys Lett 69:2701
27. Hong CC, Ahn H, Wu CY, Gwo S (2009) Strong green PL from InGaN/GaN nanorod arrays. Opt Express 17:17337
28. Huang HW et al (2006) Improvement of InGaN/GaN LED performance with a nono-roughened p-Gan surface by excimer laser-irradiation. Mater Chem Phys 99:414
29. Huang C et al (2008) Enhanced efficiency and reduced spectral shift of green LED epitaxial structure with prestrained growth. J Appl Phys 104:123106
30. Jung W, Jang J, Choi S, Kim J (2008) Growth behavior of InGaN/GaN quantum dots structure via MOCVD. Korean J Mater Res 18:535
31. Kapon E, Hwang DM, Bhat R (1989) Stimulated emission in semiconductor quantum wire heterostructures. Phys Rev Lett 63:430
32. Kar A, Alexson D, Dutta M, Stroscio MA (2008) Evidence of compositional inhomogeneity in $In_xGa_{1-x}N$ alloys using ultraviolet and visible Raman spectroscopy. J Appl Phys 104:073502
33. Karpov SYu (1998) Suppression of phase separation in InGaN due to elastic strain. MRS Internet J Nitride Semicond Res 3:16
34. Keiser G (1991) Optical fiber communications, 2nd edn. McGraw-Hill, Inc, New York, Chap. 4
35. Kish FA, Fletcher RM (1997) AlGaInP LEDs. In: Stringfellow GB, Craford MG (eds) High brightness LEDs. Academic Press, New York, Chap. 5

36. Krames M et al (1999) High-power truncated-inverted-pyramid AlGaInP/GaP LEDs exhibiting >50% external quantum efficiency. Appl Phys Lett 75:2365
37. Krames MR et al (2000) High brightness AlGaInN light emitting diodes. Proc SPIE 3938:2
38. Krames MR et al (2007) Status and future of high-power light-emitting diodes for solid-state lighting. J Display Technol 3:160
39. Krysa AB et al (2007) Low threshold InP/AlGaInP on GaAs QD laser emitting at 740 nm. J Cryst Growth 298:663
40. Kukta RV, Freund LB (1997) J Mech Phys Solids 45:1835
41. Kuykendall T, Ulrich P, Aloni S, Yang P (2007) Complete composition tunability of InGaN nanowires using a combinatorial approach. Nat Mater 6:951
42. Lai Y, Liu C, Chen Z (2006) Tuning the emitting wavelength of InGaN/GaN superlattices from blue, green to yellow by controlling the size of InGaN quasi-quantum dot. Thin Solid Films 498:128
43. Lauhon L, Gudiksen M, Wang D, Lieber CM (2002) Epitaxial core-shell and core-multishell nanowire heterostructures. Nature 420:57
44. Ledentsov NN et al (2000) Quantum-dot heterostructure lasers. IEEE J Sel Top Quantum Electron 6:439
45. Ledentsov NN, Bimberg D, Alferov ZhI (2008) Progress in epitaxial growth and performance of quantum dot and quantum wire lasers. J Lightwave Tech 26:1540
46. Leonard D et al (1993) Direct formation of quantum-sized dots from uniform coherent islands of InGaAs on GaAs surfaces. Appl Phys Lett 63:3203
47. Lester SD, Ponce FA, Craford MG, Steigerwald DA (1995) High dislocation densities in high efficiency GaN-based LEDs. Appl Phys Lett 66:1249
48. Lu C, Huang C, Chen Y, Yang CC (2008) Dependence of spectral behavior in an InGaN/GaN quantum-well light-emittingdiode on the prestrained barrier thickness. J Appl Phys 104:043108
49. Luo JW, Franceschetti A, Zunger A (2008) Quantum-size-induced electronic transitions in quantum dots: indirect band-gap GaAs. Phys Rev B 78:035306
50. Mathews JW, Blakesley AE (1976) Defects in epitaxial multilayers: III. Preparation of almost perfect multilayers. J Cryst Growth 32:265
51. Mishra U (2008) Group III nitride optoelectronics, Invited talk at electronic materials conference, Santa Barbara California
52. Mohseni PK et al (2009) Structural and optical analysis of GaAsP/GaP core-shell nanowires. J Appl Phys 106:124306
53. Moison JM et al (1994) Self-organized growth of regular nanometer-scale InAs dots on GaAs. Appl Phys Lett 64:196
54. Moustakis TD et al (2008) Growth of III-nitride QDs and their applications to blue-green LEDs. Phys Stat Sol (a) 205:2560
55. Mukai T (2002) Recent progress in group-III nitride LEDs. IEEE J Sel Top Quantum Electron 8:1077
56. Nakamura S (1997) Group III–V nitride-based ultraviolet blue–green–yellow LED and laser diodes. In: Stringfellow GB, Craford MG (eds) High brightness LEDs. Academic Press, New York, Chap. 8
57. Nakamura S (2009) Current status of GaN-based solid-state lighting. MRS Bull 34:101
58. Narukawa Y et al (2006) Ultra-high efficiency white LEDs. Jpn J Appl Phys 45:L1084
59. Novotny CJ, Yu ET, Yu PKL (2008) InP nanowire/polymer hybrid photodiode. Nano Lett 8:775
60. Park I et al (2005) Enhancement of phase separation in the InGaN layer for self-assembled In-rich quantum dots. Appl Phys Lett 87:061906
61. Park I et al (2008) Effect of InGaN quantum dot size on the recombination process in light-emitting diodes. Appl Phys Lett 92:253105
62. Petroff PM, Gossard AC, Wiegmann W (1984) Structure of AlAs-GaAs interfaces grown on (100) vicinal surfaces by molecular beam epitaxy. Appl Phys Lett 45:620

63. Piner EL, El-Mastry NA, Liu SX, Bedair SM (1998) Phase separation in InGaN grown by metalorganic chemical vapor deposition. Mater Res Soc Proc 482:125
64. Ponce FA et al (2003) Microstructure and electronic properties of InGaN alloys. Phys Stat Sol (b) 2:273
65. Potin V et al (2004) Comparison of the In distribution in InGaN/GaN quantum well structures grown by MBE and MOVPE. J Cryst Growth 262:145
66. Pristovsek M, Stellmach J, Leyer M, Kneissl M (2009) Phys Stat Sol C 6:5565
67. Quian F, Gradecak S, Li Y, Wen CY, Lieber CM (2005) Core/multishell nanowire heterostructures ad multicolor, high-Efficiency LEDs. Nano Lett 5:2287
68. Rao M, Kim D, Mahajan S (2004) Compositional dependence of phase separation in InGaN layers. Appl Phys Lett 85:1961
69. Reynard J, Kandaswarmy PK, Monroy E, Gayral B (2009) Suppression of nonradiative processes in long-lived polar GaN/AlN quantum dots. Appl Phys Lett 95:131903
70. Ross FM, Tersoff J, Tromp RM (1998) Coarsening of self-assembled Ge quantum dots on Si (001). Phys Rev Lett 80:984
71. Schubert EF (2006) Light-emitting diodes, 2nd edn. Cambridge Press, Cambridge
72. Schultz WM et al (2009) Optical and structural properties of InPquantum dots embedded in $(Al_xGa_{1-x})_{0.51}In_{0.49}P$. Phys Rev B 79:035329
73. Siefert W et al (1997) In situ growth of nanostructures by MOVPE. J Cryst Growth 170:39
74. Siefert W et al (2004) Growth of one dimensional nanostructures in MOVPE. J Cryst Growth 272:211
75. Skold N et al (2005) Growth and optical properties of strained GaAs-GaInP core-shell nanowires. Nano Lett 5:1943
76. Soh CB et al (2008) Cool white III-nitride LEDs based on phosphor-free indium-rich InGaN nanostructures. Appl Phys Lett 92:261909
77. Stringfellow GB (1972) The importance of lattice mismatch in the growth of GaInP epitaxial crystals. J Appl Phys 43:3455
78. Stringfellow GB (1974) Calculation of ternary and quaternary III–V phase diagrams. J Cryst Growth 27:21
79. Stringfellow GB (1978) VPE growth of III/V semiconductors. Annu Rev Mater Sci 8:73–98
80. Stringfellow GB (1982) Spinodal decomposition and clustering in III/V alloys. J Electron Mater 11:903
81. Stringfellow GB (1997) Materials issues in high-brightness LEDs. In: Stringfellow GB, Craford MG (eds) High brightness LEDs. Academic Press, New York, Chap. 1
82. Stringfellow GB (1998) Ordering in III/V semiconductor alloys. In: Santos M, Liu WK (eds) Thin films: heteroepitaxial systems. World Scientific Publishing, Hackensack, pp 64–116
83. Stringfellow GB (1999) Organometallic vapor phase epitaxy: theory and practice, 2nd edn. Academic Press, Boston
84. Stringfellow GB (2002) Effects of the surface on CuPt ordering during OMVPE growth. In: Mascarenhas A (ed) Spontaneous ordering in semiconductor alloys. Kluwer Academic Publishers, New York, Chap. 3
85. Stringfellow GB (2010) Microstructures produced during the epitaxial growth of InGaN alloys. J Cryst Growth 312:735
86. Stringfellow GB, Greene PE (1969) Dislocations in GaAsP. J Appl Phys 40:502
87. Su LC, Ho IH, Kobayashi N, Stringfellow GB (1994) Order/disorder heterostructures in GaInP with $\Delta E = 160$ meV. J Cryst Growth 145:140
88. Swalin RA (1972) Thermodynamics of solids. Wiley, New York
89. Sze SM, Ng KK (2007) Physics of semiconductor devices, 3rd edn. Wiley, New York
90. Tachibana K, Someya T, Arakawa A (1999) Nanometer-scale InGaN self-assembled quantum dots grown by metalorganic chemical vapor deposition. Appl Phys Lett 74:383
91. Takagahara T, Takeda K (1992) Theory of the quantum confinement effect on excitons in quantum dots of indirect-gap materials. Phys Rev B 46:15578

92. Tanoto H et al (2009) Electroluminescence and structural characteristics of InAs/InGaAs QDs grown on graded SiGe/Si substrate. Appl Phys Lett 95:141905
93. Tran CA et al (1998) Phase separation in InGaN/GaN MQWs and its relation to brightness of blue and green LEDs. J Cryst Growth 195:397
94. Tsao JY (1993) Materials fundamentals of molecular beam epitaxy. Academic Press, Boston
95. Vampola KJ et al (2008) Highly efficient broad-area blue and white LEDs on bulk GaN substrates. Phys Stat Sol (a) 206:200
96. Venables JA (2000) Introduction to surface and thin film processes. Cambridge University Press, Cambridge, pp 145–146
97. Wang XL, Ogura M, Matsuhata H (1995) Flow rate modulation epitaxy of AlGaAs/GaAs quantum wires on nonplanar substrate. Appl Phys Lett 66:1506
98. Wang XH et al (2007) White LEDs based on a single InGaN emission layer. Appl Phys Lett 91:161912
99. Weisbuch C, Nagle J (1990) Science and engineering of 1D and 0D semiconductor systems, ser. NATO ASI series. Plenum, New York, p 319
100. Weisbuch C, Vinter B (1991) Quantum semiconductor structures: fundamentals and applications. Academic Press, Boston
101. Yamaguchi T et al (2006) Two to three dimensional transitions of InGaN and the impact of GaN overgrowth. Phys Stat Sol (c) 3:1396
102. Yuan J, Wang H, van Veldhoven PJ, Notzel R (2009) Impact of base size and shape on formation control of multifaceted InP nanopyramids by selective area MOVPE. J Appl Phys 106:124304

Dielectric Physics Approach for Improvement of Organic-Field Effect Transistors Performance

Martin Weis and Mitsumasa Iwamoto

Abstract We present a brief review on charge transport in organic field-effect transistors (OFETs), which is necessary to further design nanostructured devices. Dielectric physics is used to explain charge transport of these organic devices in the steady and transient states. We clearly show the influence of internal fields on charge accumulation and transport, and propose models for potential distributions across the OFET channel. Potential drop on the electrodes (the contact resistance) is also discussed and its control is described. Improvement of OFET performance is explained in terms of the design of device dimensions, materials and operation regime.

1 Introduction

Semiconductor devices based on organic materials [1], such as thin film transistors [2] and light emitting diodes [3], have attracted much research interest for their promising applications in various fields. With the development of organic materials with high mobilities, a recent trend in the research on the OFET has been concentrated on applied research, mostly focused on increasing carrier mobility and many experimental approaches have been exerted. Among them are the

M. Weis
Institute of Physics, Slovak Academy of Sciences, Dubravska Cesta 9, 845 11 Bratislava, Slovakia

M. Iwamoto (✉)
Department of Physical Electronics, Tokyo Institute of Technology, 2-12-1 O-Okayama, Meguro-ku, 152-8552 Tokyo, Japan
e-mail: iwamoto@ome.pe.titech.ac.jp

L. Zang (ed.), *Energy Efficiency and Renewable Energy Through Nanotechnology*, Green Energy and Technology, DOI: 10.1007/978-0-85729-638-2_24,

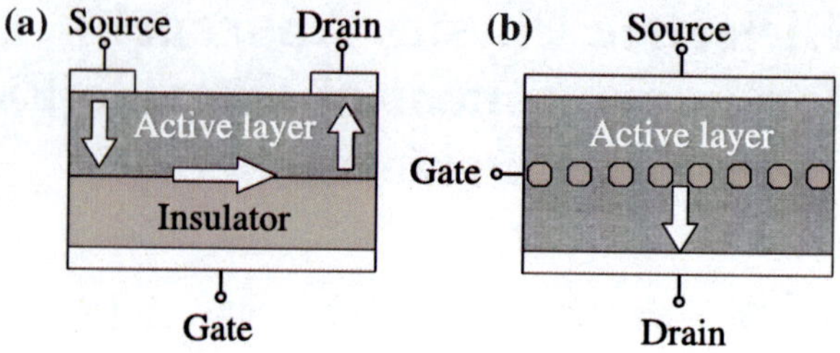

Fig. 1 Simplified view of organic electronic devices: **a** OFET and **b** OSIT. The *arrows* depict charge carrier paths from the source to the drain electrodes

development of modified surface gate insulators and the use of a single crystal semiconductor layer [4, 5]. Along with these endeavors, basic research, such as injection, accumulation and transfer mechanisms, is being utilized to improve OFET performance. Even though there has been significant study, the device physics of OFETs is not yet clear, in comparison with that of inorganic FET structures. Various theoretical studies have been carried out to clarify the device physics of OFETs [6–8], but attention was focused on transport phenomenon only. Recently it has been shown that carriers injected from a source electrode dominate OFET operation. However, owing to the ambiguities of energetics at the organic-metal and organic–organic interfaces, device performance is not fully understood. Therefore deep understanding of injection and transport processes is crucial for further application of OFETs. It should be mentioned here that organic electronic devices are promising candidates for steady-state application and also for high-frequency devices. Therefore, improvement of device performance requires an increase in output current (for a specific applied voltage) as well as reduced response time. However, the successful design of organic device requires a suitable device model.

In contrast to inorganic semiconductors, organic materials have different mechanical, optical and, to some extent, electrical properties. However, electronic devices based on organic molecular materials exhibit similar behavior to inorganic analogies; hence, the name organic semiconductor is used [1]. On the other hand, the low charge- carrier mobilities reported by many research groups suggests classification of these materials as dielectrics, i.e., materials in which the thermodynamic equilibrium is not established. The simplest structure used for investigating material properties is represented by an organic semiconductor sandwiched between two electrodes. For a low intrinsic mobility, it is usually denoted as a metal–insulator–metal (MIM) structure. The OFET, which is most common object of study, is a planar device ruled by injected excess charges (Fig. 1a). Although OFET is well-studied in academic research, it is not widely used and its application requires further improvement. It is necessary to note that OFETs with alternative structures, like vertical OFETs [9] or organic static induction transistor (OSIT) [10], have been proposed. Here, the OSIT is based on the idea of the organic semiconductor film sandwiched between the source and drain electrodes with the gate electrode in the middle of the film (Fig. 1b). The gate electrode is usually a grid [11] or semitransparent (non-continuous) film [10] of metal with blocking contact, which reduces the leakage current. Note that original idea of inorganic SIT lies in the regulation of slit width by controlling depletion layer with applied gate voltage. However, this situation is doubtful for organic SIT and development of new model is needed, as explained in Sect. 2.3.

Interestingly, for the various device structures reported, different models based on different physical background are used for the same organic material [8]. The MIM and OFET structures are discussed on the basis of insulators and semiconductors, respectively. In addition, due to ill-defined differences, the designation "semi-insulators" was proposed. Moreover, the charge transport in the OSIT device was not described clearly and suggested models [12] cannot explain experimental results [13]. Hence, there is a need to find an alternative model that can describe all the aforementioned structures.

2 Steady-State Current Flow

In the following discussion, the electrodes for injection and collection of the charge are denoted as a source and drain, respectively. Even though these names are commonly used for transistor devices only, they hold for the MIM structure, too, and can provide us with a better understanding of the carrier transport mechanism. In this study, the organic semiconductor is treated as a dielectric material with negligible intrinsic carrier density and a single type of charge carrier can be injected and transported (e.g., *p*-type conductivity). In addition, the carrier mobility μ and the dielectric constant ε are assumed constant throughout the active layer and internal field (e.g., due to dissimilar electrodes) is neglected. The current density j through the device can be described in accordance to Ohm's Law as

$$j = en\mu E \tag{1}$$

where en, μ, and E are charge density (e is elementary charge), free carrier mobility and electric field, respectively. Hence, the aim of this discussion is the proposal of an appropriate model, which can provide the charge density and average electric field.

2.1 Metal–Insulator–Metal Structure

The current–voltage characteristics are common sources of information on charge transport through MIM devices [14]. The analytical description of the charge transport is usually based on the drift–diffusion equation solved for the Poisson field [15]. However, the propagation of charges can also be solved along with the equivalent electric circuit with distributed parameters. For analysis of the structures using circuit theory, we break the problem into small parts so that the circuit element dimensions will be infinitesimal small. Thus, the parameters of equivalent circuits spread throughout the line are expressed by distributed parameters

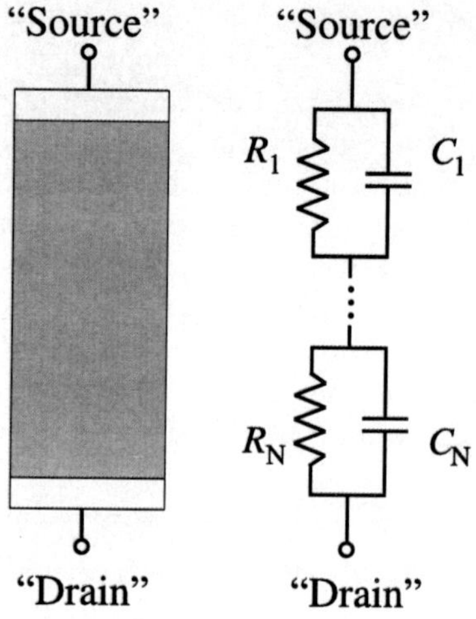

Fig. 2 Schematic view of MIM device and its equivalent circuit

(units per unit length). Here, the equivalent circuit stems from the Maxwell's basic idea that a current flowing through the active layer is given by sum of the conduction and displacement currents. Therefore, the resistor, R, and capacitor, C, components of equivalent circuit represent these two contributions and they are connected in parallel, as shown in Fig. 2.

On the other hand, these elements of the equivalent circuit stand for the physical phenomenon: R and C illustrate the distribution of the electric field and charge, respectively. In detail, for low voltages the constant electric field, i.e., linear potential profile through dx, is modeled by a constant resistor representing potential drop across the distance, dx. Hence, the Maxwell relaxation time $\tau = \mathrm{RC}$ for the region between injection (source) and charge collecting (drain) electrodes represent the spreading of the charge:

$$\tau = \frac{\varepsilon}{\sigma}, \tag{2}$$

where ε and σ represent organic material's dielectric constant and conductivity. The relaxation time, τ, is constant and is a material property. According to Maxwell's electromagnetic field theory, charge, Q, is accumulated at the interface between two dielectric materials with different relaxation times when current with density j flows across the two-material interface. In other words, the current density induces an accumulation of the charge in accordance with the relation:

$$\nabla \cdot D = \nabla \cdot \frac{\varepsilon}{\sigma} j = Q_s, \tag{3}$$

where D represents electric flux density and Q_s is surface charge density ($Q_s = Q/S$, where S is area). Hence, in the modeling of a MIM device by a series of RC loops illustrates a distribution of parameters appearing to be dependent upon charge accumulation. However, although the model of MIM device consists of multiple interfaces, their relaxation times are identical because of constant material parameters, σ and ε. Thus, there is no charge stored in the device and the current density in the low electric field region is:

$$j = \sigma \frac{V}{d}, \tag{4}$$

where V is the voltage applied on the device with film thickness d.

Subsequently, in the high electric field regime, the current follows the well-known space-charge limited conditions (SCLC) [14]:

$$j = \frac{9}{8} \varepsilon \mu \frac{V^2}{d^3}. \tag{5}$$

This tendency has also been observed experimentally, which verifies the dielectric approach for MIM devices. Equation 3 suggests that charge accumulates steadily in the film while the voltage is applied, indicating the Maxwell relaxation time given by Eq. 2 is spatially changing due to injected carriers that contribute to carrier transport (see Sect. 3.1). Further noting that Eq. 2 also accounts for charge trapping in MIM devices under the assumption that the trapping time differs from the dielectric relaxation time of the active layer, ε/σ. Thus, discussion based on the relaxation time can be carried out without the loss of generality.

2.2 *Organic-Field Effect Transistor Structure*

In contrast to the MIM devices, where the charges are transported throughout the bulk of the active layer, the charge transport across the channel in OFET devices is limited to the active layer–gate insulator interface only [8, 16]. This charge transport mechanism induces high-carrier density at the interface, which causes the drain–source current saturation for higher drain–source voltages (Fig. 3). The potential and charge distribution across the channel is simulated by another equivalent circuit (Fig. 4). Here the current flows from the source to the drain electrode through series-connected capacitors and resistors, per unit length. Therefore $C\mathrm{d}x$ and $R\mathrm{d}x$ are the capacitance and resistance, respectively, across the distance dx. Note that $R\mathrm{d}x$ represents the conductivity of surface of active layer and $C\mathrm{d}x$ the gate insulator capacitance. In accordance with the model proposed for the MIM device, the active layer capacitance C_a and gate insulator resistance R_g are connected in parallel with the active layer resistance, R_a, and the gate capacitance, C_g. However, for the OFET electrode setup, the drain–source capacitance represented by C_a is negligible. In addition, we assume ideal gate insulator ($R_g \to \infty$). Thus C_a and R_g elements of electrical equivalent circuit are neglected.

Application of the drain–source and gate–source voltage in accordance with Eq. 2 induces charge accumulation on the active layer-gate insulator interface due

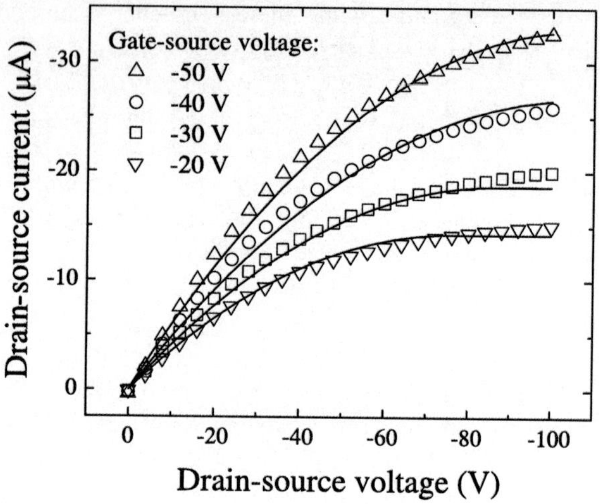

Fig. 3 The output characteristics of pentacene OFET [17] for various gate–source voltages. *Open symbols* represent experimental data, *solid lines* depict calculation result of Eq. 8

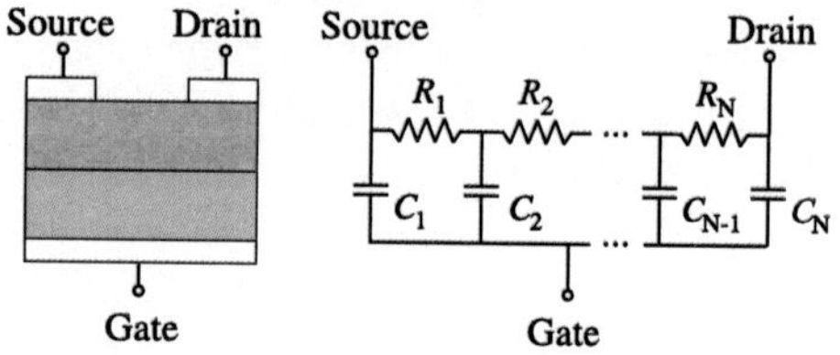

Fig. 4 Schematic view of OFET device and its equivalent circuit

to a discrepancy in relaxation times. Hence, the charge density, Q_s (=*en*), is described by the relation:

$$Q_s = C_g V_{\text{eff}} = C_g\left(V_{\text{gs}} - \frac{1}{2}V_{\text{ds}}\right)\left(1 - \frac{\tau}{\tau_g}\right), \quad (6)$$

where τ and τ_g are relaxation times of active layer and gate insulator, respectively, C_g is gate capacitance per unit of area and V_{eff} is average potential voltage across the channel. To approximate the linear potential profile through the channel region, $V_{\text{eff}} = (V_{\text{gs}} - V_{\text{ds}})x/L$, where x varies from zero to L and V_{gs} and V_{ds} are gate–source and drain–source voltages, respectively. Even with the limitation of this approximation for the saturated region [16], it is widely accepted. This assumption leads to identical resistances per unit of length across the channel. Note that although the relaxation times are again conserved, the charge Q_s is accumulated due to charge spreading on the interface, i.e., capacitors representing the gate insulator capacitance are charged through the active layer. This charge moves the electric field:

$$E_{\text{ds}} = -\nabla V_{\text{eff}} = \frac{V_{\text{ds}}}{L}. \quad (7)$$

In accordance with our previous work [16, 18] we substitute Eqs. 6 and 7 into Eq. 1 to evaluate the drain–source current:

$$I_{ds} = WQ_s\mu E_{ds} = C_g\frac{W}{L}\mu\left(V_{gs} - \frac{1}{2}V_{ds}\right)V_{ds}. \tag{8}$$

Equation 8 has a form that is identical with already derived relations [8, 18], and its fit with experimental data was reported [8, 19]. Note that Eq. 8 is derived using assumption of an ideal gate insulator, $\tau_g \gg \tau$.

It is interesting to compare OFETs with the MIM device characteristics. The drain–source current for low drain–source voltages follows Ohm's Law ($I_{ds} = C_g(W/L)\mu V_{gs}V_{ds}$), i.e., the linear region, which behaves similar to the MIM device. In this voltage region, a continuous charge sheet in the channel region is established. However, increasing the drain–source voltage over $V_{ds} \geq V_{gs}$, i.e., the saturated region, induces a limitation of the interface charge represented by the *pinch-off* position, where $V(x) = 0$ V. This spatial limitation of accumulated charges leads to a saturation of the drain–source current and follows the SCLC model, with a dependence on the square of the applied voltage. Thus, the drain–source current reaches a value of

$$I_{ds} = C_g\frac{W}{2L}\mu V_{gs}^2 \tag{9}$$

and is no longer dependent on the drain–source voltage. This is shown in an experiment (see Fig. 3), where the drain–source current is linearly proportional to the drain–source voltage at first and then, at higher applied voltages, saturates (but its value depends on the gate–source voltage). Equation 8 is suitable for modeling the device behavior (see solid lines in Fig. 3) and carrier mobility evaluation from experimental data.

In addition, although the charge transport in both OFET and MIM devices is described by Ohm's Law and SCLC, the charge in the OFET is transported along the active layer-gate insulator interface only. Hence, the charge transport mechanism is the interface charge propagation, charging of the interface due to difference in relaxation times of the active layer and gate insulator.

2.2.1 Deviation from Linear Potential Profile

As it was already mentioned in the above discussion that all common models since Shockley's famous *gradual channel approximation* [20] assume a linear potential profile across the channel region. However, there still remain questions on limits of this model and how to explain the nonlinear behavior. Electric fields across the OFET channel are crucial for device design; thus, various experimental techniques have been developed to map electric fields and potential. Interestingly, a nonlinear potential (non-constant electric field) has been reported [16, 21], as depicted in Fig. 5a. Here, the optical second-harmonic generation experiment is used to

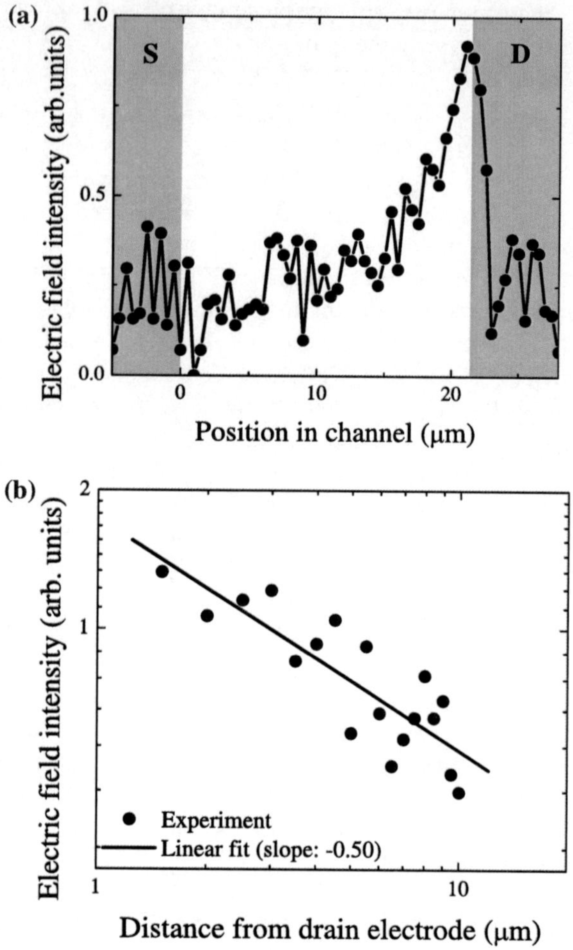

Fig. 5 **a** Electric field intensity obtained from the second-harmonic generation experiment (Manaka) and **b** re-plotted experiment in $\log_{10}$–$\log_{10}$ scale. The grey regions represent source and drain electrode positions

directly visualize electric field of pentacene OFET in the saturated region $\left(V_{\mathrm{ds}} = V_{\mathrm{gs}}\right)$.

Solving the Poisson equation is difficult if charge carriers are injected to the material. The origin of the electric field intensity distribution is influenced by various factors, such as conductive carriers, accumulated charge layer, and trapped carriers. As a result, the total electric field represents the sum of the space–charge field of conduction (mobile) carriers and the accumulated charge layer, $E_m(x) + E_a(x)$. If the contribution of accumulated and trapped charges is small, its influence on the conduction carriers can be neglected and solved independently.

Again, we can apply the concept of transmission line approximation (TLA) with distributed parameters. Here the current flows from the source to the drain electrode through series-connected circuit elements. Like before, Cdx and Rdx are capacitance and resistance, respectively (Fig. 6). The capacitance per unit of

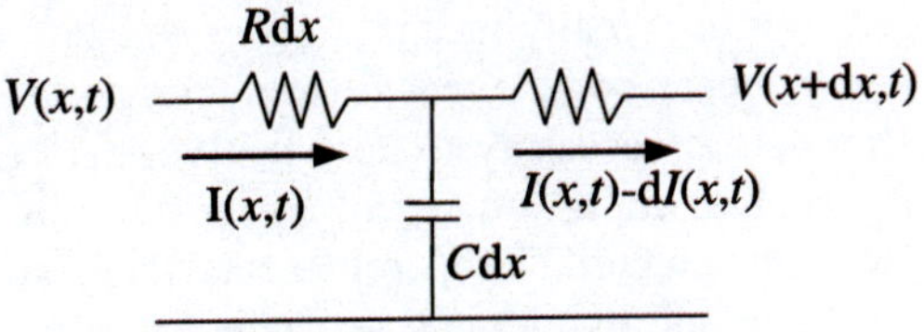

Fig. 6 Detail view of equivalent circuit part, where after local potential $V(x,t)$ and current $I(x,t)$ are denoted

length represents the gate insulator capacitance ($C = \varepsilon_0 \varepsilon_r W/d_g$, where W is the channel width and d_g is the gate insulator thickness). For the resistance, its reciprocal value, the conductance, G, can be used as follows:

$$R^{-1} = G = en\mu\,\mathrm{d}x = CV(x,t)\mu, \tag{10}$$

where $V(x, t)$ is the local potential, $en = \mathrm{C}V(x, t)$. The decrease of current in the channel region due to the charging of capacitors (i.e., the creation of an accumulation layer on the active layer-gate insulator interface) can be written in the form:

$$\begin{aligned} -\partial I &= \frac{\partial \rho}{\partial t}\partial x, \\ -\frac{\partial I}{\partial x} &= C\frac{\partial V(x,t)}{\partial t}, \end{aligned} \tag{11}$$

where ρ is charge density ($\partial Q = \rho \partial x$). Simultaneously, the current flowing through the resistor decreases the potential as follows:

$$\begin{aligned} -\partial V(x,t) &= -I(x,t)R\partial x, \\ \frac{\partial V(x,t)}{\partial x} &= -I(x,t)R, \end{aligned} \tag{12}$$

which represent the potential (carrier) propagation along the conductive channel.

The current $I(x, t)$ can be expressed in its usual way as

$$I(x,t) = en\mu E(\mathrm{LW}) = -CV(x,t)\mu\frac{\partial V(x,t)}{\partial x}, \tag{13}$$

where LW is channel area. Substitution of Eqs. 11 and 12 into Eq. 13 gives us

$$\frac{\partial}{\partial x}\left(V(x,t)\mu\frac{\partial V(x,t)}{\partial x}\right) = \frac{\partial V(x,t)}{\partial t} \tag{14}$$

for a gate capacitance independent of time and position in the channel. Here, the initial and boundary conditions are as follows: $V(0,t) = V_0, V(L,t) = 0, V(x,0) = 0$, which represent the applied voltage on the source and drain electrodes, and the empty channel. Because this solves for the charge propagation in the channel region, the boundary conditions of Eq. 14 represent the transport limited conditions, i.e., smooth injection with small charge carrier injection barrier. In more detail, our model does not have an element, such as diode, that represents the carrier injection process, indicating the carrier behavior after injection. In this sense, $V(0, t)$ really means $V(0^+, t)$ and $V(L, t)$ similarly means $V(L^-, t)$, where + and − represent the position of electrode just inside the active-organic layer. Hence, it is necessary to note that voltage below the source electrode is equal to the applied drain–source voltage reduced by the potential drop, $V_0 = V_{ds} - V_{drop}$. Here the potential drop represents the potential difference between the source electrode and the organic semiconductor. Moreover, the potential drops illustrate the decrease in potential due to contact resistance caused by the carrier injection process from the electrode ($x = 0^-$) to the active-organic layer ($x = 0^+$). Therefore, V_{drop} represents the effect of carrier injection and can be included in our model. On the other hand, Eq. 14 is the continuity equation describing charge carrier propagation if no diffusion is present. Therefore, Eq. 14 for steady state reaches a trivial solution for a potential $V(x)$ and the electric field intensity $E(x)$ $(= -\nabla V)$. $V(x)$ and $E(x)$ can be expressed as:

$$V(x) = V_0\sqrt{1 - \frac{x}{L}}, \tag{15}$$

$$E(x) = \frac{V_0}{2\sqrt{L(L - x)}} \tag{16}$$

where V_0 is the potential below the source electrode. Distribution of the electric field is illustrated in Fig. 5a; the theoretical results are compared with experimental data. Figure 5b shows the decay of the electric field intensity near the drain electrode. A linear curve fitting shows a slope of 0.5, which indicates the field generated by conduction carriers is as described by Eq. 16. If a constant field is built across the source and drain when a steady-state current flows, then $E(0^+)$ should equal $-V_0/L$. However, according to Eq. 16, $E(0^+) = -V_0/2L$. This deviation suggests a space-charge effect caused by mobile charges in the three-electrode system.

Note that in comparison to the common SCLC electric field distribution [14], here the electric field does not vanish on the injection electrode. The reason is the propagation of mobile carriers under the aforementioned boundary conditions in the calculation, and the effect of accumulated charge is discarded, that is, $E_a(x)$ is neglected. Accumulated charge will give rise to a space–charge field, i.e., $E_a(x)$, that effectively decreases the electric field at the injection electrode and regulates the carrier injection process. Actually, as depicted in Fig. 5a, the electric field intensity is low around the injection electrode. Moreover, the field on the source electrode vanishes in the limit of channel length ($L \to \infty$); hence this model describes the influence of finite channel size in OFET.

2.3 Organic Static Induced Transistor Structure

For the OSIT devices, depending on the choice of organic semiconductor material, two different behaviors of output characteristics are reported. Generally, a diode-like tendency without presence of saturation is observed, as illustrated in Fig. 7a.

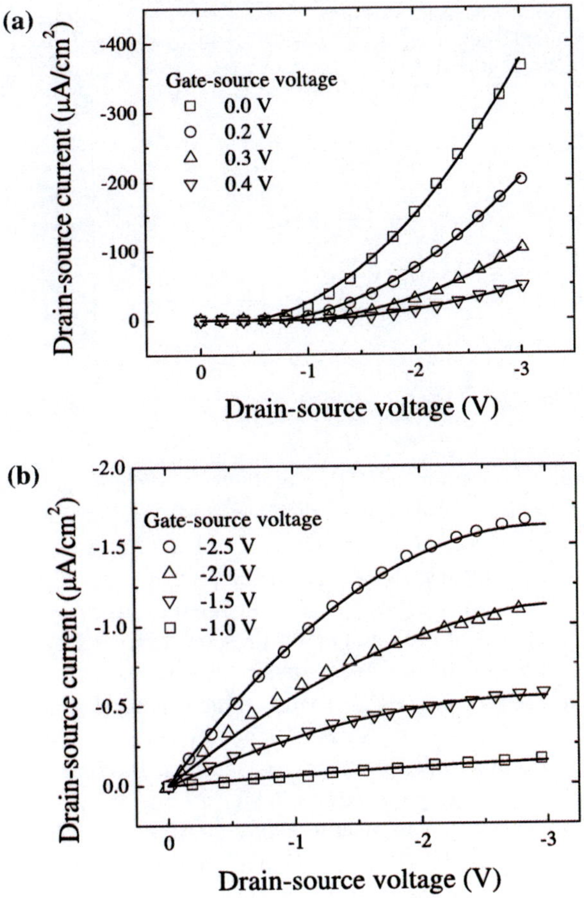

Fig. 7 The output characteristics of an OSIT device with active layer of **a** phthalocynaine and **b** BCQBT, adapted from [23] and [13]. *Open symbols* represent experimental data and *solid lines* depict the results of Eqs. 4, 5 and 19

However, for some materials, OFET-like behavior with saturated current, as depicted in Fig. 7b, is typical. This discrepancy forces us to study the charge transfer mechanism in OSIT. Surprisingly, the principle of OSIT has not been discussed in detail. The theory originally proposed for inorganic SIT devices [22] is based on electrostatics and semiconductor physics, and a similar approach is discussed for OSITs [12]. Nevertheless, this approach cannot easily explain the experimental observations for different materials. This leads us to study the charge transport phenomenon in OSIT and the origin of different behaviors dependent upon the active layer material.

In accordance with our analysis, we suggest an equivalent electrical circuit depicted in Fig. 8. Here, the OSIT problem is divided into areas between (i) source and gate (ii) gate and source and (iii) the region close to the gate electrode (Fig. 8). Again, in the low electric field region, the constant electric field is modeled by a resistor representing the potential drop over the distance, dx, where x varies from zero to $d/2$ (d is the source-drain separation distance). Hence, the Maxwell

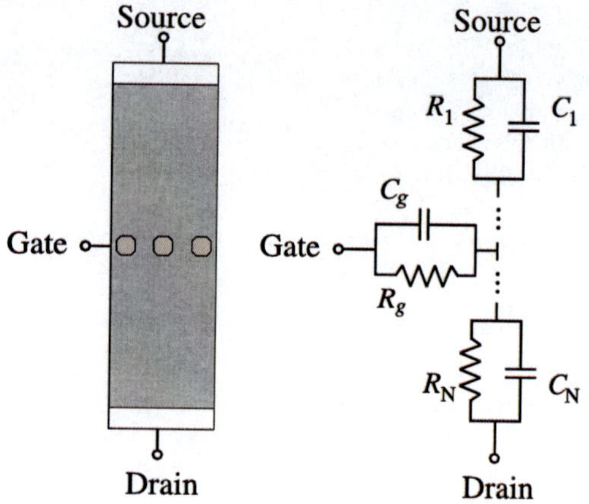

Fig. 8 Schematic view of OSIT device and its equivalent circuit

relaxation time for the region between the source (drain) and gate electrodes again depicts the spreading of the charge in active layer. The relaxation time is constant and a material parameter; the relaxation times $R_1C_1 = \cdots = R_nC_n = RC$ are independent on device geometry.

The gate electrode is also modeled by its resistance and capacitance. Here, the resistance, R_g, illustrates the leakage current ($R_g \approx V_{gs}/I_{gs}$) and its capacitance, C_g, describes the accumulated charge on gate electrode. In other words, the parameters R_g and C_g depend on the electrical and geometric properties of the gate electrode; these parameters depend on the electrode type, e.g., grid or semitransparent electrode, blocking contact, etc,. In contrast to the OFET device, here the absence of gate insulator leads to a finite value of R_g.

Here, we again meet a relaxation time, $\tau_g = R_gC_g$, representing the charging of the gate electrode region. Hence, in accordance with the solution of the Maxwell equations for the interface of two dielectrics, the charge is stored in the capacitor C_g only for unequal relaxation times, $\tau \neq \tau_g$. Therefore, the following discussion is divided into two parts with respect to relaxation times.

First we discuss the case of similar relaxation times. As already mentioned, no charge is accumulated close to the gate electrode if $\tau \approx \tau_g$. Hence, the OSIT is simplified to the MIM structure, where the gate electrode changes the effective applied potential V'_{ds}. In other words, the charge transport is driven by the drift of the carriers in the source–gate field, $E_{gs} = V_{gs}/(d/2)$, and afterward in the gate–drain field, $E_{dg} = \left(V_{ds} - V_{gs}\right)/(d/2)$ (here the drain–gate voltage is evaluated as $V_{dg} = V_{ds} - V_{gs}$). Hence, the drain–source current density, j_{ds}, is expected to follow Eqs. 4 and 5, but using of effective applied potential V'_{ds} instead of V. Interestingly, this ohmic and SCLC behavior is recorded in experimental data, as illustrated in Fig. 5a by solid lines. Here we again see ohmic conductivity ($j \propto V$) in the low electric field region, $V_{ds} < 1$ V, with a SCLC behavior ($j \propto V^2$) in the high electric field region.

Now, we will discuss the case of dissimilar relaxation times. A discrepancy in the relaxation times of the organic semiconductor film, τ, and the gate electrode region, τ_g, causes accumulation of charge close to the gate electrode. The amount of accumulated interface charge can be estimated by the MW model:

$$\begin{aligned} Q_s = \frac{G_g G}{G_g + G} = V'_{\mathrm{ds}}(\tau - \tau_g) \\ = (C_g V_g - C_1 V_1) + (C_g V_g - C_2 V_2), \end{aligned} \quad (17)$$

where $G = 1/R$ and $G_g = 1/R_g$ are the conductivities of the active layer and gate electrode, respectively, and $V_1, \ldots, V_n$ and V_g are the potential drops throughout the resistive elements of the equivalent circuit ($R_1, \ldots, R_N$ and R_g) (see Fig. 8). Assuming a low leakage current ($R_g \gg R$), we neglect the potential drop across the active layer (i.e., $V_g \approx V_{\mathrm{gs}}$) and the accumulated charge is approximately given by the relation:

$$Q_s \approx 2C_g\left(V_{\mathrm{gs}} - \frac{1}{2}V'_{\mathrm{ds}}\right), \quad (18)$$

where we assumed the middle position of the gate electrode, $V_s = V_{\mathrm{ds}}/2$, and effective drain–source voltage V'_{ds} which includes geometry effect, $V'_{\mathrm{ds}} = (C/C_g)V_{\mathrm{ds}}$.

Therefore, the gate electrode separating two conductive organic films limits the amount of transported charge. This phenomenon represents interface-limited current. As already discussed above, we estimate the drain–source current as:

$$j_{\mathrm{ds}} = Q_s \mu E_{\mathrm{ds}} = C_g \frac{2}{L}\left(V_{\mathrm{gs}} - \frac{1}{2}V'_{\mathrm{ds}}\right)V_{\mathrm{ds}}. \quad (19)$$

In a high drain–source, we again see saturation of the drain–source current to a value of

$$j_{\mathrm{ds}} = C_g \frac{1}{L} V_{\mathrm{gs}}^2. \quad (20)$$

Surprisingly, we derived a relation almost identical with relations describing OFET behavior. The drain–source current increases with applied drain–source voltage and saturates at $V_{\mathrm{ds}} = V_{\mathrm{gs}}$. This result corresponds to experimental records (Fig. 7b), where saturation of the drain–source current is observed for higher drain–source voltages. Again, Eq. 12 is suitable for the modeling of device behavior (the solid lines in Fig. 7b) and the carrier mobility can be extracted.

Moreover, due to the interface charge limitation, it is expected that a drain–source current is much smaller than in case of similar relaxation constants for the case of $\tau \ll \tau_g$. This is also observable in Fig. 5, where the interface-limited current is two orders smaller than the drift current. However, a detailed discussion of the origin of the "bottleneck", the charge transport limitation, is beyond the scope of this model and strongly depends on the OSIT gate electrode preparation.

3 Charge Propagation in the Transient State

Carrier transport through the OFET as a transient phenomenon has only received attention recently [24]. The difficulty came mostly from experimental problems as well as a focus on the steady-state only. However, recently there has been research focused on the time-of-flight (TOF) method or time-resolved microscopy second-harmonic generation (TRM-SHG) technique applied to the OFET structures [24–26] and the transit time, t_{tr}, was obtained. Hence, as a first approximation, TOF analysis adapted from MIM structures was used. However, for the three-electrode system of the OFET, the charge carrier propagation can differ due to the space-charge field.

Therefore, in the following discussion, two physical models of charge carrier transport are presented and discussed separately: the MW model and the TLA. The MW model is well-known for its physical explanation of charge accumulation at the interface and relies on the electrical properties of materials. On the other hand, the TLA is common for signal propagation analysis and relies on equivalent circuits. Both models are used to explain the charge propagation in two- and three-electrodes systems represented by MIM and OFET structures. We show that both models approach to the same result. In other words, the TLA can model the charge propagation across the channel even though the conductivity distribution changes with time. A comparison of the models points out advantages as well as limitations of both evaluations.

3.1 The Maxwell–Wagner Model

Two macroscopic physical parameters characterize the organic materials used in MIM and OFET devices. These are the dielectric constant, ε, and conductivity, σ. The ratio ε/σ gives a relaxation time and represents the spreading time of the excess charge carriers in the materials. That is, a steady-state charge distribution is established after an elapsed time of around $\tau = \varepsilon/\sigma$. Note that conductivity is proportional to the carrier density n_0 and is given by $\sigma = en_0\mu$. The carrier density is generally the intrinsic carrier density of materials at thermodynamic equilibrium. According to electromagnetic field theory, the total current flowing across the organic materials is the sum of the conduction current and the displacement current. The densities of the conduction current and displacement current are given by εE and $\partial D/\partial t$ with $D = \varepsilon E$ (D is the electric flux density), respectively.

3.1.1 The MIM Device

For the MIM structures, when the injected carrier density, n, is low but is continuously supplied from the electrodes and the space–charge field caused by the injected carriers is negligible in comparison to the applied external electric field (linear potential through the insulator), E and σE are replaced by $\sigma\ V/L$ and $\varepsilon\ V/L$, respectively. Here, L is the thickness of the insulator. This means that the MIM structure is represented merely as a parallel electrode system, $R = L/\sigma S$ and $C = \varepsilon S/L$, in carrier transport of injected carriers. We find a relationship between the time constant of the equivalent circuit, RC, and the relaxation time of the insulator ε/σ, as RC $= \varepsilon/\sigma$. That is, the time constant is free from the geometry of electrode configuration and is given only by the material parameters, σ and ε. From the equivalent circuit consideration, we find the equivalent circuit is converted into N series-connected parallel RC circuits, as illustrated in Fig. 2, suggesting the potential drop across each resistance R/N and charging of each capacitance NC must be the same, i.e., there is no charge accumulation at the connection point between segments. According to the Maxwell–Wagner effect, charge accumulation happens at the interface between two materials with different relaxation times. Hence, in an insulator represented by constant material parameters, ε and σ, there is no charge accumulation over the whole region and carriers are supplied from one electrode and are conveyed to the counter electrode across the insulator. This is actually consistent with the result of the N-series RC circuit model. The transit time of carriers across the MIM structure is given by

$$t_{\mathrm{tr}} = \frac{L}{\mu E} = \frac{L^2}{\mu V}, \tag{21}$$

and a steady-state current flows at $t = t_{\mathrm{tr}}$ after applying a step voltage, V, at $t = 0$ [14]. Obviously, we may consider this transit time gives a charge spreading time $\varepsilon/\sigma = \varepsilon/en\mu$, where n is the average carrier density of the insulator caused by injected carriers and intrinsic carriers, n_0. Therefore we obtain the relation $en = (\varepsilon V/L)/L$, representing a constant carrier distribution in the MIM after time, t_{tr}. In other words, the carrier density of the dielectric changes from n_0 to n at $t = t_{\mathrm{tr}}$.

This discussion easily extends to the case where carriers transport across I series-connected RC segments ($i < N$), i.e., from electrode to the ith connection point in the equivalent circuit shown in Fig. 2. In that situation, the time required for carriers crossing the i segments is given by

$$\begin{aligned} t'_{\mathrm{tr}} &= \frac{L'}{\mu E} = \frac{L'^2}{\mu V'}, \\ \text{with } L' &= \frac{i}{N}L, \\ \text{and } V' &= \frac{i}{N}V. \end{aligned} \tag{22}$$

Since the charge spreading into i segments should be the same as $\varepsilon/\sigma = \varepsilon/en\mu$, we obtain $en = (\varepsilon V'/L')/L'$. That is, the carrier density of the insulator changes from n_0 to n, along with the evolution of the region of injected carriers in the presence of the electric field, E.

Therefore we may conclude that charge transport can be simply described by the drift of carriers in the average electric field, satisfying the time dependence along the direction of the electric field. This is a sketch of carrier propagation for the case of a MIM device with two electrodes and by using TOF we can estimate carrier motion indirectly.

3.1.2 The OFET Device

The situation is quite different in the case of the OFET, a three-electrode system. As described above, at the interface between two materials with different relaxation times, charge is accumulated at the interface (the Maxwell–Wagner effect). This situation happens at the active-organic layer-gate insulator interface. The relaxation time of the gate insulator material is longer than that of the active layer, $\tau_g > \tau$. Therefore, charge is accumulated at the interface in a manner similar to trapped charges while a current flows and this situation is quite different from that of MIM, suggesting that carrier motion must be described using a model that considers interface charge accumulation.

In more detail, for the OFETs, carriers are injected from the source electrode in a manner similar to the case of the MIM structures, but they flow along the gate insulator-active layer interface in the direction from the source to the drain electrode, accompanying the interface charge accumulation caused by the MW effect. The amount of charge accumulated at the interface is regulated by the gate voltage and is given by

$$Q_s = C_g\left(V_{gs} - \frac{1}{2}V_{ds}\right), \tag{23}$$

where in the limit linear potential is built along the interface by the spreading of accumulated charge along the organic semiconductor-gate insulator interface. Here, C_g and V_{gs} are the gate insulator capacitance per unit area and the gate–source voltage, respectively. The carrier density at the active layer-gate insulator interface changes from n_0 to n, caused by Q_s. Therefore, similar to the case of the MIM structure, we can estimate the spreading time of the charge carrier at the interface. Since the carrier density is given by $en = C_g(V_{gs} - V_{ds}/2)/h$ (h is the channel thickness), the conductance along the interface is given by

$$G = en\mu Wh/L. \tag{24}$$

On the other hand, charge accumulation is regulated by the potential across gate insulator as described by Eq. 23, the capacitance along the channel is given by

$$C = C_g WL. \tag{25}$$

Hence, the response time is given as C/G and represents a carrier transit time, t_{tr} across the interface from the source to drain. That is,

$$t_{tr} = \frac{L^2}{\mu\left(V_{gs} - \frac{1}{2}V_{ds}\right)}. \tag{26}$$

We should note that the t_{tr} is also valid for the case when $V_{ds} = 0$ V and represents the charge accumulation condition at the interface only. Hence, it is reasonable to say that the interface charge propagation process regulates the transit time of OFETs. Furthermore, we should note that in the derivation of Eq. 23, we did not assume $|V_{gs}| \gg |V_{ds}|/2$, as in the most simple case, but the above discussion can be simplified without loss of the underlying physics [18]. In addition, although this analysis is based on a commonly-used steady-state potential distribution across the channel, it can be extended to the time-dependent accumulation of charges at the interface.

It is instructive to note that this situation can be modeled using the equivalent circuit shown in Fig. 4, where the resistance corresponds to the conductance, G, along the channel and the capacitance represents the capacitance, C, of Eq. 25. There, the potential distribution along the channel is considered and the equivalent circuit is extended to a ladder model. The transmission line model is based on this equivalent circuit, as will be discussed in following section. Furthermore, we note that in derivation of Eq. 21, we assumed the carrier injection to the interface is only from the source electrode, but t_{tr} should be reduced in case when carrier injection is also allowed from the drain. For instance, t_{tr} should be half of the t_{tr} in Eq. 21 when $V_{ds} = 0$ V instead of $V_{ds} = V_{gs}$ and we use similar electrodes as the source and drain.

3.2 *Transmission Line Approximation*

3.2.1 The MIM Device

The TOF method was originally designed for metal–semiconductor–metal (MIM) structures [27–29], where the carrier transport in the two-electrode system can be described as a one-dimensional problem (see Fig. 2a). Interestingly, the propagation of charges can also be solved using an equivalent circuit with distributed parameters. The equivalent circuit stems from Maxwell's basic idea that a current flowing inside a material is given by the sum of conduction and displacement currents. R and C represent these two contributions, respectively. On the other hand, these elements stand for physical phenomenon: a network of R's and C's illustrates the distributions of electric field and charge. In detail, a constant electric field (i.e., a linear potential through the semiconductor part of the MIM structure) is modeled by equal resistors representing a potential drop per distance, dx. In this model, the charge propagates from one electrode with area, S, to the opposite electrode situated a distance, d, through the resistors and the charge carrier distribution is depicted by the capacitors, C. Here, it is assumed that there is no electric field inside a metal electrode ($C_{metal} = 0$) and there is negligible metal resistance compared to semiconductor ($R_{metal} = 0$). Hence, if e and n are the elementary charge and carrier density, respectively, the distributed conductance σ can be estimated from the current density as follows:

$$j = en\mu E = \frac{CV}{d}\mu\frac{V}{d} \tag{27}$$

and by the differentiation of the current density with respect to E, defined as $\sigma = \partial j/\partial E$. Subsequently, the distributed resistance $R = d/(S\sigma)$ also extracted:

$$R = \frac{d^2}{CS\mu V}. \tag{28}$$

Although excess mobile charges are injected into the device, in this evaluation it is assumed that the effect of a space–charge field is negligible. That is why we derive the σ using $en\ E$, not the last term of Eq. 27. Thus, the electric field across the MIM structure is constant, $E = V/d$. If a voltage pulse with an amplitude, V, is applied to the electrodes, the transit time, which is represented by the relaxation time of the semiconductor in the MIM structure, can be simply estimated by the relation t_{tr} = RC and therefore:

$$t_{tr} = \frac{1}{\mu}\frac{L^2}{V}. \tag{29}$$

Note that here the product of RC represents the relaxation of the injected carriers, and t_{tr} = RC with $R = \Sigma R_i$ and $C = \Sigma C_i$ (and $R_i = R_j, C_i = C_j$ for all i, j) represents the carrier transport time across all RC segments illustrated in Fig. 2. In limit of infinitesimally small elements, the sum is replaced by the integration with identical results. As such, the transit time of Eq. 29 expresses the situation where the charge carriers are conveyed through a series of RC segments in the whole MIM structure. This is described by the transit time's dependence on the voltage and the film thickness. Since there is no difference between the relaxation times of RC loops, no excess charge (like trapped charge) is stored between the segments. In more detail, the R of Eq. 28 is defined considering injected mobile carriers. This means that all charge carriers are transported across the MIM structure, even though they contribute to the space-charge field formation. The presented model with distributed parameters describes this situation, and accounts for no charge accumulation corresponding to charge trapping in the insulator layer in the MIM structure [18]. Therefore, the charge transport can simply be described by the drift of carriers in the average electric field. Furthermore, Eq. 29 suggests that the carriers will propagate with the square root of time in a drift field along the direction of the electric field intensity. In summary, we conclude that the transit time is dependent on the geometric parameter as well as the applied voltage. In the following text, we discuss the case of the OFET structures and derive the transit time.

It is instructive to note that in the state of the thermodynamic equilibrium, the dielectric relaxation time, defined by RC, is a material parameter defined by ε/σ, where ε and σ are the dielectric constant and conductivity, respectively. The dielectric relaxation time becomes independent of the geometric parameter as well as the applied voltage. Here, σ is proportional to the product of carrier density in equilibrium state ($n = n_0$) and mobility, μ. We should note that n_0 is different from that defined in Eq. 27 due to the presence of injected carriers, $n = CV/d$. Hence, the value of RC for MIM structures and OFETs should be the same.

3.2.2 The OFET Device

The electric field propagation along the channel in an OFET (Fig. 4) was recently modeled by the TLA [16]. The TLA is based on solving the equivalent circuit consisting of infinitesimally small resistors and capacitors connected in series as a ladder. In this model, charges propagate through the channel at the semiconductor-gate insulator interface. Therefore, again, the resistance and capacitance are related to the distributions of the electric field and accumulated charge $(\rho(x) = C\partial V(x)\mathrm{d}x)$. In other words, the successive charging of the capacitors represents the migration of charge carriers and resistors describe the potential drop across the channel. Note that in the presented model of the three-electrode system (OFET structure), we do not assume the influence of injection properties, i.e., the potential drop due to an injection barrier, which causes insufficient charge accumulation. Also, the displacement current between the source and drain electrodes is assumed negligible by taking into account the electrode size and separation used in this experiment. In summary, with consideration of the device parameters, the parallel segmental capacitance, found in the equivalent circuit for the MIM structure and employed to express carrier transport as displacement current, is discarded in the equivalent circuit for the OFET channel. Furthermore, since the conductivity of the gate insulator of our OFET device is extremely low, the parallel resistance element to express the carrier transport across the gate insulator is omitted from the equivalent circuit.

In the following text, analogous to the analysis of the MIM structure, the charge transport can again be solved by the relaxation times of the RC loops with distributed parameters, after calculating R in the manner we derived (Eq. 28). In the linear region (the drain–source voltage is smaller than the gate–source voltage, $|V_{ds}| \ll |V_{gs}|$), the constant electric field condition should be satisfied and all resistors of the equivalent circuit have identical values. Therefore, the transit time can be written as a product of the channel resistance, R_{ch}, and capacitance, C_{ch}, i.e., $t_{tr} = R_{ch}C_{ch}$. The drain–source current I_{ds} can be expressed [8, 30] as Eq. 8. However, in contrast to the small signal analysis, here we apply a large signal that propagates across the channel. In other words, charges carry the electric field and therefore the voltage drop varies with charge carrier (i.e., potential) distribution. Therefore, the channel resistance is not constant anymore and depends on time and position, i.e., $R_{ch} = R_{ch}(x, t)$. For a linear approximation of the potential distribution between the source electrode edge ($x = 0$) and the charge carrier sheet edge in the channel region ($x = x^*$) is derived:

$$R_{ch} = \frac{V_{ds}}{I_{ds}} = \frac{x^*}{C_g W \mu (V_{gs} - \frac{1}{2}V_{ds})}. \tag{30}$$

The channel capacitance follows:

$$C_{ch} = C_g WL. \tag{31}$$

Therefore, the transit time can be written as a product of Eqs. 29 and 30 in the form:

$$t_{tr} = \frac{1}{L}\int_0^L R_{ch} C_{ch} dx^* = \frac{1}{2\mu} \frac{L^2}{V_{gs} - \frac{1}{2}V_{ds}}. \tag{32}$$

It is interesting to note that an identical result from the definition of group velocity can be also obtained:

$$v_g = \frac{dx}{dt} = \mu E. \tag{33}$$

Here, we assume a linear potential profile, $E = V'/x^*(\text{for } x \leq x^*)$ or $E = 0\,(\text{for } x > x^*)$, with an effective voltage, V'. Trivial integration of Eq. 33 for $x \in (0,L)$ and $t \in (0,t_{tr})$ provide us with a simple relation:

$$t_{tr} = \frac{1}{2\mu}\frac{L^2}{V'}. \tag{34}$$

Note that the MW model and TLA give identical result for the MIM structure (see Eqs. 21 and 29). On the other hand, the transit time estimated by MW model (Eq. 26) and TLA (Eq. 32) differs by factor 1/2. This deviation has its origin in time-dependent conductivity, which reflects the changes of the electric field during the charge propagation. In addition, this confirms the trivial analysis of the transit time, Eq. 34.

Also, it should be pointed out that the electric propagation field is transient in an OFET. In contrast to the MIM case, where the electric field was conserved during the charge propagation, here the field depends on the position of the carrier sheet front edge ($E = V'/x^*$ (for $x \leq x^*$)). Hence, the electric field evolves in time with respect to the charge location. This is clearly illustrated in the electric field visualization prepared by the TRM-SHG technique [31, 32] (Fig. 9a). The electric field obtained from the TLA model presented for comparison (Fig. 9b).

As mentioned above, we reached the same conclusion starting from the MW and TLA models. One of the important findings from these two approaches is that the transit time, defined by Eqs. 25 and 30, is valid even when $V_{ds} = 0$. This suggests that interface charge propagation regulates transient carrier transport in the OFET channel. In the following discussion, we show such a situation by visualizing the transient electric field.

Figure 10a shows a typical transient electric field imaging at various delay times under $V_{gs} = V_{ds} = -100$ V. With an increasing delay time, the electric field moves from the source to the drain electrode and represents the edge of the carrier sheet [26]. Here, it should be noted that the electric field edge position versus the measured time on a $\log_{10} - \log_{10}$ scale reveals a time dependence proportional to the square root of time (a linear fit with slope of 0.47).

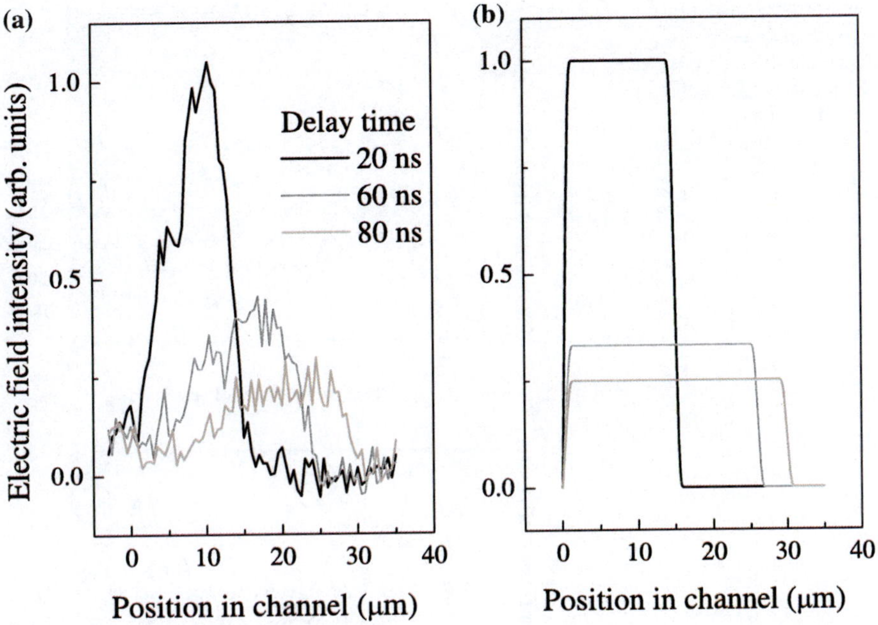

Fig. 9 Time evolution of the transient electric field propagation across the OFET channel for three different delay times, *t*, after application of voltage pulse ($V_{gs} = V_{ds}$): **a** TRM-SHG experiment [32] and **b** ideal theoretical case

Note that the carriers are predicted to migrate through the device proportionally to the square root of time ($x \propto t^{1/2}$). This result is in accordance with the experimental data shown in Fig. 10. However, our calculation also has other important consequences: (i) the charge is propagating through the OFET channel not only due to an electric field between source and drain electrodes, but also because of an interface charging phenomenon, and (ii) the charge is redistributing and accumulating within the channel even when no drain–source voltage is applied. In detail, in the three-electrode system, we found a different mechanism for carrier transport. In contrast to the two-electrode system represented by the MIM structure, where the transport mechanism is limited by charge carrier drift in the electric field, in the three-electrode system the charge is transported due to the propagation of interface charges (i.e., propagation of the accumulated charge layer). In other words, in the MIM structure with a single layer of dielectric material (organic semiconductor), carriers are transported directly in the direction of the electric field. On the other hand, in the OFET structure, a two-layer system (organic semiconductor and gate insulator), charge transport is due to the charging of the semiconductor-gate insulator interface and carrier motion is nearly perpendicular to the external gate field [26]. Note that the motive force, which conveys carriers within the channel, is due to the local electric field generated by injected excess charges. The lateral component of local electric of the excess charges causes a

Fig. 10 Time evolution of the transient electric field propagation in the pentacene OFET channel for different delay times *t* after application of voltage pulse of $V_{gs} = -100$ V and **a** $V_{ds} = V_{gs}$ and **b** $V_{ds} = 0$ V

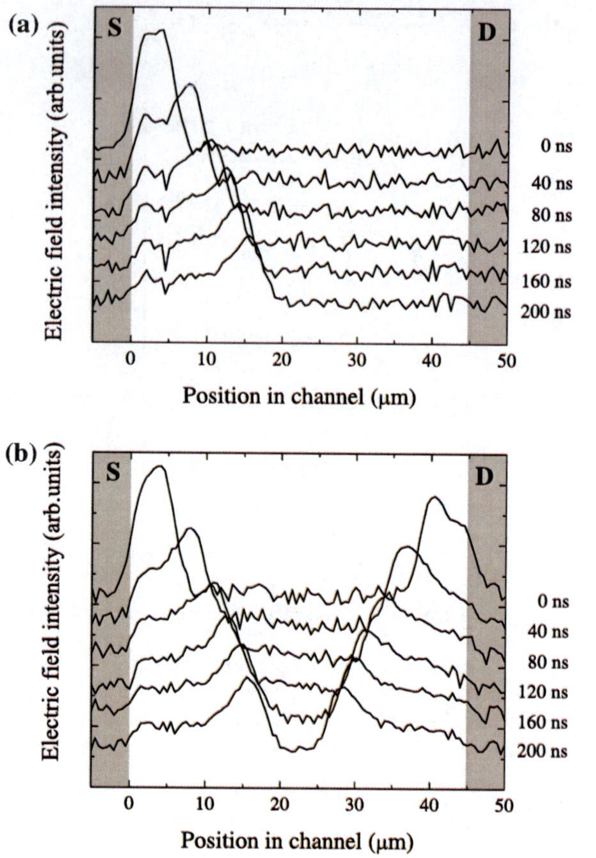

redistribution and migration within the channel toward the steady-state condition. Additionally, interface charging, which is very similar to carrier trapping but originally caused by the MW effect between the semiconductor and gate insulator layers [18], is a characteristic phenomenon originating in the OFET system. Although carrier migration through both configurations (i.e., $V_{ds} = V_{gs}$ or $V_{ds} = 0$ V) follows the square root of time law the physical reason is different. This conclusion is in accordance with the electric field visualization experiment, the electric field of the injected carriers migrates when the source and drain electrodes have the same potential (are electrically shorted, i.e., $V_{ds} = 0$ V), see Fig. 10b. The electric field records also show that the charge migration starts out symmetrically from either electrode toward the channel center due to the absence of a directional drift force between source and drain. As observed from Fig. 10, carriers migrate from both source and drain electrodes similarly. Moreover, analysis of the carrier sheet migration again shows a proportionality to the square root of time. Here we must point out that, although no drift field between source and drain electrode is established, carrier propagation is not related to the diffusion process, which is significantly slower in organic semiconductors. Intriguingly, this experimental study is in agreement with the proposed carrier transport based on interface charging, whereas the carrier drift approach contradicts this result.

Additionally, we can discuss the relation with the Maxwell–Wagner (MW) model. Recently, the MW model was used to evaluate the pentacene-gate insulator interface under the source electrode [33, 34]. In the MW model, contact resistance has been treated by a relation for charge transport similar to Eq. 32. However, the meaning is different; the MW model describes the steady-state and is based on the presence of an interface charge caused by an electric field across the interface and a difference in relaxation times, whereas the TLA model explains charge propagation in the channel region and does not require contact resistance.

Note that a model for pentacene film with distributed R and C elements (see Fig. 4), a simple resistance R can be used instead. However, in this case the relaxation times of all repeat units (RC loops) are identical; thus, no charge is accumulated in the pentacene film, unlike the case of the MIM device. Nevertheless, the charge is accumulated at the pentacene-gate insulator interface due to the MW effect and this interface charging is in conjunction with charge transport. Therefore, the proposed TLA model describes a more realistic situation in comparison with the standard steady-state MW model, where at first charge is accumulated according to the MW effect and transported thereafter. However, we must point out that also the MW model can be extended to the case of time-dependent fields with the same result as TLA because the MW effect always causes charge accumulation at the interface between two different materials with unequal relaxation times, without a time dependence on the fields. Curiously, the transit time through the OFET device is also the relaxation time of the investigated device. In other words, the charging of the OFET represents how charge is transported in a three-electrode system.

4 How can we Improve Device Performance?

As discussed above, application of organic electronics devices requires various device properties, like thermal and temporal stability, low-cost fabrication, high output current and fast response. Even though material research and chemical engineering can provide cheap and stable organic semiconductors, device design requires understanding of underlying physics. Following discussion concludes consequences of device models and suggests road map for device improvement.

4.1 Charge Accumulation

The above analysis of an OFET based on the MW model pointed out importance of the charge accumulated at the organic semiconductor-gate insulator interface. It was shown that amount of accumulated charge is proportional to the drain–source current. Hence, the main aim of various new device designs is increasing the accumulated charge, Q_s. Although the most common case is the enlargement of the gate insulator capacitance, it is not only opportunity for improvement. To recapitulate the possible ways, we follow a modified Eq. 6 as follows:

$$Q_s = C_g V_{\text{eff}} = C_g \left(V_{\text{gs}} + V_{\text{dip}} - \frac{1}{2} V_{\text{ds}} \right) \left(1 - \frac{\tau}{\tau_g} \right), \tag{35}$$

where V_{dip} is potential change due to internal fields. As a result we find three different methods to increase the accumulated charge:

1. Gate insulator capacitance,
2. Internal fields, and
3. Relaxation times.

In detail, the first way is based on increasing the gate insulator capacitance ($C_g = \varepsilon_0 \varepsilon_r / d_g$) by changing its dielectric constant (ε_r) or reducing the insulating film thickness (d_g). The second way of improvement employs modification of the internal fields. This usually involves modification of gate insulator by self-assembled layers with electrostatic dipoles [35, 36], or gate insulator films with dipoles [37]. Note that the internal field has other contributions, e.g., trapping charges; thus, the total internal field is created by the superposition of all components. Interestingly, the latest way for raising accumulated charge includes the organic semiconductor properties. In other words, appropriate selection of organic semiconductor dielectric constant and conductivity with respect to the gate insulator leads to a higher accumulation of charge at the interface.

4.2 Charge Transport

After the charges accumulate at the organic semiconductor–gate insulator interface, charge transport becomes crucial for device performance. Note that this is true not only for the current's magnitude, but also for the response time, represented by the transit time. Recently, improvements of organic semiconductor mobility have achieved a level comparable to that of amorphous silicon and have brought the OFET closer to commercial applications. However, there are still other ways to reduce of the transit time. It is instructive to note that the transit time has a direct impact on the drain–source current:

$$I_{\text{ds}} = \int_0^{V_{\text{ds}}} \frac{C_{\text{ch}}}{t_{\text{tr}}} \text{d}V = C_g \frac{W}{L} \mu \left(V_{\text{gs}} - \frac{1}{2} V_{\text{ds}} \right) V_{\text{ds}}. \tag{36}$$

That is, the total charge accumulated in the channel region is transported in time t_{tr}, i.e., $I_{ds} = Q_s/t_{tr}$. Therefore, improvement of the charge transport stands for a reduction of the transit time. In accordance with the above models, the transit time is influenced by:

1. Carrier mobility,
2. Channel length, and
3. Applied voltage.

The carrier mobility is a material parameter that depends on the organic semiconductor and the preparation technique. Note that although it is microscopic parameter related to charge transport on the molecular level, the macroscopic measurements estimate the effective value, which includes additional effects like traps and potential drops. Thus, the rise of carrier mobility depends mostly on the progress of material science and engineering.

More interesting is reduction of the channel length. The quadratic dependence of the transit time on the channel length gives us a powerful tool to reduce the response time and increase the current. This inspired researchers to design the vertical channel OFET [9] and OSIT devices. However, reducing the channel length decreases the amount of accumulated charge and the space-charge field. In other words, in the case of OSIT, the gate electrode capacitance becomes smaller than the capacitance of the source and drain electrodes. As a result the OSIT device losses its transistor behavior. Thus, the channel length reduction is limited by the preparation techniques available and the vanishing space–charge field [23].

4.3 Contact Resistance

Improvement of the carrier transport in OFETs revealed another bottleneck: charge injection. This barrier is expressed by the contact resistance, R_c, and in OFETs, it is a serious problem for practical applications [38]. The contact resistance has many origins, such as the non-uniformity of organic semiconductors, the presence of dipole layers at the metal-organic interface, electrode resistance and the interfacial energy states [39, 40]. Note that the device contact effect influences the device operation conditions, such as the potential distribution across the OFET channel [21]. The carrier mechanism is influenced by the R_c at the metal-organic material interface when the energy difference between the Fermi level of metal electrode and the highest occupied molecular orbital (HOMO) of organic semiconductor dominates the hole injection. This understanding gives us a powerful tool for contact resistance engineering. Hence, we present a brief summary of possible approaches to modify the contact resistance.

As mentioned above, the contact resistance for a specific organic semiconductor depends on the Fermi energy of electrode metal. Hence, fine-tuning the interface energetics plays a key role for smooth carrier injection. Rough adjustment of the injection barrier can be accomplished through selection of an electrode metal with a work function similar to HOMO level (for hole injection) [41]. However, the effective work function of metal can also be changed. It was shown that exposure to the UV and ozone is helpful for surface cleaning [42], which lowers the work function [43]. Moreover, this treatment creates metal oxides on the electrode surface and thus increases the work function. Here, it should be noted that for UV/ozone surface cleaning, a decrease [42] and an increase [44] of injection barrier were both reported. Unfortunately, this simple change of the injection barrier can only be applied to bottom-contact OFETs, where the electrode is formed prior to the organic semiconductor film deposition.

Another common approach to modify injection properties is introducing a dipolar layer. Usually, a self-assembled monolayer (SAM) is created on metal-organic interface in a bottom-contact OFET [45] or on the gate insulator surface for top-contact OFET [46]. In the case of SAM grown on the electrode surface, the injection barrier is changed by the interfacial dipole [45]. In other words, the electric field of the dipole can decrease the injection barrier due to the Schottky injection mechanism. In a similar way, the SAM grown on the gate insulator surface induces a field in the organic semiconductor film and metal/organic interface [46]. However, surface modification of the gate insulator has a great effect on the crystallinity of organic semiconductor [4]. An increase of the grain boundaries (i.e., decreasing grain size) is related to the rise of trapped carriers due to interface traps and grain boundary resistance [47]. Trapped (immobile) carriers are an additional source of the electric field, which compensates the applied electric field. As a result, the contact resistance increases together with the amount of traps. A similar effect was also reported for traps at the metal-organic interface of a top-contact OFET [48].

These studies drive us to include, in the field, all excess charges in the device. A high density of injected carriers accumulated on the organic semiconductor–gate insulator can induce a local field. In more detail, an electric field, E_i, induced on the injection electrode (source) is proportional to the potential drop, V_i, created by the charge, Q_i, situated on the gate insulator surface as follows:

$$E_i \propto \frac{Q_i}{C_{\mathrm{si}} + C_{\mathrm{di}} + C_{\mathrm{gi}}} \frac{1}{d} \approx \frac{Q_i}{C_{\mathrm{si}} d \left(1 + C_{\mathrm{gi}}/C_{\mathrm{si}}\right)}. \tag{37}$$

Here, the capacitances C_{si}, C_{di} and C_{gi}, are related to the induced electric field on the source, drain and gate electrodes, respectively, and d is the thickness of the organic semiconductor. Therefore, detailed evaluation of the electric field effect requires charge integration across the channel. However, for a simplified problem when charges below the source electrode are considered, only the induced electric field decreases with an increasing film thickness. Figure 11 depicts an example of the voltage dependence of the contact resistance for three different film thicknesses. It is obvious that an increase of film thickness increases the contact resistance. Moreover, in case of injection slower than charge transport from the source to the drain electrode, charge accumulation is not fully established and the potential drop increases, i.e., the contact resistance rises. Note that in a

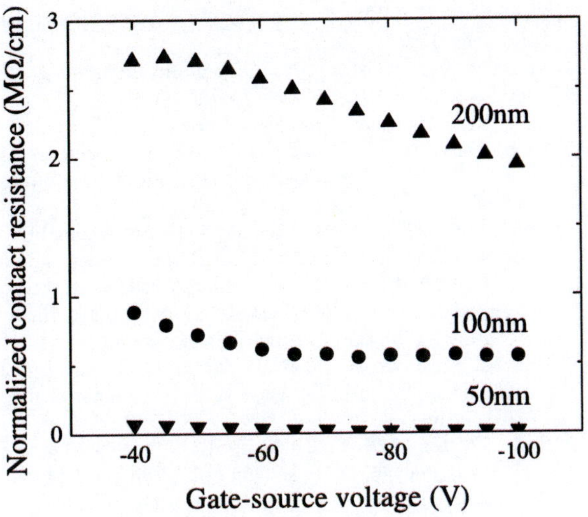

Fig. 11 Voltage dependence of the contact resistance for three selected organic semiconductor film thicknesses [50]

bottom-contact OFET, the charge cannot be accumulated below the electrode and contact resistance is minimized. In summary, a weak electric field's influence on the contact resistance for this device geometry was reported [49].

5 Concluding Remarks

In this brief summary we explained OFET models based on dielectric physics. Two different models were employed to discuss the charge propagation along the organic semiconductor-gate insulator interface. The MW model provides deep physical insight to the charge accumulation phenomenon and the discussion based on TLA is more suitable for charge propagation in the channel region with a variable electric field. In other words, the device can be explained by the MW charge accumulation and propagation, where charge transport is realized by the accumulation of charge on the organic semiconductor–gate insulator interface. The steady-state and transient state were explained separately to point out the importance of charge accumulation and transit time. This was later employed for device design and the improvement of device performance; possible methods were discussed and compared with each other. Hence, it results to the road map for device performance improvement.

References

1. Katz HE (1997) Organic molecular solids as thin film transistor semiconductors. J Mater Chem 7:369–376

2. Sirringhaus H (2005) Device physics of solution-processed organic field-effect transistors. Adv Mater 17:2411–2425
3. Tang CW, VanSlyke SA (1987) Organic electroluminescent diodes. Appl Phys Lett 51:913–915
4. Kim C, Facchetti A, Marks TJ (2007) Polymer gate dielectric surface viscoelasticity modulates pentacene transistor performance. Science 318:76–80
5. Knipp D, Street RA, Völkel A, et al (2003) Pentacene thin film transistors on inorganic dielectrics: Morphology, structural properties, and electronic transport. J Appl Phys 93:347/1–347/9
6. Tecklenburg R, Paasch G, Scheinert S (1998) Organic FET Current characteristics: extraction of unusual field dependences of hopping mobilities. Adv Mater Opt Electron 8:285–294
7. Zaumseil J, Baldwin KW, Rogers JA (2003) Contact resistance in organic transistors that use source and drain electrodes formed by soft contact lamination. J Appl Phys 93:6117/1–6117/8
8. Horowitz G (1998) Organic field-effect transistors. Adv Mater 10:365–377
9. Uno M, Tominari Y, Takeya J (2008) Three-dimensional organic field-effect transistors: charge accumulation in the vertical semiconductor channels. Appl Phys Lett 93:173301/1–173301/3
10. Kudo K, Wang DX, Iizuka M et al (1998) Schottky gate static induction transistor using copper phthalocyanine films. Thin Solid Films 331:51–54
11. Watanabe Y, Kudo K (2005) Flexible organic static induction transistors using pentacene thin films. Appl Phys Lett 87:223505/1–223505/3
12. Xue Y-B, Wang D-X (2006) Experimental analysis of operating characteristics of organic semiconductor static induction transistor. J Shanghai Univ 10:352–356
13. Fukagawa H, Watanabe Y, Kudo K, et al (2009) Ext Absr (70th Autumn Meet), Japan Society of Applied Physics, 8p-K-8 (in Japanese)
14. Lampert MA, Mark P (1970) Current injection in solids. Academic Press, New York
15. Kao KC (2004) Dielectric phenomenon in solids. Elsevier, Amsterdam
16. Weis M, Manaka T, Iwamoto M (2009) Origin of electric field distribution in organic field-effect transistor: experiment and analysis. J Appl Phys 105:24505/1–24505/7
17. Weis M, Manaka T, Iwamoto M, Effect of traps on carrier injection and transport in organic field-effect transistor, IEEJ Trans Elect Electronic Eng (in press)
18. Tamura R, Lim E, Manaka T, et al (2006) Analysis of pentacene field effect transistor as a Maxwell–Wagner effect element. J Appl Phys 100:114515/1–114515/7
19. Dimitrakopoulos CD, Purushothaman S, Kymissis J et al (1999) Low-voltage organic transistors on plastic comprising high-dielectric constant gate insulators. Science 283:822–824
20. Shockley W (1952) A unipolar "field-effect" transistor. Proceedings of the IRE 40:1365–1376
21. Silveira WR, Marohn JA (2004) Microscopic view of charge injection in an organic semiconductor. Phys Rev Lett 93:116104/1–116104/4
22. Bulucea C, Rusu A (1987) A first-order theory of the static induction transistor. Solid-State Electron 30:1227–1242
23. Wang DX, Tanaka Y, Iizuka M et al (1999) Device characteristics of organic static induction transistor using copper phthalocyanine films and al gate electrode. Jpn J Appl Phys 38:256–259
24. Dunn L, Basu D, Wang L, Dodabalapur A (2006), Organic field effect transistor mobility from transient response analysis. Appl Phys Lett 88:063507/1–063507/3
25. Basu D, Wang L, Dunn L, et al (2006) Direct measurement of carrier drift velocity and mobility in a polymer field-effect transistor. Appl Phys Lett 89:242104/1–242104/3
26. Manaka T, Lim E, Tamura R et al (2007) Direct imaging of carrier motion in organic transistors by optical second-harmonic generation. Nat Photonics 1:581–584
27. Kepler RG (1960) Charge carrier production and mobility in anthracene crystals. Phys Rev 119:1226–1229
28. Spear WE, Mort J (1963) Electron and hole transport in CdS crystals. Proc Phys Soc 81:130–140
29. Scher H, Montroll W (1975) Anomalous transit-time dispersion in amorphous solids. Phys Rev B 12:2455–2477

30. Horowitz G (2006) Organic transistors. In: Klauk H (ed) Organic electronics, Wiley-WCH, Weinheim
31. Manaka T, Liu F, Weis M, et al (2009) Mobility measurement based on visualized electric field migration in organic field-effect transistors. Appl Phys Express 2:061501/1–061501/3
32. Manaka T, Liu F, Weis M, et al (2010) Influence of traps on transient electric field and mobility evaluation in organic field-effect transistors. J Appl Phys 107:043712/1–043712/7
33. Lin J, Weis M, Taguchi D et al (2009) Carrier injection and transport in organic field-effect transistor investigated by impedance spectroscopy. Thin Solid Films 518:448–551
34. Lim E, Manaka T, Iwamoto M (2008) Analysis of pentacene field-effect transistor with contact resistance as an element of a Maxwell–Wagner effect system. J Appl Phys 104:054511/1–054511/5
35. Aswal DK, Lenfant S, Guerin D et al (2006) Self assembled monolayers on silicon for molecular electronics. Anal Chim Acta 568:84–108
36. Possanner SK, Zojer K, Pacher P et al (2009) Threshold voltage shifts in organic thin-film transistors due to self-assembled monolayers at the dielectric surface. Adv Funct Mater 19:958–967
37. Yoshita S, Tamura R, Taguchi D, et al (2009) Displacement current analysis of carrier behavior in pentacene field effect transistor with poly(vinylidene fluoride and tetrafluoroethylene) gate insulator. J Appl Phys 106:024505/1–024505/4
38. Wang SD, Minari T, Miyadera T, et al (2007) Contact-metal dependent current injection in pentacene thin-film transistors. Appl Phys Lett 91:203508/1–203508/3
39. Miyadera T, Nakayama M, Ikeda S et al (2007) Investigation of complex channel capacitance in C_{60} field effect transistor and evaluation of the effect of grain boundaries. Curr Appl Phys 7:87–91
40. Minari Y, Nemoto T, Isoda S (2004) Fabrication and characterization of single-grain organic field-effect transistor of pentacene. J Appl Phys 96:769–772
41. Diao L, Frisbie CD, Schroepfer DD, et al (2007) Electrical characterization of metal/pentacene contacts. J Appl Phys 101:14510/1–14510/8
42. Vig JR (1985) UV/ozone cleaning of surfaces. J Vac Sci Technol. A 3:1027–1034
43. Suzue Y, Manaka T, Iwamoto M (2005) Current–voltage characteristics of pentacene films: effect of UV/ozone treatment on Au electrodes. Jpn J Appl Phys 44:561–565
44. Wan A, Hwang J, Amy F et al (2005) Impact of electrode contamination on the α-NPD/Au hole injection barrier. Org Electron 6:47–54
45. Campbell IH, Rubin S, Zawodzinski TA et al (1996) Controlling Schottky energy barriers in organic electronic devices using self-assembled monolayers. Phys Rev B 54:R14321–R14324
46. Jang Y Cho JH, Kim DH et al (2007) Effects of the permanent dipoles of self-assembled monolayer-treated insulator surfaces on the field-effect mobility of a pentacene thin-film transistor. Appl Phys Lett 90:132104/1–132104/3
47. Kelley TW, Frisbie CD (2001) Gate voltage dependent resistance of a single organic semiconductor grain boundary. J Phys Chem B 105:4538–4540
48. Wang SD, Minari T, Miyadera T, et al (2008) Bias stress instability in pentacene thin film transistors: contact resistance change and channel threshold voltage shift. Appl Phys Lett 92: 63305/1–63305/3
49. Necliudov PV, Shur MS, Gundlach DJ et al (2003) Contact resistance extraction in pentacene thin film transistors. Solid-State Electron 47:259–262
50. Weis M, Nakao M, Lin J et al (2009) Thermionic emission model for contact resistance in organic field-effect transistor. Thin Solid Films 518:795–798

Aerogels for Energy Saving and Storage

Te-Yu Wei and Shih-Yuan Lu

Abstract Aerogels have drawn a great deal of research attention in recent years because their unique and advantageous structural characteristics of high porosity, high specific surface area, and mesopores, find a wide range of potential applications. Among them, their timely and imperative applications in energy saving and energy storage are particularly important to respond to the ever worsening issues of fossil energy depletion and global warming. In this chapter, aerogels serving as thermal insulation materials for energy saving and as electrode materials for supercapacitors and lithium ion batteries for energy storage are reviewed and discussed.

1 Introduction

1.1 Introduction of Aerogels

Aerogels are three-dimensional (3D) network structures, constructed via connection of nanoparticles. The first aerogel was prepared by Kistler in 1931 [1]. Most of the space in aerogels is occupied by air, and only a very limited amount of solid phase is present as illustrated in Fig. 1. The pore sizes of aerogels usually fall in the range of 2–50 nm, a pore size range defined for mesopores.

Aerogels possess several remarkable structural characteristics, including high porosity (>80%), low density (0.003–0.35 g cm^{-3}), high specific surface area (100–3,000 m^2 g^{-1}), and a large amount of percolating mesopores. These unique

T.-Y. Wei · S.-Y. Lu (✉)
Department of Chemical Engineering, National Tsing-Hua University,
30013 Hsinchu, Taiwan (R.O.C)
e-mail: sylu@mx.nthu.edu.tw

L. Zang (ed.), *Energy Efficiency and Renewable Energy Through Nanotechnology*,
Green Energy and Technology, DOI: 10.1007/978-0-85729-638-2_25,

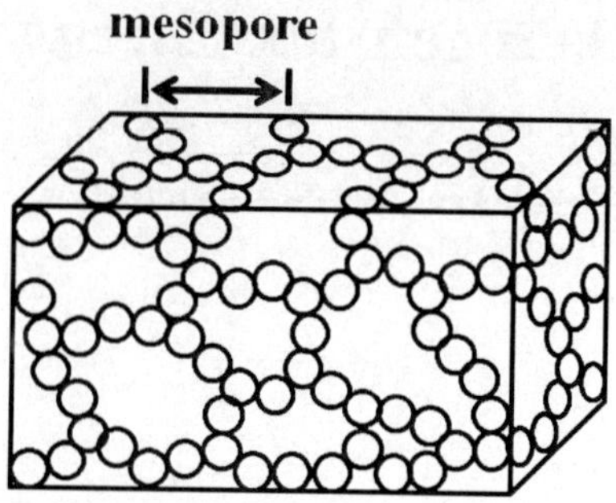

Fig. 1 Illustrative structure of an aerogel

structural features foster a wide range of applications for aerogels. For example, the high porosity makes aerogels a promising candidate for thermal insulators in energy saving applications. With the high specific surface area and mesoporous structure, aerogels can work as excellent electrode materials in supercapacitors and lithium ion batteries. Traditionally, sol–gel processes are used to synthesize these mesoporous structures in which nanoparticles are cross-linked to form 3D networks. The wet gels obtained from the sol–gel process normally require supercritical fluid drying to remove the trapped solvent without causing excessive structural shrinkage and even collapse of the brittle 3D network structure.

1.2 Synthesis of Aerogels

According to the type of precursor used, the sol–gel process for the synthesis of aerogels can be divided into two main categories, alkoxide and non-alkoxide-based sol–gel processes. The so-called sol is a stable colloidal suspension in which nanoparticles are well-dispersed in a solvent. As for the gel, it is a solid state with the solvent trapped inside and the solid phase is composed of cross-linked nanoparticles. The solvent in a wet gel has to be removed to afford the aerogel. In alkoxide-based syntheses, metal alkoxides are used as the precursor for the preparation of wet gels through hydrolysis and condensation. As in non-alkoxide-based routes, metal salts are used as the precursor. To initiate the necessary hydrolysis for metal salts, catalysts other than acids and bases are needed, and epoxides are a popular choice for the purpose. For the removal of the trapped solvent, supercritical fluid drying is normally applied. Supercritical fluid drying however requires high pressures and high temperatures, which are energy intensive and dangerous. In recent years, surface modification assisted ambient pressure drying has gained increasing popularity.

1.2.1 Alkoxide-Based Processes

In alkoxide-based processes, metal alkoxides are used as the precursor. For example, tetraethoxysilane (TEOS) and tetramethoxysilane (TMOS) are the two

Hydrolysis:

$$Si(OC_2H_5)_4 + 4H_2O \xrightarrow{HCl} Si(OH)_4 + 4C_2H_5OH$$

Condensation:

$$Si(OH)_4 + (OH)_4Si \xrightarrow{NH_4OH} (OH)_3Si\text{-}O\text{-}Si(OH)_3 + H_2O$$

$$n(OH)_3Si\text{-}O\text{-}Si(OH)_3 \longrightarrow \text{3-D network structure}$$

Scheme 1 A two-step catalysis process with TEOS as precursor for preparation of silica wet gels

most acclaimed precursors for silica aerogel preparation. The alkoxide proceeds with the hydrolysis and condensation reactions to eventually form the wet gel, which is a 3D network structure of nanoparticles containing trapped solvents. A two-step catalysis process for the silica wet gel preparation is shown below in Scheme 1 as an example.

Often, acids, such as HCl, HPO_3, and HF, are used for the hydrolysis reaction, whereas bases, such as NH_4OH and NaOH, are used for the condensation reaction. Consequently, pH value is an important parameter influencing the wet gel formation. In addition, the type of precursor also affects the final properties of aerogels.

1.2.2 Non-Alkoxide-Based Procedure

Metal alkoxides are generally expensive and sensitive to heat and moisture, and great care is needed when handling them. On the contrary, metal salts are not only cost-effective but also stable to heat and moisture. Non-alkoxide-based sol–gel routes for syntheses of aerogels have thus drawn much research attention in recent years. In the process, metal nitrides or metal chlorides are commonly used as the precursor, with water or alcohols as the solvent. An illustrative example of a non-alkoxide-based route is presented in Scheme 2. Here, propylene oxide is added to promote the hydrolysis and condensation of the metal salt.

Epoxide is commonly used as an acid scavenger in organic syntheses. An example is illustrated in Scheme 3 [3]. In a typical process, an epoxide removes protons from a strong acid and becomes protonated. This protonated epoxide would later proceed with the ring-opening reaction if suitable nucleophiles are present. Hydrated metal salts are strong acids and become hydrated metal hydroxide by losing protons to the epoxide, which can then further proceed with the condensation reaction for formation of the wet gels. These reactions are exothermic and the solution pH increases as the reaction proceeds since the protons continue to be removed by the epoxide.

Solvents also play an important role in the epoxide addition synthesis. Take the preparation of iron oxide aerogels as an example. In this synthesis, polar protic solvents, such as water, ethanol, and methanol, are necessary for the wet gel formation. The hydrogen bonding offered by the polar protic solvent can promote the cluster growth of the metal oxide. It was found difficult to obtain wet gels

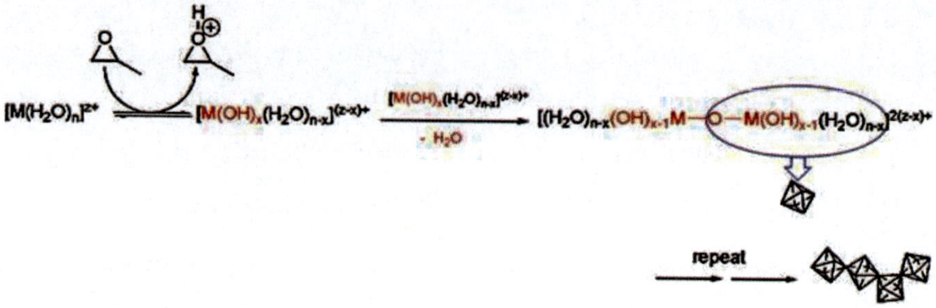

Scheme 2 Epoxide addition procedure for preparation of metal oxide wet gels (propylene oxide for example). Reprinted with permission from Ref. [2]. Copyright 2005 American Chemical Society

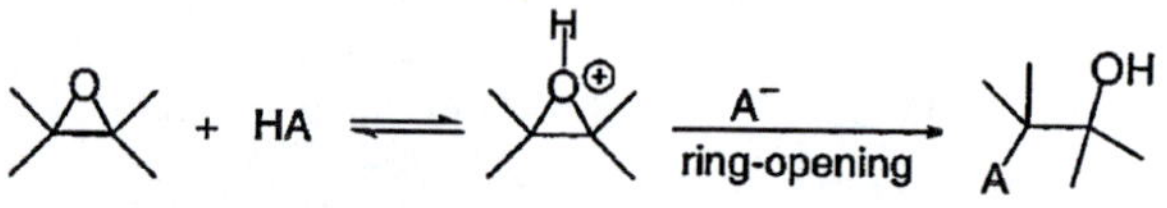

Scheme 3 Reaction of propylene oxide as an acid scavenger. Reprinted with permission from Ref. [3]. Copyright 2001 American Chemical Society

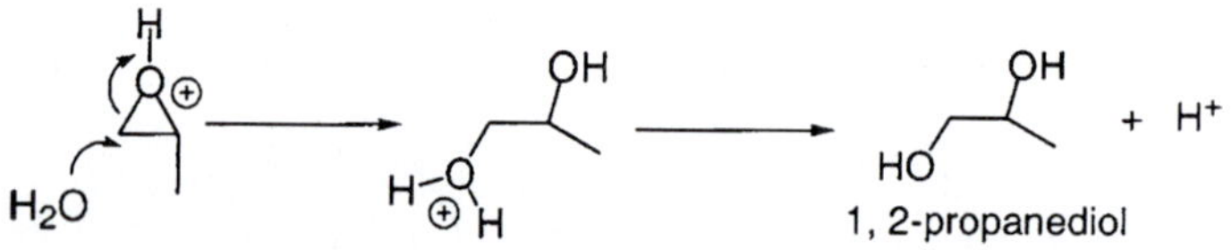

Scheme 4 Ring opening of propylene oxide in water achieved with the water attack. Reprinted with permission from Ref. [3]. Copyright 2001 American Chemical Society

if polar aprotic or nonpolar solvents were used in the process [3]. Often, precipitation occurs instead of formation of wet gels.

A slow and smooth increase in the pH value of the reaction solution was found to be a distinct feature of the epoxide addition process for successful wet gel formation. Wet gels are not likely to be obtained if the solution pH increases too fast or too slow. To control the pH increase, the interplay between the solvent and precursor ions should be considered. The protonated epoxide would proceed with the ring-opening reaction if attacked by nucleophiles, solvent, or precursor ions. For example, water is more nucleophilic than nitrate ions and thus dominates the ring-opening reaction if nitrate precursors are used with water as the solvent in an epoxide addition process [3]. As shown in Scheme 4, H^+ is formed when the propylene oxide is attacked by water for the ring-opening reaction. Consequently, the solution pH may be suppressed, leading to difficulty in wet gel formation.

On the other hand, Cl^- is more nucleophilic than water. The chloride ion would dominate the ring-opening reaction if chlorides are dissolved in water for the epoxide addition process. The attack of Cl^- on the epoxide however produces no

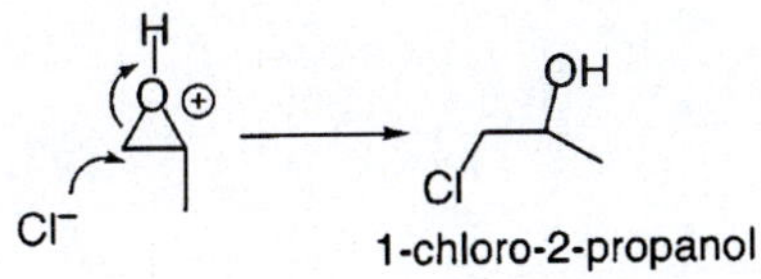

Scheme 5 Ring opening of propylene oxide in water achieved with the Cl^- attack. Reprinted with permission from Ref. [3]. Copyright 2001 American Chemical Society

H^+ as shown in Scheme 5. Therefore, the solution pH may increase at a suitable rate for wet gel formation.

Gash et al. suggested that it is easier to form wet gels by using metal chlorides than by using metal nitrates as the precursor [3]. Nevertheless, several examples of metal oxide wet gel formation did not follow this speculation. The pairing of the precursor ion and solvent influences the solution pH increase, and thus the formation of wet gels. Up to now, several transition metal oxide aerogels, including Al_2O_3 [4], SnO_2 [5], ZnO [6], CeO_2 [7], Fe_2O_3 [8], and Co_3O_4 [9], have been successfully prepared via the non-alkoxide-based procedure, using either metal chlorides or metal nitrates as the precursor.

1.2.3 Supercritical Drying

Wet gels will experience a large amount of volume shrinkage if dried at ambient pressures because of the large surface tension generated when solvent leaves the structure at the drying step. This results in the collapse of the 3D structure of the wet gel, leading to high apparent densities and low specific surface areas of the final products. Products thus obtained are called xerogels. To preserve the 3D structure, the wet gels can be dried with supercritical fluids such as the supercritical carbon dioxide, since there is no surface tension involved in the supercritical state. The products so obtained are called aerogels.

Supercritical carbon dioxide is commonly used to dry aerogels. The pressure and temperature of the drier at the drying condition should be maintained above 7.37 MPa and 31°C (the critical pressure and temperature), respectively to attain the supercritical state of the carbon dioxide as illustrated in Fig. 2.

1.2.4 Non-Supercritical Drying

Because of the high cost to achieve and difficulty in controlling the critical pressure and temperature of the drying medium for the supercritical fluid drying process, surface modification assisted ambient pressure drying is gaining popularity in recent years. The surfaces of the wet gel backbone normally are hydrophilic because of the presence of the exposing hydroxyl groups. These hydrophilic surfaces adsorb a significant amount of moisture. The moisture will later generate a large amount of surface tension during the drying process to damage the fragile 3D structure of the wet gel. To avoid the above mentioned situation, surface

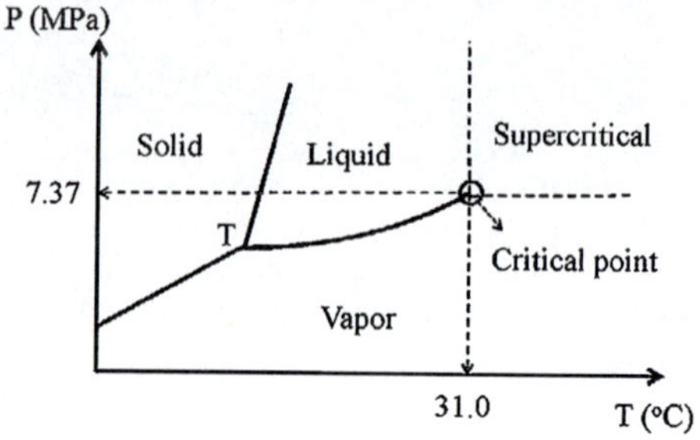

Fig. 2 Phase diagram of carbon dioxide

modification should be performed to replace the hydroxyl groups of the wet gel with less polar or even non-polar functional groups, such as –OSi$(CH_3)_3$. The wet gels then turn hydrophobic after the surface modification and the surface tension involved at the ambient pressure drying drastically reduces. It is worth mentioning that the supercritically dried aerogels, although retaining their 3D structure, remain hydrophilic and tend to absorb moisture in later use. The absorbed moisture would inevitably create surface tensions during the use of the aerogel. Consequently, the long-term usage of such products cannot be guaranteed. With the surface modification treatment, not only ambient pressure drying becomes possible, but also the hydrophobic characteristics of the product favor a long-term usage.

Many modification reagents, such as trimethylchlorosilane (TMCS) and 1H,1H,2H,2H-perfluorooctyldimethylchlorosilane [10–12], have been developed in recent years for surface modification treatment in aerogel production. Take TMCS as an example. The hydroxyl group distributed on the surface of the wet gel backbone is replaced by reacting with TMCS to form the –OSi$(CH_3)_3$ group, releasing HCl as the by-product. The –OSi$(CH_3)_3$ group is much less polar than –OH and the treated wet gel exhibits good hydrophobicity.

The extent of surface modification was found to play an important role in the quality of the resulting aerogel [13]. Recently, Wei et al. studied the effect of the extent of surface modification on the quality of silica aerogels dried at ambient pressure conditions. Silica aerogels obtained from processes involving no surface modification (NSM), single surface modification (SSM), and multiple surface modifications (MSM) were compared [13]. It was found that monolithic silica aerogels can only be obtained with the MSM process, as shown in Fig. 3. The SSM process achieved only a very limited improvement on the volume shrinkage as compared with the NSM process, as evident in Table 1. A drastic improvement in volume shrinkage, apparent density, and porosity was achieved with the MSM process, enabling formation of monolithic silica aerogels.

The successfulness of the surface modification can be judged with the FTIR spectra shown in Fig. 4. The alkyl groups coming from the surface modification reagent TMCS were detected by the peaks located at 2,980 and 845 cm^{-1}, and evidently the two peak intensities increased with increasing extent of surface modification. The peak located at 1,632 cm^{-1} was due to the presence of the H-OH bond, mainly responsible for the moisture adsorption experienced by

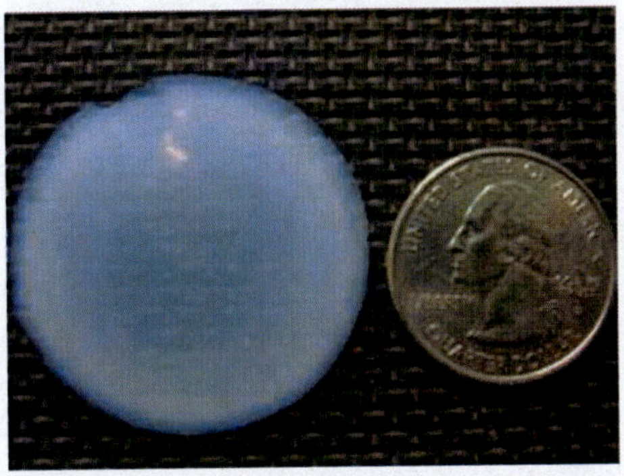

Fig. 3 Monolithic silica aerogel prepared with multiple surface modification. Reprinted with permission from Ref. [13]. Copyright 2007 John Wiley Sons

Table 1 Physical properties of as-prepared silica aerogels with different extent of surface modification

	Volume shrinkage (%)	Density (g cm^{-3})	Porosity (%)
NSM	87.4 ± 0.3	0.624 ± 0.022	72.0 ± 01.5
SSM	80.0 ± 0.3	0.502 ± 0.007	77.2 ± 0.3
MSM	<1.0 ± 0.9	0.069 ± 0.004	96.8 ± 0.2

Reprinted with permission from Ref. [13]. Copyright 2007 John Wiley & Sons

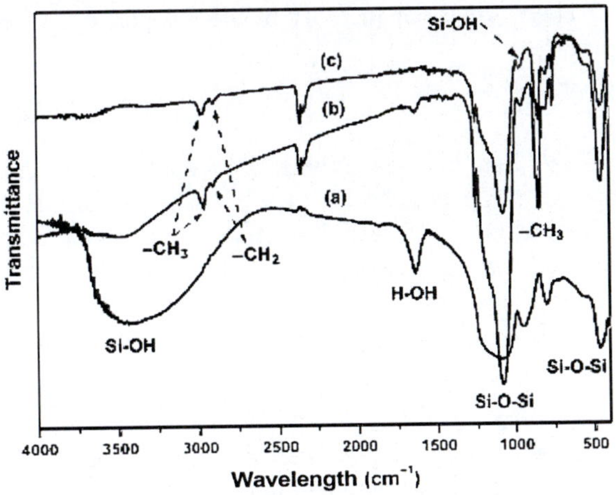

Fig. 4 FTIR of silica aerogels prepared with different extent of surface modification: **a** NSM, **b** SSM, and **c** MSM. Reprinted with permission from Ref. [13]. Copyright 2007 John Wiley & Sons

the products from the NSM and SSM processes. Note that this particular peak almost disappeared for the case of MSM, revealing the excellent hydrophobicity of the product from the MSM process [13].

2 Aerogels for Energy Saving

One of the extraordinary properties of aerogels that Kistler discovered was their very low-thermal conductivities. The renaissance of the aerogel technology around 1980 coincided with an increasing concern for energy efficiency and the environmental effects of chlorofluorocarbons (CFC's). It was then apparent that silica aerogels were an attractive alternative to traditional thermal insulation materials because of their low thermal conductivities and environment-friendly production processes.

2.1 Principle of Thermal Conductivity

The transfer of thermal energy through an insulating material occurs via three mechanisms: solid conduction, gaseous conduction, and radiative transmission. The sum of these three components gives the total thermal conductivity of the material [14].

$$k_t = k_s + k_g + k_r, \tag{1}$$

where k_t is the total thermal conductivity and k_s, k_g, and k_r are solid, gaseous, and equivalent radiation conductivities, respectively. The thermal conductivity of the solid part can be expressed as a function of its density (ρ) and thermal conductivity coefficient (α), as shown below [14].

$$k_s = \rho^{\alpha} \tag{2}$$

The thermal conductivity of the gas part can be expressed as a function of the porosity,

$$k_g = k_{g0}\Pi/(1 + 2\beta k_n) \tag{3}$$

where k_g and k_{g0} are the thermal conductivities of the gas in the material and in the ambient, respectively. Π is the porosity of the material, and β and K_n are the gas constant in the material (2 for air in silica aerogel) and Knudsen number, respectively. Equations 2 and 3 show that the thermal conductivities of the solid and gas parts are weak functions of temperature. However, the radiation term (k_r) is strongly related to temperature, as evident from Eq. 4 [14].

$$k_r = (16/3)n^2\sigma T_r^3/[(T_r)e\rho] \tag{4}$$

where n is the refractive index of the material, σ is the Stefan-Boltzmann constant, T_r is the radiation temperature, and $e\rho$ is the extinction coefficient. At room temperature, the contribution of k_r to the total thermal conductivity is almost negligible. But it increases rapidly with the increase in temperature and becomes the dominant contributor at high temperatures.

2.2 Thermal Insulators at Room Temperature

Among the many available aerogels, silica aerogels are particularly favored for thermal insulation applications because of the following three reasons. (1) The thermal conductivity of silica in bulk form (1.3 W/m-K) is lower than those of metal oxides traditionally used for thermal insulation, such as Al_2O_3 (25.08 W/m-K), ZrO_2 (1.8–2.2 W/m-K), and ZnO (116 W/m-K). (2) Air is well-known for its good thermal insulation ability. The high porosity of silica aerogels provides ample space for air and limited space for silica skeleton, favoring thermal insulation. (3) The extremely small pore sizes, typically 1–100 nm, lead to a partial suppression of the gaseous thermal conductivity [14].

Because of the above mentioned three reasons, the thermal conductivities of silica aerogels are lower than those of most commercial insulators. Thermal conductivities as low as 0.008 W/m-K have been achieved for silica aerogels, whereas the thermal conductivities of commercial insulators, such as air, polyurethane foam, polyurethane foam with CFC, and silica powders are 0.025, 0.04, 0.021, and 0.025 W/m-K, respectively [15]. Here, we discuss two different types of silica aerogels: (1) pure silica aerogels prepared with supercritical and ambient pressure drying and (2) composite and organic aerogels.

Compared with ambient pressure drying, supercritical fluid drying provides faster drying and better quality products of lower thermal conductivity, higher porosity, and higher specific surface area. However, as mentioned earlier, supercritical fluid drying is an expensive and a safety-concerning procedure, and the often hydrophilic products are unfavorable for long-term usage. As for ambient pressure drying, the resulting hydrophobic products favor long-term usage with the product thermal conductivity and porosity, although not as good but close to those obtainable from supercritical fluid drying. The ambient pressure drying procedure however needs longer processing time.

On the other hand, because of the fragile nature of pure silica aerogels, composite aerogels and organic aerogels have been developed to improve the mechanical strength of the products while preserving the low-thermal conductivity, high porosity, and high specific surface area.

2.2.1 Silica Aerogels as Thermal Insulators at Room Temperature

The thermal conductivities of pure silica aerogels prepared using different precursors and supercritical fluid drying conditions have been studied by several research groups. TEOS, polyethoxydisiloxane (PEDS), and TMOS have been used as the precursors for silica aerogel preparation [16]. The pore sizes of PEDS-derived silica aerogels were found to be larger than those of the TEOS-derived and TMOS-derived silica aerogels, as shown in Fig. 5. Large pores often lead to large pore volumes and thus large porosities. Large porosities in turn favor low-thermal conductivities. The thermal conductivity of PEDS-derived aerogels was found to be as low as 0.015 W/m-K at room temperature as reported by Wagh et al. [16].

As mentioned earlier, if wet gels without surface modification are dried at ambient pressures, the large surface tension developed at the solvent removal stage will damage the 3D structure of the wet gel and large volume shrinkages result. Large volume shrinkages give low porosities, and thus high thermal conductivity products. One of the direct results of surface modification is the reduction in

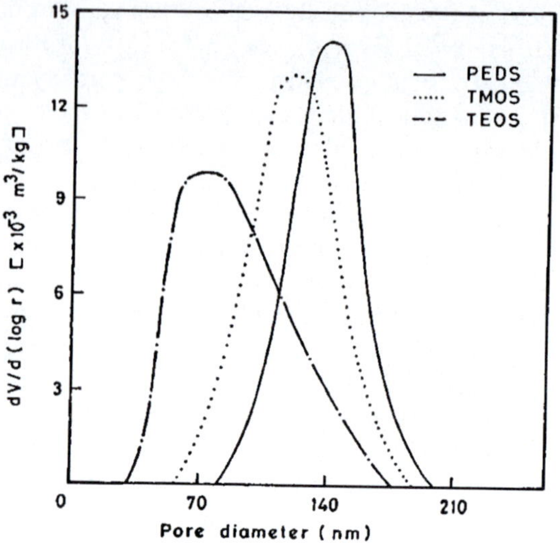

Fig. 5 Pore size distributions of pure silica aeorgels derived from PEDS, TMOS, and TEOS [16]. Reprinted with permission from Wagh et al. [16]. Copyright 1999, Elsevier

volume shrinkage, which improves the thermal insulation of the resulting aerogel. With the multiple surface modification treatment with TMCS and ambient pressure drying, Wei et al. produced silica aerogels with thermal conductivities close to those prepared with supercritical fluid drying [13]. The highest porosity and lowest thermal conductivity of the silica aerogels from their study were 97% and 0.036 W/m-K at room temperature, respectively. In addition, long-term usage of the product was made favorable because of the high hydrophobicity imparted by the surface modification.

There is one disadvantage associated with the surface modification by TMCS. That is the by-product HCl generated during the surface modification may damage the product structure, and thus needs to be removed with intensive washing, thus increasing the production cost. Besides TMCS, hexamethyldisilazane (HMDZ) can also be used as a surface modification reagent, and the by-product is NH_3 instead of HCl, which does not affect the product structure during the ambient pressure drying procedure. Gurav et al. prepared TEOS-derived silica aerogels with HMDZ serving as the surface modification reagent, and studied the effect of the HMDZ/TEOS ratio on the product thermal conductivity [17]. The –NH group of HMDZ reacts with the surface hydrophilic -OH group to form the hydrophobic $–OSi(CH_3)_3$ group as illustrated in Scheme 6 [17]. Here, NH_3 is produced as a by-product of the reaction.

Figure 6 shows the thermal conductivities (0.08–0.11 W/m-K) of the resulting silica aerogels as a function of the HMDZ/TEOS ratio. The lowest thermal conductivity (0.080 W/m-K) was obtained at the HMDZ/TEOS ratio of 0.68, with a corresponding porosity of 96%.

Direct syntheses of hydrophobic silica aerogels are possible provided that alkoxide precursors possessing non-polar groups are available. This was demonstrated by Nadargi and Roa by using methyltriethoxysilane (MTES) as the

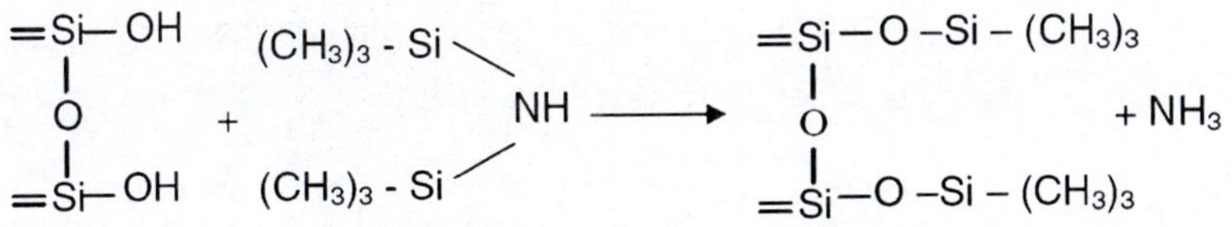

Scheme 6 Mechanism of HMDZ surface modification. Reprinted with permission from Ref. [17]. Copyright 2007 Springer

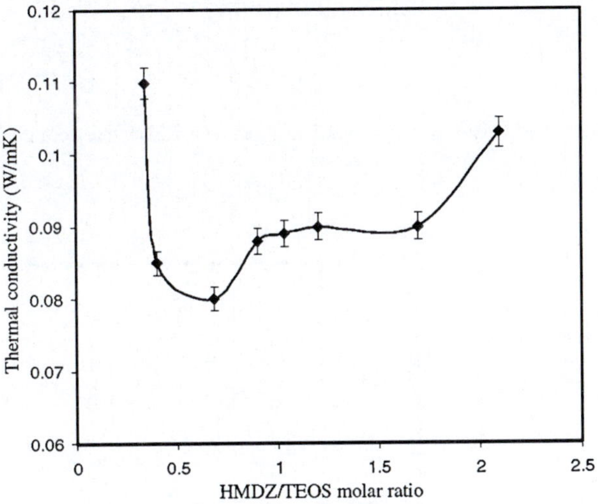

Fig. 6 Thermal conductivity versus HMDZ/TEOS ratio. Reprinted with permission from Ref. [17]. Copyright 2007 Springer

precursor [18]. The reaction mechanism of this sol–gel process is shown in Scheme 7. The exposed -CH_3 groups make the product hydrophobic.

The hydrophobicity of the product silica aerogels was characterized with the contact angle measurements [18]. The largest contact angle obtained by the authors was 163°, revealing the successfulness of the approach.

2.2.2 Organic Aerogels as Thermal Insulators at Room Temperature

Aerogels, composed of 3D network of nanoparticles, are in fact quite fragile and are difficult to process for application purposes. One way to deal with this drawback is to introduce some soft components, such as polymers, into the skeleton of the aerogel or even to prepare organic aerogels. Albert et al. successfully synthesized phenolic-furfural (PF) aerogels, a pure organic aerogel, and measured their thermal conductivities at room temperature at different system pressures [19]. Figure 7 shows the thermal conductivities of the PF aerogels at a series of system

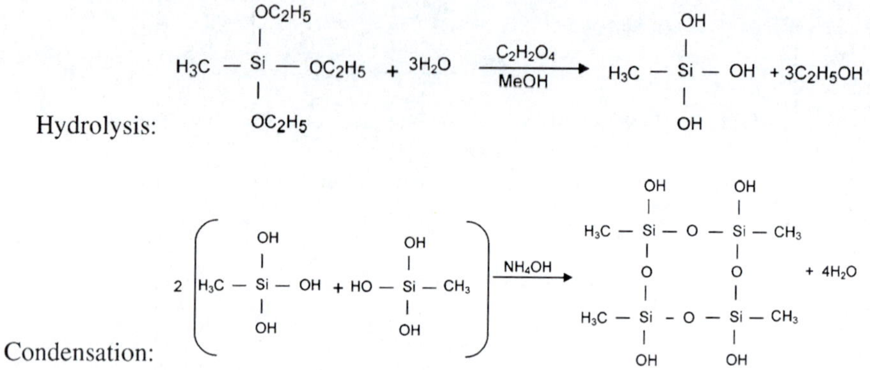

Scheme 7 Sol–gel mechanism by using MTES as precursor. Reprinted from Nadargi et al. [18]. Copyright 2009, with permission from Elsevier

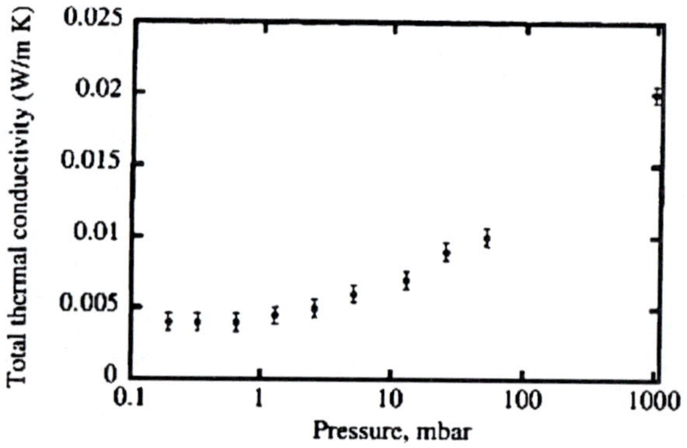

Fig. 7 Total thermal conductivity of PF aerogels at various pressures at room temperature. Reprinted with permission from [19]. Copyright 2001, Elsevier

pressures at room temperature [19]. The thermal conductivity of the PF aerogels at one atmosphere was 0.02 W/m-K, and was decreased to 0.005 W/m-K when the system pressure was reduced to below 1 mbar [19]. As the system pressure is reduced, the contribution of gaseous conductions diminishes and the total thermal conductivity obtained basically represents the thermal conductivity of the solid phase of the aerogel and the PF backbone, since the radiative contribution is negligible at room temperature.

Polyurethane and polyurea aerogels were also developed in recent years. Figure 8 shows the appearance of monolithic polyurea and polyurethane aerogels. They were opaque and yellow in color. The thermal conductivity of the polyurethane aerogel (0.022 W/m-K) was lower than that of the polyurethane foams (0.030 W/m-K) at ambient conditions mainly because of its higher porosity [20, 21]. The thermal conductivity of polyurea aerogels was 0.018 W/m-K at one atmosphere and room temperature, and reduced to 0.005 W/m-K at the system pressure of 0.075 torr [21].

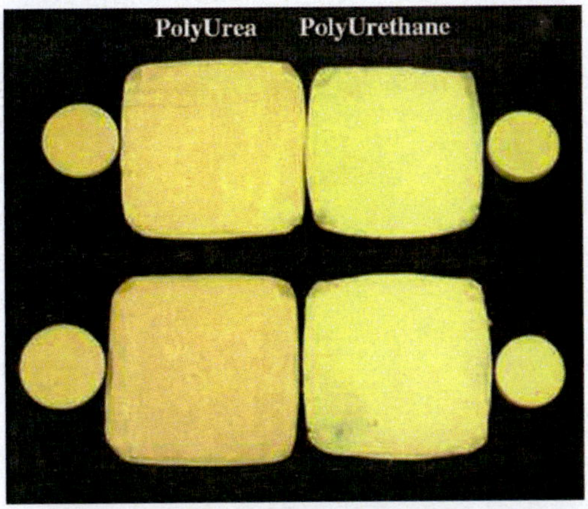

Fig. 8 Photographs of polyurea and polyurethane aerogels. Reprinted with permission from Ref. [21]. Copyright 2009, Springer

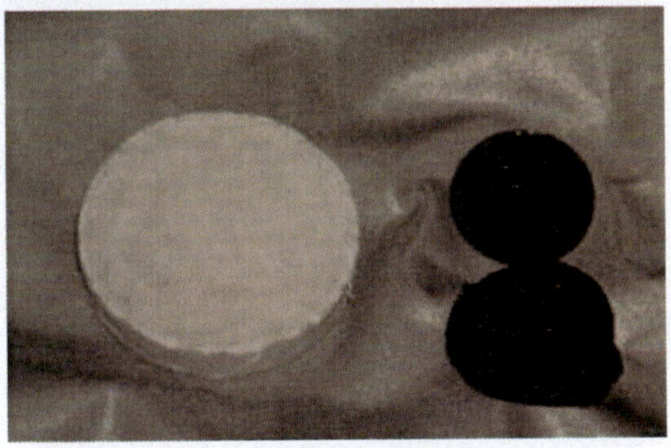

Fig. 9 Photographs of polydicyclopentadiene aerogels and xerogels. Reprinted with permission from Ref. [22]. Copyright 2007 Springer

Lee et al. developed polydicyclopentadiene aerogels via ring-opening metathesis polymerizations catalyzed by homogeneous ruthenium complexes [22]. Polydicyclopentadiene aerogels with thermal conductivities as low as 0.014 W/m-K were obtained. Both polydicyclopentadiene aerogels and xerogels were prepared by using the supercritical carbon dioxide and ambient pressure drying with the resulting volume shrinkage factors of 1.35 and 5.56, respectively. Both aerogels, as shown in Fig. 9, were opaque and monolithic. The thermal conductivities of xerogels were larger than those of aerogels, as expected since aerogels possessed higher porosities [22].

2.2.3 Composite Aerogels as Thermal Insulators at Room Temperature

Soft matters, mainly polymers, have been introduced into inorganic aerogels to improve the mechanical strength of the parent aerogels, and thus their processability for application purposes. Katti et al. synthesized isocyanate cross-linked, amine-modified silica aerogels with the skeletons coated with polyurea layers [23]. The detailed preparation flowchart is shown in Fig. 10. The diisocyanate reacted with the -NH_2 group of the amine-modified silica to form the polyurea coating.

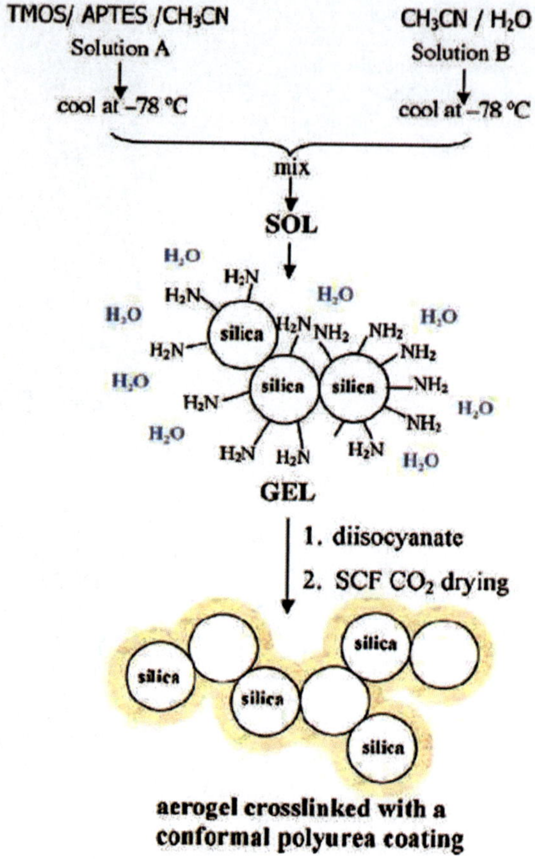

Fig. 10 Preparation flowchart of isocyanate cross-linked, amine-modified silica aerogels. Reprinted with permission from Ref. [23]. Copyright 2006 American Chemical Society

The presence of this polyurea coating improved the Young's modulus of the silica aerogel from below the detection limit to 129 MPa, while maintaining a low-thermal conductivity of 0.041 W/m-K [23, 24]. This nanocasting technology was also applied to polystyrene-enhanced silica aerogels and achieved a similar level of success, with the thermal conductivity of the composite aerogel being as low as 0.041 W/m-K.

In addition to the described chemical route to obtain composite aerogels, facile one-step polymer-incorporation sol–gel processes have also been developed for the purpose. Wei et al. incorporated poly(vinylpyrrolidone) (PVP) into pure silica aerogels via a simple polymer addition procedure during the sol–gel process to synthesize silica-PVP composite aerogels [25]. This is a much easier and greener process for composite aerogel preparation than those involving amine-modification and toxic reagents, such as acetonitrile. Figure 11 shows the photographs of the transparent, blue-toned monolithic silica-PVP composite aerogels. Multiple hydrophobic surface modification treatments along with ambient pressure drying were used to produce these composite aerogels.

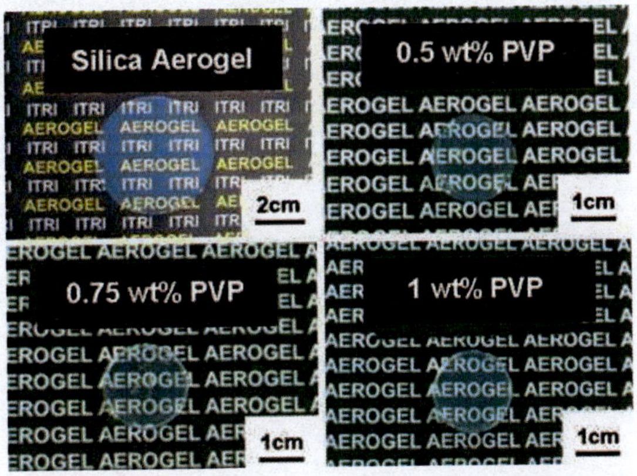

Fig. 11 Silica-PVP composite aerogels. Reprinted with permission from Ref. [25]. Copyright 2008 American Chemical Society

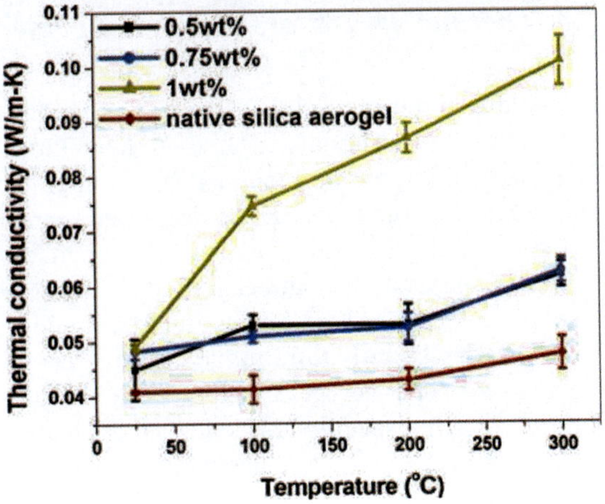

Fig. 12 Thermal conductivities of silica-PVP composite and pristine silica aerogels at increasing temperatures. Reprinted with permission from Ref. [25]. Copyright 2008 American Chemical Society

The thermal conductivities of these silica-PVP composite aerogels were maintained low at around 0.045 W/m-K, while the Young's modulus was improved to a value as high as 39 MPa. The authors also investigated the high temperature thermal conductivities of the composite aerogels, since the PVP remained stable at temperatures up to 300°C [25]. Figure 12 shows the thermal conductivities of the silica-PVP composite aerogels and pristine silica aerogel as functions of system temperature [25].

Because of the low extinction coefficients of PVP and silica, the thermal conductivities of the composite aerogels increased appreciably with increasing temperature because of the more pronounced radiative heat transfer at high temperatures. The thermal conductivity of the composite aerogel of 0.5 wt% PVP, however, achieved a low value of 0.063 W/m-K at 300°C, which was lower than that of glass fibers (0.08 W/m-K, at 300°C), a traditional high temperature thermal insulation material.

Table 2 Specific extinction coefficients of nine materials at 300 K

Material	Specific extinction coefficient at 300 K
SiO_2	22.7
SiO_2 (opacified)	84.2
TiO_2	32.6
ZrO_2	38.9
Carbon black	>1,000
Resorcinol–formaldehyde	50.1
Melamine–formaldehyde	47.2
Polyurethane	47.6
Polystyrene	47.8

2.3 Thermal Insulators at High Temperatures

It was already mentioned that thermal radiations play a prime role in the thermal conductivity of materials at high temperatures. It is important to reduce the thermal radiation for thermal insulators at high temperatures. Equation 4 indicates that thermal conductivity can be decreased by increasing the extinction coefficient of the material. Plain silica aerogels are a suitable choice for thermal insulators at moderate temperatures, but because of their low extinction coefficient, they are not a good candidate for high temperature thermal insulation applications. One way to deal with this issue is to introduce opaque components of high extinction coefficients to the silica aerogel to suppress the thermal radiation at high temperatures. Table 2 lists the specific extinction coefficients of nine materials at 300 K [26]. Among them, carbon black possesses the highest specific extinction coefficient at 300 K.

Carbon black is therefore commonly used to opacify aerogels. Unfortunately, it remains thermally stable at temperatures up to only around 300°C at ambient conditions [27], and thus cannot be used for applications requiring system temperatures above 300°C. Titania has a good thermal stability and higher extinction coefficient than silica, and has been investigated as an enhancement component for silica aerogels. Wang et al. introduced TiO_2 into pure silica aerogels and studied the extinction coefficients of the composites [27]. The measured effective specific extinction coefficient spectra of the composite aerogels are compared in Fig. 13.

Evidently, samples with a higher enhancement component loading give higher specific extinction coefficients, particularly in the wavelength range of 3.75–5.06 μm, corresponding to the thermal radiations generated at 300–500°C. Besides TiO_2, several other metal oxides have also been studied as the enhancement components for silica aerogels, such as Fe_3O_4, limenite, B_4C, and SiC [28]. Figure 14 shows the specific extinction coefficient spectra of these composite aerogels. In the wavelength range of 3.75–5.06 μm, Fe_3O_4 showed the highest specific extinction coefficient, indicating the high efficiency of radiation reduction.

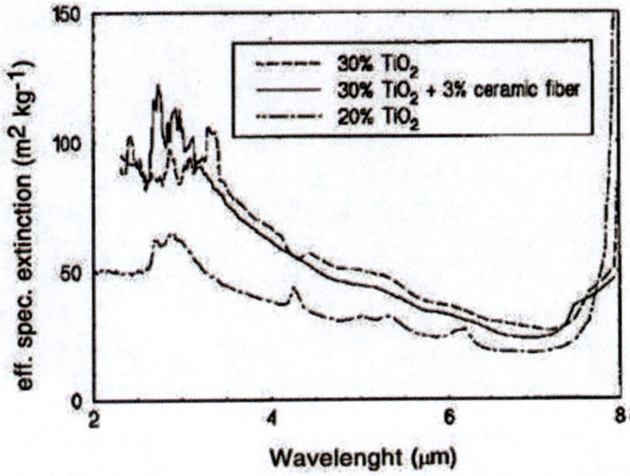

Fig. 13 Specific extinction coefficient spectra of TiO_2-silica composite aerogels with different concentrations of enhancement inclusion. Reprinted with permission from Wang et al. [27]. Copyright 1995, Elsevier

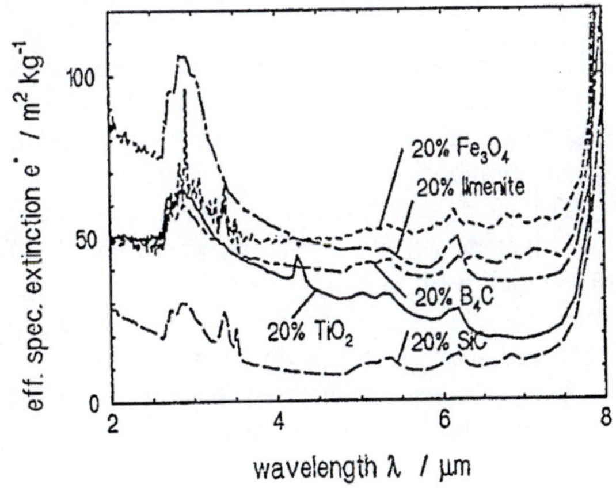

Fig. 14 Specific extinction coefficient spectra of silica composite aerogels with different enhancement components. Reprinted from Kuhn et al. [28]. Copyright 1995, with permission from Elsevier

Carbon nanofibers possess a thermal stability of above 500°C, a low-thermal conductivity of 0.07 W/m-K, and a high extinction coefficient of carbon materials. They are potentially a good opacifier for silica aerogels. Wei et al. studied the thermal conductivities and specific extinction coefficients of carbon nanofiber opacified silica aerogels [29]. Figure 15 shows the photograph and SEM image of the composite aerogels.

The length and diameter of the carbon nanofibers were 10 μm and 30–100 nm, respectively. These carbon nanofibers remained stable up to 530°C. The thermal conductivities of the carbon nanofiber opacified silica aerogels were measured at 500°C and are shown in Fig. 16.

It is evident that the thermal conductivities of pristine silica aerogels increased sharply with increasing temperature, whereas those of the carbon nanofiber opacified silica aerogels remained low even at the high temperature of 500°C. This clearly showed the effect of the incorporated carbon nanofibers in suppressing the thermal radiations at high temperatures. A small amount of carbon nanofibers,

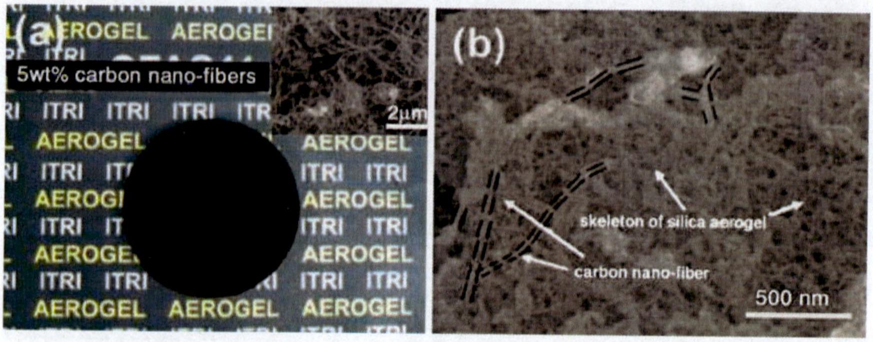

Fig. 15 **a** Photograph of carbon nanofiber opacified silica aerogels with inset showing an SEM image of carbon nanofibers. **b** SEM image of the composite aerogel with 20 wt% carbon nanofibers. Reprinted with permission from Ref. [29]. Copyright 2009 American Chemical Society

about 1 wt%, was enough to produce pronounced effects, as suggested by Fig. 16a. The low-thermal conductivity of the composite aerogels at 500°C, as shown in Fig. 16b, persisted for 600 h. This demonstrated the excellent long-term stability of the composite aerogels.

The specific extinction coefficients of the composite aerogels were further investigated, with the results shown in Fig. 17. The specific extinction coefficient of the high loading samples was found to be much greater than that of the pristine silica aerogel in the wavelength range of 3.75–5.06 μm. Because of the much improved specific extinction coefficient, the thermal radiations dominant at high temperatures were suppressed, leading to the low high-temperature thermal conductivity. The low-thermal conductivity value of 0.05 W/m-K was significantly lower than those of most of the commercial thermal insulators, such as glass fibers (0.3 W/m-K at 500°C), alumina fused brick (0.1 W/m-K at 527°C), and sillimante (1.7 W/m-K at 527°C) [27].

Carbon aerogels were also used for thermal insulations at high temperatures because of their high specific extinction coefficients. Thermal conductivities of carbon aerogels prepared at different pyrolysis temperatures were measured in vacuum by Wiener et al. [30, 31]. The pyrolysis temperature was found to be a dominant parameter in determining the thermal conductivity of the resulting carbon aerogel. As shown in Fig. 18, the thermal conductivities at 300°C under vacuum of the carbon aerogels increased with increasing pyrolysis temperature, from 0.05 W/m-K at 800°C to 0.42 W/m-K at 2,500°C.

3 Aerogels for Energy Storage

Supercapacitors and lithium ion batteries are two important energy storage devices for electronics and power systems. Because of the high specific surface area, aerogels can accommodate large amounts of electric double layers for capacitances

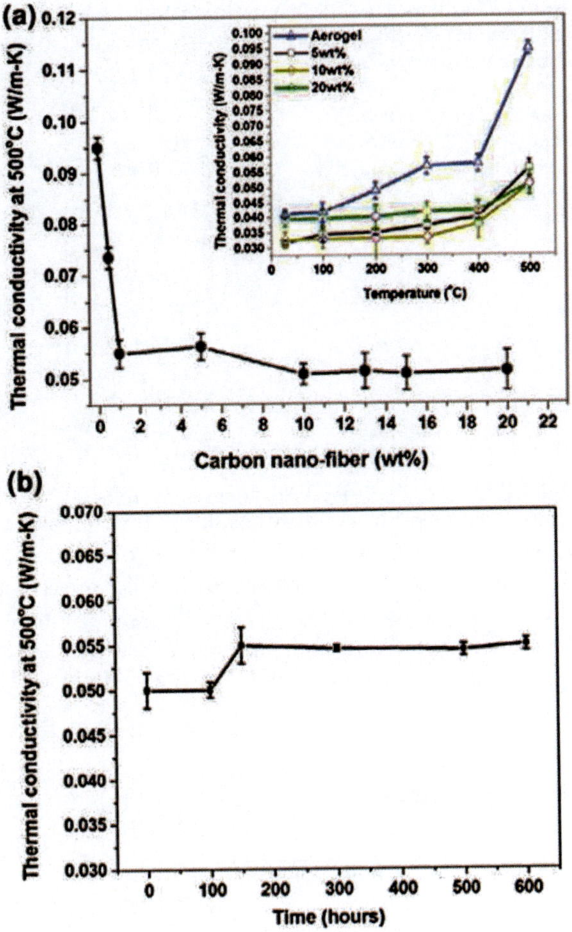

Fig. 16 **a** Thermal conductivity at 500°C as a function of carbon nanofiber loading, with the inset showing thermal conductivities of composite aerogels with various carbon nanofiber loadings as functions of temperature. **b** The stability test of thermal conductivity at 500°C. Reprinted with permission from Ref. [29]. Copyright 2009 American Chemical Society

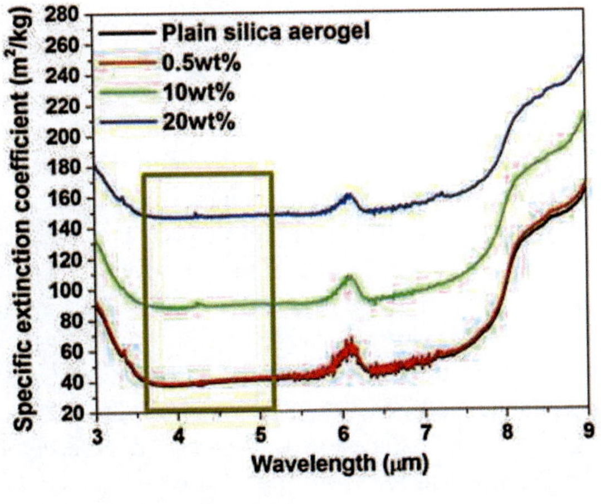

Fig. 17 Specific extinction coefficient spectra of pristine and composite silica aerogels of various carbon nanofiber loading. Reprinted with permission from Ref. [29]. Copyright 2009 American Chemical Society

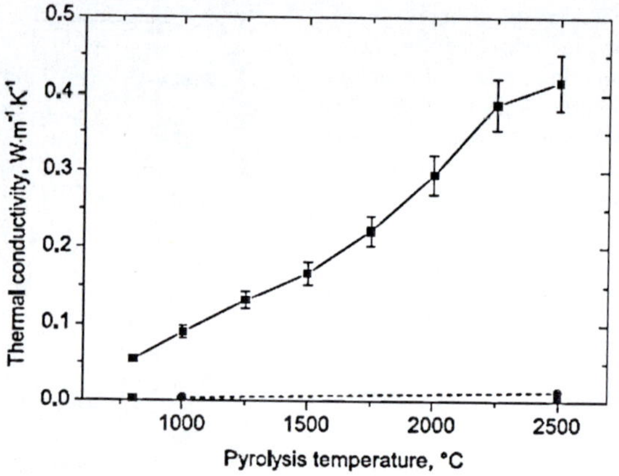

Fig. 18 Thermal conductivities of carbon aerogels measured at 300°C under vaccum vs. pyrolysis temperature (*solid line*). Reprinted with permission from Ref. [30, 31]. Copyright 2006, Springer

and superficial electroactive species to participate in pseudocapacitive faradaic redox reactions. For applications in lithium ion batteries, the high specific surface area of aerogels provides a large amount of reaction sites for lithium ions to store electrons. In addition to high specific surface areas, aerogels are highly porous and offer low mass transfer resistance pathways for ion diffusions. Miller et al. studied the storage capacities of aerogels and demonstrated fast charge/discharge cycles because of the low mass transfer resistance to ion diffusions [32, 33].

3.1 Supercapacitor

Supercapacitors are an important energy storage device. They offer transient but extremely high powers for the time-dependent power needs of modern electronics and power systems. According to the storage mechanism, they can be categorized into two types, namely, electric double-layer capacitors and pseudocapacitors.

Electric Double-Layer Capacitors. A schematic of an electric double-layer capacitor (EDLC) is shown in Fig. 19.

EDLCs are composed of a pair of polarizing electrodes and collectors separated by an electrolyte. Carbon materials are generally used as the polarizing electrodes. The charge storage in EDLCs is due to the adsorption of ions through electrostatic forces. The double layer capacitances are generated via separation of charges at the interfaces of electrodes and electrolytes. The charges are stored on the surfaces of the carbon materials, therefore the specific capacitance is related to the specific surface area of the electrode [34–37]. The electrochemical process occurring in EDLCs is shown in Scheme 8 [38].

Here, E_s, C^+, and A^- are the carbon electrode surface, cation, and anion of the electrolyte, respectively. The symbol // represents the accumulation of charges on the electrode surface to form electric double layers [38]. In the charging process, the electrons move from the positive electrode to negative electrode through the

Fig. 19 A typical structure of electric double-layer capacitor

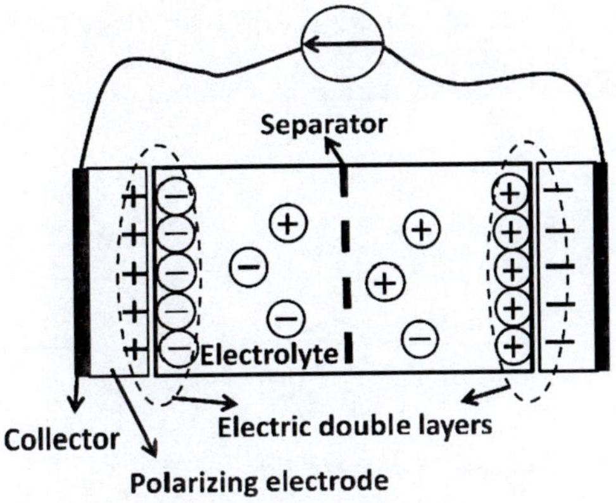

Scheme 8 Electrochemical process in EDLCs

Positive electrode:

$$E_s + A^- \underset{\text{discharge}}{\overset{\text{charge}}{\rightleftharpoons}} E_s^+//A^- + e^-$$

Negative electrode:

$$E_s + C^+ + e^- \underset{\text{discharge}}{\overset{\text{charge}}{\rightleftharpoons}} E_s^-//C^+$$

Overall reaction:

$$E_s + E_s + C^+A^- \underset{\text{discharge}}{\overset{\text{charge}}{\rightleftharpoons}} E_s^+//A^- + E_s^-//C^+$$

external power sources. The cation and anion are separated from the bulk electrolyte and move to electrode surface at the same time. During the discharging process, electrons move from the negative to positive electrode through the load. Ions are released from the electrode surface and move back to the bulk electrolyte. The double layers that accumulate electrons during the charging process are formed at the interface between the positive electrode and anion [38].

Pseudocapacitors. Pseudocapacitors store charges via the fast faradaic redox reactions in the electrodes. Transition metal oxides are commonly used as the electrodes for pseudocapacitors since they possess several oxidation states and offer rich redox reactions. Among them, ruthenium oxide is currently the most popular electrode material for supercapacitor research, holding the record of the highest specific capacitance [39]. The redox reaction that takes place in the electrode is shown in Scheme 9 The oxidation state of Ru ion changed around from 2^+ to 7^+ during the redox reaction.

3.1.1 Carbon Aerogel-Based Supercapacitors

High electrical conductivity, low diffusion resistance to protons/cations, and high electroactive area of the electrode are three key factors for the good performance of a supercapacitor. Carbon materials are commonly used as electrodes in

$$RuO_a(OH)_b + \delta H^+ + \delta e^- \longleftrightarrow RuO_{a-\delta}(OH)_{b+\delta}$$

Scheme 9 Redox reaction of ruthenium oxide

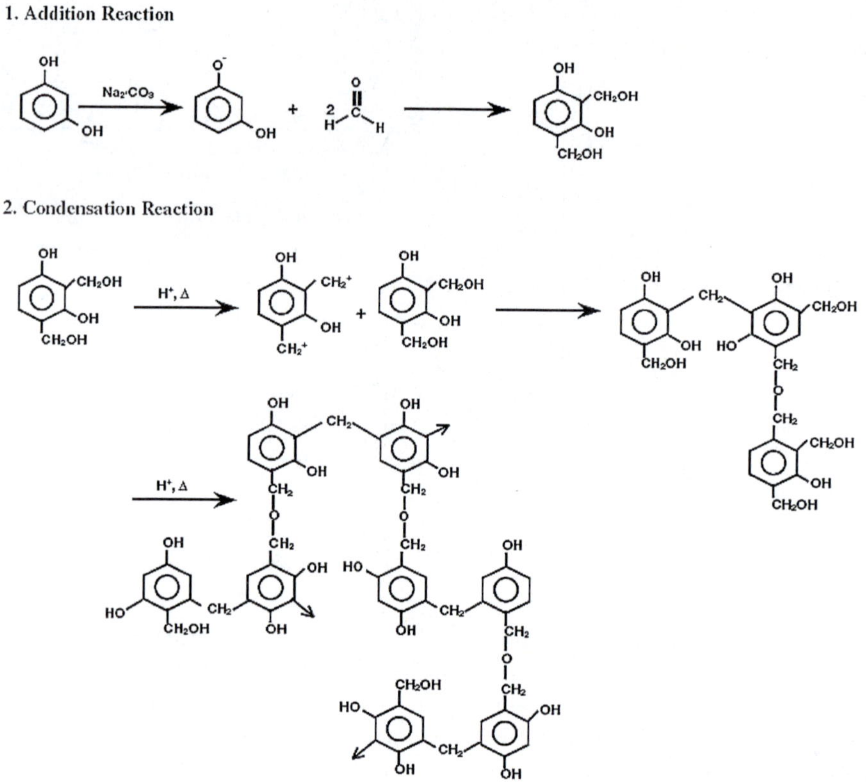

Fig. 20 Polymerization mechanism of resorcinol and formaldehyde to form RF gels [40]. Copyright 1997, with permission from Elsevier

supercapacitors due to their high electrical conductivity, and the mechanism for charge storage is based on electric double-layers. Because of the low specific surface area and the microporous structure of commercial carbon materials, carbon aerogels with high specific surfaces and mesoporous structure became a promising candidate for supercapacitor applications. Carbon aerogels can be prepared via carbonization of resorcinol (R)-formaldehyde (F) organic aerogels [40]. The RF organic aerogels in turn are synthesized by polycondensation of R and F, followed by supercritical drying. The polymerization mechanism of the RF gel is shown in Fig. 20. Sodium carbonate as a base catalyst and the proton as an acid catalyst play an important role in the addition and condensation reactions, respectively.

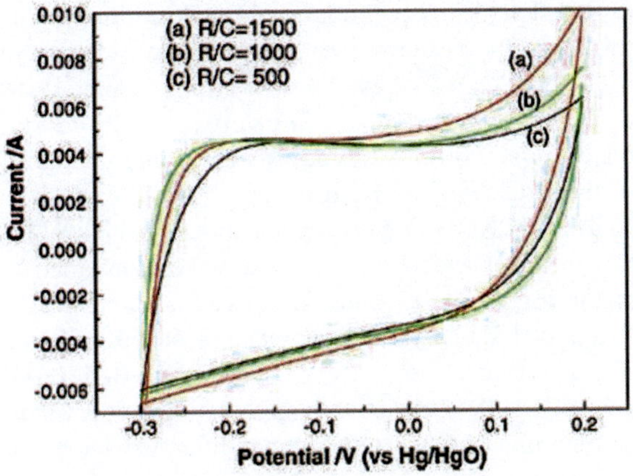

Fig. 21 Cyclic voltammograms of carbon aerogels prepared at three R/C ratios and measured in 6 M KOH at a scan rate of 1 mV/s. Reprinted with permission from Li et al. [42]. Copyright 2006, Elsevier

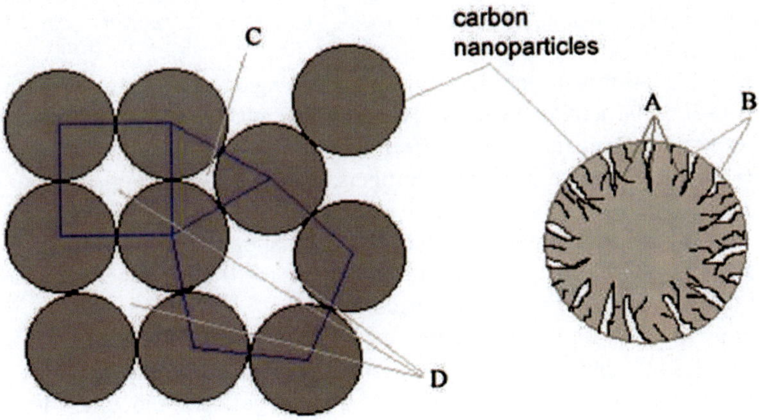

Fig. 22 Hierarchical pore structure in carbon aerogel [43]. Reprinted with permission from Copyright 2007, Elsevier.

They affect the pH of the reaction, and the gelation occurs within a narrow pH window (~1–2 pH unit).

Carbon aerogels possess high porosities, high specific surface areas, and high electrical conductivities, which make them a promising material for applications in supercapacitors. Many preparation parameters, including solvent, pH value, catalyst, as well as the carbonization and activation methods adopted in the preparation process, affect the supercapacitive properties of the carbon aerogels, as discussed below [41].

1. Reactant and pH value. When the concentration of the reactants is decreased or the pH value of the reacting solution is increased, the specific surface area, pore volume, and specific capacitance of the product carbon aerogel increase.

2. *Carbonization*. The specific capacitance increases with increasing carbonization temperature/time and activation time because of the higher electrical

conductivity of the product [41]. In addition, the specific surface area can be increased by reducing the oxygen content during the activation procedure.

The effects of various preparation parameters are further discussed below.

Effects of reactant, carbonization, and activation. Na_2CO_3 is commonly used as the catalyst (C) for the synthesis of RF organic aerogels. Li et al. studied the effect of the R/C ratio on the resulting specific capacitance of the product [42]. In the study, 6 M KOH, 1 M Na_2SO_4, and 2 M $(NH_4)_2SO_4$ were used as the electrolytes. Much higher specific capacitances were obtained for the 6 M KOH case because of the higher ionic conductivities of K^+ and OH^-. The measured cyclic voltammogram (CV) curves for the case of 6 M KOH and three R/C ratios are shown in Fig. 21. A maximum specific capacitance of 110.06 F/g was achieved with an R/C ratio of 1,500. These curves appeared to be very symmetric with respect to the zero current line and are quite rectangular in shape, indicating reversible redox processes at the scan rate of 1 mV/s. The specific capacitance, however, decreased from 110.06 to 67.2 F/g, and the symmetry and rectangular shape of the CV curves deteriorated on increasing the scan rate from 1 to 10 mV/s, mainly because of the poor ion diffusion within the porous electrodes [42].

Besides activation with thermal energy by heating at high temperatures, carbon aerogels can also be activated with KOH (K) treatments. The resulting pore structure of the activated carbon aerogel was affected by the KOH concentration. Four types of pores can be identified in carbon aerogels, as illustrated in Fig. 22 [43]. Here, types A and B are micropores (less than 2 nm) and small mesopores (2–5 nm), respectively, which are formed within the carbon nanoparticles,. Type C and D are big mesopores (5–40 nm) and macropores/channels (>40 nm), respectively, and are formed from the carbon nanoparticle packing.

Appropriate activation with KOH can increase the numbers of type A and type B pores and thus the specific surface area, leading to an increase in the specific capacitance of the activated carbon aerogel. On the other hand, the presence of type D macropores/channels plays an important role in maintaining a good specific capacitance at high charge/discharge rates for easing the ion diffusion. Activated carbon aerogels with a specific surface area as high as 3,247 m^2/g, together with a corresponding specific capacitance of 244 F/g, were obtained with this KOH activation process [43].

Effects of modification. Poor wettability and the subsequent high internal resistance of activated carbon aerogels are disadvantageous to their electrochemical capacitive performances in organic electrolytes. Fang et al. introduced chemical modification in the aerogel synthesis process to deal with this issue [44]. They treated activated carbon aerogels with surfactant sodium oleate to improve the wettability of the aerogels with organic electrolytes to ease the mass transfer within the electrode. This practice improved the electrochemical capacitive performance of the activated carbon aerogel at high discharge rates.

Carbon aerogels from different types of organic aerogels. Besides RF organic aerogels, carbon aerogels were also prepared from the carbonization of pitch-furfural (PF) gels [45]. Toluene (T) and acetic acid (A) were used as the solvent and catalyst, respectively, for the polycondensation of pitches and furfural to form

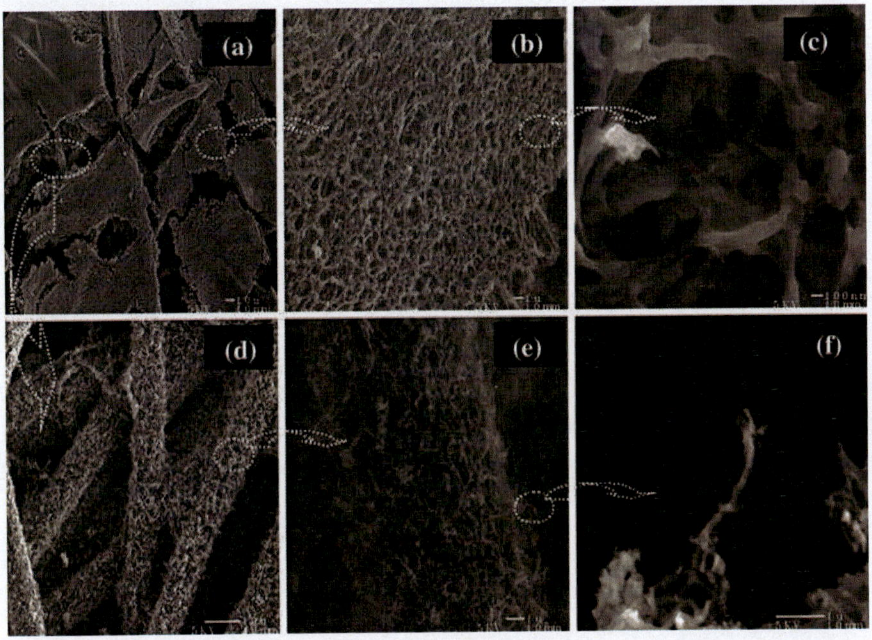

Fig. 23 SEM images of carbon aerogel-modified carbon nanotube electrodc. Reprinted with permission from Ref. [47]. Copyright 2009 John Wiley & Sons

the PF gels. The volume shrinkage of the carbon aerogel increased with increasing T/A ratio, resulting in a decrease in the specific surface area. A maximum specific capacitance of 131.9 F/g and a corresponding specific surface area of 440 m^2/g were obtained at a T/A ratio of 2.23. Upon KOH (K) activation, the specific surface area of the carbon aerogel (CA) drastically increased to as high as 2,375 m^2/g at a K/CA ratio of 5. The maximum specific capacitance of 187.2 F/g, however, was obtained for the product activated at a K/CA ratio of 2 with a corresponding specific surface area of 1,185 m^2/g [46].

Because of the high electrical conductivity of carbon nanotubes (CNTs), carbon aerogel-modified MWCNTs (multi-wall carbon nanotubes) also has drawn research attention in recent years [47]. MWCNTs alone did not show impressive specific capacitances because of their low specific surface area of 92.5 m^2/g. The idea is to combine the advantages of the high electric conductivity of CNTs with the high specific surface area of carbon aerogels. Microfibrous carbon papers were used as the substrate, with the MWCNTs deposited on top, forming the electrode to be used in supercapacitors. Finally, carbon aerogels were introduced to coat the MWCNTs, increasing the total specific surface area. Figure 23 shows the SEM images of the composites. Most of the substrate was covered with CA (a–c). Beneath the CA layer are the microfibers of the carbon paper substrate (d) uniformly coated with MWCNTs (e), which in turn were nicely covered with CA (f) [47].

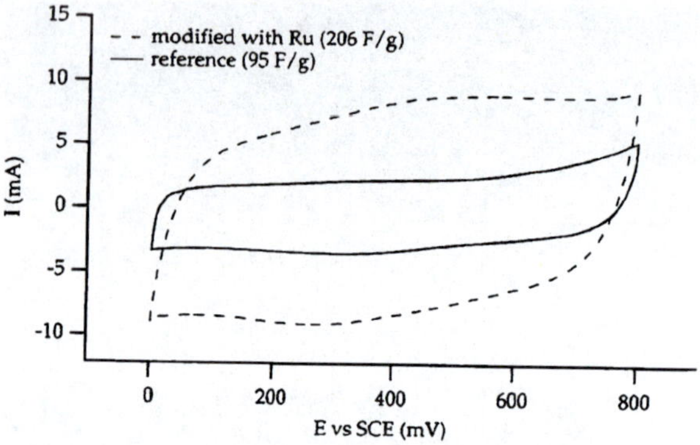

Fig. 24 Cyclic voltammograms of Ru incorporated carbon aerogel and reference carbon aerogel. Reprinted with permission from Ref. [48, 49]. Copyright 1997 Electro-chemical Society

The total specific surface area and pore volume of the final products were 1,059 m^2/g and 2.37 cc/g, respectively. The specific capacitance of the sample reached values as high as 524 F/g, four times higher than that of plain carbon aerogels [47].

Metal-doped carbon aerogels. Carbon aerogels alone offer only capacitances contributed by the electric double layer. To further boost the specific capacitances of carbon aerogels, metals or metal oxides are introduced into the framework of carbon aerogels to render pseudocapacitances coming from the faradaic redox reactions.

Ru, because of its high electronic conductivity and electrochemical activity, was introduced into carbon aerogels for applications in supercapacitors, achieving higher specific capacitances than plain carbon aerogels [48, 49]. The diameter of the Ru nanoparticles was $\sim$2 nm. These Ru nanoparticles were produced within the host carbon aerogel via a chemical vapor impregnation route. The specific surface area of the Ru incorporated carbon aerogel decreased from 724 to 434.5 m^2/g with increasing Ru content from 0 to 34.93%. The corresponding specific capacitance however increased from 85 to 206.4 F/g. The CV curves of the Ru incorporated carbon aerogel and reference carbon aerogel are shown in Fig. 24 [48, 49]. Both curves appeared symmetric and rectangular in shape, indicating reversible redox reactions at the charge/discharge cycles. The enclosed area of the CV curve of the Ru incorporated carbon aerogel was much larger than that of the plain carbon aerogel, leading to the much higher specific capacitance of the Ru incorporated carbon aerogel.

3.1.2 Transition Metal Oxide Aerogel-Based Supercapacitors

Unlike carbon materials, transition metal oxide-based supercapacitors possess not only electric double-layer capacitances but also pseudocapacitances. Most transition metal oxides have several oxidation states, thus showing rich redox reactions to generate pseudocapacitances. Aerogels, as mentioned earlier, possess high specific surface areas to offer high electrochemically active surface areas for relevant redox reactions. Also, their high porosities together with the percolating mesoporous structure are advantageous for ion diffusions. Consequently, transition metal oxide aerogels are likely to offer much higher specific capacitances than carbon aerogels. An increasing amount of research efforts have thus been directed to the study of transition metal oxide aerogel-based supercapacitors.

RuO_2-based aerogels. RuO_2 has been extensively used in electrochemistry because of its high electronic conductivity and electrochemical activity. In the area of supercapacitors, the surface sites of RuO_xH_y can be reversibly oxidized and reduced with the simultaneous exchange of protons with the contacting solution, resulting in changes of the oxidation state of the Ru ions. Since RuO_2 possesses several oxidation states (II–XII), it can offer rich redox reactions for pseudocapacitance. Its high specific capacitances have been reported in the literature. Ruthenium oxide-based aerogels (hydrous ruthenium oxide and ruthenium-titanium oxide) were synthesized via an alkoxide-based procedure and investigated as a supercapacitor material [50]. Figure 25 shows the CV curves of several hydrous RuO_2 aerogels and Ru-Ti mixed oxide aerogels. The capacitive currents decreased in the order of $RuO_2.0.05H_2O >> RuO_2.0.03H_2O > (Ru_{0.32}-Ti_{0.68})O_x$ aerogels> anhydrous RuO_2. The specific surface area and specific capacitance of the $RuO_2.0.5H_2O$ aerogel were 75 m^2/g and 900 F/g, respectively. The specific capacitance of the $RuO_2.0.5H_2O$ aerogel was about 1,000 times that of the anhydrous RuO_2 aerogel, even though there existed only a two-fold difference in specific surface area. The drastic difference in specific capacitance mainly came from the difference in the degree of hydration and degree of crystallinity. The capacitance per area of the $RuO_2.0.5H_2O$ aerogel reached a value as high as 1,200 $\mu F/cm^2$, and the corresponding symmetric and rectangular shaped CV curve was demonstrated in an electrolyte of 0.5 M H_2SO_4, as shown in Fig. 25, superior to the other three cases.

Nickel oxide-based aerogels. RuO_2, although demonstrated to be an excellent supercapacitor material, suffers from its high cost, which hinders its potential for commercialization. Therefore, research efforts to seek cost-effective alternatives to RuO_2 continue. Several candidates, such as nickel oxide, cobalt oxide, and manganese oxide, have been investigated. Nickel oxide is a low cost transition metal oxide, and has also been extensively used in electrochemisty. Aerogel-like mesoporous nickel oxides were synthesized and studied by Wu et al. via an alkoxide-based procedure followed by supercritical drying [51].

$Ni(OH)_2$ aerogels were initially obtained with the sol–gel reaction, and were transformed to crystalline NiO aerogels upon calcination at 250°C. The maximum specific surface area was 325 m^2/g, with a high porosity of 90%, after calcination at 300°C [51]. The oxidation of $Ni(OH)_2$ and NiO, as shown in Scheme 10, changes the oxidation states of Ni ions from 2^+ to 3^+. The highest specific capacitance of the NiO aerogels was determined to be 125 F/g. The corresponding CV curves were quite symmetric and rectangular in shape, indicating a good reversibility in the involved redox reaction.

Cobalt oxide-based aerogels. Cobalt oxide is another promising candidate, particularly for its higher electronic conductivity of 2×10^{-4} compared to that of

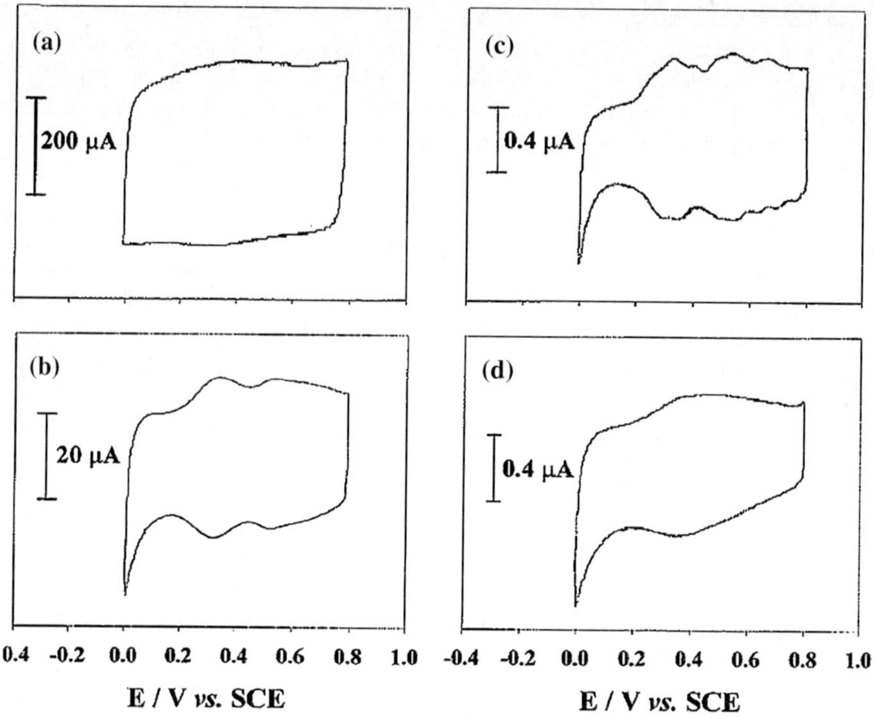

Fig. 25 Cyclic voltammograms of **a** $RuO_2.0.5H_2O$, **b** $RuO_2.0.03H_2O$, **c** anhydrous RuO_2, **d** $(Ru_{0.32}-Ti_{0.68})O_x$ aerogels. Reprinted with permission from Ref. [50]. Copyright 1999 American Chemical Society

Scheme 10 Oxidation of nickel oxide

$$Ni(OH)_2 + OH^- \longrightarrow NiOOH + H_2O + e^-$$

$$NiO + OH^- \longrightarrow NiOOH + e^-$$

nickel oxide ($8 \times 10^{-5}\ \Omega^{-1}\ cm^{-1}$). Traditionally, transition metal oxide aerogels were prepared via alkoxide-based sol–gel processes. As mentioned earlier, alkoxides are generally expensive and sensitive to moisture and heat, requiring careful handling. With that, non-alkoxide-based sol–gel processes were developed in recent years to prepare transition metal oxide aerogels. Cobalt oxide aerogels were first synthesized via a non-alkoxide-based procedure, with their electrochemical properties studied by Wei et al. [9]. The authors also prepared cobalt oxides containing micropores (termed PCO-micro) to study the pore size effect (mesopore vs. micropore) in supercapacitors. $Co(OH)_2$ was initially produced via a non-alkoxide based sol–gel process and transformed to crystalline Co_3O_4 aerogels upon calcination at 200°C. The specific surface area of cobalt oxide aerogels clacined at 200°C (termed A-200) was as high as 235 m^2/g with an average pore size of 3.8 nm. This pore size was within the 2–5 nm optimal size for

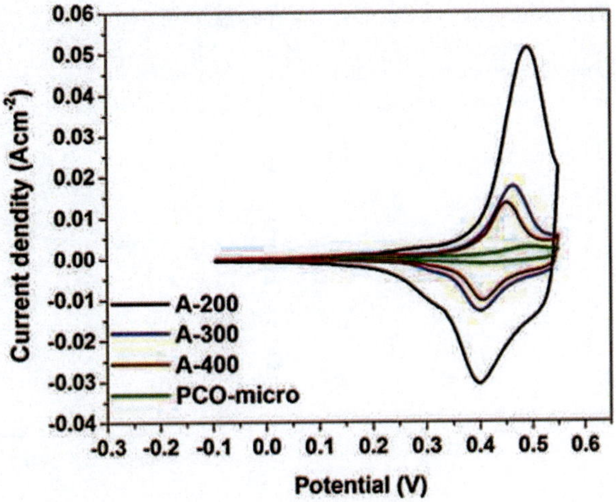

Fig. 26 Cyclic voltammograms of cobalt oxide aerogels produced with different calcination temperatures compared with cobalt oxides containing micropores (PCO-micro). Reprinted with permission from Ref. [9]. Copyright 2009 American Chemical Society

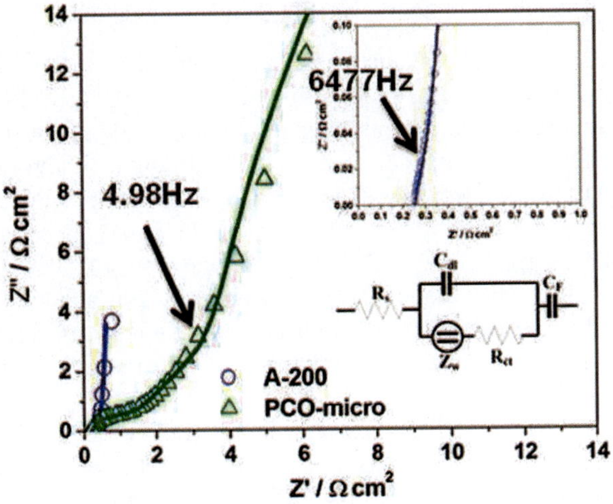

Fig. 27 Electrochemical impedance spectroscopy of cobalt oxide aerogel calcined at 200°C and cobalt oxide containing micropores. Reprinted with permission from Ref. [9]. Copyright 2009 American Chemical Society

supercapacitor applications. The CV curves of the resulting cobalt oxide samples are shown in Fig. 26 [9].

The curves were symmetric, but not rectangular in shape, signifying reversible redox reactions with a narrow operating window. The highest specific capacitance was obtained for sample A-200 with a value of 623 F/g, much higher than that of the PCO-micro sample of 41 F/g. It is the highest specific capacitance ever reported for cobalt oxide as a supercapacitor material. The pore size effect on the supercapacitive properties of the products was investigated by electrochemical impedance spectroscopy (EIS). The results are displayed in Fig. 27 [9]. Here, samples A-200 and PCO-micro were compared. Sample A-200 contained solely mesopores, whereas sample PCO-micro contained micropores of about 50% in terms of its contribution to the total surface area. Mesopores offer easier

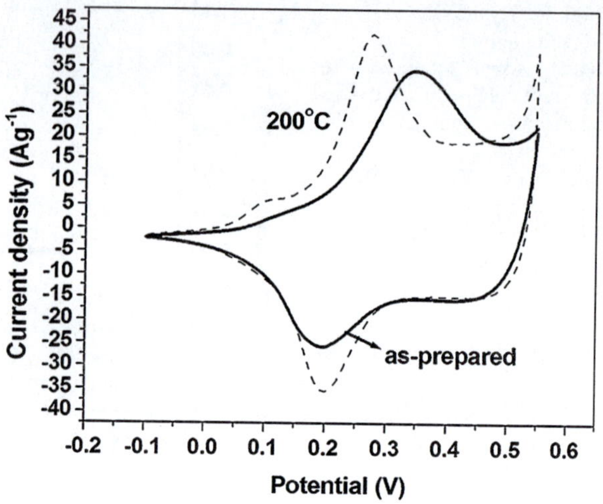

Fig. 28 Cyclic voltammograms of nickel cobaltite aerogels. Reprinted with permission from Ref. [52]. Copyright John Wiley & Sons

transportation channels for ion diffusions, whereas the mass transfer within micropores is much more difficult. As a result, the EIS curve for sample A-200 was almost vertical, mimicking the behavior of an ideal RC circuit and implying a negligible composite interfacial impedance (including mass transfer resistance) within the electrode. The EIS curve for sample PCO-micro however showed an apparent development of a semicircle, implying an appreciable composite interfacial impedance within the electrode.

The onset frequencies of samples A-200 and PCO-micro were 6,477 Hz and 4.98 Hz, respectively. High onset frequencies mean high powers. Therefore, the ease in mass transfer within the electrode also affects the power performance of the supercapacitor.

Nickel Cobaltite-based Aerogel. Nickel cobaltite was found to have a higher electronic conductivity (0.6 Ω^{-1} cm^{-1}) than nickel oxide (8×10^{-5} Ω^{-1} cm^{-1}) and cobalt oxide (2×10^{-4} Ω^{-1} cm^{-1}). Meanwhile, its electrochemical activity is also better than those of nickel oxide and cobalt oxide. Also, it is expected to offer richer redox reactions, including contributions from both nickel and cobalt ions, than the two corresponding single component oxides. Furthermore, nickel cobaltite is cheap, making it a potentially cost-effective alternative for RuO_2. Nickel cobaltite aerogels were prepared with a non-alkoxide sol–gel process by Wei et al. [52]. Surprisingly, the crystalline spinel structure of $NiCo_2O_4$ formed in the as-prepared sample without the need for heat treatment. The recorded CV curve of the product aerogel is shown in Fig. 28.

The CV curve for the aerogel calcined at 200°C showed better symmetry than that of the as-prepared aerogel, possibly because of the improved crystallinity. Its operating window is much wider than the nickel oxide and cobalt oxide aerogels, indicating the benefit of having two types of metal ions, Ni and Co, in the product. The specific capacitance of the calcined aerogel without a full activation was

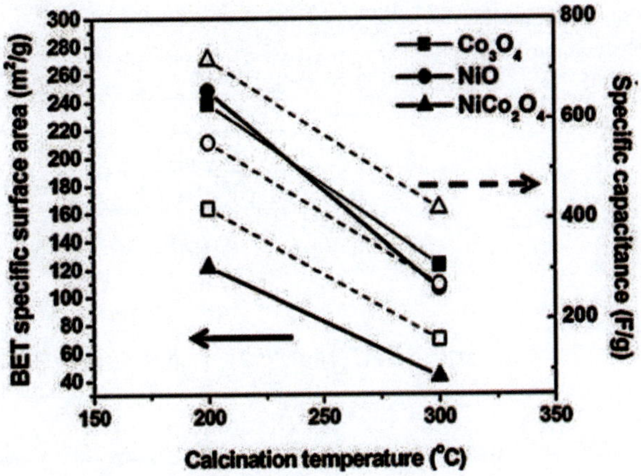

Fig. 29 Specific capacitance and specific surface area of nickel cobaltite aerogels vs. calcination temperature. Reprinted with permission from Ref. [52]. Copyright 2010 John Wiley & Sons

determined to be 719 F/g, already a high value higher than those of nickel oxide and cobalt oxide aerogels.

Figure 29 shows the BET specific surface area (solid lines) and corresponding specific capacitance (dashed lines) of the cobalt oxide, nickel oxide, and nickel cobaltite aerogels obtained at 200 and 300°C as a function of the calcination temperature. The plot shows: (1) the trend of the specific surface area that goes side by side with that of the specific capacitance, indicating the importance of the specific surface area in specific capacitances, and (2) the nickel cobaltite aerogel, although possessing the lowest specific surface area, exhibits the highest specific capacitance. When the nickel cobaltite aerogels were fully activated, the specific capacitance reached 1,400 F/g, close to the highest value ever reported for RuO_2, showing that nickel cobaltite aerogels are a promising, cost-effective alternative to RuO_2 as a supercapacitor material.

3.2 Li–Ion Battery

3.2.1 Principle of Li–Ion Batteries

Lithium–ion batteries are commonly used for consumer electronics and are gaining popularity in automotive, defense, and aerospace applications because of their high energy densities. By comparison, supercapacitors offer high power densities, whereas lithium–ion batteries offer high energy densities. It has become popular to integrate supercapacitors and lithium–ion batteries together as a hybrid energy storage and supply system to satisfy simultaneous demands of power and energy. The basic working mechanism of the lithium–ion battery is shown in Fig. 30. In the charging process, an electron is released from the cathode and transported through the external circuit to the anode. Lithium ions diffuse from the cathode to anode via the electrolyte. Intercalations of lithium ions in the electrodes occur

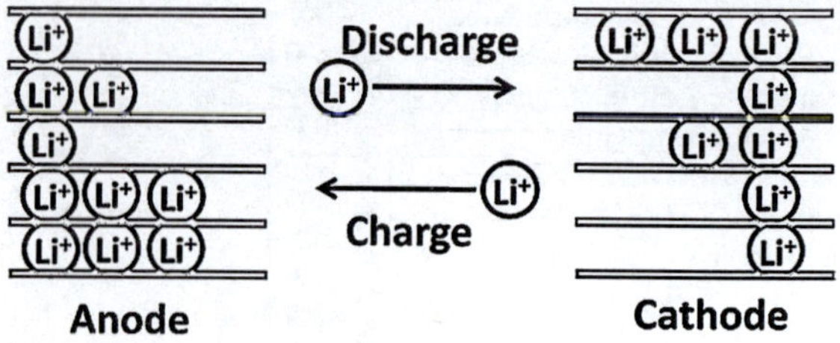

Fig. 30 Working mechanism of lithium–ion battery

$$\text{Cathode:}\quad Li_{1-x}MaMbO_2 + xLi^+ + xe^- \underset{\text{discharge}}{\overset{\text{charge}}{\rightleftharpoons}} LiMaMbO_2$$

$$\text{Anode:}\quad C_6Li_x \underset{\text{discharge}}{\overset{\text{charge}}{\rightleftharpoons}} C_6 + xLi^+ + xe^-$$

Scheme 11 Redox reactions of lithium–ion battery

during the redox reactions at the charge/discharge steps. The redox reaction is shown in Scheme 11 [53–56].

The charge storage in lithium–ion batteries is directly related to the specific surface area and pore structure of the electrode. High specific surface areas of the electrode offer a large amount of electroactive surfaces for lithium–ion intercalation and redox reactions, and highly porous structure reduces the resistance to ion diffusions. Therefore, intercalation materials in aerogel form are attractive, not only because of the high specific surface area, but also due to the small diffusion resistances in the electrode. We review below several important applications of metal oxide aerogels in lithium–ion batteries.

3.2.2 Aerogel-Based Li–Ion Batteries

Vanadium oxide-based aerogels. Vanadium oxide is well-known for its high capacity toward lithium insertion in non-aqueous electrolytes and, thus, is a popular cathode material for lithium–ion batteries. Vanadium pentoxide (V_2O_5) aerogels were prepared by Salloux et al. via an alkoxide-based procedure by using vanadyl triisopropoxide as the precursor [57]. The lithium–ion intercalation in this aerogel was determined to be up to 1.9 mol of Li per mole of aerogel. Le et al. prepared amorphous V_2O_5 aerogels via an ion exchange processing of sodium metavanadate, which possessed a high specific surface area of 450 m^2/g and a high pore volume of 2.3 cc/g [58]. The amount of lithium–ion intercalation was determined to be at least 4 mol of Li per mole of aerogel [58]. The specific energies of the V_2O_5 aerogels were found to be far higher than those of the V_2O_5 xerogels and amorphous V_2O_5 samples, as shown in Fig. 31, and reached a maximum value of 1,500 Wh/kg [58, 59].

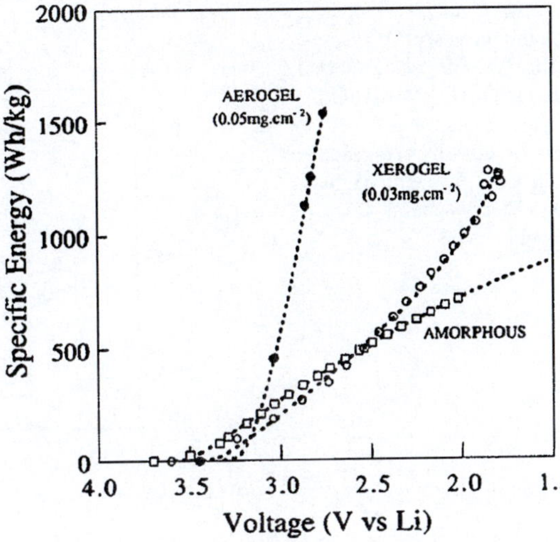

Fig. 31 Specific energy versus equilibrium voltage curves for different V_2O_5 samples. Reprinted with permission from Ref. [58]. Copyright 1996 Electrochemical Society

Molybdenum Oxide-based aerogel. Molybdenum oxide possesses a stable, 2D layered structure. These layers can be propped open by an intercalating species, such as protons, solvated lithium, sodium ions, as well as larger molecules. Consequently, molybdenum oxide is a promising candidate for lithium–ion battery applications. Dong et al. prepared molybdenum oxide aerogels by using molybdenum alkoxides as the precursor [60]. Both aerogels and xerogels were obtained. The density, specific surface area, and pore volume of the as-prepared molybdenum oxide aerogel through supercritical drying were 0.17 g/cm^3, 180 m^2/g, and 3.5 cc/g, respectively. As for the xerogel production, low surface tension solvents such as pentane, cyclohexane, and acetone were used in the ambient pressure drying step. The density and specific surface area of the as-prepared xerogel from pentane were 0.7 g/cm^3 and 270 m^2/g, respectively. The amorphous molybdenum oxide aerogels were electrochemically active materials with reversible Li intercalation up to 1.2 Li/Mo [60].

Manganese oxide-based aerogels. MnO_2 is a well-studied cathode material for its usage in the familiar 1.5 V commercial Zn/MnO_2 alkaline cells. MnO_2 is also low cost and low toxicity relative to other metal oxides, such as V_2O_5, and drew much research attention as a lithium–ion intercalation host for lithium–ion batteries. MnO_2 aerogels, ambigels, and xerogels were obtained via a $KMnO_4$-based sol–gel process and applied in lithium batteries by Long et al. [61, 62]. The specific surface area and pore volume of the MnO_2 aerogel were 210 m^2/g and 0.8 cc/g, respectively. The measured CV curves of these MnO_2 aerogels, xerogels, and ambigels in 1 M $LiClO_4$/propylene carbonate, as shown in Fig. 32, exhibited well-defined faradaic features for the lithium–ion insertion/deinsertion process.

Figure 33 compares the discharge capacities of the aerogel and xerogel samples at increasing discharge cycle. Evidently, the aerogel samples outperformed the

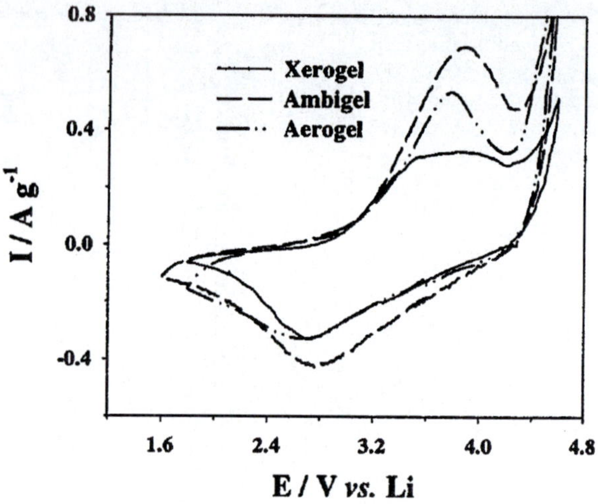

Fig. 32 Cyclic voltammgrams of MnO_2 aerogels, ambigels, and xerogels in 1.0 M $LiClO_4$/ propylene carbonate. Reprinted with permission from Ref. [62]. Copyright 2000 Electro-chemical Society

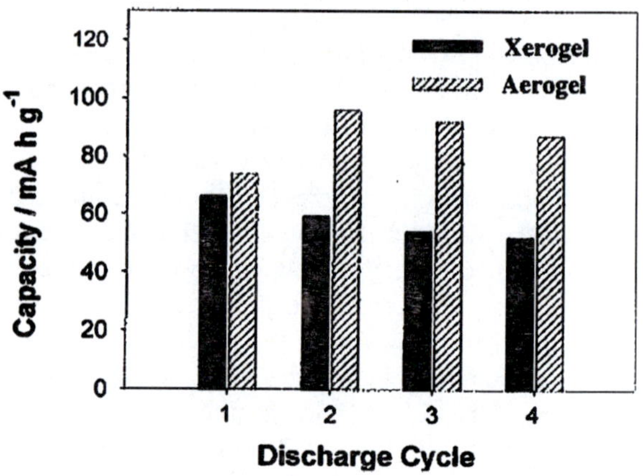

Fig. 33 Discharge capacities of aerogels and xerogels at increasing discharge cycle Reprinted with permission from Ref. [62]. Copyright 2000 Electro-chemical Society

xerogel samples, with the highest discharge capacity of almost 100 mAh/g achieved at the second discharge cycle [62].

MnO_2 aerogels, ambigels, and xerogels with birnessite structure were also synthesized via a sol–gel process by Long et al. [63]. The specific surface areas and pore volumes of the resulting aerogel, ambigel, and xerogel were (250 m^2/g, 1.8 cc/g), (210 m^2/g, 1.6 cc/g), and (140 m^2/g, 0.36 cc/g), respectively. The discharge capacities of the MnO_2 aerogel, ambigel, and xerogel were characterized at 500 mA/g and are compared in Fig. 34. Evidently, the decay rate in discharge capacity with increasing discharge/charge cycle was the greatest for the xerogel sample, attributable to its smaller specific surface area and pore volume. Apparently, the open pore structure and high surface area are beneficial to the relevant electrochemical performance.

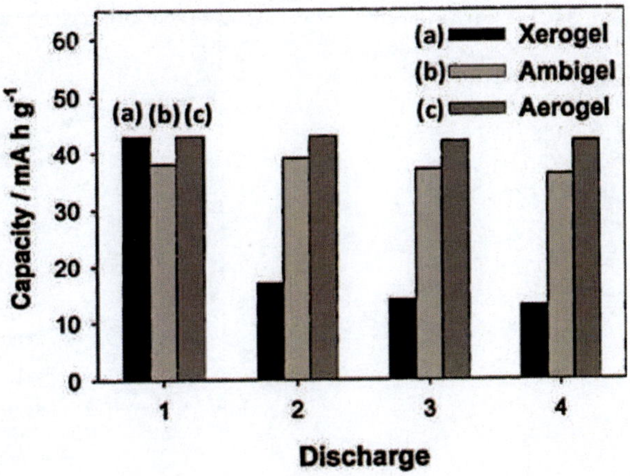

Fig. 34 Discharge capacities of MnO_2 aerogels, ambigels, and xerogels at increasing discharge cycle. Reprinted from Long et al. [63]. Copyright 2001, with permission from Elsevier

Composite aerogels. For lithium–ion batteries, high electronic conductivities for electrodes are one of the key factors to obtain high energy capacities. Because of this, RuO_2,which possesses a higher electronic conductivity than V_2O_5, was introduced into V_2O_5 aerogels in the form of nanoparticles for lithium–ion battery applications [64]. Aerogel-like V_2O_5 and RuO_2 nanoparticle-modified V_2O_5 aerogels, $(RuO_2)_{0.1}V_2O_5$, were synthesized by Zhang et al. The active material networks were coated and interconnected by in situ-formed RuO_2 nanoparticles ($\sim$10%) that ensured good mechanical stability and electronic conductivity. RuO_2 nanoparticles did not block the free volume and surface area of the V_2O_5 aerogel, and the electronic conductivity increased, leading to improved specific charge capacity over plain V_2O_5 aerogels. The electrochemical data of both aerogel samples, as shown in Fig. 35, were measured under the constant-current insertion condition at discharge rates of 2C, C, C/5, and C/10. Both samples showed smooth voltage decays upon lithium insertion, and the specific discharge capacities of the composite aerogel were larger than the corresponding ones of the plain aerogel at all discharge rates. For example, the specific discharge capacities of $(RuO_2)_{0.1}V_2O_5$ and V_2O_5 aerogels were 400 and 310 mAh/g, respectively at the discharge rate of C/10. The large differences in specific discharge capacity were attributed to the smaller ohmic effect in $(RuO_2)_{0.1}V_2O_5$ aerogels [64].

Carbon black has been used to improve the electronic conductivities of lithium intercalation electrodes. Unfortunately, carbon black, although produced as fine particles, often forms sub-micron sized aggregates during the fabrication of the composite electrode, blocking the pathway for ion diffusion. To improve the situation, Sakamoto and Dunn investigated how single-wall carbon nanotubes (SWCNT) can work with V_2O_5 aerogels as a lithium intercalation electrode [65]. They synthesized V_2O_5/SWCNT composite aerogels with different loadings of SWCNTs. SWCNTs ($\sim 10^4$ S/cm) were introduced into V_2O_5 aerogels, making good contact with the aerogel backbone without blocking the pore structure, and the composite aerogels showed good performance, even at high discharge rates.

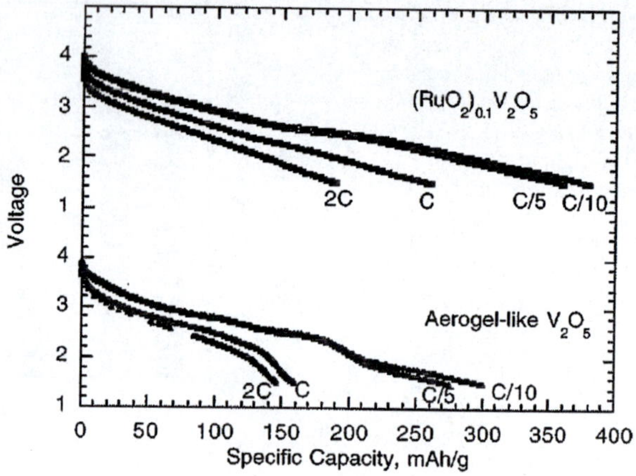

Fig. 35 Electrochemical data of $(RuO_2)_{0.1}V_2O_5$ and V_2O_5 aerogels at four different discharge rates. Reprinted with permission from Ref. [64]. Copyright 2001 Electro-chemical Society

Table 3 Specific discharge capacities for V_2O_5/KJB and V_2O_5/SWCNT composite aerogels

Specific current(mA/g)	Specific discharge capacities (mAh/g)					
	V_2O_5/KJB			V_2O_5/SWCNT		
	5wt%	9wt%	17wt%	5wt%	9wt%	17wt%
112	364	425	452	382	449	452
560	274	326	382	314	373	418
1,120	197	227	323	280	342	395
2,800	81	85	155	210	257	296

Reprinted with permission from Ref. [65]. Copyright 2002 Electro-chemical Society

The specific discharge capacities obtained at increasing specific currents for V_2O_5/SWNT composite aerogels of three different SWCNT loadings are listed in Table 3. Also included in Table 3 are the data from the carbon black-loaded V_2O_5 aerogels, V_2O_5/KJB, for comparison. At low specific currents, there was not much difference between the two types of composite aerogels of the same additive loading, although the specific discharge capacities of the V_2O_5/SWCNT composite aerogels were slightly higher than those of the corresponding V_2O_5/KJB composite aerogels. At high specific currents, there was a significant decay in the specific discharge capacity of the V_2O_5/KJB composite aerogels, whereas those of the V_2O_5/SWCNT composite aerogels showed only minor to moderate decays. This may be attributed to better contact between the SWCNTs and the aerogel backbone and fewer structural disturbances to the aerogel from the SWCNT incorporation.

4 Conclusions

Aerogels find a wide range of applications because of their unique and advantageous structural features of high porosity, high specific surface area, and percolating mesopores. Any application requiring high functional surfaces and easy transport for participating species, will find aerogels useful. More applications for aerogels are bound to be developed in the future. To meet different application requirements, aerogels need to be produced in suitable processes to acquire necessary application characteristics. We review preparation routes of alkoxide vs. non-alkoxide precursor and supercritical fluid vs. ambient pressure drying. The choice depends on the product requirements for aerogels, and the trend is to develop multi-component aerogels to further boost targeted performances and to tackle more complicated tasks.

Aerogels have been demonstrated superior in energy saving as the thermal insulation material and in energy storage as the electrode materials for supercapacitors and lithium–ion batteries. The trend is to develop composite aerogels that take advantages from individual components to suit different needs of the applications. Furthermore, aerogels of compositions other than metal oxides and carbon deserve more research attention. There exist ample opportunities for even wider applications for aerogels.

References

1. Kistler SS (1931) Coherent expanded aerogels and jellies. Nature 227:741–741
2. Chervin CN, Clapsaddle BJ, Chiu HW et al (2005) Aerogel synthesis of yttria-stabilized zirconia by a non-alkoxide sol-gel route. Chem Mater 17:3345–3351
3. Gash AE, Tillotson TM, Satcher JH Jr et al (2001) Use of epoxides in the sol-gel synthesis of porous iron(III) oxide monoliths from Fe(III) salts. Chem Mater 13:999–1007
4. Baumann TF, Gash AE, Chinn SC et al (2005) Synthesis of high-surface-area alumina aerogels without the use of alkoxide precursors. Chem Mater 17:395–401
5. Baumann TF, Kucheyev SO, Gash AE et al (2005) Facile synthesis of a crystalline, high-surface-area SnO_2 aerogel. Adv Mater 17:1546–1548
6. Gao YP, Sisk CN, Hope-Weeks LJ (2007) A sol-gel route to synthesize monolithic zinc oxide aerogels. Chem Mater 19:6007–6011
7. Laberty-Robert C, Long JW, Lucas EM et al (2006) Sol-gel-derived ceria nanoarchitectures: synthesis, characterization, and electrical properties. Chem Mater 18:50–58
8. Long JW, Logan MS, Rhodes CP et al (2004) Nanocrystalline iron oxide aerogels as mesoporous magnetic architectures. J Am Chem Soc 126:16879–16889
9. Wei TY, Chen CH, Chang KH et al (2009) Cobalt oxide aerogels of ideal supercapacitive properties prepared with an epoxide synthetic route. Chem Mater 21:3228–3233
10. Prakash SS, Brinker CJ, Hurd AJ (1995) Silica aerogel films at ambient pressure. J Non-Crystal Solids 190:264–275
11. Prakash SS, Brinker CJ, Hurd AJ et al (1995) Silica aerogel films prepared at ambient pressure by using surface derivatization to induce reversible drying shinkage. Nature 374:439–443
12. Roig A, Molins E, Rodriguez E et al (2004) Superhydrophobichydrophobic silica aerogels by fluorination at the gel stage. Chem Comm 2316–2317
13. Wei TY, Chang TF, Lu SY et al (2007) Preparation of monolithic silica aerogel of low thermal conductivity by ambient pressure drying. J Am Ceram Soc 90:2003–2007
14. Lu X, Arduini-Schuster MC, Kuhn J et al (1992) Thermal conductivity of monolithic organic aerogels. Science 255:971–972
15. Yoldas BE, Annen MJ, Bostaph J (2000) Chemical engineering of aerogel morphology formed under nonsupercriticalsupercritical conditions for thermal insulation. Chem Mater 12:2475–2484

16. Wagh PB, Begag R, Pajonk GM et al (1999) Comparison of some physical properties of silica aerogel monoliths synthesized by different precursors. Mater Chem Phys 57:214–218
17. Gurav JL, Rao AV, Nadargl DY (2009) Study of thermal conductivity and effect of humidity on HMDZ modified TEOS based aerogel dried at ambient pressure. J Sol-Gel Sci Technol 50:275–280
18. Nadargi DY, Roa AV (2009) Methyltriethixysilane: new precursor for synthesizing silica aerogels. J Alloy Compd 467:397–404
19. Albert DF, Andrews GR, Mendenhall RS et al (2001) Supercritical methanol drying as a convenient route to phenolic-furfural aerogels. J Non-Cryst Solids 296:1–9
20. Rigacci A, Marechal JC, Repoux M et al (2004) Preparation of polyurethane-based aerogels and xerogels for thermal superinsulation. J Non-Cryst Solids 350:372–378
21. Lee JK, Gould GL, Rhine W (2009) Polyurea based aerogel for a high performance thermal insulation material. J Sol-Gel Sci Technol 49:209–220
22. Lee JK, Gould GL (2007) Polydicyclopentadiene based aerogelaerogel: a new insulation material. J Sol-Gel Sci Technol 44:20–40
23. Katti A, Shimpi N, Roy S et al (2006) Chemical, physical, and mechanical characterization of isocyanate cross-linked amine-modified silica aerogels. Chem Mater 18:285–296
24. Meador MAB, Capadona LA, McCorkle L et al (2007) Structure-property relationships in porous 3D nanostructures as a function of preparation condition isocyanate cross-linked silica aerogels. Chem Mater 19:2247–2260
25. Wei TY, Lu SY, Chang YC (2008) Transparent, hydrophobic composite aerogels with high mechanical strength and low high-temperature conductivities. J Phys Chem B 112: 11881–11886
26. Hrubesh LW, Pekala RW (1994) Thermal properties of organic and inorganic aerogels. J Mater Res 9:731–738
27. Wang J, Kuhn J, Lu X (1995) Monolithic silica aerogel insulation doped with TiO2 powder and ceramic fibers. J Non-Cryst Solids 186:296–300
28. Kuhn J, Gleissner T, Arduini-Schuster MC et al (1995) Integration of mineral powders into SiO_2 aerogels. J Non-Cryst Solids 186:291–295
29. Wei TY, Lu SY, Chang YC (2009) A new class of opacified monolithic aerogels of ultralow high-temperature thermal conductivies. J Phys Chem C 113:7424–7428
30. Wiener M, Relchenauer G, Hemberger F et al (2006) Thermal conductivity of carbon aerogels as a function of pyrolysis temperature. Int J Thermophys 27:1826–1843
31. Wiener M, Reichenauer G, Braxmeier S et al (2009) Carbon aerogel-based high-temperature thermal insulation. Int J Thermophys 30:1372–1385
32. Miller JR, Burke AF (2008) Electrochemical capacitors: challenges and opportunities for real-world applications. Electrochem Soc Inter 17:53–57
33. Jayalakshmi M, Balasubramanian K (2008) Simple capacitors to supercapacitors-an overview. Int J Electrochem Sci 3:1196–1217
34. Endo M, Maeda T, Takeda T et al (2001) Capacitance and pore-size distribution in aqueous and nanaqueous electrolytes using various actived carbon electrodes. J Electrochem Soc 148:A910–A914
35. Simon P, Gogotsi Y (2008) Materials for electrochemical capacitors. Nature Mater 7: 845–854
36. Frackowiak E (2007) Carbon materials for supercapacitor application. Phys Chem Chem Phys 9:1774–1785
37. Miller JR, Simon P (2008) Electrochemical capacitors for energy management. Science 321:651–652
38. Zheng JP, Huang J, Jow TR (1997) The limitations of energy density fro electrochemical capacitors. J Electorchem Soc 114:2026–2031
39. Hu CC, Chen WC, Chang KH (2004) How to achieve maximum utilization of hydrous ruthenium oxide for supercapacitors. J Electrochem Soc 151:A281–A290
40. Lin C, Ritter JA (1997) Effect of synthesis pH on the structure of carbon xerogels. Carbon 35:1271–1278

41. Al-Muhtaseb SA, Ritter JA (2003) Preparation and properties of resorcinol-formaldehyde organic and carbon gels. Adv Mater 15:101–114
42. Li J, Wang X, Huang Q et al (2006) Studies on preparation and performances of carbon aerogel electrodes for the application of supercapacitor. J Power Sour 158:784–788
43. Wang J, Yang X, Wu D et al (2008) The porous structures of activated carbon aerogels and their effects on electrochemical performance. J Power Sour 185:589–594
44. Fang B, Wei YZ, Maruyama K et al (2005) High capacity supercapacitors based on modified activated carbon aerogel. J Appl Electrochem 35:229–233
45. Zeng X, Wu D, Fu R et al (2008) Structure and EDLC characteristics of pitch-based carbon aerogels. Mater Chem Phys 112:1074–1077
46. Zeng X, Wu D, Fu R et al (2008) Preparation and electrochemical properties of pitch-based actived carbon aerogels. Electrochim Acta 53:5711–5715
47. Bordjiba T, Mohamedi M, Dai LH (2009) New class of carbon-nanotube aerogel electrodes for electrochemical power sources. Adv Mater 20:815–819
48. Miller JM, Dunn B, Tran TD et al (1997) Deposition of ruthrnium nanoparticles on carbon aerogels for high energy density supercapacitor electrodes. J Electrochem Soc 144:L309–L311
49. Miller JM, Dunn B (1999) Morphology and electrochemistry of ruthenium/carbon aerogel nanostructures. Langmuir 15:799–806
50. Long JW, Swider KE, Merzbacher CI et al (1999) Voltammetric characterization of ruthenium oxide-based aerogels and other RuO2 solids: the nature of capacitance in nanostructured materials. Langmuir 15:780–785
51. Wu M, Gao J, Zhang S et al (2006) Synthesis and charcterization of aerogel-like mesoporous nickel oxide for electrochemical supercapacitors. J Porous Mater 13:407–412
52. Wei TY, Chen CH, Chien HC et al (2010) A cost-effective supercapacitor material of ultrahigh specific capacitances: spinel nickel cobaltite aerogels from an expoxide-drived sol-gel process. Adv Mater 22:347–351
53. Arora P, White RE, Doyle M (1998) Capacity fade mechanisms and side reactions in lithium-ion batteries. J Electrochem Soc 145:3647–3667
54. Wakihara M (2001) Recent developments in lithium ion batteries. Mater Sci Eng R33: 109–134
55. Tarascon JM, Armand M (2001) Issues and challenges facing rechargeable lithium batteries. Nature 414:359–367
56. Shukla AK, Kumar TP (2008) Materials for next-generation lithium batteries. Current Sci 94:314–331
57. Salloux K, Chaput F, Wong HP et al (1995) Lithium intercalation in vanadium pentoxide aerogels. J Electrochem Soc 142:L191–L192
58. Le DB, Passerini S, Guo J et al (1996) High surface area V_2O_5 aerogel intercalation electrodes. J Electrochem Soc 143:2099–2104
59. Passerini S, Le DB, Smyrl WH et al (1997) XAS and electrochemical characterization of lithiated high surface area V_2O_5 aerogels. Solid State Ion 104:195–204
60. Dong W, Dunn B (1998) Sol-gel synthesis and characterization of molybdenum oixde gels. J Non-Cryst Solids 225:135–140
61. Passerini S, Coustier F, Giorgetti M et al (1999) Li-Mn-O aerogels. Electrochem Solid State Lett 2:483–485
62. Long JW, Swider-Lyons KE, Stroud RM et al (2000) Design of pore and matter architectures in manganese oxide charge-storage materials. Electrochem Solid State Lett 3:453–456
63. Long JW, Stroud RM, Rolison DR (2001) Controlling the pore-solid architecture of mesoporous, high surface area manganese oxides with the birnessite structure. J Non-Cryst Solids 285:288–294
64. Zhang F, Passerini S, Owens BB et al (2001) Nanocomposites of V_2O_5 aerogel and RuO_2 as cathode materials for lithium intercalation. Electrochem Solid State Lett 4:A221–A223
65. Sakamoto JS, Dunn B (2002) Vanadium oxide-carbon nanotube composite electrodes for use in secondary lithium batteries. J Electrochem Soc 149:A26–A30

Window Glass Coatings

Jitka Mohelníková

Abstract Thin film coatings modulate optical and thermal and other properties of window glass. Coated window glasses influence indoor climate and energy efficiency in buildings. They can be classified into groups of spectrally selective glazings, chromogenic glazed systems for switchable smart window technologies and light-enhancing transparent materials. Low-emissivity glazings are typical of spectrally selective window glasses. Chromogenics have applications in light control and switching technologies that are optically, thermally, chemically or electrically activated. Transparent materials with micro-structured or holographic films and antireflective coatings serve to direct light and control light functions. An overview of several types of window glazings and coatings will be presented.

1 Introduction

Energy savings and motivation toward solar energy utilization in buildings brought forth the development of advanced glazing materials [1–3]. These glazings with thin film coatings and selective surfaces have ability to modulate the properties of window glass. Special windows glazing influences the climate in the buildings in which they are used. This is why they have found wide applications in architecture [4, 5]. They can be used for:

J. Mohelníková (✉)
Faculty of Civil Engineering, Brno University of Technology,
Veveří 95, 602 00 Brno, Czech Republic
e-mail: mohelnikova.j@fce.vutbr.cz

L. Zang (ed.), *Energy Efficiency and Renewable Energy Through Nanotechnology*,
Green Energy and Technology, DOI: 10.1007/978-0-85729-638-2_26,

- limiting heat loses and maximizing solar gains and natural lighting in buildings to reduce energy consumption for heating and artificial lighting,
- reducing glare and lowering the energy demands for cooling and air-conditioning,
- reducing the amount of cleaning water and solvents due to reduction in maintenance requirements in glazed windows.

Glazings can be classified into three main groups with respect to their applications in buildings [6–8]:

- spectrally selective glazings (e.g., glazings with modulated transmittance, reflectance and absorbance in the visible or infrared spectral range),
- chromogenic glazings (i.e., switchable glazings for dynamic light control),
- light-enhancing glazings (e.g., light reflectors, collectors or wave-guiding transparent materials, antireflective glass and glazings with self-cleaning surfaces).

2 Spectrally Selective Glazings

Spectrally selective coatings modulate the spectral properties of glass substrates. For window glazing applications they are represented mainly by thin films that reduce absorption of infrared radiation within glass panes. They have very low emissivity (between 0.2 and 0.05) compared to the emissivity of common window glass 0.84 [9–11]. These thin films are called low-emissivity (low-e) coatings. The major requirements for these coatings are:

- high transmittance in visible spectral range,
- high reflectance in infrared spectral region.

Low-emissivity coatings are classified into two main groups [9–15]:

- Coatings reflective in infrared spectral range (between 2 and 10 μm). This type of low-e glazing is also called a heat mirror or a winter film. They serve to eliminate heat radiation losses of windows or collect thermal energy.
- Coatings reflective in near-infrared part of the solar radiation spectrum in spectral range between 0.78 and 2.5 μm. They are called solar controlling low-emissivity coatings or summer films. Their main application is reduction of solar thermal loads.

2.1 Low-Emissivity Glazings

Low-emissivity (low-e) glazings have special coatings deposited on their surfaces. Low-e coatings operate as transparent heat mirrors. They can be deposited as [11, 14, 15]:

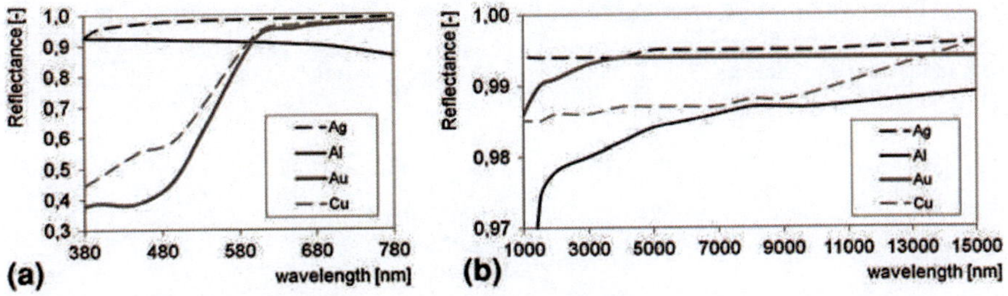

Fig. 1 Spectral reflectance of *silver* (Ag), *aluminum* (Al), *gold* (Au) and *copper* (Cu) thin films on glass, **a** visible spectral range 380–780 nm, **b** infrared spectral range 1,000–15,000 nm

- doped oxides semiconductor coatings,
- conducting micro-grid coatings,
- multilayer thin films consisting of metal and transparent dielectric layers.

Doped oxide semiconductor coatings consist of dielectric layers with mobile charge carriers. The coatings have high reflectance in the infrared spectral range and good transmittance of visible wavelengths. Materials such as SnO_2:F, In_2O_3:Sn, SnO_2:Sb, ZnO:Al or Cd_2SnO_4 are used for these coatings [11, 16–19].

The transparent heat mirror coatings can be also designed as micro-grid thin conducting films with small openings (around 2.5 μm). These openings transmit solar radiation and the rest parts of the coating reflect infrared radiation. This type of coatings has not found wide applications for window glass panes [19].

Multilayer coatings consisting of a metal film sandwiched between transparent dielectric layers are recommended for window glazings. Metals have high reflectance and low-emissivity in the infrared spectral range [20]. Figure 1 shows spectral reflectance of selected thin metal films (Aluminium-Al, Silver-Ag, Gold-Au and Copper-Cu) of equal thicknesses deposited on a glass substrate [21]. The reflectance curves are compared in the visible spectral range (380–780 nm) and in the part of infrared range between 1,000 and 15,000 nm.

It is obvious that the most apt materials are silver and gold layers. The thin silver and gold layers both have high visible transmittance and high reflectance in the infrared spectral range. Silver is the most adept metal for low-e coatings because of its high infrared reflectance and low light absorbance [21]. Aluminium is ill suited as a glazing component due to its higher reflectance and reduced transmittance in the visible range.

Dielectric coatings protect the thin metal layer and influence light transmittance of the low-e coatings. Spectral transmittance of a thin metal film (M) compared with transmittance of the same metal film with single and double dielectric (D) layers is presented in Fig. 2 (computer simulation [22]). Dielectric layers have an anti-reflective property. The top dielectric layer on the metal film has higher influence on visible transmittance.

Dielectric materials with high refractive indices, such as TiO_2, SnO_2, $SnBO_2$, In_2O_3, ZnO, ZnS, Si_3N_4 and Bi_2O_3 [23, 24], can be used in low-emissivity

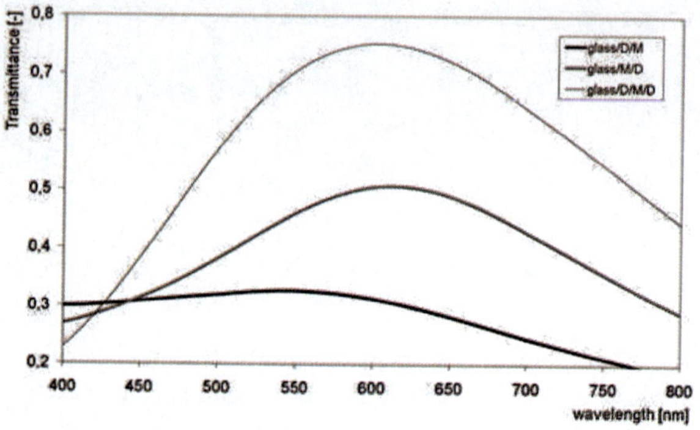

Fig. 2 Spectral transmittance of thin films with metal (M) and dielectric (D) layers deposited on glass

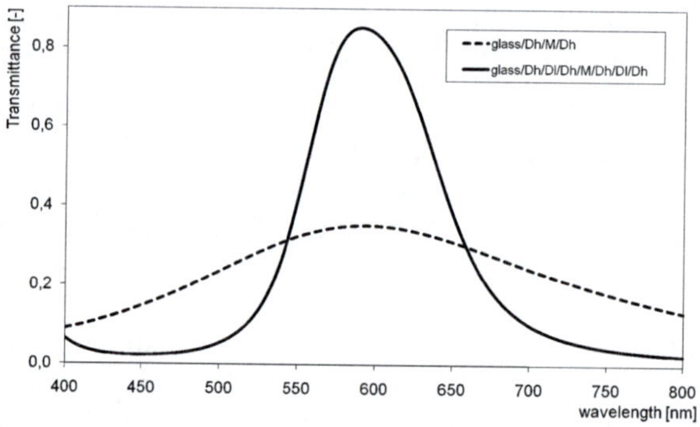

Fig. 3 Influence of dielectric layers on transmittance of a thin low-e coating (M-metal layer, Dh-dielectric layer with high refractive index, Dl-dielectric layer with low refractive index)

coatings. Multilayer low-e coatings use combinations of dielectric materials with high and low refractive indices, for example, TiO_2 and SiO_2.

The materials selected as layers in a low-e coating influence its visible transmittance. Figure 3 presents a spectral transmittance of a low-e coating with metal (M) layer and dielectric layers of materials with high (Dh) and low (Dl) refractive indices (computer simulation [22]).

The light transmittance of the five-layer Dh/Dl/Dh/M/Dh/Dl/Dh coating is very high but relatively narrow-band compared to the three-layer Dh/M/Dh coating of the same material composition, Fig. 3.

An optimized design of the multilayer composition of dielectric and metal layers yields visible transmittance and infrared reflectance coatings [23–26]. Examples of low-e coatings include [14, 27–33]:

- a central silver layer between two dielectric layers:
  glass/ZnO/Ag/ZnO
  glass/ZnS/Ag/ZnS
  glass/TiO_2/Ag/TiO_2
  glass/SnO_2/Ag/SnO_2
  glass/Bi_2O_3/Ag/Bi_2O_3

- a silver layer and three dielectric layers of two different materials: glass/TiO_2/Ag/TiO_2/SiO_2
- a copper layer between two dielectric layers: glass/SnO_2/Cu/SnO_2
- a titanium nitride layer between two dielectric layers: glass/TiO_2/TiN/TiO_2
- a zirconium nitride layer between two dielectric layers: glass/ZrO_2/ZrN/ZrO_2
- a gold layer between two dielectric layers: glass/In_2O_3/Au/In_2O_3.

The thin gold film is inert but metal coatings of silver or copper are not chemically stable. Their optical properties degrade by thermal oxidation and corrosion caused mainly by atmospheric pollutants such as chlorine and sulphur [33]. The metal layer's chemical stability and durability is increased with a protective layer [34]. The metal layer is protected by a barrier layer or blocker layer [23]. A barrier layer should be deposited on one or both metal-dielectric interfaces. Materials that can be used to protect the metal layers against corrosion include Si, Ti, TiN, $TiAO_x$, NiCr, $NiCrO_x$, Cr, Zr, Mo, W and ZrSi [23, 35–37].

Examples of characteristic compositions of low-e window coatings with protective layers:

- glazings with a single silver and a single blocker layer [35–37]: glass/SnO_2/Ag/$NiCrO_x$/SnO_2 or glass/ZnO/Ag/$TiAO_x$/SnO_2, glass/$SnBO_2$/ZnO/Ag/$NiCrO_x$/$SnBO_2$ or glass/TiO_2/ZnO/Ag/$NiCrO_x$/Si_3N_4
- glazings with a single silver and double blocker layer: glass/SnO_2/$TiAO_x$/Ag/$TiAO_x$/SnO_2 [37].

Low-e window coatings can be designed as single, double or even triple metal layer films [38, 39]. Multi-layered dielectric-metal coatings serve as the broadband infrared (IR) reflectors for low-emissivity glazings. It means these glazings can eliminate thermal radiation losses, provide near-IR solar control and substitute convenient transmittance in the whole visible spectrum [40]. Compositions of such multifunctional coatings include:

- glass/SnO_2/ZnO/Ag/$NiCrO_x$/SnO_2/Ag/$NiCrO_x$/SnO_2 [37]
- glass/ZnO/Ag/Ti/ZnO/Ag/Ti/ZnO/TiO_2/ZnO [40]
- glass/TiO_x/ZnO_x/Ag/$NiCrO_x$/TiO_x/ZnO_x/Ag/$NiCrO_x$/SiN_x [41, 42], the better thermal and mechanical stability of this coating is achieved by dividing the first layer and middle layer of titanium oxide and also two silver layers in the following way glass/TiO_x/$NiCrO_x$/TiO_x/ZnO_x/Ag/$NiCrO_x$/Ag/$NiCrO_x$/TiO_x/NiCrOx/TiO_x/ZnO_x/Ag/$NiCrO_x$/Ag/$NiCrO_x$/SiN_x [41, 42]
- glass/Si_3N_4/Ni:Cr/Ag/Ni:Cr/Si_3N_4/Ni:Cr/Ag/Ni:Cr/Si_3N_4 [43]
- glass/TiO_x/Si_3N_4/NiCr/Ag/NiCr/Si_3N_4/SnO_2/ZnO/Ag/$NiCrO_x$/SnO_2/Si_3N_4 [44]

Figure 4 shows spectral transmittance of three different compositions of the low-e coatings with single, double and triple silver layer and TiO_2/SiO_2 layers in the following compositions (computer simulation [22]):

- single silver coating: glass/TiO_2/Ag/TiO_2
- double silver coating: glass/TiO_2/Ag/TiO_2/SiO_2/TiO_2/Ag/TiO_2
- triple silver coating: glass/TiO_2/Ag/TiO_2/SiO_2/Ag/SiO_2/TiO_2/Ag/TiO_2

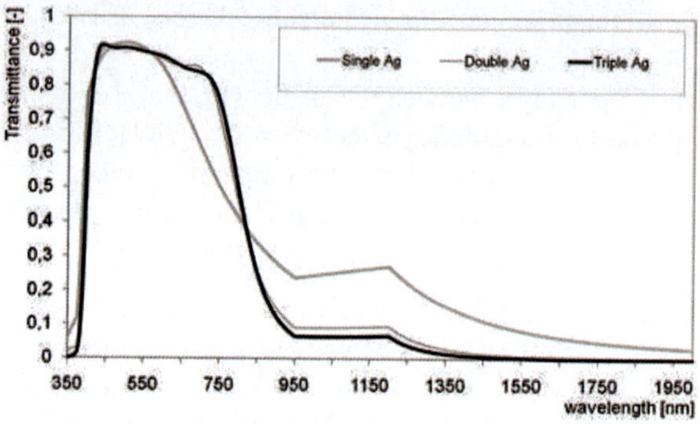

Fig. 4 Spectral transmittance of single, double and triple *silver* low-e coatings

2.2 Spectrally and Angularly Selective Solar Control Glass

Solar control glazings reduce overheating and glare [45]. Glazings with spectrally selective coatings are reflective or absorptive in the near-IR range of solar radiation spectrum. Special coatings can also influence angular selectivity of window glass. General requirements for the design of solar control reflective coatings are visible transmittance and high near-IR reflectance. These properties are achieved by fabricating thin dielectric/metal/dielectric films with metal layers, such as Ag, Cu, Au, or TiN and dielectric layers of Bi_2O_3, In_2O_3, SnO_2, TiO_2, ZnO or ZnS [9, 14].

Solar control coatings can also consist of absorbing and reflecting materials, such as CrN_x, TiN_x, and FeN_x, embedded in layers of a material with a high refractive index [46]. A thin film of Cu_xS can be also used for solar control purposes [47].

The near-IR absorbing thin film and a low-e layer within one coating is another possibility [48, 49]. The coating consists of SnO_2 layer (deposited on the glass substrate) with a dopant of antimony, tungsten, vanadium, iron, chromium, molybdenum, niobium, cobalt or nickel and the low-e layer of SnO_2 containing a dopant of fluorine or phosphorus [48, 49].

Special near-IR absorbing glazings include a light absorptive resin laminated between two glass panes [50]. The transparent acrylic resin contains copper ions, which reduce solar transmittance in the near-IR spectrum and allow for high transmittance of visible light.

Angular selective coatings can also be designed for solar control glazings. Angle selective coatings allow for seasonal self-regulation of window glazings. They reflect part of direct solar rays affecting glass panes at high angles of incidence, which happens in the summer, and transmit diffusive and low-angle incidence direct solar radiation characteristic of winter sunlight [51–54]. Coatings with angularly selective oblique columnar microstructures are used for this purpose. These coatings are based on materials such as Al, Cr, Ta and cernet [54–57]. Affected solar radiation is scattered or directly transmitted by the columnar microstructured coating, dependent upon the angle of incidence. The scattering of direct solar radiation eliminates glare. Scattering surface patterns and thin films are applied for special glare protection glazings [58, 59]. Such coatings reduce the need for cooling in summer seasons.

3 Chromogenic Glazings

Chromogenic glazing materials offer solar control switching activated optically, thermally, chemically or electrically [7, 60, 61]. Switchable glazings have many applications including architectural glazed roofs and façades, automotive windscreens and sunroofs or aircraft windows and displays [61].

Window chromogenic glazings allow dynamic changes in solar transmittance dependent upon external conditions, such as solar radiation intensity or temperature variations. Chromogenic glazings fall into two categories:

- non-electrically activated devices: photochromic (optical activation), thermochromic and thermotropic (thermal activation) and gasochromic glazings (chemical activation). Some of the glazings can also be connected to an external switch, allowing for user-defined operation with electronic control.
- electrically activated devices, such as electrochromic, electrochromic-photovoltaic glazings, photoelectrochromic glazings (they can also be applied in an electrically non-activated variation) and glazing devices with liquid crystals or suspended particles.

3.1 Photochromic Glazings

The photochromic phenomenon is a reversible change between two energy states of a material due to light absorption [62]. Photochromism has been known since 1880s [62–64]. Photochromic glass color centers are activated under radiation exposure, causing glass coloration (darkening). The destruction of the color centers of the photochromic glass occurs in the state without radiation present, when the glass is in the initial transparent state. Color centers are activated in response to exposure to ultraviolet and short-wave visible radiation. The destruction of color centers is influenced by long-wave visible and short-wave infrared radiation.

Photochromic materials are metal halides, such as silver, copper halides, cadmium halides, europium or cerium. Chromium, molybdenum or tungsten and the mineral hackmanit can be also used [62–65]. Investigations in the kinetics of photochemical reactions of photochromic glasses confirm desirability of silver halides [64, 65] as AgCl or AgBr. Another group of photochromics is organic materials. Examples include certain dyes, stereoisomers and polynuclear aromatic hydrocarbons [66, 67].

The best-known variation is ophthalmic photochromic glass. Photochromic glasses have few applications for large-area glazings in buildings [62–68]. A glazed photochromic element with special thin film based on magnesium–nickel alloy has been developed [68] and is expected to be implementable as a window coating [69]. The diffuse-reflection type of photochromic thin film (e.g. SnO_2) forms pyramidal crystal grain projections. This irregular surface is coated with Mg_6Ni layer. In a double glass unit configuration, the photochromic film faces an internal cavity. The cavity is filled with diluted hydrogen which influences reflective and transparent state variations of the unit.

The aforementioned photochromic material, SnO_2, implemented in the following device, glass/SnO_2:F/electrolyte/Pd/Mg_6Ni/irregularly-shaped SnO_2:F/glass, can be activated electrically when connected to an appropriate controller [69].

Another special photochromic device features user-controllable light transmittance. This device bleaches at the user's discretion [70]. Such a device is comprised of a radiation sensitive electrode/interconnecting medium/ion intercalative layer, embedded between two panes of glass with deposited transparent conductive oxide (TCO) layers. An electrical connection exists between the two TCO electrodes. Radiation sensitive electrode materials include zinc–cadmium sulfide and titanium oxide with dopants Ta, Nb, Sb, V and Ru [70].

3.2 *Thermochromic and Thermotropic Glazings*

Thermochromic glass alters its optical properties with response to temperature changes [71–73]. Thermochromic materials' colors are modified by a thermally induced chemical reaction or by a phase transformation. The phase transformation from semiconductor state (high visible transmittance) to metallic state (high infrared reflectance) occurs in transition metal oxides. Thermochromic coatings in the metallic state have similar properties as low-emissivity coatings.

Organic and inorganic compounds are known for their thermochromic properties [72, 73]. Such organic compounds are anil, spiropyrans, polyvinyl acetal resins and hydroxide groups. Examples of inorganic thermochromic compounds include AgI, Ag_2HgI_4, Cd_3P_3Cl, HgI, HgI_2 and $SrTiO_3$. Vanadium dioxide VO_2 has been used for thermochromic window applications [7, 74–78]. Reversible metal-to-semiconductor phase transformation can be achieved by doping VO_2. Tungsten is a frequently used dopant [76, 77]. Thin films VO_2:Mg [78] and multilayer coatings $TiO_2/VO_2/TiO_2/VO_2/TiO_2$ [7] have also been investigated for window coating applications.

Thermotropic glazings change their translucence with temperature changes from clear states to light-scattering, non-transparent states. Thermotropic devices contain special gels embedded between two glass panes. The gels influence the radiation and conductive heat transfer through the glazing. They consist of thermotropic materials of two components with different refractive indices [79–83]. These components are homogenous for low temperatures (transparent state). If temperature rises above the limit value (between 20 and 50°C) the components are phase-separate, forming very small nanoparticles. The separation causes scattering of solar radiation, leaving a non-transparent, white-colored glazing [83].

3.3 Gasochromic Glazings

Dynamic transmittance changes of gasochromic glazings are caused by chemical reactions of hydrogen with wolfram oxide [83–88]. Gasochromic devices consist of a double-glazed unit with connections to a gas supply unit and an electronic control unit. An example of a gasochromic double-glazed unit composition is: glass/WO_3/Pt/cavity filled with gas (e.g. H_2, O_2)/glass [86, 87]. The optically active component of the gasochromic glazing is a thin porous WO_3 film. The Pt layer serves as a catalyst. The glazed unit changes color when the gasochromic film is exposed to a low concentration of H_2. The bleaching process that produces a transparent state occurs with exposure of diluted O_2. The following chemical reaction WO_3 (colorless) + xH_2(Cat) → $2xH^+$ + WO_3 → $H_{2x}WO_3$ (blue) causes blue coloration of the gasochromic glazing [88]. Color intensity depends on the WO_3 film thickness and hydrogen concentration. The effect of the gasochromic coloration is also used in the aforementioned photochromic reflective light control element [69].

A switchable gasochromic mirror device based on Pd/Ni-WO_3 anodic double layer coating deposited on the glass has been developed [89]. The hydrogen gas dissociates from the Pd catalyst into H atoms, which diffuse into the Ni-WO_3 film, causing coloration.

3.4 Electrochromic Glazings

The electrochromic phenomenon has been the subject of many scientific research programs and investigations [90–110]. Electrochromic (EC) smart window glazings are just one possible application of this phenomenon.

EC glazings reversibly change their optical properties under an applied voltage. The presence of electric current causes variation of the material's chemical composition and its optical properties. EC glazings require only a small voltage (between 1 and 5 V) for switching. They are transparent when no voltage is applied. A typical composition of EC glass is the following [93–95]: glass/transparent electrode (TCO)/EC layer/central electrolyte-ion conductor/ion-storage layer/transparent electrode (TCO)/glass.

The switchable process of glazing coloration and bleaching is caused by ion migration between the EC and ion-storage layer. Ions of the EC layer move toward the ion-storage layer through the central electrolyte when voltage is applied between two transparent electrodes. The ions transport to the ion-storage layer causes EC glazing coloration. A change in electric polarity causes ion flow in the opposite direction, causing bleaching of the EC device.

Many materials have been tested for EC applications. Typical materials which are used for EC devices are [90–105], for example:

- EC layer: inorganic materials, such as WO_3, Li_xWO_3, H_xWO_3, NiO or organic materials, such as viologen,
- ion conduction layer: $LiClO_4$ + PC, Ta_2O_5, PMMA + organic,
- ion-storage layer: WO_3, WO_3:MO, Nb_2O_5, NiO, Prussian Blue, $Li_yV_2O_5$,
- transparent electrode: In_2O_3:Sn or SnO_2:F.

The coloration process is achieved by the oxidation–reduction reaction in organic EC materials [103]. The EC effect occurs in inorganic metal oxides due to the dual injection (cathodic) or ejection (anodic) of ions and electrons (e^-). A typical reaction for the cathodic coloring material is:

WO_3 (transparent) + yH^+ + ye^- $\leftrightarrow$ $HyWO_3$ (blue) [103, 104]

Ions such as Li^+, Na^+ and Ag^+ can also cause EC coloration.

A typical coloring anodic reaction is, for example:

$$Li_yV_2O_5(\text{light yellow}) - yLi^+ - ye^- \leftrightarrow V_2O_5(\text{blue}) \qquad [103]$$

$$\text{or } Ir(OH)_x(\text{transparent}) \leftrightarrow IrOx\ (\text{black}) + XH^+ + xe^- \qquad [104]$$

Possibilities of enhancing an EC glazing visible transmittance in its transparent state can be achieved through adding Al or Mg to the Ni oxide [106] or antireflective layers [107]. EC safety glass laminated with ion-conducting PVB (Polyvinyl butyral) sheets [108] and flexible polyester-based foils with EC coatings [109] have also been developed.

3.5 Photoelectrochromic Glazings

Photoelectrochromic glazings combine electrochromic and photochromic principles [110–114]. Such glazing units consist of an electrochromic material (WO_3, IrO_x, V_2O_5 or NiO) and a semiconductor film (TiO_2, CdS, ZnS, ZnO or WO) [110]. The composition of the photoelectrochromic glazing can be [111, 112]: glass/transparent electrode/WO_3 layer/nanoporous TiO_2 layer/dye monolayer/electrolyte LiI/Pt layer/transparent electrode/glass.

The thin film of TiO_2 is porous and covered with a dye monolayer. The dye is excited under illumination and donates electrons to the TiO_2, which conduct them to the WO_3 layer. This reduces the tungsten and changes its color from transparent to blue. The Pt thin film catalyzes the reverse reaction. The process takes place under illumination. This means that the whole system operates as a passive photochromic element. The device can also have transparent electrodes connected to an external switch. Electrons from the WO_3 layer can flow to the counter-electrode in case of the closed external switch (darkening state). The device is transparent in the electrically non-activated state.

3.6 Electrochromic and PV Glazing

Photovoltaics can be used as power sources for switchable glazings, creating a complete smart window system [115]. Another interesting application of photovoltaic (PV) technology for windows is the integration of PV thin films of special colors into the glazings [116]. An electrochromic (EC) device powered by an integrated amorphous silicon carbide (a-SiC:H) PV film is an example of a photovoltaic/electrochromic (PV/EC) glazing [117–119]. The PV part serves as a semitransparent power supply and the EC thin film serves as a modulator of transmittance. These two parts of the PV/EC coating consists of:

- PV part: TCO/a-SiC:H-n/a-SiC:H-i/a-SiC:H-p
- EC part: TCO/V_2O_5 (ion-storage layer)/$LiAlF_4$ (ion conductor layer)/Li_y-WO_3(EC layer)/TCO

The whole composition of a PV/EC glazing can be as follows: glass/TCO/a-SiC:H-n/a-SiC:H-i/a-SiC:H-p/TCO/V_2O_5/$LiAlF_4$/Li_yWO_3/TCO. The device is darkened when the top and bottom TCO layers are electrically connected via external switch. The middle TCO layer is used for battery charging and user control [119].

3.7 Liquid Crystal and Suspended Particle Glazings

Liquid crystals (LC) are materials comprised of thin, needle-shaped organic molecules that are randomly distributed and flow like liquids. Under certain conditions they could be aligned and ordered. They exist in several liquid crystalline phases, such as [120]:

- nemanic phase (LC orientation order but no position order),
- cholesteric phase (LC local orientation in a helical or spiral configuration),
- smetic phase (LC orientation and position order).

LC orientation alters with electric field. Orientation influences the light transmittance of the LC glazing. There are several types of LC used in switchable devices. Polymer dispersed liquid crystals (PDLC) and nemanic curvilinear aligned phase of encapsulated liquid crystals (NCAP) are used in large-area glazing applications [63, 120, 121]. Guest-host LC represents another possibility for windows [63]. PDLC have wide applications for architectural and automotive glazings [122].

These window glazings have LC within an index matched polymer matrix between two transparent conductive oxides TCO electrodes. The composition of the LC glazing can be: glass/TCO/dielectric layer/PDLC layer/dielectric layer/TCO/glass [14, 124]. LC molecules are dispersed between two conductive electrodes. Activation of the electric field causes alignment of the LC. This allows for enhanced light transmittance. The device appears translucent and white in the off-state. LC glazings require electrical activation to be in the transparent state [14, 121–126].

A new LC glazing technology which does not require constant power has been developed. A reverse mode of PDLC glazing is achieved by doping with photoconductive molecules in the polymer LC matrix [127, 128, 129].

Suspended particle glazings also require electrical activation to reach transparency. They are not transparent in the non-activated state [130–136]. A suspended particle (SP) switchable device consists of glass/TCO/dielectric layer/SP layer/dielectric layer/TCO/glass [130, 131]. Polyester sheets can be also used instead of the glass panes.

The active SP layer has needle-shaped dipole particles (less 1 μm long). They are suspended randomly in an organic fluid or gel in the electrically non-activated state, causing absorption and scattering of incident light. The suspended particles align if an electric field is applied, causing bleaching of the device.

SP glazing with modified particles could provide darkening in several colors. SP glazings with encapsulated particles in polymers and laminated the composite between polyester sheets show promise for large glazed areas applications [136].

4 Light Enhancing and Controlling Glazings

Modern solar technologies offer light enhancement and control systems. These systems serve to transport and redirect light into the interiors of buildings to improve indoor visual comfort. Optical fibers, transparent materials with microstructured surfaces or holographic films, fluorescent concentrators and mirrors can guide, redirect and concentrate light [137–140].

These materials integrated into façades and glazed roofs can eliminate glare and overheating problems in building interiors. They can also be used for solar conversion systems [140, 141]. Combinations of prismatic materials and switchable technology provide smart solar control transparent devices [116, 142–149].

Enhancement of light and solar transmittance in window glass can be achieved through antireflective coatings [150–158]. Special self-cleaning functional coatings also positively influence light transmittance of windows [159–169].

4.1 Switchable Devices and Prismatic Surfaces

Microprismatic glass coatings and retro-reflecting prisms serve for angle selective solar control. They allow variation of transmittance dependent upon the angle of solar radiation incidence [138–140]. High-angle incident light is retro-reflected but low-angle incident and diffuse light is transmitted.

The microprismatic surface patterns are produced by the interference holography or micro-machining technology. The microstructure can be periodic

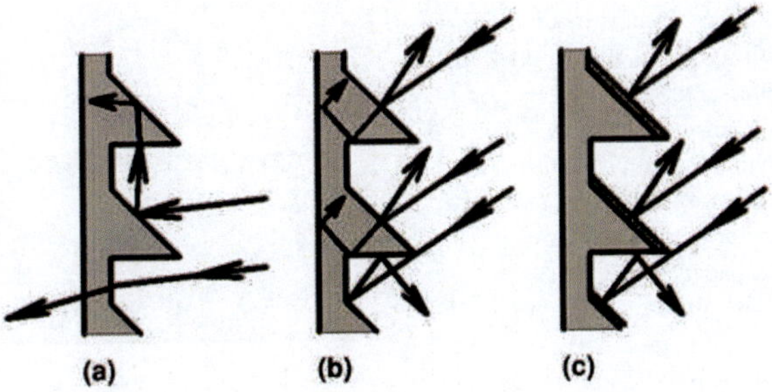

Fig. 5 Prismatic transparent materials, **a** transmission of low-angle incident solar rays, **b** retro-reflection of high-angle incident solar rays, **c** retro-reflection and absorption of solar radiation on switchable coatings (deposited on the oblique parts of the prisms) [116, 142–144]

(triangular, sinusoidal and parabolic) or random (periods between 0.2 and 50 μm) [140, 141]. The surface is laminated on the inner surfaces of double-glazed units in building applications.

Transparent materials with micro and/or macro-structured surfaces can be combined with switchable devices to achieve a user-controlled system, Fig. 5c [116, 142–144]. This combination provides dynamic regulation of solar transmittance. Parts of the prismatic surfaces exposed to high solar incidence have thin chromogenic coatings. These coatings reduce transmittance due to a switchable coloration process. Glare protection is achieved through prismatic surface retro-reflecting and light-scattering effects.

Another possible application of switchable glazing technology is the combination of laminated glass with light emitting diodes (LEDs), with a liquid crystal interlayer embedded between two glass panes with transparent conductive oxide layers [145–147].

A switchable glazing system with surface micro-blinds has been developed, as shown in Fig. 6. The device controls light transmittance in response to applied voltage. Thin metal micro-blinds are deposited on a glass substrate with a transparent conductive oxide layer and an insulator layer. The micro-blinds are rolled in the transparent electrically non-activated state. When electric field is activated a potential difference between the thin rolled metal and transparent conductive layer causes the micro-blinds to stretch and limit light transmittance (shading effect) [148, 149].

4.2 Antireflective Glazings

Special low-iron glasses and glazings with antireflective coatings can be used to improve visible and solar transmittance.

Generally an antireflective (AR) coating consists of alternating dielectric layers of materials with high (as ZrO_2, Ta_2O_5 or TiO_2) and low (as MgF_2, SiO_2 and

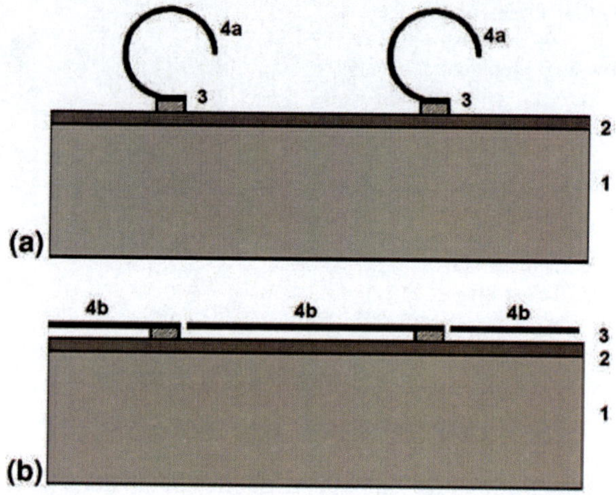

Fig. 6 Function of micro-blinds, **a** in the electrically non-activated state (*transparent*), **b** in the electrically activated state (shading), 1-glass substrate, 2-adhesion layer and transparent conductive layer, 3-insulator, 4-reflective resilient metal micro-blind (4a-rolled blind, 4-stretched blind) [148, 149]

Al_2O_3) refractive indices [150, 151]. Devices comprised of TiO_2 and SiO_2 layers are commonly used [152]. Multilayer dielectric AR coatings can enhance the transmittance of visible light. A single AR layer with low refractive index and high porosity is recommended for solar energy applications. A third type of the AR coating can improve both visible and solar near-IR transmittance [152]. Non-porous materials can also achieve the AR effect through surface etching, nano-patterns or gratings [152–155].

An example of a practical application of AR coatings for windows is the enhancement of the visible transmittance of low-e glazings. An AR coating of porous SiO_2 can be deposited on both sides of a glazing with a pyrolytic low-e tin-oxide coating [156]. Antireflective coatings are also useful in switchable glazing technologies, such as electrochromic glazings [157, 158].

4.3 Self-Cleaning Glazings and Coatings

Special coatings can create self-cleaning or easy-to-clean surfaces [159–161]. These coatings can provide:

- photocatalytic function (decomposition of organic materials on the glass surface),
- hydrophilic function (strongly wettable surface which allows water sheeting, which cleans the surface),
- hydrophobic function (weakly wettable surface).

A porous TiO_2 thin film coating can exhibit both the photocatalytic and hydrophilic effects [159–161]. Incident solar radiation (UV part) generates the catalytic activity within the TiO_2 surface coating. The excited titanium transfers its energy to the oxygen molecules, which loses an electron. Electrons migrate to the

Fig. 7 Self-cleaning surfaces, **a** water drop rolled on the hydrophobic surface, **b** water drop spread on the hydrophilic surface [164]

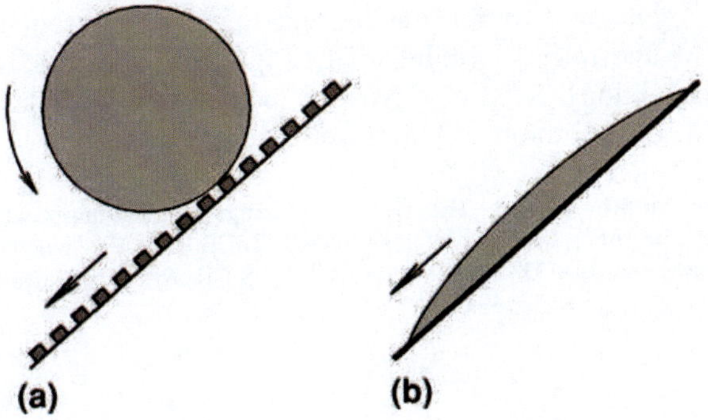

organic material on the glass surface and decompose it. This process also changes the glass surface to hydrophilic by creating oxygen vacancies, which react with water. Due to this activity water droplets wet the surface. The water washes away the decomposed organic materials. The thin TiO_2 photo-catalytically active coating should be covered by a special corrosion protective coating [162, 163] for large façade or roof glazing applications.

Another self-cleaning system is a hydrophobic surface coating. It consists of a water polymer matrix which creates a surface with nanoscale pores. The self-cleaning effect is achieved due to water surface tension. Water drops form almost spherical droplets on the hydrophobic surface. They fall off the surface due to gravity and roll dust and small dirt with them. Water drops are removed before they can evaporate on the surface [164], as shown in Fig. 7.

Special smart switchable coatings provide protective, hydrophobic or hydrophilic and other properties. Such smart coatings can also reversibly alter the wetting properties of surfaces [165–169] in response to external stimuli.

5 Conclusion

An overview of several types of window glazings and coatings has been presented. Implementations of thin film coatings for multifunctional glazed window units provide integrated systems with spectrally selective glazings of high visible transmittance and low-emissivity, which can be completed with solar control and switchable glazings with light retro-reflecting and guiding units.

The development of glazing systems continues to interest researchers. There is a trend toward applications of chromogenic glazings for large glazed areas of building envelopes. Improvements in durability, maintenance-free service and low-cost production will be the focal points of future work.

Photovoltaics offer self-powered glazed devices. Systems based on nanotechnology and advanced materials provide new types of self-activated switchable coatings for multifunctional smart glazing devices with reversibly modified properties.

Integrated multi-function glazings are representatives of systems that are able to dynamically modulate their properties in the response to external climatic conditions. Such changes can increase energy efficiency, and indoor visual and thermal comfort in buildings.

Acknowledgments The window glazings and coatings overview was elaborated within the frame of research projects MŠMT MEB 080804 and GAČR 101/09/H050. The author acknowledges Dr. Pavel Pokorný, ISI AS CR, Brno for consultations on the thin film design and computer simulations.

References

1. IEA SH&C: Task 10 solar materials, Task 18 advanced glazing materials, Task 27 energy performance of switchable glazing. http://www.iea-shc.org
2. Hutchins MG (1998) Advanced glazing materials. Sol Energy 62:145–147
3. Lampert CM (1992) Advanced glazing technology. Fenestration 2000, phase III-glazing materials. LBL-31616, Berkeley
4. Robinson PD, Hutchins MG (1994) Advanced glazing technology for the low energy buildings in the UK. Renew Energy 5:298–309
5. Bauchot M (2001) Energy, environmental and economic benefits from advanced double glazing in EU dwellings. Glass performance days, Tampere 2001
6. O'Shaughnessy D (2009) TH18-global megatrends and next-generation architectural glass. Proc Construct 2009, Indianopolis
7. Granqvist CG et al (2010) Advances in chromogenic materials and devices. Thin Solid Films 518:3046–3053
8. Lee ES, Selkowitz SE et al (2006) Active load management with advanced window wall systems: research and industry perspectives. Final project report CEC-500-2006-052-AT1. LBNL, California
9. Johnson ET (1991) Low-e glazing design guide. Butterworth Architecture, Boston
10. Pulker HK (1998) Coatings on glass. Elsevier, Amsterdam
11. Lampert CM (1981) Heat mirror coatings for energy conserving windows. Sol Energy Mater 6:1–41
12. Fan JC (1981) Sputtered films for wavelength-selective applications. Thin Solid Films 80:125–136
13. Berning PH (1983) Principles of design of architectural coatings. Appl Opt 22:4127–4141
14. Terry Hollands KG et al (2001) Glazings and coatings. In: Gordon J (ed) Solar energy: the state of the art, ISES position papers. James & James, London
15. Karlsson B (1981) Materials for solar transmitting heat reflecting coatings. Thin Solid Films 86:91–98
16. Stjerna B et al (1994) Optical and electrical properties of radio frequency sputtered tin oxide films doped with oxygen vacancies, F, Sb, or Mo. J Appl Phys 76:3797–3817
17. Hamberg I, Granqvist CG (1986) Evaporated Sn-doped In_2O_3 film: basic optical properties and applications to energy efficient windows. J Appl Phys 60:R123–R160
18. Jin ZC et al (1988) Optical properties of sputter-deposited ZnO:Al film. J Appl Phys 64:5117–5131
19. Terry Hollands KG et al (2001) Glazings and coatings. High transmittance in the visible region and reflectance in the IR. In: Gordon J (ed) Solar energy: the state of the art, ISES position papers. James & James, London, pp 56–70
20. Valkonen E et al (1984) Solar optical properties of thin films of Cu, Ag, Au. Sol Energy 32:211

21. Palik (ed) (1991) Handbook of optical constants of solids. Academic Press, New York
22. Computer program FILM*CALC 3.03—advanced optical thin film technology, FTG software associates, Princeton
23. Bräuer G (1999) Large area glass coating. Surf Coat Technol 112:358–365
24. Schaefer C et al (1997) Low emissivity coatings on architectural glass. Surf Coat Technik 93:37–45
25. Smith GB et al (2004) Energy-efficient coatings in nanohouse^{TM} initiative. Curr Appl Phys 4:381–384
26. Granqvist CG (1990) Window coatings for the future. Thin Solid Films 193:730–741
27. Gläser HJ The European history of coatings on architectural glazing. http://www.glassfiles.com
28. Fan et al (1985) Transparent heat-mirror. US Patent 4556277
29. Kim D (2010) Low temperature deposition of transparent conducting ITO/Au/ITO films by reactive magnetron sputtering. Appl Surf Sci 256:1774–1777
30. Martin-Palma RJ et al (1998) Silver-based low-emissivity coatings for architectural windows: optical and structural properties. Sol Energy Mater Sol Cells 53:55–66
31. Miyazaki M, Ando E (1994) Durability improvement of Ag-based low-emissivity coatings. J Non-Cryst Solids 178:245–249
32. Andersson KE et al (1994) Zirconium nitride based transparent heat mirror coatings preparation and characterisation. Sol Energy Mater Sol Cells 32:199–212
33. Kusano E et al (1986) Thermal stability of heat-reflective films consisting of oxide-Ag-oxide deposited by dc magnetron sputtering. J Vac Sci Technik A: Vacuum, Surfaces, and Films 4:2907–2910
34. Lampert CM (1981) Heat Mirror Coatings for Energy Conserving Windows. Solar Energy Mater 6:1–41
35. Szczyrbowski J et al (1999) New low emissivity coating based on TwinMag® sputtered TiO_2 and Si_3N_4 layers. Thin Solid Films 351:254–259
36. Beister G et al (1995) Progress in large-area glass coatings by high-rate sputtering. Surf Coat Technol 76–77:776–785
37. Schaefer C et al (1997) Low emissivity coatings on architectural glass. Surf Coat Technik 9:37–45
38. Lu Y et al (2009) Method of making low-e coating using ceramic zinc inclusive target and target used in same. USPTO Patent Application 20090205956
39. Boor WP (2008) Studying low-e glass performance. Constr Specif, http://www.ppg.com/corporate/ideascapes/SiteCollectionDocuments/274404_final.pdf
40. Steven J et al (2007) Durable low-e coated glass for use in warm temperate climates. Glass performance days 2007, Tampere
41. Glenn D et al (2009) Double silver low emissivity and solar control coating. US Patent 7632572 B2
42. Glenn D et al (2003) Double silver low emissivity and solar control coating. US Patent Application Publication 2003/0049464 A1
43. Hartig KW et al (1996) Dual silver low-e glass coating system and insulating glass made thereform. US Patent 5557462
44. Neuman G et al (2008) Coated article with low-e coating including IR reflecting layer(s) and corresponding method. US Patent 7419725
45. Manfrè G (2005) The need of innovation for solar control technologies, Glass performance days 2005, Tamper
46. Ochs D et al (2007) Progressive power supplies for architectural glass coating. Glass performance days 2007, Tampere
47. Nair PK et al (1991) Optimisation of chemically deposited Cu_xS solar control coatings. J Appl Phys 24:441–449
48. Russo D et al (2003) Solar control coated glass. US Patent 6596398
49. McKown C et al (2001) Solar control coated glass. US Patent 6218018

50. Ujiie T, Katono H (2005) High performance solar control material. Glass performance days 2005, Tampere
51. Smith GB (1990) Theory of angular selective transmittance in oblique columnar thin films containing metal and voids. Appl Opt 29:3685–3693
52. Elkadi H (2006) Cultures of glass architecture. Ashgate, Hampshire
53. Reppel J, Edmonds IR (1998) Angle-selective glazing for radiant heat control in buildings: Theory. Sol Energy 62:245–253
54. Smith GB et al (1998) Thin film angular selective glazing. Sol Energy 62:229–244
55. Dligatch S (1998) An analysis of Ag/Al_2O_3 angular selective films by X-ray reflectivity. Thin Solid Films 312:4–6
56. Smith GB et al (1998) Angular selective thin film glazing. Renew Energy 15:183–188
57. Bellac DL et al (1995) Angular selective optical transmittance through Cr-based films made by oblique angle sputtering experiment and theory. J Phys D Appl Phys 28:600
58. Ishizuka S (1992) Optical properties of angle dependent light control film. Proc SPIE 1727:241
59. Gombert A (2006) Optically functional surfaces for solar applications. ISES, Eurosun 2006, Glasgow
60. Lampert CM (2004) Chromogenic smart materials. Mater Today 7:28–35
61. Lampert CM (1995) Chromogenic switchable glazing: towards the development of the smart window. In: Proceedings of window innovations '95, Toronto
62. Hoffmann HJ (1990) Photochromic glass. In: Lampert CM, Granqvist CG (eds) Large-area chromogenics: materials and devices for transmittance control, vol IS4. SPIE, Bellingham, pp 86–101
63. Wiggington M (1996) Glass in architecture. Phaidon Press, London
64. Fanderlík I (1996) Vlastnosti skel. Informatorium, Prague
65. Araujo RJ (1980) Photochromism in glasses containing silver halides. Contemp Phys 21:77
66. Chu N (1990) Photochromic plastics. In: Lampert CM, Granqvist CG (eds) Large-area chromogenics: materials and devices for transmittance control, vol IS4. SPIE Bellingham, pp 102–121
67. Chu N (1986) Photochromic performance of spiroindolinonaphthoxazines in plastics. Sol Energy Mater 14:215
68. Richardson TJ et al (2001) Switchable mirrors based on nickel–magnesium films. Appl Phys Lett 78:3047
69. Yoshimura K, Okada M (2007) Reflective light control element with diffusible reflecting surface. US Patent 7259902
70. Teowee G et al (2001) Photochromic devices. US Patent 6246505
71. Day J, Willet R (1990) Science and technology of thermochromic materials. In: Lampert CM, Granqvist CG (eds) Large-area chromogenics: materials and devices for transmittance control, vol IS4. SPIE, Bellingham, pp 122–147
72. Jorgenson GV, Lee JC (1990) Thermochromic materials and devices: inorganic systems. In: Lampert CM, Granqvist CG (ed) Large-area chromogenics: materials and devices for transmittance control, vol IS4. SPIE, Bellingham, pp 142–159
73. Sone K, Fukuda Y (1987) Inorganic thermochromism. Springer, Berlin
74. Babulanam SM et al (1987) Thermochromic VO_2 films for energy efficient windows. Sol Energy Mat 16:347
75. Jorgenson GV, Lee JC (1986) Doped vanadium oxide for optical switching films. Sol Energy Mat 14:205
76. Parkin I, Manning T (2007) Thermochromic coatings. US Patent 0048438
77. Blackman Ch et al (2009) Atmospheric pressure chemical vapour deposition of thermochromic tungsten doped vanadium dioxide thin films for use. Thin Solid Films 517:4565–4570
78. Mlyuka NR et al (2009) Mg doping of thermochromic VO_2 films enhances the optical transmittance and decreases the metal-insulator transition temperature. Appl Phys Lett 95:171909

79. Haldimann M et al (2008) Structural use of glass. International Association for Bridge and Structural Engineering, Zürich
80. Wilson HR (1994) Optical properties of thermotropic layers. Proc SPIE 2255:473
81. Seeboth A et al (2004) Chromogenic polymer gels for reversible transparency and color control. In: Samson A et al (ed) Chromogenic phenomena in polymers, vol 888, chapter 80. ACS Symposium Series, Washington DC, pp 110–121
82. Nitz P, Hartwig H (2005) Solar control with thermotropic layers. Sol Energy 79:573–582
83. Georg A et al (1998) Switchable glazing with a large dynamic range in total solar energy transmittance (TSET). Sol Energy 62:215–228
84. Wilson HR et al (2002) The optical properties of gasochromic glazing. In: Proceedings of the 4th international conference on coating on glass, Braunschweig
85. Schwarz M (2008) Smart materials. CRC Press, Taylor & Francis, Boca Raton
86. Wittwer V et al (2004) Gasochromic windows. Sol Energy Mater Sol Cells 84:305–314
87. Wittwer V, Graf W (2001) Gaschromic glazings with a large dynamic range in total solar energy transmittance. Glass performance days 2001, Tampere
88. Lampert CM (2002) Electrochromics-history, technology, and the future, 6.1 gas-chromics. In: Chowdari B et al (ed) Solid state ionics: trends in the new millenium. Proceedings of the 8th Asian conference on world scientific, London
89. Se-hee L et al (2004) Pd/Ni-WO_3 anodic double layer gasochromic device. US Patent 6723566
90. Lampert CM, Granqvist CG (1990) Large-area chromogenics: materials and devices for transmittance control, vol IS 4. SPIE Institutes for Advanced Optical technologies, Bellingham
91. Granqvist CG (1995) Handbook of inorganic electrochromic materials. Elsevier, Amsterdam
92. Monk P et al (2007) Electrochromism and electrochromic devices. Cambridge University Press, London
93. Deb SK et al (1978) Electrochromic cell with protective overcoat layer. US Patent 4120568
94. Granqvist CG (2005) Electrochromic device. J Eur Ceram Soc 25:2907–2912
95. Granqvist CG (1992) Electrochromism and smart window design. Solid State Ionics 53–56:479–489
96. Granqvist CG (2008) Oxide electrochromics: why, how, and whither. Sol Energy Mater Sol Cells 92:203–208
97. Granqvist CG et al (2007) Nanomaterials for benign indoor environments: electrochromics for "smart windows", sensors for air quality, and photo-catalysts for air cleaning. Sol Energy Mater Sol Cells 91:355–365
98. Granqvist CG et al (2003) Electrochromic coating devices: survey of some recent advances. Thin Solid Films 442:201–211
99. Granqvist CG (2000) Electrochromic tungsten oxide films: review of progress 1993–1998. Sol Energy Mater Sol Cells 60:201–262
100. Granqvist CG et al (1997) Towards the smart window: progress in electrochromics. J Non-Cryst Solids 218:273–279
101. Granqvist CG (1990) Window coatings for the future. Thin Solid Films 193–194:730–741
102. Lampert CM (2002) Electrochromism-history, technology and the future. In: Chowdari, B et al (ed) Solid state ionics: trends in the new millenium. Proceedings of the 8th Asian conference on world scientific, London
103. Lampert CM (1998) Smart switchable glazing for solar energy and daylight control. Sol Energy Mater Sol Cells 52:207–221
104. Lampert CM (1993) Optical switching technology for glazing. Thin Solid Films 236:6–13
105. Lampert CM (1984) Electrochromic materials and devices for energy efficient windows. Sol Energy Mat 11:1–27
106. Avendaño E et al (2004) Electrochromism in nickel oxide films containing Mg, Al, Si, V, Zr, Nb, Ag, or Ta. Sol Energy Mater Sol Cells 84:337–350

107. Jonson A, Roos A (2006) Influence of the performance of antireflective coatings in electrochromic windows. In: Proceedings of ISES EuroSun 2006, Glasgow
108. Rottmann M et al (2005) Large area electrochromic safety glass; switching behaviour and transmission control of solar radiation. Glass performance days 2005, Tampere
109. Granqvist CG (2006) Solar energy materials of the future: electrochromic foils for energy efficiency and indoor comfort. In: Proceedings of ISES EuroSun 2006, Glasgow
110. Bechinger CS et al (2002) Self bleaching photoelectrochemical-electrochromic device. US Patent 6369934
111. Pichot F et al (1999) Flexible solid-state photoelectrochromic windows. J Electrochem Soc 146:4324–4326
112. Georg A, Opara Krašovec U (2006) Photoelectrochromic window with Pt catalyst. Thin Solid Films 502:246–251
113. Hauch A et al (2001) New photoelectrochromic device. Electrochim Acta 46:2131–2136
114. Gombert A (2007) New development in glazing for a better use of solar energy in buildings. Glass performance days 2007, Tampere
115. Lampert CM (2003) Large-area smart glass and integrated photovoltaics. Sol Energy Mater Sol Cells 76:489–499
116. Gombert A (2007) New developments in glazing for better use of solar energy in buildings. Glass performance days 2007, Tampere
117. Benson DK, Branz HM (1995) Design goals and challenges for a photovoltaic-powered electrochromic window covering. Sol Energy Mater Sol Cells 39:204–211
118. Gao W et al (2000) Approaches for large-area a-SiC:H photovoltaic-powered electrochromic window coatings. J Non-Cryst Solids 266–269:1140–1144
119. Gao W et al (1999) First a-SiC:H photovoltaic-powered monolithic tandem electrochromic smart window device. Sol Energy Sol Cells 59:243–254
120. Lampert CM (1990) Introduction to liquid crystals. Large-area chromogenics: materials and devices for transmittance control, vol IS 4. SPIE, Bellingham
121. Amstock JS (1997) Liquid crystals and suspended-particle device. Handbook of glass in construction. McGraw-Hill, New York
122. Sucheol P, Hong JW (2009) Polymer dispersed liquid crystal film for variable-transparency glazing. Thin Solid Films 517:3183–3186
123. Lampert CM (1999) Advances in materials and technology for switchable glazing. Glass performance days 1999
124. Lampert CM (1994) Glazing materials for solar and architectural applications, IEA SH&C Task 10 C, LBL-34436
125. Sixou P et al (2001) Switchable liquid-crystal/polymer micro-composite glazings. Glass processing days 2001
126. Baughmann et al (1990) Dual-pane thermal window with liquid crystal shade. US Patent 4964251
127. Garnier DJ et al (2009) High-efficiency multistable switchable glazing using smetic A liquid crystals. Sol Energy Mater Sol Cells 93:301–306
128. Cupelli D et al (2009) Self-adjusting smart windows based on polymer-dispersed liquid crystals. Sol Energy Mater Sol Cells 93:2008–2012
129. Cupelli D et al (2004) Fine adjustment of conductivity in polymer-dispersed liquid crystals. Appl Phys Lett 85:3292–3294
130. Lampert CM (1995) Chromogenic switchable glazing: towards the development of the smart window. In: Conference proceedings of window innovations '95, Toronto
131. Amstock JS (1997) Liquid crystals and suspended-particle device. Handbook of glass in construction, McGraw-Hill, New York
132. Vergaz R et al (2008) Modelling and electro-optical testing of suspended particle device. Sol Energy Mater Sol Cells 92:1483–1487
133. Check JA (1995) Light modulating film of improved clarity for a light valve. US Patent 5463492

134. Saxe RL (1981) Light valve containing improved light valve suspension. US Patent 4247175
135. Thompson et al (1978) Light valve. US Patent 4078856
136. Lampert CM (1999) Advances in materials and technology for switchable glazing. Glass processing days 1999, Tampere
137. Riccobono J, Ludman J (2002) Solar holography. In: Luxman J et al (eds) Holography for the new millenium. Springer, New York
138. Hans DT et al (1993) Design optimization and manufacturing of holographic windows for daylighting applications in buildings. In: Lampert CM (ed) Optical materials technology for energy efficiency and solar energy conversion XII. SPIE, Bellingham 2017, pp 35–45
139. Stojanoff CG (2006) Engineering applications of HOEs manufactured with enhanced performance DCG films. In: Bjelkhagen HI, Lessard RA (eds) Practical holography XX: materials and applications. SPIE Proceedings 6136 613601
140. Hoßfeld W et al (2003) Application of microstructured surfaces in architectural glazings. In: Proceedings of ISES solar world congress 2003, Göteborg
141. Gunther W et al (2005) Combination of microstructures and optically functional coatings for solar control glazing. Sol Energy Mater Sol Cells 89:233–248
142. Wilson V et al (2002) The optical properties of gasochromic glazings. In: Proceedings of 4th international conference coating on glass, Braunschweig
143. Gombert A (2006) Optically functional surfaces for solar applications. In: Proceedings of ISES EuroSun 2006, Glasgow
144. Graf et al (2003) Device for guiding light. WO Patent WO/2003/071079
145. Shavit D (2007) LED-and SMD-polyester film embedded in glass: history, current and future developments. Glass performance days 2007, Tampere
146. Lefèvre H (2009) Laminated glass with embedded LEDs: the use of specific power supplies able to provide continuous high voltage allows new lighting applications and colour changes in decorative applications. Glass performance days 2009, Tampere
147. http://www.glassonweb.com
148. Lamontagne B et al (2006) Microblinds and methods of fabrication thereof. US Patent 0196613
149. Lamontagne B et al (2009) The next generation of switchable glass: the micro-blinds. Glass performance days 2009, Tampere
150. Ochs D, Rettich T (2007) Progressive power supplies for architectural glass coating. Glass performance days 2007, Tampere
151. Boire P et al (2000) Glazing pane having an anti-reflection Coating. US Patent 6086914
152. Gombert A et al (1998) Glazing with very high solar transmittance. Sol Energy 62:177–178
153. Hofmann T, Kursawe M (2003) Antireflective coating on glass for solar applications glass. Glass performance days 2003, Tampere
154. Olsson G (2003) Low cost industrial manufacturing of a thin single layer antireflective surface on sheet glass. Glass performance days 2003, Tampere
155. Southwell WH (1991) Pyramid-array surface relief structures producing anti-reflection index matching on optical surfaces. J Opt Soc Am A 8:549–553
156. Hammarberg E, Roos A (2003) Antireflection treatment of low-emitting glazings for energy efficient windows with high visible transmittance. Thin Solid Films 442:222–226
157. Jonsson A, Roos A (2006) Antireflective coatings on different window surfaces. In: Proceedings of the 6th international conference on coatings on glass and plastics, Dresden
158. Roos A et al (2009) Applications of coated glass in high performance energy efficient windows. Glass performance days 2009, Tampere
159. Armand P (2003) Self-cleaning coatings for architectural application. Glass performance days 2003, Tampere
160. Hüber M (2003) TiO_2-coatings from inorganic soles: a new approach to hydrophilic and photocatalytically active glasses. Glass performance days 2003, Tampere
161. Gläser HJ The effects of weather onto glazing and their influence, Part II. http://www.glassfiles.com

162. Hohenstein H (2003) Coatings with nano-particles for windows and façades. Glass performance days 2003, Tampere
163. Nakamura M et al (2004) Hydrophilic property of SiO_2/TiO_2 double layer films. In: Plütz J et al (ed) Proceedings of ICCG5, 2004, Saarbrücken
164. Overs M (2005) Coatings for decorative glass surfaces based on chemical nanotechnology. Glass performance days 2005, Tampere
165. Lahann J et al (2006) Switchable surfaces. US Patent 7020355
166. Shen L et al (2009) Mechanism of sliding friction on a film-terminated fibrillar interface. Langmuir 25:2772–2780
167. Autumn K et al (2000) Adhesive force of a single gecko foot-hair. Nature 405:681–685
168. Provder T, Baghdachi J (2007) Smart Coatings. ACS Symposium Series 957, American Chemical Society, Oxford University Press, Washington
169. Provder T, Baghdachi J (2009) Smart Coatings II. ACS Symposium Series 1002, American Chemical Society, Oxford University Press, Washington

Editor Biography

Ling Zang, Ph.D., is a USTAR professor of Nanotechnology at the University of Utah, with expertise in the fields of nanomaterials and molecular devices. Dr. Zang's current research focuses on nanoscale imaging and molecular probing, organic semiconductors and nanostructures, optoelectronic sensors and nanodevices, with the long-term goal to achieve real applications in the areas of national security, renewable energy, and clean environment. Before moving to Utah in 2008, Dr. Zang was an associate professor at Southern Illinois University. He was previously an Alexander von Humboldt Fellow at Erlangen-Nuremberg University in Germany, a NSF CAREER Award winner, and a K. C. Wong Foundation Research Fellow. He also holds an adjunct professorship at the Institute of Chemistry, Chinese Academy of Sciences, Beijing, and serves on the editorial board of several scientific journals including Journal of Nanoengineering and Nanosystems, Journal of Nanoscience Letters, and Imaging Science and Photochemistry. Dr. Zang has been awarded various federal grants to support his broad range of research in nanoscience and nanotechnology. Beyond the regular faculty duty on campus, Dr. Zang also remains active in organizing and chairing the nanotechnology sessions of various national and international conferences, e.g., Beckman Frontiers of Science Symposium, National Academy of Sciences, AIChE Annual Meeting, NanoUtah Annual Conference, etc. Dr. Zang earned his B.S. in Physical Chemistry from Tsinghua University in 1991 and Ph.D. in Chemistry from the Chinese Academy of Sciences in 1995. More details about Dr. Zang and his research can be found at his group website at the University of Utah: http://www.eng.utah.edu/~lzang/.

L. Zang (ed.), *Energy Efficiency and Renewable Energy Through Nanotechnology*, Green Energy and Technology, DOI: 10.1007/978-0-85729-638-2,

Index

L. Zang (ed.), *Energy Efficiency and Renewable Energy Through Nanotechnology*, Green Energy and Technology, DOI: 10.1007/978-0-85729-638-2,

T